ORGANIC CHEMICALS

An Environmental Perspective

ORGANIC CHEMICALS

An Environmental Perspective

Alasdair H. Neilson

CRC Press
Taylor & Francis Group
Boca Raton London New York

CRC Press is an imprint of the
Taylor & Francis Group, an **informa** business

CRC Press
Taylor & Francis Group
6000 Broken Sound Parkway NW, Suite 300
Boca Raton, FL 33487-2742

First issued in paperback 2019

ISBN-13: 978-1-56670-376-5 (hbk)
ISBN-13: 978-0-367-39909-2 (pbk)
Library of Congress Card Number   99-052061

Library of Congress Cataloging-in-Publication Data

Neilson, Alasdair H.
    Organic chemicals : an environmental perspective / Alasdair Neilson
        p.   cm.
    Rev. Ed. of: Organic chemicals in the aquatic environment. c1994
    Includes bibliographical references and index.
    ISBN 1-56670-376-X
    1. Aquatic organisms—Effect of water pollution on. 2. Organic water
pollutants—Environmental aspects.   I. Neilson, Alasdair H. Organic chemicals in the
aquatic environment. II. Title.
    QH545.W3 N45 1999
    557.6′ 27—dc21   99-052061
                                                                                                          CIP

Visit the Taylor & Francis Web site at
http://www.taylorandfrancis.com

and the CRC Press Web site at
http://www.crcpress.com

Preface

This book stems from the author's experience with a variety of problems on the fate, distribution, and toxicity of organic compounds in the aquatic environment. It became increasingly clear that the procedures for investigating these problems crossed the traditional boundaries of organic and analytical chemistry, microbiology, and biology, and after many years this resulted in the idea of selecting the relevant aspects of these and writing the present book. Environmental problems have become increasingly complex and environmental impact studies should be based as far as possible on incontrovertible scientific facts. In this book the basic issues of chemical analysis, distribution, persistence, and ecotoxicology have therefore been discussed although the emphasis has been placed on microbial reactions with which the author is most familiar. Throughout the book an attempt has been made to include a wide range of structurally diverse compounds as illustration, and a mechanistic approach to degradation and transformation has been adopted. At the same time, the limitations in this book should be clearly appreciated: it is not designed for the specialist in any of the traditional disciplines, although it is hoped that the level of detail is acceptable to those who seek discussions of a range of environmental issues. The book is by no means comprehensive but a list of references for those seeking further detail is provided at the end of each chapter.

This volume may be considered a new edition of a previous one *Organic Chemicals in the Aquatic Environment*. It differs from that volume not only in attempting to bring up to date the contents of the original chapters and correcting some errors, but in incorporating extensive new material. This includes more extensive discussion in Chapter 2 of recent analytical procedures, a more thorough discussion in Chapter 4 of chemical and photochemical reactions including those in the troposphere, a presentation in Chapter 7 of more recent aspects of ecotoxicological assays, and a new chapter on bioremediation that makes extensive use of principles introduced in previous chapters. More extensive discussions have been given to the terrestrial and tropospheric environments, and this justifies the widening of the title from the original one.

Acknowledgments

When all is said and done, it remains to thank all those who have contributed in many different ways to the making of this book. It is a pleasure to express my deep gratitude to the following.

My teachers in the Universities of Glasgow, Cambridge, Oxford and the University of California, Berkeley — including many who are no longer with us — for illustrating by example the rigors of scientific inquiry.

The late Percy W. Brian FRS for patiently directing my faltering footsteps into microbiology many years ago.

Östen Ekengren, Director, Environmental Technology and Toxicology for allowing me a number of privileges including free access to library and copying facilities.

My collaborators, Ann-Sofie Allard, Per-Åke Hynning, Marianne Malmberg, and Mikael Remberger not only for their scientific contributions but also for their friendship and tolerance over many years.

Ann-Sofie Allard for her expertise in producing the numerous figures from sometimes erratic drafts, and for allowing me to expand and update our earlier review that has become Chapter 8.

Mirja S. Salkinoja-Salonen and her colleagues in the Department of Applied Microbiology and Applied Chemistry, University of Helsinki for friendship and stimulation during many years.

The Knut and Alice Wallenberg Foundation and its executive director Professor Gunnar Hoppe for providing instrumentation that opened new horizons to our research.

My good and long-standing friends the Søndmør families in Øversjødalen, Norway for providing the tranquility during which most of the revision of the original text was undertaken.

Eva Berg, without whose continued support and encouragement this book would never have materialized.

To A.N. and G.N.M. who no longer tread the Highland hills

Contents

3. Partition: Distribution, Transport, and Mobility... 103

4.   Persistence: General Orientation ....

8. Microbiological Aspects of Bioremediation 763

In collaboration with Ann-Sofie Allard

1

Introduction

1.1 Orientation

Although the synthesis of 1,1,1-trichloro-bis(4-chlorophenyl)-ethane (DDT) was achieved in 1874 by Othmar Zeidler, the compound elicited little interest until its potent toxic effect toward insects was discovered by P.H. Müller in 1939. The compound fulfilled only too well his criteria that a suitable compound should be chemically stable and persist for long periods of time. Although its use over the next few years was extensive and effective, DDT was to become the paradigm for an appreciation of the complex network of adverse effects brought about by the introduction of synthetic chemicals into the environment (Dunlap 1982).

It is historically convenient to date the beginning of popular concern over the environment with the publication of Rachel Carson's *Silent Spring* in 1962. Although this book made no pretense at a scientific exposure of the potential dangers of pesticides, the date happily coincided with the coming of age of the analytical instrumentation needed to provide a firm scientific foundation for the unease which was expressed.

The succeeding years have seen an increasing degree of sophistication in the analysis of environmental issues and a clear appreciation of the central role of chemical analysis. Indeed, since the birth of quantitative chemistry there has been an almost continuous search for more sensitive and selective methods of analysis. For example, it was the invention of the humble Bunsen burner almost 150 years ago that provided a nonluminous flame, which made possible atomic emission spectroscopy, and before the turn of the 19th century this had contributed to the discovery of no fewer than 12 new elements. Within the last 40 years or so, the revolution in organic analysis has been no less spectacular: application of gas chromatographic, high-resolution liquid chromatography, and capillary electrophoresis has become routine, and structural identification using mass spectrometry and nuclear magnetic resonance commonplace.

For organic compounds, evaluation of their environmental impact based on data for acute toxicity to fish and estimation of biochemical oxygen demand has been steadily replaced by the results of experiments on subacute and chronic toxicity, sophisticated investigations of microbial metabolism, and an appreciation of the true complexity of partitioning processes. It may conservatively be stated that during the 1960s, there slowly emerged a new interdisciplinary activity — environmental science.

This book attempts to provide discussions on the analysis, partitioning, metabolism, and toxicity of organic compounds. Most of the principles are applicable to both the aquatic and terrestrial environments, and the presentations are restricted to organic compounds. Although the most-detailed analysis has been given to aspects of microbiology which has been the main interest of the author during many years, these discussions that extend over three chapters have been buttressed by a brief section dealing with chemical analysis which is a cornerstone of all aspects of environmental activity, and short sections devoted to problems of partitioning and ecotoxicology. Yet none of the chapters stands in isolation; analysis, partitioning, and mechanisms of association are intimately bound to metabolism while assessment of toxicity requires consideration of bioavailability, partition, and metabolism.

There are certain threads that run throughout this series of essays: the importance of the transformation of xenobiotics, the association between xenobiotics and components of environmental matrices, and the dynamics of ecosystems. By implication, therefore, environmental hazard assessments should take into account not only the original xenobiotic but also its possible transformation products together with the actual state — free or associated — in which the compound exists in the environment. It should be clearly appreciated that both the persistence and the toxicity of xenobiotics are critically determined by the extent and reversibility of the interactions between low-molecular-weight xenobiotics and polymeric material in the environment. Indeed, the importance of dynamic interactions at both the physicochemical and the molecular levels cannot be too strongly emphasized.

The present study is essentially chemical and mechanistic in approach. Attention is drawn to a number of books dealing with the more general issue of hazard assessment even though the emphasis of some is on radiological hazard and carcinogenesis in humans (Fischoff et al. 1981; Ottoboni 1984; Lewis 1990; Rodricks 1992), to a provocative sociological book that deals with perceived risk and fear (Furedi 1997), and to a valuable book on randomness and numerical significance (Bennett 1998).

1.2   Literature Cited

One aim of this book is to provide a forum for presenting an overview of the issues that are basic to producing an environmental hazard assessment of organic compounds discharged into the environment. This evaluation requires the application of expertise in the analysis of the compounds, knowledge of their distribution among the various environmental compartments and their dissemination, and appreciation of the factors that determine their persistence and toxicology. An attempt has been made throughout to illustrate these principles with concrete examples, even though no effort has been made to provide a comprehensive account of any single group of compounds. It was hoped at the same time that the reader would obtain some appreciation of the complexities in the design and interpretation of the relevant experiments, which are being carried out in the laboratory. This has therefore necessitated some degree of compromise both in the depth of the presentation and in the number of references to the literature.

It was decided at the outset that it was clearly impossible to provide citations for every statement; on the other hand, controversial points or possibly less well-known facts which have not yet reached the textbooks deserve citation of their source. References are almost invariably given to the primary literature, which has been subjected to the scrutiny of peer review. It is therefore assured that even when the present author's interpretations should prove faulty — and this is inevitable — a solid and reproducible basis of fact is available to the critical reader. As a result, although the number of references is considerably greater than had been visualized at the outset, they represent merely an eclectic selection from a vast literature. The choice of references may, however, seem quixotic — on the one hand, unduly historical, and on the other, recent, though incomplete — relevant references have no doubt been omitted, but the writer can assure the authors of these that there is no malice in the selection. Some older work has been cited when this has led to lasting concepts, although other early work may be difficult to evaluate by the standards of today, and no doubt work at the cutting edge of current research will rightly require modification and extension in the future. When appropriate, reference has been made to reviews which present detailed coverage that lies beyond the scope of this volume. The author has therefore exercised the privilege of selection. It is too much to hope that an adequate compromise has been reached, and, without doubt, important studies have been omitted and errors of interpretation have been made; in the final analysis, all that can be done is to offer humble apologies and hope that the damage done is not too serious. The author hopes he may take refuge in the reply given to a lady inquiring why, in his great dictionary, Dr. Johnson had defined pastern as the knee of a horse: "Ignorance, madam, pure ignorance."

1.3 Limitations

It is inevitable that the coverage of the areas included for discussion is uneven. The selection of material for inclusion is therefore substantially determined by personal bias and some of these prejudices should be taken into account. To quote again Dr. Johnson discussing the state of learning in the present author's native country: "Their learning is like bread in a besieged town: every man gets a little, but no man gets a full meal." This is certainly true of the discussions presented in this book.

1.3.1 Organochlorines

There is a huge literature on organochlorine compounds since these have awakened serious and well-merited concern. Although in many cases, investigations concerned with these compounds have yielded principles of general application, an attempt has been made to redress the balance by including illustrative examples using nonchlorinated compounds: the coverage of, for example, DDT, polychlorinated biphenyls (PCBs), and chlorinated dibenzo[1,4]dioxins does not therefore reflect the substantial research effort that has been directed to these compounds. It is hoped, however, that — apart altogether from the specific interest in individual compounds — the general principles that have been developed in the course of such investigations have clearly emerged. The other large class of compounds that have been extensively studied are the polycyclic aromatic hydrocarbons (PAHs), but it is pointed out in several places that conclusions drawn from studies with neutral compounds such as PAHs and many of the organochlorines may not be directly applicable to compounds with polar substituents such as hydroxyl, amino, or ketonic groups. It is particularly important to underscore the great structural diversity of agrochemicals and pharmaceutical products that have been developed against specific biological targets using the increasing sophistication of methods available to the synthetic organic chemist.

Whereas agrochemicals have attracted attention as a result of their potential adverse effects on nontarget organisms and concern with their persistence, the same intensity of effort has not been directed so far to pharmaceutical compounds for human use. Although this presumably reflects the much smaller quantities that are involved, the widespread distribution of some of them including, for example, antibiotics and steroids merits attention. Indeed, recent interest in the possible estrogenic effects of xenobiotics has revealed low levels of the pharmaceutical diclofenac in lake water in Switzerland (Buser et al. 1998) and the contraceptive 17α-ethynlyestrone in treated domestic sewage effluent (Desbrow et al. 1998). On the other hand, large quantities of veterinary chemotherapeutic agents are used in large-scale husbandry and may be eliminated into the terrestrial environment: an illustrative example is provided by the fluoroquinolone carboxylic acids. Enrofloxacin, which is eliminated largely unchanged and as the N-deethylated metabolite, is strongly sorbed to clay minerals (Nowara et al. 1997), so that its degradation is photochemically mediated (Burhenne et al. 1997).

1.3.2 Dictates for Inclusion

Areas that have attracted intense research effort may not have done so exclusively because of their scientific interest, but as a result of economic — or even social — pressure. Two illustrative examples are the 2,3,7,8-tetrachlorodibenzo[1,4]dioxins and the phthalates. The long-standing uncertainty surrounding the hazard associated with exposure to 2,3,7,8-tetrachlorodibenzo[1,4]dioxin (U.S. EPA 1993) in spite of massive international effort illustrates both the influence of public opinion and the intrinsic difficulty of the hazard assessment process. Enormous attention has been directed to phthalate esters, which is not reflected in this book, although the biodegradation of these is addressed in Sections 6.2.1 and 6.7.3. It should be noted that *o*-phthalate is an intermediate in several degradation pathways: for example, the bacterial degradation of phenanthrene (Section 6.2.3) and quinoline (Section 6.3.1), and the fungal degradation of anthracene (Section 6.2.2).

1.3.3 Modeling

There is at least one major area of activity pertaining directly to the environment for which the reader will seek in vain. The complexity of environmental problems and the availability of personal computers have led to extensive studies on models of varying sophistication. A discussion and evaluation of these lie well beyond the competence of an old-fashioned experimentalist; this gap is left for others to fill but attention is drawn to a review that covers recent developments in the application of models to the risk assessment of xenobiotics (Barnthouse 1992), a book (Mackay 1991) that is devoted to the problem of partition in terms of fugacity — a useful term taken over from classical thermodynamics — and a chapter in the book by Schwarzenbach et al. (1993). Some superficial comments are, however, presented in Section 3.5.5 in an attempt to provide an overview of the dissemination of xenobiotics in natural ecosystems. It should also be noted that pharmacokinetic models have a valuable place in assessing the dynamics of uptake and elimination of xenobiotics in biota, and a single example (Clark et al. 1987) is noted parenthetically in another context in Section 3.1.1. In similar vein, statistical procedures for assessing community effects are only superficially noted in Section 7.4. Examples of the application of cluster analysis to analyze bacterial populations of interest in the bioremediation of contaminated sites are given in Section 8.2.6.2.

1.3.4 Enzymes and Relatedness

The discussion of enzymology is both limited and uneven. An area of increasing interest is the relation between enzymes based on sequences of amino acids or nucleotides. This has been noted in a few examples, but a satisfactory discussions lies beyond both the competence of the author and the limits of this volume.

1.4 Biotechnology

In Chapters 4 and 6, considerable emphasis is placed on the environmental significance of biotransformation as opposed to biodegradation. It therefore seems appropriate to draw brief attention also to their use in biotechnology. Microorganisms play important roles in widely diverse technical processes ranging from the leaching of ores and the preservation of food, to the production of antibiotics and food additives. It is worth drawing attention to some features of these applications.

1. Some of these processes are carried out by specific groups of organisms: for example, by acid-tolerant thiobacilli in the leaching of ores, by lactobacilli in food preservation, or by fungi and actinomyces in the production of antibiotics.
2. The production of food additives such as fructose, glutamate, lysine, citric acid, ascorbic acid, and vanillin implies that these compounds are not used as substrates for growth. This is generally accomplished by use of mutant strains that, as a bonus, produce high yields of the desired products.

Increasing attention has been directed to the use of microbial transformations in the production of valuable compounds or chemical intermediates. A survey has been given of biochemical transformations in organic chemistry (Faber 1997) and covers many reactions (including the formation of C–C bonds) that have made use of microorganisms to accomplish an important step in organic syntheses. Some of the salient reactions are discussed in Section 6.11.

Biotechnology is often interpreted to include biological waste treatment, so that one important application of studies in biodegradation is in the design and optimization of biological systems for the treatment of liquid industrial waste. Attention is therefore directed to a review that draws attention to the possible application of organisms not traditionally considered in this context (Kobayashi and Rittman 1982). Biological waste treatment has not been discussed at all here since most of the fundamental physicochemical and engineering principles are not included in this book. On the other hand, many critical — and in some cases unresolved — issues that are directly relevant have been taken up in the appropriate chapters of this book: partition into the solid phase, into biota, and into the atmosphere together with the reversibility of these processes (Chapter 3), persistence to biotic and abiotic attack and the formation of metabolites (Chapters 4 and 6), and toxic effects (Chapter 7). Microbiological issues relevant to the bioremediation or biorestoration of contaminated areas are discussed in Chapter 8, and the contamination of groundwater by leachate from landfills clearly interfaces with that of the aquatic environment. One important issue which may be of dominant significance in some of these artificial terrestrial environments — and which is clearly not so in the aquatic environment — is the water concentration, which may be a factor that limits the growth and metabolism of the relevant organisms.

Some other aspects of biological treatment systems — both aqueous and terrestrial — are clearly germane to the principles discussed in this book and attention is drawn to the following issues.

1. In most situations a number of structurally diverse substrates are simultaneously present, and because of toxicity or metabolic incompatibility, this may give rise to problems in assessing their toxicity and degradability.

2. Care must be exercised not to confuse the aims of the investigation: for example, anaerobic reactors that are developed to treat wastewater with the object of producing methane may not degrade recalcitrant xenobiotics for which inoculation with specific microorganisms may be necessary (Ahring et al. 1992).

3. There exist unresolved issues in the application of genetically engineered organisms for biological treatment (a) from the potential risk of dissemination of the organisms and (b) from the problem of maintaining these organisms in a mixed bacterial population in competition with other organisms.

4. Addition of nutrients such as nitrogen and phosphorus must be carefully adjusted so that excess is not discharged into receiving waters, while for effluents with high concentrations of readily degraded substrates, advantage might be taken of nitrogen-fixing bacteria to diminish or even eliminate additions of combined nitrogen (Neilson and Allard 1986).

1.5 Terrestrial Systems

There is considerable interest and concern over the fate of agrochemicals in the terrestrial environment and over their possible effect on nontarget organisms. These problems are considered only parenthetically here, although the transport of agrochemicals is governed by the principles of distribution discussed in Chapter 3, their fate by those given in Chapters 4 and 6, and their toxicology by those in Chapter 7. Extensive results on the persistence of agrochemicals have also provided gratuitous support for the principles of elective enrichment that are discussed in Chapter 5.

Considerable interest has arisen in the environmental problems associated with the disposal of solid waste; again this interfaces closely with the aquatic environment through leaching of organic compounds from landfills — both as solutes and as particulate material — into watercourses, rivers, and lakes. There has been considerable interest in the bioremediation of sites that have been contaminated with both municipal and industrial solid waste. A chapter on bioremediation is therefore provided for two additional reasons: (1) in important respects, bioremediation involves an extension of the principles outlined in Chapters 3, 4, 5 and 6, and (2) it illustrates that many of the principles developed within the aquatic environment that were the subject of previous chapters can be applied with suitable and relatively minor modification to the terrestrial environment.

1.6   The Atmosphere

It is apparent that separate consideration of the atmospheric, terrestrial, and aquatic phases is artificial since there are strong interactions among them. Two examples, which support this view, will be used as illustration.

1. It has been repeatedly demonstrated that organic compounds — even those of apparently low volatility — may be transported via the atmosphere over great distances and may therefore be encountered in samples of precipitation and particulate deposition collected from areas remote from the point of initial discharge. This is discussed further in Section 3.5.3.

2. The atmosphere is not an inert medium, and compounds discharged into it may be transformed by photochemical reactions or by interaction with atmospheric constituents such as oxygen, or the oxides of sulfur and nitrogen; this is discussed briefly in Section 4.1. These transformation products may subsequently enter the aquatic and terrestrial environments in the form of both particulate matter and precipitation. This important issue has been addressed in Section 4.1.3.

For these reasons, a thorough discussion of the role of atmospheric processes ought to be presented if the aquatic environment is to be adequately discussed. This is, however, a monumental task far beyond the competence of the author; all that can be offered instead are brief comments at the appropriate places in the various chapters. Reference should be made to the comprehensive discussion of principles given by Finlayson-Pitts and Pitts (1986) that is referred to in Sections 3.5.3 and 4.1.2, and to the range of compounds produced by higher plants including hydrocarbons (Cao and Hewitt 1995) and oxygenated derivatives (König et al. 1995) which may enter the atmospheric environment.

1.7 Natural Products and Microbial Metabolites

This account is directed almost exclusively to low-molecular-mass xenobiotics so that a discussion of biosynthetic reactions mediated by microorganisms and of biopolymers lies beyond its assigned limits. A few brief comments seem justified, however, because of an upsurge of interest in various groups of naturally occurring compounds, the structure of these biopolymers, and their role in carbon cycling in the environment. Attention is therefore drawn to a few of these, some of which may be persistent.

1.7.1 Organohalogen Compounds

There has been considerable interest in these due to widespread concern over possibly adverse effects of some anthropogenic organohalogens, but since the quantitative significance of almost none of these microbial metabolites has been evaluated and the toxicology of only very few has been investigated — invariably in other contexts — it seems justifiable to accord a low priority to such problems in the light of more urgent issues.

Organohalogen metabolites are produced by bacteria, fungi, and algae by *haloperoxidases* that contain protoporphyrin IX, vanadium (V), or are without metals, and catalyze their synthesis in the presence of H_2O_2, formally by production of Hal^+ (Neidleman and Geigert 1986).

1. Biosynthetic reactions carried out by microorganisms produce a structurally diverse range of chlorinated and brominated compounds (Strunz 1984; Gribble 1996; van Pée 1996; de Jong and Field 1997), while some higher plants produce fluorinated compounds. Brominated compounds are widely distributed in marine environments, and bromoform is the major haloform produced by marine macroalgae (Nightingale et al. 1995). An unusual 1,1'-dimethyl-2,2'-bipyrrole containing four bromine and two chlorine atoms has been found in the eggs of seabirds from both the Atlantic and Pacific Oceans — but not from the Great Lakes (Tittlemeier et al. 1999). Comparable highly halogenated compounds of marine origin may therefore have considerable bioconcentration potential. A plethora

of organobromo compounds is produced by higher organisms — in particular, sponges — and their possible therapeutic application has resulted in intense activity (Munro et al. 1987). Attention has also been drawn to the possibility of exploiting the synthetic capability of marine bacteria for producing compounds of pharmaceutical interest (Austin 1989), and to the use of chlorinated phenylpyrroles produced by *Pseudomonas cepacia* for controlling fruit spoilage fungi (Roitman et al. 1990).

2. Halogenated compounds are produced both by *de novo* synthesis involving direct incorporation of halide ion and by transmethylation reactions (Section 6.11.4). The former include reactions mediated by haloperoxidases in the presence of hydrogen peroxide and halide ion, and these enzyme systems have wide biosynthetic capability (Neidleman and Geigert 1986; van Pée 1996) including oxidative dimerization resulting, for example, in the formation of 2,3,7,8-tetrachlorodibenzo[1,4]dioxin from 2,4,5-trichlorophenol (Svenson et al. 1989).

3. A few naturally occurring compounds such as methyl chloride and halogenated phenols are important industrial products, although the relative quantitative contribution of the natural products on a global basis cannot be estimated reliably. In addition, it seems unlikely that the more highly chlorinated phenols — and in particular, for example, 2,4,5-trichlorophenol — are produced in significant concentrations by such reactions (Wannstedt et al. 1990).

4. It has been hypothesized on the basis of the formation of trichloroacetate from aliphatic compounds, especially acetate, by the action of chloroperoxidase in the presence of hydrogen peroxide and chloride that this might be a naturally occurring metabolite (Haiber et al. 1996). Plausible mechanisms for the formation of trichloroacetic acid by atmospheric reactions involving trichloroethane and tetrachloroethene are discussed in Section 4.1.2.

5. Although the distribution, persistence, and toxicity of these halogenated compounds — such as the chlorinated grisans, chloramphenicol, 7-chlorotetracycline, and clindamycin are antibiotics — are subject to the same principles as those outlined in this book, this aspect has seldom been examined. One exception that may serve as an illustration is the debromination of naturally occurring bromophenols by bacteria under anaerobic conditions (King 1988).

6. 3-Chloro- and 3,5-dichloroanisyl alcohols produced by the white-rot fungus *Bjerkandera* sp. may play a physiological role by producing H_2O_2 during oxidation to the aldehydes by aryl alcohol

oxidase; the chlorinated metabolites are poor substrates for lignin peroxidases so that the H_2O_2 may therefore be used for oxidation of lignin while the anisyl aldehydes are readily recycled by reduction to the alcohols (de Jong et al., 1994). The metabolite 2-chloro-1,4-dimethoxybenzene is a cofactor in the oxidation of anisyl alcohol by lignin peroxidase, and is superior to veratryl alcohol although it does not protect the enzyme against inactivation by H_2O_2 (Teunissen and Field 1998).

7. Chlorinated orcinols have been identified from the bulbs of the edible lily *(Lilium maximowiczii)* infected with the fungus *Fusarium oxysporum* f.sp. *lilii*. They were not, however, fungal metabolites; these could also be induced by ultraviolet radiation of bulb scales, and the more highly chlorinated metabolites inhibited conidial germination of *Bipolaris leersiae* (Monde et al. 1998).

1.7.2 Polymeric Compounds

There are many important groups of macromolecules. A discussion of these lies beyond the expertise of the author, but it seems appropriate to note briefly a few of these groups due to their environmental relevance and since all of them make a contribution to organic carbon cycling in the environment.

1. Lignin is an important constituent of many higher plants although the amounts vary with the taxon, the part of the plant, and the time of year. A great deal of attention has been given to the degradation of these compounds (Jeffries 1994), and to attempts to recover useful monomeric products including methane. A few aspects of the enzymes involved are given in Section 4.4.4, while the role of the lignin-degrading white-rot fungi including *Phanerochaete chrysosporium* in degrading xenobiotics is noted in Section 4.3.5 and in Chapter 6 (Sections 6.2.2, 6.5.2, and 6.8.2). The greater metabolic potential of other white-rot fungi is, however, emerging.

2. The empirically defined terms "humic acids" and "fulvic acids" have been used to denote important macromolecular constituents of aquatic systems. Even though their structures are incompletely resolved (Section 3.2.3 and 4.5.3), their association with low-molecular-mass organic compounds — and with metal cations — is important in determining the bioavailability and hence biodegradability and toxicity of organic compounds. Their role in the generation of OH radicals under anaerobic conditions is noted in Section 4.1.1, and as intermediates in electron-transfer reactions under anaerobic conditions in Section 5.5.5.

3. Considerable attention has been directed to particulate organic matter (POM) in the marine environment that is plausibly derived from organisms in the euphotic zone. This is supported by MS comparison of pyrolysis products from samples collected in sediment traps in the Mediterranean Sea with that from the diatom *Biddulphia sinensis*. Although the structures are unknown, a wide variety of compounds have been identified in the pyrolysates including aliphatic hydrocarbons and nitriles, pyrroles, indoles, and aromatic hydrocarbons (Peulvé et al. 1996).

4. Although polysaccharide, lipid, and polypeptide components of microorganisms may be presumed to be readily degraded in a variety of ways, it has been shown that algae synthesize a number of more resistant compounds. The algaenans represent an important group of aliphatic macromolecules; these have been examined in a number of marine algae and consist of linear C_{28}–C_{34} methylene groups linked by oxygen atoms (Gelin et al. 1996), and have plausibly been suggested as an important sink for organic carbon in the marine environment.

5. Biphytanes derived from membrane ether lipids of archaea have been found in water column particulates, and sedimentary organic matter (Hoefs et al. 1997).

In addition, a number of anthropogenic oligomers are produced in substantial quantities.

1. Important groups of detergents consist of alcohol ethoxylates, alkylphenyl ethoxylates, and alkylethoxy sulfates, and their biodegradation has been reviewed (White et al. 1996). Only parenthetical comments on this group of compounds are given in Chapter 6.

2. Silicones have found diverse and increasing use and are generally considered highly stable. Although the levels in effluents from municipal treatment plants were generally below the limits of quantification, they have been found in substantial amounts in the sludges, in sediments in the vicinity of the outfalls, and in agricultural soil that was amended with sewage sludge (Fendinger et al. 1997). Polydimethylsiloxanes are, however, at least partially degraded in soil to dimethylsilanediol (Carpenter et al. 1995; Fendinger et al. 1997), and the mineralization in soil of the monomer (dimethylsilanediol) has been shown (Sabourin et al. 1996). The interesting observations were made that in liquid cultures degradation could be obtained with *F. oxysporium* growing concurrently with propan-2-ol or acetone, and by a strain of *Arthrobacter* sp. growing concurrently with dimethylsulfone.

1.7.3 Polypyrroles

Chlorophylls are produced by all photosynthetic organisms — and even by some nonphotosynthetic bacteria — and details of their structures depend on their source. Collectively they represent a considerable reserve of organic carbon and nitrogen, although little seems to have been established on their persistence. A wide range of transformation products of chlorophylls has been recovered from the sediments of a freshwater eutrophic lake, and these included the unusual sterol esters of pyrophaeophorbides (Eckardt et al. 1995). It is also presumable that such chlorophyll transformation products produce the pyrroles and indoles that have been described in sediment pyrolysates noted above.

1.7.4 Polycyclic Aromatic Hydrocarbons

These have been encountered in a wide variety of environmental samples ranging from sedimentary rocks to lake sediments and street dust. Although thermal processes during incineration of fossil fuels represent a major input into the environment, it seems plausible that many of these compounds are transformation products of naturally occurring steroids and terpenoids, and it is possible that these reactions are microbially mediated. A summary of many of the structures involved has been given (Simoneit 1998), and of plausible reactions for their formation (Neilson and Hynning 1998). The occurrence of these in environmental samples could seriously compromise nonspecific analysis for aromatic compounds; this is noted again in Section 2.4.2. Detailed reviews of many aspects of PAHs including their sources, dissemination, mechanisms of their carcinogenic activity, their uptake and toxicity in aquatic biota, and their microbial metabolism have been given in a multiauthored book (Neilson 1998).

1.8   The Effect of Xenobiotics on Microbial-Mediated Processes

The present discussion is directed to the degradation and transformation of xenobiotics by microorganisms. The converse issue — the inhibitory effect of xenobiotics toward microbial processes that are not directly involved in degradation, but may play an important role in natural geochemical cycles — is not explored. This is of enormous importance, particularly in the terrestrial environment where nontarget organisms may be subjected to exposure to agrochemicals. Some of these organisms have an autotrophic way of life and utilize CO_2 as principal, or exclusive, source of carbon — rather than a heterotrophic metabolism that is characteristic of those organisms that are primarily responsible for degradation of xenobiotics. The autotrophic organisms include algae, ammonia-oxidizing bacteria, and many thiobacilli. In the context of degradation, however, the relevant microorganisms are generally sufficiently resistant to the xenobiotics to preclude serious inhibition. Some useful examples of the reactions that may be involved are found in a review (Smit et al. 1992) that discusses the hazard of genetically manipulated organisms released into the terrestrial environment.

References

Ahring, B.K., N. Christiansen, I. Mathrani, H.V. Henriksen, A.J.L. Macario, and E.C. de Macario. 1992. Introduction of a de novo bioremediation ability, aryl reductive dechlorination, into anaerobic granular sludge by inoculation of sludge with *Desulfomonile tiedjei. Appl. Environ. Microbiol.* 58: 3677–3682.

Austin, B. 1989. Novel pharmaceutical compounds from marine bacteria. *J. Appl. Bacteriol.* 67: 461–470.

Barnthouse, L.W. 1992. The role of models in ecological risk assessment: a 1990s perspective. *Environ. Toxicol. Chem.* 11: 1751–1760.

Bennett, D.J. 1998. *Randomness.* Harvard University Press, Cambridge, MA.

Burhenne, J., M. Ludwig, and M. Spiteller. 1997a. Photolytic degradation of fluoroquinolone carboxylic acids in aqueous solution. Primary photoproducts and half-lives. *Environ. Sci. Pollut. Res.* 4: 10–15.

Burhenne, J., M. Ludwig, and M. Spiteller. 1997b. Photolytic degradation of fluoroquinolone carboxylic acids in aqueous solution. Isolation and structural elucidation of polar photometabolites. *Environ. Sci. Pollut. Res.* 4: 61–67.

Buser, H.-R., T. Poiger, and M.D. Müller. 1998. Occurrence and fate of the pharmaceutical drug diclofenac in suface waters: rapid photodegradation in lake. *Environ. Sci. Technol.* 32: 3449–3456.

Cao, W.-L. and C.N. Hewitt. 1995. Detection methods for the analysis of biogenic non-methane hydrocarbons in air. *J. Chromatogr.* A, 710: 39–50.

Carpenter, J.C., Cella, J.A., and Dorn, S.B. 1995. Study of the degradation of polydimethylsiloxanes on soil. *Environ. Sci. Technol.* 29: 864–868.

Clark, T.P., R.J. Norstrom, G.A. Fox, and H.T. Won. 1987. Dynamics of organochlorine compounds in herring gulls (*Larus argentatus*): II. A two-compartment model and data for ten compounds. *Environ. Toxicol. Chem.* 6: 547–559.

de Jong, E., A.E. Cazemier, J.A. Field, and J.A.M. de Bont. 1994. Physiological role of chlorinated aryl alcohols biosynthesized de novo by the white rot fungus *Bjerkandera* sp. strain BOS55. *Appl. Environ. Microbiol.* 60: 271–277.

de Jong, E. and J.A. Field. 1997. Sulfur tuft and turkey tail: biosynthesis and biodegradation of organohalogens by Basidiomycetes. *Annu. Rev. Microbiol.* 51: 375–414.

Desbrow, C., E.J. Routledge, G.C. Brighty, J.P. Sumpter, and M. Waldock. 1998. Identification of estrogenic chemicals in STW effluent. 1. Chemical fractionation and *in vitro* biological screening. *Environ. Sci. Technol.* 32: 1549–1558.

Dunlap, T.R. 1982. *DDT. Scientists, Citizens, and Public Policy.* Princeton University Press, Princeton, NJ.

Eckardt, C.B., B.J. Keely, and J.R. Maxwell. 1995. Identification of chlorophyll transformation products in a lake sediment by combined liquid chromatography-mass spectrometry. *J. Chromatogr.* 557: 271–288.

Faber, K. 1997. *Biotransformations in Organic Chemistry,* 3rd. ed. Springer-Verlag, Berlin.

Fendinger, N.J., D.C. Mcavoy, W.S. Eckhoff, and B.B. Price. 1997. Environmental occurrence of polymethylsiloxane. *Environ. Sci. Technol.* 31: 1555–1563.

Finlayson-Pitts, B. and J. Pitts, Jr. 1986. *Atmospheric Chemistry.* John Wiley & Sons, New York.

Fischhoff, B., S. Lichtenstein, P. Slovic, S.L. Derby, and R.L. Keeney. 1981. *Acceptable Risk.* Cambridge University Press, Cambridge.

Furedi, F. 1997. *Culture of Fear.* Cassell, London.

Gelin, F., I. Boodgers, A.A.M. Noordeloos, J.S.S. Damsté, P.G. Hatcher, and J.W. de Leeuw. 1996. Novel, resistant microalgal polyethers: an important sink of organic carbon in the marine environment? *Geochim. Cosmochim. Acta* 60: 1275–1280.

Gribble, G.W. 1996. *Naturally Occurring Organohalogen Compounds — A Comprehensive Survey.* Springer-Verlag, Berlin.

Haiber, G., G. Jacob, V. Niedan, G. Nkusi, and H.F. Schöler. 1996. The occurrence of trichloroacetic acid TCAA — indications of a natural production? *Chemosphere* 33: 839–849.

Hoefs, M.J.L., S. Schouten, J.W. de Leeuw, L.L. King, S.G. Wakeham, and J.S.S. Damsté. 1997. Ether lipids of planktonic archaea in the marine water column. *Appl. Environ. Microbiol.* 63: 3090–3095.

Jeffries, T.W. 1994. Biodegradation of lignin and hemicelluloses pp. 233–277. In *Biochemistry of Microbial Degradation* (Ed. C. Ratledge). Kluwer Academic Publishers, Dordrecht, The Netherlands.

King, G.M. 1988. Dehalogenation in marine sediments containing natural sources of halophenols. *Appl. Environ. Microbiol.* 54: 3079–3085.

Kobayashi, H. and B.E. Rittman 1982. Microbial removal of hazardous organic compounds. *Environ. Sci. Technol.* 16: 170A–183A.

König, G., M. Brunda, H. Puxbaum, C.N. Hewitt, S.C. Duckham, and J. Rudolph. 1995. Relative contribution of oxygenated hydrocarbons to the total biogenic VOC emissions of selected mid-European agricultural and natural plant species. *Atmos. Environ.* 29: 861–874.

Lewis, H.W. 1990. *Technological Risk.* W.H. Norton and Company, New York.

Mackay, D. 1991. *Multimedia Environmental Models. The Fugacity Approach.* Lewis Publishers, Chelsea, MI.

Monde, K., H. Satoh, M. Nakamura, M. Tamura, and M. Takasugi. 1998. Organochlorine compounds from a terrestrial plant: structures and origin of chlorinated orcinol derivatives from diseased bulbs of *Lilium maximowiczii. J. Nat. Prod.* 61: 913–921.

Munro, M.H.G., R.T. Luibrand, and J.W. Blunt. 1987. In *Bioorganic Marine Chemistry,* Vol. 1 (Ed. P.J. Scheuer), pp. 93–176. Springer-Verlag, Berlin.

Neidleman, S.L. and J. Geigert. 1986. *Biohalogenation: Principles, Basic Roles and Applications.* Ellis Horwood, Chichester, England.

Neilson, A.H. (Ed.). 1998. *PAHs and Related Compounds. Handbook of Enviromental Chemistry* (Ed. O. Hutzinger). Vol. 3I and 3J. Springer-Verlag, Berlin.

Neilson, A.H. and A.-S. Allard. 1986. Acetylene reduction (N_2-fixation) by *Enteobacteriaceae* isolated from industrial wastewaters and biological treatment systems. *Appl. Microbiol. Biotechnol.* 23: 67–74.

Neilson, A.H. and P.-Å. Hynning. 1998. PAHs: products of chemical and biochemical transformation of alicyclic precursors, pp. 224–269. In *PAHs and Related Compounds* (Ed. A.H. Neilson). *Handbook of Environmental Chemistry* (Ed. O. Hutzinger) Vol. 3, Part I. Springer-Verlag, Berlin.

Nightingale, P.D., G. Maline, and P.S. Liss. 1995. Production of chloroform and other low molecular weight halocarbons by some species of macroalgae. *Limnol. Oceanogr.* 40: 680–689.

Nowara, A., J. Burhenne, and M. Spiteller. 1997. Binding of fluoroquinolone carboxylic acid derivatives to clay minerals. *J. Agric. Food Chem.* 45: 1459–1463.

Ottoboni, M.A. 1984. *The Dose Makes the Poison*. Vincente Books, Berkeley, CA.

Peulvé, S., J.W. de Leeuw, M.-A. Sicre, M. Baas, and A. Saliot. 1996. Characterization of organic matter in sediment traps from the northwestern Mediterranean Sea. *Geochim. Cosmochim. Acta* 60: 1239–1259.

Rodricks, J.V. 1992. *Calculated Risks*. Cambridge University Press, Cambridge.

Roitman, J.N., N.E. Mahoney, W.J. Janisiewicz, and M. Benson. 1990. A new chlorinated phenylpyrrole antibiotic produced by the antifungal bacterium *Pseudomonas cepacia*. *J. Agric. Food Chem.* 38: 538–541.

Sabourin, C.L., J.C. Carpenter, T.K. Leib, and J.L. Spivack. 1996. Biodegradation of dimethylsilanediol in soils. *Appl. Environ. Microbiol.* 62: 4352–4360.

Schwarzenbach, R.P., P.M. Gschwend, and D.M. Imboden. 1993. *Environmental Organic Chemistry*. John Wiley, New York.

Simoneit, B.T.R. 1998. Biomarker PAHs in the Environment pp. 176–221. In *PAHs and Related Compounds* Vol. 3J (Ed. A.H. Neilson) Springer-Verlag, Berlin.

Smit, E., J.D. van Elsas, and J.A. van Veen. 1992. Risks associated with the application of genetically modified microorganisms in terrestrial ecosystems. *FEMS Microbiol. Rev.* 88: 263–278.

Strunz, G.M. 1984. Microbial chlorine-containing metabolites, pp. 674–773. In *CRC Handbook of Microbiology*, Vol. V, 2nd. ed. (Eds. A.I. Laskin and H.A. Lechavalier). CRC Press, Boca Raton, FL.

Svenson, A., L.-O. Kjeller, and C. Rappe. 1989. Enzyme-mediated formation of 2,3,7,8-tetrasubstituted chlorinated dibenzodioxins and dibenzofurans. *Environ. Sci. Technol.* 23: 900–902.

Teunissen, P.J.M. and J. A. Field. 1998. 2-Chloro-1,4-dimethoxybenzene as a novel catalytic cofactor for oxidation of anisyl alcohol by lignin peroxidase. *Appl. Environ. Microbiol.* 64: 830–835.

Tittlemeier, S.A., M. Simon, W.M. Jarman, J.E. Elliott, and R.J. Norstrom. 1999. Identification of a novel $C_{10}H_6N_2Br_4Cl_2$ heterocyclic compound in seabird eggs. A bioaccumulating marine natural product? *Environ. Sci. Technol.* 33: 26–33.

U.S. EPA. 1993. Interim Report on Data and Methods for Assessment of 2,3,7,8-Tetrachlorodibenzo-*p*-Dioxin Risks to Aquatic Life and Associated Wild Life. EPA/600/R-93/055. Office of Research and Development, Washington, D.C.

van Pée, K.-H. 1996. Biosynthesis of halogenated metabolites by bacteria. *Annu. Rev. Microbiol.* 50: 375–399.

Wannstedt, C., D. Rotella, and J.F. Siuda. 1990. Chloroperoxidase mediated halogenation of phenols. *Bull. Environ. Contam. Toxicol.* 44: 282–287.

White, G.F., N.J. Russell, and E.C. Tidswell. 1996. Bacterial scission of ether bonds. *Microbiol. Rev.* 60: 216–232.

2

Analysis

SYNOPSIS Chemical analysis is an integral part of all environmental investigations. Brief attention is directed to aspects of sampling in order to avoid interference from possible artifacts, to various procedures for extraction and concentration of analytes, and to the importance of cleanup procedures before identification and quantification. Procedures for the analysis of water, soil, and sediment samples and biota are outlined, and some important reactions for derivatizing functional groups are summarized. Methods for the identification of components of environmental samples are briefly summarized, and attention is directed to developments in ionization procedures for mass spectrometry (MS) analysis, high-performance liquid chromatography (HPLC) — MS interfaces, and the application of nondestructive procedures such as nuclear magnetic resonance (NMR) spectroscopy. The importance of access to authentic reference compounds is emphasized. Gas chromatographic (GC) procedures for quantification are briefly discussed, and attention is directed to recent developments including the use of chiral support phases and a range of detector systems. The use of supercritical fluids for the extraction of samples and as mobile phases for chromatography is briefly noted, and developments in liquid chromatography (LC) and capillary electrophoresis (CE) are discussed. The potential for application of immunologically based assays is noted, and attention is drawn to the potential for application of stable isotope fractionation. The range of analytes — including transformation products — which may be encountered in environmental samples is discussed including the problems in the analysis of commercial mixtures and complex effluents. There is an inherent indeterminacy in assessing the recoverability of analytes from naturally aged sediment samples as a result of dynamic associations between analytes and macromolecular components in environmental matrices.

Introduction

It is appropriate to begin this book with a discussion of chemical analysis which lies at the heart of almost all environmental investigations whether they are devoted to monitoring the distribution of a xenobiotic, evaluating its persistence and toxicity, or determining its partition among environmental matrices. Analytical support will be incorporated into all of these programs and its central role should be clearly appreciated at the planning stage; otherwise, the conclusions that are drawn from the results of the investigation may be equivocal. These comments should not, however, be interpreted as negating the fundamental contributions that analytical chemistry itself has made, not the least of which is in illuminating the role of metabolites and transformation products, and in revealing the global distribution of hitherto unsuspected compounds.

There have been revolutionary instrumental developments during the last 35 years or so, and these have completely altered the scope and possibilities of environmental research. The following examples may be given as illustration.

1. After early concern over the accumulation of DDT in biota, analysis of DDT and its metabolites was carried out by nitration and colorometric measurement after treatment with alkali. This was superseded by GC some 15 years later, and this development facilitated more rapid analysis, greater precision, and simultaneous unambiguous analysis of DDT metabolites such as DDE and DDD.

2. Although polychlorinated biphenyls (PCBs) were first synthesized in 1881 and introduced into industrial use as electrical insulators in the 1920s, they were detected in environmental samples only some 35 years later in 1966 by Jensen (Jensen 1972; Jensen et al. 1972). This discovery was facilitated by the development of the electron-capture (EC) detector in GC, and it also illustrates an early example of the application of MS to the tentative identification of environmental contaminants. Since that date, PCBs have become recognized as virtually universal contaminants of environmental samples from all parts of the world.

3. The interest of the synthetic organic chemist in stereospecific synthesis has resulted in the need for methods for the analysis of chirons. This has led to the development of both chiral reagents and of chiral supports for GC analysis, and the use of both HPLC and CE. The application of these methods to environmental samples is beginning to draw attention to the environmental input of enantiomers without established biological activity as well as to their possible persistence. The inherent complexity when the asymmetric atom carries an electron-attracting group such as carboxyl is illustrated by the example of phenoxypropionates (Buser and Müller 1997) is discussed in Section 4.2.2.

4. The identification of xenobiotics at the concentrations existing in environmental contaminants and of metabolites formed during laboratory experiments in biodegradation and transformation has been completely revolutionized by the availability of modern

instrumentation for structure determination. For example, infrared (IR) and NMR spectroscopy and MS have been widely used, and their sensitivity has been facilitated by Fourier transform (FT) techniques for signal processing.

5. The detection and quantification of extremely low concentrations of chlorinated dibenzo[1,4]dioxins in environmental samples would not have been possible without the development of high-resolution mass spectrometers. As a dividend, the instrumentation thereby developed has been adapted to other analytes and has significantly lowered both their limits of detection and the level of quantification.

6. The increased sensitivity of NMR instrumentation has made possible the application of ^{13}C NMR to study the pathways and kinetics of microbial reactions in cell suspensions. These procedures have provided direct evidence for the structure of intermediate metabolites without the need for their isolation. Determination of the structure of environmental analytes has also become possible using natural levels of ^{13}C. Techniques have also been developed to overcome the negative magnetogyric ratio of nuclei such as ^{15}N and ^{29}Si (Morris and Freeman 1979), and these procedures have been applied to a number of environmental problems that were previously inaccessible.

Today, few modern laboratories engaged in environmental studies lack facilities for carrying out analysis using GC, HPLC, CE, or MS, and many have access to high-field NMR instrumentation. Most of these instruments are coupled to systems for automatic injection of samples and to data systems for processing the output signals and reducing background interference. This makes possible the analysis of extremely small amounts of material, and this single advantage can hardly be overestimated since it is seldom possible to isolate environmental contaminants in sufficient quantity for identification by conventional chemical procedures.

At the outset, brief mention should possibly be made of sum parameters since these have traditionally dominated environmental analysis. Some of these, such as total organic carbon (TOC), dissolved organic carbon (DOC), particulate organic carbon (POC), and their nitrogen and phosphorus analogues have been included in conventional water quality criteria, and are still useful in wide-ranging monitoring, as in oceanography for example. With the upsurge of interest in halogenated xenobiotics, extensive effort has therefore been directed to devising comparable methods for organohalogen compounds. Analysis for total PCBs which depends on estimation of decachlorobiphenyl after chlorination (Lin and Hee 1985) or by dechlorination with LiAlH$_4$ (de Kok et al. 1981), or estimation of halogenated compounds in general by reduction with dispersed sodium followed by potentiometric analysis (Ware et al. 1988) have been used, but are now primarily of historic interest although they might find application to other chemically stable analytes. Methods have also been developed for analyzing total organic chlorine, bromine, or iodine by neutron activation analysis (Gether et al. 1979) or for nonspecific halogen by coulometry or potentiometric titration after combustion (Sjöström et al. 1985). None of these procedures merits inclusion in contemporary environmental hazard assessment programs since it has been consistently demonstrated that differences in the persistence, the toxicity, and the phase partitioning among, PCB congeners for example, and polycyclic aromatic hydrocarbons (PAHs) make the use of sum parameters quite inappropriate; such parameters should be replaced by analysis for the concentrations of specific compounds. Although sum parameters have been used quite extensively in some monitoring programs (Martinsen et al. 1988), they are really justifiable only when it can be established that a restricted range of structurally similar compounds is present.

The discussion presented here makes no attempt at being comprehensive, nor does it provide experimental details of standardized methodologies. This account is therefore not directed to the professional analyst, nor is it intended to serve as a handbook although some general comments on laboratory practice are made in Section 2.5. The needs of the professional are fulfilled by several complementary books (Keith 1988; 1992; Heftmann 1992) and by references to reviews cited in the text. A few examples may be given to illustrate the substantial limitations in the present account:

1. There is no systematic presentation of analytical procedures developed by the U.S. EPA, nor of those suggested by the OECD although attention is directed to a valuable critique of EPA procedures (Hites and Budde 1991).

2. No attempt is made to evaluate, for example, the range of columns which may advantageously be used in gas–liquid or high-resolution liquid chromatography, those used for solid-phase concentration of analytes from water samples, or commercially available columns for cleaning up samples.

3. There is no discussion of technical aspects of MS or NMR, nor details of the various procedures that are finding increasing application. Attention is directed to a review (Barceló 1992) that contains references to a number of books dealing with these specialized topics.

Experimental details may be found in reviews dealing with specific groups of compounds including polycyclic aromatic compounds (Bartle et al. 1981; Colmsjö 1998; Poster et al. 1998), chlorinated hydrocarbons (Hale and Greaves 1992), PCBs (Duinker et al. 1991; Lang 1992; Schmidt and Hesselberg 1992; Creaser et al. 1992), and pesticides in general (Barceló 1991). Increased interest in aromatic nitro compounds has led to development of procedures for their analysis (Lopez-Avila et al. 1991), and some of the material presented here is covered in a review dealing specifically with marine organic pollutants (Hühnerfuss and Kallenborn 1992). Attention is also directed to special issues of the *Journal of Chromatography* (1993: 642 and 643) that are devoted to a broad spectrum of procedures that have been applied to the analysis of a wide range of analytes in environmental samples.

In broad outline, the following steps will be incorporated into virtually all environmental investigations: (1) sampling from a predetermined site, (2) extraction and cleanup of the sample, (3) identification of the components, and (4) their quantification. The objective of the investigation will naturally determine the degree of sophistication of the procedures that will be applied. It should also be appreciated that unforeseeable difficulties may arise during the investigation, and it is important to resolve these at as early a stage as possible. Attempts will be made in this chapter to discuss all of the steps outlined above, although no effort has been made to provide a comprehensive account of any single aspect. Instead, this chapter will attempt to erect signposts along the wayside, and to provide an overview of analytical problems and a perspective on their application to specific problems in partitioning, estimating persistence, and evaluating toxicity. It is hoped, however, that some of the pitfalls awaiting the unwary have been revealed, and that at least some of the major unresolved issues have emerged. The available literature is both specialized and extensive, and no attempt has been made to provide either a complete or a systematic coverage.

2.1 Sampling

Sampling from laboratory experiments, for example, on partitioning, toxicity, or on degradation and transformation generally presents few problems: the concentrations of analytes are relatively high, only small volumes of samples are required, and these systems are homogeneous. Samples may be frozen after collection if they are not analyzed immediately. On the other hand, the magnitude of the problem with field material is substantially greater since the systems are less homogeneous and the concentrations of the analytes may be low. Although no attempt at a comprehensive treatment will be made here, it cannot be emphasized too strongly that the quality of an investigation is critically determined by the care devoted to the selection of samples, their collection, and their preservation before analysis. The relative costs of the sampling and the subsequent analyses should be kept in proportion. The following issues merit brief notice.

1. Attention should be paid not only to the cleanliness of the sample containers but also to their composition; use of plastic containers is attractive due to their physical inertness and robustness, but their

use may inevitably contaminate the sample with plasticizers such as phthalate esters. Glass containers are therefore generally preferred for samples of water and sediment. Problems arising from sorption of analytes to glass surfaces may exist, and may be significant for particular groups of compounds; in the case of sterol analysis, this has been circumvented by silylation of glassware to which the analyte is exposed (Fenimore et al. 1976). Samples of biota should be wrapped in aluminum foil rather than any kind of plastic material.

2. Water samples present fewer problems than other matrices, although if low concentrations of xenobiotics are to be analyzed, it may be necessary to use high-volume samplers and process the samples on board the ship. One issue which should be recognized — even if it cannot be resolved — is that presented by samples having particulate material, and this is discussed in greater detail in Sections 2.2.4 and 2.2.5. Freezing of such samples may bring about alterations in their chemical composition and hence in their toxicity (Schuytema et al. 1989).

3. Sediment sampling may present a number of problems including the patchiness of the area being investigated (Downing and Rath 1988; Brandl et al. 1993), and the difficulty of obtaining a truly representative subsample from a possibly heterogeneous bulk sample. These problems may be particularly severe where monitoring is directed to providing a historical record of deposition: ambiguities may arise (Sanders et al. 1992) where the sediment surface is nonuniform or where diffusion and bioturbation are significant. There is probably no single optimal procedure for preserving sediment samples after collection; for chemical analysis, although freezing may eliminate or at least minimize the possibility of chemical reactions occurring after sampling, analytes may be released from sediment–xenobiotic associations during subsequent thawing. Preservation at 4°C will generally minimize microbial alterations of the analyte, but the addition of inhibitors of microbial activity such as mercuric chloride or sodium azide may introduce problems during subsequent analysis (Section 2.2.4) and thus should be avoided.

4. Particular care should be exercised in the collection of samples of fish and attempts should be made to minimize the inevitable stress of capture; this is particularly critical for analysis of compounds such as steroid hormones or for assays of enzymatic activity (Munkittrick et al. 1991; Hontela et al. 1992).

5. There has been increased interest in samples from remote areas such as the Arctic and Antarctic. There areas are, however, increasingly visited both by tourists and by scientific expeditions so that

particular care should be exercised in the conclusions drawn from the results of such sampling. Atmospheric input may not be the only source of xenobiotics, and truly undisturbed localities are becoming increasingly rare throughout the world.

6. Attention is drawn briefly to a few examples of procedures for passive sampling:

   a. Xenobiotics may be concentrated from soil by passive diffusion into solid adsorbents or into organic solvents in dialysis membranes (Zabik et al. 1992). Desorption of the solid supports such as C-18 bonded silica or XAD resins is then carried out for analysis of the xenobiotics.

   b. Xenobiotics have been concentrated from aquatic systems into dialysis membranes containing solvents such as hexane or triolein, and this procedure has been suggested as simulating uptake by biota (Södergren 1987; Huckins et al. 1990a; Meadows et al. 1998). Although these procedures are valuable for demonstrating the presence of compounds, they are of restricted applicability to hazard assessment for which the concentrations of the analyte are clearly needed.

   c. Xenobiotics may be concentrated from the air using a passive sampler that uses a thin film of polythene containing triolein, and this has been evaluated for monitoring of PCBs and gave good agreement with conventional high-volume sampling methods (Ockenden et al. 1998).

   d. Application of laser-desorption from membrane introduction MS is noted in Section 2.4.1.2 for the analysis of PAHs in water samples.

2.2 Extraction and Cleanup

After collection and transfer to the laboratory, the samples have to be analyzed. Two factors combine to magnify the problem: (1) the low concentration of the analyte which generally necessitates concentration and (2) interference from other compounds that may occur in much greater concentration and which necessitates purification or cleanup procedures to reduce their concentration. These factors have resulted in increasing demands upon the skill of the analyst, leading to the need for the application of sophisticated procedures for the pretreatment of samples before identification and quantification. The analysis will therefore generally encompass at least the following steps: extraction, concentration, and cleanup with or without derivatization, identification, and quantification. A range of matrices including samples of water, sediment, and biota ranging from microbial cultures to higher organisms, and samples of atmospheric deposition may be submitted for analysis. Each of these presents its own special problem, so that procedures for their extraction and cleanup will vary considerably. An attempt will be made to present only an outline of the methods that have been widely used and to summarize some of the problems that may be encountered.

2.2.1   Solvents and Reagents

Organic solvents are used at virtually all stages of the analytical procedures — for extraction, during cleanup, and in identification and quantification. It is therefore appropriate to preface this section with general comments on some of the important issues that should be addressed. The purity of the solvents is of cardinal importance: the initial extracts from the samples which may have volumes of tens of milliliters are generally concentrated to volumes of the order of hundreds of microliters, and volumes in the microliter range are then used for the final analysis. Concentration procedures should be designed to minimize the loss of compounds with low boiling points and appreciable volatility; for this reason, solvents such as dichloromethane, pentane, or diethyl ether or *t*-butyl methyl ether are frequently used, while a less volatile solvent such as isooctane may be added as a "keeper" to retain a liquid phase. A gas atmosphere of N_2 is generally maintained during concentration to minimize oxidation of sensitive components in the sample. There is, unfortunately, no single extraction solvent which is optimal for all analytes, and the choice depends both upon the nature of the matrix and upon the structure of the analyte. Solvents with limited capacity for dissolving water and ranging in polarity from hexane and dichloromethane through toluene or benzene to diethyl and *t*-butyl methyl ethers and ethyl acetate may be usefully exploited, although no single one of these is invariably optimal. For some purposes, such as extraction of wet sediment samples, water-miscible solvents such as methanol, propan-2-ol, tetrahydrofuran, acetonitrile, or dimethylformamide may be applicable; the analyte is then subsequently partitioned into a water-immiscible phase. The major drawback to the use of solvents such as ethyl acetate, and to a lesser extent the dialkyl ethers for extracting aqueous solutions, is the high solubility of water in these solvents: such extracts should therefore be dried, for example, with anhydrous sodium sulfate or preferably by azeiotropic distillation, before attempting derivatization with water-sensitive reagents such as acid anhydrides, acid chlorides, or isocyanates.

It is pointed out later (Sections 3.2.3 and 4.6.3) that substantial amounts of xenobiotics become associated with components of the soil and sediment with time, and therefore only a fraction of the total is recoverable by solvent extraction.

Most common solvents are now available commercially in a high state of purity; these have extremely low residue levels and they can be purchased as glass-distilled quality to obviate this step in the laboratory. Care should be exercised in their storage after purchase to prevent contamination with volatile laboratory chemicals, and especially with standards used for analysis. Particular caution should be exercised in the use of solvents which contain stabilizers: for example, 4-*t*-butyl-2-methylphenol in diethyl ether, or cyclohexene in dichloromethane. Dichloromethane is recommended for many extraction procedures because of its volatility and the simplification in extraction resulting from its greater density than water; reactions of the analyte with the cyclohexene inhibitor may, however, result in the production of artifacts such as halogenated cyclohexanes and cyclohexanols (Campbell et al. 1987; Fayad 1988), and the presence of iodocyclohexanol has been associated with low recoveries of phenolic compounds using EPA standard procedures (Chen et al. 1991). Attention is also drawn to the reaction of methanol with carbonyl compounds that may result in the formation of acetals or ketals (Hatano et al. 1988) or of esters from chlorinated acetic acids (Xie et al. 1993). Care should therefore be exercised in the analysis of such compounds particularly by HPLC for which methanol is a widely used mobile phase. For the extraction and analysis of compounds containing reactive carbonyl groups — especially aldehydes — their chemical reactivity clearly precludes the use of solvents with active methylene groups such as acetone, acetonitrile, dimethylsulfoxide, or nitromethane.

As a result of the increase in the number of laboratories carrying out environmental analysis and the corresponding increase in the volumes that are used primarily for extraction and have to be disposed of, there has been increasing use of solid-phase extraction (SPE) methods including solid-phase microextraction (SPME) procedures using fiber-coated poly(dimethylsiloxane), methyl silicone, or polyacrylate. These procedures are discussed in Sections 2.2.4 and 2.2.7.

Although most widely used reagents are available in an acceptable degree of purity, for some analytical applications interfering impurities must be removed. For solids, this can usually be accomplished by successive recrystallization from a suitable solvent. Solid phases for open chromatography may have been activated by ignition so that organic impurities have been removed. It is worth pointing out, however, that analytically pure Na_2SO_4 that is widely used as a drying agent may contain interfering organic residues that should also be removed by ignition.

2.2.2 Cleanup Procedures

Extracts from environmental samples are seldom sufficiently free of contaminants that they can be analyzed directly, since the high-resolution columns used for GC or HPLC analysis must not be overloaded with compounds other than the analyte. Pretreatment of the sample before GC, GC-MS, or HPLC analysis is therefore often a critical step in the analysis of extracts from complex matrices, and this is particularly important when the concentrations of the interfering compounds greatly exceed that of the analyte. These compounds may then seriously compromise the analytical procedure. The cleanup procedures that will be used subsequent to the initial extraction depend both on the nature of the matrix and on that of the analyte — in particular the sensitivity of the analyte to chemical reagents such as strong acids (Bernal et al. 1992) or bases (Vassilaros et al. 1982) or oxidizing agents including molecular oxygen (Wasilchuk et al. 1992). All of these may bring about significant chemical changes in the analytes and should therefore be avoided unless the stability of the analyte toward them has been unequivocally established. For example, although the quantification of chlorinated dibenzo[1,4]dioxins and some of the PCB congeners is facilitated by taking advantage of the great stability of these compounds to chemical reagents such as sulfuric acid for removing interfering compounds, comparable cleanup procedures are clearly not applicable to more sensitive compounds. Less drastic cleanup procedures should therefore be developed, and these should be adapted to the chemical structure and to the reactivity of the analyte.

The validation of analytical methods for quantification (Section 2.4.2) should be made by spiking relevant environmental samples that may often contain seriously interfering compounds. Cleanup of the samples will reveal the continued existence of such complications, although the question of accessibility of analytes in aged samples is an entirely different issue that cannot be addressed by experiment. Assays of samples containing authentic mixtures are obligatory, but they are not sufficient.

A number of different cleanup procedures for environmental samples have been effectively employed and the following may be used as illustration. A valuable summary of their application to samples of various matrices containing PAHs has been given (Colmsjö 1998).

1. Advantage may profitably be taken of classical chromatographic separation methods using open columns of silica gel, alumina, florisil, or activated charcoal although it should be realized that these do not generally provide complete separation of structurally similar compounds. Gel-permeation chromatography has been effectively used for separating classes of compounds such as toxaphene (polychlorinated bornanes and bornenes) and PCBs (Brumley et al. 1993), and has increasingly been used for the removal of interfering lipid material; this is discussed further below.

2. Preparative HPLC is an invaluable procedure that makes available substantial quantities of the analyte in a high state of purity, and may effectively be incorporated into the final stages of purification, for example, of samples that will be subjected to NMR analysis.

3. Countercurrent chromatography (CCC) has been extensively used for the isolation of natural products (Foucault 1991), and has many features applicable to the fractionation of environmental samples before analysis. A good example is the resolution of intractable

mixtures of polyphenolic compounds in plant extracts (Okuda et al. 1990), and this procedure might be effective in the analysis of organic constituents of soils and organic-rich sediments. One additional advantage is the elimination of ambiguities resulting from possible alteration in the structure of the analyte during GC separation (Nitz et al. 1992). Many features of CCC are identical to those involved in HPLC, and the main limitation probably lies in the availability of suitable immiscible solvent systems although ternary systems such as hexane/acetonitrile/*t*-butyl methyl ether or even quaternary systems such as hexane/ethyl acetate/methanol/water have been used effectively.

4. Separation of the desired analyte may be achieved by taking advantage of the formation of specific complexes. Examples include those between Ag^+ and $C=C$ bonds for separating alkenes or alkenoic acids from their saturated analogues (Christie 1989), and of clathrate complexes between urea and normal saturated aliphatic hydrocarbons for separating these from branched-chain and other hydrocarbons (Richardson and Miller 1982).

5. Specific derivatives may be used for removing interfering compounds, and ideally the desired analyte should be recoverable, for example, by hydrolysis of semicarbazones formed between aldehydes or ketones and the Girard reagents (Wheeler 1968). This reagent has been employed for the effective removal of high concentrations of interfering chlorovanillins during the analysis of low concentrations of chlorocatechols (Neilson et al. 1988).

Two serious problems frequently arise: the presence of elementary sulfur in sediments samples and the presence of lipids in samples of both sediments and biota. Fortunately, satisfactory solutions have been found for both problems.

1. Anaerobic sediments frequently contain large amounts of elementary sulfur (S_8) which may cause serious interference during GC analysis. The traditional method that is applicable to chemically stable analytes is treatment with activated copper, but an elegant and milder procedure in which elementary sulfur is converted into water-soluble thiosulfate by reaction with tetrabutyl ammonium sulfite has been developed (Jensen et al. 1977). This method cannot, however, be applied to samples containing sensitive compounds such as phenols; in such cases, the sulfur may be removed from the phenol acetates by silica gel chromatography and elution with cyclohexane (Allard et al. 1991).

2. Samples of biota — and to a lesser extent sediment — will almost invariably contain substantial amounts of lipid material which is often dominated by relatively high-molecular-weight triglycerides

that may seriously interfere with analysis of the desired analyte. The choice of procedure for the removal of lipids depends on a number of factors including the nature of the analyte, the amounts of lipid that have to be removed, and the number of samples that have to be processed. For highly stable compounds such as PAHs and related compounds, samples have been submitted to alkaline hydrolysis before further cleanup (Vassilaros et al. 1982). This is, of course, not possible with more sensitive analytes, and a number of gentler procedures have been used for the removal of lipid material. These procedures generally take advantage of the differences either in the molecular weight of the analytes and the lipids or in the sorbent properties of the analytes and the lipids. Some of the most widely used procedures include the following:

- Gel-permeation chromatography (Stalling et al. 1972) which is essentially a fractionation dependent on differences in molecular weight: Instrumental improvements to avoid the problem of tailing triglyceride peaks have been described, and may be particularly valuable for application to online systems (Grob and Kälin 1991).

- Dialysis using membranes of polyethylene and methylcyclopentane as solvent (Huckins et al. 1990b): This general procedure has been examined for samples containing 50 g of lipid or more and has proved an attractive simplification of conventional procedures (Meadows et al. 1993).

- Online columns of silica gel provided care is taken not to overload the columns (Grob et al. 1991).

- Use of HPLC and columns of silica gel with gradient elution after saponification of the glycerides (Hennion et al. 1983) or of HPLC using a Nucleosil 100-5 column (Petrick et al. 1988).

- Use of supercritical CO_2 or CO_2/methanol and silica gel or aluminum oxide (France et al. 1991): The use of supercritical fluids is discussed in a more general context in Section 2.2.5.

2.2.3   Toxicity-Directed Fractionation

The foregoing fractionation procedures are essentially empirical, and the individual fractions must be analyzed and their components identified and quantified. Clearly, this could be a substantial undertaking, and the results might not necessarily justify the expenditure of effort. Many investigations are directed to determining the toxic components in the samples. By analogy with the use of biological assays in isolation procedures for biologically active substances such as vitamins and steroid hormones, efforts have therefore been made to direct the fractionation of environmental extracts on the basis of the results of assays for toxicity.

1. The Ames test using mutant strains of *Salmonella typhimurium* has been used most widely both due to the ease and rapidity with which it may be carried out, and also because concentrates may be assayed as solutions in, for example, dimethyl sulfoxide so that problems with solubility are readily overcome. This procedure may be illustrated by its use in an investigation of marine sediments (Fernández et al. 1992), or combined with assays for aryl hydrocarbon hydroxylase activity in a study of the organic components in automobile exhausts (Alsberg et al. 1985). A wider range of toxicity assays to marine organisms has been used to direct the fractionation of high-molecular-weight components of bleachery effluents, although a direct correlation of toxicity with molecular weight was apparently not possible (Higashi et al. 1992).

2. EROD induction in rainbow trout (Section 7.6.1) was used to separate active compounds from commercial samples of the lampricide 3-trifluoromethyl-4-nitrophenol, and the products were identified as isomeric 2-trifluoro-3-nitro-7-(8)chlorodibenzo[1,4]dioxins that were, however, relatively weak inducers of MFO activity (Hewitt et al. 1998).

3. Estrogenic activity assayed by the yeast assay (Section 7.3.5) has been used to fractionate samples of domestic sewage treatment effluent (Desbrow et al. 1998). It was shown that the major estrogenic activity could be accounted for by the presence of 17β-estradiol (2.7 to 48 $\pm$ 16 ng/l) and estrone (1.4 to 76 $\pm$ 10 ng/l), whereas concentrations of 17α-ethynylestrone were much lower (0.2 to 7.0 $\pm$ 3.7 ng/l). Assays for the adverse effect of these compounds on fish is discussed in Section 7.3.5.

4. Toxicity assays using the oligochaete *Lumbriculus variegatus* in which comparison was made between cultures kept in the dark and those irradiated with ultraviolet (UV) light were used in the fractionation of pore water from a highly contaminated site to establish the presence of phototoxic compounds (Kosian et al. 1998a,b). The active components comprised a structurally heterogeneous group of aromatic compounds containing substituted quinones, stilbenes, and thia- and azaarenes.

Other types of parameters could also be used to direct the fractionation and separate the compounds of primary interest: for example, use may be made of reversed-phase HPLC to separate and isolate compounds with P_{ow} values indicative of significant bioconcentration potential (Hynning 1996).

2.2.4 Specific Matrices: Water Samples

For homogeneous water samples, there are generally few serious problems and several procedures for concentration may be used:

1. Solvent extraction of the analytes may be carried out with suitable solvents using either batch or continuous procedures.

2. For compounds with appreciable vapor pressure, purging systems may be used followed by collection of the compound on Tenax columns from which the analyte may subsequently be thermally desorbed or — alternatively, for less volatile compounds — eluted with methanol. A procedure has been developed in which the water or soil samples are purged with N_2 and directed onto a dimethylpolysiloxane membrane from which diffusing molecules are directly inserted into the ion source of the mass spectrometer (Kostinainen et al. 1998). The procedure was evaluated for a range of halogenated alkanes and alkanes, aliphatic esters and ethers, and monocyclic aromatic hydrocarbons. Although the procedure is sensitive, since no chromatographic separation is involved, it has only limited selectivity. Further comments on purge-and-trap methods for gas-phase samples are given in Section 2.2.7, and on laser desorption membrane MS in Section 2.4.1.2.

3. For less volatile compounds, closed-loop stripping has been used in which the analytes are collected on a column of activated carbon from which they are then eluted with carbon disulfide. Attention should especially be drawn to an evaluation of this system for a range of analytes (Coleman et al. 1983) and in particular to the conclusion that for quantification, recovery efficiencies must be evaluated using matrices spiked with the analyte of specific interest.

4. The analyte may be sorbed directly from the aqueous phase, for example, on XAD columns (Dietrich et al. 1988) which are then eluted with a suitable solvent such as methanol, acetonitrile, dichloromethane, or diethyl ether.

5. A range of bonded-phase silica adsorbents were evaluated for analysis of priority pollutants (Chladek and Marano 1984) and have been used increasingly. Solvents for elution of the analytes have included dichloromethane, acetonitrile, ethyl acetate, hexane, and tetrahydrofuran (Chladek and Marano 1984), and ethyl acetate or benzene (Junk and Richard 1988). Use of disks incorporating C18-bonded silica facilitates the processing of large sample volumes (Barceló et al. 1993). Alternatively, advantage may be taken of ion-exchange resins followed by elution with salt solutions or phosphate buffer (Arjmand et al. 1988) and a combination of reversed-phase C18 and ion-exchange columns has been incorporated into an automated system for the analysis of triazine metabolites in samples of soil and water (Mills and Thurman 1992). A wide range of ion-exchange columns is commercially available, and all of these systems have the advantage of using only small volumes of organic solvents. This aspect has achieved increasing prominence in attempts to diminish the use in analytical laboratories of solvents,

particularly chlorinated ones. Cartridges filled with graphitized carbon black have been used for the concentration of low concentrations of triazine, anilides, and phenoxyacids in water (Bucheli et al. 1997), metabolites from nonylphenol ethoxylates in wastewater samples (Di Corcia et al. 1998), and organotin compounds from water and sediment samples (Arnold et al. 1998).

6. SPME has been used for a range of analytes. Two procedures have been used, one in which sampling takes place in the aquatic phase, and the other in the gas phase. In addition, derivatization has been carried our either on preformed fibers before exposure to the analyte or by gas-phase exposure after sorption of the analyte. It is important to appreciate that in the aqueous phase, both adsorption and absorption may occur, and it has been shown for PCBs that the former is more important (Yang et al. 1998). In addition, true equilibrium may not be reached during the relatively short exposure times, so that reproducibility must be assured at the shortened times employed. Diverse illustrations are given in the following examples; application to evaluating the concentration of freely dissolved as opposed to total concentrations of analytes is given in Section 3.2.3.

   a. A microsystem using optical fibers coated with methyl silicone which are then used for direct injection into GC-MS systems has been suggested for concentrating and analyzing volatile aromatic hydrocarbons (Arthur et al. 1992). Fibers coated with polydimethylsiloxane have been used for sorption of PAHs containing up to five rings, and the fibers were analyzed after direct coupling to a cyclodextrin-modified capillary electrophoresis system (Nguyen and Luong 1997).

   b. PCBs have been analyzed in water samples by exposure of a polydimethylsiloxane fiber both to the aqueous phase and in the headspace (Llompart et al. 1998). Although equilibrium was not achieved, the use of headspace sorption offered greater sensitivity, and the procedure was considered as effective as liquid/liquid extraction.

   c. SPME fibers coated with polyacrylate were used for the extraction from seawater of 2,4,6-trinitrotoluene (TNT), monoaminodinitrotoluenes, and hexahydro-1,3,5-trinitro-1,3,5-triazine (RDX), and although equilibrium was not achieved, a 10-min extraction was deemed to be adequate (Barshick and Griest 1998). It was found by comparison with solid-phase extraction methods that, for unknown reasons, concentrations assayed by SPME were generally higher both for both TNT and RDX.

   d. Several different fibers were evaluated for the analysis of carcinogenic aromatic amines (Wu and Huang 1999), and the important issues of desorption discussed.

   e. A number of acidic herbicides including various phenoxyacetic
     and phenoxypropionic acids, and 2-methoxy-3,6-dichlorobenzo-
     ic acid were concentrated optimally on poly(dimethylsiloxane)
     fibers in acid media containing NaCl (Lee et al. 1998): the com-
     pounds were methylated on the fibers by gas-phase diaz-
     omethane and analyzed by GC/MS.

   f. Headspace sorption on dimethylsiloxane/divinyl benzene fibers
     was used for analysis of aniline, 2-chloroaniline, and methylan-
     lines in biological fluids after adjustment of the pH to 13 and
     addition of NaCl (DeBruin et al. 1998).

Other extraction procedures have been developed. For example, a system
for volatile compounds is noted in Section 2.2.7 and one for extraction of aromatic amines from rainwater used sorption to rayon bearing covalently
bound trisulfo-copper phthalocyanine followed by elution with an ammoni-
acal solution (Wu et al. 1995).

Samples with particulate matter may present quite serious problems, and
it may be desirable to remove particles, for example, by centrifugation, and
examine this fraction by procedures applicable to solid phases which are discussed in Section 2.2.5. Tangential-flow high-volume filtration systems have
been used for analysis of particulate fractions (>0.45 µm) where the analytes
occur in only low concentration (Broman et al. 1991). Attention has already
been drawn to artifacts resulting from reactions with cyclohexene added as
an inhibitor to dichloromethane. It has also been suggested that under basic
conditions, Mn^{2+} in water samples may be oxidized to Mn(III or IV) which in
turn oxidized phenolic constituents to quinones (Chen et al. 1991). Serious
problems may arise if mercuric chloride is added as a preservative after collection of the samples (Foreman et al. 1992) since this has appreciable solubility in many organic solvents, and its use should therefore be avoided.

Water samples may contain appreciable amounts of particulate matter, dissolved organic carbon, or colloidal material and all of these may form associations with the analytes and affect their recoverability. For these reasons,
discrepancies may arise between the concentrations of analytes determined
by liquid extraction and those obtained by sorption on polyurethane or XAD
resins (Gómez-Belinchón et al. 1988). Empirical procedures have been developed (Landrum et al. 1984) for fractionating samples to assess the relative
contribution of the associations of xenobiotics with the various organic components, while sediment traps for collection of particulate matter have been
extensively used in investigations in the Baltic Sea where appreciably turbid
water may be present (Näf et al. 1992).

2.2.5　Specific Matrices: Sediments and Soils

The analysis of solid matrices such as sediments and soils almost invariably
presents more serious problems than those encountered with water samples,

and incorporation of cleanup steps or fractionation, for example, by silica gel or alumina chromatography will generally be needed to remove high concentrations of interfering compounds that are almost always present in such samples.

Conventional Procedures

Before extraction, soil and sediment samples may be dried, for example, by freeze-drying — provided that volatile compounds are not to be analyzed — or by mixing with anhydrous sodium sulfate and extraction in a Soxhlet apparatus. It should, however, be noted that it has frequently been found advantageous to add low concentrations of water, and this is consistent with the finding that addition of water to dry soils inhibits sorption of PAHs (Karimi-Lotfabad et al. 1996). If wet samples are to be analyzed directly, acetonitrile, propan-2-ol, or ethanol may be employed first, and these may be valuable in promoting the chemical accessibility of substances sorbed onto components of the matrices; the analyte may then be extracted into water-immiscible solvents and the water phase discarded. Alternatively, if the analyte is sufficiently soluble in, for example, benzene, the water may be removed azeotropically in a Dean & Stark apparatus and the analyte then extracted with the dry solvent. Analytes may, however, be entrapped in micropores in the soil matrix so that, for example, recovery of even the volatile 1,2-dibromoethane required extraction with methanol at 75°C for 24 h (Sawhney et al. 1988).

If a large number of samples are to be processed, Soxhlet-based procedures may be unduly cumbersome. As an alternative, attention has been given to the use of sonication in a variety of solvents although this procedure should clearly be carried out with care when the analytes are sensitive to oxidation which may be exacerbated by cavitation. Sonication has been effectively used for recovering neutral priority organic pollutants from a number of matrices (Ozretich and Schroeder 1986), and for four nitrated explosives from soil (Jenkins and Grant 1987). It should be recognized, however, that the recovery efficiencies for some compounds may be quite low compared with alternative procedures.

Supercritical Fluids

Increased effort has been directed to the application of supercritical fluids for extraction of environmental samples. The range of fluids that has been examined, details of the extraction and collection procedures, and a comparison with conventional procedures for extraction have been given (Böwadt and Hawthorne 1995). An extensive review that covers both supercritical fluid chromatography and supercritical fluid extraction has been given (Chester et al. 1998) and includes 533 references. The following is therefore a highly selective and compromised summary.

1. Both static and dynamic extraction procedures have been developed, and supercritical fluids, such as methanol for the extraction of bound pesticide resides (Capriel et al. 1986) or of carbon dioxide/methanol for the cleanup of organochlorine pesticides in fatty tissue (France et al. 1991), have been used. In the latter study, attention was drawn to the importance of using carbon dioxide of suitable purity.

2. It has emerged from several investigations that only poor recovery of some analytes can be achieved using carbon dioxide without the addition of polar additives — including such apparently drastic reagents as formic acid for the extraction of resin and fatty acids from sediment samples (Lee and Peart 1992). Somewhat conflicting results on the recoverability of a number of analytes have been reported, so that caution should be exercised against uncritical application of this attractive procedure. For example, whereas one extensive study in which a total of 88 pesticides were analyzed in spiked samples of sand showed considerable variation in recoverability and the recovery of polycyclic hydrocarbons in certified samples was generally poor (Lopez-Avila et al. 1990), another in which CO_2 modified with 3% methanol was used showed good recoveries that were comparable with those from conventional procedures using sonication treatment or Soxhlet extraction (Snyder et al. 1992). A detailed analysis of the effectiveness of supercritical fluids using different matrices and different fluids revealed some important factors (Benner 1998):

   • For urban dust, the most effective fluid was chlorodifluoromethane;

   • For marine sediments, substantial differences were observed between different samples and CO_2 — wet-dichloromethane was uniformly effective for only one the samples;

   • For diesel particulate matter, none of the fluids examined was effective.

These results therefore suggest a careful evaluation of the pros and cons of supercritical fluids. Another study that included PCBs showed good agreement between Soxhlet extraction and dynamic supercritical fluid extraction of soil samples with the added advantage that no cleanup step was needed with the latter (van der Velde et al. 1992). A combined extraction using CO_2 and acetylation with acetic anhydride and triethylamine has been used for the analysis of chlorophenolic compounds in soil (Lee et al. 1992) and air-dried sediment samples (Lee et al. 1993). This is both rapid and gives results comparable to those using conventional methods of steam distillation.

It therefore appears that some procedural details remain incompletely resolved and that, as with all extraction procedures, due attention should be directed both to the nature of the analyte and to that of the matrix that is being analyzed.

Use of acidic CO_2 is clearly undesirable for the extraction of basic compounds, and N_2O has been used effectively for a number of amines (Mathiasson et al. 1989; Ashraf-Khorassani and Taylor 1990). Similarly, CO_2 was ineffective for extraction of 1-nitropyrene from diesel exhaust particulate matter although this could be accomplished effectively using the freon $CHClF_2$ (Paschke et al. 1992). In conclusion, it seems safe to state the obvious: no single supercritical fluid is likely to be optimal for the extraction of structurally diverse analytes.

The combination of supercritical fluid chromatography and supercritical fluid extraction with standard GC and LC techniques offers an attractive extension to conventional procedures (Greibrokk 1992), and this will probably receive increased attention in the analysis of samples containing a large number of components.

Association — The Recoverability of "Bound" Analytes

In light of increasing evidence for the existence of "bound" residues, which is discussed more fully in Sections 3.2.2 and 3.2.3, relatively drastic procedures have been used for the analysis of both soil and sediment samples. Difficulties experienced in the recovery of even such apparently unreactive compounds such as γ-hexachlorocyclohexane (lindane) have led to the suggestion of a procedure using BF_3 in methanol (Westcott and Worobey 1985). Extraction of glyphosate (*N*-phosphonomethylglycine) from contaminated soils was facilitated by the use of alkaline extraction (Miles and Moyne 1988), which appears to be a viable procedure for soils up to 56 days after spiking (Aubin and Smith 1992). These observations are relevant to the issue of "aging" that is particularly important for agrochemicals applied to the terrestrial environment. The principles are, however, equally relevant to aquatic sediments containing xenobiotics that have been deposited over many years. For example, application of chemical treatment has been used to make an empirical division between "free" (extractable with simple solvents) and "bound" (accessible only after chemical treatment such as methanolic alkali) concentrations of chloroguaiacols in contaminated sediments (Remberger et al. 1988). This distinction is, of course, both relative and pragmatic. In addition, it has emerged that there were clear differences in the relative recoverability of chloroguaiacols and chlorocatechols, and this was plausibly attributed to different types of association between the analytes and the sediment phase.

It cannot therefore be emphasized sufficiently that there is no such thing as the absolute concentration of a given analyte in an environmental sample.

Significant alterations in recoverability occur after deposition (aging): measured concentrations are empirical and subject to an inherent indeterminacy.

2.2.6 Specific Matrices: Biota

Procedures for the extraction of aquatic biota — particularly fish — have been extensively developed, and generally involve grinding with anhydrous sodium sulfate followed by extraction with a suitable solvent. An attractive alternative (Birkholz et al. 1988) has used homogenization of fish tissue with solid carbon dioxide; after removal of CO_2 by sublimation at –20°C, a finely divided powder is obtained and this may either be extracted directly or freeze-dried and subjected to Soxhlet extraction. As is the case for samples of soil and sediment, supercritical fluid extraction has also been used for samples of biota and the results have been compared with those using standard procedures; supercritical fluid extraction has found substantial application in the analysis of food products. After extraction, advantage may be taken of any of the procedures noted in Section 2.2.2 for removing interfering contaminants such as lipids which will generally be present in high concentration. For highly stable compounds such as chlorinated dibenzo[1,4]dioxins, quite drastic cleanup stages including, for example, treatment with sulfuric acid, may be incorporated. In contrast to this, and to the situation with bound residues in soil or sediment samples, analytes in samples of biota may be present as conjugates — generally as metabolites of the compounds to which biota were initially exposed. Many of these are sensitive to acidic or basic reagents so that mild procedures for their analysis are obligatory. Further details of some of the reactions involved in the metabolism of xenobiotics by higher organisms are given in Section 7.5; illustrative examples are the glycoside and sulfate conjugates of phenols which may be cleaved with acid or enzymatically, or glutathione conjugates which may be reduced with Raney nickel. Even neutral compounds may be tightly associated with lipid components of fish, and the extractability of both PAHs and related compounds (Vassilaros et al. 1982) and PCBs (de Boer 1988) is significantly improved by preliminary alkali treatment. Problems of accessibility may also be encountered during the analysis of cultures containing microbial cells which may effectively bind the analyte; this may be especially severe with Gram-positive bacteria for which treatment with acetonitrile combined with sonication has been found to be highly effective (Allard et al. 1985). It has been shown (Dietrich et al. 1995) that appreciable amounts of ^{14}C-labeled PCB congeners were associated with the hyphae of *Phanerochaete chrysosporium*. For 4,4'-dichlorobiphenyl that was mineralized partly to CO_2, this amounted to ~40% at the beginning of the experiments but decreased as biodegradation and biotransformation proceeded; for 3,3',4,4'-tetrachloro- and 2,2',4,4',5,5'-hexachlorobiphenyls that were resistant to biodegradation, however, over 60% was associated with the hyphae. Such associations clearly should be taken into consideration in interpreting the results of experiments on biodegradation and biotransformation.

2.2.7 Volatile Analytes

These may occur in water, sediment, or air samples, and several procedures have been used for analytes including halogenated aliphatic compounds, low-molecular-weight aliphatic ethers such as *t*-butyl methyl ether, and monocyclic aromatic hydrocarbons such as benzene, toluene, ethylbenzene, and xylenes (BTEX).

Many of the procedures have already been outlined in Section 2.2.4, so that only summarizing comments are justified.

1. Solvent extraction of water and sediment samples has already been noted in Section 2.2.4.
2. For gas-phase samples, sorption on, for example, Tenax or Carbo-pack, followed by desorption either thermally or with a suitable solvent has been used.
3. Static headspace analysis has been used for liquid samples in closed containers after equilibration between the liquid and gas phases.
4. Purge-and-trap systems for water and sediment samples followed by sorption and desorption. Both open- and closed-loop stripping have been used.
5. Solid-phase extraction has been noted in Section 2.2.4, and SPME has been developed for a range of analytes in water samples. Some further illustrations are given here.
   a. Glass fibers have been coated with the solid phases such as poly(dimethylsiloxane) methyl silicone or polyacrylate, and the analytes (benzene and alkyl benzenes) desorbed directly in the GC injector system (Arthur et al. 1992).
   b. Films of Parafilm have been used as the sorbent for analytes including benzene, chlorobenzene, and chloroform, and trans-mission FT-IR used for direct analysis (Heglund and Tilotta 1996). Specific bands were used to increase the specificity of the method, and although formal equilibrium could not be achieved in times < 200 min, reproducible levels of extraction could be obtained after 30 min.
   c. Formaldehyde in the gas phase has been analyzed by reaction with 2,3,4,5,6-pentafluorobenzylhydroxylamine sorbed onto fibers of poly(dimethylsiloxane)/divinyl benzene that were ana-lyzed directly by GC (Martos and Pawliszyn 1998).

2.2.8 Atmospheric Sampling and Analysis

This is a highly sophisticated discipline and only a superficial discussion is presented. Since, however, organic compounds in the atmosphere are scavenged by rain and by sorption onto particulate matter, some of these compounds will ultimately enter the aquatic and terrestrial environments.

A wide range of structurally diverse compounds is produced during incineration. These include PAHs and related compounds, azaarenes, and chlorinated PAHs from combustion of fossil fuels and natural wildfires. Organic compounds in the atmosphere may exist both in the free (gaseous) state or on particles of various dimensions. Recent concern has been directed to the occurrence in aerosols both of the compounds themselves and of their transformation products (secondary aerosols) (1) for their role in atmospheric chemistry and as determinants of climate (Andreae and Crutzen 1997) and (2) due to health risks since aerosol formation facilitates the transport into and sorption by the lungs.

Although it is not intended to enter into detail here, a few examples of the classes of analytes that have aroused concern are given as illustration.

- PAHs in both the particulate and gaseous phase from the incineration of polystyrene (Durlak et al. 1998) — the presence of pyrene and 5,12-dihydronaphthacene in the particulate fraction and their distribution throughout the size fractions are worth noting;

- PAHs, ketones, and quinones in particulates < 2 μm from automotive exhaust (Rogge et al. 1993), and oxygenated PAHs including ketones and quinones from an urban aerosol-size fractionation (Allen et al. 1997);

- Aliphatic and aromatic carbonyl compounds from ambient air (Kölliker et al. 1998);

- Secondary aerosols from monocyclic aromatic hydrocarbons — production and analysis of products formed under simulated conditions in smog chambers (Forstner et al. 1997);

- Azaarenes from urban atmospheric samples — fractionation and speciation (Chen and Preston 1998);

- Nitroazabenzo[*a*]pyrenes in airborne particulate matter and automotive exhaust (Sera et al. 1994);

- Chlorinated PAHs in urban air including the mutagenic 1- and 4-chloropyrene, dichloropyrenes, and 6-chlorobenzo[*a*]pyrene (Nilsson and Östman 1993);

- Chlorinated dibenzo[1,4]dioxins in urban air particulates (Czuczwa and Hites 1986);

- Nitronaphthalenes and alkylated congeners in ambient air (Gupta et al. 1996) and dinitropyrenes in urban air particulates (Hayakawa et al. 1995);

- Salts of α,ω-dicarboxylic acids in marine aerosols (Stephanou 1992).

Brief comments on procedures for collecting wet and dry deposition are relevant since considerable alteration in the distribution of congeners of, for example, chlorinated dibenzo[1,4]dioxins may occur after emission into the atmosphere and before reaching the ultimate sinks (Czuczwa and Hites 1986; Koester and Hites 1992). Samples of rain may be filtered to remove particulate matter and then analyzed by procedures used for water samples, although a considerable degree of concentration may be necessary if the concentrations of the analytes are low. Alternative procedures for the collection of dry deposition have been described using either inverted Frisbee samplers or flat plates, both of which are coated with a thin layer of mineral oil to improve collection efficiency. Samples were then removed from the collectors by wiping with glass wool, and extracted by procedures used for dried soil or sediment samples (Koester and Hites 1992). Air sampling is generally carried out using high-volume samplers that collect particulate material on quartz filters followed by sorption of volatile compounds on polyurethane, Tenax, or XAD resins. A passive sampler that uses a thin film of polythene containing triolein has been evaluated for monitoring of PCBs and gave good agreement with conventional high-volume sampling methods (Ockenden et al. 1998).

For some compounds such as peroxyacyl nitrates, the sample may be introduced directly into the GC system (Grosjean et al. 1993). Otherwise, sampling of the gas phase may be carried out by sorption on Tenax cartridges or on polyurethane foam (PUF) plugs. The analytes may then be recovered by thermal desorption of the Tenax cartridges, or by solvent extraction of the PUF plugs. Many organic compounds in the atmosphere, however, are partitioned onto particulate matter of varying dimensions. Probably most sampling of particulates is carried out using high-volume samplers in which particulate matter is collected on quartz or glass-fiber filters that have been exhaustively freed from organic contaminants. The filters are then extracted by any of the procedures outlined below and analyzed by GC or LC. It is convenient to include both types of system, first a filter for collecting particulates followed by sorbents for the volatile components. In addition, it may be necessary to fractionate particles by size, and this is carried out using cascade impactors combined with a backup for fine particulates. In all sampling, a backup sorbent is often included to preclude loss of volatile components.

The following illustrate other sampling procedures.

1. For carbonyl compounds and carboxylic acids, specific sorbents that reacted chemically with the analytes have been used. Sampling of aldehydes in a forest atmosphere used silanized porous glass impregnated with naphthalene-1-dimethylamino-5-sulfonhydrazine (Nondek et al. 1992), and carbonyl compounds from the irradiation of terpenes in the presence of NO were adsorbed on C_{18} cartridges coated with 2,4-dinitrophenylhydrazine (Grosjean et al. 1992).

2. Sampling of halogenated acetic acids was carried out using glass tubes coated with alkaline glycerol (Frank et al. 1995).

2.3 Procedures Involving Chemical Reactions: Derivatization

2.3.1 Introduction

For GC or HPLC quantification, or GC-MS identification, neutral compounds are of necessity analyzed without further chemical treatment. For compounds with functional groups such as hydroxyl, amino, carboxylic acid, or reactive carbonyl groups, however, it may be convenient to prepare suitable derivatives. This has a long tradition in organic chemistry highlighted by the use of crystalline phenylosazones of carbohydrates to prepare and separate otherwise intractable mixtures of noncrystalline carbohydrates (Fischer 1909). In the present context, there are several advantages in using such procedures:

- A degree of selectivity — and thereby purification — is automatically introduced.
- It is possible to prepare volatile compounds with shorter GC retention times than their precursors.
- By preparation of halogenated derivatives, advantage may be taken of the enhanced sensitivity of the electron capture (EC) detection system in GC analysis.
- The lifetime of GC columns is prolonged by avoiding exposure to reactive compounds such as phenols, carboxylic acids, or reactive amines.
- Reactive compounds such as phenyl ureas are thermally unstable and are advantageously derivatized before GC analysis (Karg 1993).
- For HPLC, advantage may be taken of the high sensitivity of fluorescent detection systems.

For GC analysis, several of these advantages may be combined by preparing, for example, heptafluorobutyrate esters of phenols or trichloroethyl esters of carboxylic acids. Although trimethylsilylation is widely used and gives derivatives with structurally valuable mass spectra, the trimethylsilyl ethers and particularly esters are often relatively unstable. The choice of derivatives should therefore take into consideration their suitability for the cleanup procedures that may subsequently be employed.

Derivatization may be carried out at any stage of the analysis: on the initial sample, on sample extracts, or on the GC column itself. Generally, cleaner samples for analysis are obtained when extracts are used, since direct derivatization of a soil or sediment sample, for example, will generally result in large numbers of interfering peaks on the chromatograms due to reaction of the reagent with organic components of the solid matrix as well as with the desired analytes.

Derivatization of the untreated sample may not be universally applicable, but attention is drawn to examples where this has been examined and successfully applied.

1. Methylation of 2,4-dichlorophenoxyacetic acid and 3,6-dichloro-2-methoxybenzoic acid in soil samples was examined using trimethylphenylammonium hydroxide prior to supercritical extraction with CO_2 (Hawthorne et al. 1992). Although recoveries of these compounds were apparently limited by competition for the reagent of components of the soil matrix, this procedure possesses obvious advantages for polar analytes such as carboxylic acids and is capable of extension to other analytes.

2. Analysis of chlorophenolic compounds in sediment samples has been carried out by *in situ* acetylation followed by solvent extraction (Xie et al. 1985) or by combined supercritical fluid extraction with carbon dioxide and derivatization (Lee et al. 1992).

3. Low concentrations of volatile aldehydes in air samples have been analyzed by collection on porous glass impregnated with 5-dimethylaminonaphthalene-1-sulfohydrazide with which they react to form hydrazones that were analyzed online using HPLC with fluorescence detection (Nondek et al. 1992).

2.3.2 Specific Procedures for Derivatization

There is a voluminous literature on procedures for derivatization, though in practice, only relatively few have achieved extensive application. The most valuable derivatives combine a number of features including chemical stability, physical properties which enable their incorporation into cleanup steps such as open column chromatography before analysis, and a high response to the detector system. Extensive compilations are available (Knapp 1979; Blau and Halket 1993) to which reference may be made. Attention should also be drawn to an extensive range of reagents with varying selectivity for silylating functional groups (Pierce 1979). Only a few illustrative examples of commonly used reactions are given here:

- Acylation of phenols, amines, and alcohols;
- Esterification of carboxylic acids and sulfonic acids;
- Formation of silyl ethers of alcohols and phenols;
- Formation of ureas from alcohols by reaction with isocyanates;
- Formation of oximes from aldehydes and ketones.

Some additional procedures merit attention:

1. Anilide formation of highly water-soluble carboxylic acids including fluoroacetate, mono-, di-, and trichloroacetate has been carried out using dicyclohexylcarbodiimide (Ozawa and Tsukioka 1990).

2. Single-stage reduction and trimethylsilylation of a number of hydroxylated naphthoquinones and anthraquinones have been carried out using N-methyl-N-trimethylsilyltrifluoroacetamide in the presence of NH_4I (Bakola-Christianopoulou et al. 1993).

3. To take advantage of ^{19}F NMR, carbonyl groups including quinones have been derivatized by reaction with trifluoromethyltrimethylsilane in the presence of tetramethylammonium fluoride followed by hydrolysis of the trimethylsilyl ethers (Ahvasi et al. 1999). The method was applied to analysis of the various carbonyl groups in lignin.

4. Bifunctional reagents to derivatize several functional groups simultaneously have been used:

   - Di-t-butyldichlorosilane derivatization of bifunctional analytes such as 1:2-diols, 2-hydroxybenzoates, and catechols (Brooks and Cole 1988): formation of cyclic boronates from aromatic *cis*-1,2-dihydrodiols that has been applied to determine the absolute configuration of the dihydrodiols (Resnick et al. 1995) and is noted later as an example of the use of NMR;

   - Phosgene for 2-hydroxyamines related to catechol amines (Gyllenhaal and Vessman 1988);

   - Formation of Diels–Alder adducts between cyclohexa-1,3-dienes and 4-phenyl-1,2,4-triazoline-3,5-dione (Kobal et al. 1973) — This reaction is noted later as exemplification of the use of X-ray analysis and the absolute configuration of the dihydrodiols;

   - Trimethylphosphite formed cyclic trimethylphosphate derivatives of the hydroquinones formed from 1:2 quinones (Argyropoulos and Zhang 1998), and is noted again in the context of analysis of 1:2 quinones.

5. The use of chiral reagents. The use of these greatly facilitates the analysis of the enantiomers of compounds that contain functional groups and is essentially an application of the traditional procedure for resolving optically active compounds. Most of the reagents are chiral compounds with reactive groups such as acyl chlorides, chloroformate esters, isocyanates, or amines that are suitable for the preparation of derivatives by conventional procedures. The advantage of such procedures is that standard chromatographic columns may be used. Their application to the analysis of carboxylic acids may be illustrated by two examples:

- The analysis of chiral propionic acid anti-inflammatory drugs and herbicides was carried out by converting the acids into their acid chlorides with thionyl chloride and then reacting them with R-1-phenylethylamine (Blessington et al. 1993).

- For the analysis of isomers of the synthetic pyrethroid insecticde permethrin, enantiomeric amides were prepared using carbon-yldiiminazole and a range of chiral amines of which R(–) - or S(+)-amphetamine gave baseline separation of the four isomers (Taylor et al. 1993).

There has been considerable interest in determining the configuration of dihydrodiols produced by bacterial dioxygenation of aromatic compounds, and use has been made of (+)-(R)-α-methoxy-α-(trifluoromethyl)phenylacetyl chloride. In some cases, however, the dihydrodiols undergo aromatization during the reaction, and for monocyclic aromatic compounds a Diels-Alder reaction with 4-phenyl-1,2,4-triazoline-3,5-dione has been used to produce stable adducts with the cyclohexadienes before reaction with the acyl chloride (Boyd et al. 1991). For aromatic heterocyclic compounds, the problem of aromatization has been circumvented by initial hydrogenation of the dihydrodiols to the tetrahydrodiols (Boyd et al. 1992). Enantiomerically pure (–)-S-and (+)-R- 2-(1-methoxy-ethyl)phenyl boronic acids have been prepared and used for direct NMR determination of the absolute stereochemistry of a series of bicyclic *cis*-1,2-dihydrodiols. For cyclohexadienes, however, it was necessary to convert these first by reaction with 4-phenyl-1,2,4-triazoline-3,5-dione (Resnick et al. 1995). Further details of the use of NMR in establishing the configurations of the products are given in Section 2.4.1.3.

6. Derivatives for electrospray ionization mass spectrometry. To enhance the range of analytes that are suitable, neutral derivatives of ferrocene have been examined; these were designed to be "electrochemically ionizable" and are therefore inherently compatible with the operation of the electrospray source. Ferrocenoyl carbamates of alkanols were prepared by reaction with ferrocenoyl azide which, on heating, forms the reactive isocyanate (van Berkel et al. 1998).

It should also be appreciated that problems may arise even when functional groups are to be derivatized using apparently straightforward reactions. For example:

1. The susceptibility to hydrolysis of derivatives of nitrophenolic compounds (Hynning et al. 1989);

2. The difficulty of acetylating phenolic groups which have carbonyl groups in the *para* position, for example, vanillin without the use of pyridine as a catalyst (Neilson et al. 1988);

3. The difficulty of esterifying sterically hindered carboxylic acids such as the C_{18} carboxylic acid group in dehydroabietic acid by conventional procedures using alcohols and an acid catalyst: diazoalkanes may be used, or esterification carried out with pentafluorobenzyl bromide (Lee et al. 1990);

4. Diazomethane reacts rapidly and quantitatively with acidic hydroxyl groups such as carboxylic acids though less readily even with phenols containing electron-attracting groups such as chlorine, and hardly at all with unsubstituted phenols. 2,2,2-trifluorodiazomethane is even more selective, and does not react even with carboxylic acids although it forms esters with the much more acidic sulfonic acids (Meese 1984).

Procedures using ion-pair alkylation and esterification have been effectively used in a number of cases and a few illustrative examples may be given.

1. Esterification of chlorophenoxy acids and of pentachlorophenol has been assessed using tetrabutylammonium hydroxide and methyl iodide in methanol, and has been shown to be as effective as the standard procedure using diazomethane (Hopper 1987). A further development has examined the application of methanol/-benzyltrialkylammonium for simultaneous extraction and methylation of 2,4-dichlorophenoxyacetic acid in soil (Li et al. 1991); the quaternary ammonium hydroxide may, in this case, also play a significant role in releasing the "bound" analyte.

2. Tetrabutyl ammonium hydroxide and dimethyl sulfate in dichloromethane have been used to methylate a number of aromatic hydroxyacids and hydroxyaldehydes (Ramaswamy et al. 1985).

3. An on-column methylation of dimethylphosphorothioate in urine samples used trimethylanilinium hydroxide in acetone (Moody et al. 1985).

Finally, attention is directed to procedures for less commonly encountered compounds for which specific procedures may advantageously be applied; one example is the analysis of substituted benzoquinones.

1. For halogenated benzo[1:2]quinones, an analytical procedure was used in which these compounds reacted specifically with diazomethane to give methylenedioxy compounds of the corresponding hydroquinones (Remberger et al. 1991b). Similarly, analysis of some benzo[1,4]quinones (Remberger et al. 1991a) required reduction

with dithionite since conventional reducing agents such as ascorbate were insufficiently active. In neither case would conventional procedures have been satisfactory.

2. ^{31}P NMR was used to examine the products from the reaction of both 1:2 and 1:4 benzoquinones with trimethylphosphite (Argyropoulos and Zhang 1998). The former formed cyclic trimethylphosphate derivatives of the hydroquinones in a reaction comparable to that with diazomethane, and were hydrolyzed to 2-hydroxyphenyldimethylphosphates. The 1:4 quinone formed 1-methoxy-4-phenyldimethyl phosphate.

These examples illustrate the value of using chemical methods directed to specific functional groups as an alternative to exclusive reliance on traditional procedures that have been used for alcohols, phenols, or carboxylic acids.

It should be recognized that some derivatives such as the O-acetates of phenols give mass spectra with extremely weak — or even lacking — parent ions, so that it may be useful to prepare a range of derivatives some of which may provide more informative mass spectra (Wretensjö et al. 1990). In addition, the possible thermal instability of compounds during passage through GC columns may present serious limitations to application of this technique: at the expense of resolution, shorter columns may sometimes be used to overcome this problem.

The same principles have been applied — although to a considerably lesser extent — to HPLC analysis for which underivatized samples are frequently used. For compounds without UV absorption, however, it may be convenient to prepare derivatives with suitable absorption or fluorescence for the detector system. A few illustrative examples are given:

1. Alcohols may be converted into benzoate esters (Fitzpatrick and Siggia 1973) or into fluorescent derivatives by reaction with 9-fluorenylmethyl chloroformate (Miles and Moyne 1988).

2. Amines may be condensed with 1,2-phthaldialdehyde (Lindroth and Mopper 1979) or with 2-methyl-3-keto-4-phenyl-2,3-dihydrofuran-2-yl acetate (Figure 2.1) (Chen and Novotny 1997). The latter has been applied to analysis of a range of peptides and is applicable also to capillary electrophoresis that is discussed in Section 2.4.2.3.

3. Thiols may be alkylated with 3,6,7-trimethyl-4-bromomethyl-1,5-diazabicyclo-[3.3.0]-octa-3,6-diene-2,8-dione (Figure 2.2) (Velury and Howell 1988).

4. Phenolic compounds with a free position vicinal to the hydroxyl group have been derivatized by reaction with 4-dimethylaminocinnaldehyde in acidic medium followed by detection at 640 nm. This has been illustrated with procyanidins (de Pascual-Teresa et al. 1998), and could be valuable for a greater range of analytes.

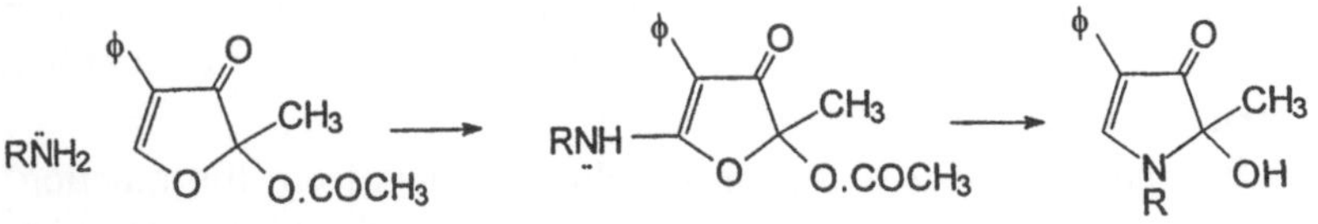

FIGURE 2.1
Derivatization of amines with 2-methyl-3-keto-4-phenyl-2,3-dihydrofuran-2-yl acetate.

5. For application to the analysis of aldehydes and ketones where the presence of NO_x during gas-phase sampling may cause side-reactions with 2,4-dinitrophenylhydrazine to produce 5-nitrobenzfurazan-3-oxide by formation of the azide and elimination of N_2; this has been obviated by use of the N-methyl compound (Buidt and Karst 1997). The analysis of 20-ketosteroids by a combination of pre- and postderivatization is described below.

6. There has been considerable interest in the development of analytical procedures for hydroxyl radicals in view of their potentially adverse biological effects (Section 4.6.1.2), and their role in transformation reactions (Section 4.1.1). A number of procedures have been developed.

   a. The reaction of the radical with dimethyl sulfoxide was used to produce methyl radicals that react with nitroxides to form N-methyl derivatives; these were analyzed by reversed phase HPLC and quantified fluorimetrically. Spin-labels have been used in different procedures; in one (Li et al. 1997), a fluorescent spin-label nitroxide is used, whereas in the other (Vaughan and Blough 1998), the spin-label contained an alkyl amino group that reacted with fluorescamine in a second step. Alternatively, the methyl radicals formed by the reaction with dimethyl sulfoxide reacted with aromatic derivatives of t-butylnitrones to produce nitroxides that could be reduced to hydroxylamines by ascorbate. Both the nitroxides and the hydroxylamines could be analyzed by HPLC using electrochemical or UV detection, and the method was applied to analysis of hydroxyl radicals in HepG2 cells and in rat hepatocytes (Stoyanovsky et al. 1999).

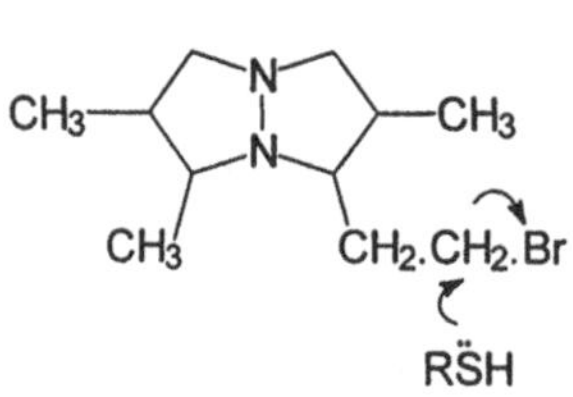

FIGURE 2.2
Derivatization of thiols with 3,6,7-trimethyl-4-bromomethyl-1,5-diazabicyclo-[3.3.0]-octa-3,6-diene-2,8-dione.

b. In a study directed to the analysis of the role of Fe and the generation of H_2O_2 in *Escherichia coli* (Section 5.5.5), the analytical procedure used EPR and a stable adduct between α-(4-pyridyl-1-oxide)-*N*-*tert*-butyl nitroxide and the α-hydroxyethyl radical formed by reaction of ethanol with hydroxyl radicals (stable adducts are not formed either by superoxide or hydroxyl radicals) (McCormick et al. 1998). The same procedure using this probe was also used for examining the formation and kinetics of Fenton reaction–derived OH radicals in soil slurries (Huling et al. 1998).

Chemiluminescence has been used as the detection system for derivatives produced by reaction of the analyte with hydrogen peroxide in the presence of an activated derivative of oxalic acid. Two applications are given as illustration:

1. For the analysis of low concentrations of aromatic nitro compounds in airborne particulate matter, the nitro compounds were reduced by pretreatment with sodium hydrosulfide, and after HPLC the fractions were treated with bis(2,4,6-trichlorophenyl)oxalate and hydrogen peroxide to produce chemiluminescent derivatives (Hayakawa et al. 1995).

2. The analysis of ketosteroids (3-keto, 20-keto- and 3,20-diketo) was carried out by precolumn conversion to the hydrazides using 5-dimethylaminonaphthalenesulfonic acid hydrazide and trifluoromethansulfonic acid. Chemiluminescent derivatives were formed by postcolumn reaction with 1,1′-oxalyldiiminazole in phosphate buffer (Appelblad et al. 1998).

In summary, almost all reactions that have traditionally been used for preparing derivatives in organic chemistry have been effectively applied — modified as necessary to increase their sensitivity to the appropriate detector systems.

2.4 Identification and Quantification: Basic Definitions

It is highly desirable to avoid any misunderstanding of these operational terms. They are therefore defined as they will be used in this account.

Identification — Use of the term clearly implies that it has been unambiguously established that the structure of an unknown compound is identical to that of an authentic standard. Comparisons are most frequently made using infra-red, mass, or nuclear magnetic resonance spectroscopy. These procedures are discussed in more detail below.

Quantification — Quantification is the determination of the concentration of a compound in a given matrix. For water samples, concentration units in terms of volume are universally used, but for other matrices, alternatives to volume have to be used for normalization. For soils and sediments, results are usually expressed relative to the total wet (or dry) weight — or preferably to organic carbon determined, for example, by loss after combustion of acidified samples: for biota, the total wet (or dry) weight may be used — or the fat or lipid content. Limits of detection are determined by the background levels, i.e., noise levels in the instrumentation used. As a working rule, the detection limit may be taken as three times, and the quantification limit as ten times the standard deviation of background levels (Keith et al. 1983).

2.4.1 Experimental Techniques: Identification

Several identification procedures are available, but two cardinal criteria should always be fulfilled:

1. Comparison should be made using several parameters; this practice is hallowed by historical application — combustion analysis of three independent derivatives for the establishment of the empirical formula of unknown compounds, or the establishment of identity by comparison of the R_f values in paper chromatographic systems using three different solvent systems.

2. The obvious fact is that samples of authentic compounds must be available. In many cases — and particularly for metabolites — these may not be commercially available and must therefore be synthesized by the laboratory carrying out the investigation. This is discussed in more detail below. Although all the PCB and chlorinated dibenzo[1,4]dioxins congeners are available, the identification of PAHs with seven or more fused rings remains a serious problem since these are not readily accessible and the structures of many depart from planarity (Biggs and Fetzer 1996). Indeed, the problem is reminiscent of that in the early 1930s for four- and five-ring PAHs.

Identification most often relies upon comparison of the GC or HPLC retention time together with the mass spectrum, or of the NMR spectrum with those of authentic reference compounds. Use of GC or LC retention times alone is not adequate, although for GC this may be an acceptable compromise procedure if several derivatives are available. Use of only the mass spectrum is clearly unacceptable since isomers generally provide essentially identical spectra. Naturally, compounds whose structures are unknown may be encountered; this presents a much greater challenge since comparison is not then possible. Determination of the structure of such compounds must therefore rely upon determination of the molecular mass by high-resolution mass spectrometry (HRMS), interpretation of the mass spectrum, and the NMR spectrum or ultimately the results of an X-ray diffraction study.

2.4.1.1 X-Ray Diffraction

X-ray structure analysis has become significantly more accessible with the advent of modern high-speed computing facilities so that its application is possibly limited mainly by the need for suitable single crystals of the pure compounds. X-ray diffraction has been effectively applied to the resolution of some important problems including the following:

- The absolute sterochemistry of the *cis*-toluene dihydrodiol produced from toluene by *Pseudomonas putida* F1 (Kobal et al. 1973);
- The structure of 3,4,5-trichloroguaiacol (Lindström and Österberg 1980);
- The structure of important congeners in technical chlordane (Dearth and Hites 1991);
- The structure of a photochemically modified $C_{10}H_9C_7$ component of toxaphene (Hainzl 1995).

2.4.1.2 Mass Spectrometry

Identification of components in environmental samples and in samples from laboratory studies of biodegradation and biotransformation is generally based on the application of MS coupled to either GC or LC systems. For environmental samples which may contain only small amounts of the relevant compounds, MS is particularly attractive in view of the extremely small amounts of samples — of the order of nanograms — which are required. An important additional advantage is that since the mass spectrometer can be interfaced with GC, LC, or capillary electrophoresis (CE) systems which incorporate separation procedures, pure samples are not required. Some salient issues in MS in the context of environmental application are summarized briefly as an introduction. Reference should be made to an exhaustive review (Burlingame et al. 1998) for instrumental details and aspects that are not covered here, such as MS of synthetic and natural polymers.

1. Instrumentation
    a. Two types of mass spectrometers have been most widely used for environmental application: low-resolution quadrupole instruments and high-resolution magnetic sector instruments, while both single-stage and tandem configurations have been developed. Some examples of the application of HRMS to quantification of low concentrations of analytes in complex mixtures are noted in this section.

b. There has been revived interest in time-of-flight (ToF) instruments whose principles have been discussed in depth (Guilhaus 1995). Examples of its application are given later in this section, and in Section 5.8.

c. In hydrocarbon-type analysis, the unfractionated sample is evaporated into a reservoir from which the vapor is introduced into the mass spectrometer. This type of analysis requires access to a high-resolution instrument, and Fourier transform ion cyclotron resonance mass spectrometry offers this degree of ultrahigh resolution (Rodgers et al. 1998). It has been used to separate thiaarenes from related arenes in diesel fuels before and after processing. The instrument was a home-built 5.6-T instrument, and some examples include the resolution of an alkylated benzothiophene $C_{10}H_{10}S^+$ 162.038 from the arene $C_{12}H_{18}^+$ 162.140, $C_{14}H_{11}O^+$ 195.083 from $C_{14}H_{13}N^+$ 195.103 and $C_{15}H_{15}^+$ 195.123, and naturally occurring ^{13}C from ^{12}CH. A 4.7-T Fourier transform ion cyclotron resonance mass spectrometer with infra-red laser desorption has been used to analyze phosphatidyethanolamine and phosphatidylglycerol lipids in whole cells of some Gram-positive and Gram-negative bacteria (Ho and Fenselau 1998), and the advantage of a suspension of cobalt powder in glycerol has been demonstrated.

d. Attention is also drawn to instruments that have been developed for measuring stable isotope ratios that are discussed in Section 2.4.4.

2. **Ionization** — Although electron impact (EI) and chemical ionization (CI) both in the negative and positive mode have traditionally been used, these have been supplemented by other procedures that are suitable for liquid samples including fast atom bombardment (FAB), electrospray (ES), and matrix-assisted laser desorption ionization (MALDI), which is particularly valuable for analysis of high-molecular-mass samples. Additional comments are provided below.

3. **Chromatographic interfaces** — Interfaces have been developed for GC, LC, and CE, and have achieved wide application.

Probably most investigations have used EI ionization, but the greater sensitivity of CI is in some instances attractive and may enable use of underivatized samples (de Witt et al. 1988); an application of negative-ion CI for quantification is noted below together with other ionization procedures. Ideally, comparison should be made between the complete mass spectrum of the analyte and that of the reference compound. If very low amounts of the analyte are available, it may not be feasible to obtain a complete spectrum and it may be necessary to carry out the comparison by monitoring selected ions determined from the mass spectrum of the authentic compound. Many compilations of suitable ions for a range of widely encountered environmental contaminants have been provided. For some compounds, it may be difficult to obtain a parent ion, and several alternative ionization systems have been used; some of these are briefly noted below for LC–MS interfaces.

Brief attention is drawn to some specific issues.

1. Isomers generally yield essentially identical EI mass spectra although halogenated biphenyls carrying substituents at both the 2- and 2'-positions yield characteristic spectra in which loss of one halogen atom yields an intense ion fragment (Sovocool et al. 1987), and aromatic dinitro compounds substituted in the *ortho-* or *peri-* positions display unique fragmentation pathways (Ramdahl et al. 1988). The spectra of the isomeric diphenylacetylene, phenanthrene, anthracene, 9-methylenefluorene, and benzazulene are identical due to the intrusion of rearrangement reactions (Ramana and Krishna 1989).

2. Interest in bromine-containing fire retardants and contaminants such as brominated dibenzo[1,4]dioxins and dibenzofurans has led to increased application of both negative CI (Buser 1986) and EI (Donnelly et al. 1987) procedures.

3. A computer program has been developed to use the mass-selective detector as a chlorine-selective detector (LaBrosse and Anderegg 1984; Johnsen and Kolset 1988), but it does not appear to have been widely evaluated. Attention should be directed to three situations which may give rise to error in their application:

   • Compounds from which HCl is eliminated during electron impact so that no characteristic isotope pattern is observed;

   • Compounds containing elements other than chlorine such as bromine or sulfur which compromise the isotope pattern;

   • Compounds lacking chlorine atoms but which fortuitously give clusters of peaks with distributions resembling or identical to those expected for molecules with chlorine substituents.

Whereas none of these problems is unique to the use of such detector systems, they clearly underscore the care that should be exercised by the conscientious mass spectroscopist.

Current developments in MS have been directed to the application of alternative ionization methods for sensitive molecules, to the use of tandem systems (Durand et al. 1992), and to the development of interfaces with chromatographic systems (Arpino 1990; Niessen et al. 1991; Arpino et al. 1993). Some examples include the following.

1. Particular attention has been directed to developments in LC–MS (Niessen and Tinke 1995; Slobodnik et al. 1995). Particle beam LC–MS systems (PB) which provide spectra compatible with conventional EI spectra and enable analysis of compounds unsuitable for GC analysis seem promising (Behymer et al. 1990; Kim et al. 1991). Two widely different applications are given as illustrations.

   • Examination of the degradation products of PAHs (Gremm and Frimmel 1994) — Increased sensitivity was obtained by including precolumn concentration using a microbore C_{18} column.

   • Application to a series of triphenyl methane dyes (Turnipseed et al. 1995) — These are used in aquaculture and concern has arisen over their accumulation in edible fish; it is difficult, however, to evaluate the application of this procedure to fish exposed to the dyes since the PB mass spectra of the dye and the reduced compounds are identical, while GC–MS procedures are available for the reduced forms.

2. For some types of compounds, LC–MS systems using atmospheric-pressure chemical ionization (APCI) are highly effective, and the following illustrations are given.

   • Determination of the components of industrial effluents (Castillo et al. 1997) that included phthalate esters, 4-nonylphenol, pentachlorophenol, 2-methylbenzenesulfonamide, and tetramethylthiourea.

   • Analysis of 2,4-dinitrophenylhydrazones of carbonyl compounds in ambient air that included aliphatic and aromatic aldehydes and ketones (Kölliker et al. 1998).

3. ES may be highly effective for determining the molecular mass of samples with low volatility, and has been combined with collision-activated dissociation (CAD) to yield structural information. The sample must be free from interfering compounds and then provide improved mass resolution over that obtained for ion fragments using MS–MS. It is most structurally useful using MS–MS and may be illustrated by its application to the detection and identification of the benzo[*a*]pyrene diol epoxide adduct with guanosine phosphate (Barry et al. 1996). ES collision-induced decomposition (CID) MS was used to identify recalcitrant metabolites from branched alkyl side-chain nonylphenol ethoxylates after fractionation by LC on a graphitized carbon black column (Di Corcia et al. 1998).

   A comparison of various LC interfaces and ionization systems including ES, TS, APCI, and FAB was made with plant extracts that contained glycosides of xanthones, flavones, secoiridoid, and triterpenoid glycoside; it clearly emerged that no single one was optimal for various classes of compounds (Wolfender et al. 1995).

It is, therefore, probably safe to state that no single ionization procedure is universally applicable.

4. Although NH$_3$ or methane is generally used for chemical ionization, attention is drawn to the use of alternatives:

   - Acetone — and most valuably [^{2}H]$_6$-acetone — to provide acyl adducts with alkenes without interference from reaction with cycloalkanes (Roussis and Fedora 1997);
   - A mixture of pentafluorobenzyl alcohol and methane (1:1) for a range of airborne alkanoic acids, and phenols (Chien et al. 1998);
   - Dichloromethane added to the HPLC eluate before introduction to an APCI interface produced adducts with bacteriopropanepolyols (Fox et al. 1998). Although the method was applied to extracts of a lake sediment, interference from other analytes made identification more difficult.

5. A system has been described in which the analytes from water samples are desorbed from a 0.010-in. silicone membrane by a low-energy CO$_2$ laser, and interfaces were developed for carrying out desorption inside and outside the vacuum manifold of an ion-trap mass spectrometer (Soni et al. 1998). The procedure was illustrated for the analysis of two-, three-, four-, five-, and six-ring PAHs.

6. There has been substantial success in the development of ionization procedures that can be applied to compounds of high molecular mass. Application of procedures such as MALDI MS has revealed the existence of high-molecular-mass PAHs and azaarenes (Herod 1998), and has achieved spectacular mechanistic insight into interaction between PAH metabolites and polynucleotides (Ramanathan and Gross 1998). Application of this procedure seems likely to provide new insights into the association between xenobiotics and macromolecules.

7. There has been considerable interest in ToF MS, and laser ToF has been used to analyze PAHs in the form of their picrate complexes (Hankin et al. 1997). This overcomes problems of loss of volatile analytes in conventional procedures.

   For analysis of compounds with high-mass MALDI–ToF has been extensively used, although its direct application to environmental problems has so far been restricted. It has been used for analysis of PCR products, and for monitoring expression of protein expression in whole cells of bacteria (Easterling et al. 1998). For analysis of PCR products, the use of delayed extraction MALDI–ToF has been shown to result in increased resolution and precision (Ross et al. 1998). Negative–ion MALDI–ToF has been applied to analysis of the PCR-amplified gene sequences for the particulate methane monooxygenases of two taxa of methanotrophis (Hurst et al. 1998).

In *Methylosinus trichosporium* OB3b a type II methanotroph, the 50-mer amplified product could readily be identified after the samples had been purified by standard procedures before analysis. *Methylomicrobium albus* BG8, a type I methanotroph with a 99-mer could also be identified, and no cross-amplification products were observed. The application of this procedure is noted in Sections 5.8 and in 8.1.2.

8. Capillary zone electrophoresis interfaced with ES ionization and quadrupole ToF spectroscopy has been used to characterize DNA oligonucleotides in the range of 20 to 120 bases (Deforce et al. 1998): the spectra of the larger nucleotides often display peaks resulting from intramolecular reactions, so that care would have to be exercised in interpreting the results from unknown sequences.

Ideally, authentic samples of pure compounds should be examined under identical operating conditions, but frequent use is made of published spectra or of those contained in libraries of spectra that are directly coupled to the data system of the mass spectrometer. Although these compilations contain mass spectral data for a large number of compounds, care in their use should be exercised since operational conditions such as the amount of the compound used for analysis and the accelerating voltage may differ from those used in a given laboratory. Where library MS spectra are not available, the relevant compounds must be prepared by the laboratory itself, and this may be a demanding exercise. Examples of the effort required may be illustrated by the synthesis of the 209 PCB congeners (Mullin et al. 1984), the 75 congeners of chlorinated dibenzo[1,4]dioxins, or the complete set of chlorinated guaiacols and catechols. Without access to these reference compounds, none of the studies dependent on their availability could have been accomplished. On the other hand, the mass spectra of even fewer microbial or other metabolites or transformation products currently exist in library catalogs so that these compounds must almost always be synthesized. A few examples may be given as illustration:

1. Various hydroxylated tri- and tetrachlorodibenzofurans have been synthesized (Burka and Overstreet 1989) since, as conjugates, these could plausibly be expected to be mammalian metabolites of 2,3,7,8-tetrachlorodibenzofuran. A number of methoxy, dimethoxy, and dihydroxy derivatives of 2,3,7,8-tetrachlorodibenzo[1,4]dioxin have also been synthesized for similar reasons (Singh and Kumar 1993).

2. A series of dibenzo[1,4]dioxins bearing nitro and trifluoromethyl substituents in one ring and chloro in the other was synthesized for comparison with impurities in commercial samples of the lampricide 3-trifluoromethyl-4-nitrophenol, and to assess their ability to induce MFO activity (Hewitt et al. 1998).

3. To facilitate identification and quantification, putative metabolites of dichlofop-methyl (methyl 2-[4-(2,4-dichlorophenoxy)phenoxy]propanoate) produced by plant-resistant species have been synthesized (Tanaka et al. 1990).

4. Decarboxylation products of dehydroabietic acid and 12,14-dichlorodehydroabietic acid have been synthesized (Hynning et al. 1993) to enable the conclusive identification of the 18-*nor* and 19-*nor* compounds in environmental samples of fish and sediment.

5. A series of hydroxy-, nitrooxy-, and dinitrooxyalkanes that are produced by reactions between alkenes and hydroxy and nitrate radicals were synthesized to develop procedures both for their collection from atmospheric samples and to facilitate their identification and analysis (Muthuramu et al. 1993).

6. Three trichlorodiphenyl sulfides were synthesized and their structures determined by ^1H and ^{13}C NMR, and these were used for comparison with compounds in pulp mill effluents and stack gas samples (Sinkkonen et al. 1993).

7. Methyl sulfones are ultimate mammalian metabolites of PCBs produced by a series of reactions involving arene oxides and glutathione conjugates and have been synthesized to enable their conclusive identification and quantification (Bergman and Wachmeister 1978; Haraguchi et al. 1987). This is discussed further in Section 7.5.2.

8. Sulfonic acids are produced as metabolites of alachlor and metolachlor, and the metabolite from metolachlor has been synthesized for FAB–MS comparison with the compound isolated from soil (Aga et al. 1996).

9. A series of alkyl tin compounds and the corresponding perdeuterated analogues were used to identify and quantify organotin contaminants in samples of water and sediment (Arnold et al. 1998).

Purely analytical work may therefore profitably be supplemented with competent — and sometimes extensive — organic synthetic activity. As noted in Section 2.5, however, great care must be taken to ensure that samples for analysis do not become contaminated in the laboratory with compounds synthesized in very much larger — in many cases gram — quantities.

2.4.1.3 Nondestructive Procedures: NMR

In contrast to GC–MS procedures, NMR methods are nondestructive although they require access to essentially pure samples. In addition, relatively large amounts of samples may be required especially if advantage is to be taken of natural levels of ^{13}C in the samples. NMR has been effectively used for many years in identifying microbial metabolites where relatively large amounts of sample can readily be obtained. Its application to environmental samples is more restricted although good illustrations are provided by the following structural determinations:

1. The two principal components of toxaphene congeners isolated from biota (Stern et al. 1992): ^{1}H NMR showed these to be octachloro and nanochloro congeners both of which display chirality although the enantiomers were not resolved in this study (Figure 2.3).

2. The structure of a transformation product of a triterpene that was established on the basis of HRMS, MS and ^{13}C NMR (Figure 2.4) (Hynning et al. 1997).

Some important issues in the development and application of NMR are summarized in the following paragraphs.

1. Over the years since its introduction there have, however, been substantial increases in the magnetic field strengths that are attainable, so that greater sensitivity and resolution are achieved with modern instruments and so that sample size has become less significant — provided that the sample is of adequate purity. Application to naturally occurring levels of ^{13}C has become possible, and examination of compounds with ^{15}N, ^{17}O, and ^{19}F has been reported. NMR has been increasingly used for *in vivo* investigations of microbial degradation since samples containing cell suspensions and the ^{13}C-labeled substrates can be examined directly, and the kinetics of the appropriate resonance signals monitored. Further details of the application of NMR to metabolic problems using ^{13}C and ^{19}F are given in Chapter 5 (Sections 5.5.4 and 5.5.5), and an example of the application of ^{13}C NMR to study the interaction of 2,4-dichlorophenol with humic acid is given in Chapter 3 (Section 3.2.4). A ^{13}C NMR method has been developed to estimate the capacity of various environmental samples of soil and water to carry out transformation and degradation of chloroacetonitrile and related compounds (Castro et al. 1996), and would be amenable to extension to a much wider range of compounds and reactions.

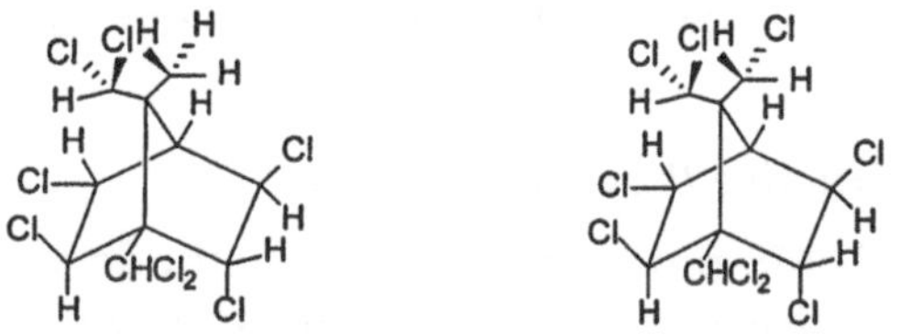

FIGURE 2.3
Structures of octachloro- and nonachlorocamphanes.

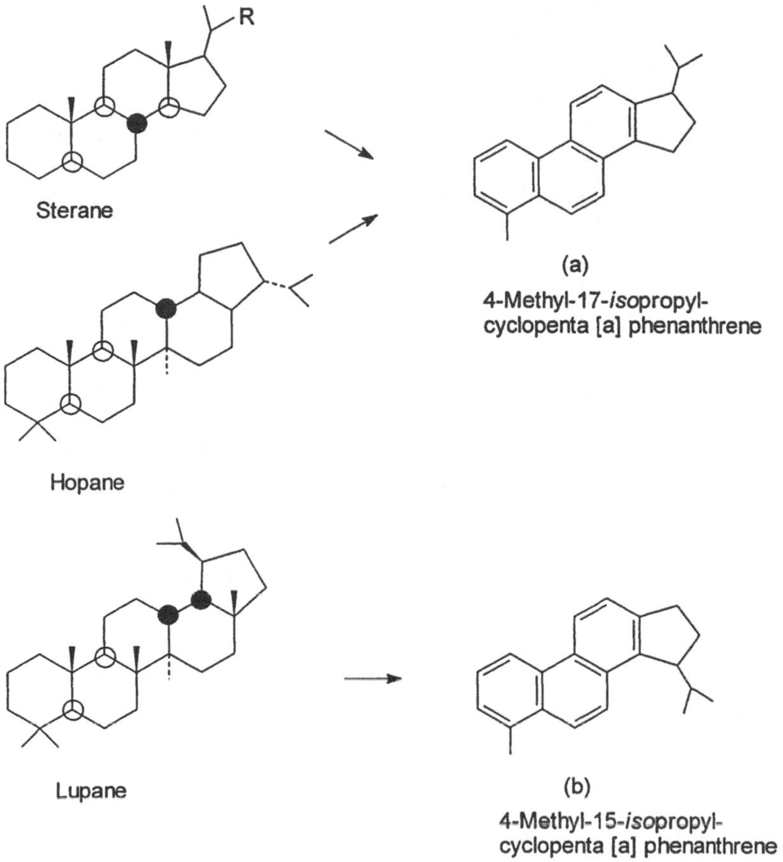

FIGURE 2.4
Transformation of a pentacyclic triterpene.

2. The size of sample required has been reduced by a number of technical developments including microinverse probes and microcells (references in Martin et al. 1998), and have been reduced even further using a 1.7-mm submicroinverse-detection gradient probe (Martin et al. 1998). Combined use of inverse detection probes with solenoid microcoils has also been developed to reduce sample volumes for ^{13}C NMR (Subramanian and Webb 1998).

3. Multidimensional spectra have been increasingly used as well as techniques including DEPT (distortionless enhancement by polarization transfer), COSY (correlated spectroscopy), and ROESY (rotating-frame Overhauser enhancement spectroscopy).

4. Since mass spectra after HPLC are generally obtained using soft ionization techniques, they provide data on the molecular mass, but there is little fragmentation so that little structural information is obtained. The combined application of an interface to NMR and to MS provides a powerful structural tool, and for MS methods that provide limited structural information, this is a particularly attractive combination. Identification by NMR provides detailed structural information, and a system for the analysis of peptides involves initial HPLC followed by splitting of the fractions for subsequent MS and NMR analysis in a flow-through system (Holt et al. 1997). Application to environmental samples after fractionation is attractive, and a configuration has been developed for combining capillary zone electrophoresis, capillary HPLC, and capillary electrochromatography with NMR including its use in stopped-flow experiments (Pusecker et al. 1998). Direct coupling to LC systems has been achieved, and application to whole cells has been possible in a number of applications. Both the continuous-flow and stop-flow modes have proved valuable, and a combination of continuous-flow HPLC interfaced with ^{1}H NMR and HPLC interfaced to thermospray mass spectrometry has been used to identify the products from the photolysis of 2,4,6-trinitrotoluene (Godejohann et al. 1998). The application of NMR analysis coupled to HPLC and supercritical fluid procedures has been made possible by a radical change in the design of the NMR instrument (Albert 1995), and further technical developments are likely to offer a wider range of applications. The procedure has been valuable for identifying urinary metabolites of pharmaceuticals and has used both continuous-flow and stopped-flow methods (Sidelmann et al. 1997).

5. Where the chirality of the product is of interest, optical rotatory dispersion and circular dichroism may be effectively used, for example, in the determination of the configuration of the *cis*-dihydrodiol metabolites formed during fungal metabolism of 7,12-dimethylbenz[*a*]anthracene (McMillan et al. 1987), and in combination with HPLC using a chiral stationary phase (Section 2.4.2.2) to resolve details of the metabolism of fungi by several fungi (Sutherland et al. 1993). This is discussed further in Section 6.2.2. Resolution of enantiomers by NMR may be achieved using a range of chiral shift reagents that are complexes of the rare earth trivalent cations europium, praseodymium, and ytterbium with enantiomeric *R*- and *S*-camphor. However, when only small quantities of sample are available and where several compounds are present, the use of chromatographic methods that have already been discussed is clearly advantageous. Increasing use has, however, been made of NMR procedures. The absolute configuration

of *cis*-dihydrodiols that are produced by bacterial dioxygenation of aromatic compounds has been established from the ¹H NMR chemical shifts after conversion to (+)-(R)-α-methoxy-α-(trifluoromethyl)phenylacetates (Boyd et al. 1991, 1992). Both the absolute configuration and the enantiomeric purity of *cis*-dihydrodiols produced by reaction of naphthalene dioxygenase with fluorene, dibenzofuran, and dibenzothiophene (Resnick and Gibson 1996) was established by ¹H NMR of the esters formed with (−)-S-1-methoxyethylphenylboronic acid (Resnick et al. 1995). This has been noted in Section 2.3, and it should be noted that it is possible to assign absolute configurations using the consistent trends in the NMR signals (Resnick et al. 1995).

More restricted has been use of the ¹⁵N, ¹⁷O, ²⁹Si, and ³¹P nuclei but some illustrative examples may be cited:

1. The application of ¹⁵N NMR to the structural determination of humic material is noted in Chapter 3 (Section 3.2.4)

2. Limited application of ¹⁷O NMR has been made to a restricted range of chlorinated aromatic compounds (Kolehmainen et al. 1992), and has been used to establish the source of oxygen in the metabolites produced from ¹⁷O acetate and ¹⁷O₂ by *Aspergillus melleus* (Staunton and Sutkowski 1991).

3. ³¹P NMR has been used to examine the effect of pentachlorophenol on the energy metabolism of abalone (*Haliotis rufescens*) (Tjeerdema et al. 1991), and the structures of the compounds formed by reaction of 1:2 and 1:4 benzoquinones with trimethylphosphite (Argyropoulos and Zhang 1998).

4. ²⁹Si NMR of hexamethyldisiloxane has been examined using a 750-MHz ¹H resonance frequency, and a number of technical issues discussed (Knight and Kinrade 1999): the use of chromium[III] acetylacetonate to reduce the ²⁹Si relaxation time and of a DEPT-45 pulse sequence.

2.4.1.4 Fluorescence Line-Narrowing Spectroscopy (FLN Spectroscopy)

The theory of this has been presented in detail by Jankowiak and Small who are the pioneers in the development and application of this procedure. This was exemplified in elegant studies on the structures of the adducts formed between PAH metabolites and DNA (Jankowiak and Small 1998). This was carried out using both *in vivo* and *in vitro* studies, and provided a valuable complement to the use of MS and NMR. Conventional analysis for PAHs has been supplemented by using Shpol`skii spectroscopy that necessitates compatibility with *n*-hexane to provide a crystalline matrix. Although FLN spectroscopy is less sensitive than NMR, the sample requires no pretreatment and polar metabolites may be determined. This has been applied to establishing the metabolites of pyrene produced in various organs of the isopod *Porcellio scaber* (Larsen et al. 1998). The presence of the metabolite 1-hydroxypyrene was established, and that of the glycoside after specific enzymatic hydrolysis. The potential of this technique seems attractive for application to polar metabolites of other xenobiotics that contain fluorophores.

2.4.2 Experimental Techniques: Separation and Quantification

The following comments are directed primarily to the issue of quantification. Clearly, however, this presupposes that the analyte has been identified, and can be separated, so that these techniques — especially preparative HPLC — have found wide application in the cleanup of environmental samples. It is worth pointing out that the problem of separating and identifying natural products that may exist as complex mixtures in low concentrations has taken advantage of many of these techniques, and indeed was the motivation for their development. The combination of complex mixtures and low concentration of analytes is characteristic of most environmental samples, so that advantage should be taken of these methods on a wider basis.

In many cases, the identity of the analyte will be known; nonetheless, it is highly desirable that this be confirmed to avoid the possibility that an interfering compound fortuitously has, for example, the same GC or HPLC retention time as that of the desired analyte. Indeed, many protocols that are now advocated use mass spectrometric systems so that this control is automatically incorporated. Samples may be spiked with internal standards to simplify calculation and eliminate small errors in pipetting and injection, or surrogate standards may be employed where, for example, incomplete extraction of the analyte is unavoidable. When MS is used as the detection system, analytes labeled with suitable isotopes have been widely used: for PAHs, fully deuterated standards, and for PCBs and agrochemicals, ^{13}C-labeled compounds. For partially labeled standards of analytes, care must be exercised in their choice if it is intended to analyze for metabolites of a substrate in which the label may have been lost.

Any measuring system may in principle be used for quantification provided that it produces a linear response between the output signal and the concentration of the analyte over a suitable range. Greatest use has probably been made of gas–liquid chromatography (GC) and HPLC and capillary electrophoresis (CE) systems interfaced with appropriate detector systems. Chromatographic systems may incorporate temperature gradients (GC) or gradients in the composition of mixed mobile phases (HPLC). Use of supercritical fluids (Section 2.2.5) incorporating, for example, carbon dioxide, offers a promising alternative technique (Schoenmakers and Uunk 1987) to conventional GC or HPLC. For GC analysis, it should be appreciated, however, that the use of a single type of column packing or of a single detector system can never be optimal for all kinds of compounds. For quantification, translation of the detector response into amounts and thereby concentrations clearly requires access to authentic compounds. These will, however, also be needed for unambiguous identification in the first place so that they will generally be available.

2.4.2.1 GC Systems

It is convenient to note briefly developments in the two basic components — the column and the detector system.

GC Columns

It is convenient to draw together various developments. Packed columns have been largely replaced by capillary columns that are available for a wide range of applications. Attention is drawn to the decomposition of DDT at the GC inlet that could result in unacceptable errors in the concentration of DDT metabolites and therefore in the persistence of DDT (Foreman and Gates 1997). The generality of this phenomenon is unknown.

1. Dual columns — The introduction of capillary columns brought about a revolution in the level of resolution, and these columns are now almost universally used. Nonetheless, no single column is able, for example, to resolve all of the 209 PCB congeners so that the application of two columns in tandem with separate ovens and detector systems has been advocated. Fraction cuts from one column are quantitiatively transferred to the other (Duinker et al. 1988). A few examples are given as illustration.

   a. The value of the procedure has been graphically illustrated by the possibility of unambiguously quantifying toxic congeners such as 3,3′,4,4′-tetrachlorobiphenyl (IUPAC no. 77) or 2′,3,4,4′,5-pentachlorobiphenyl (IUPAC no. 123) which otherwise coelute with other congeners. Analytical protocols for the analysis of the toxic PCB congeners IUPAC 77, 126 (3,3′,4,4′,5-pentachlorobiphenyl) and 169 (3,3′,4,4′,5,5′-hexachlorobiphenyl) using conventional cleanup procedures which are compatible with those used for chlorinated dibenzo[1,4]dioxins and dibenzofurans have, however, been developed (Harrad et al. 1992).

   b. The procedure has been applied successfully to separating the large number of congeners in commercial samples of toxaphene and in environmental samples of fish and seal blubber (de Boer et al. 1997).

   c. A number of column combinations were evaluated, and parallel columns after splitting the sample prior to injection have been used for a range of aromatic nitro compounds (Lopez-Avila et al. 1991).

   d. This procedure with a combination of an SE 52 column interfaced with a 2,3-di-O-ethyl-6-O-(t-butyldimethylsilyl)-β-cyclodextrin in a PS 086 capillary column was used to separate enantiomers of a number of monoterpenes including β-pinene, sabinene, limonene, linalool, and α-terpineol (Mondello et al. 1998).

2. High-molecular-mass PAHs — There has been increased interest in the analysis of PAHs with molecular mass >328. This has resulted in the development of GC columns that withstand temperatures up to 370°C, and these have been used for resolution of PAHs in carbon black and in coal-tar extracts with seven rings (benzo[c]picene, M_r 328), eight rings (benzo[a]coronene, M_r 350), and nine rings (dibenzo[a,j]coronene, M_r 400) (Bemgård et al. 1993).

3. Chiral support phases — Determination of the chirality of analytes is extremely important, since the generally significant difference in the biological activity of enantiomers is well established. Striking examples are provided by (R)-asparagine that is sweet whereas the (S)- enantiomer is bitter, or D-penicillamine that is antiarthritic whereas the L-enantiomer is toxic. In general, only one of the enantiomers of many compounds displays the desired biological activity, and this means that up to half of many compounds that have traditionally been used as rodenticides, insecticides, fungicides, and herbicides is biologically inactive toward the target organism and that this therefore contributes unnecessarily to the environmental burden of xenobiotics. Chiral support phases have therefore found important applications in a number of investigations. Detailed accounts have been provided of the methods for selected compounds, together with valuable chromatographic details (Armstrong et al. 1993). A few examples will be used to illustrate the potentially great scope for the application of these columns.

a. A stationary phase consisting of permethylated heptakis (2,3,6-tri-O-methyl)-β-cyclodextrin diluted with polysiloxane was used for the analysis of hexachlorocyclohexane isomers (Müller et al. 1992). Although the α isomer (*aaaaee*) is optically active, the synthetic product is racemic; nonetheless, the ratio of the enantiomers in environmental samples may deviate considerably from unity (Mössner et al. 1992). Further comments are given in Section 4.2.2. Application of the same chiral support phase revealed similar deviations in the composition of epimeric octachloro- and nonachloro-chlordanes from samples of biota including an Adélie penguin (*Pygoscalis adeliae*) from Ross Island, Antarctica (Buser et al. 1992a; Buser and Müller 1992). Preparative resolution of enantiomers using a number of different alkylated cyclodextrin columns has been carried out, and in some cases products of high enantiomeric purity were obtained in milligram quantities: ~95% for *cis*-heptachloro epoxide, ~93% for *trans*-chlordane, and >99% for α–hexachloro[*aaaaee*]cyclohexane (König et al. 1994). It seems plausible that biological activity mediated either by bacteria or by higher organisms is the explanation for these deviations from a ratio of 1:1, although care should be exercised in the interpretation of enantiomeric ratios close to unity since there is evidence that some of the products from the chlorination of camphene are not racemic and contain a slight enantiomeric excess of some congeners (Buser and Müller 1994). A range of chiral stationary phases based on cyclodextrins has also been used for the resolution of enantiomeric hydrocarbon components which are of interest as biomarkers (Armstrong et al. 1991). Parenthetically, it is useful to summarize the conformation of the various hexachlorocyclohexane isomers: α-isomers — *aaaaee*; β-isomer — *eeeeee*; γ-isomer — *aaaeee*; δ-isomer — *aeeeee*; ε-isomer — *aeeaee*; η-isomer — *aaeaee*; φ-isomer — *aeaeee* (Willett et al. 1998).

b. Stereospecific synthesis is of enormous importance in biotechnology in view of the generally divergent biological effects of enantiomers. A few examples of this are given in Chapter 6 (Sections 6.11.1 and 6.11.2), and only one illustration will be given here. The indanols formed by microbiological oxidation of indan with a strain of *E. coli* containing the toluene dioxygenase genes from *Pseudomonas putida* F1 were separated after conversion into their *iso*propylurethanes on an XE-60-(*S*)-valine-(*S*)-phenylethylamide-fused silica column (Brand et al. 1992).

c. A PCB mixture was separated by HPLC into fractions having different numbers of chlorine atoms in the *ortho* positions, and nine enantiomeric pairs with three or four *ortho* chlorine atoms (IUPAC 84, 91, 95, 132, 135, 136, 149, 174 and 176) were separated using a Chirasil-Dex 30 m column (Ramos et al. 1999). The separation of PCB enantioners by capillary electrophoresis is noted in Section 2.4.2.3.

GC Detector Systems

A variety of detection systems has been used including flame ionization (FID), electron capture (EC) — which is particularly sensitive to most halogenated compounds — N/P-sensitive, atomic emission, and mass selective detectors. It should be emphasized that the EC response of a compound is dependent not only upon the number of halogen atoms but also upon their position in the molecule; estimates of response based on different isomers may therefore be totally unreliable. Illustrative examples may be found among isomers of PCBs, trichlorophenols, and monochlorocatechols. Increasingly, Hall detector systems have been used on account of their combined sensitivity to organochlorine compounds and their selectivity. In a convenient splitting system, this detector may be combined with a traditional FID system (Dahlgran 1981).

Although FID and EC detection systems have been by far the most widely used, the mass selective detector is highly attractive and is becoming incorporated into many standardized protocols; mass spectral comparison with authentic compounds — and thereby identification — is simultaneously made available. These quantification procedures may make use of labeled standards incorporating either ^{13}C or ^{2}H. Suitable compounds may not, however, be commercially available and may have to be synthesized; in such cases, synthesis of ^{2}H-labeled compounds may be more readily and economically accomplished. High-resolution MS has many advantages including its specificity and sensitivity, and a graphic illustration is its use in the analysis of the complex of arenes, oxaarenes, and thiaarenes in gas oil (Guan et al. 1996). Selected ion monitoring has also been increasingly used — particularly for the quantification of chlorinated dibenzo[1,4]dioxins in environmental samples in which the concentrations may be extremely low — and its application with negative-ion chemical ionization has been effectively applied to analysis of PCB congeners in fish (Schmidt and Hesselberg 1992) and toxaphene in arctic biota (Bidleman et al. 1993).

Use of infrared spectra for identification has declined, possibly because of the need for pure and relatively large amounts of material. With the availability of interface systems for GC and FTIR, and increasing interest in supercritical fluid FTIR interfaces (Bartle et al. 1989), the situation may change. Use of FTIR methods is particularly attractive for compounds containing carbonyl groups such as aldehydes, ketones, esters, and lactones, and quinones which have highly characteristic C=O stretching frequencies well separated from other absorptions. A good example of the combined use of MS and FTIR is provided by a study of the bacterial metabolites produced from methylbenzothiophenes (Saflic et al. 1992). This is a particularly favorable case since the relevant absorptions are strong compared with others in the spectra so that effective use could be made of IR absorptions both for the S–O stretching frequencies of the sulfoxides and for the C=O stretching frequencies of the benzothiophene-2,3-diones. GC–FTIR has been used for the analysis of PCBs (Bush and Barnard 1995), and the spectra were sufficiently differentiated that identification was effectively carried out using a search program. Combined with high-resolution GC, it also offers the possibility of quantification.

As with mass spectra, substantial libraries of IR spectra are available, although, once again, these should be used with due attention to the conditions under which the spectra were obtained.

Other detection systems have proved valuable in particular applications:

1. N/P-specific detectors have been used for the analysis of azaarenes (references in Herod 1998), and the *n*-butyl esters of NTA (Schaffner and Giger 1984).

2. Atomic emission detectors (AEM) in the carbon- and sulfur-selective modes have been used for the analysis of thiaarenes (references in Herod 1998; Mössner and Wise 1999), and to determine the contents of benzothiophene, dibenzothiophene, naphthothiophenes, phenanthro[4,5-*bcd*]thiophene, and benzo[*b*]naphtho[2,1-*d*]thiophene in NIST samples of coal tar, crude oil, and shale oil (Schmid and Andersson 1997). The sulfur-selective detector has also proved valuable for the analysis of sulfones in environmental samples (Janák et al. 1998).

3. The analysis of dialkyl ethers used as octane enhancers in automobile fuel (oxygenates) was carried out by catalytic cracking to CO followed by reduction to methane (Verga et al. 1988).

4. The analysis of N_{ox}, nitric acid, or n-propyl nitrate was carried out by reduction to nitric oxide using carbon monoxide and gold catalyst (Bollinger et al. 1983).

5. A detector system (SID) based on hyperthermal negative surface ionization has been used for the analysis of alkanols and phenols (Kishi et al. 1998), and was ≈100 times more sensitive than FID.

6. Procedures for the analysis of organic compounds of elements such as As and Sn lie beyond the scope of this book. For the sake of completeness, however, it may be noted that although analysis of organic As compounds has widely used atomic absorption detection systems, these have been complemented for organic As compounds by MS procedures (Christakopolous et al. 1988). For organic Sn compounds, analytical procedures have taken advantage of both flame photometry and MS systems (Jackson et al. 1982), and MS using perdeuterated standards (Arnold et al. 1998).

2.4.2.2 HPLC Systems

An extensive review (LaCourse and Dasenbrook 1998) is devoted to column liquid chromatography (LC), and includes sections that examine in detail columns and a wide range of detectors. Attention is particularly directed here to developments in the use of tandem systems including LC–GC and LC–CE. The following is therefore merely a brief overview.

Quantification may use HPLC systems, although the sensitivity is probably somewhat less than that attained by optimal GC procedures using EC detector systems. A wide range of detection systems have, however, been developed some of which are highly selective for specific groups of compounds. Probably GC and HPLC systems should therefore be regarded as complementary, and both are invaluable for appropriate applications. UV or fluorescence detector systems are most generally used although, for specific groups of compounds such as catechols, electrochemical detectors have proved valuable. Fluorescence has been extensively used in monitoring PAHs in environmental samples. Caution should be exercised since it has been shown that the major polycyclic aromatic components of polychaete samples were not anthropogenic. These fluorescent compounds have been identified as alkylated octahydrochrysenes putatively originating from triterpenes by microbial dehydrogenation of rings A and B (Farrington et al. 1986). Conclusive identification of the analytes should therefore be incorporated into a more specific protocol for HPLC analysis. This is only one example of a much wider spectrum of transformations of di- and triterpenes to aromatic compounds that take place in sediments after deposition. As a result, a structurally wide range of polycyclic compounds containing one or more aromatic rings has been isolated from sediments, oils, coal, and amber (Wakeham et al. 1980; Tan and Heit 1981; ten Haven et al. 1992; Simoneit 1998). Plausible mechanisms for the formation of such compounds have been given (Neilson and Hynning 1998). Conclusive identification of such compounds in samples of sediment and biota by MS or NMR is therefore necessary. Attention is drawn to an authoritative review of chromatographic procedures for the analysis of PAHs that discusses important issues including the differences between mono- and polyphasic packings and column temperatures (Poster et al. 1998). It is also worth noting that hydrophobic interaction chromatography has been used as an alternative to reverse-phase LC for the analysis of polar natural products (Strege 1998). This requires that the analytes be soluble in the mobile phase that may contain 75% acetonitrile containing 6.5 mM concentrations of ammonium acetate. HPLC has also been applied to the separation of the enantiomers of 2-(2,4-dichlorophenoxy)propionic acids after microbial degradation (Ludwig et al. 1992) using a_1-AGP as the chiral stationary phase. Increasing interest in the separation of enantiomers that has been noted above will certainly encourage the further application of such methods.

There has been increasing interest in the development of HPLC–MS interfaces to which attention has been drawn in Section 2.4.1.2. It should be clearly appreciated, however, that the instrumental requirements for identification and quantification are substantially different, and that a single interface is unlikely to be equally suitable for both. For identification of unknown compounds, a full-scan mass spectrum is normally required, whereas selected ion monitoring is generally adequate for quantification of established analytes (Niessen and Tinke 1995). In addition, the polarity of the analyte and its molecular weight play an important part in determining the most suitable interface. Some examples are given that illustrate experimental complexities.

- The quantification of dibenzothiophene, phenanthrothiophenes, and benzo[*b*]naphtho[2,3-*d*] thiophene in extracts of a sediment sample is complicated by the presence of higher concentrations of PAHs. This was overcome by reversed-phase LC fractionation, and identification and quantification were based on atmospheric pressure chemical ionization MS, and MS/MS using multiple reaction monitoring (Thomas et al. 1995).

- A comparison of various LC–MS systems for the analysis of complex mixtures of PAHs showed that (1) the moving belt interface was mechanically awkward and is compatible only with a limited range of mobile phases, (2) particle-beam interface had low sensitivity, and the response was nonlinear, (3) a heated nebulizer interface that uses atmospheric pressure chemical ionization (APCI) was the preferred procedure (Anacleto et al. 1995).

2.4.2.3 CE Systems

There is increasing interest in capillary zone electrophoresis (CZE) and micellar electrokinetic chromatography (MEKC) (Song et al. 1997) that is applicable both to ionizable compounds, including carboxylic acids and quaternary nitrogen compounds, and to neutral compounds such as PAHs and PCBs. Although interfaces with MS have been developed, due attention should be given to the use of compatible ionization modes. Reference should be made to an extensive review (Beale 1998) with 535 references, so that only a very superficial account is given here. The diversity of its potential is summarized briefly in the following examples of its application.

1. Separation of PAHs has been achieved using various electrophoretic systems including CZE with the addition of sodium dioctyl sulfosuccinate to the acetonitrile/water phase (Shi and Fritz 1995), and cyclodextrin modified MEKC (Terabe et al. 1993). A range of PAHs with two to six rings was separated using polysodium undecylenic sulfate and sodium phosphate/sodium borate buffer (Shamsi et al. 1998), and baseline separation of all 16 was achieved under optimal conditions. Rapid separation of the 16 priority pollutant PAHs was achieved using capillary electrochromatography in which the mobile phase is driven by electroosmosis using a column that was packed with 1.5 μm nonporous octadecylsilia particles (Dadoo et al. 1998).

2. A range of triazine herbicides in water has been determined by MEKC using sodium dodecyl sulfate, and was applied to water samples after solid-phase extraction (Martínez et al. 1996).

3. Explosive residues containing 2,4,6-trinitrotoluene transformation products have been examined (Kleiböhmer et al. 1993), and a range of explosives including TNT and related compounds, and nitramines analyzed using amperometric detection with a silver-on-gold electrode (Hilmi et al. 1999).

4. Amino azaarenes from rainwater have been analyzed (Wu et al. 1995).

5. Ethylenediamine tetraacetate in human plasma and urine have been analyzed using ion spray tandem MS for detection (Sheppard and Henion 1997).

6. Chlorinated phenols have been analyzed using amperometric detection (Hilmi et al. 1997), as well as a range of amines at pH 2.35 in phosphate buffer containing 1,3-diamino propane (Cavallo et al. 1995).

7. Separation of 15 bile acids was performed using photometric detection (Yarabe et al. 1998).

8. Alkyl phosphonates were separated in a buffered medium containing Triton X-100, as well as the electroosmotic modifiers didodecyldimethylammonium hydroxide or cetyltrimethylammonium hydroxide and conductivity detection (Nassar et al. 1998).

9. Capillary electrophoresis procedures have become firmly established for the analysis of enantiomers:

   a. Phenoxyacids, and the effective separation of the enantiomers of phenoxypropionic acids (Nielen 1993);

   b. For separation of enantiomeric phenoxypropionates, tri-*O*-methyl-β-cyclodextrin (Garrison et al. 1994) was successfully used to determine the relative degradability of *S*(–)- and *R*(+)-2,4-dichlorophenoxy-2-propionate (Garrison et al. 1994 a,b);

   c. A series of 6-sulfato-β-cyclodextrins was evaluated for the resolution of a range of neutral, acidic, and basic mixtures of enantiomers (Vincent et al. 1997a,b);

   d. γ-Cyclodextrin-modified MEKC using 2-(*N*-cyclohexaylamino)ethansulfonate and sodium dodecyl sulfate micelles was used to separate enantiomers of the nonplanar PCB congeners (IUPAC nos. 45, 88, 91.95,136, 139, 149, 183, and 198) (Marina et al. 1996).

   e. Chiral selection of a wide range of carboxylic acids was achieved using *d*(+)-tubocurarine chloride (Nair et al. 1998).

2.4.3 Application of Immunological Assays

The preparation of environmental samples for analysis is often complex, time-consuming, and expensive. Attempts have therefore been made to circumvent this by using immunological assays. Enzyme-linked immunosorbent assays (ELISA) have been extensively applied to screening a number of environmental matrices and samples of various foods for the presence of agrochemical residues (Vanderlaan et al. 1988; Van Emon and Lopez-Avila 1992), and both ELISA and radioimmunoassays (RIA) have been summarized for a wide range of agrochemicals, although many fewer have been applied to field samples (Meulenberg et al. 1995). Nonetheless, although these have been applied increasingly to a number of pesticides and to chlorinated dibenzo[1,4]dioxins, there are a number of technical problems including their relatively low sensitivity, their lack of specificity, and interference from other compounds.

- An ELISA system has been evaluated using a polyclonal antibody system for the analysis of triazine herbicides and this was effective at the low concentrations ($< \sim 2$ μg/l) that are encountered in field situations. Confirmation of the results by GC–MS is still necessary, however (Thurman et al. 1990). This was clearly illustrated in this case since the structurally related triazines ametryn, prometon, and related compounds displayed significant cross-reactivity. On the other hand, atrazine metabolites give much lower response.

- The ELISA system has been effectively applied to monitoring a group of triazine herbicides in watercourses over a wide area in the corn- and soybean-growing areas of the United States (Thurman et al. 1992).

- Immunoassays have been used for the anilide herbicide alachlor, although initial assays were marred by false positives from the sulfonate metabolite; in view of its greater water solubility and apparent persistence, a specific ELISA method was developed and used to carry out a widespread screening for both alachlor and the sulfonate metabolite (Thurman et al. 1996). Although sample preparation is considerably less complex than that used for conventional chemical analysis, confirmation of the identity of the analytes is obligatory.

- A monoclonal immunoassay has been developed for coplanar PCBs (Chiu et al. 1995) and has been carefully evaluated for selectivity and sensitivity. The sensitivity was high by comparison to GC–MS assays in which concentration and cleanup of samples are required, and there was no cross-reaction from 2,3,7,8-tetrachlorodibenzo[1,4]dioxin or 2,3,7,8-tetrachlorodibenzofuran. There were certain problems, however, with cross-reactivity among congeners and this was strongly dependent on the solvent in which the assay was carried out.

It is certain that further development will make immunoassays even more attractive for monitoring concentrations of a range of xenobiotics in large numbers of samples.

2.4.4 Stable Isotope Fractionation

Elements such as C, N, O, S, Cl that are components of many organic compounds exist naturally as mixtures of stable isotopes. The ratios of these in a compound reflect the different rates of reaction at isotopically labeled positions and therefore reflect the fractionation — biotic or abiotic — to which the compound has been subjected or by which it was synthesized. Techniques have been developed whereby the ratios $^{13}C/^{12}C$ ($\delta\ ^{13}C$), $^{15}N/^{14}N$ ($\delta\ ^{15}N$), $^{18}O/^{16}O$ ($\delta\ ^{18}O$), $^{34}S/^{32}S$ ($\delta\ ^{34}S$), $^{37}Cl/^{35}Cl$ ($\delta\ ^{37}Cl$) can be accurately measured by MS: the differences are expressed as per mil (o/oo) deviations with respect to a standard:

$$\delta = 1000\ [(\text{ratio for sample}/\text{ratio for standard}) - 1]$$

Standard samples are used for calibration, and procedures for C and O have been given in detail by Craig (1957) and include details of the standards, of the appropriate correction factors that should be applied, and of the MS techniques. The standard for carbon is the Vienna Pee Dee Belemnite (VPDM), and for oxygen the Vienna Standard Mean Ocean Water (VMOW). Secondary standards have included graphite and atmospheric oxygen that have been related to the primary standards. Atmospheric air has been used for nitrogen, and ocean water chloride for chlorine. These values may differ slightly from those derived from the relative isotope contributions given in compilations of the elements.

The analyte must be converted into a volatile compound suitable for MS analysis. Procedures for C, N, and O follow those developed for conventional organic microanalysis: oxidation of organic C to CO_2, reduction of organic N to N_2, and conversion of O_2 to CO or CO_2. In most procedures, cryogenic purification of the products is carried out before MS, and both offline and online procedures have been developed.

1. Determination of $\delta\,^{13}C$ has been carried out by GC separation of the analytes followed by high-temperature oxidation. Although oxidation with CuO at 850°C has been used, use of NiO at 1050°C supplemented by O_2 is preferable, and it is important to remove H_2O before isotopic analysis (Merritt et al. 1995).

2. For determination of $\delta\,^{18}O$, the compound is pyrolyzed with graphite embedded with platinum wire at 800°C (Aggarwal et al. 1997) or glassy carbon at 1080°C (Saurer et al. 1998) for conversion to CO_2 and CO, respectively.

3. For $\delta\,^{15}N$, the sample is heated at 850°C with an excess of Cu (to reduce NO_x to N_2) and CuO followed by cryogenic purification of the N_2 (Kucklick et al. 1996).

4. A method has been developed for converting volatile organochlorine compounds to CO_2 and CH_3Cl for measuring isotope ratios for ^{13}C and ^{37}Cl (Holt et al. 1997a). Organic carbon was converted by reaction at 550°C with CuO to CO_2 and organic chloride to CuCl. The latter was then converted into CH_3Cl by reaction with CH_3I at 300°C.

Determination of $\delta\,^{13}C$ has been used extensively in geochronology, and $\delta\,^{18}O$ in glaciology and climatology, and these techniques have also been used in determining biosynthetic pathways of, for example, lipids and the various processes whereby CO_2 is incorporated into and disseminated in biota (Hayes 1993). They have, so far, been applied more sporadically in environmental research. A few examples are given to illustrate the potential of this methodology.

Two different applications of this procedure have been used: (1) analysis of biota in a food chain to establish their trophic level and (2) analysis of individual compounds to establish their sources and alterations during transformation or dissemination. Both rely upon the isotope effect and the relative abundance of the isotopes.

1. Values of δ ^{13}C have been measured for a number of PCB congeners and applied to a number of commercial PCB mixtures. Both the number and the position of the chlorine substituents affected the depletion of ^{13}C and this reflected the manufacturing procedures involving kinetic isotope effects as well as the source of the biphenyl starting material (Jarman et al. 1998). It was suggested that this could be applied to determine the source of PCBs in the environment.

2. The values for δ ^{13}C and δ ^{18}O were examined during degradation of diesel oil by a mixed microbial culture under aerobic conditions. Oxygen in the gas phase of closed samples was analyzed by conversion to CO_2 after cryogenic separation. The values for oxygen were particularly valuable in correlating production of carbon dioxide and loss in substrate concentration (Aggarwal et al. 1997), and it was therefore suggested that this methodology could be used to provide rates of *in situ* biodegradation.

3. The degradation of 9-[^{13}C]-anthracene in soil was studied (Richnow et al. 1998), and during 588-day incubation values of δ ^{13}C in the soil changed discontinuously from −27.0 to 18.5. Further details of this experiment are given in Sections 3.2.4 and 5.5.4.

4. Organic chloride was converted to CuCl and then into CH_3Cl by reaction with CH_3I at 300°C (Holt et al. 1997a). For C_1 and C_2 chloroalkanes and chloroalkenes δ ^{13}C values ranged from −25.58 (trichloroethane) to −58.77 (CH_3Cl) and δ ^{37}Cl values from −2.86 (trichloroethane) to +1.56 (CH_2Cl_2). Although for analysis of environmental samples the method has the disadvantage that water must first be removed from the samples, it has been used to determine the distribution of trichloroethene in a contaminated aquifer (Sturchio et al. 1998).

5. Values of both δ ^{13}C and δ ^{15}N have been given for biota in the Baltic Sea ranging from phytoplankton, through seston, zooplankton, blue mussels (*Mytilus edulis*), eider duck (*Somateria mollisima*), herring (*Clupea harengus*), and cod (*Gadus morrhua*) (Broman et al. 1992). Values of δ ^{13}C increased through the trophic levels from −22.6 to −19.3 and of δ ^{15}N from 1.5 to 12.1. Values of δ ^{15}N have been correlated with organochlorine concentrations in biota at various trophic levels (Kucklick and Baker 1998), and were used to demonstrate that, for total PCB concentrations, there was a correlation of δ ^{15}N with *both* lipid content and trophic level. This important conclusion was supported by the results of extensive studies of fish in three subarctic lakes in the Yukon Territory. It was established that for fish, linear-normalized concentrations of Σ DDT and

Σ PCB were constant with increasing values of δ ^{15}N, but that this was not the case for less hydrophobic contaminants such as Σ HCH and Σ CB (Kidd et al. 1998). This is discussed again in Section 3.5.4.

6. Determinations of δ ^{34}S have been extensively used in studies on the sulfur cycle, including reactions involving microbial anaerobic reduction of sulfate and thiosulfate (Smock et al. 1998).

2.5 General Comments

2.5.1 Introduction

In spite of the fact that this is not intended to be a handbook, some very brief comments on what may be considered organizational aspects of laboratory practice may not be out of place. These may be considered within the wider context of precautions that should be exercised in all analytical laboratories. First of all, attention is drawn to the important principles of quality control, to procedures for numerical analysis of data, and to the important issue of documentation, all of which have been covered succinctly in a review (Keith et al. 1983) and more extensively in books (Keith 1988; 1992). It cannot be too strongly emphasized that the analyst is part of a team, and that he or she should play an active part in both the planning and execution of the proposed investigation; thereby, many pitfalls — and unnecessary irritation — may be avoided. For example, some conflict may arise over the number of samples required to answer the specific questions that are posed, and resolution of this issue should take priority in planning discussions. In addition, the level of accuracy should be decided at the outset, and care taken that sufficient samples are available for duplicates to be preserved for reanalysis if necessary — and that these are preserved in an acceptable manner. It is worth emphasizing that analytical results may be used in social or political contexts in which numbers may be readily misused. The level of accuracy and interpretation of the data are therefore of cardinal importance in such circumstances.

It has already been noted in Section 2.4.1 that authentic compounds may need to be synthesized by the laboratory which also carries out analysis. Whereas this is a convenient — and indeed desirable — activity, the possibility of cross-contamination should not be underestimated. It is clearly hazardous for an analyst working with possibly nanograms of a compound to be in the close vicinity of a synthetic chemist producing gram quantities of the same substance. These two operations should therefore be separated, and particular care should be given to the design of, and access to the facility in which synthesis is carried out (Alexander et al. 1986). The operation of the ventilation system should be foolproof, and residues from synthetic activity must not enter communicating waste systems either for water or air. A good example of the effort which may be needed to trace the source of contaminants is provided by the occurrence of 1,3,6,8- and 1,3,7,9-tetrachlorodibenzo[1,4]dioxin in a laboratory for dioxin analysis. These compounds had not been synthesized in the laboratory, but gained access through the use of a phenol-based cleaning solution in which these compounds were a minor impurity (Alexander et al. 1986).

2.5.2 Laboratory Practice

A valuable review (Duinker et al. 1991) has summarized some problems encountered during analysis of PCBs and these are of sufficient general importance to merit a brief summary.

1. The purity of standards should be checked, and for GC this should include not only EC detection systems but FID systems to detect the presence of nonhalogenated impurities.

2. The purity of solvents should not be taken for granted, and if redistillation is practiced, this should be carried out in a stream of N_2 to eliminate contamination by volatile laboratory contaminants; cross-contamination is a potentially serious problem and steps should be taken to ensure that samples, standards, and solvents are not exposed to this hazard.

3. Procedures for presenting numerical data should be critically reviewed and attempts made to determine the error in the analytical results.

4. Optimum conditions in operating GC equipment including technical aspects of injection procedures should be maintained and checked periodically. Although this may seem self-evident, clear evidence of its importance — and neglect — emerged from the results of an interlaboratory study of the analysis of PCBs and agrochemicals in environmental samples (Alford-Stevens et al. 1988).

5. Attention should be directed to the important question of toxic hazards associated with some reagents that may be used. If the user is not familiar with the hazards, or how to cope with accidental spillage and emergency procedures, these compounds should not be used. With proper knowledge, good experimental practice and adequate ventilation systems, the careful experimenter may, however, safely handle such compounds. Without attempting any numerical comparison of hazard, the most dangerous compounds commonly encountered are aliphatic diazo compounds such as

diazomethane and diazoethane, dimethyl and diethyl sulfate, dicy-clohexylcarbodiimide, and phosgene. Exposure to all solvents — particularly aromatic solvents including benzene — should be min-imized, and solvents such as diethyl ether and dioxan which may form explosive peroxides under some conditions should be stored in darkness. In addition, these solvents generally contain peroxide inhibitors. The synthetic chemist is exposed to a much wider range of hazardous compounds but such discussion lies well beyond the scope of this brief summary.

2.5.3 Flexibility in Operation: An Open Approach

Standardized procedures have been developed for a number of classes of analyte, but these should not be followed slavishly. It has been clearly dem-onstrated that problems of recovery may be encountered due to the occur-rence of unprecedented chemical reactions, for example, in the analysis of phenols in water samples (Chen et al. 1991), and disappointing agreement between laboratories in an intercalibration study in which PCBs in water samples were analyzed could be attributed both to numerical errors in calcu-lation and to unacceptable operation of gas chromatographs (Alford-Stevens et al. 1988). Similar variability has been encountered during analysis of fish samples for a number of organochlorine compounds, and it has been sug-gested that for regulatory purposes a range of concentrations should be set rather than absolute values (Miskiewicz and Gibbs 1992).

Problems encountered in the analysis of soils and sediments have already been noted, and it is clear that after deposition many xenobiotics are bound to organic constituents and are therefore not accessible by simple solvent extraction. The problem of alteration in the degree of accessibility after dep-osition (aging) is unresolvable, and plausible chemical reactions between the xenobiotic and organic components of the matrix are discussed in Section 3.2.4. The extent of their reversibility is generally unpredictable so that increasing attention has been directed to purely empirical procedures for the analysis of these "bound" residues. It should be emphasized that the degree of recoverability of an analyte from a given matrix cannot validly be assessed from the results of spiking experiments, and that an inherent degree of inde-terminacy in the concentrations of xenobiotics in such matrices must be accepted. For this reason, reference samples of various matrices and contain-ing a variety of analytes including chlorophenols (Marsden et al. 1986) have been prepared to provide more realistic material.

Automation has truly revolutionized analysis, but the contribution of the skilled analyst has increased rather than diminished. The occurrence of extra-neous peaks in chromatograms due to unexpected compounds should not be merely accepted blindly but their structures should be verified. A good exam-ple is the occurrence in sediment samples of a compound which had the same GC retention time as PCB 77 (3,3′,4,4′-tetrachlorobiphenyl) but which on further examination was revealed as a C_{22} alkane (Mudroch et al. 1992). Pitfalls in interpreting mass spectra should be realized, and identification that relies solely on library searches should be accepted with caution (Swallow et al. 1988). The meaning of the term *identity* should be critically interpreted, and the qualifier *tentative* added if necessary to avoid possible misunderstanding.

2.5.4 The Spectrum of Analytes

Attention should always be directed to the presence in environmental samples of compounds not previously encountered. The occurrence of metabolites and transformation products may require the development of specific procedures for their quantification and these may be radically different from those used for their precursors. This may also necessitate substantial synthetic organic activity to provide authentic compounds (Section 2.4.1). In addition, the occurrence of such transformation products in environmental matrices complicates the evaluation of data from interlaboratory analysis of samples if appropriate attention has not been directed to this possibility. Among the more important sources of these novel analytes are the following:

- Chemical transformations during combustion whereby the products are released into the atmosphere and then enter the aquatic and terrestrial environment in the form of precipitation or particulate matter, e.g., soot;
- Reactions in the troposphere including photochemical transformations;
- Biotransformation reactions in aquatic and terrestrial systems;
- Abiotic and biotic transformations in the sediment phase.

Further details of reactions mediated by microorganisms will be found in Chapter 4, throughout Chapter 6, and by higher biota in Chapter 7 (Section 7.5). Only a brief summary will be given here to illustrate some of the widely occurring chemical, thermal, and photochemical reactions together with a few examples of transformation. Some examples in the context of monitoring are given in Section 3.6.3.

The transformation of naturally occurring terpenes has already been referred to in Section 2.4.2.2, and the occurrence of aromatized compounds and plausible mechanisms for their formation from terpenoids and steroids have been presented (Neilson and Hynning 1998).

1. The formation during combustion of a range of chlorinated aromatic compounds including polychlorinated benzenes, styrenes, and naphthalenes (Yasahura and Morita 1988) and chlorobenzoic acids (Mowrer and Nordin 1987). The formation of chlorinated dioxins and 4,4'-diphenoquinones is noted further in Chapter 3, Section 3.6.3.

2. The production of nitroaromatic compounds by both combustion and by photochemical reactions of polycyclic hydrocarbons with oxides of nitrogen and oxygenating free radicals (Nielsen et al. 1983). See also Section 4.1.2.

3. The formation of ketonic and quinonoid derivatives of aromatic hydrocarbons during combustion (Alsberg et al. 1985; Levsen 1988). The environmental distribution of benzanthrone and related compounds such as benzo[c,d]pyrene-6-one and cyclopenta[def]phenanthrene-4-one has been reviewed (Spitzer and Takeuchi 1995), and such transformation products have been identified in, for example, marine sediment samples (Fernandéz et al. 1992), fish from a contaminated river (Vassilaros et al. 1982), and urban aerosols (Galceran et al. 1995; Allen et al. 1997).

4. The analysis of azaarenes, some of which such as dibenz[a,j]acridine, dibenz[a,h]acridine, and 7H-dibenzo[c,g]carbazole are suspected carcinogens, has been discussed in detail (Herod 1998).

5. The identification of sterane and triterpene hydrocarbons in vehicle particulate emissions and that probably arise from lubricants (Rogge et al. 1993).

6. Unsaturated carboxylic acids are photooxidized to C_8 and C_9 dicarboxylic acids that have been detected in atmospheric particle samples and in recent sediments (Stephanou 1992).

7. *tris*(4-Chlorophenyl)methanol has been shown to have a global distribution in birds and marine mammals, (Jarman et al. 1992), and in samples of fish, marine mammals, and sediments from the North Sea and the Wadden Sea (de Boer et al. 1996); further comments are given in Section 3.5.3.

8. Biotransformation reactions of the original xenobiotic may result in compounds with fundamentally different physical properties and toxicities. The apparently ubiquitous distribution of halogenated anisoles (Wittlinger and Ballschmiter 1990; Führer and Ballschmiter 1998) attests to the importance of O-methylation reactions that are discussed in Section 6.11.4.

9. The demonstration and partial identification of a wide range of methyl sulfone–substituted PCBs as metabolites in samples from gray seals (*Halichoerus grypus*) serves as further confirmation (Buser et al. 1992b) of a previously established (Jensen and Jansson 1976) metabolic pathway. Further comments are given in Sections 3.6.3 and 7.5.2. Both sulfone and sulfoxide metabolites of pentachlorobenzene have been identified in samples of parsnips and were presumed to have been translocated from the soil (Cairns et al. 1987). An integrated procedure for the analysis in biological samples of aryl methyl sulfones that are metabolites of aromatic

halogenated compounds in mammals (Section 7.3.5) has been developed (Letcher et al. 1995).

10. Metabolites of the herbicides alachlor and metolachlor in which the chlorine of the –CO. $CH_2 \cdot Cl$ group is replaced by $-SO_3-$ by glutathione conjugation and oxidation have been recovered from a number of water courses in the United States Two aspects are worth noting: (a) the analytical problems of identification and (b) the fact that the sulfonate of alachlor has low bioconcentration potential and is nonmutagenic in contrast to its precursor (Field and Thurman 1996).

11. Trichloroacetic acid is formed during chlorination of wastewater and the production of pulp bleached with molecular chlorine, and analytical procedures for the analysis of a number of haloacetates in samples of water and air have been developed (Frank et al. 1995). The main source of trichloroacetic in Switzerland has been shown to be rain although its ultimate source remains unresolved (Müller et al. 1996). Plausible mechanisms for the formation of trichloro-acetic acid by atmospheric reactions involving trichoroethane and tetrachloroethene are discussed in Chapter 4, Section 4.1.3. It has also been hypothesized on the basis of its formation from aliphatic compounds, especially acetate, by the action of chloroperoxidase in the presence of hydrogen peroxide and chloride that it might be a naturally occurring metabolite (Haiber et al. 1996).

2.5.5 Multicomponent Commercial Products and Effluents

Greatest attention has been directed in the preceding sections to single compounds since these represent an analytically accessible problem and serve to illustrate most of the principles to which this chapter is devoted. It is frequently the results of such analyses that are used in environmental hazard assessments which require data on distribution, toxicity, and persistence. In practice, however, many commercial products are complex mixtures of which PCBs are probably the most conspicuous. These are, however, by no means the only examples and the following may be noted: polychlorinated naphthalenes and terphenyls, chlorinated paraffins, and the chlorinated monoterpenes toxaphene and chlordane — all of these present serious problems for their quantification and raise the specter of sum vs. congener-specific analysis which has aroused so much controversy in the case of PCBs. Attention may also be directed to creosote and to alkylphenol polyethoxylates and their halogenated analogues that have been the object of extensive studies (Ahel et al. 1987; Stephanou et al. 1988; Wahlberg et al. 1990). Many of these compounds have apparently global distributions (Chapter 3, Section 3.5) and therefore present potentially serious environmental problems.

The cardinal issue in their identification and quantification in environmental samples is, of course, the availability of authentic standard compounds, and this has been discussed in Sections 2.4.1 and 2.4.2. Whereas authentic standards of PCBs, PCDDs, and PCDFs are commercially available, increased interest in the distribution of toxaphene has drawn attention to the need for authentic samples both of chlorinated bornanes and bornenes since — as with PCBs — substantial alteration in the distribution of the various congeners has generally occurred after discharge into the environment (Bidleman et al. 1993). There is no single method for analysis that is equally suitable for all the congeners — even those generally present in commercial mixtures — and both selected- ion monitoring using electron-impact high-resolution MS (HRMS) and EC negative-ion MS are complementary rather than exclusive (Lau et al. 1996). In addition, there are many enantiomers, and resolution has been achieved using chiral columns (Buser and Müller 1994a,b) and the results used to indicate selective transformation after discharge. A particularly serious problem is represented by polychlorinated alkanes that have many congeners and enantiomers, and for which authentic standards of individual components are not available. An interlaboratory study (Tomy et al. 1999) on the quantification of commercial C_{10}–C_{13} products containing 60 and 70% chlorine revealed several serious and unresolved problems, and in the words of the authors, "the results of the study met with mixed success."

Industrial effluents almost always contain a wide range of compounds, often by-products of manufacture, and their analysis may present formidable problems both in identification and quantification. An illustrative example of site-specific contaminants is provided by the recovery from samples of sediment and fish in the Niagara River–Lake Ontario system of a benzophenone, a difluorodiphenylmethane, and several biphenyls — by-products from the manufacture of 4-chloro(trifluoromethyl)benzene — all of which contained chloro and trifluoromethyl substituents (Jaffe and Hites 1985). In addition, these effluents may contain polymeric material with which monomers are associated. A classic — and possibly unique — example is the plethora of compounds produced during the manufacture of bleached pulp by conventional technologies using molecular chlorine. These effluents contain both low-molecular-weight components and high-molecular-weight "chlorolignin" (references in Neilson et al. 1991). A recent investigation (Jokela and Salkinoja-Salonen 1992) has revealed a new facet of the composition of such effluents. Analysis of the molecular weight distribution of these compounds by size exclusion chromatography (SEC) and by ultrafiltration has revealed significant differences in the results obtained by the two procedures. Micelle formation during ultrafiltration which is broken by dilution, and association phenomena in aqueous solutions that are analyzed by SEC led to significantly revised estimates of the molecular weight distribution of the components. In contrast to previous estimates, approximately 85 to 95% of the organochlorine constituents have molecular weights <1000. The importance of these observations, and of associations in general, should therefore be critically evaluated in analyzing complex effluents.

2.6 Conclusions

Quantitative data are essential components of virtually all environmental investigations, but the range of chemical compounds involved and the different matrices in which xenobiotics may be encountered makes generalizations on analytical procedures extremely hazardous. Whereas it is relatively straightforward to take advantage of modern instrumentation and methodology, and to apply widely accepted procedures for detection, identification, and quantification, there are several fundamental difficulties to which specific attention should be directed and possible solutions sought:

1. Procedures for preparation of samples before analysis are of cardinal importance and particular emphasis should be placed on procedures for removal of interfering substances without destruction or alteration of the analyte.

2. Attention should be given to the possibility that the compounds are not present in the "free" state but are associated with macromolecules, both soluble and insoluble organic matter, and inorganic material. Analytical procedures should attempt to take this into account using the appropriate empirical fractionation protocols.

3. There are inherent dangers in assessing recoverability from the results of spiking experiments, and the significance of aging in natural sediments should be appreciated even if it cannot be quantitatively evaluated.

4. Attention should be directed to developments in structural determination including new techniques in mass spectrometry, the increasing accessibility of X-ray methods, and particularly the value of nondestructive methods such as NMR.

5. All natural systems are dynamic, and appropriate attention should be directed to kinetic processes such as sorption and desorption, and to the significance and complexities arising from abiotic transformation and the metabolism of xenobiotics by biota.

References

Aga, D.S., E.M. Thurman, M.E. Yockel, L.R. Zimmerman, and T.D. Williams. 1996. Identification of a sulfonic acid metabolite of metolachlor in soil. *Environ. Sci. Technol.* 30: 592–597.

Aggarwal, P.K., M.E. Fuller, M.M. Gurgas, J.F. Manning, and M.A. Dillon. 1997. Use of stable oxygen and carbon isotope analyses for monitoring the pathways and rates of intrinsic and enhanced *in situ* biodegradation. *Environ. Sci. Technol.* 31: 590–596.

Ahel, M., T. Conrad, and W. Giger. 1987. Persistent organic chemicals in sewage effluents. 3. Determinations of nonylphenoxy carboxylic acids by high-resolution gas chromatography/mass spectrometry and high-performance liquid chromatography. *Environ. Sci. Technol.* 21: 697–703.

Ahvazi, B.C., C. Crestini, and D.S. Argyropoulos. 1999. ^{19}F nuclear magnetic resonance spectroscopy for the quantitative detection and classification of carbonyl groups in lignin. *J. Agric. Food Chem.* 47: 190–201.

Albert, K. 1995. On-line use of NMR detection in separation chemistry. *J. Chromatogr. A* 703: 123.

Alexander, L.R, D.G. Patterson, G.L. Myers, and J.S. Holler. 1986. Safe handling of chemical toxicants and control of interferences in human tissue analysis for dioxins and furans. *Environ Sci. Technol.* 20: 725–730.

Alford-Stevens, A.L., J.W. Eichelberger, and W.L. Budde. 1988. Multilaboratory study of automated determinations of polychlorinated biphenyls and chlorinated pesticides in water, soil, and sediment by gas chromatography/mass spectrometry. *Environ. Sci. Technol.* 22: 304–312.

Allard, A.-S., M. Remberger, and A.H. Neilson. 1985. Bacterial O-methylation of chloroguaiacols: effect of substrate concentration, cell density, and growth conditions. *Appl. Environ. Microbiol.* 49: 279–288.

Allard, A.-S., P.-Å. Hynning, C. Lindgren, M. Remberger, and A.H. Neilson. 1991. Dechlorination of chlorocatechols by stable enrichment cultures of anaerobic bacteria. *Appl. Environ. Microbiol.* 57: 77–84.

Allen, J.O., N.M. Dookeran, K. Taghizadeh, A.L. Lafleur, K.A. Smith, and A.F. Sarofim. 1997. Measurement of oxygenated polycyclic aromatic hydrcarbons associated with a size-segregated urban aerosol. *Environ. Sci. Technol.* 31: 2064–2070.

Alsberg, T., U. Stenberg, R. Westerholm, M. Strandell, U. Rannug, A. Sundvall, L. Romert, V. Bernson, B. Petterson, R. Toftgård, B. Franzén, M. Jansson, J.Å. Gustafsson, K.E. Egebäck, and G. Tejle. 1985. Chemical and biological characterization of organic material from gasoline exhaust particles. *Environ. Sci. Technol.* 19: 43–50.

Anacleto, J.F., L. Ramaley, F.M. Benoit, R.K. Boyd, and M.A. Quilliam. 1995. Comparison of liquid chromatography/mass spectrometry interfaces for the analysis of polycyclic aromatic compounds. *Anal. Chem.* 67: 4145–4154.

Andreae, M.O. and P.J. Crutzen. 1997. Atmospheric aerosols: biogeochemical sources and role in atmospheric chemistry. *Science* 276: 1052–1058.

Appelblad, P., T. Jonsson, T. Bäckström, and K. Irgum. 1998. Determination of C-21 ketosteroids in serum using trifluromethansulfonic acid catalyzed precolumn dansylation and 1,1′-oxalyldiimidazole postcolumn peroxyoxalate chemiluminescent detection. *Anal. Chem.* 70: 5002–5009.

Argyropoulos, D.S., and L. Zhang. 1998. Semiquantitative determination of quinonoid structures in isolated lignins by ^{31}P nuclear magnetic resonance. *J. Agric. Food Chem.* 46: 4626–4634.

Arjmand, M., T.D. Spittler, and R.O. Mumma. 1988. Analysis of Dicamba from water using solid-phase extraction and ion-pair high-performance liquid chromatography. *J. Agric. Food Chem.* 36: 492–494.

Armstrong, D.W., Y. Tang, and J. Zukowski. 1991. Resolution of enantiometic hydrocarbon biomarkers of geochemical importance. *Anal. Chem.* 63: 2858–2861.

Armstrong, D.W., G.L. Reid, M.L. Hilton, and C.-D. Chang. 1993. Relevance of enantiomer separations in environmental science. *Environ. Pollut.* 79: 51–58.

Arnold, C.G., M. Berg, S.R, Müller, U. Dommann, and R.P. Schwartzenbach. 1998. Determination of organotin compounds in water, sediments, and sewage sludge using perdeuterated internal standards, accelerated solvent extraction, and large-volume-injection GC/MS. *Anal. Chem.* 70: 3094–3101.

Arpino, P. 1990. Coupling techniques in LC/MS and SFC/MS. *Fresenius Z. Anal. Chem.* 337: 667–685.

Arpino, P.J., F. Sadoun, and H. Virelizier. 1993. Reviews on recent trends in chromatography/mass spectrometry coupling. Part IV. Reasons why supercritical fluid chromatography is not so easily coupled with mass spectrometry as originally assessed. *Chromatographia* 36: 283–288.

Arthur, C.L., L.M. Killam, S. Motlagh, M. Lim, D.W. Potter, and J. Pawliszyn. 1992. Analysis of substituted benzene compounds in groundwater using solid-phase microextraction. *Environ. Sci. Technol.* 26: 979–983.

Ashraf-Khorassani, M. and L.T. Taylor. 1990. Nitrous oxide versus carbon dioxide for supercritical fluid extraction and chromatography of amines. *Anal. Chem.* 62: 1177–1180.

Aubin, A.J. and A.E. Smith. 1992. Extraction of [^{14}C]glyphosate from Saskatchewan soils. *J. Agric. Food Chem.* 40: 1163–1165.

Bakola-Christianopoulou, M.N., V.P. Papageorgiou, and K.K. Apazidou. 1993. Gas chromatographic–mass spectrometric study of the reductive silylation of hydroxyquinones. *J. Chromatogr.* 645: 293–301.

Barceló, D. 1991. Occurence, handling, and chromatographic determination of pesticides in the aquatic environment. *Analyst* 116: 681–689.

Barceló, D. 1992. Mass spectrometry in environmental organic analysis. *Anal. Chim. Acta* 263: 1–19.

Barceló, D., G. Durand, V. Bouvot, and M. Nielen. 1993. Use of extraction disks for trace enrichment of various pesticides from river water and simulated seawater samples followed by liquid chromatography-rapid-scanning UV-visible and thermospray-mass spectrometry detection. *Environ. Sci. Technol.* 27: 271–277.

Barry, J.P., C. Norwood, and P. Vouros. 1996. Detection and identification of benzo[a]pyrene diol epoxide adducts to DNA utilizing capillary electrophoresis-electrospray mass spectrometry. *Anal. Chem.* 68: 1432–1438.

Barshick, S.-A. and W.H. Griest. 1998. Trace analysis of explosives in seawater using solid-phase microextraction and gas chromatography/ion trap mass spectrometry. *Anal. Chem.* 70: 3015–3020.

Bartle, K.D., M.L. Lee, and S.A. Wise. 1981. Modern analytical methods for environmental polycyclic aromatic compounds. *Chem. Soc. Rev.* 10: 113–158.

Bartle, K.D., M.W. Raynor, A.A. Clifford, I.L. Davies, J.P. Kithinji, G.F. Shilstone, J.M. Chalmers, and B.W. Cook. 1989. Capillary supercritical fluid chromatography with Fourier transform infrared detection. *J. Chromatogr. Sci.* 27: 283–292.

Beale, S.C. 1998. Capillary electrophoresis. *Anal. Chem.* 70: 279R–300R.

Behymer, T.D., T.A. Bellar, and W.L. Budde. 1990. Liquid chromatography/particle beam/mass spectrometry of polar compounds of environmental interest. *Anal. Chem.* 62: 1686–1690.

Bemgård, A., Colmsjö, A., and Lundmark, B.-O. 1993. Gas chromatographic analysis of high-molecular-mass polycyclic aromatic hydrocarbons II. Polycyclic aromatic hydrocarbons with relative molecular masses exceeding 328. *J. Chromatogr.* 630: 287–295.

Benner, B.A. 1998. Summarizing the effectiveness of supercritical fluid extraction of polycyclic aromatic hydrocarbons from natural matrix environmental samples. *Anal. Chem.* 70: 4594–4601.

Bergman, Å. and C.A. Wachtmeister. 1978. Synthesis of methylthio- and methylsulphonyl-polychlorobiphenyls via nucleophilic aromatic substitution of certain types of polychlorobiphenyls. *Chemosphere* 7: 949–956.

Bernal, J.L., M.J. Del Nozal, and J.J. Jiménez. 1992. Some observations on clean-up procedures using sulphuric acid and florisil. *J. Chromatogr.* 607: 303–309.

Bidleman, T.F., M.D. Walla, D.C.G. Muir, and G.A. Stern. 1993. Selective accumulation of polychlorocamphenes in aquatic biota from the Canadian Arctic. *Anal. Chim. Acta* 12: 701–709.

Biggs, W.R. and J.C. Fetzer. 1996. Analytical techniques for large polycyclic aromatic hydrocarbons: a review. *Trends Anal.Chem.* 15: 196–205.

Birkholz, D.A., R.T. Coutts, and S.E. Hrudey. 1988. Determination of polycyclic aromatic compounds in fish tissue. *J. Chromatogr.* 449: 251–260.

Blau, K. and J.M. Halket (Eds.) 1993. *Handbook of Derivatives for Chromatography.* John Wiley & Sons, New York.

Blessington, B., N. Crabb, S. Karkee, and A. Northage. 1993. Chromatographic approaches to the quality control of chiral propionate anti-inflammatory drugs and herbicides. *J. Chromatogr.* 469: 183–190.

Bollinger, M.J., R.E. Sievers, D.W. Fahey, and F.C. Fehsenfeld. 1983. Conversion of nitrogen dioxide, nitric acid, and *n*-propyl nitrate to nitric oxide by gold-catalyzed reduction with carbon monoxide. *Anal. Chem.* 55: 1980–1986.

Böwadt, S. and S.B. Hawthorne. 1995. Supercritical fluid extraction in environmental analysis. *J. Chromatogr.* A 703: 549–571.

Boyd, D.R., M.R.J. Dorrity, M.V. Hand, J.F. Malone, N.D. Sharma, H. Dalton, D.J. Gray, and G.N. Sheldrake. 1991. Enantiomeric excess and absolute configuration of *cis*-dihydrodiols from bacterial metabolism of monocyclic arenes. *J. Am. Chem. Soc.* 113: 666–667.

Boyd, D.R., N.D. Sharma, R. Boyle, R.A.S. McMordie, J. Chima, and H. Dalton. 1992. A ¹H-NMR method for the determination of enantiomeric excess and absolute configuration of *cis*-dihydrodiol metabolites of polycyclic arenes and heteroarenes. *Tetrahedron Lett.* 33: 1241–1244.

Brand, J.M., D.L. Cruden, G.J. Zylstra, and D.T. Gibson. 1992. Stereospecific hydroxylation of indan by *Escherichia coli* containing the cloned toluene dioxygenase genes from *Pseudomonas putida* F1. *Appl. Environ. Microbiol.* 58: 3407–3409.

Brandl, H., K.W. Hanselmann, R. Bachofen, and J. Piccard. 1993. Small-scale patchiness in the chemistry and microbiology of sediments in Lake Geneva, Switzerland. *J. Gen. Microbiol.* 139: 2271–2275.

Broman, D., C. Näf, C. Rolff, and Y. Zebühr. 1991. Occurrence and dynamics of polychlorinated dibenzo-*p*-dioxins and dibenzofurans and polycyclic aromatic hydrocarbons in the mixed surface layer of remote coastal and offshore waters of the Baltic. *Environ. Sci. Technol.* 25: 1850–1863.

Brooks, C.J.W. and W.J. Cole. 1988. Cyclic di-*tert*-butylsilylene derivatives of substituted salicylic acids and related compounds. A study by gas chromatography–mass spectrometry. *J. Chromatogr.* 441: 13–29.

Brumley, W.C., C.M. Brownrigg, and A.H. Grange. 1993. Determination of toxaphene in soil by electron-capture negative ion mass spectrometry after fractionation by high-performance gel permeation chromatography. *J. Chromatogr.* 633: 177–183.

Bucheli, T.D., F.C. Grübler, S.R. Müller, and R.P. Schwarzenbach. 1997. Simultaneous determination of neutral and acidic pesticides in natural waters at the low nanogram per liter level. *Anal. Chem.* 69: 1569–1576.

Büldt, A. and U. Karst. 1997. 1-methyl-1-(2,4-dinitrophenyl)hydrazine as a new reagent for the HPLC determination of aldehydes. *Anal. Chem.* 69: 3617–3622.

Burka, L.T. and D. Overstreet. 1989. Synthesis of possible metabolites of 2,3,7,8-tetrachlorodibenzofuran. *J. Agric. Food Chem.* 37: 1528–1532.

Burlingame, A.L., R.K. Boyd, and S.J. Gaskell. 1998. Mass spectrometry. *Anal. Chem.* 70: 647R–716R.

Buser, H.-R. 1986. Selective detection of brominated aromatic compounds using gas chromatography/negative chemical ionization mass spectrometry. *Anal. Chem.* 58: 2913–2919.

Buser, H.-R. and M.D. Müller. 1992. Enantiomer separation of chlordane components and metabolites using chiral high resolution gas chromatography and detection by mass spectrometric techniques. *Anal. Chem.* 64: 3168–3175.

Buser, H.-R. and M.D. Müller. 1994a. Isomeric and enantiomeric composition of different commercial toxaphenes and of chlorination products of (+)- and (–)-camphenes. *J. Agric. Food Chem.* 42: 393–400.

Buser, H.-R. and M.D. Müller. 1994b. Isomer- and enantiomer-selective analyses of toxaphene components using chiral high-resolution gas chromatography and detection by mass spectrometry/mass spectrometry. *Environ. Sci. Technol.* 28: 119–128.

Buser, H.-R. and M.D. Müller. 1997. Conversion reactions of various phenoxyalkanoic acid herbicides in soil. 2. Elucidation of the enantiomerization process of chiral phenoxy acids from incubation in a D_2O/soil system. *Environ. Sci. Technol.* 31: 1960–1967.

Buser, H.-R., D.R. Zook, and C. Rappe. 1992b. Determination of methyl sulfone-substituted polychlorobiphenyls by mass spectrometric techniques with application to environmental samples. *Anal. Chem.* 64: 1176–1183.

Buser, H.-R., M.D. Müller, and C. Rappe. 1992a. Enantioselective determination of chlordane components using chiral high-resolution gas chromatography-mass spectrometry with application to environmental samples. *Environ. Sci. Technol.* 26: 1533–1540.

Bush, B. and E.L. Barnard. 1995. Gas phase infrared spectra of 209 polychlorinated biphenyl congeners using gas chromatography with Fourier transform infrared detection: internal standardization with a ^{13}C-labelled congener. *Arch. Environ. Contam. Toxicol.* 29: 322–326.

Cairns, T., E.G. Siegmund, and F. Krick. 1987. Identification of several new metabolites from pentachloronitrobenzene by gas chromatography/mass spectrometry. *J. Agric. Food Chem.* 35: 433–439.

Campbell, J.A., M.A. LaPack, T.L. Peters, and T.A. Smock. 1987. Gas chromatography/mass spectroscopy identification of cyclohexene artifacts formed during extraction of brine samples. *Environ. Sci. Technol.* 21: 110–112.

Capriel, P., A. Haisch, and S.U. Khan. 1986. Supercritical methanol: an efficaceous technique for the extraction of bound pesticides from soil and plant samples. *J. Agric. Food Chem.* 34: 70–73.

Castillo, M., M.F. Alpendurada, and D. Barceló. 1997. Characterization of organic pollutants in industrial effluents using liquid chromatography-atmospheric pressure chemical ionization-mass spectrometry. *J. Mass Spectrom.* 32: 1100–1110.

Castro, C.E., S.K. O'Shea, W. Wang, and E.W. Bartnicki. 1996. ^{13}C-NMR reactivity probes for the environment. *Environ. Sci. Technol.* 30: 1185–1191.

Cavallo, A., V. Piangerelli, F. Nerini, S. Cavalli, and C. Reschiotto. 1995. Selective determination of aromatic amines in water samples by capillary zone electrophoresis and solid-phase extraction. *J. Chromatogr.* A 709: 361–366.

Chen, H.-Y. and M.R. Preston. 1998. Azaarenes in the aerosol of an urban atmosphere. *Environ. Sci. Technol.* 32: 577–583.

Chen, P. and M.V. Novotny. 1997. 2-Methyl-3-oxo-4-phenyl-2,3-dihydrofuran-2-yl acetate: a fluorogenic reagent for detection and analysis of primary amines. *Anal. Chem.* 69: 2806–2811.

Chen, P.H., W.A. VanAusdale, and D.F. Roberts. 1991. Oxidation of phenolic acid surrogates and target analytes during acid extraction of neutral water samples for analysis by GC/MS using EPA method 625. *Environ. Sci. Technol.* 25: 540–546.

Chester, T.L., J.D. Pinkston, and D.E. Raynie. 1998. Supercritical fluid chromatography and extraction. *Anal. Chem.* 70: 301R–319R.

Chien, C.-J., M.J. Xharles, K.G. Sexton, and H.E. Jeffries. 1998. Analysis of ariborne carboxylic acids and phenols as their pentafluorobenzyl derivatives: gas chromatography/ion trap mass spectrometry with a novel chemical ionization treagent, PFBOH. *Environ. Sci. Technol.* 32: 299–309.

Chiu, Y.-W., R.E. Carlson, K.L. Marcus, and A.E. Karu. 1995. A monoclonal immunoassay for the coplanar polychlorinated biphenyls. *Anal. Chem.* 67: 3829–3939.

Chladek, E. and R.S. Marano. 1984. Use of bonded phase silica sorbents for the sampling of priority pollutants in wastewaters. *J. Chromatogr. Sci.* 22: 313–320.

Christakopolous, A., B. Hamasur, H. Norin, and I. Nordgren. 1988. Quantitative determination of arsenocholine and acetylarsenocholine in aquatic organisms using pyrolysis and gas chromatography-mass spectrometry. *Biomed. Environ. Mass Spectrom.* 15: 67–74.

Christie, W.W. 1989. Silver ion chromatography using solid-phase extraction columns packed with a bonded-sulfonic acid phase. *J. Lipid Res.* 30: 1471–1473.

Coleman, W.E., J.W. Munch, R.W. Slater, R.G. Melon, and F.C. Kopfler. 1983. Optimization of purging efficiency and quantification of organic contaminants from water using a 1-L closed-loop stripping apparatus and computerized capillary column GC/MS. *Environ. Sci. Technol.* 17: 571–576.

Colmsjö, A. 1998. Concentration and extraction of PAHs from environmental samples pp. 55–76. In *PAHs and Related Compounds* (Ed. A.H. Neilson). *Handbook of Environmental Chemistry* (Ed. O. Hutzinger) Vol. 3, Part I. Springer-Verlag, Berlin

Craig, H. 1957. Isotopic standards for carbon and oxygen and correction factors for mass-spectrometric analysis of carbon dioxide. *Geochim. Cosmochim. Acta* 12: 133–149.

Creaser, C.S., F. Krokos, and J.R. Startin. 1992. Analytical methods for the determination of non-*ortho*-substituted chlorobiphenyls: a review. *Chemosphere* 25: 1981–2008.

Czuczwa, J.M. and R.A. Hites. 1986. Airborne dioxins and dibenzofurans: sources and fates. *Environ. Sci. Technol.* 20: 195–200.

Dadoo, R., R.N. Zare, C. Yan, and D.S. Anex. 1998. Advances in capillary electrochromatography: rapid and high-efficiency separations of PAHs. *Anal. Chem.* 70: 4787–4792.

Dahlgran, J.R. 1981. Simultaneous detection of total and halogenated hydrocarbons in complex environmental samples. *J. High Resolution Chromatogr. Commun.* 4: 393–397.

de Boer, J. 1988. Chlorobiphenyls in bound and non-bound lipids of fishes: comparison of different extraction methods. *Chemosphere* 17: 1803–1810.

de Boer, J., P.G. Wester, E.H.G. Evers, and U.A.T. Brinkman. 1996. Determination of *tris*4-chlorophenylmethanol and *tris*4-chlorophenylmethane in fish, marine mammals and sediment. *Environ. Pollut.* 93: 39–47.

de Boer, J., H.-J. de Geus, and U.A.T. Brinkman. 1997. Multidimensional gas chromatographic analysis of toxaphene. *Environ. Sci. Technol.* 31: 873–879.

de Kok, A., R.B. Geerdink, R.W. Frei, and U.A.Th. Brinkman. 1981. The use of dechlorination in the analysis of polychlorinated biphenyls and related classes of compounds. *Int. J. Environ. Anal. Chem.* 9: 301–318.

de Pascual-Teresa, S., D. Treutter, J.C. Rivas-Gonzalo, and C. Santos-Buelga. 1998. Analysis of flavanols in beverages by high-performance liquid chromatography with chemical reaction detection. *J. Agric. Food Chem.* 46: 4209–4213.

De Witt, J.S.M., C.E. Parker, K.B. Tomer, and J.W. Jorgenson. 1988. Separation and identification of trifluralin metabolites by open-tubular liquid chromatography/negative chemical ionization mass spectrometry. *Biomed. Environ. Mass Spectrom.* 17: 47–53.

Dearth, M.A. and R.A. Hites. 1991. Complete analysis of technical chlordane using negative ionization mass spectrometry. *Environ. Sci. Technol.* 25: 245–254.

deBruin, L.S., P.D. Josephy, and J.B. Pawliszyn. 1998. Solid-phase microextraction of monocyclic aromatic amines from biological fluids. *Anal. Chem.* 70: 1986–1992.

Deforce, D.L.D., J. Raymackers, L. Meheus, F. Van Wijnendaele, A. De Leenheer, E.G. Van den Eeckhout. 1998. Characterization of DNA oligonucleotides by coupling of capillary zone electrophoresis to electrospray ionization Q-TOF mass spectrometry. *Anal. Chem.* 70: 3060–3068.

Desbrow, C., E.J. Routledge, G.C. Brighty, J.P. Sumpter, and M. Waldock. 1998. Identification of estrogenic chemicals in STW effluent. 1. Chemical fractionation and *in vitro* biological screening. *Environ. Sci. Technol.* 32: 1549–1558.

Di Corcia, A., A. Costantino, C. Crescenti, E. Marinoni, and R. Samperi. 1998. Characterization of recalcitrant intermediates from biotransformation of the branched alkyl side chain of nonylphenol ethoxylate surfactants. *Environ. Sci. Technol.* 32: 2401–2409.

Dietrich, A.M., D.S. Millington, and Y.-H. Seo. 1988. Specific identification of synthetic organic chemicals in river water using liquid–liquid extraction and resin adsorption coupled with electron impact, chemical ionization and accurate mass measurement gas chromatography–mass spectrometry analysis. *J. Chromatogr.* 436: 229–241.

Dietrich, D., W.J. Hickey, and R. Lamar. 1995. Degradation of 4,4'-dichlorobiphenyl, 3,3',4,4'-tetrachlorobiphenyl, and 2,2',4,4',5,5'-hexachlorobiphenyl by the white rot fungus *Phanerochaete chrysosporium*. *Appl. Environ. Microbiol.* 61: 3904–3909.

Donnelly, J.R., W.D. Munslow, T.L. Vonnahme, N.J Nunn, C.M. Hedin, G.W. Svocool, and R.K. Mitchum. 1987. The chemistry and mass spectrometry of brominated dibenzo-*p*-dioxins and dibenzofurans. *Biomed. Environ. Mass Spectrom.* 14: 465–472.

Downing, J.A. and L.C. Rath. 1988. Spatial patchiness in the lacustrine environment. *Limnol. Oceanogr.* 33: 447–458.

Duinker, J.C., D.E. Schulz, and G. Petrick. 1988. Multidimensional gas chromatography with electron capture detection for the determination of toxic congeners in polychlorinated biphenyl mixtures. *Anal. Chem.* 60: 478–482.

Duinker, J.C., D.E. Schulz, and G. Petrick. 1991. Analysis and interpretation of chloro-biphenyls: possibilities and problems. *Chemosphere* 23: 1009–1028.

Durand, G., P. Gille, D. Fraisse, and D. Barceló. 1992. Comparison of gas chromato-graphic–mass spectrometic methods for screening of chlorotriazione pesticides in soil. *J. Chromatogr.* 603: 175–184.

Durlak, S.K., P. Biswas, J. Shi, and M.J. Bernhard. 1998. Characterization of polycyclic aromatic hydrocarbon particulate and gaseous emissions from polystyrene com-bustion. *Environ. Sci. Technol.* 32: 2301–2307.

Easterling, M.L., C.M. Colangelo, R.A. Scott, and I.J. Amster. 1998. Monitoring protein expression in whole bacterial cells with MALDI time-of-flight mass spectrometry. *Anal. Chem.* 70: 2704–2709.

Farrington, J.W., S.G. Wakeham, J.B Livramenta, B.W. Tripp, and J.M. Teal. 1986. Aromatic hydrocarbons in New York Bight polychaetes: ultraviolet fluorescence analysis and gas chromatography/gas chromatography–mass spectrometry analysis. *Environ. Sci. Technol.* 20: 69–72.

Fayad, N.M. 1988. Gas chromatography/mass spectroscopy identification of artifacts formed in methylene chloride extracts of saline water. *Environ. Sci. Technol.* 22: 1347–1348.

Fenimore, D.C., C.M. Davis, J.H. Whitford, and C.A. Harrington. 1976. Vapor phase silylation of laboratory glassware. *Anal. Chem.* 48: 2289–2290.

Fernández, P., M. Grifoll, A.M. Solanas, J. M. Bayiona, and J. Albalgés. 1992. Bioassay-directed chemical analysis of genotoxic compounds in coastal sediments. *Environ. Sci. Technol.* 26: 817–829.

Field, J.S. and E.M. Thurman. 1996. Glutathione conjugation and contaminant trans-formation. *Environ. Sci. Technol.* 30: 1413–1418.

Fischer, E. 1909. *Untersuchungen über Kohlenhydrate und Fermente (1884–1908)*. pp. 138–238. Verlag Von Julius Springer, Berlin.

Fitzpatrick, F.A. and S. Siggia. 1973. High resolution liquid chromatography of derivatized non-ultraviolet-absorbing hydroxy steroids. *Anal. Chem.* 45: 2310–2314.

Foreman, W.T. and P.M. Gates. 1997. Matrix-enhanced degradation of p,p'-DDT during gas chromatographic analysis: a consideration. *Environ. Sci. Technol.* 31: 905–910.

Foreman, W.T., S.D. Zaugg, L.M. Faires, M.G. Werner, T.J. Leiker, and P.F. Rogerson. 1992. Analytical interferences of mercuric chloride preservative in environmental water samples: determination of organic compounds isolated by continuous liquid–liquid extraction or closed-loop stripping. *Environ. Sci. Technol.* 26: 1307–1312.

Forstner, H.J.L., R.C. Flagan, and J.H. Seinfeld. 1997. Secondary organic aerosol from the photooxidation of aromatic hydrocarbons: molecular composition. *Environ. Sci. Technol.* 31: 1345–1358.

Foucault, A.P. 1991. Countercurrent chromatography. *Anal. Chem.* 63: 569A–579A.

Fox, P.A., J. Carter, and P. Farrimond. 1998. Analysis of bacteriohopanepolyols in sediment and bacterial extracts by high performance liquid chromatography/atmospheric pressure chemical ionization mass spectrometry. *Rapid Comm. Mass Spectrom.* 12: 609–612.

France, J.E., J.W. King, and J.M. Snyder. 1991. Supercritical fluid-based cleanup tech-nique for the separation of organochlorine pesticides from fats. *J. Agric. Food Chem.* 39: 1871–1874.

Frank, H., D. Renschen, A. Klein, and H. Scholl. 1995. Trace analysis of airborne haloacetates. *J. High Resol. Chromatogr.* 18: 83–88.

Führer, U. and K. Ballschmiter. 1998. Bromochloromethoxybenzenes in the marine troposphere of the Atlantic Ocean: a group of organohalogens with mixed biogenic and anthropogenic origin. *Environ. Sci. Technol.* 32: 2208–2215.

Galceran, M.T., E. Moyano, and J.M. Poza. 1995. Pentafluorobenzyl derivatives for gas chromatographic determination of hydroxy-polycyclic aromatic hydrocarbons in urban aerosols. *J. Chromatogr.* A 710: 139–147.

Garrison, A.W., P. Schmitt, and A. Kettrup. 1994a. Enantiomeric selectivity in the environmental degradation of dichloroprop as determined by high-performance capillary electrophoresis. *Environ. Sci. Technol.* 30: 2449–2455.

Garrison, A.W., P. Schmitt, and A. Kettrup. 1994b. Separation of phenoxy acid herbicides and their enantiomers by high-performance capillary electrophoresis. *J. Chromatogr.* A: 317–327.

Gether, J., G. Lunde, and E. Steinnes. 1979. Determination of the total amount of organically bound chlorine, bromine and iodine in environmental samples by instrumental neutron activation analysis. *Anal. Chim. Acta* 108: 137–147.

Godejohann, M., M. Astratov, A. Preiss, K. Levsen, and C. Mügge. 1998. Application of continuous-flow HPLC-proton-nuclear magnetic resonance spectroscopy and HPLC-thermospray spectroscopy for the structrual elucidation of phototransformation products of 2,4,6-trinitrotoluene. *Anal. Chem.* 70: 4104–4110.

Gómez-Belinchón, J.I., J.O. Grimalt, and J. Albalgés. 1988. Intercomparison study of liquid–liquid extraction and adsorption on polyurethane and amberlite XAD-2 for the analysis of hydrocarbons, polychlorobiphenyls, and fatty acids dissolved in seawater. *Environ. Sci. Technol.* 22: 677–685.

Greibrokk, T. 1992. Recent developments in the use of supercritical fluids in coupled systems. *J. Chromatogr.* 626: 33–40.

Gremm, T.J. and F.H. Frimmel. 1994. Application of liquid chromatography–particle beam mass spectrometry and gas chromatography–mass spectrometry for the identification of metabolites of polycyclic aromatic hydrocarbons. *Chromatographia* 38: 781–788.

Grob, K. and I. Kälin. 1991. Towards on-line SEC-GC pesticide residues? The problem of tailing triglyceride peaks. *J. High Resol. Chromatogr.* 14: 451–454.

Grob, K., I. Kälin, and A. Artho. 1991. Coupled LC-GC: the capacity of silica gel (HP)LC columns for retaining fat. *J. High Resol. Chromatogr.* 14: 373–376.

Grosjean, D., E.L. Williams II, and J.H. Seinfeld. 1992. Atmospheric oxidation of selected terpenes and related carbonyls: gas-phase carbonyl products. *Environ. Sci. Technol.* 26: 1526–1533.

Grosjean, D., E.L. Williams, and E. Grosjean. 1993. A biogenic precursor of peroxypropionyl nitrate: atmospheric oxidation of *cis*-3-hexen-1-ol. *Environ. Sci. Technol.* 27: 979–981.

Guan, S., A.G. Marshall, and S.E. Scheppele. 1996. Resolution and chemical formula identification of aromatic hydrocarbons and aromatic compounds containing sulfur, nitrogen, or oxygen in petroleum distillates and refinery streams. *Anal. Chem.* 68: 46–71.

Guilhaus, M. 1995. Principles and instrumentation in time-of-flight mass spectrometry. *J. Mass Spectrom.* 30: 1519–1532.

Gupta, P., W.P. Harger, and J. Arey. 1996. The contribution of nitro- and methylnitronaphthalenes to the vapour-phase mutagenicity of ambient air samples. *Atmos. Environ.* 30: 3157.

Gyllenhaal, O. and J. Vessman. 1988. Phosgene as a derivatizing reagent prior to gas and liquid chromatography. *J. Chromatogr.* 435: 259–269.

Haiber, G., G. Jacob, V. Niedan, G. Nkusi, and H.F. Schöler. 1996. The occurrence of trichloroacetic acid TCAA — indications of a natural production? *Chemosphere* 33: 839–849.

Hainzl, D. 1995. Spectroscopic behaviour and X-ray analysis of the toxaphene component 2,2,3-*exo*,8b,8c,9c,10α-heptachlorocamphene. *J. Agric. Food Chem.* 43: 277–280.

Hale, R.C. and J. Greaves. 1992. Methods for the analysis of persistent chlorinated hydrocarbons in tissues. *J. Chromatogr.* 580: 257–278.

Hankin, S.M., P. John, and G.P. Smith. 1997. Laser time-of-flight mass spectrometry of PAH-picrate complexes. *Anal. Chem.* 69: 2927–2930.

Haraguchi, K., H. Kuroki, and Y. Masuda. 1987. Synthesis and characterization of tissue-retainable methylsulfonyl polychlorinated biphenyl isomers. *J. Agric. Food Chem.* 35: 178–182.

Harrad, S.J., A.S. Sewart, R. Boumphrey, R. Duarte-Davidson, and K.C. Jones.1992. A method for the determination of PCB congeners 77, 126 and 169 in biotic and abiotic matrices. *Chemosphere* 24: 1147–1154.

Hatano, T., T. Yoshida, and T. Okuda. 1988. Chromatography of tannins III. Multiple peaks in high-performance liquid chromatography of some hydrolyzable tannins. *J. Chromatogr.* 435: 285–295.

Hawthorne, S.B., D.J. Miller, D.E. Nivens, and D.C. White. 1992. Supercritical fluid extraction of polar analytes using *in situ* chemical derivatization. *Anal. Chem.* 64: 405–412.

Hayakawa, K., T. Murahashi, M. Butoh, and M. Miyazaki. 1995. Determination of 1,3-, 1,6-, and 1,8-dinitropyrenes in urban air by high-performance liquid chromatography using chemiluminescence detection. *Environ. Sci. Technol.* 29: 928–932.

Hayes, J.M. 1993. Factors controlling ^{13}C contents of sedimentary organic compounds: principles and evidence. *Mar. Geol.* 113: 111–125.

Heftmann, E. (Ed.). 1992. Chromatography. In *Fundamentals and Applications of Chromatography and Related Differential Migration Methods. Part A: Fundamentals and Techniques.* 5th ed. Elsevier Science Publishers, Amsterdam.

Heglund, D.L. and D.C. Tilotta. 1996. Determination of volatile organic compounds in water by solid phase microextraction and infrared spectroscopy. *Environ. Sci. Technol.* 30: 1212–1219.

Hennion, M.C., J.C. Thieblement, R. Rosset, P. Scribe, J.C. Marty, and A. Saliot. 1983. Rapid semi-preparative class separation of organic compounds from marine lipid extracts by high-performance liquid chromatography and subsequent quantitative analysis by gas chromatography. *J. Chromatogr.* 280: 351–362.

Herod, A.A. 1998. Azaarenes and thiaarenes, pp. 271–323. In *PAHs and Related Compounds* (Ed. A.H. Neilson). *Handbook of Environmental Chemistry* (Ed. O. Hutzinger) Vol. 3, Part I. Springer-Verlag, Berlin.

Hewitt, L.M., J.H. Carey, K.R. Munkittrick, J.L. Parrott, K.R. Solomon, and M.R. Servos. 1998. Identification of chloro-nitro-trifluoromethyl-substituted dibenzo-p-dioxins in lampricide formulations of 3-trifluoromethyl-4-nitrophenol: assessment to induce mixed function oxidase activity. *Anal. Chim. Acta* 17: 941–950.

Higashi, R.M., G.N. Cherr, J.M. Shenker, J.M. Macdonald, and D.G. Crosby. 1992. A polar high molecular mass constituent of bleached kraft mill effluent is toxic to marine organisms. *Environ. Sci. Technol.* 26: 2413–2420.

Hilmi, A., J.H.T. Luong, and A.-L. Nguyen. 1997. Application of capillary electrophoresis with amperometric detection to study degradation of chlorophenols in contaminated soil. *Environ. Sci. Technol.* 31: 1794–1800.

Hilmi, A., J.H.T. Luong, and A.-L. Nguyen. 1999. Development of electrokinetic capillary electrophoresis equipped with amperometric detection for analysis of explosive compounds. *Anal. Chem.* 71: 873–878.

Hites, R.A. and W.L. Budde. 1991. EPA's analytical methods for water: the next generation. *Environ. Sci. Technol.* 25: 998–1006.

Ho, Y.-P. and C. Fenselau. 1998. Application of 1.06-μm IR laser desorption on a Fourier transform mass spectrometer. *Anal. Chem.* 70: 4890–4895.

Holt, B.D., N.C. Sturchio, T.A. Abrajano, and L.J. Heraty. 1997a. Conversion of chlorinated volatile organic compounds to carbon dioxide and methyl chloride for isotopic analysis of carbon and chlorine. *Anal. Chem.* 69: 2727–2733.

Holt, M.J. Newman, F.S. Pullen, D.S. Richards, and A.G. Swanson. 1997b. High-performance liquid chromatography/NMR spectrometry/mass spectrometry: further advances in hyphenated technology. *J. Mass Spectrom.* 32: 64–70.

Hontela, A., J.B. Rasmussen, C. Audet, and G. Chevalier. 1992. Impaired cortisol stress response in fish from environments polluted by PAHs, PCBs, and mercury. *Arch. Environ. Contam. Toxicol.* 22: 278–283.

Hopper, M.L. 1987. Methylation of chlorophenoxy acid herbicides and pentachlorophenol residues in foods using ion-pair alkylation. *J. Agric. Food Chem.* 35: 285–289.

Huckins, J.N., M.W. Tubergen, and G.K Manuweera 1990a. Semipermeable membrane devices containing model lipid: a new approach to monitoring the biavailability of lipophilic contaminants and estimating their bioconcentration potential. *Chemosphere* 20: 533–552.

Huckins, J.N., M.W. Tubergen, J.A. Lebo, R.W. Gale, and T.R. Schwartz. 1990b. Polymeric film dialysis in organic solvent media for cleanup of organic contaminants. *J. Assoc. Off. Anal. Chem.* 73: 290–293.

Hühnerfuss, H. and R. Kallenborn. 1992. Chromatographic separation of marine organic pollutants. *J. Chromatogr.* 580: 191–214.

Huling, S.G., R.G. Arnold, R.A. Sierka, and M.R. Miller. 1998. Measurement of hydroxyl radical activity in a soil slurry using the spin trap α-(4-pyridyl-1-oxide)-N-*tert*-butyl nitrone. *Environ. Sci. Technol.* 32: 3436–3441.

Hurst, G.B., K. Weaver, M.J. Doktycz, M.V. Buchanan, A.M. Costello, and M.E. Lidstrom. 1998. MALDI-TOF analysis of polymerase chain reaction products from methanotrophic bacteria. *Anal. Chem.* 70: 2693–2698.

Hynning, P.-Å. 1996. Separation, identification, and quantification of components of industrial effluents with bioconcentration potential. *Water Res.* 30: 1103–1108.

Hynning, P.-Å., M. Remberger, and A.H. Neilson. 1989. Synthesis, gas-liquid chromatographic analysis, and gas chromatographic-mass spectrometric identification of nitrovanillins, chloronitrovanillins, nitroguaiacols, and chloronitroguaiacols. *J. Chromatogr.* 467: 99–110.

Hynning, P.-Å., Remberger, M., A.H. Neilson, and P. Stanley.1993. Identification and quantification of 18-nor- and 19-norditerpenes and their chlorinated analogues in samples of sediments and fish. *J. Chromatogr.* 643: 439–452.

Hynning, P.-Å., A.H. Neilson, M. Remberger, M. Kipps, and P. Stanley. 1997. Broad spectrum analysis of a contaminated sediment: exemplification of a protocol. *J. Chromatogr. A* 774: 311–319.

Jackson, J.-A.A., W.R. Blair, F.E. Brinckman, and W.P. Iverson. 1982. Gas-chromatographic speciation of methylstannanes in the Chesapeake Bay using purge and trap sampling with a tin-selective detector. *Environ. Sci. Technol.* 16: 110–119.

Jaffe, R. and R.A. Hites. 1985. Identification of new, fluorinated biphenyls in the Niagara River-Lake Ontario area. *Environ. Sci. Technol.* 19: 736–740.

Jankowiak, R. and G.J. Small. 1998. Analysis of PAH-DNA adducts — fluorecence line-narrowing spectroscopy, pp. 119–145. In *PAHs and Related Compounds*, Vol. 3. (A.H. Neilson ed.) Springer-Verlag, Berlin.

Jarman, W.M., M. Simon, R.J. Norstrom, S.A. Burns, C.A. Bacon, B.R.T. Simoneit, and R.W. Risebrough. 1992. Global distribution of tris(4-chlorophenyl)methanol in high trophic level birds and mammals. *Environ. Sci. Technol.* 26: 1770–1774.

Jarman, W.M., A. Hilkert, C.E. Bacon, J.W. Collister, K. Ballschmiter, and R.W. Risebrough. 1998. Compound-specific carbon isotopic analysis of Arochlors, Clophens, Kanechlors, and Phenoclors. *Environ. Sci. Technol.* 32: 833–836.

Jenkins, T.F. and C.L. Grant. 1987. Comparison of extraction techniques for munitions residues in soil. *Anal. Chem.* 59: 1326–1331.

Jensen, S. 1972. The PCB story. *Ambio* 1: 123–131.

Jensen, S. and B. Jansson. 1976. Methyl sulfone metabolites of PCB and DDE. *Ambio* 5: 257–260.

Jensen, S., A.G. Johnels, M. Olsson, and G. Otterlind. 1972. DDT and PCB in herring and cod from the Baltic, the Kattegat and the Skagerak. *Ambio Spec. Rep.* 1: 71–85.

Jensen, S., L. Renberg, and L. Reutergård. 1977. Residue analysis of sediment and sewage sludge for organochlorines in the presence of elemental sulfur. *Anal. Chem.* 49: 316–318.

Johnsen, S. and K. Kolset. 1988. The mass-selective detector as a chlorine-selective detector. *J. Chromatogr.* 438: 233–242.

Jokela, J.K. and M. Salkinoja-Salonen. 1992. Molecular weight distributions of organic halogens in bleached kraft pulp mill effluents. *Environ. Sci. Technol.* 26: 1190–1197.

Junk, G.A., and J.J. Richard. 1988. Organics in water: solid phase extraction on a small scale. *Anal. Chem.* 60: 451–454.

Karg, F.P.M. 1993. Determination of phenylurea pesticides in water by derivatization with heptafluorobutyric anhydride and gas chromatography-mass spectrometry. *J. Chromatogr.* 634: 87–100.

Karimi-Lotfabad, S., M.A. Pickard, and M.R. Gray. 1996. Reactions of polynuclear aromatic hydrocarbons on soil. *Environ. Sci. Technol.* 30: 1145–1151.

Keith, L.H. (Ed). 1988. *Principles of Environmental Sampling.* American Chemical Society, Washington, D.C.

Keith, L.H. 1992. *Environmental Sampling and Analysis: A Practical Guide.* Lewis Publishers, Chelsea, MI.

Keith, L.H., W. Crimmett, J. Deegan, R.A. Libby, J.K. Taylor, and G. Wentler. 1983. Principles of environmental analysis. *Anal. Chem.* 55: 2210–2218.

Kidd, K.A., D.W. Schindler, R.H. Hesslein, and D.C.G. Muir. 1998. Effect of trophic position and lipid on organochlorine concentrations in fishes from subarctic lakes in Yukon Territory. *Can. J. Fish. Aquat. Sci.* 55: 869–881.

Kim, I.S., F.I. Sasins, R.D. Stephens, J. Wang, and M.A. Brown. 1991. Determination of chlorinated phenoxy acid and ester herbicides in soil and water by liquid chromatography particle beam mass spectrometry and ultraviolet spectrophotometry. *Anal. Chem.* 63: 819–823.

Kishi, H., H. Arimoto, and T. Fujii. 1998. Analysis of alcohols and phenols with a newly designed gas chromatographic detector. *Anal. Chem.* 70: 3488–3492

Kleiböhmer, W., K. Cammann, J. Robert, and E. Mussenbrock. 1993. Determination of explosives residues in soils by micellar electrokinetic capillary chromatography and high performance liquid chromatography. *J. Chromatogr.* 638: 349–356.

Knapp, D.R. 1979. *Handbook of Analytical Derivatization Reactions.* John Wiley & Sons, New York.

Knight, C.T.G. and S.D. Kinrade. 1999. Silicon-29 nuclear magnetic resonance spectroscopy detection limits. *Anal. Chem.* 71: 265–267.

Kobal, V.M., D.T. Gibson, R.E. Davis, and A. Garza. 1973. X-ray determination of the absolute stereochemistry of the initial oxidation product formed from toluene by *Pseudmonas putida* 39/D. *J. Am. Chem. Soc.* 95: 4420–4421.

Koester, C.J. and R.A. Hites. 1992. Wet and dry deposition of chlorinated dioxins and furans. *Environ. Sci. Technol.* 26: 1375–1382.

Kolehmainen, E., K. Laihia, J. Knuutinen, and J. Hyötyläinen. 1992. ^{1}H, ^{13}C and ^{17}O NMR study of chlorovanillins and some related compounds. *Magn. Reson. Chem.* 30: 253–258.

Kölliker, S., M. Oehme, and C. Dye. 1998. Structure elucidation of 2,4-dinitrophenyl-hydrazone derivatives of carbonyl compounds in ambient air by HPLC/MS and multiple MS/MS using atmospheric chemical ionization in the negative mode. *Anal. Chem.* 70: 1979–1985.

König, W.A., I. H. Hardt, B. Gehrcke, D.H. Hochmuth, H. Hühnerfuss, B. Pfaffenberger, and G. Rimkus. 1994. Optically active reference compounds for environmental analysis obtained by preparative enantioselective gas chromatography. *Angew. Chem. Int. Ed. Engl.* 33: 2085–2087.

Kosian, P.A.,E.A. Makynen, P.D. Monson, D.R. Mount, A. Spacie, O.G. Mekenyan, and G.T. Ankley. 1998b. Application of toxicity-based fractionation techniques and structure-activity relationship models for the identification of phototoxic polycyclic aromatic hydrocarbons in sediment pore water. *Anal. Chim. Acta* 17: 1021–1033.

Kostinainen, R., T. Kotiaho, I. Mattila, T. Mansikka, M. Ojala, and R.A. Ketola. 1998. Analysis of volatile organic compounds in water and soil samples by purge-and-membrane mass spectrometry. *Anal. Chem.* 70: 3028–3032.

Kucklick, J.R. and J.E. Baker. 1998. Organochlorines in Lake Superior's food web. *Environ. Sci. Technol.* 32: 1192–1198.

Kucklick, J.R., H.R. Harvey, P.H. Ostrom, N.E. Ostrom, and J.E. Baker. 1996. Organochlorine dynamics in the pelagic food web of Lake Baikal. *Anal. Chim. Acta* 15: 1388–1400.

LaBrosse, J.L. and R.J. Anderegg. 1984. The mass spectrometer as a chlorine-selective detector I. Description and evaluation of the technique. *J. Chromatogr.* 314: 83–92.

LaCourse, W.R. and C.O. Dasenbrock. 1998. Column liquid chromatography: equipment and instrumentation. *Anal. Chem.* 70: 37R–52R.

Landrum, P., S.R. Nihart, B.J. Eadie, and W.S. Gardner. 1984. Reverse-phase separation method for determining pollutant binding to Aldrich humic acid and dissolved organic carbon of natural waters. *Environ. Sci. Technol.* 18: 187–192.

Lang, V. 1992. Polychlorinated biphenyls in the environment. *J. Chromatogr.* 595: 1–43.

Larsen, O.F.A., I.S. Kozin, A.M. Rija, G.J. Stroomberg, J.A. de Knecht, N.H. Velthorst, and C. Gooijer. 1998. Direct identification of pyrene metabolites in organs of the isopod *Porcello scaber* by fluorescence line narrowing spectroscopy. *Anal. Chem.* 70: 1182–1185.

Latourte, L., J.-C. Blais, J.-C. Tabet, and R.B. Cole. 1997. Desorption behaviour and distributions of fluorinated polymers in MALDI and electrospray ionization mass spectrometry. *Anal. Chem.* 69: 2742–2750.

Lau, B., D. Weber, and P. Andrews. 1996. GC/MS analysis of toxaphene: a comparative study of different spectrometric techniques. *Chemosphere* 32: 1021–1041.

Lecher, R.J., R.J. Norstrom, and D.C.G. Muir. 1998. Biotransformation versus bioaccumulation: sources of methyl sulfone PCB and 4,4′-DDE metabolites in polar bear food chain. *Environ. Sci. Technol.*, 32: 1656–1661.

Lee, H.-B. and T.E. Peart. 1992. Supercritical carbon dioxide extraction of resin and fatty acids from sediments at pulp mill sites. *J. Chromatogr.* 594: 309–315.

Lee, H.-B., T.E. Peart, and J.M. Carron. 1990. Gas chromatographic and mass spectrometric determination of some resin and fatty acids in pulpmill effluents as their pentafluorobenzyl ester derivatives. *J. Chromatogr.* 498: 367–379.

Lee, H.-B., T.E. Peart, and R.L. Hong-You. 1992. *In situ* extraction and derivatization of pentachlorophenol and related compounds from soils using a supercritical fluid extraction system. *J. Chromatogr.* 605: 109–113.

Lee, H.-B., T.E. Peart, and R.L. Hong-You. 1993. Determination of phenolics from sediments of pulp mill origin by *in situ* supercritical carbon dioxide extraction and derivatization. *J. Chromatogr.* 636: 263–270.

Lee, M.-R., R.-J. Lee, Y.-W. Lin, C.-M. Chen, and B.-H. Hwang. 1998. Gas-phase postderivatization following solid-phase microextraction for determining acidic herbicides in water. *Anal. Chem.* 70: 1963–1968

Letcher, R.J., R.J. Norstrom, and Å. Bergman. 1995. An integrated analytical method for determination of polychlorinated aryl methyl sulfone metabolites and polychlorinated hydrocarbon contaminants in biological matrices. *Anal. Chem.* 67: 4155–4163.

Levsen, K. 1988. The analysis of diesel particulate. *Fresenius Z. Anal. Chem.* 331: 467–478.

Li, B., P.L. Gutierraz, and N.V. Bloiugh. 1997. Trace determination of hydroxyl radical in biological systems. *Anal. Chem.* 69: 4295–4302.

Li, C., L.M. Markovec, R.J. Magee, and B.D. James. 1991. A convenient method using methanol/benzyltrialkylammonium reagents for simultaneous extraction and methylation of 2,4-dichlorophenoxyacetic acid in soil, with subsequent analysis via gas chromatography. *J. Agric. Food Chem.* 39: 1110–1112.

Lin, J.M. and S.S. Que Hee. 1985. Optimization of perchlorination conditions for some representative polychlorinated biphenyls. *Anal. Chem.* 57: 2130–2134.

Lindroth, P. and K. Mopper. 1979. High performance liquid chromatographic determination of subpicomole amounts of amino acids by precolumn fluorescence derivatization with o-phthaldialdehyde. *Anal. Chem.* 51: 1667–1674.

Lindström, K. and F. Österberg. 1980. Synthesis, X-ray structure determination, and formation of 3,4,5-trichloroguaiacol occurring in kraft mill spent bleach liquors. *Can. J. Chem.* 58: 815–822.

Llompart, M., K. Li, amd M. Fingas. 1998. Solid-phase microextraction and headspace solid-phase microextraction for the determination of polychlorinated biphenyls in water samples. *Anal. Chem.* 70: 2510–2516.

Lopez-Avila, V., N.S. Dodhiwala, and W.F. Beckert. 1990. Supercritical fluid extraction and its application to environmental analysis. *J. Chromatogr. Sci.* 28: 468–476.

Lopez-Avila, V., J. Benedicto, E. Baldin, and W.F. Beckert. 1991. Analysis of classes of compounds of environmental concern: I. nitroaromatic compounds. *J. High Resol. Chromatogr.* 14: 601–607.

Ludwig, P., W. Gunkel, and H. Hühnerfuss. 1992. Chromatographic separation of the enantiomers of marine pollutants. Part 5: Enantioselective degradation of phenoxycarboxylic acid herbicides by marine organisms. *Chemosphere* 24: 1423–1429.

Marina, M.L., I. Benito, J.C. Díez-Masa, and M.J. González. 1996. Chiral separation of polychlorinated biphenyls by micellar electrokinetic chromatography with γ-cyclodextrin as modifier in the separation buffer. *Chromatographia* 42: 269–272.

Marsden, P.J., E.N. Amick, F.L Shore, L.R. Williams, V.R. Bohman, and C.R. Blincoe. 1986. Characterization of bovine urine and adipose interlaboratory evaluation samples containing biologically incorporated chlorophenols. *J. Agric. Food Chem.* 34: 795–800.

Martin, G.E., J.E. Guido, R.H. Robins, M.H.M. Sharaf, P.L. Schiff, and A.N. Tackie. 1998. Submicro inverse-detection gradient NMR: a powerful new way of conducting structure elucidation studies with <0.05 μmol samples. *J. Nat. Prod.* 61: 555–559.

Martinez, R.C., E.R. Gonzalo, A.I.M. Domingues, J.D. Alvarez, and J.H. Mendez. 1996. Determination of triazine herbicides in water by micellar electrokinetic capillary chromatography. *J. Chromatogr.* A 733: 349–360.

Martinsen, K., A. Kringstad, and G.E. Carlberg. 1988. Methods for determination of sum parameters and characterization of organochlorine compounds in spent bleach liquors from pulp mills and water, sediment and biological samples from receiving waters. *Water Sci. Technol.* 20[2]: 13–24.

Martos, P.A. and J. Pawliszyn. 1998. Sampling and determination of formaldehyde using solid-phase microextraction with on-fiber derivatization. *Anal. Chem.* 70: 2311–2320.

Mathiasson, L., J.Å. Jönsson, and L. Karlsson. 1989. Determination of nitrogen compounds by supercritical fluid chromatography using nitrous oxide as the mobile phase and nitrogen-sensitive detection. *J. Chromatogr.* 467: 61–74.

McCormick, M.L., G.R. Buettner, and B.E. Britigan. 1998. Endogenous superoxide dismutase levels regulate iron-dependent hydroxyl radical formation in *Escherichia coli* exposed to hydrogen peroxide. *J. Bacteriol.* 180: 622–625.

McMillan, D. C., P.P. Fu, and C.E. Cerniglia. 1987. Stereoselective fungal metabolism of 7,12-dimethylbenz[a]anthracene: identification and enantiomeric resolution of a K-region dihydrodiol. *Appl. Environ. Microbiol.* 53: 2560–2566.

Meadows, J., D. Tillitt, J. Huckins, and D. Schroeder. 1993. Large-scale dialysis of sample lipids using a semipermeable membrane device. *Chemosphere* 26: 1993–2006.

Meese, C.O. 1984. 2,2,2-Trifluorodiazoethane: a highly selective reagent for the protection of sulfonic acids. *Synthesis* 1041–1042.

Merritt, D.A., K.H. Freeman, M.P. Ricci, S.A. Studley, and J.M. Hayes. 1995. Performance and optimization of a combustion interface for isotope ratio monitoring gas chromatography/mass spectrometry. *Anal. Chem.* 67: 2461–2473.

Meulenberg, E.P., W.H. Mulder, and P.G. Stoks. 1995. Immunoassays for pesticides. *Environ. Sci. Technol.* 29: 553.

Miles, C.J. and H.A. Moyne. 1988. Extraction of glyphosate herbicide from soil and clay minerals and determination of residues in soils. *J. Agric. Food Chem.* 36: 486–491.

Mills, M.S. and E.M. Thurman. 1992. Mixed-mode islation of triazine metabolites from soil and aquifer sediments using automated solid-phase extraction. *Anal. Chem.* 64: 1985–1990.

Miskiewicz, A.G. and P.J. Gibbs. 1992. Variability in organochlorine analysis in fish: an interlaboratory study and its implications for environmental monitoring and regulatory standards. *Arch. Environ. Contam. Toxicol.* 23: 45–53.

Mondello, L., A. Verzera, P. Previti, F. Crispo, and G. Dugo. 1998. Multidimensional capillary GC-GC for the analysis of complex samples. 5. Enantiomeric distribution of monoterpene hydrocarbons, monoterpene alcohols, and linalyl acetate of bergamot (*Citrus bergamia* Risso et Poiteau) oils. *J. Agric. Food Chem.* 46: 4275–4282.

Moody, R.P., C.A. Franklin, D. Riedel, N.I. Muir, R. Greenhalgh, and A. Hladka. 1985. A new GC on-column methylation procedure for analysis of DMTP (*O,O*-dimethyl phosphorothioate) in urine of workers exposed to fenitrothion. *J. Agric. Food Chem.* 33: 464–467.

Morris, G.A. and R. Freeman. 1979. Enhancement of nuclear magnetic resonance signals by polarization transfer. *J. Am. Chem. Soc.* 101: 760–762.

Mössner, S.G. and S.A. Wise. 1999. Determination of polycyclic aromatic sulfur heterocycles in fossil fuel-related samples. *Anal. Chem.* 71: 58–69.

Mössner, S., T.R. Spraker, P.R. Becker, and K. Ballschmiter. 1992. Ratios of enantiomers of alpha-HCH and determination of alpha-, beta-, and gamma-HCH isomers in brain and other tissues of neonatal northern fur seals (*Callorhinus ursinus*). *Chemosphere* 24: 1171–1180.

Mowrer, J. and J. Nordin. 1987. Characterization of halogenated organic acids in flue gases from municipal waste incinerators. *Chemosphere* 16: 1181–1192.

Mudroch, A., R.J. Allen, and S.R. Joshi. 1992. Geochemistry and organic contaminants in the sediments of Great Slave Lake, Northwest Territories, Canada. *Arctic* 45: 10–19.

Müller, M.D., M. Schlabach, and M. Oehme. 1992. Fast and precise determination of alpha-hexachlorocyclohexane enantiomers in environmental samples using chiral high-resolution gas chromatography. *Environ. Sci. Technol.* 26: 566–569.

Müller, S.R., H.-R. Zweifel, D.J. Kinnison, J.A. Jacobsen, M.A. Meier, M.M. Ulrich, and R.P. Schwarzenbach. 1996. Occurrence, sources, and fate of trichloroacetic acid in Swiss waters. *Anal. Chim. Acta* 15: 1470–1478.

Mullin, M., C.M. Pochini, S. McCrindle, M. Romkes, S.H. Safe, and L.M. Safe. 1984. High-resolution PCB analysis: synthesis and chromatographic properties of all 209 PCB congeners. *Environ. Sci. Technol.* 18: 468–476.

Munkittrick, K.R., C.B. Portt, G.J. Van Der Kraak, I.R. Smith, and D. A. Rokosh. 1991. Impact of bleached kraft mill effluent on population characteristics, liver MFO activity, and serum steroid levels of a Lake Superior white sucker (*Catostomus commersoni*) population. *Can. J. Fish. Aquat. Sci.* 48: 1371–1380.

Muthuramu, K., P.B. Shepson, and J.M. O'Brien. 1993. Preparation, analysis, and atmospheric production of multifunctional organic nitrates. *Environ. Sci. Technol.* 27: 1117–1124.

Näf, C., D. Broman, H. Pettersen, and Y. Zebühr. 1992. Flux estimates and pattern recognition of particulate polycyclic aromatic hydrocarbons, polychlorinated dibenzo-*p*-dioxins, and dibenzofurans in the waters outside various emission sources on the Swedish Baltic Coast. *Environ. Sci. Technol.* 26: 1444–1457.

Nair, U.B., D.W. Armsttrong, and W.L. Hinze. 1998. Characterization and evaluation of d(+)-tubocurarine chloride as a chiral selectror for capillary electrophoretic enantioseparation. *Anal. Chem.* 70: 1059–1065.

Nassar, A.-E.F., S.V. Lucas, W.R. Jones, and L.D. Hoffland. 1998. Separation of chemical warfare agent degradation products by the reversal of electroosmotic flow in capillary electrophoresis. *Anal. Chem.* 70: 1085–1091.

Neilson, A.H. and P.-Å. Hynning. 1998. PAHs: products of chemical and biochemical transformation of alicyclic precursors pp. 224–269. In *PAHs and Related Compounds* (Ed. A.H. Neilson). *Handbook of Environmental Chemistry* (Ed. O. Hutzinger) Vol. 3, Part I. Springer-Verlag, Berlin.

Neilson, A.H., A.-S. Allard, P.-Å. Hynning, and M. Remberger. 1988. Transformations of halogenated aromatic aldehydes by metaboliocally stable anaerobic enrichment cultures. *Appl. Environ. Microbiol.* 54: 2226–2236.

Neilson, A.H., A.-S. Allard, P.-Å Hynning, and M. Remberger. 1991. Distribution, fate and persistence of organochlorine compounds formed during production of bleached pulp. *Toxicol. Environ. Chem.* 30: 3–41.

Nguyen, J.A.-L. and J.H.T. Luong. 1997. Separation and determination of polycyclic aromatic hydrocarbons by solid phase microextraction/cyclodextrin-modified capillary electrophoresis. *Anal. Chem.* 69: 1726–1731.

Nielen, M.W.F. 1993. (Enantio-)separation of phenoxy acid herbicides using capillary zone electrophoresis. *J. Chromatogr.* 637: 81–90.

Nielsen, T., T. Ramdahl, and A. Björseth. 1983. The fate of airborne polycyclic organic matter. *Environ Health Perspect.* 47: 103–114.

Niessen, W.M.A. and A.P. Tinke. 1995. Liquid chromatography-mass spectrometry. General principles and instrumentation. *J. Chromatogr.* A 703: 37–57.

Niessen, W.M.A, U.R. Tjaden, and J. van der Greef. 1991. Strategies in developing interfaces for coupling liquid chromatography and mass spectrometry. *J. Chromatogr.* 534: 3–26.

Nilsson, U.L. and C.E. Östman. 1993. Chlorinated polycyclic aromatic hydrocarbons: method of analysis and their occurrence in urban air. *Environ. Sci. Technol.* 27: 1826–1831

Nitz, S., M.H. Spraul, F. Drawert, and M. Spraul. 1992. 3-Butyl-5,6-dihydro-4H-isobenzofuran-1-one, a sensorial active phthalide in parsley roots. *J. Agric. Food Chem.* 40: 1038–1040.

Nondek, L., D.R. Rodler, and J.W. Birks. 1992. Measurement of sub-ppbv concentrations of aldehydes in a forest atmosphere using a new HPLC technique. *Environ. Sci. Technol.* 26: 1174–1178.

Ockenden, H.F. Prest, G.O.Thomas, A. Sweetman, and K.C. Jones. 1998. Passive air sampling of PVCs: field calculation of atmospheric sampling rates by triolein-containing semipermeable membrane devices. *Environ. Sci. Technol.* 32: 1538–1543.

Okuda, T., T. Yoshida, T. Hatano, K. More, and T. Fukuda. 1990. Fractionation of pharmacologically active plant polyphenols by centrifugal partition chromatography. *J. Liq. Chromatogr.* 13: 3637–3650.

Ozawa, H. and T. Tsukioka. 1990. Gas chromatographic separation and determination of chloroacetic acids in water by a difluoroanilide derivatization method. *Analyst* 115: 1343–1347.

Ozretich, R.J. and W.P. Schroeder. 1986. Determination of selected neutral priority organic pollutants in marine sediment, tissue, and reference materials utilizing bonded-phase sorbents. *Anal. Chem.* 58: 2041–2048.

Paschke, T., S.B. Hawthorne, and D.J. Miller. 1992. Supercritical fluid extraction of nitrated polycyclic aromatic hydrocarbons and polycyclic aromatic hydrocarbons from diesel exhaust particulate matter. *J. Chromatogr.* 609: 333–340.

Petrick, G., D.E. Schulz, and J.C. Duinker. 1988. Clean-up of environmental samples by high-performance liquid chromatography for analysis of organochlorine compounds by gas chromatography with electron-capture detection. *J. Chromatogr.* 435: 241–248.

Pierce, A.E. 1979. Silylation of Organic Compounds. Pierce Chemical Company, Rockford, IL.

Poster, D., L.C. Sander, and S.A. Wise. 1998. Chromatographic methods of analysis for the determination of PAHs in environmental samples pp.77–135. *In PAHs and Related Compounds* (Ed. A.H. Neilson). *Handbook of Environmental Chemistry* (Ed. O. Hutzinger) Vol. 3, Part I. Springer-Verlag, Berlin.

Pusecker, K., J. Schewitz, P. Gfrörer, L.-H. Tseng, K. Albert, and E. Bayer. 1998. On-line coupling of capillary electrochromatography, capillary electrophoresis, and capillary HPLC with nuclear magnetic resonance spectroscopy. *Anal. Chem.* 70: 3280–3285.

Ramana, D.V. and N.V.S. Krishna. 1989. Extensive substituent scrambling in substituted diphenylacetylenes on electron impact. *Org. Mass Spectrom.* 24: 903–908.

Ramanathan, R. and M.L. Gross. 1998. Mass spectroscopy techniques: DNA adducts of PAHs and related compounds pp. 147–210. In *PAHs and Related Compounds* (Ed. A.H. Neilson). Handbook of Environmental Chemistry (Ed. O. Hutzinger) Vol. 3, Part J. Springer-Verlag, Berlin.

Ramaswamy, S., M. Malalyandi, and G.W. Buchanan. 1985. Phase-transfer-catalyzed methylation of hydroxyaromatic acids, hydroxyaromatic aldehydes, and aromatic polycarboxylic acids. *Environ. Sci. Technol.* 19: 507–512.

Ramdahl, T., B. Zielinska, J. Arey, and R.W. Kondrat. 1988. The electron impact mass spectra of di- and trinitrofluoranthrenes. *Biomed. Environ. Mass Spectrom.* 17: 55–62.

Ramos, E.U., S.N. Meijer, W.H.J. Vaes, H.J.M. Verhaar, and J.L.M. Hermens. 1998. Using solid-phase microextraction to determine partition coefficients to humic acids and bioavailable concentrations of hydrophobic chemicals. *Environ. Sci. Technol.* 32: 3430–3435.

Ramos, L., L.M. Hernández. and M.-J. González. 1999. Simultaneous separation of coplanar and chiral polychlorinated biphenyls by off-line pyrenyl-silica liquid chromatography and gas chromatography. Enantiomeric ratios of chiral congeners. *Anal. Chem.* 71: 70–77.

Remberger, M., P.-Å. Hynning, and A.H. Neilson. 1988. Comparison of procedures for recovering chloroguaiacols and chlorocatechols from contaminated sediments. *Anal. Chim. Acta* 7: 795–805.

Remberger, M., P.-Å. Hynning, and A.H. Neilson. 1991a. 2,5-Dichloro-3,6-dihydroxybenzo-1,4-quinone: identification of a new organochlorine compound in kraft mill bleachery effluents. *Environ. Sci. Technol.* 25: 1903–1907.

Remberger, M., P.-Å. Hynning, and A.H. Neilson. 1991b. Chlorinated benzo-1,2-quinones: an example of chemical transformation of toxicants during tests with aquatic organisms. *Ecotoxicol. Environ. Saf.* 22: 320–336.

Resnick, S.M. and D.T. Gibson. 1996. Regio- and stereospecific oxidation of fluorene, dibenzofuran, and dibenzothiophene by naphthalene dioxygenase from *Pseudomonas* sp. strain NCIB-4. *Appl. Environ. Microbiol.* 62: 4073–4080.

Resnick, S.M., D.S. Torok, and D.T. Gibson. 1995. Chemoenzymatic synthesis of chiral boronates for the ^{1}H NMR determination of the absolute configuration and enantiomeric excess of bacterial and synthetic *cis*-diols. *J. Org. Chem.* 60: 3546–3549.

Richardson, J.S. and D.E. Miller 1982. Identification of dicyclic and tricyclic hydro-carbons in the saturate fraction of a crude oil by gas chromatography/mass spectrometry. *Anal. Chem.* 54: 765–768.

Richnow, H.H., A. Eschenbach, B. Mahro, R. Seifert, P. Wehrung, P. Albrecht, and W. Michaelis. 1998. The use of ^{13}C-labelled polycyclic aromatic hydrocarbons for the analysis of their transformation in soil. *Chemosphere* 36: 2211–2224.

Rodgers, R.P., F.M. White, C.L. Hendrickson, A.G. Marshall, and K.V. Andersen. 1998. Resolution, elementary composition, and simultaneous monitoring by Fourier transform ion cyclotron resonance mass spectrometry of organosulfur species before and after diesel fuel processing. *Anal. Chem.* 70: 4743–4750.

Rogge, W.F., L.M. Hildemann, M.A. Mazurek, G.R. Cass, and B.R.T. Simoneit. 1993. Sources of fine organic aerosol. 2. Noncatalyst and catalyst-equipped automo-biles and heavy-duty diesel trucks. *Environ. Sci. Technol.* 27: 636–651.

Ross, P.L., P.A. Davis, and P. Belgrader. 1998. Analysis of DNA fragments from conventional and microfabricated PCR devices using delayed extraction MALDI-TOF mass spectrometry. *Anal. Chem.* 70: 2067–2073.

Roussis, S.G. and J.W. Fedora. 1997. Determination of alkenes in hydrocarbon ma-trices by acetone chemical ionization mass spectrometry. *Anal. Chem.* 69: 1550–1556.

Saflic, S., P.M. Fedorak, and J.T. Andersson. 1992. Diones, sulfoxides, and sulfones from the aerobic cometabolism of methylbenzothiophenes by *Pseudomonas* strain BT1. *Environ. Sci. Technol.* 26: 1759–1764.

Sanders, G., K.C. Jones, J. Hamilton-Taylor, and H. Dörr. 1992. Historical inputs of polychlorinated biphenyls and other organochlorines to a dated lacustrine sed-iment core in rural England. *Environ. Sci. Technol.* 26: 1815–1821.

Saurer, M., I. Robertson, R. Siegwolf, and M. Leuenberger. 1998. Oxygen isotope analysis of cellulose: an interlaboratory comparison. *Anal. Chem.* 70: 2074–2080.

Sawhney, B.L., J.J. Pignatello, and S.M. Steinberg. 1988. Determination of 1,2-dibro-moethane (EDB) in field soils: implications for volatile organic compounds. *J. Environ. Qual.* 17: 149–152.

Schaffner, C. and W. Giger. 1984. Determination of nitrilotriacetic acid in water by high resolution gas chromatography. *J. Chromatogr.* 312: 413–421.

Schmid, B. and J.T. Andersson. 1997. Critical examination of the quantification of aromatic compounds in three standard reference materials. *Anal. Chem.* 69: 3476–3481.

Schmidt, L.J. and R.J. Hesselberg. 1992. A mass spectroscopic method for analysis of AHH-inducing and other polychlorinated biphenyl congeners and selected pes-ticides in fish. *Arch. Environ. Contam. Toxicol.* 23: 37–44.

Schoenmakers, P.J. and L.G.M. Uunk. 1987. Supercritical fluid chromatography — recent and future developments. *Eur. Chromatogr. News* 1: 14–22.

Schuytema, G.S., A.V. Nebeker, W.L. Griffis, and C.E. Miller. 1989. Effects of freezing on toxicity of sediments contaminated with DDT and endrin. *Anal. Chim. Acta* 8: 883–891.

Sera, N., K. Fukuhara, N. Miyata, and H. Tokiwa. 1994. Detection of nitroazaben-zo(a)pyrene derivatives in the semi-volatile phase originating from airborne particulate matter, diesel and gasoline vehicles. *Mutagenesis* 9: 347.

Shamsi, S.A., C. Akbay, and I.M. Warner. 1998. Polymeric anionic surfactant for electrokinetic chromatography: separation of 16 priority polycyclic aromatic hy-drocarbon pollutants. *Anal. Chem.* 70: 3078-3083.

Sheppard, R.L. and J. Henion. 1997. Quantitative capillary electrophoresis/ion spray tandem mass spectrometry determination of EDTA in human plasma and urine. *Anal. Chem.* 69: 2901–2907.

Shi, Y. and J.S. Fritz. 1995. HPCZE of nonionic compounds using a novel anionic surfactant additive. *Anal. Chem.* 67: 3023–3027.

Sidelmann, U. Braumann, M. Hofmann, M. Spraul, J.C. Lindon, J.K. Nicholson, and S.H.Hansen. 1997. Directly coupled 800 MHz HPLC-NMR spectroscopy of urine and its application to the identification of the major phase II metabolites of tolfenamic acid. *Anal. Chem.* 69: 607–612.

Simoneit, B.T.R. 1998. Biomarker PAHs in the environment pp. 176–221. In *PAHs and Related Compounds* Vol. 3. Part J (A.H. Neilson Ed.) Springer-Verlag, Berlin.

Singh, S.K. and S. Kumar. 1993. Synthesis of oxygenated derivatives of 2,3,7,8-tetra-chlorodibenzo-*p*-dioxin. *J. Agric. Food Chem.* 41: 1511-1516.

Sinkkonen, S., E. Kolehmainen, K. Lalhia, J. Kolstinen, and T. Rantio. 1993. Polychlo-rinated diphenyl sulfides: preparation of model compounds, chromatography, mass spectrometry, NMR, and environmental analysis. *Environ. Sci. Technol.* 27: 1319–1326.

Sjöström, L., R. Rådeström, G.E. Carlberg, and A. Kringstad. 1985. Comparison of two methods for the determination of total organic halogen (TOX) in receiving waters. *Chemosphere* 14: 1107–1113.

Slobodnik, J., B.L.M. van Baar, and U.A.Th. Brinkman. 1995. Column liquid chroma-tography-mass spectrometry: selected techniques in environmental applications for polar pesticides and related compounds. *J. Chromatogr.* A 703: 81–121.

Smock, A.M., M.E. Böttcher, and H. Cypionka. 1998. Fractionation of sulfur isotopes during thiosulfate reduction by *Desulfovibrio desulfuricans*. Arch. Microbiol. 169: 460–463.

Snyder, J.L., R.L. Grob, M.E. McNally, and T.S. Oostdyk. 1992. Comparison of super-critical fluid extraction with classical sonication and Soxhlet extractions for se-lected pesticides. *Anal. Chem.* 64: 1940–1946.

Södergren, A. 1987. Solvent-filled dialysis membranes simulate uptake of pollutants by aquatic organisms. *Environ. Sci. Technol.* 21: 855–859.

Song, L., Z. Xu, J. Kang, and J. Cheng. 1997. Analysis of environmental pollutants by capillary electrophoresis with emphasis on micellar electrokinetic chromato-graphy. *J. Chromatogr.* A 780: 297–328.

Soni, M.H., J.H. Callahan, and S.W. McElvany. 1998. Laser desorption-membrane introduction mass spectrometry. *Anal. Chem.* 70: 3103–3113.

Sovocool, G.W., R.K. Mitchum, and J.R. Donnelly. 1987. Use of the "*ortho* effect" for chlorinated biphenyl and brominated biphenyl isomer identification. *Biomed. Environ. Mass Spectrom.* 14: 579–582.

Spitzer, T. and T. Takeuchi. 1995. Determination of benzanthrone in environmental samples. *J. Chromatogr.* A 710: 109–116.

Stalling, D.L., R.C. Tindle, and J.L. Johnson. 1972. Cleanup of pesticide and polychlo-rinated biphenyl residues in fish extracts by gel permeation chromatography. *J. Assoc. Off. Anal. Chem.* 55: 32–38.

Staunton, J. and A.C. Sutowski. 1991. ^{17}O NMR in biosynthetic studies: aspyrone, asperolactone isoasperoloactone, metabolites of *Aspergillus melleus*. *J. Chem. Soc. Chem. Commun.* 1106–1108.

Stephanou, E.G. 1992. α,ω–Dicarboxylic acid salts and α,ω-dicarboxylic acids. Pho-tooxidation products of unsaturated fatty acids, present in marine aerosols and marine sediments. *Naturwissenschaften.* 79: 128–131.

Stephanou, E., M. Reinhard, and H.A. Ball. 1988. Identification and quantification of halogenated and non-halogenated octylphenol polyethoxylate residues by gas chromatography/mass spectrometry using electron ionization and chemical ionization. *Biomed. Environ. Mass Spectrom.* 15: 275–282.

Stern, G.A., D.C.G. Muir, C.A. Ford, N.P. Grift, E. Dewailly, T.F. Bidleman, and M.D. Walla. 1992. Isolation and identification of two major recalcitrant toxaphene congeners in aquatic biota. *Environ. Sci. Technol.* 26: 1838–1840.

Stoyanovsky, D.A., Z. Melnikov, and A.I. Cederbaum. 1999. ESR and HPLC-EC analysis of the interaction of hydroxyl radicals with DMSO: rapid reduction and quantification of POBN and PBN nitroxides. *Anal. Chem.* 71: 715–721.

Strege, M.A. 1998. Hydrophobic interaction chromatography-electrospray mass spectrometry analysis of polar compounds for natural products drug discovery. *Anal. Chem.* 70: 2439–2445.

Sturchio, N.C., J.L. Clausen, I.J. Heraty, L. Huang, B.D. Holt., and T.A. Abrajano. 1998. Chlorine isotope investigation of natural attenuation of trichloroethene in an aerobic aquifer. *Environ. Sci. Technol.* 32: 3037–3042.

Subramanian, R. and A.G. Webb. 1998. Design of solenoidal microcoils for high-resolution ^{13}C NMR spectroscopy. *Anal. Chem.* 70: 2454–2458.

Sutherland, J.B., P.P. Fu, S.K. Yang, L.S. von Tungeln, R.P. Casillas, S.A. Crow, and C.E. Cerniglia. 1993. Enantiomeric composition of the *trans*-dihydrodiols produced from phenanthrene by fungi. *Appl. Environ. Microbiol.* 59: 2145–2149.

Swallow, K.C., N.S. Shifrin, and P.J. Doherty. 1988. Hazardous organic compound analysis. *Environ. Sci. Technol.* 22: 136–142.

Tan, Y.L. and M. Heit. 1981. Biogenic and abiogenic polynuclear aromatic hydrocarbons in sediments from two remote Adirondack lakes. *Geochim. Cosmochim. Acta* 45: 2267–2279.

Tanaka, F.S., R.G. Wien, R.G. Zaylskie, and B.L. Hoffer. 1990. Synthesis of possible ring-hydroxylated metabolites of Diclofop-methyl. *J. Agric. Food Chem.* 38: 553–559.

Taylor, W.G., D.D. Vedres, and J.L. Elder. 1993. Capillary gas chromatographic separation of some diastereomeric amides from carbonyldiimidazole-mediated microgram-scale derivatizations of the acid moiety of permethrin insecticide. *J. Chromatogr.* 645: 303–310.

Ten Haven, H.L., T.M. Peakman, and J. Rullkötter. 1992. Early diagenetic transformation of higher-plant triterpenoids in deep-sea sediments from Baffin Bay. *Geochim. Cosmochim. Acta* 56: 2001–2024.

Terabe, S., Y. Miyashita, Y. Ishama, and O. Shibata. 1993. Cyclodextrin-modified micellar electrokinetic chromatography: separation of hydrophobic and enantiomeric compounds. *J. Chromatogr.* 636: 47–55.

Thomas, D., S.M. Crain, P.G. Sim, and F.M. Benoit. 1995. Application of reversed phase liquid chromatography with atmospheric pressure chemical ionization tandem mass spectrometry to the determination of polycyclic aromatic sulfur heterocycles in environmental samples. *J. Mass Spectrom.* 30: 1034–1040.

Thurman, E.M., M. Meyer, M. Pomes, C.A. Perry, and A.P. Schwab. 1990. Enzyme-linked immunosorbent assay compared with gas chromatography/mass sopectrometry for the determination of triazine herbicides in water. *Anal. Chem.* 62: 2043–2048.

Thurman, E.M., D.A. Goolsby, M.T. Meyer, M.S. Mills, M.L. Pomes, and D.W. Kolpin. 1992. A reconnaisance study of herbicides and their metabolites in surface water of the midwestern United States using immunoassay and gas chromatography/mass spectrometry. *Environ. Sci. Technol.* 26: 2440–2447.

Thurman, E.M., D.A. Goolsby, D.S. Aga, M.L. Pomes, and M.T. Meyer. 1996. Occurrence of alachlor and its sulfonated metabolite in rivers and reservoirs of the midwestern United States: the importance of sulfonation in the transport of chloroacetaniline herbicides. *Environ. Sci. Technol.* 30: 569–574.

Tjeerdema, R.S., T.W.-M. Fan, R.M. Higashi, and D.G. Crosby. 1991. Effects of pentachlorophenol on energy metabolism in the abalone (*Haliotis rufescens*) as measured by *in vivo* ^{31}P NMR spectroscopy. *J. Biochem. Toxicol.* 6: 45–56.

Tomy, G.T., J.B. Westmore, G.A. Stern, D.C.G. Muir, and A.T. Fisk. 1999. Interlaboratory study on quantitative methods of analysis of C_{10}-C_{13} polychloro-*n*-alkanes. *Anal. Chem.* 71: 446–451.

Turnipseed, S.B., J.E. Roybal, H.S. Rupp, J.A. Hurlbut, and A.R. Lomg. 1995. Particlebeam liquid chromatography-mass spectrometry of triphenylmethane dyes: application to confirmation of malachite green in incurred catfish tissue. *J. Chromatogr.* 670: 55–62.

Van Berkel, G.J., J.M.E. Quirke, R.A. Tigani, A.S. Dilley, and T.R. Covey. 1998. Derivatization for electrospray ionization mass spectroscopy. 3. Electrochemically ionizable derivatives. *Anal. Chem.* 70: 1544–1554.

Van der Velde, E.G., W. de Haan, and A.K.D. Liem. 1992. Supercritical fluid extraction of polychlorinated biphenyls and pesticides from soil. Comparison with other extraction methods. *J. Chromatogr.* 626: 135–143.

Van Emon, K.M., and Lopez-Avila, V. 1992. Immunochemical methods for environmental analysis. *Anal. Chem.* 64: 79A–88A.

Vanderlaan, M., B.E. Watkins, and L. Stanker. 1988. Environmental monitoring by immunoasssy. *Environ. Sci. Technol.* 22: 247–254.

Vassilaros, D.L., P.W. Stoker, G.M. Booth, and M.L. Lee. 1982. Capillary gas chromatographic determination of polycyclic aromatic compounds in vertebrate fish tissue. *Anal. Chem.* 54: 106–112.

Vaughan, P.P. and N.V. Blough. 1998. Photochemical formation of hydroxyl radicals by constituents of natural waters. *Environ. Sci. Technol.* 32: 2947–2953.

Velury, S. and S.B. Howell. 1988. Measurements of plasma thiols after derivatization with monobromobimane. *J. Chromatogr.* 424: 141–146.

Verga, G.R., A. Sironi, W. Schneider, and J.Ch. Frohne.1988. Selective determination of oxygenates in complex samples with the O-FID analyzer. *J. High Resol. Chromatogr. Commun.* 11: 248–252.

Vincent, J.B., A.D. Sokolowski, T.V. Nguyen, and G. Vigh. 1997a. A family of single-isomer chiral resolving agents for capillary electrophoresis. 1. Heptakis(2,3-diacetyl-6-sulfato)-β-cyclodextrin. *Anal. Chem.* 69: 4226–4233.

Vincent, J.B., D.M. Kirby, T.V. Nguyen, and G. Vigh. 1997b. A family of single-isomer chiral resolving agents for capillary electrophoresis. 2. Hepta-6-sulfato-β-cyclodextrin. *Anal. Chem.* 69: 4419–4428.

Wahlberg, C., L. Renberg, and U. Wideqvist. 1990. Determination of nonylphenol and nonylphenol ethoxylates as their pentafluorobenzoates in water, sewage sludge and biota. *Chemosphere* 20: 179–195.

Wakeham, S.G., C. Schaffner, and W. Giger. 1980. Polycyclic aromatic hydrocarbons in recent lake sediments — II. Compounds derived from biogenic precursors during early diagenesis. *Geochim. Cosmochim. Acta* 44: 415–429.

Ware, M.L., M.D. Argentine, and G.W. Rice. 1988. Potentiometric determination of halogen content in organic compounds using dispersed sodium reduction. *Anal. Chem.* 60: 383–384.

Wasilchuk, B.A., P.W. Le Quesne, and P. Vouros. 1992. Monitoring cholesterol autoxidation processes using multideuteriated cholesterol. *Anal. Chem.* 64: 1077–1087.

Westcott, N.D. and B.L. Worobey. 1985. Novel solvent extraction of lindane from soil. *J. Agric. Food Chem.* 33: 58–60.

Wheeler, O.H. 1968. The Girard reagents. *J. Chem. Ed.* 45: 435–437.

Willett, K.L., E.M. Ulrich, and R.A. Hites. 1998. Differential toxicity and environmental fates of hexachlorocyclohexane isomers. *Environ. Sci. Technol.* 32: 2197–2207.

Wittlinger, R. and K. Ballschmiter 1990. Studies of the global baseline pollution XIII. C_6-C_{14} organohalogens (a and g [sic]-HCH, HCB, PCB, 4,4'-DDT, 4,4-DDE, *cis*- and *trans*-chlordane, *trans*-nonachlor, anisols) in the lower troposphere of the southern Indian Ocean. *Fresenius Z. Anal. Chem.* 336: 193–200.

Wolfender, J.L., S. Rodriguez, K. Hostettmann, and W. Wagner-Redeker. 1995. Comparison of liquid chromatography/electrospray, atmospheric pressure chemical ionization, thermospray and continuous-flow fast atom bombardment mass spectrometry for the determination of secondary metabolitres in crude plant extracts. *J. Mass Spectrom. Rapid Commun. Mass Spectrom.* S35–S46.

Wretensjö, I., L. Svensson, and W.W. Christie. 1990. Gas chromatographic-mass spectrometric identification of the fatty acids in borage oil using the picolinyl ester derivatives. *J. Chromatogr.* 521: 89–97.

Wu, J., M.K. Wong, H.K .Lee, and C.N. Ong. 1995. Capillary zone electrophoretic determination of heterocyclic aromatic amines in rain. *J. Chromatogr. Sci.* 33: 712–716.

Wu, Y.-C. and S.-D. Huang. 1999. Solid-phase microextraction coupled with high performance liquid chromatography for the determination of aromatic amines. *Anal. Chem.* 71: 310–318.

Xie, T.-M., E. Abrahamson, E. Fogelqvist, and B. Josefsson. 1986. Distribution of chlorophenols in a marine environment. *Environ. Sci. Technol.* 20: 457–463.

Xie, Y., D.A. Reckhow, and R.V. Rajan. 1993. Spontaneous methylation of haloacetic acids in methanolic stock solutions. *Environ. Sci. Technol.* 27: 1232–1234.

Yang, Y., S.B. Hawthorne, D.J. Miller, Y. Liu, and M.L. Lee. 1998. Adsorption versus absorption of polychlorinated biphenyls onto solid-phase microextraction coatings. *Anal. Chem.* 70: 1866–1869.

Yarabe, H.H., S.A. Shamsi, and I.M. Warner. 1998. Capillary zone electrophoresis of bile acids with indirect photometric detection. *Anal. Chem.* 70: 1412–1418.

Yasahura, A. and M. Morita. 1988. Formation of chlorinated aromatic hydrocarbons by thermal decomposition of vinylidene chloride polymer. *Environ. Sci. Technol.* 22: 646–650.

Zabik, J.M., L.S. Aston, and J.N. Seiber. 1992. Rapid characterization of pesticide residues in contaminated soils by passive sampling devices. *Anal. Chim. Acta* 11: 765–770.

3

Partition: Distribution, Transport, and Mobility

SYNOPSIS The dissemination of a xenobiotic after discharge into the environment is determined by its partition between the water, the soil and sediment, and the atmospheric phases, and its potential for concentration in biota. These processes determine both the biological impact of the xenobiotic and the extent of its dissemination. Procedures for determining the partition of xenobiotics into biota are discussed, and attention is drawn to complicating factors, including the association of xenobiotics with macromolecules, and to the important interdependence of metabolism and bioconcentration. Surrogate procedures for evaluating bioconcentration potential that use physicochemical partition coefficients are outlined and their intrinsic limitations are pointed out. Such systems are unable to take into account the important issues of metabolism in biota and the structure of biological lipid membranes. Procedures for determining the distribution of xenobiotics between aqueous and solid phases are presented. The desorption of xenobiotics from the soil and sediment phases is discussed, and a brief account is given of interaction mechanisms between xenobiotics and components of solid matrices. Attention is drawn to the phase heterogeneity of the water mass in many natural systems and to the role of both particulate and dissolved matter in the distribution and dissemination of xenobiotics in lakes and rivers. Brief comments are devoted to the partitioning of xenobiotics between the aquatic phase and the atmosphere and to the significance of atmospheric transport on a global scale. A discussion of monitoring strategies is presented together with brief comments on the complexities in evaluating biomagnification. It is emphasized throughout that partitioning involves a complex set of molecular interactions, that these are reversible to varying degrees, and that attention should be directed both to the structure of the xenobiotic and to the ecosystem to which the results are to be applied. Equations used for correlating partition coefficients with physicochemical parameters have been presented and some of their limitations have been noted.

Introduction

With the availability of suitable analytical procedures, the next question that should be addressed is the distribution of xenobiotics among the various phases after their discharge into the environment. This information provides a basis for deciding upon the ultimate fate of these compounds — particularly those that are not readily degradable — whose dissemination, persistence, and toxicity have aroused the greatest environmental concern. The distribution of xenobiotics is determined on the one hand by physicochemical equilibria and on the other by chemical or biologically mediated reactions, some of which may result in essentially irreversible associations between the xenobiotic and organic or inorganic components of the aquatic and sediment phases. The distribution of xenobiotics is therefore a function of many interacting factors and it is to a discussion of these that this chapter is devoted. The most-detailed discussions are devoted to the aquatic, and soil and sediment phases, although attention is also directed to the atmosphere because of its established significance in the global dissemination of many xenobiotics. In this chapter, the term *aquatic phase* will be taken to include the water phase together with biota (e.g., algae and fish) and particulate material (seston), while the term *aqueous phase* will be applied in a more restricted sense to the water phase alone.

A valuable overview of the global dissemination of persistent organic compounds has been given (Wania and Mackay 1996), and application of fugacity models to the distribution of PAHs (Mackay and Callcott 1998). Attention should also be directed to the different physiology and biochemistry of the organisms as well as to their trophic level; important details of food chains are, however, noted only tangentially in this account.

The partitioning of organic compounds between the aqueous and the sediment phases and between the aqueous phase and particulate matter including algae is important for a number of rather different reasons:

1. It determines the exposure of biota to a potential toxicant initially discharged into the aqueous phase (Section 3.2) and the extent to which it is justifiable to correlate observed biological effects with measured concentrations of the toxicant.

2. It has a significant bearing on the persistence of a xenobiotic which is discussed in greater detail in Section 4.6.3.

3. An assessment of the ultimate fate of xenobiotics — and of putative metabolites — requires estimates of their concentration and distribution in all environmental compartments.

4. The dissemination of xenobiotics (Section 3.5) initially discharged, for example, into the aquatic phase may take place in several of the phases — within the water mass including suspended particulate

matter, in the sediment phase to which the compound is sorbed, or via the atmosphere — and alterations in the structure of the xeno- biotic may take place during transport within all of these phases.

Possibly the greatest attention, however, traditionally has been directed to the concentration of organic compounds from the aqueous phase into biota. This effort has been motivated by the consistent recovery of many com- pounds of industrial interest such as PCBs and the more persistent agro- chemicals such as DDT (and its metabolite DDE), mirex, and aldrin from samples of fish, birds, and marine mammals such as seals, whales, and polar bears. In a few cases, a plausible correlation has been established between injury to biota and exposure to a toxicant, and this is discussed in a wider per- spective in Chapter 7, Section 7.7.2. Only two examples of such correlations will therefore be given here as illustration:

1. Exposure of bottom-dwelling fish to concentrations of PAHs in contaminated sediments in Puget Sound, Washington and the inci- dence of disease including hepatic neoplasms (Malins et al. 1984). This is discussed, with emphases on karien flatfish by de Maagd and Vethaak (1998).
2. Exposure of fish-eating herring gulls (*Larus argentatus*) in the Great Lakes and the incidence of porphyria in the gulls (Fox et al. 1988).

However, even though exposure of biota to xenobiotics does not necessarily result in toxification of these organisms, the possibility that such compounds could enter the food chain and could therefore ultimately be consumed by the final predator — humans — has awakened serious concern over the dis- semination of such compounds. A good example is provided by the concern over possible adverse effects on human health, including reproduction, that could result from the consumption of fish from the Great Lakes that may be heavily contaminated with organochlorine compounds including PCBs (Swain 1991).

It should be appreciated that the concentration of a xenobiotic in biota is a dynamic process and represents a balance between uptake and elimination and that, as discussed in Sections 3.1.2 and 3.1.3, elimination may involve both the unchanged xenobiotic and its metabolites. Depuration therefore pro- vides both a mechanism for the detoxification of the xenobiotic and its return — either unaltered or in the form of metabolites — to the aquatic phase, and a means of its dissemination within the water mass; this aspect is discussed in Section 3.5.2.

Aquatic ecosystems are highly heterogeneous and comprise at least three apparently distinct phases: the aqueous phase, seston, the sediment phase and the biota. None of these phases should, however, be considered as an independent entity: for example, probably most sediments have a rich biota consisting of microorganisms together with a spectrum of higher organisms such as oligochaetes and amphipods, and the exposure of sediment-dwelling biota to toxicants is significantly determined by exposure to the interstitial water in the sediment phase. The distribution of an organic compound initially discharged into the aquatic environment is therefore exceedingly complex and is determined by the dynamics of a number of partition processes between (1) the aquatic phase and biota including, for example, microalgae, higher plants, invertebrates, and fish; (2) the aquatic phase and the sediment phase; and (3) the sediment and sediment-dwelling biota.

Almost all of these involve potentially reversible partitions all of which should be taken into consideration; they may be mediated, for example, by chemical desorption processes from the sediment phase or by depuration and elimination from biota. It should also be appreciated that few — if any — of these distributions are in true equilibrium; this fact should be borne in mind especially in extrapolating the results of laboratory experiments to natural ecosystems. In addition, the situation is complicated by the fact that none of these phases is truly homogeneous. Even the aquatic phase is heterogeneous and often contains particulate matter including inorganic material, and both soluble and insoluble organic matter originating from aquatic biota and terrestrial plants. In addition, components of the sediment phase may have originated from atmospheric transport and deposition; the quantitative importance of all of these distribution processes has therefore received increasing attention. As a result, intensive investigations have increasingly been directed to factors whose quantitative significance had not been fully appreciated previously. A few examples may be given to illustrate some of the important issues:

1. The sorption to particulate matter in the water column, and the dynamics and resuspension of surficial sediments;

2. The role of dissolved organic matter in the water column, accompanied by an increased appreciation of the important distinction between truly dissolved and finely divided particulate matter that may be colloidal;

3. The significance of interstitial water both in mediating exposure particularly to sediment-dwelling biota and in diffusion of xenobiotics into the water mass;

4. The importance of partitioning between the aquatic phase and the atmosphere even for compounds with relatively low volatility, and the role of the atmosphere in mediating the long-distance transport of xenobiotics.

These factors have focused attention on important new aspects of the phase partitioning of organic substances, and have indeed often revealed complexities that have merited intensive investigation and resulted in new perspectives. It is appropriate to note an increased awareness of possible limitations in extrapolating data from laboratory studies to the natural environment. Two simple examples may be used as illustration — both involving PCBs.

1. Partitioning between the aquatic and particulate phase in New Bedford Harbor was not strongly correlated with values of P_{ow} and revealed the importance of temperature (Bergen et al. 1993).

2. Studies of partitioning between the aquatic phase and algae have revealed that in natural ecosystems equilibrium is not reached in growing populations of algae so that use of P_{ow} values is not justified (Swackhamer and Skoglund 1993); this may indeed have wider implications and is discussed in greater detail subsequently.

It is important not to be left with the impression that biota and sediments function solely or primarily as sinks for xenobiotics. A number of mechanisms exist for their elimination from these phases including metabolism and depuration in biota (Section 7.5), and desorption from the sediment phase (Section 3.2.2). Elimination from biota may also depend on diffusion mechanisms when the biota are in intimate contact with another phase. Two illustrative example are given:

1. Elimination of 2,3,3′-trichlorobiphenyl, DDE, and γ-hexachloro[*aaaeee*]cyclohexane from larvae of the midge *Chironomus riparius* was generally greater in sediments with higher organic content, and a significant correlation was found between the rate of elimination and the octanol / water partition coefficient of the compounds (Lydy et al. 1992).

2. Mayflies (*Hexagenia* sp.) were chosen for monitoring PCBs in the upper Mississippi River on account of their long intrinsic need to be in contact with substrates at the base of the food chain (Steingraeber et al. 1994).

It is appropriate to make some comments here on the similarities between the soil and the sediment phases, since this is relevant both to the contents of this chapter and to the issues that are taken up in Chapters 4, 6, and 8. At first sight the soil and sediment phases appear to be totally different, but closer examination reveals important similarities and many of the principles set forth for aquatic systems are directly applicable, or with minor modification, to the terrestrial systems.

1. In both there may be substantial amounts of organic carbon, and sorption to and desorption from both mineral and organic components are essentially comparable.

2. Except on the surface of tropical deserts and arid lands, there is a subsurface water component of soils, and interstitial water is

important in the dynamics of sorption and desorption, and as a determinant of the bioavailability, toxicity, and persistence of xenobiotics to biota.

3. Both aerobic and anaerobic processes are important, and at greater depths the environment becomes essentially anaerobic.

4. Bioturbation is important in both environments.

5. Atmospheric deposition occurs directly onto terrestrial and aquatic environments, and from both by further partitioning may enter the sediment phase.

There are important reasons for including discussions of the atmosphere since this interfaces with terrestrial plants, soil surfaces, and the aquatic environment. A holistic view must therefore take into account all of these partitions. Some specific reasons for considering the atmospheric environment include the following:

a. The discharge of xenobiotics during incineration involves both "free" and particulate components, and partitions involving these in the atmosphere are extremely important (Mackay and Callcott 1998)

b. The atmospheric environment is a dynamic one, and transformation products may reenter the terrestrial and aquatic environments. This is discussed in detail in Section 4.1.2.

Considerable effort has been given to correlations between physicochemical properties and the various partitions. These properties themselves may also be of environmental significance. For example, quadricyclane that has been suggested as a high-performance aviation fuel, has the propensity to form microemulsions that could play a significant role in the dissemination of the compound in groundwater (Hill et al. 1997). There has also been concern that the water-soluble *t*-butyl methyl ether might act as a cosolvent for aromatic hydrocarbons such as BTEX that would probably occur at the same site (Poulsen et al. 1992). Field measurements in the United States suggest, fortunately, that this is not likely to pose a serious threat (Squillace et al. 1996).

3.1   Partitioning into Biota: Uptake of Xenobiotics from the Aqueous Phase

3.1.1   Direct Measurements of Bioconcentration Potential

3.1.1.1   *Outline of Experimental Procedures*

For aquatic organisms, *bioconcentration* is the accumulation of a chemical from the aqueous phase; exposure takes place only via the water although the compound may exist either in the dissolved form or associated with dissolved organic material. It is therefore distinguished from *bioaccumulation* that includes all modes of uptake including that of particulate matter; this is discussed more fully in a later section.

Bioconcentration factors (BCF) may be calculated by either of two procedures: the basic assumption in both is that the uptake and depuration are governed by first-order kinetics although deviations may occur that may be accounted for by the induction of enzymes for metabolism of the xenobiotic. In practice, a number of additional factors may be involved including the toxicokinetics of different organs and possible interference from growth of the test organism if the compound is only slowly accumulated. In one method, concentrations in the biota and in the surrounding medium are measured after a steady state has been reached, and the ratio of the two concentrations is used to obtain the BCF value. In the other, rates of uptake and elimination of the xenobiotic are measured and the ratio is used to calculate concentrations in the biota: the BCF is then calculated from these values. The experimental difficulties of maintaining a constant substrate concentration and achieving a steady state have been overcome by using a procedure based on iterative integration of the experimental data (Gobas and Zhang 1992). The possible complications resulting from metabolism of the test compound are discussed in Section 3.1.5, and more fully in Chapter 7, Section 7.5.

In laboratory experiments using fish, exposure takes place primarily by uptake through the gills directly from the aquatic phase (Pärt 1990), and the bioconcentration factor may be estimated by either or both of the procedures outlined above. Both procedures have been evaluated in experiments in which guppy (*Poecilia reticulata*) were exposed to a series of organophosphorus pesticides that are metabolized only slowly. It was shown that there was a linear relation between the BCFs and the ratios of the uptake and elimination rates within the logarithmic range of 2.6 and 4.7 (de Bruijn and Hermens 1991). Although in this case the two procedures produced essentially identical results, some discrepancy would be expected if the compounds were metabolized to a significant extent and the metabolites were subsequently eliminated from the fish. This is discussed more fully in Section 3.1.5.

Exposure to the xenobiotic generally extends over a period of days or weeks and even for up to several months; either semistatic or flow-through systems may be used, and analytical control of the concentrations of the test substrate should be maintained. After exposure, fish are generally maintained in a xenobiotic-free environment to allow excretion of toxicants or their metabolites to take place. A variety of different fish including rainbow trout (*Oncorhynchus mykiss* syn. *Salmo gairdneri*), fathead minnows (*Pimephales promelas*), guppy (*Poecilia reticulata*), zebra fish (*Brachydanio rerio*), and medaka (*Oryzias latipes*) have been employed, even though it has been clearly established that fish have highly effective metabolic potential for a wide range of compounds (Sections 3.1.5 and 7.5.1) and that this metabolic potential varies with the species. Different BCF values may therefore be found in experiments using different fish; for example, for a restricted range of chlorobenzenes using fathead minnows, green sunfish (*Lepomis cyanellus*), and rainbow trout, experimental values for rainbow trout were the lowest (Veith et al. 1979a), and this might plausibly be correlated with their established metabolic capability. It should also be appreciated that the disposition of xenobiotics within the organisms may differ significantly; for example, a number of neutral organochlorine compounds are accumulated in the central nervous system (CNS) of cod (*Gadus morhua*) but not in that of rainbow trout, and for hexachlorobenzene it has been shown that whereas it is the xenobiotic itself that is present in the CNS system, it is metabolites that are found in cerebrospinal fluid (Ingebrigtsen et al. 1992). Low concentrations of the test compound are generally employed, and particular care should be exercised with compounds displaying even subliminal toxic effects at the concentrations used during exposure; for example, the value of 39,000 for the BCF of 2,3,7,8-tetrachlorodibenzo[1,4]dioxin at the concentration where rainbow trout were least affected may well be too low, since the corresponding value of the less toxic 2,3,7,8-tetrachlorodibenzofuran increased from 2455 at a concentration of 3.93 ng/l to 6049 at a concentration of 0.41 ng/l (Mehrle et al. 1988).

Virtually any aquatic organism may, of course, be used and, for example, common mussels (*Mytilus edulis*) have been used for investigating the uptake of a restricted range of neutral organochlorine compounds (Ernst 1979), the crustacean *Daphnia pulex* for the uptake of azaarenes (Southworth et al. 1980), and freshwater mussels (*Anodonta anatina*) for the uptake of chlorophenolic compounds (Mäkelä et al. 1991). Attention is drawn (Section 3.1.5) to the differences that may be observed between fish and bivalves, and this may plausibly be attributed to the relatively lower metabolic capacity of bivalves (Livingstone and Farrar 1984). This is noted in the wider context of toxicity and metabolism in Chapter 7, Section 7.5.2.

The design of the uptake experiments themselves and the analytical determinations are straightforward: specific analysis may be carried out for the compounds being examined (together with their metabolites) or advantage may be taken of, for example, ^{14}C-labeled substrates. A number of important limitations in the numerical significance of the values obtained in laboratory studies have been pointed out (Oliver and Niimi 1985) and these are worth emphasizing:

1. Uptake of the xenobiotic may be so slow that the length of exposure is insufficient to attain a steady state.

2. The molecules may be too large for uptake, for example, via the gills of fish so that BCF values are negligible, and uptake in the environment is dominated by uptake via the food; this is discussed in greater detail below.

3. The xenobiotic is metabolized by the biota and this results in erroneously low concentrations in the biota and hence low BCF values; the interdependence of bioconcentration and metabolism in fish is considered in Section 3.1.5, and in a wider context in

Section 3.5.2, while additional details of metabolism by fish are given in Section 7.5.1.

These limitations have been systematically explored for 34 halogenated compounds in rainbow trout, and they were shown to be particularly relevant to making realistic predictions of the concentrations in wild biota. Indeed, the striking incidence of DDE (the principal transformation product of DDT) in environmental samples is consistent with its bioconcentration in field samples to a degree greatly exceeding that predicted from laboratory measurements (Oliver and Niimi 1985).

Significant differences in measured BCF values may also result from the design of the experiments and from the inevitable biological variability in the test organism. For example, BCF values for 2,3,4,5-tetrachloroaniline in guppy (*P. reticulata*) increased with increasing exposure time or increasing concentration of the test compound (de Wolf et al. 1992b), and log BCF values on a lipid basis for the same trichloroaniline isomer obtained in the same laboratory over a period of time using different strains of guppy ranged from 2.61 to 3.21 for 2,3,4-trichloroaniline and from 2.88 to 3.40 for 2,4,5-trichloroaniline (de Wolf et al. 1993). The significant role of the lipid content of the test organism is discussed in detail later.

It cannot therefore be too strongly emphasized that all of these considerations should be critically evaluated in discussions of bioconcentration potential.

3.1.1.2 The Molecular Size and Shape of Xenobiotics and the Role of Lipid Content of Biota

Increasing evidence points to the specific role of lipids in determining bioconcentration potential, and two different kinds of situation may be clearly distinguished. It should be clearly appreciated, however, that the term *lipid* is used for a class of structurally diverse compounds united by a single physicochemical property (solubility in organic solvents). They include, for example, neutral glyceryl triesters and glyceryl galactosides, zwitterionic phosphate diesters of glycerol and ethanolamine, and diesters of glycerol and inositol.

Some compounds such as hexabromobenzene, octachlorodibenzo[1,4]dioxin, and tetradecachloroterphenyl are accumulated by fish only to a minor extent, presumably due to the size and configuration of the molecules (Bruggeman et al. 1984). Such compounds have been termed *superhydrophobic* since they have values of log $P_{ow} > 6$, but it has been shown on the other hand that many of these compounds have — possibly unexpectedly — only low lipid solubility and that this decreases with increasing P_{ow} (Chessells et al. 1992). In the case of decachlorobiphenyl, it has been suggested that only 3% of the substrate in the aqueous phase is available to guppy (*P. reticulata*) and this results in a BCF value that is between 10- and 100-fold lower than would be predicted on the basis of the P_{ow} value of the compound (Gobas et al. 1989). Consistent with the overall role of values of P_{ow} the tetra- and pentabrominated diphenyl ethers (log P_{ow} < 6.97) have been recovered from fish, whereas the decabromodiphenyl compound (log P_{ow} = 9.97) was essentially absent (Sellström et al. 1998). It has, however, been shown (Kierkegaard et al. 1999) that this can be transported into muscle and liver of *Oncorhynjchus mykiss* by consumption of food (cod chips) spiked with decabromodiphenyl ether that is noted in Section 3.1.2.

By definition, hydrophobic compounds would be expected to accumulate in lipid material so that variations in the lipid content of biota may be an important determinant of uptake both in laboratory experiments and in feral populations. For example, in a study on the uptake of hexachlorocyclohexanes and of pentachlorophenol by the mussel *M. edulis* and by the polychaete *Lanice conchilega*, variations in the lipid content of the test organism could have introduced serious errors (Ernst 1979), and the BCF values for a number of organochlorine compounds increased linearly with lipid content to maxima at 5% (Tadokoro and Tomita 1987). The lipid content of fish is also a function of their age, and in lake trout, for example, the lipid content increased from 7% in the age group 3 to 5 years to 16% in those in the age group 7 to 10 years (Thomann and Connolly 1984). The lipid content may also be subject to seasonal variation, and its significance in determining the half-lives of a number of organochlorine compounds in herring gulls (*Larus argentatus*) has been examined: analyses for lipid content and for plasma concentrations were then incorporated into a two-compartment model to describe the dynamics of clearance of the compounds (Clark et al. 1987). Lipids serve as reserves for migrating fish such as catadromous eels and anadromous salmonids so that their concentrations are far from constant. For example, in a study of lipid content of sockeye salmon (*Oncorhynchus nerka*) during migration from the ocean to spawn, the lipid content diminished continuously over a 400 km stretch of the Copper River, Alaska (Ewald et al. 1998); as a result, the concentrations based on lipid content of total PCBs and DDT in both muscle tissue and gonads increased. It may therefore be generally valuable to express BCF values based on lipid concentration (Mackay 1982) as well as on wet weight. These observations have directed attention to the whole question of the role of lipids and this is discussed further in Section 3.1.1 in the context of surrogate procedures for estimating bioconcentration potential, and in Section 3.5.4 in the context of biomagnification. In particular, the structure of biological lipid membranes should be taken into consideration.

In summary, all of these results illustrate the care that must be exercised in predicting the concentrations of xenobiotics in natural biota from values of bioconcentration factors assuming that uptake takes place exclusively from the water mass. The various factors that may seriously compromise the interpretation of measurements of bioconcentration potential are as follows:

- Uptake may occur by uptake of particulate material—bioaccumulation.

- The compound may be "bound" to dissolved components of the aquatic or sediment phases so that it is not freely accessible to biota.

- Transport into biota is not passive and must be evaluated in its relation to metabolism (Section 3.1.5).

- There may exist intrinsic limitations to transport into cells due to steric effects or the mere size and shape of the xenobiotic molecule.

The preceding discussion has placed emphasis on the role of lipid in the sequestering of essentially hydrophobic xenobiotics in biota. It should, however, be noted that other biomolecules may be involved. For example, in a microcosm study using relatively high concentrations of ^{14}C-labeled trifluoroacetate, it was shown that in leaves of jewelweed (*Impatiens capensis*), in oligochaetes, and in the microbial flora the substrate was distributed more extensively in protein than in lipid (Standley and Bott 1998). It seems plausible to attribute these results to specific reaction with amino acid side chains in the proteins.

3.1.2 The Role of Particulate Matter and Uptake via Food

The inhomogeneity of many water masses is well established, so that attention has been directed to the role of particulate matter in binding xenobiotics and to its significance in determining their uptake and their bioavailability. The term *bioaccumulation* includes all transport routes including exposure to the xenobiotic in food and particulate matter, although the numerical difference between factors for bioconcentration and bioaccumulation will generally not be large except for highly hydrophobic compounds. The influence of organic matter in any form — dissolved or particulate — should, however, be kept in mind. For example, the BCFs of chlorobenzenes were reduced when guppy (*P. reticulata*) were exposed to these compounds in sediment suspensions (Schrap and Opperhuizen 1990), and even dissolved organic carbon may also form associations with xenobiotics and thereby diminish their bioavailability (Landrum et al. 1985).

It has become increasingly realized that the exposure of biota to xenobiotics may occur not only from the dissolved state, but also to a significant degree through consumption of particulate matter in sediments or in the water mass: indeed, this exposure route may be dominant for demersal fish and for sediment-dwelling organisms. Examples that support the role of particulate matter in determining exposure to xenobiotics may be given as illustration.

1. It has been estimated that in New Bedford Harbor, MA, sediment is responsible for 83% of the body burden of tetrachlorinated PCBs in winter flounder, and for 42% in lobster (Connolly 1991).

2. In the clam *Macoma nasuta*, mass balance studies showed that the major route of uptake of hexachlorobenzene was via the gut from ingested solids (60 to 80%), with contributions of around 10% for other routes including interstitial and overlying water (Boese et al. 1990).

3. It has been shown in laboratory experiments that the dominant transport of decabromodiphenyl ether into rainbow trout occurs by ingestion of contaminated food (Kierkegaard et al. 1999).

Biota Sediment Accumulation — The biota sediment accumulation factor (BSAF) is defined as the ratio of the tissue concentration based on lipid concentration and the sediment concentration based on the concentration of total organic carbon. It has been examined in the same organism with a range of PCB congeners, and it has been suggested that even though variations with sediment type were found, this could provide a suitable criterion for assessing sediment quality (Boese et al., 1995). As the following illustrative examples show, however, a number of important determinants should be taken into consideration including the type of matrix used for assay, the length of time to which the matrix has been exposed to the toxicant, which is discussed in Section 3.2.3, the role of interstitial water (Sections 3.3.2), the intrusion of metabolism (Section 3.1.5 and 7.5), and the structure of the xenobiotic.

1. Experimental values of BSAF for 2,3,7,8-tetrachlorodibenzo[1,4]dioxin and 2,3,7,8-tetrachlorodibenzofuran in contaminated sediments were obtained from exposure of the polychaete *Nereis virens* in a flow-through system for 28 day followed by a 24-h depuration period. The results were used to calculate the tissue concentrations of these contaminants at another site (Schrock et al. 1997). Although there was good agreement between the measured and predicted levels, it was pointed out that validation of this procedure would depend on the study of a more extensive data set.

2. A study on the biaccumulation of dieldrin in the oligochaete *Lumbriculus variegatus* showed significant dependence on the type of sediment (Standley 1997). Although bioaccumulation was best correlated with solvent-extractable organic matter, it was pointed out that prediction of biological response from chemical measures of bioavailability should be carried out with caution.

3. The results of a study using earthworms (*Eisenia foetida*) and atrazine, phenanthrene, and naphthalene that had been incubated for increasing times in sterile soil (Kelsey and Alexander 1997) illustrated the effect of length of exposure to the toxicant, and were consistent with the increasing significance of aging.

4. The bioaccumulation of ^{14}C–γ-hexachloro[*aaaeee*]cyclohexane and ^{14}C-hexachlorobenzene in the tubificid oligochaetes *Tubifex tubifex*

and *Limnodrilus hoffmeisteri* were examined in a standardized artificial sediment (Egeler et al. 1997). There was evidence for the formation of metabolites, and values of the bioaccumulation factor (BAF) were derived from rates of uptake and elimination. Values of BASF were comparable, although values for both toxicants were somewhat lower for *L. hoffmeisteri*. It was pointed out that it may be unacceptable to extrapolate values of BCF for fish to BAF for sediment-dwelling organisms, although it may be noted that exposure concentrations were very different due to the low water solubilities of the toxicants in fish assays.

5. Bioaccumulation of PAHs has been examined in a number of organisms including marine mollusks and polychaetes, a terrestrial oligochaete and crustaceans (references in van Brummelen et al. 1998). Values for accumulation of benzo[*a*]pyrene by the terrestrial isopod *Porcellio scaber* were low (van Brummelen et al. 1996), and this may plausibly be attributable to biotransformation since it has been established that this organism metabolizes pyrene to 1-hydroxypyrene (Stroomberg et al. 1996). Low values for the polychaete *N. virens* may also be the result of metabolism since sediment-dwelling polychaetes have the potential for metabolizing some xenobiotics. For example, *N. virens* is able to metabolize both PCBs (McElroy et al. 1988) and a number of PAHs (McElroy 1990), while *N. diversicolor* and *S. viridis* are able to metabolize benzo[*a*]pyrene (Driscoll and McElroy 1996). Consistent with this, values of BSAF for *N. diversicolor* and *S. viridis* were 0.028 and 0.163 based on the parent PAH, compared with values of 0.58 and and 0.37 based on total benzo[*a*]pyrene equivalents (parent + metabolites). On the other hand, the values for *Leitoscopolos fragilis* were essentially similar.

6. Values of BSAF for highly chlorinated congeners of PCBs with 7 to 10 chlorine substituents were examined in a range of fish collected from the vicinity of the discharge from a former chloroalkali factory. The congeners had values of P_{ow} ranging from 6.7 to > 9, and values of BASF were negatively correlated both with P_{ow} and lipid-normalized values of the trophic transfer factor (Maruya and Lee 1998).

It is important to examine in a little more detail the significance of uptake via food, although biomagnification in a wider context in field situations is discussed in Section 3.5.4. As a general rule, it has been accepted that uptake via food rather than in the dissolved state directly from the water mass is the dominant exposure route for compounds with log P_{ow} > 5 (Connolly and Pedersen 1988): (Thomann 1989). This alternative exposure route is not, however, generally taken into consideration in laboratory studies since it is difficult — and indeed may be impossible — to distinguish between direct uptake via the food and simultaneous desorption from the food that results in direct uptake from the aquatic phase. Evidence that bioaccumulation via contaminated food is the principal route of uptake for poorly water soluble compounds has, however, been clearly demonstrated. Three simple illustrations will be given here:

1. Lake trout (*Salvelinus namaycush*) in Lake Michigan have concentrations of PCBs which are three to two times higher than in alewife (*Alosa pseudoharengus*) which is preyed upon by older fish (Thomann and Connolly 1984).

2. PCB concentrations in the food chain of a small freshwater lake in Holland increase in the higher trophic levels (van der Oost et al. 1988). In the latter study, attention is also directed to the important issue of differences in the distribution of PCB congeners at the various trophic levels. An exhaustive study undertaken over a complete life cycle of guppy (*Poecilia reticulata*) has revealed important details of such processes (Sijm et al. 1992), while the wider issue of biomagnification including additional factors and the issue of cotransport with lipid material are discussed briefly in Section 3.5.4.

Whereas similar conclusions on the significance of uptake by food have been drawn from the results of a laboratory study with chlorinated dibenzo[1,4]dioxins using juvenile rainbow trout and fathead minnows (Muir and Yarechewski 1988), laboratory experiments with PAHs using rainbow trout did not reveal significant accumulation through dietary exposure apparently as a result of poor absorption efficiency from the diet and rapid elimination of the xenobiotics (Niimi and Dookhran 1989). The results of these experiments with PAHs probably do not, however, exclude the significance of this exposure route for demersal fish that are exposed to high concentrations of these compounds in contaminated sediments.

Collectively, the results of these studies clearly illustrated that attention should be directed both to the feeding habits and to the physiology of specific organisms and that there may exist serious limitations in the application of models attempting universal application and which fail to take these into account.

3.1.3   Concentration of Xenobiotics into Algae and Higher Plants

The preceding discussion has dealt almost exclusively with bioconcentration by fish or invertebrates, but transport into photosynthetic organisms also merits attention in the context of dissemination of toxicants into higher trophic levels.

Algae

These are primary producers in aquatic systems and therefore play a key role in the food chain and in the transport of xenobiotics into higher trophic levels and — after their death — into the sediment phase. An interesting study (Swackhamer and Skoglund 1993) investigated the bioaccumulation of a range of PCB congeners into a strain of *Scenedesmus* sp. under different laboratory conditions, and into field phytoplankton at different seasons of the year. Two important conclusions could be drawn:

1. In laboratory experiments at 11°C when growth was slow and when the length of exposure was 3 days or more, the lipid-normalized bioaccumulation factor was a linear function of P_{ow} with a slope of unity for congeners with log P_{ow} < 7. When comparable experiments were carried out at 20°C in growing cells the linearity was observed only for congeners with log P_{ow} < 5.5.

2. In the experiments using field material, for samples collected in both summer and winter there was linearity for all congeners up to those with log P_{ow} of 8, but whereas the slope for winter samples was 0.93, that for summer samples was only 0.4.

These results show that equilibrium conditions do not prevail in growing populations of algae. The use of P_{ow} values to assess the bioaccumulation of hydrophobic xenobiotics into phytoplankton under growth conditions is therefore not justified.

Higher Aquatic Plants

Concentration of xenobiotics into aquatic plants may also be important and presents another redistribution pathway for xenobiotics. The uptake of a few agrochemicals has been investigated using the aquatic plant *Hydrilla verticillata* (Hinman and Klaine 1992), although only low levels of atrazine, chlordane, and lindane were accumulated. These plants have moreover only low levels of lipids and this is consistent with the role of lipid material in determining uptake. Even these low levels of bioconcentration should, however, be taken into consideration in lakes with high densities of such plants. An illustrative example is the bioconcentration of a number of chlorinated aromatic hydrocarbons into the submerged macrophyte *Myriophyllum spicatum* (Gobas et al. 1991): log BCF values ranged from 1.52 for 1,3,5-trichlorobenzene to 3.79 for octachlorostyrene, and 5.79 for 2,2′,3,3′,4,4′,5,5′-octachlorobiphenyl, and it was estimated that submerged macrophytes could play a small, although important, role in the removal of chlorinated aromatic hydrocarbons from rivers and lakes. As for fish, metabolism of the xenobiotic by plants may take place after uptake; for example, pentachlorophenol is metabolized by *Eichhornia crassipes* to a number of metabolites including chlorocatechols, chlorohydroquinones, pentachloroanisole, and tetrachloroveratroles (Roy and Hänninen 1994). These are formed by hydroxylation (Figure 3.1a), O-methylation (Figure 3.1b), and dechlorination (Figure 3.1c).These products should be compared with the phenolic compounds formed during the photochemical stage (Figure 4.4) and the initial stage in the microbiological metabolism of

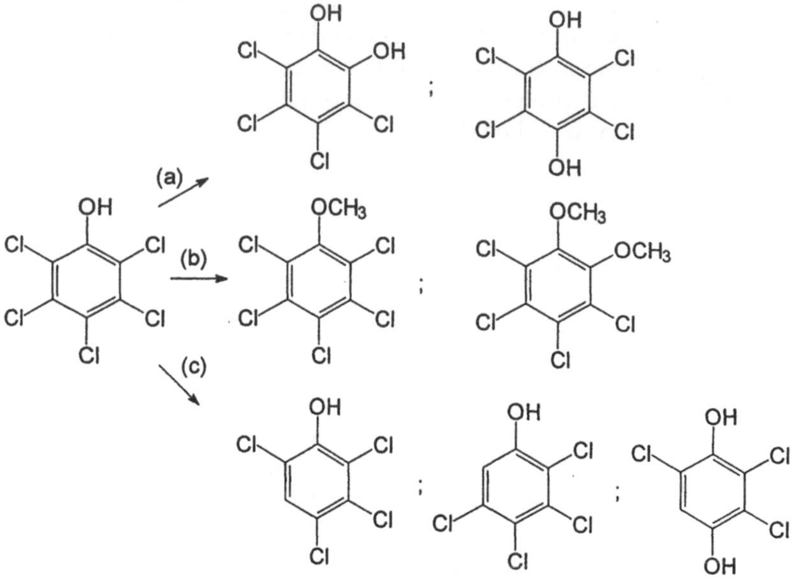

FIGURE 3.1
Metabolism of pentachlorophenol by *E. crassipes*.

pentachlorophenol (Chapter 6, Section 6.5.1.2) with subsequent *O*-methyla-tion (Section 6.11.4). Quite complex transformations may be mediated by higher plants, and the metabolism of phoxim by plant organs and cell suspen-sion of soybean (*Glycine max*) may be given as illustration (Höhl and Barz 1995) (Figure 3.2). The cardinal issue is that the metabolites may have parti-tions and toxicity quite different from those of their precursors.

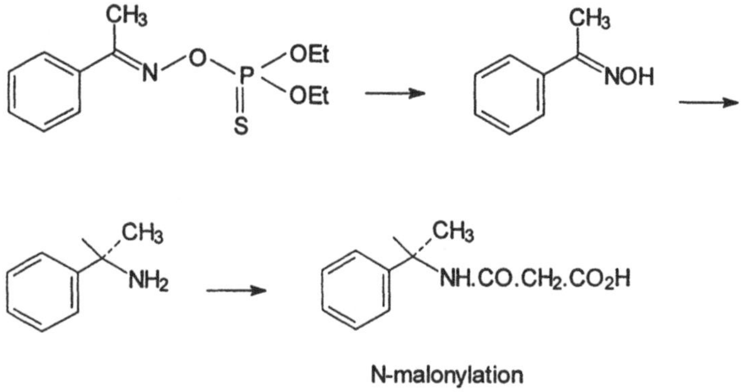

FIGURE 3.2
Metabolism of phoxim by *G. max* 4.300.

Terrestrial Plants

The situation for terrestrial plants is complicated by the fact that exposure may be mediated by a number of partitioning processes including soil/water, water/plant roots, and atmosphere/leaves, and by soil/atmosphere partitioning. After uptake, the toxicant may be translocated within the plant and be metabolized. For all these reasons, a simple relation between BCFs factors and P_{ow} is not to be expected, and this is confirmed by results using different plants and different xenobiotics (Scheunert et al. 1994). The possible role of plants is noted again in the context of assays for terrestrial toxicity in Chapter 7, Sections 7.3.1.3 and 7.5.5, and of bioremediation programs in Chapter 8, Section 8.1.1. A number of cardinal issues are summarized here and these lead to a better understanding of the global dissemination of xenobiotics.

1. It has been suggested that a suitable surrogate parameter for atmosphere/plant partition is the octanol/air partition coefficient (Harner and Mackay 1995). Direct measurements of this coefficient for a number of chlorinated aromatic hydrocarbons revealed, however, its sensitivity to temperature. The results of a study in which a number of PAHs were analyzed throughout the year in samples of tree bark, leaves, pine needles, and in the atmosphere also underscore the importance of temperature, since there was a cyclical partition between the atmosphere and the tree canopy (Simonich and Hites 1994). The contribution to the atmosphere through direct volatilization from the terrestrial environment was not resolved in this study.

2. Plant/air partition coefficients for a range of PCBs that were measured in ryegrass (*Lolium multiflorum*), clover (*Trifolium repens*), plantain (*Plantago lanceolata*), Hawk's beard (*Crepis biennis*), and yarrow (*Achillea millefolium*) varied widely and suggested that the lipophilic plant components were not well simulated by octanol (Kömp and McLachlan 1997). The sorption of volatile organics onto plant surfaces has been examined for a number of compounds (Welke et al. 1998). Values of the partitions, cuticular matrix/air (K_{ma}) cuticular matrix/water (K_{mw}) and air/water (K_{aw}) were determined and regressions of K_{ma} with the boiling points (T_b) and saturation vapor pressure (p^0) were given,

$$\log K_{ma} = 1.343 + 0.017\, T_b = 6.290 - 0.892 \log p^0$$

and were used to estimate on the basis of measured atmospheric concentrations the amounts of compounds in the cuticular matrix. Values ($\mu g/kg$) ranged from 17 for toluene to 0.22 for tetrachloromethane and 0.04 for freon-113.

3. The uptake of xenobiotics by a range of agricultural plants including barley, lettuce, carrots, and radish illustrated the different modes of uptake and translocation (Schroll et al. 1994).

    a. Hexachlorobenzene was translocated neither from roots to shoots nor vice versa.

    b. For the triazine terbuthylazine, uptake from the roots predominated over foliar uptake.

    c. Trichloroacetate was highly mobile within the plant.

    d. Trichloroethene transport is dominated by foliar uptake and is highly mobile within the plant.

These results clearly showed the relative significance of different routes of exposure, and the varying degrees of mobility after uptake. This is consistent with results for the uptake of polychlorinated dibenzo[1,4]dioxins and dibenzofurans from soils. In zucchini and pumpkin belonging to the genus *Cucurbita*, root uptake was dominant, whereas for cucumber (*Cucumis sativus*) foliar uptake was the primary mode and was much lower (Hülster et al. 1994).

Metabolism after uptake has been observed and should be taken into account. Some illustrative examples are given here, and in a wider context in Section 4.3.7.

1. It has been shown that the leaves of trees along a heavily trafficked road in Japan contained not only pyrene and 1-hydroxypyrene, but also the β–O–glucoside and β–O–glucuronide conjugates (Nakajima et al. 1996). The total concentrations of conjugates in the leaves exceeded that of free 1-hydroxypyrene, and the results suggest that the total PAH burden transferred into plants from the atmosphere may be even more significant than considered hitherto. The formation of 1-hydroxypyrene as a metabolite of pyrene is noted again in Section 6.2.2, and its toxicity in Section 7.3.6.

2. The uptake and biotransformation of benzene from soil and from the atmosphere have been studied in a number of plants, and it was shown that in leaves of spinach (*Spinacia oleracea*), the label in ^{14}C-benzene was found in muconic, fumaric, succinic, malic, and oxalic acids as well as in specific amino acids, and that an enzyme preparation in the presence of NADH or NADPH produced phenol (Ugrekhelidze et al. 1997).

3. Hybrid poplars are able to transport and metabolize various xenobiotics: (a) trichloroethene is metabolized to trichloroethanol and trichloroacetate (Newman et al. 1997), (b) atrazine is metabolized by reactions involving dealkylation and hydrolytic dechlorination to yield 2-hydroxy-4,6-diamino-1,3,5-triazine (Burken and Schnoor 1997).

3.1.4 Surrogate Procedures for Evaluating Bioconcentration Potential

Introduction

The concept of bioconcentration is derived from that of distribution coefficients in physical chemistry: in these, the equilibrium concentrations of a compound distributed between two phases are measured, for example, between water and a water-immiscible solvent such as hexane. If partitioning were a passive reaction, direct physicochemical measurements of the partition between an aquatic phase and a suitable model for the biological membrane would be possible. It would therefore be attractive to measure distribution coefficients in a chemically defined system and to seek a correlation between the values found and those obtained by direct measurements in biota.

The Octan-1-ol – Water Partition as a Surrogate

A commonly used system measures — directly or otherwise — partition between water and octan-1-ol to derive the distribution coefficient (P_{ow}), and then applies an empirical formula to translate these values into bioconcentration factors (BCF) using a range of benchmark compounds. As would be expected, the numerical relationships depend on the organism used so that different equations result. Some equations that have been used for different organisms are the following (Mackay 1982; Hawker and Connell 1986):

- Fish $\qquad \log BCF = \log P_{ow} - 1.320$
- Mollusks $\qquad \log BCF = 0.844 \log P_{ow} - 1.235$
- Daphnids $\qquad \log BCF = 0.898 \log P_{ow} - 1.315$

As implied in Sections 3.1.1 and 3.1.3, it should be noted that such equations cannot, however, be applied to uptake by aquatic plants — or indeed other biota — with low lipid content (Hinman and Klaine 1992).

Accurate values of P_{ow} are clearly necessary for the ultimate calibration of all surrogate systems, but, in practice, direct measurements of P_{ow} by the traditional shake-flask method are seldom used. Particularly for compounds with low water solubility, experimental difficulties may arise from problems in phase separation without carryover, sorption to glass surfaces, or from formation of emulsions. All of these introduce serious uncertainties into the concentrations in the appropriate phases, and may consequently lead to substantial errors in the estimates of partition coefficients. The problem is particularly acute for compounds with extremely low solubility in water such as the chlorinated dibenzo[1,4]dioxins for which widely varying values have been reported (Marple et al. 1986; Shiu et al. 1988). For such compounds, use of a generator column has been advocated (De Voe et al. 1981; Woodburn et al. 1984). In essence, the following steps are carried out: (1) a solution of the test substance in octanol is equilibrated with water and the concentration in the octanol phase is determined, (2) the octanol phase is loaded onto a column packed with Chromosorb W, and (3) octanol-saturated water is pumped through the column and the solute collected in a Sep-Pak cartridge for analysis.

A possibly more expedient procedure is the slow-stirring method that has been applied to a structurally diverse range of hydrophobic xenobiotics (De Bruijn et al. 1989), and it was shown that there was generally good agreement with the values obtained from HPLC or the generator-column procedures.

The octanol–water partition coefficient — and hence the bioconcentration potential — has also been correlated with the aqueous solubility, although the experimental determination of the latter for poorly water soluble compounds also presents some problems. A dialysis procedure that is applicable to a wide range of water solubilities ranging from cyclohexanol (37.5 g/l) to anthracene (0.0488 mg/l) has been developed (Etzweiler et al. 1995). The following relations have been proposed (Mackay 1982):

Liquids: $\log P_{ow}$ = 3.25 - $\log X_l$ where X_l is the molar solubility (mol.m^{-3}) in water

Solids: $\log P_{ow}$ = 3.25 - $\log X_s$ + 2.95 (1 -T_m/T) where X_s is the molar solubility of the solid, T_m is the melting point, and T the ambient temperature

A detailed thermodynamic discussion has been presented (Miller et al. 1985), and a valuable critique of the measurement and use of P_{ow} values has been given (Franke 1996). This draws attention to a number of important issues including (1) the relevance of the cutoff value of $\log P_{ow} < 3$ for assessing the existence of bioconcentration potential (2) the necessity for using test concentrations relevant to environmental situations, and (3) the important role of metabolism and excretion that is discussed below.

There are a number of basic questions which must also be addressed in the application of such surrogate procedures; among the most fundamental are the choice of the water-immiscible solvent and the neglect of metabolic transformation of the test compound.

Glycerol trioleate has been used in an attempt to simulate lipid membranes and to take into account some of the solvent associations plausibly occurring in biota; an impressive direct correlation was observed between $\log P_{tw}$ (P_{tw} is the partition coefficient between glycerol trioleate and water) and BCF values in rainbow trout expressed on a lipid basis, and these results were used to support the view that the bioconcentration of nonpolar hydrophobic xenobiotics is significantly determined by their lipid solubility (Chiou 1985). This conclusion is further supported by the results of an extensive examination of a series of highly hydrophobic compounds which do not demonstrate a high potential for bioconcentration (Chessells et al. 1992).

Xenobiotics have been concentrated from aquatic systems into dialysis membranes containing solvents such as hexane, and this procedure has been suggested as simulating uptake by biota (Södergren 1987; Huckins et al.

1990a, b; Meadows et al. 1993). The kinetics of uptake of PCB congeners from the aqueous phase during 28 days have been compared (Meadows et al. 1998) for triolein-filled membranes and brown trout (*Salmo trutta*). The pattern of congener uptake was similar, and rates for both were comparable over a 500-fold range of P_{ow} values provided that uptake was mediated only through solution and not via ingestion of food. It was suggested that use of a permeability reference compound be used to adjust kinetic values for a range of other environments. A word of caution: attention has already been drawn to the low lipid solubility of some superhydrophobic compounds (Chessells et al. 1992).

Application of Liposome–Water and Biomembrane–Water Systems

The importance of lipids in bioconcentration is emphasized several times in this chapter, and various devices have been explored to take this into account. Experiments using liposomes prepared with L-α-dipalmitoyl and L-α-dioleylphosphatidylcholine, and membranes prepared from *Rhodobacter sphaeroides* were therefore studied in an attempt to produce more realistic models of the lipid phase in partition experiments (Escher and Schwartzenbach 1996). The system was evaluated using a number of phenols of varying pK_a and log K_{ow} values, and it was shown that both systems provided good models for all species of phenolic compounds. An extremely important observation that has wide implications for ecotoxicology emerged: not only the neutral phenols partitioned into the liposomes but also the anionic species.

Alternative Surrogate Procedures

A number of other surrogate procedures have been developed. These include the use of reverse-phase thin-layer chromatography (TLC) (Bruggeman et al. 1982; Renberg et al. 1985) to measure relative mobilities (R_m) or reverse-phase high-performance liquid chromatography (HPLC) to measure capacity factors (Veith et al. 1979b). These values are then correlated with experimentally established P_{ow} values for standard compounds, and the correlation is then used to calculate values of P_{ow} for the unknown compounds.

The TLC procedure is extremely easy to carry out but is essentially restricted to neutral compounds, and correlations for a range of structurally diverse compounds must be carried out with caution since appreciably different relations between R_m and P_{ow} exist for different classes of compounds such as PAHs and chlorinated compounds.

Use of the reverse-phase HPLC system is highly flexible since it can also be applied to ionizable compounds such as carboxylic acids, phenols, and amines. The partition coefficients relate to the unionized compounds that are generally assumed to be the principal forms in which these compounds are transported into biota, even though their concentration may be low in comparison with the dissociated states at physiological pH values: acidic compounds such as highly chlorinated phenols or many carboxylic acids have

pK_a values < 7 and aqueous solutions of compounds with pK_a values < 6 contain only approximately < 10% of the free acid at neutral pH values. Although the influence of toxicity on pH has been examined (Neilson et al. 1990), its effect on bioconcentration has been less extensively explored. There is evidence, however, (Pärt 1990; Escher and Schwartzenbach 1996) that for some compounds both the unionized (free) and the dissociated forms may be accumulated. In the application of these methods, difficulties may emerge due to the absence of suitable detection methods; for example, in quantification of compounds lacking ultraviolet absorbance, fluorescence, or groups suitable for electrochemical detection.

A solid-phase microextraction method for estimating P_{ow} values < 3.5 has been used (Dean et al. 1996), and a procedure based on microemulsion electrokinetic chromatography has been described. This can be carried out at pH values of 1.19 or 12 at which most compounds will exist in their unionized state, and is applicable to P_{ow} values < 4.4 (Gluck et al. 1996). These values must then be corrected to take into account the degree of ionization at the pH encountered in the environment.

Clearly, surrogate systems cannot take into account the metabolic activity of biological systems. Although the extent of biotransformation may be restricted for some classes of compounds, it is unlikely to be totally absent, and it has been suggested that for highly lipophilic compounds which have a low rate of physicochemical elimination, the total rate of elimination may be significantly affected even by low rates of biological elimination (de Wolf et al. 1992a). The intrusion of metabolism results in lower concentrations in biota than would be predicted on the basis of the linear relationships between values of BCF and P_{ow} that have been noted above (de Wolf et al. 1992a). Such discrepancies have been observed for fish with compounds of diverse structure including trichloroanilines (de Wolf et al. 1993), chloronitrobenzenes (Niimi et al. 1989), and azaarenes (Southworth et al. 1980, 1981). For the azaarenes, the expected correlation was shown to hold for *Daphnia pulex* that apparently did not metabolize the compounds (Southworth et al. 1980).

Care should therefore be exercised in extrapolation of the results from all surrogate methods to assessing the uptake of xenobiotics into natural biota from the water phase.

3.1.5   Interdependence of Bioconcentration and Metabolism

It appears plausible to extrapolate to biological systems the concept of partitioning between two phases — representing the aquatic phase to which biota are exposed and a water-immiscible phase representing the lipid membranes. This simplification fails, however, to take into account a number of significant factors. The limitations concerning lipids have been briefly noted in Section 3.1.1, but there is an additional and frequently invalid assumption that is not always sufficiently appreciated. In fish, although the structure of the compound will generally remain unaltered during the partitioning between the aquatic phase and outer biological membanes (gills and skin), this will seldom be the case after transport into the organisms. Most of them have developed the capability of metabolizing many — if not most — xenobiotics and some striking examples exist: for example, even compounds such as 1,3,6,8-tetra-chlorodibenzo[1,4]dioxin can be metabolized by fish (Muir et al. 1986) although this is only one facet that may account for the apparently low bioconcentration potential of this compound. The same situation prevails in more complex natural systems. For example, whereas mirex may pass through the food chain, *D. magna* – *Lepomis macrochirus* without apparent metabolic change (Skaar et al. 1981), alterations in the relative distribution of the PCB congeners as they are transported through the food chain cod–seal–polar bear (Muir et al. 1988) clearly suggest metabolism by the terminal predator. Many xenobiotics are toxic to biota so that their metabolism in higher organisms generally serves as a mechanism for their detoxification and elimination. Some of these reactions are described in more detail in Chapter 7, Section 7.5, and for fish at least, metabolism frequently results in the transformation of the xenobiotic to water-soluble compounds which are then excreted. The concentrations of the xenobiotic in the organism are therefore determined by the rates of metabolic processes as well as by the kinetics of bioconcentration. For example, elimination of the hydrophobic organophosphate insecticide chlorpyrifos (*O,O*-diethyl-*O*- [3,5,6-trichloropyridyl]phosphorothioate) from guppy (*Poecilia reticulata*) is accomplished almost exclusively by metabolism (Welling and de Vrise 1992). In channel catfish (*Ictalurus punctatus*), the major metabolite in urine and bile is the glucuronide conjugate of 3,5,6-trichloropyridinol while the parent chloropyrifos is strongly bound to blood proteins (Barron et al. 1993). The significant role of metabolism has already been discussed in Section 3.1.4 in the context of surrogate procedures for assessing bioconcentration potential. There is, however, no sharp line dividing compounds that may be metabolized and those that are more persistent; these classes of compounds differ only in the magnitude of their rates of transformation. Some groups of compounds are only slowly eliminated from fish: (1) polychlorinated benzenes are only slowly metabolized by fish, and have therefore been suggested as a suitable example of "inert compounds" (references in de Wolf et al. 1992a); (2) lake trout (*Salvelinus namayvcush*) retained ~ 80% of the PCB burden that are essentially constant (Madenjian et al. 1998). It is probably true, however, that, in the absence of evidence to the contrary, most xenobiotics can be metabolized by the fish species that are widely used for evaluating bioconcentration potential. A striking illustration is provided by the fact that even compounds such as 1,3,6,8-tetrachlorodibenzo[1,4]dioxin can be metabolized by fish (Muir et al. 1986). Although details of the metabolism of xenobiotics by higher biota are given in Section 7.5, it may be useful to summarize here some of the groups of compounds for which metabolism should be taken into consideration in interpreting the results of experiments on bioconcentration: (1) PAHs and azaarenes; (2) phenols, anilines, and benzoates; and (3) chloronitrobenzenes.

The metabolic capacity of fish is generally greater than that of bivalves, and this may plausibly explain the substantial differences in BCF values for a number of tetrachlorinated phenyltolylmethanes (tetrachlorobenzyltoluenes) that have been measured (van Haelst et al. 1996) in guppy (*P. reticulata*), and in the zebra mussel (*Dreissena polymorpha*). This observation is particularly relevant to the choice of organisms for use in monitoring (Section 3.6) and in establishing toxicity (Sections 7.5.1 and 7.5.2).

Further consideration suggests that it may indeed be inappropriate to assess independently the apparently separate issues of bioconcentration and metabolism. Experiments on bioconcentration are generally designed so that exposure is continued until an apparent steady state is achieved; the test organisms are then generally maintained in the absence of the test compound to evaluate clearance by depuration. It would therefore be particularly attractive to combine these investigations with metabolic studies in which the nature of the metabolites is identified although this seems only seldom to have been carried out.

Two general conclusions may be drawn from the foregoing discussion. First, the interpretation of data from experiments on bioconcentration of xenobiotics should recognize possible complications from the effects of metabolism and excretion, and second that even when aquatic organisms, such as leaches, mussels, or crustaceans that are assumed to display limited metabiolic potential for xenobiotics, are used for monitoring purposes, interpretation of the data should consider the possibility of metabolism and excretion after initial exposure to the toxicant.

The role of metabolism in the wider context of the detoxification and elimination of xenobiotics from biota is discussed in Chapter 7, Section 7.5, and its potential role in the dissemination of xenobiotic metabolites in Section 3.5.

3.1.6 Cautionary Comments

It may be questioned whether, in a dynamic system, the concept of bioconcentration as currently defined is experimentally accessible, although pragmatically the concept is certainly valuable provided that its limitations are sufficiently appreciated. Although it has been shown that there are some inherent ambiguities in the concept of bioconcentration potential, direct estimates can be made using, for example, fish, and surrogate procedures are well developed, although care should be exercised in the interpretation of the results, and especially in extrapolating these to calculate the concentrations of xenobiotics in natural biota where uncertainties about the exposure route may exist. Correlations between bioconcentration potential and various parameters such as water solubility, octanol–water partition, and relative mobility on reversed-phase chromatographic systems have been demonstrated, and may be considered satisfactory if agreement within a power of 10 is achieved. For more refined analysis of field material and for an assessment of potential public health hazard, however, a much greater degree of certainty may be required. It should also be pointed out that there appears to be a striking difference in the relative concentrations of different types of xenobiotics in sediments and fish; for example, in samples collected in the same area, the sediment/fish ratio was 200 for PAHs, but only 0.05 for the organochlorine compounds (Malins et al. 1984). It would be plausible to attribute the difference to the fact that the PAHs are more readily metabolized by fish than the relatively recalcitrant organochlorine compounds. These issues are discussed in the context of biomagnification in Section 3.5.4.

An experimentally and interpretatively serious problem emerges with complex mixtures which are probably typical of many industrial effluents. The question immediately arises: if the nature of the compounds is unknown, how can measurements be made of partition? There are several strategies — each with inherent difficulties. Probably the least objectionable is the obvious one of first identifying the components of the mixture and then determining their partition coefficients. An equally acceptable — and possibly more realistic procedure — would be to fractionate the mixture into groups of compounds with putative bioconcentration potential having values of log P_{ow} > 3. This could be carried out for example using HPLC, followed by identification of the relevant compounds (Hynning 1996). The least attractive method is to quantify unknown compounds using a surrogate with no established relation to the compounds in question. This procedure has been used because of the ease with which it can be carried out, but its adoption seriously increases the number of links in the chain between measurements of partition coefficients and estimations of bioconcentration potential and thereby seriously jeopardizes estimates of bioconcentration potential.

3.2 Partition between the Aquatic and Sediment Phases

The partitioning of compounds from the aqueous phase into biota is not the only significant process that occurs after the initial discharge of xenobiotics into aquatic systems. Partitioning of xenobiotics from the aqueous phase into the sediment phase may be of equal significance, and its significance is attested by the structural range of organic compounds that have been recovered from contaminated sediments. Many of these compounds such as PAHs, PCBs, and PCDDs are widely distributed and only selected — and more or less random references — have been provided here. Some of these compounds certainly enter ecosystems as a result of long-distance transport but, irrespective of their origin, the sediment phase clearly functions as a highly effective — though not sole — sink for these compounds.

Examples of classes of xenobiotics recovered from sediment samples:

Hydrocarbons	
Polycyclic aromatic hydrocarbons	Prahl and Carpenter 1983
Alkylated aromatic hydrocarbons	Peterman and Delfino 1990
Chlorinated aromatic compounds	
Chlorobenzenes	Pereira et al. 1988
Polychlorinated biphenyls	Swackhamer and Armstrong 1986
Polychlorinated dibenzo[1,4]dioxins	Czuczwa and Hites 1986; Macdonald et al. 1992
Chlorinated guaiacols and catechols	Remberger et al. 1988
Nitrogen-containing aromatic compounds	
Azaarenes and aromatic nitriles	Krone et al. 1986
Oxygenated aromatic compounds	
2,4-Dipentyl phenol	Carter and Hites 1992
Polycyclic quinones and ketones	Fernandez and Bayona 1992
Aliphatic carboxylic acids	
C_8 and C_9 dicarboxylic acids	Stephanou 1992

There are a number of important environmental consequences resulting from the partitioning of xenobiotics into the sediment phase: (1) sediment-dwelling biota and demersal fish may be exposed to these compounds and the recovery of, for example, PAHs, polycyclic thiaarenes, and azaarenes (Vassilaros et al. 1982) from fish clearly demonstrates their progress through the food chain; and (2) since the sediment phase is not static but is subjected to the effect of currents and tides, the sediment phase may act as an effective transport system. These are discussed in greater detail in Section 3.5, and the important issue of the bioavailability of sediment-sorbed xenobiotics is discussed in Section 3.2.2.

3.2.1 Outline of Experimental Procedures

Natural sediments vary widely both in their physical structure and in their chemical composition. In addition to the inorganic matter that is universally present in the sediment phase, many shallow-water sediments contain appreciable amounts of humic material together with other organic matter originating from terrestrial plants. Furthermore, sediments in the neighborhood of industrial discharge often contain high concentrations of organic matter originating from manufacturing processes. This heterogeneity should be carefully evaluated in interpreting the results of experiments on partitioning, and on the degree of recoverability of xenobiotics from natural sediments. Although it is customary to normalize partition data to the organic content of the sediment, using a relation

$$K_p = f_{oc} \cdot K_{oc}$$

where f_{oc} is the fractional organic carbon in the sediment and K_{oc} represents the partition to "generic" organic matter, it should be appreciated that the chemical structure of components of the sediment may play a critical role; the use of total organic carbon may therefore be misleading. For example, the sorption of toluene and trichloroethene to soil was dependent on the nature of specific organic components (Garbarini and Lion 1986), and the partition of pyrene to dissolved organic humic material was influenced by its structure and was dependent on factors other than the total content of organic carbon (Gauthier et al. 1987). The same structural dependence holds for association between xenobiotics and the organic constituents of interstitial water, and for marine samples, unexpectedly high sorption may be due to the high lipid content (Chin and Gschwend 1992). On the other hand, for aquifer samples with a low content of organic carbon, there was no correlation with the organic carbon content (Stauffer et al. 1989). Care should therefore be exercised in comparing the results of partitioning using sediments that have predominantly mineral components with those containing substantial amounts of structurally diverse organic components.

Direct Measurements of Sediment/Water Partition

The experimental determination of partition coefficients in laboratory experiments is, in principle, straightforward: it involves mixing samples of the sediment and of the aqueous phase until a (pseudo) steady state is reached — generally within 24 h — followed by analyses of the phases after separation. Azide may conveniently be added to inhibit bacterial transformation of the xenobiotic during the experiment. Attention should, however, be directed to an important factor that may seriously compromise the results: after equilibration, the aqueous phase may contain dissolved organic material from the sediment phase, and this may compromise estimates of the truly dissolved concentrations of the xenobiotic. This problem may be especially acute in determining the partition of compounds with extremely low water solubility such as 1,3,6,8-tetrachlorodibenzo[1,4]dioxin (Servos and Muir 1989), and it should also be clearly appreciated that the values of partition coefficients obtained in this way cannot take into account the significant alterations (aging) that occur after deposition and that may be of cardinal significance (Section 3.2.3).

Surrogate Procedures

Extensive use of surrogate procedures has been used for estimating BCFs, and this is also the case for K_{oc}. Similarly, it should be appreciated that some implicit assumptions are made: first, that the compounds are neutral and hydrophobic and do not react with the sediment phase and, second, that the partitioning is determined by the organic carbon content of the sediment. Detailed descriptions of two surrogate procedures have been provided (Karickhoff 1984) so that only an outline of these is required here.

1. On the basis of water solubility and, for solids, thermodynamic parameters such as the entropy of fusion and the gas constant, two

equations have been suggested that take into account the gross chemical structures of the compounds:

$$\log K_{oc} = -0.9211 \log X_s - 1.405 - 0.00953 (T_m - 25): \text{PAHs}$$

$$\log K_{oc} = -0.83 \log X_s - 0.93 - 0.01 (T_m - 25): \text{organochlorines}$$

where X_s is the molar solubility in water and T_m is the melting point in °C.

2. Cogent arguments have been presented to support the existence of a correlation between K_{oc} and P_{ow} and again substantially different equations had to be applied to different structural classes of compound:

$$\log K_{oc} = 0.72 \log P_{ow} + 0.49: \text{simple benzenoid compounds}$$

$$\log K_{oc} = \log P_{ow} - 0.317: \text{PAHs}$$

Application of either procedure generally yields acceptable predictions for values for K_{oc} (Karickhoff 1981), although critical attention should be directed to the important limitations inherent in the method particularly when using empirically derived values of P_{ow}. As is generally the case, the correlations are least reliable for extreme values of P_{ow} (> ca.10^6).

3.2.2   Reversibility: Sorption and Desorption

It is well established that many compounds after introduction into the environment are not readily accessible to chemical recovery. This does not necessarily imply, however, that they are of no environmental significance: the degree to which they are desorbed and therefore become accessible to biota is a central issue that has implications both for the toxicity of xenobiotics and for their resistance to microbial attack. Several general considerations are worth noting.

1. There is a substantial literature showing that a significant fraction of agrochemicals introduced into the terrestrial environment is not recoverable by standard chemical procedures, and is apparently bound irreversibly to either organic or inorganic components of the soil matrix (Bollag et al. 1983; Lee 1985; Ou et al. 1985; Smith 1985) or physically inaccessible through inclusion in micropores (Steinberg et al. 1987).

2. Laboratory experiments on sorption have shown that even over a short period of time sorption may be irreversible or exhibit hysteresis: illustrative examples are provided by chlorophenols in sediment fractions (Isaacson and Frink 1984), naphthalene and

phenanthrene in sediments (Fu et al. 1994), chlorinated alkanes and alkenes (Pignatello 1990), and trichloroethene in soil (Pavlostathis and Jaglal 1991), and agrochemicals (references in Celis and Koskinen 1999).

These experiments often display two-stage kinetic processes of which the second may be associated with irreversible binding (Karickhoff 1984; Pavliostathis and Mathavan 1992). Some important details and the implication of slow desorption are given below.

a. The desorption from contaminated sediments of a range of compounds including chlorobenzenes, PCBs, and PAHs revealed the presence of two slowly desorbing fractions, and that the rate of slow-desorption is considerably increased with increasing temperature (Cornelissen et al. 1997). It was suggested that this might provide a readily accessible means of assessing the feasibility of bioremediation of samples from naturally contaminated sites.

b. The sorption of naphthalene and 2,2',5,5'-tetrachlorobiphenyl has been examined in detail (Kan et al. 1997), and several important and novel features emerged from the results:

   i. The amounts in the irreversibly sorbed compartments increased linearly with the number of adsorption steps until a maximum was attained;

   ii. Thereafter the adsorption was reversible;

   iii. The irreversibly sorbed compartment was in equilibrium with the aqueous phase at concentrations of 2-5 µg/l for naphthalene and 0.05 to 0.8 µg/L for 2,2',5,5'-tetrachlorobiphenyl.

   These results do not appear to be readily rationalized on the basis of conventional models of sorption.

c. The irreversible sorption of toluene, naphthalene, phenanthrene, 1,2-dichlorobenzene, 4,4'-dichlorobiphenyl, 2,2',5,5'-tetrachlorobiphenyl, and DDT was examined by cyclic sorption–desorption experiments (Kan et al. 1998). Although values of log K_{oc} (normalized to the organic content of the soil) ranged from 2.17 to 4.84, the values of the irreversible partition coefficient log K_{oc}^{irr} were essentially constant at a value of 5.53 ml/g for all soils and all sorbents. The results were consistent with a biphasic model with a linear reversible phase followed by a term that can be rearranged to a Langmuir isotherm. The model was used to predict the difference between the volumes of interstitial water that would be required to reduce concentrations in a soil by a factor of 10^4. On the assumption of reversible desorption ca. 22 pore volumes would be required as opposed to ca. 3300 based on the model that took irreversibility into account. This result

is of enormous practical significance in the context of bioremediation (Section 8.2.1).

d. Experiments with tetrachloro-, pentachloro-, and hexachlo-
robenzenes and PCBs (IUPAC 28, 65, and 118) using both mineral and organic sorbents have shown that slow desorption except from soils with low organic matter is determined by organic
content rather than from mineral micropores (Cornelissen et al.
1998).

e. The desorption of a range of hydrophobic compounds including
chlorobenzenes with 2 to 6 substituents, PCBs (IUPAC 28 and
118), and PAHs (2-methylnaphthalene, biphenyl, phenanthrene,
and fluoranthene) was examined in contaminated sediments
from Lake Ketelmeer in The Netherlands (ten Hulscher et al.
1999). Rapid, slow, and very slow desorption phases could be
distinguished, and attention was drawn to the large amounts of
very slowly (10^{-4} to 10^{-5} h^{-1}) desorbing material in decade-long
contaminated sediments. These results underscore clearly the
limitation of laboratory experiments.

3. The mobilization of sorbed xenobiotics is of serious concern in
areas subjected to historical pollution and this has motivated extensive investigations on desorption. For example, studies with sediment from New Bedford Harbor, Massachusetts revealed both the
significant role of organic carbon and that increased desorption of
PCB congeners occurred in distilled water than in saline water;
such data clearly support the concept of three phases in partitioning models (Brannon et al. 1991).

4. Probably most laboratory studies on sorption/desorption have
used single substrates, although this is almost certainly an over-
simplification of most natural situations. An example of the significance of interactions is afforded by a study with poly(*N*,*N*-
dimethylaminoethyl methacrylate) (Figure 3.3) of which a substantial fraction was irreversibly adsorbed on a sediment with
high ion-exchange capacity; in addition, presorption of the polymer to the sediment significantly increased the subsequent sorption of naphthalene (Podoll and Irwin 1988); (Takimoto et al. 1998).

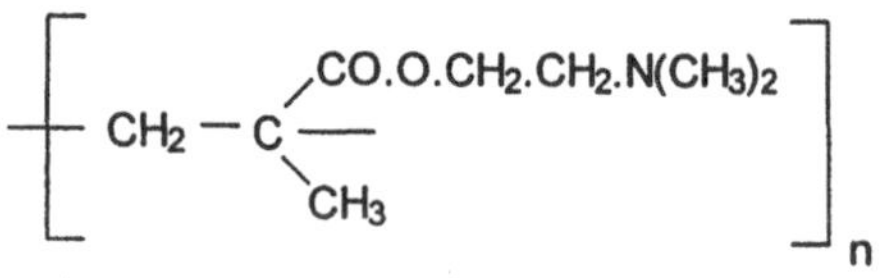

$$\left[CH_2 - C \substack{\diagup CO.O.CH_2.CH_2.N(CH_3)_2 \\ \diagdown CH_3} \right]_n$$

FIGURE 3.3
Poly (*N*,*N*-dimethylaminoethylmethylacrylate).

5. The most relevant sorbents both for the aquatic and terrestrial phases are humic and fulvic acids, and it has been shown that for pyrene the binding with a range of naturally occurring and "synthetic" humic substances was dependent on their molecular weight and their degree of aromaticity (Chin et al. 1997).

6. Soils also contain a number of low-molecular-mass water-soluble aromatic acids including benzoic, 4-hydroxybenzoic, cinnamic, coumaric (4-hydroxycinnamic), caffeic (3,4-dihydroxycinnamic), ferulic (3-methoxy-4-hydroxycinnamic), and vanillic (3-methoxy-4-hydroxybenzoic) acids. It has been shown that at plausible concentrations of these (< 100 μg/l) and at pH 5.6, there is competition between these and 2,4-dichlorophenol for sorption sites on soil organic matter (Xing and Pignatello 1998).

7. Probably greatest attention has been directed to hydrophobic compounds especially PAHs and PCBs. For example, the desorption of ^{14}C-phenanthrene and ^{14}C-chrysene preloaded onto previously contamined soils has been examined and correlated with the kinetics of mineralization (Carmichael et al. 1997).

8. Attention has been directed in Chapter 2, Section 2.5 and Chapter 4, Section 4.1.2 to the occurrence of trichloroacetic acid in rainwater and its plausible production from the photolysis of chloroalkenes. A comparable issue in the formation of trifluoracetic acid (TFA) and this has prompted a study of its sorption to and retention in soils. Although many soils did not retain TFA, retention was observed in those with a high organic content or with high concentrations of Fe and Al (Richey et al. 1997). In view of the extreme toxicity of TFA and its probably recalcitrance to microbial and chemical degradation, this is a disturbing issue.

3.2.3 Aging and Bioavailability

Results that indicate decreasing recoverability of xenobiotics from the sediment and soil phases with increasing time from deposition may be accommodated under the general description of "aging." This is the result of interactions between the xenobiotics and organic and inorganic components of the soil or sediment matrix, and the mechanisms of some of these associations are discussed in Section 3.2.4. This important issue has many ramifications including the influence on chemical recoverability, toxicity, and degradability; reference should therefore be made to Sections 3.1.2 and 3.2, Chapter 4, Section 4.6.3, and Chapter 8, Sections 8.1.1 and 8.2.1, that address the role of surfactants.

The aging of soils and sediments introduces a potentially serious indeterminacy into estimates both of the chemical concentration and into the degree of bioavailability of xenobiotics, and may therefore result in a serious ambiguity in correlating the exposure of biota to xenobiotics and the effects that are observed. Its significance has been illustrated for example with PAHs and with 2,4,5, 2′,4′,5′-hexachlorobiphenyl in the amphipod *Pontoporeia hoyi* (Landrum 1989), and may be even greater for compounds such as chloroguaiacols since a substantial fraction of these are chemically inaccessible in naturally aged sediments (Remberger et al. 1988). Aging has been demonstrated conclusively in the terrestrial environment, and the critical effect of residence time on the degree of recovery has been evaluated (Capriel et al. 1985; McCall and Agin 1985; Winkelmann and Klaine 1991). It seems unlikely, however, that these "bound" residues are merely static reserves (sinks), so that the critical issue is the extent to which they are desorbed and thereby become directly available to biota (Knezovich et al. 1987).

Soils vary greatly in their type and composition, and a study using phenanthrene and atrazine with soils differing in content of organic carbon and clay, and different pH showed that organic solvent extractability does not provide a good measure of the bioavailabity to microorganisms (Chung and Alexander 1998). Such procedures provide rather a measure of the total concentration of the analyte including both sorbed and freely dissolved fractions. The application of negligible depletion solid-phase microextraction using SPME fibers coated with poly(dimethylsiloxane) has been used to determine the freely dissolved fraction of the hydrophobic polychlorobenzenes, PCB 77 and DDT (Ramos et al. 1998). Increasing concentration of Aldrich humic acid resulted in an appropriate decrease in the freely dissolved concentrations of the analytes, and the results were used to calculate values of log K_{DOC} that were in good agreement with values reported in the literature. It is, however, widely appreciated that Aldrich humic acid is a rather poor surrogate for natural humic acids.

A more extensive discussion of this important topic is given in Chapter 4, Section 4.6.3 and the issues of sorption/desorption and bioavailability to microorganisms have achieved increasing prominence in the context of the bioremediation of contaminated terrestrial sites that is discussed in Chapter 8.

The degree of bioavailability of organic compounds depends critically on their chemical structure which determines the kinds of interaction that may take place within the solid phase. For example, linear alkylbenzenesulfonates are readily desorbed from sediments, so that their biodegradability and potential toxicity is largely unaffected by the presence of sediments (Hand and Williams 1987). On the other hand, benzo[*a*]pyrene even though accessible to chemical extraction appears to be available to biota only to a limited extent (Varanasi et al. 1985). This is consistent with the view noted earlier that chemical extractability is not a useful measure of bioavailability for naturally aged samples as opposed to those which have been spiked with the contaminant (Kelsey et al. 1997).

It is possible to distinguish two apparently opposing environmental effects:

1. Sorption of toxicants may result in diminished exposure of biota to deleterious concentrations of xenobiotics. This has been clearly demonstrated in experiments under clearly defined laboratory systems involving dissolved organic carbon in the aquatic phase. These experiments showed reduced bioavailability and hence diminished toxicity. A number of organisms and a range of toxicants have been examined: for example, *Salmo salar* and a range of organochlorine compounds including both chlorophenolic and neutral compounds (Carlberg et al. 1986), *Oncorhynchus mykiss* (syn. *S. gairdneri*) and benzo[a]pyrene and 2,2',5,5'-tetrachlorobiphenyl (Black and McCarthy 1988), and *Diporeia* sp. (syn. *P. hoyi*) and PAHs and 2,2',4,4'-tetrachlorobiphenyl (Landrum et al. 1987). Reduced toxicity during aging has also been shown for terrestrial systems in laboratory experiments. For example, it was shown that the toxicity to the house fly, fruit fly, and German cockroach of DDT and dieldrin spiked into sterile soil decreased with aging during long-time exposure and was particularly marked for dieldrin (Robertson and Alexander 1998).

2. On the other hand, the persistence of xenobiotics may be increased if they are not accessible to the relevant degradative microorganisms. This is discussed more extensively in Chapter 4, Section 4.6.3, so that only a few illustrative examples will be given here. Experiments on the biodegradation of compounds as diverse as *iso*propylphenyl diphenyl phosphate (Heitkamp et al. 1984), aliphatic esters of 4-aminobenzoate (Flenner et al. 1991) in spiked sediments, or substituted phenols in the presence of naturally occurring humic acids (Shimp and Pfaender 1985) support the view that increased persistence of xenobiotics is to be expected when the substrates are not freely available to microorganisms in the aquatic phase. Conclusions concerning the influence of sediment redox potential and pH on the degradation of pentachlorophenol (DeLaune et al. 1983) therefore appear to be equally consistent with the role of binding of pentachlorophenol to the sediment phase. In all cases, the key issue is probably the rate of transport into the microbial cells, since rates of chemical hydrolysis of phosphorothioate esters under neutral conditions were apparently unaffected by sediment sorption (Macalady and Wolfe 1985); on the other hand, the presence of humic material reduced the rate of alkaline hydrolysis of the octyl ester of 2,4-dichlorophenoxyacetic acid (Perdue and Wolfe 1982). A similar apparent contradiction emerges from the results of studies in which the role of dissolved organic carbon either facilitated or retarded the transport of xenobiotics through porous material such as sand (Magee et al. 1991). These differences may, on the other hand, merely reflect significant differences in the structure of the humic material used in these experiments.

The issue of bioavailability to deposit-feeding organisms has been examined (Weston and Mayer 1998a,b) using digestive fluid from the subsurface deposit-feeding marine polychaete *Arenicola brasiliensis* and the filter-feeding echurian *Urechis caupo*. Unlabeled and ^{3}H-labeled phenanthrene and benzo[*a*]pyrene were used, and it was shown that compared with seawater, greater amounts of the substrates were released in the presence of the fluids. The increased desorption was related to the organic carbon content of the sediment, was not due to lipase, protease activity, or esterase activities, and was attributed to surfactant activity. It was correlated with traditional estimates of bioavailability based on uptake clearance. A number of important unresolved issues merit further examination: (1) extension to sediments with greater concentrations of organic carbon than was used in these experiments that used sandy sediment with a low content of organic carbon, (2) whether the extractability of unlabeled substrates from aged sediment that was greater than that for sediment spiked with ^{3}H-labeled substrates was an effect of the different concentrations.

Regardless of the details, it may be concluded that the mechanisms whereby xenobiotics become associated with particulate matter and the degree to which these interactions are reversible are of cardinal importance in assessing the environmental impact of xenobiotics.

3.2.4 Mechanisms of Interaction between Xenobiotics and Components of Solid Matrices

It is now appropriate to consider in more detail the mechanisms of interaction between xenobiotics and the components of soils and sediments. It may plausibly be assumed that the principles are applicable equally to both of these matrices — with one important exception: interactions in the terrestrial environment catalyzed by fungal enzymes will probably play at most a minimal role in most aquatic systems. Details of the mechanisms by which xenobiotics are "bound" to components of the sediment phase have not been fully established although several plausible hypotheses have been put forward. Proposed mechanisms of interaction include ionic and covalent binding, long-range (van der Waals) forces, or sorption by undefined mechanisms, although these are pragmatic and probably conceal the complexity of the molecular processes. Three major components of the solid matrix have generally been considered:

1. Inorganic minerals dominated by the most abundant elements such as aluminum, iron, and silicon;
2. Organic constituents such as lignin-derived compounds and undefined compounds similar to humic and fulvic acids originating from the terrestrial environment;

3. Detrital material resulting from the decomposition of aquatic and sediment biota comprising both lipid and proteinaceous material.

It is worth pointing out that humic material in addition to its role in associations with xenobiotics may also play an important role in redox systems in the sediment phase. This is discussed later in this section.

Essentially three broad mechanisms of interaction may be discerned: (1) sorption involving interaction with inorganic components, (2) covalent reactions involving the organic constituents both by chemical and biologically-mediated processes, and (3) physical entrapment.

Interactions with Inorganic Components

Extensive studies (Hayes et al. 1978a,b) have been directed to the sorption onto clay minerals of pyridinium and bipyridinium compounds which are valuable agrochemicals. The mechanisms were clearly different in a number of respects from those noted below for other types of compound: adsorption correlated with the cation exchange capacity of the clays, and when this was saturated, sorption was attributed to van der Waals interactions between the pyridinium rings. As might be expected for these compounds, quaternization with methyl groups reduced the degree of adsorption. The sorption of a wide range of nitroaromatic compounds to mineral surfaces has been examined (Haderlein and Schwarzenbach 1993) and it has been proposed that interaction involves the formation of electron donor–acceptor complexes that are particularly strong for compounds containing several electron-withdrawing groups such as nitro. The results suggested the possibly significant role of such interactions in the transport of such compounds in aquifers. It has also been hypothesized that partitioning of chlorinated catechols from the aquatic phase into the sediment phase took place through formation of complexes with Fe and Al components, and this has been correlated with simultaneous desorption of chlorocatechols, Fe and Al (Remberger et al. 1993). Collectively, these observations clearly demonstrate the importance of interactions between organic compounds and mineral surfaces.

Interactions Involving Organic Components

The detailed chemical structures of the organic components of soil and sediments are largely unknown, and terms such as *humic acid* and *fulvic acid* are primarily descriptive rather than representing chemically defined entities. A brief summary is given of studies aimed at providing information on the structural components of humic and fulvic acids. Further studies using specific chemical reactions are given below.

1. Studies on the structure of fulvic acid using ^{13}C NMR, IR and UV spectroscopy, and titrimetry have resulted in proposals for the environment of the carboxyl groups in the polymer: measurement

of pK$_a$ values with those of model compounds were consistent with a structure for fulvic acid in which the carboxyl groups were in the proximity of cyclic ether, carbonyl, or aromatic structures (Leenheer et al. 1995).

2. Application of ^{13}C NMR using ramped amplitude cross-polarization–magic angle spinning (Ramp-CP-MAS) to extensively characterized solid state samples of fulvic and humic acid revealed that carbohydrate entities are a significant part of the structure of fulvic acid (Cook and Langford 1998). Whereas the humic acid contained substituted aromatic units that account for its functionality and may be presumed to be of major significance in associations, the fulvic acid contained unfunctionalized aliphatic entities and carbohydrate structures that accounted for the presence of both weakly acidic hydroxyl and carboxyl groups.

3. Derivatization of preparations of humic acid with trimethyl phosphite (Section 2.3) followed by ^{31}P NMR have been used to provide approximate quantification of quinone groups whose concentrations lay between 0.020 and 0.055 m mol/g (Argyropoulos and Zhang 1998). Analysis of the various carbonyl groups in lignin used ^{19}F NMR of derivatives prepared with trifluoromethyltrimethylsilane in the presence of tetramethylammonium fluoride (Ahvasi et al. 1999).

4. (Dixon et al. 1999)

Chemical Reactions

The most detailed studies on the mechanisms of interaction have been directed to interactions between hypothetical structures of humic material and xenobiotics such as phenols, aromatic amines, and carboxylic acids; two essentially different mechanisms for binding of xenobiotics to organic components of solid matrices have been considered. They are (1) chemical reaction between the xenobiotic and functional groups on the humus structure and (2) biologically mediated reactions of incorporation which merit attention as representing plausible models particularly in the terrestrial environment.

Some examples of different types of chemical reactions may be used as illustration. Most of these are dependent on reaction between the analyte and carbonyl groups in lignin.

1. It has been hypothesized that carbonyl and quinone groups occur in humic material and the presence of these has now been confirmed by a study in which diverse fulvic and humic acid samples were derivatized with ^{15}N hydroxylamine and the products examined by ^{15}N and ^{13}C-NMR (Thorn et al. 1992) (Figure 3.4). These observations underscore the relevance of the results from an earlier

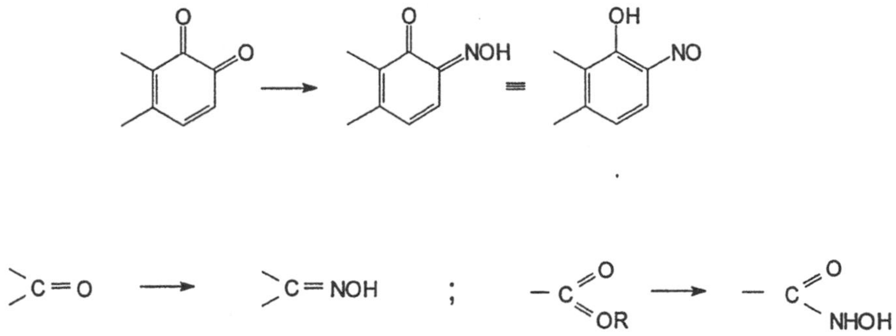

FIGURE 3.4
Structural entities identified in humic and fulvic acids by ^{15}N NMR after reaction with [^{15}N] hydroxylamine.

investigation (Parris 1980) in which the interaction of aromatic amines with carbonyl and quinone groups was studied: it was shown that after a rapid reversible reaction, a slow irreversible reaction took place probably involving addition of the amines to quinones followed by tautomerism and oxidation (Figure 3.5).

2. ^{15}N aniline was used in a study of the reactions of aniline with humic and fulvic acids (Thorn et al. 1996), and the detection of resonances attributed to anilinoquinone, imines, and N-heterocyclic compounds are fully consistent with the foregoing hypotheses. Structures in which phenols are covalently linked to C_3-guaiacyl residues have been examined as models for interaction between chlorophenols and lignin residues in humic acids (Zitzelsberger et al. 1987).

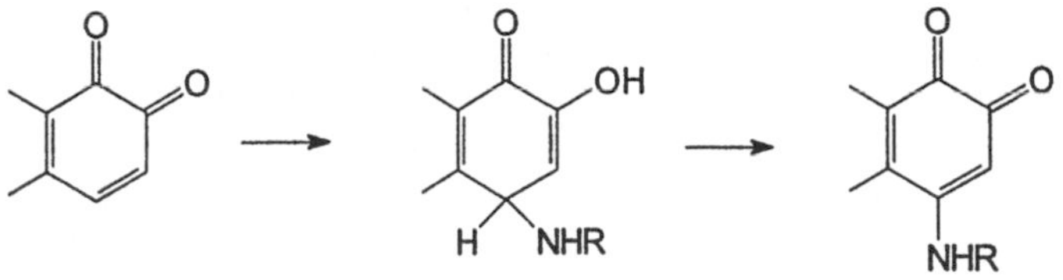

FIGURE 3.5
Reaction of amines with quinone groups.

Biologically Mediated Reactions

1. Formation of Associations with Organic Components of Soil and Sediment

These may involve both the original compound and its metabolites produced by biological reactions. This mechanism has wide implications and has been most extensively documented in the terrestrial environment.

1. Naphth-1-ol is an established fungal metabolite of naphthalene and may play a role in the association of naphthalene with humic material (Burgos et al. 1996).

2. ^{13}C-labeled metabolites of 9-[^{13}C])-anthracene including 2-hydroxy-anthracene-3-carboxylate and phthalate that were not extractable with acetone or dichloromethane could be recovered after alkaline hydrolysis (Richnow et al. 1998).

3. The nonextractable fraction of ^{14}C-labeled pyrene that had been introduced into pristine soil and incubated with and without the addition of azide was substantially greater in the latter (Guthrie and Pfaender 1998): microbial activity produced a number of unidentified polar metabolites that might plausibly be involved in the association.

4. The metabolism of ^{14}C-labeled BTX has been examined in soil cultures and a mass balance constructed after 4 weeks aerobic incubation (Tsao et al. 1998). Mineralization of all substrates was ca. 70% but ca. 20% of the label in toluene and ca. 30% in *o*-xylene were found in humus. It was suggested that the alkylated catechol metabolites were responsible for this association.

5. The mechanism of the interaction of cyprodinil (4-cyclopropyl-6-methyl-2-phenylaminopyrimidine) with soil organic matter has been examined. The association with soil organic carbon was biologically mediated, and it was shown that this increased during incubation for up to 180 days (Dec et al. 1997a). After 169 days of incubation, the fractions obtained by methanol extraction, and the humic acid and fulvic acid fractions after alkali extraction were examined by ^{13}C NMR (Dec et al. 1997b). Both the phenyl and the pyrimidine rings were associated with humic material, although only partly in the form of intact cyprodinil.

6. Considerable attention has been directed to enzymatic reactions mediated by fungal oxidoreductase enzymes such as phenol oxidase, peroxidase, and laccase. These systems have been used to copolymerize structurally diverse xenobiotics including substituted anilines (Bollag et al. 1983) and benzo[*a*]pyrene quinone (Trenck and Sandermann 1981) to lignin-like structures. One great advantage of the use of these model systems is that it has been possible to isolate the products of the reactions and determine their chemical structure. Some examples may be given to illustrate the different substrates involved and the types of products that may be produced.

   a. Incubation of pentachlorophenol with a crude supernatant from *Phanerochaete chrysosporium* in the presence of a lignin precursor (ferulic acid), and H_2O_2 produced a high-molecular-mass polymer (Rüttimann-Johnson and Lamar 1996). It was suggested that this might mimic the association of pentachlorophenol with humic material and the formation of heteropolymers between pentachlorophenol and lignin monomers.

b. Reaction between halogenated phenols and syringic acid in the presence of laccase from the fungus *Rhizoctonia praticola* resulted in the formation of a series of diphenyl ethers containing one ring originating from the chlorophenol together with 1,2-quinonoid products resulting from partial *O*-demethylation and oxidation (Bollag and Liu 1985) (Figure 3.6); comparable reactions have also been postulated to occur between 2,4-dichlorophenol and fulvic acid (Sarkar et al. 1988).

c. It has been shown that oligomerization of 4-chloroaniline mediated by oxidoreductases may produce 4,4'-dichloroazobenzene and 4-chloro-4'-aminodiphenyl as well as trimers and tetramers (Simmons et al. 1987). A study using guaiacol and 4-chloroaniline and a number of oxidoreductases has demonstrated the synthesis of oligomeric quinonimines together with compounds resulting from the reaction of the aniline with diphenoquinones produced from guaiacol (Simmons et al. 1989) (Figure 3.7).

d. Direct evidence of the existence of covalent bonding between 2,4-dichlorophenol and peat humic acid in the presence of horseradish peroxidase has been provided by the results of an NMR study using 2,6-[^{13}C]-2,4-dichlorophenol (Hatcher et al. 1993). In the absence of suitable model compounds, interpretation of the results was based on estimated chemical shifts for a range of plausible structures. The most important contributions came from those with an ester linkage with the phenol group, and covalent bonds between carbon atoms of the humic acid and C_4 (with loss of chlorine) and C_6 of the chlorophenol.

e. Laccase-catalyzed reactions between bentazon (3-isopropyl-*H*-2,1,3-benzothiadiazine-4(3*H*)-one 2,2-dioxide) and various humic acid monomeric components have been studied, and the products from reactions with catechol examined in detail by both ^{1}H and ^{13}C NMR (Kim et al. 1997). Products with masses of 348 and 586 were isolated and were produced by reactions between the N-atom of bendazon and the 1,2-quinone formed by the laccase.

f. Coniferyl alcohol that is the monomeric precursor of lignin was copolymerized by peroxidase and H_2O_2 in the presence of ^{15}N aniline and 3,4-dichloroaniline in various ratios and the products examined by ^{1}H, ^{13}C, and ^{15}N NMR (Lange et al. 1998). The conjugates were formed by reaction at the benzylic carbon atom of the coniferyl alcohol polymer. Although the anilines could be recovered by acid hydrolysis, it was pointed out that this may be the result of the high molar ratio of anilines used for the copolymerization.

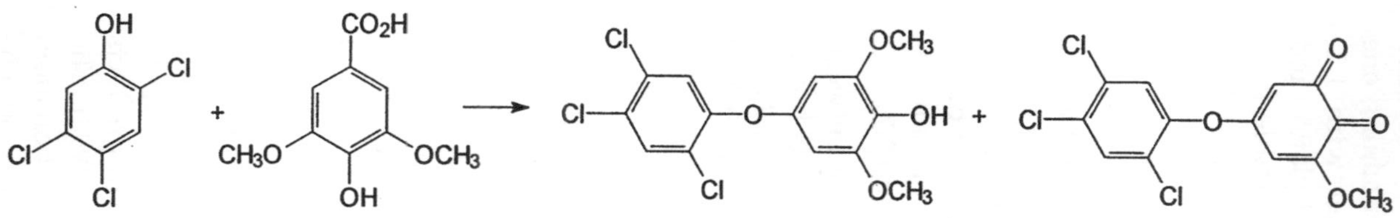

FIGURE 3.6
Reaction between 2,4,5-trichlorophenol and syringic acid catalyzed by laccase.

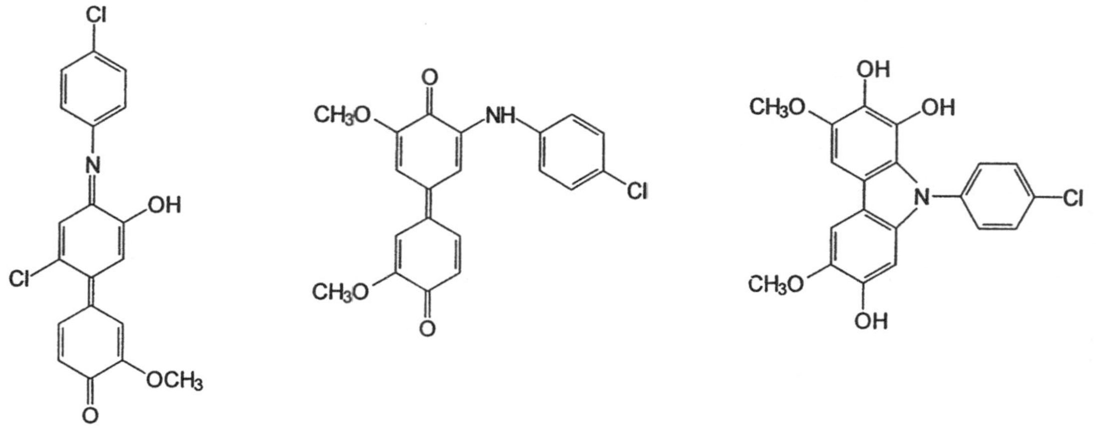

FIGURE 3.7
Products from the enzymatic copolymerization of guaiacol and 4-chloroaniline.

There is therefore extensive evidence that may be used to rationalize the occurrence of "bound" residues in soils, and this phenomenon is of particular significance for agrochemicals. Such processes influence not only their recovery by chemical procedures but also their biological effect and their biodegradability (Calderbank 1989); the latter is discussed further in Chapter 4, Section 4.6.3. The extent to which these principles are applicable to aquatic systems — in which fungal oxidoreductases are less well established — appears not to have been established although it is clearly possible that comparable mechanisms exist.

2. Reactions Involving the Redox System in Humic Acids

Humic acids are widely distributed in aquatic and terrestrial systems, and it has been shown above that they contain a number of reactive groups including electrophilic quinones that form redox systems with the corresponding hydroquinones, and are able to react with nucleophiles such as amines and phenols. Increasing attention has been directed to the significance of redox reactions involving humic acids, the Fe(III)/Fe(II) system, organic substrates, and bacterial systems.

1. It has been shown that iron-reducing bacteria are able to reduce humic acids using acetate, lactate, or H_2 as reductants (Lovley et al. 1996, 1998) thereby mediating the oxidation of organic substrates such as acetate with the reduction of Fe(III).

2. In the presence of high concentrations of humic acids, the redox balance in the products of bacterial fermentation may be altered; the fermentation of lactate by *Propionibacterium freudenreichii* is altered in favor of increasing amounts of acetate relative to propionate, and the fermentation of glucose by *Enteroccus cecorum* produces increased amounts of acetate without the production of ethanol (Benz et al. 1998).

Physical Entrapment

Physical entrapment may also be significant for some molecules and attention is drawn to the occurrence of micropores in soils and the role of these in retaining xenobiotics. This has been demonstrated for 1,2-dibromoethane which is notoriously persistent in agricultural soils (Steinberg et al. 1987). Although this mechanism has seldom been considered as a quantitatively significant phenomenon in a wider context, the chemical stability of a wide range of clathrate compounds (Chapter 2, Section 2.2.2) may be worth examining (Hagan 1962), and the existence of complexes between the pyrazole phenyl ether herbicide and cyclodextrin (Garbow and Gaede 1992) could provide a viable model for such essentially physical interactions.

An important conclusion from all these studies is that the mechanisms of interaction are specifically related to the structure of the xenobiotic, and that exclusive concentration of effort on neutral hydrophobic compounds may divert attention from important principles that are relevant to groups of structurally different compounds.

3.3   Phase Heterogeneity: Dissolved Organic Carbon, Interstitial Water, and Particulate Matter

Partitioning of a xenobiotic from the aquatic phase into biota or into the sediment phase is not a terminal process; indeed, from an environmental point of view these represent merely the introduction of the xenobiotic into a complex network of interactions. At least three important factors should be evaluated: (1) direct desorption mechanisms from the sediment phase into the aquatic phase, (2) resuspension of particulate matter from the sediment phase into the water column, and (3) the role of interstitial water in the sediment phase.

3.3.1   The Inhomogeneity of the Water Column

The preceding discussion has taken into account only partitioning between two single phases. In natural ecosystems, however, the situation is almost invariably more complex and significant complications are introduced as a result of the inhomogeneity of the water column. This may often contain organic carbon in various states of aggregation, and the distinction between these is empirical rather than theoretical. A useful pragmatic procedure has used SepPak C-18 columns from which humus-bound xenobiotics are eluted directly, whereas the dissolved components are retained and may then be eluted with methanol (Landrum et al. 1984). There are therefore at least the following three states — sometimes referred to as phases — in which xenobiotics may be found in the water column: (1) truly dissolved in the aqueous phase, (2) associated with dissolved (including high-molecular-weight) organic carbon, and (3) bound to particulate material.

It should be emphasized that these divisions are purely empirical, that dynamic interactions between them occur continuously, and that the relative concentrations in the various fractions will depend on the specific nature and the geographic location of the water mass. For example, in samples of water from sites in Lake Michigan, although the freely dissolved concentration of organochlorine compounds was dominant and the contribution from compounds bound to dissolved organic carbon was generally < 5% (Eadie et al. 1990), this may not necessarily be the case for other kinds of lakes or for brackish or marine systems. The details of exposure may not, however, always be unequivocally defined. For example, the burden of freshwater mussels exposed to natural contaminants in Lake St. Clair in the presence of contaminated or uncontaminated sediments clearly showed that uptake was mediated principally via the water column — although the possible role of uptake via compounds sorbed to dissolved or particulate material was not unambiguously resolved (Muncaster et al. 1990). Attention has already been directed to the fact that the uptake of highly hydrophobic compounds into biota may take place via food rather than by direct uptake from the water mass (Section 3.1.1), and such compounds may be associated with organic compounds in any of the three states distinguished above. Reference has already been made (Section 3.2.3) to the possible detoxification which may be exerted by dissolved organic carbon in the aqueous phase.

3.3.2 The Role of Interstitial Water

There has been extensive interest in interstitial water in view of its significance in determining the exposure of sediment-dwelling biota to xenobiotics and its role in the dissemination of xenobiotics. It should be appreciated, however, that the term *interstitial* is operational rather than absolute. Although various methods have been advocated including pressure filtration, centrifugation at moderate centrifugal force (ca. $1000 \times g$) is probably the preferred procedure. Interstitial water is not generally a homogeneous phase and xenobiotics may occur in any of the states described in Section 3.3.2. At least for hydrophobic xenobiotics such as PCBs in coastal marine sediments, these are associated in the interstitial water with organic carbon that is probably colloidal though of undetermined structure (Brownawell and Farrington 1986). In an evaluation of the role of colloidal phase in the partitioning of PAHs in a contaminated freshwater aquifer, it was shown that (1) the PAHs associated principally with those > 100 nm and (2) the K_{oc} values were linearly related on a logarithmic scale to the P_{ow} values of the individual PAHs (Villholth 1999).

The role of interstitial water in mediating the exposure of biota to xenobiotics is discussed in this Section while its significance in the dissemination of xenobiotics is discussed briefly in section 3.5.1.

For sediment-dwelling organisms, one important factor that determines the degree of exposure to xenobiotics in the sediment phase is the partitioning from the true sediment phase into interstitial water from which the xenobiotic may then be accumulated by biota. Exposure of sediment biota to xenobiotics is, however, a complex process, since uptake may proceed either via particulate material or via interstitial water, or by both routes. In the equilibrium partition model the concentration of a xenobiotic in the interstitial water (C_{iw}) is given by the following relation:

$$C_{iw} = C_s / K_{oc} \cdot f_{oc}$$

where C_s is the concentration in the sediment, K_{oc} the partition coefficient between water and "generic" organic carbon, and f_{oc} the fraction of organic carbon in the sediment. As an example, this relation has been verified for fluoranthene at interstitial concentrations less than 50 mg/l, although it was not valid at higher concentrations (Swartz et al. 1990). There may, however, be significant differences between the concentrations observed in interstitial water and those predicted by equilibrium partitioning from concentrations in the sediment. This has been demonstrated with PAHs in sediment cores from Boston Harbor: the fraction of PAHs in the sediment phase that was available for partitioning into the interstitial water varied from 0.2 to 5% for phenanthree and from 5 to 70% for pyrene (McGroddy and Farrington 1995). There are a number of important conclusions that may be drawn from this study:

- The degree of equilibration depended on the the structure of the PAHs.

- PAHs were probably derived from the deposition of particular material. Soot from combustion contains a high fraction of aromatic structures, and its porosity will facilitate association with PAHs.

- The discrepancies observed with PAHs were not found for PCBs so that the structure of the xenobiotic should be taken into consideration.

These results have been confirmed in laboratory experiments on desorption (McGroddy et al. 1996), and underscore the limitations of equilibrium partition models for PAHs — although not apparently for PCBs.

It has been shown that the uptake of a number of halogenated xenobiotics is mediated by interstitial water. Examples include the transport of chlorobenzenes into larvae of the midge *Chironomus decorus* (Knezovich and Harrison 1988) and of a range of chlorinated compounds into oligochaete worms (Oliver 1987). In addition, its significance may be inferred from the results of experiments on the uptake of PCB by the polychaete *Nereis diversicola* (Fowler et al. 1978). Interstitial water has therefore been widely used to assess the toxicity of sediments using a number of aquatic organisms (Carr et al. 1989; Ankley et al. 1991) and the equilibrium partition model justifies this application even for organisms such as the marine amphipods *Rhepoxynius abronius* and *Corophium spinicorne* which have potentially different routes of exposure to xenobiotics in the sediment phase (Swartz et al. 1990). A number of important facts have emerged which illustrate the caution that should be exercised in the application and interpretation of the results of assays using interstitial water:

1. Care should be exercised to ensure that the test organisms realistically represent the situation near the water–sediment interface (Ankley et al. 1991).

2. The rate of accumulation of PAHs from spiked sediments by *Diporeia* sp. could not be predicted from measurements of partitioning between interstitial water and sediment particles (Landrum et al. 1991).

3. Although the toxicity of DDT to the amphipod *Hyalella azteca* decreased with increasing carbon content of the sediment, this was not the case for endrin (Nebeker et al. 1989) so that specific mechanisms of interaction even between neutral xenobiotics and the organic carbon in the sediment phase may be of determinative significance. The results with DDT are, in fact, consistent with evidence from equilibrium dialysis experiments of its association with dissolved humic material (Carter and Suffet 1982).

Interstitial water may also have an important influence on rates of biodegradation. For example, the effect of interstitial water velocity and residence time on the degradation of 2,4-dichlorophenoxyacetate has been examined in batch conditions using soil columns (Langer et al. 1998). Although it was shown that interstitial water velocity had an influence on the degradation rate constants, the complex of responsible factors could not be determined: among these are changes in the microbial community, the residence time of the substrate, and effects on transport rates of nutrients. These results underscore the difficulty of predicting transport and degradation of xenobiotics in soils that have different rates of flow of interstitial water.

There are therefore unresolved factors which restrict the extrapolation of laboratory-determined partitioning parameters to field situations, and the results underscore the limitation of models for partitioning which encompass compounds with significantly different physicochemical properties. There therefore remains a need for the application of direct assays for toxicity using true sediment organisms such as oligochaetes (Wiederholm et al. 1987), although the results even from such tests should be interpreted with caution in view of the complication of aging which has been briefly noted above.

3.3.3 The Role of Sediment and Particulate Matter in the Aquatic Phase

The dynamics of transport from the aqueous phase into biota, into sediments, and into the atmosphere are cardinal determinants of the dissemination of xenobiotics after discharge. For the aquatic phase, the complexity of the situation that may prevail in natural ecosystems may be illustrated by two apparently conflicting results from laboratory experiments using the same compound. The results of experiments in which fathead minnows (*Pimephales promelas*), the worm (*Lumbriculus variegatus*), and two amphipods were exposed to hexachlorobenzene in water and in spiked sediment suggested that the sediment was a more efficient sink than the biota (Schuytema et al. 1990). On the other hand, hexachlorobenzene was apparently desorbed from contaminated sediments and accumulated in algae (Autenrieth and DePinto 1991). The critical issues are therefore the relative rates of the various partition processes and these may be determined specifically by the biota present in the system and by the lipid content of the relevant biota (Section 3.1.1); attention has already been directed (Section 3.2.2) to the degree of reversibility of these xenobiotic associations, and these factors underscore the care that should be exercised in extrapolating to natural systems the results of laboratory experiments on partition.

Association between xenobiotics and particulate — although suspended — material should also be viewed in a broader perspective. On the one hand, stratification of the water mass of lakes may result in xenobiotic-associated particulate material from the upper layers of the lakes entering the superficial layers of the sediment surface; on the other hand, resuspension of sediment material at the sediment/water interface may bring about reentry of such material into the water column. The extent of the various processes depends critically on the structures of the compounds so that, for example, more highly chlorinated PCBs in Lake Superior were lost from the water column whereas the less highly chlorinated congeners entered the water phase from the sediments (Baker et al. 1985). Simplistic views that sediments function exclusively as "sinks" for highly hydrophobic organic compounds must therefore be viewed with caution, and the details of the dynamics of the water masses in lakes should be taken into consideration.

As illustrated by the following example, the term *particulate* should also be defined carefully. A sequential protocol for the fractionation of particulate material was applied to PAHs containing three to six rings (Leppard et al. 1998). It was shown that ca. 80% of all of them were associated with particles >0.45 μm, and that the size distribution among particles from 2 to >80 μm varied with the molecular size of the PAHs: phenanthrene was found among the larger particles whereas benzo[*ghi*]perylene was more evenly distributed among particles in the same range.

It may therefore be concluded that understanding of the basic mechanisms and, in particular, the dynamics of the partitioning of xenobiotics between all phases — including truly dissolved organic matter, colloidal material, and particulate matter — and their relative importance in determining the exposure of biota to xenobiotics is strictly limited and that refinement of many unresolved details is highly desirable.

3.4   Partitions Involving the Atmospheric Phase

3.4.1   Partitioning between the Aquatic Phase and the Atmosphere

It is obvious that volatile organic compounds such as low-molecular-weight hydrocarbons and chlorinated hydrocarbons may readily be partitioned into the atmosphere from the aquatic phase, and indeed this may be the major sink for such compounds (Smith et al. 1980). Attention has been directed especially to volatile chlorinated compounds in the light of their role in atmospheric chemical reactions, in particular the destruction of ozone. Increasing evidence of the long-distance transport of apparently nonvolatile compounds such as PCBs (Swackhamer and Armstrong 1986) which is discussed in greater detail in Section 3.5.3 illustrates, however, the need for assessing the magnitude of partitioning between the aquatic phase and the atmosphere even of quite large molecules that are presumably considerably less volatile than the chlorinated alkanes. For example, it has been estimated that in Lake Michigan loss of PCBs by volatilization from the aqueous phase makes a contribution that is approximately one third that due to sedimentation (Hornbuckle et al. 1995).

The rate of transfer from the aquatic phase to the air is a complex function of several mass-transfer parameters including Henry's law constant (H): this is defined as the ratio of the vapour pressure of a compound to its solubility in water, and the value of H is of particular significance in many natural situations. Due to experimental difficulties, values of H are frequently calculated rather than measured experimentally, although this may result in substantial errors in the estimated values. An extensive compilation of values of H has been provided by Mackay and Shiu (1981). Three experimental procedures have been used:

1. A direct method using a wetted-wall column has been used for several pesticides with values of H between 10^{-5} and 10^{-7} (dimensionless units) (Fendinger and Glotfelty 1988) and for selected pesticides, PAHs, and PCBs (Fendinger and Glotfeldy 1990).

2. A dynamic headspace gas-partitioning method has been used to assess the effect of dissolved organic carbon on the value of H for mirex (Yin and Hassett 1986), and with a gas chromatographic detection system for analysis, this is applicable to native water samples containing mirex.

3. A gas-purge system has been used for a number of chlorinated aromatic hydrocarbons and PAHs (ten Hulscher et al. 1992).

Particular attention should be directed to the units used for reporting values of H and those for both vapor pressure and solubility. Dimensionless values of H may be reported as exemplified above, and are the least ambiguous; to convert them into conventional $kPa \cdot m^3 \cdot mol^{-1}$ they should be multiplied by RT which has a value of 0.246 in this case. Vapor pressures may be expressed in atmospheres or torr (1 kPa = 7.5 torr = 9.87×10^{-3} atm), and solubilities in $g \cdot m^{-3}$ or $mol \cdot m^{-3}$.

Estimates of the transfer rates for a number of aquatic environments have been made (Smith et al. 1981), from which it emerges clearly that the half-lives of compounds in the aqueous phase will be less than 10 days for compounds with H values greater than $\sim 10^{-1}$ $kPa \cdot m^3 \cdot mol^{-1}$ in lakes and greater than $\sim 10^{-2}$ $kPa \cdot m^3 \cdot mol^{-1}$ in rivers. This is consistent with the prediction (Mackay and Yuen 1980) that significant rates of volatilization occur for compounds having values of H > 10^{-1} $kPa \cdot m^3 \cdot mol^{-1}$, and may still be significant at values of 10^{-3} $kPa \cdot m^3 \cdot mol^{-1}$: a lower limit may be set for water which has a value of 3×10^{-5} $kPa \cdot m^3 \cdot mol^{-1}$. The values of H (ten Hulscher et al. 1992) suggest appreciable rates of volatilization for all the chlorinated aromatic compounds (ranging from 0.192 $kPa \cdot m^3 \cdot mol^{-1}$ for 1,3,5-trichlorobenzene to 0.016 $kPa \cdot m^3 \cdot mol^{-1}$ for 2,2',5,5'-tetrachlorobiphenyl) but not for the PAHs that had values < 5×10^{-5} $kPa \cdot m^3 \cdot mol^{-1}$. Partition from the aqueous phase into the atmosphere is particularly significant for compounds with appreciable vapor pressure and low water solubility (Mackay and Yuen 1980).

As for the other partition processes discussed above, empirical relations have been sought to relate the vapor pressure (P) to simple molecular parameters. Correlation between P and the boiling point (T_b) and melting point (T_m) has resulted in the proposal of the following equation (Mackay and Yuen 1980) which holds for a wide range of neutral compounds:

$$\ln P = 10.6 \, (1 - T_b/T) + 6.8 \, (1 - T_m/T)$$

where T is the ambient temperature. Correlations have also been used to esti-mate P for a few compounds including chlorinated guaiacols from gas-chro-matographic retention times (Bidleman and Renberg 1985). Since values of P_{ow} have been correlated with aqueous solubility (S), it is simple to express H_c as a function of T_b and P_{ow}:

$$\ln H_c = 10.6 \, (1 - T_b/T) + \ln P_{ow} - 12.1$$

It is thus clear that there is a network of relations between the functions deter-mining the folllowing partitions: octanol – water (P_{ow}), water – biota (BCF), water – generic organic carbon in sediment (K_{oc}), water – atmosphere (H_c) — and between each of them and thermodynamic properties such as aqueous solubility (S), vapor pressure (P), and melting point (T_m).

3.4.2   Partition between Solid Phases and the Atmosphere

This is an important issue since a range of xenobiotics may exist on the sur-face of soils and dredged sediments, although some components in contam-inated sediments may have undergone transformation — though not degradation. The large volumes of such sediments that are treated involve a number of important partitions including volatilization during drying.

1. Values of the soil – air partitioning (K_{sa}) were determined for a range of "semivolatile" compounds (Hippelein and McLachlan 1998) including PCBs and the PAHs phenanthrene, fluorene, and pyrene. Relations with the octanol – water (K_{ow}), octanol – air (K_{oa}), and air –water (K_{aw}) partitions were proposed:

   $$\log K_{sa} = 0.987 \log K_{ow}/K_{aw} - 1.696 = 0.951 \log K_{oa} - 1.754$$

2. The emission of naphthalene, phenanthrene, and pyrene from con-taminated sediments has been examined and its extent depended on a number of factors including the moisture content and the relative humidity of the air (Valsaraj et al. 1999). Reworking of the sediment and displacement by water in high moisture air increased the emission.

3. A study was made of sediments that had been contaminated with Arochlor 1248 and that were subjected to drying cycles that simu-late natural weathering (Bushart et al. 1998). A number of PCB congeners were lost by volatilization, and were the *ortho*-substi-tured di- and trichlorinated congeners (2,2'-dichloro, and 2,2',6-, 2,2'4-, 2,6,4', and 2,6,4'-trichloro) most often produced during par-tial microbial dechlorination of the original Arochlor 1248.

3.5 Dissemination of Xenobiotics

It is now appreciated that environmental hazards cannot be evaluated solely on the basis of the effects observed in the immediate neighborhood of point discharges. Xenobiotics may be transported over long distances, and may then be recovered from samples remote from the source of initial discharge. This is particularly significant for the most persistent compounds such as organochlorines, in particular, PCBs, PCCs, PCDDs, and PCDFs, and some agrochemicals — and indeed the recovery of these from environmental matrices confirms the view that these compounds are not readily degraded and are therefore recalcitrant. The effects of these compounds may therefore be exerted in pristine areas remote from the initial discharge that may be particularly sensitive to such perturbations. The important role of metabolites produced by microorganisms and of conjugates of metabolites excreted by higher biota should also be taken into consideration.

In the discussions presented in earlier sections attention has been directed to the partition between pairs of phases: between the aqueous phase and biota including fish, algae, and higher plants; between the aqueous phase and sediments including particulate matter; and between the aqueous phase and the atmosphere. In most natural ecosystems all of these are simultaneously present, and most of the partition processes are reversible to a greater or lesser extent. The distribution of a xenobiotic is therefore determined by the dynamics of sorption/desorption, bioconcentration/elimination after metabolism, and atmospheric deposition/evaporation from the aquatic phase. All of these are interconnected and the transport of xenobiotics associated with both particulate matter and with algae may be mediated by mechanisms as different as those for uptake by biota or deposition onto and incorporation into the sediment phase. Some examples of these partitioning processes that have been revealed in studies using mesocosms are discussed in Chapter 7, Section 7.4.3. It is also important to take into account the geographical location, the morphometry, and the topographical surroundings of the water mass. For example, although the range of organochlorine compounds identified in samples of water and biota from Lake Baikal, Siberia was similar to that found in the Great Lakes of North America, the atmosphere–water partitioning in the lake and the input from snowmelt from the mountains that surround Lake Baikal will be very different (Kucklick et al. 1994). The spectrum of biota that may be exposed to the toxic effects of xenobiotics is therefore wide and is the subject of Chapter 7. It is important also to appreciate the progress of a xenobiotic through the food chain. Although few of these are so truncated as that in Antarctic waters involving phytoplankton → krill → baleen whales and crabeater seals (Knox 1970), all involve a primary producer (algae) and a succession through crustaceans and fish to sea mammals such as seals, dolphins, and whales and, in many cases, ultimately to humans. For example, in Arctic waters the succession: phytoplankton →

copepod → cod → ringed seal → polar bear is important (Welch et al. 1992). In addition, the possible intrusion of insects should not be underestimated (Section 3.5.2). For example, mayflies (*Hexagenia* sp.) that were analyzed for PCBs in a survey of the Mississippi River (Steingraeber et al. 1994) were chosen for monitoring since, during development, they have a long-time exposure to sediment detritus and are important diet for fish. At each stage, the partitions noted above will occur — even though true equilibrium will seldom be achieved. In the following sections, some general remarks are presented on the factors that determine the dissemination of xenobiotics in natural systems, and an attempt is made in Section 3.5.5 to present the role of models in providing an overview of all of these processes.

The potential ambiguity in experiments on bioconcentration resulting from metabolism by biota has already been noted (Section 3.1.5), but it is important to appreciate other significant consequences of metabolism by fish and higher biota:

1. Metabolism may serve as an effective mechanism for the elimination of the xenobiotic: examples of a number of important reactions are given in Chapter 7, Section 7.5.

2. The metabolites may themselves be toxicants. A good example is provided by the diol epoxides of some PAHs that are putatively the causative agents of liver carcinomas in fish (Chapter 7, Section 7.7.2).

3. The metabolites may be excreted from the organism and thereby the transformation products of the xenobiotic are introduced into the environment; an example of this is given in Section 3.5.2.

3.5.1   Transport within Aquatic Systems: The Role of Water and Sediment

The physical transport of dissolved xenobiotics within aquatic systems may obviously take place under the influence of currents and tides. It has also been shown, however, that particulate matter and river sediment may be important vehicles, and that the dissemination of xenobiotics may also take place by subsequent diffusion from these transported sediments into the water column. An attempt will be made to illustrate the operation of these mechanisms.

A study conducted in the St. Clair and Detroit Rivers illustrated that, for hexachlorobenzene, octachlorostyrene, and PCBs, transport by the water mass and on suspended solids were of equal importance although at least in these systems, river bed sediments were of lesser significance (Lau et al. 1989). The extensive evidence for binding of lipophilic xenobiotics to particulate material in lakes (Baker et al. 1991) indicates that this may play an important role in the transport of xenobiotics; compounds sorbed to particulate matter with a low density may be effectively transported over long distances before deposition onto sediment surfaces. The dynamics of resuspension should also be considered since it has been shown by the use of sediment traps that resuspension of superficial sediments may be quantitatively important and result in the relatively uniform distribution of xenobiotics in large sedimentation basins (Oliver et al. 1989). The possible geographical extent of dissemination by rivers may be illustrated from the results of measurements in the Mackenzie River system which extends for 1770 km and which transports ~ 10^8 t/year sediment into the floodplain (Carson et al. 1998). During the summer, a substantial fraction of this is in the form of flocs with a mean size of 22.6 μm containing up to 2.45 mg/l particulate organic carbon (Droppo et al. 1998). A study identified PCB congeners in sediment samples from Great Slave Lake, Northwest Territories, Canada (Mudroch et al. 1992), and showed that these compounds probably enter the lake by atmospheric deposition. It was suggested that resuspension of these sediments into the Mackenzie River drainage system could result in their transport into the Arctic Ocean in the Beaufort Sea. An investigation of organic material in river particulate and shelf sediment revealed, however, that total concentration of PCBs was less than ~ 4.3 μg/kg sediment, and that concentrations of combustion-related PAHs from the atmosphere were less than those from petrogenic and biogenic sources (Yunker and Macdonald 1995). Contamination of rivers by leaching of agrochemicals from terrestrial systems may also be significant, and attention has been drawn to the significance of overland flow during heavy spring rainfall in addition to transport of these compounds by contaminated groundwater (Squillace and Thurman 1992).

Xenobiotics may have become associated with particulate matter including algae by transportation within the water mass and at any stage this may be deposited onto the sediments, processes that have been discussed in Section 3.2. The partition of the xenobiotic from the bulk sediments into interstitial water may then result in reexposure of biota to the xenobiotic; this has been discussed in Section 3.3.2. Other important processes are diffusion within the sediment phase and from interstitial water into the water column, and both of these are important in the dissemination of xenobiotics. For example, diffusion of PCBs within sediment cores in Lake Ontario has been established (Eisenreich et al. 1989), and there is good evidence to support the reentry of PCBs into the water column by diffusion from interstitial water (Baker et al. 1985). The state of these xenobiotics is not unequivocally established although they are probably associated with colloidal material (Brownawell and Farrington 1986); the degree of their bioavailability therefore remains unresolved. It may also be noted that, for reactive compounds, their chemical stability may be increased by the association. For example, chlorocatechols that are associated with particulate matter in interstitial water are thereby protected from atmospheric oxidation that would result in rapid transformation of the freely dissolved compounds (Remberger et al. 1993).

In the foregoing examples, there is no doubt that the organochlorine compounds are anthropogenic. The situation for PAHs is, however, more complex since, in putatively pristine sediments, the major components may be of plant origin and the contribution from atmospheric deposition of anthropogenic compounds by long-distance transport minimal. A good example is provided by sediments from the Mackenzie River delta and shallow-water deposits from the Beaufort Sea that were dominated by PAHs derived from biotic material (Yunker and Macdonald 1995). Attention has already been directed in Chapter 2, Sections 2.4.2 and 2.5 to the wider issues, and to mechanisms for the formation of such biogenic and petrogenic PAHs.

The main factors that operate may be summarized as (1) transport of dissolved compounds within the aqueous phase, (2) transport of sediment- and particle-associated compounds within the water mass, (3) diagenesis within the sediment phase, and (4) diffusion via the interstitial water from the sediment phase into the water mass. The relative quantitative significance of these will depend on a number of factors including the hydrological conditions, the organic components in the system, and the structure of the xenobiotic.

3.5.2 Transport within Aquatic Systems: the Role of Biota

Extensive discussion has already been devoted to the concentration of xenobiotics from the aquatic phase into biota (Section 3.1) and this may contribute significantly to the dissemination of the compounds. A number of illustrative examples include the following, some of which have already been noted in the context of bioconcentration:

1. The accumulation of xenobiotics into planktonic algae or plants (Section 3.1.3) may introduce the compound into higher organisms in the food chain such as fish, while the detrital material may eventually enter the sediment phase and be dispersed by any of the mechanisms noted in Section 3.5.1.

2. After deposition into sediments, particulate-associated xenobiotics may be desorbed into sediment interstitial water and thereby mediate exposure of the xenobiotic to sediment-dwelling organisms such as oligochaetes, amphipods, or chironomids and thence into higher organisms in the food chain; this has been discussed in Section 3.3.2. It is important to appreciate that such processes may occur at sites remote from those at which the initial sorption to the particulate matter took place, and that their extent will depend both on hydrological conditions in the water mass as well as on the specific nature of the association.

3. Xenobiotics may be accumulated from the aquatic phase into fish, many of which move over substantial distances, and some of which serve as prey to larger fish (Thomann and Connolly 1984). This may

result in biomagnification, which is discussed in Section 3.5.4. Migrating fish may also introduce xenobiotics into regions distant from their exposure. For example, it has been shown that arctic grayling (*Thymallus arcticus*) that occupy the highest trophic level in lakes used by sockeye salmon (*Oncorhynchus nerka*) for spawning, have a pattern of organochlorine compounds resembling those of salmon, whereas those in lakes not used for spawning had a different spectrum that suggested atmospheric transport (Ewald et al. 1998). In addition, fish may metabolize the xenobiotic and excrete the conjugates into the aquatic system. This is discussed in greater detail in Chapter 7 (Section 7.5.1), and will often function as an effective mechanism for the elimination of the xenobiotic. A single example will suffice to illustrate its role in the dissemination of the transformation products of compounds accumulated by fish from the aquatic phase: 3,4,5-trichloroveratrole (a bacterial metabolite of 3,4,5- and 4,5,6-trichloroguaiacol) is accumulated in zebra fish (*Brachydanio rerio*) where it can be metabolized with the formation of 3,4,5- and 4,5,6-trichloroguaiacol and 3,4,5-trichlorocatechol that are conjugated to sulfate and glucuronate. These water-soluble metabolites are then excreted into the aquatic phase resulting in dispersal not of .the original xenobiotic but of its metabolites (Neilson et al. 1989).

4. More complex interacting systems that mediate dissemination deserve brief mention. One interesting possibility is illustrated by the demonstration that in a mescocosm system, 2,3,7,8-tetrachlorodibenzofuran may be accumulated from spiked sediments into the larval stages of insects. After emergence as flying insects, these may then be consumed by a range of aquatic and terrestrial species including birds (Fairchild et al. 1992). Even though the magnitude of the transfer may be small, it should not be neglected in mass balance studies of lakes, and in a wider perspective, this suggests an additional process for the widespread dissemination of xenobiotics.

5. It has been suggested that bacteria may play a role in the transport of hydrophobic compounds in soils (Lindqvist and Enfield 1992), and it has been shown in batch experiments with *Bacillus subtilis* that sorption of 2,4,6-trichlorophenol may involve both neutral and anionic species (Daughney and Fein 1998). This mechanism could potentially apply also to aquatic systems where such processes could reasonably be included under particulate transport. There are, however, obvious unresolved issues concerning the subsequent desorption and bioavailability of these sorbed compounds.

6. Attention has already been drawn earlier in this section and in Section 3.1.1 to the similarity in the spectrum of organochlorines in sockeye salmon and arctic grayling in a lake chosen for spawning (Ewald et al. 1998). In such cases, trophic transfer may be quite

complex involving direct consumption of contaminated salmon eggs by predators, accumulation in insect larvae, and further transfer both to fish such as grayling, and the possibility of long-distance transport to other ecosystems. Coho salmon (*O. kisutch*) that were introduced into Lake Michigan have a diet varying with the time of year: in early spring, terrestrial insects and benthic invertebrates replaced by alewife (*Alosa pseudoharengus*) during the summer and early fall. The sum of PCB congeners that were analyzed were used to calculate the net trophic transfer efficiency from prey to predator (Madenjian et al. 1998) and showed that this was the major source of PCBs in coho salmon with a value of 0.5.

There are, therefore, a number of important roles that biota play in the dissemination of xenobiotics including (1) deposition of algal-associated xenobiotics onto the sediment phase, (2) progressive dissemination of xenobiotics via predators through the food chain, (3) elimination of xenobiotics as metabolites, and all of these may take place at any point in aquatic systems.

3.5.3   The Role of Atmospheric Transport

Introduction

It is clear from the preceding discussion that many xenobiotics will be recovered from matrices differing from those into which they were initially discharged. An important vector over long distances is the atmosphere, and the compounds may then reach remote areas both in the form of precipitation and as particulate deposition. The following example may be used to illustrate the magnitudes that may be involved in the atmospheric transport of particulate matter. Transport via dust particles that were subsequently deposited onto the surface of snow in the Canadian Arctic amounted to some thousands of tons and was attributed to atmospheric transport of dust from agricultural land in China (Welch et al. 1991). Increasing concern has been expressed over the contamination of hitherto pristine regions such as the Arctic and Antarctic. A general review has examined in detail the situation in the Arctic for a number of compounds including organochlorines and PAHs (Barrie et al. 1992), and others have been devoted specifically to the arctic marine ecosystem (Muir et al. 1992), the arctic terrestrial ecosystem (Thomas et al. 1992), and freshwaters in the Canadian Arctic (Lockhart et al. 1992). Airborne xenobiotics may also be deposited onto or accumulated in higher plants that may be consumed by herbivores and thus eventually transmitted to humans (Riederer 1990).

The processes whereby xenobiotics enter the atmosphere in the first place have received increasing attention; the principles have already been discussed in Section 3.4 and a model describing the various processes has been developed (Mackay et al. 1986). It has emerged that it is particularly compounds with low water solubility and appreciable vapor pressure that are effectively retransported from the aquatic phase into the atmosphere. This accounts for the recovery of compounds such as PCBs, PCCs (polychlorinated camphene consisting of chlorinated bornanes and bornenes), DDT, and some of the higher PAHs from widely distributed environmental samples. On the other hand, volatilization of α-hexachloro[*aaaaee*]cyclohexane in the Great Lakes is appreciable only during the summer months and the overall flux is from the atmosphere to the aquatic phase (McConnell et al. 1993).The complex dynamics of the equilibria between the phases are clearly demonstrated by extensive data from studies of PCBs at sites remote from possible direct discharge (Swackhamer and Hites 1988; Swackhamer et al. 1988). Some of the most significant conclusions from these elegant studies are of general significance and are worth noting. First, the distribution among the aqueous phase, the biota, and the sediment was complex and depended critically on the PCB congener involved, and, second, although the sole source of PCB input was from the atmosphere and originated from remote sites of discharge, a substantial part of the burden from the atmosphere was returned from the aquatic phase after deposition, and this reentry was accompanied by significant changes in the distribution of the various PCB congeners.

The latter conclusion is clearly supported by the results of an investigation that measured the concentrations of PCBs and PAHs in samples of air and water in a number of samples collected in Lake Superior. These were used to calculate fugacity gradients across the atmosphere–water interface (Baker and Eisenreich 1990) and the results clearly showed that during the summer months these xenobiotics were transferred from the aquatic phase into the atmosphere. Additional investigations in Lake Michigan revealed the significant roles of wind velocity and water concentration (Achman et al. 1993). In addition, all of these results clearly underscore the necessity of considering the different behavior of individual components of commercial products such as PCBs or of complex mixtures such as PAHs and chlorinated dibenzo[1,4]dioxins.

Indeed, the atmospheric input of PCBs into Lake Michigan has been found to exceed that from landfills (Hornbuckle et al. 1995), so that important interphase partitions should be taken into account. Atmospheric deposition and transport of toxaphene into sediments in the Great Lakes are dominant, although during the more recent 10-year period there is some increase in the hepta- and hexachlorinated congeners that suggests a slow transformation by dechlorination with an estimated half-life > 50 years (Pearson et al. 1997). Attention has been drawn in Chapter 2, Sections 2.4.1.3, 2.4.2.1, and 2.5 to analytical problems with toxaphene, and specifically to the presence of isomeric enantiomers.

Partition between vegetation and the atmosphere and between the soil and the atmosphere should also be considered; its significance in the accumulation of xenobiotics in higher plants and its application to monitoring have been discussed (Simonich and Hites 1995). A number of significant issues have been discussed in Section 3.1.3 and these investigations add further weight to those already enunciated for water – biota partition: (1) the unsuitability of octanol as a surrogate for biological lipids and (2) the significance of metabolism after uptake into biota. The possible role of plants is noted again in the context of assays for terrestrial toxicity in Chapter 7 (Section 7.5.5) and of bioremediation programs in Chapter 8 (Section 8.1.1).

It is important to appreciate that partition processes also take place in the atmosphere; scavenging by rain and sorption to particulate matter are of particular importance. Their significance has been demonstrated for the chlorinated dibenzo[1,4]dioxins (Koester and Hites 1992) and accounts for the enhancement of octachlorodibenzo[1,4]dioxin which has consistently been observed to dominate the other congeners in environmental samples (Czuczwa and Hites 1986) except fish (Zacharewski et al. 1989) possibly due to its limited uptake (Bruggeman et al. 1984).

Long-Distance Dissemination of Xenobiotics

Although complex chemical transformations — mainly photochemical — take place in the atmosphere, many chemically stable compounds may be transported intact via the atmosphere and subsequently enter the aquatic and terrestrial environments in the form of precipitation. Although the whole issue of chemical reactions in the troposphere lies outside the scope of this account, some comments are given in Chapter 4, Section 4.1.2, and reference should be made to the comprehensive account of principles given by Finlay-son-Pitts and Pitts (1986). The persistence in the troposphere of xenobiotics — even those of moderate or low volatility — is determined by the rates of transformation processes. These involve reactions with hydroxyl radicals, nitrate radicals, and ozone, or direct photolysis. Reactions with hydroxyl radicals are generally the most important. Illustrative values are given for the rates of reaction (cm^3 s^{-1} $molecule^{-1}$) with hydroxyl radicals, nitrate radicals, and ozone (Atkinson 1990).

	Hydroxyl Radicals	Nitrate Radicals	Ozone
n-Butane	2.54×10^{-12}	6.5×10^{-17}	9.8×10^{-24}
Acetaldehyde	15.8×10^{-12}	2.7×10^{-15}	$<10^{-20}$

Rates of reaction with hydroxyl radicals of a range of chlorinated dibenzo[1,4]- dioxins. dibenzofurans, and biphenyls extrapolated from experimental measurements of the less highly chlorinated congeners are sufficiently low to (1) suggest appreciable persistence in the atmosphere, (2) underscore the importance of atmospheric dissemination of the unaltered compounds, and (3) account for observed ambient concentrations (Kwok et al. 1995). There has been concern over the fate of halogenated aliphatic compounds in the atmosphere, and a single illustration of the diverse consequences is noted here. The initial reaction of 1,1,1-trichloroethane with hydroxyl radicals produces the $Cl_3C.CH_2$ radical which then undergoes a complex series of further reactions including the following:

$$Cl_3C \cdot CH_2 + O_2 \rightarrow Cl_3C \cdot CH_2 \cdot O_2$$

$$2Cl_3 \, C \cdot CH_2 \cdot O_2 \rightarrow 2Cl_3C \cdot CH_2 \cdot O$$

$$Cl_3C \cdot CH_2 \cdot O + O_2 \rightarrow Cl_3C \cdot CHO$$

In addition, the alkoxy radical $Cl_3C.CH_2.O$ produces highly reactive phosgene $(COCl_2)$ (Nelson et al. 1990, Platz et al. 1995) . Interest in the formation of trichloroacetaldehyde is heightened by the observation that the occurance of trichloroacetic acid in rainwater is a major source of this contaminant (Müller et al. 1996).

Once again, complications from binding of these compounds to particulate matter including for example, soot particles, must be taken into consideration, and this is particularly important for PAHs that are bound to particulate matter before emission. The virtual global distribution of some xenobiotics attests to the probable importance of atmospheric transport and provides evidence that this is by no means restricted to highly volatile compounds. Nonetheless, some evidence suggests that for compounds including PCBs, hexachlorobenzene, and hexachlorocyclohexanes, although transport within the troposphere is important, its role in transport between the hemispheres is marginal (Wittlinger and Ballschmiter 1990). A few examples of long-range transport will be used to illustrate both the geographical range covered and the diversity of the compounds involved.

1. Persistent organochlorine compounds have been identified in rainfall (Strachan 1988) and compounds such as hexachlorocyclohexanes, a-endosulfan, and dieldrin have been detected in samples of snow even in remote areas such as the Canadian Arctic (Gregor and Gummer 1989).

2. Analysis of pine (*Pinus sylvestris*) needles collected over large areas of Europe showed the presence of a number of halogenated compounds including DDT, pentachlorophenol, and PCBs (Jensen et al. 1992). Whereas the origin of the DDT, which has been banned for many years in Europe, could be attributed to its application in an area of southern Germany, the source of pentachlorophenol which was highest in samples from Northern Sweden could not be unequivocally established.

3. Pine needles from North America have also been used for monitoring polychlorinated dibenzodioxins (Safe et al. 1992); the spectrum of the congeners which was dominated by octachlorodibenzodioxin and the most highly chlorinated dibenzofurans suggested their ori-

gin from impurities in commercial pentachlorophenol used for wood impregnation rather as a result of the combustion of chlorinated aromatic compounds. This is consistent with a similar conclusion drawn from previous results of the analysis of air particulates and sediments (Czuczwa and Hites 1986).

4. Analysis of lichens from the Great Lakes region has been used to assess the dissemination of a number of organochlorine compounds and has drawn attention to the significance of atmosphere–plant partitioning (Muir et al. 1993): the pattern of compounds that were found differed from those found in atmospheric or rainfall samples, and this fact illustrates clearly the specific influence of both the chemical structure of plant surfaces and of their surface area.

5. Samples of seal blubber from the high-arctic archipelago Svalbard contained levels of polychlorinated camphenes comparable with those found in biota in the Baltic (Andersson et al. 1988), and the same group of compounds was dominant in samples of narwhal (*Monodon monoceros*) blubber from Pond Inlet, Baffin Island in the Canadian high arctic (Muir et al. 1992).

6. Residues of organochlorine compounds including such ubiquitous representatives as hexachlorobenzene, hexachlorocyclohexanes, DDT, and its metabolites have been found in samples of lichens and moss from several localities in the Antarctic (Focardi et al. 1991).

7. A variety of samples from an ice island in the Canadian Arctic contained a range of organochlorine compounds, and the role of atmospheric delivery was supported by their occurrence in samples of air, water, snow, zooplankton, and benthic amphipods (Bidleman et al. 1989). The significance of arctic haze was inferred from comparable data collected in May and September from the same ice island (Hargrave et al. 1988).

8. The occurrence of PCBs in samples of cod, seal, and polar bear in the Canadian Arctic (Muir et al. 1988) attests to the ubiquitous distribution of these compounds, and the existence of biomagnification within this food chain is discussed further in Section 3.5.4.

9. The carcinogenic amino-a-carbolines, 2-amino-9H-pyrido[2,3-b]-indole and its 3-methyl derivative (Figure 3.8) that are pyrolysis products of tryptophan have been identified in environmental samples including airborne particles and rainwater as well as cigarette-polluted indoor air (Manabe et al. 1992). A number of other azaarenes including a substituted imidazo [4,5-b]pyridine and an imidazo[4,5-f]quinoxaline have been found in rainwater and their presence was attributed to forest burning (Wu et al. 1995). If generally confirmed, this finding might suggest the existence of a potentially serious threat to human health.

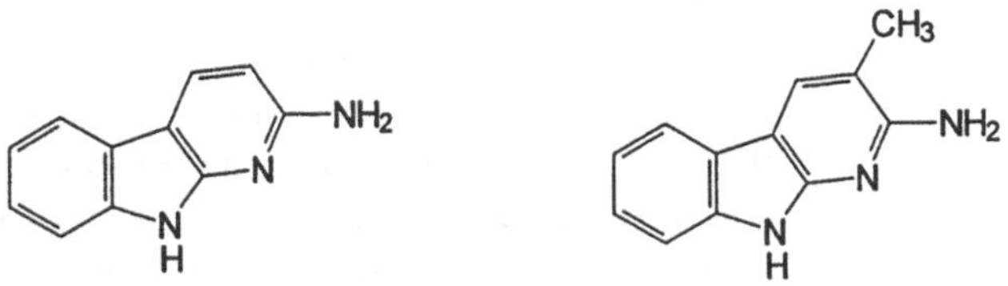

FIGURE 3.8
2-Amino-9*H*-pyrido[2,3*b*]indole and 2-amino-3-methyl-9*H*-pyrido[2,3*b*]indole.

10. Tris(4-Chlorophenyl)methanol has been recovered from samples of marine mammals and bird eggs from the Arctic, the Antarctic, Australia, and North America (Jarman et al. 1992), and both this and the corresponding methane in a range of biota from the Baltic Sea coast (Falandysz et al. 1999). Tris(4-Chlorophenyl)methanol is probably a transformation product of tris (4-chlorophenyl)methane that has been prepared in the laboratory as a by-product of the procedure used commercially for DDT, and that has been detected in samples of technical DDT as well as in samples of biota (Buser 1995).

It is clear from this summary that polar regions may be substantially exposed to certain groups of organochlorine compounds, and it has been suggested on the basis of an analysis of their physicochemical properties that polar regions may function as an effective condensation trap for xenobiotics of intermediate volatility (Wania and Mackay 1993). This is clearly supported by analysis of PCB congeners in lake sediments from 49° 45′N to 81° 45′N (Muir et al. 1996). Fluxes of di- and trichlorobiphenyls were not dependent upon latitude, in contrast to those for the less volatile tetra- to octachlorobiphenyls that decreased with latitude; the contribution of the former to the total concentration of PCBs increased with latitude whereas that of octachlorobiphenyl decreased.

3.5.4 Biomagnification

Biomagnification is the process whereby a compound enters a food chain at a lower trophic level and, via a series of predators, is increasingly concentrated in higher levels. Biomagnification may be considered as sequential bioaccumulation, and some general comments have been given in the begining of this section. There has been considerable discussion of the mechanism of this process, and field observations are not always readily interpretable on the basis of laboratory experiments.

Simplified fugacity models may not predict the existence of biomagnification, although bioenergetic models allow for biomagnification which may be especially significant for compounds with values of log $P_{ow} > 4$ (Connolly and Pedersen 1988). Considerable attention has been directed to theoretical considerations, and the data appear to suggest only moderate levels of biomagnification (Thomann 1989; Bierman 1990). It is probably fair to state, however, that there are very substantial approximations involved in such estimates since they inherit all the existing ambiguities of those for predicting bioconcentration potential. In addition, few results are reported for the concentrations of xenobiotics normalized to the same base — of which lipid weight is probably the most relevant and widely applicable.

The role of lipids has already been discussed (Section 3.1.1), and its significance in biomagnification is supported by evidence that the increasing bioconcentration with increasing trophic level may be correlated with increasing lipid concentration. It has been suggested that, in organisms at higher trophic levels, observed concentrations of xenobiotics in biota are the result of two additive factors: (1) decreased biotransformation and elimination of the xenobiotic and (2) increased lipid concentration.

1. On the basis of the somewhat limited data that were available for analysis, and for values of log P_{ow} > 5, a relation between BCF values for organisms at lower ($BCF_<$) and higher trophic levels ($BCF_>$) was proposed: $BCF_> = BCF_< [8.2 \log P_{ow} - 40]$. True biomagnification will only be significant for compounds having values of log P_{ow} >~6.3 (Leblanc 1995).

2. It has been shown that lipid content played a dominant role in determining the concentration of a wide range of organochlorines in Lake Superior biota at all trophic levels (Kucklick and Baker 1998). There was a linear relation between $\delta\ ^{15}N$ and both lipid content and total PCB concentration for the range of biota, so that both lipid and trophic level play a determinative role. The analytes included PCBs, toxaphene, α-hexachloro[*aaaaee*]cyclohexane, hexachlorobenzene, and dieldrin, and the biota, planktonic *Mysis relicta,* copepods *Limnocalanus* sp.; amphipods, *Diporeia hoyi;* deepwater sculpin, *Myoxocephalus thomsoni;* smelt, *Osmerus mordax;* herring, *Coregonis artedii;* bloater, *Coregonus hoyi;* slimy sculpin, *Cottus congatus;* spoonhead sculpin, *C. ricei.* In addition, it was concluded that settling of particulate matter onto sediments was an important source of organochlorines to benthic organisms. Extensive studies of fish samples from three subarctic lakes in the Yukon Territory are consistent with these results in showing that $\delta\ ^{15}N$ values depended both on the trophic level and on the lipid content (Kidd et al. 1998). In addition, it was shown that, whereas lipid-normalized concentrations of DDT + DDE + DDD (Σ DDT) in the three lakes were constant for increasing values of $\delta\ ^{15}N$, this was not observed for less hydrophobic contaminants such as total hexachlorocyclohexane isomers (Σ HCH). Lipid-adjusted concentrations in lake trout (*Salvelinus namaycush*) remained significantly different between lakes and this was attributed to variable lengths in the food chain.

Attention has already been drawn to a laboratory study (Sijm et al. 1992) in which successive generations of guppy (*Poecilia reticulata*) were exposed to PCB-contaminated food. A study with guppy (*P. reticulata*) and goldfish (*Carassius auratus*) and food contaminated with a mixture of 1,2,4,5-tetra-, penta-, and hexachlorobenzene, 2,2',4,4',6,6'-hexachlorobiphenyl, and mirex has illustrated the importance in biomagnification of food digestion and food absorption in the gastrointestinal tract (Gobas et al. 1993a). This has been examined in detail (Gobas et al. 1999) in laboratory experiments using rainbow trout *Oncorhynchus mykiss*, and it was shown that, for 2,2',4,4',6,6'-hexachlorobiphenyl, the fugacity of the toxicant in the gastrointestinal tract may be up to eight times that in the consumed food. This increase is the result of the greater absorption efficiency of lipid compared with that for the toxicant, and the several additional factors in interpretating the results were carefully discussed. The magnitude of gastrointestinal tract biomagnification will be greater when the diet in the prey is rich in lipid, and care should be exercised in attempting to produce widely applicable values.

Biomagnification has been examined extensively in field samples from the marine food chain. The following examples involve not only different organisms but also different experimental methodologies.

1. Biomagnification using biota at different trophic levels has been carefully examined in the arctic marine food chain — arctic cod (*Boreogadus saida*), ringed seal (*Phoca hispida*), and polar bear (*Ursus maritimus*) (Muir et al. 1988)—and biomagnification factors (BMFs) were calculated. The existence of biomagnification was demonstrated, and a number of other important facts emerged:

   - BMFs for fish/seal and seal/bear were dependent on the sex of the seals.

   - In the fish/seal system, BMFs for DDE were greater than for total DDT though for seal/bear values were < 1 for both compounds.

   - The highest BCFs were observed for the most highly chlorinated PCB congeners with 7 to 9 chlorine atoms; this is consistent with the unique pattern of PCB congeners in polar bears which have relatively high concentrations of these congeners, but apparently successfully metabolize those with lower degrees of chlorination (Norstrom et al. 1988).

   Levels of biomagnification of highly chlorinated PCBs, of DDT, DDE, and chlordane-related compounds were disturbingly high, and all of these compounds together with α-hexachloro[*aaaaee*]cyclohexane have been identified in the diet of Inuit who consume and are essentially dependent on the harvest of ring seal, beluga, eider, and arctic char (Cameron and Weis 1993). The more controversial issue of the risks as opposed to the benefits of such consumption has been addressed (Kinloch et al. 1992).

Dietary exposure to organochlorine compounds probably exists for all circumpolar inhabitants that pursue a traditional way of life, and attention has already been drawn (Section 3.5.3) to the suggestion that polar regions may be an effective condensation trap for such compounds.

2. Biomagnification has been investigated in a marine food chain in the western Pacific Ocean (Tanabe et al. 1984), and this study examined samples of zooplankton, myctophids (lanternfish, *Diaphus suborbitalis*), squid (*Todarodes pacificus*), and striped dolphin (*Stenella coeruleoalba*) that were analyzed for a number of organochlorine compounds. In contrast to the results from the fish/seal/bear study, the spectrum of PCB congeners in myctophid, squid, and dolphin was essentially similar, although in comparison with the seawater, it was markedly enriched in the more highly chlorinated congeners. On the other hand, compared with squid, myctophid, and zooplankton, hexachlorocyclohexane isomers were completely different in striped dolphin with a dominance of the β-isomer [*eeeeee*] and almost none of the α-isomer [*aaaaee*]. Bioconcentration factors were presented in terms of total body weight, although concentrations were also given in terms of whole-body lipid which ranged from ~ 3% in mycophid and squid to over 13% for striped dolphin. Once again, there is no doubt that biomagnification existed — even though detailed kinetic analysis is not possible — and that the level of bioconcentration depended on the trophic level of the biota.

3. On the basis of $\delta^{15}N$ values for biota in the Baltic Sea — phytoplankton, seston, zooplankton, blue mussels, eider duck, herring, and cod — it was concluded that biomagnification occurred for 2,3,7,8-tetrachlorodibenzo[1,4]dioxin and 2,3,7,8-tetrachlorodibenzofuran, although not for the octachloro congeners (Broman et al. 1992). Apart from the possible intrusion of different levels of lipid content that has been noted previously in Section 3.1.1, it should be noted that the octachloro congeners have been shown to exhibit only low levels of lipid solubility and bioconcentration potential (Section 3.1.1) so that their levels in the bile of eider ducks remain unexplained.

4. Other systems that have been examined include fish/cormorant (*Phalacrocorax carbo sinensis*) in Schleswig-Holstein on the German North Sea coast. These data were not presented on a comparable basis for the two components, however, so that numerical comparison with other systems cannot be made. PCB concentrations in the subcutaneous fat of cormorants were from 10 to 100 times those in muscle tissue of marine fish with tenfold values for freshwater fish (Scharenberg 1991). There is evidence that birds, and especially those consuming fish, have only low levels of microsomal monooxygenase activity (Walker 1983), although it is important to distinguish the differential activity of 3-methylcholanthrene-induced and

phenobarbital-induced monooxygenase systems in the metabolism of different xenobiotics (Braune and Norstrom 1989), and this could have an important bearing on these observations on cormorants.

5. A number of studies have been directed to freshwater systems in North America. An example from the Great Lakes nicely illustrates the differences that may be encountered at different trophic levels and for different organochlorine compounds. On the basis of a limited number of analyses, there appeared to be little biomagnification between zooplankton and forage fish for 2,3,7,8-tetrachlorodibenzo[1,4]dioxin, (Whittle et al. 1992), whereas values of 32 between herring gulls (*Larus argentatus*) and alewife (*Alosa pseudoharengus*) have been reported for 2,4,3',4'-tetrachlorobiphenyl (Braune and Norstrom 1989). For PCBs in this system, the BMFs depended critically on both the position and the number of chlorine atoms with a maximum value of 219 for 2,3,4,5,2',3',4',5'-octachlorobiphenyl, but were consistently higher than the values for biomagnification from alewife to salmonids calculated on the basis of data from a previous study (Oliver and Niimi 1988). This was attributed to the different energy demands on fish and birds, which are much greater in the latter.

In summary, whereas there is no reason for doubting the existence of evidence that may be interpreted as biomagnification, four general comments are worth inserting in addition to those at the beginning of this section:

1. Experiments using goldfish (*Carassius auratus*) and a range of neutral organochlorine compounds including chlorinated benzenes, biphenyls, and octachlorostyrene showed that uptake was not dependent on the lipid content of the diet and was therefore determined primarily by passive diffusion into the gastrointestinal tract rather than by cotransport with lipids (Gobas et al. 1993). Essentially similar conclusions may be drawn from experiments (Larsen et al. 1992) on the bioavailability of PCB congeners to the earthworm *Lumbricus rubellus*.

2. The investigations cited above demonstrated the selective accumulation in various organs, and that for the PCBs this was congener specific; for example, decachlorobiphenyl was found only in the brain tissue of cormorants (Scharenberg 1991). This provides further support for the critical comments that have been raised against sum parameters.

3. Although the transport of PAHs from particulate matter in the aquatic phase through blue mussels (*Mytilus edulis*) to the common eider (*Somateria mollissima*) has been demonstrated in the Baltic Sea, there was a successive change in the relative concentrations of the various individual hydrocarbons (Broman et al. 1990). This may be attributed to increasing metabolic activity at the higher trophic

levels that could result in the production of putatively mutagenic compounds. This is an issue of general importance that merits detailed evaluation (Braune and Norstrom 1989).

4. Investigations have been largely limited to neutral organochlorine compounds with a high potential for bioconcentration, and it would clearly be desirable to extend this to other groups of compounds including those containing polar substituents such as carboxylic acid, phenolic, or amino groups.

In conclusion, the following should be critically evaluated in discussion of biomagnification.

- Normalization to a relevant base of the concentration of the xeno-biotic at the various trophic levels;
- Physiological differences in the organisms at the various trophic levels and their bioenergetics;
- The metabolism of the xenobiotic as it is transported through the various trophic levels.

The metabolism of xenobiotics by birds and their possible role in the dissemination is discussed in Chapter 7, Section 7.5.3.

3.5.5   The Role of Models in Evaluating the Distribution of Xenobiotics

It will have become apparent from the preceding discussions that xenobiotics after discharge from a point source may enter any of the various environmental compartments: aquatic systems including biota and sediment, the atmosphere, terrestrial systems including soils, biota, and in the long run possibly the ultimate predator — humans. Considerable effort has therefore been devoted to the development and application of models to evaluate this dissemination in quantitiative terms. These involve the concept of fugacity, and it seems appropriate at the beginning to examine this concept briefly.

In classical thermodynamics, which deals with the behavior of ideal reversible systems, the term fuqaaty is used to preserve the relationship between the free energy and the temperature and pressure by replacing the pressure with the fugacity (Partington 1950), and it may be shown that the concept can be applied also to other phases and to mixtures (Tolman 1938). It is important to realize, however, that the additional terms included in the van der Waals' equation for real gases derive from interaction mechanisms between molecules that lie beyond the valid application of reversible thermodynamics (Jeans 1952). The application of the term to natural systems that are — by definition — seldom or never in equilibrium should therefore be constantly kept in mind, and the reasons for departure from predictions sought in molecular mechanisms of interaction. Models are generally classified as belonging to Levels I and II which depict equilibrium conditions whereas Level III takes into account nonequilibrium steady-state conditions; a clear presentation of the relevant assumptions as well as of Levels IV and V models has been given (Paterson and Mackay 1985). All of the models require the input of physico-chemical parameters for the compound(s) and, for Level III estimated rates of degradation. Estimation of these probably presents the greatest uncer-tainty. Inevitably, the predictive value of all the models depends critically on the accuracy of the input data as well as on the environmental realism of the assumptions that are incorporated.

As pointed out in Chapter 1, modeling lies far beyond the expertise of the author, but in view of its importance a few inevitably superficial remarks are offered by way of illustration.

1. A Level III study using data for 2,3,7,8-tetrachlorodibenzo[1,4]-dioxin showed that the low water solubility, low vapor pressure, and high P_{ow} resulted in partitioning of the compound mainly into soil (69%) and sediment (29%) and that the major removal process is air advection. On the basis of reasonable assumptions, it was concluded that the major exposure route for humans is via food, mainly meat and fruit and vegetables (Mackay et al. 1985).

2. A comprehensive application of Level III was made to the accu-mulation and elimination of a range of neutral halogenated xeno-biotics by fish from food and water and took into account transport across the gills, activity in the gastrointestinal tract including metabolism, and the bioavailability in the water mass (Clark et al. 1990). The results illustrated the importance of the uptake of xeno-biotics via the food that may result in biomagnification, and that this process is most significant for hydrophobic compounds that are poorly metabolized.

3. Concern has been expressed over human exposure to xenobiotics transported in the atmosphere and accumulated in higher plants that are consumed both directly or indirectly, for example, as milk from herbivores. A simple model for assessing the transport of xeno-biotics from the atmosphere to vegetation has been developed (Ried-erer 1990) and was used to calculate equilibrium concentrations in plant tissues and bioconcentration factors from air to vegetation. The important issue of revolatilization was addressed since this deter-mines the persistence of xenobiotics in plants. A detailed illustration of the processes that determine the dissemination of PAHs has been given (Mackay and Caldcott 1998), which draws attention to the complexity of atmospheric partitioning when particulate matter is present.

4. Two different models were used to examine the dispersion of a limited number of chlorophenolic compounds in a segment of the Gulf of Bothnia (Kolset and Heiberg 1988); one of these, EXAMS, takes into account ionized and nonionized forms separately,

whereas FEQUM does not. The application of models to such complex situations is extremely difficult in the absence of the relevant hydrological data, and the use of chloroform in this case to calibrate the model is somewhat questionable since this compound differs so significantly in its physicochemical properties from the chloroguaiacols and chlorocatechols that were the subject of study. In spite of this, generally reasonable distributions were obtained for the chloroguaiacols — although less so for tetrachlorocatechol.

The further successful application of such models would appear to depend on more explicit understanding of the mechanism of transport processes and metabolism, as well as more precise numerical definition of cardinal environmental parameters.

3.5.6 Leaching and Recovery from Other Solid Phases

The preceding discussion has focused attention on aquatic systems including sediments. Some xenobiotics are, however, deliberately applied to terrestrial systems; for example, modern agriculture is dependent on the use of substantial amounts of agrochemicals for control of weeds, insects, and pathogenic microorganisms. Concern has been expressed both about the residues of agrochemicals in food for human consumption and about the persistence and mobilization of these compounds in the soil. Extensive investigations have therefore been directed to the analysis of these residues. What is, of course, of equal significance is the extent to which nontarget organisms are exposed to these potential toxicants, and this was the concern to which Rachel Carson originally drew attention.

It has already been pointed out that the exposure of biota to xenobiotics is critically determined by the dynamics of desorption processes and that these depend on the nature of the solid matrix: agricultural soils throughout the world are extremely variable both in texture, organic content, and mineral composition. Standard extraction procedures, some of which have been outlined in Chapter 2, Section 2.2, have been widely adopted even though the level of recovery may be quite low and this is plausibly attributable to association between the xenobiotic and humic components of the soil. The significance of physical entrapment has been illustrated by the extreme persistence of low-molecular-weight and normally volatile compounds such as 1,2-dibromoethane (Steinberg et al. 1987; Sawhney et al. 1988). An illustrative example of other issues involved with agrochemicals is provided by the transformation products of chloroacetanilide herbicides (Field and Thurman 1996). These react with glutathione to form conjugates which, after scission and oxidation, form aliphatic sulfonates that have environmentally significant properties.

- They are water soluble and are therefore readily disseminated in watercourses.

- They contain an element not present in their precursors and may present a problem of identification in water and soil samples (Aga et al. 1996).

The same also applies to other *S*-containing metabolites, for example, the sulfonate of alachlor which has a low bioconcentration potential and is nonmutagenic in contrast to its precursor (Section 2.5).

The whole issue of recoverability of xenobiotics from solid phases, of their desorption, and of their bioavailability has emerged increasingly in a wider context: the problem of solid waste disposal (Suflita et al. 1992) and the effectiveness of bioremediation of contaminated sediments (Harkness et al. 1993). Leaching from what are euphemistically termed *landfills* could result in the pollution of both ground and surface waters by a wide range of organic chemicals, most of which are probably recalcitrant and many of which are toxic (Chapter 8, Sections 8.2.8 and 8.2.9). This will almost certainly absorb substantial activity in the future, and will claim increasing attention from chemists, microbiologists, and toxicologists to provide a basis for the rational application of control and recovery strategies. The important issue of emission of xenobiotics from solid phases into the atmosphere has been discussed in Section 3.4.2.

3.6 Monitoring

Many xenobiotics initially discharged into restricted geographicl areas now have a worldwide distribution as a result of the various transport processes that have been discussed in this chapter. This knowledge has accumulated from extensive monitoring studies that have achieved increasing prominence and encompassed an ever-widening range of compounds. Monitoring may be used to provide evidence for evaluating the persistence of a xenobiotic in the environment, and at the same time determining the extent to which it has entered environments geographically distant from the initial point of discharge. If, of course, a compound is continuously introduced into the environment, its recovery does not necessarily provide evidence of its persistence; an example of this is the phthalates whose ubiquity is clearly established even though they are degradable and cannot therefore be regarded as recalcitrant. On the other hand, an ingenious application of monitoring to demonstrate the microbial degradation of PCBs in contaminated natural sediments may be given. The sediments were analyzed for the presence of chlorobenzoates that had previously been established in laboratory experiments as microbial metabolites (Flanagan and May 1993). These results demonstrate both the necessity of laboratory experiments on biodegradation for interpreting the results of an environmental monitoring program, and the value of carefully selected analytes for inclusion.

Since the results will be interpreted on a comparative basis, it is important to take into account physiological and biochemical aspects of the organisms that are used, and the physical environment from which they were collected. An illustrative example may be taken from a study using two different shellfish from the North Island of New Zealand (Hickey et al. 1995). The salient features that emerged may be summarized.

- The role of exposure and uptake: *Macoma liliana* was a deposit feeder, whereas *Austroventus stutchburyi* was a filter feeder.
- The health of the organisms should be taken into consideration, although in this case there was no correlation between their health and lipid content.
- The habitat of the organisms, including salinity and temperature tolerence, should be taken into consideration.
- The organisms may differ significantly in metabolic potential (Section 3.1.5 and Chapter 7, Section 7.5), so that compounds that are originally taken up may be depurinated as water-soluble metabolites. The organisms may have a specificity toward a specific xenobiotic, so that it should not be assumed that all — even neutral compounds such as organochlorines or PAHs — are equally accessible or persistent after uptake.
- In their natural habitat, the organisms may display a patchiness so that it may be difficult to obtain a representative sample.

Even a cursory examination of the literature shows that analysis of virtually every environmental sample reveals contamination from polycyclic aromatic hydrocarbons — resulting from incomplete combustion processes — and a range of the more recalcitrant organohalogen compounds such as DDT (together with its degradation product DDE), PCBs, hexachlorocyclohexanes, compounds related to aldrin, and mixtures present in commercial toxaphene preparations. Possibly the most disturbing fact, which has already been noted, is the occurrence of these compounds in samples from remote and largely isolated locations in the Arctic and Antarctic.

3.6.1   Choice of Samples

Although monitoring has traditionally been used to determine whether, in the final analysis, xenobiotics are persistent in the environment, the question of what kinds of samples to monitor and what compounds to analyze cannot easily or universally be answered. Samples from virtually all possible matrices have been used including the atmosphere, snow, water, sediments, and a diverse range of biota. Among biota, the U.S. National Contaminant Biomonitoring Program uses fish collected from a network of stations (Schmitt et al. 1990) and the Mussel Watch Program uses bivalves (Goldberg 1986; Sericano et al. 1993). A much greater diversity of organisms has, however, been less extensively used and includes aquatic leeches (Metcalfe et al. 1984, 1988); herring gulls (Norstrom et al. 1986) and their eggs (Hebert et al. 1994) in the Great Lakes; cormorants in the North Sea coast of Germany (Scharenberg 1991); polar bears in the high arctic (Muir et al. 1988); insect larvae in the Hudson River (Novak et al. 1988); lamprey ammocoetes in the St. Lawrence River basin (Renaud et al. 1995); female mayflies along a 1250-km stretch of the upper Mississippi River (Steingraeber et al. 1994).

These programs will not necessarily provide comparable results since they vary, for example, in feeding habits and lipid content. Monitoring programs using fish should take into account the season at which sampling is made, and their age, size, sex, and fat content, and analyze individuals rather than pooled samples to avoid serious errors in interpretation of the results (Bignert et al. 1993). A study of PCB concentrations in the Great Lakes between 1972 and 1990 has underscored the importance not only of using the appropriate statistical procedures for describing variations, but also of biological factors such as food chain dynamics (Smith 1995). Attention is directed to the use of blood plasma in the analysis of a range of organochlorine compounds in falcons from Greenland (Jarman et al. 1994). This is clearly attractive for wild populations in which only low volumes of sample are available, and since only simple cleanup procedures are needed. Precautions exercised with aquatic biota such as fish and mussels should be applied also to other indicator species such as plants and trees. The organisms used should clearly be identified to genus and species although in a study using pine needles, this was not carried out (Calamari et al. 1994).

Considerable care should be exercised in the choice of biota and this should take into account their metabolic potential for the compounds that are being monitored (Sections 3.1.5 and Chapter 7, Section 7.5). Indeed, advantage of this capability has been used for monitoring exposure to chlorophenolic compounds by analyzing fish bile for conjugates (Oikari and Kunnamo-Ojala 1987; Wachtmeister et al. 1991; Brumley et al. 1996). Two apparently extreme situations may be used to illustrate potential problem areas. One apparent advantage of using seabirds is the apparently low level of liver mixed-function oxidase enzymes involved in the metabolism of most xenobiotics (Walker 1983); the xenobiotic may therefore be recovered metabolically intact and this considerably simplifies the analysis of the samples. There is, however, evidence that illustrates the potential of birds to metabolize xenobiotics (Braune and Norstrom 1989). On the other hand, the unique pattern of PCB congeners found in polar bears attests to their ability to metabolize at least the less highly chlorinated congeners (Norstrom et al. 1988).

When sediments are employed, there are a number of critical experimental factors to which careful attention should be given. These include the size of the sample, the patchiness of natural sediments (Swartz et al. 1989), and the conditions for transport and preservation of the sample after collection. Concentrations of xenobiotics will generally be normalized to dry weight, amounts of organic carbon, or to specific compound groups such as lipids, but even so comparison between samples from different geographical regions or using different biota is often extremely difficult. It is therefore essential to clearly define at the outset the objective of the investigation. Unfortunately the importance of this is not always appreciated and may result in unnecessary cost and avoidable frustration. Although monitoring sediment-dwelling biota is attractive, attention should be given to their metabolic capability. For example, polychaete worms belonging to the genera *Nereis* and *Scolecolepides* have extensive metabolic potential (Chapter 7, Section 7.5.4) so that these taxa are less suitable for monitoring than taxa with limited metabolic potential.

3.6.2 Temporal Record of Input

Monitoring has also been carried out to provide a historical record of input to ecosystems and has generally been correlated in time with industrial production; this might be termed environmental archeology and has been most widely applied to organochlorine compounds particularly PCBs and chlorinated dibenzo[1,4]dioxins. The results of such analyses generally provide at the same time convincing support for the widespread dissemination of many xenobiotics via atmospheric transport (Sections 3.4 and 3.5), but analysis for a specific compound may be invaluable in determining the details of dissemination process after discharge from a single point. An almost unique example is the use of 2,4-dipentylphenol to map its distribution in Lake Erie (Carter and Hites 1992). This was possible since the compound was produced virtually exclusively by a single plant and is an otherwise unusual xenobiotic. Over relatively short time spans (< 100 years), peat bogs have been used to assess the magnitude of atmospheric input of xenobiotics including PCBs, DDT, hexchlorobenzene, and toxaphene (Rapaport and Eisenreich 1988). The advantage of these environments for such monitoring lies in the relatively low rates of degradation and minimal mobilizing of the xenobiotics in the organic matrix.

One of the cardinal issues is the dating of the samples. Ideally, advantage should be taken of archived samples collected at established dates, but this is possible only for samples of soils and some kinds of biota — and its unambiguous application depends critically on how adequately the samples have been preserved.

Virtually unique examples are provided by the analysis of soil samples from the Broadbalk Experiment at Rothamstead, England. This experiment began in 1843 and samples collected during more than a century have been used to investigate the deposition of PAHs which result from combustion processes of various kinds (Jones et al. 1989), and of polychlorinated dibenzo[1,4]dioxins and furans (Alcock et al. 1998). The latter study confirmed previous results (Kjeller et al. 1991) showing the full range of congeners, and the increase in all of them including the presumptively toxic ones since 1881 until the present time. It is worth noting that concentrations (ng/kg dry weight) of Σ PCDD/PCDF increased from 1881 to 1986 from 50 to 84, and values of Σ TEF from 0.7 to 1.4 ng / kg dry weight, increases that must be considered rather insubstantial. Importantly, contamination either from dust during subsampling or from laboratory air was apparently excluded (Baker and Hites 1999; Alcock et al. 1999). The source of these compounds cannot be unambiguously determinate in England which had used substantial volumes of wood and coal for iron smelting and glass production, and where chlorine for bleaching had been introduced during the late 18th century. A survey of PCB levels carried out by the same groups using a wider range of samples that included other areas in England provided interesting details of temporal trends and the alterations in congener composition that have taken place probably mediated by volatilization (Alcock et al. 1993).

More commonly, sediment samples have been investigated. In this case, cores have been collected, and segments used simultaneously for radiodating using ^{210}Pb, ^{134}Cs, and for recent sediments ^{137}Cs, or for older ones ^{14}C, and for chemical analysis. A striking example of its application is the finding of various chlorinated dibenzo[1,4]dioxins in sediment core samples from Japan (Hashimoto et al. 1990). Especially notable is the finding of the 1,2,3,4,6,7,9-heptachloro and octachloro congeners in the deepest samples dated at ~ 8000 years old. In this study, the possible complicating circumstances were apparently eliminated, but it may be valuable to summarize briefly the most important of them since they are relevant to all monitoring studies that seek to provide an unequivocal record of input. Analytical artifacts must, of course, be eliminated, and the levels of analytes in blanks must be essentially zero. Other less obvious factors may seriously compromise the interpretation of the results so that their significance should be appreciated and taken into account:

1. Sediments may not remain undisturbed due to diffusion or bioturbation within the sediment phase; of these, bioturbation is probably dominant in many situations (Karickhoff and Morris 1985).

2. Sedimentation may not be uniform so that some degree of inhomogeneity may exist within the sediment layers; this may introduce potentially serious errors especially if pooled samples are analyzed (Sanders et al. 1992).

3. It cannot validly be assumed that even apparently recalicitrant compounds remain unaltered over very long periods of time, and the possibility of their transformation should be taken into consideration; this includes both abiotic reactions that are discussed in Chapter 4, Sections 4.1.1 and 4.1.3 as well as those mediated by microorganisms that are the subject of detailed analysis in Chapter 6.

4. It has already been pointed out in Section 3.3 that resuspension of surficial sediments may result in reentry of a part of the burden of the sediment phase into the aquatic phase.

Even data from the analysis of snow and ice samples from ice caps in polar regions should be interpreted with caution in light of possible compromising factors including ablation and percolation of meltwater during the summer months.

3.6.3 Choice of Analytes

The compounds chosen for analysis will generally belong to established groups of xenobiotics such as organochlorines, PAHs, or the structurally diverse range of agrochemicals, although halogenated anisoles that are transformation products of the corresponding phenols (Wittlinger and Ballschmiter 1990) and organobromine compounds (Sellstrom et al. 1998) have begun to be included. In addition, attention should be drawn to the diversity of abiotic transformation products that have been noted in Chapter 2 (Section 2.5) and are discussed in greater detail in Chapter 4 (Sections 4.1.1 and 4.1.3), together with those mediated by microorganisms (Chapter 6) and higher organism including fish (Section 7.5). An illustrative example of the use of established bacterial metabolites to provide evidence for the degradation of the lower congeners of PCBs in sediments is given in Chapter 5 (Section 5.3.3), while the care that must be exercised in the interpretation of results from monitoring is exemplified by the results of an elegant study of chlorinated dibenzo[1,4]dioxins (Albrecht et al. 1999). In microcosm experiments using a sediment that was dominated by the octachloro congener, it was shown that treatment with 2-bromodibenzo[1,4]dioxin resulted in a doubling in the concentration of the 2,3,7,8 congener at the expense of the original octachloro congener, and that hydrogen amendments resulted in the production of the 2-chloro congener without increase in the 2,3,7,8 congener. The absolute concentration of the 2,3,7,8-congener therefore provides no unequivocal evidence of its persistence, since there is a dynamic situation in which both abiotic and biotic reactions are involved. The role of brominated compounds in promoting anaerobic dechlorination is noted in Chapter 8, Section 8.2.4.2.

All of these factors necessitate care both in the design and in the interpretation of the results of monitoring studies; particular care should be exercised in the matter of conclusive identification of analytes. It may, indeed, not be possible to identify even tentatively all the components in the samples; for example, a study of organic compounds in Florida sediments revealed three unidentified compounds in addition to PAHs, benzonaphthothiophene, and terpene-related compounds (Garcia et al. 1993).

Increasing attention has been devoted to the enantiomeric composition of xenobiotics isolated from environmental samples. Even when the analyte is released in an optically inactive form, a preponderance of one anentiomer may occur after discharge, mediated by both chemical and microbiolgical activity. Considerable effort has been directed to the analysis of highly chlorinated bornanes and bornenes produced by the chlorination of camphenes.

Other compounds that may deserve attention include the following which illustrate the vigilance which may profitably be exercised in the analysis of environmental samples.

1. Tris(4-chlorophenyl) methanol is widely distributed among marine mammals and bird eggs from widely separated geographicl localities (Jarman et al. 1992). Its occurrence as a by-product in the manufacture of DDT has already been noted in Section 3.5.3, and the results suggest a substantial input, and its widespread distribution and persistence.

2. Substantial attention has been directed (Section 2.5) to the occurrence of sulfonated sulfur-containing metabolites including sulfones and sulfonates. These include both methylsulfonyl derivatives of PCBs and of 2,2′-bis(4-chlorophenyl)-1,1-dichloroethene (DDE) (Janák et al. 1998). These are noted further in Section 7.5.2 in the context of the metabolism of PCBs by seals and polar bears.

3. A number of brominated compounds including hexabromocyclododecane and a series of polybrominated diphenyl ethers have been found in samples of sediment and fish from a Swedish river (Sellström et al. 1998).

4. Chlorinated and methylated 4,4′-diphenoquinones have been found in a number of samples that have been subjected to pyrolysis, and occur concurrently with chlorinated dioxins and related compounds (Otto et al. 1998). It should be noted that both the quinones and the corresponding hydroquinones occur simultaneously.

5. In addition to PCBs and polychlorinated dioxins and furans, there are other groups of widely distributed polychlorinated aromatic compounds:

    a. Chlorinated PAHs have been found in a range of environmental samples, and studies have been directed to the reactions under which these are synthesized (Nilsson and Colmsjö 1990).

    b. Tetra- to decachlorinated diphenyl ethers of unestablished origin have been found in fish at the higher trophic levels in the Great Lakes (Niimi et al. 1994).

    c. Tetra- to heptachloronaphthalenes have been found in samples of fish and sediment in Sweden (Järnberg et al. 1993).

Commercial mixtures of compounds or effluents containing large numbers of compounds present particularly difficult choices in monitoring programs. Analytical problems for the analysis of a number of such groups of compounds have been discussed in Chapter 2, Section 2.5.

1. If a commercial product contains several components, it is clearly desirable to provide analytical data for each of them. For example, in the case of PCBs, most current studies provide data for specific congeners, and since all 209 PCB congeners have been synthesized and gas-chromatographic analytical methods for their analysis have been developed (Mullin et al. 1984), there seems little justification for not incorporating such data into all investigations. The value of doing so is clearly motivated by several considerations including the following.

   • The difference in volatility of the various congeners determines their redistribution among environmental compartments after deposition.

   • Some of the PCB congeners are more persistent than others, so that changes take place in the composition of the mixture after discharge. This may be due to microbial reactions such as anaerobic dechlorination which is discussed more fully in Chapter 6, Section 6.6, or the result of metabolism by biota: an illustration in the cod–seal–polar bear food chain has been cited above.

   • The toxicity of the congeners varies considerably so that attention has been directed particularly to the planar congeners that do not contain two or more chlorine atoms in positions adjacent to the phenyl–phenyl ring junction, and which are significantly more toxic (De Voogt et al. 1990).

2. On the other hand, the complexity of some industrial effluents may make it difficult to propose a restricted number of specific compounds for analysis. An unusually difficult choice is presented by effluents from the production of bleached pulp with conventional technologies using molecular chlorine. Investigations on the distribution of organochlorine compounds in samples of sediment and biota from the Gulf of Bothnia and the Baltic Sea have made extensive use of measurements of the concentration of cyclohexane-extractable organically bound chlorine (Martinsen et al. 1988). The interpretative limitations of this procedure are, however, at least twofold. First, the origin of the source remains undefined in the absence of data for specific compounds related to the supposed industrial operation, and, second, biological effects cannot validly be correlated with such measurements since there may exist other compounds inducing the observed effects. Analytical studies of extracts from sediments have substantiated the limitations of such parameters, since the greater part of the sample contained unchlorinated compounds, and a substantial fraction (>85%) of the organochlorine compounds remains unidentified (Remberger et al. 1990). In this case, additional complications have emerged from the established molecular weight distribution of the organochlorine

components in the effluents (Jokela and Salkinoja-Salonen 1992) and the role of high-molecular-weight fractions in determining toxicity (Higashi et al. 1992).

3.6.4 Monitoring and Ecoepidemiology

Monitoring may also be incorporated into ecoepidemiological investigations that are discussed in greater detail in Chapter 7, Section 7.7.2. Some general comments may, however, be provided here. For example, above-background concentrations of PCBs have been associated with impaired reproduction in several species of seals. Unless, however, it is shown that the compounds in question do indeed bring about the postulated biological effect, the correlation may be fortuitous. The complexity of the possible correlation of impaired reproduction in marine mammals and exposure to organochlorine compounds (Addison 1989) suggests a wider application of the principle of Koch's postulates for determining causal pathogenicity of bacteria to animals or plants: (1) isolation of the putatively causative organism (substance), (2) demonstration of its pathogenicity (toxicity), and (3) recovery from biota infected with the organism isolated (affected by the putative toxicant).

The crux lies in fulfillment of the third criterion since, in natural ecosystems, biota are seldom exposed to only a single potential toxicant. This whole issue of association between biological effects and putative causes is discussed again in Chapter 7, Section 7.7.2.

3.7 Conclusions

It has been shown that phase partitioning involves complex interactions the details of which have not been completely resolved. A number of important conclusions may, however, be drawn from this discussion, and at least the following factors should be quantitatively evaluated particularly in interpreting the results from a monitoring program.

1. The dynamics of partitioning between phases, and the role of interstitial water in sediments and soils. Specific attention should be directed to the quantitative significance of associations between xenobiotics and both organic and inorganic components of the soil and sediment phases and the extent to which these processes are reversible.

2. The importance of alterations in the structure of the xenobiotic after release should be taken into consideration. These changes may be mediated by both abiotic and biotic reactions, and may produce

compounds with physical and chemical properties significantly different from the original xenobiotic.

3. It has been demonstrated that many organic compounds are effectively partitioned into the gas phase, and attention should be directed to this in light of its significance in the global distribution of xenobiotics.

References

Achman, D.R., K.C. Hornbuckle, and S.J. Eisenreich. 1993. Volatilization of polychlorinated biphenyls from Green Bay, Lake Michigan. *Environ. Sci. Technol.* 27: 75–87.

Addison, R.F. 1989. Organochlorines and marine mammal reproduction. *Can. J. Fish. Aquat. Sci.* 46: 360–368.

Aga, D.S., E.M. Thurman, M.E. Yockel, L.R. Zimmerman, and T.D. Williams. 1996. Identification of a new sulfonic acid metabolite of metolachlor in soil. *Environ. Sci. Technol.* 30: 592–597.

Ahvazi, B.C., C. Crestini, and D.S. Argyropoulos. 1999. [19]F nuclear magnetic resonance spectroscopy for the quantitative detection and classification of carbonyl groups in lignin. *J. Agric. Food Chem.* 47: 190–201.

Albrecht, I.D., A.L. Barkovskii, and P. Adriaens. 1999. Production and dechlorination of 2,3,7,8-tetrachlorodibenzo-*p*-dioxin in historically contaminated estuarine sediments. *Environ. Sci. Technol.* 33: 737–744.

Alcock, R.E., A.E. Johnston, S.P. McGrath, M.L. Barrow, and K.C. Jones. 1993. Longterm changes in the polychlorinated biphenyl content of United Kingdom soils. *Environ. Sci. Technol.* 27: 1918–1923.

Alcock, R.E., M.S. Mclachlan, A.E. Johnston, and K.C. Jones. 1998. Evidence for the presence of PCDD/Fs in the environment prior to 1900 and further studies on their temporal trends. *Environ. Sci. Technol.* 32: 1580–1587.

Alcock, R.E., K.C. Jones, M.S. Mclachlan, and A.E. Johnston. 1999. Response to comment on "Evidence for the presence of PCDD/Fs in the environment prior to 1900 and further sudies on their temporal trends." *Environ. Sci. Technol.* 33: 206–207.

Andersson, Ö., C.-E. Linder, M. Olsson, L. Reutergårdh, U.B. Uvemo, and U. Wideqvist. 1988. Spatial differences and temporal trends of organochlorine compounds in biota from the northwestern hemisphere. *Arch. Environ. Contam. Toxicol.* 17: 755–765.

Ankley, G.T., M.K. Schubauer-Berigan, and J.R. Dierkes. 1991. Predicting the toxicity of bulk sediments to aquatic organisms with various test fractions: pore water vs. elutriate. *Environ.Toxicol. Chem.* 10: 1359–1366.

Argyropoulos, D.S. and L. Zhang. 1998. Semiquantitative determination of quinonoid structures in isolated lignins by ^{31}P nuclear magnetic resonance. *J. Agric. Food Chem.* 46: 4626–4634.

Arjmand, M. and H. Sandermann. 1985. Mineralization of chloroaniline/lignin conjugates and of free chloroanilines by the white rot fungus *Phanerochaete chrysosporium*. *J. Agric. Food Chem.* 33: 1055–1060.

Atkinson, R. 1990. Gas-phase tropospheric chemistry of organic compounds: a review. *Atmos. Environ.* 24A: 1–41.

Autenrieth, R.L. and J.V. DePinto. 1991. Desorption of chlorinated hydrocarbons from phytoplankton. *Environ. Toxicol. Chem.* 10: 857–872.

Baker, J.E. and S.J. Eisenreich. 1990. Concentrations and fluxes of polycyclic aromatic hydrocarbons and polychlorinated biphenyls across the air-water interface of Lake Superior. *Environ. Sci. Technol.* 24: 342–352.

Baker, J.E., S.J. Eisenreich, T.C. Johnson, and B.M. Halfman. 1985. Chlorinated hydrocarbon cycling in the benthic nepheloid layer of Lake Superior. *Environ. Sci. Technol.* 19: 854–861.

Baker, J.E., S.J. Eisenreich, and B.J. Eadie. 1991. Sediment trap fluxes and benthic recycling of organic carbon, polycyclic aromatic hydrocarbons, and polychlorobiphenyl congeners in Lake Superior. *Environ. Sci. Technol.* 25: 500–509.

Baker, J.I. and R.A. Hites. 1999. Comment on "Evidence for the presence of PCDD/Fs in the environment prior to 1900 and further studies on their temporal trends." *Environ. Sci. Technol.* 33: 205.

Barrie, L.A., D. Gregor, B. Hargrave, R. Lake, D. Muir, R. Shearer, B. Tracey, and T. Bidelman. 1992. Arctic contaminants: sources, occurrence and pathways. *Sci. Total Environ.* 122: 1–74.

Barron, M.G., S.M. Plakas, P.C. Wilga, and T. Ball. 1993. Absorption, tissue distribution and metabolism of chloropyrifos in channel catfish following waterborne exposure. *Environ. Toxicol. Chem.* 12: 1469–1476.

Benz, M., B. Schink, and A. Brune. 1998. Humic acid reduction by *Propionibacterium freudenreichii* and other fermenting bacteria. *Appl. Environ. Microbiol.* 64: 4507–4512.

Bergen, B.J., W.G. Nelson, and R.J. Pruell. 1993. Partitioning of polychlorinated biphenyl congeners in the seawater of New Bedford Harbor, Massachusetts. *Environ. Sci. Technol.* 27: 938–942.

Bidleman, T.F. and L. Renberg. 1985. Determination of vapour pressures for chloroguaiacols, chloroveratroles and nonylphenol by gas chromatography. *Chemosphere* 14: 1475–1481.

Bidleman, T.F., G.W. Patton, M.D. Walla, B.T. Hargrave, W.P. Vass, P. Erickson, B. Fowler, V. Scott, and D.J. Gregor. 1989. Toxaphene and other organochlorines in Arctic Ocean fauna: evidence for atmospheric delivery. *Arctic* 42: 307–313.

Bierman, V.J. 1990. Equilibrium partitioning and biomagnification of organic chemicals in benthic animals. *Environ. Sci. Technol.* 24: 1407–1412.

Bignert, A., A. Göthberg, S. Jensen, K. Litzén, T. Odsjö, M. Olsson, and L. Reutergard. 1993. The need for adequate biological sampling in ecotoxicological investigations: a retrospective study of twenty years pollution monitoring. *Sci. Tot. Environ.* 128: 121.139.

Black, M.C. and J.F. McCarthy. 1988. Dissolved organic macromolecules reduce the uptake of hydrophobic organic contaminants by the gills of rainbow trout (*Salmo gairdneri*). *Environ. Toxicol. Chem.* 7: 593–600.

Boese, B.L., H. Lee, D.T. Specht, and R.C. Randall. 1990. Comparison of aqueous and solid-phase uptake for hexachlorobenzene in the tellinid clam *Macoma nasuta* (Conrad): a mass balance approach. *Environ. Toxicol. Chem.* 9: 221–231.

Boese, B.L., M. Winsor, H. Lee II, S. Evchols, J. Pelletier, and R. Randall. 1995. PCB congeners and hexachlorobenzene biota sediment accumulation factors for *Macoma nasuta* exposed to sediments with different total organic carbon content. *Environ. Toxicol. Chem.* 14: 303–310.

Bollag, J.-M. and S.-Y. Liu. 1985. Copolymerization of halogenated phenols and syringic acid. *Pestic. Biochem. Physiol.* 23: 261–272.

Bollag, J.-M., R.D. Minard, and S.-Y. Liu. 1983. Cross-linkage between anilines and phenolic humus constituents. *Environ. Sci. Technol.* 17: 72–80.

Brannon, J.M., T.E. Myers, D. Gunnison, and C.B. Price. 1991. Nonconstant polychlorinated biphenyl partitioning in New Bedford Harbor sediment during sequential batch leaching. *Environ. Sci. Technol.* 25: 1082–1087.

Braune, B.M. and R.J. Norstrom. 1989. Dynamics of organochlorine compounds in herring gulls: III. Tissue distribution and bioaccumulation in Lake Ontario gulls. *Environ. Toxicol. Chem.* 8: 957–968.

Broman, D., C. Näf, I. Lundbergh, and Y. Zebühr. 1990. An *in situ* study on the distribution, biotransformation and flux of polycyclic aromatic hydrocarbons (PAHs) in an aquatic food chain (seston-*Mytilus edulis* L. – *Somateria mollissima* L.) from the Baltic: an ecotoxicological perspective. *Environ. Toxicol. Chem.* 9: 429–442.

Broman, D., C. Näf, C. Rolff, Y. Zebühr, B. Fry, and J. Hobbie. 1992. Using ratios of stable nitrogen isotopes to estimate bioaccumulation and flux of polychlorinated dibenzo-*p*-dioxins (PCDDs) and dibenzofurams (PCDFs) in two food chains in the Northern Baltic. *Environ. Toxicol. Chem.* 11: 331–345.

Brownawell, B.J. and J.W. Farrington. 1986. Biogeochemistry of PCBs in interstitial waters of a coastal marine sediment. *Geochim. Cosmochim. Acta* 50: 157–169.

Bruggeman, W.A., J. Van der Stenen, and O. Hutzinger. 1982. Reversed-phase thin-layer chromatography of polynuclear aromatic hydrocarbons and chlorinated biphenyls. Relationship with hydrophobicity as measured by aqueous solubility and octanol–water partition coefficient. *J. Chromatogr.* 238: 335–346.

Bruggeman, W.A., A. Opperhuizen, A. Wijbenga, and O. Hutzinger. 1984. Bioaccumulation of super-lipophilic chemicals in fish. *Toxicol. Environ. Chem.* 7: 173–189.

Brumley, C., V.S. Haritos, J.T. Ahokas, and D.A. Holdway. 1996. Metabolites of chlorinated syringols in fish bile as biomarkers of exposure to bleached eucalypt pulp effluents. *Ecotoxicol. Environ. Saf.* 33: 253–260.

Burgos, W.D., J.T. Novak, and D.F. Berry. 1996. Reversible sorption and irreversible binding of naphthalene and α-naphthol to soil: elucidation of processes. *Environ. Sci. Technol.* 30: 1205–1211.

Burken, J.G. and J.L. Schnoor. 1997. Uptake and metabolism of atrazine by poplar trees. *Environ. Sci. Technol.* 31: 1399–1406.

Buser, H.-R. 1995. DDT, a potential source of environmental *tris*(4-chlorophenyl)methane and *tris*(4-chlorophenyl)methanol. *Environ. Sci. Technol.* 29: 2133–2139.

Bushart, S.P., B. Bush, E.L. Barnard, and A. Bott. 1998. Volatilization of extensively dechlorinated polychlorinated biphenyls from historically contaminated sediments. *Environ. Toxicol. Chem.* 17: 1927–1933.

Calamari, D., P. Tremolada, A. di Guardo, and M. Vighi. 1994. Chlorinated hydrocarbons in pine needles in Europe: fingerprint for the past and recent use. *Environ. Sci. Technol.* 287: 429–434.

Calderbank, A. 1989. The occurrence and significance of bound pesticide residues in soil. *Rev. Environ. Contam. Toxicol.* 108: 1–1103.

Cameron, M. and I.M. Weis. 1993. Organochlorine contaminants in the country food diet of the Belcher Island Inuit, Northwest Territories, Canada. *Arctic* 46: 42–48.

Capriel, P., A. Haisch, and S.U. Khan.1985. Distribution and nature of bound (non-extractable) residues of atrazine in a mineral soil 9 years after the herbicide application. *J. Agric. Food Chem.* 33: 567–569.

Carlberg, G.E., K. Martinsen, A. Kringstad, E. Gjessing, M. Grande, T. Källqvist, and J.U. Skåre. 1986. Influence of aquatic humus on the bioavailability of chlorinated micropollutants in Atlantic salmon. *Arch. Environ. Contam. Toxicol.* 15: 543–548.

Carmichael, L.M., R.F. Christman, and F.K. Pfaender. 1997. Desorption and mineralization kinetics of phenanthrene and chysene in contaminated soils. *Environ. Sci. Technol.* 31: 126–132.

Carr, R.S., J.W. Williams, and C.T.B. Fragata. 1989. Development and evaluation of a novel marine sediment pore water toxicity test with the polychaete *Dinophilus gyrociliatus*. *Environ. Toxicol. Chem.* 8: 533–543.

Carson, M.A., J.N. Jasper, and F.M. Conly. 1998. Magnitude and sources of sediment input to the Mackenzie delta, Northwest Territories, 1974–94. *Arctic* 51: 116–124.

Carter, C.W. and I.H. Suffet. 1982. Binding of DDT to dissolved humic materials. *Environ. Sci. Technol.* 16: 735–740.

Carter, D.S. and R.A. Hites. 1992. Fate and transport of Detroit River derived pollutants throughout Lake Erie. *Environ. Sci. Technol.* 26: 1333–1341.

Celis, R. and W.C. Koskinen. 1999. An isotope exchange method for the characterization of the irreversibility of pesticide sorption-desorption in soil. *J. Agric. Food Chem.* 47: 782–790.

Chessells, M., D.W. Hawker, and D.W. Connel. 1992. Influence of solubility in lipid on bioconcentration of hydrophobic compounds. *Ecotoxicol. Environ. Saf.* 23: 260–273.

Chin, Y.-P. and P.M. Gschwend. 1992. Partitioning of polycyclic aromatic hydrocarbons to marine porewater organic colloids. *Environ. Sci. Technol.* 26: 1621–1626.

Chin, Y.-P., G.R. Aitken, and K.M. Danielsen. 1997. Binding of pyrene to aquatic and commercial humic substances: the role of molecular weight and aromaticity. *Environ. Sci. Technol.* 31: 1630–1635.

Chiou, C.T. 1985. Partition coefficients of organic compounds in lipid-water systems and correlations with fish bioconcentration factors. *Environ. Sci. Technol.* 19: 57–62.

Chung, N. and M. Alexander. 1998. Differences in sequestration and bioavailability of organic compounds aged in different soils. *Environ. Sci. Technol.* 32: 855–860.

Clark, K.E., F.A.P.C. Gobas, and D. Mackay. 1990. Model of organic chemical uptake and clearance by fish from food and water. *Environ. Sci. Technol.* 24: 1203–1213.

Clark, T.P., R.J. Norstrom, G.A. Fox, and H.T. Won. 1987. Dynamics of organochlorine compounds in herring gulls (*Larus argentatus*): II. A two-compartment model and data for ten compounds. *Environ. Toxicol. Chem.* 6: 547–559.

Connolly, J.P. 1991. Application of a food chain model to polychlorinated biphenyl contamination of the lobster and winter flounder food chains in New Bedford Harbor. *Environ. Sci. Technol.* 25: 760–770.

Connolly, J.P. and C.J. Pedersen. 1988. A thermodynamic-based evaluation of organic chemical accumulation in aquatic organisms. *Environ. Sci. Technol.* 22: 99–103.

Cook, R.L. and C.H. Langford. 1998. Structural characterization of a fulvic acid and a humic acid using solid-state Ramp-CP-MAS ^{13}C nuclear magnetic resonance. *Environ. Sci. Technol.* 32: 719–725.

Cornelissen, G., P.C.M. van Noort, J.R. Parsons, and H.A.J. Govers. 1997. Temperature dependence of slow adsorption and desorption kinetcs of organic compounds in sediments. *Environ. Sci. Technol.* 31: 454–460.

Cornelissen, G., P.C.M. van Noort, and H.A.J. Govers. 1998. Mechanism of slow desorption of organic compounds from sediments: a study using model sorbents. *Environ. Sci. Technol.* 32: 3124–3131.

Czuczwa, J.M. and R.A. Hites. 1986. Airborne dioxins and dibenzofurans: sources and fates. *Environ. Sci. Technol.* 20: 195–200.

Daughney, C.J., and J.B. Fein. 1998. Sorption of 2,4,6-trichlorophenol by *Bacillus subtilis*. *Environ. Sci. Technol.* 32: 749–752.

de Boer, J., P.G. Wester, E.H.G. Evers, and U.A.T. Brinkman. 1996. Determination of *tris*4-chlorophenylmethanol and *tris*4-chlorophenylmethane in fish, marine mammals and sediment. *Environ. Pollut.* 93: 39–47.

De Bruijn, J., and J. Hermens. 1991. Uptake and elimination kinetics of organophosphorous [sic] pesticides in the guppy (*Poecilia reticulata*): correlations with the octanol/water partition coefficient. *Environ. Toxicol. Chem.* 10: 791–804.

De Bruijn, J., F. Busser, W. Seinen, and J. Hermens. 1989. Determination of octanol / water partition coefficients for hydrophobic organic chemicals with the slow-stirring method. *Environ. Toxicol. Chem.* 8: 499–512.

De Voe, H., M.M. Miller, and S.P. Wasik. 1981. Generator column and high pressure liquid chromatography for determining aqueous solubilities and octanol–water partition coefficients of hydrophobic substances. *J. Res. Natl. Bur. Stand. (U.S.)* 86: 361–366.

De Voogt, P., D.E. Wells, L. Reutergård, and U.A.T. Brinkman. 1990. Biological activity, determination and occurrence of planar, mono- and di-ortho PCBs. *Int. J. Environ. Anal. Chem.* 40: 1–46.

De Wolf, W., J.H.M. de Bruijn, W. Seinen, and J.L.M. Hermens. 1992a. Influence of biotransformation on the relationship between bioconcentration factors and octanol–water partition coefficients. *Environ. Sci. Technol.* 26: 1197–1201.

De Wolf, W., W. Seinen, A. Opperhuizen, and J.L.M. Hermens. 1992b. Bioconcentration and lethal body burden of 2,3,4,5-tetrachloroaniline in guppy, *Poecilia reticulata*. *Chemosphere* 25: 853–863.

De Wolf, W., W. Seinen, and J.L.M. Hermens. 1993. Biotransformation and toxicokinetics of trichloroanilines in fish in relation to their hydrophobicity. *Arch. Environ. Contam. Toxicol.* 25: 110–117.

Dean, J.R., W.R. Tomlinson, V. Makovskaya, R. Cumming, M. Hetheridge, and M. Comber. 1996. Solid-phase microextraction as a method for estimating the octanol–water partition coefficient. *Anal. Chem.* 68: 130–133.

Dec, J., K. Haider, A. Benesi, V. Rangaswamy, A. Schäffer, E. Fernandes, and J.-M. Bollag. 1997a. Formation of soil-bound residues of cyprodinil and their plant uptake. *J. Agric. Food Chem.* 45: 514–520.

Dec, J., K. Haider, A. Benesi, V. Rangaswamy, A. Schäffer, U. Plücken, and J.-M. Bollag. 1997b. Analysis of soil-bound residues of ^{13}C-labelled fungicide cyprodinil by NMR spectroscopy. *Environ. Sci. Technol.* 31: 1128–1135.

DeLaune, R.D., R.P. Gambrell, and K.S. Reddy. 1983. Fate of pentachlorophenol in estuarine sediment. *Environ. Pollut. Ser. B.* 6: 297–308.

Driscoll, S.K. and A.E. McElroy. 1996. Bioaccumulation and metabolism of benzo[a]pyrene in three species of polychaete worms. *Environ. Toxicol. Chem.* 15: 1401–1410.

Droppo, I.G., D. Jeffries, C. Jaskot, and S. Backus. 1998. The prevalence of freshwater flocculation in cold regions: a case study from the Mackenzie River delta, Northwest Territories, Canada. *Arctic* 51: 155–164.

Eadie, B.J., N.R. Morehead, and P.F. Landrum. 1990. Three-phase partitioning of hydrophobic organic compounds in Great Lakes waters. *Chemosphere* 20: 161–178.

Egeler, P., J. Römbke, M. Meller, Th. Knacker, C. Franke, G. Studinger, and R. Nagel. 1997. Bioaccumulation of lindane and hexachlorobenzene by tubificid sludgeworms (*Oligochaeta*) under standardized laboratory conditions. *Chemosphere* 35: 835–852.

Eisenreich, S.J., P.D. Capel, J.A. Robbins, and R. Bourbonniere. 1989. Accumulation and diagenesis of chlorinated hydrocarbons in lake sediments. *Environ. Sci. Technol.* 23: 1116–1126.

Ernst, W. 1979. Factors affecting the evaluation of chemicals in laboratory experiments using marine organisms. *Ecotoxicol. Environ. Saf.* 3: 90.98.

Escher, B.I. and R.P. Schwartzenbach. 1996. Partition of substituted phenols in liposome–water, biomembrane–water, and octanol–water systems. *Environ. Sci. Technol.* 30: 260–270.

Etzweiler, F., E. Senn, and H.W.H. Schmidt. 1995. Method for measuring aqueous solubilities of organic compounds. *Anal. Chem.* 67: 655–658.

Ewald, G., P. Larsson, H. Linge, L. Okla, and N. Szarzi. 1998. Biotransport of organic pollutants to an inland Alaska lake by migrating sockeye salmon (*Oncorhynchus nerka*). *Arctic* 51: 40–47.

Fairchild, W.L., D.C.G. Muir, R.S. Curri, and A.L. Yarechewski. 1992. Emerging insects as a biotic pathway for movement of 2,3,7,8-tetrachlorodibenzofuran from lake sediments. *Environ.Toxicol. Chem.* 11: 867–872.

Falandysz, J., B. Strandberg, L. Strandberg, and C. Rappe. 1999. Tris(4-chlorophenyl)methane and tris(4-chlorophenyl)methanol on sediment and food webs from the Baltic south coast. *Environ. Sci. Technol.* 33: 517–521.

Fendinger, N.J. and D.E. Glotfelty. 1988. A laboratory method for the experimental determination of air–water Henry's law constants for several pesticides. *Environ. Sci. Technol.* 22: 1289–1293.

Fendinger, N.J. and D.E. Glotfelty. 1990. Henry's law constants for selected pesticides, PAHs and PCBs. *Environ. Toxicol. Chem.* 9: 731–735.

Fernandez, P. and J.M. Bayona. 1992. Use of off-line gel permeation chromatography-normal–phase liquid chromatography for the determination of polycyclic aromatic compounds in environmental samples and standard reference materials (air particulate matter and marine sediment). *J. Chromatogr.* 625: 141–149.

Field, J.A. and E.M. Thurman. 1996. Glutathione conjugation and contaminant transformation. *Environ. Sci. Technol.* 30: 1413–1418.

Finlayson-Pitts, B. and J. Pitts, Jr. 1986. *Atmospheric Chemistry.* John Wiley & Sons, New York.

Flanagan, W.P. and R.J. May. 1993. Metabolite detection as evidence for naturally occurring aerobic PCB degradation in Hudson River sediments. *Environ. Sci. Technol.* 27: 2207–2212.

Flenner, C.K., J.R. Parsons, S.M. Schrap, and A. Opperhuizen. 1991. Influence of suspended sediment on the biodegradation of alkyl esters of *p*-aminobenzoic acid. *Bull. Environ. Contam. Toxicol.* 47: 555– 560.

Focardi, S., C. Gaggi, G. Chemello, and E. Bacci. 1991. Organochlorine residues in moss and lichen samples from two Antarctic areas. *Polar Rec.* 162: 241–244.

Fowler, S.W., G.G. Polikarpov, D.L. Elder, P. Parsi, and J.-P. Villeneuve.1978. Polychlorinated biphenyls: accumulation from contaminated sediments and water by the polychaete *Nereis diversicolor*. *Mar. Biol.* 48: 303–309.

Fox, G.A., S.W. Kennedy, R.J. Norstrom, and DC. Wigfield. 1988. Porphyria in herring gulls: a biochemical response to chemical contamination of Great Lakes food chains. *Environ. Toxicol. Chem.* 7: 831–839.

Franke, C. 1996. How meaningful is the bioconcentration factor for risk assessment? *Chemosphere* 32: 1897–1905.

Fu, G., A.T. Kan, and M. Tomson. 1994. Adsorption and desorption hysteresis of PAHs in surface sediment. *Environ. Toxicol. Chem.* 13: 1559–1567.

Garbarini, D.R. and L.W. Lion. 1986. Influence of the nature of soil organics on the sorption of toluene and trichloroethylene. *Environ. Sci. Technol.* 20: 1263–1269.

Garbow, J.R. and B.J. Gaede. 1992. Analysis of a phenyl ether herbicide-cyclodextrin inclusion complex by CPMAS ^{13}C NMR. *J. Agric. Food Chem.* 40: 156–159.

Garcia, K.L., J.J. Delfino, and D.H. Powell. 1993. Non-regulated organic compounds in Florida sediments. *Water Res.* 27: 1601–1613.

Gauthier, T.D., W.R. Selz, and C.V.L. Grant. 1987. Effects of structural and compositional variations of dissolved humic materials on pyrene K_{oc} values. *Environ. Sci. Technol.* 21: 243–248.

Gluck, S.J., M.H. Benko, R.K. Hallberg, and K.P. Steele. 1996. Indirect determination of octanol–water partition coefficients by microemulsion electrokinetic chromatography. *J. Chromatogr.* A 744: 141–146.

Gobas, F.A.P.C. and X. Zhang. 1992. Measuring bioconcentration factors and rate constants of chemicals in aquatic systems under conditions of variable water concentrations and short exposure time. *Chemosphere* 25: 1961–1971.

Gobas, F.A.P.C., K.E. Clark, W.Y. Shiu, and D. Mackay. 1989. Bioconcentration of polybrominated benzenes and biphenyls and related superhydrophobic chemicals in fish: role of bioavailability and elimination into the feces. *Environ. Toxicol. Chem.* 8: 231–245.

Gobas, F.A.P.C., E.J. McNeil, L. Lovett-Doust, and G.D. Haffner. 1991. Bioconcentration of chlorinated aromatic hydrocarbons in aquatic macrophytes. *Environ. Sci. Technol.* 25: 924–929.

Gobas, F.A.P.C., J.R. McCorquodale, and G.D. Haffner. 1993. Intestinal absorption and biomagnification of organochlorines. *Environ. Toxicol. Chem.* 12: 567–576.

Gobas, F.A.P.C., J.B. Wilcockson, R.W. Russell, and G.D. Haffner. 1999. Mechanism of biomagnification in fish under laboratory and field conditions. *Environ. Sci. Technol.* 33. 133–141.

Goldberg, E.D. 1986. The mussel watch concept. *Environ. Monit. Assess.* 7: 91–103.

Gregor, D.J. and W.D. Gummer. 1989. Evidence of atmospheric transport and deposition of organochlorine pesticides and polychlorinated biphenyls in Canadian Arctic snow. *Environ. Sci. Technol.* 23: 561–566.

Guthrie, E.A. and F.K. Pfaender. 1998. Reduced pyrene bioavailability in microbially active soils. *Environ. Sci. Technol.* 32: 501–508.

Haderlein, S.B. and R.P. Schwarzenbach. 1993. Adsorption of substituted nitrobenzenes and nitrophenols to mineral surfaces. *Environ. Sci. Technol.* 27: 316–326.

Hagan, M. 1962. Clathrate Inclusion Compounds. Reinhold, New York.

Hand, V.C. and G.K. Williams. 1987. Structure–activity relationships for sorption of linear alkylbenzenesulfonates. *Environ. Sci. Technol.* 21: 370–373.

Hargrave, B.T., W.P. Vass, P.E. Erickson, and B.R. Fowler. 1988. Atmospheric transport of organochlorines to the Arctic Ocean. *Tellus* 40B: 480–493.

Harkness, M.R., J.B. McDermott, D.A. Abramowicz, J.J. Salvo, W.P. Flanagan, M.L. Stephens, F.J. Mondello, R.J. May, J.H. Lobos, K.M. Carroll, M.J. Brennan, A.A. Bracco, K.M. Fish, G.L. Warner, P.R. Wilson, D.K. Dietrich, D.T. Lin, C.B. Morgan, and W.L.Gately. 1993. *In situ* stimulation of aerobic PCB biodegradation in Hudson River sediments. *Science* 259: 503–507.

Harner, T. and D. Mackay. 1995. Measurement of octanol–air partition coefficients for chlorobenzenes, PCBs, and DDT. *Environ. Sci. Technol.* 29: 1599–1606.

Hashimoto, S., T. Wakimoto, and R. Tatsukawa. 1990. PCDDs in the sediments accumulated about 8120 years ago from Japanese coastal areas. *Chemosphere* 21: 825–835.

Hatcher, P.G., J.M. Bortiatynski, R.D. Minard, J. Dec, and J.-M. Bollag. 1993. Use of high-resolution ^{13}C NMR to examine the enzymatic covalent binding of ^{13}C-labeled 2,4-dichlorophenol to humic substances. *Environ. Sci. Technol.* 27: 2098–2103.

Hawker, D.W. and D.W. Connell. 1986. Bioconcentration of lipophilic compounds by some aquatic organisms. *Ecotoxicol. Environ. Saf.* 11: 184–197.

Hayes, M.H.B., M.E. Pick, and B.A. Thoms. 1978a. The influence of organocation structure on the adsorption of mono- and of bipyridinium cations by expanding lattice clay minerals. I. Adsorption by Na^+-montmorillonite. *J. Colloid Interface Sci.* 65: 254–265.

Hayes, M.H.B., M.E. Pick, and B.A. Thoms. 1978b. The influence of organocation structure on the adsorption of mono- and of bipyridinium cations by expanding lattice clay minerals. II. Adsorption by Na^+-vermiculite. *J. Colloid Interface Sci.* 65: 266–275.

Hebert, C.E., R.J. Norstrom, M. Simon, B.M. Braune, D.V. Weseloh, and C.R. Macdonald. 1994. Temporal trends and sources of PCDDs and PCDFs in the Great Lakes: herring gull egg monitoring, 1981–1991. *Environ. Sci. Technol.* 28: 1268–1277.

Heitkamp, M.A., J.N. Huckins, J. D. Pettie, and J.L. 1984. Fate and metabolism of isopropylphenyl diphenyl phosphate in freshwater sediments. *Environ. Sci. Technol.* 18: 434–439.

Hickey, C.W., D.S. Roper, P.T. Holland, and T.M. Trower.1995. Accumulation of organic contaminants in two sediment-dwelling feeding modes: deposit *Macoma liliana* and filter-feeding *Austroventus stuchburyi*. *Arch. Environ. Contam. Toxicol.* 29: 221–231.

Higashi, R.M., G.N. Cherr, J.M. Shenker, J.M. Macdonald, and D.G. Crosby. 1992. A polar high molecular mass constituent of bleached kraft mill effluent is toxic to marine organisms. *Environ. Sci. Technol.* 26: 2413–2420.

Hill, W.E., J. Szechi, C. Hofstee, and J.H. Dane. 1997. Fate of a highly strained hydrocarbon in aqueous soil environment. *Environ. Sci. Technol.* 31: 651–655.

Hinman, M.L. and S.J. Klaine. 1992. Uptake and translocation of selected organic pesticides by the rooted aquatic plant *Hydrilla verticillata* Royle. *Environ. Sci. Technol.* 26: 609–613.

Hippelein, M. and M.S. McLachlan. 1998. Soil/air partitioning of semivolatile organic compounds. 1. Method development and influence of physical-chemical properties. *Environ. Sci. Technol.* 32: 310–316.

Höhl, H.-U. and W. Barz. 1995. Metabolism of the insecticide phoxim in plants and cell suspension cultures of soybean. *J. Agric. Food Chem.* 43: 1052–1056.

Hornbuckle, K.C., C.W. Sweet, R.F. Pearson, D.L. Swackhamer, and S.J. Eisenreich. 1995. Assessing annual water–air fluxes of polychlorinated biphenyls in Lake Michigan. *Environ. Sci. Technol.* 29: 869–877.

Hülster, A., J.F. Müller, and H. Marschner. 1994. Soil-plant transfer of polychlorinated dibenzo-*p*-dioxins and dibenzofurans to vegetables of the cucumber family (Cucurbitaceae). *Environ. Sci. Technol.*28: 1110–1115.

Ingebrigtsen, K., H. Hektoen, T. Andersson, E.K. Wehler, Å. Bergman, and I. Brandt. 1992. Enrichment of metabolites in the cerebrospinal fluid of cod (*Gadus morhua*) following oral administration of hexachlorobenzene and 2,4',5-trichlorobiphenyl. *Pharmacol. Toxicol.* 71: 420–425.

Isaacson, P.J. and C.R. Frink. 1984. Nonreversible sorption of phenolic compounds by sediment fractions: the role of sediment organic matter. *Environ. Sci. Technol.* 18: 43–48.

Janák, K., G. Becker, A. Colmsjö, C. Östman, M. Athanasiadou, K. Valters, and Å. Bergman. 1998. Methyl sulfonyl polychlorinated biphenyls and 2,2-bis(4-chlorophenyl)-1,1-dichloroethene in gray seal tissues determined by gas chromatography with electron capture detection and atomic emission detection. *Environ. Toxicol. Chem.* 17: 1046–1055.

Jarman, W.M., M. Simon, R.J. Norstrom, S.A. Burns, C.A. Bacon, B.R.T. Simoneit, and R.W. Risebrough. 1992. Global distribution of *tris*(4-chlorophenyl)methanol in high trophic level birds and mammals. *Environ. Sci. Technol.* 26: 1770–1774.

Jarman, W.M., S.A. Burns, W.G. Mattox, and W.S. Seegar. 1994. Organochlorine compounds in the plasma of peregrine falcons and gyrfalcons nesting in Greenland. *Arctic* 47: 334–340.

Järnberg, U., L. Asplund, C. de Witt, A.-K. Grafström, P. Haglund, B. Jansson, K. Lexén, M. Strandell, M. Olsson, and B. Jonsson. 1993. Polychlorinated biophenyls and polychlorinated naphthalene in Swedish sediment and biota: levels, patterns, and time trends. *Environ. Sci. Technol.* 27: 1364–1374.

Jeans, J. 1952. *An Introduction to the Kinetic Theory of Gases*, pp. 63–68. Cambridge University Press, Cambridge.

Jensen, S., G. Eriksson, H. Kylin, and W.M.J. Strachan. 1992. Atmospheric pollution by persistent organic compounds: monitoring with pine needles. *Chemosphere* 24: 229–245.

Jokela, J.K. and M. Salkinoja-Salonen. 1992. Molecular weight distributions of organic halogens in bleached kraft pulp mill effluents. *Environ. Sci. Technol.* 26: 1190–1197.

Jones, K.C., J.A. Stratford, K.S. Waterhouse, E.T. Furlong, W. Giger, R.A. Hites, C. Schaffner, and A.E. Johnston. 1989. Increases in the polynuclear aromatic hydrocarbon content of an agricultural soil over the last century. *Environ. Sci. Technol.* 23: 95–101.

Kan, A.T., G. Fu, M.A. Hunt, and M.B. Tomson. 1997. Irreversible adsorption of naphthalene and tetrachlorobiphenyl to Lul and surrogate sediments. *Environ. Sci. Technol.* 31: 2176–2185.

Kan, A.T., G. Fu, M. Hunter, and M.B. Tomson. 1998. Irreversible sorption of neutral hydrocarbons to sediments: experimental observations and model predictions. *Environ. Sci. Technol.* 32: 892–902.

Karickhoff, S.W. 1981. Semi-empirical estimation of sorption of hydrophobic pollutants on natural sediments and soils. *Chemosphere* 10: 833–846.

Karickhoff, S.W. 1984. Organic pollutant sorption in aquatic systems. *J. Hydraul. Eng.* 100: 707–735.

Karickhoff, S.W. and K.R. Morris. 1985. Impact of tubificid oligochaetes on pollutant transport in bottom sediments. *Environ. Sci. Technol.* 19: 51–56.

Kelsey, J.W. and M. Alexander. 1997. Declining bioavailability and inappropriate estimation of risk of persistent compounds. *Environ. Toxicol. Chem.* 16: 582–585.

Kelsey, J.W., B.D. Kottler, and M. Alexander. 1997. Selective chemical extractants to predict bioavailability of soil-aged organic chemicals. *Environ. Sci. Technol.* 31: 214–217.

Kidd, K.A., D.W. Schindler, R.H. Hesslein, and D.C.G. Muir. 1998. Effect of trophic position and lipid on organochlorine concentrations in fishes from subarctic lakes in Yukon Territory. *Can. J. Fish. Aquat. Sci.* 55: 869–881.

Kim, J.-E., E. Fernandes, and J.-M. Bollag. 1997. Enzymatic coupling of the herbicide bentazon with humus monomers and characterization of reaction products. *Environ. Sci. Technol.* 31: 2392–2398.

Kinloch, D., H.V. Kuhnlein, and D. Muir. 1992. Inuit foods and diet. A preliminary assessment of benefits and risks. *Sci. Total Environ.* 122: 245–276.

Kjeller, L.-O., K.C. Jones, A.E. Jonnston, and C. Rappe. 1991. Increases in the poly-chloroinated dibenzo-*p*-dioxin and -furan content of soils and vegetation since the 1840s. *Environ. Sci. Technol.* 25: 1619–1627.

Knezovich, J.P. and F.L. Harrison. 1988. The bioavailability of sediment-sorbed chlo-robenzenes to larvae of the midge, *Chironomus decorus. Ecotoxicol. Environ. Saf.* 15: 226–241.

Knezovich, J.P., F.L. Harrison, and R.G. Wilhelm. 1987. The bioavailability of sedi-ment-sorbed organic chemicals: a review. *Water Air Soil Pollut.* 32: 233–245.

Knox, G.A. 1970. Antarctic marine ecosystems, pp. 69–96. In *Antarctic Ecology* (Ed. M.W. Holdgate). Vol. 1. Academic Press, London.

Koester, C.J. and R.A. Hites. 1992. Wet and dry deposition of chlorinated dioxins and furans. *Environ. Sci. Technol.* 26: 1375–1382.

Kolset, K. and A. Heiberg. 1988. Evaluation of the "Fugacity" (FEQUM) and the "EXAMS" chemical fate and transport models: a case study on the pollution of the Norrsundet Bay (Sweden). *Water Sci. Technol.* 20 (2): 1–12.

Kömp, P. and M.S. McLachlan. 1997. Interspecies variability of the plant/air parti-tioning of polychlorinated biphenyls. *Environ. Sci. Technol.* 31: 2944–2948.

Krone, C.A., D.G. Burrows, D.W. Brown, P.S. Robisch, A.J. Friedman, and D.C. Malins. 1966. Nitrogen-containing aromatic compounds in sediments from a polluted harbor in Puget Sound. *Environ. Sci. Technol.* 20: 1144–1150.

Kucklick, J.R. and J.E. Baker. 1998. Organochlorines in Lake Superior's food web. *Environ. Sci. Technol.* 32: 1192–1198.

Kucklick, J.R., T.F. Bidleman, L.L. McConnell, M.D. Walla, and G.P. Ivanov. 1994. Organochlorines in water and biota of Lake Baikal, Siberia. *Environ. Sci. Technol.* 28: 31–37.

Kwok, E.S.C., R. Atkinson, and J. Arey. 1995. Rate constants for the gas-phase reactions of the OH radical with dichlorobiphenyls, 1-chlorodibenzo-*p*-dioxin, 1,2-dimethoxybenzene, and diphenyl ether: estimation of OH radical reaction rate constants for PCBs, PCDDs, and PCDFs. *Environ. Sci. Technol.* 29: 1591–1598.

Landrum, P.F. 1989. Bioavailability and toxicokinetics of polycyclic aromatic hydro-carbons sorbed to sediments for the amphipod *Pontoporeia hoyi. Environ. Sci. Technol.* 23: 588–595.

Landrum, P.F., S.R. Nihart, B.J. Eadie, and W.S. Gardner. 1984. Reverse-phase sepa-ration for determining pollutant binding to Aldrich humic acid and dissolved organic carbon of natural waters. *Environ. Sci. Technol.* 18: 187–192.

Landrum, P.F., M.D. Reinhold, S.R. Nihart, and B.J. Eadie. 1985. Predicting the bio-availability of organic xenobiotics to *Pontoporeia hoyi* in the presence of humic and fulvic materials and natural dissolved organic matter. *Environ. Toxicol. Chem.* 4: 459–467.

Landrum, P.F., S.R. Nihart, B.J. Eadie, and L.R. Herche. 1987. Reduction in bioavailability of organic contaminants to the amphipod *Pontoporeia hoyi* by dissolved organic matter of sediment interstitial waters. *Environ. Toxicol. Chem.* 6: 11–20.

Landrum, P.F., B.J. Eadie, and W.R. Faust. 1991. Toxicokinetics and toxicity of a mixture of sediment-associated polycyclic aromatic hydrocarbons to the amphipod *Diporeia* sp. *Environ. Toxicol. Chem.* 10: 35–46.

Lange, B.M., N. Hertkorn, and H. Sandermann. 1998. Chloroaniline/lignin conjugates as model system for nonextractable pesticide residues in crop plants. *Environ. Sci. Technol.* 32: 2113–2118.

Langner, H.W., W.P. Inskeep, H.M. Gaber, W.L. Jones, B.S. Das, and J.M. Wraith. 1998. Pore water velocity and residence time effects on the degradation of 2,4-D during transport. *Environ. Sci. Technol.* 32: 1308–1315.

Larsen, B., F. Pelusio, H. Skejö, and A. Paya-Perez. 1992. Bioavailability of polychlorinated biphenyl congeners in the soil to earthworm (*L. rubellus*) system. *Int. J. Environ. Anal. Chem.* 46: 149–162.

Lau, Y.L., B.G. Oliver, and B.G. Krishnappan. 1989. Transport of some chlorinated contaminants by the water, suspended sediments, and bed sediments in the St. Clair and Detroit Rivers. *Environ. Toxicol. Chem.* 8: 293–301.

Leblanc, G.A. 1995. Trophic-level differences in the bioconcentration of chemicals: implications in assessing environmental biomagnification. *Environ. Sci. Technol.* 29: 154–160.

Lee, P.W. 1985. Fate of fenvalerate (pydin insecticide) in the soil environment. *J. Agric. Food Chem.* 33: 993–998.

Leenheer, J.A., R.L. Wershaw, and M.M. Reddy. 1995. Strong-acid, carboxyl-group structures in fulvic acid from the Suwannee River, Georgia. 2. Major structures. *Environ. Sci. Technol.* 29: 399–405.

Leppard, G.G., D.T. Flannigan, C.H. Marvin, D.W. Bryant, and B.E. McCarry. 1998. Binding of polycyclic aromatic hydrocarbons by size classes of particulate in Hamilton Harbor water. *Environ. Sci. Technol.* 32: 3633–3639.

Lindqvist, R. and C.G Enfield. 1992. Biosorption of dichlorodiphenyltrichloroethane and hexachlorobenzene in groundwater and its implications for facilitated transport. *Appl. Environ. Microbiol.* 58: 2211–2218.

Livingstone, D.R. and S.V. Farrar. 1984. Tissue and subcellular distribution of enzyme activities of mixed-function oxygenase and benzo[*a*]pyrene metabolism in the common mussel *Mytilis edulis* L. *Sci. Total Environ.* 39: 209–235

Lockhart, R., Wagemann, B. Tracey, D. Sutherland, and D.J. Thomas. 1992. Presence and implications of chemical contaminants in the freshwaters of the Canadian Arctic. *Sci. Total Environ.* 122: 165–243.

Lovley, D.R., J.D. Coates, E.L. Blunt-Harris, E.J.P. Philipps, and J.C. Wodward. 1996. Humic substances as electron acceptors for microbial respiration. *Nature* (London) 382: 445–448.

Lovley, D.R., J.L. Fraga, E.L. Blunt-Harris, L.A. Hayes, E.J.P. Philipps, and J.D. Coates. 1998. Humic substances as a mediator for microbially catalyzed metal reduction. *Acta Hydrochim. Hydrobiol.* 26: 152–157.

Lydy, M.J., J.T. Oris, P.C. Baumann, and S.W. Fisher. 1992. Effects of sediment organic carbon content on the elimination rates of neutral lipophilic compounds in the midge (*Chironomus riparius*). *Environ. Toxicol. Chem.* 11: 347–356.

Macalady, D.L. and N.L. Wolfe. 1985. Effects of sorption and abiotic hydrolyses. 1. Organophosphorothioate esters. *J. Agric. Food Chem.* 33: 167–173.

Macdonald, R.W., W.J. Cretney, N. Crewe, and D. Paton. A history of octachlorodibenzo-*p*-dioxin, 2,3,7,8-tetrachlorodibenzofuran, and 3,3′,4,4′-tetrachlorobiphenyl contamination in Howe Sound, British Columbia. *Environ. Sci. Technol.* 26: 1544–1550.

Mackay, D. 1982. Correlation of bioconcentration factors. *Environ. Sci. Technol.* 16: 274–278.

Mackay, D. and D. Callcott. 1998. Partitioning and physical chemical properties of PAHs, pp. 325–246. In *Handbook of Environmental Chemistry* (Ed. A.H. Neilson) Vol. 3, Part I. Springer-Verlag, Berlin.

Mackay, D. and W.Y. Shiu. 1981. Vapor pressure of organic compounds. *J. Phys. Chem. Ref. Data* 10: 220–243.

Mackay, D. and T.K. Yuen. 1980. Volatilization rates of organic contaminants from rivers. *Water Pollut. Res. J. Can.* 15: 83–98.

Mackay, D., S. Paterson, and B. Cheung. 1985. Evaluating the environmental fate of chemicals. The fugacity — Level III approach as applied to 2,3,7,8, TCDD. *Chemosphere* 14: 859–863.

Mackay, D., S. Paterson, and W.H. Schroeder. 1986. Model describing the rates of transfer processes of organic chemicals between atmosphere and water. *Environ. Sci. Technol.* 20: 810–816.

Madenjian, C.P., R.F. Elliot, L.J. Schmidt, T.J. Desorcie, R.J. Hesselberg, R.T. Quintal, L.J. Begnoche, P.M. Bouchard, and M.E. Holey. 1998. Net trophic transfer efficiency of PCBs to Lake Michigan coho salmon from their prey. *Environ. Sci. Technol.* 32: 3063–3067.

Magee, B.R., L.W. Lion, and A.T. Lemley. 1991. Transport of dissolved organic macromolecules and their effect on the transport of phenanthrene in porous media. *Environ. Sci. Technol.* 25: 323–331.

Mäkelä, T.P., T. Petänen, J. Kukkonen, and A.O. Oikari. 1991. Accumulation and depuration of chlorinated phenolics in the freshwater mussel (*Anodonta anatina* L.). *Ecotoxicol. Environ. Saf.* 22: 153–163.

Malins, D.C., B.B. McCain, D.W. Brown, S.-L. Chan, M.S. Meyers, J.T. Landahl, P.G. Prohaska, A.J. Friedman, L.D. Rhodes, D.G. Burrows, W.D. Gronlund, and H.O. Hodgins. 1984. Chemical pollutants in sediments and diseases of bottom-dwelling fish in Puget Sound. *Environ. Sci. Technol.* 18: 705–713.

Manabe, S., O. Wada, M. Morita, S. Izumikawa, K. Asakuno, and H. Suzuki. 1992. Occurrence of carcinogenic amino-α-carbolines in some environmental samples. *Environ. Pollut.* 75: 301–305.

Marple, L., B. Berridge, and L. Throop. 1986. Measurement of the water–octanol partition coefficient of 2,3,7,8-tetrachlorodibenzo-*p*-dioxin. *Environ. Sci. Technol.* 20: 397–399.

Martinsen, K., A. Kringstad, and G.E. Carlberg. 1988. Methods for determination of sum parameters and characterization of organochlorine compounds in spent bleach liquors from pulp mills and water, sediment and biological samples from receiving waters. *Water Sci. Technol.* 20(2): 13–24.

Maruya, K.A. and R.A. Lee. 1998. Biota-sediment accumulation and trophic transfer factors for extremely hydrophobic polychlorinated biphenyls. *Environ. Toxicol. Chem.* 17: 2463–2469.

McCall, P.J. and G.L. Agin. 1985. Desorption kinetics of picloram as affected by residence time in the soil. *Environ. Toxicol. Chem.* 4: 37–44.

McConnell, L.L., W.E. Cotham, and T.F. Bidleman. 1993. Gas exchange of hexachlorocyclohexane in the Great Lakes. *Environ. Sci. Technol.* 27: 1304–1311.

McElroy, A.E. 1990. Polycyclic aromatic hydrocarbon metabolism in the polychaete *Nereis virens. Aquat. Toxicol.* 18: 35–50.

McElroy, A.E. and J.C. Means. 1988. Uptake, metabolism, and depuration of PCBs by the polychaete *Nereis virens. Aquat. Toxicol.* 11: 416–417.

McGroddy, S.E. and J.W. Farrington. 1995. Sediment porewater partitioning of polycyclic aromatic hydrocarbons in three cores from Boston Harbor, Massachusetts. *Environ. Sci. Technol.* 29: 1542–1550.

McGroddy, S.E., K.W. Farrington, and P.M. Gschwend. 1996. Comparison of the *in situ* and desorption sediment–water partitioning of polycyclic aromatic hydrocarbons and polychlorinated biphenyls. *Environ. Sci. Technol.* 30: 172–177.

Meadows, J.C., K.R. Echols, J.N. Huckins, F.A. Borsuk, R.F. Carline, and D.E. Tillitt. 1998. Estimation of uptake rate constants for PCB congeners accumulated by semipermeable membrane devices and brown trout (*Salmo trutta*). *Environ. Sci. Technol.* 32: 1847–1852.

Mehrle, P.M., D.R. Buckler, E.E. Little, L.M. Smith, J.D. Petty, P.H. Peterman, D.L. Stalling, G.M. De Graeve, J.J. Coyle, and W.J. Adams. 1988. Toxicity and bioconcentration of 2,3,7,8-tetrachlorodibenzodioxin and 2,3,7,8-tetrachlorodibenzofuran in rainbow trout. *Environ. Toxicol. Chem.* 7: 47–62.

Metcalfe, J.L., M.E. Fox, and J.H. Carey. 1984. Aquatic leeches (Hirudinea) as bioindicators of organic chemical contaminants in freshwater ecosystems. *Chemosphere* 13: 143–150.

Metcalfe, J.L., M.E. Fox, and J.H. Carey. 1988. Freshwater leeches (Hirudinea) as a sceening tool for detecting organic contaminants in the environment. *Environ. Monit. Assess.* 11: 147–169.

Miller, M.M., S.P. Wasik, G.-L. Huang, W.-Y. Shiu, and D. Mackay. 1985. Relationships between octanol–water partition coefficients and aqueous solubility. *Environ. Sci. Technol.* 19: 522–529.

Mudroch, A., R.J. Allan, and S.R. Joshi. 1992. Geochemistry and organic contaminants in the sediments of Great Slave Lake, Northwest Territories, Canada. *Arctic* 45: 10–19.

Muir, D.C.G. and A.L. Yarechewski. 1988. Dietary accumulation of four chlorinated dioxin congeners by rainbow trout and fathead minnows. *Environ. Toxicol. Chem.* 7: 227–236.

Muir, D.C.G., A.L. Yarechewski, and A. Knoll. 1986. Bioconcentration and disposition of 1,3,6,8- tetrachlorodibenzo-*p*-dioxin and octachlorodibenzo-*p*-dioxin by rainbow trout and fathead minnows. *Environ. Toxicol. Chem.* 5: 261–272.

Muir, D.C.G., R.J. Norstrom, and M. Simon. 1988. Organochlorine contaminants in arctic marine food chains: accumulation of specific polychlorinated biphenyls and chlordane-related compounds. *Environ. Sci. Technol.* 22: 1071–1079.

Muir, D.C.G., C.A. Ford, N.P. Grift, and R.E.A. Stewart. 1992. Organochlorine contaminants in narwhal (*Monodon monoceros*) from the Canadian arctic. *Environ. Pollut.* 75: 307–316.

Muir, D.C.G., R. Wagemann, B.T. Hargrave, D.J. Thomas, D.B. Peakall, and R.J. Norstrom. 1992. Arctic marine ecosystem contamination. *Sci. Total Environ.* 122: 75–134.

Muir, D.C.G., M.D. Segstro, P.M. Welbourn, D. Toom, S.J. Eisenreich, C.R. Macdonald, and D.M. Whelpdale. 1993. Pattern of accumulation of airborne organochlorine contaminants in lichens from the upper Great Lakes region of Ontario. *Environ. Sci. Technol.* 27: 1201–1210.

Muir, D.C.G., A. Omelchenko, N.P. Grift, D.A. Savoie, W.L. Lockhart, P. Wilkinson, and G.J. Brunskill. 1996. Spatial trends and historical deposition of polychlorinated biphenyls in Canadian midlatitude and arctic lake sediments. *Environ. Sci. Technol.* 30: 3609–3617.

Müller, S.R., H.-R. Zweifel, D.J. Kinnison, J.A. Jacobsen, M.A. Meier, M.M. Ulrich, and R.P. Schwarzenbach. 1996. Occurrence, sources, and fate of trichloroacetic acid in Swiss waters. *Environ. Toxicol. Chem.* 15: 1470–1478.

Mullin, M., C.M. Pochini, S. McCrindle, M. Romkes, S.H. Safe, and L.M. Safe. 1984. High-resolution PCB analysis: synthesis and chromatographic properties of all 209 PCB congeners. *Environ. Sci. Technol.* 18: 468–476.

Muncaster, B.W., P.D.N. Herbert, and R. Lazar. 1990. Biological and physical factors affecting the body burden of organic contaminants in freshwater mussels. *Arch. Environ. Contam. Toxicol.* 19: 25–34.

Nakajima, D., E. Kojima, S. Iwaya, J. Suzuki, and S. Suzuki. 1996. Presence of 1-hydroxypyrene conjugates in woody plant leaves and seasonal changes in their concentration. *Environ. Sci. Technol.* 30: 1675–1679.

Nebeker, A.V., G.S. Schuytema, W.L. Griffis, J.A. Barbitta, and L.A. Carey. 1989. Effect of sediment organic carbon on survival of *Hyalella azteca* exposed to DDT and endrin. *Environ. Toxicol. Chem.* 705–718.

Neilson, A.H., H. Blanck, L. Förlin, L. Landner, P. Pärt, A. Rosemarin, and M. Söderström. 1989. Advanced hazard assessment of 4,5,6-trichloroguaiacol in the Swedish environment. In *Chemicals in the Aquatic Environment*, pp. 329–374. Ed. L. Landner, Springer-Verlag, Berlin.

Neilson, A.H., A.-S. Allard, S. Fischer, M. Malmberg, and T. Viktor. 1990. Incorporation of a subacute test with zebra fish into a hierarchical system for evaluating the effect of toxicants in the aquatic environment. *Ecotoxicol. Environ. Saf.* 20: 82–97.

Nelson, L., I. Shanahan, H.W. Sidebottom, J. Treacy, and O.J. Nielsen. 1990. Kinetics and mechanism for the oxidation of 1,1,1-trichloroethane. *Int. J. Chem. Kinet.* 22: 577–590.

Newman, L.A., S.E. Strand, N. Choe, J. Duffy, G. Ekuan, M. Ruiszai, B.B. Shurtleff, J. Wilmoth, and M.P. Gordon. 1997. Uptake and biotransformation of trichloroethylene by hybrid poplars. *Environ. Sci. Technol.* 31: 1062–1067.

Niimi, A.J. and G.P. Dookhran. 1989. Dietary absorption efficiencies and elimination rates of polycyclic aromatic hydrocarbons (PAHs) in rainbow trout (*Salmo gairdneri*). *Environ.Toxicol. Chem.* 8: 719–722.

Niimi, A.J., H.B. Lee, and G.P. Kissoon. 1989. Octanol/water partition coefficients and bioconcentration factors of chloronitrobenzenes in rainbow trout (*Salmo gairdneri*). *Environ. Toxicol. Chem.* 8: 817–823.

Niimi, A.J., C.D. Metcalfe, and S.Y. Huestis. 1994. Chlorinated diphenyl ethers in Great Lakes fish and their environmental implication. *Environ. Toxicol. Chem.* 13: 1133–1138.

Nilsson, U.L. and A.L. Colmsjö. 1990. Formation of chlorinated polycyclic aromatic hydrocarbons in different chlorination reactions. *Chemosphere* 21: 939–951.

Norstrom, R.J., T.P. Clark, D.A. Jeffrey, H.T. Won, and A.P. Gilman. 1986. Dynamics of organochlorine compounds in herring gulls (*Larus argentatus*): 1. Distribution and clearance of [^{14}C]DDE in free-living herring gulls (*Larus argentatus*). *Environ. Toxicol. Chem.* 5: 41–48.

Norstrom, R.J., M. Simon, D.C.G Muir, and R.E. Schweinsberg. 1988. Organochlorine contaminants in arctic marine food chains: identification, geographical distribution, and temporal trends in polar bears. *Environ. Sci. Technol.* 22: 1063–1071.

Novak, M.A., A.A. Reilly, and S.J. Jackling. 1988. Long-term monitoring of polychlorinated biphenyls in the Hudson River (New York) using caddisfly larvae and other macroinvertebrates. *Arch. Environ. Contam. Toxicol.* 17: 699–710.

Oikari, A. and T. Kunnamo-Ojala. 1987. Tracing of xenobiotic contamination in water with the aid of fish bile metabolites: a field study with caged rainbow trout (*Salmo gairdneri*). *Aquat. Pollut.* 9: 327–341.

Oliver, B.G. 1987. Biouptake of chlorinated hydrocarbons from laboratory-spiked and field sediments by oligochaete worms. *Environ. Sci. Technol.* 21: 785–790.

Oliver, B.G. and A.J. Niimi. 1985. Bioconcentration of some halogenated organics for rainbow trout: limitations in their use for prediction of environmental residues. *Environ. Sci. Technol.* 19: 842–849.

Oliver, B.G. and A.J. Niimi. 1988. Trophodynamic analysis of PCB congeners and other chlorinated hydrocarbons in the Lake Ontario ecosystem. *Environ. Sci. Technol.* 22: 388–397.

Oliver, B.G., M.N. Charlton, and R.W. Durham. 1989. Distribution, redistribution, and geochronology of polychlorinated biphenyl congeners and other chlorinated hydrocarbons in Lake Ontario sediments. *Environ. Sci. Technol.* 23: 200–208.

Opperhuizen, A., P. Serné, and J.M.D. van Steen. 1988. Thermodynamics of fish/water and octan-1-ol/water partitioning of some chlorinated benzenes. *Environ. Sci. Technol.* 22: 286–292.

Otto, F., G. Leupold, H. Pariar, R. Rosemann, M. Bahadir, and H. Hopf. 1998. Chlorinated diphenoquinones: a new class of dioxin isomeric compounds discovered in fly ashes, slags, and pyrolysis oil samples by using HPLC/ELCD and HRGC/MS. *Anal. Chem.* 70: 2831–2838.

Ou, L., K.S Edvardsson, and P.S.C. Rao. 1985. Aerobic and anaerobic degradation of Aldicarb in soils. *J. Agric. Food Chem.* 53: 72–78.

Parris, G.E. 1980. Covalent binding of aromatic amines to humates. 1. Reactions with carbonyls and quinones. *Environ. Sci. Technol.* 14: 1099–1106.

Pärt, P. 1990. The perfused fish gill preparation in studies of the bioavailability of chemicals. *Ecotoxicol. Environ. Saf.* 19: 106–115.

Partington, J.R. 1950. *Thermodynamics*, pp. 112–114. Constrable & Company, London.

Paterson, S. and D. Mackay. 1985. The fugacity concept in environmental modelling, pp. 121–140. In *Handbook of Environmental Chemistry* (Ed. O. Hutzinger), Vol. 2 Part C. Springer-Verlag, Berlin.

Pavliostathis, S.G. and K. Jaglal. 1991. Desorptive behavior of trichloroethylene in contaminated soil. *Environ. Sci. Technol.* 25: 274–279.

Pavliostathis, S.G. and G.N. Mathavan. 1992. Desorption kinetics of selected volatile organic compounds from field contaminated soils. *Environ. Sci. Technol.* 26: 532–538.

Pearson, R.F., D.L. Swackhamer, S.J. Eisenreich, and D.T. Long. 1997. Concentrations, accumulations and inventories of toxaphene in sediments of the Great Lakes. *Environ. Sci. Technol.* 31: 3523–3529.

Perdue, E.M. and N.L. Wolfe. 1982. Modification of pollutant hydrolysis kinetics in the presence of humic substances. *Environ. Sci. Technol.* 16: 847–852.

Pereira, W.E., C.E. Rostad,, C.T. Chiou, T.I. Brinron, L.B. Barber, D.K. Demcheck, and C.R. Demas. 1988. Contamination of estuarine water, biota, and sediment by halogenated organic compounds: a field study. *Environ. Sci. Technol.* 22: 772–778.

Peterman, P.H. and J.J. Delfino. 1990. Identification of isopropylbiphenyl, alkyl diphenylmethanes, diisopropylnaphthalene, linear alkyl benzenzenes and other polychlorinated biphenyl replacement compounds in effluents, sediments and fish in the Fox River system, Wisconsin. *Biomed. Environ. Mass Spectrom.* 19: 755–770.

Phale, P.S., H.S. Savithri, N.A. Rao, and C.S. Vaidyanathan. 1995. Production of biosurfactant "Biosur-Pm" by *Pseudomonas maltophilia* CSV89: characterization and role in hydrocarbon uptake. *Arch. Microbiol.* 163: 424–431.

Pignatello, J.J. 1990. Slowly reversible sorption of aliphatic halocarbons in soils. I. Formation of residual fractions. *Environ. Toxicol. Chem.* 9: 1107–1115.

Platz, J., O.J. Nielsen, J. Sehested, and T.J. Wallington. 1995. Atmospheric chemistry of 1,1,1-trichloroethane: UV absorption spectra and self-reaction kinetics of CCl_3CH_2 and $CCl_3CH_2O_2$ radicals, kinetics of the reactions of the $CCl_3CH_2O_2$ radical with NO and NO_2, and the fate of alkoxy radical CCl_3CH_2O. *J. Phys. Chem.* 99: 6570–6579.

Podoll, R.T. and K.C. Irwin. 1988. Sorption of cationic oligomers on sediments. *Environ. Toxicol. Chem.* 7: 405–415.

Poulsen, M., L. Lemon, and J.F. Barker. 1992. Dissolution of monoaromatic hydrocarbons into groundwater from gasoline-oxygenate mixtures. *Environ. Sci. Technol.* 26: 2483–2489.

Prahl, F.G. and R. Carpenter. 1983. Polycyclic aromatic hydrocarbon (PAH)-phase associations in Washington coastal sediment. *Geochim. Cosmochim. Acta* 47: 1013–1023.

Rapaport, R.A. and S.J. Eisenreich. 1988. Historical atmospheric inputs of high molecular weight chlorinated hydrocarbons to Eastern North America. *Environ. Sci. Technol.* 22: 931–941.

Remberger, M., P.-Å. Hynning, and A.H. Neilson. 1988. Comparison of procedures for recovering chloroguaiacols and chlorocatechols from contaminated sediments. *Environ. Toxicol. Chem.* 7: 795–805.

Remberger, M., P.-Å. Hynning, and A.H. Neilson.1990. Gas chromatographic analysis and gas chromatographic–mass spectrometric identification of components in the cyclohexane-extractable fraction from contaminated sediment samples. *J. Chromatogr.* 508: 159–178.

Remberger, M., P.-Å. Hynning, and A.H. Neilson.1993. Release of chlorinated catechols from a contaminated sediment. *Environ. Sci. Technol.* 27:

Renaud, C.B., K.L.E. Kaiser, M.E. Comba, and J.L. Metcalfe-Smith. 1995. Comparison between lamprey ammocoetes and bivalve molluscs as biomonitors of organochlorine contaminants. *Can. J. Fish. Aquat. Sci.* 52: 276–282.

Renberg, L. O., S.G. Sundström, and A.-C. Rosén-Olofsson. 1985. The determination of partition coefficients of organic compounds in technical products and waste waters for the estimation of their bioaccumulation potential using reverse phase thin layer chromatography. *Toxicol. Environ Chem.* 10: 333–349.

Richey, D.G., C.T. Driscoll, and G.E. Likens. 1997. Soil retention of trifluoroacetate. *Environ. Sci. Technol.* 31: 1723–1727.

Richnow, H.H., A. Eschenbach, B. Mahro, R. Seifert, P. Wehrung, P. Albrecht, and W. Michaelis. 1998. The use of ^{13}C-labelled polycyclic aromatic hydrocarbons for the analysis of their transformation in soil. *Chemosphere* 36: 2211–2224.

Riederer, M. 1990. Estimating partitioning and transport of organic chemicals in the foliage/atmosphere system: discussion of a fugacity-based model. *Environ. Sci. Technol.* 24: 829–837.

Roy, S. and O. Hänninen. 1994. Pentachlorophenol: uptake/elimination kinetics and metabolism in an aquatic plant. *Eichhornia crassipes. Environ. Toxicol. Chem.* 13: 763–773.

Rüttimann-Johnson, C. and Lamer, R.T. 1996. Polymerization of pentachlorophenol and ferulic acid by fungal extracellular lignin-degrading enzymes. *Appl. Environ. Microbiol.* 62: 38909–3893.

Safe, S., K.W. Brown, K.C. Donnelly, C.S. Anderson, K.V. Markiewicz, M.S. McLachlan, A. Reischi, and O. Hutzinger. 1992. Polychlorinated dibenzo-*p*-dioxins and dibenzofurans associated wirth wood-preserving chemical sites: biomonitoring with pine needles. *Environ. Sci. Technol.* 26: 394–396.

Sanders, G., K.C. Jones, J. Hamilton-Taylor, and H. Dörr. 1992. Historical inputs of polychlorinated biphenyls and other organochlorines to a dated lacustrine sediment core in rural England. *Environ. Sci. Technol.* 26: 1815–1821.

Sarkar, J.M., R.L. Malcolm, and J.-M. Bollag. 1988. Enzymatic coupling of 2,4-dichlorophenol to stream fulvic acid in the presence of oxidoreductases. *Soil Sci. Am. J.* 52: 688–694.

Sawhney, B.L., J.J. Pignatello, and A.M. Steinberg. 1988. Determination of 1,2-dibromoethane (EDB) in field soils: implications for volatile organic compounds. *J. Environ. Qual.* 17: 149–152.

Scharenberg, W. 1991. Cormorants (*Phalacrocorax carbo sinensis*) as bioindicators for polychlorinated biphenyls. *Arch. Environ. Contam. Toxicol.* 21: 536–540.

Scheunert, I., E. Topp, A. Attar, and F. Korte. 1994. Uptake pathways of chlorobenzenes in plants and their corrrelation with *N*-octanol/water partition coefficients. *Ecotoxicol. Environ. Saf.* 27: 90–104.

Schmitt, C.J., J.L. Zajick, and P.H. Peterman. 1990. National contaminants biomonitoring program: residues of organochlorine chemicals in U.S. freshwater fish, 1976–1984. *Arch. Environ. Contam. Toxicol.* 19: 748–781.

Schrap, S.M. and A. Opperhuizen. 1990. Relationship between bioavailability and hydrophobicity: reduction of the uptake of organic chemicals by fish due to sorption on particles. *Environ. Toxicol. Chem.* 9: 715–724.

Schrock, M.E., E.S. Barrows, and L.B. Rosman. 1997. Biota-to-sediment accumulation factors for TCDD and TCDF in worms from 28 d bioaccumulation tests. *Chemosphere* 34: 1333–1339.

Schroll, R., B. Bierling, G. Cao, U. Dörfer, M. Lahaniati, T. Langenbach, I. Scheunert, and R. Winkler. 1994. Uptake pathways of organic chemicals from soil by agricultural plants. *Chemosphere* 28: 297–303.

Schuytema, G.S., D.F. Krawczyk, W.L. Griffis, A.V. Nebeker, and M.L. Robideaux. 1990. Hexachlorobenzene uptake by fathead minnows and macroinvertebrates in recirculating sediment/water systems. *Arch. Environ. Contam. Toxicol.* 19: 1–9.

Sellström, U., A. Kierkegaard, C. de Witt, and B. Jansson. 1998. Polybrominated diphenyl ethers and hexabromocyclododecane in sediment and fish from a Swedish river. *Environ. Toxicol. Chem.* 17: 1065–1072.

Sericano, J.L., T.L. Wade, J.M. Brooks, E.L. Atlas, R.R. Fay, and D.L. Wilkinson. 1993. National status and trends Mussel Watch Program: chlordane-related compounds in Gulf of Mexico oysters, 1986–90. *Environ. Pollut.* 82: 23–32.

Servos, M. and D.C.G. Muir. 1989. Effect of suspended sediment concentration on the sediment to water partition coefficient for 1,3,6,8-tetrachlorodibenzo-*p*-dioxin. *Environ. Sci. Technol.* 23: 1302–1306.

Shimp R. and F.P. Pfaender.1985. Influence of naturally occurring humic acids on biodegradation of monosubstituted phenols by aquatic bacteria. *Appl. Environ. Microbiol.* 49: 402–407.

Shiu, W.Y., W. Doucette, F.A.P.C. Gobas, A. Andren, and D. Mackay. 1988. Physical-chemical properties of chlorinated dibenzo-*p*-dioxins. *Environ. Sci. Technol.* 22: 651–658.

Sijm, D.T.H.M., W. Seinen, and A. Opperhuizen. 1992. Life-cycle biomagnification study in fish. *Environ. Sci. Technol.* 26: 2162–2174.

Simmons, K.E., Minard, R.D., and Bollag, J.-M. 1987. Oligomerization of 4-chloroaniline by oxidoreductases. *Environ. Sci. Technol.* 21: 999–1003.

Simmons, K.E., R.D. Minard, and J.-M. Bollag. 1989. Oxidative co-oligomerization of guaiacol and 4-chloroaniline. *Environ. Sci. Technol.* 23: 115–121.

Simonich, S.L. and R.A. Hites. 1994. Vegetation-atmosphere partitioning of polycyclic aromatic hydrocarbons. *Environ. Sci. Technol.* 28: 939–943.

Simonich, S.L. and R.A. Hites. 1995. Organic pollutant accumulation in vegetation. *Environ. Sci. Technol.* 29: 2905–2913.

Skaar, D.R., B.T. Johnson, J.R. Jones, and J.N. Huckins. 1981. Fate of kepone and mirex in a model aquatic environment: sediment, fish, and diet. *Can. J. Fish. Aquat. Sci.* 38: 931–938.

Smith, A.E. 1985. Persistence and transformation of the herbicides ^{14}C Fenoxaprop-ethyl and ^{14}C Fenthiaprop-ethyl in two prairie soils under laboratory and field conditions. *J. Agric. Food Chem.* 33: 483–488.

Smith, D.W. 1995. Are PCBs in the Great Lakes approaching a "new equilibrium"? *Environ. Sci. Technol.* 29: 42A–46A.

Smith, J.H., D.C. Bomberger, and D.I. Haynes. 1980. Prediction of the volatilization rates of high-volatility chemicals from natural water bodies. *Environ. Sci. Technol.* 14: 1332–1337.

Smith, J.H., D.C. Bomberger, and D.I. Haynes. 1981. Volatilization rates of intermediate and low volatility chemicals from water. *Chemosphere* 10: 281–289.

Southworth, G.R., C.C. Keffer, and J.J. Beauchamp. 1980. Potential and realized bioconcentration. A comparison of observed and predicted bioconcentration of azaarenes in the fathead minnow (*Pimephales promelas*). *Environ. Sci. Technol.* 14: 1529–1531.

Southworth, G.R., C.C. Keffer, and J.J. Beauchamp. 1981. The accumulation and disposition of benz(a)acridine in the fathead minnow, *Pimephales promelas. Arch. Environ. Contam. Toxicol.* 10: 561–569.

Squillace, P.J. and E.M. Thurman. 1992. Herbicide transport in rivers: importance of hydrology and geochemistry in nonpoint-source contamination. *Environ. Sci. Technol.* 26: 538–545.

Squillace, P.J., J.S. Zogorski, W.G. Wilber, and C.V. Price. 1996. Preliminary assessment of the occurrence and possible sources of MTBE in groundwater in the United States, 1993–1994. *Environ. Sci. Technol.* 30: 1721–1730.

Standley, L.J. 1997. Effect of sedimentary organic matter composition on the partitioning and bioavailability of dieldrin to the oligochaete *Lumbriculus variegatus. Environ. Sci. Technol.* 31: 2577–2583.

Standley, L.J. and T.L. Bott. 1998. Trifluoroacetate, an atmospheric breakdown product of hydrofluorocarbon refrigerants: biomolecular fate in aquatic organisms. *Environ. Sci. Technol.* 32: 469–475.

Stauffer, T.B., W.C. MacIntyre, and D.C. Wickman. 1989. Sorption of nonpolar organic chemicals on low-carbon-content aquifer materials. *Environ. Toxicol. Chem.* 8: 845–852.

Steffensen, W.S. and M. Alexander. 1995. Role of competition for inorganic nutrients in the biodegradation of mixtures of substrates. *Appl. Environ. Microbiol.* 61: 2859–2862.

Steinberg, S.M., J.J. Pignatello, and B.L. Sawhney. 1987. Persistence of 1,2-dibromo-ethane in soils: entrapment in intraparticle micropores. *Environ. Sci. Technol.* 21: 1201–1208.

Steingraeber, M.T., T.R. Schwartz, J.G. Wiener, and J.A. Lebo. 1994. Polychlorinated biphenyl congeners in emergent mayflies from the upper Mississippi River. *Environ. Sci. Technol.* 28: 707–714.

Stephanou, E.G. 1992. α,ω-Dicarboxylic acid salts and α,ω-dicarboxylic acids. Photo-oxidation products of unsaturated fatty acids, present in marine aerosols and marine sediments. *Naturwissenschaften* 79: 128–131.

Strachan, W.M.J. 1988. Toxic contaminants in rainfall in Canada: 1984. *Environ. Toxicol. Chem.* 7: 871–877.

Stroomberg, G.J., C. Reuther, I. Konin, T.C. van Brummelen, C.A.M. van Gestel, C. Gooijer, and W.P. Cofino. 1996. Formation of pyrene metabolites by the terrestrial isopod *Porcello scaber*. *Chemosphere* 33: 1905–1914.

Suflita, J.M., C.P. Gerba, R.A. Ham, A. C. Palmisano, W.L. Rathje, and J.A. Robinson. 1992. The world's largest landfill. A multidisciplinary investigation. *Environ. Sci. Technol.* 26: 1486–1495.

Swackhamer, D.L. and D.E. Armstrong. 1986. Estimation of the atmospheric and nonatmospheric contributions and losses of polychlorinated biphenyls for Lake Michigan on the basis of sediment records of remote lakes. *Environ. Sci. Technol.* 20: 879–883.

Swackhamer, D.L. and R.A. Hites. 1988. Occurrence and bioaccumulation of orga-nochlorine compounds in fishes from Siskiwit Lake, Isle Royale, Lake Superior. *Environ. Sci. Technol.* 22: 543–548.

Swackhamer, D.L. and R.S. Skoglund. 1993. Bioaccumulation of PCBs by algae: ki-netics versus equilibrium. *Environ. Toxicol. Chem.* 12: 831–838.

Swackhamer, D.L., B.D. McVeety, and R.A. Hites. 1988. Deposition and evaporation of polychlorobiphenyl congeners to and from Siskiwit Lake, Isle Royale, Lake Superior. *Environ. Sci. Technol.* 22: 664.–672.

Swain, W.R. 1991. Effects of organochlorine chemicals on the reproductive outcome of humans who consumed contamined Great Lakes fish: an epidemiologic con-sideration. *J. Toxicol. Environ. Health* 33: 587–639.

Swartz, R.C., P.F. Kemp, D.W. Schults, G.R. Ditsworth, and R.J. Ozretich. 1989. Acute toxicity of sediment from Eagle Harbor, Washington, to the infaunal amphipod *Rhepoxynius abronius*. *Environ. Toxicol. Chem.* 8: 215–222.

Swartz, R.C., D.W. Schults, T.H. Dewitt, G.R. Ditsworth, and J.O. Lamberson. 1990. Toxicity of fluoranthene in sediments to marine amphipods: a test of the equi-librium partitioning approach to sediment quality criteria. *Environ. Toxicol. Chem.* 9: 1071–1080.

Tadokoro, H. and Y. Tomita. 1987. The relationship between bioaccumulation and lipid content in fish. pp. 363–373. In *QSAR in Environmental Toxicology - II.* (Ed. K.L.E. Kaiser). Reidel Publishers, Dordrecht.

Tanabe, S., H. Tanaka, and R. Tatsukawa. 1984. Polychlorobiphenyls, Σ DDT, hexachlorocyclohexane isomers in the western North Pacific ecosystem. *Arch. Environ. Contam. Toxicol.* 731–738.

Ten Hulscher, Th. E.M., L.E. van der Velde, and W.A. Bruggeman. 1992. Temperature dependence of Henry's law constants for selected chlorobenzenes, polychlorinated biphenyls and polycyclic aromatic hydrocarbons. *Environ. Toxicol. Chem.* 11: 1595–1603.

Ten Hulscher, Th. E. M., B.A. Vrind, H. van den Heuvel, L.E. van der Velde, P.C.M. van Noort, J.E.M. Beurskens, and H.A.J. Govers. 1999. Triphasic desorption of highly resistant chlorobenzenes, polychlorinated biphenyls, and polycyclic aromatic hydrocarbons in field contaminated sediment. *Environ. Sci. Technol.* 33: 126–132.

Thomann, R.V. 1989. Bioaccumulation model of organic chemical distributiion in aquatic food chains. *Environ. Sci. Technol.* 23: 699–707.

Thomann, R.V. and J.P. Connolly. 1984. Model of PCB in the Lake Michigan lake trout food chain. *Environ. Sci. Technol.* 18: 65–71.

Thomas, D.J., B. Tracey, H. Marshall, and R.J. Norstrom. 1992. Arctic terrestrial ecosystem contamination. *Sci. Total Environ.* 122: 135–164.

Thorn, K.A., J.B. Arterburn, and M.A. Mikita. 1992. ^{15}N and ^{13}C NMR investigation of hydroxylamine-derivatized humic substances. *Environ. Sci. Technol.* 26: 107–116.

Thorn, K.A., Pettigrew, P.J., Goldenberg, W.S., and Weber, E.J. 1996. Covalent binding of aniline to humic substances. 2. ^{15}N NMR studies of nucleophilic addition reactions. *Environ. Sci. Technol.* 30: 1764–1775.

Tolman, R.C.1938. *The Principles of Statistical Mechanics* p. 602. Oxford University Press, Oxford.

Trenck, T.v.d. and H. Sandermann. 1981. Incorporation of benzo[a]pyrene quinones into lignin. *FEBS Lett.* 125: 72–76.

Tsao, C.-W., H.-G. Song, and R. Bartha. 1998. Metabolism of benzene, toluene, and xylene hydrocarbons in soil. *Appl. Environ. Microbiol.* 64: 4924–4929.

Ugrekhelidze, D., Kortc, F., and G. Kvesitadze. 1997. Uptake and transformation of benzene and toluene by plant leaves. *Ecotoxicol. Environ. Saf.* 37: 24–29.

Valsaraj, K.T., R. Ravikrishna, B. Choy, D.D. Reible, L.J. Thibodeaux, C.B. Price, S. Yosy, J.M. Brannon, and T.E. Myers. 1999. Air emissions from exposed contaminated sediments and dredged material. *Environ. Sci. Technol.* 33: 142–149.

Van der Oost, R., H. Heida, and A. Opperhuizen. 1988. Polychlorinated biphenyl congeners in sediments, plankton, molluscs, crustaceans, and eel in a freshwater lake: implications of using reference chemicals and indicator organisms in bioaccumulation studies. *Arch. Environ. Contam. Toxicol.* 17: 721–729.

Van Haelst, A.G., H. Loonen, F.W.M. van der Wielen, and H.A.J. Govers. 1996. Comparison of bioconcentration factors of tetrachlorobenzyltoluenes in the guppy *Poecilia reticulata* and zebramussel *Dreissena polymorpha*. *Chemosphere* 32: 1117–1122.

Varanasi, U., W.L. Reichert, J.E. Stein, D.W. Brown, and H.R. Sanborn. 1985. Bioavailability and biotransformation of aromatic hydrocarbons in benthic organisms exposed to sediment from an urban estuary. *Environ. Sci. Technol.* 19: 836–841.

Vassilaros, D.L., P.W. Stoker, G.M. Booth, and M.L. Lee. 1982. Capillary gas chromatographic determination of polycyclic aromatic compounds in vertebrate fish tissue. *Anal. Chem.* 54: 106–112.

Veith, G.D., D.L. DeFoe, and B.V. Bergstedt. 1979a. Measuring and estimating the bioconcentration factor of chemicals in fish. *J. Fish. Res. Board Can.* 36: 1040–1048.

Veith, G.D., N.M. Austin, and R.T. Morris. 1979b. A rapid method for estimating log P for organic chemicals. *Water Res.* 13: 43–47.

Villholth, K.G. 1999. Colloid characterization and colloidal phase partitioning of polycyclic aromatic hydrocarbons in two creosote-contaminated aquifers in Denmark. *Environ. Sci. Technol.* 33: 691–699.

Wachtmeister, C.A., L. Förlin, K.C. Arnoldsson, and J. Larsson. 1991. Fish bile as a tool for monitoring aquatic pollutants: studies with radioactively labelled 4,5,6-trichloroguaiacol. *Chemosphere* 22: 39–46.

Walker, C.H. 1983. Pesticides and birds—mechanisms of selective toxicity. *Agric. Ecosyst. Environ.* 9: 211–226.

Wania, F. and D. Mackay. 1993. Global fractionation and cold condensation of low volatility organochlorine compounds in polar regions. *Ambio* 22: 10–18.

Wania, F. and Mackay, D. 1996. Tracking the distribution of persistent organic pollutants. *Environ. Sci. Technol.* 30: 390A–396A.

Welch, H.E., D.C.G. Muir, B.N. Billeck, W.L. Lockhart, G.J. Brunskill, H.J. Kling, M.P. Olson, and R.M. Lemoine. 1991. Brown snow: a long-range transport event in the Canadian Arctic. *Environ. Sci. Technol.* 25: 280–286.

Welch, H.E., M.A. Bergmann, T.D. Siferd, K.A. Martin, M.F. Curtis, R.E. Crawford, R.J. Conover, and H. Hop. 1992. Energy flow through a marine ecosystem of the Lancaster Sound Region, Arctic Canada. *Arctic* 45: 343–357.

Welke, B., K. Ettlinger, and M. Riedewrer. 1998. Sorption of volatile organic chemicals in plant surfaces. *Environ. Sci. Technol.* 32: 1099–1104.

Welling, W. and J.W. de Vries. 1992. Bioconcentration kinetics of the organophosphate insecticide chlorpyrifos in guppies (*Poecilia reticulata*). *Ecotoxicol. Environ. Saf.* 23: 64–75.

Westcott, N.D. and B.L. Worobey. 1985. Novel solvent extraction of lindane from soil. *J. Agric. Food Chem.* 33: 58–60.

Weston, D.P. and L.M. Mayer. 1998a. *In vitro* digestive fluid extraction as a measure of the bioavailability of sediment-associated polycyclic aromatic hydrocarbons: sources of variation and implication for partitioning models. *Environ. Toxicol. Chem.* 17: 820–829.

Weston, D.P. and L.M. Mayer. 1998b. Comparison of *in vitro* digestive fluid extraction and traditional *in vivo* approaches as measures of polycyclic aromatic hydrocarbon bioavailability from sediments. *Environ. Toxicol. Chem.* 17: 830– 840.

Whittle, D.M., D.B. Sargent, S.Y. Huestis, and W.H. Hyatt. 1992. Foodchain accumulation of PCDD and PCDF isomers in the Great Lake aquatic community. *Chemosphere* 25: 181–184.

Wiederholm, T., A.-M. Wiederholm, and G. Milbrink. 1987. Bulk sediment bioassays with five species of fresh-water oligochaetes. *Water Air Soil Pollut.* 36: 131–154.

Willett, K.L., E.M. Ulrich, and R.A. Hites. 1998. Differential toxicity and environmental fates of hexachlorocyclohexane isomers. *Environ. Sci. Technol.* 32: 2197–2207.

Winkelmann, D.A. and S.J. Klaine. 1991. Degradation and bound residue formation of four atrazine metabolites, deethylatrazine, deisopropylatrazine, dealkylatrazine and hydroxyatrazine, in a western Tennessee soil. *Environ. Toxicol. Chem.* 10: 347–354.

Wittlinger, R. and K. Ballschmiter 1990. Studies of the global baseline pollution XIII. C6-C14 organohalogens (a and g [sic]-HCH, HCB, PCB, 4,4'-DDT, 4,4-DDE, *cis*- and *trans*-chlordane, *trans*-nonachlor, anisols) in the lower troposphere of the southern Indian Ocean. *Fresenius J. Anal. Chem.* 336: 193–200.

Woodburn, K.B., W.J. Doucette, and A.W. Andren. 1984. Generator column determination of octanol/water partition coefficients for selected polychlorinated biphenyl congeners. *Environ. Sci. Technol.* 18: 457–459.

Wu, J., M.K. Wong, H.K.Lee, and C.N. Ong. 1995. Capillary zone electrophoretic determination of heterocyclic aromatic amines in rain. *J. Chromatogr.* Sci. 33: 712–716.

Xing, B. and J.J. Pignatello. 1998. Competitive sorption between 1,3-dichlorobenzene or 2,4-dichlorophenol and natural aromatic acids in soil organic matter. *Environ. Sci. Technol.* 32: 614–619.

Yin, C. and J.P. Hassett. 1986. Gas-partitioning approach for laboratory and field studies of mirex fugacity in water. *Environ. Sci. Technol.* 20: 1213–1217.

Yunker, M.B. and R.W. Macdonald. 1995. Composition and origins of polycyclic aromatic hydrocarbons in the Mackenzie River and on the Beaufort Sea Shelf. *Arctic* 48: 118–129.

Zacharewski, T., L. Safe, S. Safe, B. Chittim, D. DeVault, K. Wiberg, P.-A. Bergqvist, and C. Rappe. 1989. Comparative analysis of polychlorinated dibenzo-*p*-dioxin and dibenzofuran congeners in Great Lakes fish extracts by gas chromatography–mass spectrometry and *in vitro* enzyme induction activities. *Environ. Sci. Technol.* 23: 730–735.

Zitzelsberger, W., W. Ziegler, and P.R. Wallnöfer. 1987. Stereochemistry of the degradation of veratrylglycerol-β–2,4-dichlorophenyl-ether, a model compound for lignin bound xenobiotic residues by *Phenerochaete chrysosporium, Corynebacterium equi* and photosensitized rioboflavin. *Chemosphere* 16: 1137–1142.

4

Persistence: General Orientation

SYNOPSIS An overview is presented of the factors that determine the persistence of xenobiotics including the role of both abiotic and biotic reactions. Examples are given of photochemical reactions including those that take place in the troposphere and transformation products that may subsequently enter aquatic or terrestrial systems, and of chemical transformations including hydrolysis, dehalogenation, oxidation, and reduction. It is pointed out that combinations of abiotic and biotic processes may be of determinative significance, and the significance of these reactions in determining the analytes that may be included in monitoring programs is emphasized. Biotic reactions are discussed in detail and the important distinction between biodegradation and biotransformation is emphasized. Attention is directed to the metabolic potential of groups of microorganisms that have been less extensively examined; these include enteric bacteria, ammonia oxidizers, marine and lithotrophic bacteria, algae, and anaerobic phototrophic bacteria. The significance of electron acceptors other than oxygen is noted, and examples are illustrated with organisms using nitrate and related compounds, and those growing anaerobically by reduction of Fe(III), Mn(IV), or U(VI). Some important reactions mediated by yeasts, fungi, and algae are outlined. The mechanisms whereby oxygen is introduced into xenobiotics are discussed, and brief accounts of the enzymology are included. Attention is directed to metabolic interactions where several organisms and a single substrate are present, or where several substrates and a single organism occur. Examples are given of metabolic limitations imposed by enzyme regulatory mechanisms and of metabolic situations where a single readily degraded substrate is present in addition to a more recalcitrant xenobiotic. Factors that may critically determine the biodegradability of xenobiotics in natural systems are summarized; these include temperature, the oxygen concentration, the substrate concentration, the synthesis of natural emulsifying agents, the nature of transport mechanisms, and the cardinal issue of the bioavailability of the xenobiotic. A number of incompletely resolved issues are discussed including biodegradation in pristine environments, natural enrichment in contaminated environments, estimation of the rates of metabolic reactions both in laboratory and natural systems, and the significance of toxic metabolites. Brief comments are given on the role of catabolic plasmids.

Introduction

Procedures for the analysis of environmental samples have been outlined in Chapter 2, and the processes that determine the dissemination of xenobiotics after discharge from point sources have been discussed in Chapter 3. In the next three chapters, the factors that determine the ultimate fate of xenobiotics will be discussed. This chapter attempts to present an overview of the factors that determine the persistence of xenobiotics, while Chapter 5 will be devoted to experimental procedures for carrying out the relevant investigations, and Chapter 6 to a detailed examination of the pathways taken for the degradation and transformation of a wide range of structurally diverse xenobiotics. Attention will be focused on microorganisms, and in particular on bacteria that are the most important degradative organisms in virtually all aquatic ecosystems. A certain degree of overlap between this chapter and Chapter 6 is inevitable, but an attempt has been made to minimize this by inclusion of cross-references.

It was the persistence of DDT which raised the greatest alarm over its extensive use during the years 1940 to 1968. Although levels since its banning have decreased dramatically, those of its metabolite DDE may still be appreciable and serve to sustain the initial concern. Many organic compounds have become environmentally suspect, but it is especially the highly chlorinated ones such as the polychlorinated biphenyls, polychlorinated camphenes, and mirex which have acquired the reputation of being unacceptable due to their apparent persistence. As a result of these fears, there has emerged a general concern with all synthetic chlorinated organic compounds (Hileman 1993) which may possibly have deflected interest from other groups which merit comparable attention. It should, of course, be appreciated that on the other hand a number of compounds and products such as modern plastics have been developed for their stability under a variety of conditions — and are produced with this end in view.

For these reasons, studies on biodegradation began to occupy a central position in discussions on the environmental impact of organic chemicals, and the complexities have been clearly presented (Landner 1989). It should be appreciated at the outset that the terms *persistent* and *recalcitrant* are relative rather than absolute since probably most chemical structures can be degraded or transformed by microorganisms. The crucial issue is the rate at which the reactions occur, and the area between slowly degradable compounds and truly persistent ones is often unresolved. For example, in spite of the fact that degradation of some PCB congeners has been demonstrated under aerobic conditions, and biotransformation (dechlorination) under anaerobic conditions, these compounds are still recoverable from many environmental samples; they should therefore be regarded as persistent. The critical questions are both what reactions take place and the rate at which they occur in the environment into which the compound is discharged. Both of these should be addressed in investigations aimed at incorporating environmental relevance.

Two essentially different processes determine the persistence of an organic compound in the aquatic environment. The first are abiotic reactions, and for some groups of compounds these reactions may be dominant in determining their fate. The second are biotic reactions mediated by a wide range of organisms. Only microorganisms will be discussed in the next three chapters, although a brief discussion of the metabolism of xenobiotics by higher organisms is given in Chapter 7 (Section 7.5).

4.1 Abiotic Reactions

Virtually any of the plethora of reactions known in organic chemistry may be exploited for the abiotic degradation of xenobiotics. Hydrolytic reactions may convert compounds such as esters, amides, or nitriles into the corresponding carboxylic acids, or ureas and carbamides into the amines. These abiotic reactions may therefore be the first step in the degradation of such compounds; the transformation products may, however, be resistant to further chemical transformation so that their ultimate fate is dependent upon subsequent microbial reactions. For example, for urea herbicides the limiting factor is the rate of microbial degradation of the chlorinated anilines which are the initial hydrolysis products. The role of abiotic reactions should, therefore, always be taken into consideration, and should be carefully evaluated in all laboratory experiments on biodegradation and biotransformation (Section 5.3). It should be appreciated that the results of experiments directed to microbial degradation are probably discarded if they show substantial interference from abiotic reactions. A good illustration of the complementary roles of abiotic and biotic processes is offered by the degradation of tributyl tin compounds. Earlier experiments (Seligman et al. 1986) had demonstrated the transformation of tributyltin to dibutyltin primarily by microbial processes. It was subsequently shown, however, that an important abiotic reaction mediated by fine-grained sediments resulted in the formation also of monobutyltin and inorganic tin (Stang et al. 1992). It was therefore concluded that both processes were important in determining the fate of tributyl tin in the marine environment.

A study of the carbamate biocides, carbaryl and propham, illustrates the care that should be exercised in determining the relative importance of chemical hydrolysis, photolysis, and bacterial degradation (Figure 4.1) (Wolfe et al. 1978a). For carbaryl, the half-life for hydrolysis increased from 0.15 day at pH 9 to 1500 day at pH 5, while that for photolysis was 6.6 day: biodegradation was too slow to be significant. On the other hand, the half-lives of propham for hydrolysis and photolysis were $>10^4$ and 121 day - so greatly exceeding the half-life of 2.9 day for biodegradation that abiotic processes would be considered to be of subordinate significance. Close attention to structural features of xenobiotics is therefore clearly imperative before making generalizations on the relative significance of alternative degradative pathways.

O.CO.NHCH₃

NH.CO.O.CH(CH₃)₂

A   B

FIGURE 4.1
Carbaryl (A) and propham (B).

4.1.1 Photochemical Reactions in Aqueous and Terrestrial Environments

Photochemical reactions may be important especially in areas of high solar irradiation, or on the surface of soils, or in aquatic systems containing ultraviolet (UV) absorbing humic and fulvic acids (Zepp et al. 1981), and they may be especially relevant for otherwise recalcitrant compounds. It has also been shown (Zepp and Schlotqhauer 1983) that, although the presence of algae may enhance photometabolism, this is subservient to direct photolysis at the cell densities likely to be encountered in rivers and lakes. It should be noted that different products may be produced in natural river water and in buffered medium; for example, photolysis of triclopyr (3,5,6-trichloro-2-pyridyloxy-acetic acid) in sterile medium at pH 7 resulted in hydrolytic replacement of one chlorine atom, whereas in river water the ring was degraded to form oxamic acid as the principal product (Woodburn et al. 1993). Particular attention has understandably therefore been directed to the photolytic degradation of biocides — including agrochemicals — that are applied to terrestrial systems. There has been increased interest in the phototoxicity toward a range of biota (references in Monson et al. 1999), and this may be attributed to some of the reactions and transformations that are discussed later in this chapter. It should be emphasized that photochemical reactions may produce molecules structurally more complex and less susceptible to degradation than their precursors, even though the deep-seated rearrangements induced in complex compounds such as the terpene santonin during UV irradiation (Figure 4.2) are not likely to be encountered in environmental situations.

The Diversity of Photochemical Transformations

In broad terms, the following types of reactions are mediated by the homolytic fission products of water (formally, hydrogen and hydroxyl radicals) and molecular oxygen or its excited states: hydrolysis, elimination, oxidation, reduction, and cyclization.

The Role of Hydroxyl Radicals

The hydroxyl radical plays two essentially different roles: (1) as a reactant mediating the transformations of xenobiotics and (2) as a toxicant operating by damaging DNA. Hydroxyl radicals are important in a number of

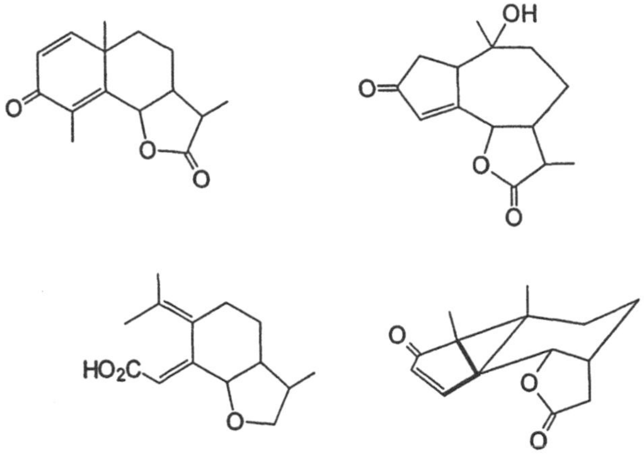

FIGURE 4.2
Photochemical transformation of santonin.

environments: (1) in aquatic systems under irradiation, (2) in the troposphere that is discussed later in this section, and (3) in biological systems that are noted in the context of superoxide dismutase and the role of Fe in Section 4.6.1.2 and in Sections 5.2.4 and 5.5.5. Hydroxyl radicals in aqueous media may be generated by (1) photolysis of nitrite and nitrate (Brezonik and Fulkerson-Brekken 1998), (2) the Fenton reaction with H_2O_2 and Fe^{2+} in the presence of light that is noted later, and (3) photolysis of fulvic acids under anaerobic conditions (Vaughan and Blough 1998), and (d) reaction of Fe(III) or Cu(II) complexes of humic acids with hydrogen peroxide (Paciolla et al. 1999).

For the sake of completeness, attention is drawn to the following: (1) the interactive role of hydroxyl radicals, superoxide, and Fe levels in wild and mutant strains of *Escherichia coli* lacking Fe and Mn superoxide dismutase is discussed in Sections 5.2.4 and 5.5.5 and (2) the possible role of hydroxyl radicals in mediating the transformations accomplished by the brown-rot fungus *Gleophyllum striatum* which is supported by the overall similarity in the structures of the fungal metabolites with those produced with Fenton's reagent (Wetzstein et al. 1997).

Analytical procedures for hydroxyl radicals noted in Section 2.3 and have been used to demonstrate the role of the anticancer drug 2,5-bis(1-azacyclopropyl)-3,6-bis(carboethoxyamino)benzo-1,4-quinone in mediating the production of hydroxyl radicals in JB6 mouse epidermal cells (Li et al. 1997).

Illustrative Examples of Photochemical Transformations in Aqueous Solutions

1. Atrazine is successively transformed to 2,4,6-trihydroxy-1,3,5-triazine (Pelizzetti et al. 1990) by dealkylation of the alkylamine

side chains and hydrolytic displacement of the ring chlorine and amino groups (Figure 4.3). A comparison has been made between direct photolysis and nitrate-mediated hydroxyl radical reactions (Torrents et al. 1997). The rates of the latter were much greater under the conditions of this experiment, and the major difference in the products was the absence of ring hydroxylation with loss of chloride.

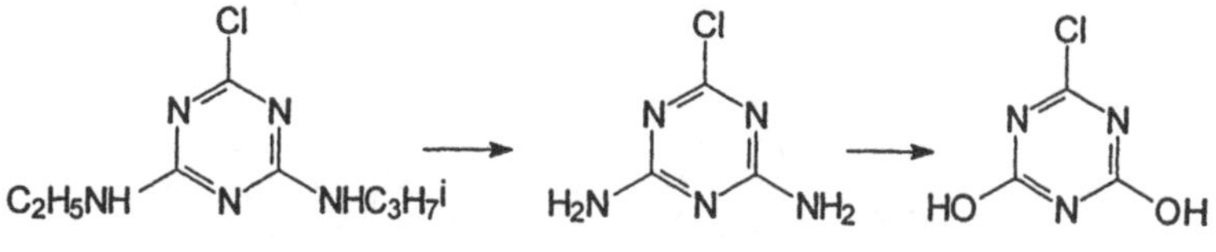

FIGURE 4.3
Photochemical transformation of atrazine.

2. Pentachlorophenol produces a wide variety of transformation products including chloranilic acid (2,5-dichloro-3,6-dihydroxy-benzo-1,4-quinone) by hydrolysis and oxidation, a dichlorocyclopentanedione by ring contraction, and dichloromaleic acid by cleavage of the aromatic ring (Figure 4.4) (Wong and Crosby 1981).

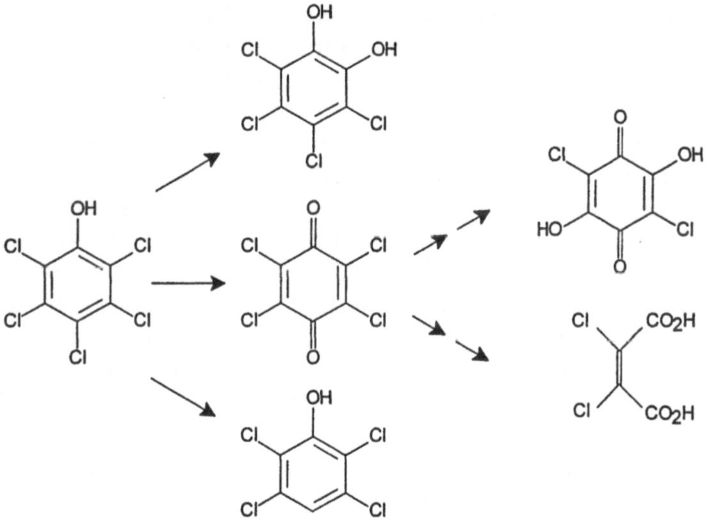

FIGURE 4.4
Photochemical transformation of pentachlorophenol.

3. The main products of photolysis of 3-trifluoromethyl-4-nitrophenol are 2,5-dihydroxybenzoate produced by hydrolytic loss of the nitro group and oxidation of the trifluoromethyl group, together with a compound identified as a condensation product of the original compound and the dihydroxybenzoate (Figure 4.5) (Carey and Cox 1981).

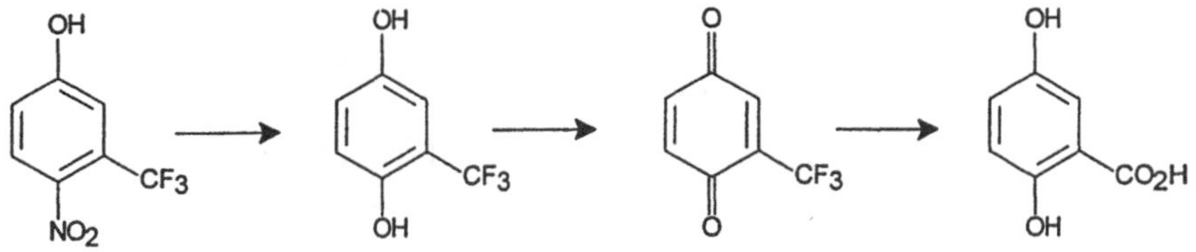

FIGURE 4.5
Photochemical transformation of 3-trifluoromethyl-4-nitrophenol.

4. The potential insecticide that is a derivative of tetrahydro-1,3-thi-azine undergoes a number of reactions resulting in some 43 products of which the dimeric azo compound is the principal one in aqueous solutions (Figure 4.6) (Kleier et al. 1985).

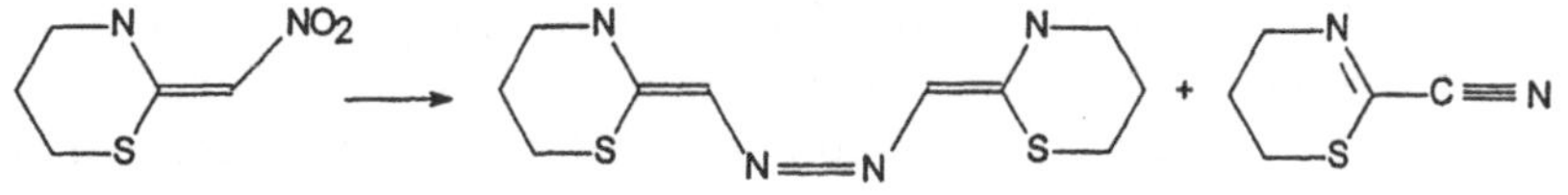

FIGURE 4.6
Photochemical transformation of a tetrahydro-1,3-thiazine.

5. The herbicide trifluralin undergoes a photochemical reaction in which the n-propyl side chain of the amine reacts with the vicinal nitro group to form the benzopyrazine (Figure 4.7) (Soderquist et al. 1975).

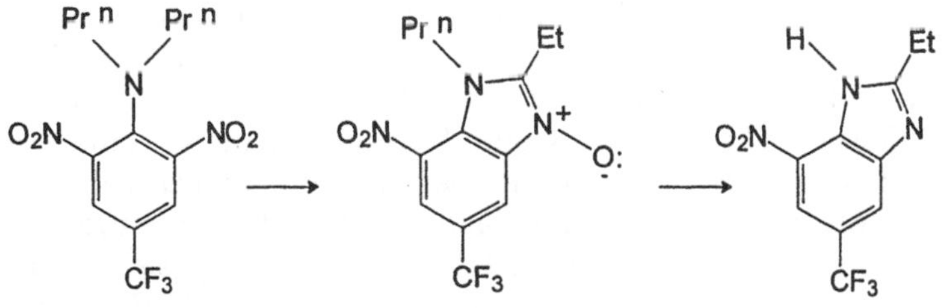

FIGURE 4.7
Photochemical transformation of trifluralin.

6. Heptachlor and *cis*-chlordane both of which are chiral form caged or half-caged structures (Figure 4.8) on irradiation and these products have been identified in biota from the Baltic, from the Arctic, and from the Antarctic (Buser and Müller 1993).

7. Methylcyclopentadienyl manganese tricarbonyl that has been suggested as a fuel additive is decomposed in aqueous medium primarily by photolysis. This resulted in the formation of methylcyclopentadiene that may plausibly be presumed to polymerize, and a manganese carbonyl that decomposed to Mn_3O_4 (Garrison et al. 1995).

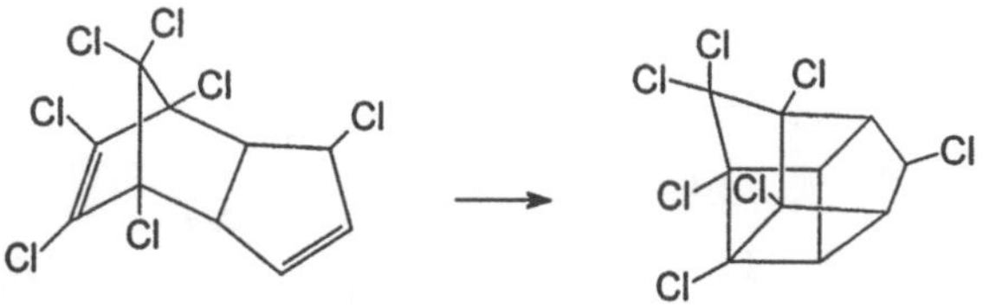

FIGURE 4.8
Photochemical transformation of chlordane.

8. Stilbenes that are used as fluorescent whitening agents are photolytically degraded by reactions involving *cis-trans* isomerization followed by hydration of the double bond or oxidative fission of the double bond to yield aldehydes (Kramer et al. 1996).

9. The photolysis of chloroalkanes and chloroalkenes has received attention and results in the formation of phosgene as one of the final products. The photodegradation of 1,1,1-trichloroethane proceeds by hydrogen abstraction and oxidation to trichloroacetaldehyde that is degraded by a complex series of reactions to phosgene (Nelson et al. 1990; Platz et al. 1995).Tetrachloroethene is degraded by reaction with chlorine radicals and oxidation to pentachloropropanol radical which also forms phosgene (Franklin 1994). Attention has already been drawn to the significance of these reactions in the context of environmental analytes (Section 2.5), and the atmospheric dissemination of xenobiotics (Section 3.5.3).

10. Although EDTA is biodegradable under specific laboratory conditions (Belly et al. 1975; Lauff et al. 1990; Nörtemann 1992; Witschel et al. 1997), the primary mode of degradation in the natural aquatic environment involves photolysis of the Fe complex (Lockhart and Blakeley 1975; Kari and Giger 1995). Other metal complexes are relatively resistant, so that its persistence is critically determined not only by the degree of insolation but by the concentration of Fe in the environment. The available evidence suggests that, in contrast to NTA that is more readily biodegradable, EDTA is likely to be persistent except in environments in which concentrations of Fe greatly exceed those of other cations.

11. The photolytic degradation of the fluoroquinolone antibiotic enrofloxacin involves a number of reactions that produce 6-fluoro-7-amino-1-cyclopropylquinolone 2-carboxylic acid that is then degraded to CO_2 via reactions involving fission of the benzenoid ring with loss of fluoride, dealkylation, and decarboxylation (Burhenne et al. 1997a,b) (Figure 4.9).

12. Photolysis of the oxime group in the pyrazole miticide fenpyroximate resulted in the formation of two principal transformation products: the nitrile via an elimination reaction and the aldehyde by hydrolysis (Swanson et al. 1995).

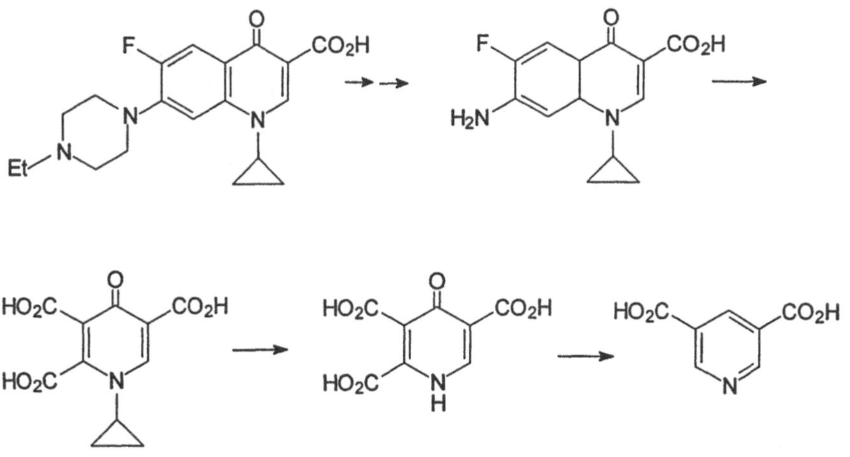

FIGURE 4.9
Photochemical degradation of enrofloxin.

13. Photochemical transformation of pyrene in aqueous media produced the 1,6- and 1,8-quinones as stable end products after initial formation of 1-hydroxypyrene (Sigman et al. 1998). Irrespective of mechanism, these reactions are formally comparable to those operating during the transformation of benzo[*a*]pyrene by *Phanerochaete chrysosporium* (Chapter 6, Section 6.2.2).

14. The transformation of isoquinoline has been studied both under photochemical conditions with hydrogen peroxide, and in the dark with hydroxyl radicals (Beitz et al. 1998). The former resulted in fission of the pyridine ring with formation of phthalic dialdehyde and phthalimide whereas the major product from the latter involved oxidation of the benzene ring with formation of the 5,8-quinone and a hydroxylated quinone.

15. In the presence of both light and hydrogen peroxide, 2,4-dinitrotoluene is oxidized to the corresponding carboxylic acid; this is then decarboxylated to 1,3-dinitrobenzene which is degraded further by hydroxylation and ring fission (Figure 4.10) (Ho 1986). Comparable reaction products were formed from 2,4,6-trinitrotoluene and hydroxylated to various nitrophenols and nitrocatechols before cleavage of the aromatic rings, and included the dimeric 2,2'carboxy-3,3',5,5'-tetranitroazoxybenzene (Godejohann et al. 1998).

Hydroxyl Radicals in the Destruction of Contaminants

The use of hydroxyl-radical mediated reactions has attracted interest in the context of destruction of contaminants, and two are provided as illustration. These reactions should be viewed against those with hydroxyl radicals that occur in the troposphere that are considered in Section 4.1.2.

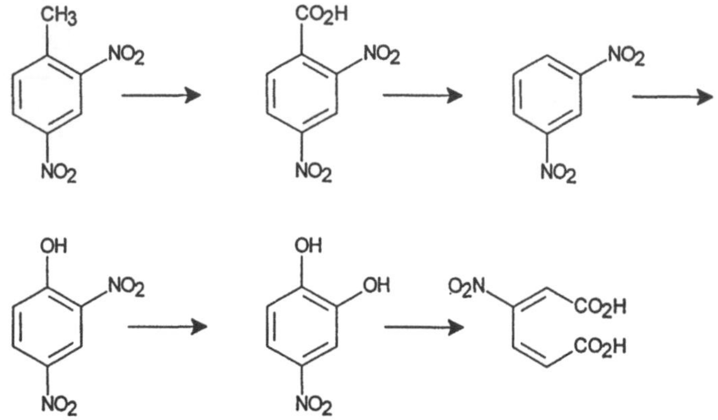

FIGURE 4.10
Photochemical transformation of 2,4-dinitrotoluene.

1. Fenton's reagent — hydrogen peroxide in the presence of Fe^{2+} or Fe^{3+} — both in the presence of oxygen and under the influence of irradiation. The reaction involves hydroxyl radicals and has been studied particularly intensively for the destruction of chlorinated phenoxyacetic acid herbicides (Sun and Pignatello 1993). Systematic investigations have been carried out on the effect of pH, the molar ratio of H_2O_2/substrate, and the possible complications resulting from the formation of iron complexes. Although this reaction may have limited environmental relevance except under rather special circumstances, an example of its use in combination with biological treatment of PAHs is given in Chapter 8, Section 8.2.1. Attention is drawn to it here since, under conditions where the concentration of oxidant is limiting, intermediates may be formed that are stable and that may possibly exert adverse environmental effects. Some examples that illustrate the formation of intermediates are given, although it should be emphasized that total destruction of the relevant xenobiotics under optimal conditions can be successfully accomplished. The structure of the products that are produced by the action of Fenton's reagent on chlorobenzene are shown in Figure 4.11a (Sedlak and Andren 1991), and those from 2,4-dichlorophenoxyacetate in Figure 4.11b (Sun and Pignatello 1993). Whereas the degradation of azo dyes by Fenton's reagent produced water-soluble and $CHCl_2$-soluble transformation products including nitrobenzene from Disperse Orange 3 that contains a nitro group, benzene was tentatively identified among volatile products from Solvent Yellow 14 (Spadaro et al. 1994).

2. Photolytic degradation on TiO_2.

   a. In the presence of slurries of TiO_2 that served as a photochemical sensitizer, methyl *t*-butyl ether was photochemically

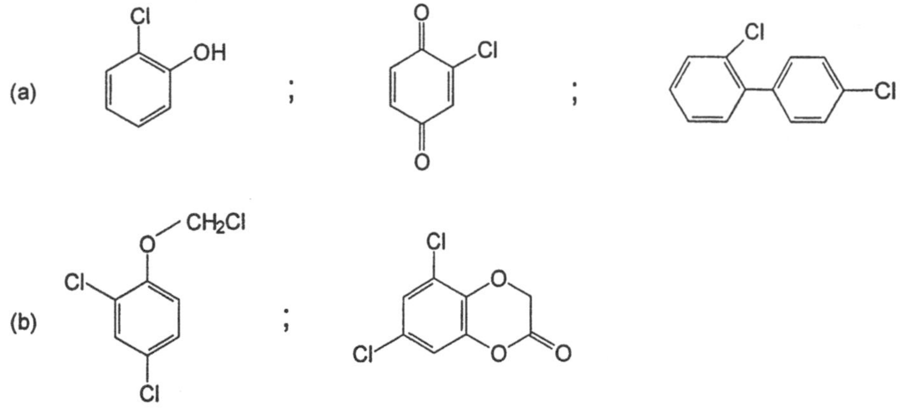

FIGURE 4.11
Transformation products from (a) chlorobenzene and (b) 2,4-dichlorophenoxyacetate.

decomposed at wavelengths < 290 nm. The mechanism involves photochemical production of a free electron in the conduction band (e_{cb}^-) and a corresponding hole (h_{vb}^+) in the valence band. Both of these produce H_2O_2, and thence hydroxyl radicals, and the products were essentially the same as those produced by hydroxyl radicals under atmospheric conditions (Barreto et al., 1995): *t*-butyl formate and *t*-butanol were rapidly formed and further degraded to formate, acetone, acetate, and but-2-ene.

b. The degradation of haloalkanes has been extensively studied and involves the same principles that have been noted earlier. For these substrates, the initial reaction is abstraction of a hydrogen atom, and this is followed by a complex series of reactions. From trichloroethene, a number of products are formed including tetrachloromethane, hexachloroethane, pentachloroethane, and tetrachlororethene, although the last two were shown to be degradable in separate experiments (Hung and Marinas 1997). In TiO_2 slurries, the photochemical degradation of chloroform, bromoform, and tetrachloromethane involves initial formation of the trihalomethyl radicals. In the absence of oxygen, these are further decomposed via dihalocarbenes to CO. Dichlorocarbene was found as an intermediate in the degradation of trichloroacetate (Choi and Hoffmann 1997).

c. The photocatalytic oxidation of various EDTA complexes has been examined (Madden et al. 1997). The rates and efficiencies were strongly dependent on the metal and the reactions are generally similar to those involved in electrochemical oxidation (Pakalapati et al. 1996).

Other Photochemically Induced Reactions

1. Two groups of reactions are important in the photochemical transformation of PAHs: those with molecular oxygen, and those involving cyclization. Illustrative examples are provided by the
   photooxidation of 7,12-dimethylbenz[a]anthracene-3,4-dihy-
   drodiol (Lee and Harvey 1986) (Figure 4.12a) and benzo[a]pyrene
   (Lee-Ruff et al. 1986) (Figure 4.12b), and the cyclization of *cis*-
   stilbene (Figure 4.12 c).

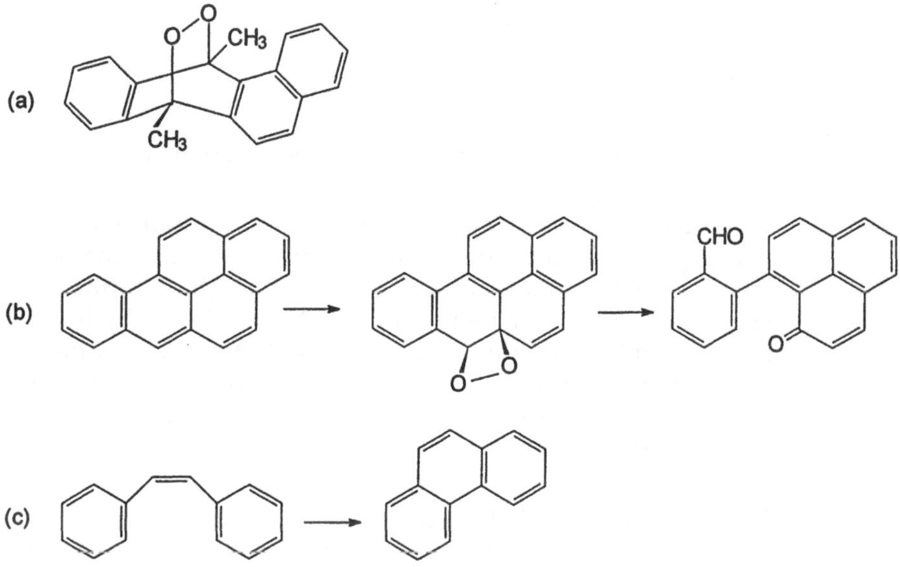

FIGURE 4.12
Photooxygenation of (a) 7,12-dimethylbenz[a]anthracene, (b) benzo[a]pyrene, and (c) photocyl-
ization of *cis*-stilbene.

2. In nonaqueous solutions, two other groups of reactions have been
   observed with polycyclic arenes: condensation via free-radical reactions and oxidative ring fission.

   a. Irradiation of benz[a]anthracene in benzene solutions in the
      presence of xanth-9-one or vanillin produced a number of trans-
      formation products tentatively identified as the result of oxida-
      tion and cleavage of ring A, ring C, ring D, and rings C and D,
      and rings B, C, and D (Jang and McDow 1997).

   b. 1-Nitropyrene is a widely distributed contaminant produced
      in the troposphere by reaction of nitrate radicals with pyrene
      that is discussed in Section 4.1.2. A solution in benzene was

photochemically transformed into 9-hydroxy-1-nitropyrene that is less mutagenic than its precursor (Koizumi et al. 1994).

3. The photochemical transformation of phenanthrene sorbed on silica gel (Barbas et al. 1996) resulted in a variety of products including *cis*-9,10-dihydrodihydroxyphenanthrene and phenanthrene-9,10-quinone, and a number of ring fission products including biphenyl-2,2′-dicarboxaldehyde, naphthalene-1,2-dicarboxylic acid, and benzo[c]coumarin. This may be compared with the products from the activated solution photooxidation of benz[a]anthracene that have already been noted.

4. The photooxidation of naphthylamines adsorbed on particles of silica and alumina produced products putatively less toxic than their precursors (Hasegawa et al. 1993) (Figure 4.13).

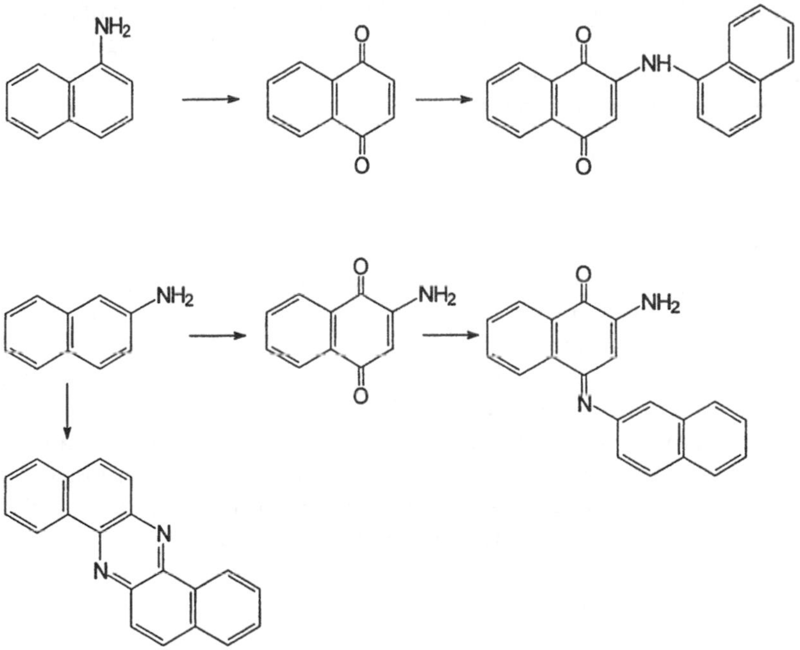

FIGURE 4.13
Photooxidation of sorbed naphthylamines.

5. It has been suggested that photochemically induced reactions may take place between biocides and biomolecules of plant cuticles: laboratory experiments have examined addition reactions between DDT and methyl oleate and have been used to illustrate reactions which result in the production of "bound" DDT residues (Figure 4.14) (Schwack 1988).

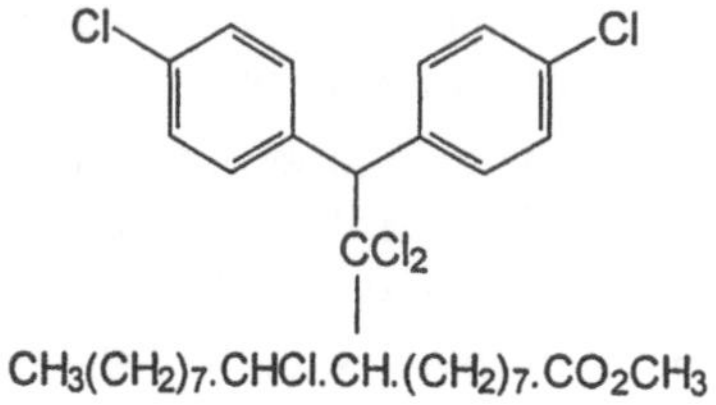

FIGURE 4.14
Product of reaction between DDT and methyl oleate.

Interactions Between Photochemical and Other Reactions

It has been shown that a combination of photolytic and biotic reactions may result in enhanced degradation of xenobiotics in municipal treatment systems, for example, of chlorophenols (Miller et al. 1988a) and benzo[*a*]pyrene (Miller et al. 1988b). Two examples may be used to illustrate the success of a combination of microbial and photochemical reactions in accomplishing the degradation of widely different xenobiotics in natural ecosystems: both of them involved marine bacteria and it therefore seems plausible to assume that such processes might be especially important in warm-water marine environments.

1. The degradation of pyridine dicarboxylates (Amador and Taylor 1990).
2. The degradation of 3- and 4-trifluoromethylbenzoate: the microbial transformation resulted in the formation of catechol intermediates that were converted into 7,7,7-trifluoro-hepta-2,4-diene-6-one carboxylate. This was subsequently degraded photochemically with the loss of fluoride (Taylor et al. 1993) (Figure 4.15). This degradation may be compared to the purely photochemical degradation of 3-trifluoromethyl-4-nitrophenol that has already been noted and contrasted with the resistance to microbial degradation of trifluoromethylbenzoates that is noted in Section 6.10.

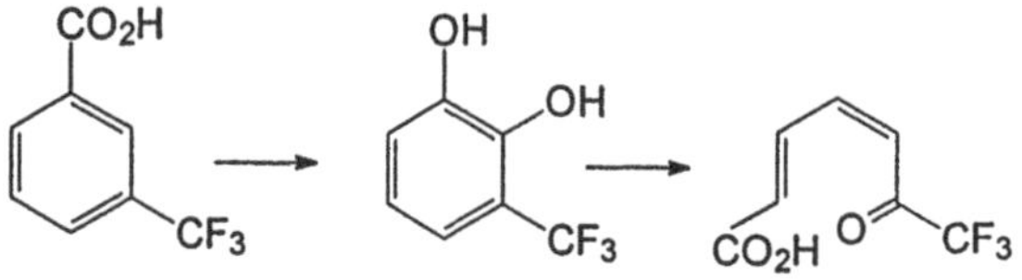

FIGURE 4.15
Microbial followed by photochemical degradation of 3-trifluorobenzoate.

Collectively, these examples illustrate the diversity of transformations of xenobiotics that are photochemically induced in aquatic and terrestrial systems. Photochemical reactions in the troposphere are also extremely important in determining the fate and persistence not only of xenobiotics but also of naturally occurring compounds. These are discussed more fully with mechanistic details in Section 4.1.2.

4.1.2 Reactions in the Troposphere

Although chemical transformations in the troposphere may seem peripheral to this discussion, these reactions should be kept in mind since their products may subsequently enter the aquatic and terrestrial environments. The persistence and the toxicity of these secondary products are therefore directly relevant to this discussion. Details of the relevant principles and details of the methodology are covered in the comprehensive treatise by Finlayson-Pitts and Pitts (1986), and reference should be made to reviews on tropospheric air pollution (Finlayson-Pitts and Pitts 1997) and atmospheric aerosols (Andreae and Crutzen 1997). The reactions are dominated by those involving free radicals.

There are several important reasons for discussing the reactions of organic compounds in the troposphere.

1. The partitioning of compounds between the various phases has been discussed in Chapter 3, and those of sufficient volatility or associated with particles may be transported over long distances. This is not a passive process, however, since important transformations may take place in the troposphere so that attention should also be directed to their transformation products.

2. Considerable attention has been given to the persistence and fate of organic compounds in the troposphere, and this has been increasingly motivated by their possible role in the production of ozone by reactions involving NO_x.

3. Concern has been expressed over the destruction of ozone in the stratosphere brought about by its reactions with chlorine atoms produced from chlorofluoroalkanes that are persistent in the troposphere and that may contribute to radiatively acting gases other than CO_2.

By way of introduction, a few examples are given here.

1. The occurrence of C_8 and C_9 dicarboxylic acids in samples of atmospheric particles and in recent sediments (Stephanou 1992; Stephanou and Stratigakis 1993) has been attributed to photochemical degradation of unsaturated carboxylic acids that are widespread in almost all biota.

2. The formation of peroxyacetyl nitrate from isoprene (Grosjean et al. 1993a) and of peroxypropionyl nitrate (Grosjean et al. 1993b)

from *cis*-3-hexen-1-ol that is derived from higher plants, may be given as illustration of important contributions to atmospheric degradation (Seefeld and Kerr 1997).

3. Attention has been given to possible adverse effects of incorporating *t*-butyl methyl ether into automobile fuels, and it has been shown that photolysis of *t*-butyl formate (that is an established product of photolysis) in the presence of NO can produce the relatively stable *t*-butoxyformyl peroxynitrate. This has a stability comparable to that of peroxyacetyl nitrate and may therefore increase the potential for disseminating NO_x (Kirchner et al. 1997).

Reactions in the troposphere are mediated by reactions involving hydroxyl radicals produced photochemically during daylight, by nitrate radicals that are significant during the night (Platt et al. 1984), by ozone, and in some circumstances by $O(^3P)$.

The overall reactions involved in the production of hydroxyl radicals are

$$O_3 + hv \rightarrow O_2 + O\ (^1D); O(^1D) + H_2O \rightarrow 2OH$$

$$O(^1D) \rightarrow O(^3P); O(^3P) + O_2 \rightarrow O_3$$

Note that Roman capitals (S, P, D) are used for the states of atoms and Greek capitals (Σ, Π, Δ) for those of molecules, and that the ground state of O_2 is a triplet $O_2(^3\Sigma)$. The reaction $O_3 + hv \rightarrow O(^1D) + O_2(^1\Delta)$ has an energy threshold at 310 nm, and the other possible reaction $O_3 + hv \rightarrow O(^1D) + O_2(^3\Sigma)$ is formally forbidden by conservation of spin. Increasing evidence has, however, accumulated to show that the rate of production of $O(^1D)$ and therefore hydroxyl radicals at wavelengths >310 nm is significant, and that, therefore, in contrast to previous assumptions, the latter reaction makes an important contribution (Ravishankara et al. 1998).

Nitrate radicals are formed from NO which is produced during combustion processes and are significant only during the night in the absence of photochemically produced OH radicals. They are formed by the reactions:

$$NO + O_3 \rightarrow NO_2 + O_2; NO_2 + O_3 \rightarrow NO_3 + O_2$$

The concentrations of all these depend on local conditions, the time of day, and both altitude and latitude. Values of $\sim 10^6$ molecules·cm^{-3} for OH, 10^8 to 10^{10} molecules.cm^{-3} for NO_3 and $\sim 10^{11}$ molecules·cm^{-3} for ozone are representative. Not all of these reactants are equally important, and the rates of reaction of a substrate vary considerably; reactions with hydroxyl radicals are generally the most important and some illustrative values are given for the rates of reaction (cm^3 s^{-1} molecule^{-1}) with hydroxyl radicals, nitrate radicals, and ozone (Atkinson 1990; summary of PAHs by Arey 1998),

	Hydroxyl Radicals	Nitrate Radicals	Ozone
n-Butane	2.54×10^{-12}	6.5×10^{-17}	9.8×10^{-24}
Acetaldehyde	15.8×10^{-12}	2.7×10^{-15}	$<10^{-20}$
Naphthalene	23.16×10^{-11}	$3.6 \times 10^{-28}[NO_2]$	$<2 \times 10^{-19}$

Survey of Reactions

The major reactions carried out by hydroxyl and nitrate radicals may be represented for a primary alkane RH or a secondary alkane R_2CH; in both, hydrogen abstraction is the initiating reaction.

1. Hydrogen abstraction:

    $$RH + HO \rightarrow R + H_2O$$

    $$RH + NO_3 \rightarrow R + HNO_3$$

2. Formation of alkylperoxy radicals:

    $$R + O_2 \rightarrow R \cdot O \cdot O$$

3. Reactions of alkylperoxy radicals with NO_x:

    $$R \cdot O \cdot O + NO \rightarrow R \cdot O + NO_2$$

    $$R \cdot O \cdot O + NO \rightarrow R \cdot O \cdot NO_2$$

4. Reactions of alkyloxy radicals:

    $$R_2CH \cdot O + O_2 \rightarrow R_2CO + HO_2$$

    $$R_2CH \cdot O + NO \rightarrow R_2CH \cdot O \cdot NO_2 \rightarrow R_2CO + HNO$$

The concentration of NO determines the relative importance of reaction 3, and the formation of NO_2 is particularly significant since this is readily photolyzed to produce $O(^3P)$ that reacts with oxygen to produce ozone. This alkane–NO_x reaction may produce O_3 at the troposphere/stratosphere interface:

$$NO_2 \rightarrow NO + O(^3P); \qquad O(^3P) + O_2 \rightarrow O_3$$

This is the main reaction for the formation of ozone, but under equilibrium conditions, the concentrations of NO_2, NO, and O_3 are interdependent and no net synthesis of O_3 occurs. When, however, the equilibrium is disturbed and NO is removed by reactions with alkylperoxy radicals (reactions 1 + 2 + 3),

$$RH + OH \rightarrow R + H_2O; R + O_2 \rightarrow RO_2; RO_2 + NO \rightarrow RO + NO_2$$

synthesis of O_3 may take place. The extent to which this occurs depends on a number of factors (Finlayson-Pitts and Pitts 1997), including the reactivity of the hydrocarbon which is itself a function of many factors. It has been proposed that the possibility of ozone formation is best described by a reactivity index incremental hydrocarbon reactivity (Carter and Atkinson 1987; 1989) that combines the rate of formation of O_3 with that of the reduction in the concentration of NO. The method has been applied, for example, to oxygenate additives to automobile fuel (Japar et al. 1991), and both anthropogenic compounds and naturally occurring hydrocarbons may be reactive.

Clearly, whether or not ozone is formed depends also on the rate at which it is destroyed, for example, by reaction with unsaturated hydrocarbons. Rates of reactions with alkanes are, as noted above, much slower than for reaction with OH radicals, and reactions with ozone are of the greatest significance with unsaturated aliphatic compounds. The pathways plausibly follow those involved in chemical ozonization (Hudlicky 1990), and some of these are noted later.

Details of the kinetics of the various reactions have been explored in detail using large-volume chambers that can be used to simulate the reactions in the troposphere, and have frequently used hydroxyl radicals formed by photolysis of methyl (or ethyl) nitrite, with the addition of NO to inhibit photolysis of NO_2. This would result in the formation of $O(^3P)$ atoms, and subsequent reaction with O_2 to produce ozone and hence NO_3 radicals from NO_2. Nitrate radicals are produced by the thermal decomposition of N_2O_5, and in experiments with O_3, a scavenger for hydroxyl radicals is added (Chapter 5, Section 5.1). Details of the different experimental procedures for the measurement of absolute and relative rates have been summarized, and attention drawn to the often considerable spread of values for experiments carried out at room temperature (~298 K) (Atkinson, 1986). It should be emphasized that in the real troposphere, both the rates — and possibly the products — of transformation will be determined by seasonal differences both in temperature and the intensity of solar radiation. These are determined both by latitude and altitude.

The kinetics of the reactions of many xenobiotics with hydroxyl and nitrate radicals have been examined under simulated atmospheric conditions and include (1) aliphatic and aromatic hydrocarbons (Tuazon et al. 1986) and substituted monocyclic aromatic compounds (Atkinson et al. 1987c); (2) terpenes (Atkinson et al. 1985a); (3) amines (Atkinson et al. 1987a); (4) heterocyclic compounds (Atkinson et al. 1985b); and (5) chlorinated aromatic hydrocarbons (Kwok et al. 1995). For PCBs (Anderson and Hites 1996), rate constants were highly dependent on the number of chlorine atoms, and calculated atmospheric lifetimes varied from 2 days for 3-chlorobiphenyl to 34 days for 2,2′,3,5′,6-pentachlorbiphenyl. It was estimated that loss by hydroxylation in the atmosphere was a primary process for removal of PCBs from the environment. It was later shown that the products were chlorinated benzoic acids produced by initial reaction with a hydroxyl radical at the 1-position followed by transannular dioxygenation at the 2- and 5-positions followed by ring fission (Brubaker and Hites 1998). Reactions of hydroxyl radicals with polychlorinated dibenzo[1,4]dioxins and dibenzofurans also play an important role for their removal from the atmosphere (Brubaker and Hites 1997). The gas phase and the particulate phase are in equilibrium, and the results show that gas-phase reactions with hydroxyl radicals are important for the compounds with fewer numbers of chlorine atoms whereas for those with larger numbers of substituents particle-phase removal is significant.

Considerable attention has been directed to determining the products from reactions of aromatic compounds and unsaturated compounds including biogenic terpenes that exhibit appreciable volatility. These studies have been conducted both in simulation chambers that have been noted and using natural sunlight in the presence of NO.

Aromatic Hydrocarbons

Ring fission of aromatic hydrocarbons may take place; for example, *o*-xylene forms diacetyl, methylglyoxal, and gloxal (Tuazon et al. 1986) which are also the products of ozonolysis (Levine and Cole 1932), while naphthalene forms 2-formylcinnamaldehyde (Arey 1998). The photooxidation of alkyl benzenes that are atmospheric contaminants with high volatility has been studied in detail and the reaction pathways have been delineated (Yu et al. 1997). Products from alkyl benzenes included both those with the ring intact such as aromatic aldehydes and quinones together with a wide range of aliphatic compounds containing alcohol, ketone, and epoxy functional groups resulting from ring fission. The significance of epoxide intermediates (Yu and Jeffries 1997) is noted in the next section. Attention is drawn later to the important reactions of arenes that result in the production of nitroarenes.

Biogenic Terpenes

Monoterpenes are appreciably volatile and are produced in substantial quantities by a range of higher plants and trees. Only some summary remarks are given here.

1. The photochemical reactions of isoprene (references in Grosjean et al. 1993a);
2. The products from reaction of α-pinene with ozone that produced a range of cyclobutane carboxylic acids (Kamens et al. 1999);
3. The rapid reactions of linalool with OH radicals, NO_3 radicals and ozone in which the major products were acetone and 5-ethenyldihydro-5-methyl-2(3*H*)-furanone (Shu et al. 1997);
4. The plant metabolite *cis*-hex-3-ene-1-ol that is the precursor of peroxypropionyl nitrate (Grosjean et al. 1993b) analogous to peroxyacetyl nitrate;

5. The degradation of many other terpenes has been examined including the β-pinene, D-limonene, and *trans*-caryophyllene (Grosjean et al. 1993b).

6. The products formed by reaction of NO$_3$ radicals with α-pinene have been identified and include pinane epoxide, 2-hydroxypinane-3-nitrate, 3-ketopinan-2-nitrate formed by reactions at the double bond, and pinonaldehyde that is produced by ring fission between C2 and C3 (Wängberg et al. 1997). These reactions should be viewed in the general context of "odd nitrogen" to which alkyl nitrates belong (Schneider et al. 1998).

7. Gas-phase products from the reactions of ozone with the monoterpenes (–)-β-pinene and (+)-sabinene which include the ketones formed by oxidative fission of the exocyclic C=C bonds as well as ozonides from the addition of ozone to this bond (Griesbaum et al. 1998).

Reentry of Tropospheric Transformation Products

Some important illustrative examples are given in which the tropospheric transformation products enter aquatic or terrestrial ecosystems by deposition on particles.

1. Halogenated Alkanes and Alkenes

The stability of perchlorofluoroalkanes is due to the absence of hydrogen atoms that may be abstracted in reaction with hydroxyl radicals. Attention has therefore been directed to chlorofluoroalkanes containing at least one hydrogen atom (Hayman and Derwcut 1997). Considerable effort has been directed to the reactions of chloroalkanes and chloroalkenes, and this deserves a more-detailed examination in the light of interest in the products formed.

a. There has been concern over the fate of halogenated aliphatic compounds in the atmosphere, and a single illustration of the diverse consequences is noted here. The initial reaction of 1,1,1-trichloroethane with hydroxyl radicals produces the $Cl_3C.CH_2$ radical by abstraction of H and then undergoes a complex series of reactions including the following:

$$Cl_3C.CH_2 + O_2 \rightarrow Cl_3C.CH_2O_2$$
$$2Cl_3C.CH_2O_2 \rightarrow 2Cl_3C.CH_2O$$
$$Cl_3C.CH_2O + O_2 \rightarrow Cl_3C.CHO$$

In addition, the alkoxy radical $Cl_3C.CH_2O$ produces highly reactive phosgene ($COCl_2$) (Platz et al. 1995; Nelson et al. 1990) that has been identified in atmospheric samples and was attributed to the transformation of gem-dichloro aliphatic compounds (Grosjean 1991).

b. The atmospheric degradation of tetrachloroethene produces trichloroacetyl chloride as the primary intermediate which is formed by an initial reaction with Cl radicals followed by the following reactions (Franklin 1994):

$$Cl_3CCl_2 + O_2 \rightarrow Cl_3C.CCl_2O_2$$
$$Cl_3C.CCl_2O_2 + NO \rightarrow Cl_3C.CCl_2O + NO_2$$
$$Cl_3C.CCl_2O \rightarrow Cl_3C.COCl + Cl$$
$$\rightarrow COCl_2 + CCl_3.$$
$$CCl_3 + O_2 + NO \rightarrow COCl_2 + NO_2 + Cl$$

An overview of the reactions involving $X_3C.CHYZ$, where X, Y, and Z are halogen atoms, has been given in the context of ozone depletion (Hayman and Derwent 1997). Interest in the formation of trichloroacetaldehyde formed from trichloroethane and tetrachloroethene is heightened by the phytotoxicity of trichloroacetic acid (Frank et al. 1994), and by its occurrence in rainwater which seems to be a major source of this contaminant (Müller et al. 1996). The situation in Japan seems, however, to underscore the possible significance of other sources including chlorinated wastewater (Hashimoto et al. 1998).

Low concentrations of trifluoroacetate have been found in lakes in California and Nevada (Wujcik et al. 1998). It is formed by atmospheric reactions from 1,1,1,2-tetrafluoroethane, and from the chlorofluorocarbon replacement compound $CF_3.CH_2F$ (HFC-134a) in an estimated yield of 7 to 20% (Wallington et al. 1996); CF_3OH formed from CF_3 in the stratosphere is apparently a sink for its oxidation products (Wallington and Schneider 1994).

$$CF_3.CH_2F + OH \rightarrow CF_3.CHF$$
$$CF_3.CHF + O_2 \rightarrow CF_3.CHFO_2$$
$$CF_3.CHFO_2 + NO \rightarrow CF_3.CHFO + NO_2$$
$$CF_3.CHFO + O_2 \rightarrow CF_3.COF$$
$$CF_3.CHFO \rightarrow CF_3 + H.COF$$

Although trifluoroacetate is accumulated by a range of biota through incorporation into biomolecules (Standley and Bott 1998), unlike trichloroacetate it is only weakly phytotoxic and there is no evidence for its inhibitory effect on methanogenesis (Emptage et al. 1997).

2.   *Arenes and Nitroarenes*

The transformation of arenes in the troposphere has been discussed in detail (Arey 1998). Destruction can be mediated by reaction with hydroxyl radicals, and from naphthalene, a wide range of compounds is produced, including 1- and 2-naphthols, 2-formylcinnamaldehyde, phthalic anhydride, and, with less certainty, 1,4-naphthoquinone and 2,3-epoxynaphthoquinone. Both

1- and 2-nitronaphthalene were formed through the intervention of NO_2 (Bunce et al. 1997). Attention has also been directed to the composition of secondary organic aerosols from the photooxidation of monocyclic aromatic hydrocarbons in the presence of NO_x (Forstner et al. 1997); the main products from a range of alkylated aromatics were 2,5-furandione and the 3-methyl and 3-ethyl congeners.

Considerable attention has been directed to the formation of nitroarenes which may be formed by two different mechanisms: (a) initial reaction with hydroxyl radicals followed by reactions with nitrate radicals or NO_2 and (b) direct reaction with nitrate radicals. The first is important for arenes in the troposphere, whereas the second is a thermal reaction that occurs during combustion of arenes. The kinetics of formation of nitroarenes by gas-phase reaction with N_2O_5 has been examined for naphthalene (Pitts et al. 1985a) and methylnaphthalenes (Zielinska et al. 1989); biphenyl (Atkinson et al. 1987b); acephenanthrylene (Zielinska et al. 1988), and for adsorbed pyrene (Pitts et al. 1985b). Both 1- and 2-nitronaphthalene were formed through OH-radical-initiated reactions with napthhalene by the intervention of NO_2 (Bunce et al. 1997). From naphthalene the major product from the first group of reactions is 2-nitronaphthalene, and a number of other nitroarenes have been identified including nitropyrene and nitrofluoranthenes (Arey 1998). The tentative identification of hydroxylated nitroarenes in air particulate samples (Nishioka et al. 1988) is consistent with operation of this dual mechanism. Reaction of methyl arenes with nitrate radicals in the gas phase gives rise to a number of products. From toluene, the major product was benzaldehyde with lesser amounts of 2-nitrotoluene > benzyl alcohol nitrate > 4-nitrotoluene > 3-nitrotoluene (Chlodini et al. 1993). An interesting example is the formation of the mutagenic 2-nitro- and 6-nitro-6*H*-dibenzo[*b,d*]pyran-6-ones (Figure 4.16) from the oxidation of phenanthrene in the presence of NO_x and methyl nitrite as a source of hydroxyl radicals (Helmig et al. 1992a). These compounds have been identified in samples of ambient air (Helmig et al. 1992b), and analogous compounds from pyrene have been tentatively identified (Sasaki et al., 1995). These compounds add further examples to the list of mononitroarenes that already include 2-nitropyrene and 2-nitrofluoranthene, and it appears plausible to suggest that comparable reactions are involved in the formation

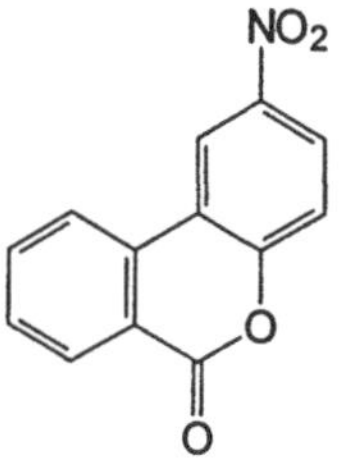

FIGURE 4.16
Product from the photochemical reaction of phenanthrene and NO_x.

of the 1,6- and 1,8-dinitroarenes that have been identified in diesel exhaust. 3-nitrobenzanthrone that is formally analogous to the dibenzopyrones noted earlier has also been identified in diesel exhaust and is also highly mutagenic to *Salmonella typhimurium* strain TA 98 (Enya et al. 1997).

Many nitroarenes are direct-acting frame-shift mutagens in the Ames test (Rosenkranz and Mermelstein 1983) and, although the mechanism has not been finally resolved, it appears to involve metabolic participation of the test organisms. 2-nitronaphthalene is a potent mutagen. In addition, nitroarenes may be reduced microbiologically (Chapter 6, Section 6.8.2) in terrestrial and aquatic systems to the amino compounds that have two undesirable properties: (a) some, including 2-aminonaphthalene are carcinogenic to mammals and (b) they react with components of humic and fulvic acids (Section 3.2.4) which makes them more recalcitrant to degradation and therefore more persistent in ecosystems.

It should also be noted that a wide range of azaarenes are produced during combustion (Herod 1998) and may enter the troposphere, so that formation of the corresponding nitro derivatives may occur.

3. Alkylated Arenes

The products from the oxidation of alkylbenzenes under simulated atmospheric conditions have been noted earlier. Both ring epoxides that were highly functionalized and aliphatic epoxides from ring fission were tentatively identified (Yu and Jeffries 1997), and formation of the latter, many of which are mutagenic, may cause further concern over transformation products from monocyclic aromatic hydrocarbons in the atmosphere.

4. Sulfides and Disulfides

An example in which formation of a carbon radical is not the initial reaction is provided by the atmospheric reactions of organic sulfides and disulfides. They also provide an example in which rates of reaction with nitrate radicals exceed those with hydroxyl radicals. 2-dimethylthiopropionic acid is produced by algae and by the marsh grass *Spartina alternifolia*, and may then be metabolized in sediment slurries under anoxic conditions to dimethyl sulfide (Kiene and Taylor 1988), and by aerobic bacteria to methyl sulfide (Taylor and Gilchrist 1991). It should be added that methyl sulfide can be produced by biological methylation of sulfide itself (HS^-) (Section 6.11.4). Dimethyl sulfide — and possibly also methyl sulfide — is oxidized in the troposphere to sulfur dioxide and methanesulfonic acids.

$$CH_3 \cdot SH \rightarrow CH_3 \cdot SO_3H$$

$$CH_3 \cdot S \cdot S \cdot CH_3 \rightarrow CH_3 \cdot SO_3H + CH_3 \cdot SO$$

$$CH_3 \cdot SO \rightarrow CH_3 + SO_2$$

It has been suggested that these compounds may play a critical role in promoting cloud formation (Charleson et al. 1987) so that the long-term effect of the biosynthesis of methyl sulfides on climate alteration may be considerable — and yet at first glance this seems far removed from the production of an osmolyte by higher plants, its metabolism in aquatic systems, or microbial methylation. The occurrence of methyl sulfates in atmospheric samples (Eatough et al. 1986) should be noted although the mechanism of its formation appears not to have been established fully. These reactions provide a good example of the long chain of events which may bring about environmental effects through the subtle interaction of biotic and abiotic reactions in both the aquatic and atmospheric environments.

Appreciation of the interactive processes outlined earlier has been able to illuminate discussion on mechanisms of problems as diverse as acidification of water masses, climate alteration, ozone formation, and destruction, and the possible environmental roles of trichloroacetic acid and nitroarenes. The analysis and distribution of these—and other—transformation products is therefore clearly motivated (Sections 2.5 and 3.6).

4.1.3 Chemically Mediated Transformation Reactions

Only a limited number of the plethora of known chemical reactions are involved in the transformations of xenobiotics. An attempt is made merely to present some examples of chemical degradation or transformation on the basis of a classification of the reactions that take place.

4.1.3.1 Hydrolytic Reactions

Organic compounds containing carbonyl groups flanked by alkoxy groups (esters) or by amino or substituted amino groups (amides, carbamates, and ureas) may be hydrolyzed by purely abiotic reactions under appropriate conditions of pH; the generally high pH of seawater (~8.2) may be noted so that chemical hydrolysis may be quite important in this environment. On the other hand, although very few natural aquatic ecosystems have pH values sufficiently low for acidic hydrolysis to be of major importance, this may be important in terrestrial systems. It is therefore important to distinguish between alkaline or neutral, and acidic hydrolytic mechanisms. It should also be appreciated that both hydrolytic and photolytic mechanisms may operate simultaneously and that the products may not necessarily be identical.

Substantial numbers of important agrochemicals contain the carbonyl groups noted earlier, so that abiotic hydrolysis may be the primary reaction in their transformation; the example of carbaryl has already been cited (Wolfe et al. 1978a). The same general principles may be extended to phosphate and thiophosphate esters, although in these cases, it is important to bear in mind the stability to hydrolysis of primary and secondary phosphate esters under neutral or alkaline conditions that prevail in most natural ecosystems. On the other hand, sulfate esters and sulfamides are generally quite resistant to chemical hydrolysis except under rather drastic conditions so that their hydrolysis is generally mediated by sulfatases and sulfamidases. A few examples will be given to illustrate the diversity of hydrolytic reactions; these involve structurally diverse agrochemicals that may enter aquatic systems by leaching.

1. The cyclic sulfite of a- and b-endosulfan (Singh et al. 1991);
2. The carbamate phenmedipham that results in the intermediate formation of m-tolyl isocyanate (Figure 4.17) (Bergon et al. 1985);

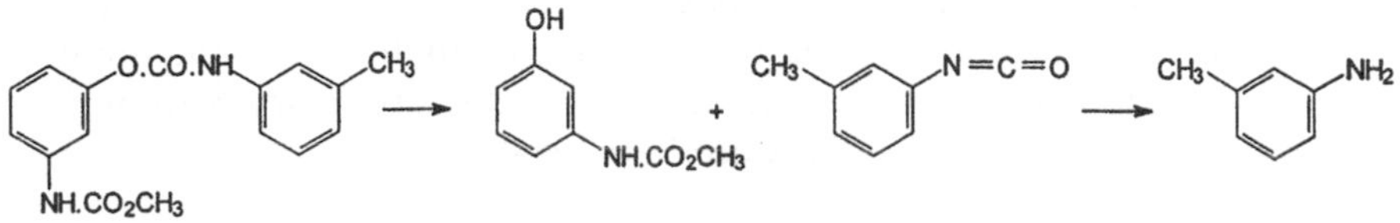

FIGURE 4.17
Hydrolysis of phenmedipham.

3. 2-(thiocyanomethylthio)benzthiazole with initial formation of 2-thiobenzthiazole; this metabolite is then rapidly degraded photochemically to benzthiazole and 2-hydroxybenzthiazole (Brownlee et al. 1992);
4. Aldicarb that undergoes simple hydrolysis at pH values above 7 whereas at pH values below 5, an elimination reaction intervenes (Figure 4.18) (Bank and Tyrrell 1984).

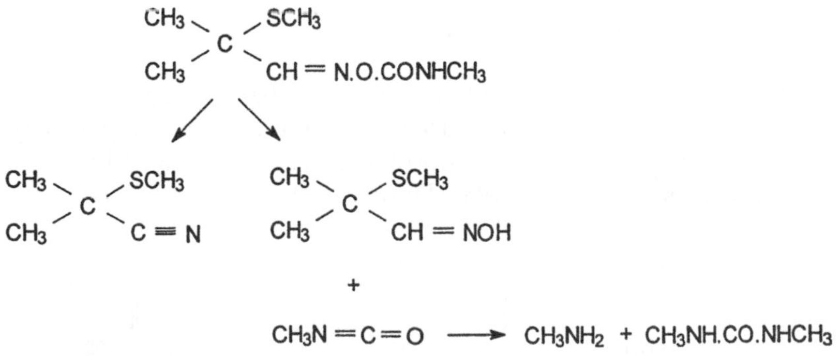

FIGURE 4.18
Hydrolysis of aldicarb.

5. The sulfonyl urea sulfometuron methyl that is stable at neutral or alkaline pH values but is hydrolyzed at pH 5 to methyl 2-aminosulfonylbenzoate that is cyclized to saccharin (Figure 4.19) (Harvey et al. 1985). The original compound is completely degraded to CO_2 by photolysis.

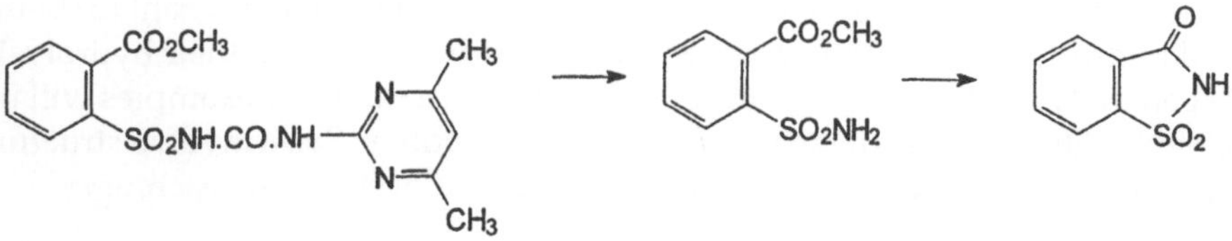

FIGURE 4.19
Hydrolysis of sulfometuron.

6. The pyrethroid insecticides fenvalerate and cypermethrin that are hydrolyzed under alkaline conditions at low substrate concentrations, but at higher concentrations the initially formed 3-phenoxybenzaldehyde reacts further with the substrate to form dimeric compounds (Figure 4.20) (Camilleri 1984).

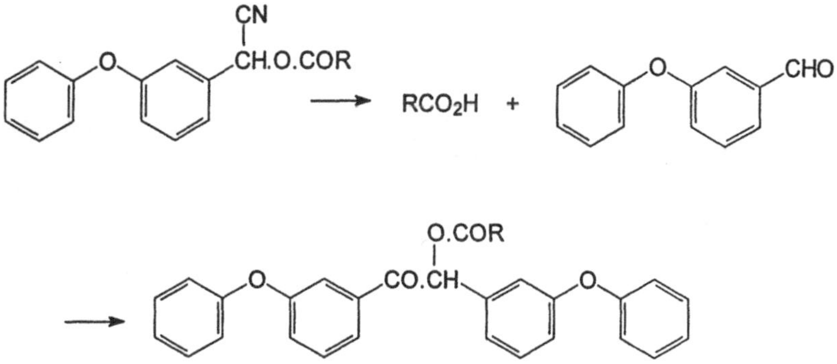

FIGURE 4.20
Hydrolysis of the pyrethroid insecticides fenvalerate and cypermethrin.

7. The sulfonyl urea herbicide rimsulfuron that is degraded increasingly rapidly at pHs from 5 to 9; the main degradation pathway is by rearrangement of the sulfonyl urea group followed by hydrolysis (Schneiders et al. 1993) (Figure 4.21).

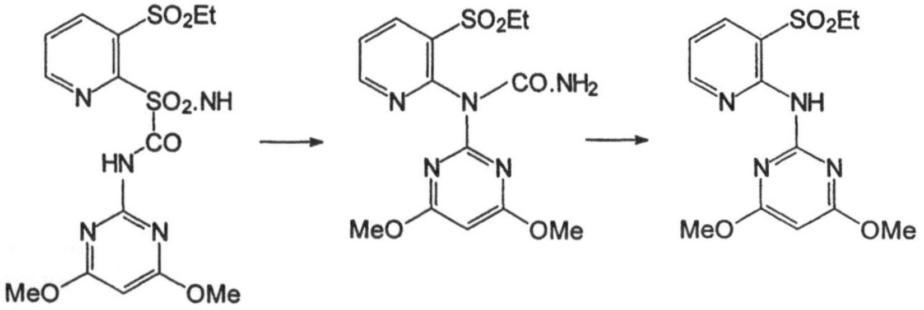

FIGURE 4.21
Hydrolysis of rimsulfuron.

8. The thiophosphate phorate that is degraded in aqueous solutions at pH 8.5 to yield diethyl sulfide and formaldhyde that are formed by nucleophilic attack either at the P=S atom or the methylene dithioketal carbon atom (Hong and Pehkonen 1998).

Three important comments should be added:

1. It should be emphasized that abiotic hydrolysis generally accomplishes only a single step in the ultimate degradation of the compounds that have been used for illustration. The intervention of subsequent biotic reactions is therefore almost invariably necessary for their complete mineralization; these reactions are discussed more fully in Chapter 6.

2. The operation of these hydrolytic reactions is independent of the oxygen concentration of the system so that — in contrast to biotic degradation and transformation — these reactions may occur effectively under both aerobic and anaerobic conditions.

3. Rates of hydrolysis may be influenced by the presence of dissolved organic carbon or sediment and the effect is determined by the structure of the compound and by the kinetics of its association with these components. For example, whereas the neutral hydrolysis of chlorpyrifos was unaffected by sorption to sediments, the rate of alkaline hydrolysis was considerably slower (Macalady and Wolf 1985); humic acid also reduced the rate of alkaline hydrolysis of 1-octyl 2,4-dichlorophenoxyacetate (Perdue and Wolfe 1982). Conversely, sediment sorption had no effect on the neutral hydrolysis of 4-chlorostilbene oxide although the rate below pH 5 where acid hydrolysis dominates was reduced (Metwally and Wolfe 1990).

4.1.3.2 Dehalogenation Reactions

Reductive Processes

The mechanism of chemical dechlorination of a range of organochlorine compounds has received increasing prominence. Attention has been directed to the role of corrins and porphyrins in the absence of biological systems, and a number of structurally diverse compounds have been shown to be dechlorinated including DDT (Zoro et al. 1974), lindane (Marks et al. 1989), mirex (Holmstead 1976), C_1 chloroalkanes (Krone et al. 1989), C_2 chloroalkenes (Gantzer and Wackett 1991), and C_2 chloroalkanes (Schanke and Wackett 1992). Detailed mechanistic examination of the dehydrochlorination of pentachloroethane to tetrachloroethene reveals, however, the potential complexity of this reaction, and the possibly significant role of pentachloroethane in the abiotic transformation of hexachloroethane (Roberts and Gschwend 1991). Considerable attention has been directed to dehalogenation mediated by corrinoids and porphyrins in the presence of a chemical reductant (references. in Workman et al. 1997), and an illustration is provided by the dechlorination and elimination reactions carried out by titanium(III) citrate and hydroxocobalamin (Bosma et al. 1994); hexachlorobuta-1,3-diene was dechlorinated to the pentachloro compound, and by dechlorination and elimination successively to trichloro-but-1-ene-3-yne (probably the 1,2,2-trichloro isomer) and but-1-ene-2-yne. The specificity of corrins and porphyrins is of particular interest since it seems to be significantly less than that of the enzymes generally implicated in microbial dechlorination. At the same time, however, it should be appreciated that both porphyrins and corrins are constituents of many bacteria — the porphyrins as prosthetic groups of cytochromes and the corrins as the chromophore in vitamin B_{12} coenzyme and related compounds. The interesting — though possibly philosophical rather than scientific — question then arises whether reactions carried out by cells containing these pyrrolic compounds are biochemically or chemically mediated. The study of these reactions may, however, help to elucidate the mechanism of microbial dechlorination reactions and this is illustrated further in Chapter 6 (Sections 6.4.4 and 6.6). Interest in the adverse environmental effects of chlorofluoroalkanes has stimulated interest in their anaerobic degradation which is probably mediated by abiotic reactions possibly involving porphyrins (Lovely and Woodward 1992; Lesage et al. 1992) although it should be noted that the C–F bond is apparently retained in the products (Lesage et al. 1992).

An additional aspect of these dehalogenations that elucidates the role of vitamin B_{12} is provided by experiments with *Shewanella alga* strain BrY (Workman et al. 1997). This organism carries out reduction of Fe(III) and Co(III) during growth with lactate and H_2, and was used to reduce vitamin B_{12a} anaerobically in the presence of an electron donor. The biologically reduced vitamin B_{12} was then able to transform tetrachloromethane to CO.

Nucleophilic Reactions

The foregoing reactions involve reductive dechlorination or elimination, but nucleophilic displacement of chloride may also be important in some circumstances. This has been examined with dihalomethanes using HS⁻ at concentrations that might be encountered in environments where active anaerobic sulfate reduction is taking place. The rates of reaction with HS⁻ exceeded those for hydrolysis, and at pH values above 7 in systems in equilibrium with elementary sulfur, the rates with polysulfide exceeded those with HS⁻. The principal product from dihalomethanes is the polythiomethylene HS–$(CH_2.S)_nH$ (Roberts et al. 1992). Two examples of the potential role of HS⁻ in formally reductive dechlorination are provided by (1) the formation of tetrachloroethene from hexachloroethane in the presence of 5-hydroxy-naphtho-1,4-quinone (Perlinger et al. 1996) and (2) the reduction of hexachloroethane by both HS⁻ and polysulfides (S_4^{2-}) to tetrachloroethene and pentachloroethane (Miller et al. 1998).

Attention is briefly drawn to procedures that have been considered for the destruction of xenobiotics. Although these are carried out under conditions that are not relevant to the aquatic environment, they may be useful as a background to alternative remediation programs that are considered in Chapter 8. Two examples involving CaO and related compounds may be used as examples of important and unprecedented reactions.

1. The destruction of DDT by ball-milling with CaO resulted in substantial loss of chloride and produced a graphitic product containing some residual chlorine. In addition, an interesting rearrangement occurred with the formation of bis(4-chlorophenyl)ethyne that was identified by ^{1}H NMR (Hall et al. 1996) (Figure 4.22).

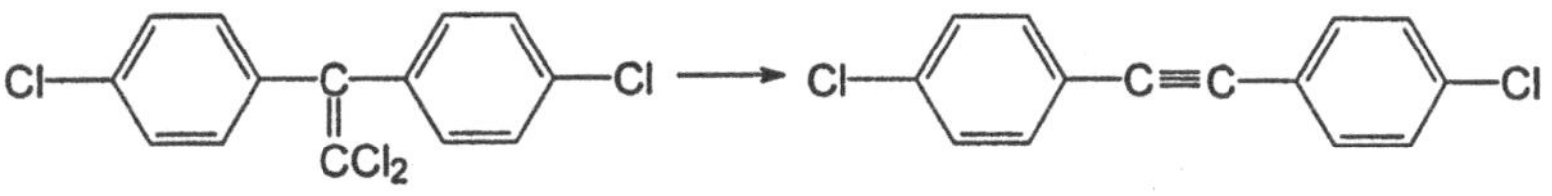

FIGURE 4.22
Alkaline destruction of DDT.

2. Treatment of 1,2,3,4-tetrachlorodibenzo[1,4]dioxin on Ca-based sorbents at 160 to 300°C resulted in the conversion to products with molecular masses of 302 and 394 that were tentatively identified as chlorinated benzofurans and 1-phenylnaphthalene or anthracenes (Gullett et al. 1997).

4.1.3.3 Oxidation Reactions

These have already been noted in the context of hydroxyl radical–initiated oxidations, and reference should be made to an extensive review by Worobey (1989) that covers a wider range of abiotic oxidations. Some have attracted interest in the context of the destruction of xenobiotics, and reference has already been made to photochemically - induced oxidations. Their combination with biological treatment of PAHs is noted again in Chapter 8, Section 8.2.1.

An interesting study examined the anodic oxidation of EDTA at alkaline pH on a smooth platinum electrode (Pakalapati et al. 1996). Degradation is initiated by removal of the acetate side chains as formaldehyde, followed by deamination of the ethylene diamine that is formed to glyoxal and oxalate. Oxalate and formaldehyde are oxidized to CO_2, and adsoption was an integral part of the oxidation.

4.1.3.4 Reduction Reactions

Reduction of monocyclic aromatic nitro compounds with reduced sulfur compounds mediated by a naphthoquinone or an iron porphyrin

(Schwarzenbach et al. 1990), and by Fe(II) and magnetite produced by the action of the anaerobic bacterium *Geobacter metallireducens* (Heijman et al. 1993) have been demonstrated, and it has been suggested that such reactions may be significant in determining the fate of aromatic nitro compounds in the environment. The reduction of nitrobenzenes in the presence of sulfide and natural organic matter from a variety of sources has also been demonstrated and could be expected to be a significant abiotic process in some natural systems (Dunnivant et al. 1992).

Cell-free supernatants may catalyze reductions; (1) the reduction of aromatic nitro compounds by the filtrate from a strain of *Streptomyces* sp. that is known to synthesize cinnaquinone (2-amino-3-carboxy-5-hydroxybenzo-1,4-quinone) and the 6,6'–diquinone (dicinnaquinone) as secondary metabolites (Glaus et al. 1992), and (2) the dechlorination of tetrachloro- and trichloromethane by extracellular products from *Methanosarcina thermophila* grown with Fe^0 (Novak et al. 1998).

The aerobic biodegradation of N-heterocylic aromatic compounds frequently involves a reductive step (Section 6.3.1.3), but purely chemical reduction may take place under highly anaerobic conditions and has, for example, been encountered with the substituted 1,2,4-triazolo[1,5a]pyrimidine Flumetsulam (Wolt et al. 1992) (Figure 4.23).

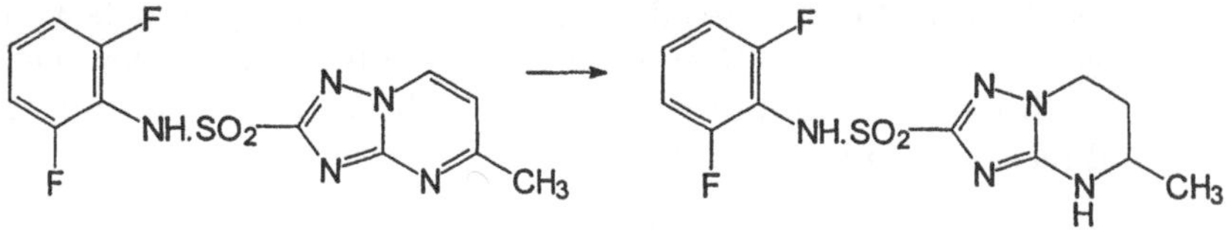

FIGURE 4.23
Reductive degradation of 1,2,4-triazolo[1,5a]pyrimidine.

4.1.3.5 Thermal Reactions during Incineration

The products of incomplete combustion may be associated with particulate matter before their discharge into the atmosphere, and these may ultimately enter the aquatic and terrestrial environments in the form of precipitation and dry deposition. The spectrum of compounds involved is quite extensive and a number of them are formed by reactions between hydrocarbons and inorganic sulfur or nitrogen constituents of air. Some illustrative examples involving other types of reaction include the following:

1. The pyrolysis of vinylidene chloride produces a range of chlorinated aromatic compounds including polychlorinated benzenes, styrenes, and naphthalenes (Yasahura and Morita 1988), and a series of chlorinated acids including chlorobenzoic acids has been identified in emissions from a municipal incinerator (Mowrer and Nordin 1987).

2. Nitroaromatic compounds have been identified in diesel engine emissions (Salmeen et al. 1984) and attention has been directed particularly to 1,8- and 1,6-dinitropyrene since these compounds are mutagenic and possibly carcinogenic (Nakagawa et al. 1983).

3. A wide range of azaarenes including acridines and benzoacridines, 4-azafluorene, and 10-azabenzo[*a*]pyrene (Figure 4.24) has been identified in particulate samples of urban air and some of these have been recovered from contaminated sediments (Yamauchi and Handa 1987).

FIGURE 4.24
Azaarenes identified in particulate samples of urban air.

4. Ketonic and quinonoid derivatives of aromatic hydrocarbons have been identified in automobile (Alsberg et al. 1985) and diesel exhaust particulates (Levsen 1988), and have been recovered from samples of marine sediments (Fernandéz et al. 1992).

5. Halogenated phenols particularly 2-bromo-, 2,4-dibromo-, and 2,4,6-tribromophenol have been identified in automotive emissions and are the products of thermal reactions involving dibromoethane fuel additive (Müller and Buser 1986). It can therefore no longer be assumed that such compounds are exclusively the products of biosynthesis by marine algae.

6. Complex reactions occur during high-temperature treatment of aromatic hydrocarbons. An important class of reactions involve the cyclization and condensation of simpler PAHs to form highly condensed polycyclic compounds. This is discussed more fully by Zander (1995).

   a. A number of pentacyclic aromatic hydrocarbons have been identified as products of the gas-phase pyrolysis of methyl naphthalenes. These, from 1-methyl- and 2-methylnaphthalene, were formed by dimerization (Lang and Buffleb 1958) at various positions, whereas direct coupling with loss of the methyl group was found to be dominant with 2-methylnaphthalene (Lijinsky and Taha 1961) (Figure 4.25).

   b. A hypothetical scheme involving 2-carbon and 4-carbon additions has been used to illustrate the formation of coronene (circumbenzene) and ovalene (circumnaphthalene) from phenanthrene (Figure 4.26).

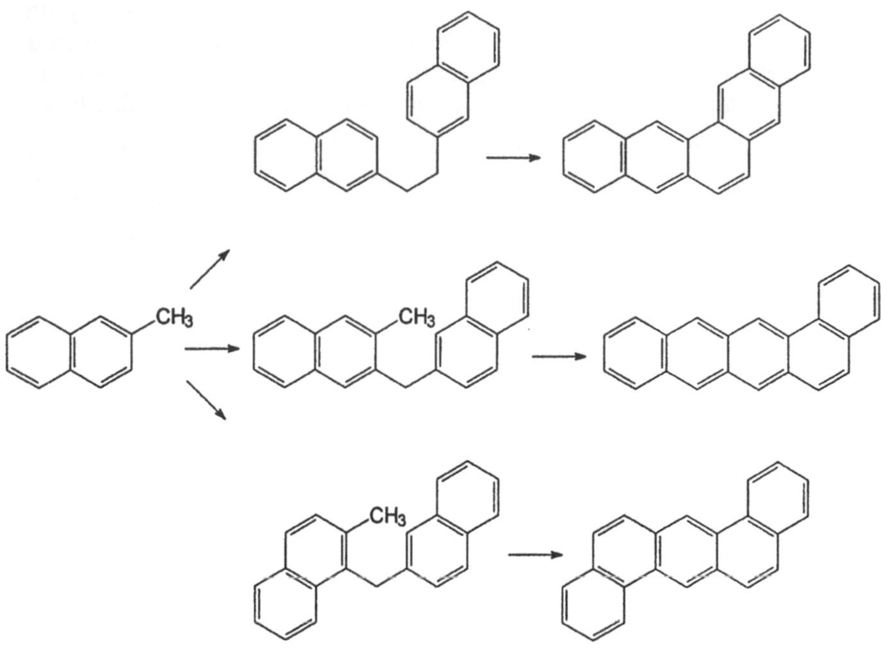

FIGURE 4.25
Products from the pyrolysis of 2-methylnaphthalene.

4.2 Biotic Reactions

It is generally conceded that biotic reactions are of great significance in determining the fate and persistence of organic compounds in most natural aquatic ecosystems. By way of providing continuity with the abiotic reactions that have been discussed earlier, one example may be given of the long chain of events which may bring about environmental effects through the subtle interaction of biotic and abiotic reactions: 2-dimethylthiopropionic acid is produced by algae and by the marsh grass *Spartina alternifolia*, and may then be metabolized in sediment slurries under anoxic conditions to dimethyl sulfide (Kiene and Taylor 1988), and by aerobic bacteria to methyl sulfide (Taylor and Gilchrist 1991).

$$(CH_3)_2S \cdot CH_2 \cdot CO_2H \rightarrow CH_3 \cdot S \cdot CH_2 \cdot CH_2 \cdot CO_2H \rightarrow HS \cdot CH_2 \cdot CH_2 \cdot CO_2H + CH_3SH$$

Dimethyl sulfide — and possibly also methyl sulfide — is oxidized in the troposphere to sulfuric and methanesulfonic acids, and it has been suggested that these compounds may play a critical role in promoting cloud formation

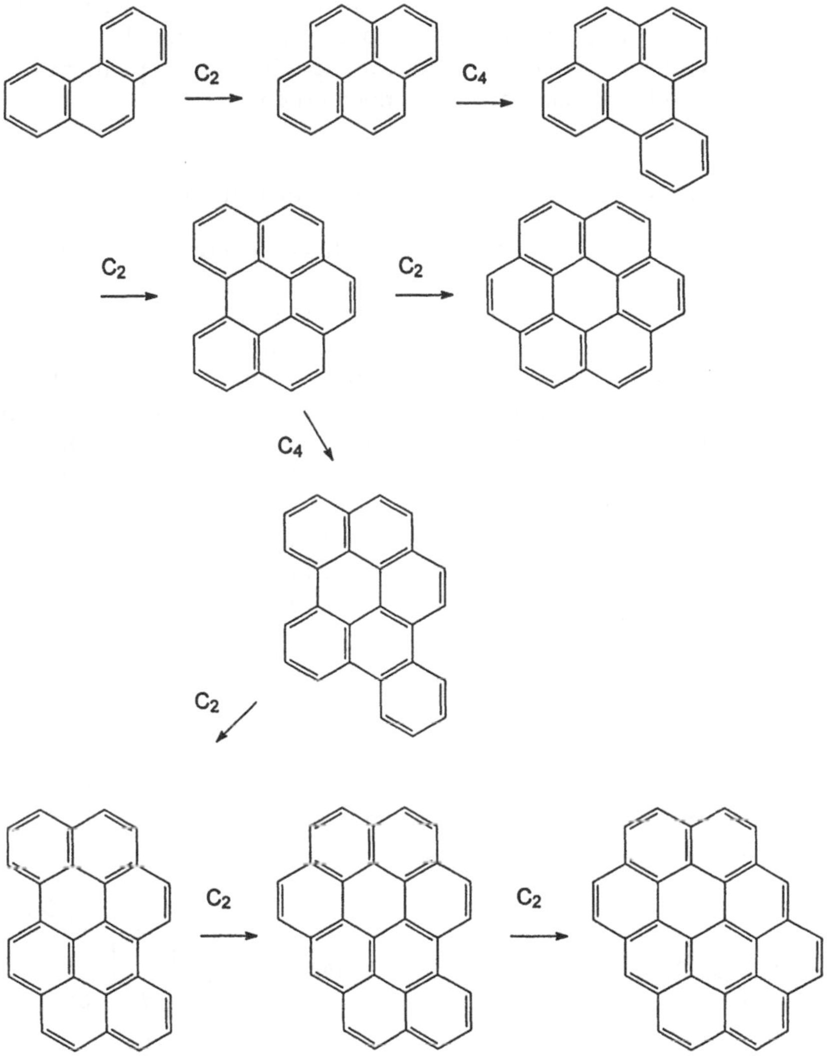

FIGURE 4.26
Successive C_2 and C_4 addition reactions.

(Charleson et al. 1987). It should be added that sulfide itself can be biologically methylated to methyl sulfide, and this is noted in Chapter 6, Section 6.11.4. The long-term effect of the biosynthesis of methyl sulfides on climate alteration may be considerable — and yet at first glance, this seems far removed from the production of an osmolyte by higher plants, its metabolism in aquatic systems, or microbial methylation.

Bacteria, cyanobacteria, fungi, yeasts, and algae comprise a large and diverse number of taxa. Only a relatively small number of even the genera have, however, been been examined, and there is no way of determining how representative of the groups these are. Care should therefore be exercised in drawing conclusions about the metabolic capability of the plethora of taxa included within these major groups of microorganisms.

4.2.1   Definitions — Degradation and Transformation

It is essential at the start to make a clear distinction between biodegradation and biotransformation.

Biodegradation under aerobic conditions results in the mineralization of an organic compound to carbon dioxide and water and — if the compound contains nitrogen, sulfur, phosphorus, or chlorine — with the release of ammonium (or nitrite), sulfate, phosphate, or chloride. These inorganic products may then enter well-established geochemical cycles. Under anaerobic conditions, methane may be formed in addition to carbon dioxide, and sulfate may be reduced to sulfide.

During biotransformation, on the other hand, only a restricted number of metabolic reactions is accomplished, and the basic framework of the molecule remains essentially intact. Some illustrative examples of biotransformation reactions include the following:

1. The hydroxylation of dehydroabietic acid by fungi (Figure 4.27) (Kutney et al. 1982);

2. The epoxidation of alkenes by bacteria (Patel et al. 1982; van Ginkel et al. 1987); this is discussed again in Chapter 6, Section 6.1.3;

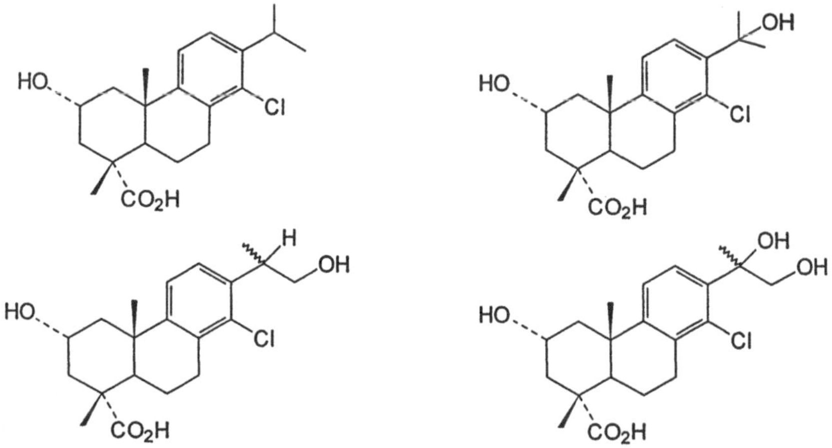

FIGURE 4.27
Biotransformation of dehydroabietic acid by *Mortierella isabellina.*

3. The formation of 16-chlorohexadecyl-16-chlorohexadecanoate from hexadecyl chloride by *Micrococcus cerificans* (Kolattukudy and Hankin 1968)

$$CH_3[CH_2]_{14} \cdot CH_2Cl \rightarrow ClCH_2[CH_2]_{14} \cdot CH_2 \cdot O \cdot CO \cdot [CH_2]_{14} \cdot CH_2Cl$$

4. The *O*-methylation of chlorophenols to anisoles by fungi (Cserjesi and Johnson 1972; Gee and Peel 1974) and by bacteria (Suzuki 1978; Rott et al. 1979; Neilson et al. 1983; Häggblom et al. 1988);

5. The formation of glyceryl-2-nitrate from glyceryl trinitrate by *Phanerochaete chrysosporium* (Servent et al. 1991).

The initial biotransformation products may, in some cases, be incorporated into cellular material. For example, the carboxylic acids formed by the oxidation of long-chain *n*-alkyl chlorides were incorporated into cellular fatty acids by strains of *Mycobacterium* sp.(Murphy and Perry 1983), and metabolites of metolachlor that could only be extracted from the cells with acetone were apparently chemically bound to unidentified sulfur-containing cellular components (Liu et al. 1989). More extensive details of a wider range of microbial transformation reactions will be found throughout Chapter 6.

Biodegradation and biotransformations are, of course, alternatives, but they are not mutually exclusive. For example, it has been suggested that for chlorophenolic compounds, the *O*-methylation reaction may be an important alternative to reactions that bring about their degradation (Allard et al. 1987). Apart from the environmental significance of biotransformation reactions, many of them have enormous importance in biotechnology: for example, in the synthesis of sterol derivatives, and in reactions that take advantage of the oxidative potential of methanotrophic bacteria (Lidstrom and Stirling 1991) and of rhodococci (Finnerty 1992). A few of these and related reactions are discussed further in Chapter 6, Section 6.11.

It is important also to consider the degradation of xenobiotics in the wider context of metabolic reactions carried out by the cell. The cell must obtain energy to carry out essential biosynthetic (anabolic) reactions for its continued existence, and to enable growth and cell division to take place. The substrate cannot therefore be degraded entirely to carbon dioxide or methane, for example, and a portion must be channeled into the biosynthesis of essential molecules. Indeed, many organisms will degrade xenobiotics only in presence of a suitable more readily degraded growth substrate that supplies both cell carbon and the energy for growth; this is discussed later (Section 4.5.2) in the context of "cometabolism" and "concurrent metabolism." Growth under anaerobic conditions is demanding both physiologically and biochemically since the cells will generally obtain only low energy yields from the growth substrate, and must additionally maintain a delicate balance between oxidative and reductive processes. Only a few examples are given of mechanism for ATP generation in anaerobes.

- Clastic reactions from 2-keto acid CoA esters produced in a number of degradations;

- Reactions involving carbamyl phosphate in the degradation of arginine in clostridia and the fermentation of allantoin by *Streptococcus allantoicus;*

- The activity of formyl THF synthase during the fermentation of purines by clostridia (Chapter 6, Section 6.7.4.1);

- The reductive dechlorination of 3-chlorobenzoate by *Desulfomonile tiedjei* DCB-1 (Chapter 6, Section 6.6);

- The proton pump in *Oxalobacter formigenes* (Section 4.6.2);

- The biotin-dependent carboxylases that couple the decarboxylation of malonate to acetate in *Malonomonas rubra* to the transport of Na$^+$ across the across the cytoplasmic membrane (Section 4.6.2).

True fermentation implies that a single substrate is able to provide carbon for cell growth and at the same time, satisfy the energy requirements of the cell: a simple example of fermentation is the catabolism of glucose by facultatively anaerobic bacteria to pyruvate which is further transformed into a variety of products including acetate, butyrate, propionate, or ethanol by different organisms. On the other hand, a range of electron acceptors may be used under anaerobic conditions to mediate oxidative degradation of the carbon substrate at the expense of the reduction of the electron acceptors. For example, the following reductions may be coupled to oxidative degradation: nitrate to nitrogen (or nitrous oxide), sulfate to sulfide, carbonate to methane, fumarate to succinate, trimethylamine-*N*-oxide to trimethylamine, or dimethylsulfoxide to dimethyl sulfide (Styrvold and Ström 1984) (Figure 4.28). The environments required by the relevant organisms are directly related to

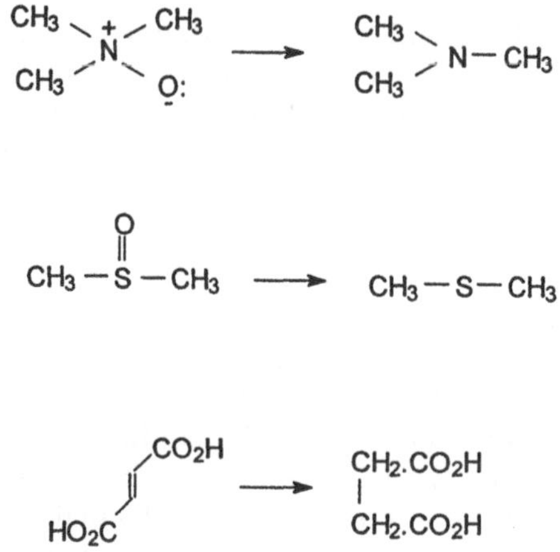

FIGURE 4.28
Examples of alternative electron acceptors and their reduction products.

the redox potential of the prevailing reactions, so that increasingly reducing conditions are required for reduction of nitrate, sulfate, and carbonate. Further discussion and examples of the degradation of xenobiotics using alternative electron acceptors is given in Section 4.3.3.

Attention may also be drawn to dechlorination by anaerobic bacteria of both chlorinated ethenes and chlorophenolic compounds that serve as electron acceptors using electron donors including formate, pyruvate, and acetate. This is more fully discussed in Chapter 6 (Sections 6.4.4 and 6.6).

Probably most of the microbial degradations and transformations that are discussed in this book are carried out by heterotrophic microorganisms that use the xenobiotic as a source of both carbon and energy. Examples in which the xenobiotics are used only as sources of N, S, or P are, however, given in Chapter 5, Section 5.2.3. Attention is briefly drawn here to groups of organisms, many of whose members are autotrophic or lithotrophic; discussion of the complex issue of the organic nutrition of chemolithotrophic bacteria and the use of the term *autotroph* is given in a review (Matin 1978). The groups of organisms that are discussed here in the context of biotransformation include (1) ammonia-oxidizing bacteria, for example, *Nitrosomonas europeae* (Section 4.3.2); (2) the facultatively heterotrophic thiobacilli that use a number of organic sulfur compounds as energy sources (Chapter 6, Section 6.9.3); and (3) photolithotrophic algae and cyanobacteria (Section 4.3.5). It is important to underscore the fact that carbon dioxide is required not only for the growth of strictly phototrophic and lithotrophic organisms: many organisms which are heterotrophic have an obligate requirement for carbon dioxide for their growth. Some illustrative examples are anaerobic bacteria such as the acetogens, methanogens, and the propionic bacteria, and aerobic bacteria that degrade propane (MacMichael and Brown 1987), the branched hydrocarbon 2,6-dimethyloct-2-ene (Fall et al. 1979), or oxidize carbon monoxide (Meyer and Schlegel 1983). The lag after diluting glucose-grown cultures of *E. coli* into fresh medium may indeed be eliminated by the addition of $NaHCO_3$, and this is consistent with the requirement of this organism for low concentrations of CO_2 for growth (Neidhardt et al. 1974). The role of CO_2 in determining the products formed from propene oxide by a strain of *Xanthobacter* sp. is noted in Chapter 6, Section 6.1.3, and its significance in the anaerobic biotransformation of aromatic compounds is discussed in Chapter 6, Section 6.7.3. A review (Ensign et al. 1998) provides a brief summary of the role of CO_2 in the metabolism of epoxides by *Xanthobacter* sp. strain Py2, and of acetone by both aerobic and anaerobic bacteria.

4.2.2 Biodegradation of Enantiomers

The structures of many compounds do not possess any element of symmetry, so that they may exist as pairs of mirror-image enantiomers. In the context of their analysis, important examples in which the concentration of one enantiomer exceeds that of the other have been noted in Chapter 2, Section 2.4.2.

This may plausibly be attributed to preferential destruction or transformation of one enantiomer, and is consistent with significant differences in the biodegradability of the enantiomers. Different strategies for biodegradation of racemates may be used, and are illustrated in the following examples:

1. Both enantiomers of mandelate were degraded through the activity of a mandelate racemase (Hegeman 1966), and the racemase (mdlA) is encoded in an operon that includes the following two enzymes in the pathway of degradation, S-mandelate dehydrogenase (mdlB) and benzoylformate decarboxylase (mdlC) (Tsou et al. 1990).

2. Only the $R(+)$ enantiomer of the herbicide 2-(2-methyl-4-chlorophenoxy)propionic acid was degraded (Tett et al. 1994), although cell extracts of *Sphingomonas herbicidovorans* grown with the (R) or (S) enantiomer, respectively, transformed selectively the (R) or (S) substrates to 2-methyl-4-chlorophenol (Nickel et al. 1997).

3. Cells of *Acinetobacter* sp. NCIB 9871 grown with cyclohexanol carried out enantiomerically specific degradation of a racemic substituted norbornanone to a single ketone having >95% enantiomeric excess (Levitt et al. 1990).

4. The specific degradation of epoxides of *cis*- and *trans*-pent-2-enes is discussed in Chapter 6, Section 6.1.3.

Stereospecific biotransformation is frequently observed. *Bauveria sulfurescens* stereospecifically hydroxylated an azabrendane at the quaternary carbon atom (Archelas et al. 1988; Chapter 6, Section 6.1.2), while steroid and terpenoid hydroxylations are discussed in Chapter 6, Section 6.11.2.

In natural systems, the situation may be quite complex. For example, the enantiomerization of phenoxyalkanoic acids containing a chiral side chain has been studied in soil using ^2H_2O (Buser and Müller 1997). It was shown that there was an equilibrium between the R- and S- enantiomers of 2-(4-chloro-2-methylphenoxy)propionic acid (MCPP) and 2-(2,4-dichlorophenopxy)propionic acid (DCPP) with an equilibrium constant favoring the herbicidally active R-enantiomers. The exchange reactions proceeded with both retention and inversion of configuration at the chiral sites. This important issue is discussed further in Chapter 8, Section 8.2.3 and will certainly attract increasing attention in the context of the preferential microbial synthesis of intermediates of specific configuration. Some examples for aromatic compounds have been summarized in a review (Neilson and Allard 1998).

4.2.3 Sequential Microbial and Chemical Reactions

Microbial activity may produce a reactive intermediate which undergoes spontaneous chemical transformation to a terminal metabolite. This is not an unusual occurrence, and its diversity is illustrated by the following examples.

1. The formation of nitro-containing metabolites during degradation of 4-chlorobiphenyl and 2-hydroxybiphenyl is consistent with the intermediate formation of arene oxides that react with nitrate or nitrite in the medium (Chapter 5, Section 5.2.4 and Chapter 6, Section 6.3.1.2).

2. A bacterial strain BN6 oxidizes 5-aminonaphthalene-2-sulfonate by established pathways to 6-amino-2-hydroxybenzalpyruvate that undergoes spontaneous cyclization to 5-hydroxyquinoline-2-carboxylate (Figure 4.29a) (Nörtemann et al. 1993).

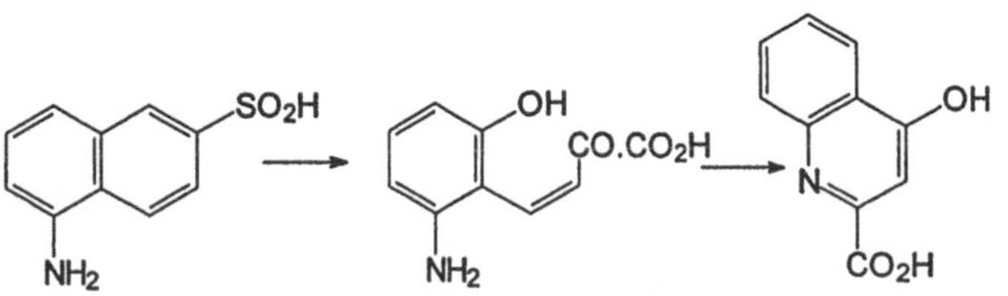

FIGURE 4.29
(a) Transformation of 5-aminonaphthalene-2-sulfonate, (b) benzo[*b*]thiophene, (c) 4-chlorobiphenyl, (d) 4-nitrotoluene, (e) 3,5-dichlor-4-methoxybenzyl alcohol, (f) 2,3-diaminonaphthalene in presence of nitrate, (g) 3,4-dichloroaniline presence of nitrate.

3. Oxidation of benzo[*b*]thiophene by strains of pseudomonads produces the sulfoxide that undergoes an intramolecular Diels–Alder reaction followed by further transformation to benzo[*b*]naphtho[1,2-*d*]thiophene (Figure 4.29b) (Kropp et al. 1994).

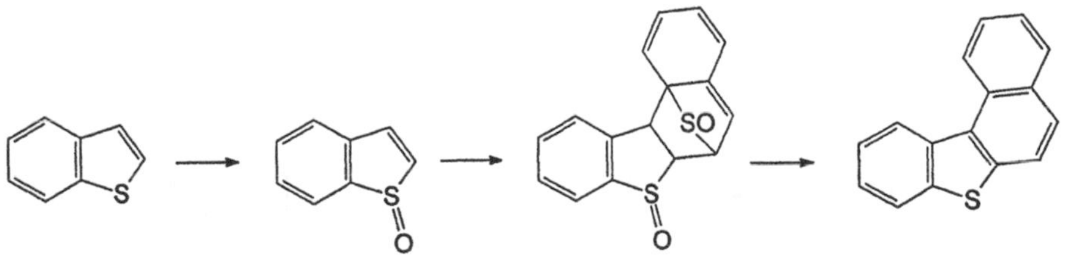

FIGURE 4.29 b (continued)

4. Transformation of 4-chlorobiphenyl by *S. paucimobilis* strain BPSI-3 produced chloropyridine carboxylates by reaction of intermediate 4-chlorocatechol fission products with NH_4^+ (Davison et al. 1996) (Figure 4.29c).

5. 4-Nitrotoluene is degraded by a strain of *Mycobacterium* sp. via the corresponding 4-amino-3-hydroxytoluene (Spiess et al. 1998); this is dimerized abiotically to form a dihydrophenoxazinone, and after extradiol cleavage to 5-methylpyridine-2-carboxylate (Figure 4.29d).

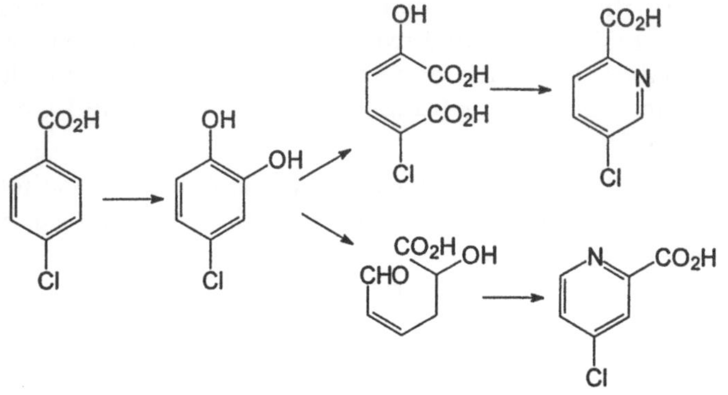

FIGURE 4.29 c (continued)

6. Incubation of 3,5-dichloro-4-methoxybenzyl alcohol with methanogenic sludge produced the de-*O*-methylated compound that was transformed to 2,6-dichloro-4-hydroxymethylphenol, and abiotically dimerized to bis(3,5-dichloro-4-hydroxyphenyl)methane (Verhagen et al. 1998) (Figure 4.29e).

7. The transformation of aromatic amino acids to the 2-ketoacids was mediated by *Morganella morganii*, and these subsequently underwent a hemin-dependent chemical transformation with the production of CO (Hino and Tauchi 1987).

Another situation arises when nitrite (or compounds at the same oxidation level) that are produced microbiologically from nitrate react with the xenobiotic to produce stable end products: the production of nitrite is the sole metabolic function of the bacteria.

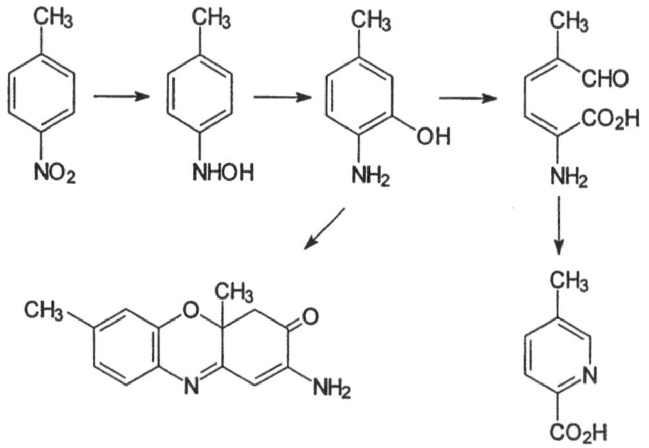

FIGURE 4.29 d (continued)

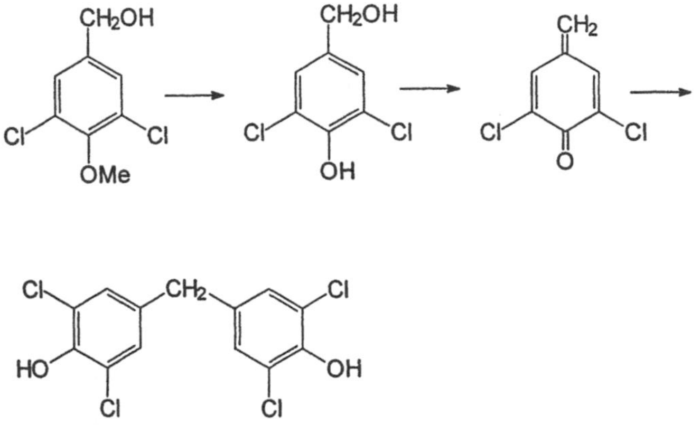

FIGURE 4.29 e (continued)

1. A strain of *E. coli* produces a naphthotriazole from 2,3-diaminon-aphthalene and nitrite that is formed from nitrate by the action of nitrate reductase. The initial product is NO which is converted into the active nitrosylating agent by reactions with oxygen; this then reacts chemically with the amine (Ji and Hollocher 1988). A comparable reaction may plausibly account for the formation of dimethylnitrosamine by *Pseudomonas stutzeri* during growth with the amine in the presence of nitrite (Mills and Alexander 1976). On the other hand, for other organisms including *E. coli*, enzymatic reactions may be involved although their nature has not been clearly delineated (Figure 4.29f).

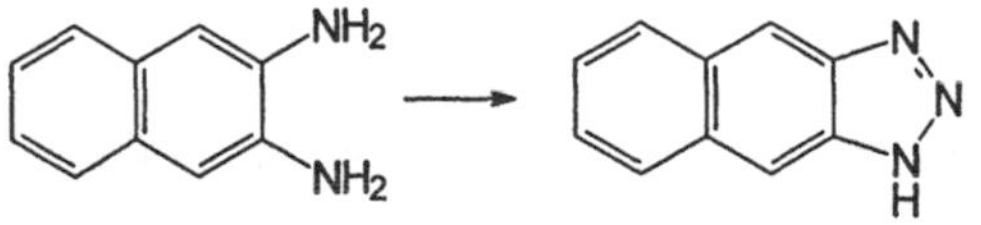

FIGURE 4.29 f (continued)

2. The formation of 3,3′,4,4′-tetrachloroazobenzene, 1,3-bis(3,4-dichlorophenyl)triazine, and 3,3′,4,4′-tetrachlorobiphenyl from 3,4-dichloroaniline and nitrate by *E. coli* presumably involves intermediate formation of the diazonium compound by reaction of the amine with nitrite (Corke et al. 1979) (Figure 4.29g).

In view of concern over the presence of nitrate in groundwater, the possible environmental significance of these or analogous reactions should not be overlooked.

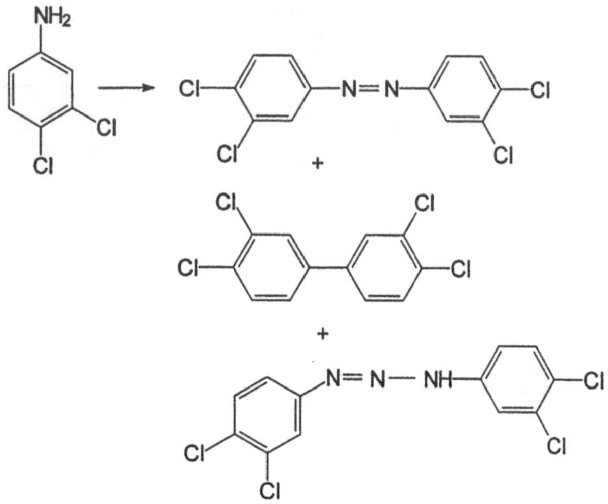

FIGURE 4.29 g (continued)

4.3   The Spectrum of Organisms

4.3.1   Introduction

Before discussing some of the larger groups of microorganisms that have been implicated in degradation and transformation, some general comments are made on groups of organisms that seem to be less well represented.

Marine and Oligotrophic Bacteria

Although most illustrations in this book have been taken from investigations of freshwater environments — lakes and rivers — fewer relate to the marine environment. In view of the area of the globe that is covered by open sea this may seem surprising, and the degradation of xenobiotics by marine bacteria has received only limited attention. There are important examples that are briefly summarized here.

1. The isolation of strains of *Marinobacter* (Gauthier et al. 1992) that are able to degrade aliphatic hydrocarbons and related compounds: this is discussed in a wider context in Chapter 6, Section 6.1.1.

2. The demonstration that a halophilic Archaeon assigned to the genus *Haloferax* is able to use a restricted number of aromatic substrates including benzoate as sole carbon and energy source (Emerson et al. 1994) is noted as an extension of the range of substrates for this genus.

3. Organisms belonging to the genus *Cycloclasticus* have been isolated from the Gulf of Mexico and Puget Sound, and were able in artificial seawater to degrade at concentrations ranging from 1 to 5 ppm a range of PAHs including alkylated naphthalenes, phenanthrene, anthracene, and fluorene (Geiselbrecht et al. 1998). Strains from both localities were numerically important and were similar based both on 16S rDNA sequences and phylogenetic relationships based on dioxygenases.

4. Marine bacteria were isolated from a creosote-contaminated sediment in Puget Sound by enrichment with naphthalene (Hedlund et al. 1999). It was shown that the gene containing the naphthalene dioxygenase ISP was not closely related to those from naphthalene-degrading strains of *Pseudomonas* or *Burkholderia*, and although analysis of 16S ribosomal DNA suggested a close relation to the genus *Oceanospirillum*, the differences were considered sufficient to assign these strain to a new genus *Neptunomonas*.

It is necessary to take into account critical aspects of the physiology and biochemistry of these marine organisms, and although parenthetical comments on marine bacteria are made in various sections of this book (Sections 4.3 and 4.5.2, and Chapter 5, Section 5.2.1), it seems convenient to bring together some of their salient features. It should also be appreciated that terrestrial organisms that have high salinity tolerance may be isolated from inshore seawater samples; among these are the yeasts that have been isolated from coastal marine sediments and that are noted in Section 4.3.6. Such organisms are excluded from this discussion, which is restricted to oceanic water.

- It is experimentally difficult to obtain numerical estimates of the total number of bacteria present in seawater, and the contribution of ultramicroorganisms having a small cell volume and low concentrations of DNA may be seriously underestimated. Although it is possible to evaluate their contribution to the uptake and mineralization of readily degraded compounds such as amino acids and carbohydrates, it is more difficult to estimate their potential for degrading xenobiotics at realistic concentrations. This is especially the case in the absence of appropriate commercially available radiolabeled substrates.

- Use of conventional plating procedures may result in the isolation only of fast-growing organisms that outgrow others — which may be numerically dominant and which are unable to produce colonies on such media; the substrate concentrations used for isolation may have been unrealistically high so that obligately oligotrophic organisms were outnumbered during attempted isolation.

- There is an indeterminacy in the term *oligotroph*, and the dilemma is exacerbated by the fact that it may be impossible to isolate obligate oligotrophs by established procedures; the application of DNA probes should, however, contribute to an understanding of the role of these "non-cultivable" organisms. The issue of oligotrophic bacteria growing at low substrate concentrations in the marine environment is briefly discussed in Section 4.6.2 (Schut et al. 1993), and from pristine environments in Section 4.6.4 (Schut et al. 1995).

- During prolonged storage in the laboratory under conditions of nutrient starvation, facultatively oliogotrophic bacteria may be isolated, and these display transport systems for the uptake of amino acids and glucose that are coregulated.

- Organisms in natural ecosystems may not be actively dividing but may, nonetheless, be metabolically active. This may be particularly important for ultramicro marine bacteria in their natural habitat.

Lithotrophic Bacteria

These are major groups of microorganisms that have achieved restricted prominence in discussions on biodegradation and biotransformation, and include both photolithotrophs and chemolithotrophs. Some brief comments are therefore justified; photolithotrophs are described in greater detail in Section 4.3.5.

These organisms use CO_2 as their principal, or exclusive, source of carbon, and this is incorporated into cellular material generally by the Benson–Calvin cycle. Energy for growth is obtained either from photochemical reactions (photolithotrophs), or by chemical oxidation of inorganic substrates such as reduced forms of nitrogen or sulfur (chemolithotrophs). In some organisms, organic carbon can be taken up and incorporated during growth even in organisms that are obligately chemolithotrophic or photolithotrophic, although in some circumstances organic carbon has an inhibitory effect on growth. Some species and strains of these organisms may also grow heterotrophically using organic carbon as sources both of energy and cell carbon. Attention is directed to reviews that cover the sometimes controversial aspects of lithotrophy and autotrophy (Kelly 1971; Rittenberg 1972; Whittenbury and Kelly 1977; Matins 1978). It is worth noting that many aerobic bacteria that belong to groups with well-established heterotrophic activity are also chemolithoautotrophic and use the oxidation of hydrogen as their source of energy (Bowien and Schlegel 1981). Attention is drawn to the reassignment of the *Pseudomonas* strains *P. flava*, *P. pseudoflava*, and *P. palleroni* to the genus *Hydrogenophaga* (Chapter 5, Section 5.9), and the degradative and transformation activity of *Xanthobacter* species (Chapter 6, Sections 6.1.3, 6.3.1.1, 6.4.2, and 6.5.1.1).

It has become clear that representatives of chemolithotrophic microorganisms may be effective in carrying out the transformation of xenobiotics. As illustration, attention is directed to three groups of organisms: (1) the ammonia oxidizers, (2) the thiobacilli, and (3) algae and cyanobacteria. Habitats to which these organisms are physiologically adapted should therefore be considered in discussions on biodegradation and biotransformation. Some examples of the reactions carried out by these groups of organisms are given in Sections 4.3.2 and 4.3.5.

Bacteria in Their Natural Habitats

Illustrations of the plethora of pathways used by bacteria for the degradation and biotransformation of xenobiotics will be provided in Chapter 6, but it is appropriate here to say something of the spectrum of metabolic potential of specific organisms, especially those that have hitherto achieved less prominence in discussions of biodegradation and biotransformation. It should be appreciated that, in natural situations, bacteria may be subjected to severe nutrient limitation so that they are compelled to reproduce at extremely low rates in order to conserve their metabolic energy (Kjellberg et al.1987; Siegele and Kolter 1992). This does not necessarily mean, however, that these organisms have negligible metabolic potential toward xenobiotics. Other slow-growing organisms may be well adapted to the natural environment (Poindexter 1981) although they may not be numerically dominant among organisms isolated by normal procedures (Schut et al. 1993). An unusual situation has been observed in *P. putida* strain mt-2 which contains the TOL plasmid pWWO. After growth with 3-methylbenzoate, cells were exposed to concentrations of toluene from 4 mg/l (growth supporting) to 130 mg/l (inhibitory) to 267 mg/l (lethal). Protein synthesis was rapidly inhibited with the concomitant production of new proteins which were characteristic of cell starvation, and which could be suppressed by addition of 3-methylbenzoate as carbon source. Cells exposed to 4 mg/l ceased to produce "starvation proteins" within 3 h and growth was initiated. At the higher concentrations these proteins persisted for increasing lengths of time, and at 267 mg/l there was a rapid loss of viability (Vercellone-Smith and Herson 1997). A discussion of methods for analyzing populations of bacteria is given in Chapter 5, Section 5.8.

The Role of Fungi

Although the cardinal importance of fungi in the terrestrial environment is unquestioned and some comments in the context of natural organochlorine metabolites are given in Chapter 1, Section 1.4, most attention in the aquatic environment has been directed to bacteria. An important exception is provided by studies in Lake Bonney, South Australia. The microflora of the lake contained a population of fungi including *Trichoderma herzianum* that was able to degrade the "free" but not the associated chloroguaiacols (van Leeuwen et al. 1996), and *Epicoccum* sp., *Mucor circinelloides*, and *Penicillium expansum* — soil fungi that are widely dispersed and may have entered the lake from runoff) — were capable of bringing about association of tetrachloroguaiacol with organic components in the aqueous phase, so that this material could subsequently enter the sediment phase (van Leeuwen et al. 1997).

Due to the similarity of the metabolic systems of fungi to those of mammalian systems, it has been suggested that fungi could be used as models for screening purposes and several interesting examples have been provided (Ferris et al. 1976; Smith and Rosazza 1983; Griffiths et al. 1992). The fungus *Cunninghamella elegans* has attracted particular attention (Section 4.3.6), and it may be noted that the reactions involved in the transformation of alachlor by this organism (Section 4.3.6; Pothuluri et al. 1993) are similar to those carried out by mammalian systems, and to the biotransformation of the analogous metalochlor by bluegill sunfish (*Lepomis macrochirus*) and by a soil actinomycete that are noted in Chapter 7, Section 7.5.1. A wide range of taxonomically diverse fungi has also been used for the synthesis of less readily available compounds and a few examples are given in Chapter 6, Section 6.11.2. Fungi — especially white-rot fungi — have attracted considerable attention in the context of bioremediation of contaminated terrestrial sites, and several examples are given in Chapter 6, Sections 6.2.2, 6.3.2, and 6.5.2, and in Chapter 8.

4.3.2   Aerobic and Facultatively Anaerobic Bacteria

The well-established metabolic versatility of groups such as the pseudomonads and their relatives, and the methanotrophs has possibly deflected attention from other groups which may be present in aquatic systems and which may play an important role in determining the fate of xenobiotics. Although the potential of other Gram-negative groups such as the acinetobacters, moraxellas, and species of *Alcaligenes* is well established, Gram-positive groups seem to have generally achieved less prominence in aquatic systems. In the succeeding paragraphs, some examples of the metabolic importance of a few of these important groups of organisms are presented.

Gram-Positive Aerobic Bacteria

The metabolic versatility of organisms belonging to the genera *Mycobacterium* and *Rhodococcus* is becoming well established and some examples will be given as illustration.

Mycobacteria can oxidize short-chain alkenes (DeBont et al. 1980), and organisms known under the invalid specific name "*Mycobacterium paraffinicum*" (Wayne et al. 1991) are able to degrade a number of alkanes. It has also been shown that strains of mycobacteria grown with propane are able to oxidize the apparently unrelated substrate trichloroethylene (Wackett et al. 1989). There has also been a revival of interest in the role of mycobacteria in the degradation of aromatic polycyclic hydrocarbons including naphthalene (Kelley et al. 1990), phenanthrene (Guerin and Jones 1988), and pyrene

(Heitkamp et al. 1988; Grosser et al. 1991); both of the pyrene-degrading strains belong to the group of fast-growing scotochromogenic mycobacteria (Govindaswami et al. 1995). A strain of *Mycobacterium* sp. that is able to use all of these as a sole source of carbon and energy has also been isolated (Boldrin et al. 1993). This seems appropriate in view of the historical importance of similar organisms, even though those were designated as mycobacteria (Gray and Thornton 1928), were not acid-fast, and some at least would be currently assigned to the genus *Rhodococcus*. This illustrates a potentially serious taxonomic pitfall which should be avoided, since the distinction between the genera *Mycobacterium* and *Rhodococcus* has not always been unequivocal in the older literature (Finnerty 1992). Indeed, the organism that is discussed later and initially described as *R. chlorophenolicus* has now been transferred to the genus *Mycobacterium* (Häggblom et al. 1994; Briglia et al. 1994). There has been increasing interest in rhodococci, and this has been sustained by their potential application in biotechnology (Finnerty 1992). Rhodococci have a wide metabolic potential and, for example, an organism capable of degrading acetylene and assigned to the genus *Mycobacterium* almost certainly belongs to the genus *Rhodococcus* (DeBont 1980). Considerable interest has been expressed in the chlorophenol-degrading organism *R. chlorophenolicus* (Apajalahti al. 1986) (*M. chlorophenolicum*) partly motivated by its potential for application to bioremediation of chlorophenol-contaminated industrial sites, and a strain of *Rhodococcus* sp. is capable of degrading a number of chlorinated aliphatic hydrocarbons including vinyl chloride and trichloroethene as well as the aromatic hydrocarbons benzene, naphthalene, and biphenyl (Malachowsky et al. 1994). A strain of *R. opacus* isolated by enrichment with chlorobenzene was able to use a wide range of halogenated compounds including 1,3- and 1,4-dichlorobenzene, 1,3- and 1,4-dibromobenzene, 2-, 3-, and 4-fluorophenol, 2-, 3-, and 4-chlorophenol, 4-nitrophenol, 3- and 4-fluorobenzoate, and 3-chlorobenzoate (Zaitsev et al. 1995). Several rhodococci have attracted interest for their ability to degrade PCBs: an organism (*Acinetobacter* sp. strain P6) now assigned to *R. globerulus* (Asturias and Timmis 1993), *R. erythropolis* (Maeda et al. 1995), and *Rhodococcus* sp. strain RHA1 (Seto et al. 1995). A number of biotransformations have also been accomplished using rhodococci, and these include, for example, the hydrolysis of nitriles which is discussed in Chapter 6, Section 6.11.1, and the reduction of the conjugated C=C double bond in 2-nitro-1-phenylprop-1-ene (Sakai et al. 1985). Hydroxylations of alkyl groups by a strain of *Rhodococcus* sp. and by *Streptomyces griseolus*, and of phenanthrene by *S. flavovirens* are noted in Section 4.4.1.2 in the context of cytochrome P-450 mediated reactions.

A satisfying evaluation of the metabolic potential of microorganisms in natural ecosystems should not therefore fail to consider these organisms which are certainly widespread, and to distinguish between rates of degradation and metabolic potential: slow-growing organisms may be extremely important in degrading xenobiotics in natural ecosystems.

Gram-Negative Aerobic Bacteria

Attention is briefly drawn to groups that are widely distributed but that have attracted attention primarily in the context of nitrogen fixation.

1. Azotobacters — It was established in the 1930s by Burk and Winogradsky that these could be readily obtained from soil samples by elective enrichment with benzoate. The degradative pathway for benzoate has been elucidated (Hardisson et al. 1969), and the range of substrates extended to 2,4,6-trichlorophenol (Li et al., 1992; Latus et al. 1995).

2. Rhizobia — Taxa belonging to both the genera *Rhizobium* and *Bradyrhizobium* are capable of degrading simple aromatic compounds including benzoate (Chen et al. 1984) and 4-hydroxybenzoate (Parke and Ornston 1986; Parke et al. 1991), and it has been shown that 4-hydroxybenzoate hydroxylase is required for its transport into the cell (Wong et al. 1994). In strains of *R. trifolium*, metabolism of benzoate is mediated either by protocatechuate 3,4-dioxygenase (Chen et al. 1984) or by catechol 1,2-dioxygenase (Chen et al. 1985). Rhizobia do have, however, a much broader metabolic capability as illustrated by the following examples:

   a. The degradation of PCB congeners (Damaj and Ahmad 1996);

   b. The degradation of flavones (Rao et al. 1991; Rao and Cooper 1994);

   c. The dechlorination — although not the degradation — of atrazine (Bouquard et al. 1997);

   d. The use of phosphonomycin (1,2-epoxypropylphosphonate) as a source of carbon, energy, and phosphorus (McGrath et al. 1998).

Gram-Negative Facultatively Anaerobic Bacteria

1. *Enterobacteriaceae*

Facultatively anaerobic bacteria, and especially those belonging to the family Enterobacteriaceae, have a long history as agents of disease in humans although there is extensive evidence for their occurrence in a wide variety of environmental samples. Methods for their identification and classification have therefore been extensively developed, and have traditionally used the ability to ferment a wide range of carbohydrates as taxonomic characters. This has had the possibly unfortunate effect of deflecting attention from the capability of these organisms to degrade other classes of substrates, although their ability to utilize substrates such as 3- and 4-hydroxybenzoates (Véron and Le Minor 1975) and nicotinate (Grimmond et al. 1977) under aerobic conditions has been quite extensively used for taxonomic classification. A few examples illustrating their ability to degrade and transform diverse xenobiotics may be used to illustrate the metabolic capabilities of these somewhat neglected organisms:

- The degradation of DDT by organisms designated *Aerobacter aerogenes* (possibly currently *Klebsiella aerogenes*) (Wedemeyer 1967) (Figure 4.30), and the partial reductive dechlorination of methoxychlor by *K. pneumoniae* (Baarschers et al. 1982).

- The biotransformation of methyl phenyl phosphonate to benzene by *K. pneumoniae* (Cook et al. 1979) (Figure 4.31a). Further examples of the cleavage of the carbon–phosphorus bond by other members of the Enterobacteriaeae and a discussion of the metabolism of phosphonates are given in Chapter 6, Section 6.9.4.

- The biotransformation of 2,4,6-trihydroxy-1,3,5-triazine and atrazine under anaerobic conditions by an unidentified facultative anaerobe (Jessee et al. 1983).

- The biotransformation of γ-hexachloro[*aaaeee*]cyclohexane to tetrachlorocyclohexene by *Citrobacter freundii* (Figure 4.31b) (Jagnow et al. 1977).

- The biotransformation of 3,5-dibromo-4-hydroxybenzonitrile to the corresponding acid by a strain of *K. pneumoniae* ssp. *ozaenae* which uses the substrate as sole source of nitrogen (Figure 4.31c) (McBride et al. 1986).

- The decarboxylation of 4-hydroxycinnamic acid to 4-hydroxystyrene, and of ferulic acid (3-methoxy-4-hydroxycinnamic acid) to 4-vinylguaiacol by several strains of *Hafnia alvei* and *H. protea*, and by single strains of *Enterobacter cloacae* and *K. aerogenes* (Figure 4.31d) (Lindsay and Priest 1975). The enzyme has been purified from *Bacillus pumilis* (Degrassi et al. 1995). Several taxa of Enterobacteriaceae including *K. pneumoniae*, *Ent. aerogenes*, and *Proteus mirabilis* are able to decarboxylate the amino acid histidine that is abundant in the muscle tissue of scombroid fish (Yoshinaga and Frank 1982), and the histamine produced has been associated with an incident of scombroid fish poisoning (Taylor et al. 1979).

- The metabolism of ferulic acid (3-methoxy-4-hydroxycinnamic acid) by *Ent. cloacae* to a number of products including phenylpropionate and benzoate (Figure 4.32) (Grbic-Galic 1986).

- Utilization of uric acid as a nitrogen source by strains of *Aer. aerogenes*, *K. pneumoniae*, and *Serratia kiliensis* (Rouf and Lomprey 1968).

- The sequential reduction of one of the nitro groups of 2,6-dinitrotoluene by *Salmonella typhimurium* (Sayama et al. 1992) — a taxon that is not generally noted for its degradative capability — is noted again in Chapter 6, Section 6.8.2.

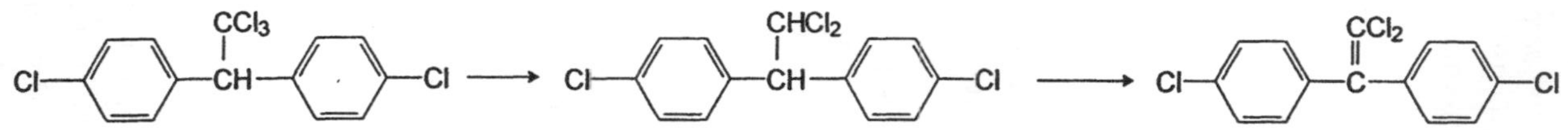

FIGURE 4.30
Degradation of DDT by *Aerobacter aerogenes*.

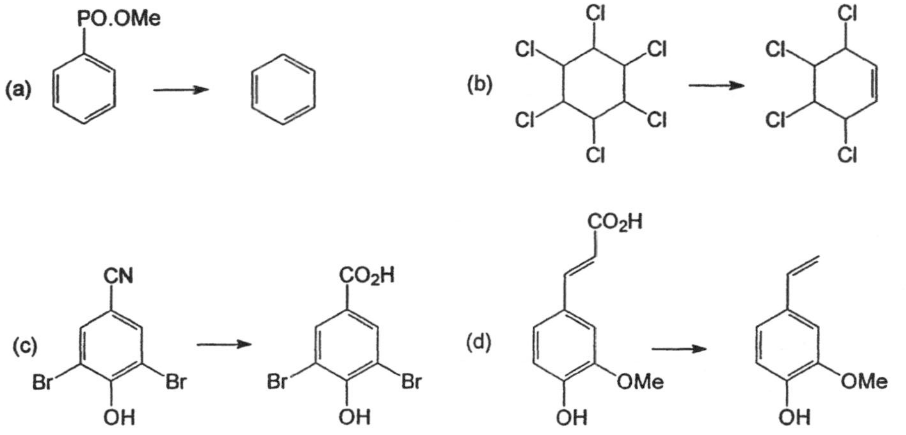

FIGURE 4.31
Examples of biotransformations carried out by Enterobacteriaceae: (a) methylphenyl phosphonate, (b) γ-hexachlorocyclohexane, (c) 3,5-dibromo-4-hydroxybenzonitrile, (d) decarboxylation.

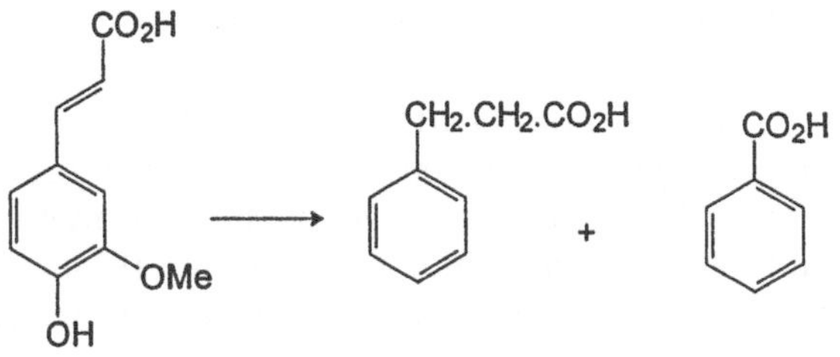

FIGURE 4.32
Metabolism of ferulic acid by *Enterobacter cloacae*.

- Under conditions of oxygen limitation, strains of *Morganella morganii* and *Providencia rettgeri* degraded hexahydro-1,3,5-trinitro-1,3,5-triazine after initial reduction to the nitroso compounds (Kitts et al. 1994).

- After growth in a medium containing suitable reductants such as glucose, a strain of *Ent. agglomerans* was able to reduce tetrachloroethene successively to trichloroethene and *cis*-1,2- dichloroethene (Sharma and McCarty 1996).

The taxonomic application of the ability of enteric organisms to grow with hydroxylated 4-hydroxyphenylacetate (Cooper and Skinner 1980) and of 3-hydroxyphenylpropionic acid (Burlingame and Chapman 1983) has been established, and it has been demonstrated that the enzyme that carries out the hydroxylation has a wide substrate range extending to 4-methylphenol and even to 4-chlorophenol (Prieto et al. 1993).

2. *Vibrionaceae*

Relatively few examples have emerged on the degradative capability of these organisms, and these include the following:

- An organism tentatively identified as *Vibrio* strain that was able to degrade the 2-carboxylates of furan, pyrrole, and thiophene (Evans and Venables 1990).

- A strain of *Aeromonas* sp. that was able to degrade phenanthrene through *o*-phthalate (Kiyohara et al. 1976).

3. *Gram-Positive Facultatively Anaerobic Organisms*

The genus *Staphylococcus* is traditionally associated with disease in humans, but the demonstration (Monna et al. 1993) that a strain of *S. auriculans* — isolated by enrichment with dibenzofuran and with no obvious clinical association — could degrade this substrate and carry out limited biotransformation of fluorene and dibenzo-1,4-dioxin may serve to illustrate the unsuspected metabolic potential of facultatively anaerobic Gram-positive organisms.

Bacterial Metabolism of C-1 Compounds: Methanotrophs, Methylotrophs, and Related Organisms

Methane monooxygenase and related systems — A few of the reactions carried out by the monooxygenase system of methanotrophic bacteria are summarized in Figure 4.33, and it is because of this that methylotrophs have received attention for their technological potential (Lidstrom and Stirling 1990). An equally wide metabolic potential has also been demonstrated for cyclohexane monooxygenase which has been shown to accomplish two broad types of reaction: one in which formally nucleophilic oxygen reacts with the substrate and the other in which formally electrophilic oxygen is involved (Figure 4.34); (Branchaud and Walsh 1985). In addition, it has emerged that the monooxygenase system in methanotrophs is similar to that in the nitrite-oxidizing bacteria and that the spectrum of biotransformations is equally wide. Additional comments on the soluble three-component methane monooxygenase are given in Chapter 6, Section 6.4.3. There are also important groups of facultatively C_1-utilizing bacteria including members of the genus *Xanthobacter* (Padden et al. 1997).

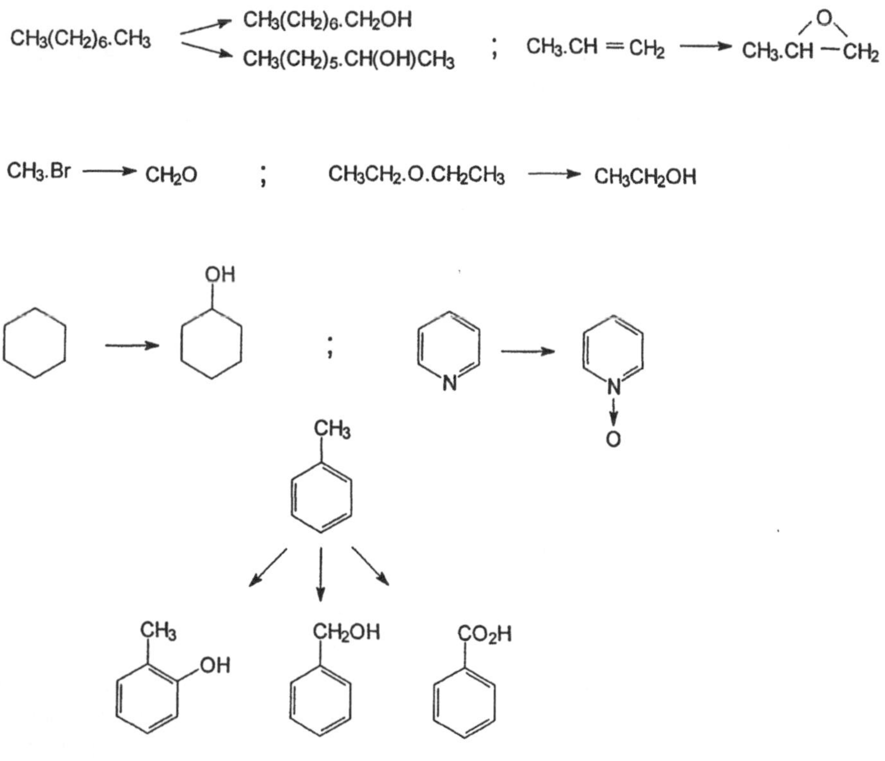

FIGURE 4.33
Examples of reactions mediated by the monooxygenase system of methanotrophic bacteria.

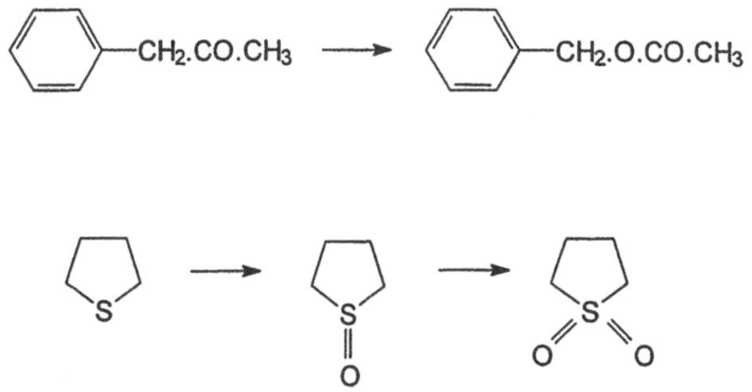

FIGURE 4.34
Classes of reactions mediated by cyclohexane monooxygenase.

The following illustrate some of the biotransformations which have been observed with *Nitrosomonas europaea* and these are particularly intriguing since this organism has an obligate dependency on CO_2 as carbon source, and has traditionally been considered to be extremely limited in its ability to use organic carbon for growth:

- Oxidation of benzene to phenol and 1,4-dihydroxybenzene (Figure 4.35a) (Hyman et al. 1985), and others including both side chain and ring oxidation of ethyl benzene, and ring-hydroxylation of halogenated benzenes and nitrobenzene (Keener and Arp 1994).

- Oxidation of alkanes (C_1 to C_8) to alkanols, and alkenes (C_2 to C_5) to epoxides (Hyman et al. 1988).

- Oxidation of methyl fluoride to formaldehyde (Hyman et al. 1994), and of chloroalkanes at carbon atoms substituted with a single chlorine atom to the corresponding aldehyde (Rasche et al. 1991).

- Oxidation of a number of chloroalkanes and chloroalkenes including dichloromethane, chloroform, 1,1,2-trichloroethane, and 1,2,2-trichloroethene (Vannelli et al. 1990).

- Oxidation of the trichloromethyl group in 2-chloro-6-(trichloromethyl)-pyridine to the corresponding carboxylic acid (Vannelli and Hooper 1992) occurs at high oxygen concentrations during cooxidation of ammonia or hydrazine. At low oxygen concentrations in the presence of hydrazine, reductive dechlorination to 2-chloro-6-dichloromethylpyridine occurs (Vannelli and Hooper 1992) (Figure 4.35b).

- A range of sulfides including methylsulfide, tetrahydrothiophene, and phenylmethylsulfide are oxidized to the corresponding sulfoxides (Juliette et al. 1993).

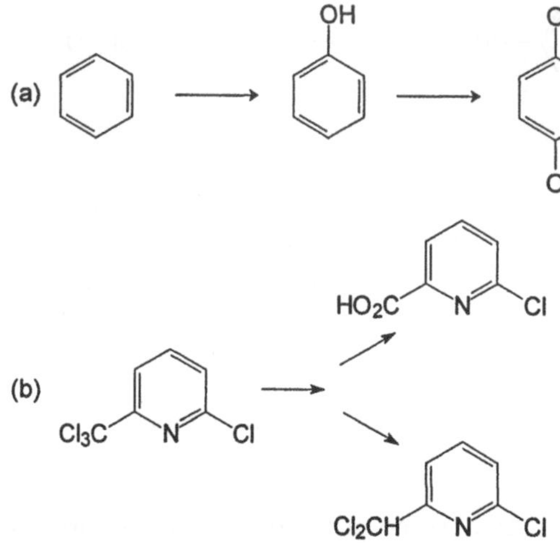

FIGURE 4.35
Metabolism by *Nitrosomonas europaea* (a) benzene and (b) 2-chloro-6-trichloromethylpyridine.

As is the case for methanotrophic bacteria, such transformations are probably confined neither to a single organism nor to strains of specific taxa within the group. For example, both *N. oceanus* and *N. europaea* are able to oxidize methane to CO_2 (Jones and Morita 1983; Ward 1987). The versatility of this group of organisms clearly motivates a reassessment of their ecological significance particularly in the marine environment where they are widely distributed. The overall similarity of these transformations to those carried out by eukaryotic cytochrome P-450 systems (Guengerich 1990) is striking.

4.3.3 Organisms Using Electron Acceptors Other than Oxygen

A number of facultatively anaerobic bacteria are able to carry out a respiratory metabolism in the absence of oxygen using alternative inorganic electron acceptors. For example, the following reductions may be coupled to oxidative degradation: nitrate to nitrogen (or nitrous oxide), sulfate to sulfide, carbonate to methane, fumarate to succinate, trimethylamine-N-oxide to trimethylamine, or dimethylsulfoxide to dimethyl sulfide (Styrvold and Ström 1984) (see Figure 4.28). An attempt will be made here to illustrate the metabolic potential of organisms under these conditions. Some introductory comments are given.

1. The conditions under which these function and their regulation depend on the organism. For example, whereas in *E. coli*, oxygen represses the synthesis of the other reductases, and under anaerobic conditions the reductases for fumarate, DMSO, and TMAO are repressed by nitrate, this is not the case for *Wolinella succinogenes*

in which sulfur represses the synthesis of the more positive electron acceptors nitrate and fumarate (Lorenzen et al. 1993).

2. The dimethyl sulfoxide reductase from *E. coli* (Weiner et al. 1988) has a very broad substrate specificity and is able to reduce a range of sulfoxides and *N*-oxides.

Anaerobic sulfate reduction is not discussed here, and many examples of the widespread metabolic potential of these organisms is given in Chapter 6.

Nitrate and Related Compounds

The degradation of organic compounds with nitrate in the absence of oxygen — denitrification or nitrate dissimilation — has been known for a long time and has been used as a valuable character in bacterial classification; the products are either dinitrogen or nitrous oxide and the reaction is generally inhibited by oxygen so that it occurs to a significant extent only under anoxic conditions. For example, although it has been reported that *Thiosphaera pantotropha* (*Paracoccus denitrificans*) is capable of both denitrification and nitrification under aerobic conditions (Robertson and Kuenen 1984), it has been shown that the suite of enzymes necessary for denitrification is not expressed constitutively and that rates of denitrification under aerobic conditions are very much slower than under anaerobic conditions (Moir et al. 1995). Intermediates in denitrification involve sequential formation of nitrite, nitric oxide and nitrous oxide. The chemical oxidation of biogenic nitric oxide, to N_{ox} has been considered in the context of increased ozone formation (Stohl et al. 1996). Nitric oxide may also be produced microbiologically in two completely different reactions:

- Oxidation of L-arginine by a strain of *Nocardia* sp. produces nitric oxide and L-citrulline (Chen and Rosazza 1995). The enzyme (nitric oxide synthase) carries out two distinct reactions: (1) hydroxylation of L-arginine catalyzed by an enzyme analogous to cytochrome P-450 and (2) a one-electron oxidation of N^{ω}-hydroxy-L-arginine to NO and L-citrulline.
- During metabolism of glycerol trinitrate by *Phanerochaete chrysosporium* (Servent et al. 1991).

These reactions are of interest in view of the physiological role of nitric oxide in mammalian systems (Feldman et al. 1993; Crane et al. 1997), in parasitic infections (James 1995), and in the inhibition of respiration in *Helicobacter pylovi* (Nagata et al. 1998).

Renewed interest has been focused on degradation of xenobiotics under anaerobic conditions in the presence of nitrate — possibly motivated by the extent of leaching of nitrate fertilizer from agricultural land into groundwater. In studies with such organisms, a clear distinction should be made between degradation of the substrate under three conditions which may or may not be biochemically equivalent: (1) aerobic conditions, (2) anaerobic conditions in the presence of nitrate, and (3) strictly anaerobic conditions in the absence of any electron acceptor where fermentation is predominant. The last of these is briefly discussed in Section 4.3.4. Some examples are given as illustration of the degradations that have been observed with facultatively anaerobic organisms using nitrate as electron acceptor.

- The degradation of carbon tetrachloride to CO_2 by a *Pseudomonas* sp. (Criddle et al. 1990). The complexity of this degradation is discussed in Chapter 6, Section 6.4.2.

- The nonstoichiometric production of trichloromethane from tetrachloromethane by *Shewanella putrefaciens* (Picardal et al. 1993).

- The degradation of benzoate (Taylor and Heeb 1972; Williams and Evans 1975; Ziegler et al. 1987) and *o*-phthalate (Afring and Taylor 1981; Nozawa and Maruyama 1988).

- The degradation of alkyl benzenes (Hutchins, 1991; Evans et al. 1991a,b; Altenschmidt and Fuchs 1991). In these studies, some of the organisms referred to the genus *Pseudomonas*, have been transferred to the genus *Thauera* (Anders et al. 1995).

- The degradation of pristane in microcosms and in enrichment cultures (Bregnard et al. 1997).

- The mineralization of cholesterol by an organism related to *Rhodocyclus, Thauera*, and *Azoarcus* (Harder and Probian 1997).

The mechanism of many of these reactions remains incompletely resolved and some of them are discussed further in Chapter 6, Section 6.7.3: the potential complexities of the pathways for the aromatic compounds may be illustrated by the demonstration that phenol undergoes a reversible exchange reaction with CO_2 to produce 4-hydroxybenzoate before dehydroxylation to benzoate (Tschech and Fuchs 1989).

It is also important to emphasize that an organism that can degrade a given substrate under conditions of nitrate dissimilation may not necessarily display this potential under aerobic conditions. For example, a strain of *Pseudomonas* sp. could be grown with vanillate under anaerobic conditions in the presence of nitrate but was unable to grow under aerobic conditions with the same substrate. On the other hand, cells grown anaerobically with nitrate and vanillate were able to oxidize vanillate under both aerobic and anaerobic conditions; the cells were also able to demethylate a much wider spectrum of aromatic methoxy compounds under anaerobic conditions than under aerobic conditions (Taylor 1983). Such subtleties should be clearly appreciated and taken into consideration in evaluating the degradative potential of comparable organisms under different physiological conditions.

Oxyanions: Chlorate, Selenate and Arsenate

1. Chlorate has been shown to support the growth of an anaerobic community growing at the expense of acetate (Malmqvist et al. 1991), and a pure culture designated *Ideonella dechlorans* has been isolated (Malmqvist et al. 1994). A bacterium, strain GR-1 is able to carry out the sequential reduction of perchlorate to chlorate, chlorite, and chloride at the expense of acetate (Rikken et al. 1996), and the chlorite dismutase has been purified and characterized (Van Ginkel et al. 1996).

2. A facultatively anaerobic organism *Thauera selenatis* is able to use selenate as electron acceptor for anaerobic growth (Macy et al. 1993), while both selenate and arsenate are used by a microaerophilic organism for growth with lactate under anaerobic conditions (Laverman et al. 1995). These reductions should be carefully distinguished from the situation in which selenate is gratuitously reduced during aerobic growth (Maiers et al. 1988). A strictly anaerobic strain of *Desulfotomaculum auripigmentum* is able to use either sulfate or — preferably arsenate — as electron acceptors for growth with lactate (Newman et al. 1997).

Iron(III), Manganese(IV), Chromium(VI), and Uranium(VI)

Many bacteria that ferment organic substrates are able to reduce Fe(III) to Fe(II) gratuitously, but these organisms are apparently unable to use the energy of this reduction for growth and couple this to the oxidation of organic substrates. There are, however, bacteria that can accomplish this, and some of them can also effectively use Mn(IV) and U(VI). There are two broadly different groups of organisms with this metabolic potential: one is facultatively anaerobic, the other strictly anaerobic.

1. Strains of *Shewanella putrefaciens* (*Alteromonas putrefaciens*) are widely distributed in environmental samples and are generally considered as aerobic organisms with the capability of reducing thiosulfate in complex media to sulfide. They are also able, however, to grow anaerobically using Fe(III) as electron acceptor and to oxidize formate, lactate, or pyruvate. These substrates cannot, however, be completely oxidized to CO_2 since the acetate produced from the C_2 compounds is not further metabolized; Mn(IV) may also function in a similar way (Lovely et al. 1989). The bioenergetics of the system have been examined in cells of another strain of this organism grown anaerobically on lactate with either fumarate or nitrate as electron acceptors, and respiration-linked proton translocation in response to Mn(IV), fumarate, or oxygen was clearly demonstrated (Myers and Nealson 1990). Levels of Fe(III) reductase,

nitrate reductase, and nitrite reductase are elevated by growth under microaerophilic conditions and the organism probably possesses three reductase systems, each of which apparently consists of low-rate and high-rate components (DiChristina 1992).

Another organism with hitherto unknown taxonomic affinity has been isolated (Caccavo et al. 1992) and is able to couple the oxidation of lactate to the reduction of Fe(III), Mn(IV), and U(VI).

2. A strictly anaerobic organism designated GS-15 and assigned to the taxon *Geobacter metallireducens* (Lovley et al. 1993) is able to use Fe(III) as electron acceptor for the oxidation of a number of compounds including acetate (Lovley and Lonergan 1990) and toluene and phenols under anaerobic conditions (Champine and Goodwin 1991). This organism is also able to oxidize acetate by reduction of Mn(IV) and U(VI) (Lovley et al. 1993). Organisms belonging to the genus *Geobacter* are widely distributed in anaerobic environments in which Fe(III) occurs, and all of the isolates are able to use acetate as electron donor (Coates et al.1996). In addition, a number of sulfate-reducing anaerobic bacteria can oxidize S^0 to sulfate at the expense of Mn(IV) (Lovley and Phillips 1994). The phylogenetically distinct *Geovibrio ferrireducens* strain PAL-1 is able to use a wide range of organic compounds as electron donors including acetate, propionate, succinate, and proline (Caccavo et al. 1996).

3. An undefined mixed flora of microorganisms is able to oxidize benzoate anaerobically at the expense of reducing Cr(VI) to Cr(III) (Shen et al. 1996). This was achieved after addition of nitrate to the initial anaerobic cultures containing chromate and benzoate, and oxygen was also an inducer.

These transformations illustrate important processes for the cycling of organic carbon in sediments where Fe(III) may have been precipitated, and it seems likely that comparable geochemical cycles involving manganese (Lovley and Phillips 1988) will also achieve greater prominence (Lovely 1991; Nealson and Myers 1992). It has also been suggested that such organisms could be used for the immobilization of soluble U(VI) in wastewater containing both U(VI) and organic compounds by conversion to insoluble U(IV) (Lovely et al. 1991).

4.3.4 Anaerobic Bacteria

There are a number of reasons for the incre2ased interest in transformations carried out by anaerobic bacteria. One is that many xenobiotics are partitioned from the aquatic phase into the sediment phase after discharge into the aquatic environment (Section 3.2). Another is that sediments in the vicinity of industrial discharge often contain in addition readily degraded organic matter; the activity of aerobic and facultatively anaerobic bacteria then renders these sediments effectively anaerobic.

The fate of xenobiotics in the environment is therefore significantly determined by the degradative activity of anaerobic bacteria.

The terms *anaerobic* and *anoxic* are purely operational and imply merely the absence of air or oxygen, and the absolute distinction between aerobic and anaerobic organisms is becoming increasingly blurred; the problem of defining anaerobic bacteria is therefore best left to philosophers. Possibly the critical issue is the degree to which low concentrations of oxygen are either necessary for growth or toxic. During growth of bacteria in the absence of externally added electron acceptors, the term *fermentation* implies that a redox balance is achieved between the substrate (which may include CO_2) and its metabolites. A few examples are given to illustrate the apparently conflicting situations that may be encountered, and the gradients of response that may exist.

1. Although strictly anaerobic bacteria do not generally grow in the presence of high-potential electron acceptors such as oxygen or nitrate, an intriguing exception is provided by an obligately anaerobic organism that uses nitrate as electron acceptor during the degradation of resorcinol (Gorny et al. 1992). This isolation draws attention to the unknown extent to which such organisms exist in natural systems, since strictly anaerobic conditions are seldom used for the isolation of organisms using nitrate as electron acceptor; other formally similar organisms are either facultatively anaerobic or have a fermentative metabolism. On the other hand, the normally strictly anaerobic sulfate-reducing organisms *Desulfobulbus propionicus* and *Desulfovibrio desulfuricans* may grow by reducing nitrate or nitrite to ammonia using hydrogen as electron donor (Seitz and Cypionka 1986). *Desulfovibrio vulgaris* Hildeborough is capable of growth at oxygen concentrations of 0.24 to 0.48 μM, and it was suggested that this organism may protect anoxic environments from adverse effects resulting from intrusion of oxygen (Johnson et al. 1997). In the presence of nitrate the acetogen *Clostridium thermoaceticum* oxidizes the *O*-methyl groups of vanillin or vanillate to CO_2 without production of acetate that is the normal product in the absence of nitrate (Seifritz et al. 1993).

2. Some anaerobic organisms such as clostridia are appreciably tolerant of exposure to oxygen, and indeed others such as *Wolinella succinogenes* that have hitherto been classified as anaerobes are in fact microaerophilic (Han et al. 1991). On the other hand, organisms such as *Nitrosomonas europaea* that normally obtain energy for growth by oxidation of ammonia to nitrite may apparently bring about denitrification of nitrite under conditions of oxygen stress

(Poth and Focht 1985), or under anaerobic conditions in the presence of pyruvate (Abeliovich and Vonshak 1992).

3. Attention is drawn in Section 4.6.1.2 to the role of oxygen concentration so that it is appropriate to note the existence of microaerotolerant or microaerophilic organisms. The example of *W. succinogenes* has been noted earlier, and another is provided by *Malonomonas rubra* that uses malonate as sole source of carbon and energy (Dehning and Schink 1989a). On the other hand, *Propionigenium modestum* that obtains its energy by decarboxylating succinate to propionate is a strictly anaerobic organism (Schink and Pfennig 1982).

Attention may be briefly directed to two groups of anaerobic bacteria that display metabolic versatility to structurally diverse compounds — clostridia and sulfate-reducers.

1. The classical studies on the anaerobic metabolism of amino acids, purines, and pyrimidines by clostridia not only set out the relevant experimental procedures and thereby laid the foundations for virtually all future investigations, but also brought to light the importance and range of coenzyme-B_{12}-mediated rearrangements. These reactions were mainly carried out by species of clostridia and indeed some of these degradations belong to the classical age of microbiology in Delft (Liebert 1909). The number of different clostridia investigated may be gained from the following selected examples: the classic purine-fermenting organisms *Clostridium acidurici* and *Cl. cylindrosporum* ferment several purines including uric acid, xanthine, and guanine, while the degradation of pyrimidines such as orotic acid is accomplished by *Cl. oroticum*. A large number of clostridia including *Cl. perfringens, Cl. saccharobutyricum, Cl. propionicum, Cl. tetani, Cl. sporogenes,* and *Cl. tetanomorphum* ferment a range of single amino acids, while some participate in the Stickland reaction involving two amino acids. In Chapter 6, the pathways employed for the degradation of aminoacids are discussed in Chapter 6, Section 6.7.1 and those for the degradation of N-containing heterocyclic compounds in Chapter 6, Section 6.7.4.1.

2. The spectrum of compounds degraded by anaerobic sulfate-reducing bacteria is continuously widening and now includes, for example, alkanes (Aeckersberg et al. 1991; 1998), alkanols, alkylamines, and alkanoic acids (Chapter 6, Section 6.7.1), nicotinic acid (Imhoff-Stuckle and Pfennig 1983), indoles (Bak and Widdel 1986), methoxybenzoates (Tasaki et al. 1991), benzoate, hydroxybenzoate, and phenol (Drzyga et al. 1993), and catechol (Szewzyk and Pfennig 1987). Further details of some of the pathways involved are provided in Chapter 6, Section 6.7. It should also be noted that S^0

may serve as oxidant for organisms belonging to a range of genera including *Desulfomicrobium, Desulfurella, Desulfuromonas,* and *Desulfuromusa* (references in Liesack and Finster 1994). Syntrophic associations are briefly discussed in Section 4.5.1, and it is worth noting that anaerobic oxidation of propionate may be accomplished in pure cultures of *Syntrophobacter wolinii* and *S. pfennigi* at the expense of sulfate reduction (Wallrabenstein et al. 1995).

It is appropriate to mention briefly a few other aspects of anaerobic organisms:

1. Compounds such as oxalate (Dehning and Schink 1989b) and malonate (Dehning et al. 1989; Janssen and Harfoot 1992) that are degraded by decarboxylation with only a modest energy contribution are nevertheless able to support the growth of the appropriate organisms. The bioenergetics of the anaerobic *Oxalobacter formigenes* (Anantharam et al. 1989) and the microaerophilic *Malonomonas rubra* have been elucidated (Hilbi et al. 1993; Dimroth and Hilbi 1997). Further comment on decarboxylation of dicarboxylic acids by strictly anaerobic bacteria is given in Chapter 6, Section 6.7.1.

2. Extensive effort has been directed to the anaerobic dechlorination of aromatic compounds and alkenes although hitherto only relatively few pure cultures have been isolated: (a) a sulfate-reducing organism *Desulfomonile tiedjei* (DeWeerd et al. 1990), (b) a spore-forming organism (Madsen and Licht 1992), (c) a sulfite-reducing organism *Desulfitobacterium dehalogenans* (Utkin et al. 1994), (d) an organism within the myxobacteria (Cole et al. 1994), (e) *Dehalospirillum multivorans* that accomplishes the sequential reduction of tetrachloroethene to trichloroethene and *cis*-1,2-dichloroethene (Neumann et al. 1994). Further comments on these reactions are given in Chapter 6, Section 6.6.

4.3.5 Phototrophic Organisms

The metabolic significance of oxygenic algae and cyanobacteria has received relatively limited attention in spite of the fact that they are important components of many ecosystems and may, for example, in the marine environment be of particular significance. Whereas the heterotrophic growth of algae at the expense of simple carbohydrates, amino acids, lower aliphatic carboxylic acids, and simple polyols is well documented (Neilson and Lewin 1974), the potential of algae for metabolism of xenobiotics has been much less extensively explored. Among these metabolic possibilities which have received less attention than they possibly deserve, the following may be used as illustration:

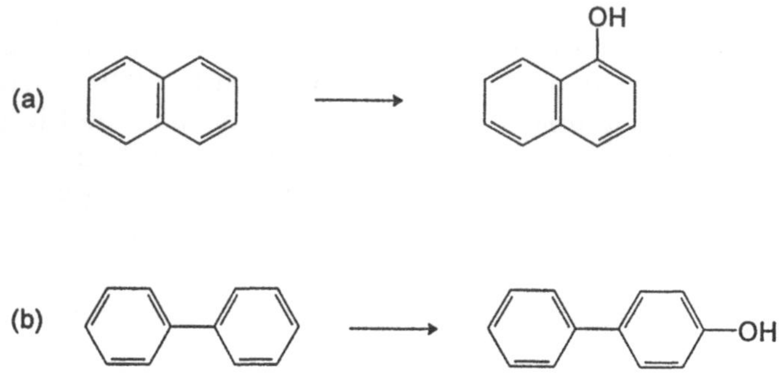

FIGURE 4.36
Examples of the biotransformation of aromatic compounds by algae.

1. The transformation — although not apparently the degradation — of naphthalene has been examined in cyanobacteria and microalgae including representatives of green, red, and brown algae and diatoms (Cerniglia et al. 1980a; 1982), and the transformation of biphenyl (Cerniglia et al. 1980b), aniline (Cerniglia et al. 1981) and methyl naphthalenes (Cerniglia et al. 1983) by cyanobacteria (Figure 4.36). Phenanthrene is metabolized by *Agmenellum quadruplicatum* to the *trans*-9,10-dihydrodiol by a monooxygenase system, and the transiently formed 1-hydroxyphenanthrene is *O*-methylated (Narro et al. 1992). The biotransformation of benzo[*a*]pyrene has been demonstrated in a number of green algae, although this was not metabolized by a chlamydomonad, a chrysophyte, a euglenid, or a cyanobacterium (Warshawsky et al. 1995); the relative amounts of the products depended on the light sources and their intensity, and included dihydrodiols (9,10-, 4,5-, 11,12-, and 7,8-), the toxic 3,6-quinone, and phenols. The dihydrodiols (11,12-, 7,8-, and 4,5-) produced by *Selenastrum capricornutum* all had the *cis* configuration that suggests their formation by the mediation of a dioxygenase (Warshawsky et al. 1988). It is worth noting that *Ochromonas danica* is able to degrade phenol by extradiol fission of the catechol that is initially formed (Semple and Cain 1996).

2. A number of green algae are able to use aromatic sulfonic acids (Figure 4.37a) (Soeder et al. 1987) and aliphatic sulfonic acids (Figure 4.37b) (Biedlingmeier and Schmidt 1983) as sources of sulfur. Cultures of *Scenedesmus obliquus* under conditions of sulfate limitation metabolized naphthalene-1-sulfonate to 1-hydroxynaphthalene-2-sulfonate and the glucoside of naphth-1-ol (Kneifel et al. 1997). These results are consistent with formation of a 1,2-epoxide followed by an NIH shift.

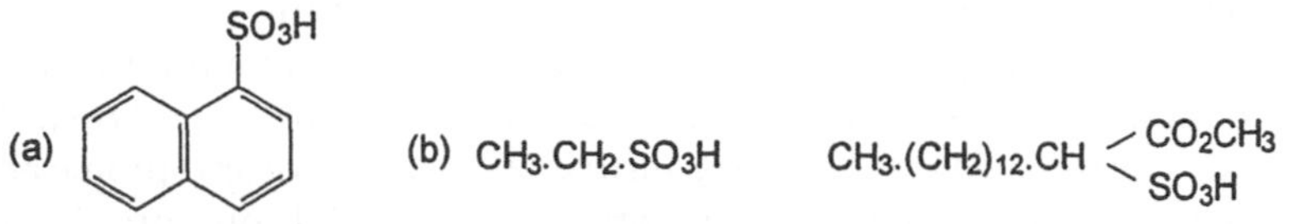

FIGURE 4.37
Organosulfur compounds used as S-sources by algae.

3. The cyanobacteria *Anabaena* sp. strain PCC 7120 and *Nostoc ellipsosporum* dechlorinated γ-hexachloro[*aaaeee*]cyclohexane in the light in presence of nitrate to γ-pentachlorocyclohexene (Figure 4.38), and to a mixture of chlorobenzenes (Kuritz and Wolk 1995), and the reaction is dependent on the functioning of the *nir* operon involved in nitrite reduction (Kuritz et al. 1997).

4. Representatives of the major groups of algae are able to use a range of amino acids as nitrogen sources (Neilson and Larsson 1980).

5. The transformation of DDT to DDE— albeit in rather low yield — by elimination of one molecule of HCl has been observed in several marine algae (Rice and Sikka 1973).

There has been a revival of interest in the metabolic potential of anaerobic phototrophic bacteria, particularly the purple nonsulfur organisms that can degrade aromatic compounds (Khanna et al. 1992). Such organisms are widely distributed in appropriate ecosystems, and may therefore play a significant role in the degradation of xenobiotics. Less appears to be known of

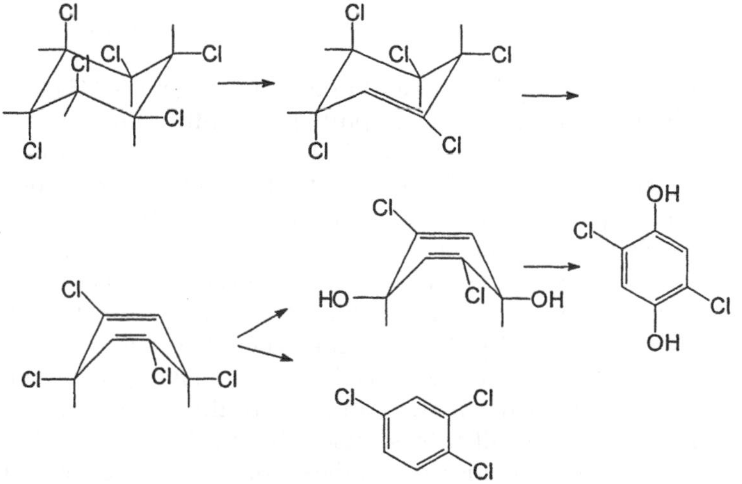

FIGURE 4.38
Transformation of lindane by cyanobacteria.

the potential of other anaerobic phototrophs such as the purple and green sulfur bacteria to degrade xenobiotics. Dehalogenation of a number of halogenated alkanoic acids has been observed with *Rhodospirillum rubrum, R. photometricum,* and *Rhodopseudomonas palustris*: 2- and 3-chloropropionic acid by all of them, chloroacetic acid by *R. photometricum,* and 2-bromopropionic acid by *R. rubrum* and *Rh. palustris* (McGrath and Harfoot 1997).

In the context of Fe(III) metabolism that has been discussed earlier, the converse reaction whereby Fe(II) serves as an electron donor for anaerobic phototrophic growth of purple bacteria may be noted (Ehrenreich and Widdel 1994).

4.3.6 Eukaryotic Microorganisms: Fungi and Yeasts

Matabolism by Fungi

In the terrestrial environment — and possibly also in a few specialized aquatic ecosystems — fungi and yeasts play a cardinal role in biodegradation and biotransformation. The role of yeasts in the coastal marine environment is illustrated from results of their frequency and metabolic potential for transformation of phenanthrene and benz[*a*]anthracene (MacGillivray and Shiaris 1993). The transformation of PAHs by fungi is comparable to that in mammalian systems so that fungal metabolism has been explored as a model for higher organisms, and extensive studies have been carried out with *Cunninghamella elegans*. This is discussed further in Chapter 6, Section 6.2.2.

It has been suggested that the transformations accomplished by the brown-rot fungus *Gleophyllum striatum* may involve hydroxyl radicals, and this is supported by the overall similarity in the structures of the fungal metabolites with those produced with Fenton's reagent (Wetzstein et al. 1997). Yeasts are able to carry out a number of oxidations that are noted in Sections 4.3.6, 4.4.1.2, and 4.4.4.

Biotransformation (hydroxylation) of a wide range of PAHs and related compounds including biphenyl, naphthalene, anthracene, phenanthrene, 4- and 7-methylbenz[*a*]anthracene, 7,12-dimethylbenz[*a*]anthracene has been examined in a number of fungi, most extensively in species of *Cunninghamella* especially in *C. elegans* (McMillan et al. 1987). Hydroxylation of benzimidazole (Seigle-Murandi et al. 1986) by *Absidia spinosa* and of biphenyl ether by *C. echinulata* (Seigle-Murandi et al. 1991) has also been studied due to the industrial interest in the metabolites. The biotransformation of alachlor (2-chloro-*N*-methoxymethyl-*N*-[2,6-diethylphenyl]acetamide) by *C. elegans* has already been mentioned (Section 4.3.6) and involves primarily hydroxylation at the benzylic carbon atom and loss of the methoxymethyl group (Pothuluri et al. 1993). In all of these cases, the reactions are purely biotransformations since the aromatic rings of these compounds are not destroyed.

Particular attention has been focused on the white-rot fungus *Phanerochaete chrysosporium* because of its ability to degrade lignin and to metabolize a wide range of unrelated compounds including PAHs (Bumpus 1989) and organochlorine compounds such as DDT (Bumpus and Aust 1987), PCBs (Eaton 1985), lindane and chlordane (Kennedy et al. 1990), pentachlorophenol (Mileski et al. 1988), and 2,7-dichlorobenzo-1,4-dioxin (Valli et al. 1992a). The novel pathways for the degradation of 2,4-dichlorophenol (Valli and Gold 1991) and 2,4-dinitrotoluene (Valli et al. 1992b) are discussed in a wider context in Chapter 6 (Sections 6.5.2 and 6.8.2). Degradation of all these compounds is apparently mediated by two peroxidase systems — lignin peroxidases and manganese-dependent peroxidases — and a laccase that is produced by several white-rot fungi though not by *P. chrysosporium*. A potentially serious interpretative ambiguity has, however, emerged from the observation that lignin peroxidase is able to polymerize a range of putative aromatic precursors to lignin, but that it is not the functional enzyme in the depolymerization of lignin (Sarkanen et al. 1991). The regulation of the synthesis of these oxidative enzymes is complex and is influenced by nitrogen limitation, the growth status of the cells, and the concentration of manganese in the medium (Perez and Jeffries 1990). In addition, it seems clear that monooxygenase and epoxide hydrolase activities are also involved since the biotransformation of phenanthrene takes place even in the absence of peroxidase systems (Sutherland et al. 1991). The ability of *P. chrysosporium* to degrade anthracene by oxidation to anthra-9,10-quinone followed by ring fission to yield *o*-phthalate is noted in Chapter 6, Section 6.2.2. Further comments on the synthesis of these enzyme systems is given in Chapter 5, Section 5.2.4. Clearly, therefore, a number of important unresolved issues remain. In addition, attention has already been drawn in Chapter 3, Section 3.2.3 to the role of fungal redox systems in the covalent linking of xenobiotics to aromatic components of humus in soils.

The metabolic activity of other white-rot fungi including *P. chrysosporium* and *Pleurotus ostreacus* is discussed in Chapter 6, Section 6.2.2 in the context of PAHs, and the mineralization capacity of the manganese peroxide system from *Nematoloma frowardii* for a number of substrates has been demonstrated (Hofrichter et al. 1998). The formation of CO_2 from labeled substrates ranged from 7% from pyrene to 36% for pentachlorophenol, 42% for 2-amino-4,6-dinitrotoluene and 49% for catechol.

Several strains of white-rot fungi have been examined for their ability to degrade and mineralize selected PCB congeners (Beaudette et al. 1998). Mineralization of 2,4',5 [U-^{14}C]trichlorobiphenyl as a fraction of the substrate added, ranged from ~ 4.2% for a strain of *Pleurotus ostreacus*, 4.9% and 6.9% for two strains of *Bjerkandera adusta*, to 11% for a strain of *Trametes versicolor*, whereas *Phanerochaete chrysosporium* produced only ~ 2% of [14]CO_2. There was no apparent correlation among levels of lignin peroxidase, manganese peroxidase and degradative ability.

The importance of *P. chrysosporium* that is able to degrade simultaneously chlorobenzene and toluene (Yadav et al. 1995) is noted in Sections 4.8.1 and 8.1.2.

The degradation of phenolic compounds by fungi may involve rather unusual features of which two are given as illustration.

- *Aspergillus fumigatus* degrades phenol using simultaneously two pathways: first, *ortho*-hydroxylation to catechol followed by ring cleavage to β-ketoadipate and, second, successive hydroxylation to hydroquinone and 1,2,4-trihydroxybenzene before ring cleavage (Jones et al. 1995). The metabolism of 2-aminobenzoate (anthranilate) to 2,3-dihydroxybenzoate by *A. niger* is accomplished by apparent incorporation of one atom of oxygen from each of O_2 and H_2O, and is not a flavoprotein (Subramanian and Vaidyanathan 1984).

- The degradation of phloroglcinol by *Fusarium solani* involves rearrangement to pyrogallol followed by ring cleavage to 2-keto-hex-3-ene-1,6-dicarboxylate (Walker and Taylor 1983). This rearrangement is the opposite of that involved in the anaerobic degradation of 3,4,5-trihydroxybenzoate (Chapter 6, Section 6.7.3.1).

Metabolism by Yeasts

The metabolic capabilities of yeasts have attracted attention in different contexts and it has emerged that, in contrast to many fungi, they are able to bring about fission of aromatic rings. Some examples that illustrate the various possibilities are given.

1. Ring cleavage clearly occurs during the metabolism of phenol (Walker 1973) by the yeast *Rhodotorula glutinis*, and of aromatic acids by various fungi (Cain et al. 1968; Durham et al. 1984; Gupta et al. 1986).

2. Comparable ring cleavage reactions have also been found in studies on the metabolism of aromatic compounds by the yeast *Trichosporon cutaneum* whose metabolic versatility is indeed quite comparable to that of bacteria; examples of this may be found in the degradation of phenol and resorcinol (Gaal and Neujahr 1979), of tryptophan and anthranilate (Anderson and Dagley 1981a), and of aromatic acids (Anderson and Dagley 1981b).

3. The ability to grow at the expense of 4-hydroxy- and 3,4-dihydroxybenzoate has been used for classification of medically important yeasts including *Candida parapsilosis* (Cooper and Land 1979), and this organism degrades these by oxidative decarboxylation catalyzed by a flavoprotein monooxygenase (Eppink et al. 1997).

4. The phenol-assimilating yeast *C. maltosa* degraded a number of phenols even though these were unable to support growth: hydroxylation of 3-chloro- and 4-chlorophenol produced initially 4-chlorocatechol and then 5-chloropyrogallol (Polnisch et al. 1992). The yeast *Rhodotorula cresolica* was able to assimilate phenol, 3- and 4-methylphenol, catechol and 3- and 4-methylcatechol, resorcinol

and hydroquinone, and a wide range of phenolic carboxylic acids (Middelhoven and Spaaij 1997).

5. Diphenyl ether is transformed by *Trichosporon beigelii* through successive hydroxylation and *exo* ring cleavage of the resulting catechol at the 4 to 5 position (Schauer et al. 1995). In a formally comparable way, metabolism of dibenzofuran by the yeast *T. mucoides* involves initial hydroxylation of one of the rings followed by ring fission at the 2,3-position (Hammer et al. 1998).

Yeasts are also able to degrade long-chain alkanes. This is accomplished in two subcellular organelles: in microsomes, cytochrome P-450 and the associated NADH reductase (Käppeli 1986) carry out hydroxylation, while the alkanols are dehydrogenated and undergo β-oxidation in peroxisomes that are induced during growth with alkanes (Tanaka and Ueda 1993). Peroxisomes are discussed further in Section 4.4.4. Clearly then, yeasts possess metabolic potential for the degradation of xenobiotics little inferior to that of many bacterial groups, so that their role in natural ecosystems justifies the greater attention that has been directed to them (MacGillivray and Shiaris 1993).

4.3.7 Other Organisms

The metabolic potential of fish has already been briefly noted in the context of bioconcentration (Section 3.1.3) and is discussed more extensively in Chapter 7, Section 7.5.1 in the context of toxicology. It is therefore sufficient here merely to state that fish may effectively metabolize a wide range of xenobiotics even though complete degradation is seldom achieved; instead, simple biotransformations are carried out, such as introduction of hydroxyl groups into aromatic rings followed by conjugation with sulfate or glucuronic acid.

Limited investigations have revealed the metabolic potential of a taxonomically diverse eukaryotic organisms:

1. The apochlorotic alga (protozoan) *Prototheca zopfii* that is able to degrade aliphatic hydrocarbons (Walker and Pore 1978; Koenig and Ward 1983);

2. *Tetrahymena thermophila* that transforms pentachloronitrobenzene to the corresponding aniline and pentachlorothioanisole (Figure 4.39) (Murphy et al. 1982);

3. *Daphnia magna* that has been shown to bring about dechlorination and limited oxidation of heptachlor (Figure 4.40) (Feroz et al. 1990).

A number of other groups of biota are also able to bring about transformation of structurally diverse compounds. A few examples suffice as illustration.

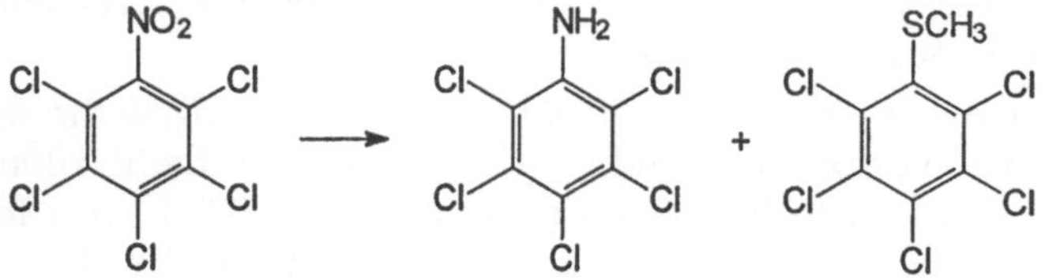

FIGURE 4.39
Biotransformation of pentachloronitrobenzene by *Tetrahymena thermophila*.

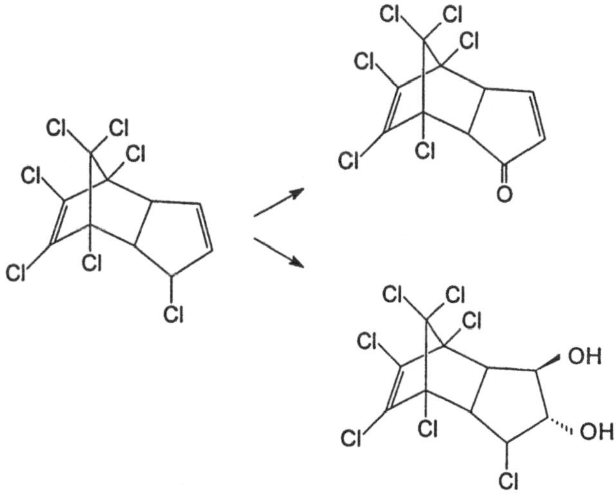

FIGURE 4.40
Biotransformation of heptachlor by *Daphnia magna*.

Metabolism by Higher Plants

The detoxification and metabolism of higher plants has been reviewed (Sandermann 1994) (see also Chapter 8, Section 8.1.1), and a few examples are given here as illustration.

1. Pentachlorophenol is metabolized by the aquatic plant *Eichhornia crassipes* to a number of metabolites including di-, tri-, and tetra-chlorocatechol, 2,3,5-tri- and tetrachlorohydroquinone, pentachloroanisole, and tetrachloroveratrole (Roy and Hänninen 1994). The phenolic compounds should be compared with those produced during the photochemical (see Figure 4.4) and the initial stages in the microbiological metabolism of pentachlorophenol (Chapter 6, Section 6.5.1.2), followed by *O*-methylation (Section 6.11.4).

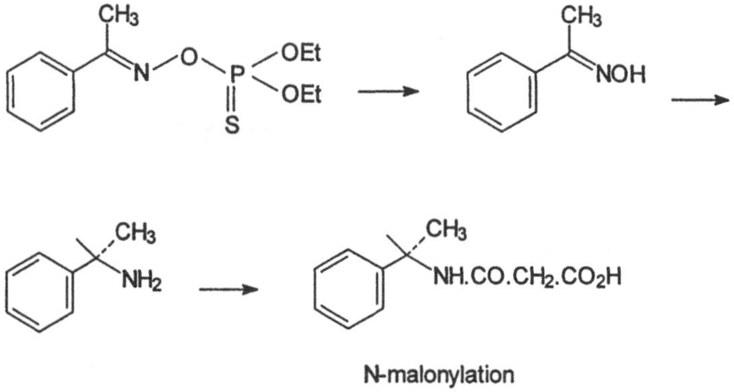

FIGURE 4.41
Metabolism of phoxim.

2. Quite complex transformations may be mediated, and the metabolism of phoxim by plant organs and cell suspension of soybean (*Glycine max*) may be given as illustration (Höhl and Barz 1995) (Figure 4.41).

3. The uptake and biotransformation of benzene from soil and from the atmosphere have been studied in a number of plants and it was shown that in leaves of spinach (*Spinacia oleracea*) the label in ^{14}C-benzene was found in muconic, fumaric, succinic, malic, and oxalic acids as well as in specific amino acids, and that an enzyme preparation in the presence of NADH or NADPH produced phenol (Ugrekhelidze et al. 1997).

4. Hybrid poplars are able to transport and metabolize diverse xenobiotics: (a) trichloroethene is metabolized to trichloroethanol and trichloroacetate (Newman et al. 1997) and (b) atrazine is metabolized by reactions involving dealkylation and hydrolytic dechlorination to yield 2-hydroxy-4,6-diamino-1,3,5-triazine (Burken and Schnoor 1997).

5. 2,4,6-Trinitrotoluene is reduced by the aquatic plant *Myriophyllum spicatum* to aminodinitrotoluenes (Pavlostathis et al. 1998), and in axenic root cultures of *Catharanthus roseus* the initial metabolites 2-amino-4,6-dinitrotoluene and 4-amino-2,6-dintrotoluene are conjugated probably with C6-units (Bhadra et al. 1999). There are then several important unresolved issues including the phytotoxicity of these metabolitres before phytoremediation can be exploited.

Metabolism by Insects

The herbicide alachlor is transformed by chironomid larvae, and proceeds by *O*-demethylation followed by loss of the chloroacetyl group to produce ultimately 2,6-diethylaniline (Figure 4.42) (Wei and Vossbrinck 1992).

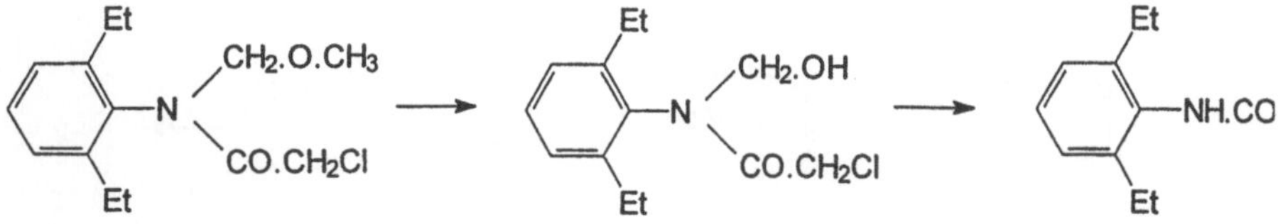

FIGURE 4.42
Transformation of alachlor by chironomid larvae.

Metabolism by Isopods

The uptake and elimination of benzo[a]pyrene by the terrestrial isopod *Porcellio scaber* have been investigated (van Brummelen and van Straalen 1996), and 1-hydroxypyrene was identified among the metabolites of pyrene in this organism (Stroomberg et al. 1996).

Metabolism by Oligochaetes

Both (+) and (−) limonene were transformed by larvae of the cutworm *Spodoptera litura* (Miyazawa et al. 1998): the reactions for both enantiomers involved (1) dihydroxylation between C-8 and C-9 and (2) oxidation of the C-1 methyl group to carboxyl. These transformations were not dependent on the intestinal microflora in comparison with the transformation of α–terpinene to *p*-mentha-1,3-dien-7-ol and *p*-cymene whose formation could be attributed to the intestinal flora.

Metabolism by Polychaetes

Polychaete worms belonging to the genera *Nereis* and *Scolecolepides* have extensive metabolic potential. *N. virens* is able to metabolize PCBs (McElroy et al. 1988) and a number of PAHs (McElroy 1990), while *N. diversicolor* and *S. viridis* are able to metabolize benzo[a]pyrene (Driscoll and McElroy 1996).

4.4 Mechanisms for the Introduction of Oxygen

There are a variety of strategies used by microorganisms for the introduction of oxygen. There are discussed in the following sections. By way of introduction, the formal reactions are summarized for monooxygenases in Figure 4.43a,b, and an outline of bacterial dioxygenases in Figure 4.43c,d. Reference should be made to reviews (Mason and Cammack 1992; Harayama et al. 1992) for detailed discussion of the mechanism and enzymology of prokaryotic oxygenases.

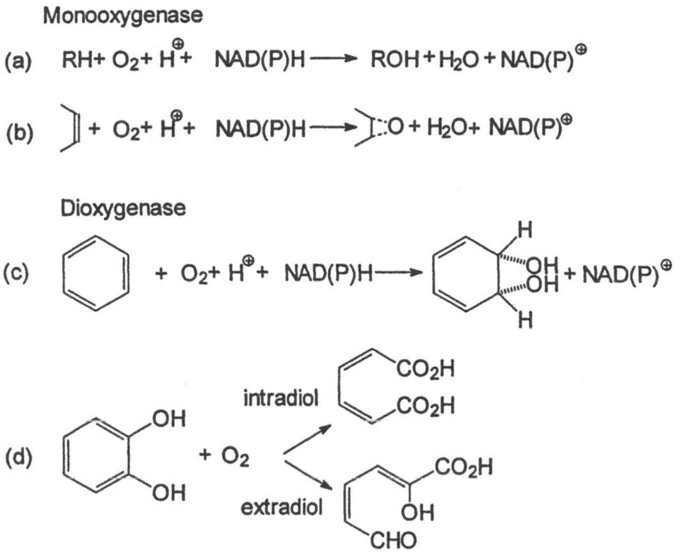

Monooxygenase

(a) $RH + O_2 + H^{\oplus} + NAD(P)H \longrightarrow ROH + H_2O + NAD(P)^{\oplus}$

(b) $\rrbracket + O_2 + H^{\oplus} + NAD(P)H \longrightarrow \rrbracket{::}O + H_2O + NAD(P)^{\oplus}$

Dioxygenase

(c) $+ O_2 + H^{\oplus} + NAD(P)H \longrightarrow$ $+ NAD(P)^{\oplus}$

(d) $+ O_2$

intradiol → (cis,cis-muconate with CO₂H, CO₂H)

extradiol → (with CO₂H, OH, CHO)

FIGURE 4.43
(a,b) Monooxygenase and (c,d) dioxygenase reactions.

4.4.1 Microbial Monooxygenase and Hydroxylase Systems

Microbial monooxygenases — both prokaryotic and eukaryotic — are not confined to a single taxonomic group of organisms. Dioxygenases are discussed in Sections 4.4.2 and 4.4.3 and, as noted later, some of them also possess monooxygenase activity.

Monooxygenases that carry out formal hydroxylation are discussed in greater detail in later sections and only some illustrative examples are given here.

1. Methane mooxygenase — and formally related systems — are involved in the bacterial hydroxylation of alkanes and cycloalkanes, and the epoxidation of alkenes. This has already been discussed in the context of C_1-metabolism by bacteria (Section 4.3.2) and in greater detail in Chapter 6, Section 6.1.1.

2. Monooxygenases carry out a Baeyer–Villiger-type reaction in the introduction of a single atom of oxygen into cycloalkanones with the production of lactones. Examples are given in Chapter 6, Section 6.1.2.

3. Bacterial monooxygenases that are flavoproteins requiring NADPH are involved in the hydroxylation of a number of phenolic compounds (a) phenol and chlorophenols, (b) salicylate, (c) 4-hydroxybenzoate, (d) 4-hydroxyphenylacetate, and (e) anthranilate in fungi. Further details of some of these are given later in this chapter and in Chapter 6 (Sections 6.2.1 and 6.5.1.2).

4. Cytochrome P-450 systems in bacteria, yeasts, and fungi carry out hydroxylation of a number of diverse compounds and are discussed later.

Attention is also drawn to other reactions in which a single atom of oxygen from H_2O is involved. Hydratases that catalyze the addition of the elements of H_2O to activated bonds are involved in the β-oxidation of long-chain alkanoates by yeasts (Section 4.4.4), in aerobic degradation of N-heterocyclic compounds (Chapter 6, Section 6.3.1.1), and both the aerobic and the anaerobic degradation of alkynes (Chapter 6, Sections 6.1.4 and 6.7.1).

4.4.1.1 Monooxygenases

Monooxygenation, or hydroxylation as it is often termed (compare hydroxylation of azaarenes by hydration–dehydrogenation, Chapter 6, Section 6.3.1.1) is involved in the aerobic degradation of both aliphatic and aromatic compounds.

Hydroxylation of Alkanes

The oxidation of the simplest alkane methanol is carried out by methanotrophs which may be obligate or facultative; only brief mention is made here of methane monooxygenase that catalyzes the introduction of oxygen. The enzyme exists in both a soluble (sm) and a particulate (ammo) form of which the former has been more extensively studied. The enzyme consists of three components: a hydroxylase, a regulatory protein that is not directly involved in electron transfer between the hydroxylase, and a third protein that is a reductase containing FAD and an [2Fe–2S] cluster. Details of the structure of the hydroxylase and the mechanism of its action involving the $Fe^{III} - O - Fe^{III}$ at the active site are given in a review (Lipscomb 1994). The metabolic versatility of the enzyme has already been noted in Section 4.3.2. The particulate enzyme contains copper (Nguyen et al. 1994), and both copper and iron (Zahn and DiSpirito 1996), and the concentration of copper determines the catalytic activity of the enzyme (Sontoh and Semrau 1998; see Chapter 5, Section 5.2.4).

The metabolism of higher alkanes and alkanoic acids is considered in Chapter 6, Section 6.1.1 and only brief comments are given here. Hydroxylation of *n*-alkanes and *n*-alkanoic acids differ only in detail.

1. Cytochrome P-450 systems are involved in *Pseudomonas oleovorans* and strains of *Acinetobacter* sp. (Section 4.4.1.2). In *P. oleovorans*, hydroxylation is carried out by a three-component system comprising a hydroxylase, a rubredoxin and a rubredoxin reductase. The *alkBFGHKL* and *alkST* genes are clustered on a plasmid (van Beilin et al. 1994), and the hydroxylase is encoded by the *alkB* gene, the rubredoxin reductase by *alkT*, the rubredoxin by *alkG*, an alcohol dehydrogenase by *alkJ*, the aldehyde dehydrogenase by *alkH*, and the acyl-CoSA synthase by *alkK*.

2. In *Acinetobacter* sp. strain ADP1, the hydroxylase gene encoded by *alkM* is chromosomal and the hydroxylation system is completely differed from that in *P. oleovorans* (Ratajczac et al. 1998).

The unusual involvement of a dioxygenase in the hydroxylation of *n*-alkanes by *Acinetobacter* sp. strain M-1 is noted later in Section 4.4.2.

Monooxygenation of Aromatic Compounds

Although the ring-cleavage reactions involved in the bacterial degradation of aromatic compounds are generally initiated by dioxygenation (Section 4.4.2), monooxygenases may be involved additionally or alternatively in the introduction of oxygen into the ring. Further details of toluene monooxygenation are given in Chapter 6, Section 6.2.1, and of oxygenating dehalogenation of chlorophenols in Chapter 6, Section 6.5.1.2.

1. Side-chain oxidation of alkylated aromatic hydrocarbons — Although, as noted later, and in Chapter 6, Sections 6.2.1 and 6.2.2, degradation of aromatic hydrocarbons generally proceeds by initial dioxygenation, an alternative pathway for the degradation of alkyl benzenes involves oxidation of methyl substituents to carboxylates. This is also used for the degradation of 4-nitrotoluene by *Pseudomonas* sp. strain TW3, and the *ntnWCMAB** genes that encode the enzymes that convert the substrate to 4-nitrobenzoate are similar to those in the upper pathway of the TOL plasmid (Harayama et al. 1989): an alcohol dehydrogenase, benzaldehyde dehydrogenase, a two-component monooxygenase, and part of a benzyl alcohol dehydrogenase (James and Williams 1998). In this strain, however, the enzymes are chromosomal, and the benzyl alcohol dehydrogenase ntnB* differs from the corresponding xylB protein.

The oxidation by strains of *P. putida* of the methyl group in compounds containing a hydroxyl group in the *para* position is, however, carried out by a different mechanism. The initial step is a dehydrogenation to a quinone methide followed by hydration (hydroxylation) to the benzyl alcohol (Hopper 1976) (Figure 4.44). The reaction with 4-ethylphenol is partially stereospecific and (McIntire et al. 1984), and the enzymes that catalyze the first two steps are flavocytochromes (McIntire et al. 1985). The role of hydroxylation in the degradation of azaarenes is discussed later.

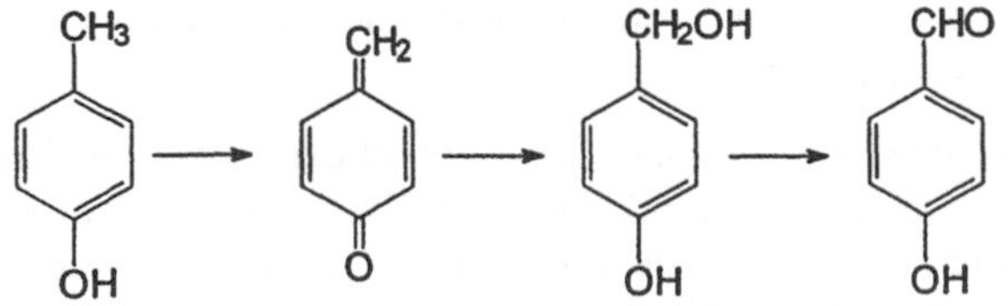

FIGURE 4.44
Degradation of 4-methylphenol by hydroxylation.

2. *Aromatic hydrocarbons* — Concurrent synthesis of *both* mono- and dioxygenases has been shown in a number of instances.

a. A number of strains of *Pseudomonas* sp. that were induced for toluene dioxygenase activity catalyzed the enantiomeric monooxygenation of indan to indan-1-ol and indene to inden-1-ol and *cis*-indan-1,2-diol (Wackett et al. 1988), while purified naphthalene dioxygenase from a strain of *Pseudomonas* sp. catalyzed the enantiomeric monooxygenation of indan to indan-1-ol and the dehydrogenation of indan to indene (Gibson et al. 1995).

b. By cloning genes for benzene/toluene degradative enzymes in *Pseudomonas* (*Burkholderia*) sp. strain JS150, it has been found that this strain also carries genes for a monooxygenase (Johnson and Olsen 1995) in addition to those for the dioxygenase. Initial products from the metabolism of toluene are therefore 3-methyl catechol produced by 2,3-dioxygenation, 4-methylcatechol by 4-monooxygenation, and both 3- and 4-methylcatechols by 2-monooxygenation (Johnson and Olsen 1997).

c. *Sphingomonas yanoikuyae* (*Beijerinckia* sp. strain B1 metabolizes biphenyl by initial dioxygenation followed by dehydrogenation to 2,3-dihydroxybiphenyl. Cells of a mutant (strain B8/36) that lacks *cis*-biphenyl dihydrodiol dehydrogenase induced with 1,3-dimethylbenzene transformed dihydronaphthalene by three reactions: (i) monooxygenation to (+)-(R)-2-hydroxy-1,2-dihydronaphthalene, (ii) dioxygenation to (+)-(1R,2S)-*cis*-naphthalene dihydrodiol, and (iii) dehydrogenation to naphthalene followed by dioxygenation to (+)-(1R,2S)-*cis*-naphthalene dihydrodiol (Eaton et al. 1996).

d. The degradation of pyrene by a *Mycobacterium* sp. involves both dioxygenase and monooxygenase activities (Heitkamp et al. 1988), and is discussed again in Chapter 6, Section 6.2.3.

e. Arene oxides may be intermediates in the bacterial transformation of aromatic compounds and may result in rearrangements (NIH shifts) (Dalton et al. 1981; Cerniglia et al. 1984; Adriaens 1994). These reactions are noted again in Chapter 6 (Sections 6.2.2, 6.5.1.1, and 6.5.2). As already noted, arene oxides may plausibly account for the formation of nitro-substituted products during degradation of aromatic compounds when nitrate is present in the medium (Sylvestre et al. 1982; Omori et al. 1992; Chapter 5, Sections 5.2.4 and Chapter 6, 6.3.1.2).

f. Toluene monooxygenases provide alternatives to dioxygenation (Chapter 6, Section 6.2.1), and the *ortho*-monooxygenase in *Burkholderia* (*Pseudomonas*) *cepacia* G4 has been shown to be carried on a plasmid TOM (Shields et al. 1995). In *P. mendocina* KR1, the degradation of toluene is initiated by a three-component toluene-4-monooxygenase that converts toluene to 4-methylphenol. The formation of an intermediate arene oxide is consistent with the observation of an NIH shift in which 68% of the deuterium in 4-[^{2}H]-toluene is retained in the 4-methyl phenol (Whited and Gibson 1991). On the other hand, the enzyme that oxidizes toluene-4-sulfonate to the benzyl alcohol in the first step of its degradation is a monooxygenase that consists of only two components: an Fe–S flavoprotein that serves as a reductase and an oxygenase (Locher et al. 1991).

g. *Xanthobacter* sp. strain Py2 was isolated by enrichment on propene which is metabolized by initial metabolism to the epoxide. The monooxygenase that is closely related to aromatic monooxygenases is able to hydroxylate benzene to phenol before degradation, and toluene to a mixture of the 2-, 3-, and 4-methylphenols that are not further metabolized (Zhou et al. 1999).

h. The degradation of quinol-2-one in *P. putida* 86 is initiated by introduction of oxygen at the 8 position. The two component enzyme consists of a reductase containing FAD and a [2Fe–2S] and an oxygenase containing Rieske-type (2Fe–2S) clusters (Rosche et al. 1997). The amino acid sequence of the reductase is similar to class 1B reductases whereas that for the oxygenase suggests an affinity with class 1A oxidases (Rosche et al. 1997). The formation of 3-hydroxycarbazole from carbazole (Resnick et al. 1993) may be carried out by comparable monooxygenation and is discussed in Section 6.3.1.1.

3. *Phenols* — Monohydroxylation by a monooxygenase is generally the first step in the degradation of phenols and their ethers.

a. The 4-methoxybenzoate monooxygenase from *P. putida* shows low substrate specificity: although it introduces only a single atom of oxygen into 3-hydroxy-, and 4-hydroxybenzoate, it accomplishes the conversion of 4-vinylbenzoate into the corresponding side-chain diol (Wende et al. 1989).

b. The metabolism of 4-hydroxybenzoate involves conversion to 3,4-dihydroxybenzoate by a hydroxylase that has been purified and characterized from a strain of *P. fluorescens* (Howell et al. 1972) and from *P. aeruginosa* strain PAO1 (Entsch and Ballou 1989). The enzyme is a flavoprotein containing FAD and requires NADPH for activity.

c. The enzymes that introduce an oxygen atom into 4-hydroxyphenylacetate in a strain of *P. putida* (Arunachalam et al. 1992) and in *E. coli* W (Prieto and Garcia 1994) are flavoprotein monooxygenases that require a further protein component (coupling protein) for activity. In the absence of this protein, oxidation of NADH produces H_2O_2 (Arunachalam et al. 1992). This phenomenon has also been observed with other oxygenases including salicylate hydroxylase, 2,4-dichlorophenol hydroxylase, and 2,4,6-trichlorophenol monooxygenase.

d. Salicylate is an intermediate in the metabolism of PAHs including naphthalene and phenanthrene, and its degradation involves oxidation to catechol. The hydroxylase (monooxygenase) has been extensively studied (references in White-Stevens and Kamin 1972) and, in the presence of an analogue that does not serve as a substrate NADH, is oxidized with the production of H_2O_2 (White-Stevens and Kamin 1972). This "uncoupling" is characteristic of flavoenzymes and is exemplified also by the chlorophenol hydroxylase from an *Azotobacter* sp. that is noted later.

4. *Chlorophenols* — Monooxygenases have been implicated in the introduction of oxygen to several chlorophenols. There are two situations that may conveniently be distinguished: formation of a catechol that undergoes ring fission without loss of chlorine, and as a prerequisite to further reactions that result in dechlorination before ring fission.

a. The hydroxylase that converts 2,4-dichlorophenol to 3,5-dichlorocatechol has been purified from a strain of *Acinetobacter* sp. (Beadle and Smith 1982) and from *Alcaligenes eutrophus* JMP 134 (Don et al. 1985; Perkins et al. 1990). The reductant is NADPH, the enzyme is a flavoprotein containing FAD, and in the presence of compounds that are not substrates, NADPH and O_2 are consumed with the production of H_2O_2.

i. The first step in the degradation of pentachlorophenol by *Flavobacterium* sp. 39732 involves introduction of oxygen at C4 to produce tetrachlorohydroquinone. The enzyme is not a hydroxylase, but a flavin-containing monooxygenase (Xun et al. 1992).

ii. An NADH-requiring chlorophenol monooxygenase from *Burkholderia cepacia* AC1100 successively dehalogenates 2,4,5-trichlorophenol to 2,5-dichloro-hydroquinone and 5-chloro-1,2,4-trihydroxybenzene (Xun 1996). It is a two component enzyme, and component A contains FAD and an NADH reductase.

iii. The enzyme from *Azotobacter* sp. strain GP1 that catalyzes the formation of 2,6-dichlorohydroquinone from 2,4,6-trichlorophenol is also a monooxygenase that requires NADH, FAD, and O_2 (Wieser et al. 1997). The enzyme is able to accept other chlorophenols with consumption of NADH including 2,4-, 2,6-, 3,4-dichlo-, 2,4,5-trichloro-, and 2,3,4,5- and 2,3,4,6-tetrachlorophenol, and in the absence of a substrate results in unproductive formation of H_2O_2.

5. *Nitrophenols* — Monooxygenation is involved in the degradation of 4-nitrophenol by a strain of *Moraxella* sp. (Spain and Gibson 1991), and the enzyme appears to be a particulate flavoprotein. In the Gram-positive organisms *Arthrobacter* sp. (Jain et al. 1994) and *Bacillus sphaericus* strain JS905

(Kadiyala and Spain 1998), 4-nitrophenol degradation involves initial hydroxylation to 4-nitrocatechol followed by elimination of nitrite with the formation of benzene-1,2,4-triol before ring fission to β-ketoadipate. In *B. sphaericus*, the first two reactions are carried out by a single two-component enzyme composed of a flavoprotein reductase and an oxygenase.

6. *Formal hydroxylation of azaarenes by oxidoreductases* — In contrast to the foregoing reactions that are true oxygenases involving one atom of O_2, the degradation of azaarenes frequently involves true hydroxylation using the oxygen atom in H_2O. The aerobic degradation of quinoline is accomplished by a number of organisms and the initial product is the 2-hydroxy compound (quinol-2-one). This reaction has been studied extensively by Lingens and his co-workers (Schach et al. 1995), and involves introduction of the oxygen atom in water. The initial reaction is therefore a hydration followed by a dehydrogenation, and the enzymes that have been purified from a number of taxa have been termed *oxidoreductases*.

7. Finally, an important example is given whereby a monooxygenase induced by an aromatic substrate is capable of bringing about degradation of a structurally unrelated compound. Trichloroethene is not able to support growth of single organisms but it may be cooxidized by a number of monooxygenases during growth, for example, of *Alcaligenes eutrophus* JMP134 with phenol or 2,4-dichlorophenoxyphenoxyacetate (Harker and Kim 1990), or with the cells of a number of organisms grown with toluene, phenol, or various short-chain aliphatic hydrocarbons. Further details may be found in Shields and Reagin (1992). This is discussed also in Chapter 6, Section 6.4.3.

4.4.1.2 Cytochrome P-450 Systems

Reference should be made to a number of reviews that discuss mechanistic aspects of cytochrome P-450 reactions (Guengerich 1990), the reactions mediated by both eukaryotic and prokaryotic cytochrome P-450 systems (Sariaslani 1991), their widespread role in the transformation of xenobiotics (Smith and Davis 1968), and their occurrence and activities in yeasts (Käppeli 1986). The functioning of cytochrome P-450 involves both oxygenation and reduction reactions (Tyson et al. 1972), and this is consistent with its role in both hydroxylation and reductive transformations. The structures of prokaryotic and eukaryotic cytochrome P-450 systems are compared in Figure 4.45.

Prokaryotic Organisms

This activity is widely distributed among both Gram-positive and Gram-negative organisms, and mediates a number of important degradations and transformations. Cytochrome P-450 systems in actinomyces have been reviewed (O'Keefe and Harder 1991), and the systematic nomenclature of some important bacterial cytochrome P-450 systems has been given

Classification of microbial cytochrome P-450 monooxygenases

Class I   Two component enzyme of eukayrotic organisms
   (1) NADPH reductase containing FAD and FMN
   (2) Monooxygenase

Class II   Three component system of prokaryotic systems
   (1) NADP reductase containg FAD
   (2) Ferredoxin
   (3) Monooxygenase

Class III Single monooxygenase component system of *Bacillus megaterium*

FIGURE 4.45
Cytochrome P-450 systems in prokaryotes and eukaryotes.

(Munro and Lindsay 1996). The diversity of reactions that are catalyzed by this system is illustrated by the following examples.

1. The initial hydroxylation in the degradation of some terpenes: the ring methylene group of camphor by *Pseudomonas putida* (Katagiri et al. 1968; Tyson et al. 1972; Koga et al. 1986), and the *isopropylidene* methyl group of linalool by a strain of *P. putida* (Ullah et al. 1990).

2. When the 5- and 6-positions of camphor are blocked by substituents, hydroxylation at other positions may take place. For example, the quaternary methyl group of 5,5-difluorocamphor is hydroxylated to the 9-hydroxymethyl compound (Eble and Dawson 1984; Figure 4.46a).

3. The cytochrome P-450$_{cam}$ is able to bring about the stereoselective epoxidation of *cis*-methylstyrene to the 1*S*,2*R* epoxide (Ortiz de Montellano et al. 1991).

4. Adamantane (A) and adamantan-4-one (B) were specifically hydroxylated at the quaternary C1 by the bacterial cytochrome P450$_{cam}$ to produce C and D, whereas the eukaryotic cytochrome P-450$_{LM2}$ formed also the C2 compound from adamantane, and both 5-hydroxyadamantan-1-one (D) and the 4-*anti*-hydroxyadamantan-1-one (E) from adamantan-4-one (White et al. 1984; Figure 4.46b).

5. The hydroxylation of cyclohexane by a strain of *Xanthobacterium* sp. (Trickett et al. 1991). In cell extracts, a range of other substrates were oxidized including cyclopentane, pinane, and toluene (Warburton et al. 1990).

6. Hydroxylation of *n*-octane by cell extracts of *Corynebacterium* sp. strain 7E1C (Cardini and Jurtshuk 1968), and by some strains of *Acinetobacter calcoaceticus* induced with *n*-hexadecane (Asperger et

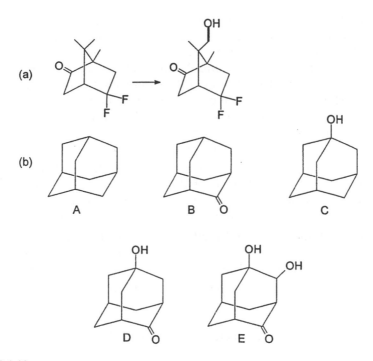

FIGURE 4.46
Hydroxylation of (a) 5,5-difluorocamphor and (b) adamantane (A) and adamantan-4-one (B).

al. 1981). The genes *alkBFGHKL* involved in the degradation of long-chain alkanes by *P. oleovorans* are tightly clustered on a plasmid and encode the alkane hydroxylase, rubredoxin, rubredoxin reductase, aldehyde dehydrogenase, alcohol dehydrogenase, acyl-CoA synthase, and a membrane protein of unknown function.

7. Hydroxylation of long-chain carboxylic acids, amides, and alcohols — but not the esters or the corresponding alkanes — at the ω-1, ω-2, and ω-3 positions by a soluble enzyme system from *Bacillus megaterium* (Miura and Fulco 1975; Narhi and Fulco 1987).

8. Hydroxylation of progesterone and closely related compounds at the 15-β position by cell extracts of *B. megaterium* (Berg et al. 1976).

9. The transformation of benzene, toluene, naphthalene, biphenyl, and benzo[*a*]pyrene to the corresponding phenols (Trower et al. 1989) by *Streptomyces griseus*, and phenanthrene by *S. flavovirens* to (-)*trans*-[9*S*,10*S*]-9,10-dihydrodihydroxyphenanthrene and minor amounts of 9-hydroxyphenanthrene (Sutherland et al. 1990).

10. The transformation of sulfonylureas by *S. griseolus* grown in a complex medium containing glucose in which the methyl group of the heterocyclic moeities is hydroxylated, and for some substrates subsequently oxidized to the carboxylic acid (Romesser and O'Keefe 1986; O'Keefe et al. 1988).

11. The dealkylation of S-ethyl dipropylthiocarbamate and atrazine by a strain of *Rhodococcus* sp. (Nagy et al. 1995a,b), and of 2-ethoxyphenol and 4-methoxybenzoate by *Rhodococcus rhodochrous* (Karlson et al. 1993). In *R. erythropolis* NI86/21, however, the *thcF* gene involved in the degradation of thiocarbamate herbicides is a non-heme haloperoxidase that does not occur in other strains of rhodococci that can degrade thiocarbamates (de Schrijver et al. 1997); this specificity occurs also for the cytochrome P-450 system encoded by the *thcRBCD* genes (Nagy et al. 1995a,b).

12. The reductive dehalogenation of polyhalogenated methanes (Castro et al. 1985) and polyhalogenated ethanes (Li and Wackett 1993) by *Pseudomonas* strain PpG786. This is noted again in Chapter 6, Section 6.10.1 in the context of the metabolism of chlorofluoroalkanes.

13. The oxidation of *t*-butyl methyl ether to *t*-butanol (Steffan et al. 1997) which is mediated by the cytochrome P-450 from camphor-grown *P. putida* CAM, but not by that from *R. rhodochrous* strain 116.

14. The degradation of morpholine by a strain of *Mycobacterium* sp. (Poupin et al. 1998).

Some unusual aspects of bacterial cytochrome P-450 systems are worth pointing out:

1. The induction in *B. megaterium* by barbiturates, and the fact that the enzyme is unusual in having a single 119-kD protein coupling NADH reduction to oxygenation (Narhi and Fulco 1986).

2. The induction of the enzyme in *S. griseus* by genistein (5,7,4'-trihydroxyisoflavone) contained in the soybean flower that is used as the growth medium (Trower et al. 1989).

3. Cytochrome P-450 systems may be both substrate- and organism-specific, and multiple forms of the enzyme may be produced in the same organism.

Eukaryotic Organisms

Cytochrome P-450 hydroxylation activity is well established in eukaryotic yeasts and some fungi, and this activity has attracted attention in various contexts.

1. The initial oxidation of aliphatic hydrocarbons and amines by yeasts (references in Käppeli 1986). The enzyme (designated cytochrome P-450$_{aO}$) is induced in organisms including species of *Candida* and *Endomycopsis* during growth with alkanes, and therefore differs from the enzyme (P-450$_{14DM}$)synthesized during growth of saccharophilic yeasts under conditions of low oxygen

concentration. Details of the following metabolic steps are noted later and in Section 4.4.4.

2. The aryl hydroxylase of *Saccharomyces cerevisiae* that transforms benzo[*a*]pyrene to the 3- and 9-hydroxy compounds and the 7,8-dihydrodiol (King et al. 1984)

3. In fungi that serve as models for mammalian transformation of xenobiotics (Ferris et al. 1976; Smith and Rosazza 1983; Section 4.3.6).

4. The hydroxylation of steroids at various positions by a range of fungi together with some representatives of other polycyclic compounds is discussed in Chapter 6, Section 6.11.2, and probably involves the cytochrome P-450 system (Breskvar and Hudnik-Plevnik, 1977).

5. The oxidative removal of the 14α–methyl group in lanosterol by *S. cerevisiae* with concomitant production of the $\Delta^{14\text{-}15}$ bond (Aoyama et al. 1984) is plausibly facilitated by formation of the conjugated 8-9,14-15-diene. The enzyme, designated P-450$_{14DM}$ is induced under "semi-anaerobic" conditions during growth with glucose and is different from both the P450$_{aO}$ enzyme noted earlier, and from P-450$_{22DS}$ that brings about Δ^{22}-dehydrogenation (Hata et al. 1983).

It is important to note that most eukaryotic organisms, yeasts, and fungi contain peroxisomes. The degradation of long-chain alkanes by yeasts is carried out in two separate organelles: hydroxylation to alkanols in microsomes (Käppeli 1986), and oxidation to the alkanoic acid CoA esters that are further metabolized in peroxisomes. Peroxisomes are induced during growth with alkanes and are multipurpose organelles. Some of the salient features include the following (Tanaka and Ueda 1993), which differ from the pathway for bacterial degradation of alkanoic acids.

1. In *Candida* sp., degradation of the CoA alkanoic esters to the alk-enoic acid esters is catalyzed by an acyl-CoA oxidase and results in the production of H_2O_2 that is converted to O_2 by catalase activity. The enzyme from *C. tropicalis* contains FAD (Jiang and Thorpe 1983), and in *C. lipolytica* carries out a stereospecific *anti*-elimination of hydrogen (Kawaguchi et al. 1980).

2. The CoA alkenoic esters are degraded further by a bifunctional enol-CoA hydratase and 3-hydroxyacetyl-CoA dehydrogenase to the β-keto acid which is then degraded to acetyl CoA and the lower alkanoate ester by 3-ketoacetyl CoA thiolase and acetyl CoA thiolase activities.

3. With the exception of the acetyl CoA thiolase, all of these enzymes are located exclusively in the peroxisomes, whereas the enzymes that are involved in lipid synthesis are found in the microsomes and the mitochondrion.

4.4.2 Bacterial Dioxygenase Systems

The following brief comments are devoted to reactions that involve dioxygenation, although no details of the enzymology are provided. Dioxygenases are involved in the bacterial degradation of many aromatic compounds, and carry out two kinds of reaction: (1) the initial introduction of O_2 to produce *cis*-1,2-dihydrodiols and (2) ring cleavage mediated by catechol oxygenases that carry out either intradiol (1:2) or extradiol (2:3) fission. Examples of these are given in Chapter 6 (Sections 6.2.1 and 6.2.3), and further details of the aromatic hydrocarbon dioxygenases have been summarized (Neilson and Allard 1998). Examples are given in Chapter 5, Section 5.7 of their application to screening cells for a desired enzymatic using the activity toward analogue substrates that produce colored metabolites.

Dioxygenases for the Initial Introduction of Oxygen into Aromatic Substrates

The dioxygenases that catalyze the initial introduction of molecular oxygen into aromatic hydrocarbons have been extensively studied, and are multicomponent enzymes generally carrying out three functions: reduction of NAD by a reductase, electron transport by a ferredoxin, and transfer of electrons onto the substrate simultaneously with introduction of oxygen by a terminal oxygenase. The classes of bacterial dioxygenases are outlined in Figure 4.47.

Classification of bacterial dioxygenases

Class I   Two component enzyme
  (1)   Reductase - flavin + [2Fe-2S] redox center
  (2)   Oxygenase
  Class IA   FMN as flavin : e.g., 4-sulfobenzoate 3,4-dioxygenase
  Class IB   FAD as flavin : e.g. benzoate dioxygenase

Class II   Three component enzyme
  (1)   Reductase - flavoprotein
  (2)   Ferredoxin
  (3)   Oxygenase
  Class IIA   Plant-type ferredoxin : e.g. pyrazon dioxygenase
  Class IIB   Rieske Fe-S clusters in ferredoxin : e.g. benzene dioxygenase and toluene
                  dioxygenase

Class III   Three component type e.g. naphthalene dioxygenase
  (1)   Reductase- flavin + [2Fe-2S] redox center
  (2)   [2Fe-2S] ferredoxin
  (3)   Oxygenase

FIGURE 4.47
Bacterial dioxygenase systems.

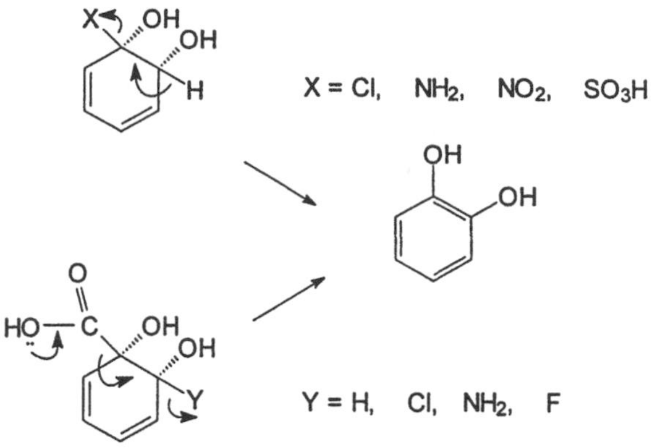

FIGURE 4.48
Dioxygenation of substituted benzenes and benzoic acids.

During dioxygenation, elimination of a substituent may occur simultaneously. Illustrative examples of simultaneous elimination and dioxyygenation during degradation are discussed in the appropriate sections in Chapter 6, but may usefully be summarized here. They are represented formally by the reactions given in Figure 4.48.

- Amine groups in the conversion of aniline to catechol (Bachofer and Lingens 1975; Fukumori and Saint 1997), and of both amine and carboxyl groups in the conversion of anthranilic acid to catechol (Taniuchi et al. 1964).

- Carboxyl groups in the conversion of benzoate to catechol (Neidle et al. 1991), and during the degradation of 4-methyl-*o*-phthalate by *P. fluorescens* strain JT701 that takes place by analogous oxidative decarboxylation with the formation of 4-methyl-2,3-dihydroxybenzoate (Ribbons et al. 1984).

- Chloride during the conversion of 1,2,4,5-tetrachlorobenzene to 3,4,6-trichlorocatechol (Sander et al. 1991), and both chloride and carboxyl during the degradation of 2-chlorobenzoate (Fetzner et al. 1992; Romanov and Hausinger 1994).

- Fluoride during the conversion of 3-fluorotoluene to 3-methylcatechol (Renganathen 1989), and of both fluoride and carboxyl during the degradation of 2-fluorobenzoate (Engesser et al. 1980).

- Nitrite during conversion of 2-nitrotoluene to catechol (An et al. 1994), 2,4-dinitrotoluene to 4-nitrocatechol (Spanggord et al. 1991) and 1,3-dinitrobenzene to 4-nitrocatechol (Dickel and Knackmuss 1991). Further details of the enzymes are given in Chapter 6, Section 6.8.2.

- Sulfite during the conversion of naphthalene-1-sulfonate to 1,2-dihydroxynaphthalene (Kuhm et al. 1991).

Attention is drawn to two important issues:

1. When both CO_2 and a substituent such as halogen are eliminated, no dehydrogenation is required to produce the catechol. This is also the case in the degradation of 3,4-dichlorobenzoate where spontaneous loss of HCl from the initially produced 4,5-*cis*-diol formed 3-chloro-4,5-dihydroxybenzoate without intervention of a dehydrogenase that is necessary for 3-chlorobenzoate (Nakatsu et al. 1997). These examples stand in contrast to the situation for aromatic hydrocarbons in which a dehydrogenase is necessary.

2. There are a number of two-component dioxygenases consisting of only a reductase and an oxygenase, containing a flavin cofactor and a [2Fe–2S] center but lacking the ferredoxin component that is invariably required for aromatic hydrocarbon dioxygenation. Examples of these dioxygenases include (a) 4-sulfobenzoate-3,4-dioxygenase (Locher et al. 1991), (b) *o*-phthalate 4,5-dioxygenase (Batie et al. 1987a), (c) benzoate 1,2-dioxygenase (Yamaguchi and Fujisawa 1982), (d) 4-chlorophenylacetate-3,4-dioxygenase from *Pseudomonas* sp. strain CBS32 (Schweizer et al. 1987), (e) 3-chlorobenzoate 3,4-dioxygenase (Nakatsu et al. 1995), (f) 2-halobenzoate 1,2-dioxygenase (Fetzner et al. 1992); the corresponding dioxygenase from *P. aeruginosa* 142 is, however, a three-component enzyme (Romanov and Hausinger 1994).

3. The degradation of *o*-phthalate by *Burkholderia cepacia* DBO1 (syn. *P. fluorescens* PHK, syn. *P. putida*) takes place by dioxygenation to the *cis*-4,5-dihydrodiol, dehydrogenation to the catechol, and decarboxylation to 3,4-dihydroxybenzoate (Chapter 6, Section 6.2.1). The reductase component of the dioxygenase coded by *ophA1* and the decarboxylase coded by *ophC* are adjacent, and the genes for the oxygenase component of the dioxygenase coded by *ophA2* and for the dihydrodiol dehydrogenase by *ophB* are linked (Chang and Zylstra 1998).

4. The metabolism of dehydroabietic acid by *Pseudomonas abietaniphila* BKME-9 is initiated by hydroxylation and dehydrogenation at C-8 followed by dioxygenation at C-11 and C-12. The dioxygenase contains a ferredoxin that is different from the [2Fe-2S] that is generally found in ring hydroxylating dioxygenases, and the protein sequence of the α-subunit of the oxygenase is distinct from those of classes I, II, or III (Martin and Mohn 1999).

It is important to appreciate that dioxygenases are capable of carrying out a number of other reactions than the introduction of both atoms of oxygen into the substrate. Some illustrative examples are given here.

1. Dioxygenases may also accomplish monooxygenation — Although naphthalene dioxygenase is enantiomer specific producing the (+)-*cis*-(1*R*,2*S*)-dihydrodiol, it also possesses dehydrogenase and monooxygenase activities (Gibson et al. 1995) so that monooxygenase and dioxygenase activities are not exclusive.

   - Naphthalene dioxygenase carried out dioxygenation of indene to *cis*-(1*R*,2*S*)-indandiol, monooxygenation to 1*S*-indenol, and in addition dehydrogenation of indan to indene. Further details of this and other reactions of indanol are given in Chapter 6, Section 6.2.3.

   - Toluene is oxidized by purified naphthalene dioxygenase from *Pseudomonas* sp. strain NCIB 9816-4 to benzyl alcohol and benzaldehyde, and ethylbenzene to (*S*)-1-phenylethanol and acetophenone. Whereas the initial reactions involve monooxygenation, oxidation to the aldehyde and ketone are dioxygen-dependent (Lee and Gibson 1996).

   - Toluene dioxygenase from *P. putida* strain F1, and chlorobenzene dioxygenase from *Burkholderia* sp. strain PS12 that are involved in degradation of these substrates by dioxygenation bring about monooxygenation of 2- and 3-chlorotoluene by side-chain oxidation to produce the corresponding benzyl alcohols that were slowly oxidized further (Lehning et al. 1997).

2. Naphthalene dioxygenase from *Pseudomonas* sp. strain NCIB 9816-4 is able to oxygenate a number of substrates in addition to naphthalene: fluorene to fluorene-3,4- *cis*-dihydrodiol, dibenzofuran to dibenzofuran-1,2- *cis*-dihydrodiol, and dibenzothiophene to dibenzothiophene-1,2- *cis*-dihydrodiol (Resnick and Gibson 1996). A comparable diversity has been shown with the carbazole 1,9a-dioxygenase from *Pseudomonas* sp. strain CA10 (Chapter 6, Section 6.3.1.2): this is able to dioxygenate carbazole, dibenzofuran dibenzo[1,4] dioxin, xanthene, and phenoxathiin with concomitant fission of the bond between the oxygenated ring and the nitrogen or oxygen heteroatom, but also naphthalene, biphenyl, anthracene and fluoranthene to *cis*-dihydrodiols (Nojiri et al. 1999

3. Naphthalene dioxygenase from *P. putida* strain F1 is able to oxidize a number of halogenated ethenes, propenes, and butenes, and *cis*-2-heptene and *cis*-2-octene (Lange and Wackett 1997). Alkenes with halogen and methyl substituents at double bonds form allyl alcohols, whereas those with only alkyl or chloromethyl groups form diols.

   It is also worth drawing attention to some metabolic similarities between naphthalene dioxygenase and 2,4-dinitrotoluene dioxygenase from *Burkholderia* sp. strain DNT, and it has been suggested that these enzymes may have a common ancestor (Suen et al. 1996).

4. Protein sequences among biphenyl dioxygenases may be similar or identical in spite of distinct differences in the range of PCB congeners that are attacked (Kimura et al. 1997; Mondello et al. 1997). The roles of the α- and β-subunits of the terminal dioxygenase (ISP) have been examined in more detail with divergent conclusions.

- Using enzymes from hybrid naphthalene dioxygenase and 2,4-dinitrotoluene dioxygenase genes introduced into *E. coli* (Parales et al. 1998). Although the rates were different for the wild-type and hybrid enzymes, the products and the enantiomeric specificities were the same for each. It was shown that whereas the β–subunit of the dioxygenase was necessary for activity, it was the large α-subunit containing a Rieske [2Fe–2S] center that determined substrate specificity.

- Chimeras were constructed from the α- and β-subunits of the terminal dioxygenase proteins ISP$_{BPH}$ (Hurtubise et al. 1998) from two strains of PCB-degrading strains, *Comamonas testosteroni* strain B-356 and *Pseudomonas* sp. strain LB400. The enzymes were purified, and it was shown from the substrate specificities of the purified enzymes that the structures of both α and β subunits influenced the specificities to various substrates.

- The function of the β-subunit that contains no detectable prosthetic groups is not fully understood, although the Tod C2 subunit is needed to obtain catalytic activity of the α-subunit (Tod C1) in *P. putida* F1 (Jiang et al. 1999).

Ring-Cleaving Dioxygenases

The ring-cleavage enzymes are of two broad classes, and only some illustrative examples are given.

1. 1,2-Catechol (intradiol: *ortho*) cleavage enzymes
   - Catechol-1,2-dioxygenase contains Fe(III), and lacks heme and sulfur–iron components. The protein consists of α- and β- subunits with masses of 30 and 32 in the combinations (α,β) or (α,α) or (β,β), and the molecular mass of the enzyme is 63 kD.
   - 3,4-Dihydroxybenzoate-3,4-dioxygenase is an Fe(III) protein and consists of an aggregate of α- and β-subunits with masses of 22.2 and 26.6 in a (α,β)$_{12}$ structure with a molecular mass of 587.

2. 2,3-Catechol (extradiol: *meta*) cleavage enzymes that have been divided into three classes (Peng et al. 1998).
   - Catechol-2,3-dioxygenase contains Fe(II) and consists of subunits with a mass of 35 in an (α)$_4$ structure with a molecular mass of 140.

- 3,4-Dihydroxybenzoate-4,5-dioxygenase is an Fe(II) protein with α- and β-subunits of masses 17 and 33 in a $(\alpha,\beta)_2$ structure.
- 3,4-Dihydroxyphenylacetate-2,3-dioxygenase from *Arthrobacter globiformis* (Boldt et al. 1995) and from *Bacillus brevis* (Que et al. 1981) is an Mn(II) enzyme, and is neither activated by Fe(II) nor rapidly inhibited by H_2O_2.

The dioxygenase that brings about ring fission of 1-hydroxynaphthalene-2-carboxylate is different from the dioxygenases that catalyze fission of catchols (Iwabuchi and Harayama 1998).

2,5-Dihydroxybenzoate (gentisate) 1,2-dioxygenase that produces maleylpyruvate without loss of the carboxyl group belongs to a different class of ring-cleaving dioxygenases, since the protein sequences of the gene product from *Sphingomomas* sp. strain RW5 revealed little or no similarities to those of intradiol or extradiol dioxygenases (Wergath et al. 1998). By contrast, the degradation of the analogous salicylate (Section 4.4.1.1) proceeds by hydroxylation to catechol with loss of carboxyl, followed by ring fission of the catechol.

In addition to these reactions, dioxygenases are involved in a number of other reactions although the enzymes appear to be quite distinct from those involved in the metabolism of aromatic compounds.

1. The oxidation of long-chain alkanes by *Acinetobacter* sp. strain M-1 (Maeng et al. 1996). This does not involve hydroxylation, but initial formation of an alkyl hydroperoxide that forms the aldehyde directly and does not require NAD(P)H.

2. The oxidation of propane to propan-1-ol by *Rhodococcus rhodochrous* may formally involve a reaction comparable to that for the eukaryotic oxidation of 2-nitropropane (Section 4.4.3) since one molecule of oxygen is incorporated into 2 mol of propane (Babu and Brown 1984). Details of the latter reaction suggest, however, a different mechanism for the propane oxidation.

3. The degradation of 2,4-dichlorophenoxyacetate by *Alcaligenes eutrophus* strain JMP134 involves initial formation of 2,4-dichlorophenol and glyoxylate. This is accomplished by a dioxygenase that couples this reaction to the conversion of 2-ketoglutarate to succinate and CO_2 (Fukumori and Hausinger 1993; Figure 4.49 a). The sequence of the chromosomal gene *tfdA* in *Burkholderia* spp. that degrades 2,4-dichlorophenoxyacetate by this pathway was only 77.2% homologous to the plasmid-borne gene of *A. eutrophus* strain JMP 134, but shared a 99.5% identity with the chromosomal gene from another strain of *Burkholderia* sp. from a geographically distinct area (Matheson et al. 1996). In *S. herbicidovorans*, the 2-ketoglutarate-dependent dioxygenase converted 4-chloro-2-methylphenoxypropionate (mecoprop) to 4-chloro-2-methylphenol, pyruvate and succinate each of which were formed with incorporation of one atom of O_2 (Nickel et al. 1997).

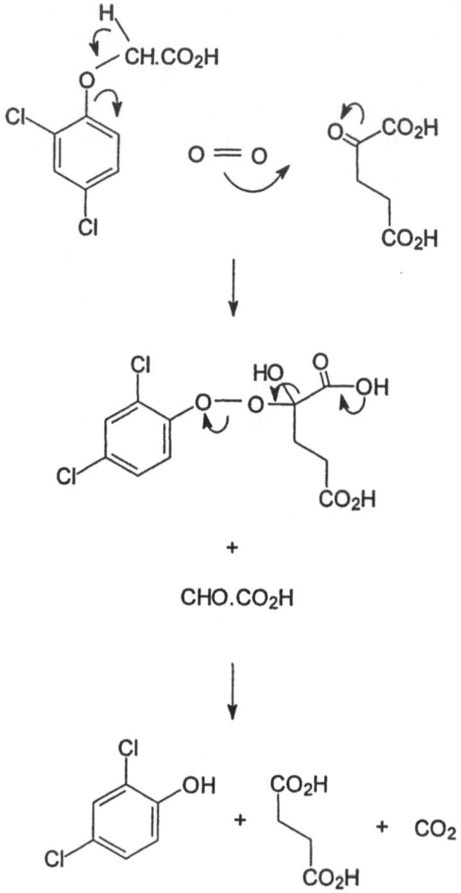

FIGURE 4.49 a
Dioxygenation of (a) 2,4-dichlorophenoxyacetate .

4. An analogous reaction mediated by a dioxygenase is the intramolecular dioxygenation of 4-hydroxyphenylpyruvate to 2,5-dihydroxyphenylacetate (Lindblad et al. 1970), in which both the 2-hydroxyl and the carboxylate oxygen atoms are derived from O_2 (Figure 4.49b). The degradation of 4-hydroxyphenylpyruvate formed from tyrosine by *Pseudomonas* sp. strain PJ 874 grown at the expense of tyrosine is mediated by 4-hydroxyphenylpyruvate dioxygenase that has been purified (Lindstedt et al. 1977). The metabolism of L-tyrosine may proceed by transamination, dioxygenation to 4-phenylpyruvate followed by either ring fission or oxidation to 2-carboxymethyl 1,4-benzoquinone and polymerization (Denoya et al. 1994).

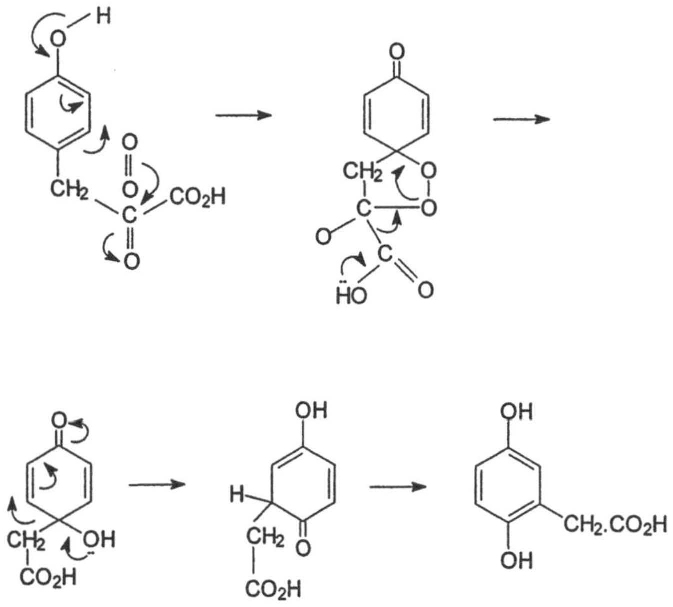

FIGURE 4.49 b
Dioxygenation of (b) 4-hydroxyphenylpyruvate.

5. The oxidative fission of the C=C bond in a substituted stilbene by *Pseudomonas* sp. strain TMY 1009 to a mixture of aromatic aldehydes involves 1 mol each of substrate and oxygen (Habu et al. 1989).

4.4.3 Eukaryotic Dioxygenases

There are a few eukyotic dioxygenases that are important for the degradation of xenobiotics.

1. The oxidation of 2-nitropropane by the yeast *Hansenula mrakii* is carried out by a flavoenzyme and produces 2 mol of acetone from 2 mol of substrate and one molecule of O_2 that is activated by conversion to superoxide (Kido et al. 1978b; Figure 4.50). A similar enzyme has been purified and characterized from the heterothallic ascomycete *Neurospora crassa* (Gorlatova et al. 1998); 2-nitropropane is the optimal substrate.

2. 1,2,4-Trihydroxybenzene is an intermediate in the degradation by *Phanerochaete chrysosporium* of vanillate, 2,4-dichlorophenol, 2,4-dinitrophenol, and 2,7-dichlorodibenzo[1,4]dioxin (Chapter 6, Sections 6.5.2, 6.8.2 and 6.3.2, respectively), and its degradation is mediated by a dioxygenase that carries out intradiol ring fission (Rieble et al. 1994).

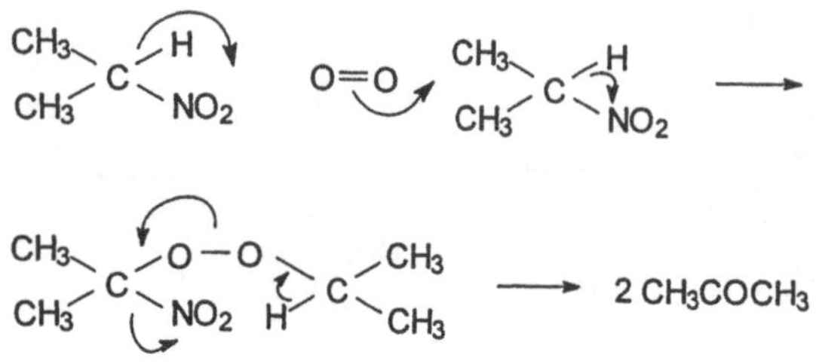

FIGURE 4.50
Dioxygenation of 2-nitropropane.

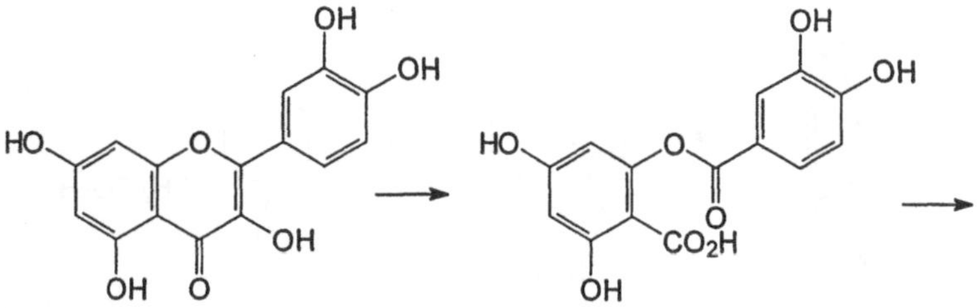

FIGURE 4.51
Dioxygenation of rutin.

3. The degradation of rutin by *Aspergillus flavus* proceeds by hydrolysis to the aglycone followed by degradation to a depside with release of the unusual metabolite carbon monoxide (Figure 4.51) and is accomplished by dioxygenation (Krishnamurty and Simpson 1970).

4.4.4 Oxidases, Peroxidases, and Haloperoxidases

Oxidases

These are produced by both prokaryotes and eukaryotes and catalyze a number of important reactions. They are flavoproteins that produce potentially destructive H_2O_2 which is removed by catalase or peroxidase activity. The reactions are formally outlined in Figure 4.52.

(1a) $R_2CHX + H_2O + O_2 \longrightarrow R_2CO + HX + H_2O_2$

(1b) $RCH\underset{CO_2H}{\overset{NH_2}{<}} + H_2O + O_2 \longrightarrow RCO.CO_2H + NH_3 + H_2O_2$

(2) $RCH_2CH_2CO_2H + O_2 \longrightarrow RCH=CH.CO_2H + H_2O_2$

FIGURE 4.52
Overview of reactions carried out by oxidases.

1. Primary amines are widely used as a nitrogen source by bacteria, and the first step of the degradation pathway involves formation of the corresponding aldehyde. Although this may be accomplished by dehydrogenation, some organisms use an amine oxidase. For example, whereas *Pseudomonas putida* (ATCC 12633) and *P. aeruginosa* (ATCC 17933) employed a dehydrogenase, *Klebsiella oxytoca* (ATCC 8724) and *E. coli* (ATCC 9637) used a copper quinoprotein amine oxidase (Hacisalohoglu et al. 1997). The alternative flavopotein oxidase is, however, used for deamination of tyramine by *Sarcina lutea*. The amine oxidase functions by oxidation of the amine to the aldehyde concomitant with the reduction of 2,4,5-trihydroxyphenylalanine quinone (TPQ) to an aminoquinol which, in the form of a Cu(I) radical, reacts with O_2 to form H_2O_2, Cu(II) and the imine. Methylamine dehydrogenase, however, involves tryptophane tryptophylquinone (TTQ) but does not result in formation of H_2O_2 (References in Stubbe and van der Donk 1998).

2. *Hyphomicrobium* sp. strain EG is able to grow at the expense of dimethyl sulfide, or dimethyl sulfoxide and produces initially methanethiol. This is then further oxidized to formaldehyde, sulfide and H_2O_2 by an oxidase that has been purified (Suylen et al. 1987).

3. The oxidation of cholesterol to cholest-4-ene-3-one is carried out by an oxidase in several bacteria. This activity has been found in *Brevibacterium sterolicum* and *Streptomyces* sp. strain SA-COO (Ohta et al. 1991), and the extracellular enzyme that has been purified from *Pseudomonas* sp. strain ST-200 (Doukyu and Aono 1998) has a preference for 3β-hydroxy compounds.

4. D-Amino acid oxidase has been isolated from a number of yeasts, and the nucleotide sequence of the enzyme from *Rhodotorula gracilis* ATCC 26217 has been established (Alonso et al. 1998). The gene could be overexpressed in *E. coli,* and levels of the enzyme were greater under conditions of low aeration; the enzyme isolated from the recombinant organisms was apparently the apoenzyme since maximum activity required the presence of FAD. The biotechnological interest in this enzyme is noted in Chapter 6, Section 6.11.2.

5. Yeasts belonging to the genera *Candida* and *Endomycopsis* are able to degrade alkanes. The initial hydroxylation is carried out by cytochrome P-450 that is found in the microsome (references in Käppeli 1986) (Section 4.4.1.2). The degradation of the alkanoic acids is carried out by enzymes that are contained in peroxisomes and are induced during growth with alkanes. Acyl CoA oxidase carries out the first step in the degradation of the alkanoic acid CoA esters, and although this is formally a dehydrogenase, acyl CoA dehydrogenase activity is absent (Kawamoto et al. 1978; Tanaka and Ueda 1993). The three subsequent enzymatic activities are apparently contained in a single protein.

6. Nitroalkanes are degraded by some yeasts and fungi. The dioxygenase pathway noted in Section 4.4.3 is not, however, the only one. The fungus *Fusarium oxysporum* is able to degrade a number of nitroalkanes including 1- and 2-nitropropane, nitroethane, and nitrocyclohexane by an inducible oxidase, and the reaction rate of the enzyme is enhanced by addition of catalase and inhibited by both NADH and NADPH (Kido et al. 1978a). The oxidase is a flavoprotein that occurs in an inactive form as 5-nitrobutyl-FAD that is convertible to the active form of the enzyme (Gadda et al. 1997). This enzyme is clearly different from the dioxygenase.

Extracellular H_2O_2 is required for the activity of peroxidases in white-rot fungi, and is produced by a several fungal reactions.

1. Glyoxal oxidase is produced from *Phanerochaete chrysosporium* under high concentrations of oxygen, is stimulated by Cu^{2+}, and oxidizes a range of substrates with the production of H_2O_2 from O_2. Substrates include methylglyoxal, glyoxylic acid, and glycolaldehyde, and the pure enzyme requires activation by lignin peroxidase; under these conditions, in the presence of catalytic amounts of H_2O_2, pyruvate, and veratraldehyde are produced, respectively, from methylglyoxal and 3,4-dimethoxybenzyl alcohol (veratryl alcohol) (the lignin peroxidase substrate) (Kersten 1990).

2. An aryl-alcohol oxidase produced optimally under carbon limitation from *Bjerkandera adusta* oxidized a number of benzyl alcohols including 4-methoxybenzyl alcohol, 3,4-dimethoxybenzyl alcohol (veratryl alcohol), and 4-hydroxy-3-methoxybenzyl alcohol, with the production of H_2O_2 from O_2; monosaccharides were not oxidized (Muheim et al. 1990). An aryl-alcohol oxidase from *Pleurotus eryngii* is a flavoprotein with range of substrates comparable to that from *B. adusta* (Guillén et al. 1992).

Peroxidases

Under specific growth conditions that include carbon, nitrogen or sulfur limitation, manganese concentration, and increased oxygen concentration, two distinct extracellular enzymes are produced by white-rot fungi — lignin peroxidase (LiP) and manganese peroxidase (MnP). These are involved in the degradation of lignin and in the biotransformation of xenobiotics: examples are given in Sections 4.3.6, Chapter 6, Sections 6.2.2, 6.5.2, and 6.8.2, and Chapter 8, Section 8.2.1. These enzymes require extracellular H_2O_2 that is produced by a number of these organisms as a result of oxidase activity, which was noted earlier. Details of the relevant reactions have emerged from a study of the degradation of a model lignin substrate by *Phanerochaete chrysosporium* (Hammel et al. 1994). 1-(3,4-Dimethyoxyphenyl)-2-phenoxypropane-1,3-diol was metabolized in the presence of H_2O_2 to glycolaldehyde that was identified by ^{13}C-NMR. The oxidase activity of glyoxal oxidase that was synthesized simultaneously with lignin peroxidase produced oxalate and 3 mol H_2O_2 that could then be recycled.

Lignin peroxidase functions by generating cation radicals from aromatic rings. In the degradation of lignin and model compounds, this results in fission of the alkyl chain between C-1 and C-2 (Kirk and Farrell 1987) or, for substrates such as PAHs, may be followed by nonenzymatic nucleophilic reactions of the cation radical (Hammel et al. 1986; Haemmerli et al. 1986). Manganese peroxidase oxidizes Mn(II) to Mn(III) which is the active oxidant and, in the presence of H_2O_2, is capable of oxidizing a number of PAHs (Bogan and Lamar 1996) and mineralizing substituted aromatic compounds (Hofrichter et al. 1998). Both activities may bring about oxidation of PAHs in reactions that are mimicked by oxidation with manganic (III) acetate (Cremonesi et al. 1989) or electrochemical oxidation (Jeftic and Adams 1970).

Haloperoxidases

Although these are of primary significance in the biosynthesis of organohalogen compounds and are distributed among mammals, marine biota, bacteria, and fungi, a few brief comments are given. These enzymes catalyze the reaction

$$S\text{-}H + H_2O_2 + HalH \rightarrow S\text{-}Hal + 2H_2O$$

and Hal may be chloride, bromide, or iodide (references in Neidleman and Geigert 1986). There is, however, one example of a degradative enzyme that is a haloperoxidase. In *Rhodcoccus erythropolis* NI86/21, the *thcF* gene involved in the degradation of thiocarbamate herbicides is a nonheme haloperoxidase that does not occur in other strains of rhodococci that can degrade thiocarbamates (de Schrijver et al. 1997).

1. The chloroperoxidase from *Caldaromyces fumago* has been isolated in pure form. It is a glycoprotein containing ferroprotoporphyrin IX and displays, in addition, peroxidase and catalase activities. In the absence of organic substrates it catalyzes the oxidation of chloride and bromide to Cl_2 and Br_2 (Morris and Hager 1966; Libby et al. 1982).

2. Bromoperoxidase has been isolated from *Pseudomonas aureofaciens* ATCC 15926, also displays peroxidase and catalase activities, and contains ferriprotoporphyrin IX (van Pée and Lingens 1985). Four different bromoperoxidases have been isolated from *Streptomyces griseus* (Zeiner et al. 1988). Only one of them, however, contains ferriprotoporphyrin IX and displays peroxidase and catalase activities. This illustrates that there are two different groups of enzymes, one of which lacks heme prosthetic groups.

4.5 Interactions

4.5.1 Single Substrates: Several Organisms

Cultures of a single microorganism will occur naturally only in circumstances where extreme selection pressure operates, for example, in hydrothermal vents or in the sediments of Antarctic lakes. Generally, many different organisms with diverse metabolic potential will exist side by side, so that metabolic interactions are probably the rule rather than the exception in natural ecosystems. Both the nature and the tightness of the association may vary widely, and in some cases degradation of a single compound may necessitate the cooperation of two (or more) organisms. Some well-defined interactions and the different mechanisms underlying the cooperation are illustrated by the following examples:

1. One of the organisms fulfills the need for a growth requirement by the other. Examples are provided by vitamin requirements of one organism that is provided by the other: biotin in cocultures of *Methylocystis* sp. and *Xanthobacter* sp. (Lidstrom-O'Connor et al. 1983) and thiamin in cocultures of *Pseudomonas aeruginosa* and an undefined *Pseudomonas* sp. that degrade the phosphonate herbicide glyphosate (Moore et al. 1983).

2. One organism may be able to carry out only a single step in the biodegradation. Many examples among aerobic organisms have been provided (Reanney et al. 1983; Slater and Lovatt 1984) so that this is probably a widespread situation; three examples will therefore suffice as illustration:

   - The degradation of parathon is carried out by a mixed culture of *P. stutzeri* and *P. aeruginosa* (Daughton and Hsieh 1977) in which the 4-nitrophenol initially formed by the former is metabolized by the latter (Figure 4.53).

   - The degradation of 4-chloroacetophenone is accomplished by a mixed culture of an *Arthrobacter* sp. and a *Micrococcus* sp.: the first organism is able to carry out all the degradative steps except the conversion to 4-chlorocatechol of the intermediate 4-chlorophenol that is toxic to the first organism (Havel and Reineke 1993). The details of the pathway are discussed again in Chapter 6, Section 6.2.1.

   - The aerobic degradation of polyethylene glycol can be carried out by a consortium of a *Flavobacterium* sp. and a *Pseudomonas* sp. in which the latter is required for the degradation of the glycollate produced by the former (Kawai and Yamanaka 1986).

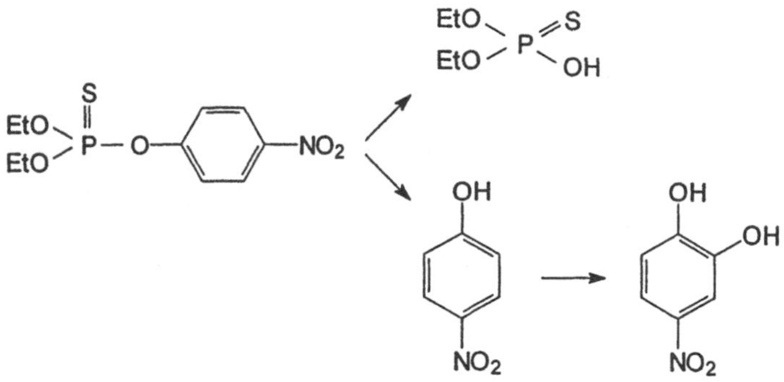

FIGURE 4.53
Degradation of parathion by a mixed culture of two pseudomonads.

$$HO\cdot(CH_2\cdot CH_2O)_n\cdot CH_2\cdot CH_2\cdot O\cdot CH_2\cdot CH_2OH \rightarrow$$
$$HO\cdot(CH_2\cdot CH_2O)_n\cdot CH_2\cdot CH_2\cdot O\cdot CH_2\cdot CO_2H \rightarrow$$
$$HO\cdot(CH_2\cdot CH_2O)_n\cdot CH_2\cdot CH_2\cdot OH + HO\cdot CH_2\cdot CO_2H$$

3. Two organisms are required to maintain the redox balance. Among anaerobic bacteria, hydrogen transfer is important since the redox balance must be maintained, and the hydrogen concentration in the mixed cultures may be critical. Interspecies hydrogen transfer has been demonstrated especially among populations of rumen bacteria containing methanogens where the concentration of hydrogen must be limited for effective functioning of the consortia. Illustrative examples have been summarized (Wolin 1982), and a few additional comments on degradative reactions that are dependent on hydrogen transfer mediated by one of the cooperating organisms are added here.

Stable metabolic associations generally between pairs of anaerobic bacteria have been termed *syntrophs,* and these are effective in degrading a number of aliphatic carboxylic acids or benzoate under anaerobic conditions; these reactions have been discussed in reviews (Schink 1991; 1997; Lowe et al. 1993) that provide lucid accounts of the role of syntrophs in the degradation of complex organic matter, while a number of important degradations and transformations of aromatic compounds by mixed cultures are discussed in Chapter 6, Section 6.7.3.2 . Two examples may be used here to illustrate the experimental intricacy of the problems besetting the study of syntrophic metabolism under anaerobic conditions.

1. Oxidation under anaerobic conditions of long-chain aliphatic carboxylic acids was established in syntrophic cultures of *Clostridium bryantii/Desulfovibrio* sp. (Stieb and Schink 1985), *Syntrophomonas sapovorans/Methanospirillum hungatei* (Roy et al. 1986), and *S. wolfei* in coculture with H_2-utilizing anaerobic bacteria (McInerney et al.

1981). The role of the second syntroph was to metabolize the reducing equivalents produced by oxidation of the carboxylic acids. *S. wolfei* was subsequently, however, adapted to grow with crotonate in pure culture (Beaty and McInerney 1987), and this procedure was also used for *Cl. bryantii*: 16S rRNA sequence analysis was then used to show the close relationship of these two organism and to assign them to a new genus *Syntrophospora* (Zhao et al. 1990). On the other hand, anaerobic oxidation of carboxylic acids with chain lengths of up to ten has been demonstrated in pure cultures of species of *Desulfonema* (Widdel et al. 1983), and even aliphatic hydrocarbons may be completely oxidized to CO_2 by a sulfate-reducing bacterium (Aeckersberg et al. 1991).

2. There has been considerable interest in the anaerobic degradation of propionate that is a fermentation product of many complex substrates, and syntrophic associations of acetogenic and methanogenic bacteria have been obtained.

   - As in the example noted earlier, during metabolism of propionate in a syntrophic culture (Houwen et al. 1991), the methanogens serve to remove hydrogen produced during the oxidation of propionate to acetate. Growth of syntrophic propionate-oxidizing bacteria in the absence of methanogens has, however, been accomplished using fumarate as sole substrate (Plugge et al. 1993). Fumarate plays a central role in the metabolism of this organism since it is produced from propionate via methylmalonate and succinate, and fumarate itself is metabolized by the acetyl-CoA cleavage pathway via malate, oxalacetate, and pyruvate.

   - Pure cultures of organisms that may oxidize propionate either in the presence of a methanogen or using sulfate as electron acceptor have been obtained; these include both *Syntrophobacter wolinii* and *S. pfenigii* (Wallrabenstein et al. 1995).

The interaction of two organisms is therefore clearly not obligatory for the ability to degrade these carboxylic acids under anaerobic conditions.

4.5.2 Cometabolism and Related Phenomena

In natural ecosystems, it is seldom indeed that either a pure culture or a single substrate exists. In general, several substrates will be present and these will include compounds of widely varying susceptibility to microbial degradation. The phenomenon where degradation occurs in the presence of two substrates has been termed *cooxidation* or less specifically *cometabolism* or *concurrent metabolism*. Unfortunately, the term cometabolism (Horvath 1972) has been used in different and conflicting ways; since the prefix *co* implies *together*, it should not therefore be used when only a single substrate is present and the term *biotransformation* is unambiguous and seems appropriate and entirely adequate. Detailed discussions have been presented (Dalton and Stirling 1982) and some of the conflicting aspects have been briefly summarized (Neilson et al. 1985).

Whereas probably most of the aerobic bacteria that have been used in experiments on biodegradation were selected by elective growth using the chosen substrate as sole source of both carbon and energy, anaerobic bacteria are often more fastidious in their nutritional demands. Addition of nutrient supplements — such as yeast extract or even ruminant fluid — may be needed to stimulate or maintain growth. Although the metabolic conclusions from such experiments are unambiguous, care should be exercised in uncritically extrapolating the results to natural ecosystems that are unlikely to provide such a nutritious environment. In such experiments, unequivocal results may often be advantageously obtained by using suspensions of washed cells (Chapter 5, Section 5.3.1).

Cometabolism merits careful analysis since important metabolic principles lie behind most of the experiments, even though confusion may have arisen as a result of ambiguous terminology. An attempt is therefore made to ignore semantic implications and to adopt a wide perspective in discussing this environmentally important issue. A pragmatic point of view has been adopted, and the following examples attempt to illustrate the kinds of experiments that have been carried out under various conditions.

Experiments in Which Only a Single Substrate Is Present

Organisms may be obtained after elective enrichment with a given substrate but are subsequently shown to be unable to use the substrate for growth although they are able to accomplish its partial metabolism; such situations may be adequately classified as biotransformation. A typical example is the partial oxidation of 2,3,6-trichlorobenzoate to 3,5-dichlorocatechol (Figure 4.54) (Horvath 1971). In some cases, strains unselected by enrichment have been used, for example, oxidation of 4-chlorophenol or 3-chlorobenzoate to 3-chlorocatechol (Figure 4.55) by several strains of bacteria grown with non-chlorinated substrates such as phenol, naphth-2-ol, or naphthalene, and that were unable to use the chlorinated compounds for growth (Spokes and Walker 1974), and oxidation of nitrobenzene to 3-nitrocatechol by strains of *Pseudomonas* sp. grown with toluene or chlorobenzene (Haigler and Spain 1991).

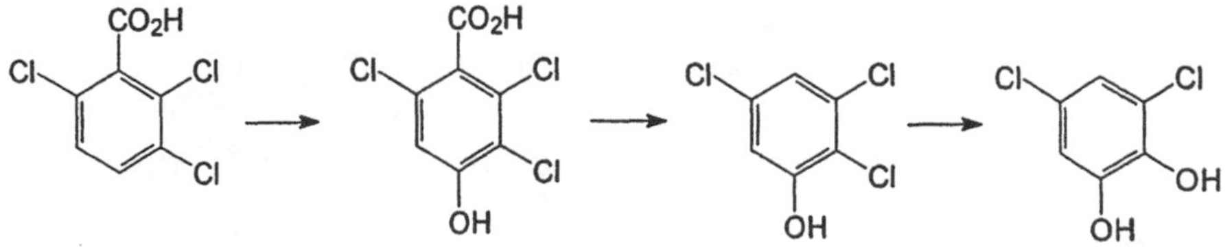

FIGURE 4.54
Biotransformation of 2,3,6-trichlorobenzoate.

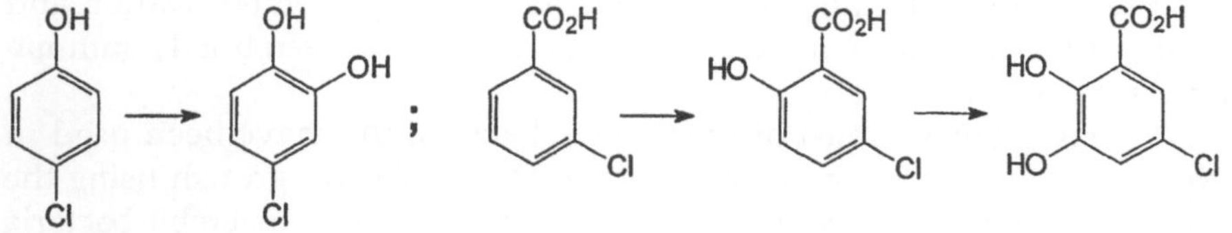

FIGURE 4.55
Biotransformation of 2-chlorophenol and 3-chlorobenzoate.

The Induction of Catabolic Enzymes by Preexposure to an Analogue Substrate

Cells may be grown before exposure to the xenobiotic on an analogue substrate; although the xenobiotic is extensively degraded, it cannot be used alone to support growth of the cells. When both substrates are simultaneously present, the term *cometabolism* is appropriate, but this term is quite unjustifiable for the situation in which cells grown with a given inducing substrate are then used for studying the metabolism of a single xenobiotic. The following examples are given as illustration:

- The oxidation of 2-chloronaphthalene to chloro-2-hydroxy-6-keto-hexa-2,4-dienoic acids (Figure 4.56) by cells in which biotransformation of the substrate was induced by growth with naphthalene (Morris and Barnsley 1982).

- The degradation of polychlorinated biphenyls has been established in a number of organisms enriched with biphenyl as carbon source, for example, *Pseudomonas* sp. strain LB400 (Bopp 1986), *Alcaligenes eutrophus* strain H850 (Bedard et al. 1987), and *P. pseudoalcaligenes* (Furukawa and Arimura 1987). Further details are discussed in Chapter 6, Section 6.5.1.1.

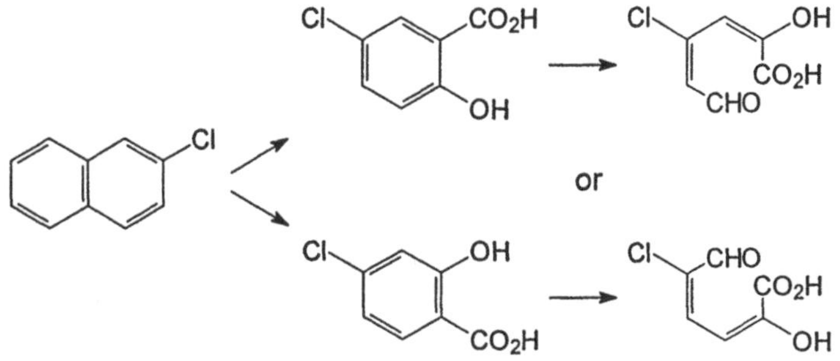

FIGURE 4.56
Biotransformation of 2-chloronaphthalene.

- The use of brominated biphenyls has been examined to induce dechlorination of highly chlorinated biphenyls including the hepta-, hexa-, and pentachloro congeners (Bedard et al. 1998). Di- and tri-bromo congeners were the most effective and were themselves reduced to biphenyl. This is noted again in Chapter 8, Section 8.2.4.2.

- The oxidation of methylbenzothiophenes by cells grown with 1-methylnaphthalene (Saflic et al. 1992).

- The mineralization of benz[*a*]anthracene, chrysene, and benzo[*a*]pyrene by organisms isolated by enrichment and growth with phenanthrene (Aitken et al. 1998).

- A low level of stimulation of the degradation by *Burkholderia cepacia* of benz[*a*]anthracene and dibenz[*a,h*]anthracene by addition of phenanthrene to the cultures (Juhasz et al. 1997).

Comparable situations could be encountered during experiments on bioremediation in which it may be experimentally expedient to grow cells on a suitable analogue (Klecka and Maier 1988) or to introduce the organisms into the contaminated site (Harkness et al. 1993). There are, however, inherent dangers in this procedure since, for example, cells able to degrade 5-chlorosalicylate (Crawford et al. 1979), 2,6-dichlorotoluene (Vandenbergh et al. 1981), and pentachlorophenol (Stanlake and Finn 1982) were unable to degrade the nonhalogenated analogues, and the degradation of 4-nitrobenzoate was inhibited by benzoate even though the strain could use either substrate separately (Haller and Finn 1978).

Enzyme Induction by Growth on Unrelated Compounds

Enzymes necessary for the metabolism of a substrate may be induced by growth on structurally unrelated compounds. In the examples used for illustration, monooxygenases play a cardinal role. The versatility of methane monooxygenase has already been noted in Section 4.3.2, while the monooxygenases that may be involved in toluene degradation are discussed in Section 4.4.1.1 and Chapter 6, Section 6.2.1.

1. Trichloroethene and aromatic compounds — A striking example is the degradation of trichloroethene by different strains of *Pseudomonas* sp. grown with phenol (Folsom et al. 1990) or with toluene. This capability has already been noted in Section 4.4.1.1 in the context of monooxygenase reactions, and has attracted attention for the bioremediation of contaminated sites (Hopkins and McCarty 1995). Conversely, toluene degradation is induced (a) by trichloroethene in a strain of *P. putida* (Heald and Jenkins 1994) and (b) in *P. mendocina* — although not in *Burkholderia* (*Pseudomonas*) *cepacia* or *P. putida* strain F1 — by trichloroethene, pentane, and

hexane (McClay et al. 1995). This metabolic versatility may be consistent with the different pathways that may be followed in toluene degradation (Figure 4.57): *P. putida* F1 by the classical toluene dioxygenase system (Zylstra et al. 1989), *B. cepacia* G4 by monooxygenation to 2-methylphenol (Shields et al. 1989), *P. mendocina* KR by monooxygenation to 4-methylphenol (Whited and Gibson 1991), and *P. pickettii* PKO1 by monooxygenation to 4-methylphenol (Olsen et al. 1994). The last is mediated by a monooxygenase that can be induced by benzene, toluene, and ethylbenzene, and also by xylenes and styrene. A plausibly comparable situation exists for strains of *Pseudomonas* sp. and *Rhodococcus erythropolis* that were obtained by enrichment with *iso*propylbenzene, and that could be shown to oxidize trichloroethene (Dabrock et al. 1992). In addition, one of the pseudomonads could oxidize 1,1-dichloroethene, vinyl chloride, trichloroethane, and 1,2-dichloroethane.

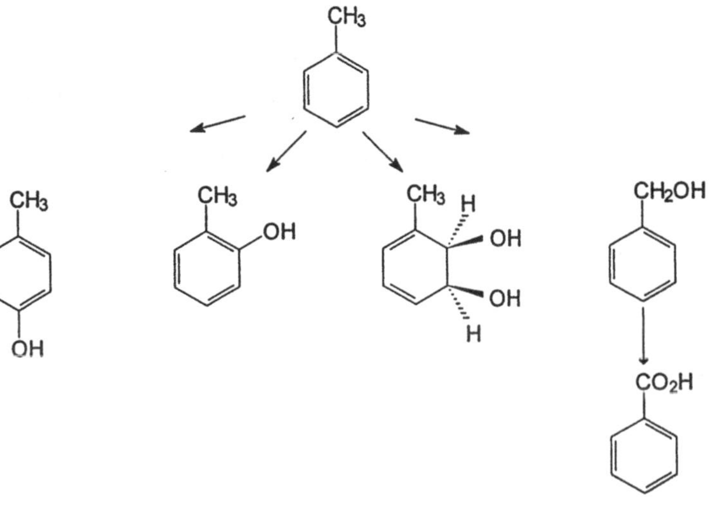

FIGURE 4.57
Pathways for the biotransformation of toluene.

2. Methane, butane, and chloroform — Cells of *Methylosinus trichosporium* grown with methane and of *P. butanovora* and *Mycobacterium vaccae* grown with butane were able to partially degrade chloroform (Hamamura et al. 1997). Again, this may be the result of the induction of monooxygenase activity.

3. Strain G4/PR1 of *B. cepacia* that is constitutive for the synthesis of toluene-2-monooxygenase is able to degrade a number of ethers including diethyl ether and *n*-butyl methyl ether but not *t*-butyl methyl ether (Hur et al. 1997).

4. A good example is provided by DDE, the first metabolite in the conventional degradation pathway of DDT. Pure cultures of aerobic and anaerobic bacteria that are able to degrade 1,1-dichloroethene and 4,4'-dichlorobiphenyl that were considered to represent important structural features were unable to degrade DDE even during incubation with dense cell suspensions (Megharaj et al. 1997). Cell extracts of the aerobic organisms were also ineffective, and it was therefore concluded that recalcitrance lay in the structural entity 1,1-diphenyl-2,2-dichloroethene since 1,1-diphenylethene could be used as sole substrate for growth of styrene-degrading strains of *Rhodococcus* sp. (Megharaj et al. 1998). On the other hand, cells of *P. acidovorans* strain M3GY during growth with biphenyl have been shown to degrade DDE with the fission of one ring and production of 4-chlorobenzoate (Hay and Focht 1998).

This situation may be of widespread occurrence and further examples of its existence will be facilitated by insight into the mechanisms of pathways for biodegradation.

The Role of Readily Degraded Substrates

Whereas the results from experiments on biodegradation in which readily degraded substrates such as glucose are added have probably restricted relevance to natural ecosystems in which such substrates exist in negligible concentration, situations in which readily degraded substrates are present in addition to those less readily degraded undoubtedly occur in biological waste treatment systems. In these circumstances, at least three broadly different metabolic situations may exist:

1. The presence of glucose may suppress degradation of a recalcitrant compound. Two examples are (a) strains of *P. pickettii* degrade 2,4,6-trichlorophenol, and its degradation is induced by several other chlorophenols but is repressed in the presence of glucose or succinate (Kiyohara et al. 1992), and (b) the presence of glucose decreased the rate of degradation of phenol in a natural lake-water community, although the rate was increased by arginine (Rubin and Alexander 1983).

2. The presence of glucose may on the other hand enhance the degradation of a recalcitrant compound. Several different metabolic situations may be discerned, each representing a different mechanism for the stimulatory effect.

   • Experiments in which degradation of fluorobenzoates by a mixed bacterial flora was enhanced by the presence of glucose might plausibly be attributed to an increase in the cell density of the appropriate organism(s) (Horvath and Flathman 1976). A comparable conclusion could also be drawn from the data for the degradation of 2,4-dichlorophenoxyacetate and O,O-dimethyl-O-[3-methyl-4-nitrophenyl] phosphorothioate in cyclone fermentors (Liu et al. 1981).

- The presence of readily degraded substrates such as glutamate, succinate, or glucose had a stimulatory effect on the degradation of pentachlorophenol by a *Flavobacterium* sp. possibly by ameliorating the toxic effects of pentachlorophenol (Topp et al. 1988), and enhanced the ability of natural communities to degrade a number of xenobiotics (Shimp and Pfaender 1985a).

- The presence of glucose facilitated the anaerobic dechlorination of pentachlorophenol and may plausibly be attributed to the increased level of reducing equivalents (Henriksen et al. 1991). A comparable phenomenon is the enhancement of the dechlorination of tetrachloroethene in anaerobic microcosms by the addition of carboxylic acids including lactate, propionate, butyrate, and crotonate (Gibson and Sewell 1992).

3. The xenobiotic may be degraded in preference to glucose that is not a universal growth substrate. This situation is encountered in a phenol-utilizing strain of the yeast *Trichosporon cutaneum* that possesses a partially constitutive catechol 1,2-dioxygenase (Shoda and Udaka 1980) and illustrates the importance of regulatory mechanisms in determining the degradation of xenobiotics. Constitutive synthesis of the appropriate enzyme systems may indeed be of determinative significance in many natural ecosystems.

In some circumstances, therefore, the presence of readily degraded substrates may clearly facilitate degradation of more recalcitrant xenobiotics, although this is neither universally the case nor has a generally valid mechanism for these positive effects emerged. Whereas the addition of metabolizable analogues may increase the overall rates of degradation (Klecka and Maier 1988), it should be emphasized that the presence of readily degraded substrates in enrichments would generally be expected to be counterselective to the development of organisms degrading a given xenobiotic and that the observed enhancements summarized earlier were generally observed during relatively short time intervals.

Simultaneous metabolism of two structurally related substrates in which only one of them serves as growth substrate during biotransformation of the other may exist. A simple example is the O-demethylation of 4,5,6-trichloroguaiacol to 3,4,5-trichlorocatechol followed by successive O-methylation during growth of a strain of *Rhodococcus* sp. at the expense of vanillate (Figure 4.58) (Allard et al. 1985). The results of experiments with mixtures of benzoate and 2,5-dichlorobenzoate using variants of a strain of *P. aeruginosa* grown in chemostat cultures have revealed important aspects of environmental significance (van der Woude et al. 1995):

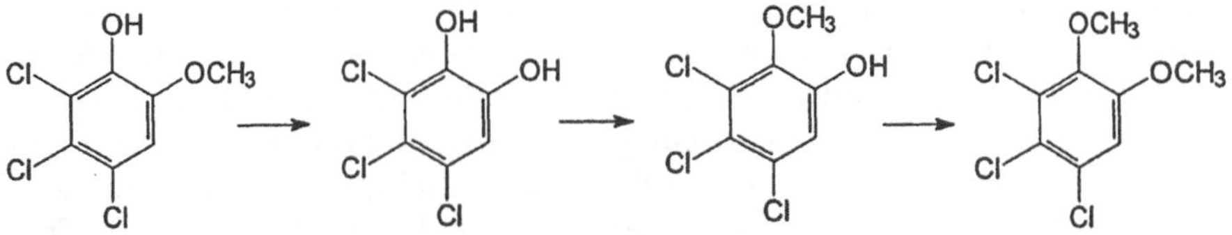

FIGURE 4.58
Biotransformation of 4,5,6-trichloroguaiacol by *Rhodococcus* sp. during growth with vanillate.

- Variants that formed stable cultures could be obtained by growth limitation with both benzoate and 2,5-dichlorobenzoate.

- One of these variants was capable of 2,5-dichlorobenzoate-limited growth at an oxygen concentration of 11 µM, but the presence of benzoate increased the residual concentration of 2,5-dichlorobenzoate from 0.05 to 1.27 µM.

A number of factors may be responsible for these observations including the oxygen gradient within the cell and the oxygen concentration required for synthesis of the degradative oxygenases. It was concluded that, under the low oxygen tensions that may exist in natural ecosystems, the presence of the more readily degraded benzoate necessitated high oxygen affinity for organisms to achieve complete degradation of the 2,5-dichlorobenzoate. In this case, it is relevant to note that the organism was originally isolated after enrichment at high oxygen concentrations. The issue of oxygen concentration is discussed in a wider context in Section 4.6.1.2.

4.6 Determinative Parameters

4.6.1 Physical Parameters

Physical parameters such as temperature, salinity, pH, and oxygen concentration may critically determine the persistence or otherwise of a xenobiotic under natural conditions and these should therefore be critically evaluated. Experiments can be carried out under any of the conditions that simulate the natural environment, and these can be imposed both during isolation of the organisms by enrichment and incorporated into the design of subsequent experiments on biodegradation and biotransformation. In practice, most experiments are carried out with mesophiles and at pH values in the vicinity of pH 7, presumably motivated by the fact that these are — or are assumed to be — prevalent in natural ecosystems. It should also be emphasized that water temperatures during the winter in high latitudes in both the Northern and Southern Hemispheres are low — probably well under 10°C, so that greater advantage should be taken of psychrophiles particularly in investigations aimed at providing realistic estimates of rates for microbial transformation under natural conditions. With the notable exception of the anaerobic sulfate-reducing bacteria, surprisingly few investigations have used truly marine bacteria in spite of the fact that substantial quantities of xenobiotics are discharged into the sea either directly or via the input from rivers. These issues are discussed in the next section.

4.6.1.1 Temperature

The role of temperature may be of particular significance for mixed cultures of organisms in the natural environment, or when they are used in laboratory studies. Temperature may effect important changes both in the composition of the microbial flora and on the rates of reaction. Greatest attention has hitherto been directed to hydrocarbons and PCBs. Some illustrative examples of important issues that are especially relevant to bioremediation (Chapter 8) include the following.

1. An anaerobic sediment sample was incubated with 2,3,4,6-tetra-chlorobiphenyl at various temperatures between 4 and 66°C (Wu et al. 1997a). The main products were 2,4,6- and 2,3,6-trichlorobiphenyl and 2,6-dichlorobiphenyl: the first was produced maximally and discontinuously at 12 and 34°C, the second maximally at 18°C and the third was dominant from 25 to 30°C. Dechlorination was not observed above 37°C. In a further comparable study with Arochlor 1260, hexa- to nonachlorinated congeners were dechlorinated with a corresponding increase in tri- and tetrachloro congeners, and four dechlorination patterns associated with different temperature ranges could be distinguished (Wu et al. 1997b).

2. Sediment samples from a contaminated site were spiked with Arochlor 1242 and incubated at 4°C for several months (Williams and May 1997). Degradation by aerobic organisms in the upper layers of the sediment — but not in those at >15 mm from the surface — occurred with the selective production of di- and trichlorobiphenyls. Some congeners were not found including 2,6- and 4,4-dichlorobiphenyls, and a wider range of trichlorobiphenyls.

3. Enrichment of arctic soils from Northwest Territories, Canada with biphenyl yielded organisms that were assigned to the genus *Pseudomonas* (Master and Mohn 1998). Rates of removal of individual congeners of Arochlor 1242 were examined at 7°C and compared with those for the mesophilic *Burkholderia cepacia* strain LB400. The spectrum of rates for all congeners was similar for the arctic strains, and for some of the trichlorinated congeners was considerably greater than for *B. cepacia* strain LB400.

4. Mineralization of ^{14}C-ring-labeled toluene was examined in contaminated and uncontaminated samples from an aquifer in Alaska and one in South Carolina (Bradley and Chapelle 1995). A number of interesting facts emerged:

   a. Rates for organisms from South Carolina were greater for the contaminated sample than for the uncontaminated sample, and for the uncontaminated sample were zero at 5°C.

   b. Rates for organisms from the contaminated site in Alaska were highly sensitive to temperature and showed a distinct maximum at 20°C that was not apparent for those from the uncontaminated site, whereas rates were virtually identical at 5 and 35°C for both groups of organisms.

5. A number of psychotrophic organisms isolated from various sites in northern Canada were capable of mineralizing naphthalene, toluene, and linear dodecane and hexadecane at 5°C. Although some of the positive strains possessed genes homologous with those required for metabolism of the substrates by established pathways — *nahB* for naphthalene, and *xylE* and *todC1* for toluene — others showed only low homologies and may have possessed novel catabolic genes (Whyte et al. 1996). Two organisms assigned to the genus *Pseudomonas* degraded alkanes (C_5 to C_{12}), toluene, and naphthalene at both 5 and 25°C (Wayne et al. 1997). It was shown by PCR and DNA sequence analysis that the plasmid-borne catabolic genes were comparable to those for the *alk* pathway in *P. oleovorans* and the *nah* pathway in *P. putida* G7.

6. Experiments using a fluidized-bed reactor showed that removal of chlorophenols could be accomplished by organisms adapted to a temperature of 5 to 7°C (Järvinen et al. 1994).

4.6.1.2 *Oxygen Concentration*

The question of oxygen concentration is a good deal more complicated: whereas the extreme conditions in which oxygen concentration is high, for example, in well-mixed surface waters, or essentially absent, for example, in organic-rich deep sediments, are readily visualized and realized in laboratory experiments, it is worth drawing attention to a number of less obvious situations. It should also be pointed out that under conditions of oxygen limitation in continuous culture, mixed populations of aerobic bacteria and the strictly anaerobic *Methanobacterium formicicum* and *Methanosarcina barkeri* may be maintained (Gerritse and Gottschal 1993a).

The role of oxygen concentration has already been noted in a variety of contexts: (1) in the regulation of the synthesis of peroxidases in white-rot fungi (Section 4.4.4), (2) in determining the outcome of the degradation of the trichloromethyl group in a substituted pyridine by *Nitrosomonas europeae*

(Section 4.3.2), (3) in the induction of cytochrome P-450 $_{14DH}$ in *Saccharomyces cerevisiae* (Section 4.4.1.2). The oxygen tolerance among anaerobic bacteria has been discussed in Section 4.3.4. Further comments on other organisms are given here.

1. The existence of microaerophilc organisms such as *Wolinella succinogenes* that was formerly considered an anaerobe or, conversely, the oxygen tolerance of many clostridia suggests that such organisms may occupy an ecological niche between the two extremes noted earlier.

2. Attention has already been drawn to facultatively anaerobic organisms (Section 4.3.2) that can employ either fermentative or oxidative modes for the metabolism of appropriate substrates such as carbohydrates, and to the existence of others (Section 4.3.3) that can use nitrate as an alternative electron acceptor in the absence of oxygen; such organisms have therefore two metabolic options, although they are generally mutually exclusive. An illustrative example of the metabolic flexibility of facultatively anaerobic organisms is provided by the type strains of all the species of the enteric genus *Citrobacter*. These organisms are able to degrade a number of amino acids including glutamate using either respiratory or fermentative metabolism and they can rapidly switch between these alternatives (Gerritse and Gottschal 1993b). Under anaerobic conditions, the initial steps of glutamate degradation involve formation of 3-methylaspartate, mesaconate, and citramalate that are typical of clostridial fermentations (Chapter 6, Section 6.7.1). Although the oxygen concentration in natural environments may be highly variable, organisms with such a high degree of metabolic flexibility may reasonably be presumed to be at an advantage.

3. There are a number of aspects of oxygen toxicity under aerobic conditions that are the result of the synthesis of compounds including superoxide, hydrogen peroxide, and hydroxyl radicals. Further details emerge from observations that the Fe isoenzyme of superoxidase dismutase is synthesized by *E. coli* under anaerobic conditions in contrast to the Mn isoenzyme that is produced only under aerobic conditions. The synthesis of the former facilitates the transition from anaerobic to aerobic conditions by destroying the superoxide radical generated after exposure to oxygen (Kargalioglu and Imlay 1994). This flexibility is important in environments where fluctuating oxygen concentrations prevail. Studies with mutants of *P. aeruginosa* in which the synthesis of Mn superoxide dismutase and Fe superoxide were impaired showed that growth in a complex medium or in a defined glucose medium was adversely affected

by the latter but only insignificantly by the former (Hassett et al. 1995). The transport of Fe into the cell is subject to complex regulatory control since excess within the cell would result in the production of deleterious hydroxyl radicals from molecular oxygen (Touati et al. 1995). This is noted again in Chapter 5, Section 5.5.5.

4. There is substantial evidence that organisms that are strictly dependent on the aerobic metabolism of substrates for growth and replication may nonetheless accomplish biodegradations and biotransformations under conditions of low oxygen concentration. Indeed, such conditions may inadvertently prevail in laboratory experiments using dense cell suspensions. It is, of course, important to remember that the *growth* of these aerobic organisms is strictly dependent on the availability of oxygen. A number of examples are given as illustration of the environmental role of putatively aerobic organisms under conditions of low oxygen concentration.

    a. The rate of biodegradation of hexadecane in a marine enrichment culture was unaffected until oxygen concentrations were lower than 1% saturation (Michaelsen et al. 1992).

    b. It has been shown that the degradation of pyrene by a strain of *Mycobacterium* sp. can take place at low oxygen concentrations (Fritzsche 1994).

    c. *Alcaligenes* sp. strain L6 was obtained after enrichment with 3-chlorobenzoate and 2% oxygen in the gas phase and possessed both 2,5- and 3,4-dihydroxybenzote dioxygenase activities but lacked catechol dioxygenase activity (Krooneman et al. 1996). The substrate affinity of cells grown under oxygen limitation was three times greater than under excess oxygen, and exceeded values for other bacteria growing with benzoate or 2,5-dichlorobenzoate. This strain that metabolized 3-chlorobenzoate by the gentisate pathway was, under low oxygen concentrations, able to compete successfully with a *Pseudomonas* sp. strain A that used the catechol pathway (Krooneman et al. 1998).

    d. Further details emerged from the results of experiments that compared the kinetic properties of catechol-2,3-dioxygenases from toluene-degrading pseudomonads isolated from aquifer sands or groundwater (strains W 31 and CFS 215) with other strains (*P. putida* F1). This revealed a number of significant differences that may plausibly be associated with the low oxygen concentrations pertaining in the environment from which the strains were isolated: for strains that degraded toluene via *cis*-toluene-2,3-dihydrodiol and 3-methylcatechol, lower values of K_m were observed for O_2 and 3-methylcatechol, and higher values of V_{max} for O_2 and 3-methylcatechol (Kukor and Olsen 1996).

Strain	K_m		V_{max}	
	O_2	3-Methylcatechol	O_2	3-Methylcatechol
F1	9.7	16.9	4.3	17
W31	2.0	0.3	293	125
CFS 215	0.9	0.5	400	180

e. The growth of *P. putida* strains KT2442 and mt-2 on aromatic carboxylates decreased with O_2 concentrations < 10 μM, and under these conditions strain KT2442 excreted catechol during growth with benzoate, or 3,4-dihydroxybenzoate from 4-hydroxybenzoate (Arras et al. 1998). In these experiments, the rate of oxygenation rather than the pO_2 was the limiting factor since the K_m for catechol 1,2-dioxygenase was 20 μM O_2.

f. The synthesis of cytochrome P-450 in yeasts that assimilate long-chain alkanes has been examined in facultatively anaerobic species of *Candida*. In *C. guilliermondii*, levels of cytochrome P-450 that were induced during growth with hexadecane — although not with glucose — increased at dissolved O_2 concentrations < 20% (Mauresberger et al. 1980). For cells of *C. tropicalis*, growth with hexadecane levels of cytochrome P-450 increased when the oxygen concentration was <2 kPa (14% dissolved O_2). On the other hand, the levels of alcohol dehydrogenase, aldehyde dehydrogenase, catalase, and cytochrome c were not influenced (Gmünder et al. 1981).

5. An issue of particular relevance in the context of bioremediation is illustrated by the increased rate of cell death of an established naphthalene-degrading *P. putida* G7 brought about by naphthalene under conditions of oxygen (or combined nitrogen) limitation (Ahn et al. 1998).

6. Growth of *Bacillus subtilis* may take place not only under aerobic conditions, but using nitrate as electron acceptor and even under fermentative conditions with glucose provided, pyruvate is available as an electron acceptor since the organism lacks pyruvate formate hydrogen lyase (Nakano and Zuber 1998).

It should also be appreciated that the oxygen concentration required for accomplishing different stages of a biodegradation pathway may differ significantly.

• The degradation of pentachlorophenol by *Rhodococcus chlorophenolicus* proceeds by initial formation of tetrachlorohydroquinone; whereas its formation is oxygen-dependent, its subsequent degradation can be accomplished in the absence of oxygen (Apajalahti and Salkinoja-Salonen 1987).

- The biodegradation of halogenated alkanes by *P. putida* containing the CAM plasmid and genes for toluene dioxygenase proceeds by an initial anaerobic dehalogenation followed by oxidation; further details are given in Chapter 6, Sections 6.4.4 and 6.10.1.

- Biotransformation as opposed to biodegradation may in fact be favored by limited oxygen concentration: a good example is provided by the synthesis of $7\alpha,12\beta$-dihydroxy-1,4-androstadien-3,17-dione from cholic acid by a strain of *Pseudomonas* sp. (Smith and Park 1984), in which oxygen limitation restricts the rate of C-9α hydroxylation that precedes degradation.

4.6.1.3 Redox Potential

For anaerobic bacteria, it is usually not sufficient that the medium is anoxic: it must also be poised at the correct redox potential. It is for this reason that prereduced media (Chapter 5, Section 5.2.6) are used. This is accomplished by addition of, for example, sulfide, dithionite, or titanium(III) citrate, and the media generally contain a redox indicator such as resazurin. Examples have been given earlier of reactions carried out by aerobic organisms at low oxygen concentrations and the outcome of reactions carried out by such organisms may be influenced by the redox potential. For example, cells of the aerobic *P. cepacia* carried out the degradation of carbon tetrachloride only when a negative potential was maintained, and the maximum rate occurred at a potential of approximately -150 mV (Jin and Englande 1997).

4.6.1.4 The Role of Association of Bacteria with Particulate Material

Increasing attention has been directed to the degradation of xenobiotics in aquifers, and it has been shown that most of the relevant bacteria are associated with fine particles rather than existing as free entities; it is therefore important to include such material in laboratory experiments using unenriched communities that attempt to simulate natural conditions (Holm et al. 1992). The interdependence of surfactant sorption and biodegradability was proposed from the results of laboratory experiments which showed that *Pseudomonas* sp. strain DES1 was able to degrade sodium dodecyl sulfate attached to sediment particles more effectively than organisms that were unable to degrade analogous (nondegradable) substrates (Marchesi et al. 1994). Although this was attributed to the effect of the metabolite dodecan-1-ol, further study has revealed a more complex situation (Owen et al. 1997) as a result of differences in response of different strains and toward different surfactants. Bacterial associations with particulate material should also be evaluated in the context of the bioavailability of the substrate that is discussed in Section 4.6.3 and in bioremediation strategies (Chapter 8, Section 8.1.2).

4.6.2   Substrates, Concentration, Transport into Cells, and Toxicity

Range of Substrates for Growth of Organisms

The range of substrates used for growth of organisms used in studies of bio-degradation may be pragmatically divided into four groups: (1) complex media containing peptones or yeast extract for organisms that have undeter-mined and complex nutritional requirements, including, for example, some fungi and anaerobic bacteria; (2) the substance whose biodegradation is being examined; (3) analogues of this such as, for example, biphenyl for PCB-degrading organisms; (4) substrates that are related to those naturally avail-able to the degrading organism. The last appears to have been exploited only occasionally, and only three illustrations are provided.

a. The use of a range of coumarin (esculetin) and flavanone (quercitin, narignenin) glycones and aglycones, and a flavane (catechin) (Figure 4.59) as substrates for the anaerobic dechlorination of chlo-rocatechols (Allard et al. 1991; 1992).

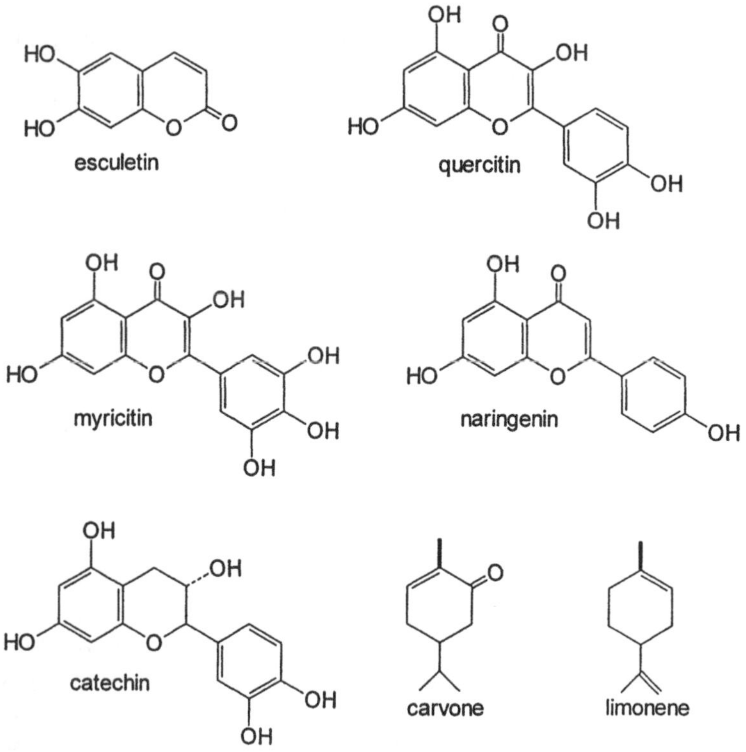

FIGURE 4.59
Structures of natural products used as substrates or inducers.

b. The growth of established PCB-degrading bacteria on a range of flavones, flavanols, and flavanones (catechin, naringenin, and myricitin) (Figure 4.59) (Donnelly et al. 1994).

c. Reference is made in Section 4.8.1 to the induction by terpenoids, including carvone, and limonene (Figure 4.59) of the degradation of PCB congeners (Gilbert and Crowley 1997).

Utilization of Low Substrate Concentrations

Probably most laboratory experiments on biodegradation and biotransformation have been carried out using relatively high concentrations of the appropriate substrates, even though these may be far in excess of those that are likely to be encountered in natural ecosystems (Subba-Rao et al. 1982; Alexander 1985). This limitation is particularly severe during conventional tests for biodegradability and seriously restricts the degree to which the results of these experiments are environmentally relevant. Except in the immediate neighborhood of industrial discharge, low concentrations of xenobiotics are almost certainly the rule rather than the exception in natural situations, and a number of significant experimental observations should be taken into consideration.

Investigations into the flora of natural waters have revealed the presence of bacteria able to grow with extremely low substrate concentrations, and these have been termed *oligotrophs* (Poindexter 1981). It is, of course, well established that organisms such as *Aeromonas hydrophila* (Van der Kooij et al. 1980), *P. aeruginosa* (Van der Klooij et al. 1982), and a species of *Spirillum* (Van der Kooij and Hijnen 1984) may proliferate in natural waters supplemented with low concentrations of additional organic carbon. It has been suggested that there exist bacteria specially adapted to such conditions, although doubt has been cast on the absolute distinction between eutrophic and oligotrophic bacteria (Martin and MacLeod 1984). This distinction is blurred by the fact that after prolonged nutrient starvation under laboratory conditions, initially oligotrophic marine bacteria can be isolated on media containing high substrate concentrations (Schut et al. 1993); these organisms are therefore facultatively oligotrophic. The authors point out that obligate oligotrophy may be determined rather by life history than by invariant physiological characteristics. One critical issue seems to be the effectiveness and regulation of substrate transport into the cells (Schut et al. 1995). The significance of such organisms in natural marine systems burdened with xenobiotics is less clearly resolved, although an organism designated as *Cycloclasticus oligotrophus* strain RB1 harbors genes with a high degree of homology to those encoding xylene degradation in terrestrial pseudomonads, and is able to grow at the expense of xylenes, toluene, biphenyl, naphthalene, and phenanthrene (Wang et al. 1996).

Possible Existence of Threshold Concentrations

An issue which has received some — although almost certainly insufficient — attention exists when bacteria are exposed to extremely low concentrations of a xenobiotic, of the order of nanogram per liter or less. For example, although the rates of biodegradation of phenol, benzoate, benzylamine, 4-nitrophenol, and di(2-ethylhexyl)phthalate in natural lake water were linear over a wide range of substrate concentrations between ng/l and µg/l (Rubin et al. 1982), it has been shown that the rates of degradation of 2,4-dichlorophenoxyacetate at concentrations of the order of µg/l were extremely low (Boethling and Alexander 1979). This observation has subsequently been extended to a greater range of compounds (Hoover et al. 1986). These and other data could be interpreted as supporting the concept of a threshold concentration below which growth and degradation does not take place — or occurs at insignificant rates.

Although the reasons for the existence of such threshold concentrations have not been clearly resolved, a number of plausible hypotheses may be suggested:

1. The substrate concentrations may be too low for effective transport into the cells.

2. There may be a limiting substrate concentration required for induction of the appropriate catabolic enzymes; at low substrate concentrations, the necessary enzymes would simply not be synthesized and this could be the determining factor in some circumstances (Janke 1987).

Two contrasting results may be instructive. Experiments with chlorinated benzenes in which the effect of substrate concentration was examined in batch cultures and in recirculating fermentors showed that although substrates could be degraded completely in the former, a residual concentration of the substrate persisted in the latter (van der Meer et al. 1992). On the other hand, experiments using *Burkholderia* sp. strain PS14 failed to detect residual concentrations >0.5 nM after mineralization of 1,2,4,5-tetra- and 1,2,4-trichlorobenzene at concentrations of 500 nM (Rapp and Timmis 1999). These results are highly relevant in the context of bioremediation (Chapter 8, Section 8.1.2).

All these observations emphasize that tests for biodegradability carried out at high substrate concentrations may not adequately predict the rates of degradation occurring in natural ecosystems where only low concentrations of xenobiotics are encountered (Alexander 1985). This phenomenon is therefore of enormous environmental importance since it would imply the possibility of extreme persistence of low concentrations in natural ecosystems. The further exploration of this phenomenon is probably only limited in practice by the access to analytical methods for measuring sufficiently accurately substrate concentrations at the level of ng/l or lower. Most studies that have been carried out have therefore used ^{14}C-labeled substrates which necessarily limits the range of compounds accessible and restricts the elucidation of metabolic pathways in which only biotransformation or partial mineralization has occurred.

Strategies Used by Cells for Substrates with Low or Negligible Water Solubility

Although this issue will be discussed in Chapter 5 in the context of the design of laboratory experiments on biodegradation and biotransformation and in

Chapter 8 (Section 8.2.1) in the context of bioremediation of PAH-contaminated terrestrial systems, a more formal approach will be adopted here. Organisms have developed their own strategies to circumvent the low water solubility of their substrates and a few examples may be given as illustration:

1. They may produce extracellular enzymes which attack the substrate without the need for transport into the cell; examples are cellulase, DNAse, or gelatinase.

2. They may synthesize surface-active emulsifying compounds during growth. This problem has been extensively investigated on account of its commercial application (Gerson and Zajic 1979), and details of the structural aspects of biosurfactants elaborated for degradation of water-insoluble substrates have been given (Hommel 1994). Some of the key conclusions from a wide range of studies may be briefly summarized.

   - Glycolipids consisting of long-chain carboxylic acids and rhamnose (Itoh and Suzuki 1972; Rendell et al. 1990) or trehalose (Suzuki et al. 1969; Singer et al. 1990) have been isolated during growth of a number of different bacteria on *n*-alkanes. The rhamnolipid surfactant produced by a strain of *Pseudomonas* sp. was effective in enhancing degradation of octadecane (Zhang and Miller 1992), although the concentration used in these experiments was rather high (300 mg/l) to encourage its practical application. A model that includes the effect of rhamnolipids on solubilization biodegradation, and bioavailability within surfactant micelles has been presented to rationalize the data from batch studies on the dissolution, bioavailability, and biodegradation of phenanthrene (Zhang et al. 1997).

   - A polyanionic heteropolysaccharide (emulsan) is produced during growth of a strain of *Acinetobacter calcoaceticus* with hydrocarbon mixtures, and the high-molecular-weight polymer is necessary for emulsifying activity (Shoham and Rosenberg 1983). The value of emulsan for treating oil spills seems, however, equivocal in light of results that demonstrate reduced biodegradation in its presence (Foght et al. 1989).

   - A *Rhodococcus* sp. synthesizes a glycolipid during growth with *n*-alkanes and *n*-alkanols but not with carboxylic acids, triglycerides, or carbohydrates, and its formation was favored by nitrogen limitation (Singer and Finnerty 1990); these conditions may, of course, be counterproductive to optimal growth of the organisms.

   - *P. maltophilia* produces an extracellular surfactant during growth with naphth-1-oate, and this organisms displays much greater emulsifying activity toward one- and two-ring aromatic hydrocarbons than toward aliphatic hydrocarbons (Phale et al. 1995).

- A strain of *Pseudomonas* sp. produces a surfactant in the presence of high concentrations of glucose or mannitol and naphthalene or phenanthrene (Déziel et al. 1996).

- The synthesis of an emulsifying agent produced by *Candida lipolytica* is inducible during growth with a number of *n*-alkanes, but is not synthesized during growth with glucose (Cirigliano and Carman 1984).

A number of different mechanisms have therefore clearly emerged and it seems premature to draw general conclusions especially with respect to the application of these natural surfactants to bioremediation that is discussed in greater detail in Chapter 8, Section 8.2.1. It is important to note that production of biosurfactants may not be the only mechanism for facilitating the uptake of substrates with low water solubility. For a strain of *Rhodococcus* sp. that did not produce surfactants, the rates of degradation of pyrene dissolved in water in the presence of insoluble, nondegradable 2,2,4,4,6,8,8-heptamethylnonane exceeded those predicted for physicochemical transfer from the solvent to the aqueous phase, but could be accounted for on the basis of uptake both from the interface and from the aqueous solution (Bouchez et al. 1997).

Transport Mechanisms

The mechanisms whereby carbohydrates, carboxylic acids, and glycerol are transported across the bacterial cell membrane prior to metabolism have been elucidated in great detail. Attention should, however, be drawn to an investigation of the regulation of the transport of amino acids and glucose into a facultatively oligotrophic marine bacterium that revealed the lack of specificity of the constitutive system for amino acids, and the interaction of the regulatory systems for amino acids and glucose (Schut et al. 1995). It should also be noted that different transport systems for C_4 dicarboxylic acids into *Escherichia coli* operate under aerobic or anaerobic conditions (Six et al. 1994), so that uptake is dependent on the physiological state of the cells.

By contrast, less effort has apparently been directed to the transport of xenobiotics, and there is an intrinsic difficulty that, in contrast to organisms that utilize carbohydrates or amino acids, mutants defective in the metabolism of the substrate are not so widely available. This makes it impossible to determine directly whether active transport is involved.

It has been suggested that active transport systems for benzoate (Thayer and Wheelis 1982) and for mandelate (Higgins and Mandelstam 1972) are involved. In *Rhizobium leguminosarum*, 4-hydroxybenzoate hydroxylase activity is required for the uptake of 4-hydroxybenzoate (Wong et al. 1994), while in *P. putida* a gene cluster *pcaRKF* is involved in the transport of 4-hydroxybenzoate into the cells, their chemotactic response to the substrate and its degradation by ring hydroxylation (Harwood et al. 1994; Nichols and Harwood 1997). The situation for phenylacetate transport into *P. putida* U is apparently different (Schleissner et al. 1994) since degradation of the substrate by this strain is distinct from that used for 4-hydroxyphenylacetate.

Investigations have, however, examined the transport systems for a restricted range of xenobiotics.

1. A detailed investigation (Groenewegen et al. 1990) has examined the uptake of 4-chlorobenzoate by a coryneform bacterium that degraded this compound. The uptake was inducible and occurred in cells grown with 4-chlorobenzoate but not with glucose. A proton motive force (Δp)-driven mechanism was almost certainly involved, and uptake could not take place under anaerobic conditions unless an electron acceptor such as nitrate was present.

2. The transport of toluene-4-sulfonate into *Comamonas testosteroni* has been examined (Locher et al. 1993), and rapid uptake required growth of the cells with toluene-4-sulfonate or 4-methylbenzoate. From the results of experiments with various inhibitors, it was concluded that a toluenesulfonate anion/proton symport system operates rather than transport driven by a difference in electrical potential ($\Delta\psi$).

3. The uptake of benzoate was examined in two strains of *Alcaligenes denitrificans*: the transport system was inducible, was carrier mediated, was energy dependent, and involved a proton symport system. On the other hand, the uptake of 2,4-dichlorobenzoate by one of the strains was constitutive, displayed no saturation kinetics, and appeared to occur by passive diffusion (Miguez et al. 1995). The uptake of 2,4-dichlorophenoxyacetate has been studied in *Ralstonia eutropha* JMP134(pJP4) in which the degradative genes are plasmid borne (Sections 4.4.2 and 4.9). Uptake was inducible by the substrate, did not take place in fructose-grown cells, and was inhibited by cyanide, which prevents development of a protomotive force, and by the protophore carbonylcyanide-3-chlorophenylhydrazone (Leveau et al. 1998). The protein involved was encoded by an open reading frame on the plasmid designated *tfdK*.

4. *Sphingomonas herbicidovorans* strain MH is able to grow at the expense of both enantiomers of 2-(2,4-dichlorophenoxy)propionate with a preference for the S enantiomer, and the uptake of both enantiomers and 2,4-dichlorophenoxyacetate is inducible. Although the ATPase inhibitor *N,N'*-dicyclohexylcarbodiimide had only slight effect on intracellular levels of ATP, uptake was inhibited by the protophore carbonylcyanide-3-chlorophenylhydrazone and by nigericin that dissipates ΔpH in the presence of high concentration of K^+, but not by valinomycin. It was suggested that uptake is driven by a protomotive force (Δp), and that ΔpH rather than $\Delta\psi$ is the determinant of uptake (Zipper et al. 1998).

5. In *Oxalobacter formigenes*, oxalate and its decarboxylation product formate form a one-to-one antiport system which involves the consumption of an internal proton during decarboxylation and serves as a proton pump to generate ATP by decarboxylative phosphorylation (Anantharam et al. 1989).

6. The transport of EDTA into a bacterial strain capable of its degradation has been examined (Witschel et al. 1999). Inhibition was observed with DCCD (ATPase inhibitor), nigericin (dissipates ∆pH), but not valinomycin (dissipates ∆ψ), and was dependent on the stability constant of metal-EDTA complexes. This has implications both for its biodegradability and its toxicity in natural environments.

All of these important results illustrate the potential complexity of transport systems in bacteria. This has hitherto been a rather neglected aspect of the degradation of xenobiotics, and extension to other organisms and to a wider range of xenobiotics is clearly merited.

Resistance to Toxicity

It is probably no exaggeration to state that most xenobiotics can be metabolized — albeit to varying degrees — under the appropriate conditions, providing, of course, that the lethal toxicity is not exceeded. As noted in Section 4.9, many pathogenic bacteria have, however, developed enzymes for detoxifying antibiotics including β-lactamases, chloroamphenicol acetyltransferases, and the phosphorylases, adenylases, and acetylases for aminoglycosides. Bacteria may also become resistant to the toxicity of heavy metals in the form both of their cations and oxyanions, and to the oxyanions of Se, As, and Te. Nonetheless, even such normally toxic compounds as carbon monoxide (Meyer and Schlegel 1983; Mörsdorf et al. 1992), cyanide (Harris and Knowles 1983), toluene (Claus and Walker 1964), and fluoroacetate (Meyer et al. 1990) can be metabolized by bacteria. Although this is generally accomplished by using low substrate concentrations, changes in the lipid structure of the cells may occur when strains are exposed to lipophilic substrates.

1. A strain of *Pseudomonas putida* that tolerates high concentrations of toluene circumvents toxicity by altering membrane fluidity through synthesis of *trans* unsaturated fatty acids in place of the *cis* compounds (Weber et al. 1994).

2. A strain of *Rhodococcus* sp. that degrades benzene is able to tolerate concentrations of 2% (v/v) per day in continuous culture (Paje et al. 1997), and strains of rhodococci that degrade aromatic compounds including phenols, 4-chlorophenol, and benzene synthesized concentrations of 10-methyl-fatty acids apparently at the expense of unsaturated fatty acids (Tsitko et al. 1999).

4.6.3 Bioavailability: "Free" and "Bound" Substrates

In the discussion of partition in Chapter 3 (Sections 3.2.2 and 3.2.3), it has been pointed out that xenobiotics exist not only in the "free" state but also in association with organic and mineral components of the particles in the water mass and in the soil and sediment phases. This association is a central determinant of the persistence of xenobiotics in the environment, since the extent to which such residues are accessible to microbial attack is largely unknown, and its magnitude is critical also to the effectiveness of bioremediation (Harkness et al. 1993). Attention is also drawn to the critical role of complexes between xenobiotics and Fe in determining biodegradability, and this is discussed in Chapter 5, Section 5.2.4. Only a brief attempt is made to place this important subject in perspective, and to illustrate some of the major unresolved issues. Although the most persuasive evidence for the significance of reduced bioavailability comes from data on the persistence of agrochemicals in terrestrial systems (Calderbank 1989), there seems no reason to doubt the validity of translating the principles to aquatic sediments that contain organic matter structurally resembling that of soils.

Many experiments on the recoverability, persistence, and toxicity of xenobiotics have used spiked samples and therefore do not take into account the cardinal issue of aging after deposition that has been discussed in Chapter 3, Section 3.2.3 and that should be critically evaluated. Some examples are given as illustrative.

1. Experiments on the effect of aging on biodegradability using sterilized samples of soils that were spiked and aged under laboratory conditions have shown that the rates and extent of degradation of phenanthrene and 4-nitrophenol by an added strain of *Pseudomonas* sp. decreased markedly with prolonged aging (Hatzinger and Alexander 1995).

2. The sorption–desorption of PAHs has been extensively investigated, and the role of desorption in determining their biodegradability in aged sediments has been widely accepted (references in Carmichael et al. 1997). A definitive study using ^{14}C-phenanthrene and ^{14}C-chrysene showed that, in contaminated soils, their rates of mineralization were much lower than the rates of desorption from spiked sediments. By contrast, for aged substrates desorption rates were essentially comparable to rates of mineralization. This suggests that the indigenous microflora may have become adapted to the low substrate concentrations available by desorption (Carmichael et al. 1997).

3. It was shown that the toxicity to the house fly, fruit fly, and German cockroach of DDT and dieldrin spiked into sterile soil decreased with aging during long time exposure and was particularly marked for dieldrin. After 270-day incubation when no mortality remained toward any test organism, 92% of the toxicant could be recovered

by Soxhlet extraction (Robertson and Alexander 1998). Clearly then, chemical extractability would have seriously overestimated the toxicity of the samples. This has already been discussed in Section 3.2.3.

Examples that illustrate the generality of the concept include the following.

1. Montmorillonite complexes with benzylamine at concentrations below 200 µg/l decreased the extent of mineralization in lake water samples, although a similar effect was not noted with benzoate (Subba-Rao and Alexander 1982). Even in apparently simple systems, general conclusions cannot therefore be drawn even for two not entirely dissimilar aromatic compounds, both of which are readily degradable under normal circumstances in the dissolved state.

2. Suspensions of 2,4-dichlorophenoxyacetate sorbed to sterile soils were completely protected from degradation by either free or sorbed bacteria, and degradation of the substrate required access by the bacteria to the free compound in solution (Ogram et al. 1985). Rates of degradation in a high organic soil were lower than for a low organic soil (Greer and Shelton 1992), and this further supports the significance of desorption of the xenobiotic in determining its biodegradability.

3. ^{14}C-labeled 2,4-dichlorophenol bound to synthetic or natural humic acids, or polymerized by H_2O_2 and peroxidase was mineralized to CO_2 only to a limited extent (<10%) and the greater part remained bound to the polymers (Dec et al. 1990).

4. ^{14}C-labeled 3,4-dichloroaniline–lignin conjugates were degraded to [14]CO_2 by *Phanerochaete chrysosporium* as effectively as the free compound (Arjmand and Sandermann 1985) and it was therefore concluded that these "bound" residues were not persistent in the environment. This may, however, represent a special case for several reasons: (a) although this organism is able to degrade lignin, the relevance of such organisms in most aquatic environments is possibly marginal, (b) the lignin peroxidases implicated in lignin degradation are generally extracellular so that soluble substrates are probably not necessary.

5. The presence of humic acids had a detrimental effect of the degradation of substituted phenols by a microbial community after lengthy adaption to the humic acids and was not alleviated by the addition of inorganic nutrients (Shimp and Pfaender 1985b). The diminished number of organisms with degradative capability was responsible for the reduced degree of degradation so that the predominant effect was probably the toxicity of the humic acids even toward adapted microorganisms.

6. In short-term experiments with carbofuran (2,3-dihydro-2,2-dimethyl-7-benzofuranyl-*N*-methylcarbamate), degradation was accomplished by organisms in an enrichment culture obtained from soils with a low carbon content where sorption of the substrate is low, but was essentially absent in cultures obtained from soils with a high organic matter (Singh and Sethunathan 1992).

7. Experiments using a chlorocatechol-contaminated sediment and interstitial water prepared from it showed that the concentrations of total 3,4,5-tri- and tetrachlorocatechols (i.e., including the fraction that is released only after alkaline extraction) were apparently unaltered during prolonged anaerobic incubation even after addition of cultures with established dechlorinating capability for the soluble chlorocatechols (Allard et al. 1994).

All of the preceding investigations have been concerned with polar compounds for which plausible mechanisms for their association with organic components of water, soil, and sediment may be more readily conceptualized. It may therefore be valuable to provide a wider perspective by giving some examples for neutral compounds additional to those discussed at the beginning of this section.

1. The aerobic mineralization of α-hexachloro[*aaaaee*]cyclohexane by endemic bacteria in the soil is limited by the rate of its desorption and by intraparticle mass transfer (Rijnaarts et al. 1990).

2. Whereas degradation of the readily extractable toluene in spiked soil by *Pseudomonas putida* was rapidly accomplished, there was a residue that was degraded much more slowly at a rate that was apparently dependent on its desorption (Robinson et al. 1990).

3. The extent of bioremediation of sediments contaminated with PCBs appears to be limited by the association of a significant fraction with organic components of the sediment phase (Harkness et al. 1993).

4. Immobilization of neutral xenobiotics in soils by quaternary ammonium cations has been established, and its significance on the bioavailability of naphthalene to bacteria has been examined: bioavailability was determined by rates of desorption and these differed between a strain of *P. putida* and one of *Alcaligenes* sp. (Crocker et al. 1995). This is noted in Chapter 8, Section 8.2.1.

The results of these experiments in both aquatic systems and terrestrial systems may profitably be viewed against the extensive evidence for the persistence of agrochemicals in the terrestrial environment. Considerable effort has been directed to the issue of bound residues of agrochemicals (Calderbank 1989) and to its significance in determining both their biological effects and their persistence; this is now fully accepted in contemporary thinking. At the same time, it should be appreciated that from an economic point of view, enhanced rates of degradation of agrochemicals in the terrestrial environment may be highly undesirable (Racke and Coats 1990). Nonetheless, care should be exercised in making generalizations. For example, whereas it has been established that soil microorganisms may significantly increase the evolution of $^{14}CO_2$ from ^{14}C parathion (Racke and Lichtenstein 1985), this was not observed with chlorpyrifos (Racke et al. 1990).

Analysis of these diverse observations clearly demonstrates that the persistence of a xenobiotic in the aquatic or in the terrestrial environment may be significantly increased if the xenobiotic is bound either to inorganic minerals or to any of a range of complex natural polymers such as humic and fulvic acids. One of the key issues is the rate of desorption of the xenobiotic from the matrices; this has already been discussed in Chapter 3, Section 3.2.2 and this may depend critically on the mechanism of the association (Chapter 3, Section 3.2.4). The degree of bioavailability may also depend on the nature of the relevant organisms, since, for example, soil-sorbed naphthalene was degraded at markedly different rates by two naphthalene-degrading organisms (Guerin and Boyd 1992) and, at low substrate concentrations, 2,4-dichlorophenoxyacetate was degraded at different rates by the two strains that were examined (Greer and Shelton 1992). Although a number of important unresolved issues remain, it is clear that the degree of bioavailability of xenobiotics in natural systems introduces an important additional uncertainty in extrapolating the results of studies on biodegradation and biotransformation of "free" xenobiotics in laboratory experiments to processes and rates in natural ecosystems.

This is important not only in field investigations. Even in laboratory experiments on the metabolism of xenobiotics, problems of association between the substrate and the microbial cells may occur. Two illustrative examples that have been noted in Chapter 2, Section 2.2.6 may be given.

- Chloroguaiacols with Gram-positive bacteria (Allard et al. 1985);
- PCB congeners with the hyphae of *Phanerochaete chrysosporium* (Dietrich et al. 1995).

If this were not quantitatively evaluated or eliminated, the results and interpretation of such experiments would be seriously compromised.

4.6.4 Preexposure: Pristine and Contaminated Environments

Experimental aspects of the elective enrichment procedure are discussed in some detail in Chapter 5, but the question of its existence and significance in natural populations already exposed to xenobiotics is conveniently addressed here. It is important to distinguish between induction (or derepression) of catabolic enzymes and selection for a specific phenotype. The former is a relatively rapid response, so that exposure of samples from uncontaminated areas to xenobiotics for a period of weeks or months would be expected to result in the selection of organisms with degradative potential rather than merely be the result of low rates of enzyme induction. Caution should, however, be exercised in establishing a correlation between exposure to xenobiotics and the existence of organisms with the relevant degradative capacity. Diverse examples are given as illustration.

1. It has been shown (Kamagata et al. 1997) that bacteria isolated from a pristine site with no established contamination were capable of degrading 2,4-dichlorophenoxyacetate (2,4-D) and differed from those traditionally isolated from contaminated sites. The new isolates grew slowly, and although one of them could be assigned to the genus *Variovorax* and carried the *tfdA* gene, the other five did not and had no 2,4-D-specific 2–ketoglutarate-dependent dioxygenase activity.

2. These results are consistent with previous evidence (Fulthorpe et al. 1996) from a study of pristine soils that, although populations existed that could degrade both 3-chlorobenzoate and 2,4-dichlorophenoxyacetate, isolation of the latter strains was generally unsuccessful by the methods used. These results should be viewed against the general comments on oligotrophs and bradytrophs in Section 4.3.

3. Soils from putatively pristine areas in Southwest Australia, South Africa, California, Central Chile, Saskatchewan, and Russia were enriched with 3-chlorobenzoate and assayed for their capability to mineralize the substrate (Fulthorpe et al. 1998). Genetic procedures were used to show that 91% of the genotypes were unique to the sites from which the organisms were isolated. These results suggest that the genotypes were endemic and were not the result of global dispersion of a single genotype.

4. By comparison of organisms isolated from a contaminated and a pristine system in Canada, it was shown that, although the established genotypes for the degradation of 3-chlorobenzoate *clc-*, *cba-*, and *fcb-*, encoding enzymes for 3-chlorocatechol 2,3-dioxygenase (Frantz and Chakrabarty 1987), 3,4-(4,5)-dioxygenase (Nakatsu and Wyndmam 1993), and the hydrolytic pathway for 4-chlorobenzoate (Chang et al. 1992; see Section 6.5.1.2) were present in strains from the contaminated site, these were absent in strains isolated from the pristine site. Based on substrate utilization patterns, these traits were distributed among phenotypically distinct groups (Peel and Wyndham 1999).

Experiments by Tattersfield (1928) clearly showed that the biocidal effect of naphthalene decreased during successive applications and that this was due to biodegradation. Since then, many results have accumulated that support the view that preexposure increases the number of organisms capable of degrading a given xenobiotic, although fewer attempts have been made to quantify the number of organisms involved. There is convincing evidence that exposure to unusual substrates in laboratory experiments elicits the synthesis of genes for their degradation (Mortlock 1982), and increasing support for the view that exposure to xenobiotics increases the probability of mutations that are favorable to the degradation of these substrates (Hall 1990; Thomas et al. 1992). The following examples attempt to illustrate the spectrum of responses that have been observed.

1. Rates of mineralization of the more readily degraded PAHs such as naphthalene and phenanthrene were greater in samples from PAH-contaminated areas than in those from pristine sediments, although even in the former, the rates for benz[*a*]anthracene and benzo[*a*]pyrene were extremely low (Herbes and Schwall 1978). Examination of the sequences of 16S rRNA and of naphthalene dioxygenase Fe–S protein genes (*nahAc*) of nine strains of bacteria capable of degrading naphthalene and isolated from the same site showed that whereas in seven strains the former differed by as much as 7.9%, they had a single *nahAc* allele. All strains contained plasmids of different sizes that contained the gene for naphthalene degradation, and it was suggested that horizontal transfer of plasmids may play a role in the adaption of microbial communities to xenobiotics (Herrick et al. 1997).

2. Experiments using marine sediment slurries have examined the effect of preexposure to various aromatic hydrocarbons on the rate of subsequent degradation of the same or other hydrocarbons. The results clearly illustrated the complexity of the selection process; for example, whereas preexposure to benzene, naphthalene, anthracene, or phenanthrene enhanced the rate of mineralization of naphthalene, similar preexposure to naphthalene stimulated the degradation of phenanthrene, but had no effect on that of anthracene (Bauer and Capone 1988).

3. In experiments using soil samples from a pristine aquifer exposed in the laboratory to a range of compounds, widely diverse responses were observed (Aelion et al. 1987; 1989):
   - The bacterial population was apparently already adapted to some of the compounds such as phenol, 4-chlorophenol, and 1,2-dibromoethane at the start of the experiment, and these substrates were therefore rapidly degraded.
   - No adaptation was found with chlorobenzene or 1,2,4-trichlorobenzene, and only slight mineralization was observed.
   - A linear increase in the rate of degradation with increasing length of exposure was noted for some generally readily degraded compounds such as aniline.
   - True adaptation by selection of the appropriate organisms was observed only for 4-nitrophenol.

It therefore seems premature to draw general conclusions on the influence of preexposure on the biodegradability of structurally different substrates.

4. Systematic studies on the degradation of 4-nitrophenol (Spain et al. 1984) showed that the rates of adaptation in a natural system were comparable to those observed in a laboratory test system and were associated with an increase in the number of degrading organisms by up to 1000-fold.

5. It has been consistently observed that a wide range of agrochemicals applied successively to the same plots are increasingly readily degraded presumably due to enrichment of the appropriate degradative microorganisms (Racke and Coats 1990): examples include compounds as diverse as naphthalene (Gray and Thornton 1928), γ-hexachloro[*aaaeee*]cyclohexane (Wada et al. 1989), and triazines (Cook and Hütter 1981).

6. Dehalogenation of polychlorinated or polybrominated biphenyls was more rapid in cultures using inocula prepared from sediments contaminated with the chlorinated or brominated biphenyl, respectively (Morris et al. 1992).

These results are particularly relevant to bioremediation and suggest that organisms originally isolated from areas either contaminated naturally (Fredrickson et al. 1991) or as a result of industrial activity (Grosser et al. 1991) may be particularly attractive. The discussion hitherto has been devoted to the issue of selection, but for the sake of completeness, it should be noted that the alternative approach of deliberately adding an inducer has also been examined. Salicylate is an inducer of the enzymes for degradation of naphthalene, and its addition to soil has been shown to result in a modest increase in the number of organisms degrading naphthalene (Ogunseitan et al. 1991), and in the mineralization of some four- and five-ring PAHs (Chen and Aitken 1999). Since, however, naphthalene and phenanthrene are readily degradable compounds, the wider evaluation of this interesting idea would have to be explored before its application to practical situations could be justified (Chapter 8, Section 8.2.1).

This discussion may also be viewed against the background of studies on genetic transfer of catabolic activity toward a given xenobiotic. Considerable care should be exercised in the interpretation even of data which seem supportive of this process, since selection and enrichment from a small population of the organisms initially present must be excluded. One interesting investigation on the biodegradation of aniline revealed the existence of two genotypes differing in their tolerance to the substrate. The dominant organism which was originally present assimilated aniline at micromolar concentrations but was inhibited at higher concentrations; a mutant could, however, be isolated from a population of several hundred cells or by continuous culture and this organism tolerated millimolar concentrations. Populations of the two organisms in a natural system were apparently regulated by the prevailing concentration of aniline (Wyndham 1986).

In attempting to synthesize the results of all of these experiments, it is hardly possible to escape the conclusion that our understanding of the processes regulating the population dynamics of microorganisms in natural systems is limited.

4.7   Rates of Metabolic Reaction

4.7.1   Kinetic Aspects

The question of the rates at which xenobiotics are degraded or transformed is of cardinal importance, since it is upon their quantification that a given compound can, in the final analysis, be designated persistent or otherwise (Battersby 1990). It should be appreciated at the outset that, even if acceptable rates of degradation are observed in laboratory experiments, the final assessment of persistence depends upon the demonstration that the compound is indeed degraded under natural conditions. And in the long run, it is the latter fact that is of primary environmental significance. As will have been appreciated from earlier discussions, however, persistence is determined not only by rates of biotic and abiotic degradation but also by the accessibility of the substrate which may have been concentrated into biota or associated with organic or inorganic components in the water mass or in the oil and sediment phases. There are a number of important issues which should be addressed in discussions of rates and some of them are briefly discussed in the following paragraphs.

1. Even if rate constants can be measured in laboratory experiments, these must be normalized to the number of microbial cells. This may pose only minor problems with an axenic culture in the laboratory and has been consistently used in some investigations where well-defined kinetics prevail (Allard et al. 1987), but this becomes a major problem in natural situations: How many of the organisms are metabolically active in accomplishing the given reaction? Specific DNA probes have been used for detection of genes coding for heavy metal resistance (Diels and Mergeay 1990), and for the detection of pathogens (Samadapour et al. 1990). This procedure has, however, been less extensively applied to organisms of catabolic significance (Sayler et al. 1985; Holben et al. 1992) although the values obtained for 2,4-dichlorophenoxyacetic-degrading populations agreed well with

those using conventional most-probable-number methods (Holben et al. 1992). This is an attractive approach that is discussed further in Chapter 5, Section 5.9.

2. Although some plausible models have been assembled (Simkins and Alexander 1984) and their application has been evaluated (Simkins et al. 1986), microbial reactions may not follow well-defined kinetics. They may, for example, exhibit multiphase kinetics which has been illustrated during the transformation of methyl parathion by a *Flavobacterium* sp. (Lewis et al. 1985): system I was a high-affinity, low-capacity system, whereas system II was the opposite. In experiments with mixed microbial cultures on the degradation of phenol, 4-chlorophenol, 4-methylphenol, acetone, and methanol, multiphase kinetics were encountered. Failure to take this into account would have resulted in serious errors in evaluating the rates of degradation (Hwang et al. 1989). The results of these studies illustrate only one of the factors that may invalidate predictions which fail to take into account multiphase kinetics.

3. Attempts have been made to apply the structure–activity concept (Hansch and Leo 1979) to environmental problems, and this has been successfully applied to the rates of hydrolysis of carbamate pesticides (Wolfe et al. 1978b) and of esters of chlorinated carboxylic acids (Paris et al. 1984). This has been extended to correlating rates of biotransformation with the structure of the substrates and has been illustrated with a number of single-stage reactions. Clearly, this approach could be extensively refined if knowledge of the structure and function of the relevant enzyme was available. Some examples are sufficient to illustrate the application of this procedure.

   • Rates of bacterial hydroxylation of substituted phenols to catechols by *Pseudomonas putida* correlated well with the van der Waals' radii of the substituents (Paris et al. 1982), and this was also demonstrated for the biotransformation of anilines to catechols both by this strain and by a natural population of bacteria (Paris and Wolfe 1987).

   • Rates of hydrolysis of substituted aromatic amides by the bacterial population of pond water correlated well with the infrared C=O stretching frequencies of the substrates (Steen and Collette 1989).

   • Rates of anaerobic dechlorination of aromatic hydrocarbons (Peijnenburg et al. 1992) and of the hydrolysis of aromatic nitriles under anaerobic conditions (Peijnenburg et al. 1993) have been correlated with a number of parameters including Hammett σ-constants, inductive parameters, and evaluations of the soil/water partitioning.

There are, however, some important limitations to a general extension of this principle for a number of reasons: biodegradation — as opposed to biotransformation — of complex molecules necessarily involves a number of sequential reactions, each of whose rates may be determined by complex regulatory mechanisms. For novel compounds containing structural entities that have not previously been investigated, the level of prediction is necessarily limited by the lack of relevant data.

4. The significant and incompletely resolved issue of the possible occurrence of threshold substrate concentrations below which rates of biodegradation may become extremely low or even nonexistent has been noted in Section 4.6.2.

Too Olympian a view of the problem of rates should not, however, be adopted; an overly critical attitude should not be allowed to pervade the discussions — provided that the limitations of the procedures that are used are clearly appreciated and set forth. In view of the great practical importance of quantitative estimates of persistence to microbial attack any procedure — even if it provides merely orders of magnitude — should not be neglected.

4.7.2 Metabolic Aspects: Nutrients

In many natural ecosystems, microbial growth and metabolism are limited by the concentrations of inorganic nutrients such as nitrogen, phosphorus, or even iron. Systematic investigation of such limitations on the biodegradation of xenobiotics has seldom been carried out except when these contain N or P that is used for growth of the cells. The role of Fe, Mn, and Al as determinants of biodegradability is discussed in Chapter 5, Section 5.2.4. Only two examples of the importance of nutrient limitation in determining the realization of biodegradation are noted here.

1. Expression of lignin peroxidases in *Phanerochaete chrysosporium* is induced by N-limitation and also by the concentration of Mn(II) in the medium (Perez and Jeffries 1990); this is discussed further in Chapter 5, Section 5.2.4.

2. Under conditions of selenium starvation, *Clostridium purinilyticum* degrades uric acid by an unusual pathway involving cleavage of the iminazole ring to produce 5,6-diaminouracil which is then degraded to formate, acetate, glycine, and CO_2 (Dürre and Andreesen 1982).

In experiments using field samples containing natural assemblies of microorganisms, two effects of supplementation with nitrogen or phosphorus have been encountered: (1) a decreased lag phase was observed before transformation of 4-methylphenol (Lewis et al. 1986) and (2) the bacterial population increased, although there was no effect on the second-order rates of transformation of a number of compounds, including phenol and the agrochemicals propyl 2,4-dichlorophenoxyacetate, methyl parathion, and methoxychlor (Paris and Rogers 1986). Limitation in the concentrations of these inorganic nutrients does not therefore appear to have a dramatic effect on the persistence of the relatively few compounds that have been examined systematically. On the other hand, it has been suggested that competition between organisms for inorganic phosphorus may account for differences in biodegradability when several carbon sources are present (Steffensen and Alexander 1995).

Consistent with the preceding comments on the metabolism of xenobiotics in the presence of additional carbon substrates (Section 4.5.2), the situation with carbon additions may be quite complex and need not be addressed in detail again. Two simple examples may be given — in these, addition of glucose apparently elicited two differing responses although it should be emphasized that, since the concentration of such readily degradable substrates in natural aquatic systems will be extremely low, the environmental relevance of such observations will inevitably be restricted — (1) the rate of mineralization of phenol by the flora of natural lake water decreased (Rubin and Alexander 1983) and (2) the rate of mineralization of 4-nitrophenol was enhanced in lake water inoculated with a *Corynebacterium* sp. when rates of mineralization were low (Zaidi et al. 1988).

4.8 Regulation and Toxic Metabolites

4.8.1 Regulation

Whereas the following discussion is directed primarily to the role of metabolites, it may be noted that environmental factors may also be important (Section 4.6.1). Some examples are given as illustration, and these should be seen in the context of the comments in Section 4.6.1 and Chapter 8, Section 8.1.2.

1. In facultatively anaerobic yeasts such as *Candida* sp. that can use long-chain alkanes as growth substrates, the synthesis of cytochrome P-450 that is involved in hydroxylation of alkanes is induced by hexadecane, and the levels are regulated by the oxygen concentration and are maximal under oxygen limitation. Under these conditions, the synthesis of levels of alcohol dehydrogenase and aldehyde dehydrogenase is derepressed, and their levels and that of cytochrome P-450 are very much higher than during growth

with glucose (Gmünder et al. 1981). Induction of cytochrome P-450$_{14DH}$ under low oxygen concentration has been noted for *Saccharomyces cerevisiae* in Section 4.4.1.2. The positive effect of low oxygen concentration is also observed with facultatively anaerobic Enterobacteriaceae (Section 4.3.2), and is obligatory for organisms that use nitrate as an electron acceptor under anaerobic conditions (Section 4.3.3).

2. *Alcaligenes* sp. strain L6 was obtained after enrichment with 3-chlorobenzoate and 2% oxygen in the gas phase, and possessed 2,5- and 3,4-dihydroxybenzoate dioxygenase activities but lacked catechol dioxygenase (Krooneman et al. 1996). The affinity for O$_2$ and 3-methylcatechol of cells grown under oxygen limitation was three times greater than under excess oxygen, and exceeded values for other bacteria growing with benzoate or 2,5-dichlorobenzoate. Further details are given in Section 4.5.2.

3. Methanotrophs synthesize both soluble (sMMO) and particulate (pMMO) monooxygenases and the synthesis of the particulate methane monooxygenase is determined by the concentration of Cu in the medium; this is noted again in Chapter 5, Section 5.2.4.

Classic studies were devoted to the regulation of the enzymes for conversion of catechol and protocatechuate to β-ketoadipate by *Pseudomonas putida* (Ornston 1966), the β-ketoadipate pathway in *Moraxella calcoacetica* (Cánovas and Stanier 1967), and the mandelate pathway in *P. aeruginosa* (Rosenberg 1971). All of these organisms cleave the catechol by intradiol fission catalyzed by a 1,2-dioxygenase, and the induction patterns may be briefly summarized. Synthesis of the enzymes for the primary oxidations is induced by the growth substrates, whereas synthesis of those for later transformations is induced by their products — either by *cis,cis* muconate or by β–ketoadipate. On the other hand, for *Alcaligenes eutrophus* that metabolizes phenol and 4-methylphenol by extradiol fission catalyzed by a catechol 2,3-dioxygenase, all six enzymes that result in the production of 4-hydroxy-2-ketovalerate are inducible by the initial growth substrate (Hughes and Bayly 1983). The induction patterns may, however, be quite complex and strongly dependent on the growth substrate. For example, *P. paucimobilis* strain Q1 that was isolated by enrichment with biphenyl is capable of utilizing biphenyls, xylene and toluene, salicylate, and benzoates (Furukawa et al. 1983); all the enzymes for degradation of the first four of these that are degraded by a 2,3-dioxygenase ring cleavage were induced by growth on these substrates, whereas growth with benzoate induced only the 1,2-dioxygenase rather than the 2,3-dioxygenase. It is instructive to note that whereas the enzymes of the β-ketoadioate pathway are generally inducible, this is not always the case. Those in slow-growing species of *Bradyrhizobium* spp. are constitutive (Parke and Ornston 1986). Further details of the experimental procedures are given in Chapter 5, Section 5.5.2, and of the pathways in Chapter 6, Section 6.2.1. By comparison, relatively fewer studies have been directed primarily to the genetics and the regulation of the enzymes for the degradation of xenobiotics with the important exception of aromatic hydrocarbons. A few examples are merely noted as illustration.

* In *P. putida* strain F1 that degrades toluene with the methyl group intact (Finette et al. 1984);
* For the *nah* operon in *P. putida* strain G1064 involved in degrading naphthalene to salicylate (Eaton and Chapman 1992), the enzymes are generally induced by growth with salicylate (Austen and Dunn 1980);
* The initial steps in the degradation of biphenyls by *Pseudomonas* sp. strain LB400 (Mondello 1989), by *P. pseudoalcaligenes* strain KF707 (Taira et al. 1992), and by *Rhodococcus globerulus* (Asturias and Timmis 1993).

Further details of the pathways for the degradation of PAHs are described in Chapter 6, Section 6.2.1 and in a review (Neilson and Allard 1998). It seems that most of the degradative enzymes are inducible, and this is consistent with the fact that most strains have been isolated after specific enrichment with the xenobiotic. The case of the partially constitutive synthesis of catechol 1,2-dioxygenase in the yeast *Trichosporon cutaneum* (Shoda and Udaka 1980) has been noted (Section 4.5.2). In the case of biotransformation, however, there are sporadic examples of the constitutive synthesis of enzymes. For example, the system carrying out the *O*-methylation of halogenated phenolic compounds was apparently constitutive (Neilson et al. 1988); this observation is consistent with the isolation of the strains by enrichment with C_1 compounds structurally unrelated to the halogenated substrates. The *O*-methylation reaction may function primarily as a detoxification system, so that in this case constitutive synthesis of the enzyme would clearly be advantageous to the survival of the cells.

Since it is only seldom that organisms are exposed to a single substrate, attention has been directed to the metabolism of mixtures of benzoate and aliphatic carboxylic acids. In *Ralstonia eutropha* (*Alcaligenes eutrophus*), diauxic growth is exhibited when both benzoate and actetate are supplied as substrates but not when benzoate and succinate are used (Ampe et al. 1997). In the latter situation, the growth rate is increased by the presence of both substrates and under these circumstances, there was probably a more optimal distribution of metabolites among the various anabolic and catabolic pathways. By contrast, the simultaneous presence of acetate and benzoate would result in the production of energy that would exceed the requirements of the cell. On the other hand, during growth with both acetate and phenol, synthesis of both phenol hydroxylase and the extradiol dioxygenase involved in the metabolism of catechol decreased, and benzoate blocked the diauxic growth with phenol (Ampe et al. 1998).

An interesting and ecologically relevant observation has been made on the induction of metabolism of PCBs by *Arthrobacter* sp. strain B1B (Gilbert and Crowley 1997). Cells were grown in a mineral medium with fructose and carvone (50 mg/l). Effective degradation of a number of congeners in Arochlor 1242 was induced by carvone that could not, however, be used as a growth substrate and was toxic at high concentrations (>500 mg/l). Other structurally related compounds including limonene, *p*-cymene, and isoprene were also effective, and such results may be relevant to bioremediation programs for PCBs. It has also been shown by protein analysis sequencing in *Comamonas testosteroni* that testosterone induces the synthesis of both steroid-catabolizing enzymes and those for the degradation of aromatic compounds (Möbus et al. 1997).

The induction of the monooxygenases for the degradation of trichloroethene by aromatic substrates, and vice versa, has already been discussed (Sections 4.4.1.1 and 4.5.2).

4.8.2   Toxic or Inhibitory Metabolites

There are a number of examples in which metabolites are produced which themselves toxify the organism responsible for their synthesis. The classic example is fluoroacetate (Peters 1952), which enters the TCA cycle and is thereby converted into fluorocitrate. This compound effectively inhibits aconitase — the enzyme involved in the next metabolic step — so that cell metabolism itself is inhibited with the resulting death of the cell. It should be noted, however, that bacteria able to degrade fluoroacetate to fluoride exist so that some organisms have developed the capability for overcoming this toxicity (Meyer et al. 1990).

The significance of toxic metabolites is important in a number of diverse metabolic situations: (1) when a pathway results in the synthesis of a toxic or inhibitory metabolite and (2) when pathways for the metabolism of two (or more) analogous substrates supplied simultaneously are incompatible due to the production of a toxic metabolite by one of the substrates. A number of examples are provided to illustrate these possibilities that have achieved considerable attention in the context of the biodegradation of chloroaromatic compounds, and further discussion is provided in Chapter 6, Section 6.5.1.2. The issue of toxic or inhibitory metabolites also merits attention in a wider context (Chapter 7, Section 7.5).

1. Biotransformation of chlorobenzene by *P. putida* grown on toluene or benzene resulted in the formation of 3-chlorocatechol; this inhibited further metabolism by catechol 2,3-dioxygenase, so that its presence resulted in the formation of catechols even from benzene and toluene (Gibson et al. 1968; Klecka and Gibson 1981; Bartels et al. 1984). The same situation has emerged in the degradation of 3-chlorobenzoate: the inhibition that would result from

the inhibitory effect of 3-chlorocatechol on catechol 2,3-dioxygen-
ase is lifted by the synthesis of catechol 1,2-dioxygenase (Reinecke
et al. 1982). This is indeed a general strategy that is discussed again
in Chapter 6, Section 6.5.1.2. Inactivation of *extra*diol dioxygenases
— reversible or otherwise — by other substituted catechols has
also been reported: (a) 2,3-dihydroxybiphenyl 1,2-dioxygenase
from a biphenyl-degrading strain of *Pseudomonas* sp. by 4-phenyl
catechol that is not a substrate for 2,3-dihydroxybiphenyl-1,2-diox-
ygenase (Lloyd-Jones et al. 1995) and (b) catechol-2,3-dioxygenase
by 4-ethyl catechol that is a metabolite of 4-ethylbenzoate (Ramos
et al. 1987). Although this inactivation may present a limitation to
the degradative potential of the relevant strains, *P. putida* strain
GJ31 degrades chlorobenzene via 3-chlorocatechol and extradiol
fission (Kaschabek et al. 1998). This is accomplished by a chloro-
catechol-2,3-dioxygenase that hydrolyses the initially formed
cis,cis-hydroxymuconic acid chloride to 2-hydroxymuconate;
thereby the irreversible reaction of the acid chloride with nucleophiles or the formation of pyr-2-one-6-carboxylate as a terminal
metabolite is avoided. The catechol-2,3-dioxygenase from this
strain encoded by *cbzE* is plasmid- borne and is capable metabolizing both 3-chlorocatechol and 3-methylcatechol: it belongs to the
2.C subfamily of type 1 extradiol dioxygenases (Mars et al. 1999).
The alternative extradiol fission of 3-chlorocatechol may take place
between the 1 and 6 positions, and this has been shown for the 2,3-
dihydroxybiphenyl-1,2-dioxygenase from the naphthalene sulfonate–degrading *Sphingomonas* sp. strain BN6 (Riegert et al. 1998).

The converse situation occurs during the degradation of 4-chlo-
robenzoate in which the synthesis of catechol-1,2-dioxygenase circumvents the production of chloroacetaldehyde by the action of
catechol-2,3-dioxygenase (Reinecke et al. 1982). It has been shown
that protoanemonin may be produced during the degradation of
4-chlorobenzoate that is formed by ring fission of 4-chlorobiphenyl
(Blasco et al. 1995), and that this may inhibit the growth of PCB-
degrading organisms in the soil (Blasco et al. 1997). Although protoanemonin is produced from chloro-*cis-cis*-muconates, its formation may be circumvented by the action of chloromuconate
cycloisomerases to form dienlactones, and by conversion by
dienelactone hydrolase to *cis*-acetylacrylate (Brückmann et al.
1998). (Figure 4.60).

2. Ring fission during degradation of chloroaromatic compounds is
generally catalyzed by catechol-1,2-dioxygenases, whereas for the
corresponding methyl compound a catechol-2,3-dioxygenase is
involved. These pathways are generally incompatible due to the
inactivation of catechol-2,3-oxygenase by 3-chlorocatechol, although
simultaneous degradation of chloro- and methyl-substituted

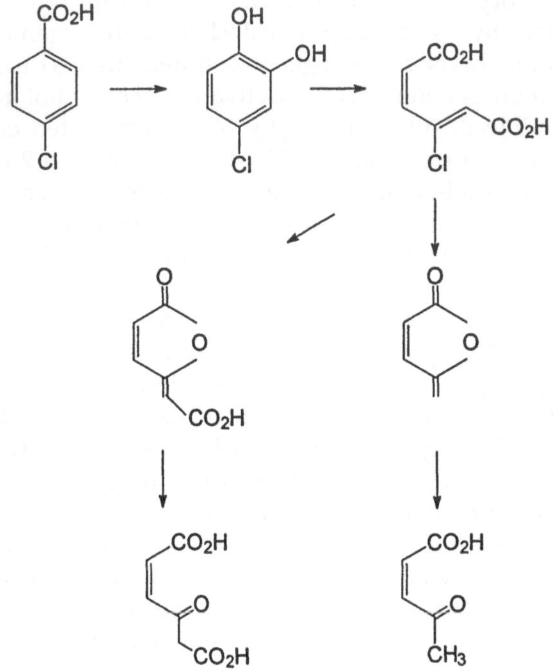

FIGURE 4.60
Biodegradation of 4-chlorobenzoate.

aromatic compounds has been shown to occur in some strains (Taeger 1988; Pettigrew et al. 1991), and a strain of *Ralstonia* sp. strain
JS705 carries genes for both chlorocatechol degradation and toluene
dioxygenation (van der Meer et al. 1998). Although *Comamonas testosteroni* strain JH5 degraded 4-chlorophenol and 2-methylphenol
simultaneously, 4-chlorophenol and 4-methylphenol were degraded
sequentially; a catechol-2,3-dioxygenase for both 4-methylcatechol
and 4-chlorocatechol was induced by growth with 4-chlorophenol
or 4-methylphenol (Hollender et al. 1994). *Phanerochaete chrysosporium* is able to degrade simultaneously chlorobenzene and toluene
(Yadav et al. 1995). This phenomenon is highly relevant to the biological treatment of industrial effluents since most of these consist
of complex mixtures of substrates, and to bioremediation that is
discussed in Chapter 8, Section 8.1.2.

3. Total degradation of PCBs necessitates degradation of the chlorobenzoates produced by dioxygenation, dehydrogenation, and
ring cleavage reactions. Metabolism of chlorobenzoate may produce chlorocatechols and thence muconic acids, and it has been
shown with *P. testosteroni* strain B-356 that the metabolites from
3-chlorobenzoate strongly inhibited the activity of 2,3-dihydroxy-

biphenyl 1,2-dioxygenase, and therefore the degradation of the original substrates (Sondossi et al. 1992). This inhibition is reminiscent of the inhibition of catechol-2,3-dioxygenase by 3-chlorocatechol that is noted earlier.

4. Simultaneous metabolism of closely related substrates may be restricted by the synthesis of inhibitory metabolites. For example, cells of an *Acinetobacter* sp. grown with 4-chlorobenzoate could dehalogenate 3,4-dichlorobenzoate although the organism cannot use this as sole growth substrate, and the dichlorobenzoate inhibited both its own metabolism and that of the growth substrate 4-chlorobenzoate (Adriaens and Focht 1991a,b).

5. A strain of *Pseudomonas* sp. grew with a wide range of aromatic compounds including phenol, benzoate, benzene, toluene, naphthalene, and chlorobenzene (Haigler et al. 1992), and mixtures of these substrates were degraded in continuous cultures without evidence for accumulation of metabolites. In this case, degradation depended on the presence of a nonspecific toluene dioxygenase, and of induction of enzymes for the intradiol (*ortho*) fission, extradiol (*meta*) fission, and the modified intradiol fission (*ortho*) pathways for degradation of the catechols. The results implied the existence of a metabolic situation additional to that normally encompassed by the terms *biodegradation* and *biotransformation*: enzymes induced by the presence of one substrate facilitated the degradation of another substrate that is not normally used as sole source of carbon and energy. These results have obvious implications for implementing bioremediation programs.

6. There may be several reasons an analogue substrate cannot be metabolized by an organism. This is well illustrated by 4-ethylbenzoate which could not be used by strains that degraded 3- and 4-methylbenzoate (Ramos et al. 1987). There were two reasons for this: (a) 4-ethylcatechol that is a metabolite inactivates catechol-2,3-dioxygenase that is obligatory for its degradation and (b) 4-ethylcatechol, in contrast to 3- and 4-methylcatechol, does not activate the xylS protein whose gene is the positive regulator of the promoter of the TOL plasmid extradiol fission pathway. These limitations could, however, be overcome by construction of mutant strains.

7. The degradation of trichloroethene by methylotrophic bacteria involves the epoxide as intermediate (Little et al. 1988); further transformation of this results in the production of CO which may toxify the bacterium, both by competition for reductant and by enzyme inhibition (Henry and Grbic-Galic 1991). The inhibitory effect of CO may, however, be effectively overcome by adding a reductant such as formate.

8. The growth of a strain of *Alcaligenes* sp. that could degrade
   2-aminobenzenesulfonate was inhibited by equimolar concentrations of 3-methylbenzoate, but an exconjugant after insertion of the
   TOL catabolic genes was able to carry out sequential degradation
   of both substrates (Jahnke et al. 1993).

4.9 Catabolic Plasmids

Plasmids may be defined as fragments of DNA which replicate outside the
bacterial chromosome, and they are important in a number of different
contexts:

- As carriers of antibiotic resistance: The emergence of antibiotic-
  resistant strains has had serious repercussions in the application
  of antibiotic therapy and has seriously increased the danger of
  nosicomial infections.
- The presence of unusual carbohydrate fermentation patterns (particularly for lactose) and the ability to use citrate among Enterobacteriaceae has hindered — and in some cases jeopardized — the
  identification of pathogenic strains including *Salmonella typhi*.
- Resistance to heavy metals such as Hg may be mediated by plasmid-borne genes.
- Genes coding for the catabolism of a large number of diverse xenobiotics are carried on plasmids: This is highly germane to the
  present discussion, and some examples are given in Table 4.1.

It should be noted, however, that many bacteria harbor plasmids with no
hitherto established function.

It is worth pointing out that genes for specific substrates may be borne either
on plasmids or chromosomally. The following examples serve as illustration.

1. The degradation of toluene: In this example, different pathways
   are used by the plasmid and chromosomally borne genes — side-
   chain oxidation to benzoate and ring 2,3-dioxygenation, respectively (Chapter 6, Section 6.2.1).
2. The degradation of trichloroethene by *Alcaligenes eutrophus* JMP134
   is carried out by either a chromosomal phenol-dependent pathway
   or by the plasmid-borne 2,4-dichlorophenoxyacetate pathway
   (Harker and Kim 1990).
3. The degradation of 2,4-dichlorophenoxyacetate: The initial 2–ketoglutarate-dependent dioxygenase that results in the production of

2,4-dichlorophenol (Section 4.4.2) is generally plasmid borne, although chromosomally borne genes have been identified in *Burkholderia* sp. strain RASC (Suwa et al. 1996), and the transfer of both the plasmid-borne and chromosomally borne *tfdA* gene may contribute to the dissemination of the capability to degrade 2,4-dichlorophenoxyacetate (Matheson et al. 1996; McGowan et al. 1998).

TABLE 4.1

Examples of Plasmids Carrying Genes Coding for the Biodegradation or Biotransformation of Xenobiotics

Hydrocarbons and Related Compounds

Octane	Chakrabarty et al. 1973
Camphor	Rheinwald et al. 1973
Citronellol, geraniol	Vandenbergh et al. 1983
Linalool	Vandenbergh et al. 1986
Benzene	Tan and Mason 1990
Toluene and xylene	Williams and Worsey 1976
Naphthalene	Connors and Barnsley 1982
Naphthalene, phenanthrene, and anthracene	Sanseverino et al. 1993
Dibenzothiophene	Monticello et al. 1985

Halogenated compounds

Chloroalkanoates	Kawasaki et al. 1981; Hardman et al. 1986
Dichloromethane	Gälli and Leisinger 1988
Chlorobenzoates	Chatterjee et al. 1981
Chlorophenoxyacetates	Don and Pemberton 1985; Chaudry and Huang 1988
Chlorobiphenyls	Furukawa and Chakrabarty 1982; Shields et al. 1985
Chloridazon	Kreiss et al. 1981
Bromoxynil	Stalker and McBride 1987
Chlorobenzenes	van der Meer et al. 1991

Other Structural Groups

6-Aminohexanoate cyclic dimer	Negoro et al. 1980; Kanagawa et al. 1989
Parathion	Serdar et al. 1982; Mulbry et al. 1986
Nicotine	Brandsch et al. 1982
Cinnamic acid	Andreoni and Bestetti 1986
Ferulic acid	Andreoni and Bestetti 1986
S-Ethyl-*N,N*-dipropylthiocarbamate	Tam et al. 1987
Aniline	Anson and Mackinnon 1984; McClure and Venables 1987; Karns and Eaton 1997
2-Aminobenzenesulfonate	Jahnke at al. 1990
1,1'-dimethyl-4,4'-bipyridinium dichloride	Salleh and Pemberton 1993
Arylsulfonates	Junker and Cook 1997

Probably the greatest interest in catabolic plasmids stems from the possibility of constructing strains having increased metabolic potential toward xenobiotics, and from the potential application of such strains to waste treatment systems. Considerable effort has been devoted to aromatic chlorinated compounds, and conspicuous success has been achieved in overcoming problems with, for example, synthesis of toxic metabolites such as 3-chlorocatechol during the degradation of chlorobenzenes by natural strains. There are several important issues which should, however, be addressed before considering the application of such artificially constructed strains to biological treatment systems:

1. Competition with endemic strains which may eventually outnumber and eliminate the introduced strains;
2. Genetic instability if selection pressure is removed through fluctuations in the loading of the xenobiotic;
3. Concern over the discharge of such strains into the environment; attempts have therefore been made to incorporate safeguards so that the strains are unlikely to survive in competition with natural strains in the ecosystem.

The questions of plasmid transmission and the stability of plasmids in natural ecosystems have received somewhat greater attention, but caution should be exercised in drawing general conclusions on the basis of the sometimes fragmentary evidence from laboratory experiments. Some important principles may be illustrated with the following observations.

1. A study with a strain of plasmid-borne antibiotic resistant strain of *Escherichia coli* indicated that the strain did not transmit these plasmids to indigenous strains after introduction into the terrestrial environment (Devanas et al. 1986).
2. In enteric bacteria carrying thermosensitive plasmids coding for the utilization of citrate and for resistance to antibiotics, it has been shown (Smith et al. 1978) that rates of transmission are negligible at 37°C but appreciable at 23°C — a temperature more nearly approaching that which prevails in some natural ecosystems.
3. There is evidence that, even in the absence of selective pressure exerted by the presence of a xenobiotic, bacterial populations may retain a small number of organisms carrying the relevant degradative plasmids. For example, strains of *Pseudomonas putida* in which the degradation of toluene is mediated by genes on the nonconjugative TOL plasmid, maintain a small population of cells carrying the plasmid even in the absence of toluene (Keshavarz et al. 1985). It has also been shown (Duetz et al. 1994) that cultures of *P. putida* grown with growth-limiting concentrations of succinate express TOL catabolic genes in both the upper and lower pathways in

response to *o*-xylene (that is not metabolized) but fail to do so during nonlimited growth with succinate.

4. Strains of indigenous groundwater bacteria, and a strain of *P. putida* carrying a TOL plasmid were introduced into microcosms prepared from a putatively pristine aquifer. DNA-specific probes were used to monitor the numbers of organisms carrying the genotypes, and it was found that the stability of genotypes for the degradation of toluene was maintained in the absence of selective pressure over the 8-week period of the experiment (Jain et al. 1987).

5. A strain of *Alcaligenes* sp. carrying a plasmid bearing the genes for the degradation of 3-chlorobenzoate was introduced into a freshwater microcosm system, and a specific DNA probe was used to enumerate the organisms bearing this gene (Fulthorpe and Wyndham 1989). A number of generally important results were obtained:

   • The presence of 3-chlorobenzoate was needed to maintain the catabolic genotype.

   • The number of probe-positive organisms often greatly outnumbered that of the original organism determined by plate counts.

   • The nature of these "additional" probe-positive organisms is unknown.

6. It has been noted in Section 4.6.4 that detailed analysis of bacteria isolated from the sediment of a coal-tar contaminated site demonstrated horizontal transfer of genes involved in naphthalene degradation, and that identical alleles NahAc were shared among seven taxonomically diverse hosts (Herrick et al. 1997). Analysis of restriction fragment length polymorphism (RFPL) (Chapter 5, Section 5.8) showed the existence of two catabolic plasmids in 12 of the isolates and that these were closely related both to each other and to the plasmid pDGT1 of *P. putida* strain NCIB 9816-4. The plasmids were stable in this environment, and it was therefore suggested that these plasmids were involved in the development of the naphthalene-degrading community as a result of the selective pressure exerted by the coal-tar contamination (Stuart-Keil et al. 1998).

Clearly, therefore, further investigation is required before generalizations can be made on the cardinal issues of the stability of plasmids in natural ecosystems, the extent to which these plasmids are transmissible, and the stability of the genotypes in the absence of selective pressure. At least some of the apparently conflicting views may be attributed to the different organisms that have been used including their nutritional demands — compounded by the widely varying environments in which their stability has been examined (Sobecky et al. 1992). Currently, the greatest volume of research is apparently being devoted to purely genetic aspects of these problems.

4.10   Conclusions

It may be valuable to summarize the broader conclusions that may be drawn from the extensive literature.

1. It is imperative to distinguish clearly between the degradation and the transformation of xenobiotics: this applies equally to biotic and abiotic processes. Failure to do so may result in erroneous estimates of the fate of a compound and of its toxicity that may be mediated by its metabolites.

2. Conventional assays for determining the biodegradability of a xenobiotic do not provide a rational basis for extrapolating the results to conditions that occur in the natural environment.

3. Abiotic reactions including those induced photochemically may be important in degradation and in addition may act in concert with microbial processes.

4. A wide range of taxonomically distinct organisms is able to degrade or transform xenobiotics, and reactions carried out under anaerobic conditions merit continued attention. Sequential transformations reactions carried out by different organisms may be necessary to accomplish degradation of the initial substrate, and attempts to mimic such situations in laboratory experiments have been illuminating.

5. The extent to which a xenobiotic is susceptible to microbial attack depends on a number of environmental factors including the concentration of the substrate, its bioavailability, and the presence of other substrates as well as on physical parameters such as temperature, salinity, and pH.

6. Attention should be directed to the complex reactions that take place in the natural environment and in particular to the situation that exists when microorganisms are exposed to several substrates at different concentration and to the complex growth kinetics that may result.

References

Abeliovich, A. and A. Vonshak. 1992. Anaerobic metabolism of *Nitrosomonas europaea*. *Arch. Microbiol.* 158: 267–270.

Adriaens, P. 1994. Evidence for chlorine migration during oxidation of 2-chlorobiphenyl by a type II methanotroph. *Appl. Environ. Microbiol.* 60: 1658–1662.

Adriaens, P. and D.D. Focht. 1991a. Evidence for inhibitory substrate interactions during cometabolism of 3,4-dichlorobenzoate by *Acinetobacter* sp. strain 4-CB1. *FEMS Microbiol. Ecol.* 85: 293–300.

Adriaens, P. and D.D. Focht. 1991b. Cometabolism of 3,4-dichlorobenzoate by *Acinetobacter* sp. strain 4-CB1. *Appl. Environ. Microbiol.* 57: 173–179.

Aeckersberg, F., F. Bak, and F. Widdel. 1991. Anaerobic oxidation of saturated hydrocarbons to CO_2 by a new type of sulfate-reducing bacterium. *Arch. Microbiol.* 156: 5–14.

Aeckersberg, F., F.A. Rainey, and F. Widdel. 1998. Growth, natural relationships, cellular fatty acids and metabolic adaptation of sulfate-reducing bacteria that utilize long-chain alkanes under anoxic conditions. *Arch. Microbiol.* 170: 361–369.

Aelion, C.M., C.M. Swindoll, and F.K. Pfaender. 1987. Adaptation to and biodegradation of xenobiotic compounds by microbial communities from a pristine aquifer. *Appl. Environ. Microbiol.* 53: 2212–2217.

Aelion, C.M., D.C. Dobbins, and F.K. Pfaender. 1989. Adaptation of aquifer microbial communities to the biodegradation of xenobiotic compounds: influence of substrate concentration and preexposure. *Environ. Toxicol. Chem.* 8: 75–86.

Afring, R.P. and B.F. Taylor. 1981. Aerobic and anaerobic catabolism of phthalic acid by a nitrate-respiring bacterium. *Arch. Microbiol.* 130: 101–104.

Ahn, I.-S., W.C. Ghiorse, L.W. Lion, and M.L. Shuler. 1998. Growth kinetics of *Pseudomonas putida* G7 on naphthalene and toxicity during nutrient deprivation. *Biotechnol. Bioeng.* 59: 587–594.

Aitken, M.D., W.T. Stringfellow, R.D. Nagel, C. Kazuga, and S.-H. Chen. 1998. Characteristics of phenanthrene-degrading bacteria isolated from soils contaminated with polycyclic aromatic hydrocarbons. *Can. J. Microbiol.* 44: 743–752.

Alexander, M. 1985. Biodegradation of organic chemicals. *Environ. Sci. Technol.* 18: 106–111.

Allard, A.-S., M. Remberger, and A.H. Neilson. 1985. Bacterial O-methylation of chloroguaiacols: effect of substrate concentration, cell density and growth conditions. *Appl. Environ. Microbiol.* 49: 279–288.

Allard, A.-S., M. Remberger, and A.H. Neilson. 1987. Bacterial O-methylation of halogen-substituted phenols. *Appl. Environ. Microbiol.* 53: 839–845.

Allard, A.-S., P.-Å. Hynning, C. Lindgren, M. Remberger, and A.H. Neilson. 1991. Dechlorination of chlorocatechols by stable enrichment cultures of anaerobic bacteria. *Appl. Environ. Microbiol.* 57: 77–84.

Allard, A.-S., P.-Å. Hynning, M. Remberger, and A.H. Neilson. 1992. Role of sulfate concentration in dechlorination of 3,4,5-trichlorocatechol by stable enrichment cultures grown with coumarin and flavanone glycones and aglycones. *Appl. Environ. Microbiol.* 58: 961–968.

Allard, A.-S., P.-Å. Hynning, M. Remberger, and A.H. Neilson. 1994. Bioavailability of chlorocatechols in naturally contaminated sediment samples and of chloroguaiacols covalently bound to C_2-guaiacyl residues. *Appl. Environ. Microbiol.* 60: 777–784.

Alonso, J., J.L. Barredo, B. Díez, E. Mellado, F. Salto, J.L. García, and E. Cortés. 1998. D-Amino-acid oxidase gene from *Rhodotorula gracilis* (*Rhodosporidium toruloides*) ATCC 26217. *Microbiology* (U.K.) 144: 1095–1101.

Alsberg, T., U. Stenberg, R. Westerholm, M. Strandell, U. Rannug, A. Sundvall, L. Romert, V. Bernson, B. Petterson, R. Toftgård, B. Franzén, M. Jansson, J.Å. Gustafsson, K.E. Egebäck, and G. Tejle. 1985. Chemical and biological characterization of organic material from gasoline exhaust particles. *Environ. Sci. Technol.* 19: 43–50.

Altenschmidt, U. and G. Fuchs. 1991. Anaerobic degradation of toluene in denitrifying *Pseudomonas* sp.: indication for toluene methylhydroxylation and benzoyl-CoA as central aromatic intermediate. *Arch. Microbiol.* 156: 152–158.

Amador, J.A. and B.F. Taylor. 1990. Coupled metabolic and photolytic pathway for degradation of pyridinedicarboxylic acids, especially dipicolinic acid. *Appl. Environ. Microbiol.* 56: 1352–1356.

Ampe, F., J.-L. Uribelarrea, G.M.F. Aragao, and N.D. Lindley. 1997. Benzoate degradation via the *ortho* pathway in *Alcaligenes eutrophus* is perturbed by succinate. *Appl. Environ. Microbiol.* 63: 2765–2770.

Ampe, F., D. Léonard, and N.D. Lindley. 1998. Repression of phenol catabolism by organic acids in *Ralstonia eutropha*. *Appl. Environ. Microbiol.* 64: 1–6.

An, D., D.T. Gibson, and J.C. Spain. 1994. Oxidative release of nitrite from 2-nitrotoluene by a three component enzyme system from *Pseudomomas* sp. strain JS42. *J. Bacteriol.* 176: 7462–7467.

Anantharam, V., M.J. Allison, and P.C. Maloney. 1989. Oxalate: formate exchange. *J. Biol. Chem.* 264: 7244–7250.

Anders, H.-J., A. Kaetzke, P. Kämpfer, W. Ludwig, and G. Fuchs. 1995. Taxonomic position of aromatic-degrading denitrifying pseudomonad strains K 172 and KB 740 and their description as new members of the genera *Thauera*, as *Thauera aromatica* sp.nov., and *Azoarcus*, as *Azoarcus evansii* sp. nov., respectively, members of the beta subclass of the *Protobacteria*. *Int. J. Syst. Bacteriol.* 45: 327–333.

Anderson, J.J. and S. Dagley. 1981a. Catabolism of aromatic acids in *Trichosporon cutaneum*. *J. Bacteriol.* 141: 534–543.

Anderson, J.J. and S. Dagley. 1981b. Catabolism of tryptophan, anthranilate, and 2,3-dihydroxybenzoate in *Trichosporon cutaneum*. *J. Bacteriol.* 146: 291–297.

Anderson, P.N. and R.A. Hites. 1996. OH radical reactions: the major removal pathway for polychlorinated biphenyls from the atmosphere. *Environ. Sci. Technol.* 30: 1756–1763.

Andreae, M.O. and Crutzen. 1997. Atmospheric aerosols: biogeochemical sources and role in atmospheric chemistry. *Science* 276: 1052–1058.

Andreoni, V. and G. Bestetti. 1986. Comparative analysis of different *Pseudomonas* strains that degrade cinnamic acid. *Appl. Environ. Microbiol.* 52: 930–934.

Anson, J.G. and G. Mackinnon. 1984. Novel plasmid involved in aniline degradation. *Appl. Environ. Microbiol.* 48: 868–869.

Aoyama, Y., Y. Yoshida, and R. Sato. 1984. Yeast cytochrome P-450 catalyzing lanosterol 14α-demethylation II. Lanosterol metabolism by purified P-450$_{14DM}$ and by intact micrososmes. *J. Biol. Chem.* 259: 1661–1666.

Apajalahti, J.H. and M.S. Salkinoja-Salonen. 1987. Dechlorination and *para*-hydroxylation of polychlorinated phenols by *Rhodococcus chlorophenolicus*. *J. Bacteriol.* 169: 675–681.

Apajalahti, J.H., P. Kärpänoja, and M.S. Salkinoja-Salonen. 1986. *Rhodococcus chlorophenolicus* sp. nov., a chlorophenol-mineralizing actinomycete. *Int. J. Syst. Bacteriol.* 36: 246–251.

Archelas, A., J.-D. Fourneron, R. Furstoss, M. Cesario, and C. Pascard. 1988. Microbial transformations. 8. First example of a highly enentioselective microbiological hydroxylation process. *J. Amer. Chem. Soc.* 53: 1797–1799.

Arey, J. 1998. Atmospheric reactions of PAHs oncluding formation of nitroarenes, pp. 347–385. In *PAHs and Related Compounds* (Ed. A.H. Neilson). *Handbook of Environmental Chemistry* (Ed. O. Hutzinger) Vol. 3, Part I. Springer-Verlag, Berlin.

Arino, S., R. Marchal, and J.-P. Vandecasteele. 1998. Involvement of a rhamnolipid-producing strain of *Pseudomonas aeruginosa* in the degradation of polycycic aromatic hydrocarbons by a bacterial community. *J. Appl. Microbiol.* 84: 769–776.

Arjmand, M. and H. Sandermann. 1985. Mineralization of chloroaniline/lignin conjugates and of free chloroanilines by the white rot fungus *Phanerochaete chrysosporium. J. Agric. Food Chem.* 33: 1055–1060.

Arras, T., J. Schirawski, and G. Unden. 1998. Availability of O_2 as a substrate in the cytoplasm of bacteria under aerobic and microaerobic conditions. *J. Bacteriol.* 180: 2133–2136.

Arunachalam, U., V. Massey, and C.S. Vaidyanathan. 1992. *p*-Hydroxyphenacetate-3-hydroxylase. A two-component enzyme. *J. Biol. Chem.* 267: 25848–25855.

Asperger, O., A. Naumann, and H.-P. Kleber. 1981. Occurrence of cytochrome P-450 in *Acinetobacter* strains after growth on *n*-hexadecane. *FEMS Microbiol. Lett.* 11: 309–312.

Asturias, J.A. and K.N. Timmis. 1993. Three different 2,3-dihydroxybiphenyl-1,2-dioxygenase genes in the Gram-positive polychlorobiphenyl-degrading bacterium *Rhodococcus globerulus* P6. *J. Bacteriol.* 175: 4631–4640.

Atkinson, R. 1986. Kinetics and mechanisms of the gas-phase reactions of the hydroxyl radical with organic compounds under atmospheric conditions. *Chem. Rev.* 86: 29–201.

Atkinson, R. 1990. Gas-phase troposphere chemistry of organic compounds: a review. *Atmos. Environ.* 24A: 1–41.

Atkinson, R. and W.P.L. Carter. 1984. Kinetics and mechanisms of the gas-phase reactions of ozone with organic compounds under atmospheric conditions. *Chem. Rev.* 84: 437–470.

Atkinson, R., S.M. Aschmann, A.M. Winer, and J.N. Pitts. 1985a. Kinetics and atmospheric implications of the gas-phase reactions of NO_3 radicals with a series of monoterpenes and related organics at 294 ± 2 K. *Environ. Sci. Technol.* 19: 159–163.

Atkinson, R., S.M. Aschmann, A.M. Winer, and W.P.L. Carter. 1985b. Rate constants for the gas-phase reactions of NO_3 radicals with furan, thiophene, and pyrrole at 295 ± 1 K and atmospheric pressure. *Environ. Sci. Technol.* 19: 87–90.

Atkinson, R., E.C. Tuazon, T.J. Wallington, S.M. Aschmann, J. Arey, A.M. Winer, and J.N. Pitts. 1987a. Atmospheric chemistry of aniline, *N,N*-dimethylaniline, pyridine, 1,3,5-triazine and nitrobenzene. *Environ. Sci. Technol.* 21: 64–72.

Atkinson, R., J. Arey, B. Zielinska, and S.M. Aschmann. 1987b. Kinetics and products of the gas-phase reactions of OH radicals and N_2O_5 with naphthalene and biphenyl. *Environ. Sci. Technol.* 21: 1014–1022.

Atkinson, R., S.M. Aschmann, and A.W. Winer. 1987c. Kinetics of the reactions of NO_3 radicals with a series of aromatic compounds. *Environ. Sci. Technol.* 21: 1123–1126.

Austen, R.A. and N.W. Dunn. 1980. Regulation of the plasmid-specified naphthalene catabolic pathway of *Pseudomonas putida. J. Gen. Microbiol.* 117: 521–528.

Baarschers, W.H., A.I. Bharaty, and J. Elvish. 1982. The biodegradation of methoxychlor by *Klebsiella pneumoniae. Can. J. Microbiol.* 28: 176–179.

Babu, J.P. and L.R. Brown. 1984. New type of oxygenase involved in the metabolism of propane and isobutane. *Appl. Environ. Microbiol.* 48: 260–264.

Bachofer, R. and F. Lingens. 1975. Conversion of aniline into pyrocatechol by a *Nocardia* sp.: incorporation of oxygen-18. *FEBS Lett.* 50: 288–290.

Bak, F. and F. Widdel. 1986. Anaerobic degradatiion of indolic compounds by sulfate-reducing enrichment cultures, and description of *Desulfobacterium indolicum* gen. nov., sp. nov. *Arch. Microbiol.* 146: 170–176.

Bank, S. and R.J. Tyrrell. 1984. Kinetics and mechanism of alkaline and acidic hydrolysis of aldicarb. *J. Agric. Food Chem.* 32: 1223–1232.

Barbas, J.T., M.E. Sigman, and R. Dabestani. 1996. Photochemical oxidation of phenanthrene sorbed on silica gel. *Environ. Sci. Technol.* 30: 1776–1780.

Baretto, R.D., K.A. Gray, and K. Anders. 1995. Photocatalytic degradation of methyl-*tert*-butyll ether in TiO$_2$ slurries: a proposed reaction scheme. *Water Res.* 29: 1243–1248.

Bartels, I., H.-J. Knackmuss, and W. Reineke. 1984. Suicide inactivation of catechol 2,3-dioxygenase from *Pseudomonas putida* mt-2 by 3-halocatechols. *Appl. Environ. Microbiol.* 47: 500–505.

Batie, C.J., E. LaHaie, and D.P. Ballou. 1987a. Purification and characterization of phthalate oxygenase and phthalate oxygenase reductase from *Pseudomonas cepacia*. *J. Biol. Chem.* 262: 1510–1518.

Batie, C.J., E. LaHaie, and D.P. Ballou. 1987b. Purification and characterization of phthalate oxygenase reductase from *Pseudomonas cepacia*. *J. Biol. Chem.* 262: 766–769.

Battersby, N.S. 1990. A review of biodegradation kinetics in the aquatic environment. *Chemosphere* 21: 1243–1284.

Bauer, J.E. and D.G. Capone. 1988. Effects of co-occurring aromatic hydrocarbons on degradation of individual polycyclic aromatic hydrocarbons in marine sediment slurries. *Appl. Environ. Microbiol.* 54: 1649–1655.

Beadle, T.A. and A.R.W. Smith. 1982. The purification and properties of 2,4-dichlorophenol hydroxylase from a strain of *Acinetobacter* species. *Eur. J. Biochem.* 123: 323–332.

Beaty, P.S. and M.J. McInerney. 1987. Growth of *Syntrophomonas wolfei* in pure culture on crotonate. *Arch. Microbiol.* 147: 389–393.

Beaudette, L.A., S. Davies, P.M. Fedorak, O.P. Ward, and M.A. Pickard. 1998, Comparison of gas chromatography and mineralization experiments for measuring loss of selected polychlorobiphenyl congeners in cultures of white rot fungi. *Appl. Environ. Microbiol.* 64: 2020–2025.

Beitz, T., W. Bechmann, and R. Mitzner. 1998. Investigations of reactions of selected azaarenes with radicals in water.1. Hydroxyl and sulfate radicals. *J. Phys. Chem.* A 102: 6760–6765.

Belly, R.T., J.J. Lauff, and C.T. Goodhue. 1975. Degradation of ethylenediaminetetraacetic acid by microbial populations from an aerated lagoon. *Appl. Microbiol.* 29: 787–794.

Berg, A., J.Å. Gustafsson, M. Ingelman-Sundberg, and K. Carlström. 1976. Characterization of a cytochrome P-450-dependent steroid hydroxylase present in *Bacillus megaterium*. *J. Biol. Chem.* 251: 2831–2838.

Bergon, M., N.B. Hamida, and J.-P. Calmon. 1985. Isocyanate formation in the decomposition of phenmedipham in aqueous media. *J. Agric. Food Chem.* 33: 577–583.

Berndt, T., O. Böge, and W. Rolle. 1997. Products of gas-phase reactions of NO$_3$ radicals with furan and tetramethylfuran. *Environ. Sci. Technol.* 31: 1157–1162.

Bhadra, R., D.G. Wayment, J.B. Hughes and J.V. Shanks. 1999. Confirmation of conjugation processes during TNT metabolism by axenic plant roots. *Environ. Sci. Technol.* 33: 446–452.

Biedlingmeier, J.J. and A. Schmidt. 1983. Arylsulfonic acids and some S-containing detergents as sulfur sources for growth of *Chlorella fusca. Arch. Microbiol.* 136: 124–130.

Boethling, R.S. and M. Alexander. 1979. Effect of concentration of organic chemicals on their biodegradation by natural microbial communities. *Appl. Environ. Microbiol.* 37: 1211–1216.

Bogan, B.W. and R.T. Lamar. 1996. Polycyclic aromatic hydrocarbon-degrading capabilities of *Phanerochaete laevis* HHB-1625 and its extracellular lignolytic enzymes. *Appl. Environ. Microbiol.* 62: 1597–1603.

Boldrin, B., A. Tiehm, and C. Fritzsche. 1993. Degradation of phenanthrene, fluorene, fluoranthene, and pyrene by a *Mycobacterium* sp. *Appl. Environ. Microbiol.* 59: 1927–1930.

Boldt, Y.R., M.J. Sadowsky, L.B.M. Ellis, L. Que, and L.P. Wackett. 1995. A manganese-dependent dioxygenase from *Arthrobacter globiformis* CM-2 belongs to the major extradiol dioxygenase family. *J. Bacteriol.* 177: 1225–1232.

Bosma, T.N.P., F.H.M. Cottaar, M.S. Posthumus, C.J. Teunis, A. van Veidhuizen, G. Schraa, and A.J.B. Zehnder. 1994. Comparison of reductive dechlorination of hexachloro-1,3-butadiene in Rhine sediments and model systems with hydroxocobalamin. *Environ. Sci. Technol.* 28: 1124–1128.

Bouchez, M., D. Blanchet, and J.-P. Vandecasteele. 1997. An interfacial uptake mechanism for the degradation of pyrene by a *Rhodococcus* strain. *Microbiology* (U.K.) 143: 1087–1093.

Bouquard, C., J. Ouazzani, J.-C. Promé, Y. Michel-Briand, and P. Plésiat. 1997. Dechlorination of atrazine by a *Rhizobium* sp. isolate. *Appl. Environ. Microbiol.* 63: 762–866.

Bowien, B. and H.G. Schlegel. 1981. Physiology and biochemistry of hydrogen-oxidizing bacteria. *Annu. Rev. Microbiol.* 35: 405–452.

Bradley, P.M. and F.H. Chapelle. 1995. Rapid toluene mineralization by aquifer microorganisms at Adak, Alaska: implications for intrinsic bioremediation in cold envrionments. *Environ. Sci. Technol.* 29: 2778–2781.

Branchaud, B.P. and C.T. Walsh. 1985. Functional group diversity in enzymatic oxygenation reactions catalyzed by bacterial flavin-containing cyclohexanone oxygenase. *J. Am. Chem. Soc.* 107: 2153–2161.

Brandsch, R., A.E. Hinkkanen, and K. Decker. 1982. Plasmid-mediated nicotine degradation in *Arthrobacter oxidans. Arch. Microbiol.* 132: 26–30.

Bregnard, T.P.-A., A. Häner, P. Höhener, and J. Zeyer. 1997. Anaerobic degradation of pristane in nitrate-reducing microcosms and enrichment cultures. *Appl. Environ. Microbiol.* 63: 2077–2081.

Breskvar, K. and T. Hudnik-Plevnik. 1977. A possible role of cytochrome P-450 in hydroxylation of progesterone by *Rhizopus nigricans. Biochem. Biophys. Res. Commun.* 74: 1192–1198.

Brezonik, P.L. and J. Fulkerson-Brekken. 1998. Nitrate-induced photoysis in natural waters: controls on concentrations of hydroxyl radical photo-intermediates by natural scavenging agents. *Environ. Sci. Technol.* 32: 3004–3010.

Briglia, M., R.I.L. Eggen, D.J. van Elsas, and W.M. de Vos. 1994. Phylogenetic evidence for transfer of pentachlorophenol-mineralizing *Rhodococcus chlorophenolicus* PCP-I^T to the genus *Mycobacterium. Int. J. Syst. Bacteriol.* 44: 494–498.

Brownlee, B.G., J.H. Carey, G.A. MacInnes, and I.T. Pellizzari. 1992. Aquatic environmental chemistry of 2-(thiocyanomethylthio)benzothiazole and related benzothiazoles. *Environ. Toxicol. Chem.* 11: 1153–1168.

Brubaker, W.W. and R.A. Hites. 1997. Polychlorinated dibenzo-*p*-dioxins and dibenzofurans: gas-phase hydroxyl radical reactions and related atmospheric removal. *Environ. Sci. Technol.* 31: 1805–1810.

Brubaker, W.W. and R.A. Hites. 1998. Gas-phase oxidation products of biphenyl and polychlorinated biphenyls. *Environ. Sci. Technol.* 32: 3913–3918.

Brückmann, M., R. Blasco, K.H. Timmis, and D.H. Pieper. 1998. Detoxification of protoanemonin by dienelactone hydrolase. *Appl. Environ. Microbiol.* 64: 400–402.

Bumpus, J.A. 1989. Biodegradation of polycyclic aromatic hydrocarbons by *Phanerochaete chrysosporium*. *Appl. Environ. Microbiol.* 55: 154–158.

Bumpus, J.A. and S.D. Aust. 1987. Biodegradation of DDT [1,1,1-trichloro-2,2-bis(4-chlorophenyl)ethane] by the white rot fungus *Phanerochaete chrysosporium*. *Appl. Environ. Microbiol.* 53: 2001–2008.

Bunce, N.J., L. Liu, J. Zhu, and D.A. Lane. 1997. Reaction of naphthalene and its derivatives with hydroxyl radicals in the gas phase. *Environ. Sci. Technol.* 31: 2252–2259.

Burhenne, J., M. Ludwig, and M. Spiteller. 1997a. Photolytic degradation of fluoro-quinolone carboxylic acids in aqueous solution. Primary photoproducts and half-lives. *Environ. Sci. Pollut. Res.* 4: 10–15.

Burhenne, J., M. Ludwig, and M. Spiteller. 1997b. Photolytic degradation of fluoro-quinolone carboxylic acids in aqueous solution. Isolation and structural eluci-dation of polar photometabolites. *Environ. Sci. Pollut. Res.* 4: 61–67.

Burken, J.G. and J.L. Schnoor. 1997. Uptake and metabolism of atrazine by poplar trees. *Environ. Sci. Technol.* 31: 1399–1406.

Burlingame, R. and P.J. Chapman. 1983. Catabolism of phenylpropionic acid and its 3-hydroxy derivative by *Escherichia coli*. *J. Bacteriol.* 155: 113–121.

Buser, H.-R. and M.D. Müller. 1993. Enantioselective determination of chlordane components, metabolites, and photoconversion products in environmental sam-ples using chiral high-resolution gas chromatography and mass spectrometry. *Environ. Sci. Technol.* 27: 1211–1220.

Buser, H.-R. and M.D. Müller. 1997. Conversion reactions of various phenoxyalkanoic acid herbicides in soil. 2. Elucidation of the enantiomerization process of chiral phenoxy acids from incubation in a D_2O/soil system. *Environ. Sci. Technol.* 31: 1960–1967.

Caccavo, F., R.P. Blakemore, and D.R. Lovley. 1992. A hydrogen-oxidizing, Fe(III)-reducing microorganism from the Great Bay estuary, New Hampshire. *Appl. Environ. Microbiol.* 58: 3211–3216.

Caccavo, F., J.D. Coates, R. Rossello-Mora, W. Ludwig, K.H. Schleifer, D.R. Lovley, and M.J. McInerney. 1996. *Geovibrio ferrireducens*, a phylogenetically distinct dissimilatory FeIII-reducing bacterium. *Arch. Microbiol.* 165: 370–376.

Cain, R.B., R.F. Bilton, and J.A. Darrah. 1968. The metabolism of aromatic acids by microorganisms. *Biochem. J.* 108: 797–832.

Calderbank, A. 1989. The occurrence and significance of bound pesticide residues in soil. Rev. Environ. Contam. Toxicol. 108: 1–1103.

Camilleri, P. 1984. Alkaline hydrolysis of some pyrethroid insecticides. *J. Agric. Food Chem.* 32: 1122–1124.

Cánovas, J.L. and R.Y. Stanier. 1967. Regulation of the enzymes of the β-ketoadipate pathway in *Moraxella calcoacetica*. *Eur. J. Biochem.* 1: 289–300.

Cardini, G. and P. Jurtshuk. 1968. Cytochrome P-450 involvement in the oxidation of *n*-octane by cell-free extracts of *Corynebacterium* sp. strain 7E1C. *J. Biol. Chem.* 243: 6070–6072.

Carey, J.H. and M.E. Cox. 1981. Photodegradation of the lampricide 3-trifluoromethyl-4-nitrophenol (TFM) 1. Pathway of the direct photolysis in solution. *J. Great Lakes Res.* 7: 234–241.

Carmichael, L.M., R.F. Christman, and F.K. Pfaender. 1997. Desorption and mineralization kinetics of phenanthrene and chysene in contaminated soils. *Environ. Sci. Technol.* 31: 126–132.

Carter, W.P.L. and R. Atkinson. 1987. An experimental study of incremental hydrocarbon reactivity. *Environ. Sci. Technol.* 21: 670–679

Carter, W.P.L. and R. Atkinson. 1989. Computer modeling of incremental hydrocarbon reactivity. *Environ. Sci. Technol.* 23: 864–880.

Castro, C.E., R.S. Wade, and N.O.Belser. 1985. Biodehalogenation reactions of cytochrome P-450 with polyhalomethanes. *Biochemistry* 24: 204–210.

Cerniglia, C.E., D.T. Gibson, and C. van Baalen. 1980a. Oxidation of naphthalene by cyanobacteria and microalgae. *J. Gen. Microbiol.* 116: 495–500.

Cerniglia, C.E., C. van Baalen, and D.T. Gibson. 1980b. Oxidation of biphenyl by the cyanobacterium, I sp., strain JCM. *Arch. Microbiol.* 125: 203–207.

Cerniglia, C.E., J.P. Freeman, and C. van Baalen. 1981. Biotransformation and toxicity of aniline and aniline derivatives in cyanobacteria. *Arch. Microbiol.* 130: 272–275.

Cerniglia, C.E., D.T. Gibson, and C. van Baalen. 1982. Naphthalene metabolism by diatoms isolated from the Kachemak Bay region of Alaska. *J. Gen. Microbiol.* 128: 987–990.

Cerniglia, C.E., J.P. Freeman, J.R. Althaus, and C. van Baalen. 1983. Metabolism and toxicity of 1- and 2-methylnaphthalene and their derivatives in cyanobacteria. *Arch. Microbiol.* 136: 177–183.

Cerniglia, C.E., J.P. Freeman, and F.E. Evans. 1984. Evidence for an arene oxide–NIH shift pathway in the transformation of naphthalene to 1-naphthol by *Bacillus cereus*. *Arch. Microbiol.* 138: 283–286.

Chakrabarty, A.M., G. Chou, and I.C. Gunsalus. 1973. Genetic regulation of octane dissimilation plasmid in *Pseudomonas*. *Proc. Natl. Acad. Sci. U.S.A.* 70: 1137 1140.

Champine, J.E. and S. Goodwin. 1991. Acetate catabolism in the dissimilatory iron-reducing isolate GS-15. *J. Bacteriol.* 173: 2704–2706.

Chang, H.-K. and G.J. Zylstra. 1998. Novel organization of the genes for phthalate degradation from *Burkholdeia cepacia* DBO1. *J. Bacteriol.* 180: 6529–6537.

Charleson, R.J., J.E. Lovelock,, M.O. Andreae, and S.G. Warren. 1987. Oceanic phytoplankton, atmospheric sulfur, cloud albedo and climate. *Nature* (London) 326: 655–661.

Chatterjee, D.K., S.T. Kellog, S. Hamada, and A.M. Chakrabarty. 1981. Plasmid specifying total degradation of 3-chlorobenzoate by a modified *ortho* pathway. *J. Bacteriol.* 146: 639–646.

Chaudry, G.S. and G.H. Huang. 1988. Isolation and characterization of a new plasmid from a *Flavobacterium* sp. which carries the genes for degradation of 2,4-dichlorophenoxyacetate. *J. Bacteriol.* 170: 7897–3902.

Chen, S.-W. and M.D. Aitken. 1999. Salicylate stimulates the degradation of high-molecular-weight polycyclic aromatic hydrocarbons by *Pseudomonas saccharophila* P15. *Environ. Sci. Technol.* 33: 435–439.

Chen, Y. and J.P.N. Rosazza. 1995. Purification and characterization of nitric oxide synthase NOS_{Noc} from a *Nocardia* sp. *J. Bacteriol.* 177: 5122–5128.

Chen, Y.P., M.J. Dilworth, and A.R. Glenn. 1984. Aromatic metabolism in *Rhizobium trifolii* — protocatechuate 3,4-dioxygenase. *Arch. Microbiol.* 138: 1897–190.

Chen, Y.P., A.R. Glenn, and M.J. Dilworth. 1985. Aromatic metabolism in *Rhizobium trifolii* — catechol 1,2-dioxygenase. *Arch. Microbiol.* 141: 225–228.

Chlodini, G., B. Rindone, F. Cariati, S. Polesello, G. Restelli, and J. Hjorth. 1993. Comparison between the gas-phase and the solution reaction of the nitrate radical and methylarenes. *Environ. Sci. Technol.* 27: 1659–1664.

Choi, W. and M.R. Hoffman. 1997. Novel photocatalytic mechanisms for $CHCl_3$, $CHBr_3$, and $CCl_3CO_2^-$ degradation and fate of photogenerated trihalomethyl radicals on TiO_2. *Environ. Sci. Technol.* 31: 89–95.

Cirigliano, M.C., and G.M. Carman. 1984. Isolation of a bioemulsifier from *Candida lipolytica*. *Appl. Environ. Microbiol.* 48: 747–750.

Claus, D. and N. Walker. 1964. The decomposition of toluene by soil bacteria. *J. Gen. Microbiol.* 36: 107–122.

Coates, J.D., E.J.P. Phillips, D.J. Lonergan, H. Jenter, and D.R.Lovley. 1996. Isolation of *Geobacter* species from diverse sedimentary environments. *Appl. Environ. Microbiol.* 62: 1531–1536.

Cole, J.R., A.L. Cascarelli, W.W. Mohn, and J.M. Tiedje. 1994. Isolation and characterization of a novel bacterium growing via reductive dehalogenation of 2-chlorophenol. *Appl. Environ. Microbiol.* 60: 3536–3542.

Connors, M.A. and E.A. Barnsley. 1982. Naphthalene plasmids in pseudomonads. *J. Bacteriol.* 149: 1096–1101.

Cook, A.M. and R. Hütter. 1981. Degradation of s-triazines: a critical view of biodegradation pp. 237–249. In *Microbial Degradation of Xenobiotics and Recalcitrant Compounds*. T. Leisinger, A.M. Cook, R. Hütter, and J. Nüesch (Eds.). Academic Press, London.

Cook, A.M., C.G. Daughton, and M. Alexander. 1979. Benzene from bacterial cleavage of the carbon–phosphorus bond of phenylphosphonates. *Biochem. J.* 184: 453–455.

Cooper, B.H. and G.A. Land. 1979. Assimilation of protocatechuate acid and *p*-hydroxybenzoic acid as an aid to laboratory identification of *Candida parapsilosis* and other medically important yeasts. *J. Clin. Microbiol.* 10: 343–345.

Cooper, R.A. and M. A. Skinner. 1980. Catabolism of 3- and 4-hydroxyphenylacetate by the 3,4-dihydroxyphenylacetate pathway in *Escherichia coli*. *J. Bacteriol.* 143: 302–306.

Corke, C.T., N.J. Bunce, A.-L. Beaumont, and R.L. Merrick. 1979. Diazonium cations as intermediates in the microbial transformations of chloroanilines to chlorinated biphenyls, azo compounds and triazenes. *J. Agric. Food Chem.* 27: 644–646.

Crawford, R.L., P.E. Olson, and T.D. Frick. 1979. Catabolism of 5-chlorosalicylate by a *Bacillus* isolated from the Mississippi River. *Appl. Environ. Microbiol.* 38: 379–384.

Cremonsei, P., E.L. Cavalieri, and E.G. Rogan. 1989. One-electron oxidation of 6-substituted benzo(a)pyrenes by manganic acetate. A model for metabolic activation. *J. Org. Chem.* 54: 3561–3570.

Criddle, C.S., J.T. DeWitt, D. Grbic-Galic, and P.L. McCarty. 1990. Transformation of carbon tetrachloride by *Pseudomonas* sp. strain KC under denitrifying conditions. *Appl. Environ. Microbiol* 56: 3240–3246.

Crocker, F.H., W.A. Guerin, and S.A. Boyd. 1995. Bioavailability of naphthalene sorbed to cationic surfactant-modified smectite clay. *Environ. Sci. Technol.* 29: 2953–2958.

Cserjesi, A.J. and E.L. Johnson. 1972. Methylation of pentachlorophenol by *Trichoderma virginatum*. *Can. J. Microbiol.* 18: 45–49.

Dabrock, B., J. Riedel, J. Bertram, and G. Gottschalk. 1992. Isopropylbenzene (cumene) — a new substrate for the isolation of trichloroethene-degrading bacteria. *Arch. Microbiol.* 158: 9–13.

Dalton, H. and D.I. Stirling. 1982. Co-metabolism. *Philos. Trans. R. Soc. London.* B 297: 481–496.

Dalton, H., B.T. Golding, B.W. Waters, R. Higgins, and J.A. Taylor. 1981. Oxidation of cyclopropane, methylcyclopropane, and arenes with the mono-oxygenase systems from *Methylococcus capsulatus. J. Chem. Soc. Chem. Commun.* 482–483.

Damaj M. and D. Ahmad. 1996. Biodegradation of polychlorinated biphenyls by rhizobia: a novel finding. *Biochem. Biophys. Res. Commun.* 218: 908–915.

Daughton, C.G. and D.P.H. Hsieh. 1977. Parathion utilization by bacterial symbionts in a chemostat. *Appl. Environ. Microbiol.* 34: 175–184.

Davison, A.D., P. Karuso, D.R. Jardine, and D.A. Veal. 1996. Halopicolinic acids, novel products arising through the degradation of chloro- and bromobiphenyl by *Sphingomonas paucimobilis* BPS!—3. *Can. J. Microbiol.* 42: 66–71.

de Schrijver, A., I. Nagy, G. Schoofs, P. Proost, J. Vanderleyden, K.-H. van Pée, and R. de Mot. 1997. Thiocarbamate herbicide-indicible nonmheme haloperoxidase of *Rhodococcus erythropolis* NI86/21. *Appl. Environ. Microbiol.* 63: 1811–1916.

DeBont, J.A.M., S.B. Prtimrose, M.D. Collins, and D. Jones. 1980. Chemical studies on some bacteria which utilize gaseous unsaturated hydrocarbons. *J. Gen. Microbiol.* 117: 97–102.

Dec, J., K.L. Shuttleworth, and J.-M. Bollag. 1990. Microbial release of 2,4-dichlorophenol bound to humic acid or incorporated during humification. *J. Environ. Qual.* 19: 546–551.

Degrassi, G., P. P. de Laureto, and C.V. Bruschi. 1995. Purification and characterization of ferulate and *p*-coumarate decarboxylase from *Bacillus pumilis. Appl. Environ. Microbiol.* 61: 326–332.

Dehning, I. and B. Schink. 1989a. *Malonomonas rubra* gen. nov. sp. nov., a microaerotolerant anaerobic bacterium growing by decarboxylation of malonate. *Arch. Microbiol.* 151: 427 433.

Dehning, I. and B. Schink. 1989b. Two new species of anaerobic oxalate-fermenting bacteria, *Oxalobacter vibrioformis* sp. nov. and *Clostridium oxalicum* sp. nov. from sediment samples. *Arch. Microbiol.* 153: 79–84.

Dehning, I., M. Stieb, and B. Schink. 1989. *Sporomusa malonica* sp. nov., a homoacetogenic bacterium growing by decarboxylation of malonate or succinate. *Arch. Microbiol.* 151: 421–426.

Denoya, C.D., D.D. Skinner, and M.R. Morgenstern. 1994. A *Streptomyces avermitilis* gene encoding a 4-hydroxyphenylpyruvic acid dioxygenase-like protein that directs the production of homogentisic acid and an ochronotic pigment in *Escherichia coli. J. Bacteriol.* 176: 5312–5319.

Devanas, M.A., D. Rafaeli-Eshkol, and G. Stotsky. 1986. Survival of plasmid-containing strains of *Escherichia coli* in soil: effect of plasmid size and nutrients on survival of hosts and maintenance of plasmids. *Curr. Microbiol.* 13: 269–277.

DeWeerd, K.A., L. Mandelco, R.S. Tanner, C.R. Woese, and J.M. Suflita. 1990. *Desulfomonile tiedjei* gen. nov. and sp. nov., a novel anaerobic, dehalogenating, sulfate-reducing bacterium. *Arch. Microbiol.* 154: 23–30.

Déziel, É., G. Paquette, R. Villemur, F. Lépine, and J.-G. Bisaillon. 1996. Biosurfactant production by a soil *Pseudomonas* strain growing on polycyclic aromatic hydrocarbons. *Appl. Environ. Microbiol.* 62: 1908–1912.

DiChristina, T.J. 1992. Effects of nitrate and nitrite on dissimilatory iron reduction by Shewanella putrefaciens 200. *J. Bacteriol.* 174: 1891–1896.

Dickel, O.D. and H.-J. Knackmuss. 1991. Catabolism of 1,3-dinitrobenzene by *Rhodococcus* sp. QT-1. *Arch. Microbiol.* 157: 76–79.

Diels, L. and M. Mergeay. 1990. DNA probe-mediated detection of resistant bacteria from soils highly polluted by heavy metals. *Appl. Environ. Microbiol.* 56: 1485–1491.

Dietrich, D., W.J. Hickey, and R. Lamar. 1995. degradation of 4,4'-dichlorobiphenyl, 3,3',4,4'-tetrachlorobiphenyl, and 2,2',4,4',5,5'hexachlorobiphenyl by the white rot fungus *Phanerochaete chrysosporium*. *Appl. Environ. Microbiol.* 61: 3904–3909.

Dimroth, P. and H. Hilbi. 1997. Enzymatic and genetic basis for bacterial growth on malonate. *Mol. Microbiol.* 25: 3–10.

Don, R.H. and J.M. Pemberton. 1985. Genetic and physical map of the 2,4-dichlorophenoxyacetic acid-degradative plasmid pJP4. *J. Bacteriol.* 161: 466–468.

Don, R.H., A.J. Wightman, H.H. Knackmuss, and K.N. Timmis. 1985. Transposon mutagenesis and cloning analysis of the pathways of degradation of 2,4-dichlorophenoxyacetic acid and 3-chlorobenzoate in *Alcaligenes eutrophus* JMP134(pJP4). *J. Bacteriol.* 161: 85–90.

Donnelly, P.K., R.S. Hegde, and J.S. Fletcher. 1994. Growth of PCB-degrading bacteria on compounds from photosynthetic plants. *Chemosphere.* 28: 981–988.

Doukyu, N. and R. Aono. 1998. Purification of extracellular cholesterol oxidase with high activity in the presence of organic solvents from *Pseudomonas* sp. strain ST-200. *Appl. Environ. Microbiol.* 64: 1929–1932.

Drzyzga, O., J. Küver, and K.-H. Blottevogel. 1993. Complete oxidation of benzoate and 4-hydroxybenzoate by a new sulfate-reducuing bacterium resembling *Desulfoarculus*. *Arch. Microbiol.* 159: 109–113.

Duetz, W.A., S. Marqués, C. de Jong, J.L. Ramos, and J. G. van Andel. 1994. Inducibility of the TOL catabolic pathway in *Pseudomonas putida* (pWWO) growing on succinate in continuous culture: evidence of carbon catabolite repression control. *J. Bacteriol.* 176: 2354–2361.

Dunnivant, F.M., R.P. Schwarzenbach, and D.L. Macalady. 1992. Reduction of substituted nitrobenzenes in aqueous solutions containing natural organic matter. *Environ. Sci. Technol.* 26: 2133–2141.

Durham, D.R., C.G. McNamee, and D.P. Stewart. 1984. Dissimilation of aromatic compounds in *Rhodotorula graminis*: biochemical characterization of pleiotrophically negative mutants. *J. Bacteriol.* 160: 771–777.

Dürre, P. and J.R. Andreesen. 1982. Anaerobic degradation of uric acid via pyrimidine derivatives by selenium-starved cells of *Clostridium purinolyticum*. *Arch. Microbiol.* 131: 255–260.

Eaton, D.C. 1985. Mineralization of polychlorinated biphenyls by *Phanerochaete chrysosporium*, a lignolytic fungus. *Enzyme Microbiol. Technol.* 7: 194–196.

Eaton, R.W. and P.J. Chapman. 1992. Bacterial metabolism of naphthalene: construction and use of recombinant bacteria to study ring cleavage of 1,2-dihydroxynaphthalene and subsequent reactions. *J. Bacteriol.* 174: 7542–7554.

Eatough, D.J., V.F. White, L.D. Hansen, N.L. Eatough, and J.L. Cheney. 1986. Identification of gas-phase dimethyl sulfate and monomethyl hydrogen sulfate in the Los Angeles atmosphere. *Environ. Sci. Technol.* 20: 867–872.

Eble, K.S. and J.H. Dawson. 1984. Novel reactivity of cytochrome P-450-CAM. Methyl hydroxylation of 5,5-difluorocamphor. *J. Biol. Chem.* 259: 14389–14393.

Ehrenreich, A. and F. Widdel. 1994. Anaerobic oxidation of ferrous iron by purple bacteria, a new type of phototrophic metabolism. *Appl. Environ. Microbiol.* 60: 4517–4526.

Emerson, D., S. Chauhan, P. Oriel, and J.A. Breznak. 1994. *Haloferax* sp D1227, a halophilic Archaeon capable of growth on aromatic compounds. *Arch. Microbiol.* 161: 445–452.

Emptage, M., J. Tabinowski, and J.M. Odom. 1997. Effect of fluoroacetates on methanogenesis in samples from selected methanogenic environments. *Environ. Sci. Technol.* 31: 732–734.

Engesser, K.-H., E. Schmidt, and H.-J. Knackmuss. 1980. Adaptation of *Alcaligenes eutrophus* B9 and *Pseudomonas* sp. B13 to 2-fluorobenzoate as growth substrate. *Appl. Environ. Microbiol.* 39: 68–73.

Ensign, S.A., F.J. Smakk, J.R. Allen, and M.K. Sluis. 1998. New roles for CO_2 in the microbial metabolism of aliphatic epoxides and ketones. *Arch. Microbiol.* 169: 179–187.

Entsch, B. and D.P. Ballou. 1989. Purification, properties, and oxygen reactivity of p-hydroxybenzoate hydroxylase from *Pseudomonas aeruginosa*. *Biochim. Biophys. Acta* 999: 253–260.

Enya, T., H. Suzuki, T. Watanabe, T. Hiurayama, and Y. Hisamatsu. 1997. 3-Nitrobenzanthrone, a powerful bacterial mutagen and suspected human carcinogen found in diesel exhaust and airborne particulates. *Environ. Sci. Technol.* 31: 2772–2776.

Eppink, M.H.M., S.A. Boeren, J. Vervoort, and W.J.H. van Berkel. 1997. Purification and properties of 4-hydroxybenzoate 1-hydroxylase (decarboxylating), a novel flavin adenein dinucleotide-dependent monooxygenase from *Candida parapilosis* CBS604. *J. Bacteriol.* 179: 6680–6687.

Evans, J.S. and W.A. Venables. 1990. Degradation of thiophene-2-carboxylate, furan-2-carboxylate, pyrrole-2-carboxylate and other thiophene derivatives by the bacterium *Vibrio* YC1. *Appl. Microbiol. Biotechnol.* 32: 715–720.

Evans, P.J., D.T. Mang, K.S. Kim, and L.Y. Young. 1991a. Anaerobic degradation of toluene by a denitrifying bacterium. *Appl. Environ. Microbiol.* 57: 1139–1145.

Evans, P.J., D.T. Mang, and L.Y. Young. 1991b. Degradation of toluene and *m*-xylene and transformation of *o*-xylene by denitrifying enrichment cultures. *Appl. Environ. Microbiol.* 57: 450–454.

Fall, R.R., J.I. Brown, and T.L. Schaeffer. 1979. Enzyme recruitment allows the biodegradation of recalcitrant branched hydrocarbons by *Pseudomonas citronellolis*. *Appl. Environ. Microbiol.* 38: 715–722.

Feldman, P.L., O.W. Griffith, and D.J. Stuehr. 1993. The surprising life of nitric oxide. *Chem. Eng. News* 71 (51): 26–39.

Fernández, P., M. Grifoll, A.M. Solanas, J. M. Bayiona, and J. Albalgés. 1992. Bioassay-directed chemical analysis of genotoxic compounds in coastal sediments. *Environ. Sci. Technol.* 26: 817–829.

Feroz, M., A.A. Podowski, and M.A.Q. Khan. 1990. Oxidative dehydrochlorination of heptachlor by *Daphnia magna*. *Pestic. Biochem. Physiol.* 36: 101–105.

Ferris, J.P., L.H. MacDonald, M.A. Patrie and M.A.Martin. 1976. Aryl hydrocarbon hydroxylase activity in the fungus *Cunninghamella bainieri*; evidence for the presence of cytochrome P-450. *Arch. Biochem. Biophys.* 175: 443–452.

Fetzner, S., R. Müller, and F. Lingens. 1992. Purification and some properties of 2-halobenzoate 1,2-dioxygenase, a two-component enzyme system from *Pseudomonas cepacia* 2CBS. *J. Bacteriol.* 174: 279–290.

Finette, B.A., V. Subramanian, and D.T. Gibson. 1984. Isolation and characterization of *Pseudomonas putida* PpF1 mutants defective in the toluene dioxygenase enzyme system. *J. Bacteriol.* 160: 1003–1009.

Finlayson-Pitts, B. and J. Pitts, Jr. 1986. *Atmospheric Chemistry.* John Wiley & Sons, New York.

Finlayson-Pitts, B.J. and J.N. Pitts. 1997. Tropospheric air pollution: ozone, airborne toxics, polycyclic aromatic hydrocarbons, and particles. *Science* 276: 1045–1052.

Finnerty, W.R. 1992. The biology and genetics of the genus *Rhodococcus. Annu. Rev. Microbiol.* 46: 193–218.

Foght, J.M., D.L. Gutnick, and D.W.S. Westlake. 1989. Effect of emulsan on biodegradation of crude oil by pure and mixed bacterial cultures. *Appl. Environ. Microbiol.* 55: 36–42.

Folsom, B.R., P.J. Chapman, and P.H. Pritchard. 1990. Phenol and trichloroethylene degradation by *Pseudomonas cepacia* G4: kinetics and interactions between substrates. *Acinetobacter* sp. strain 4-CB1. *Appl. Environ. Microbiol.* 56: 1279–1285.

Forstner, H.J.L., R.C. Flagan, and J.H. Seinfeld. 1997. Secondary organic aerosol from the photooxidation of aromatic hydrocarbons: molecular composition. *Environ. Sci. Technol.* 31: 1345–1358.

Fox, B.G., W.A. Froland, J.E. Dege, and J.D. Lipscomb. 1989. Methane monooxygenase from *Methylosinus trichosporium* OB3b. Purification and properties of a three-component system with a high specific activity from a type II methanotroph. *J. Biol. Chem.* 264: 10023–10033.

Frank, H., H. Scholl, D. Renschen, B. Rether, A. Laouedj, and Y. Norokorpi. 1994. Haloacetic acids, phytotoxic secondary air pollutants. *Environ. Sci. Pollut. Res.* 1: 4–14.

Frank, H., D. Renschen, A. Klein, and H. Scholl. 1995. Trace analysis of airborne haloacetates. *J. High Resol. Chromatogr.* 18: 83–88.

Franklin, J. 1994. The atmospheric degradation and impact of perchloroethylene. *Toxicol. Environ. Chem.* 46: 169–182.

Fredrickson, J.K., F.J. Brockman, D.J. Workmnan, S.W. Li, and T.O. Stevens. 1991. Isolation and characterization of a subsurface bacterium capable of growth on toluene, naphthalene, and other aromatic compounds. *Appl. Environ. Microbiol.* 57: 796–803.

Fritzsche, C. 1994. Degradation of pyrene at low defined oxygen concentrations by *Mycobacterium* sp. *Appl. Environ. Microbiol.* 60: 1687–1689

Fukumori, F. and R.P. Hausinger. 1993. *Alcaligenes eutrophus* JMP 134 2,4-chlorophenoxyacetate "monooxygenase" is an α–ketoglutarate-dependent dioxygenase. *J. Bacteriol.* 175: 2083–2086.

Fulthorpe, R.R. and R. C. Wyndham. 1989. Survival and activity of a 3-chlorobenzoate-catabolic genotype in a natural system. *Appl. Environ. Microbiol.* 55: 1584–1590.

Fukumori, F. and C.P. Saint. 1997. Nucleotide sequences and regulational analysis of genes involved in conversion of aniline to catechol in *Pseudomonas putida* UCC22 (pTDN1). *J. Bacteriol.* 179: 399–408.

Fulthorpe, R.R., A.N. Rhodes, and J.M. Tiedje. 1996. Pristine soils mineralize 3-chlorobenzoate and 2,4-dichlorophenoxyacetate via different microbial populations. *Appl. Environ. Microbiol.* 62: 1159–1166.

Fulthorpe, R.R., A.N. Rhodes, and J.M. Tiedje. 1998. High levels of endemicity of 3-chlorobenzoate-degrading soil bacteria. *Appl. Environ. Microbiol.* 64: 1620–1627.

Furukawa, K. and A. M. Chakrabarty. 1982. Involvement of plasmids in total degradation of chlorinated biphenyls. *Appl. Environ. Microbiol.* 44: 619–626.

Furukawa, K., J.R. Simon and A.M. Chakrabarty. 1983. Common induction and regulation of biphenyl, xylene/toluene, and salicylate catabolism in *Pseudomonas paucimobilis*. *J. Bacteriol.* 154: 1356–1362.

Gaal, A. and H.Y. Neujahr. 1979. Metabolism of phenol and resorcinol in *Trichosporon cutaneum*. *J. Bacteriol.* 137: 13–21.

Gadda, G., R.D. Edmondson, D.H. Russell, and P.F. Fitzpatrick. 1997. Identification of the naturally occurring flavin of nitroalkane oxidase from *Fusarium oxysporum* as a 5-nitrobutyl-FAD and conversion of the enzyme to the active FAD-containing form. *J. Biol. Chem.* 272: 5563–5570.

Gälli, R. and T. Leisinger. 1988. Plasmid analysis and cloning of the dichloromethane-utilizing genes of *Methylobacterium* sp. DM4. *J. Gen. Microbiol.* 134: 943–952.

Gantzer, C.J. and L.P. Wackett. 1991. Reductive dechlorination catalyzed by bacterial transition-metal coenzymes. *Environ. Sci. Technol.* 25: 715–722.

Garrison, A.W., M.G. Cippolone, N.L. Wolfe, and R.R. Swank. 1995. Environmental fate of methylcyclopentadienyl manganese tricarbonyl. *Environ. Toxicol. Chem.* 14: 1859–1864.

Gauthier, M.J., B.Lafay, R. Christen, L. Fernandez, M. Acquaviva, P. Bonin, and J.-C. Bertrand. 1992. *Marinobacter hydrocarbonoclasticus* gen. nov., sp. nov., a new extremely halotolerant, hydrocarbon-degrading marine bacterium. *Int. J. Syst. Bacteriol.* 42: 568–576.

Gee, J.M. and J.L. Peel. 1974. Metabolism of 2,3,4,6-tetrachlorophenol by micro-organisms from broiler house litter. *J. Gen. Microbiol.* 85: 237–243.

Geiselbrecht, A.D., B.P. Hedland, M.A. Tichi, and J.T. Staley. 1998. Isolation of marine polycyclic aromatic hydrocarbon (PAH)-degrading *Cycloclasticus* strains from the Gulf of Mexico and comparison of their PAH degradation ability with that of Puget Sound *Cycloclasticus* strains. *Appl. Environ. Microbiol.* 64: 4703–4710.

Gerritse, J. and J.C. Gottschal. 1993a. Two-membered mixed cultures of methanogenic and aerobic bacteria in O_2-limited chemostats. *J. Gen. Microbiol.* 139: 1853–1860.

Gerritse, J. and J.C. Gottschal. 1993b. Oxic and anoxic growth of a new *Citrobacter* species on amino acids. *Arch. Microbiol.* 160: 51–61.

Gerson, D.F. and J.E. Zajic. 1979. Comparison of surfactant production from kerosene by four species of *Corynebacterium*. *Antonie van Leeuwenhoek* 45: 81–94.

Gibson, D.T., J.R. Koch, C.L. Schuld, and R.E. Kallio. 1968. Oxidative degradation of aromatic hydrocarbons by microorganisms. II. Metabolism of halogenated aromatic hydrocarbons. *Biochemistry* 7: 3795–3802.

Gibson, D.T., S.M. Resnick, K. Lee, J.M. Brand, D.S. Torok, L.P. Wackett, M.J. Schocken, and B.E. Haigler. 1995. Desaturation, dioxygenation, and monooxygenation reactions catalyzed by naphthalene dioxygenase from *Pseudomonas* sp. strain 9816-4. *J. Bacteriol.* 177: 2615–2621.

Gibson, S.A. and G.W. Sewell. 1992. Stimulation of reductive dechlorination of tetrachloroethene in anaerobic aquifer microcosms by addition of short-chain organic acids or alcohols. *Appl. Environ. Microbiol.* 58: 1392–1393.

Gilbert, E.S. and D.E. Crowley. 1997. Plant compounds that induce polychlorinated biphenyl degradation by *Arthrobacter* sp. strain B1B. *Appl. Environ. Microbiol.* 63: 1933–1938.

Glaus, M.A., C.G. Heijman, R.P. Schwarzenbach, and J. Zeyer. 1992. Reduction of nitroaromatic compounds mediated by *Streptomyces* sp. exudates. *Appl. Environ. Microbiol.* 58: 1945–1951.

Gmünder, F.K., O. Käppeli, and A. Fiechter. 1981. Chemostat studies on the hexade-cane assimilation by the yeast *Candida tropicalis*. *Eur. J. Appl. Microbiol. Biotechnol.* 12: 135–142.

Godejohann, M., M. Astratov, A. Preiss, K. Levsen, and C. Mügge. 1998. Application of continuous-flow HPLC-proton-nuclear magnetic resonance spectroscopy and HPLC-thermospray spectroscopy for the structrual elucidation of phototransfor-mation products of 2,4,6-trinitrotoluene. *Anal.Chem.* 70: 4104–4110.

Gorlatova, N., M. Tchorzewski, T. Kurihara, K. Soda, and N. Esaki. 1998. Purification, characterization, and mechanism of a flavin mononucleotide-dependent 2-nitro-propane dioxygenase from *Neurospora crassa*. *Appl. Environ. Microbiol.* 64: 1029–1033.

Gorny, N., G. Wahl, A. Brune, and B. Schink. 1992. A strictly anaerobic nitrate-reducing bacterium growing with resorcinol and other aromatic compunds. *Arch. Microbiol.* 158: 48–53.

Govindaswami, M., D.J. Feldjake, B.K. Kinkle, D.P. Mindell, and J.C. Loper. 1995. Phylogenetic comparison of two polycyclic aromatic hydrocarbon-degrading mycobacteria. *Appl. Environ. Microbiol.* 61: 3221–3226.

Gray, P.H.H. and H.G. Thornton. 1928. Soil bacteria that decompose certain aromatic compounds. *Centralbl. Bakteriol. Parasitenkd. Infektionskr.* (2 Abt.) 73: 74–96.

Grbic-Galic, D. 1986. O-Demethylation, dehydroxylation, ring-reduction and cleav-age of aromatic substrates by *Enterobacteriaceae* under anaerobic conditions. *J. Appl. Bacteriol.* 61: 491–497.

Greer, L.E. and D.R. Shelton. 1992. Effect of inoculant strain and organic matter content on kinetics of 2,4-dichlorophenoxyacetic acid degradation in soil. *Appl. Environ. Microbiol.* 58: 1459–1465.

Griesbaum, K., V. Miclaus, and I.C. Jung. 1998. Isolation of ozonides from gas-phase ozonolysis of terpenes. *Environ. Sci. Technol.* 32: 647–649.

Griffiths, D.A., D.E. Brown, and S.G. Jezequel. 1992. Biotransformation of warfarin by the fungus *Beauveria bassiana*. *Appl. Microbiol. Biotechnol.* 37: 169–175.

Grimond, P.A.D., F. Grimond, H.L.C. Dulong de Rosnay, and P.H.A. Sneath. 1977. Taxonomy of the genus *Serratia*. *J. Gen. Microbiol.* 98: 39–66.

Groenewegen, P.E.J., A.J.M. Diessen, W.M. Konigs, and J.A.M. de Bont. 1990. Energy-dependent uptake of 4-chlorobenzoate in the coryneform bacterium NTM-1. *J. Bacteriol.* 172: 419–423.

Grosjean, D. 1991. Atmospheric chemistry of toxic contaminants. 4. Saturated haloge-nated aliphatics: methyl bromide, epichlorhydrin, phosgene. *J. Air Waste Manage. Assoc.* 41: 56–61.

Grosjean, D., E.L. Williams II, and J.H. Seinfeld. 1992. Atmospheric oxidation of selected terpenes and related carbonyls: gas-phase carbonyl products. *Environ. Sci. Technol.* 26: 1526–1533.

Grosjean, D., E.L. Williams II, and E. Grosjean. 1993a. A biogenic precursor of per-oxypropionyl nitrate: atmospheric oxidation of *cis*-3-hexen-1-ol. *Environ. Sci. Technol.* 27: 979–981.

Grosjean, D., E.L. Williams II, and E. Grosjean. 1993b. Atmospheric chemistry of isoprene and its carbonyl products. *Environ. Sci. Technol.* 27: 830–840.

Grosjean, D., E.L. Williams, E. Grosjean, J.M. Andino, and J.H. Seinfeld. 1993c. At-mospheric oxidation of biogenic hydrocarbons: reaction of ozone with β-pinene, D-liminene, and *trans*-caryophyllene. *Environ. Sci. Technol.* 27: 2754–2758.

Grosser, R.J., D. Warshawsky, and J.R. Vestal. 1991. Indigenous and enhanced min-eralization of pyrene, benzo[a]pyrene, and carbazole in soils. *Appl. Environ. Microbiol.* 57: 3462–3469.

Guengerich, P. 1990. Enzymatic oxidation of xenobiotic chemicals. *Crit. Rev. Biochem. Mol. Biol.* 25: 97–153.

Guerin, W.F. and S.A. Boyd. 1992. Differential bioavailability of soil-sorbed naphthalene by two bacterial species. *Appl. Environ. Microbiol.* 58: 1142–1152

Guerin, W.F. and G.E. Jones. 1988. Mineralization of phenanthrene by a *Mycobacterium* sp. *Appl. Environ. Microbiol.* 54: 937–944.

Guillén, F., A.T. Martínez, and M. J. Martínez. 1992. Substrate specificity and properties of the aryl-alcohol oxidase from the lignolytic fungus *Pleurotus eryngii. Eur. J. Biochem.* 209: 603–611

Gullett, B.K., D.F. Natschke, and K.R.E. Bruce. 1997. Thermal treatment of 1,2,3,4-tetrachlorodibenzo-*p*-dioxin by reaction with Ca-based sorbents at 23–300°C. *Environ. Sci. Technol.* 31: 1855–1862.

Gupta, J.K., C. Jebsen, and H. Kneifel. 1986. Sinapic acid degradation by the yeast *Rhodotorula graminis. J. Gen. Microbiol.* 132: 2793–2799.

Habu, N., M. Samejima, and T. Yoshimoto. 1989. A novel dioxygenase responsible for the C_α–C_β cleavage of lignin model compounds from *Pseudomonas* sp. TMY 1009. *Mokuzai Gakkaishi* 35: 26–29.

Hacisalihoglu, A., J.A. Jongejan, and J.A. Duine. 1997. Distribution of amine oxidases and amine dehydrogenases in bacteria grown on primary amines and characterization of the amine oxidase from *Klebsiella oxytoca. Microbiology* (U.K.) 143: 505–512.

Haemmerli, S.D., M.S.A. Leisola, D. Sanglard, and A. Fiechter. 1986. Oxidation of benzo(a)pyrene by extracellular ligninases of *Phanerochaete chrysosporium.* Veratryl alcohol and stability of ligninase. *J. Biol. Chem.* 261: 6900–6903.

Häggblom, M., D. Janke, P.J.M. Middeldorp, and M. Salkinoja-Salonen. 1988. O-Methylation of chlorinated phenols in the genus *Rhodococcus. Arch. Microbiol.* 152: 6–9.

Häggblom, M.M., L.J. Nohynek, N.J. Palleroni, K. Kronqvist, E.-L. Nurmiaho-Lassila, M.S. Salkinoja-Salonen, S. Klatte, and R.M. Kroppenstedt. 1994. Transfer of polychlorophenol-degrading *Rhodococcus chlorophenolicus* Apajalahti et al. 1986 to the genus *Mycobacterium* as *Mycobacterium chlorophenolicum* comb. nov. *Int. J. Syst. Bacteriol.* 44: 485–493.

Haigler, B.E. and J.C. Spain. 1991. Biotransformation of nitrobenzene by bacteria containing toluene degradative pathways. *Appl. Environ. Microbiol.* 57: 3156–3162.

Haigler, B.E., C.A. Pettigrew, and J.C. Spain. 1992. Biodegradation of mixtures of substituted benzenes by *Pseudomonas* sp. strain JS150. *Appl. Environ. Microbiol.* 58: 2237–2244.

Hall, A.K., J.M. Harrowfied, R.J. Hart, and P.G. McCormick. 1996. Mechanochemical reaction of DDT with calcium oxide. *Environ. Sci. Technol.* 30: 3401–3407

Hall, B.G. 1990. Spontaneous point mutations that occur more often when advantageous than when neutral. *Genetics* 126: 5–16.

Haller, H.D. and R.K. Finn. 1978. Kinetics of biodegradation of *p*-nitrobenzoate and inhibition by benzoate in a pseudomonad. *Appl. Environ. Microbiol.* 35: 890–896.

Hamamura, N., C. Page, T. Long, L. Semprini, and D.H. Arp. 1997. Chloroform cometabolism by butane-grown CF8, *Pseudomonas butanovora,* and *Mycobacterium vaccae* JOB5 and methane-grown *Methylosinus trichosporium* OB3b. *Appl. Environ. Microbiol.* 63: 3607–3613.

Hammel, K.E., B. Kalyanaraman, and T.K. Kirk. 1986. Oxidation of polycyclic aromatic hydrocarbons and dibenzo(*p*)dioxins by *Phanmerochaete chrysosporium* ligninase. *J. Biol. Chem.* 261: 16948–16952.

Hammel, K.E., M.D. Mozuch, K.A. Jensen, and P.J. Kersten. 1994. H_2O_2 recycling during oxidation of the arylglycerol β-aryl ether lignin structure by lignin peroxidase and glyoxal oxidase. *Biochemistry* 33: 13349–13354.

Hammer, E., D. Krpowas, A. Schäfer, M. Specht, W. Francke, and F. Schauer. 1998. Isolation and characterization of a dibenzofuran-degrading yeast: identification of oxidation and ring clavage products. *Appl. Environ. Microbiol.* 64: 2215–2219.

Han, Y.-H., R.M. Smibert, and N.R. Krieg. 1991. *Wolinella recta, Wolinella curva, Bacteroides ureolyticus,* and *Bacteroides gracilis* are microaerophiles, not anaerobes. *Int. J. Syst. Bacteriol.* 41: 218–222.

Hansch, C. and A. Leo. 1979. *Substituent Constants for Correlation Analysis in Chemistry and Biology.* John Wiley & Sons, New York.

Harayama, S., M. Rekik, M. Wubbolts, K. Rose, R.A. Leppik, and K.N. Timmis. 1989. Characterization of five genes in the upper-pathway operon of TOL plasmid pPWW0 from *Pseudomonas putida* and identification of the gene products. *J. Bacteriol.* 171: 5048–5055.

Harayama, S., M. Kok, and E.L. Neidle. 1992. Functional and evolutionary relationships among diverse oxygenases. *Annu. Rev. Microbiol.* 46: 565–601.

Harder, J. and C. Probian. 1997. Anaerobic mineralization of cholesterol by a novel type of denitrifying bacterium. *Arch. Microbiol.* 167: 269–274.

Hardisson, C., J.M. Sala-Trapat, and R.Y. Stanier. 1969. Pathways for the oxidation of aromatic compounds by *Azotobacter. J. Gen. Microbiol.* 59: 1–11.

Hardman, D.J., P.C. Gowland, and J.H. Slater. 1986. Large plasmids from soil bacteria enriched on halogenated alkanoic acids. *Appl. Environ. Microbiol.* 51: 44–51.

Harker, A.R. and Y. Kim. 1990. Trichloroethylene degradation by two independent aromatic-degrading pathways in *Alcaligenes eutrophus* JMP134. *Appl. Environ. Microbiol.* 56: 1179–1181.

Harkness, M.R., J.B. McDermott, D.A. Abramowicz, J.J. Salvo, W.P. Flanagan, M.L. Stephens, F.J. Mondello, R.J. May, J.H. Lobos, K.M. Carroll, M.J. Brennan, A.A. Bracco, K.M. Fish, G.L. Warner, P.R. Wilson, D.K. Dietrich, D.T. Lin, C.B. Morgan, and W.L. Gately. 1993. *In situ* stimulation of aerobic PCB biodegradation in Hudson River sediments. *Science* 259: 503–507.

Harris, R. and C.J. Knowles. 1983. Isolation and growth of a *Pseudomonas* species that utilizes cyanide as a source of nitrogen. *J. Gen. Microbiol.* 129: 1005–1011.

Harvey, J., J.J. Dulka, and J.J. Anderson. 1985. Properties of sulfometuroin methyl affecting its environmental fate: aqueous hydrolysis and photolysis, mobility and adsorption on soils, and bioaccumulation potential. *J. Agric. Food Chem.* 33: 590–596.

Harwood, C.S., N.N. Nichols, M.-K. Kim, J.J. Ditty, and R.E. Parales. 1994. Identification of the *pcaRKF* gene cluster from *Pseudomonas putida*: involvement in chemotaxis, biodegradation, and transport of 4-hydroxybenzoate. *J. Bacteriol.* 176: 6479–6488.

Hasegawa, K., and nine coauthors. 1993. Photooxidation of naphthalenamines adsorbed on particles under simulated atmospheric conditions. *Environ. Sci. Technol.* 27: 1819–1825.

Hashimoto, S., T. Azuma, and A. Otsuki. 1998. Distribution, sources, and stability of haloacetic acids in Tokyo Bay, Japan. *Environ. Toxicol. Chem.* 17: 798–805.

Hassett, D.J., H.P. Schweitzer, and D.E. Ohman. 1995. *Pseudomonas aeruginosa sodA* and *sodB* mutants defective in manganese- and iron-cofactored superoxide dismutase activity demonstrate the importance of the iron-cofactored form in aerobic metabolism. *J. Bacteriol.* 177: 6330–6337.

Hata, S., T. Nishino, H. Katsuki, Y. Aoyama, and Y. Yoshida. 1983. Two species of cytochrome P-450 involved in ergosterol biosynthesis in yeast. *Biochem. Biophys. Res. Commun.* 116: 162–166.

Hatzinger, P.B. and M. Alexander. 1995. Effect of aging of chemicals in soil on their biodegradability and extractability. *Environ. Sci. Technol.* 29: 537–545.

Havel, J. and W. Reineke. 1993. Microbial degradation of chlorinated acetophenones. *Appl. Environ. Microbiol.* 59: 2706–2712.

Hay, A.G. and D.D. Focht. 1998. Cometabolism of 1,1-dichloro-2,2-bis(4-chlorophenyl)ethylene by *Pseudomomas acidovorans* M3GY grown on biphenyl. *Appl. Environ. Microbiol.* 64: 2141–2146.

Hayman, G.D. and R.G. Derwent. 1997. Atmospheric chemical reactivity and ozone-forming potentials of potential CFC replacements. *Environ. Sci. Technol.* 31: 317–336.

Heald, S. and R.O. Jenkins. 1994. Trichloroethylene removal and oxidation toxicity mediated by toluene dioxygenase of *Pseudomonas putida*. *Appl. Environ. Microbiol.* 60: 4634–4637.

Hedlund, B.P., A.D. Geiselbrecht, T.J. Bair, and J.T. Staley. 1999. Polycyclic aromatic hydrocarbon degradation by a new marine becterium, *Neptunomonas naphthovorans* gen. nov., sp. nov. *Appl. Environ. Microbiol.* 65: 251–259.

Hegeman, G.D. 1966. Synthesis of the enzymes of the mandelate pathway by *Pseudomonas putida*. I. Synthesis of enzymes of the wild type. *J. Bacteriol.* 91: 1140–1154.

Heijman, C.G., C. Holliger, M.A. Glaus, R.P. Schwarzenbach, and J. Zeyer. 1993. Abiotic reduction of 4-chloronitrobenzene to 4-chloroaniline in a dissimilatory iron-reducing enrichment culture. *Appl. Environ. Microbiol.* 59: 4350–4353.

Heitkamp, M.A., J.P. Freeman, D.W. Miller, and C.E. Cerniglia. 1988. Pyrene degradation by a *Mycobacterium* sp.: identification of ring oxidation and ring fission products. *Appl. Environ. Microbiol.* 54: 2556–2565.

Helmig, D., J. Arey, W.P. Harger, R. Atkinson, and J. López-Cancio. 1992a. Formation of mutagenic nitrodibenzopyranones and their occurrence in ambient air. *Environ. Sci. Technol.* 26: 622–624

Helmig, D., J. López-Cancio, J. Arey, W.P. Harger, and R. Atkinson. 1992b. Quantification of ambient nitrodibenzopyranones: further evidence for atmospheric mutagen formation. *Environ. Sci. Technol.* 26: 2207–2213.

Henriksen, H., V., S. Larsen, and B.K. Ahring. 1991. Anaerobic degradation of PCP and phenol in fixed-film reactors: the influence of an additional substrate. *Water Sci. Technol.* 24: 431–436.

Henry, S.M. and D. Grbic-Galic. 1991. Inhibition of trichloroethylene oxidation by the transformation intermediate carbon monoxide. *Appl. Environ. Microbiol.* 57: 1770–1776

Herbes, S.E. and L.R. Schwall. 1978. Microbial transformation of polycyclic aromatic hydrocarbons in pristine and petroleum-contaminated sediments. *Appl. Environ. Microbiol.* 35: 306–316.

Herrick, J.B., K.G. Stuart-Keil, W.C. Ghiorse, and E.L. Madsen. 1997. Natural horizontal transfer of a naphthalene dioxygenase gene between bacteria native to a coal tar-contaminated field site. *Appl. Environ. Microbiol.* 63: 2330–2337.

Higgins, S.J. and J. Mandelstam. 1972. Evidence for induced synthesis of an active transport factor for mandelate in *Pseudomonas putida*. *Biochem. J.* 126: 917–922.

Hilbi, H., R. Hermann, and P. Dimroth. 1993. The malonate decarboxylase enzyme system of *Malonomonas rubra*: evidence for the cytoplasmic location of the biotin-containing component. *Arch. Microbiol.* 160: 126–131.

Hileman, B. 1993. Concerns broaden over chlorine and chlorinated hydrocarbons. *Chem. Eng. News* 71(16): 11–20.

Hino, S. and H. Tauchi. 1987. Production of carbon monoxide from aromatic acids by *Morganella morganii*. *Arch. Microbiol.* 148: 167–171.

Ho, P.C. 1986. Photooxidation of 2,4-dinitrotoluene in aqueous solution in the presence of hydrogen peroxide. *Environ. Sci. Technol.* 20: 260–267.

Hofrichter, M., K. Scheibner, I. Schneegaß, and W. Fritsche. 1998. Enzymatic combustion of aromatic and aliphatic compounds by manganese peroxidase from *Nematoloma frowardii*. *Appl. Environ. Microbiol.* 64: 399–404.

Höhl, H.-U. and W. Barz. 1995. Metabolism of the insecticide phoxim in plants and cell suspension cultures of soybean. *J. Agric. Food Chem.* 43: 1052–1056.

Holben, W.E., J.K. Jansson, B.K. Chelm, and J.M. Tiedje. 1988. DNA probe method for the detection of specific microorganisms in the soil bacterial community. *Appl. Environ. Microbiol.* 54: 703–711.

Holben, W.E., B.M. Schroeter, V.G.M. Calabrese, R.H. Olsen, J.K. Kukor, V.O. Biederbeck, A.E. Smith, and J.M. Tiedje. 1992. Gene probe analysis of soil microbial populations selected by amendment with 2,4-dichlorophenoxyacetic acid. *Appl. Environ. Microbiol.* 58: 3941–3948.

Holden, P.A., L.J. Halverson, and M.K. Firestone. 1997. Water stress effects on toluene biodegradation by *Pseudomonas putida*. *Biodegradation* 8: 143–151.

Hollender, J., W. Dott, and J. Hopp. 1994. Regulation of chloro- and methylphenol degradation in *Comamonas testosteroni* JH5. *Appl. Environ. Microbiol.* 60: 2330–2338.

Holm, P.E., P.H. Nielsen, H.-J. Albrechtsen, and T.H. Christensen. 1992. Importance of unattached bacteria and bacteria attached to sediment in determining potentials for degradation of xenobiotic organic contaminants in an aerobic aquifer. *Appl. Environ. Microbiol.* 58: 3020–3026.

Holmstead, R.L. 1976. Studies of the degradation of mirex with an iron (II) porphyrin model system. *J. Agric. Food Chem.* 24: 620–624.

Hommel, R.K. 1994. Formation and function of biosurfactants for degradation of water-insoluble substrates. In *Biochemistry of Microbial Degradation* (Ed. C. Rattledge) pp. 63–87. Kluwer Academic Publishers, Dordrecht, The Netherlands.

Hong, F. and S. Pehkonen. 1998. Hydrolysis of phorate using simulated and environmental conditions: rates, mechanisms, and product analysis. *J. Agric. Food Chem.* 46: 1192–1199.

Hoover, D.G., G.E. Borgonovi, S.H. Jones, and M. Alexander. 1986. Anomalies in mineralization of low concentrations of organic compounds in lake water and sewage. *Appl. Environ. Microbiol.* 51: 226–232.

Hopkins, G.D. and P.L. McCarty. 1995. Field evaluation of *in situ* aerobic cometabolism of trichloroethylene and three dichloroethylene isomers using phenol and toluene as primary substrates. *Environ. Sci. Technol.* 29: 1628–1637.

Hopper, D.J. 1976. The hydroxylation of p-cresol and its conversion to p-hydroxybenzaldehyde in *Pseudomonas putida*. *Biochem. Biophys. Res. Commun.* 69: 462–468.

Horvath, R.S. 1971. Cometabolism of the herbicide 2,3,6-trichlorobenzoate. *J. Agric. Food Chem.* 19: 291–293.

Horvath, R.S. 1972. Microbial cometabolism and the degradation of organic compounds in nature. *Bacteriol. Rev.* 36: 146–155.

Horvath, R.S. and P. Flathman. 1976. Cometabolism of fluorobenzoates by natural microbial populations. *Appl. Environ. Microbiol.* 31: 889–891.

Houwen, F.P., C. Dijkema, A.J.M. Stams, and A.J.B. Zehnder. 1991. Propionate metabolism in anaerobic bacteria; determination of carboxylation reactions with ^{13}C-NMR spectroscopy. *Biochim. Biophys. Acta* 1056: 126–132.

Howell, J.G., T. Spector, and V. Massey. 1972. Purification and properties of *p*-hydroxybenzoate hydroxylase from *Pseudomonas fluorescens. J. Biol. Chem.* 247: 4340–4350.

Hudlický, M. 1990. *Oxidations in Organic Chemistry.* American Chemical Society, Washington, D.C.

Hung, C.-H. and B.J. Marinas. 1997. Role of chlorine and oxygen in the photocatalytic degradation of trichloroethylene vapor on TiO_2 films. *Environ. Sci. Technol.* 31: 562–568.

Hur, H.-G., L.M. Newman, L.P. Wackett, and M.J. Sadowsky. 1997. Toluene 2-monooxygenase-dependent growth of *Burkholderia cepacia* G4/PR1 on diethyl ether. *Appl. Environ. Microbiol.* 63: 1606–1609.

Hurtubise, Y., D. Barriault, and M. Sylvestre. 1998. Involvement of the terminal oxygenase β subunit in the biphenyl dioxygenase reactivity pattern towards chlorobiphenyls. *J. Bacteriol.* 180: 5828–5835.

Hutchins, S.R. 1991. Biodegradation of monoaromatic hydrocarbons by aquifer microorganisms using oxygen, nitrate or nitrous oxide as the terminal electron acceptors. *Appl. Environ. Microbiol.* 57: 2403–2407.

Hwang, H.-M., R.E. Hodson, and D.L. Lewis. 1989. Microbial degradation kinetics of toxic organic chemicals over a wide range of concentrations in natural aquatic systems. *Environ. Toxicol. Chem.* 8: 65–74.

Hyman, M.R., A.W. Sansome-Smith, J.H. Shears, and P.M. Wood. 1985. A kinetic study of benzene oxidation to phenol by whole cells of *Nitrosomonas europaea* and evidence for the further oxidation of phenol to hydroquinone. *Arch. Microbiol.* 143: 302–306.

Hyman, M.R., I.B. Murton, and D.J. Arp. 1988. Interaction of ammonia monooxygenase from *Nitrosomonas europaea* with alkanes, alkenes, and alkynes. *Appl. Environ. Microbiol.* 54: 3187–3190.

Hyman, M.R., C.L. Page, and D.J. Arp. 1994. Oxidation of methyl fluoride and dimethyl ether by ammonia monooxygenase in *Nitrosomonas europaea. Appl. Environ. Microbiol.* 60: 3033–3035.

Imhoff-Stuckle, D. and N. Pfennig. 1983. Isolation and characterization of a nicotinic acid-degrading sulfate-reducing bacterium, *Desulfococcus niacini* sp. nov. *Arch. Microbiol.* 136: 194–198.

Itoh, S. and T. Suzuki. 1972. Effect of rhamnolipids on growth of *Pseudomonas aeruginosa* mutant deficient in *n*-paraffin-utilizing ability. *Agr. Biol. Chem.* 36: 2233–2235.

Iwabuchi, T. and S. Harayama. 1998. Biochemical and molecular characterization of 1-hydroxy-2-naphthoate dioxygenase from *Nocardioides* sp. KP7. *J. Biol. Chem.* 273: 8332–8336,

Jagnow, G., K. Haider, and P.-C. Ellwardt. 1977. Anaerobic dechlorination and degradation of hexachlorocyclohexane by anaerobic and facultatively anaerobic bacteria. *Arch. Microbiol.* 115: 285–292.

Jahnke, M., T. El-Banna, R. Klintworth, and G. Auling. 1990. Mineralization of ortha-nilic acid is a plasmid-associated trait in *Alcaligenes* sp. O-1. *J. Gen. Microbiol.* 136: 2241–2249.

Jahnke, M., F. Lehmann, A. Schoebel, and G. Auling. 1993. Transposition of the TOL catabolic genes *Tn46521* into the degradative plasmid pSAH of *Alcaligenes* sp. O-1 ensures simultaneous mineralization of sulpho- and methyl-substituted aromatics. *J. Gen. Microbiol.* 139: 1959–1966.

Jain, R.K., G.S. Sayler, J.T. Wilson, L. Houston, and D. Pacia. 1987. Maintenance and stability of introduced genotypes in groundwater aquifer material. *Appl. Environ. Microbiol.* 53: 996–1002.

Jain, R.K., J.H. Dreisbach, and J.C. Spain. 1994. Biodegradation of *p*-nitrophenol via 1,2,4-benzenetriol by an *Arthrobacter* sp. *Appl. Environ. Microbiol.* 60: 3030–3032.

James, K.D. and P.A. Williams. 1998. *ntn* genes determining the early steps in the divergent catabolism of 4-nitrotoluene and toluene in *Pseudomonas* sp. strain TW3. *J. Bacteriol.* 180: 2043–2049.

James, S.L. 1995. Role of nitric oxide in parasitic infections. *Microbiol. Rev.* 59: 533–547.

Jang, M. and S.R. McDow. 1997. Products of benz/*a*/anthracene photodegradation in the presence of known organic constituents of atmospheric aerosols. *Environ. Sci. Technol.* 31: 1046–1053.

Janke, D. 1987. Use of salicylate to estimate the threshold inducer level for *de novo* synthesis of the phenol-degrading enzynes in *Pseudomonas putida* strain H. *J. Basic Microbiol.* 27: 83–89.

Janssen, P.H. and C.G. Harfoot. 1992. Anaerobic malonate decarboxylation by *Citrobacter diversus*. *Arch. Microbiol.* 157: 471–474.

Japar, S.M., T.J. Wallington, S.J. Rudy, and T.Y. Chang. 1991. Ozone-forming potential of a series of oxygenated organic compounds. *Environ. Sci. Technol.* 25: 415–420.

Järvinen, K.T., E.S. Melin, and J.A. Puhakka. 1994. High-rate bioremediation of chlorophenol-contaminated groundwater at low temperatures. *Environ. Sci. Technol.* 28: 2387–2392.

Jeftic, L. and R.N. Adams. 1970. Electrochemical oxidation pathways of benzo(a)pyrene. *J. Am. Chem. Soc.* 92: 1332–1337.

Jessee, J.A., R.E. Benoit, A.C. Hendricks, G.C. Allen, and J.L. Neal. 1983. Anaerobic degradation of cyanuric acid, cysteine, and atrazine by a facultative anaerobic bacterium. *Appl. Environ. Microbiol.* 45: 97–102.

Ji, X.-B. and T.C. Hollocher. 1988. Mechanism for nitrosation of 2,3-diaminonaphthalene by *Escherichi coli*: enzymatic production of NO followed by O_2-dependent chemical nitrosation. *Appl. Environ. Microbiol.* 54: 1791–1794.

Jiang, H., R.E. Parales, and D.T. Gibson. 1999. The α subunit of toluene dioxygenase from *Pseudomonas putida* F1 can accept electrons from reduced ferredoxin$_{TOL}$ but is catalytically inactive in the absence of the β subunit. *Appl. Environ. Microbiol.* 65: 315–318.

Jiang, Z.-Y. and C. Thorpe. 1983. Acyl-CoA oxidase from *Candida tropicalis*. *Biochemistry* 22: 3752–3758.

Jin, G. and A.J. Englande. 1997. Biodegradation kinetics of carbon tetrachloride by *Pseudomonas cepacia* under varying oxzidation-reduction potential conditions. *Water Environ. Res.* 69: 1094–1099.

Johnson, G.R. and R.H. Olsen. 1995. Nucleotide sequence analysis of genes encoding a toluene/benzene-2-monooxygenase from *Pseudomonas* sp. strain JS150. *Appl. Environ. Microbiol.* 61: 3336–3346.

Johnson, G.R. and R.H. Olsen. 1997. Multiple pathways for toluene degradation in *Burkholderia* sp. strain JS150. *Appl. Environ. Microbiol.* 63: 4047–4052.

Johnson, M.S., I.B. Zhulin, M.-E.R. Gapuzan, and B.L. Taylor. 1997. Oxygen-dependent growth of the obligate anaerobe *Desulfovibrio vulgaris* Hildenborough. *J. Bacteriol.* 179: 5598–5601.

Jones, K.H., P.W. Trudgill, and D.J. Hopper. 1995. Evidence for two pathways for the metabolism of phenol by *Aspergillus fumigatus*. *Arch. Microbiol.* 163: 176–181.

Jones, R.D. and R.Y. Morita. 1983. Methane oxidation by *Nitrosococcus oceanus* and *Nitrosomonas europaea*. *Appl. Environ. Microbiol.* 45: 401–410.

Junker, F. and A.M. Cook. 1997. Conjugative plasmids and the degradation of arylsulfoinates in *Comamonas testosteroni*. *Appl. Environ. Microbiol.* 63: 2403–2410.

Kadiyala, V. and J.C. Spain. 1998. A two-component monoxygenase catalyzes both the hydroxylation of *p*-nitrophenol and the oxidative release of nitrite from 4-nitrocatechol in *Bacillus sphaericus* JS905. *Appl. Environ. Microbiol.* 64: 2479–2484.

Kamagata, Y., R.R. Fulthorpe, K. Tamura, H. Takami, L.J. Forney, and J.M. Tiedje. 1997. Pristine environments harbor a new group of oligotrophic 2,4-dichlorophenoxyacetic acid-degrading bacteria. *Appl. Environ. Microbiol.* 63: 2266–2272.

Kanagawa, K., S. Negoro, N. Takada, and H. Okada. 1989. Plasmid dependence of *Pseudomonas* sp. strain NK87 enzymes that degrade 6-aminohexanoate-cyclic dimer. *J. Bacteriol.* 171: 3181–3186.

Käppeli, O. 1986. Cytochromes P-450 of yeasts. *Microbiol. Rev.* 50: 244–258.

Kargalalioglu, Y. and J.A. Imlay. 1994. Importance of anaerobic superoxide dismutase synthesis in facilitating outgrowth of *Escherichia coli* upon entry into an aerobic habitat. *J. Bacteriol.* 176: 7653–7658.

Kari, F.G. and W. Giger. 1995. Modelling the photochemical degradation of ethylenediaminetetraacetate in the River Glatt. *Environ. Sci. Technol.* 29: 2814–2827.

Karlson, U., D.F. Dwyer, S.W. Hooper, E.R.B. Moore, K.N. Timmis, and L.D. Eltis. 1993. Two independently regulated cytochromes P-450 in a *Rhodococcus rhodochrous* strain that degrades 2-ethoxyphenol and 4-methoxybenzoate. *J. Bacteriol.* 175: 1467–1474.

Karns, J.S. and R.W. Eaton. 1997. Genes encoding *s*-triazine degradation are plasmidborne in *Klebsiella pneumoniae* strain 99. *J. Agric. Food Chem.* 45: 1017–1022.

Kaschabek, S.R., T. Kasberg, D. Müller, A.E. Mars, D.B. Janssen, and W. Reineke. 1998. Degradation of chloroaromatics: purification and characterization of a novel type of chlorocatechol 2,3-dioxygenase of *Pseudomonas putida* GJ31. *J. Bacteriol.* 180: 296–302.

Katagiri, M., B.N. Ganguli, and I.C. Gunsalus. 1968. A soluble cytochrome P-450 functional in methylene hydroxylation. *J. Biol. Chem.* 243: 3543–3546.

Kawaguchi, A., S. Tsubotani, Y. Seyama, T. Yamakawa, T. Osumi, T. Hashimoto, Y. Kikuchi, M. Ando, and S. Okuda. 1980. Stereochemistry of dehydrogenation catalyzed by acyl-CoA oxidase. *J. Biochem.* 88: 1481–1486.

Kawai, F. and H. Yamanaka. 1986. Biodegradation of polyethyelene glycol by symbiotic mixed culture (obligate mutualism). *Arch. Microbiol.* 146: 125–129.

Kawamoto, S., C. Nozaki, A. Tanaka, and S. Fukui. 1978. Fatty acid β–oxidation system in microbodies of *n*-alkane-grown *Candida tropicalis*. *Eur. J. Biochem.* 83: 609–613.

Kawasaki, H., H. Yahara, and K. Tonomura. 1981. Isolation and characterization of plasmid pUO1 mediating dehalogenation of haloacetate and mercury resistance in *Moraxella* sp. B. *Agric. Biol. Chem.* 45: 1477–1481.

Keener, W.K. and D.J. Arp. 1994. Transformations of aromatic compounds by *Nitrosomonas europaea. Appl. Environ. Microbiol.* 60: 1914–1921.

Kelley, I., J.P. Freeman, and C.E. Cerniglia. 1990. Identification of metabolites from degradation of naphthalene by a *Mycobacterium* sp. *Biodegradation* 1: 283–290.

Kelly, D.P. 1971. Autotrophy: concepts of lithotrophic bacteria and their organic metabolism. *Annu. Rev. Microbiol.* 25: 177–210.

Kennedy, D.W., S.D. Aust, and J.A. Bumpus. 1990. Comparative biodegradation of alkyl halide insecticides by the white rot fungus, *Phanerochaete chrysosporium* (BKM-F-1767). *Appl. Environ. Microbiol.* 56: 2347–2353.

Kersten, P.J. 1990. Glyoxal oxidase of *Phanerochaete chrysosporium*: its characterization and activation by lignin peroxidase. *Proc. Natl. Acad. Sci. U.S.A.* 87: 2936–2940.

Keshavarz, T., M.D. Lilly, and P.C. Clarke. 1985. Stability of a catabolic plasmid in continuous culture. *J. Gen. Microbiol.* 131: 1193–1203.

Khanna, P., B. Rajkumar, and N. Jothikumar. 1992. Anoxygenic degradation of aromatic substances by *Rhodopseudomonas palustris. Curr. Microbiol.* 25: 63–67.

Kido, T., K. Hashizume, and K. Soda. 1978a. Purification and properties of nitroalkane oxidase from *Fusarium oxysporium. J. Bacteriol.* 133: 53–58.

Kido, T., K. Sida, and K. Asada. 1978b. Properties of 2-nitropropane dioxygenase of *Hansenula mrakii. J. Biol.Chem.* 253: 226–232.

Kiene, R.P. and B.F. Taylor. 1988. Demethylation of dimethylsulfoniopropionate and production of thiols in anoxic marine sediments. *Appl. Environ. Microbiol.* 54: 2208–2212.

Kimura, N., A. Nishi, M. Goto, and K. Furukawa. 1997. Functional analyses of a variety of chimeric dioxygenases constructed from two biphenyl dioxyganses that are similar structurally but different functionally. *J. Bacteriol.* 179: 3936–3943.

King, D.J., M.R. Azari, and A. Wiseman. 1984. Studies on the properties of highly purified cytochrome P-448 and its dependent activity benzo[a]pyrene hydroxylase, from *Saccharomyces cerevisiae. Xenobiotica* 14: 187–206.

Kirchner, F., L.P. Thüner, I. Barnes, K.H. Becker, B. Donner, and F. Zabel. 1997. Thermal lifetimes of peroxynitrates occurring in the atmospheric degradation of oxygenated fuel additives. *Environ. Sci. Technol.* 31: 1801–1804.

Kirk, T.K. and R.L. Farrell. 1987. Enzymatic "combustion": the microbial degradation of lignin. *Annu. Rev. Microbiol.* 41: 465–505.

Kitts, C.L., D.P. Cunningham, and P.J. Unkefer. 1994. Isolation of three hexahydro-1,3,5-trinitro-1,3,5-triazine-degrading species of the family *Enterobacteriaceae* from nitramine explosive-contaminated soil. *Appl. Environ. Microbiol.* 60: 4608–4711.

Kiyohara, H., T. Hatta, Y. Ogawa, T. Kakuda, H. Yokoyama, and N. Takizawa. 1992. Isolation of *Pseudomonas pickettii* strains that degrade 2,4,6-trichlorophenol and their dechlorination of chlorophenols. *Appl. Environ. Microbiol.* 58: 1276–1283.

Kjellberg, S., M. Hermansson, P. Mårdén, and G.W. Jones. 1987. The transient phase between growth and nongrowth of heterotrophic bacteria with emphasis on the marine environment. *Annu. Rev. Microbiol.* 41: 25–49.

Klecka, G.M. and D.T. Gibson. 1981. Inhibition of catechol 2,3-dioxygenase from *Pseudomonas putida* by 3-chlorocatechol. *Appl. Environ. Microbiol.* 41: 1159–1165.

Klecka, G.M. and W.J. Maier. 1988. Kinetics of microbial growth on mixtures of pentachlorophenol and chlorinated aromatic compounds. *Biotechnol. Bioeng.* 31: 328–335.

Kleier, D., I. Holden, J.E. Casida, and L.O. Ruzo. 1985. Novel photoreactions of an insecticidal nitromethylene heterocycle. *J. Agric. Food Chem.* 33: 998–1000.

Kneifel, H., K. Elmendorff, E. Hegewald, and C.J. Soeder. 1997. Biotransformation of 1-naphthalenesulfonic acid by the green alga *Scenedesmus obliquus*. *Arch. Microbiol.* 167: 32–37.

Koenig, D.W. and H.B. Ward. 1983. Prototheca zopfii Krüger strain UMK-13 growth on acetate or *n*-alkanes. *Appl. Environ. Microbiol.* 45: 333–336.

Koga, H., H. Aramaki, E. Yamaguchi, K. Takeuchi, T. Horiuchi, and I.C. Gunsdalus. 1986. *camR*, a negative regulator locus of the cytochrome P-450$_{cam}$ hydroxylase operon. *J. Bacteriol.* 166:1089–1095.

Koizumi, A., N. Saitoh, T. Suzuki, and S. Kamiyama. 1994. A novel compound, 9-hydroxy-1-nitropyrene, is a major photodegraded compound of 1-nitropyrene in the environment. *Arch. Environ. Health.* 49: 87–92.

Kolattukudy, P.E. and L. Hankin. 1968. Production of omega-haloesters from alkyl halides by *Micrococcus cerificans*. *J. Gen. Microbiol.* 54: 145–153.

Kramer, J.B., Canonica, S., Hoigné, and Kaschig, J. 1996. Degradation of fluorescent whitening agents in sunlit natural waters. *Environ. Sci. Technol.* 30: 2227–2234.

Kreiss, M., J. Eberspächer, and F. Linmgens. 1981. Detection and characterization of plasmids in Chloridazon and Antipyrin degrading bacteria. *Zentrabl. Bakteriol. Hyg., I. Abt. Orig.* C2: 45–60.

Krishnamurty, H.G. and F.J. Simpson. 1970. Degradation of rutin by *Aspergillus flavus*. Studies with oxygen 18 on the action of a dioxygenase on quercitin. *J. Biol. Chem.* 245: 1467–1471.

Krone, U.E., R.K. Thauer, and H.P.C. Hogenkamp. 1989. Reductive dehalogenation of chlorinated C1-hydrocarbons mediated by corrinoids. *Biochemistry* 28: 4908–4914.

Krooneman, J., E.B.A. Wieringa, E.R.B. Moore, J. Gerritse, R.A. Prins, and J.C. Gottschal. 1996. Isolation of *Alcaligenes* sp. strain L6 at low oxygen concentrations and degradation of 3-chlorobenzoate via a pathway not involving (chloro)catechols. *Appl. Environ. Microbiol.* 62: 2427–2434.

Krooneman, J., E.R.B. Moore, J. C.L. van Velzen, R.A. Prins, L.J. Forney, and J.C. Gottschal. 1998. Competition for oxygen and 3-chlorobenzoate between two aerobic bacteria using different degradation pathways. *FEMS Microbiol. Ecol.* 26: 171–179.

Kropp, K.G., J. A. Goncalves, J.T. Anderson, and P.M. Fedorak. 1994. Microbially mediated formation of benzonaphthothiophenes from benzo[*b*]thiophenes. *Appl. Environ. Microbiol.* 60: 3624–3631.

Kuhm, A.E., A. Stolz, K.-L. Ngai, and H.-J. Knackmuss. 1991. Purification and characterization of a 1,2-dihydroxynaphthalene dioxygenase from a bacterium that degrades naphthalenesulfonic acids. *J. Bacteriol.* 173: 3795–3802.

Kukor, J.J. and R.H. Olsen. 1996. Catechol 2,3-dioxygenases functional in oxygen-limited hypoxic environments. *Appl. Environ. Microbiol.* 62: 1728–1740.

Kunisaki, N. and M. Hayashi. 1979. Formation of N-nitrosamines from secondary amines and nitrite by resting cells of *Escherichi coli* B. *Appl. Environ. Microbiol.* 37: 279–282.

Kuritz, T. and C.P. Wolk. 1995. Use of filamentous cyanobacteria for biodegradation of organic pollutants. *Appl. Environ. Microbiol.* 61: 234–238.

Kuritz, T., L.V. Bocanera, and N.S. Rivera. 1997. Dechlorination of lindane by the cyanobacterium *Anabaena* sp. strain PCC 7120 depends on the function of the *nir* operon. *J. Bacteriol.* 179: 3368–3370.

Kutney, J.P., E. Dimitriadis, G.M. Hewitt, P.J. Salisbury, and M. Singh. 1982. Studies related to biological detoxification of kraft mill effluent. IV — The biodegradation of 14-chlorodehydroabietic acid with *Mortierella isabellina*. *Helv. Chem. Acta* 65: 1343–1350.

Kwok, E.S.C., R. Atkinson, and J. Arey. 1995. Rate constants for the gas-phase reactions of the OH radical with dichlorobiphenyls, 1-chlorodibenzo-*p*-dioxin, 1,2-dimethoxybenzene, and diphenyl ether: estimation of OH radical reaction rate constants for PCBs, PCDDs, and PCDFs. *Environ. Sci. Technol.* 29: 1591–1598.

Landner, L. (Ed.) 1989. *Chemicals in the Aquatic Environment*. Springer-Verlag, Berlin.

Lang, K.F. and H. Buffleb. 1958. Die Pyrolyse des α- and β-methyl-naphthalins. *Chem. Ber.* 91: 2866–2870.

Lange, C.C. and L.P. Wackett. 1997. Oxidation of aliphatic olefins by toluene dioxygenase: enzyme rates and product identification. *J. Bacteriol.* 179: 3858–3865.

Latus, M., H.-J. Seitz, J. Eberspächer, and F. Lingens. 1995. Purification and characterization of hydroxyquinol 1,2-dioxygenase from *Azotobacter* sp. strain GP1. *Appl. Environ. Microbiol.* 61: 2453–2460.

Lauff, J.L., D.B. Steele, L.A. Coogan, and J.M. Breitfeller. 1990. Degradation of the ferric chelate of EDTA by a pure culture of an *Agrobacterium* sp. *Appl. Environ. Microbiol.* 56: 3346–3353.

Laverman, A.M., J.S. Blum, J.K. Schaefer, E.J.P. Phillips, D.R. Lovley, and R.S. Oremland. 1995. Growth of strain SES-3 with arsenate and other diverse electron acceptors. *Appl. Environ. Microbiol.* 61: 3556–3561.

Lee, H. and R.G. Harvey. 1986. Synthesis of the active diol epoxide metabolites of the potent carcinogenic hydrocarbon 7,12-dimethybenz[*a*]anthrene. *J. Am. Chem. Soc.* 51: 3502–3507.

Lee, K. and D.T. Gibson. 1996. Toluene and ethylbenzene oxidation by purified naphthalene dioxygenase from *Pseudomonas* sp. strain NCIB 9816-4. *Appl. Environ. Microbiol.* 62: 3101–3106

Lee-Ruff, E., H. Kazarians-Moghaddam, and M. Katz. 1986. Controlled oxidations of benzo[*a*]pyrene. *Can. J. Chem.* 64: 11297–11303.

Lehning, A., U. Fgock, R.-M. Wittich, K.N. Timmis, and D.H. Pieper. 1997. Metabolism of chlorotoluenes by *Burkholderia* sp. strain PS12 and toluene dioxygenase of *Pseudomonas putida* F1: evidence for monooxygenation by toluene and chlorobenzene dioxygenases. *Appl. Environ. Microbiol.* 63: 1974–1979.

Lesage, S., S. Brown, and K.R. Hosler. 1992. Degradation of chlorofluorocarbon-113 under anaerobic conditions. *Chemosphere* 24: 1225–1243.

Leveau, J.H.L., A.J.B. Zehnder, and J.R. van der Meer. 1998. The *tfdK* gene product facilitates uptake of 2,4-dichlorophenoxyacetate by *Ralstonia eutropha* JMP134(pJP4). *J. Bacteriol.* 180: 2237–2243.

Levine, A. A. and A.G. Cole. 1932. The ozonides of ortho-xylene and the structure of the benzene ring. *J. Am. Chem. Soc.* 54: 338–341.

Levitt, M.S., R.F. Newton, S.M. Roberts, and A.J. Willetts. 1990. Preparation of optically active 6′-fluorocarbocyclic nucleosides utilizing an enantiospecific enzyme-catalysed Baeyer-Villiger type oxidation. *J. Chem. Soc. Chem. Commun.* 619–620

Levsen, K. 1988. The analysis of diesel particulate. *Fresenius Z. Anal. Chem.* 331: 467–478.

Lewis, D.L., R.E. Hodson, and L.F. Freeman. 1985. Multiphase kinetics for transformation of methyl parathion by *Flavobacterium* species. *Appl. Environ. Microbiol.* 50: 553–557.

Lewis, D.L., H.P. Kollig, and R.E. Hodson. 1986. Nutrient limitation and adaptation of microbial populations to chemical transformations. *Appl. Environ. Microbiol.* 51: 598–603.

Li, B., P.L. Gutierraz, and N.V. Bloiugh. 1997. Trace determination of hydroxyl radical in biological systems. *Anal. Chem.* 69: 4295–4302.

Li, D.-Y., J. Eberspächer, B. Wagner, J. Kuntzer, and F. Lingens. 1992. Degradation of 2,4,6-trichlorophenol by *Azotobacter* sp. strain GP1. *Appl. Environ. Microbiol.* 57: 1920–1928.

Li, S. and L.P. Wackett. 1993. Reductive dehalogenation by cytochrome $P450_{CAM}$: substrate binding and catalysis. *Biochemistry* 32: 9355–9361.

Libby, R.D., J.A. Thomas, L.W. Kaiser, and L.P. Hager. 1982. Chloroperoxidase halogenation reactions. *J. Biol. Chem.* 257: 5030–5037.

Lidstrom, M.E. and D.I. Stirling. 1990. Methylotrophs: genetics and commercial applications. *Annu. Rev. Microbiol.* 44: 27–58.

Lidstrom-O'Connor, M.E., G.L. Fulton, and A.E. Wopat. 1983. "*Methylobacterium ethanolicum*:" a syntrophic association of two methylotrophic bacteria. *J. Gen. Microbiol.* 129: 3139–3148.

Liebert, F. 1909. The decomposition of uric acid by bacteria. *Proc. K. Ned. Akad. Wet.* 12: 54–64.

Liesack, W. and K. Finster. 1994. Phylogenetic analysis of five strains of Gram-negative, obligately anaerobic, sulfur-reducing bacteria and description of *Desulfuromusa* gen. nov., including *Desulfuromusa kysingii* sp.nov., *Desulfuromusa bakii* sp. nov., and *Desulfuromusa succinoxidan*s sp. nov. *Int. J. Syst. Bacteriol.* 44: 753–758.

Lijinsky, W. and C.R. Taha. 1961.The pyrolysis of 2-methylnaphthalene. *J. Org. Chem.* 26: 3566–3568

Lindblad, B., G. Lindstedt, and S. Lindstedt. 1970. The mechanism of enzymatic formation of homogentisate from *p*-hydroxyphenylpyruvate. *J. Am. Chem. Soc.* 92: 7446–7449.

Lindsay, R.F. and F.G. Priest. 1975. Decarboxylation of substituted cinnamic acids by enterobacteria: the influence on beer flavour. *J. Appl. Bacteriol.* 39: 181–187.

Lindstedt, S., B. Odelhög, and M. Rundgren. 1977. Purification and properties of 4-hydroxyphenylpyruvate dioxygenase from *Pseudomonas* sp. P.J. 874. *Biochemistry* 16: 3369–3377.

Lipscomb, J.D. 1994. Biochemistry of the soluble methane monooxygenase. *Annu. Rev. Microbiol.* 48: 371–399

Little, C.D., A.V. Palumbo, S.E. Herbes, M.E. Lidström, R.L. Tyndall, and P.J. Gilmour. 1988. Trichloroethylene biodegradation by a methane-oxidizing bacterium. *Appl. Environ. Microbiol.* 54: 951–956.

Liu, D., W.M.J. Strachan, K. Thomson, and K. Kwasniewska. 1981. Determination of the biodegradability of organic compounds. *Environ. Sci. Technol.* 15: 788–793.

Liu, S.-Y., Z. Zheng, R. Zhang, and J.-M. Bollag. 1989. Sorption and metabolism of metolachlor by a bacterial community. *Appl. Environ. Microbiol.* 55: 733–740.

Lloyd-Jones, G., R.C. Ogden, and P.A. Williams. 1995. Inactivation of 2,3-dihydroxybiphenyl 1,2-dioxygenase from *Pseudomonas* sp. strain CB406 by 3,4-dihydroxybiphenyl (4-phenyl caetchol). *Biodegradation* 6: 11–17.

Locher, H.H., T. Leisinger, and A.M. Cook. 1991. 4-Toluene sulfonate methyl monooxygenase from *Comamonas testosteroni* T-2: purification and some properties of the oxygenase component. *J. Bacteriol.* 173: 3741–3748.

Locher, H.H., B. Poolman, A.M. Cook, and W.N. Konings. 1993. Uptake of 4-toluene sulfonate by *Comamonas testosteroni* T-2. *J. Bacteriol.* 175: 1075–1080.

Lockhart, H.B. and R.V. Blakeley. 1975. Aerobic photodegradation of Fe(III)-(ethyl-enedinitrilo)tetraacetate (Ferric EDTA). Implications for natural waters. *Environ. Sci. Technol.* 9: 1035–1038.

Lorenzen, J.P., A. Kröger, and G. Unden. 1993. Regulation of anaerobic respiratory pathways in *Wolinellla succinogenes* by the presence of electron acceptors. *Arch. Microbiol.* 159: 477–483.

Louie, T.M. and W.W. Mohn. 1999. Evidence for a chemiosmotic model of dehalores-piration in *Desulfomonile tiedjei* DCB-1. *J. Bacteriol.* 181: 40–46.

Lovley, D.R. 1991. Dissimilatory Fe(III) and Mn(IV) reduction. *Microbiol. Rev.* 55: 259–287.

Lovley, D.R. and D.J. Lonergan. 1990. Anaerobic oxidation of toluene, phenol, and *p*-cresol by the dissimilatory iron-reducing organism, GS-15. *Appl. Environ. Microbiol.* 56: 1858–1864.

Lovley, D.R. and E.J.P. Phillips. 1988. Novel mode of microbial energy metabolism: organic carbon oxidation coupled to dissimilatory reduction of iron or manga-nese. *Appl. Environ. Microbiol.* 54: 1472–1480.

Lovley, D.R. and E.J.P. Phillips. 1994. Novel processes for anaerobic sulfate production from elemental sulfur by sulfate-reducing bacteria. *Appl. Environ. Microbiol.* 60: 2394–2399.

Lovley, D.R. and J.C. Woodward. 1992. Consumption of freons CFC-11 and CFC-12 by anaerobic sediments and soils. *Environ. Sci. Technol.* 26: 925–929.

Lovley, D.R., E.J.P. Phillips, and D.J. Lonergan. 1989. Hydrogen and formate oxidation coupled to dissimilatory reduction of iron or manganese by *Alteromonas putre-faciens*. *Appl. Environ. Microbiol.* 55: 700–706.

Lovley, D.R., E.J.P. Phillips, Y. A. Gorby, and E.R. Landa. 1991. Microbial reduction of uranium. *Nature* 350: 413–416.

Lovley, D.R., S.J. Giovannoni, D.C. White, J.E. Champine, E.J.P. Phillips, Y.A. Gorby, and S. Goodwin. 1993. *Geobacter metallireducens* gen nov. sp. nov., a microorgan-ism capable of coupling the complete oxidation of organic compounds to the reduction of iron and other metals. *Arch. Microbiol.* 159: 336–344.

Lowe, S.E., M.K. Jain, and J.G. Zeikus. 1993. Biology, ecology, and biotechnological applications of anaerobic bacteria adapted to environmental stresses in temper-ature, pH, salinity, or substrates. *Microbiol. Revs.* 57: 451–509.

Macalady and N.L. Wolfe. 1985. Effects of sediment sorption and abiotic hydrolyses.1. Organophosphorothioate esters. *J. Agric. Food. Chem.* 33: 167–173.

MacGillivray, A.R. and M.P. Shiaris. 1993. Biotransformation of polycyclic aromatic hydrocarbons by yeasts isolated from coastal sediments. *Appl. Environ. Microbiol.* 59: 1613–1618.

MacMichael, G.J. and L.R. Brown. 1987. Role of carbon dioxide in catabolism of propane by "*Nocardia paraffinicum*" (*Rhodococcus rhodochrous*). *Appl. Environ. Microbiol.* 53: 65–69.

Macy, J.M., S. Rech, G. Auling, M. Dorsch, E. Stackebrandt, and L.I. Sly. 1993. *Thauera selenatis* gen. nov., sp. nov., a member of the beta subclass of *Proteobacteria* with a novel type of anaerobic respiration. *Int. J. Syst. Bacteriol.* 43: 135–142.

Madden, T.H., A.K. Datye, M. Fulton, M.R. Prairie, S.A. Majumdar, and B.M. Stange. 1997. Oxidarion of metal–EDTA complexes by TiO_2 photocatalysis. *Environ. Sci. Technol.* 31: 3475–3481.

Madsen, T. and D. Licht. 1992. Isolation and characterization of an anaerobic chlorophenol-transforming bacterium. *Appl. Environ. Microbiol.* 58: 2874–2878.

Maeng, J.H., Y. Sakai, Y. Tani, and N. Kato. 1996. Isolation and characterization of a novel oxygenase that catalyzes the first step of *n*-alkane oxidation in *Acinetobacter* sp. strain M-1. *J. Bacteriol.* 178; 3695–3700.

Maiers, D.T., P.L. Wichlacz, D.L. Thompson, and D.F. Bruhn. 1988. Selenate reduction by bacteria from a selenium-rich environment. *Appl. Environ. Microbiol.* 54: 2591–2593.

Malachowsky, K.J., T.J. Phelps, A.B. Teboli, D.E. Minnikin, and D.C. White. 1994. Aerobic mineralization of trichloroethylene, vinyl chloride, and aromatic compounds by *Rhodococcus* species. *Appl. Environ. Microbiol.* 60: 542–548.

Malmqvist, Å., T. Welander, and L. Gunnarsson. 1991. Anaerobic growth of microorganisms with chlorate as an electron acceptor. *Appl. Environ. Microbiol.* 57: 2229–2232.

Malmqvist, Å., T. Wellander, E. Moore, A. Ternström, G. Molin, and I.-M. Stenström. 1994. *Ideonella dechlorans* gen. nov., sp. nov., a new bacterium capabale of growing anaerobically with chlorate as electron acceptor. *Syst. Appl. Microbiol.* 17: 58–64.

Marchesi, J.R., S.A. Owen, G.F. White, W.A. House, and N.J. Russell. 1994. SDS-degrading bacteria attach to riverine sediment in response to the surfactant or its primary degradation product dodecan-1-ol. *Microbiology* (U.K.) 140: 2999–3006.

Marks, T.S., J.D. Allpress, and A. Maule. 1989. Dehalogenation of lindane by a variety of porphyrins and corrins. *Appl. Environ. Microbiol.* 55: 1258–1261.

Mars, A.E., J. Kingma, S.R. Kaschabek, W. Reinke, and D.B. Janssen. 1999. Conversion of 3-chlorocatechol by various catechol 2,3-dioxygenases and sequence analysis of the chlorocatechol dioxygense region of *Pseudomonas putida* GJ31. *J. Bacteriol.* 181: 1309–1318.

Martin, P. and R.A. MacLeod. 1984. Observations on the distinction between oligotrophic and eutrophic marine bacteria. *Appl. Environ. Microbiol.* 47: 1017–1022.

Mason, J.R. and R. Cammack. 1992. The electron-transport proteins of hydroxylation bacterial dioxygenases. *Annu. Rev. Microbiol.* 46: 277–305.

Master, E.M. and W.W. Mohn. 1998. Psychrotolerant bacteria isolated from arctic soil that degrade polychlorinated biphenyls at low temperatures. *Appl. Environ. Microbiol.* 64: 4823–4829.

Matheson, V.G., L.J. Forney, Y. Suwa, C.H. Nakatsu, A.J. Sextone, and W.E. Holben. 1996. Evidence for acquisition in nature of a chromosomal 2,4-dichlorophenoxyacetic acid/α-ketoglutarate dioxygenase gene by different *Burckholderia* spp. *Appl. Environ. Microbiol.* 62: 2457–2463.

Matin, A. 1978. Organic nutrition of chemolithotrophic bacteria. *Annu. Rev. Microbiol.* 32: 433–468.

Mauersberger, S., R.N. Matyashova, H.-G. Müller, and A.B. Losinov. 1980. Influence of the growth substrate amd the oxygen concentration in thre medium on the cytochrome P-450 content in *Candida guilliermondii*. *Eur. J. Appl. Microbiol.* Biotechnol. 9: 285–294.

McBride, K.E., J.W. Kenny, and D.M. Stalker. 1986. Metabolism of the herbicide bromoxynil by *Klebsiella pneumoniae* subspecies *ozaenae*. *Appl. Environ. Microbiol.* 52: 325–330.

McClay, K., S.H. Streger and R.J. Steffan. 1995. Induction of toluene oxidation in *Pseudomonas mendocina* KR1 and *Pseudomonas* sp. strain ENVPC5 by chlorinated solvents and alkanes. *Appl. Environ. Microbiol.* 61: 3479–3481.

McGowan, C., R. Fulthorpe, A. Wright, and J.M. Tiedje. 1998. Evidence for interspecies gene transfer in the evolution of 2,4-dichlorophenoxyacetic acid degraders. *Appl. Environ. Microbiol.* 64: 4089–4092.

McGrath, J.E. and C.G. Harfoot. 1997. Reductive dehalogenation of halocarboxylic acids by the phototrophic genera *Rhodospirillum* and *Rhodopseudomonas*. *Appl. Environ. Microbiol.* 63: 3333–3335.

McGrath, J.W., F. Hammerschmidt, and J.P. Quinn. 1998. Biodegradation of phosphonomycin by *Rhizobium huakuii* PMY1. *Appl. Environ. Microbiol.* 64: 356–358.

McInerney, M.J., M.P. Bryant, R.B. Hespell, and J.W. Costerton. 1981. *Syntrophomonas wolfei* gen. nov., sp. nov., an anaerobic, syntrophic, fatty acid-oxidizing bacterium. *Appl. Environ. Microbiol.* 41: 1029–1039.

McIntire, W., D.J. Hopper, J.C. Craig, E.T. Everhart, R.V. Webster, M.J. Causer, and T.P. Singer. 1984. Stereochemistry of 1-(4'-hydroxyphenyl)ethanol produced by hydroxylation of 4-ethylphenol by *p*-cresol methylhydroxylase. *Biochem. J.* 224: 617–621.

McIntire, W., D.J. Hopper, and T.P. Singer. 1985. *p*-Cresol methylhydroxylase. Assay and general properties. *Biochem. J.* 228: 325–335.

McMillan, D. C., P.P. Fu, and C.E. Cerniglia. 1987. Stereoselective fungal metabolism of 7,12-dimethylbenz[a]anthracene: identification and enantiomeric resolution of a K-region dihydrodiol. *Appl. Environ. Microbiol.* 53: 2560–2566.

Megharaj, M., A. Jovcic, H.L. Boul, and J.H. Thiele. 1997. Recalcitrance of 1,1-dichloro-2,2-bis(*p*-chlorophenyl)ethylene (DDE) to cometabolic degradation by pure cultures of aerobic and anaerobic bacteria. *Arch. Environ. Contam. Toxicol.* 33: 141–146.

Megharaj, M., S. Hartmans, K.-H. Engesser, and J.H. Thiele. 1998. Recalcitrance of 1,1-dichloro-2,2-bis(*p*-chlorophenyl)ethylene to degradation by pure cultures of 1,1-diphenylethylene-degrading aerobic bacteria. *Appl. Microbiol. Biotechnol.* 49: 337–342.

Metwally, M.E.-S. and N.L. Wolfe. 1990. Hydrolysis of chlorostilbene oxide. II. Modelling of hydrolysis in aquifer samples and in sediment-water systems. *Environ. Toxicol. Chem.* 9: 963–973.

Meyer, J.J.M., N. Grobbelaar, and P.L. Steyn. 1990. Fluoroacetate-metabolizing pseudomonad isolated from *Dichapetalum cymosum*. *Appl. Environ. Microbiol.* 56: 2152–2155.

Meyer, O., and H.G. Schlegel. 1983. Biology of aerobic carbon monoxide-oxidizing bacteria. *Annu. Rev. Microbiol.* 37: 277–310.

Michaelsen, M., R. Hulsch, T. Höpner, and L. Berthe-Corti. 1992. Hexadecane mineralization in oxygen-controlled sediment-seawater cultivations with autochthonous microorganisms. *Appl. Environ. Microbiol.* 58: 3072–3077.

Middelhoven, W.J., and F. Spaaij. 1997. *Rhodotorula cresolica* sp. nov., a cresol-assimilating yeast species isolated from soil. *Int. J. Syst. Bacteriol.* 47: 324–327.

Miguez, C.B., C.W. Greer, J.M. Ingram, and R.A. MacLeod. 1995. Uptake of benzoic acid and chloro-substituted benzoic acids by *Alcaligenes denitrificans* BRI 3010 and BRI 6011. *Appl. Environ. Microbiol.* 61: 4152–4159.

Mileski, G., J.A. Bumpus, M.A. Jurek, and S.D. Aust. 1988. Biodegradation of pentachlorophenol by the white rot fungus, *Phanerochaete chrysosporium*. *Appl. Environ. Microbiol.* 54: 2885–2889.

Miller, P.L., D. Vasudevan, P.M. Gschwend, and A.L. Roberts. 1998. Transformation of hexachloroethane in a sulfidic natural water. *Environ. Sci. Technol.* 32: 1269–1275.

Miller, R.M., G.M. Singer, J.D. Rosen, and R. Bartha. 1988a. Sequential degradation of chlorophenols by photolytic and microbial treatment. *Environ. Sci. Technol.* 22: 1215–1219.

Miller, R.M., G.M. Singer, J.D. Rosen, and R. Bartha. 1988b. Photolysis primes bio-degradation of benzo[a]pyrene. *Appl. Environ. Microbiol.* 54: 1724–1730.

Mills, A.L. and M. Alexander. 1976. N-Nitrosamine formation by cultures of several microorganisms. *Appl. Environ. Microbiol.* 31: 892–895.

Miura, Y. and A.J. Fulco. 1975. ω-1, ω-2 and ω-3 hydroxylation of long-chain fatty acids, amides and alcohols by a soluble enzyme system from *Bacillus megaterium*. *Biochim. Biophys. Acta* 388: 305–317.

Miyazawa, M., T. Wada, and H. Kameoka. 1998. Biotransformation of (+) and (–) limonene by the larvae of common cutworm (*Spodoptera litura*). *J. Agric. Food Chem.* 46: 300–303.

Möbus, E., M. Jahn, R. Schmid, D. Jahn, and E. Maser. 1997. Testosterone-regulated expression of enzymes involved in steroid and aromatic hydrocarbon catabolism in *Comamonas testosteroni*. *J. Bacteriol.* 179: 5951–5955.

Moir, J.W.B., D.J. Richardson, and S.J. Ferguson. 1995. The expression of redox pro-teins of denitrification in *Thiosphaera pantropha* grown with oxygen, nitrate, and nitrous oxide. *Arch. Microbiol.* 164: 43–49.

Mondello, F.J. 1989. Cloning and expression in *Escherichia coli* of *Pseudomonas* strain LB400 genes encoding polychlorinated biphenyl degradation. *J. Bacteriol.* 171: 1725–1732.

Mondello, F.J., M.P. Turcich, J.H. Lobos, and B.D. Erickson. 1997. Identification and modification of biphenyl dioxygenase sequences that determine the specificity of polychlorinated biphenyl degradation. *Appl. Environ. Microbiol.* 63: 3096–3103.

Monna, L., T. Omori, and T. Kodama. 1993. Microbial degradation of dibenzofuran, fluorene, and dibenzo-p-dioxin by *Staphylococcus auriculans* DBF63. *Appl. Environ. Microbiol.* 59: 285–289.

Monson, P.D., D.J. Call, D.A. Cox, K. Liber, and G.T. Ankley. 1999. Photoinduced toxicity of fluoranthene to northern leopard frogs (*Rana pipens*). *Environ. Toxicol. Chem.* 18: 308–312.

Monticello, D.J., D. Bakker, and W.R. Finnerty. 1985. Plasmid-mediated degradation of dibenzothiophene by *Pseudomonas species*. *Appl. Environ. Microbiol.* 49: 756–760.

Moore, J.K., H.D. Braymer, and A.D. Larson. 1983. Isolation of a *Pseudomonas* sp. which utilizes the phosphonate herbicide glyphosate. *Appl. Environ. Microbiol.* 46: 316–320.

Morris, C.M. and E.A. Barnsley. 1982. The cometabolism of 1- and 2-chloronaphtha-lene by pseudomonads. *Can. J. Microbiol.* 28: 73–79.

Morris, D.R. and L.P. Hager. 1966. Chloroperoxidase I. Isolation and properties of the crystalline glycoprotein. *J. Biol. Chem.* 241: 1763–1768.

Morris, P.J., J.F. Quensen III, J.M. Tiedje, and S.A. Boyd. 1992. Reductive debromina-tion of the commercial polybrominated biphenyl mixture Firemaster BP6 by anaerobic microorganisms from sediments. *Appl. Environ. Microbiol.* 58: 3249–3256.

Mörsdorf, G., K. Frunzke, D. Gadkari, and O. Meyer. 1992. Microbial growth on carbon monoxide. *Biodegradation* 3: 61–82.

Mortlock, R.P. 1982. Metabolic acquisition through laboratory selection. *Annu. Rev. Microbiol.* 34: 37–66.

Mowrer, J., and J. Nordin. 1987. Characterization of halogenated organic acids in flue gases from municipal waste incinerators. *Chemosphere* 16: 1181–1192.

Muheim, A., R. Waldner, M.S.A. Leisola, and A. Fiechter. 1990. An extracellular aryl-alcohol oxidase from the white-rot fungus *Bjerkandera adusta*. *Enzyme Microbiol. Technol.* 12: 204–209.

Mulbry, W.W., J.S. Karns, P.C. Kearney, J.O. Nelson, C.S. McDaniel, and J.R. Wild. 1986. Identification of a plasmid-borne Parathion hydrolase gene from *Flavobacterium* sp. by Southern hybridization with *opd* from *Pseudomonas diminuta*. *Appl. Environ. Microbiol.* 51: 926–930.

Müller, M.D. and H.-R. Buser. 1986. Halogenated aromatic compounds in automotive emissions from leaded gasoline additives. *Environ. Sci. Technol.* 20: 1151–1157.

Müller, S.R., H.-R. Zweifel, D.J. Kinnison, J.A. Jacobsen, M.A. Meier, M.M. Ulrich, and R.P. Schwarzenbach. 1996. Occurrence, sources, and fate of trichloroacetic acid in Swiss waters. *Environ. Toxicol. Chem.* 15: 1470–1478.

Munro, A.W. and J.G. Lindsay. 1996. Bacterial cytochromes P-450. *Mol. Microbiol.* 20: 1115–1125.

Murphy, G.l. and J.J. Perry. 1983. Incorporation of chlorinated alkanes into fatty acids of hydrocarbon-utilizing mycobacteria. *J. Bacteriol.* 156: 1158–1164.

Murphy, S.E., A. Drotar, and R. Fall. 1982. Biotransformation of the fungicide pentachloronitrobenzene by *Tetrahymena thermophila*. *Chemosphere* 11: 33–39.

Myers, C.R. and K.H. Nealson. 1990. Respiration-linked proton translocation coupled to anaerobic reduction of manganese(IV) and iron (III) in *Shewanella putrefaciens*. *J. Bacteriol.* 172: 6232–6238.

Nagata, K., H. Yu, M. Nishikawa, M. Kashiba, A. Nakamura, E.F. Sato, T. Tamura, and M. Inoue. 1998. *Helicobacter pylori* generates superoxide radicals and modulates nitric oxide metabolism. *J. Biol. Chem.* 273: 14071–14073.

Nagy, I., F. Compernolle, K. Ghys, J. Vanderleyden, and R. De Mot. 1995a. A single cytochrome P-450 system is involved in degradation of the herbicides EPTC (*S*-ethyl dipropylthiocarbamate) and atrazine by *Rhodococcus* sp. N186/21. *Appl. Environ. Microbiol.* 61: 2056–2060.

Nagy, I., G. Schoofs, F. Compermolle, P. Proost, J. Vanderleyden, and R. De Mot. 1995b. Degradation of the thiocarbamate herbicide EPTC *S*-ethyl dipropylcarbamoylthioate and biosafening by *Rhodococcus* sp. strain N186/21 involve an inducible cytochrome P-450 system and aldehyde dehydrogenase. *J. Bacteriol.* 177: 676–687.

Nakagawa, R., S. Kitamori, K. Horikawa, K. Nakashima, and H. Tokiwa. 1983. Identification of dinitropyrenes in diesel-exhaust particles. Their probable presence as the major mutagens. *Mutat. Res.* 124: 201–211.

Nakano, N.M. and P. Zuber. 1998. Anaerobic growth of a "strict aerobe" (*Bacillus subtilis*). *Annu. Rev. Microbiol.* 52: 165–190.

Nakatsu, C.H.J., N.A. Straus, and R.C. Wyndham. 1995. The nucleotide sequence of the Tn 5271 3-chlorobenzoate 3,4-dioxygenase genes (*cbaAB*) unites the class 1A oxygenases in a single linkage. *Microbiology (U.K.)* 141: 485–595.

Nakatsu, C.H., M. Providenti, and R.C. Wyndham. 1997. The *cis*-diol dehydrogenase *cbaC* gene of Tn5271 is required for growth on 3-chlorobenzoate but not 3,4-dichlorobenzoate. *Gene* 196: 200–218.

Narhi, L.O. and A.J. Fulco. 1986. Characterization of a catalytically self-sufficient 119,000-Dalton cytochrome P-450 monooxygenase induced by barbiturates in *Bacillus megaterium*. *J. Biol. Chem.* 261: 7160–7169.

Narhi, L.O. and A.J. Fulco. 1987. Identification and characterization of two functional domains in cytochrome P-450$_{BM-3}$, a catalytically self-sufficient monooxygenase induced by barbiturates in *Bacillus megaterium*. *J. Biol. Chem.* 262: 6683–6690.

Narro, M.L., C.E. Cerniglia, C. Van Baalen, and D.T. Gibson. 1992. Metabolism of phenanthrene by the marine cyanobacterium *Agmenellum quadruplicatum* PR-6. *Appl. Environ. Microbiol.* 58: 1351–1359.

Nealson, K.H. and C.R. Myers. 1992. Microbial reduction of manganese and iron: new approaches to carbon cycling. *Appl. Environ. Microbiol.* 58: 439–443.

Negoro, S., H. Shinagawa, A. Naata, S. Kinoshita, T. Hatozaki, and H. Okada. 1980. Plasmid control of 6-aminohexanoic acid cyclic dimer degradation enzymes of *Flavobacterium* sp. KI72. *J. Bacteriol.* 143: 238–345.

Neidhardt, F.C., P.L. Bloch, and D.F. Smith. 1974. Culture medium for enterobacteria. *J. Bacteriol.* 119: 736–747.

Neidle, E.L., C. Hartnett, L.N. Ornmston, A. Bairoch, M. Rekin, and S. Harayama. 1991. Nucleotide sequences of the *Acinetobacter calcoaceticus benABC* genes for benzoate 1,2-dioxygenase reveal evolutionary relationship among multicomponent oxygenases. *J. Bacteriol.* 173: 5385–5395.

Neidleman, S.L. and J. Geigert. 1986. *Biohalogenation: Principles, Basic Roles and Applications.* Ellis Horwood, Chichester, England.

Neilson, A.H. and R.A. Lewin. 1974. The uptake and utilization of organic carbon by algae: an essay in comparative biochemistry. *Phycologia* 13: 227–264.

Neilson, A.H., A.-S. Allard, and M. Remberger. 1985. Biodegradation and transformation of recalcitrant compounds pp. 29–86. In *Handbook of Environmental Chemistry* (Ed. O. Hutzinger), Vol. 2, Part C. Springer-Verlag, Berlin.

Neilson, A.H., A.-S. Allard, P.-Å. Hynning, M. Remberger, and L. Landner. 1983. Bacterial methylation of chlorinated phenols and guiaiacols: formation of veratroles from guaiacols and high molecular weight chlorinated lignin. *Appl. Environ. Microbiol.* 45: 774–783.

Neilson, A.H. and T. Larsson. 1980. The utilization of organic nitrogen for growth of algae: physiological aspects. *Physiol. Plant.* 48: 542–553.

Neilson, A.H., C. Lindgren, P.-Å. Hynning, and M. Remberger. 1988. Methylation of halogenated phenols and thiophenols by cell extracts of Gram-positive and Gram-negative bacteria. *Appl. Environ. Microbiol.* 54: 524–530.

Nelson, L., I. Shanahan, H.W. Sidebottom, J. Treacy, and O.J. Nielsen. 1990. Kinetics and mechanism for the oxidation of 1,1,1-trichloroethane. *Int. J. Chem. Kinet.* 22: 577–590.

Neumann, A., H. Scholz-Muramatsu, and G. Diekert. 1994. Tetrachloroethene metabolism of *Dehalospirillum multivorans. Arch. Microbiol.* 162: 295–301.

Newman, D.K., E.K. Kennedy, J.D. Coates, D. Ahmann, D.J. Ellis, D.K. Lovley, and F.M.M. Morel. 1997. Dissimilatory arsenate and sulfate reduction in *Desulfotomaculum auripigmentum. Arch. Microbiol.* 168: 380–388.

Newman, L.A., S.E. Strand, N. Choe, J. Duffy, G. Ekuan, M. Ruiszai, B.B. Shurtleff, J. Wilmoth, and M.P. Gordon. 1997. Uptake and biotransformation of trichloroethylene by hybrid poplars. *Environ. Sci. Technol.* 31: 1062–1067.

Nguyen, H.-H.T., A.K. Shiemka, S.J. Jacobs, B.J. Hales, M.E. Lidstrom, and S.I. Chan. 1994. The nature of the copper ions in the membranes containing the particulate methane monooxygenase from *Methylococcus capsulatus* (Bath). *J. Biol. Chem.* 269: 14995–15005.

Nichols, N.N. and C.S. Harwood. 1997. PcaK, a high-affinity permease for the aromatic compounds 4-hydroxybenzoate and protocatechuate from *Pseudomonas putida. J. Bacteriol.* 179: 5056–5061.

Nickel, K., M.J.-F. Suter and H.-P.E. Kophler. 1997. Involvement of two α-ketoglutarate-dependent dioxygenases in enantioselective degradation of (*R*)- and (*S*)-mecoprop by *Sphingomonas herbicidovorans* MH. *Appl. Environ. Microbiol.* 63: 6674–6679.

Nishioka, M.G., C.C. Howard, D.A. Conros, and L.M. Ball. 1988. Detection of hydroxylated nitro aromatic and hydroxylated nitro polycyclic aromatic compounds in ambient air particulate extract using bioassay-directed fractionation. *Environ. Sci. Technol.* 22: 908–915.

Nörtemann, B. 1992. Total degradation of EDTA by mixed cultures and a bacterial isolate. *Appl. Environ. Microbiol.* 58: 671–676.

Nörtemann, B., A. Glässer, R. Machinek, G. Remberg, and H.-J. Knackmuss. 1993. 5-Hydroxyquinoline-2-carboxylic acid, a dead-end metabolite from the bacterial oxidation of 5-aminonaphthalene-2-sulfonic acid. *Appl. Environ. Microbiol.* 59: 1898–1903.

Novak, P.J., L. Daniels, and G.F. Parkin. 1998. Rapid dechlorination of carbon tetrachloride and chloroform by extracellular agents in cultures of *Methanosarcina thermophila*. *Environ. Sci. Technol.* 32: 3132–3136.

Nozawa, T. and Y. Maruyama. 1988. Anaerobic metabolism of phthalate and other aromatic compounds by a denitrifying bacterium. *J. Bacteriol.* 170: 5778–5784.

O'Keefe, D.P and P.A. Harder. 1991. Occurrence and biological function of cytochrome P-450 monooxygenase in the actinomycetes. *Mol. Microbiol.* 5: 2099–2105.

O'Keefe, D.P., J.A. Romesser, and K.J. Leto. 1988. Identification of constitutive and herbicide inducible cytochromes P-450 in *Streptomyces griseolus*. *Arch. Microbiol.* 149: 406–412.

Ogram, A.V., R.E. Jessup, L.T. Ou, and P.S.C. Rao. 1985. Effects of sorption on biological degradation rates of (2,4-dichlorophenoxy)acetic acid in soils. *Appl. Environ. Microbiol.* 49: 582–587.

Ogunseitan, O.A., I.L. Deklgado, Y.-L. Tsai, and B.H. Olson. 1991. Effect of 2-hydroxybenzoate on the maintenance of naphthalene-degrading pseudomonads in seeded and unseeded soils. *Appl. Environ. Microbiol.* 57: 2873–2879.

Ohta, T., K. Fujishoro, K. Yamaguchi, Y. Tamura, K. Aisaka, T. Uwajima, and M. Hasegawa. 1991. Sequence of gene *choB* encoding cholesterol oxidase of *Brevibacterium sterolicum*: comparison with *choA* of *Streptomyces* sp. SA-COO. *Gene* 103: 93–96.

Olsen, R.H., J.J. Kukor, and B. Kaphammer. 1994. A novel toluene-3-monooxygenase pathway cloned from *Pseudomonas pickettii* PKO1. *J. Bacteriol.* 176: 3749–3756.

Omori, T., L. Monna, Y. Saiki, and T. Kodama. 1992. Desulfurization of dibenzothiophene by *Corynebacterium* sp. strain SY1. *Appl. Environ. Microbiol.* 58: 911–915.

Oremland, R.S., J.S. Blum, C.W. Culbertson, P.T. Visscher, L.G. Miller, P. Dowdle, and F.E. Strohmaier. 1994. Isolation, growth, and metabolism of an obligately anaerobic, selenate-respiring bacterium, strain SES-3. *Appl. Environ. Microbiol.* 60: 3011–3019.

Ornston, L.N. 1966. The conversion of catechol and protocatechuate to β–ketoadipate by *Pseudomonas putida*. IV. Regulation. *J. Biol. Chem.* 241: 3800–3810.

Ortiz de Montellano, P.R., J.A. Fruetel, J.R. Collins, D.L. Camper, and G.H. Loew. 1991. Theoretical and experimental analysis of the absolute stereochemistry of *cis*-β-methylstyrene epoxidation by cytochrome P-450$_{cam}$. *J. Am. Chem. Soc.* 113: 3195–3196.

Owen, S.A., N.J. Russell, W.A. House, and G.F. White. 1997. Re-evaluation of the hypothesis that biodegradable surfactants stimulate surface attachment of competent bacteria. *Microbiology* (U.K.) 143: 3649–3659.

Padden, A.N., F.A. Rainey, D.P. Kelly, and A.P. Wood. 1997. *Xanthobacter tagetidis* sp. nov., an organism associated with *Tagetes* species and able to grow on substituted thiophenes. *Int. J. Syst. Bacteriol.* 47: 394–401.

Paje, M.L.F., B.A. Neilan, and I. Coupoerwhitre. 1997. A *Rhodococcus* species that thrives on medium saturated with liquid benzene. *Microbiology* (U.K.) 143: 2975–2981.

Pakalapati, S.N.R., B.N. Popov, and R.E. White. 1996. Anodic oxidation of ethylenediaminetetraacetic acid on platinum electrode in alkaline medium. *J. Electrochem. Soc.* 143: 1636–1643.

Parales, R.E., M.D. Emig, N.A. Lynch, and D.T. Gibson. 1998. Substrate specificities of hybrid naphthalene and 2,4-dinitrotioluene dioixygenase enzyme systems. *J. Bacteriol.* 180: 2337–2344.

Paris, D.F. and J. E. Rogers. 1986. Kinetic concepts for measuring microbial rate constants: effects of nutrients on rate constants. *Appl. Environ. Microbiol.* 51: 221–225.

Paris, D.F. and N.L. Wolfe. 1987. Relationship between properties of a series of anilines and their transformation by bacteria. *Appl. Environ. Microbiol.* 53: 911–916.

Paris, D.F., N.L. Wolfe, and W.C. Steen. 1982. Structure–activity relationships in microbial transformation of phenols. *Appl. Environ. Microbiol.* 44: 153–158.

Paris, D.F., N.L. Wolfe, and W.C. Steen. 1984. Microbial transformation of esters of chlorinated carboxylic acids. *Appl. Environ. Microbiol.* 47: 7–11.

Parke, D. and L.N. Ornston. 1986. Enzymes of the β-ketoadipate pathway are indicible in *Rhizobium* and *Agrobacterium* spp. and constitutive in *Bradyrhizobium* spp. *J. Bacteriol.* 165: 288–292.

Parke, D., F. Rynne, and A. Glenn. 1991. Regulation of phenolic metabolism in *Rhizobium leguminosarum* biovar trifolii. *J. Bacteriol.* 173: 5546–5550.

Patel, R.N., C.T. Hou, A.I. Laskin, and A. Felix. 1982. Microbial oxidation of hydrocarbons: properties of a soluble monooxygenase from a facultative methaneutilizing organisms *Methylobacterium* sp. strain CRL-26. *Appl. Environ. Microbiol.* 44: 1130–1137.

Pavlostathis, S.G., K.K. Comstock, M.E. Jacobson, and F.M. Saunders. 1998. Transformation of 2,4,6-trinitrotoluene by the aquatic plant *Myriophyllum spicatum*. *Environ. Toxicol. Chem.* 17: 2266–2273.

Peel, M.C. and R.C. Wyndham. 1999. Selection of *clc, cba*, and *fbc* chlorobenzoatecatabolic genotypes from groundwater and surface waters adjacent to the Hyde Park, Niagara Falls, chemical landfill. *Appl. Environ. Microbiol.* 654: 1627–1635.

Peijnenburg, W.J.G.M., M.J. 't Hart, H.A. den Hollander, D. van de Meent, H.H. Verboom, and N.L. Wolfe. 1992. QSARs for predicting reductive transformation constants of halogenated aromatic hydrocarbons in anoxic sediment systems. *Environ. Toxicol. Chem.* 11: 301–314.

Peijnenburg, W.J.G.M., K.G.M. de Beer, H.A. den Hollander, M.H.L. Stegeman, and H. Verboom. 1993. Kinetics, products, mechanisms, and QSARs for the hydrolytic transformation of aromatic nitriles in anaerobic sediment slurries. *Environ. Toxicol. Chem.* 12: 1149–1161.

Pelizzetti, E., V. Maurino, C. Minero, V. Carlin, E. Pramauro, O. Zerbinati, and M.L. Tosata. 1990. Photocatalytics degradation of atrazine and other *s*-triazine herbicides. *Environ. Sci. Technol.* 24: 1559–1565.

Peng, X., T. Egashira, K. Hanashiro, E. Masai, S. Nishikawa, Y. Katayama, K. Kimbara, and M. Fukuda. 1998. Cloning of a *Sphingomonas paucibilis* SYK-6 gene encoding a novel oxygenase that cleaves lignin-related biphenyl and characterization of the enzyme. *Appl. Environ. Microbiol.* 64: 2520–2527.

Perdue, E.M. and N.L. Wolfe. 1982. Modification of pollutant hydrolysis kinetics in the presence of humic substances. *Environ. Sci. Technol.* 16: 847–852.

Perez, J. and T.W. Jeffries. 1990. Mineralization of ^{14}C-ring-labelled synthetic lignin correlates with the production of lignin peroxidase, not of manganese peroxidase or laccase. *Appl. Environ. Microbiol.* 56: 1806–1812.

Perkins, E.J., M.P. Gordon, O. Caceres, and P.F. Lurquin. 1990. Organization and sequence analysis of the 2,4-dichlorophenol hydroxylase and dichlocatechol ox-idative operons of plasmid pJP4. *J. Bacteriol.* 172: 2351–2359.

Perlinger, J.A., W. Angst, and R.P. Schwarzenbach. 1996. Kinetics of the reduction of hexachloroethane by juglone in solutions containing hydrogen sulfide. *Environ. Sci. Technol.* 30: 3408–3417

Peters, R. 1952. Lethal synthesis. *Proc. R. Soc.* (London) B 139: 143–170.

Pettigrew, C.A., B.E. Haigler, and J.C. Spain. 1991. Simultaneous biodegradation of chlorobenzene and toluene by a *Pseudomonas* strain. *Appl. Environ. Microbiol.* 57: 157–162.

Picardal, F.W., R.G. Arnold, H. Couch, A.M. Little, and M.E. Smith. 1993. Involvement of cytochromes in the anaerobic biotransformation of tetrachloromethane by *Shewanella putrefaciens* 200. *Appl. Environ. Microbiol.* 59: 3763–3770.

Pitts, J.N., Jr., R. Atkinson, J.A. Sweetman, and B. Zielinska. 1985a. The gas-phase reaction of naphthalenes with N_2O_5 to form nitronaphthalenes. *Atmos. Environ.* 19: 701–705.

Pitts, J.N., Jr., B. Zielinska, J.A. Sweetman, R. Atkinson, and A.M. Winer. 1985b. Reactions of adsorbed pyrene and perylene with gaseous N_2O_5 under simulated atmospheric conditions. *Atmos. Environ.* 19: 911–915.

Platt, U.F., A.M. Winer, H.W. Biermann, R. Atkinson, and J.N. Pitts, Jr. 1984. Mea-surement of nitrate radical concentrations in continental air. *Environ. Sci. Technol.* 18: 365–369.

Platz, J., O.J. Nielsen, J. Sehested, and T.J. Wallington. 1995. Atmospheric chemistry of 1,1,1-trichloroethane: UV absorption spectra and self-reaction kinetics of CCl_3CH_2 and $CCl_3CH_2O_2$ radicals, kinetics of the reactions of the $CCl_3CH_2O_2$ radical with NO and NO_2, and the fate of alkoxy radical CCl_3CH_2O. *J. Phys. Chem.* 99: 6570–6579.

Plugge, C.M., C. Dijkema, and A.J.M. Stams. 1993. Acetyl-CoA cleavage pathways in a syntrophic propionate oxidizing bacterium growing on fumarate in the absence of methanogens. *FEMS Microbiol. Lett.* 110: 71–76.

Poindexter, J.S. 1981. Oligotrophy. Fast and famine existence. *Adv. Microb. Ecol.* 5: 63–89.

Polnisch, E., H. Kneifel, H. Franzke, and K.L. Hofmann. 1992. Degradation and dehalogenation of monochlorophenols by the phenol-assimilating yeast *Candida maltosa*. *Biodegradation* 2: 193–199.

Poth, M. and D.D. Focht. 1985. [15]N kinetic analysis of N_2O production by *Nitrosomonas europaea*: an examination of nitrifier denitrification. *Appl. Environ. Microbiol.* 49: 1134–1141.

Pothuluri, J.V., J.P. Freeman, F.E. Evans, T.B. Moorman, and C.E. Cerniglia. 1993. Metabolism of alachlor by the fungus *Cunninghamella elegans*. *J. Agric. Food Chem.* 41: 483–488.

Poupin, P., N. Truffaut, B. Combourieu, M. Sancelelme, H. Veschambre, and A.M. Delort. 1998. Degradation of morpholine by an environmental *Mycobacterium* strain involves a cytochrome P-450. *Appl. Environ. Microbiol.* 64: 159–165.

Prieto, M.A. and J.L. Garcia. 1994. Molecular characterization of 4-hydroxyphenylac-etate 3-hydroxylase *Escherichia coli*. *J. Biol. Chem.* 269: 22823–22829.

Prieto, M.A., A. Perez-Aranda, and J.L. Garcia. 1993. Characterization of an *Escherichia coli* aromatic hydroxylase with a broad substrate range. *J. Bacteriol.* 175: 2162–2167.

Que, L., J. Widom, and R.L. Crawford. 1981. 3,4-Dihydroxyphenylacetate 2,3-dioxy-genase: a manganese(II) dioxygenase from *Bacillus brevis*. *J. Biol. Chem.* 256: 10941–10944.

Racke, K.D. and J.R. Coats 1990. *Enhanced Biodegradation of Pesticides in the Environment*. American Chemical Society Symposium Series 426. American Chemical Society, Washington, D.C.

Racke, K.D. and E.P. Lichtenstein. 1985. Effects of soil microorganisms on the release of bound ^{14}C residues from soils previously treated with [^{14}C]parathion. *J. Agric. Food Chem.* 33: 938–943.

Racke, K.D., D.A. Laskowski, and M.R. Schultz. 1990. Resistance of chloropyrifos to enhanced biodegradation in soil. *J. Agric. Food Chem.* 38: 1430–1436.

Ramos, J.L., A. Wasserfallen, K. Rose, and K.N. Timmis. 1987. Redesigning metabolic routes: manipulation of TOL plasmid pathway for catabolism of alkylbenzoates. *Science* 235: 593–596.

Rao, J.R. and J.E. Cooper 1994. Rhizobia catabolize *nod* gene-inducing flavonoids via C-ring fission mechanisms. *J. Bacteriol.* 176: 5409–5413.

Rao, J.R., N.D. Sharma, J.T.G. Hamilton, D.R. Boyd, and J.E. Cooper. 1991. Biotrans-formation of the pentahydroxy flavone quercitin by *Rhizobium loti* and *Bradyrhizobium* strains Lotus. *Appl. Environ. Microbiol.* 57: 1563–1565.

Rasche, M.E., M.R. Hyman, and D.J. Arp. 1991. Factors limiting aliphatic chlorocarbon degradation by *Nitrosomonas europaea*: cometabolic inactivation of ammonia mo nooxygenase and substrate specificity. *Appl. Environ. Microbiol.* 57: 2986–2994.

Ratajczak, A., W. Geißdörfer, and W. Hillen. 1998. Alkane hydroxylase from *Acineto-bacter* sp. strain ADP1 is encoded by *alkM* and belongs to a new family of bacterial integral-membrane hydrocarbon hydroxylases. *Appl. Environ. Microbiol.* 64: 1175–1179.

Ravishankara, A.R., G. Hancock, M. Kawasaki, and Y. Matsumi. 1998. Photochemistry of ozone: surprises and recent lessons. *Science* 280: 60–61.

Reanney, D.C., P.C. Gowland, and J.H. Slater. 1983. Genetic interactions among mi-crobial communities. *Symp. Soc. Gen. Microbiol.* 34: 379–421.

Reinecke, W., D.J. Jeenes, P.A. Williams, and H.-J. Knackmuss. 1982. TOL plasmid pWWO in constructed halobenzoate-degrading *Pseudomonas* strains: prevention of meta pathway. *J. Bacteriol.* 150: 195–201.

Rendell, N.B., G.W. Taylor, M. Somerville, H. Todd, R. Wilson, and P.J. Cole. 1990. Characterization of *Pseudomonas* rhamnolipids. *Biochim. Biophys. Acta* 1045: 189–193.

Renganathan, V. 1989. Possible involvement of toluene-2,3-dioxygenase in defluori-nation of 3-fluoro-substituted benzenes by toluene-degrading *Pseudomonas* sp. strain T-12. *Appl. Environ. Microbiol.* 55: 330–334.

Resnick, S.M. and D.T. Gibson. 1996. Regio- and stereospecific oxidation of fluorene, dibenzofuran, and dibenzothiophene by naphthalene dioxygenase from *Pseudomonas* sp. strain NCIB-4. *Appl. Environ. Microbiol.* 62: 4073–4080.

Rheinwald, J.G., A.M. Chakrabarty, and I.C. Gunsalus. 1973. A transmissible plasmid controlling camphor oxidation in *Pseudomonas putida*. *Proc. Natl. Acad. Sci. U.S.A.* 70: 885–889.

Ribbons, D.W., P. Keyser, D.A. Kunz, B.F. Taylor, R.W. Eaton, and B.N. Anderson. 1984. Microbial degradation of phthalates pp. 371–395. In *Microbial Degradation of Organic Compounds* (D.T. Gibson ed.). Marcel Dekker, New York.

Rice, C.P. and Sikka, H.C. 1973. Uptake and metabolism of DDT by six species of marine algae. *J. Agric. Food Chem.* 21: 148–152.

Rieble, S., D.K. Joshi, and M.H. Gold. 1994. Purification and characterization of a 1,2,4-trihydroxybenzene 1,2-dioxygenase from the basidiomycete *Phanerochaete chrysosporium*. *J. Bacteriol.* 176: 4838–4844.

Riegert, U., G. Heiss, P. Fischer, and A. Stolz. 1998. Distal cleavage of 3-chlorocatechol by an extradiol dioxygenase to 3-chloro-2-hydroxymuconic semialdehyde. *J. Bacteriol.* 180: 2849–2853.

Rijnaarts, H.H.M., A. Bachmann, J.C. Jumelet, and A.J.B. Zehnder. 1990. Effect of desorption and intraparticle mass transfer on the aerobic biomineralization of alpha-hexachlorocyclohexane in a contaminated calcareous soil. *Environ. Sci. Technol.* 24: 1349–1354.

Rikken, G.B., A.G.M. Croon, and C.G. van Ginkel. 1996. Transformation of perchlorate into chloride by a newly isolated bacterium: reduction and dismutation. *Appl. Microbiol. Biotechnol.* 45: 420–426.

Rittenberg, S.C. 1972. The obligate autotroph — the demise of a concept. *Antonie van Leeuwenhoek* 38: 457–478.

Roberts, A.L. and P.M. Gschwend. 1991. Mechanism of pentachloroethane dehydrochlorination to tetrachloroethylene. *Environ. Sci. Technol.* 25: 76–86.

Roberts, A.L., P.N. Sanborn, and P.M. Gschwend. 1992. Nucleophilic substitution of dihalomethanes with hydrogen sulfide species. *Environ. Sci. Technol.* 26: 2263–2274.

Robertson, B.K. and M. Alexander. 1998. Sequestration of DDT and dieldrin in soil: disappearance of acute toxicity but not the compounds. *Environ.Toxicol. Chem.* 17: 1034–1038.

Robertson, L.A. and J.G. Kuenen. 1984. Aerobic denitrification: a controversy revived. *Arch. Microbiol.* 139: 351–354.

Robinson, K.G., W.S. Farmer, and J.T. Novak. 1990. Availability of sorbed toluene in soils for biodegradation by acclimated bacteria. *Water Res.* 24: 345–350.

Romanov, V. and R.P. Hausinger. 1994. *Pseudomonas aeruginosa* 142 uses a three-component *ortho*-halobenzoate 1,2-dioxygenase for metabolism of 2,4-dichloro- and 2-chlorobenzoate. *J. Bacteriol.* 176: 3368–3374.

Romesser, J.A. and D.P. O'Keefe. 1986. Induction of cytochrome P-450-dependent sulfonylurea metabolism in *Streptomyces griseolus*. *Biochem. Biophys. Res. Commun.* 140: 650–659.

Rosche, B., B. Tshisuaka, B. Hauer, F. Lingens, and S. Fetzner. 1997. 2-Oxo-1,2-dihydroquinoline 8-monooxygenase: phylogenetic relationship to other multicomponent nonheme iron oxygenases. *J. Bacteriol.* 179: 3459–3554.

Rosenberg, S. 1971. Regulation of the mandelate pathway in *Pseudomonas aeruginosa*. *J. Bacteriol.* 108: 1257–1269.

Rosenkranz, H.S. and R. Mermelstein. 1983. Mutagenicity and genotoxicity of nitroarenes. All nitro-containing chemicals were not created equal. *Mutat. Res.* 114: 216–267.

Rott, B., S. Nitz, and F. Korte. 1979. Microbial decomposition of sodium pentachlorophenolate. *J. Agric. Food Chem.* 27: 306–310.

Rouf, M.A., and R.F. Lomprey. 1968. Degradation of uric acid by certain aerobic bacteria. *J. Bacteriol.* 96: 617–622.

Roy, F., E. Samain, H.C. Dubourguier, and G. Albagnac. 1986. *Synthrophomonas* [sic] *sapovorans* sp. nov., a new obligately proton reducing anaerobe oxidizing saturated and unsaturated long chain fatty acids. *Arch. Microbiol.* 145: 142–147.

Roy, S. and O. Hänninen. 1994. Pentachlorophenol: uptake/elimination kinetics and metabolism in an aquatic plant. *Eichhornia crassipes. Environ. Toxicol. Chem.* 13: 763–773.

Rubin, H.E. and M. Alexander. 1983. Effect of nutrients on the rates of mineralization of trace concentrations of phenol and *p*-nitrophenol. *Environ. Sci. Technol.* 17: 104–107.

Rubin, H.E., R.V. Subba-Rao, and M. Alexander. 1982. Rates of mineralization of trace concentrations of aromatic compounds in lake water and sewage samples. *Appl. Environ. Microbiol.* 43: 1133–1138.

Saflic, S., P.M. Fedorak, and J.T. Andersson. 1992. Diones, sulfoxides, and sulfones from the aerobic cometabolism of methylbenzothiophenes by *Pseudomonas* strain BT1. *Environ. Sci. Technol.* 26: 1759–1764.

Sakai, K., A. Nakazawa, K. Kondo, and H. Ohta. 1985. Microbial hydrogenation of nitroolefins. *Agric. Biol. Chem.* 49: 2231–2236.

Salleh, M.A. and J.M. Pemberton. 1993. Cloning of a DNA region of a *Pseudomonas* plasmid that codes for detoxification of the herbicide paraquat. *Curr. Microbiol.* 27: 63–67.

Salmeen, I.T., A.M. Pero, R. Zator, D. Schuetzle, and T.L. Riley. 1984. Ames assay chromatograms and the identification of mutagens in diesel particle extracts. *Environ. Sci. Technol.* 18: 375–382.

Samadapour, M., J. Liston, J.E. Ongerth, and P.I. Tarr. 1990. Evaluation of DNA probes for detection of Shiga-like-toxin-producing *Escherichia coli* in food and calf fecal samples. *Appl. Environ. Microbiol.* 56: 1212–1215.

Sander, P., R.-M. Wittich, P. Fortnagel, H. Wilkes, and W. Francke. 1991. Degradation of 1,2,4-trichloro- and 1,2,4,5-tetrachlorobenzene by *Pseudomonas* strains. *Appl. Environ. Microbiol.* 57: 1430–1440.

Sandermann, H. 1994. Higher plant metabolism of xenobiotics: the "green liver" concept. *Pharmacogenetics* 4: 225–241.

Sanseverino, J., B.M. Applegate, J.M.H. King, and G.S. Sayler. 1993. Plasmid-mediated mineralization of naphthalene, phenanthrene, and anthracene. *Appl. Environ. Microbiol.* 59: 1931–1937.

Sariaslani, F.S. 1991. Microbial cytochromes P-450 and xenobiotic metabolism. *Adv. Appl. Microbiol.* 36: 133–177.

Sarkanen, S., R.A. Razal, T. Piccariello, E. Yamamoto, and N.G. Lewis. 1991. Lignin peroxidase: toward a clarification of its role *in vivo. J. Biol. Chem.* 266: 3636–3643.

Sasaki, J., J. Arey, and W.P. Harger. 1995. Formation of mutagens from the photooxidation of 2-4-ring PAH. *Environ. Sci. Technol.* 29: 1324–1335.

Sayama, M., M. Inoue, M.-A. Mori, Y. Maruyama, and H. Kozuka. 1992. Bacterial metabolism of 2,6-dinitrotoluene with *Salmonella typhimurium* and mutagenicity of the metabolites of 2,6-dinitrotoluene and related compounds. *Xenobiotica* 22: 633–640.

Sayler, G.S., M.S. Shields, E.T. Tedford, A. Breen, S.W. Hooper, K.M. Sirotkin, and J.W. Davis. 1985. Application of DNA-DNA colony hybridization to the detection of catabolic genotypes in environmental samples. *Appl. Environ. Microbiol.* 49: 1295–1303.

Schach, S., B. Tshisuaka, S. Fetzner, and F. Lingens. 1995. Quinoline 2-oxidoreductase and 2-oxo-1,2-dihydroquinoline-5,6-dioxygenase from *Comamonas testeteroni* 63, the first two enzymes in quinoline and 3-methylquinoline degradation. *Eur. J. Biochem.* 232: 536–544.

Schanke, C.A. and L.P. Wackett. 1992. Environmental reductive elimination reactions of polychlorinated ethanes mimicked by transition-metal coenzymes. *Environ. Sci. Technol.* 26: 830–833.

Schauer, F., K. Henning, H. Pscheidl, R.M. Wittich, P. Fortnagel, H. Wilkes, V. Sinnwell, and W. Francke. 1995. Biotransformation of diphenyl ether by the yeast *Trichosporon beigelii* SBUG 765. *Biodegradation* 6: 173–180.

Schink, B. 1991. Syntrophism among prokaryotes pp. 276–299. In *The Prokaryotes* (Eds. A. Balows, H.G. Trüper, M. Dworkin, W. Harder, and K.-H. Schleifer). Springer-Verlag, Heidelberg.

Schink, B. 1997. Energetics of syntrophic cooperation in methanogenic degradation. *Microbiol. Mol. Biol. Rev.* 61: 262–280.

Schink, B. and N. Pfennig. 1982. *Propionigenium modestum* gen. nov. sp. nov. a new strictly anaerobic, nonsporing bacterium growing on succinate. *Arch. Microbiol.* 133: 209–216.

Schleissner, C., E.R. Olivera, M. Fernández-Valverde, and J.M. Luengo. 1994. Aerobic catabolism of phenylacetic acid in *Pseudomonas putida* U: biochemical characterization of a specific phenylacetic acid transport system and formal demonstration that phenylacetyl-coenzyme A is a catabolic intermediate. *J. Bacteriol.* 176: 7667–7676.

Schneider, M., O. Luxenhofer, A. Deissler, and K. Ballschmiter. 1998. C_1–C_{15} alkyl nitrates, benzyl nitrate and bifunctional nitrates: measurements in California and South Atlantic air and global comparison using C_2C_{14} and $CHBr_3$ as marker molecules. *Environ. Sci. Technol.* 32: 3055–3062.

Schneiders, G.E., M.K. Koeppe, M.V. Naidu, P. Horne, A.M. Brown, and C.F. Mucha. 1993. Rate of rimsulfuron in the environment. *J. Agric. Food Chem.* 41: 2404–2410.

Schut, F., E.J. de Vries, J.C. Gottschal, B.R. Robertson, W. Harder, R.A. Prins, and D.K. Button. 1993. Isolation of typical marine bacteria by dilution culture: growth, maintenance, and characteristics of isolates under laboratory conditions. *Appl. Environ. Microbiol.* 59: 2150–2160.

Schut, F.M. Jansen, T.M.P. Gomes, J.C. Gottschal, W. Harder, and R.A. Prins. 1995. Substrate uptake and utilization by a marine ultramicrobacterium. *Microbiology* (U.K.) 141: 351–361.

Schwack, W. 1988. Photoinduced additions of pesticides to biomolecules. 2. Model reactions of DDT and methoxychlor with methyl oleate. *J. Agric. Food Chem.* 36: 645–648.

Schwarzenbach, R.P., R. Stierliu, K. Lanz, and J. Zeyer. 1990. Quinone and iron porphyrin mediated reduction of nitroaromatic compounds in homogeneous aqueous solution. *Environ. Sci. Technol.* 24: 1566–1574.

Schweizer, D., A. Markus, M. Seez, H.H. Ruf, and F. Lingens. 1987. Purification and some properties of component B of the 4-chlorophenylacetate 3,4-dioxygenase from *Pseudomonas* sp. strain CBS3. *J. Biol. Chem.* 262: 9340–9346.

Sedlak, D.L. and A.W. Andren. 1991. Oxidation of chlorobenzene with Fenton's reagent. *Environ. Sci. Technol.* 25: 777–782.

Seefeld, S. and J.A. Kerr. 1997. Kinetics of reactions of propionylperoxy radicals with NO and NO_2: peroxypropionyl nitrate formation under laboratory conditions related to the troposphere. *Environ. Sci. Technol.* 31: 2949–2953.

Seigle-Murandi, F., R. Steiman, F. Chapella, and C. Luu Duc. 1986. 5-Hydroxylation of benzimidazole by Micromycetes. II. Optimization of production with *Absidia spinosa. Appl. Microbiol. Biotechnol.* 25: 8–13.

Seigle-Murandi, F.M., S.M.A. Krivobok, R.L. Steiman, J.-L.A. Benoit-Guyod, and G.-A. Thiault. 1991. Biphenyl oxide hydroxylation by *Cunninghamella echinulata. J. Agric. Food Chem.* 39: 428–430.

Seitz, H.-J. and H. Cypinka. 1986. Chemolithotrophic growth of *Desulfovibrio desulfuricans* with hydrogen coupled to ammonification with nitrate or nitrite. *Arch. Microbiol.* 146: 63–67.

Seligman, P.F., A.O. Valkirs, and R.F. Lee. 1986. Degradation of tributytin in San Diego Bay, California, waters. *Environ. Sci. Technol.* 20: 1229–1235.

Semple, K.T. and R.B. Cain. 1996. Biodegradation of phenols by the alga *Ochromonas danica. Appl. Environ. Microbiol.* 62: 1265–1273

Serdar, C.M., D.T. Gibson, D. M. Munnecke, and J.H. Lancaster. 1982. Plasmid involvement in Parathion hydrolysis in *Pseudomonas diminuta. Appl. Environ. Microbiol.* 44: 246–249.

Servent, D., C. Ducrorq, Y. Henry, A. Guissani, and M. Lenfant. 1991. Nitroglycerin metabolism by *Phanerochaete chrysosporium*: evidence for nitric oxide and nitrite formation. *Biochim. Biophys. Acta* 1074: 320–325.

Seto, M., K. Kimbura, M. Shimura, T. Hatta, M. Fukuda, and K. Yano. 1995. A novel transformation of polychlorinated biphenyls by *Rhodococcus* sp. strain RHA1. *Appl. Environ. Microbiol.* 61: 3353–3358.

Sharma, P.K., and P.L. McCarty. 1996. Isolation and characterization of a facultatively aerobic bacterium that reductively dehalogenates tetrachloroethene to *cis*-dichloroethene. *Appl. Environ. Microbiol.* 62: 761–765.

Shen, H., P.H. Pritchard, and G.W. Sewell. 1996. Microbial reduction of CrVI during anaerobic degradation of benzoate. *Environ. Sci. Technol.* 30: 1667–1674.

Shields, M.S. and M.J. Reagin. 1992. Selection of a *Pseudomonas cepacia* strain constitutive for the degradation of trichloroethylene. *Appl. Environ. Microbiol.* 58: 3977–3983.

Shields, M.S., S.W. Hooper, and G.S. Sayler. 1985. Plasmid-mediated mineralization of 4-chlorobiphenyl. *J. Bacteriol.* 163: 882–889.

Shields, M.S., S.O. Montgomery, P.J. Chapman, S.M. Cuskey, and P.H. Pritchard. 1989. Novel pathway of toluene catabolism in the trichloroethylene-degrading bacterium G4. *Appl. Environ. Microbiol.* 55: 1624–1629.

Shields, M.S., M.J. Reagin, R.R. Gerger, R. Campbell, and C. Somerville. 1995. TOM, a new aromatic degradative plasmid from *Burkholderia (Pseudomonas) cepacia* G4. *Appl. Environ. Microbiol.* 61: 1352–1356.

Shimp, R.J. and F.K. Pfaender. 1985a. Influence of easily degradable naturally occurring carbon substrates on biodegradation of monosubstituted phenols by aquatic bacteria. *Appl. Environ. Microbiol.* 49: 394–401.

Shimp, R.J. and F.K. Pfaender. 1985b. Influence of naturally occurring humic acids on biodegradation of monosubstituted phenols by aquatic bacteria. *Appl. Environ. Microbiol.* 49: 402–407.

Shoda, M. and S. Udaka. 1980. Preferential utilizatiion of phenol rather than glucose by *Trichosporon cutaneum* possessing a partiallty constitutive catechol 1,2-oxygenase. *Appl. Environ. Microbiol.* 39: 1129–1133.

Shoham, Y. and E. Rosenberg. 1983. Enzymatic depolymerization of emulsan. *J. Bacteriol.* 156:161–167.

Shu, Y., E.S.C. Kwok, E.C. Tuazon, R. Atkinson, and J. Arey. 1997. Products of the gas-phase reactions of linalool with OH radicals, NO_3 radicals, and O_3. *Environ. Sci. Technol.* 31: 896–904.

Siegele, D.A. and R. Kolter. 1992. Life after log. *J. Bacteriol.* 174: 345–348.

Sigman, M.E., P.F. Schuler, M.M. Ghosh, and R.T. Dabestani. 1998. Mechanism of pyrene photochemical oxidation in aqueous and surfactant solutions. *Environ. Sci. Technol.* 32: 3980–3985.

Simkins, S. and M. Alexander. 1984. Models for mineralization kinetics with the variables of substrate concentration and population density. *Appl. Environ. Microbiol.* 47: 1299–1306.

Simkins, S., R. Mukherjee, and M. Alexander. 1986. Two approaches to modeling kinetics of biodegradation by growing cells and application of a two-compartment model for mineralization kinetics in sewage. *Appl. Environ. Microbiol.* 51: 1153–1160.

Singer, M.E.V. and W.R. Finnerty. 1990. Physiology of biosurfactant synthesis by *Rhodococcus* species H13-A. *Can. J. Microbiol.* 36: 741–745.

Singer, M.E.V., W.R. Finnerty, and A. Tunelid. 1990. Physical and chemical properties of a biosurfactant synthesized by *Rhodococcus* sp. H13-A. *Can. J. Microbiol.* 36: 746–750.

Singh, N. and N. Sethunathan. 1992. Degradation of soil-sorbed carbofuran by an enrichment culture from carbofuran-retreated *Azolla* plot. *J. Agric. Food Chem.* 40: 1062–1065.

Singh, N.C., T.P. Dasgupta, E.V. Roberts, and A. Mansingh. 1991. Dynamics of pesticides in tropical conditions. 1. Kinetic studies of volatilization, hydrolysis, and photolysis of dieldrin and a- and b-endosulfan. *J. Agric. Food Chem.* 39: 575–579.

Six, S., S.C. Andrews, G. Unden, and J.R. Guest. 1994. *Escherichia coli* possesses two homologous anaerobic C_4-dicarboxylate membrane transporters (*Dcua* and *Dcub*) distinct from the aerobic dicarboxylate transport system (*Dct*). *J. Bacteriol.* 176: 6470–6478.

Slater, J.H. and D. Lovatt. 1984. Biodegradation and the significance of microbial communities pp. 439–485. In *Microbial Degradation of Organic Compounds* (Ed. D.T. Gibson). Marcel Dekker, New York.

Smith, H.W., Z. Parsell, and P. Green. 1978. Thermosensitive antibiotic resistance plasmids in enterobacteria. *J. Gen. Microbiol.* 109: 37–47.

Smith, M.G. and R.J. Park. 1984. Effect of restricted aeration on catabolism of cholic acid by two *Pseudomonas* species. *Appl. Environ. Microbiol.* 48: 108–113.

Smith, R.V. and P.J. Davis. 1968. Induction of xenobiotic monooxygenases. *Adv. Biochem. Eng.* 4: 61–100.

Smith, R.V. and J.P. Rosazza. 1983. Microbial models of mammalian metabolism. *J. Nat. Prod.* 46: 79–91.

Sobecky, P.A., M.A. Schell, M.A. Moran, and R.E. Hodson. 1992. Adaptation of model genetically engineered microorganisms to lake water: growth rate enhancements and plasmid loss. *Appl. Environ. Microbiol.* 58: 3630–3637.

Soderquist, C.J., D.G. Crosby, K.W. Moilanen, J.N. Seiber, and J.E. Woodrow. 1975. Occurrence of trifluralin and its photoproducts in air. *J. Agric. Food Chem.* 23: 304–309.

Soeder, C.J., E. Hegewald, and H. Kneifel. 1987. Green microalgae can use naphthalenesulfonic acids as sources of sulfur. *Arch. Microbiol.* 148: 260–263.

Sondossi, M., M. Sylvestre, and D. Ahmad. 1992. Effects of chlorobenzoate transformation on the *Pseudomonas testosteroni* biphenyl and chlorobiphenyl degradation pathway. *Appl. Environ. Microbiol.* 58: 485–495.

Sontoh, S. and J.D. Semrau. 1998. Methane and trichloroethylene degradation by *Methylosinus trichosporium* OB3b expressing particulate methane monooxygenase. *Appl. Environ. Microbiol.* 64: 1106–1114.

Spadaro, J.T., L. Isabelle, and V. Renganathan. 1994. Hydroxyl radical mediated degradation of azo dyes: evidence for benzene generation. *Environ. Sci. Technol.* 28: 1389–1393.

Spain, J.C. and D.T. Gibson 1991. Pathway for biodegradation of *p*-nitrophenol in a *Moraxella* sp. *Appl. Environ. Microbiol.* 57: 812–819.

Spain, J.C., P.A. van Veld, C.A. Monti, P.H. Pritchard, and C.R. Cripe. 1984. Comparison of *p*-nitrophenol biodegradation in field and laboratory test systems. *Appl. Environ. Microbiol.* 48: 944–950.

Spiess, T., F. Desiere, P. Fischer, J.C. Spain, H.-J. Knackmuss, and H. Lenke. 1998. A new 4-nitrotoluene degradation pathway in a *Mycobacterium* strain. *Appl. Environ. Microbiol.* 64: 446–452.

Spokes, J.R. and N. Walker. 1974. Chlorophenol and chlorobenzoic acid co-metabolism by different genera of soil bacteria. *Arch. Microbiol.* 96: 125–134.

Stalker, D.M. and K.E. McBride. 1987. Cloning and expresssion in *Escherichia coli* of a *Klebsiella ozaenae* plasmid-borne gene encoding a nitrilase specific for the herbicide Bromoxynil. *J. Bacteriol.* 169: 955–960.

Standley, L.J. and T.L. Bott. 1998. Trifluoroacetate, an atmospheric breakdown product of hydrofluorocarbon refrigerants: biomolecular fate in aquatic organisms. *Environ. Sci. Technol.* 32: 469–475.

Stang, P.M., R.F. Lee, and P.F. Seligman. 1992. Evidence for rapid, nonbiological degradation of tributyltin compounds in autoclaved and heat-treated fine-grained sediments. *Environ. Sci. Technol.* 26: 1382–1387.

Stanlake, G.J. and R.K. Finn. 1982. Isolation and characterization of a pentachlorophenol-degrading bacterium. *Appl. Environ. Microbiol.* 44: 1421–1427.

Steen, W.C. and T.W. Collette. 1989. Microbial degradation of seven amides by suspended bacterial populations. *Appl. Environ. Microbiol.* 55: 2545–2549.

Steffan, R.J., K. McClay, S. Vainberg, C.W. Condee, and D. Zhang. 1997. Biodegradation of the gasoline oxygenates methyl *tert*-butyl ether, ethyl *tert*-butyl ether, and amyl *tert*-butyl ether by propane-oxidizing bacteria. *Appl. Environ. Microbiol.* 63: 4216–4222.

Stephanou, E.G. and N. Stratigakis 1993. Oxocarboxylic and α,ω-dicarboxylic acids: photooxidation products of biogenic unsaturated fatty acids present in urban aerosols. *Environ. Sci. Technol.* 27: 1403–1407.

Stephanou, E.G. 1992. α,ω-Dicarboxylic acid salts and α,ω-dicarboxylic acids. Photooxidation products of unsaturated fatty acids, present in marine aerosols and marine sediments. Naturwissenschaften. 79: 28–131.

Stieb, M. and B. Schink. 1985. Anaerobic oxidation of fatty acids by *Clostridium bryantii* sp. nov., a sporeforming, obligately syntrophic bacterium. *Arch. Microbiol.* 140: 387–390.

Stohl, A., E. Williams, G. Wotawa, and H. Kromp-Kolb. 1996 A European inventory of soil nitric oxide emissions and the effect of these emissions on the photochemical formation of ozone. *Atmos. Environ.* 30: 3741–3755.

Stuart-Keil, K.G., A.M. Hohnstock, K.P. Drees, J.B. Herrick, and E.L. Madsen. 1998. Plasmids responsible for horizontal transfer of naphthalene catabolism genes between bacteria at a coal tar-contaminated site are homologous to pDTG1 from *Pseudomonas putida* NCIB 9816-4. *Appl. Environ. Microbiol.* 64: 3633–3640.

Stubbe, J. and W.A. van der Donk. 1998. Protein radicals in enzyme catalysis. *Chem. Rev.* 98: 705–762.

Styrvold, O.B. and A.R. Ström. 1984. Dimethylsulphoxide and trimethylamine oxide respiraton of *Proteus vulgaris*. *Arch. Microbiol.* 140: 74–78.

Subba-Rao, R.V. and M. Alexander. 1982. Effect of sorption on mineralization of low concentrations of aromatic compounds in lake water samples. *Appl. Environ. Microbiol.* 44: 659–668.

Subba-Rao, R.V., H.E. Rubin, and M. Alexander. 1982. Kinetics and extent of mineralization of organic chemicals at trace levels in freshwater and sewage. *Appl. Environ. Microbiol.* 43: 1139–1150.

Subramanian, V. and C.S. Vaidyanathan. 1984. Anthranilate hydroxylase from *Aspergillus niger*: new type of NADPH-linked nonheme iron monooxygenase. *J. Bacteriol.* 160: 651–655.

Suen, W.-C., B.E. Haigler, and J.C. Spain 1996. 2,4-Dinitrotoluene dioxygenase from *Burkholderia* sp. strain DNT: similarity to naphthalene dioxygenase. *J. Bacteriol.* 178: 4926–4934.

Sun, Y. and J.J. Pignatello. 1993. Organic intermediates in the degradation of 2,4-dichlorophenoxyacetic acid by Fe^{3+}/H_2O_2 and $Fe^{3+}/H_2O_2/UV$. *J. Agric. Food. Chem.* 41: 1139–1142.

Sutherland, J.B., J.P. Freeman, A.L. Selby, P.P. Fu, D.W. Miller, and C.E. Cerniglia. 1990. Stereoselective formation of a K-region dihydrodiol from phenanthrene by *Streptomyces flavovirens*. *Arch. Microbiol.* 154: 260–266.

Sutherland, J.B., A.L. Selby, J.P. Freeman, F.E. Evans, and C.E. Cerniglia. 1991. Metabolism of phenanthrene by *Phanerochaete chrysosporium*. *Appl. Environ. Microbiol.* 57: 3310–3316.

Suwa, Y., W.E. Olben, and L.J. Forney. 1996. Characterization of chromosomally encoded 2,4-dichlorophenoxyacetic acid-α-ketoglutarate dioxygenase from *Burkholderia* sp. strain RASC. *Appl. Environ. Microbiol.* 62: 2464–2469.

Suylen, G.M.H., P.J. Large, J.P. van Dijken, and J.G. Kuenen. 1987. Methyl mercaptan oxidase, a key enzyme in the metabolism of methylated sulphur compounds by *Hyphomicrobium* EG. *J. Gen. Microbiol.* 133: 2989–2997.

Suzuki, T. 1978. Enzymatic methylation of pentachlorophenol and its related compounds by cell-free extracts of *Mycobacterium* sp. isolated from soil. *J. Pestic. Sci.* 3: 441–443.

Suzuki, T., K. Tanaka, I. Matsubara, and S. Kinoshita. 1969. Trehalose lipid and alpha-branched-beta-hydroxy fatty acid formed by bacteria grown on *n*-alkanes. *Agric. Biol. Chem.* 33: 1619–1627.

Swanson, M.B., W.A. Ivancic, A.M. Saxena, J.D. Allton, G.K. O'Brian, T. Suzuki, H. Nishizawa, and M. Nokata. 1995. Direct photolysis of fenpyroximate in a buffered aqueous solution. *J. Agric. Food Chem.* 43: 513–518.

Sylvestre, M., R. Massé, F. Messier, J. Fauteux, J.-G. Bisaillon, and R. Beaudet. 1982. Bacterial nitration of 4-chlorobiphenyl. *Appl. Environ. Microbiol.* 44: 871–877.

Szewzyk, R. and N. Pfennig. 1987. Complete oxidation of catechol by the strictly anaerobic sulfate-reducing *Desulfobacterium catecholicum* sp. nov. *Arch. Microbiol.* 147: 163–168.

Taeger, K., H.-J. Knackmuss, and E. Schmidt. 1988. Biodegradability of mixtures of chloro- and methylsubstituted aromatics: simultaneous degradation of 3-chlorobenzoate and 3-methylbenzoate. *Appl. Microbiol. Biotechnol.* 28: 603–608.

Taira, K., J. Hirose, S. Hayashida, and K. Furukawa. 1992. Analysis of *bph* operon from the polychlorinated biphenyl-degrading strain of *Pseudomonas pseudoalcaligenes* KF707. *J. Biol. Chem.* 267: 4844–4853.

Tam, A.C., R.M. Behki, and S.U. Khan. 1987. Isolation and characterization of an S-ethyl-N,N-dipropylthiocarbamate-degrading *Arthrobacter* strain and evidence for plasmid-associated S-ethyl-N,N-dipropylthiocarbamate degradation. *Appl. Environ. Microbiol.* 53: 1088–1093.

Tamaguchi, M. and H. Fujisawa. 1982. Subunit structure of oxygenase component in benzoate-1,2-dioxygenase system from *Pseudomonas arvilla* C-1. *J. Biol. Chem.* 257: 12497–12502.

Tan, H.-M. and J.R. Mason. 1990. Cloning and expression of the plasmid-encoded benzene 2,3,4,6-tetrachlorobiphenyl dioxygenase genes from *Pseudomonas putida* ML2. *FEMS Microbiol. Lett.* 72: 259–264.

Tanaka, A. and M. Ueda. 1993. Assimilation of alkanes by yeasts: functions and biogenesis of peroxisomes. *Mycol. Res.* 98: 1025–1044.

Taniuchi, H., M. Hatanaka, S. Kuno, O. Hayashi, M. Nakazima, and N. Kurihara. 1964. Enzymatic formation of catechol from anthranilic acid. *J. Biol. Chem.* 239: 2204–2211.

Tasaki, M., Y. Kamagata, K. Nakamura, and E. Mikami. 1991. Isolation and characterization of a thermophilic benzoate-degrading, sulfate-reducing bacterium, *Desulfotomaculum thermobenzoicum* sp. nov. *Arch. Microbiol.* 155: 348–352.

Taylor, B.F. 1983. Aerobic and anaerobic catabolism of vanillic acid and some other methoxy-aromatic compounds by *Pseudomonas* sp. strain PN-1. *Appl. Environ. Microbiol.* 46: 1286–1292.

Taylor, B.F. and D.C. Gilchrist. 1991. New routes for aerobic biodegradation of dimethylsulfoniopropionate. *Appl. Environ. Microbiol.* 57: 3581–3584.

Taylor, B.F. and M.J. Heeb. 1972. The anaerobic degradation of aromatic compounds by a denitrifying bacterium. Radioisotope and mutant studies. *Arch. Microbiol.* 83: 165–171.

Taylor, B.F., J.A. Amador, and H.S. Levinson. 1993. Degradation of *meta*-trifluoromethylbenzoate by sequential microbial and photochemical treatments. *FEMS Microbiol. Lett.* 110: 213–216.

Taylor, S.L., L.S. Guthertz, M. Leatherwood, and E.R. Lieber. 1979. Histamine production by *Klebsiella pneumoniae* and an incident of scombroid fish poisoning. *Appl. Environ. Microbiol.* 37: 274–278.

Thayer, J.R. and M.L. Wheelis. 1982. Active transport of benzoate in *Pseudomonas putida*. *J. Gen. Microbiol.* 128: 1749–1753.

Thomas, A.W., J. Lewington, S. Hope, A.W. Topping, A.J. Weightman, and J.H. Slater. 1992. Environmentally directed mutations in the dehalogenase system of *Pseudomonas putida* strain PP3. *Arch. Microbiol.* 158: 176–182.

Topp, E., R.L. Crawford, and R.S. Hanson. 1988. Influence of readily metabolizable carbon on pentachlorophenol metabolism by a pentachlorophenol-degrading *Flavobacterium* sp. *Appl. Environ. Microbiol.* 54: 2452–2459.

Torrents, A., B.G. Anderson, S. Bilboulian, W.E. Johnson, and C.J. Hapeman. 1997. Atrazine photolysis: mechanistic investigations of direct and nitrate-mediated hydroxyl radical processes and the influence of dissolved organic carbon from the Chesapeake Bay. 1997. *Environ. Sci. Technol.* 31: 1476–1482.

Touati, D., M.Jacques, B. Tardat, L. Bouchard, and S. Despied. 1995. Lethal oxidative damage and mutagenesis are generated by iron Δfur mutants of *Escherichia coli*: protective role of superoxide dismutase. *J. Bacteriol.* 177: 2305–2314.

Trickett, J.M., E.J. Hammonds, T.L. Worrall, M.K. Trower, and M. Griffin. 1991. Characterization of cyclohexane hydroxylase; a three-component enzyme system from a cyclohexane-grown *Xanthobacter* sp. *FEMS Microbiol. Lett.* 82: 329–334.

Trower, M.K., F.S. Sariaslani, and D.P. O'Keefe. 1989. Purification and characterization of a soybean flour-induced cytochrome P-450 from *Streptomyces griseus*. *J. Bacteriol.* 171: 1781–1787.

Tschech, A. and G. Fuchs. 1989. Anaerobic degradation of phenol via carboxylation to 4-hydroxybenzoate: *in vitro* study of isotope exchange between $^{14}CO_2$ and 4-hydroxybenzoate. *Arch. Microbiol.* 152: 594–599.

Tsitko, I.V., G.M. Zaitsev, A.G. Lobanok, and M.S. Salkinoja-Salonen. 1999. Effect of aromatic compounds on cellular fatty acid composition of *Rhodococcus opacus*. *Appl. Environ. Microbiol.* 65: 853–855.

Tsou, A.Y., S.C. Ransom, J.A. Gerlt, D.D. Buechter, P.C. Babbitt, and G.L. Kenyon. 1990. Mandelate pathway of *Pseudomonas putida*: sequence relationships involving mandelate racemase, (S)-mandelate dehydrogenase, and benzoylformate decarboxylase and expression of benzoylformate decarboxylase in *Escherichia coli*. *Biochemistry* 29: 9856–9862.

Tuazon, E.C., H. MacLeod, R. Atkinson, and W.P.L. Carter. 1986. α Dicarbonyl yields from the NO_x-air photooxidations of a series of aromatic hydrocarbons in air. *Environ. Sci. Technol.* 20: 383–387.

Tyson, C.A., J.D. Lipscomb, and I.C. Gunsalus. 1972. The roles of putidaredoxin and $P450_{cam}$ in methylene hydroxylation. *J. Biol. Chem.* 247: 5777–5781.

Ullah, A.J.H., R.I. Murray, P.K. Bhattacharyya, G.C. Wagner, and I.C. Gunsalus. 1990. Protein components of a cytochrome P-450 linalool 8-methyl hydroxylase. *J. Biol. Chem.* 265: 1345–1351.

Utkin, I., C. Woese, and J. Wiegel. 1994. Isolation and characterization of *Desulfitobacterium dehalogenans* gen. nov., sp.nov., an anaerobic bacterium which reductively dechlorinates chlorophenolic compounds. *Int. J. Syst. Bacteriol.* 44: 612–619.

Uwajima, T. and M. Hasegawa. 1991. Sequence of gene *choB* encoding cholesterol oxidase of *Brevibacterium sterolicum*: comparison with *choA* of *Streptomyces* sp. SA-COO. *Gene* 103: 93–96.

Valli, K. and M.H. Gold. 1991. Degradation of 2,4-dichlorophenol by the lignin-degrading fungus *Phanerochaete chrysosporium*. *J. Bacteriol.* 173: 345–352.

Valli, K., H. Wariishi, and M.H. Gold. 1992a. Degradation of 2,7-dichlorodibenzo-*p*-dioxin by the lignin-degrading basidiomycete *Phanerochaete chrysosporium*. *J. Bacteriol.* 174: 2131–2137.

Valli, K., B.J. Brock, D.K. Joshi, and M.H. Gold. 1992b. Degradation of 2,4-dinitrotoluene by the lignin-degrading fungus *Phanerochaete chrysosporium*. *Appl Environ. Microbiol.* 58: 221–228.

van Beilin, J.B., M.G. Wubbolts, and B. Witholt. 1994. Genetics of alkane oxidation by *Pseudomonas oleovorans*. *Biodegradation* 5: 161–174.

Van der Kooij, D. and W.A.M. Hijnen. 1984. Substrate utilization by an oxalate-consuming *Spirillum* species in relation to its growth in ozonated water. *Appl. Environ. Microbiol.* 47: 551–559.

Van der Kooij, D., A. Visser, and J.P. Oranje. 1980. Growth of *Aeromonas hydrophila* at low concentrations of substrates added to tap water. *Appl. Environ. Microbiol.* 39: 1198–1204.

Van der Kooij, D., J.P. Oranje, and W.A.M. Hijnen. 1982. Growth of *Pseudomonad aeruginosa* in tap water in relation to utilization of substrates at concentrations of few micrograms per liter. *Appl. Environ. Microbiol.* 44:1086–1095.

Van der Meer, J.R., A.R.W. van Neerven, E.J. de Vries, W.M. de Vos, and A.J.B. Zehnder. 1991. Cloning and characterization of plasmid-encoding genes for the degradation of 1,2-dichloro-, 1,4-dichloro-, and 1,2,4-trichlorobenzene of *Pseudomonas* sp. strain P51. *J. Bacteriol.* 173: 6–15.

Van der Meer, J.R., T.N.P. Bosma, W.P. de Bruin, H. Harms, C. Holliger, H.H.M. Rijnaarts, M. E. Tros, G. Schraa, and A.J.B. Zehnder. 1992. Versatility of soil column experiments to study biodegradation of halogenated compounds under environmental conditions. *Biodegradation* 3: 265–284.

Van der Meer, J.R., C. Werlen, S.F. Nishino, and J.C. Spain. 1998. Evolution of a pathway for chlorobenzene metabolism leads to natural attenuation in contaminated groundwater. *Appl. Environ. Microbiol.* 64: 4185–4193.

Van der Woude, B.J., J.C. Gottschal, and R.A. Prins. 1995. Degradation of 2,5-dichlorobenzoic acid by *Pseudomonas aeruginosa* JB2 at low oxygen tensions. *Biodegradation* 6: 39–46

Van Ginkel, C.G., H.G.J. Welten, and J.A.M. de Bont. 1987. Oxidation of gaseous and volatile hydrocarbons by selected alkene-utilizing bacteria. *Appl. Environ. Microbiol.* 53: 2903–2907.

Van Ginkel, C.G., G.B. Rikken, A.G.M. Kroon, and S.W.M. Kengen. 1996. Purification and characterization of chlorite dismutase: a novel oxygen-generating enzyme. *Arch. Microbiol.* 166: 321–326.

Van Leeuwen, B.C. Nicholson, K.P. Hayes, and D.E. Mulcahy. 1996. Resistance of bound chloroguaiacols and AOX from pulp mill effluent to degradation by *Trichoderma harzianum* isolated from Lake Bonney, South-eastern Australia. *Mar. Fresh Water Res.* 47: 961–969.

Van Leeuwen, B.C. Nicholson, G. Levay, K.P. Hayes, and D.E. Mulcahy. 1997. Transformation of free tetrachloroguaiacol to bound compounds by fungi isolated from Lake Bonney, south-eastern Australia. *Mar. Fresh Water Res.* 48: 551–557.

van Pée, K.-H., and F. Lingens. 1985. Purification of bromoperoxidase from *Pseudomonas aureofaciens. J. Bacteriol.* 161: 1171–1175.

Vandenbergh, P.A. and R.L. Cole. 1986. Plasmid involvement in linalool metabolism by *Pseudomonas fluorescens. Appl. Environ. Microbiol.* 52: 939–940.

Vandenbergh, P.A. and A.M. Wright. 1983. Plasmid involvement in acyclic isoprenoid metabolism by *Pseudomonas putida. Appl. Environ. Microbiol.* 45: 1953–1955.

Vandenbergh, P.A., R.H. Olsen, and J.E. Colaruotolo. 1981. Isolation and genetic characterization of bacteria that degrade chloroaromatic compounds. *Appl. Environ. Microbiol.* 42: 737–739.

Vannelli, T. and A.B. Hooper. 1992. Oxidation of nitrapyrin to 6-chloropicolinic acid by the ammonia-oxidizing bacterium *Nitrosomonas europaea. Appl. Environ. Microbiol.* 58: 2321–2325.

Vannelli, T., M. Logan, D.M. Arciero, and A.B. Hooper. 1990. Degradation of halogenated aliphatic compounds by the ammonia-oxidizing bacterium *Nitrosomonas europaea. Appl. Environ. Microbiol.* 56: 1169–1171.

Vercellone-Smith, P. and D.S. Herson. 1997. Toluene elicits a carbon starvation response in *Pseudomonas putida* mt-2 containing the TOL plasmid pWWO. *Appl. Environ. Microbiol.* 63: 1925–1932.

Verhagen, F.J.M., H.J. Swarts, J.B.P.A. Wijnberg, and J.A. Field. 1998. Biotransformation of the major fungal metabolite 3,5-dichloro-*p*-anisyl alcohol under anaerobic conditions and its role in formation of bis(3,5-dichloro-4-hydroxyphenyl)methane. *Appl. Environ. Microbiol.* 64: 3225–3231.

Véron, M. and L. Le Minor. 1975. Nutrition et taxonomie des *Enterobacteriaceae* et bactéries voisines. III. Caractéres nutritionnels et différenciation des groupes taxonomiques. *Ann. Microbiol. (Inst. Pasteur)* 126B: 125–147.

Wackett, L.P., L.D. Kwart, and D.T.Gibson. 1988. Benzylic monooxygenation catalyzed by toluene dioxygenase from *Pseudomonas putida. Biochemistry* 27: 1360–1367.

Wackett, L.P., G.A. Brusseau, S.R. Householder, and R.S. Hanson. 1989. Survey of microbial oxygenases: trichloroethylene degradation by propane-oxidizing bacteria. *Appl. Environ. Microbiol.* 55: 2960–2964.

Wada, H., K. Sendoo, and Y. Takai. 1989. Rapid degradation of γ-HCH in the upland soil after multiple application. *Soil Sci. Plant Nutr.* 35: 71–77.

Walker, J.D. and R. S. Pore. 1978. Growth of *Prototheca* isolates on *n*-hexadecane and mixed-hydrocarbon substrate. *Appl. Environ. Microbiol.* 35: 694–697.

Walker, J.R.L. and B.G. Taylor. 1983. Metabolism of phloroglucinol by *Fusarium solani. Arch. Microbiol.* 134: 123–126.

Walker, N. 1973. Metabolism of chlorophenols by *Rhodotorula glutinis. Soil Biol. Biochem.* 5: 525–530.

Wallington, T.J. and W.F. Schneider. 1994. The stratospheric fate of CF_3OH. *Environ. Sci. Technol.* 28: 1198–1200.

Wallington, T.J., M.D. Hurley, J.M. Fracheboud, J.J. Orlando, G.S. Tyndall, J. Sehested, T.E. Mögelberg, and O.J. Nielsen. 1996. Role of excited CF_3CFHO radicals in the atmospheric chemistry of HFC-134a. *J. Phys. Chem.* 100: 18116–18122.

Wallrabenstein, C., E. Hauschild, and B. Schink. 1995. *Syntrophobacter pfennigii* sp. nov., new syntrophically propionate-oxidizing anaerobe growing in pure culture with propionate and sulfate. *Arch. Microbiol.* 164: 346–352.

Wang, Y., P.C.K. Lau, and D.K. Button. 1996. A marine oligobacterium harboring genes known to be part of aromatic hydrocarbon degradation pathways of soil pseudomonads. *Appl. Environ. Microbiol.* 62: 2169–2173.

Wangberg, I., I. Barnes, and K.H. Becker. 1997. Product and mechanistic study of the reaction of NO_3 radicals with α-pinene. *Environ. Sci. Technol.* 31: 2130–2135.

Warburton, E.J., A.M. Magor, M.K. Trower, and M. Griffin. 1990. Characterization of cyclohexane hydroxylase; involvement of a cytochrome P-450 system from a cyclohexane-grown *Xanthobacter* sp. *FEMS Microbiol. Lett.* 66: 5–10.

Ward, B.B. 1987. Kinetic studies on ammonia and methane oxidation by *Nitrosococcus oceanus. Arch. Microbiol.* 147: 126–133.

Warshawsky, D., M. Radike, K. Jayasimhulu, and T. Cody. 1988. Metabolism of benzo(a)pyrene by a dioxygenase system of the freshwater green alga *Selenastrum capricornutum. Biochem. Biophys. Res. Commun.* 152: 540–544.

Warshawsky, D., T. Cody, M. Radike, R. Reilman, B. Schumann, K. LaDow, and J. Schneider. 1995. Biotransformation of benzo[*a*]pyrene and other polycyclic aromatic hydrocarbons and heterocyclic analogues by several green algae and other algal species under gold and white light. *Chem.- Biol. Interact.* 97: 131–148.

Wayne, L.G., and 17 coauthors. 1991. Fourth report of the cooperative, open-ended study of slowly growing mycobacteria by the international working group on mycobacterial taxonomy. *Int. J. Syst. Bacteriol.* 41: 463–472.

Weber, F.J., S. Isken, and J.A.M. de Bont. 1994. *Cis/trans* isomerization of fatty acids as a defense mechanism of *Pseudomonas putida* strains to toxic concentrations of toluene. *Microbiology* (U.K.) 140: 2013–2017.

Wedemeyer, G. 1967. Dechlorination of 1,1,1-trichloro-2,2-bis[p-chlorophenyl]ethane by *Aerobacter aerogenes. Appl. Microbiol.* 15: 569–574.

Weiner, J.H., D.P. MacIsaac, R.E. Bishop, and P.T. Bilous. 1988. Purification and properties of *Escherichia coli* dimethyl sulfoxide reductase, an iron-sulfur molybdoenzyme with broad substrate specificity. *J. Bacteriol.* 170: 1505–1510.

Wende, R., F.-H. Bernhardt, and K. Pfleger. 1989. Substrate-modulated reactions of putidamonooxin. The nature of the active oxygen species formed and its reaction mechanism. *Eur. J. Biochem.* 81: 189–197.

Wergath, J., H.-A. Arfmann, D.H. Pieper, K.N. Timmis, and R.-M. Wittich. 1998. Biochemical and genetic analysis of a gentisate 1,2-dioxygenase from *Sphingomonas* sp. strain RW 5. *J. Bacteriol.* 180: 4171–4176.

Wetzstein, H.-G., N. Schmeer, and W. Karl. 1997. Degradation of the fluoroquinolone enrofloxacin by the brown-rot fungus *Gleophyllum striatum*: identification of metabolites. *Appl. Environ. Microbiol.* 63: 4272–4281.

White, R.E., M.B. McCarthy, K.D. Egeberg, and S.G. Sligar. 1984. Regioselectivity in the cytochromes P-450: control by protein constraints and by chemical reactivities. *Arch. Biochem. Biophys.* 228: 493–502.

White-Stevens, R.H. and H. Kamin. 1972. Studies of a flavoprotein, salicylate hydroxylase. I. Preparation, properties, and the uncoupling of oxygen reduction from hydroxylation. *J. Biol. Chem.* 247: 2358–2370.

Whited, G.M. and D.T. Gibson. 1991. Separation and partial characterization of the enzymes of the toluene-4-monooxygenase catabolic pathway in *Pseudomonas mendocina* KR1. *J. Bacteriol.* 173: 3017–3020.

Whittenbury, R. and D.P. Kelly. 1977. Autotrophy: a conceptual phoenix. *Symp. Soc. Gen. Microbiol.* 27: 121–149.

Whyte, L.G., C.W. Greer, and W.E. Innis. 1996. Assessment of the biodegradation potential of psychrotrophic microorganisms. *Can. J. Microbiol.* 42: 99–106.

Whyte, L.G., L. Bourbonnière, and C.W. Greer. 1997. Biodegradation of petroleum hydrocarbons by psychotrophic *Pseudomonas* strains possessing both alkane (*alk*) and naphthalene (*nah*) catabolic pathways. *Appl. Environ. Microbiol.* 63: 3719–3723.

Widdel, F., G.-W. Kohring, and F. Mayer. 1983. Studies on dissimilatory sulfate-reducing bacteria that decompose fatty acids. *Arch. Microbiol.* 134: 286–294.

Wieser, M., B. Wagner, J. Eberspächer, and F. Lingens. 1997. Purification and characterization of 2,4,6-trichlorophenol-4-monooxygenase, a dehalogenating enzyme from *Azotobacter* sp. strain GP1. *J. Bacteriol.* 179: 202–208.

Williams, P.A. and M.J. Worsey. 1976. Ubiquity of plasmids in coding for toluene and xylene metabolism in soil bacteria: evidence for the existence of new TOL plasmids. *J. Bacteriol.* 125: 818–828.

Williams, R.J. and W.C. Evans. 1975. The metabolism of benzoate by *Moraxella* species through anaerobic nitrate respiration. *Biochem. J.* 148: 1–10.

Williams, W.A. and R.J. May. 1997. Low-temperature microbial aerobic degradation of polychlorinated biphenyls in sediment. *Environ. Sci. Technol.* 31: 3491–3496.

Witschel, M., S. Nagel, and T. Egli. 1997. Identification and characterization of the two-enzyme system catalyzing the oxidation of EDTA in the EDTA-degrading bacterial strain DSM 9103. *J. Bacteriol.* 179: 6937–6943.

Wolfe, N.L., R.G. Zepp, and D.F. Paris. 1978a. Carbaryl, propham and chlorpropham: a comparison of the rates of hydrolysis and photolysis with the rates of biolysis. *Water Res.* 12: 565–571.

Wolfe, N.L., R.G. Zepp, and D.F. Paris. 1978b. Use of structure–reactivity relationships to estimate hydrolytic persistence of carbamate pesticides. *Water Res.* 12: 561–563.

Wolin, M.J. 1982. Hydrogen transfer in microbial communities. In *Microbial Interactions and Communities*. (Eds. A.T. Bull and J.H. Slater) Vol.1 pp. 323–356.

Wolt, J.D., J.D. Schwake, F.R. Batzer, S.M. Brown, L.H. McKendry, J.R. Miller, G.A. Roth, M.A. Stanga, D. Portwood, and D.L. Holbrook. 1992. Anaerobic aquatic degradation of flumetsulam [N-(2,6-difluorophenyl)-5-methyl[1,2,4]triazolo[1,5a]pyrimidine-2-sulfonamide]. *J. Agric. Food Chem.* 40: 2302–2308.

Wong, A.S. and D.G. Crosby. 1981. Photodecomposition of pentachlorophenol in water. *J. Agric. Food Chem.* 29: 125–130.

Wong, C.M., M.J. Dilworth, and A.R. Glenn. 1994. Cloning and sequencing show that 4-hydroxybenzoate hydroxylase *Poba* is required for uptake of 4-hydroxybenzoate in *Rhizobium leguminosarum. Microbiology* (U.K.) 140: 2775–2786.

Workman, S.L. Woods, Y.A. Gorby, J.K. Fredrickson, and M.J. Truex. 1997. Microbial reduction of vitamin B_{12} by *Shewanella alga* strain BrY with subsequent transformation of carbon tetrachloride. *Environ. Sci. Technol.* 31: 2292–2297.

Worobey, B.L. 1989. Nonenzymatic biomimetic oxidation systems: theory and application to transformation studies of environmental chemicals. In *The Handbook of Environmental Chemistry* Vol. 2E (Ed. O.Hutzinger), pp. 58–110, Springer-Verlag, Berlin.

Wu, Q., D.L. Bedard, and J. Wiegel. 1997a. Effect of incubation temperature on the route of microbial reductive dechlorination of 2,3,4,6-tetrachlorobiphenyl in polychlorinated biphenyl (PCB)-contaminated and PCB-free freshwater sediments. *Appl. Environ. Microbiol.* 63: 28366–2843.

Wu, Q., D.L. Bedard, and J. Wiegel. 1997b. Temperature determines the pattern of anaerobic microbial dechlorination of Arochlor 1260 primed by 2,3,4,6-tetrachlorobiphenyl in Woods Pond sediment. *Appl. Environ. Microbiol.* 63: 4818–4825.

Wujcik, C.E., T.M. Cahill, and J.N. Seiber. 1998. Extraction and analysis of trifluoroacetic acid in environmental waters. *Anal. Chem.* 70: 4074–4080.

Wyndham, R.C. 1986. Evolved aniline catabolism in *Acinetobacter calcoaceticus* during continuous culture of river water. *Appl. Environ. Microbiol.* 51: 781–789.

Xun, L. 1996. Purification and characterization of chlorophenol 4-monooxygenase from *Burkholderia cepacia* AC1100. *J. Bacteriol.* 178: 2645–2649.

Yamaguchi, M. and H. Fujisawa. 1982. Subunit structure of oxygenase components in benzoate 1,2-dioxygenase system from *Pseudomonas arvilla* C-1. *J. Biol. Chem.* 257: 12497–12502.

Yamauchi, T. and T. Handa. 1987. Characterization of aza heterocyclic hydrocarbons in urban atmospheric particulate matter. *Environ. Sci. Technol.* 21: 1177–1181.

Yasahura, A. and M. Morita. 1988. Formation of chlorinated aromatic hydrocarbons by thermal decomposition of vinylidene chloride polymer. *Environ. Sci. Technol.* 22: 646–650.

Yoshinaga, D.H. and H.A. Frank. 1982. Histamine-producing bacteria in decomposing skipjack tuna (*Katsuwonus pelamis*). *Appl. Environ. Microbiol.* 44: 447–452.

Yu, J. and H.E. Jeffries. 1997. Atmospheric photooxidation of alkylbenzenes II. Evidence of formation of epoxide intermediates. *Atmos. Environ.* 31: 2281–2287.

Yu, J., H.E. Jeffries, and K.G. Sexton. 1997. Atmospheric photooxidation of alkylbenzenes — I. Carbonyl product analysis. *Atmos. Environ.* 31: 2261–2280.

Zahn, J.A. and A. A. DiSpirito. 1996. Membrane associated methane monooxygenase from *Methylococcus capsulatus* (Bath). *J. Bacteriol.* 178: 1018–1029.

Zaidi, B.R., Y. Murakami, and M. Alexander. 1988. Factors limiting success of inoculation to enhance biodegradation of low concentrations of organic chemicals. *Environ. Sci. Technol.* 22: 1419–1425.

Zaitsev, G.M., J.S. Uotila, I.V. Tsitko, A.G. Lobanok, and M.S. Salkinoja-Salonen. 1995. Utilization of halogenated benzenes, phenols, and benzoates by *Rhodococcus opacus* GM-14. *Appl. Environ. Microbiol.* 61: 4191–4201.

Zander, M. 1995. *Polycyclische Aromaten*, pp. 213–217. B.G. Teuber, Stuttgart, Germany.

Zeiner, R., K.-H. van Pée, and F. Lingens. 1988. Purification and partial characterization of multiple bromoperoxidases from *Streptomyces griseus*. *J. Gen. Microbiol.* 134: 3141–3149.

Zepp, R.G. and P.F. Schlotzhauer. 1983. Influence of algae on photolysis rates of chemicals in water. *Environ. Sci. Technol.* 17: 462–468.

Zepp, R.G., G.L. Baugham, and P.A. Scholtzhauer. 1981. Comparison of photochemical behaviour of various humic substances in water I. Sunlight induced reactions of aquatic pollutants photosensitized by humic substances. *Chemosphere* 10: 109–117.

Zhang, Y. and R.M. Miller. 1992. Enhanced octadecane dispersion and biodegradation by a *Pseudomonas* rhamnolipid surfactant (biosurfactant). *Appl. Environ. Microbiol.* 58: 3276–3282.

Zhang, Y., W.J. Maier, and R.M. Miller. 1997. Effect of rhamnolipids on the dissolution, bioavailability and biodegradation of phenanthrene. *Environ. Sci. Technol.* 31: 2211–2217.

Zhao, H., D. Yang, C.R. Woese, and M.P. Bryant. 1990. Assignment of Clostridium bryantii to *Syntrophospora bryantii* gen. nov., comb. nov. on the basis of a 16S rRNA sequence analysis of its crotonate-grown pure culture. *Int. J. Syst. Bacteriol.* 40: 40–44.

Zhou, N.-Y., A. Jenkins, C.K.N. Chan kwo Chion, and D.J. Leak. 1999. The alkene monooxygenase from *Xanthobacter* strain Py2 is closely related to aromatic monooxygenases and catalyzes aromatic monooxygenation of benzene, toluene, and phenol. *Appl. Environ. Microbiol.* 65: 1589–1595.

Ziegler, K., K. Braun, A. Böckler, and G. Fuchs. 1987. Studies on the anaerobic degradation of benzoic acid and 2-aminobenzoic acid by a denitrifying *Pseudomonas* strain. *Arch. Microbiol.* 149: 62–69.

Zielinska, B., J. Arey, R. Atkinson, and P.A. McElroy. 1988. Nitration of acephenanthrylene under simulated atmospheric conditions and in solution and the presence of nitroacephenanthrylenes in ambient particles. *Environ. Sci. Technol.* 22: 1044–1048.

Zielinska, B., J. Arey, R. Atkinson, and P.A. McElroy. 1989. Formation of methylnitronaphthalenes from the gas-phase reactions of 1- and 2-methylnaphthalene with hydroxyl radicals and N_2O_5 and their occurrence in ambient air. *Environ. Sci. Technol.* 23: 723–729.

Zipper, C., M. Bunk, A.J.B. Zehnder, and H.-P. E. Kohler. 1998. Enantioselective uptake and degradation of the chiral herbicide dichloroprop [(RS)-2-(2,4-dichlorophenoxy)propionic acid] by *Sphingomonas herbicidovorans* MH. *J. Bacteriol.* 180: 3368–3374.

Zoro, J.A., J.M. Hunter, G. Eglinton, and C.C. Ware. 1974. Degradation of *p,p'*-DDT in reducing environments. *Nature* 247: 235–237.

Zylstra, G.J., L.P. Wackett, and D.T. Gibson. 1989. Trichloroethylene degradation by *Escherichia coli* containing the cloned *Pseudomonas putida* F1 toluene dioxygenase genes. *Appl. Environ. Microbiol.* 55: 3162–3166.

5

Persistence: Experimental Aspects

SYNOPSIS Experimental procedures for investigating the degradation and transformation of xenobiotics are outlined, together with a critical analysis of the limitations inherent in conventional tests for determining ready biodegradability. A brief description of procedures for elective enrichment is given including the basic components of growth media together with salient features of their composition and use. Attention is directed to problems associated with low water solubility, toxicity, or volatility of the substrates. An outline is given of the various types of experiments that have been designed to assess biodegradability and biotransformation including the application of both batch- and continuous-culture procedures. Examples are provided of the use of pure cultures and stable consortia, microcosms, and larger-scale field systems. A number of procedures for investigating metabolic pathways are given including the use of mutants, the use of metabolic inhibitors, the application of stable isotopes, and *in vivo* studies using nuclear magnetic resonance. A brief discussion addresses procedures for analyzing natural populations and attempting to establish their metabolic roles. The difficulty in assigning organisms of metabolic interest to taxonomic groups is noted in addition to the generally limited existing knowledge of their genetic systems.

Introduction

A general overview of the processes that determine the fate and persistence of xenobiotics in the environment has been presented in Chapter 4, but it is important to anchor this to the experimental methods on which the conclusions from such investigations must ultimately be based. That is the aim of the present chapter.

5.1 Abiotic Reactions

The significance of these reactions has been exemplified in Chapter 4, Section 4.1, and some brief comments on experimental aspects should be given. The cardinal problem of isolating and identifying transformation products has been discussed in detail in Chapter 2.

For chemical reactions, few details are needed since their design follows those used in traditional preparative organic chemistry integrated with measurements of rates by methods that are widely used in physical organic chemistry. Two issues are worth noting and should be taken into account: (1) that substrate concentrations that are environmentally realistic are much lower than those generally used in purely chemical investigations and (2) environmental temperatures are highly variable from low summer water temperatures in northern latitudes to high temperatures on the surface of soils during periods of high insolation.

Experiments on photochemical reactions and transformations have been carried out under a number of different conditions.

1. Reactions in liquid medium may be carried out by illumination with radiation of the relevant wavelengths generally in quartz vessels of various configurations.

2. Gas-phase reactions have been carried out in 160-ml quartz vessels and the products analyzed online by mass spectrometry (Brubaker and Hites 1998). Hydroxyl radicals were produced by photolysis of ozone in the presence of water:

$$O_3 + h\nu \rightarrow O(^1D) + O_2; \quad O(^1D) + H_2O \rightarrow 2OH$$

3. Reactions that simulate tropospheric conditions have been carried out in Teflon bags with volumes of ~6 m^3 fitted with sampling ports for introduction of reactants and substrates, and removal of samples for analysis. Substrates may be added in the gas phase or as aerosols that form a surface film. The reactants have been noted in Section 4.1.2 and are the hydroxyl and nitrate radicals, and ozone. These must be prepared before use by Reactions 5.1 to 5.3.

Hydroxyl radicals by photolysis of methyl nitrite:

$$CH_3ONO + h\nu \quad \rightarrow CH_3O + NO \qquad\qquad (5.1)$$

$$CH_3O + O_2 \quad \rightarrow CH_2O + HO_2$$

$$HO_2 + NO \rightarrow OH \quad + NO_2$$

NO is added to inhibit photolysis of NO$_2$ that would produce O$_3$ and NO$_3$ radicals:

$$NO_2 + h\nu \quad \rightarrow NO \quad + O(^3P) \qquad\qquad (5.2)$$

$$O(^3P) + O_2 \quad \rightarrow O_3$$

$$O_3 + NO_2 \rightarrow NO_3 \quad + O_2$$

Nitrate radicals are prepared by photolysis of dinitrogen pentoxide in the dark:

$$N_2O_5 \rightarrow NO_3 + NO_2 \qquad (5.3)$$

Ozone by a corona discharge in O_2.

Products from the reactions are collected on Tenax cartridges and the analytes desorbed by heating, or on polyurethane form plugs from which the analytes may be recovered by elution with a suitable organic solvent.

5.2 Microbial Reactions

Like organic chemistry on which it critically depends, the study of microbial metabolism is a relatively young discipline, not much more than 100 years old. And the cardinal experimental procedures for isolation of microorganisms for studies on the metabolism of xenobiotics remain those of elective enrichment pioneered by Winogradsky, Beijerinck, Kluyver, van Niel, and their successors, backed up by the use of pure cultures using procedures developed by Koch in his classic investigations on anthrax. The major innovation has been the development of generally applicable procedures for the isolation of strictly anaerobic organisms that were introduced by Hungate (1969). There have, however, been major developments in methods for the elucidation of metabolic pathways: (1) the availability of isotopically labeled compounds — in particular ^{13}C and ^{14}C, (2) the application of genetic procedures, and (3) the application of modern analytical procedures and physical methods for structure determination that have been outlined in greater detail in Chapter 2. In studies of microbial metabolism, the advantages resulting from the requirement for only extremely small quantities of material needed for gas-chromatographic quantification and gas chromatographic–mass spectrometric identification can hardly be overestimated.

In the following sections, an attempt will be made to provide a critical outline of experimental aspects of investigations directed to biodegradation and biotransformation with particular emphasis on outstanding issues to which sufficient attention has not always been paid, and which have not therefore received ultimate resolution. Before proceeding further, it is desirable to define clearly some operational terms in addition to those that have been given in Chapter 4.

Mineralization is conventionally used for the aerobic degradation of a compound to CO_2 and H_2O.

Ready biodegradability refers to the situation in which the test compound is totally degraded (under aerobic conditions to CO_2, H_2O, etc.)

within the time span of a standardized test usually lasting 5, 7, or 28 days.

Inherent biodegradability is applied when the compound *may* be degraded, although not under the standard conditions generally used: their degradation may require, for example, preexposure to the xenobiotic.

Recalcitrant is a valuable concept (Alexander 1975) that has been applied to compounds which have not been demonstrably degraded under the conditions used for their examination.

Biotransformation is applied to situations in which even though degradation is not achieved, minor structural modifications of the test compound have occurred; illustrative examples have already been given in Chapter 4.

Rigid boundaries between these terms should not, however, be drawn, since all of them are operational rather than absolute.

5.2.1 Determination of Ready Biodegradability

Because of the central role that estimates of biodegradability play in environmental impact assessments, a great deal of effort has been devoted to developing standardized test procedures (Gerike and Fischer 1981). In spite of this, conventional tests for biodegradability under aerobic conditions retain some questionable, or even undesirable, features from an environmental point of view. Attention is therefore drawn to two valuable critiques of widely used procedures (Howard and Banerjee 1984; Battersby 1990). Some of the important issues in the design of such tests are therefore only briefly summarized here.

The Inoculum

For assessment of biodegradation in freshwater systems that have been most extensively examined, the inoculum is generally taken from municipal sewage treatment plants and is therefore dominated by microorganisms that have been subjected to selection primarily for their ability to use readily degraded substrates. Whereas these organisms are clearly valuable in evaluating the persistence of compounds that might be poorly degraded in such treatment systems and therefore might be discharged unaltered into the environment, they are not necessarily equally suited to investigations on the degradability of possibly recalcitrant substances in natural ecosystems that are dominated by microorganisms adapted to a different environment.

Concentration of the Substrate

In conventional assay systems, substrate concentrations are generally used at levels appropriate to measurements of the uptake of oxygen or the evolution of carbon dioxide, and these concentrations greatly exceed those likely to be encountered in receiving waters; attention has been drawn to this issue in Section 4.6.2. Application to poorly water-soluble (De Morsier et al. 1987) or to volatile substrates may be difficult or even impossible due to the need for the high substrate concentrations. In addition, high substrate concentrations may be toxic to the test organism and thereby provide false-negative results. Use of ^{14}C-labeled substrates and measurement of the evolution of $^{14}CO_2$ enables very much lower substrate concentrations to be used and this is particularly important if the contribution of marine oligotrophic bacteria is to be evaluated. This has been applied to populations of marine bacteria using ^{14}C-labeled acetate, aniline, 4-nitrophenol, desmethyl methylparathion, and 4-chloroaniline (Nyholm et al. 1992). The wider application of the use of isotopes is limited primarily by the availability of labeled substrates, although an increasingly wider range of industrially important compounds including agrochemicals is becoming commercially available.

The End Points

For aerobic degradation, uptake of oxygen or the evolution of carbon dioxide (labeled or unlabeled) is most widely used. Use of the concentration of dissolved organic carbon may present technical problems when particulate matter is present, although analysis of dissolved inorganic carbon in a closed system has been advocated (Birch and Fletcher 1991) and may simultaneously overcome problems with poorly soluble or volatile compounds.

For anaerobic degradation, advantage has been taken of methane production (Birch et al. 1989; Battersby and Wilson 1989). Whereas this may be valuable in the context of municipal sewage treatment plants, it is more questionable whether this is generally a valid parameter in investigations concerned with anaerobic degradation in natural ecosystems in view of the extensive evidence for anaerobic degradation by nonmethanogenic bacteria such as sulfate-reducing anaerobes; relevant examples are given in Sections 6.6 and 6.7.

Design of Experiments

Probably most investigations have been carried out in conventional batch cultures, but attention should be drawn to an attractive and flexible procedure using a cyclone fermentor (Liu et al. 1981).

Metabolic Limitations

In the most widely applied procedures, the test system is restricted in flexibility by the salinity and pH requirements of the test organisms, but probably the most serious limitation of all these test systems is that no account is taken of biotransformation reactions, nor is identification of their products routinely attempted. Examples in which this has been carried out include the following.

1. In a study of the degradation of sodium dodecyltriethoxy sulfate under mixed-culture die-away conditions using acclimated cultures (Griffiths et al. 1986), the metabolites were identified, and the kinetics of their synthesis compared with the degradation pathways elucidated in investigations using pure cultures (Hales et al. 1982, 1986).

2. The biodegradation of branched-chain alkanol ethoxyethylates was carried out by the standard OECD confirmatory tests and the metabolites fractionated after solid-phase extraction. The structures of the metabolites were determined by electrospray mass spectrometry and this made it possible to derive a scheme for the partial degradation of the compounds (Di Corcia et al. 1998a,b).

These procedures could advantageously receive more general application.

Application to Marine Systems

In view of the substantial quantities of xenobiotics that enter the marine environment, surprisingly little effort has been directed to this problem. The wider aspects of marine bacteria have been discussed in Chapter 1, Section 1.4 and Chapter 4, Section 4.3.1. The degradation of several structurally diverse substrates including nitrilotriacetate, 3-methylphenol and some chlorobenzenes was evaluated from the rates of incorporation of ^{14}C-labeled substrates into biomass and production of $^{14}CO_2$. These were used to evaluate the differences among freshwater, estuarine, and marine environments and revealed the difficulty of correlating rates with characteristics of the microbial community (Bartholomew and Pfaender 1983). Recent studies have used both dissolved organic carbon and oxygen uptake as parameters (Nyholm and Kristensen 1992), or analysis for specific ^{14}C-labeled compounds at low substrate concentration which are proposed as "simulation tests" (Nyholm et al. 1992). Both these tests used the indigenous organisms present in seawater and thereby provided a valuable degree of relevance even though they inevitably encountered the variability in the nutritional status — particularly for organic carbon — of seawater. Attention has been briefly made in Chapter 4, Sections 4.3.1 and 4.6.2 to oligotrophic marine ultramicrobacteria that are of undoubted importance in oceanic systems. There are, however, important aspects of their isolation that should be appreciated including the extent to which their physiology may be altered during maintenance under severe carbon limitation in the laboratory (Schut et al. 1993). Biodegradation of xenobiotics by these organisms has not attracted the attention it merits, and it is hazardous to extrapolate results from freshwater or brackish water systems to marine systems since all of the factors noted above coupled to the low temperatures that characterize ocean water introduce complexities that remain to be systematically investigated. Attention is drawn to the extensive metabolic potential of anaerobic sulfate-reducing bacteria isolated from marine muds and sediments. Examples are provided throughout Chapter 6.

5.2.2   Isolation and Elective Enrichment

It is only seldom that it has been possible to obtain bacteria with a desired metabolic capability directly from natural habitats. Almost always large numbers of other organisms are present so that some form of selection or enrichment is generally adopted before metabolic studies are attempted. Use of antibiotics or even more drastic procedures using alkali or hypochlorite have been used only infrequently except for the isolation of pathogenic bacteria such as *Mycobacterium tuberculosis*. Although valuable results have been obtained from experiments using metabolically stable mixed cultures, the problems of repeatability somewhat limit their application. Probably most metabolic studies on xenobiotics have therefore been carried out with pure cultures of organisms. With the possible exception of anaerobic bacteria, and in particular methanogens, relatively few of these organisms have been isolated from samples of municipal sewage sludge. Most have been obtained after elective enrichment of natural samples of water, sediments, or soils. This methodology was developed by Beijerinck and Winogradsky, and has been extensively exploited in the pioneering investigations carried out by the Delft school and their successors over many years. In the present context, one of its particularly attractive features is the inherent degree of environmental realism introduced by its application and the flexibility whereby virtually any environmental condition can be mimicked must be considered as one of the most attractive features of this procedure. There is extensive evidence for the existence of enrichment in the natural environment, and a number of examples illustrating its operation have been given in Section 4.6.4, but possibly the last word should be left to one of the pioneers of its application:

> But once an elective culture method for a particular microbe is available, it may safely be concluded that this organism will also be found in nature under conditions corresponding in detail to those of the culture, and that it will carry out the same transformations. (van Niel 1955)

5.2.3   General Procedures

In its simplest form, the procedure consists of elective enrichment of the microorganisms in an environmental sample by growth at the expense of a single compound serving as the sole source of carbon and energy, successive transfer into fresh medium after growth has occurred, followed by isolation of the appropriate organisms. Some of the experimental details are briefly described in the following paragraphs. The three successive stages used in isolating the desired organisms are outlined first, followed in Section 5.2.4 by a more extensive discussion of media.

1. An appropriate mineral medium containing the organic compound which is to be studied is inoculated with a sample of water, soil, or sediment; in studies of the environmental fate of a xenobiotic in a specific ecosystem, samples are generally taken from the area

putatively contaminated with the given compound so that a degree
of environmental relevance is automatically incorporated. Recent
attention has, in addition, been directed to pristine environments,
and to the question of adaptation or preexposure which has already
been discussed in Chapter 4, Section 4.6.4.

If the test compound is to serve as a source of sulfur, nitrogen, or phosphorus, these elements will generally be omitted from the medium, and an
appropriate carbon source must be supplied either by the xenobiotic under
investigation or by another intentionally added substrate. In such experiments, glassware must be scrupulously cleaned to remove interfering traces
of, for example, detergents which may contain residues of all of these nutrients and which could therefore compromise the outcome of the experiment.
The metabolism of phosphonates (Chapter 6, Section 6.9.4) and of sulfate and
phosphate esters (Chapter 6, Section 6.11.1) may be inducible only in the
absence of inorganic phosphorus or sulfur. The following may be given as
illustration of enrichments designed to obtain organisms using organic compounds as sources of N, S, or P — although not necessarily able to use the test
substrates as carbon sources.

- Nitrophenols as N-source using succinate as C-source (Bruhn et
  al. 1987);
- 2-Chloro-4,6-diamino-1,3,5-triazine as N-source and lactate as
  C-source (Grossenbacher et al. 1984) or 2-chloro-4-aminoethyl-
  6-amino-1,3,5-triazine as N-source and glycerol as C-source
  (Cook and Hütter 1984);
- Arylsulfonic acids as S-source and succinate or glycerol as
  C-sources (Zürrer et al. 1987);
- 2-(1-Methylethyl)amino-4-hydroxy-6-methylthio-1,3,5-triazine
  as S-source and glucose as C-source (Cook and Hütter 1982);
- Glyphosate (*N*-phosphonomethylglycine) as P-source and glu-
  cose as C-source (Talbot et al. 1984).

These procedures may of course result in the dominance of organisms that
carry out only biotransformation of the xenobiotic, although the biodegradation of many of these compounds has also been demonstrated using the same
or other organisms; relevant illustrations are given in Chapter 6.

2. The cultures are then incubated under relevant conditions of temperature, pH, salinity, and oxygen concentration, and after growth
   has occurred, successive transfer to fresh medium is carried out;
   interfering particulate matter will thereby be removed by dilution
   and a culture suitable for isolation may be obtained. Incubation is
   generally carried out in the dark although for the isolation of

phototrophic organisms, illumination at suitable wavelength and intensity must obviously be supplied. There are no rigid rules on how many transfers should be carried out, but especially for anaerobic organisms, sufficient time should elapse between transfers to allow growth of these often slow-growing organisms; as a result, the enrichment may take up to a year or even longer.

3. After metabolically stable cultures have been obtained, pure cultures of the relevant organisms may then be obtained by either of three basic procedures:

    a. By preparing serial dilutions of the culture in a suitable buffer medium and spreading portions onto solid media (agar plates, or roll tubes for anaerobic bacteria) containing the organic compound as source of carbon, sulfur, nitrogen, or phosphorus. The plates (or tubes) are then incubated under appropriate conditions; after growth has taken place, single colonies are then selected and pure cultures obtained by repeated restreaking on the original defined medium. Use of complex media or substrate analogues may introduce serious ambiguities since overgrowth of unwanted, rapidly growing organisms may occur. Considerable difficulty may be experienced when "spreading" organisms conceal the desired organism and this may make the isolation of single colonies a tedious procedure.

    b. Serial dilution may be carried out in a defined liquid growth medium and the dilutions incubated under suitable conditions; successive transfers are then made from the highest dilution showing growth. This experimentally tedious procedure may indeed be obligatory for organisms such as *Thermomicrobium fosteri* — a name no longer accepted by the authors (Zarilla and Perry 1984) — which are unable to produce colonies on agar plates (Phillips and Perry 1976), and have been quite extensively used for anaerobic bacteria in which the liquid medium is replaced by a soft-agar medium. A cyclic procedure involving an additional plating step was used to reduce the complexity of the original population and to isolate an organism utilizing high concentrations of 2,4,6-trichlorophenol (Maltseva and Oriel 1997): further details that involved DNA fingerprinting are given in Chapter 5, Section 5.8.

    c. Mechanical methods may be used for the isolation of filamentous organisms. For example, washing on the surface of membrane filters, or micromanipulation on the surface of agar medium (Skerman 1968), has occasionally been employed, and an ingenious procedure that uses an electron microscope grid for preliminary removal of other organisms has been described for use under anaerobic conditions (Widdel 1983).

Lack of repeatability in the results of metabolic studies using laboratory strains which have been maintained by repeated transfer for long periods under nonselective conditions may be encountered. These strains may have lost their original metabolic capabilities, and this may be particularly prevalent when the strains carry catabolic plasmids which may be lost under such conditions. For these reasons, strains should be maintained in the presence of a cryoprotectant such as glycerol at low temperatures ($-70°C$ or in liquid N_2) as soon as possible after isolation. Freeze-drying is also widely adopted.

In some cases, difficulty may be experienced in isolating pure cultures with the desired metabolic capability, and the mixed enrichment cultures or consortia must be used for further studies. For example, although an enrichment culture effectively degraded atrazine (2-chloro-4-ethylamine-6-isopropylamine-1,3,5-triazine) none of the 200 pure cultures isolated from this were able to use the substrate as N-source (Mandelbaum et al. 1993). A probably frequent situation that is illustrated in Chapter 4, Section 4.5.1 is when two (or more) organisms cooperate in degradation of the substrate. Considerable effort may then be required to determine the appropriate combination of organisms. For example, an enrichment with 4-chloroacetophenone yielded eight pure strains although none of these could degrade the substrate in pure culture. All pairwise combinations were then analyzed and this revealed that the degradation of the substrate was accomplished by strains of an *Arthrobacter* sp. and a *Micrococcus* sp.; these were then used to elucidate the metabolic pathway (Havel and Reineke 1993). Attention is drawn to the fact that, in this case, isolation of the pure strains from the enrichment culture was carried out using a complex medium under nonselective conditions.

These batch procedures for enrichment and successive transfer may be replaced by the use of continuous culture, and this may be particularly attractive when the test compound is toxic, when it is poorly soluble in water, or where the investigation is directed to substrate concentrations so low that clearly visible growth is not to be expected. These problems remain, however, for subsequent isolation of the relevant organisms. One considerable problem in long-term use arises from growth in the tubing that should be renewed periodically and on the surfaces of the culture vessel.

5.2.4 Basal Media

The choice of appropriate basal media is of cardinal importance and a number of important practical considerations should be taken into account.

Basal Mineral Media

A plethora of basal media for the growth of freshwater organisms has been formulated and these may differ significantly, particularly in the concentrations of phosphate, while for anaerobic bacteria the inclusion of bicarbonate and a suitable reductant is standard practice. Numerous examples of suitable media have been collected in *The Prokaryotes* (Balows et al. 1992).

Clearly, if the organic substrate is to serve as a source of nitrogen or sulfur or phosphorus, these elements must be omitted from the basal medium. Otherwise, these inorganic nutrient requirements will generally be supplied by the following: S (generally as sulfate except for organisms such as chlorobia which require reduced sulfur as S^{2-}); N (generally as ammonium or nitrate except for N_2-fixing organisms); P (as phosphate); Mg^{2+}; Ca^{2+}; and lower concentrations of Na^+ and K^+. For marine organisms, the basal medium is constructed to resemble natural seawater in the concentrations of Na^+, K^+, Mg^{2+}, Ca^{2+}, Cl^-, and SO_4^{2-} and a number of different formulations have been used (Taylor et al. 1981; Neilson 1980) although the exact composition does not seem to be critical. A few brief comments may, however, be inserted on the nitrogen source and the importance of its concentration.

1. Caution should be exercised regarding the inclusion of nitrate as a sole or supplementary nitrogen source. This may result in the synthesis of metabolites containing nitro substituents which are introduced plausibly via arene oxide intermediates (Sylvestre et al. 1982; Omori et al. 1992); further comments are given in Chapter 6 (Section 6.3.1.2).

2. The intrusion of chemical reactions between amines and reduction products of nitrate has already been noted in Chapter 4, Section 4.2.3 and again in Chapter 6, Section 6.11.3.

3. The nitrogen status of the growth medium determines the levels of lignin peroxidases and manganese-dependent peroxidases that are synthesized in *Phanerochaete chrysosporium*. The role of Mn concentration is noted below, while the metabolic consequence of nitrogen concentration on the metabolism of polycyclic aromatic hydrocarbons (PAH) is briefly discussed in Section 6.2.2.

The CO_2 requirement of various groups of bacteria has been noted in Section 4.3 and is generally satisfied by gaseous CO_2 supplemented, for anaerobes, with bicarbonate.

Trace Elements

Many enzymes and coenzymes contain metals and B, Zn, Cu, Fe, Mn, Co, Ni, and Mo, are generally provided at low concentrations. A large number of different formulations of trace elements have been published: the A4 formulation (Arnon 1938) supplemented with Co^{2+} and Mo(VI) has been widely used, or one of the SL series of formulations developed by Pfennig and his co-workers particularly for the cultivation of anaerobic bacteria. SL 9 is typical of the latter series of formulations (Tschech and Pfennig 1984). Somewhat conflicting views exist on the possibly deleterious effects resulting from the incorporation of complexing agents, particularly ETDA, so that their concentration should probably be kept to a minimum.

Anaerobic bacteria are more fastidious in their trace metal requirements and routine addition is made of selenium as selenite and of tungsten as tungstate (Tschech and Pfennig 1984). Selenium is required for the synthesis of active xanthine dehydrogenase in purine-fermenting clostridia (Wagner and Andreesen 1979) and of formate dehydrogenase in a number of organisms including methanogens (Jones and Stadtman 1981) and clostridia (Yamamoto et al. 1983). Addition of W to growth media is necessary for the growth of various methanogens (Winter et al. 1984; Zellner et al. 1987), an anaerobic cellulolytic bacterium (Taya et al. 1985), and for the synthesis of various enzymes in anaerobic bacteria:

- The carboxylic acid reductase in acetogenic clostridia such as *Clostridium thermoaceticum* (White et al. 1989; Strobl et al. 1992).

- The benzylviologen-linked aldehyde oxidoreductase in *Desulfovibrio gigas* grown with ethanol (Hensgens et al. 1995) and the corresponding enzyme in *D. simplex* (Zellner and Jargon 1998). For the latter, it was suggested that the flavins FMN or FAD were the natural cofactors.

- The acetylene hydratase of *Pelobacter acetylenicus* (Rosner and Schink 1995).

- It may be incorporated into some proteins of purinolytic clostridia (Wagner and Andreesen 1987) and into formylmethanofuran dehydrogenase in *Methanobacterium thermoautotrophicum* (Bertram et al. 1994).

For the sake of completeness, it is worth noting the role of vanadium.

a. This is required for the synthesis of alternative nitrogenase by several aerobic bacteria, especially in the genus *Azotobacter* (Fallik et al. 1991): low yields of hydrazine are produced and this suggests a different affinity of the nitrogenase for the N_2 substrate (Dilworth and Eady 1991).

b. A number of haloperoxidases including the chloroperoxidase in the fungus *Curvularia inaequalis* and the bromoperoxidases in marine red and brown algae contain vanadium at the active sites, and this is involved in the formation of V^v peroxo complexes (Almeida et al. 1998; Butler 1998).

Control of pH

Buffering of the medium is usually achieved with phosphate, although excessive concentrations should be avoided since they cause problems of precipitation during sterilization by autoclaving. A number of organic buffers have been used effectively in different applications. In studies employing organic phosphorus compounds as P-sources, phosphate has been replaced by, for example, HEPES (Cook et al. 1978) and MOPS has been incorporated into media for growth of Enterobacteriaceae (Neidhardt et al. 1974). For media requiring low pH, MES has been used in a medium with extremely low phosphate concentration (Angle et al. 1991), and TRIS has been incorporated into media for growth of marine bacteria (Taylor et al. 1981). In all of these cases, it should be established that metabolic complications do not arise as a result of the ability of the organisms to use the buffer as sources of nitrogen or sulfur. It is also possible for the buffer to react with metabolic intermediates, and this is illustrated by the isolation of a compound produced by reaction of the carbon atom (formally CO) of CCl_4 with HEPES (Lewis and Crawford 1995). This is discussed again in Section 6.4.2. Metal chelating agents such as NTA or ETDA may be used but their concentrations should be kept to a minimum in view of their potential toxicity; the former must obviously be omitted in studies of utilization of organic nitrogen as N-source since both NTA and its metal complexes are apparently quite readily degraded (Firestone and Tiedje 1975).

Vitamins

Vitamins such as thiamin, biotin, and vitamin B_{12} are often added. Once again, the requirements of anaerobes are somewhat greater, and a more extensive range of vitamins that includes pantothenate, folate, and nicotinate is generally employed. In some cases, additions of low concentrations of peptones, yeast extract, casamino acids, or rumen fluid may be used, although in higher concentrations, metabolic ambiguities may be introduced since these compounds may serve as additional carbon sources.

Sterilization

Mineral basal media may be sterilized by autoclaving, but for almost all organic compounds which are used as sources of C, N, S, or P, it is probably better to prepare concentrated stock solutions and sterilize these by filtration, generally using 0.2 μm cellulose nitrate or cellulose acetate filters. The same applies to solutions of vitamins and to solutions of bicarbonate and sulfide which are components of many media used for anaerobic bacteria.

The Role of Metal Concentration in Metabolism

Iron ·

Although Fe is required as a trace element, its uptake is critically regulated since excess leads to the generation of toxic hydroxyl radicals, and there exist complex interactions involving Fe(II) and Fe(III) within the cell (Touati et al. 1995). Details of the role of Fe and its relation to the generation of toxic hydroxyl radicals has been further explored by analysis of a strain of *Escherichia coli* and a mutant strain lacking both Fe and Mn superoxide dismutase. The mutant strain showed a marked increase in the hydroxyl radical after exposure to H_2O_2 (McCormick et al. 1998); preincubation with an Fe chelator inhibited this difference, and redox-active Fe defined as EPR-detectable ascorbyl radicals was greater in the mutant than in the wild-type strain.

Iron may also play a subtler role in determining the biodegradability of a substrate that forms complexes with Fe. Two examples may be used as illustration of probably different underlying reasons for this effect.

1. A strain of *Agrobacterium* sp. was able to degrade ferric EDTA though not the free compound (Lauff et al. 1990). This may be due either to the adverse effect of free EDTA on the cells or the inability of the cells to transport the free compound. The former is supported by the established sensitivity (Wilkinson 1968) of some Gram-negative organisms to ETDA and the increased surface permeability in enteric organisms exposed to EDTA (Leive 1968). These results are consistent with the requirement for high concentrations of Ca^{2+} in the enrichment medium used for isolating a mixed culture capable of degrading EDTA (Nörtemann 1992). On the other hand, although the degradative enzymes have been purified and characterized from cells of a pure culture grown with Mg ETDA, the enzyme complex was unable to use Fe EDTA as a substrate (Witschel et al. 1997). This is consistent with the results using an unenriched culture and ^{14}C-labeled EDTA (Allard et al. 1996). In the aquatic environment, degradation is accomplished primarily by photolysis of the Fe complex (Kari and Giger 1995).

2. A strain of *Pseudomonas fluorescens* biovar II is able to degrade citrate whose metabolism requires access to the hydroxyl group. This group is, however, implicated in the tridentate ligand with Fe(II) so that this complex is resistant to degradation in contrast to the bidentate ligand with Fe(III) that has a free hydroxyl group and is readily degraded (Francis and Dodge 1993). Similarly, the bidentate complexes containing Fe(III), Ni, and Zn are readily degraded by *P. fluorescens*, in contrast to the tridentate complexes containing Cd, Cu, and U that are not degraded (Joshi-Tope and Francis 1995).

These results may be viewed in the wider context of interactions between potential ligands of multifunctional xenobiotics and metal cations in aquatic environments and the subtle effects of the oxidation level of cations such as Fe.

The Fe status of a bacterial culture has an important influence on synthesis of the redox systems of the cell since many of the electron transport proteins contain Fe. This is not generally evaluated systematically, but the degradation of tetrachloromethane by a strain of *Pseudomonas* sp. under denitrifying conditions clearly illustrated the adverse effect of Fe on the biotransformation of the substrate (Lewis and Crawford 1993; Tatara et al. 1993); this possibility should therefore be taken into account in the application of such organisms to bioremediation programs.

Manganese

The role of manganese concentration has seldom been explicitly examined in the context of biodegradation, although it is essential for the growth of the purple nonsulfur anaerobic phototrophs *Rhodospirillum rubrum* and *Rhodopseudomonas capsulata* during growth with N_2 but not with glutamate (Yoch 1979). It does, however, play an essential role in the metabolic capability of the white-rot fungus *Phanaerochaete chrysosporium*. This organism produces two groups of peroxidases during secondary metabolism — lignin peroxidases and manganese-dependent peroxidases. Both of them are synthesized when only low levels of Mn(II) are present in the growth medium, whereas high concentrations of Mn result in repression of the synthesis of the lignin peroxidases and an enhanced synthesis of manganese-dependent peroxidases (Bonnarme and Jeffries 1990; Brown et al. 1990). Experiments with a nitrogen-deregulated mutant have shown that N-regulation of both these groups of peroxidases is independent of Mn(II) regulation (Van der Woude 1993).

Copper

In *Methylosinus trichosporium* OB3b that expresses a particulate monooxygenase, the concentration of copper plays a significant role in the kinetic parameters for the consumption of both methane and trichloroethene (Sontoh and Semrau 1998). For methane, V_{max} decreased from 300 to 82 when the concentration of Cu increased from 2.5 to 20 µm, and K_s decreased from 62 to 8.3 under these conditions. For trichloroethene, V_{max} and K_s were unmeasurable at Cu concentrations of 2.5 µM even in the presence of formate, but for concentrations of 20 µM in the presence of formate were 4.1 and 7.9.

Other Metals

Heavy metal cations (e.g., Pb^{2+}, Hg^{2+}) and oxyanions (e.g., UO_4^{2-}) are toxic to bacteria although resistance may be induced by various mechanisms after exposure. Attention has been drawn to an unusual example in which Al^{3+} may be significant since the catechol 1,2-dioxygenase and protocatechuate 3,4-dioxygenase that are involved in the metabolism of benzoate by strains of *Rhizobium trifolii* are highly sensitive to inhibition by Al^{3+} (Chen et al. 1985).

The Redox Potential of Media

Cultivation of strictly anaerobic organisms clearly requires that the medium be oxygen-free, but this is often not sufficient: the redox potential of the medium must generally be lowered to be compatible with that required by the organisms. This may be accomplished by addition of reducing agents such as sulfide, dithionite, or titanium(III) citrate although any of these may be toxic, so that only low concentrations should be employed and attention has been drawn to the fact that titanium(III) citrate-reduced medium may be inhibitory to bacteria during initial isolation (Wachenheim and Hespell 1984). Further comments on procedures are given in Section 5.2.6.

5.2.5   Organic Substrates

Although organic substrates such as carboxylic acids are sufficiently thermally stable that they may be sterilized with the basal media, many others including, for example, carbohydrates, esters, or amides are better prepared as concentrated stock solutions, sterilized by filtration through 0.22-μm filters, and added to the sterile basal medium.

Toxic Substrates

A major problem arises when the desired organic substrates are poorly soluble in water or highly volatile or toxic.

For volatile or gaseous substrates sealed systems such as desiccators or sealed ampoules may be used. In addition, it has been found convenient to supply toxic substrate such hydrogen cyanide in the gas phase (Harris and Knowles 1983). Serious problems may arise for substrates that are too toxic to be added in the free state at concentrations sufficient for growth. The toxicity of long-chain aliphatic compounds with low water solubility has been examined in yeasts (Gill and Ratledge 1972) and the principles that emerged have been applied to circumventing toxicity in liquid media by supplying substrates such as toluene in an inert hydrophobic carrier (Rabus et al. 1993) or the water-soluble hexadecyltrimethylammonium chloride adsorbed on silica (van Ginkel et al. 1992). The use of dibutyl phthalate as a cosolvent for the toxic cyclohexane epoxide has been examined, although, in this case, the essential problem was the susceptibility of the substrate to hydrolysis to the diol that was more readily degradable (Carter and Leak 1995). Some additional methods that have been applied to the preparation of solid media are described later.

Volatile Substrates

Gaseous or highly volatile substrates present a problem that may be overcome by the use of enclosed systems such as desiccators or sealed ampoules, or by closing bottles with Teflon-lined crimp caps and inverting them during incubation. Attention is drawn to the permeability to organic compounds of many types of rubber sealings. This procedure has been employed for 4-chloroacetophenone (Havel and Reineke 1993) since the 4-chlorophenol produced is toxic to one of the components of the consortium. In this case, low concentrations of the toxic intermediate could also be maintained by adding gelatin to the medium, and this procedure facilitated growth of one of the components to occur. This example has already been discussed in the context of enrichment procedures in Section 5.2.3 and as an example of metabolic interaction of organisms in Chapter 4, Section 4.5.1, while metabolic details are given in Chapter 6, Section 6.2.1. It should be noted that, particularly in sealed systems, it is important to satisfy the obligate requirement of many organisms for CO_2 (Chapter 4, Section 4.2). The result of enrichments may sometimes depend upon whether the substrate is applied in the vapor phase or in the aqueous phase; for example, enrichments with α-pinene in the vapor phase yielded predominantly Gram-positive organisms in contrast to the Gram-negative organisms obtained when the substrate was added to the liquid medium. This may plausibly be related to the greater sensitivity of Gram-positive organisms which are exposed to only low substrate concentrations in the vapor phase (Griffiths et al. 1987).

Solid Media

Although solid media have been prepared from silica gels, these have not been widely used. Agar for preparing solid medium should be of the highest quality, and as free as possible from alternative carbon sources. It is generally preferable to autoclave agar separately from the mineral base; both are prepared at double the final concentration and mixed after autoclaving (Stanier et al. 1966). The problem of preparing plates for testing the metabolic capacity of poorly water-soluble substances is a serious one for which no universal solutions are available. Some of the techniques which have been advocated include the following:

1. Liquid hydrocarbons have been adsorbed on silica powder and dispersed in the agar medium. The silica itself may be autoclaved (Baruah et al. 1967), although this may be avoided by sterilizing the silica by heating and carrying out the adsorption and removal of solvent under aseptic conditions.

2. Solutions of the substrate have been prepared, for example, in acetone or diethyl ether, and added to or spread over the surface of agar plates either before or after inoculation (Sylvestre 1980; Shiaris and Cooney 1983)

3. For compounds with relatively high volatility such as benzene, toluene, naphthalene, or other sufficiently volatile hydrocarbons, the substrate may be contained in a tube placed in the liquid medium (Claus and Walker 1964) or on the lid of a petri dish (Söhngen 1913). For benzene and toluene, this also obviates problems with toxicity since the organisms are exposed to only low concentrations of the substrate.

4. A solution of the substrate in ethanol may be mixed with the bacterial suspension in agarose and poured over agar plates of the base medium (Bogardt and Hemmingsen 1992). This is the general procedure used with top agar in the Ames test (Maron and Ames 1983) although in this case dimethyl sulfoxide is generally used as the water-miscible solvent.

Growth at the Expense of Alternative Substrates

It may be found that, after enrichment, growth does not occur on plates prepared with the compound which showed satisfactory growth in liquid medium, and some probable reasons for this have been outlined in Section 4.5. One additional possibility worth examining is that of attempting to grow the organisms with a potential metabolite, although it should be kept in mind that organisms may be unable to utilize compounds which are clearly established metabolites of the desired compound. Examples are the inability of fluorescent pseudomonads which degrade aromatic compounds via *cis,cis* muconate to use this as a substrate (Robert-Gero et al. 1969) or of alkane-degrading bacteria to grow with the corresponding carboxylic acids which are the early metabolites in alkane degradation (Zarilla and Perry 1984). One plausible reason could be the lack of an effective transport system for the metabolite; another could be the failure of the metabolite to induce the enzymes necessary for its production. For example, whereas salicylate normally induces the enzymes required for the degradation of naphthalene, this is apparently not the case for a naphthalene-degrading strain of *Rhodococcus* sp. (Grund et al. 1992).

Use of complex media for isolating organisms after elective enrichment is, on the other hand, a potentially hazardous procedure. Media that are routinely used for nonmetabolic studies in clinical laboratories generally contain high concentrations of peptones, yeast extract, or carbohydrates, and these may provide alternative carbon sources. Their use may therefore result in only low selection pressure for the emergence of the desired organisms; overgrowth by undesired microorganisms may then all too readily take place.

5.2.6 Techniques for Anaerobic Bacteria

Substantial and increasing attention has been directed to the growth and isolation of anaerobic bacteria. In addition to the nutritional requirements noted above, their general requirement for CO_2 should be taken into consideration. Broadly, three types of experimental methodologies have been used:

1. Anaerobe jars containing a catalyst for the reaction between oxygen and hydrogen that is either added to or generated within the system; these systems have limitations in the kinds of experiment which can be carried out since at some stage exposure to air cannot be avoided. In addition, many workers have experienced the unreliability of these systems, and they are not suitable for work with highly oxygen-sensitive organisms such as methanogens.

2. The classical Hungate technique (Hungate 1969) has been successfully used over many years and incorporates a number of features designed to minimize exposure to oxygen. This procedure enables incubation to be carried out under a variety of gas atmospheres and is designed to produce a redox potential in media that is suitable for growth. Roll-tubes have been used instead of petri dishes, and thereby a strictly anoxic environment may be maintained during manipulation. A modification using serum bottles has been introduced (Miller and Wolin 1974), and roll-tubes have

also been successfully used for isolation of anaerobic phycomyce-
tous fungi from rumen fluid (Joblin 1981).

3. Anaerobe chambers of varying design have achieved increasing
   popularity since they enable standard manipulations to be carried
   out under anoxic conditions; these systems maintain a gas atmosphere of N_2, H_2, and CO_2 (generally 90:5:5), and include a heated
   catalyst for the maintenance of anaerobic conditions. They may
   employ either a glove box design, or free access through wrist
   bands and enable quite sophisticated experiments to be carried out
   and cultures maintained over lengthy periods. These can therefore
   be unequivocally recommended, although their maintenance costs
   should not be underestimated.

5.3 Design of Experiments on Biodegradation and Biotransformation

There are four essentially different kinds of experiments which may be carried out. Two are laboratory based and two are field investigations:

1. Laboratory experiments using pure cultures or stable consortia;
2. Laboratory experiments using communities in microcosms simulating natural systems;
3. Field experiments in model ecosystems — pools or steams;
4. Large-scale field experiments under natural conditions.

It should be clearly appreciated that the objectives of these various investigations are rather different. The first two aim at elucidating the basic facts of
metabolism, the products formed, and the kinetics of their synthesis; studies
using pure cultures may ultimately be directed to studying more-sophisticated
aspects of the regulation and genetics of biodegradation. On the other hand,
the last two procedures are designed to obtain data of more direct environmental relevance and may profitably — and even necessarily — draw upon
the results obtained using the first two procedures. While the degree of environmental realism increases from 1 through 4, so also do the experimental difficulties and the interpretative ambiguities; all of the procedures have clear
advantages for specific objectives, and are complementary. Indeed, it is highly
desirable that several of them be combined.

There are significant differences in the control experiments that are possible
in each of these systems. Before the quantifier *bio-* can be applied, the possibility of abiotic alteration of the substrate during incubation must be eliminated. On this point, only the first design lends itself readily to rigorous

control, and even then there may be experimental difficulties. For experiments using cell suspensions, the obvious controls are incubation of the substrate in the absence of cells or using autoclaved cultures. Care should be exercised in the interpretation of the results, however, since some reactions may apparently be catalyzed by cell components in purely chemical reactions: the question may then legitimately be raised whether or not these are biochemically mediated. Two examples may be used as illustration of intrusion by apparently chemically mediated reactions:

1. Dechlorination reactions of organochlorine compounds which have been referred to in Section 4.1.3.2 involving corrinoids and porphyrins, and are discussed again in Sections 6.4.4 and 6.6;
2. Reduction of aromatic nitro compounds by sulfide and catalyzed by extracellular compounds excreted into the medium during growth of *Streptomyces griseoflavus* (Glaus et al. 1992).

In experiments where small volumes of sediment suspensions are employed, autoclaving may significantly alter the structure of the sediment as well as introducing possibly severe analytical difficulties; in such cases, there seem few alternatives to incubation in the presence of toxic agents such as NaN_3 which has been used at a concentration of 2 g/l. There remains, of course, the possibility that azide-resistant strains could emerge during prolonged incubation, and the occurrence of reactions between the substrate and azide must also be taken into consideration.

Only controls using inhibitors of microbial growth are applicable to microcosm experiments, and it seems unlikely that even these can be applied to outdoor systems, which therefore combine, and generally fail to discriminate between, abiotic and biotic reactions.

5.3.1   Pure Cultures and Stable Consortia

Different kinds of experimental procedures have been used and these should be evaluated against the background presented in Section 4.5 that is devoted to situations in which several organisms or several substrates are simultaneously present. There are no essential differences in the design of experiments using pure cultures and those employing metabolically stable consortia. It should be emphasized, however, that even in the latter case the experiments should be carried out under aseptic conditions; otherwise, interpretation of the results may be compromised by adventitious organisms.

Cell Growth at the Expense of the Xenobiotic

In the simplest case, growth of the organism that has been isolated may be studied using the test substance that fulfills the nutritional requirement of the organism as the sole source of carbon, sulfur, nitrogen, or phosphorus. For a compound used as sole source of carbon and energy, the end points could be growth and conversion of the substance into CO_2 under aerobic conditions, or growth under anaerobic conditions accompanied by, for example, production of methane or sulfide from sulfate. In some studies, however, only diminution in the concentration of the initial substrate has been demonstrated, and this alone clearly does not constitute evidence for biodegradation. Biotransformation is equally possible and should be taken into consideration. Ideally, use may be made of radiolabeled substrates followed by identification of the labeled products. ^{14}C, ^{35}S, ^{36}Cl, ^{31}P have been used, although the relevant labeled products may not always be available commercially, and the required synthetic expertise may not be available in all laboratories. Further comments on the use of these isotopes together with the application of the nonradioactive isotopes ^{13}C and ^{19}F are given in Section 5.5.4. It should be noted that CO_2 is incorporated not only into phototrophs and chemolithotrophs, but also into heterotrophic organisms. The role of CO_2 in determining the products formed from propene oxide by a strain of *Xanthobacter* sp. is given in Section 6.1.3, and its significance in the anaerobic biotransformation of aromatic compounds is discussed in Section 6.7.3. A review (Ensign et al. 1998) provides a brief summary of the role of CO_2 in the metabolism of epoxides by *Xanthobacter* sp. strain Py2, and of acetone by both aerobic and anaerobic bacteria.

Since considerable weight has been given to the environmental significance of biotransformation and the synthesis of toxic metabolites (Chapter 6, Section 6.11.5), it is particularly desirable to direct effort to the identification of such metabolites. This presents a substantially greater challenge than that of quantifying the original substrate for several reasons:

1. The structure of the metabolite will generally be unknown and must be predicted from knowledge of putative degradation pathways and confirmed by any of the methods outlined in Chapter 2, Section 2.4.1.

2. The metabolite will frequently be more polar than the initial substrate so that specific procedures for extraction and analysis must generally be developed, and the pure compound must be available for quantification.

3. The metabolite may be transient with unknown kinetics of its formation and further degradation.

In practice, there is only one really satisfactory solution: the kinetics of the transformation must be followed. The justification for this substantial increase in effort is the dividend resulting in the form of a description of the metabolic pathway including the synthesis of possibly inhibitory metabolites. On the basis of this, it may then be possible to formulate generalizations on the degradation of other structurally related xenobiotics.

Stable Enrichment Cultures

Some investigations have used metabolically stable enrichment cultures that are particle free to study biotransformation, and these are preferable to the use of sediment slurries even though these may be environmentally realistic. There are several important reasons for this preference:

- They avoid the ambiguities resulting from the presence of xenobiotics in the original soil or sediment.
- They eliminate association of the added xenobiotic with particulate matter.
- They simplify analytical procedures.
- They overcome the unresolved function of organic components in the sediment.
- They make it possible to carry out reproducible experiments under defined conditions.

Particularly for anaerobes, however, it is often difficult to obtain pure cultures and sediment or sludge slurries have frequently been used. Examples of studies that have used stable enrichment cultures that are free of sediment by successive transfer in defined media during extensive periods of time include the following.

1. The extensive use in a series of experiments on the anaerobic dechlorination of chlorocatechols (references in Allard et al. 1994).
2. Studies on the anaerobic dechlorination of 2,3,5,6-tetrachlorobiphenyl (Cutter et al. 1998). Further details of these experiments are given in Section 5.3.1

Use of Dense Cell Suspensions

Dense cell suspensions have traditionally been used for experiments on the respiration of microbial cells at the expense of organic substrates, and they are equally applicable to experiments on biodegradation and biotransformation. Cells are grown in a suitable medium generally containing the test compound (or an alternative growth substrate), collected by centrifugation, washed in a buffer solution to remove remaining concentrations of the growth substrate and its metabolites, and resuspended in fresh medium before further exposure to the xenobiotic. For aerobic organisms, there are generally few experimental difficulties with three important exceptions:

1. For organisms which grow poorly in liquid medium, it may be difficult to obtain sufficient quantities of cells; in such cases, cultures may be grown on the surface of agar plates and the cells removed by scraping.

2. Organisms that have fastidious nutritional requirements may require undefined growth additives such as peptones, yeast extract, rumen fluid, or serum. Subsequent exposure to the xenobiotic may then be used to induce synthesis of the relevant catabolic enzymes. For example, the chlorophenol-degrading bacterium *Rhodococcus chlorophenolicus* has been grown in media containing yeast extract or rhamnose, and exposure to pentachlorophenol used to induced the enzymes required for the degradation of a wide range of chlorophenols (Apajalahti and Salkinoja-Salonen 1986).

3. Organisms such as actinomycetes may not produce well-suspended growth, and shaking, for example, in baffled flasks or in flasks with coiled-wire inserts may be advantageous in partially overcoming this problem.

For anaerobic bacteria, the same principles apply except that additional attention must be directed to preparing the cell suspensions. Use of an anaerobic chamber in which cultures can be transferred to tightly capped centrifuge tubes is virtually obligatory, and addition of an anaerobic indicator should be used to ensure that subsequent entrance of oxygen does not take place inadvertently. Oak Ridge centrifuge tubes are particularly suitable for centrifugation.

Use of Immobilized Cells

Cells may be immobilized on a number of suitable matrices in a reactor and the medium containing the test substrate circulated continuously. Although this methodology has been motivated by interest in biotechnology and in bioremediation technology, it is clearly applicable to laboratory experiments on biodegradation and biotransformation which could readily be carried out under sterile conditions. This procedure has been used, for example, to study the biodegradation of 4-nitrophenol (Heitkamp et al. 1990), pentachlorophenol (O'Reilly and Crawford 1989), and 6-methylquinoline (Rothenburger and Atlas 1993). A few additional comments may be inserted.

1. It is possible to carry out experiments with immobilized cells in essentially nonaqueous media (Rothenburger and Atlas 1993), and this might prove an attractive strategy for compounds with limited water solubility provided that solvents can be found that are compatible with the solubility of the substrate and the sensitivity of the cells to organic solvents.

2. Encapsulated cells have been successfully used for the commercial biosynthesis of a number of valuable compounds such as amino acids, and this technique could readily be adapted to investigations on biodegradation. This methodology offers the advantage that the metabolic activity of the cells can be maintained over long periods

of time so that a high degree of reproducibility in the experiments is guaranteed, and the stability of such systems may be particularly attractive in studies of recalcitrant compounds.

3. With appropriate experimental modifications, the various procedures could be adapted to study biodegradation under anaerobic conditions; for example, sparging with air could be replaced by the use of a gas mixture containing appropriate concentrations of CO_2 and H_2 in an inert gas such as N_2.

Application of Continuous Culture Procedures

The use of these has been very briefly noted in Chapter 4, Sections 4.6.1.2 and 4.8.2, and in Section 5.2.3, and they have been particularly valuable in studies using low concentrations of xenobiotics and for the isolation of consortia that have been used in elucidating metabolic interactions between the various microbial components. In many cases, consortia containing several organisms are obtained even though only a few of their members are actively involved in the metabolism of the xenobiotic; it is possible that the low substrate concentrations that have been used in these experiments favor selection for organisms able to take advantage of the lysis products from such cells that do not play a direct role in the degradation of the xenobiotic. Three examples may suffice as illustration.

1. Enrichment in a two-stage chemostat with parathion (*O,O*-diethyl-*O*-(4-nitrophenyl)-phosphorothioate) as the sole source of carbon and sulfur resulted in a community which was stable for several years (Daughton and Hsieh 1977). It should be noted that, due to the toxicity of parathion to the culture, only low substrate concentrations could be used, and this methodology is ideally adapted to such situations. Degradation was accomplished by two organisms, *Pseudomonas stutzeri* and *P. aeruginosa*, whereas the third organism in the stable community had no defined function: *P. stutzeri* functioned only in ester hydrolysis which is the first step in the degradation of parathion. This was the first demonstration of degradation of parathion by a metabolically defined microbial consortium, although degradation by a culture consisting of nine organisms had already been demonstrated (Munnecke and Hsieh 1976).

2. Chemostat enrichment was carried out with a mixture of linear alkylbenzene sulfonates as the sole sources of carbon and sulfur at a concentration of 10 mg/l, and resulted in the development of a four-component consortium (Jiménez et al. 1991). Three of the organisms were apparently necessary to accomplish this apparently straightforward degradation, although the isolation procedure that used a complex medium with glucose as carbon source

is not entirely unequivocal. A similar situation arises with hexade-cyltrimethylammonium chloride from which three strains that could grow with the substrate were obtained again after streaking on yeast–glucose medium (van Ginkel et al. 1992).

3. Chemostat enrichment with 2-chloropropionamide yielded a community of at least six organisms; one of these, a *Mycoplana* sp., carried out hydrolysis of the amide, while various other components used the resulting free acid for growth. An interesting observation was that after prolonged incubation at a dilution rate of 0.01/h, a single strain of *Pseudomonas* sp. capable of growth solely on 2-chloropropionamide as carbon source could be isolated (Reanney et al. 1983).

There are some general conclusions which may plausibly be drawn from the results of all these experiments:

- The reactions necessary for degradation were apparently relatively simple ones, and would be expected to be accessible to single organisms.
- The first stages for two of these degradations were straightforward hydrolytic reactions.
- In all cases, organisms with undefined metabolic functions were present, and probably fulfilled an important role in providing complex organic substrates in the form of cell lysis products or nutritional requirements (Section 4.5.1).

Reaction sequences used for the degradation of xenobiotics in natural systems may therefore be more complex than might plausibly be predicted on the basis of studies with pure cultures using relatively high substrate concentrations.

Attention should be drawn to experiments in which solutions of the substrate in a suitable mineral medium are percolated through soil that is used as the source of inoculum. This is one of the classical procedures of soil microbiology and has been exploited to advantage in studies on the degradation of a range of chlorinated contaminants in groundwater (van der Meer et al. 1992). Apart from the fact that this mimics closely the natural situation and incorporates the features inherent in any enrichment methodology, this procedure offers a degree of flexibility that enables systematic exploration of the following:

1. The effect of varying redox conditions since, by altering the gas phase, experiments can readily be carried out under aerobic, microaerophilic, or anaerobic conditions.

2. The effect of substrate concentration, and the important issue of the existence, or otherwise, of threshold concentrations below

which degradation is not effectively accomplished; this has been discussed in Section 4.6.2.

3. The influence of sorption/desorption on biodegradation that has been discussed in a wider context in Section 4.6.3.

Apart from its application to the specific problem of groundwater contamination, this procedure offers a potentially valuable procedure for simulating bioremediation of contaminated soils.

Simultaneous Presence of Two Substrates

Analogous to the fact that pure cultures of microorganisms seldom occur in natural ecosystems, it is very rare for a single organic substrate to exist in appreciable concentrations. The relevant microorganisms under natural situations are therefore exposed simultaneously to several compounds, and this situation can be simulated in laboratory experiments. The terminology has undergone a variety of different designations that have already been discussed in Chapter 4, Section 4.5.2 so that only brief mention is justified here: although the term *cometabolism* has been used extensively, it has been applied to conflicting metabolic situations and the pragmatic term *concurrent metabolism* (Neilson et al. 1985) offers an attractive alternative when more than one substrate is present. Three examples will be used to illustrate the application of this procedure to experiments in which the pathways for the biotransformation of different xenobiotics have been established.

1. Experiments have used cells with a metabolic capability that may plausibly be predicted as relevant to that of the xenobiotic. For example, elective enrichment failed to yield organisms able to grow at the expense of dibenzo-[1,4]-dioxin, but its metabolism could be studied in a strain of *Pseudomonas* sp. capable of growth with naphthalene (Klecka and Gibson 1979). Cells were grown with salicylate (1 g/l) in the presence of dibenzo-[1,4]-dioxin (0.5 g/l), and two metabolites of the latter were isolated: *cis*-1,2-dihydro-1,2-diol and 2-hydroxydibenzo-1,4-dioxin. The former is consistent with the established dioxygenation of naphthalene and the role of salicylate as coordinate inducer of the relevant enzymes for conversion of naphthalene into salicylate.

2. An environmentally relevant situation may be simulated by the growth of an organism with a single substrate at a relatively high concentration and simultaneous exposure to a structurally unrelated xenobiotic present at a significantly lower concentration. A series of investigations has used growth substrates at concentrations of 200 mg/l and xenobiotic concentrations of 100 µg/l; it may reasonably be assumed that growth with compounds at the latter concentration is negligible. For example, during growth of a stable

anaerobic enrichment culture with 3,4,5-trimethoxybenzoate, 4,5,6-trichloroguaiacol was transformed into 3,4,5-trichlorocatechol which was further dechlorinated to 3,5-dichlorocatechol (Neilson et al. 1987).

3. There has been considerable discussion on the mineralization of DDT, and in particular of the biodegradation of the intermediate DDE. Cells of *P. acidovorans* strain M3GY have, however, during growth with biphenyl been shown to degrade DDE with the fission of one ring and production of 4-chlorobenzoate (Hay and Focht 1998).

Use of Unenriched Cultures: Undefined Natural Consortia

Laboratory experiments using natural consortia under defined conditions have particular value from several points of view. They are of direct environmental relevance, and their use minimizes the ambiguities in extrapolation from the results of studies with pure cultures. They provide valuable verification of the results of studies with pure cultures and make possible an evaluation of the extent to which the results of such studies may be justifiably extended to the natural environment.

It should be appreciated, however, that in some cases the habitats from which the inoculum was taken may already have been exposed to xenobiotics so that "natural" enrichment may already have taken place; this has been discussed briefly in Chapter 4, Section 4.6.4.

Extensive studies — some of which have already been cited in Chapter 4, Sections 4.6.2 and 4.6.3 — on the effect of substrate concentration and of the bioavailability of the substrate to the appropriate microorganisms have employed samples of natural lake water supplemented with suitable nutrients. There are few additional details that need to be added since the experimental methods are straightforward and present no particular difficulties. Considerable use has also been made of a comparable methodology to determine the fate of agrochemicals in the terrestrial environment.

Because of the difficulty of obtaining pure cultures of anaerobic bacteria, extensive use has been made of anaerobic sediment slurries in laboratory experiments. In some of these, although no enrichment was deliberately incorporated, experiments were carried out over long periods of time in the presence of contaminated sediments and adaptation of the natural flora to the xenobiotic during exposure in the laboratory may therefore have taken place. The design of these experiments may also inevitably result in interpretative difficulties. For example, although the results of experiments on the dechlorination of pentachlorophenol (Bryant et al. 1991) enabled elucidation of the pathways to be elucidated, this study also revealed one of the limitations in the use of such procedures; detailed interpretation of the kinetics of pentachlorophenol degradation using dichlorophenol-adapted cultures was equivocal due to carryover of phenol from the sediment slurries. The biodegradation of acenaphthene and naphthalene under denitrifying conditions was examined in soil-water slurries (Mihelcic and Luthy 1988), although in this case only analyses for the concentrations of the initial substrates were carried out. In both of these examples, growth of the degradative organisms was supported at least partly by organic components of the soil and sediment, so that the physiological state of the cells could not be precisely defined. It would therefore be desirable to avoid ambiguity by using cultures in which the sediment is no longer present. This approach is illustrated by extensive investigations on the anaerobic dechlorination of chlorocatechols (Allard et al. 1991; 1994) and of 2,3,5,6-tetrachlorobiphenyl (Vutter et al., 1998).

5.3.2 Microcosm Experiments

Microcosms are laboratory systems generally consisting of tanks such as fish aquaria containing natural sediment and water, or soil. In those which have been most extensively evaluated for aquatic systems, continuous-flow systems are used. In all of them, continuous measurement of $^{14}CO_2$ evolved from ^{14}C-labeled substrates may be incorporated, and recovery of both volatile and nonvolatile metabolites is possible so that a material balance may be constructed (Huckins et al. 1984). It should be pointed out that the term *microcosm* has also been used to cover much smaller scale experiments that have been carried out in flasks under anaerobic conditions (Edwards et al. 1992), and to systems for evaluating the effect of toxicants on biota (Section 7.4.2). Some examples are given to illustrate different facets of the application of microcosms to study various aspects of biodegradation.

1. Biodegradation of *t*-butylphenyl diphenyl phosphate was examined using sediments either from an uncontaminated site or from one having a history of chronic exposure to agricultural chemicals (Heitkamp et al. 1986). Mineralization was very much more extensive in the latter case, but was inhibited by substrate concentrations exceeding 0.1 mg/l. Low concentrations of diphenyl phosphate, 4-*t*-butylphenol, and phenol indicated the occurrence of esterase activity, while the recovery of triphenyl phosphate suggested dealkylation by an unestablished pathway. A comparable study of naphthalene biodegradation (Heitkamp et al. 1987) found more rapid degradation when sediments chronically exposed to petroleum hydrocarbons were used and isolation of *cis*-1,2-dihydronaphthalene-1,2-diol, 1- and 2-naphthol, salicylate, and catechol confirmed the pathway established for the degradation of naphthalene. The results of both investigations illustrate the potential for a more extensive application of the procedure and at the same time the significance of preexposure to the xenobiotic.

2. One of the key issues in bioremediation is the survival of the organisms deliberately introduced into the contaminated system. A microcosm prepared from a pristine ecosystem was inoculated with a strain of *Mycobacterium* sp. that had a wide capacity for degrading PAHs and this organisms was used to study the degradation of 2-methylnaphthalene, phenanthrene, pyrene, and benzo[*a*]pyrene (Heitkamp and Cerniglia 1989). The test strain survived in the system with or without exposure to PAHs, but the addition of organic nutrients was detrimental to its maintenance. Clearly, an almost unlimited range of parameters could be varied to enable a realistic evaluation of the effectiveness of bioremediation in natural circumstances.

3. Concern has been expressed on the potential hazards from discharge into the environment of organisms carrying catabolic genes on plasmids. Investigations to which reference has been made in Chapter 4, Section 4.9 (Jain et al. 1987; Fulthorpe and Wyndham 1989; Sobecky et al. 1992) used a set of microcosms to determine the conditions needed to preserve the genotype and its stability. Once again, the advantage of the technique is the ease of incorporating important variables that may be difficult to analyze in natural systems.

4. A sediment–water system was used to study the partition and the degradation of ^{14}C-labeled 4-nitrophenol and 3,4-dichloroaniline (Heim et al. 1994). The results clearly illustrated the importance of water-to-sediment partitioning, and that a substantial fraction of the substrates existed in the form of nonextractable residues.

5. A study using resuspended river sediment (Marchesi et al. 1991) illustrated the important interdependence of attachment of the substrate to particulate matter and its biodegradability: addition of sodium dodecyl sulfate that is degradable resulted in a relative increase in the number of particle-associated bacteria, whereas this was not observed with the nondegradable sodium tetradecyl sulfate or with sodium dodecane sulfonate.

6. A series of soil microcosms were used to study the biodegradation and bioavailability of pyrene during long-term incubation. The nonextractable fraction of ^{14}C-labeled pyrene that had been introduced into pristine soil and incubated with and without the addition of azide was substantially greater in the latter (Guthrie and Pfaender 1998). It was also shown that microbial activity produced a number of unidentified polar metabolites that might plausibly be involved in the association. This experiment is discussed in greater detail in Chapter 8, Section 8.2.1.

True field experiments on microbial reactions are extremely difficult to carry out, but a series of microcosm experiments on the substrates that may support anaerobic sulfate reduction approached this ideal situation quite closely (Parkes et al. 1989). The investigation used inhibition of sulfate reduction by molybdate to study the increase in the levels of a wide range of organic substrates endogenous in the sediments used. These included both a range of alkanoic acids and amino acids, and very considerably increased the range of organic substrates able to support sulfate reduction. Both *in situ* microcosms and laboratory systems were used to compare and evaluate first-order rates of degradation of a range of mixed substrates including aromatic hydrocarbons and phenolic compounds (Nielsen et al. 1996). The observed rates were comparable, but no systematic differences were observed with the exception of 2,6-dichlorophenol that was not degraded in the laboratory system.

5.3.3 Experiments in Models of Natural Aquatic Systems

It is extremely difficult to carry out field investigations in natural ecosystems with the rigor necessary to unravel metabolic intricacies, although such experiments have been successfully carried out in investigations aimed at determining the fate and persistence of agrochemicals in the terrestrial environment, and in the context of bioremediation. In general, however, simplified systems have been developed: these attempt to simulate critical segments of natural ecosystems in a clearly defined way. Outdoor model systems have been used, and two examples may be used to illustrate the kinds of data that can be assembled and the range of conclusions — and their limitations — that may be drawn from such experiments. Not only purely microbiological determinants of persistence may be revealed, but in addition important data on the distribution and fate of the xenobiotic may be acquired. Attention has already been drawn to the general issue of partition among various environmental compartments in Chapter 3 and to the significance of bioavailability in Chapter 4 (Section 4.6.3).

1. Studies in an artificial stream system were designed to provide confirmation in a field situation of the results from laboratory experiments that had demonstrated the biodegradability of pentachlorophenol. Pentachlorophenol was added continuously to the system during 88 days and its degradation followed (Pignatello et al. 1983; 1985). The results confirmed that pentachlorophenol was indeed degraded by the natural populations of microorganisms and, in addition, drew attention to the significance of both sediments and surfaces in the partitioning of pentachlorophenol between the phases within the system.

2. 4,5,6-Trichloroguaiacol was added continuously during several months to mesocosm systems simulating the Baltic Sea littoral zone: samples of water, sediment, and biota including algae were removed periodically for analysis both of the original substance and of metabolites identified previously in extensive laboratory experiments (Neilson et al. 1989). A complex of metabolic transformations of 4,5,6-trichloroguaiacol was identified, including *O*-methylation to 3,4,5-trichloroveratrole, *O*-demethylation to 3,4,5-trichlorocatechol, and partial dechlorination to a dichlorocatechol, and these metabolites were distributed among the various matrices in the system. Of particular significance was the fact that a material balance unequivocally demonstrated the role of the sediment phase as a sink for both the original substrate and the metabolites, so that a number of interrelated factors determined the fate of the initial substrate.

3. Mesocosms placed in shallow Finnish lakes were used to evaluate changes with extended incubation in bleachery effluent from mills using chloride dioxide that had been biologically treated. The mesocosms had a volume of ~2 m^3 and were constructed of translucent polyethene, or black polyethene to simulate dark reactions. The experiments were carried out at ambient temperatures throughout the year and sum parameters were used to trace the fate of the organically bound chlorine. In view of previous studies on the molecular mass distribution of effluents (Jokela and Salkinoja-Salonen 1992), this was measured as an additional marker. Important features were that (a) sedimentation occurred exclusively within the water mass within the mesocosm, (b) the atmospheric input could be estimated from control mesocosms, (c) the microbial flora included both indigenous organisms in lake water and those carried over from the treatment process (Saski et al. 1996a,b). There were a number of important general conclusions:

   - The atmospheric input of organically bound halogen was negligible compared with that in the effluent.

   - There was >50% loss of organohalogen in the water phase and this included compounds with masses of both <500 and >500.

   - The tetrahydrofuran-extractable organic chlorine in the *de novo* sediment had a molecular mass (average 1400) much higher than that from the water-phase extracts (average 360). There was no evidence for selective sorption of higher-molecular-mass components and transformation of lower-molecular-mass components to higher-molecular-mass hydrophobic compounds cannot be excluded.

Comparable experiments in natural aquatic ecosystems are much more difficult to design (Madsen 1991), but some examples of what may be accomplished are given as illustration, and are applicable when sufficient information is known about the degradative pathways of the xenobiotic.

1. Analysis of chlorobenzoates in sediments that had been contaminated with PCBs was used to demonstrated that the lower PCB congeners that had initially been produced by anaerobic dechlorination were subsequently degraded under aerobic conditions: the chlorobenzoates were transient metabolites and their concentrations were extremely low since bacteria that could successfully degrade them were present in the sediment samples (Flanagan and May 1993).

2. The bacterial aerobic degradation of pyrene has been shown to proceed by initial formation of *cis*-pyrene-4,5-dihydrodiol. This metabolite has been used to demonstrate the biodegradability of pyrene in an environment in which there was continuous input of the substrate when it is not possible to use any diminution in its concentration as evidence for biodegradation (Li et al. 1996). The corresponding metabolite from naphthalene — *cis*-naphthalene-1,2-dihydrodiol — has been used both in site-derived enrichment cultures and in leachate from the contaminated site to demonstrate biodegradation of naphthalene (Wilson and Madsen 1996).

3. It has been shown that under anaerobic denitrifying conditions pure cultures of bacteria may form benzylsuccinate as metabolites of toluene (Evans et al. 1992; Migaud et al. 1996; Beller et al. 1996; see also Chapter 6, Section 6.7.3): demonstration of these and the corresponding methylbenzyl succinates from xylenes has been used to demonstrate metabolism of toluene and xylene in an anaerobic aquifer (Beller et al. 1995).

Care must, however, be exercised in the interpretation of results that show the presence of putative metabolites. An illustrative example is provided by a study of the biodegradation of a range of PAHs in compost-amended soil, unamended soil, and sterilized soil (Wischmann and Steinhart 1997).

- Neither dihydrodiols formed by bacterial dioxygenation nor phenols from fungal monooxygenation followed by rearrangement or hydrolysis and elimination were found.

- On the other hand, plausible oxidation products of anthracene, acenaphthylene, fluorene, and benz[*a*]anthracene — anthracene-9,10-quinone, acenaphthene-9,10-dione, fluorene-9-one, and benz[*a*]anthracene-7,12-quinone — were found transiently in compost-amended soil. It was shown, however, that these were formed even in sterile controls by undetermined abiotic reactions.

These results clearly illustrate the care that must be exercised in interpreting the occurrence of PAH oxidation products as evidence of biodegradation.

Due to the potential health hazard, the application of radiolabeled substrates that have been the cornerstone of metabolic experiments is not generally acceptable in field experiments. Examples of experiments using stable isotopes have, however, been carried out to determine biodegradation under field conditions and to establish the source of contaminants. Further details of the procedure have been given in Chapter 2, Section 2.4.4.

1. Fully deuterated benzene, toluene, 1,4-dimethylbenzene, and naphthalene were used to determine their dissemination and biodegradation in an aquifer plume with bromide as an inert marker (Thierrin et al. 1995). Analysis of samples was readily accomplished by GC-MS, and after taking into account loss by sorption and dispersion, half-lives of the substrates were calculated. At the oxic upper surface of the plume, rapid degradation occurred, and continued at a slower rate into the anoxic zone. Although benzene was the most persistent substrate, there is evidence that it is degradable under anaerobic methanogenic, sulfate-reducing, Fe(III)-reducing (Kazumi et al. 1997), and nitrate-reducing (Burland and Edwards 1999) conditions.

2. The availability of accurate methods for measuring stable isotope ratios for oxygen ($^{18}O/^{16}O$) and carbon ($^{13}C/^{12}C$) have made it possible to use these for estimating the changes that occur during degradation of a given substrate. Isotope concentrations are expressed as per mil (o/oo) deviations δ from standard values, and the values for oxygen, nitrate, and sulfate are affected principally by microbial processes. The values for δ ^{13}C and δ ^{18}O were examined during degradation of diesel oil by a mixed culture under aerobic conditions, and the values for oxygen were particularly valuable in correlating production of CO_2 and loss in substrate concentration (Aggarwal et al. 1997). It was therefore suggested that this methodology could be used to provide rates of *in situ* biodegradation.

3. A method has been developed for converting volatile organochlorine compounds to CO_2 and CH_3Cl for measuring isotope ratios for ^{13}C and ^{37}Cl (Holt et al. 1997), and applied to chloroalkanes and chloroalkenes. For the C_1 and C_2 compounds, δ ^{13}C values ranged from −25.58 to −58.77 and δ ^{37}Cl from −2.86 to +1.56, and it was suggested that the method could be used to study the fate and distribution of such compounds. The method has, however, the disadvantage that water must first be removed from the sample.

5.4   Experimental Problems: Water Solubility, Volatility, Sampling, and Association of the Substrate with Microbial Cells

These issues have been briefly noted in Section 5.2.5, but deserve some further comment. For freely water-soluble substrates which have low volatility, there are few difficulties in carrying out the appropriate experiments described above. There is, however, increasing interest in xenobiotics such as PAHs and highly chlorinated compounds including, for example, PCBs, which have only low water solubility; in addition, attention has been focused on volatile chlorinated aliphatic compounds such as the chloroethenes, dichloromethane, and carbon tetrachloride. All of these substrates present experimental difficulties of greater or lesser severity.

1. Whereas suspensions of poorly water soluble substrates can be used for experiments on the identification of metabolites, these methods are not suitable for kinetic experiments that necessitate the quantification of substrate concentrations. In such cases, the whole sample must be sacrificed at each sampling time, and care must be taken to ensure that substrate concentrations in each incubation vessel are, as far as possible, equal. In addition, the whole sample must be extracted for analysis since representative aliquots cannot be removed. Solutions of the substrates in suitable solvents such as acetone, diethyl ether, methanol, or ethyl acetate may be prepared, sterilized by filtration, and suitable volumes dispensed into each incubation vessel. The solvent is then removed in a stream of sterile N_2 and the cell suspension added. As an alternative, solutions in any of these solvents or in others that are much less volatile such as dimethyl sulfoxide and dimethyl formamide, have been added directly to media after sterilization by filtration. Since, however, the remaining concentration of the solvents may not be negligible, care must be taken to ensure that these are neither toxic nor compromise the results of the metabolic experiments. When the test substrate is a solid, it may be preferable to prepare saturated solutions in the basal medium, remove undissolved substrate by filtration through glass-fiber filters, and sterilize the solution by filtration before dispensing and adding cell suspensions.

2. Investigation of substrates with appreciable volatility — and many organic compounds including solids have significant vapor pressures at ambient temperatures — presents a greater experimental challenge, especially if experiments are to be conducted over any length of time. Incubation vessels such as tubes or bottles may be closed with rubber stoppers or with Teflon-lined crimp caps fitted with rubber seals, and these are particularly convenient for

sampling with syringes. Even with Teflon seals, however, these cap inserts may be permeable to compounds with appreciable vapor pressure, and sorption of the test substrate may also occur to a significant extent; both of these factors result in controls that display undesirable diminution in the concentration of the test substrate. Good illustrative examples are provided by the results of a study with endosulfans and related compounds (Guerin and Kennedy 1992), and on the metabolism of 4,4-dichlorobiphenyl by *Phanerochaete chrysosporium* (Dietrich et al. 1995). These results support the importance even in laboratory experiments of taking into account gas/liquid partitioning that has been discussed in Chapter 3, Section 3.4. Completely sealed glass ampoules may be used for less volatile compounds—although clearly not for highly volatile compounds—and subsampling cannot be carried out. For aerobic organisms, one serious problem with all these closed systems is that of oxygen limitation; the volume of the vessels should therefore be very much greater than that of the liquid phase, and conclusions from the experiments should recognize that microaerophilic conditions will almost certainly prevail during prolonged incubation. Such limitations clearly do not prevail for anaerobic organisms for which this is a valuable procedure.

3. Associations between xenobiotics and higher biota have been discussed in Chapter 3, Section 3.1.1 in the context of bioconcentration. Such associations may also introduce ambiguities into the interpretation of the results of metabolic experiments using microorganisms. Three illustrative examples are given of association of xenobiotics with microorganisms.

- Chloroguaiacols with Gram-positive bacteria (Allard et al. 1985);

- PCB congeners with the hyphae of *P. chrysosporium* (Dietrich et al. 1995);

- 2,4,6-Trichlorophenol with *Bacillus subtilis* (Daughney and Fein 1998).

Analytical procedures should therefore be designed to take into account such associations. If they are not quantitatively evaluated or eliminated, the results of such experiments may be seriously compromised.

5.5 Procedures for Elucidating Metabolic Pathways

The general principles follow those used in all biochemical investigations, although some procedures are particularly well suited to microbial systems.

The structure of the various metabolites and the order in which they are formed will reveal the broad outlines of the metabolic pathways followed during the degradation of the substrate. Subsequent refinements may involve isolating and characterizing the relevant enzymes, followed by investigation of their regulation. At all of these stages, the availability of mutants blocked at specific steps of the metabolic sequence is of the greatest value. Good examples of this are provided by the investigations of the metabolism of aromatic compounds which are discussed in detail in Chapter 6, Section 6.2.1. The specific details of the investigation must also take into account problems including the stability of the relevant enzymes, the availability of suitable enzyme assays, and access to the relevant mutants. The complete elucidation of a metabolic pathway is therefore a time-consuming and intellectually demanding operation of which only a few salient features can be presented briefly here.

5.5.1   The Principle of Sequential Induction

On the basis of the intensive and extensive investigations into microbial metabolism that have been carried out painstakingly over many years, it may now be possible to propose a hypothetical degradative pathway on the basis of plausible biochemical reactions. This pathway may then be confirmed by determining the structure of the intermediate metabolites and by further metabolic experiments using, for example, suitable isotopes (Section 5.5.4). Metabolic pathways for a wide range of structurally diverse compounds are discussed in detail in Chapter 6, so that only a general outline will be given here by way of illustration. The delineation of the pathways and the role of hypothetical metabolites may be confirmed — or eliminated — by application of the principle of simultaneous or sequential induction (Stanier 1947) in any of its many forms. Good illustrations may be taken from two early studies on biodegradation:

1. Strains of bacteria able to grow with naphthalene were unable to utilize 1- or 2-naphthol, phthalic acid, or catechol. This clearly suggested that these compounds were not *directly* involved in the degradation (Tausson 1927), and this conclusion has been amply confirmed in detailed subsequent studies.

2. For strains of bacteria during growth with toluene (Claus and Walker 1964), utilization of aromatic substrates was inducible and toluene-grown cells were assayed for oxidation of a number of possible intermediates. From the results of these experiments, it could be concluded that the degradation in these strains did not proceed through benzaldehyde and benzoic acid but by oxygenation to 3-methyl catechol. Degradation via benzoate has, of course, been established as an alternative in other organisms.

In both examples, further details of the mechanism of the ring-cleavage reaction that is discussed in Chapter 4, Section 4.4.2 and Chapter 6, Section 6.2 had to await chemical characterization and structural determination of unstable intermediates.

The principle of sequential induction can be further extended to include the assay of specific enzymes when an inducer is added to cells grown with a noninducing carbon source. For example, addition of DL-mandelate to a culture of *Pseudomonas putida* grown with succinate resulted in immediate synthesis of L-mandelate dehydrogenase that is the first enzyme of the degradative pathway, but a delay before synthesis of the muconate lactonizing enzyme that is involved several stages later after cleavage of the aromatic ring (Hegeman 1966a,b).

5.5.2 Application of Mutants

Mutants have been extensively used in elucidating biosynthetic pathways in a number of widely different biological systems. The principles are equally applicable to metabolic studies using mutants defective in specific genes required for the complete degradation of the substrate, and these are essentially the same as those applied in sequential induction. Mutagenesis may be accomplished by a number of procedures which differ both in the degree of killing and in the type of mutation induced. Ethyl methanesulfonate has been used traditionally, although this has been largely superseded by N-methyl-N′-nitro-N-nitrosoguanidine which induces primarily base substitutions, or by compounds related to substituted 5-aminoacridines which induce addition or deletion of one or more bases (Miller 1977). Use has also been made of a procedure whereby organisms are grown with a substrate and a structurally related halogenated analogue (Wigmore and Ribbons 1981). Selection of mutants with the desired lesions may be accomplished by replica plating onto basal media containing the initial substrate or substrates that are suspected as metabolites: colonies which grow on the latter, but not on the former, are then purified by restreaking. An attractive procedure employed in investigations on tryptophan metabolism by pseudomonads involved the use of media with a high concentration of tryptophan (1 g/l) together with a low concentration of asparagine (0.1 g/l) which permitted production of only small colonies from mutants defective in tryptophan degradation (Palleroni and Stanier 1964).

Mention should be made of two entirely different procedures that offer a wide range of potential application. Transposon mutagenesis offers a means of obtaining insertion mutants in gene functions for which there are no means of direct selection. This has been used to study the genetics of the gentisate pathway in *P. alcaligenes* (Tham and Poh 1993), although less extensively used in studies of xenobiotics. Two widely different applications are given as illustration.

1. Mutants of *Alcaligenes eutrophus* JMP(pJP4) that can use 2,4-dichlorophenoxyacetate and 3-chlorobenzoate as sole source of carbon and energy were obtained by transposon mutagenesis using Tn5 and Tn1771, and used to localize the genes involved in the degradation of 2,4-dichlorophenoxyacetate (Don et al. 1985). It should be noted that the overall pathway had already been established by previous workers.

2. A mutant defective in the chemotaxis toward 4-hydroxybenzoate in *P. putida* was obtained by Tn5 transposon mutagenesis, and allowed identification of a gene cluster *pcaRKF* that encodes enzymes of the 3,4-dihydroxybenzoate branch of the β-ketoadipate pathway: *pcaR* encodes regulatory genes, *pcaK* the thiolase in the last step of the pathway, and *pcaF* is involved in transport of, and chemotaxis toward 4-hydroxybenzoate (Harwood et al. 1994).

Vector insertion mutagenesis has been applied to the elucidation of the pathway of nicotinate degradation in *Azorhizobium caulinodans* (Kitts et al. 1992).

The degradation of a wide range of aromatic compounds converges on catechol whose degradation therefore occupies a key position. Although details of the pathways will be presented in Chapter 6, some examples from the comprehensive investigations from the Berkeley laboratory that were directed to the elucidation of the pathways and their regulation will be given here. They illustrate the various steps in the investigations that include good applications of the use of mutants. The regulation of the pathways has been briefly noted in Section 4.7. A number of complementary procedures were used of which the following were cardinal:

1. The use of purified enzymes for the synthesis of the appropriate metabolites;

2. Analysis of induction patterns after growing cells with different substrates and using appropriate assays to determine the levels of the relevant enzymes:

3. Isolation of mutants to study enzyme regulation.

Some examples of the application of these procedures will be given as illustration.

Degradation of catechol and 3,4-dihydroxybenzoate — The key observation was that the ring-cleavage product of catechol or 3,4-dihydroxybenzoate was β-ketoadipate that is formed by a series of lactonizations and rearrangements. The various steps were elucidated using pure samples of the proposed intermediates and enzyme preparations to study induction patterns. Mutants were then used to elucidate the regulation of the pathways: *cis,cis*-muconate is the inducer for the catechol pathway, and β-ketoadipate for the 3,4-dihydroxybenzoate pathway (Ornston 1966).

Degradation of L-mandelate — L-Mandelate is degraded by successive stages to benzoate which is then metabolized by the pathways outlined above for 3,4-dihydroxybenzoate. Details of the degradation were elucidated by similar means except that the task was considerably simplified by the commercial availability of most of the intermediates (Hegeman 1966a,b).

Degradation of L-tryptophan — Mutants defective in the suite of enzymes required for degradation of the substrate were isolated: for initial cleavage of the ring (tryptophan pyrrolase), for hydrolysis of the formyl group of the cleavage product (formylkynurenine formamidase), and for conversion of the hydrolysis product, kynurenine, into anthranilic acid (kynureninase). These were used to provide a detailed picture of the regulation of the degradative pathway and to demonstrate that of the three potential inducers for the degradation of L-tryptophan, it is kynurenine that coordinately induces synthesis of all three enzymes (Palleroni and Stanier 1964).

Dioxygenation of toluene — Mutants of *P. putida* PpF1 defective in genes coding for each of the three enzymes involved in the initial dioxygenation of toluene, reductase$_{tol}$, ferredoxin$_{tol}$, and ISP$_{tol}$ were obtained after *N*-methyl-*N'*-nitro-*N*-nitrosoguanidine mutagenesis. These were selected by differentiating color on an indicator medium containing toluene and arginine, and Nitro Blue tetrazolium and 2,3,5-triphenyl-2*H*-tetrazolium chloride (Finette at al. 1984).

In studies examining regulation such as those discussed above, isolation of the appropriate enzymes and development of assays for enzymatic activity play a key role. The application of modern methods for protein fractionation together with the technique of cloning genes into *Escherichia coli* followed by gene amplification has greatly simplified the preparation and purification of catabolic enzymes. There are no simple rules for the preparation of cell extracts, and some enzymes are notoriously unstable. In these cases, use of toluene-permeabilized cells may be employed. Whole cells may be disrupted in a number of ways, for example, by grinding with abrasives such as Al_2O_3 — following the original method of Buchner who used a mixture of quartz sand and diatomite — by sonication, or by application of pressure. Probably the last procedure is generally preferable and it possesses the additional advantage that cell disruption can readily be carried out in an anaerobic atmosphere.

5.5.3 Use of Metabolic Inhibitors

As an alternative to the use of mutants, metabolic inhibitors may be used to interrupt metabolic pathways; even transient intermediates may then be accumulated and provide evidence for the details of the consecutive steps. A wide range of compounds has been used in investigations on electron transport pathways and bioenergetics, but these lie beyond the scope of this account. Examples that have been used in metabolic studies with bacteria include molybdate as an inhibitor of anaerobic sulfate reduction, methyl fluoride and difluoromethane (Miller et al. 1998) as inhibitors of the aerobic oxidation of methane, acetylene as an inhibitor of monooxygenase systems, and 2-bromoethanesulfonate as an inhibitor of methanogenesis. Some caution should be exercised in view the incomplete specificity of these—and indeed all — inhibitors. For example, methyl fluoride also inhibits methanogenesis (Janssen and Frenzel 1997) and 2-bromoethanesulfonate also inhibits the dechlorination of chloroethenes (Löffler et al. 1997). Two examples will be given to illustrate the use of inhibitors.

1. They were used to delineate an unusual pathway for the degradation of 3-methylphenol carried out by a methanogenic consortium (Londry and Fedorak 1993). Three compounds were used as selective inhibitors:

   a. 6-Fluoro-3-methylphenol that brought about the accumulation of 4-hydroxy-2-methylbenzoate;

   b. 3-Fluorobenzoate that resulted in the transient accumulation of 4-hydroxybenzoate;

   c. Addition of bromoethanesulfonate that inhibited the synthesis of methane and caused the accumulation of benzoate.

   In addition, the corresponding fluorinated metabolites were identified, and this enabled construction of the complete pathway for the degradation that included the unusual reductive loss of a methyl group from the aromatic ring; these reactions are discussed in a wider context in Section 6.7.3.2.

2. They were used in an elegant set of enrichment experiments designed to enrich for the organisms that carried out the *ortho* dechlorination of 2,3,5,6-tetrachlorobiphenyl (Holoman et al. 1998). The community structure was followed by analysis of total community genes for 16S rRNA, and it was shown that the diversity of the community could be reduced in mineral medium by addition of inhibitors for methanogens (2-bromoethanesulfonic acid), and *Clostridium* spp.(vancomycin) without eliminating dechlorination and that this was inhibited by addition of molybdate that inhibited sulfate reduction. The bacteria that actively carried out *ortho* dechlorination belonged to three groups: the δ group, the low-G+C Gram-positive group, and the *Thermotogales* subgroup which had not hitherto been implicated in anaerobic dechlorination.

5.5.4   Use of Synthetic Isotopes

The use of substrates isotopically labeled in specific positions makes it possible to follow the fate of individual atoms during the microbial degradation of xenobiotics. Under optimal conditions, both the kinetics of the degradation and of the formation of metabolites may be followed — provided, of course, that samples of the labeled metabolites are available. Many of the classical studies on the microbial metabolism of carbohydrates, carboxylic acids, and amino acids used radioactive ^{14}C-labeled substrates and specific chemical degradation of the metabolites to determine the position of the label. The method is indeed obligatory for distinguishing between degradative pathways when the same products are produced from the substrate by different pathways. A good example is provided by the β-methylaspartate and hydroxyglutarate fermentations of glutamate, both of which produce butyrate but which can be clearly distinguished by the use of [4-^{14}C]-glutamate (Buckel and Barker 1974). An attempt will be made to illustrate briefly the application of a wider range of isotopes without going into the details which are provided in the references that are cited.

Carbon (^{14}C and ^{13}C)

Traditional use has been made of the isotope ^{14}C which has the convenience of being radioactive, and details need hardly be given here. Illustrative examples include the elucidation of pathways for the anaerobic degradation of amino acids and purines that are discussed in greater detail in Sections 6.7.1 and 6.7.4.1. On the other hand, increasing application has been made of ^{13}C using high-resolution Fourier transform NMR in whole cell suspensions; this is equally applicable to molecules containing the natural ^{19}F or the synthetic ^{31}P nuclei and further details of its application are given in Section 5.5.5.

The use of ^{13}C-labeled substrates in conventional experiments on biodegradation is restricted only by the availability of the substrate. Illustrative examples of its application to elucidate the mechanism of microbiological reactions include the following.

1. The degradation of $^{13}CCl_4$ by *Pseudomonas* sp. strain KC that involves formation of intermediate $COCl_2$ trapped as a HEPES complex and by reaction with cysteine (Lewis and Crawford 1995) is noted in Section 6.4.2.

2. Use of ^{13}C[bicarbonate] and NMR to demonstrate that the first product in the metabolism of propene epoxide is acetoacetate that is then reduced to β-hydroxybutyrate (Allen and Ensign 1996).

3. Use of ^{13}C[bicarbonate] and mass spectrometry to demonstrate the formation of carboxylic acids during the sulfidogenic mineralization of naphthalene and phenanthrene (Zhang and Young 1997).

4. Use of 9[^{13}C]-anthracene to study its degradation in soil and the formation of labeled metabolites that could be released only after alkaline hydrolysis (Richnow et al. 1998). It was possible to construct a carbon balance during the 599-day incubation, and to distinguish metabolically formed phthalate from indigenous phthalate in the soil.

Sulfur (³⁵S) and Chlorine (³⁶Cl)

Although quite extensive use of ³⁵S has been made in studies on the degradation of alkyl sulfonates (Hales et al. 1986), ³⁶Cl has achieved only limited application, on account of technical difficulties resulting from the low specific activities and the synthetic inaccessibility of appropriately labeled substrates. One of the few examples of its application to the degradation of xenobiotics is provided by a study of the anaerobic dechlorination of hexachlorocyclohexane isomers (Jagnow et al. 1977), the results of which will be discussed later in Section 6.4.1.

Hydrogen (²H) and Oxygen (¹⁸O)

Although the radioactive isotope ³H has been extensively used for studies on the uptake of xenobiotics into whole cells, the intrusion of exchange reactions and the large isotope effect renders this isotope possibly less attractive for general metabolic studies. Both deuterium ²H-labeled substrates, and oxygen ¹⁸O₂ and ¹⁸OH₂ have been used in metabolic studies since essentially pure labeled compounds are increasingly readily available, and access to the necessary mass spectrometer facilities has increased over the years.

Examples of the use of ²H in different applications include the following:

1. Deuterium labeling has been invaluable in studying rearrangement reactions involving protons; for example, it has been used to reveal the operation of the NIH-shift during metabolism of [4-²H]ethylbenzene by the monooxygenase system from *Methylococcus capsulatus* (Dalton et al. 1981), of [2,2′,3,3′,5,5′,6,6′-²H]biphenyl by *Cunninghamella echinulata* (Smith et al. 1981), and of [1-²H]naphthalene and [2-²H]naphthalene by *Oscillatoria* sp. (Narro et al. 1992).

2. ²H-labeled substrates have been used to determine the dissipation and degradation of aromatic hydrocarbons in a contaminated aquifer plume (Thierrin et al. 1995). Its application is particularly appropriate since the site was already contaminated with the substrates. With suitable precautions, this procedure seems capable of extension to determining the presence — although not the complete structure — of metabolites provided that the possibility of exchange reactions were taken into account.

3. The conversion of long-chain alkanoate CoA esters to the alkenoate CoA esters by acyl-CoA oxidase involves an *anti* elimination reaction. The stereochemistry of the reaction in *Candida lipolytica* was established using stearoyl-CoA labeled with ²H at the 2(*R*)-, 3(*R*)-, and 3(*S*)-positions (Kawaguchi et al. 1980).

4. The use of ²H-labeled substrates has been used to determine details of the dehydrogenation of *cis*-dihydrodiols produced by

dioxygenases from aromatic substrates (Morawski et al. 1997), and it was possible to demonstrate the specificity of the hydrogen transfer from the dihydrodiol substrates to NAD.

5. The enantiomerization of phenoxyalkanoic acids containing a chiral side chain has been studied in soil using ^2H_2O (Buser and Müller 1997). It was shown that there was an equilibrium between the *R* and *S* enantiomers of 2-(4-chloro-2-methylphenoxy)propionic acid (MCPP) and 2-(2,4-dichlorophenopxy)propionic acid (DCPP) with an equilibrium constant favoring the herbicidally active *R* enantiomer. The exchange reactions proceeded with both retention and inversion of configuration at the chiral sites.

Application of ^{18}O has been less frequent but has been used effectively to determine the source of oxygen and the number of oxygen atoms incorporated during metabolism of xenobiotics under both aerobic and anaerobic conditions. Three typical examples are given:

1. During the biodegradation of 2,4-dinitrotoluene by a strain of *Pseudomonas* sp. two atoms of oxygen are incorporated from $^{18}O_2$ during the formation of 4-methyl-5-nitrocatechol by dioxygenation with loss of nitrite (Spanggord et al. 1991).

2. During degradation of 2-chloroacetophenone by a strain of *Alcaligenes* sp., one atom of $^{18}O_2$ is incorporated into 2-chlorophenol formed from the 2-chlorophenyl acetate that is initially formed by a Baeyer–Villiger monooxygenation (Higson and Focht 1990).

3. Benzene and toluene are anaerobically hydroxylated to phenol and 4-hydroxytoluene, and experiments with $H_2^{18}O$ showed that the oxygen atoms come from water (Vogel and Grbic-Galic 1986).

In experiments involving the use of both of these isotopes, care should be taken to exclude chemical exchange reactions involving potentially labile C–H or C–O bonds since these reactions could seriously compromise the conclusions. A good illustration of the pitfalls in such investigations is shown by a study on the dechlorination of pentachlorophenol by the dehalogenase from a strain of *Arthrobacter* sp. The initial reaction in the degradation of pentachlorophenol is mediated by a pentachlorophenol dehalogenase that produces tetrachloro-1,4-dihydroxybenzene. Experiments using the enzyme showed that ^{18}O is incorporated into this metabolite only after incubation with $H_2^{18}O$ and not with $^{18}O_2$: in fact, the labeling occurs as a result of exchange between the initially formed unlabeled metabolite and $H_2^{18}O$. An unambiguous elucidation of the mechanism of the reaction is not therefore possible since even if ^{18}O had been incorporated during the reaction with $^{18}O_2$, exchange with the excess $H_2^{16}O$ in the medium would yield an unlabeled product (Schenk et al. 1990).

5.5.5 Application of NMR Using ^{13}C, ^{15}N, ^{19}F, and ^{31}P and Electron Paramagnetic Resonance (EPR)

Application of NMR

The advantages of this technique which is nondestructive is that the unambiguous structure of metabolites is possible. This may be carried out not only in extracts after purification by conventional means, but also directly in culture supernatants and even during their synthesis in the spectrometer tubes; these may be equipped with gas inlets which enable studies to be carried out under virtually any metabolic condition. In addition, there are no experimental restrictions in handling radioactive material, although for studies of carbon metabolism ^{13}C-labeled substrates must be available; increasingly, however, many of these are no longer less accessible than their ^{14}C analogues. On the other hand, the relatively low sensitivity of the method has hitherto precluded identification of metabolites which may be formed transiently only in low concentration.

Carbon ^{13}C

Interpretative difficulties may arise from the inherent design of NMR experiments that may necessitate the use of high substrate concentrations due to the relatively low sensitivity of the procedure. A single illustrative example may be given of the occurrence of artifacts that may be encountered: during a study of the metabolism of mandelate by *P. putida* (Halpin et al. 1981) benzyl alcohol was unexpectedly identified when experiments were carried out at high substrate concentration (50 m*M*). This was, however, subsequently shown to be due to the action of a nonspecific alcohol dehydrogenase under the anaerobic conditions prevailing at the high substrate concentration used for the identification of the metabolites (Collins and Hegeman 1984).

^{13}C NMR using whole cells has been applied to the study a number of relatively straightforward metabolic reactions involving small molecules and these include the following that may be used as illustration of the potential of the procedure:

- Glycolysis in *Saccharomyces cerevisiae* (den Hollander et al. 1986);
- The reduction of dimethyl sulfoxide and trimethylamine-*N*-oxide by *Rhodobacter capsulatus* (*Rhodopseudomonas capsulata*) (King et al. 1987);
- The metabolism of acetate and methanol in *Pseudomonas* sp.(Narbad et al. 1989);
- Nicotinate and pyridine nucleotide metabolism in *Escherichia coli* and *Saccharomyces cerevisiae* (Unkefer and London 1984);
- Combined use of ^{1}H and ^{13}C NMR was used in powerful combination in deducing the pathways of degradation of vinyl chloride (Castro et al 1992a,b);

- The incorporation of $^{13}CO_2$ into poly-β-hydroxybutyrate by a strain of *Xanthomonas* sp. that metabolizes propene or its oxide (Small and Ensign 1995; Allen and Ensign 1996);

- The degradation of acetonitrile by *Methylosinus trichosporium* (Castro et al. 1996);

- The metabolism of $^{13}CCl_4$ by *Pseudomonas* sp. strain KC (Lewis and Crawford 1995) that has been noted in Section 5.2.4;

- Detection of glycolate and 2-(2-aminoethoxy)acetate as intermediates in the degradation of morpholine by *Mycobacterium aurum* strain MO1 (Combourieu et al. 1998).

^{13}C has been used as a tracer in the study of the degradation of [1-^{13}C]-acenaphthene both in pure cultures that were degrading naphthalene and phenanthrene by cometabolism, and in a mixed culture that was enriched with creosote (Selifonov et al. 1998). The degradation pathway that is initiated by benzylic monooxygenation could be determined by isolation of intermediate metabolites, and the method proved applicable to the situation where only limited biotransformation of the substrates takes place by partial oxidation.

The procedure is particularly suited to the study of anaerobic transformations since there are no problems resulting from problems with oxygen limitation. A good example is provided by the application of ^{13}C NMR to the intricate relations of fumarate, succinate, propionate, and acetate in a syntrophic organism both in the presence (Houwen et al. 1991) and in the absence of methanogens (Stams et al. 1993; Plugge et al. 1993).

A single example will be given to illustrate both the strengths and the limitations of the technique. During the metabolism of 2-^{13}C-acetate in methylotrophic strains of *Pseudomonas* sp., it was shown that the substrate was converted into α, α-trehalose in isocitrate lyase-negative strains although not in one which synthesized this enzyme. In addition, an unknown compound was revealed by the *in vitro* experiments but was not present in perchloric acid extracts of the cells. Possibly more disturbing, however, was the fact that analysis of a strain during growth with ^{13}C-methanol did not reveal the presence of the intermediates known to be part of the serine pathway that functions in this organism (Narbad et al. 1989).

Nitrogen ^{15}N

Some examples have been given in Chapter 3, Section 3.2.4, in the context of studies on humic and fulvic acids, but these illustrations may conveniently be repeated here.

1. ^{15}N NMR has been used in conjunction with ^{13}C NMR in studies on reactions of [^{15}N] hydroxylamine with fulvic and humic substances (Thorn et al. 1992).

2. The availability of ^{15}N aniline has made possible a direct study of the reactions of aniline with humic and fulvic acids (Thorn et al. 1996), and the detection of resonances attributed to anilinoquinone, imines, and N-heterocyclic compounds are fully consistent with reactions involving quinone and ketone groups.

Fluorine [19]F

Application to fluorine-containing molecules is particularly attractive since naturally occurring fluorine is monoisotopic and since the range of chemical shifts in fluorine compounds is very much greater than the proton shifts for hydrogen-containing compounds. Although only a few examples will be given as illustration, there is a vast potential for the application of ^{19}F NMR to metabolic studies of fluorine compounds.

1. Substantial attention has been devoted to the metabolism of 5-fluorouracil and related compounds: for example, both anabolic reactions involving pyrimidine nucleotides and degradation to α-fluoro-β-alanine by the fungus *Nectria haematococca* have been successfully analyzed by ^{19}F NMR (Parisot et al. 1989; 1991) both in cell extracts and in whole mycelia.

2. Fluoroacetate and 4-fluorothreonine are synthesized from fluoride by *Streptomyces cattleya*, and analysis of supernatants was used to elucidate the details of their biosynthesis: they were apparently synthesized by independent routes, and it was suggested that what is at least formally glycollate could be their precursor (Reid et al., 1995).

3. Oxidative decarboxylation of hydroxybenzoates by the yeast *Candida parapsilosis* is catalyzed by a flavin monooxygenase that is able to use a range of fluorinated hydroxybenzoates that were examined by ^{19}F NMR (Eppink et al. 1997).

4. ^{19}F NMR was used to study the effect of pH on the hydroxylation of 3-fluorophenol by the hydroxylase from *Trichosporon cutaneum* and revealed that the ratio of products, as well as the yield, was pH dependent (Peelen et al. 1993).

5. The metabolism of 3,5- and 2,5-difluorobenzoate was studied in a mutant of *P. putida* PpJT103 that is unable to aromatize the 1,2-dihydrodiol of benzoate. ^{19}F NMR was used to establish that in both 3,5- and 2,5-difluorobenzoate dioxygenation took place at C1 and C2 and that for the latter, the dihydrodiol lost fluoride to produce 4-fluorocatechol (Cass et al. 1987).

6. The metabolism of fluorophenols by phenol hydroxylase from *Trichosporium cutaneum*, catechol 1,2-dioxygenase from *P. arvilla* C-1, and by the fungus *Exophilia jeanselmei* was examined, and detailed NMR data were given for the ring fission fluoromuconates (Boersma et al. 1998).

7. The metabolism of di- and trifluorophenols by strains of *Rhodococcus* sp. that involves synthesis of catechols both by hydroxylation and by dioxygenation with elimination of fluoride (Bondar et al. 1998) is noted in Chapter 6, Section 6.10.1.

This methodology seems worthy of wider exploitation in the study of other groups of organofluorine compounds that are of industrial importance as agrochemicals and that have awakened increased environmental interest.

Phosphorus ^{31}P

Until recently, there has been only limited interest in the catabolism of organophosphorus compounds, although considerable attention has been directed to anabolic reactions. Application of ^{31}P NMR has been used to examine nucleotide pools and transmembrane potential in bacteria after exposure to pentachlorophenol, and to demonstrate the differences between *E. coli* which does not degrade the substrate and a *Flavobacterium* sp. which is able to do so (Steiert et al. 1988). ^{31}P NMR has been used to examine the effects of pentachlorophenol on the energy metablism of abalone (*Haliotis rufescens*) (Tjeerdema et al. 1991). One example may be given to illustrate the strengths and limitations of the technique. ^{31}P was used to examine the effect of ethanol on the metabolism of glucose by *Zymomonas mobilis*: whereas the sensitivity was sufficient to establish changes in nucleoside triphosphates during *in vivo* experiments, details of the changes in the various phosphorylated metabolites necessitated the use of perchloric acid cell extracts (Strohhäcker et al. 1993). Attention has already been directed (Section 3.2.4) to the application of ^{31}P NMR to quantify quinone groups in lignin (Argyropoulos and Zheng 1998).

Application of EPR

Compared with the extensive application of NMR procedures, EPR has been used only infrequently. Some examples are used to reveal its potential.

1. It has been used to elucidate the unexpected complexity of an apparently straightforward metabolic pathway: the metabolism of glycerol trinitrate by *Phanerochaete chrysosporium* does not proceed by simple hydrolysis to nitrate and glycerol, but involves formation of nitric oxide bound to both nonheme and heme proteins (Servent et al. 1991).

2. Interest in the biochemical significance of nitric oxide has led to the development of a spin-labeled EPR assay involving reaction with 1,2-bis(*iso*propylidene)cyclohexa-3,5-diene (Korth et al. 1992). The formation of NO (nitric oxide) by oxidation of L-arginine to L-citrulline by a strain of *Nocardia* sp. has been noted in Chapter 4, Section 4.3.3.

3. EPR has been used in studies of the role of hydroxyl radicals and their role in the toxicity of H_2O_2 to bacteria. In a study directed to the analysis of the role of Fe and the generation of H_2O_2 in *E. coli* (McCormick et al. 1998), hydroxyl radicals were specifically trapped by reaction with ethanol to give the α-hydroxyethyl radical that then forms a stable adduct with α-(4-pyridyl-1-oxide)-*N-t*-butyl nitroxide (stable adducts are not formed either by superoxide or hydroxyl radicals). The role of redox-reactive iron used EPR to analyze the EPR-detectable ascorbyl radicals.

4. Pyruvate formate lyase that catalyzes the conversion of pyruvate to formate and acetyl CoA is a key enzyme in the anaerobic degradation of carbohydrates in some enteric bacteria. Using an enzyme selectively ^{13}C-labeled with glycine, it was shown by EPR that the reaction involves production of a free radical at C-2 of glycine (Wagner et al. 1992). This was confirmed by destruction of the radical with O_2, and determination of part of the structure of the small protein which contained an oxalyl residue originating from gly-734.

5. *Geobacter metallireducens* is able to oxidize acetate to CO_2 under anaerobic conditions in the presence of Fe(III) (Section 4.3.3). A study of the intermediate role of humic and fulvic acids used ESR to detect and quantify free radicals produced from oxidized humic acids by cells of *G. metallireducens* in the presence of acetate. There was a substantial increase in the radical concentration after incubation with the cells, and it was plausibly suggested that these were semiquinones produced from quinone entities in the humic and fulvic structures (Scott et al. 1998).

5.6 Application of Surrogate Substrates to Establish Enzymatic Activity

Induction of appropriate degradative enzymes by their substrates or by appropriate analogues has been discussed in Sections 4.5.2 and 4.8.2, but there is need for a rapid method of assessing this. For screening a large number of colonies for a specific enzymatic activity, advantage has been taken of the ability of the appropriate enzyme to accept an analogue that produced colored metabolites. Some illustrative examples include the following.

- The broad substrate activity of naphthalene dioxygenase was used to produce indigo from indole by monooxygenation followed by chemical dimerization (Ensley et al. 1983).

- In analogous fashion, toluene 2,3-dioxygenase was able to produce indigo using indole-2-carboxylate and indole-3-carboxylate (Eaton and Chapman 1995).
- The 2,4-dichlorophenoxyacetic acid/α-ketoglutarate dioxygenase could accept 4-nitrophenoxyacetic acid to produce colored 4-nitrophenol (Sassanella et al. 1997). It was pointed out, however, that a number of strains that were able to degrade 2,4-dichlorophenoxyacetic acid were unable to accept the surrogate substrate.

5.7 Classification and Identification of Organisms

It is unfortunately true that the level of classification of bacteria illustrating important pathways for the degradation of xenobiotics is sometimes rudimentary. As a consequence, the assignment of names may be based on the slenderest evidence, and both they and the accompanying organisms may eventually find their way into national culture collections. One cannot help comparing the rigor applied to the taxonomy of major groups such as the aerobic pseudomonads and their relatives, the Enterobacteriaceae, or the mycobacteria to the somewhat peripheral descriptions of other environmentally significant and interesting organisms.

In one sense, the problem is particularly acute for aerobic organisms since there is such a wealth of information on previously accepted taxa that an attempt at placing an unknown organism within an acceptable circumscription of established taxa is indeed a formidable undertaking. By currently accepted standards, this would necessitate not only nutritional and biochemical characterization, but extensive studies of, for example, DNA/DNA hybridization, identification of fatty acids, cell wall components, ubiquinones, or 16S rRNA base sequencing. These are increasingly the field of specialists, and laboratories well equipped for studies on metabolism may lack both the facilities and the expertise for carrying out such investigations. It may not be a scientifically acceptable defense, but the fact remains that the effort required to assign an unknown organism — which might previously have been described, for example, simply as *Pseudomonas* sp. — to the correct taxon may now simply not be available. The creation of the genera *Acidovorax, Burkholderia, Comamonas, Hydrogenophaga*, and *Variovorax* for organisms that were originally assigned to the genus *Pseudomonas* attests to the level of sophistication that is being applied and its success in differentiating grossly similar genera. An overview has been given (Kersters et al. 1996), and in the context of biodegradation some important examples of new generic names that replace *Pseudomonas* include the following.

Current Designation	Former Designation
Aminobacter aminovorans	*Pseudomonas aminovorans*
Brevundimonas diminuta	*P. diminuta*
Burkolderia cepacia	*P. cepacia*
Comamonas testosteroni	*P. testosteroni*
Comamonas acidovorans	*P. acidovorans*
Ralstonia pickettii	*P. pickettii*
Sphingomonas paucimobilis	*P. paucimobilis*
Stenotrophus maltophilia	*P. maltophilia*

There are also important changes in the classification of other genera of aerobic bacteria and attention is drawn to some that have been noted in the text of Chapters 4 and 6.

- *Alcaligenes eutrophus* to *Ralstonia eutropha;*

- *P. putrefaciens* to *Alteromonas putrefaciens* to *Shewanella putrefaciens;*

- *Bacillus macerans* and *B. polymyxa* to *Paenibacillus macerans* and *Pae. polymyxa;*

- Organisms of the genus *Marinobacter* to which reference is made in Section 6.1.1;

- Those of the genus *Terrabacter* that are capable of degrading dibenzofuran and that were originally assigned to the genus *Brevibacterium* (Section 6.3.1.2); strains classified as belonging to the genus *Pimelobacter* have been transferred to the genera *Terrabacter* and *Nocardiodes* including a pyridine-degrading strain *N. pyridolyticus* (Yoon et al. 1997);

- Thiobacilli which have a complex taxonomy; some organisms originally assigned to the genus *Thiobacillus* have now been transferred to other taxa: *T. versutus* to *Paracoccus versutus,* and *Thiosphaera pantotropha* and organisms using carbon disulfide and carbonyl sulfide as energy sources to *Paracoccus denitrificans* whose description has therefore been amended (Jordan et al. 1997);

Organisms originally assigned only to the genus level may cause particular confusion since it may be difficult to relate the new name to the original. Examples include the following:

- Organisms growing anaerobically with nitrate at the expense of aromatic compounds and assigned to *Pseudomonas* sp. and now assigned to the taxa *Thauera aromatica* and *Azoarcus evansii* for organisms using benzoate and phenol, respectively (Anders et al. 1995);

- *Sphingomonas yanoikuyae* for the organism originally designated *Beijerinckia* sp. strain B1 (Khan et al. 1996).

An illustrative example of the magnitude of the problem may be taken from a study which examined 19 Gram-negative aerobic organisms that degraded a range of xenobiotics including 3-chlorobenzoate, a number of aromatic sulfonic acids, 2,6-dinitrophenol, phosphonic acids, and 1,3,5-triazines (Busse et al. 1991). All of them had DNA G+C contents in the range 61 to 68% and most of them were formally unclassified. An extensive number of characters including the presence of ubiquinones, patterns of soluble proteins, and DNA DNA hybridization enabled assignment of only 8 out of the 19 organisms to specific taxa — *Comamonas testosteroni*, *C. acidovorans*, and species of the genus *Alcaligenes*. Comparison of 16S rRNA fragments suggested possible assignment of one of the remaining organisms to *Acidovorax facilis* and six others to "*Alcaligenes eutrophus*." Clearly, there remain serious difficulties in successfully assigning such strains to established taxa, and the problem is made more difficult by the current uncertainty surrounding the nomenclature and taxonomic position of organisms previously assigned to the genera *Pseudomonas* and *Xanthomonas*. An interesting summary of the problems surrounding assignment of a single organism, "*Pseudomonas maltophilia*" provides a good illustration of the present confusion (Van Zyl and Syeyn 1992). Some of the salient issues were as follows:

1. Problems arising from including "*Pseudomonas maltophilia*" in the genus *Xanthomonas* since the organism shows important differences from other members of the genus that includes many plant pathogens;
2. The divergence of opinion in interpreting results of DNA DNA homology, analysis for ubiquinones, and oligonucleotide sequences of 16S rRNA.

The conclusion that the organism should be assigned to a new genus is plausibly argued and this has now been implemented under the name *Stenotrophomonas* (Palleroni and Bradbury 1993). Although this is a taxonomically tidy solution, the general adoption of this principle would inevitably lead to the proposal of a very large number of new genera such as *Chelatobacter* and *Chelatococcus* for nitrilotriacetate-degrading organisms (Aulin et al. 1993).

An extensive study (Collins et al. 1994) of clostridia using 16S rRNA gene sequences enabled construction of a hierarchical scheme with 14 families for clostridia and related organisms. This resulted in the creation of five new genera and in a clear recognition of the taxonomic heterogeneity of this important group of organisms.

In summary, there seem to be at least three central difficulties in the taxonomic assignment of organisms with established metabolic capabilities —
and indeed to any newly isolated organism:

1. The increasing level of sophistication necessary for classification;
2. The apparent existence of relatively few relevant strains for comparison with new taxa: in many cases, only a few or even a single
   strain exists; possible phenotypic differences within the taxon are
   not therefore available;
3. The proliferation of new names for existing taxa which make it
   increasingly difficult for the nonspecialist to be informed of the
   correct synonymies.

Perhaps one could hope that organisms with significant degradative capabilities be described in the literature with at least basic data of taxonomic relevance and that specialists be encouraged to carry out the studies — such as
the one already described — which are needed for acceptable classification.
It must, however, be accepted that with limited research funding few of those
engaged in studies of biodegradation possess either the means or the experience to take upon themselves this extra responsibility.

Possibly inevitably, a similar situation prevails for the genetics of these
organisms: detailed maps exist, of course, for the two classic enteric organisms — *E. coli* and *Salmonella typhimurium* — and for two species of *Pseudomonas*. Otherwise, systematic genetic studies encompassing a wide range of
genetic markers hardly exist. With the possibility of cloning degradative
genes for study in more suitable organisms, it seems possible that detailed
genetic studies on metabolic regulation of xenobiotic degradation may be
accomplished by alternative methods. Interest in the application of microorganisms to biotechnology has led to increased interest in the genetics of two
other important groups — the methylotrophs and the rhodococci — although
this does not currently reach the level of sophistication attained in the other
groups of bacteria. Possibly the greatest impetus for such investigations will
come from their application to various aspects of biotechnology.

5.8   Procedures for Analysis of Degradative Populations

There has been substantial interest in identifying bacteria in environmental
samples with a view to determining the components of bacterial communities and understanding their role in biodegradation and biotransformation.
Significant advances have been made in the use of appropriate genetic probes
to determine the existence of specific metabolic activity and of analysis of 16S
rRNA and 16S rDNA to assess the similarity of isolates to other taxa. These
procedures have also been used in investigations on the fate of genetically

modified microorganisms in the terrestrial environment (van Elsas et al. 1998). A valuable overview of the principles and application of a range of molecular procedures has been given (Power et al. 1998), and negative-ion MALDI-ToF has been applied to analysis of the PCR-amplified gene sequences for the particulate methane monooxygenases of two taxa of methanotrophic bacteria. In *Methylosinus trichosporium* OB3b, a type II methanotroph, the 50-mer amplified product could readily be identified after the samples had been purified by standard procedures before analysis. *Methylomicrobium albus* BG8, a type I methanotroph with a 99-mer, could also be identified, and no cross-amplification products were observed.

These studies have revealed a number of important issues for microbial ecology: (1) the complexity of bacterial ecosystems, (2) the number of organisms of hitherto undetermined taxonomy, (3) the existence of organisms that have not been cultivated, and (4) the difficulty of assigning metabolic roles to many of the sequences. A number of strategies have been proposed as potentially attractive solutions to evaluate the number of specific degradative organisms and of the appropriate degradative enzymes. A few examples will be given to illustrate the range and application of these procedures.

5.8.1 Specific Metabolic Activity

1. Application of PCR to specific degradative organisms has been reviewed (Steffan and Atlas 1991), and it has been applied to hydrocarbon-degrading *Mycobacterium* sp. (Wang et al. 1996b).

2. Probes for toluene-2-monooxygenase have been used to evaluate the potential number of trichloroethene-degrading organisms in an aquifer (Fries et al. 1997b). In this study, repetitive extragenic palindromic PCR (REP-PCR) (de Bruijn 1992) of isolates was used to classify their metabolic capability.

3. The reverse-sample genome probe procedure has been used to monitor sulfate-reducing bacteria in oil-field samples (Voordouw et al. 1991), further extended to include 16 heterotrophic bacteria (Telang et al. 1997) and applied to evaluating the effect of nitrate on an oil-field bacterial community. Denatured genomic DNAs were used in a reverse-sample genomic probe procedure to examine the effect of toluene and dicyclopentadiene on the community structure of bacteria in a contaminated soil. Hybridization of total-community DNA isolated from soil exposed to toluene showed enrichment of strains able to metabolize toluene, while DNA from the soil contaminated with cyclopentadiene indicated enrichment-revealed organisms that were able to mediate formation of products with masses of 146 and 148 that correspond to oxygenated derivatives containing epoxide or ketonic functions; no mineralization was observed (Shen et al. 1998).

4. Soil samples were enriched in medium containing 2,4,6-trichlorophenol and subjected to a procedure to reduce the complexity of the mixed flora. This involved cyclic serial dilution, plating, and growth in liquid medium. Repetitive extragenic palindromic PCR (REP-PCR) and amplified ribosomal DNA restriction analysis (ARDRA) were used to monitor the flora during the cycles. From the fourth cycle four organisms were isolated: although three of them were unable to utilize 2,4,6-trichlorophenol, an additional slow-growing organism that was able to do so was isolated and tentatively assigned to the genus *Nocardioides* (Maltseva and Oriel 1997).

5. Analysis of the populations of two phenol-degrading bacteria *P. putida* BH and *Comamonas* sp. strain E6 introduced into an activated sludge system has been carried out by extraction of DNA, PCR amplification of the *gyrB* gene fragments using strain-specific primers, and quantification after electrophoresis by densitometry (Watanabe et al. 1998). Appropriate dilutions of the extracted DNA were used to maintain linearity between the measured intensity and population density. The initial inocula of $\sim 10^8$ cells/ml fell to $\sim 10^4$ cell/ml after 10 days and thereafter remained steady for a further 18 days.

6. The occurrence of bacteria that can mineralize 3-chlorobenzoate has been examined in soil samples from widely separated regions in five continents (Fulthorpe et al. 1998). The genotypes of the isolates was examined by two procedures: (a) repetitive extragenic palindromic PCR genomic fingerprints (REP-PCR) (de Brujin 1992) and (b) analysis of restriction digests of the 16S rRNA (ARDRA) (Weisberg et al. 1991). The results showed that each genotype was generally (91%) restricted to the site from which the samples were collected and that therefore the genotypes were not derived by global dispersion.

7. DNA was extracted from soil samples that had been treated with three phenylurea herbicides during 10 years, and from an untreated control. The PCR products using two primers were analyzed by denaturing gradient gel electrophoresis (DGGE), and the patterns used to assess the quantitative similarities of the bands. PCR using two different primers followed by DGGE was used to obtain 16S rDNA for sequencing (El Fantroussi et al. 1999). The microbial diversity determined from the gel profiles had decreased in the treated soils, and the sequencing revealed that the organisms most affected belonged to uncultivated taxa. Enrichment cultures showed that dichlorinated linuron was more readily degraded in cultures using the treated soils than from those using the untreated soils. DGGE analysis and sequencing showed that one of the components that was found only in these enrichments showed a 95%

similarity to *Variovorax* sp. that was also found in enrichments using 2,4-dichlorophenoxyacetate (Kamagata et al. 1997; see also Section 4.6.4). The effect of the herbicide applications had therefore affected both the composition of the bacteria flora and the metabolic capabilities of its components.

5.8.2 Nondirected Examination of Natural Populations

1. Core samples from three redox zones at depths of 6.1, 7.3, and 9.2 m in an aquifer contaminated with hydrocarbons and chlorinated were examined (Dojka et al. 1998). Small-subunit rRNA genes from core DNA were amplified directly by PCR with Bacteria- or Archaea-specific primers and cloned, and the clones screened by restriction fragment length polymorphisms (RFLP). Analysis of the sequences showed the existence of (a) 10 types with no known taxonomic divisions, (b) 21 types with no cultured representatives, and (c) 63 types with recognized divisions. The sequences were classified into seven major groups, and it was shown that the two most abundant sequence types could be correlated with sequences of *Methanosaeta* sp. and *Syntrophus* sp., respectively. It was proposed that hydrocarbon degradation proceeded by oxidation of the hydrocarbons, fermentation of the resulting carboxylic acids by organisms with sequences related to *Syntrophus* sp., followed by aceticlastic methanogenesis by organisms related in sequence to *Methanosaeta* sp.

2. Small-subunit rRNA genes were amplified directly by PCR from DNA from a sediment sample from Yellowstone National Park and cloned: universally conserved or Bacteria-specific rDNA primers were used (Hugenholtz et al. 1998). RFLP was used to classify rDNA fragments. Most of the sequences were representatives of established bacterial division, although 30% were not so related. Database matches suggested the presence of organisms closely related to *Thermodesulfovibrio yellowstonii* that would be able to carry out anaerobic sulfate reduction, to the organotrophs *Thermus* sp. and *Dictyoglomus thermophilum*, and to the hydrogen-oxidizing *Calderiobacterium hydrogenophilum*. In contrast to previous perceptions, however, members of Bacteria dominated over Archaea in this sediment, so that the ecological boundaries between Bacteria and Archaea are becoming less clearly defined.

5.8.3 Examination for Established Metabolites or Specific Enzymes

Metabolically related parameters have also been used directly, and these include establishing any of the following.

1. The presence of metabolites determined from laboratory experiments of degradation pathways. Examples include *cis*-dihydrodiols of PAHs in a marine sediment (Li et al. 1996) and of naphthalene in leachate from a contaminated site (Wilson and Madsen 1996); chlorobenzoates from the degradation of PCBs (Flanagan and May 1993); and benzyl succinates from BTEX (Beller et al. 1995).

2. Examples have been given in Section 5.7 of enzymes with a low substrate specificity that are able to accept analogue substrates with the formation of colored metabolites. This may be used to detect enzymatic activity toward the desired substrates when a large number of colonies have to be examined. Examples for PAHs include toluene 2,3-dioxygenase (Eaton and Chapman 1995), naphthalene dioxygenase (Ensley et al. 1983), and manganese peroxidase (Bogan et al. 1996); for trichloroethene cometabolism toluene-2-monooxygenase (Fries et al. 1997b); and for degradation of 2,4-dichlorophenoxyacetate, 2,4-dichlorophenoxyacetate/2-ketoglutarate dioxygenase that can accept 4-nitrophenoxyacetate as a substrate with the formation of colored 4-nitrophenol (Sassenella et al. 1997). Care ought, however, to be exercised since it was also shown in the last example that not all strains with established degradative capability for 2,4-dichlorophenoxyacetate were able to form 4-nitrophenol from the surrogate substrate (Sassenella et al. 1997).

3. On the basis of examination of a collection of strains of *Sphingomonas* sp. from a range of localities, it has been suggested that the glutathione *S*-transferase-encoding gene might be used as a marker for PAH-degrading bacteria (Lloyd-Jones and Lau 1997).

References

Aggarwal, P.K., M.E. Fuller, M.M. Gurgas, J.F. Manning, and M.A. Dillon. 1997. Use of stable oxygen and carbon isotope analyses for monitoring the pathways and rates of intrinsic and enhanced *in situ* biodegradation. *Environ. Sci. Technol.* 31: 590–596.

Alexander, M. 1975. Environmental and microbiological problems arising from recalcitrant molecules. *Microbial Ecol.* 2: 17–27.

Allard, A.-S., C. Lindgren, P.-Å. Hynning, M. Remberger, and A.H. Neilson. 1991. Dechlorination of chlorocatechols by stable enrichment cultures of anaerobic bacteria. *Appl. Environ. Microbiol.* 57: 77–84.

Allard, A.-S., P.-Å. Hynning, M. Remberger, and A.H. Neilson. 1994. Bioavailability of chlorocatechols in naturally contaminated sediment samples and of chloroguaiacols covalently bound to C_2-guaiacyl residues. *Appl. Environ. Microbiol.* 60: 777–784.

Allard, A.-S., L. Renberg, and A.H. Neilson. 1996. Absence of $^{14}CO_2$ evolution from ^{14}C-labelled EDTA and DTPA and the sediment/water partition ratio using a sediment sample. *Chemosphere* 33: 577–583.

Allen, J.R. and S.A. Ensign. 1996. Carboxylation of epoxides to β-keto acids in cell extracts of *Xanthobacter* strain Py2. *J. Bacteriol.* 178: 1469–1472.

Almeida, M., M. Humanes, R. Melo, J.A.Silva, J.J.R. F da Silva, H. Vilter, and R. Weaver. 1998. *Sacchorhiza polyschides* (Phaeophycea: Phyllariacea) a new source for vanadium-dependent haloperoxidases. *Phytochem.* 48: 229–239.

Anders, H.-J., A. Kaetzke, P. Kämpfer, W. Ludwig, and G. Fuchs. 1995. Taxonomic position of aromatic-degrading denitrifying pseudomonad strains K 172 and KB 740 and their description as new members of the genera *Thauera*, as *Thauera aromatica* sp.nov., and *Azoarcus*, as *Azoarcus evansii* sp. nov., respectively, members of the beta subclass of the Protobacteria. *Int. J. Syst. Bacteriol.* 45: 327–333.

Angle, J.S., S.P. McGrath, and R.L. Chaney. 1991. New culture medium containing ionic concentrations of nutrients similar to concentrations found in soil solution. *Appl. Environ. Microbiol.* 57: 3674–3676.

Apajalahti, J.H.A. and M.S. Salkinoja-Salonen. 1986. Degradation of polychlorinated phenols by *Rhodococcus chlorophenolicus*. *Appl. Microbiol. Biotechnol.* 25: 62–67.

Argyropoulos, D.S. and L. Zhang. 1998. Semiquantitative determination of quinonoid structures in isolated lignins by ^{31}P nuclear magnetic resonance. *J. Agric. Food Chem.* 46: 4626–4634.

Arnon, D.I. 1938. Microelements in culture-solution experiments with higher plants. *Am. J. Bot.* 25: 322–325.

Aulin, G., H.-J. Busse, T. Egli, T. El-Banna, and E. Stackebrandt. 1993. Description of the Gram-negative, obligately aerobic, nitrilotriacetate (NTA)-utilizing bacteriia as *Chelatobacter heintzii*, gen. nov., sp. nov., and *Chelatococcus asaccharovorans*, gen. nov., sp. nov. *Syst. Appl. Microbiol.* 16: 104–112.

Balows, A., H.G. Trüper, M. Dworkin, W. Harder, and K.-H. Schleifer. (Eds.) *The Prokaryotes*. Springer-Verlag, Heidelberg.

Bartholomew, G.W. and F.K. Pfaender. 1983. Influence of spatial and temporal vari ations on organic pollutant biodegradation rates in an estuarine environment. *Appl. Environ. Microbiol.* 45: 103–109.

Baruah, J.N., Y. Alroy, and R.I. Mateles. 1967. Incorporation of liquid hydrocarbons into agar media. *Appl. Microbiol.* 15: 961.

Battersby, N.S. 1990. A review of biodegradation kinetics in the aquatic environment. *Chemosphere* 21: 1243–1284.

Battersby, N.S. and V. Wilson. 1989. Survey of the anaerobic biodegradation potential of organic chemicals in digesting sludge. *Appl. Environ. Microbiol.* 55: 433–439.

Beller, H.R., W.-S. Ding, and M. Reinhard. 1995. Byproducts of anaerobic alkylbenzene metabolism useful as indicators of *in situ* bioremediation. *Environ. Sci. Technol.* 29: 2864–2870

Beller, H.R., A.M. Spormann, P.K. Sharma, J.R. Cole, and M. Reinhard. 1996. Isolation and characterization of a novel toluene-degrading sulfate-reducing bacterium. *Appl. Environ. Microbiol.* 62: 1188–1196

Bertram, P.A., R.A. Schmitz, D. Linder, and R.K. Thauer. 1994. Tungsten can substitute for molybdate in sustaining growth of *Methanobacterium thermoautotrophicum*: identification and characterization of a tungsten isoenzyme of formylmethanofuran dehydrogenase. *Arch. Microbiol.* 161: 220–228.

Birch, R.R. and R.J. Fletcher. 1991. The application of dissolved inorganic carbon measurements to the study of aerobic biodegradability. *Chemosphere* 23: 507–524.

Birch, R.R., C. Biver, R. Campagna, W.E. Gledhill, and U. Pagga. 1989. Screening chemicals for anaerobic degradability. *Chemosphere* 19: 1527–1550.

Boersma, M.G., T.Y. Dinarieva, W.J. Middelhoven, W.J.H. van Berkel, J. Doran, J. Vervoort, and I.M.-C.M. Rietjens. 1998. [19]F nuclear magnetic resonance as a tool to investigate microbial degradation of fluorophanols to fluorocatechols and fluoromuconates. *Appl. Environ. Microbiol.* 64: 1256–1263.

Bogardt, A.H. and B.B. Hemmingsen. 1992. Enumeration of phenanthrene-degrading bacteria by an overlay technique and its use in evaluation of petroleum-contaminated sites. *Appl. Environ. Microbiol.* 58: 2579–2582.

Bonnarme, P. and T.W. Jeffries. 1990. Mn(II) regulation of lignin peroxidases and manganese-dependent peroxidases from lignin-degrading white-rot fungi. *Appl. Environ. Microbiol.* 56: 210–217.

Brown, J.A., J.K. Glenn, and M.H. Gold. 1990. Manganese regulates expression of manganese peroxidase by *Phanerochaete chrysosporium*. *J. Bacteriol.* 172: 3125–3130.

Brubaker, W.W. and R.A. Hites. 1998. OH reaction kinetics of gas-phase α- and γ-hexachlorocyclohexane and hexachlorobenzene. *Environ. Sci. Technol.* 32: 766–769.

Bruhn, C., H. Lenks, and H.-J. Knackmuss. 1987. Nitrosubstituted aromatic compounds as nitrogen source for bacteria. *Appl. Environ. Microbiol.* 53: 208–210.

Bryant, F.O., D.D. Hale, and J.E. Rogers. 1991. Regiospecific dechlorination of pentachlorphenol by dichlorophenol-adapted microorganisms in freshwater anaerobic sediment slurries. *Appl. Environ. Microbiol.* 57: 2293–2301.

Buckel, W. and H.A. Barker. 1974. Two pathways of glutamate fermentation by anaerobic bacteria. *J. Bacteriol.* 117: 1248–1260.

Burland, S.M. and E.A. Edwards. 1999. Anaerobic benzene biodegradation linked to nitrate reduction. *Appl. Environ. Microbiol.* 65: 529–533.

Buser, H.-R. and M.D. Müller. 1997. Conversion reactions of various phenoxyalkanoic acid herbicides in soil. 2. Elucidation of the enantiomerization process of chiral phenoxy acids from incubation in a D_2O/soil system. *Environ. Sci. Technol.* 31: 1960–1967.

Busse, H.-J., T. El-Banna, H. Oyaizu, and G. Auling. 1991. Identification of xenobiotic-degrading isolates from the beta subclass of the Proteobacteria by a polyphasic approach including 16S rRNA sequencing. *Int. J. Syst. Bacteriol.* 42: 19–26.

Butler, A. 1998. Vanadium haloperoxidases. *Curr. Opinion Chem. Biol.* 2: 279–285.

Castro, C.E., D.M. Riebeth, and N.O. Belser. 1992a. Biodehalogenation: the metabolism of vinyl chloride by *Methylosinus trichosporium* OB-3b. A sequential oxidative and reductive pathway through chloroethylene oxide. *Environ. Toxicol. Chem.* 11: 749–755.

Castro, C.E., R.S. Wade, D.M. Riebeth, E.W. Bartnicki, and N.O. Belser. 1992b. Biodehalogenation: rapid metabolism of vinyl chloride by a soil *Pseudmonas* sp. Direct hydrolysis of a vinyl C–Cl bond. *Environ. Toxicol. Chem.* 11: 757–764

Castro, C.E., S.K. O'Shea, W. Wang, and E.W. Bartnicki. 1996. Biodehalogenation: oxidative and hydrolytic pathways in the transformations of acetonitrile, chloroacetonitrile, chloroacetic acid, and chloroacetamide by *Methylosinus trichosporium* OB-3b. *Environ. Sci. Technol.* 30: 1180–1184.

Chen, Y.P., A.R. Glenn, and M.J. Dilworth. 1985. Aromatic metabolism in *Rhizobium trifolii* — catechol 1,2-dioxygenase. *Arch. Microbiol.* 141: 225–228.

Claus, D. and N. Walker. 1964. The decomposition of toluene by soil bacteria. *J. Gen. Microbiol.* 36: 107–122.

Collins, J. and G. Hegeman. 1984. Benzyl alcohol metabolism by *Pseudomonas putida*: a paradox resolved. *Arch. Microbiol.* 138: 153–160.

Collins, M.D., P.A. Lawson, A. Willems, J.J. Cordoba, J. Fernandez-Garayzabal, P. Garcia, J. Cai, H. Hippe, and J.A.E. Farrow. 1994. The phylogeny of the genus *Clostridium*: proposal of five new genera and eleven new species combinations. *Int. J. Syst. Bacteriol.* 44: 812–826.

Cook, A.M. and R. Hütter. 1982. Ametryne and prometryne as sulfur sources for bacteria. *Appl. Environ. Microbiol.* 43: 781–786.

Cook, A.M. and R. Hütter. 1984. Deethylsimazine: bacterial dechlorination, deamination and complete degradation. *J. Agric. Food Chem.* 32: 581–585.

Cook, A.M., C.G. Daughon, and M. Alexander. 1978. Phosphonate utilization by bacteria. *J. Bacteriol.* 133: 85–90.

Combourieu, B., P. Besse, M. Sancelme, H. Veschambre, A.M. Delort, P. Poupin, and N. Truffaut. 1998. Morpholine degradation pathway of *Mycobacterium aurum* MO1: direct evidence of intermediates by *in situ* ^{1}H nuclear magnetic resonance. *Appl. Environ. Microbiol.* 64: 153–158.

Cutter, L., K.R. Sowers, and H.D. May. 1998. Microbial dechlorination of 2,3,5,6-tetrachlorobiphenyl under anaerobic conditions in the absence of soil or sediment. *Appl. Environ. Microbiol.* 64: 2966–2969.

Dalton, H., B.T. Golding, B.W. Waters, R. Higgins, and J.A. Taylor. 1981. Oxidations of cyclopropane, methylcyclopropane, and arenes with the mono-oxygenase system from *Methylococcus capsulatus*. *J. Chem. Soc. Chem. Commun.* 482–483.

Daughney, C.J. and J.B. Fein. 1998. Sorption of 2,4,6-trichlorophenol by *Bacillus subtilis*. *Environ. Sci. Technol.* 32: 749–752.

Daughton, C.G. and D.P.H. Hsieh. 1977. Parathion utilization by bacterial symbionts in a chemostat. *Appl. Environ. Microbiol.* 34: 175–184.

de Bruijn, F.J. 1992. Use of repetitive (repetitive extragenic palindromic and enterobacterial repetitive intergeneric consensus) sequences and the polymerase chain reaction to fingerprint the genomes of *Rhizobium meliloti* isolates and other soil bacteria. *Appl. Environ. Microbiol.* 58: 2180–2187.

De Morsier, A., J. Blok, P. Gerike, L. Reynolds, H. Wellens, and W.J. Bontinck. 1987. Biodegradation tests for poorly-soluble compounds. *Chemosphere* 16: 269–277.

Den Hollander, J.A., T.R. Brown, K. Ugurbil, and R.G. Shulman. 1986. Studies of anaerobic and aerobic glycolysis in *Saccharomyces cerevisiae*. *Biochemistry* 25: 203–211.

Di Corcia, A., C. Crescenzi, A. Marcomini, and R. Samperi. 1998. Liquid-chromatography–electrospray–mass spectrometry as a valuable tool for characterizing biodegradation intermediates of branched alcohol ethoxylate surfactants. *Environ. Sci. Technol.* 32: 711–718.

Dietrich, D., W.J. Hickey, and R. Lamar. 1995. degradation of 4,4'-dichlorobiphenyl, 3,3',4,4'-tetrachlorobiphenyl, and 2,2',4,4',5,5'hexachlorobiphenyl by the white rot fungus *Phanerochaete chrysosporium*. *Appl. Environ. Microbiol.* 61: 3904–3909.

Dilworth, M.J. and R.R. Eady. 1991. Hydrazine is a product of dinitrogen reduction by the vanadium-nitrogenase from *Azotobacter chroococcum*. *Biochem. J.* 277: 465–468.

Dojka, M.A., P. Hugenholtz, S.K. Haack, and N.R. Pace. 1998. Microbial diversity in a hydrocarbon- and chlorinated–solvent–contaminated aquifer undergoing intrinsic bioremediation. *Appl. Environ. Microbiol.* 64: 3869–3877.

Don, R.H., A.J. Wightman, H.H. Knackmuss, and K.N. Timmis. 1985. Transposon mutagenesis and cloning analysis of the pathways of degradation of 2,4-dichlorophenoxyacetic acid and 3-chlorobenzoate in *Alcaligenes eutrophus* JMP134(pJP4). *J. Bacteriol.* 161: 85–90.

Doudoroff, M. 1940. The oxidative assimilation of sugars and related substances by *Pseudomonas saccharophila* with a contribution to the direct respiration of di- and poly-saccharides. *Enzymologia* 9: 59–72.

Eaton, R.W. and Chapman, P. 1995. Formation of indigo and related compounds from indolecarboxylic acids by aromatic acid-degrading bacteria: chromogenic reactions for cloning genes encoding dioxygenases that act on aromatic acids. *J. Bacteriol.* 177: 6983–6988.

Eaton, S.L., S.M. Resnick, and D.T. Gibson. 1996. Initial reactions in the oxidation of 1,2-dihydronaphthalene by *Sphingomonas yanoikuyae* strains. *Appl. Environ. Microbiol.* 62: 4388–4394.

Edwards, E.A., L.E. Williams, M. Reinhard, and D. Grbic-Galic. 1992. Anaerobic degradation of toluene and xylene by aquifer microrganisms under sulfate-reducing conditions. *Appl. Environ. Microbiol.* 58: 794–800

El Fantroussi, S., L. Verschuere, W. Verstraete, and E.M. Top. 1999. Effect of phenylurea herbicides on soil microbial communities estimated by analysis of 16S rRNA gene fingerprints and community-level physiological profiles. *Appl. Environ. Microbiol.* 65: 982–988.

Ensign, S.A., F.J. Smakk, J.R. Allen, and M.K. Sluis. 1998. New roles for CO_2 in the microbial metabolism of aliphatic epoxides and ketones. *Arch. Microbiol.* 169: 179–187.

Ensley, B.D., Ratzken, B.J., Osslund, T.D., Simon, M.J., Wackett, L.P., and. Gibson, D.T. 1983. Expression of naphthalene oxidation genes in *Escherichi coli* results in the biosynthesis of indigo. *Science* 222, 167–169.

Eppink, M.H.M., S.A. Boeren, J. Vervoort, and W.J.H. van Berkel. 1997. Purification and properties of 4-hydroxybenzoate 1-hydroxylase (decarboxylating), a novel flavin adenein dinucleotide-dependent monooxygenase from *Candida parapilosis* CBS604. *J. Bacteriol.* 179: 6680–6687.

Evans, P.J., W. Ling, B. Goldschmidt, E.R. Ritter, and L.Y. Young. 1992. Metabolites formed during anaerobic transformation of toluene and o-xylene and their proposed relationship to the initial steps of toluene mineralization. *Appl. Environ. Microbiol.* 58: 496–501

Fallik, E., Y.-K. Chan, and R.L. Robson. 1991. Detection of alternative nitrogenases in aerobic Gram-negative nitrogen-fixing bacteria. *J. Bacteriol.* 173: 365–371.

Feldman, P.L., O.W. Griffith, and D.J. Stuehr. 1993. The surprising life of nitric oxide. *Chem. Eng. News* 71 (51): 26–39.

Finette, B.A., V. Subramanian, and D.T. Gibson. 1984. Isolation and characterization of *Pseudomonas putida* PpF1 mutants defective in the toluene dioxygenase enzyme system. *J. Bacteriol.* 160: 1003–1009.

Firestone, M.K. and J.M. Tiedje. 1975. Biodegradation of metal-nitrilotriacetate complexes by a *Pseudomonas* species: mechanism of reaction. *Appl. Microbiol.* 29: 758–764.

Flanagan, W.P. and R.J. May. 1993. Metabolite detection as evidence for naturally occurring aerobic PCB degradation in Hudson River sediments. *Environ. Sci. Technol.* 27: 2207–2212.

Francis, A.J. and C.J. Dodge. 1993. Influence of complex structure on the biodegradation of iron-citrate complexes. *Appl. Environ. Microbiol.* 59: 109–113.

Fulthorpe, R.R. and R.C. Wyndham. 1989. Survival and activity of a 3-chlorobenzoate-catabolic genotype in a natural system. *Appl. Environ. Microbiol.* 55: 1584–1590.

Fulthorpe, R.R., A.N. Rhodes, and J.M. Tiedje. 1998. High levels of endemicity of 3-chlorobenzoate-degrading soil bacteria. *Appl. Environ. Microbiol.* 64: 1620–1627.

Gerike, P. and W.K. Fischer. 1981. A correlation study of biodegradability determinations with various chemicals in various tests. II. Additional results and conclusions. *Ecotoxicol. Environ. Saf.* 5: 45–55.

Gill, C.O. and C. Ratledge. 1972. Toxicity of *n*-alkanes, *n*-alkenes, *n*-alkan-1-ols and *n*-alkyl-1-bromides towards yeasts. *J. Gen. Microbiol.* 72: 165–172.

Glaus, M.A., C.G. Heijman, R.P. Schwarzenbach, and J. Zeyer. 1992. Reduction of nitroaromatic compounds mediated by *Streptomyces* sp. exudes. *Appl. Environ. Microbiol.* 58: 1945–1951.

Griffiths, E.T., S.G. Hales, N.J. Russell, G.K. Watson, and G.F. White. 1986. Metabolite production during biodegradation of the surfactant sodium dodecyltriethoxy sulphate under mixed-culture die-away conditions. *J. Gen. Microbiol.* 132: 963–972.

Grossenbacher, H., C. Horn, A.M. Cook, and R. Hütter. 1984. 2-Chloro-4-amino-1,3,5-triazine-6(5*H*)-one: a new intermediate in the biodegradation of chlorinated *s*-triazines. *Appl. Environ. Microbiol.* 48: 451–453.

Grund, E., B. Denecke, and R. Eichenlaub. 1992. Naphthalene degradation via salicylate and gentisate by *Rhodococcus* sp. strain B4. *Appl. Environ. Microbiol.* 58: 1874–1877.

Guerin, T.F. and I.R. Kennedy. 1992. Distribution and dissipation of endosulfan and related cyclodienes in sterile aqueous systems: implication for studies on biodegradation. *J. Agric. Food Chem.* 40: 2315–2323.

Guthrie, E.A. and F.K. Pfaender. 1998. Reduced pyrene bioavailability in microbially active soils. *Environ. Sci. Technol.* 32: 501–508.

Hales, S.G., K.S. Dodgson, G.F. White, N. Jones, and G.K. Watson. 1982. Initial stages in the biodegradation of the surfactant sodium dodecyltriethoxy sulfate by *Pseudomonas* sp. strain DES1. *Appl. Environ. Microbiol.* 44: 790–800.

Hales, S.G., G.K. Watson, K.S. Dodgson, and G.F. White. 1986. A comparative study of the biodegradation of the surfactant sodium dodecyltriethoxy sulphate by four detergent-degrading bacteria. *J. Gen. Microbiol.* 132: 953–961.

Halpin, R.A., G.D. Hegeman, and G.L. Kenyon. 1981. Carbon-13 nuclear magnetic resonance studies of mandelate metabolism in whole bacterial cells and in isolated, *in vivo* cross-linked enzyme complexes. *Biochemistry* 20: 1525–1533.

Harris, R., and C.J. Knowles. 1983. Isolation and growth of a *Pseudomonas* species that utilizes cyanide as a source of nitrogen. *J. Gen. Microbiol.* 129: 1005–1011.

Harwood, C.S., N.N. Nichols, M.-K. Kim, J.J. Ditty, and R.E. Parales. 1994. Identification of the *pcaRKF* gene cluster from *Pseudomonas putida*: involvement in chemotaxis, biodegradation, and transport of 4-hydroxybenzoate. *J. Bacteriol.* 176: 6479–6488.

Havel, J. and W. Reineke. 1993. Microbial degradation of chlorinated acetophenones. *Appl. Environ. Microbiol.* 59: 2706–2712.

Hay, A.G. and D.D. Focht. 1998. Cometabolism of 1,1-dichloro-2,2-bis(4-chlorophenyl)-ethylene by *Pseudomomas acidovorans* M3GY grown on biphenyl. *Appl. Environ. Microbiol.* 64: 2141–2146.

Hegeman, G.D. 1966a. Synthesis of the enzymes of the mandelate pathway by *Pseudomonas putida*. I. Synthesis of enzymes of the wild type. *J. Bacteriol.* 91: 1140–1154.

Hegeman, G.D. 1966b. Synthesis of the enzymes of the mandelate pathway by *Pseudomonas putida*. II. Isolation and properties of blocked mutants. *J. Bacteriol.* 91: 1155–1160.

Heim, K., I. Schuphan, and B. Schmidt. 1994. Behaviour of [^{14}C]-4-nitrophenol and [^{14}C]-3,4-dichloroaniline in lab sediment-water systems. 1. Metabolic fate and partitioning of radioactivity. *Environ. Toxicol. Chem.* 13: 879–888.

Heitkamp, M.A. and C.E. Cerniglia. 1989. Polycyclic aromatic hydrocarbon degradation by a *Mycobacterium* sp. in microcosms containing sediment and water from a pristine ecosystem. *Appl. Environ. Microbiol.* 55: 1968–1973.

Heitkamp, M.A., J.P. Freeman, and C.E. Cerniglia. 1986. Biodegradation of *tert*-butylphenyl diphenyl phosphate. *Appl. Environ. Microbiol.* 51: 316–322.

Heitkamp, M.A., J.P. Freeman, and C.E. Cerniglia. 1987. Naphthalene biodegradation in environmental microcosms: estimates of degradation rates and characterization of metabolites. *Appl. Environ. Microbiol.* 53: 129–136.

Heitkamp, M.A., V. Camel, T.J. Reuter, and W.J. Adams. 1990. Biodegradation of *p*-nitrophenol in an aqueous waste stream by immobilized bacteria. *Appl. Environ. Microbiol.* 56: 2967–2973.

Hensgens, C.M.H., W.R. Hagen, and T.A. Hansen. 1995. Purification and characterization of a benzylviologen-linked, tungsten-containing aldehyde oxidoreductase from *Desulfovibrio gigas*. *J. Bacteriol.* 177: 6195–6200.

Higson, F.K. and D.D. Focht. 1990. Bacterial degradation of ring-chlorinated acetophenones. *Appl. Environ. Microbiol.* 56: 3678–3685.

Holoman, T.R.P., M.A. Elberson, L.A. Cutter, H.D. May, and K.R. Sowers. 1998. Characterization of a defined 2,3,5,6-tetrachlorobiphenyl-*ortho*-dechlorinating microbial community by comparative sequence analysis of genes coding for 16S rRNA. *Appl. Environ. Microbiol.* 64: 3359–3367.

Holt, B.D., N.C. Sturchio, T.A. Abrajano, and L.J. Heraty. 1997. Conversion of chlorinated volatile organic compounds to carbon dioxide and methyl chloride for isotopic analysis of carbon and chlorine. *Anal. Chem.* 69: 2727–2733.

Houwen, F.P., C. Dijkema, A.J.M. Stams, and A.J.B. Zehnder. 1991. Propionate metabolism in anaerobic bacteria; determination of carboxylation reactions with ^{13}C-NMR spectroscopy. *Biochim. Biophys. Acta* 1056: 126–132.

Howard, P.H. and S. Banerjee. 1984. Interpreting results from biodegradability tests of chemicals in water and soil. *Environ. Toxicol. Chem.* 3: 551–562.

Huckins, J.N., J.D. Petty, and M.A. Heitkamp. 1984. Modular containers for microcosm and process model studies on the fate and effects of aquatic contaminants. *Chemosphere* 13: 1329–1341.

Hugenholtz, P., C. Pitulle, K.L. Hershberger, and N.R. Pace. 1998. Novel division level bacterial diversity in a Yellowstone hot spring. *J. Bacteriol.* 180: 366–376.

Hungate, R. E. 1969. A roll tube method for cultivation of strict anaerobes. In *Methods in Microbiology* (Eds. Norris and D.W. Ribbons), Vol. 3B p 117–132. Academic Press, New York.

Hurst, G.B., K. Weaver, M.J. Doktycz, M.V. Buchanan, A.M. Costello, and M.E. Lidstrom. 1998. MALDI-TOF analysis of polymerase chain reaction products from methanotrophic bacteria. *Anal. Chem.* 70: 2693–2698.

Jagnow, G., K. Haider, and P.-C. Ellwardt. 1977. Anaerobic dechlorination and degradation of hexachlorocyclohexane by anaerobic and facultatively anaerobic bacteria. *Arch. Microbiol.* 115: 285–292.

Jain, R.K., G.S. Sayler, J.T. Wilson, L. Houston, and D. Pacia. 1987. Maintenance and stability of introduced genotypes in groundwater aquifer material. *Appl. Environ. Microbiol.* 53: 996–1002.

Janssen, P.H. and P. Frenzel. 1997. Inhibition of methanogenesis by methyl fluoride: studies of pure and defined mixed cultures of anaerobic bacteria and archaea. *Appl. Environ. Microbiol.* 63: 4552–4557.

Jiménez, L., A. Breen, N. Thomas, T.W. Federle, and G.S. Sayler. 1991. Mineralization of linear alkylbenzene sulfonate by a four-member aerobic bacterial consortium. *Appl. Environ. Microbiol.* 57: 1566–1569.

Joblin, K.N. 1981. Isolation, enumeration, and maintenance of rumen anaerobic fungi in roll tubes. *Appl. Environ. Microbiol.* 42: 1119–1122.

Jokela, J.K. and M. Salkinoja-Salonen. 1992. Molecular weight distributions of organic halogens in bleached kraft pulp mill effluents. *Environ. Sci. Technol.* 26: 1190–1197.

Jones, J.B. and T.C. Stadtman. 1981. Selenium-dependent and selenium-independent formate dehydrogenase of *Methanococcus vannielii*. Separation of the two forms and characterization of the purified selenium-independent form. *J. Biol. Chem.* 256: 656–663.

Jordan, S.L., I.R. McDonald, A.J. Kraczkiewicz-Dowjat, D.P. Kelly, F.A. Rainey, J-.C. Murrell, and A.P. Wood. 1997. Autotrophic growth on carbon disulfide is a property of novel strains of *Paracoccus denitrificans*. *Arch. Microbiol.* 168: 225–236.

Joshi-Tope, G. and A.J. Francis. 1995. Mechanisms of biodegradation of metal-citrate complexes by *Pseudomonas fluorescens*. *J. Bacteriol.* 177: 1989–1993.

Kamagata, Y., R.R. Fulthorpe, K. Tamura, H. Takami, L.J. Forney, and J.M. Tiedje. 1997. Pristine environments harbor a new group of oligotrophic 2,4-dichlorophenoxyacetic acid-degrading bacteria. *Appl. Environ. Microbiol.* 63: 2266–2272.

Kari, F.G. and W. Giger. 1995. Modelling the photochemical degradation of ethylenediaminetetraacetate in the River Glatt. *Environ. Sci. Technol.* 29: 2814–2827.

Kawaguchi, A., S. Tsubotani, Y. Seyama, T. Yamakawa, T. Osumi, T. Hashimoto, Y. Kikuchi, M. Ando, and S. Okuda. 1980. Stereochemistry of dehydrogenation catalyzed by acyl-CoA oxidase. *J. Biochem.* 88: 1481–1486.

Kazumi, J., M.E. Caldwell, J.M. Suflita, D.R. Lovley, and L.Y. Young. 1997. Anaerobic degradation of benzene in diverse anoxic environments. *Environ. Sci. Technol.* 31: 813–818.

Kersters, K., W. Ludwig, M. Vancanneyt, P. de Vos, M. Gillis, and K.-H. Schleifer. 1996. *Syst. Appl. Microbiol.* 19: 465–477.

Khan, A.A., R.-F. Wang, W.-W. Cao, W. Franklin, and C.E. Cerniglia. 1996. Reclassification of a polycyclic aromatic hydrocarbon-metabolizing bacterium, *Beijerinckia* sp. strain B1 as *Sphingomonas yanoikuyae* by fatty acid analysis, protein pattern analysis, DNA-DNA hybridization, and 16S ribosomal DNA sequencing. *Int. J. Syst. Bacteriol.* 46: 466–469.

King, G.F., D.J. Richardson, J.B. Jackson, and S.J. Ferguson. 1987. Dimethyl sulfoxide and trimethylamine-*N*-oxide as bacterial electron acceptors: use of nuclear magnetic resonance to assay and characterize the reductase system in *Rhodobacter capsulatus*. *Arch. Microbiol.* 149: 47–51.

Kitts, C.L., J.P. Lapointe, V.T. Lam, and R.A. Ludwig. 1992. Elucidation of the complete *Azorhizobium* nicotinate catabolism pathway. *J. Bacteriol.* 174: 7791–7797.

Klecka, G.M. and D.T. Gibson. 1979. Metabolism of dibenzo[1,4]dioxan by a *Pseudomonas* species. *Biochem. J.* 180: 639–645.

Korth, H.-G., K.U. Ingold, R. Sustmann, H. de Groot, and H. Sies. 1992. Tetramethyl-*ortho*-chinodimethan (NOCT-1), das erste Mitglied einer Familie maßge-schneiderter cheletroper Spinfänger für Stickstoffmonoxid. *Angew. Chem.* 104: 915–917.

Lauff, J.J., D.B. Steele, L.A. Coogan, and J.M. Breitfeller. 1990. Degradation of the ferric chelate of EDTA by a pure culture of an *Agrobacterium* sp. *Appl. Environ. Microbiol.* 56: 3346–3353.

Leive, L. 1968. Studies on the permeability change produced in coliform bacteria by ethylenediaminetetraacetate. *J. Biol. Chem.* 243: 2373–2380.

Lewis, T.A. and R.L. Crawford. 1993. Physiological factors affecting carbon tetrachlo-ride dehalogenation by the denitrifying bacterium *Pseudomonas* sp. strain KC. *Appl. Environ. Microbiol.* 59: 1635–1641.

Lewis, T.A. and R.L. Crawford. 1995. Transformation of carbon tetrachloride via sulfur and oxygen substitution by *Pseudomonas* sp. strain KC. *J. Bacteriol.* 177: 2204–2208.

Li, X-F., X.-C. Le, C.D. Simpson, W.R. Cullen, and K.J. Reimer. 1996. Bacterial trans-formation of pyrene in a marine environment. *Environ. Sci. Technol.* 30: 1115–1119.

Liu, D., W.M.J. Strachan, K. Thomson, and K. Kwasniewska. 1981. Determination of the biodegradability of organic compounds. *Environ. Sci. Technol.* 15: 788–793.

Löffler, F.E., K.M. Ritalahti, and J.M. Tiedje. 1997. Dechlorination of chloroethenes is inhibited by 2-bromoethanesulfonate in the absence of methanogens. *Appl. En-viron. Microbiol.* 63: 4982–4985.

Londry, K.L. and P.M. Fedorak. 1993. Use of fluorinated compounds to detect aro-matic metabolites from *m*-cresol in a methanogenic consortium: evidence for a demethylation reaction. *Appl. Environ. Microbiol.* 59: 2229–2238.

Madsen, E.L. 1991. Determining *in situ* biodegradation. Facts and challenges. *Environ. Sci. Technol.* 25: 1663–1673.

Maltseva, O. and P. Oriel. 1997. Monitoring of an alkaline 2,4,6-trichlorophenol-degrading enrichment culture by DNA fingerprinting methods and isolation of the responsible organism, haloalkaliphilic *Nocardioides* sp. strain M6. *Appl. En-viron. Microbiol.* 63: 4145–4149.

Mandelbaum, R.T., L.P. Wackett, and D.L. Allan. 1993. Mineralization of the *s*-triazine ring of atrazine by stable bacterial mixed cultures. *Appl. Environ. Microbiol.* 59: 1695–1701.

Marchesi, J.R., N.J. Russel, G.F. White, and W.A. House. 1991. Effects of surfactant adsorption and biodegradability on the distribution of bacteria between sedi-ments and water in a freshwater microcosm. *Appl. Environ. Microbiol.* 57: 2507–2513.

Maron, D.M. and B.N. Ames. 1983. Revised methods for the Salmonella mutagenicity test. *Mutat. Res.* 113: 173–215.

McCormick, M.L., G.R. Buettner, and B.E. Britigan. 1998. Endogenous superoxide dismutase levels regulate iron-dependent hydroxyl radical formation in *Escher-ichia coli* exposed to hydrogen peroxide. *J. Bacteriol.* 180: 622–625.

Migaud, M.E., J.C. Chee-Sandford, J.M. Tiedje, and J.W. Frost. 1996. Benzylfumaric, benzylmaleic, and Z- and E-phenylitaconic acids: synthesis, characterization, and correlation with a metabolite generated by *Azoarcus tolulyticus* Tol-4 during anaeribic toluene degradation. *Appl. Environ. Microbiol.* 62: 974–978

Mihelcic, J.R. and R.G. Luthy. 1988. Microbial degradation of acenaphthene and naphthalene under denitrification conditions in soil-water systems. *Appl. Envi-ron. Microbiol.* 54: 1188–1198.

Miller, J.H. 1977. *Experiments in Molecular Genetics*. Cold Spring Harbor Laboratories, Cold Spring Habor, NY.

Miller, L.G., C. Sasson, and R.S. Oremland. 1998. Difluoromethane, a new and improved inhibitor of methanotrophy. *Appl. Environ. Microbiol.* 64: 4357–4362.

Miller, T.L. and M.J. Wolin. 1974. A serum bottle modification of the Hungate technique for cultivating obligate anaerobes. *Appl. Microbiol.* 27: 985–987.

Morawaski, B., G. Casy, C. Illaszewicz, H. Griengl, and D.W. Ribbons. 1997. Stereochemical course of two arene-*cis*-diol degydrogenases specifically induced in *Pseudomonas putida*. *J. Bacteriol.* 179: 4023–4029.

Munnecke, D.M. and D.P.H. Hsieh. 1976. Pathways of microbial metabolism of parathion. *Appl. Environ. Microbiol.* 31: 63–69.

Narbad, A., M.J. Hewlins, and A. G. Callely. 1989. ^{13}C-NMR studies of acetate and methanol metabolism by methylotrophic *Pseudomonas* strains. *J. Gen. Microbiol.* 135: 1469–1477.

Narro, M.L., C.E. Cerniglia, C. Van Baalen, and D.T. Gibson. 1992. Evidence for an NIH shift in oxidation of naphthalene by the marine cyanobacterium *Oscillatoria* sp. strain JCM. *Appl. Environ. Microbiol.* 58: 1360–1363.

Neidhardt, F.C., P.L. Bloch, and D.F. Smith. 1974. Culture medium for enterobacteria. *J. Bacteriol.* 119: 736–747.

Neilson, A.H. 1980. Isolation and characterization of bacteria from the Swedish west coast. *J. Appl. Bacteriol.* 49: 215–223.

Neilson, A.H., A.-S. Allard, and M. Remberger. 1985. Biodegradation and transformation of recalcitrant compounds pp. 29–86. In *Handbook of Environmental Chemistry* (Ed. O. Hutzinger), Vol. 2 Part C. Springer-Verlag, Berlin.

Neilson, A.H., A.-S. Allard, C. Lindgren, and M. Remberger. 1987. Transformations of chloroguaiacols, chloroveratroles and chlorocatechols by stable consortia of anaerobic bacteria. *Appl. Environ. Microbiol.* 53: 2511–2519.

Neilson, A.H., H. Blanck, L. Förlin, L. Landner, P. Pärt, A. Rosemarin, and M. Söderström. 1989. Advanced hazard assessment of 4,5,6-trichloroguaiacol in the Swedish environment. In *Chemicals in the Aquatic Environment*, pp. 329–374. Ed. L. Landner, Springer-Verlag, Berlin.

Nielsen, P.H., P.L. Bjerg, P. Nielsen, P. Smith, and T.H. Christensen. 1996. *In situ* and laboratory determined first-order degradation rate constants of specific organic compounds in an aerobic aquifer. *Environ. Sci. Technol.* 30: 31–37.

Nörtemann, B. 1992. Total degradation of EDTA by mixed cultures and a bacterial isolate. *Appl. Environ. Microbiol.* 58: 671–676.

Nyholm, N. and P. Kristensen. 1992. Screening methods for assessment of biodegradability of chemicals in seawater—results from a ring test. *Ecotoxicol. Environ. Saf.* 23: 161–172.

Nyholm, N., A. Damborg, and P. Lindgaard-Jörgensen. 1992. A comparative study of test methods for assessment of the biodegradability of chemicals in seawater—screening tests and simulation tests. *Ecotoxicol. Environ. Saf.* 23: 173–190.

O'Reilly, K.T. and R.L. Crawford. 1989. Degradation of pentachlorophenol by polyurethane-immobilized *Flavobacterium* cells. *Appl. Environ. Microbiol.* 55: 2113–2118.

Omori, T., L. Monna, Y. Saiki, and T. Kodama. 1992. Desulfurization of dibenzothiophene by *Corynebacterium* sp. strain SY1. *Appl. Environ. Microbiol.* 58: 911–915.

Ornston, L.N. 1966. The conversion of catechol and protocatechuate to beta-ketoadipate by *Pseudomonas putida*. IV. Regulation. *J. Biol. Chem.* 241: 3800–3810.

Palleroni, N.J. and J.F. Bradbury. 1993. *Stenotrophomonas*, a new bacterial genus for *Xanthomonas maltophilia* (Hugh 1980) Swings et al. 1983. *Int. J. Syst. Bacteriol.* 43: 606–609.

Palleroni, N.J. and R.Y. Stanier. 1964. Regulatory mechanisms governing synthesis of the enzymes for tryptophan oxidation by *Pseudomonas fluorescens. J. Gen. Microbiol.* 35: 319–334.

Parisot, D., M.C. Malet-Martino, P. Crasnier, and R. Martino. 1989. [19]F nuclear magnetic resonance analysis of 5-fluorouracil metabolism in wild-type and 5-fluorouracil-resistant *Nectria haematococca. Appl. Environ. Microbiol.* 55: 2474–2479.

Parisot, D., M.C. Malet-Martino, R. Martino, and P. Crasnier. 1991. [19]F nuclear magnetic resonance analysis of 5-fluorouracil metabolism in four differently pigmented strains of *Nectria haematococca. Appl. Environ. Microbiol.* 57: 3605–3612.

Parkes, R.J., G.R. Gibson, I. Mueller-Harvey, W.J. Buckingham, and R.A. Herbert. 1989. Determination of the substrates for sulphate-reducing bacteria within marine and estuarine sediments with different rates of sulphate reduction. *J. Gen. Microbiol.* 135: 175–187.

Peelen, S., I.M.C.M. Rietjens, W.J.H. van Berkel, W.A.T. van Workum, and J. Vervoort. 1993. ^{19}F-NMR study on the pH-dependent regioselectivity and rate of the *ortho*-hydroxylation of 3-fluorophenol by phenol hydroxylase from *Trichosporon cutaneum. Eur. J. Biochem.* 218: 345–353.

Phillips, W.E. and J.J. Perry. 1976. *Thermomicrobium fosteri* sp. nov., a hydrocarbon-utilizing obligate thermophile. *Int. J. Syst. Bacteriol.* 26: 220–225.

Pignatello, J.J., M.M. Martinson, J.G. Steiert, R.E. Carlson, and R.L. Crawford. 1983. Biodegradation and photolysis of pentachlorophenol in artificial freshwater streams. *Appl. Environ. Microbiol.* 46: 1024–1031.

Pignatello, J.J., L.K. Johnson, M.M. Martinson, R.E. Carlson, and R.L. Crawford. 1985. Response of the microflora in outdoor experimental streams to pentachlorophenol: compartmental contributions. *Appl. Environ. Microbiol.* 50: 127–132.

Plugge, C.M., C. Dijkema, and A.J.M. Stams. 1993. Acetyl-CoA cleavage pathway in a syntrophic propionate oxidizing bacterium growing on fumarate in the absence of methanogens. *FEMS Microbiol. Lett.* 110: 71–76.

Power, M., J.R. van der Meer, R. Tchelet, T. Egli, and R. Eggen. 1998. Molecular-based methods can contribute to assessment of toxicological risks and bioremediation strategies. *J. Microbiol. Methods* 32: 1078–119.

Rabus, R., R. Nordhaus, W. Ludwig, and F. Widdel. 1993. Complete oxidation of toluene under strictly anoxic conditions by a new sulfate-reducing bacterium. *Appl. Environ. Microbiol.* 59: 1444–1451.

Reanney, D.C., P.C. Gowland, and J.H. Slater. 1983. Genetic interactions among microbial communities. *Symp. Soc. Gen. Microbiol.* 34: 379–421.

Reid, K.A., J.T.G. Hamilton, R.D. Bowden, D. O'Hagan, L. Dasaradhi, M.R. Amin, and D.B. Harper. 1995. Biosynthesis of fluorinated secondary metabolites by *Streptomyces cattleya. Microbiology* (U.K.) 141: 1385–1393.

Richnow, H.H., A. Eschenbach, B. Mahro, R. Seifert, P. Wehrung, P. Albrecht, and W. Michaelis. 1998. The use of ^{13}C-labelled polycyclic aromatic hydrocarbons for the analysis of their transformation in soil. *Chemosphere* 36: 2211–2224.

Robert-Gero, M., M. Poiret, and R.Y. Stanier. 1969. The function of the beta-ketoadipate pathway in *Pseudomonas acidovorans. J. Gen. Microbiol.* 57: 207–214.

Rosner, B.M. and B. Schink. 1995. Purification and characterization of acetylene hydratase of *Pelobacter acetylenicus*, a tungsten iron-sulfur protein. *J. Bacteriol.* 177: 5767–5772.

Rothenburger, S. and R.M. Atlas. 1993. Hydroxylation and biodegradation of 6-methylquinoline by pseudomonads in aqueous and nonaqueous immobilized-cell bioreactors. *Appl. Environ. Microbiol.* 59: 2139–2144.

Saski, E.K., J.K. Jokela, and M.S. Salkinoja-Salonen. 1996a. Biodegradability of different size classes of bleached kraft mill effluent organic halogens during wastewater treatment and in lake environments pp. 179–193. In *Environmental Fate and Effects of Pulp and Paper Mill Effluents* (Eds. M.R. Servos, K.R. Munlittrick, J.H. Carey, and G.J. van der Kraak). St. Lucie Press, Delray Beach, FL.

Saski, E.K., A. Vähätalo, K. Salonen, and M.S. Salkinoja-Salonen. 1996b. Mesocosm simulation on sediment formation induced by biologically treated bleached kraft pulp mill wastewater in freshwater recipients pp. 261–270. In *Environmental Fate and Effects of Pulp and Paper Mill Effluents* (Eds. M.R. Servos, K.R. Munlittrick, J.H. Carey, and G.J. van der Kraak). St. Lucie Press, Delray Beach, FL.

Sassanella, T.M., F. Fukumori, M. Bagdasarian, and R.P. Hausinger. 1997. Use of 4-nitrophenoxyacetic acid for detection and quantification of 2,4-dichlorophenoxyacetic acid 2,4-D/α-ketoglutarate dioxygenase activity in 2,4-D-degrading microorganisms. *Appl. Environ. Microbiol.* 63: 1189–1191.

Schenk, T., R. Müller, and F. Lingens. 1990. Mechanism of enzymatic dehalogenation of pentachlorophenol by *Arthrobacter* sp. strain ATCC 33790. *J. Bacteriol.* 172: 7272–7274.

Schut, F., E.J. de Vries, J.C. Gottschal, B.R. Robertson, W. Harder, R.A. Prins, and D.K. Button. 1993. Isolation of typical marine bacteria by dilution culture: growth, maintenance, and characteristics of isolates under laboratory conditions. *Appl. Environ. Microbiol.* 59: 2150–2160.

Scott, D.T., D.M. McKnight, E.L. Blunt-Harris, S.E. Kolesar, and D.R. Lovley. 1998. Quinone moieties act as electron acceptors in the reduction of humic substances by humics-reducing microorganisms. *Environ. Sci. Technol.* 32: 2984–2989.

Selifonov, S.A., P.J. Chapman, S.B. Akkerman, J.E. Gurst, J.M. Bortiatynski, M.A. Nanny, and P.G. Hatcher. 1998. Use of ^{13}C nuclear magnetic resonance to assess fossil fuel biodegradation: fate of [1-^{13}C]acenaphthene in creosote polycyclic aromatic compound mixtures degraded by bacteria. *Appl. Environ. Microbiol.* 64: 1447–1453.

Servent, D., C. Ducrorq, Y. Henry, A. Guissani, and M. Lenfant. 1991. Nitroglycerin metabolism by *Phanerochaete chrysosporium*: evidence for nitric oxide and nitrite formation. *Biochim. Biophys. Acta* 1074: 320–325.

Shen, Y., L.G. Stehmeier, and G. Voordouw. 1998. Identification of hydrocarbon-degrading bacteria in soil by reverse sample gene probing. *Appl. Environ. Microbiol.* 64: 637–645.

Shiaris, M.P. and J.J. Cooney. 1983. Replica plating method for estimating phenanthrene-utilizing and phenanthrene-cometabolizing microorganisms. *Appl. Environ. Microbiol.* 45: 706–710.

Skerman, V.B.D. 1968. A new type of micromanipulator and microforge. *J. Gen. Microbiol.* 54: 287–297.

Small, F.J. and S.A. Ensign. 1995. Carbon dioxide fixation in the metabolism of propylene and propylene oxide by *Xanthobacter* strain Py2. *J. Bacteriol.* 177: 6170–6175.

Smith, R.V., P.J. Davis, A.M. Clark, and S.K. Prasatik. 1981. Mechanism of hydroxylation of biphenyl by *Cunninghamella echinulata*. *Biochem. J.* 196: 369–371.

Sobecky, P.A., M.A. Schell, M.A. Moran, and R.E. Hodson. 1992. Adaptation of model genetically engineered microorganisms to lake water: growth rate enhancements and plasmid loss. *Appl. Environ. Microbiol.* 58: 3630–3637.

Söhngen, N.L. 1913. Benzin, Petroleum, Paraffinöl und Paraffin als Kohlenstoff- und Energiequelle für Mikroben. *Zentralbl. Bakteriol. Paraisitenk. Infektionskr. 2 Abt.* 37: 595–609.

Sontoh, S. and J.D. Semrau. 1998. Methane and trichloroethylene degradation by *Methylosinus trichosporium* OB3b expressing particulate methane monooxygenase. *Appl. Environ. Microbiol.* 64: 1106–1114.

Spanggord, R.J., J.C. Spain, S.F. Nshino, and K.E. Mortelmans. 1991. Biodegradation of 2,4-dinitrotoluene by a *Pseudomonas* sp. *Appl. Environ. Microbiol.* 57: 3200–3205.

Stams, A.J.M., J.B. van Dijk, C. Dijkema, and C.M. Plugge. 1993. Growth of syntrophic propionate-oxidizing bacteria with fumarate in the absence of methanogenic bacteria. *Appl. Environ. Microbiol.* 59: 1114–1119.

Stanier, R.Y. 1947. Simultaneous adaptation: a new technique for the study of metabolic pathways. *J. Bacteriol.* 54: 339–348.

Stanier, R.Y., N.J. Palleroni, and M. Doudoroff. 1966. The aerobic pseudomonads: a taxonomic study. *J. Gen. Microbiol.* 43: 159–271.

Steiert, J.G., W.J. Thoma, K. Ugurbil, and R.L. Crawford. 1988. [31]P nuclear magnetic resonance studies of effects of some chlorophenols on *Escherichia coli* and a pentachlorophenol-degrading bacterium. *J. Bacteriol.* 170: 4954–4957.

Strobl, G., R. Feicht, H. White, F. Lottspeich, and H. Simon. 1992. The tungsten-containing aldehyde oxidoreductase from *Clostridium thermoaceticum* and its complex with a viologen-accepting NADPH oxidoreductase. *Biol. Chem. Hoppe-Seyler* 373: 123–132.

Strohhäcker, J., A.A. de Graaf, S.M. Schoberth, R.M. Wittig, and H. Sahm. 1993. [31]P nuclear magnetic resonance studies of ethanol inhibition in *Zymomonas mobilis*. *Arch. Microbiol.* 159: 484–490.

Sylvestre, M. 1980. Isolation method for bacterial isolates capable of growth on *p*-chlorobiphenyl. *Appl. Environ. Microbiol.* 39: 1223–1224.

Sylvestre, M., R. Massé, F. Messier, J. Faiteux, J.G. Bisaillon, and R. Beaudet. 1982. Bacterial nitration of 4-chlorobiphenyl. *Appl. Environ. Microbiol.* 44: 871–877.

Talbot, H.W., L.M. Johnson, and D.M. Munnecke. 1984. Glyphosate utilization by *Pseudomonas* sp. and *Alkaligenes* sp. isolated from environmental sources. *Curr. Microbiol.* 10: 255–260.

Tatara, G.M., M.J. Dybas, and C.S. Criddle. 1993. Effect of medium and trace metals on kinetics of carbon tetrachloride transformation by *Pseudomonas* sp. strain KC. *Appl. Environ. Microbiol.* 59: 2126–2131.

Tattersfield, F. 1928. The decomposition of naphthalene in the soil, and the effect upon its insecticidal action. *Ann. Appl. Biol.* 15: 57–80

Tausson, W.O. 1927. Naphthalin als Kohlenstoffquelle für Bakterien. *Planta* 4: 214–256.

Taya, M., H. Hinoki, and T. Kobayashi. 1985. Tungsten requirement of an extremely thermophilic cellulolytic anaerobe (strain NA 10). *Agric. Biol. Chem.* 49: 2513–2515.

Taylor, B.F., R.W. Curry, and E.F. Corcoran. 1981. Potential for biodegradation of phthalic acid esters in marine regions. *Appl. Environ. Microbiol.* 42: 590–595.

Tham, J.M.L. and C.L. Poh. 1993. Insertional mutagenesis, cloning and expession of gentisate pathway genes from *Pseudomonas alcaligenes* NCIB 9867. *J. Appl. Bacteriol.* 75: 159–163.

Thierrin, J., G.B. Davis, and C. Barber. 1995. A ground-water tracer test with deuterated compounds for monitoring *in situ* biodegradation and retardation of aromatic hydrocarbons. *Ground Water* 33: 469–475.

Thorn, K.A., J.B. Arterburn, and M.A. Mikita. 1992. ^{15}N and ^{13}C NMR investigation of hydroxylamine-derivatized humic substances. *Environ. Sci. Technol.* 26: 107–116.

Thorn, K.A., Pettigrew, P.J., Goldenberg, W.S., and Weber, E.J. 1996. Covalent binding of aniline to humic substances. 2. ^{15}N NMR studies of nucleophilic addition reactions. *Environ. Sci. Technol.* 30: 1764–1775.

Tjeerdema, R.S., T.W.-M. Fan, R.M. Higashi, and D.G. Crosby. 1991. Effects of pentachlorophenol on energy metabolism in the abalone (*Haliotis rufescens*) as measured by *in vivo* ^{31}P NMR spectroscopy. *J. Biochem. Toxicol.* 6: 45–56.

Touati, D., M. Jacques, B. Tardat, L. Bouchard, and S. Despied. 1995. Lethal oxidative damage and mutagenesis are generated by iron Δ*fur* mutants of *Escherichia coli*: protective role of superoxide dismutase. *J. Bacteriol.* 177: 2305–2314.

Tschech, A. and N. Pfennig. 1984. Growth yield increase linked to caffeate reduction in *Acetobacterium woodii*. *Arch. Microbiol.* 137: 163–167.

Unkefer, C.J. and R.E. London. 1984. *In vivo* studies of pyridine nucleotide metabolism in *Escherichia coli* and *Saccharomyces cerevisiae* by carbon-13 NMR spectroscospy. *J. Biol. Chem.* 2311–2320.

Van der Meer, J.R., T.N.P. Bosma, W.P. de Bruin, H. Harms, C. Holliger, H.H.M. Rijnaarts, M.E. Tros, G. Schraa, and A.J.B. Zehnder. 1992. Versatility of soil column experiments to study biodegradation of halogenated compounds under environmental conditions. *Biodegradation* 3: 265–284.

Van der Woude, M.W., K. Boominathan, and C.A. Reddy. 1993. Nitrogen regulation of lignin peroxidase and mangenese-dependent peroxidase production is independent of carbon and mangansese regulation in *Phanerochaete chrysosporium*. *Arch. Microbiol.* 160: 1–4.

van Elsas, J.D., G.F. Duarte, A.S. Rosada, and K. Smalla. 1998. Microbiological and molecular biological methods for monitoring microbial inoculants and their effects in the soil environment. *J. Microbiol. Methods* 32: 133–154.

Van Ginkel, C.G., J.B. van Dijl, and A.G.M. Kroon. 1992. Metabolism of hexadecyltrimethylammonium chloride in *Pseudomonas* strain B1. *Appl. Environ. Microbiol.* 58: 3083–3087.

Van Niel, C.B. 1955. Natural selection in the microbial world. *J. Gen. Microbiol.* 13: 201–217.

Van Zyl, E. and P.L. Syeyn. 1992. Reinterpretation of the taxonomic position of *Xanthomonas maltophilia* and taxonomic criteria in this genus. Request for an opinion. *Int. J. Syst. Bacteriol.* 42: 193–198.

Vogel, T.M. and D. Grbic-Galic. 1986. Incorporation of water into toluene and benzene during anaerobic fermentative transformation. *Appl. Environ. Microbiol.* 52: 200–202.

Wachenheim, D.E., and R.E. Hespell. 1984. Inhibitory effects of titanium(III) citrate on enumeration of bacteria from rumen contents. *Appl. Environ. Microbiol.* 48: 444–445.

Wagner, A.F.V., M. Frey, F.A. Neugebauer, and W. Schäfer. 1992. The free radical in pyruvate formate-lyase is located on glycine-734. *Proc. Natl. Acad. Sci. U.S.A.* 89: 996–1000.

Wagner, R. and J.R. Andreesen. 1979. Selenium requirement for active xanthine dehydrgenase from *Clostridium acidiurici* and *Clostridium cylindrosporum*. *Arch. Microbiol.* 121: 255–260.

Wagner, R. and J.R. Andressen. 1987. Accumulation and incorporation of ^{185}W-tungsten into proteins of *Clostridium acidiurici* and *Clostridium cylindrosporum*. *Arch. Microbiol.* 147: 295–299.

Watanabe, K., S. Yamamoto, S. Hino, and S. Harayama. 1998. Population dynamics of phenol-degrading bacteria in activated sludge determined by *gyrB*-targeted quantitative PCR. *Appl. Environ. Microbiol.* 64: 1203–1209.

Weisberg, W.G., S.M. Barns, D.A. Pelletier, and D.J. Lane. 1991. 16S ribosomal DNA amplification for phylogenetic study. *J. Bacteriol.* 173: 697–703.

White, H., G. Strobl, R. Feicht, and H. Simon. 1989. Carboxylic acid reductase: a new tungsten enzyme catalyses the reduction of non-activated carboxylic acids to aldehydes. *Eur. J. Biochem.* 184: 89–96.

Widdel, F. 1983. Methods for enrichment and pure culture isolation of filamentous gliding sulfate-reducing bacteria. *Arch. Microbiol.* 134: 282–285.

Wigmore, G.J. and D.W. Ribbons. 1981. Selective enrichment of *Pseudomonas* spp. defective in catabolism after exposure to halogenated substrates. *J. Bacteriol.* 146: 920–927.

Wilkinson, S.G. 1968. Studies on the cell walls of *Pseudomonas* species resistant to ethylenediaminetetra-acetic acid. *J. Gen. Microbiol.* 54: 195–213.

Wilson, M.S. and E.L. Madsen. 1996. Field extraction of a transient intermediary metabolite indicative of real time *in situ* naphthalene biodegradation. *Environ. Sci. Technol.* 30: 2099–2103

Winter, J., C. Lerp, H.-P. Zabel, F.X. Wildenauer, H. König, and F. Schindler. 1984. *Methanobacterium wolfei*, sp. nov., a new tungsten-requiring, thermophilic, autotrophic methanogen. *Syst. Appl. Microbiol.* 5: 457–466.

Wischmann, H. and H. Steinhart. 1997. The formation of PAH oxidation products in soils and soil/compost mixtures. *Chemosphere* 35: 1681–1698.

Witschel, M., S. Nagel, and T. Egli. 1997. Identification and characterization of the two-enzyme system catalyzing the oxidation of EDTA in the EDTA-degrading bacterial strain DSM 9103. *J. Bacteriol.* 179: 6937–6943.

Yamamoto, I., T. Saiki, S.M. Liu, and L.G. Ljungdahl. 1983. Purification and properties of NADP-dependent formate dehydrogenase from *Clostridium thermoaceticum*, a tungsten-selenium-iron protein. *J. Biol. Chem.* 258: 1826–1832.

Yoch, D.C. 1979. Manganese, an essential trace element for N_2 fixation by *Rhodospirilllum rubrum* and *Rhodopseudomonas capsulata*: role in nitrogenase regulation. *J. Bacteriol.* 140: 987–995.

Yoon, J.-H., S.-K. Rhee, J.-S. Lee, Y.-H. Park, and S.T. Lee. 1997. *Nocardiodes pyridinolyticus* sp. nov., a pyridine-degrading bacterium isolated from the oxic zone of an oil shale column. *Int. J. Syst. Bacteriol.* 47: 933–938.

Zarilla, K.A. and J.J. Perry. 1984. *Thermoleophilum album* gen. nov. and sp. nov., a bacterium obligate for thermophily and *n*-alkane substrates. *Arch. Microbiol.* 137: 286–290.

Zellner, G. and A. Jargon. 1997. Evidence for a tungsten-stimulated aldehyde dehydrogenase activity of *Desulfovibrio simplex* that oxidizes aliphatic and aromatic aldehydes with flavins as coenzymes. *Arch. Microbiol.* 168: 480–485.

Zellner, G., C. Alten, E. Stackebrandt, E.C. de Macario, and J. Winter. 1987. Isolation and characterization of *Methanocorpusculum parvum*, gen. nov., spec. nov., a new tungsten-requiring, coccoid methanogen. *Arch. Microbiol.* 147: 13–20.

Zhang, X. and L.Y. Young. 1997. Carboxylation as an initial reaction in the anaerobic metabolism of naphthalene and phenanthrene by sulfidogenic consortia. *Appl. Environ. Microbiol.* 63: 4759–4764.

Zürrer, D., A.M. Cook, and T. Leisinger. 1987. Microbial desulfonation of substituted naphthalenesulfonic acids and benzenesulfonic acids. *Appl. Environ. Microbiol.* 53: 1459–1463.

6

Pathways of Biodegradation and Biotransformation

SYNOPSIS An attempt is made to describe the pathways used by microorganisms to degrade or transform xenobiotics. Most of the major structural groups are considered, including aliphatic, alicyclic, aromatic, and heterocyclic compounds, including those with oxygen, sulfur, nitrogen, phosphorus, or halogen substituents. Although organochlorine compounds have received most attention, an attempt has been made to include also organobromine compounds; the degradation of organofluoro compounds is discussed separately since these compounds differ significantly even in their chemical properties from those of the other halogens. Reactions carried out by both aerobic and anaerobic bacteria and—to a somewhat lesser extent by yeasts and fungi—are considered, although no details of the enzyme systems are given. Attention is directed to the pathways that are used by different organisms for the degradation of a given xenobiotic. Investigations using aerobic bacteria have almost invariably been exemplified from the results of experiments using pure cultures, whereas for anaerobic bacteria this has been supplemented by results using mixed cultures or stable consortia. Some examples are given of the application of biotransformation reactions in biotechnology and of the environmental significance of the biotransformations of xenobiotics which may result in metabolites more toxic than their precursors. Finally, an attempt has been made to classify the reactions involved in the degradation of xenobiotics on the basis of well-established chemical transformations, and specific reference has been made to the appropriate sections in which these reactions are discussed in greater detail.

Introduction

It is desirable to explain both the motivation and the objectives of this chapter since some of the material has already been presented from a different perspective in preceding chapters: Chapter 4 attempted to provide a general background with a microbiological emphasis, while Chapter 5 filled this out with an outline of procedures for carrying out the appropriate experiments.

This chapter attempts a survey of the pathways by which a range of structurally diverse xenobiotics are degraded or transformed by microorganisms; the emphasis is on reactions mediated by bacteria which are the most effective agents in carrying out biodegradation in most natural aquatic ecosystems.

It is appropriate to begin by underscoring the two rather different—and possibly conflicting—approaches to addressing problems of biodegradation and biotransformation, and to which attention has already been briefly drawn. These concern the level at which assessments of biodegradability are carried out.

On the one hand, conventional tests for assessing ready biodegradability do not provide an adequate base for determining what occurs after release of the compound into natural ecosystems, even though they may be adequate for assessing biodegradability in the municipal treatment systems from which the inocula were taken. Indeed, the effort directed to developing standardized test systems may even have been counterproductive to environmental relevance for reasons that have been outlined in more detail in Chapter 5, Section 5.2.1.

On the other hand, the comprehensive investigations which have been pursued on the physiology, biochemistry, genetics, and regulation of biodegradation cannot realistically be incorporated even into an advanced hazard assessment except in a very few instances.

An additional problem arises from the immense structural range of organic compounds that are used industrially or have been incorporated into commercial products. The skill of the organic chemist is seemingly unlimited and with the inevitable need for new compounds, which attempt to avoid the undesirable consequences of their traditional counterparts, the number of compounds—as well as their structural diversity—seems unlikely to diminish.

There is an enormous literature on the microbial degradation and transformation of organic compounds, and it would be attractive to take advantage of this to construct generalizations on the pathways used for the degradation of broad structural classes of xenobiotics. That is the objective of this chapter. This approach has been illustrated by Alexander (1981), and this chapter attempts to provide details that were not possible within the space of that seminal review. In addition, this procedure would have a predictive capability that is not restricted to compounds which have already been investigated. Although structure–activity relationships are useful for classifying existing data, they have an inevitably restricted potential for application to completely novel chemical structures. Support for this mechanistic approach is provided by its success in assessing the biodegradability of 50 structurally diverse xenobiotics (Boethling et al. 1989).

This chapter does not attempt to encompass the enzymology of the reactions involved in the degradation of xenobiotics, so that the word *pathways* is more appropriate than *mechanisms:* however desirable, discussions of enzymology lie both beyond the scope of the present work and the competence of the author. A few parenthetical comments on the enzymology of the reactions have, however, been made if they elucidate the scope and the generality of the reactions under consideration. Attention has been drawn to the role of free radicals in enzymatic reactions (Section 4.4.4), and there is increasing appreciation of their wider significance in reactions catalyzed by enzymes (Stubbe and van der Donk 1998); examples are given in Sections 6.7.1 and 6.7.3.1. This recognition parallels the development by D.H.R. Barton of radical-mediated reactions in synthetic organic chemistry in which mild conditions, high yield, and specificity are combined.

Some important details of the reaction pathways involved in degradation have been deliberately omitted in the figures used to illustrate the various sequences: for example, (1) even when the degradation of carboxylic acids takes place through initial formation of the coenzyme-A esters, sequences have depicted the free carboxylic acids and (2) in some cases, although the structures of intermediates have not been rigorously determined, these have been included to illustrate more clearly the structural relationships between the initial substrate and the various metabolites.

The presentation is made on the basis of the chemical structure of xenobiotics and is dominated by examples of reactions carried out by aerobic and anaerobic bacteria and—to a lesser extent—aerobic fungi and yeasts; some examples of biotransformation reactions carried out by other microorganisms are given in Chapter 4, Section 4.3, and by higher organisms in Section 7.5. Although anaerobic fungi are known and are certainly important in rumen metabolism (Mountfort 1987), their existence in other habitats does not appear to have been established and their potential for degrading xenobiotics does not seem to have been explored. Since the emphasis is on degradative pathways, less attention will be devoted to the taxonomic delineation of the various organisms except where these belong to less common taxa while reference has already been made in Chapter 5, Section 5.9 to the serious problems that have been encountered in attempting to classify bacteria of established degradative importance. In addition, no attempt has been made to provide the currently acceptable taxonomic assignment of the organisms that are involved, and the designations used by the authors have been retained with only a few exceptions. Except for the simplest reaction sequences, structural representations of the various pathways are given in the form of flow diagrams rather than by using conventional chemical nomenclature. It is hoped that the reactions are thereby more clearly perceived in geometric terms, particularly to those who are not organic chemists and who are understandably repelled by the seemingly barbarous complexity and apparent incomprehensibility of systematic organic chemical nomenclature.

In the following sections, an account will be presented of the pathways by which xenobiotics are degraded by microorganisms. At the same time, it is essential to bear in mind certain fundamental aspects of the microbiology and biochemistry of the cells carrying out these reactions and, in particular, the role of metabolites that are required for biosynthesis on which continued growth and replication of the organisms ultimately depend.

1. If an organic compound is to support growth and replication of an organism, it must also provide the necessary metabolic energy and serve as the source of carbon (and, in some cases, also nitrogen or sulfur or phosphorus) for the synthesis of cell material. Details of these metabolic reactions are not given here, and a good account may be found, for example, in Mandelstam, Macquillan, and Dawes (1982). These reactions then determine the extent to which the constituent atoms of xenobiotics are incorporated into the global carbon, nitrogen, sulfur, or phosphorus cycles; these are not discussed here, and reference may be made to the valuable account of carbon cycles into which the products from the degradation of xenobiotics are incorporated (Hagedorn et al. 1988). Whereas the functional operation of these reactions is a prerequisite for biodegradation, biotransformation may be accomplished by nongrowing cells, or in cells growing at the expense of more readily degradable substrates; this has been discussed in Section 4.5.2.

2. Just as there is no single pathway universally used for the catabolism of simple substrates such as glucose, there are no unique pathways for the degradation of a given xenobiotic. The following examples may be used to illustrate the considerable differences in the pathways used for the degradation of xenobiotics by bacteria and by fungi, or even by different taxa of bacteria.

   • The degradation of DDT by *Phanerochaete chrysosporium* (Bumpus and Aust 1987) and by *Aerobacter aerogenes* (Wedemeyer 1967);

   • The degradation of 2,4-dichlorophenol by *Ph. chrysosporium* (Valli and Gold 1991) and by a strain of *Acinetobacter* sp. (Beadle and Smith 1982);

   • The degradation of quinoline by pseudomonads and by *Rhodococcus* sp. (Schwarz et al. 1989);

   • The degradation of tryptophan by *Pseudomonas fluorescens* that takes place via the β-ketoadipate pathway and by *P. acidovorans* that utilizes the quinoline pathway (Stanier 1968);

   • The plethora of pathways for the aerobic degradation of toluene (Section 6.2.1).

3. There is no absolute distinction between the degradative pathways used by aerobic and by anaerobic bacteria. Simple reductions are carried out by organisms with a strictly aerobic metabolism. These include, for example, reductive dechlorination of phenolic compounds by *R. chlorophenolicus* (Apajalahti and Salkinoja-Salonen 1987a), reduction of hydroxylated pyrimidines by *P. stutzeri* (Xu and West 1992), and degradation of anthranilate by a strain of *Pseudomonas* sp. that is able to use this as a source of both carbon and nitrogen and degrades the substrate by initial reactions

involving the reduction of the aromatic ring (Altenschmidt and Fuchs 1992b). This should not, however, be interpreted to imply that the underlying cellular metabolism of aerobic and anaerobic microorganisms is necessarily comparable.

4. Although a synopsis of the reactions used by microorganisms for the degradation and transformation of organic compounds is given in Section 6.12, it may be valuable to provide some general comments at this stage. The basic reactions known in organic chemistry provide a suitable background for rationalizing most biochemical reactions—addition, elimination, substitution, oxidation, reduction, and rearrangement—and all of these can be mediated by microorganisms although, for example, degradation involving addition reactions is rather unusual. The degradation of aliphatic (and alicyclic) and aromatic (including heterocyclic) compounds has been treated separately in this chapter, since both their chemistry and their microbial degradation pathways differ significantly. The following categorical summary may illustrate the broad types of reactions that are most commonly encountered and may serve as a prelude to the more–detailed discussions of individual groups of compounds that follow. A detailed summary is given in Section 6.12.

Oxygenation—Most organic xenobiotics are relatively highly reduced compounds so that their degradation to CO_2 and H_2O inevitably involves introduction of oxygen into the molecule either by monooxygenation or dioxygenation from O_2 or by hydroxylation from H_2O. Whereas oxygenation is clearly restricted to aerobic conditions, hydroxylation can be accomplished both aerobically and anaerobically. It should be appreciated that oxidation can occur under anaerobic conditions provided that the redox balance is preserved within the system. Methanogenesis is the terminal—although complex—step in the reduction of the precursors (CO_2 or acetate) that are produced by the degradation of more complex substrates.

Dehydrogenation or desaturation—Under aerobic conditions, dehydrogenations may be involved and these may be important under anaerobic conditions.

Dehalogenation—The degradation of compounds carrying halogenated substituents will involve loss of halogen that may occur by elimination or by displacement reactions; these may be reductive, oxidative, or hydrolytic.

Rearrangement—These are particularly important among anaerobic bacteria where they involve coenzyme-B_{12}. The unrelated rearrangement of the substituents on aromatic rings (the NIH shift) is well established particularly among fungi.

A cardinal issue for the successful biodegradation of xenobiotics is the bioenergetics of these reactions, although this aspect is not discussed here. Whereas the synthesis of ATP under aerobic conditions is at least formally straightforward, a much greater range of mechanisms operates under anaerobic conditions where ATP may be generated, for example, from intermediate acyl phosphates (generally acetyl phosphate), carbamyl phosphate, or 10-formyl tetrahydrofolate (Section 4.2.1).

As has already been emphasized, citations to the literature are eclectic rather than complete. Comprehensive reviews of many of the groups of compounds have been provided in the books and in the review articles that are given at the beginning of the reference list in Chapter 4, and these should be consulted for further details.

6.1   Aerobic Degradation of Nonaromatic Hydrocarbons

6.1.1   Alkanes

There is an enormous literature on the microbial degradation of alkanes; this has been motivated by aims as diverse as the utilization of microorganisms for the production of single-cell protein or their application to combating oil spills. Both the number and the taxonomic range of microorganisms are equally impressive, and they include many different taxa of bacteria, yeasts, and fungi. Extensive reviews that cover most aspects have been presented (Ratledge 1978; Britton 1984).

The simplest alkane is, of course, methane, but the pathways for its degradation and assimilation do not reflect this structural simplicity. In outline, the pathway of degradation is straightforward and involves successive oxidation to methanol catalyzed by methane monooxygenase, and successive dehydrogenation to formaldehyde and formate. The cells must, however, be capable of synthesizing cell material from the substrate so that some fraction of the C_1 metabolites must also be assimilated. Several distinct pathways for this have been described, but these are merely summarized here since a comprehensive and elegant presentation of the details has been given (Anthony 1982):

1. The ribulose bisphosphate pathway for the assimilation of CO_2 which is identical to the Benson–Calvin cycle used by photosynthetic organisms;

2. The ribulose monophosphate cycle for the incorporation of formaldehyde;

3. The serine pathway for the assimilation of formaldehyde.

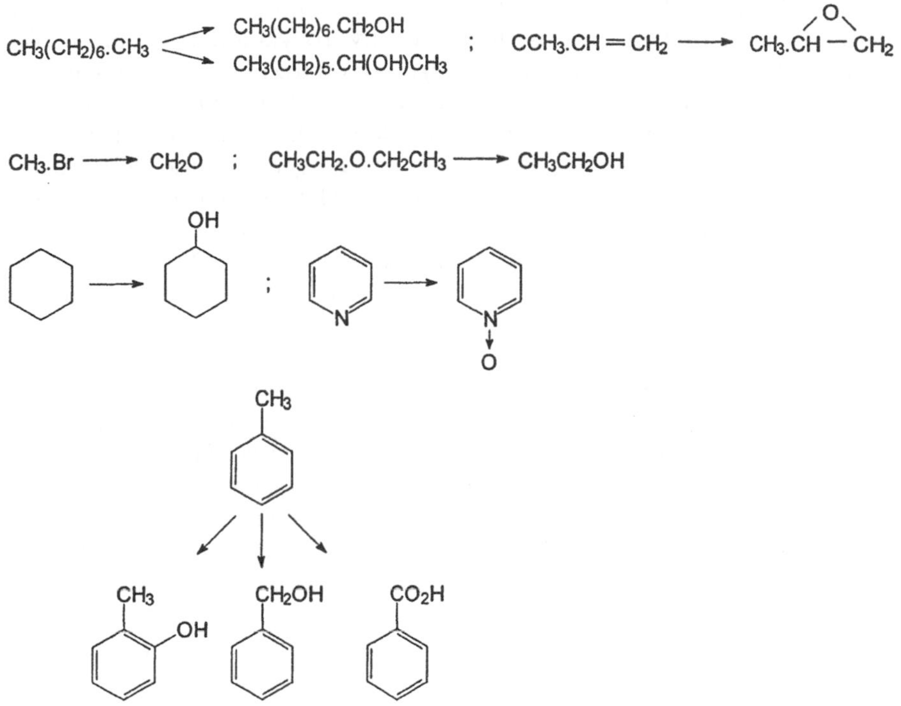

$$CH_3(CH_2)_6.CH_3 \xrightarrow{\quad} \begin{array}{l} CH_3(CH_2)_6.CH_2OH \\ CH_3(CH_2)_5.CH(OH)CH_3 \end{array} \quad ; \quad CCH_3.CH=CH_2 \longrightarrow CH_3.CH\overset{O}{\underset{}{-}}CH_2$$

$$CH_3.Br \longrightarrow CH_2O \quad ; \quad CH_3CH_2.O.CH_2CH_3 \longrightarrow CH_3CH_2OH$$

FIGURE 6.1
Examples of the reactions catalyzed by methane monooxygenase.

Methane monooxygenase may exist in either soluble (sMMO) or particulate (pMMO) forms. These display different substrate ranges and different rates of transformation, and most methanotrophs express only the latter form of the enzyme (Hanson and Hanson 1996). The role of Cu in determining whole-cell activity of pMMO is discussed in Chapter 5, Section 5.2.4. The soluble methane monooxygenase from both type I and type II methanotrophs is a three-component system comprising a nonheme iron hydroxylase containing a bridged oxo-bridged binuclear Fe cluster (A), a metal-free protein component without redox cofactors (B), and a NADH reductase (C), containing FAD and a [2Fe–2S] cluster (Fox et al. 1989).

One additional aspect is the wide spectrum of substrates which can be metabolized by the methane monooxygenase system, and some illustrative examples are given in Figure 6.1. Attention has already been drawn in Chapter 4, Section 4.3.2 to the similarity of this enzyme to that involved in the oxidation of ammonia, while the broad substrate specificity of cyclohexane oxygenase is noted again in Section 6.1.2.

The initial hydroxylation of alkanes is mediated by both membrane-bound and soluble hydroxylases, and the genetics of alkane hydroxylation and alkanol dehydrogenation in pseudomonads is complex involving in *Pseudomonas oleovorans* the loci alkA, alkB, alkC, alkD, alkE (Fennewald et al.

1979; Kok et al. 1989). In *P. putida* that carries the OCT plasmid, there is duplication of some of the loci: those for alkane hydroxylation (alkA, alkB, alkC) and for alkanol dehydrogenation (alcO) occur on the plasmid, whereas those for alcA and alcB, and for aldehyde dehydrogenation (aldA, aldB) occur in the chromosome (Grund et al. 1975). (Note the different symbols used for genetic loci in these studies.) The corresponding genes on the OCT plasmid of *P. oleovorans* and in *Acinetobacter* sp. strain ADP1 have been discussed in Chapter 4, Section 4.4.1.1. There is also some structural similarity between the nucleotide sequence for alkane hydroxylase and the subunits of the monooxygenase coded by *xylA* and *xylM* that are involved in the side-chain oxidation of toluene and xylene by *P. putida* PaW1 (Suzuki et al. 1991).

The conversion of methanol to formaldehyde is carried out by methanol dehydrogenase. A complex array of genes is involved in this oxidation, and the dehydrogenase contains pyrroloquinoline quinone (PQQ) as a cofactor (references in Ramamoorthi and Lidstrom 1995). The details of its function must, however, differ from methylamine dehydrogenase that also contains a quinoprotein–tryptophan tryptophylquinone (TTQ) (Section 4.4.4). The initial reactions involved in the metabolism of higher alkanes (> C_1) are formally similar to those used for the metabolism of methane, and the soluble alkanol dehydrogenases also contain PQQ (references in Anthony 1992). Enzymatically, however, the details may be more complex since, for example, a number of distinct alcohol and fatty acid dehydrogenases have been isolated from an *Acinetobacter* sp. during the metabolism of hexadecane (Singer and Finnerty 1985a,b).

Further degradation of the carboxylic acid following oxidation of the aldehydes involves a β-oxidation with successive loss of acetate residues (Figure 6.2). A structurally wide range of hydrocarbons may be degraded by microorganisms including linear alkanes with both even numbers of carbon atoms up to at least C_{30}, some odd numbered alkanes including the plant wax $C_{29}H_{60}$ (Hankin and Kolattukudy 1968), and branched alkanes such as pristane (2,6,10,12-tetramethylpentadecane) (McKenna and Kallio 1971; Pirnik et al. 1974). A number of details merit brief comment.

1. Yeasts are able to degrade long-chain alkanes. The initial hydroxylation is carried out in micrososmes by cytochrome P-450, while degradation of the alkanoate is carried out in peroxisomes that contain the β-oxidation enzymes: alkanoate oxidase, enoyl-CoA hydratase, and 3-hydroxyacyl-CoA dehydrogenase. Further details are given in Chapter 4, Sections 4.4.1.2 and 4.4.4.

$R.CH_2.CH_2.CH_2.CH_3 \longrightarrow R.CH_2.CH_2.CH_2.CH_2OH \longrightarrow R.CH_2.CH_2.CH_2.CHO$

$\longrightarrow R.CH_2.CH_2.CH_2.CO_2H \longrightarrow R.CH_2.CO_2H + CH_3.CO_2H \longrightarrow \longrightarrow$

FIGURE 6.2
Outline of the metabolism of alkanes.

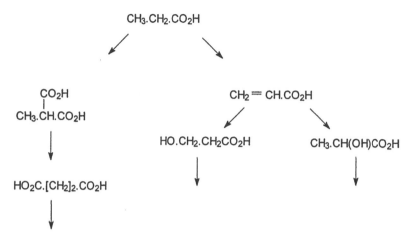

FIGURE 6.3
Pathways for the biodegradation of propionate.

2. In some cases, reaction between the initially formed alkanol and its oxidation product, the alkanoic acid, may produce esters which are resistant to further degradation (Kolattukudy and Hankin 1968).

3. For complete degradation and assimilation of the products into anabolic pathways, the cells must clearly be capable of synthesizing the appropriate enzymes. When β-oxidation results in the production of acetate, cells must be capable of synthesizing the enzymes of the glyoxylate cycle. When odd-membered alkanes are oxidized, propionate is also produced and its further degradation may follow a number of alternative pathways (Figure 6.3) (Wegener et al. 1968). A further alternative has been demonstrated in *Escherichia coli* K12 in which the initial reaction is condensation of propionyl-CoA with oxalacetate to form methylcitrate that is then converted into succinate and pyruvate (Textor et al. 1997). The enzyme was identified as citrate synthetase II which has an established role in the metabolism of propionate by the yeast *Candida lipolytica* (Uchiyama and Tabuchi 1976).

4. Oxidation of compounds such as pristane proceeds by both β-oxidation and ω-oxidation (McKenna and Kallio 1971; Pirnik et al. 1974) (Figure 6.4). Pristane may also be degraded under nitrate-reducing conditions in microcosms and enrichment cultures (Bregnard et al. 1997).

5. The existence of chain branching may present an obstacle to degradation, although this can be circumvented by a carboxylation pathway (Figure 6.5) (Fall et al. 1979) that is formally comparable to that illustrated above for the degradation of propionate. Carboxylation is also used in one of the pathways used by *Marinobacter*

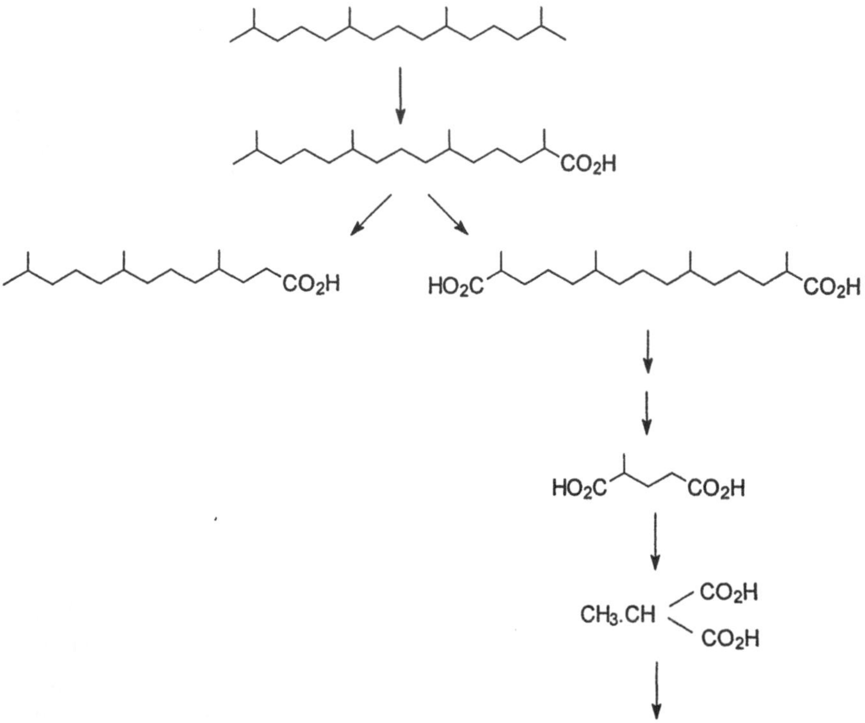

FIGURE 6.4
Pathways for the biodegradation of pristane.

sp. strain CAB for the degradation of 6,10,14-trimethylpentadecan-2-one (Rontani et al. 1997). The degradation is carried out by two independent reactions. The first involves terminal hydroxylation, loss of formaldehyde with formation of the aldehyde, dehydrogenation to the carboxylate, followed by steps comparable to those used for pristane involving carboxylation at the (original) 6-methyl group. The second involves introduction of oxygen between C-2 and C-3 by a Baeyer-Villiger-type oxidation that is noted in Section 6.1.2. Hydrolysis of the lactone is followed by dehydrogenation of the alkanol and subsequent degradation.

6. A number of substituted 2,2-bisphenylpropanes are degraded by oxidation and cleavage at the quaternary carbon atom (Figure 6.6) (Lobos et al. 1992), although this is probably facilitated by the presence of the phenyl rings. On the other hand, cytochrome P-450$_{cam}$ hydroxylated adamantane exclusively at C_1, a position that is specially accessible in this structure (White et al. 1984), and further example of hydroxylation at quaternary carbon atoms are given in Section 6.1.2. Oxidation at *t*-butyl methyl groups has also been observed.

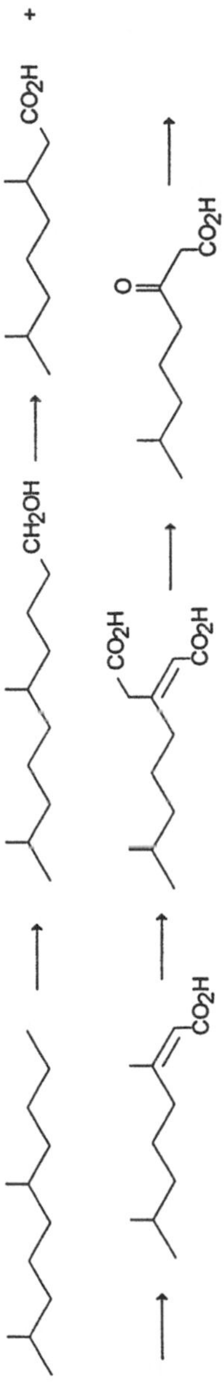

FIGURE 6.5
Carboxylation pathway for the biodegradation of a branched alkane.

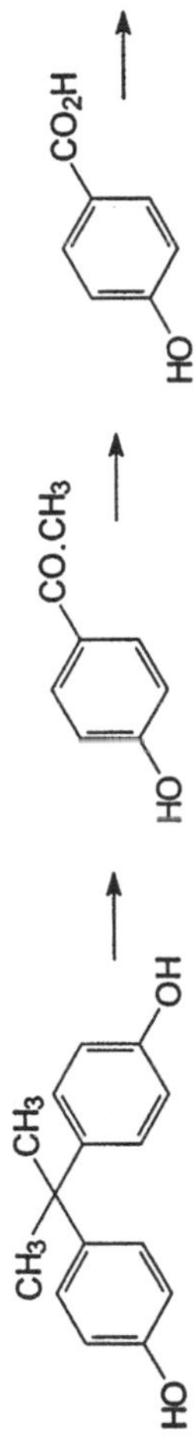

FIGURE 6.6
Biodegradation of 2,2-bisphenylpropane.

- Oxidation of the methyl group of *t*-butylphenyl diphenylphos‑ phate to the carboxylic acid by *Cuninghamella elegans* has been demonstrated (Heitkamp et al. 1985).

- Degradation of *t*-butyl methyl ether with intermediate formation of *t*-butanol has been accomplished with a mixed culture (Salan‑ itro et al. 1994), and propane-grown cells of *Mycobacterium vaccae* to JOB5 and a strain ENV425 obtained by propane enrichment degraded *t*-butyl methyl ether to *t*-butanol and further to 2-hydroxyisobutyric acid that was not, however, used as a growth substrate for the organisms (Steffan et al. 1997).

7. An unusual pathway has been proposed for the degradation of *n*-alkanes to the carboxylic acids by a *Pseudomonas* sp. under anaer‑ obic conditions; this involves initial dehydrogenation and hydrox‑ ylation followed by successive oxidations (Figure 6.7) (Parekh et al. 1977).

$$R-CH_2-CH_3 \longrightarrow R.CH=CH_2 \longrightarrow R.CH_2.CH_2OH \longrightarrow R.CH_2.CO_2H$$

FIGURE 6.7
Biodegradation of an alkane under anaerobic conditions.

The degradation of alkanoic acids by β-oxidation has been noted paren‑ thetically above, but alternative pathways may occur. For example, the metabolism of hexanoic acid by strains of *Pseudomonas* sp. may take place by ω-oxidation with subsequent formation of succinate and 2-tetrahydrofurany‑ lacetate as a terminal metabolite (Kunz and Weimer 1983). In a strain of *Corynebacterium* sp., the specificities of the relevant catabolic enzymes are consistent with the production of dodecanedioic acid by ω-oxidation of dode‑ cane but not of hexadecanedioic acid from hexadecane (Broadway et al. 1993). Hydroxylation at subterminal (ω-1, ω-2, and ω-3) positions of carbox‑ ylic acids with chain lengths of 12 to 18—and less readily of the correspond‑ ing alcohols, but not the carboxylic acids or the alkanes—has been observed (Miura and Fulco 1975) for a soluble enzyme system from a strain of *Bacillus megaterium*. Whereas in this organism ω-2 hydroxylation is carried out by a soluble cytochrome P-450 $_{BM-3}$ (Narhi and Fulco 1987), ω-hydroxylation in *P. oleovorans* that carries the OCT plasmid is mediated by a three-component hydroxylase that behaves like a cytoplasmic membrane protein (Ruettinger et al. 1974; Kok et al. 1989).

The initial reaction in the biodegradation of primary alkylamines is conver‑ sion to the aldehyde and subsequent reactions converge on those for the deg‑ radation of primary alkanes. The conversion of alkylamines to aldehydes may be accomplished by two different mechanisms: (1) by oxidases (Chapter 4, Section 4.4.4) that convert molecular oxygen to H_2O_2 or (2) by dehydroge‑ nases. Which of these is used depends on the organism: for example, the dehydrogenase is used by *P. aeruginosa* ATCC 17933, *P. putida* ATCC 12633, and the methylotroph *Paracoccus versutus* ATCC 25364, whereas *Klebsiella oxytoca* ATCC 8724, *Escherichia coli* ATCC 9637, and *Arthrobacter* sp. NCIB 11625 used a copper-quinoprotein amine oxidase (Hacisalihoglu et al. 1997). The degradation of secondary and tertiary amines is discussed in Section 6.9.1.

A number of yeasts belonging to the genera *Candida* and *Endomycopsis* are able to degrade alkanes that have chain lengths >4 (Käppeli 1986; Tanaka and Ueda 1993). The initial hydroxylation is brought about by cytochrome P-450$_{aO}$ that is located in microsomes, and further degradation of the alkanoates takes place in peroxisomes. This is discussed further in Chapter 4, Section 4.4.4.

6.1.2   Cycloalkanes

Reviews of the degradation of alicyclic compounds including monoterpenes have been given by one of the pioneers (Trudgill 1978; 1984; 1994), and these should be consulted for further details; only a bare outline with significant new developments will therefore be given here.

Even though the first two steps in the oxidation of cycloalkanes are formally similar to those used for degradation of linear alkanes, it was some time before pure strains of microorganisms were isolated that could grow with cycloalkanes or their simple derivatives. The degradation of cyclohexane has been examined in detail (Stirling et al. 1977; Trower et al. 1985) and there are two critical steps in its degradation: (1) hydroxylation of the ring and (2) subsequent cleavage of the alicyclic ring that involves insertion of oxygen in a reaction formally similar to the Baeyer–Villiger persulfate oxidation. The pathway is illustrated for cyclohexane (Figure 6.8) (Stirling et al. 1977), and a comparable one operates also for cyclopentanol (Griffin and Trudgill 1972) while the enantiomeric specificity of this oxygen-insertion reaction has been examined in a strain of camphor-degrading *Pseudomonas putida* (Jones et al. 1993). Attention has already been drawn in Section 4.3.2 to the wide metabolic versatility of cyclohexane oxygenase (Branchaud and Walsh 1985) which is reminiscent of that of methane monooxygenase. Cyclohexylacetate is degraded to cyclohexanone by elimination of the side chain after hydroxylation at the ring junction (Ougham and Trudgill 1982), but the cyclopropane ring in *cis*-11,12-methyleneoctadecanoate that is a lipid constituent of *Escherichia coli* is degraded by *Tetrahymena pyriformis* using the alternative ring opening pathway (Figure 6.9) (Tipton and Al-Shathir 1974).

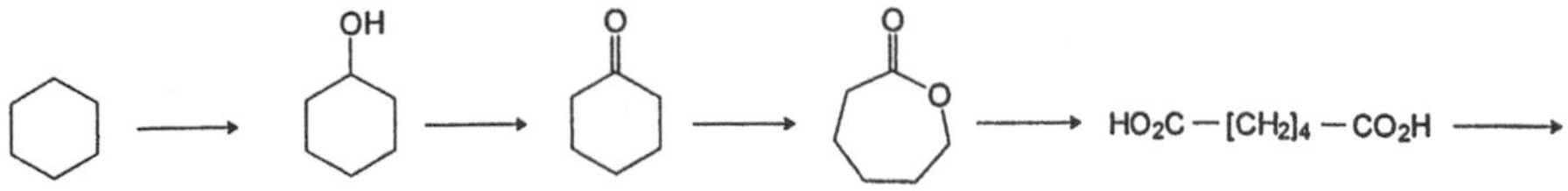

FIGURE 6.8
Biodegradation of cyclohexane.

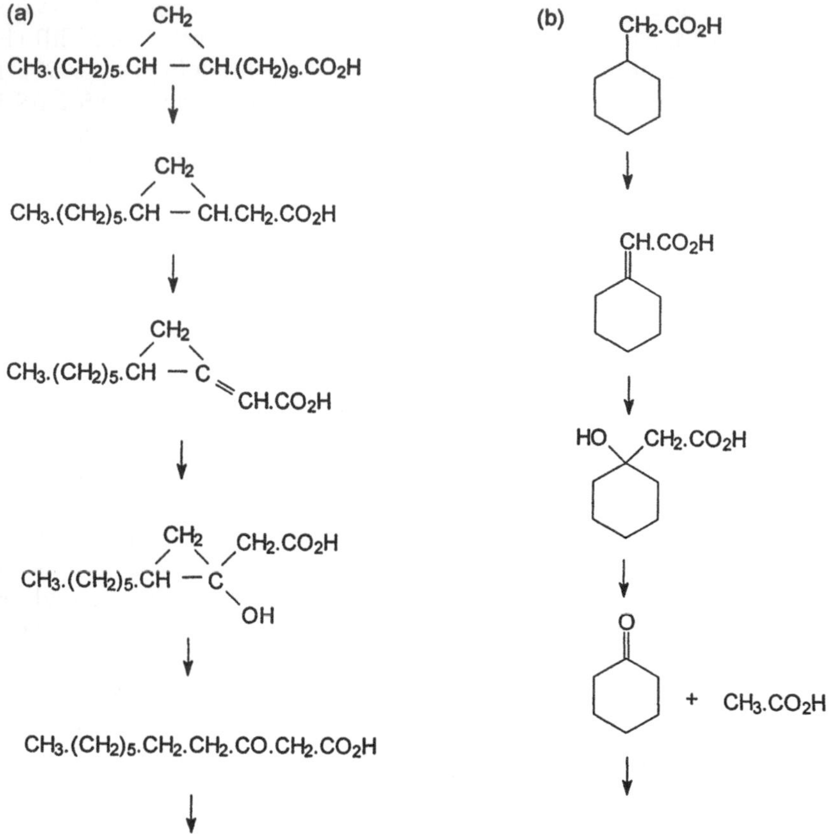

FIGURE 6.9
Biodegradation of a cycloalkylacetate.

An alternative and unusual pathway for the degradation of cyclohexane-carboxylate has also been found in which the ring is dehydrogenated to 4-hydroxybenzoate before ring cleavage (Figure 6.10) (Blakley 1974; Taylor and Trudgill 1978). The degradation of polyhydroxylated cyclohexanes such as quinate and shikimate also involves aromatic intermediates (Ingledew et

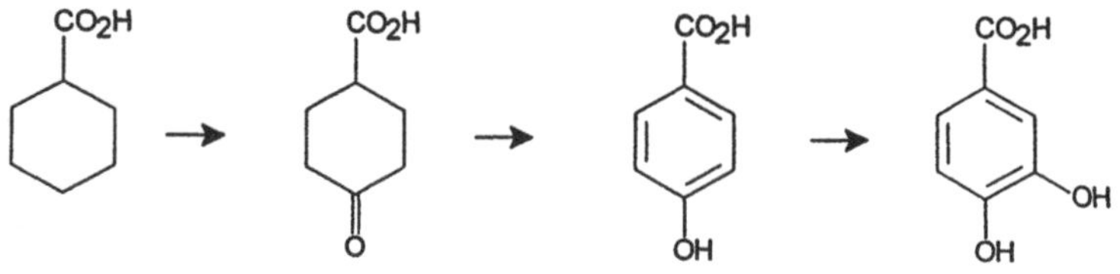

FIGURE 6.10
Alternative pathway for the biodegradation of a cyclohexane carboxylate.

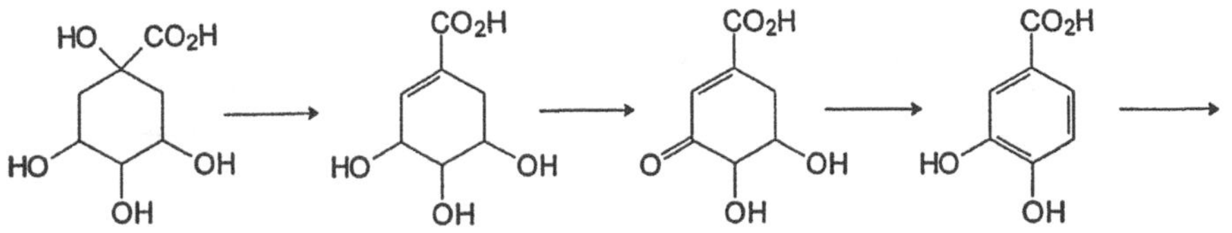

FIGURE 6.11
Biodegradation of quinate.

al. 1971), although in these examples, a mechanism for the formation of the aromatic ring by elimination reactions is more readily rationalized (Figure 6.11). The interrelation between the metabolism cyclohexane carboxylates and their benzenoid analogues may be seen in the pathways for the anaerobic degradation of cyclohexane carboxylate by *Rhodopseudomonas palustris*. This takes place by the action of a ligase (AliA) to form the coenzyme ester, followed by a dehydrogenase (BadJ) to produce cyclohex-1-ene-1-carboxy-CoA, which is then fed into the pathway used for the anaerobic degradation of benzoyl-CoA (Section 6.7.3.1; Egland and Harwood 1999).

Comparable oxidations are also used for the degradation of alicyclic compounds containing one or more rings such as terpenes and sterols. For example, some of them are part of the sequence of reactions involved in the degradation of the monocyclic monoterpene limonene (van der Werf et al. 1998) and the bicyclic monoterpene camphor, its derivatives, and structurally similar compounds. In *Pseudomonas putida* carrying the CAM plasmid, this involves an initial cytochrome P-450 hydroxylation at C-5 followed by oxidation and the introduction of an oxygen atom adjacent to the quaternary methyl group (Figure 6.12) (Ougham et al. 1983). In an organism designated *Mycobacterium rhodochrous*, hydroxylation occurs, however, at C-6, followed by ring fission of the 1,3-diketone (Figure 6.13) (Chapman et al. 1966).

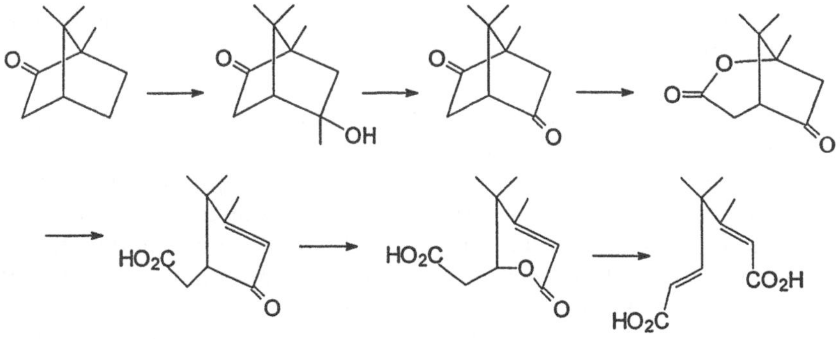

FIGURE 6.12
Degradation of camphor by *Pseudomonas putida*.

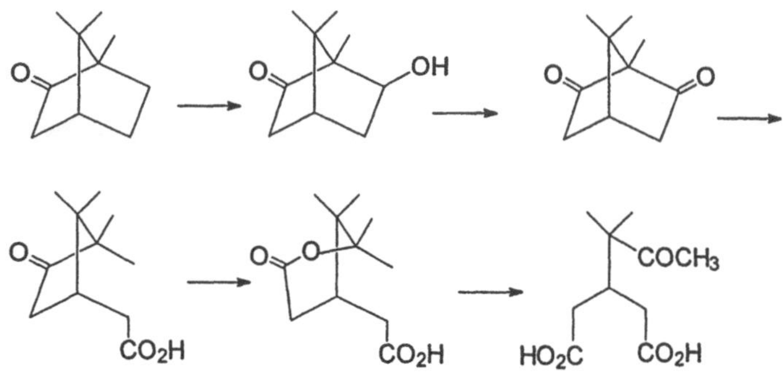

FIGURE 6.13
Degradation of camphor by *Mycobacterium rhodochrous*.

1. P-450$_{CAM}$ is able to hydroxylate the –CH$_3$ group of the quaternary methyl group of 5,5-difluorocamphor (Figure 6.14a) to the 9-hydroxymethyl compound (Eble and Dawson 1984), and adamantane and adamantan-4-one at the –CH quaternary carbon atom (Figure 6.14b) (White et al. 1984).

2. Patchoulol is transformed by *Botrytis cinerea* to a number of products principally to those involving hydroxylation at the C-5 and C-7 quaternary atoms (Aleu et al. 1999) (Figure 6.14c).

3. Cells of *Acinetobacter* sp. NCIB 9871 grown with cyclohexanol carried out enantiomerically specific degradation of 5-bromo-7-fluoronorbornanone and production of a lactone with >95% enantiomeric excess (Figure 6.15) (Levitt et al. 1990).

4. *Bauveria sulfurescens* stereospecifically hydroxylated an azabrendane at the quaternary carbon atom (Figure 6.16) (Archelas et al. 1988).

5. *Penicillium lilacinum* transformed testosterone successively to androst-4-ene-3,17-dione and testololactone (Prairie and Talalay 1963): once again, the oxygen atom is introduced into ring D at the quaternary position between C-13 and C-17.

The degradation of sterols and related compounds has been extensively studied and, for bile acids, involves a complex sequence of reactions that illustrate additional metabolic possibilities. For compounds oxygenated at C$_3$, initial reactions lead to the formation of the 1,4-diene-3-one, but the critical reaction that results in cleavage of the B-ring is hydroxylation at C$_9$ with formation of the 9,10-seco compound under the driving force of aromatization of the A-ring (Figure 6.17) (Leppik 1989).

The biodegradation of cyclic monoterpenes has been investigated under both aerobic and denitrifying conditions (Foss and Harder 1998), and may involve key reactions other than the Baeyer–Villiger type ring cleavage of ketones (Trudgill 1994). For example, in the degradation of α-pinene,

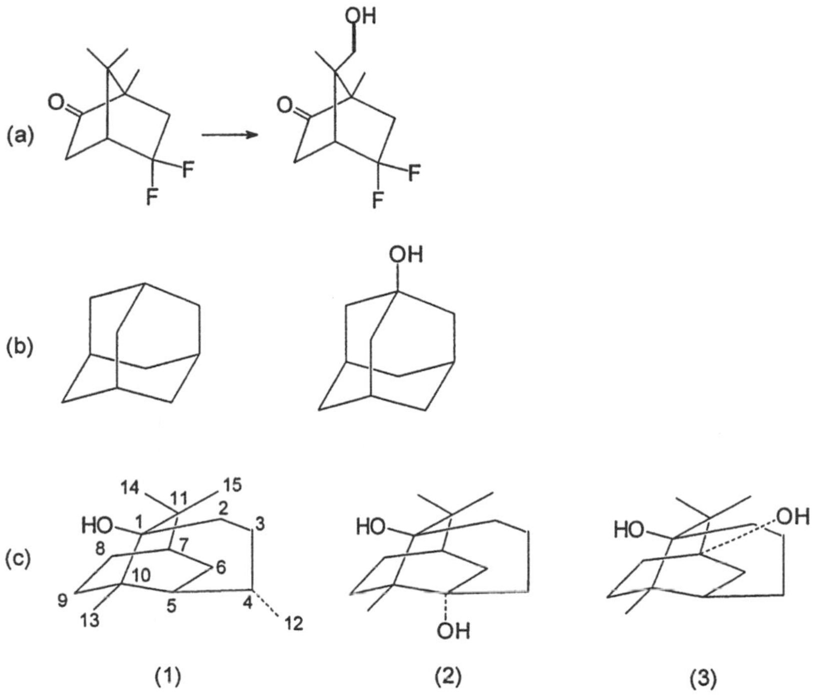

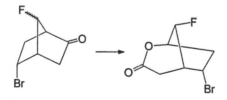

FIGURE 6.14
Hydroxylation of (a) 5,5-difluorocamphor, (b) adamantane, and (c) patchoulol.

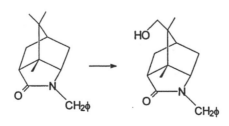

FIGURE 6.15
Hydroxylation of 5-bromo-7-fluoronorbornanone.

FIGURE 6.16
Hydroxylation of azabrendane derivative.

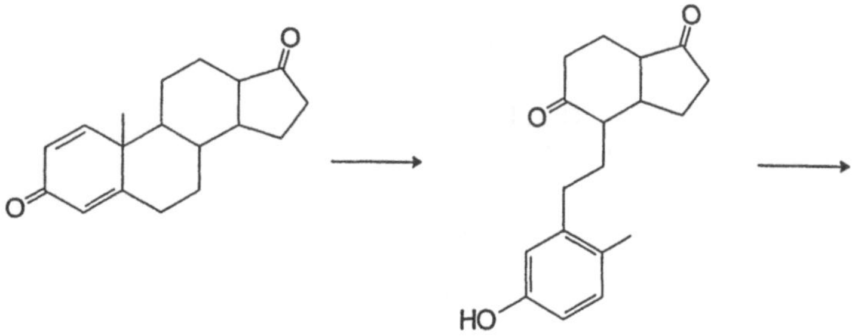

FIGURE 6.17
Biodegradation of a C_3-oxygenated bile acid.

although some strains of *Pseudomonas* sp. degrade this by rearrangement to limonene, oxidation, and ring cleavage of a β-ketoacid; in others, the initial reaction is formation of the epoxide that underwent rearrangement and ring cleavage of both the cyclohexane and cyclobutane rings to produce 2-methyl-5-isopropylhexa-2,5-dienal (Best et al. 1987; Griffiths et al. 1987) (Figure 6.18).

The degradation of atropine has been examined in *Pseudomonas* sp. strain AT3 and produces tropine as the initial metabolite. The degradation of this proceeds by oxidative loss of the *N*-methyl group, and elimination of ammonia to form 6-hydroxy-cyclohepta-1,4-dione followed by 1,3-diketone fission to 4,6-diketoheptanoate (Figure 6.19) (Bartholomew et al. 1996). Although the enzymology was unresolved, loss of ammonia presumably occurs either by successive hydroxylations at the tertiary carbon atoms adjacent to the –NH group or by successive dehydrogenations.

Attention should be directed to numerous transformation reactions—generally hydroxylations, oxidations of alcohols to ketones, or dehydrogenations—of both terpenes and sterols that have been accomplished by microorganisms especially fungi. This interest has been motivated by the great interest of the pharmaceutical industry in the products (Smith et al. 1988), and some of these reactions are summarized briefly in Section 6.11.2. An illustration of the plethora of reactions that may occur is afforded by the transformation of caryophyllene oxide by *Botrytis cinerea*. Although most of the reactions were hydroxylations or epoxidations, two involved transannular reactions (1) between the C-4-epoxide oxygen and C-7 and (2) between the C-4-epoxide and C-13 with formation of a caryolane (Figure 6.20) (Duran et al. 1999).

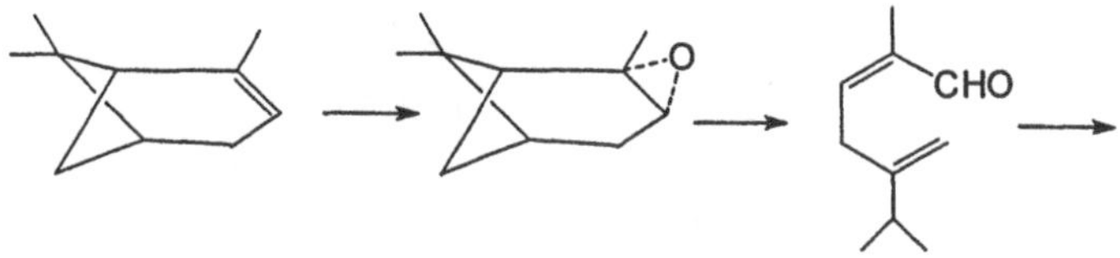

FIGURE 6.18
Degradation of α-pinene.

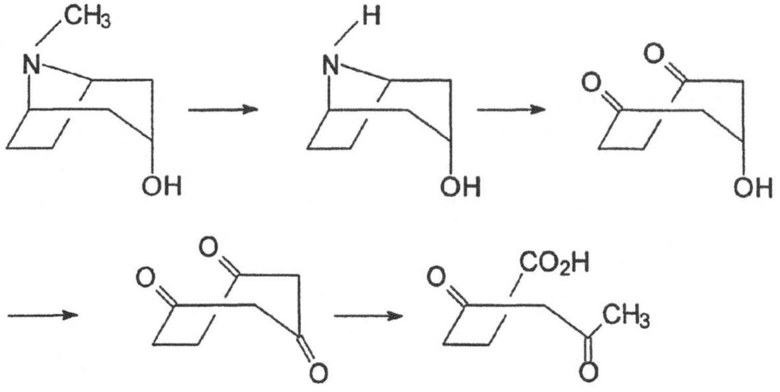

FIGURE 6.19
Degradation of tropine.

6.1.3 Alkenes

There are two different kinds of investigations which have been carried out:
(1) on growth of microorganisms at the expense of alkenes and (2) on
biotransformations resulting in the synthesis of epoxides. Studies have demonstrated growth, for example, at the expense of propene and butene (van
Ginkel and de Bont 1986), and an interesting observation is the pathway for
the degradation of intermediate *n*-alkenes produced by an aerobic organism
under anaerobic conditions (Parekh et al. 1977) that has already been noted
in Section 6.1.1. Although the generality of this pathway remains unknown,
it is clearly possible that such degradations might be accomplished even by
aerobic bacteria under anoxic conditions, and, for example, the degradation
of hexadecane may be accomplished at quite low oxygen concentrations
(Michaelsen et al. 1992). Attention should also be drawn to the possibility that
intermediate metabolites may be incorporated into biosynthetic pathways:
for example, hexadecene is oxidized by the fungus *Mortierella alpina* by ω-oxidation (Shimizu et al. 1991), but the lipids contain carboxylic acids with both
18 and 20 carbon atoms including the unusual polyunsaturated acid 5-*cis*, 8-
cis, 11-*cis*, 14-*cis*, 19-eicosapentaenoic acid.

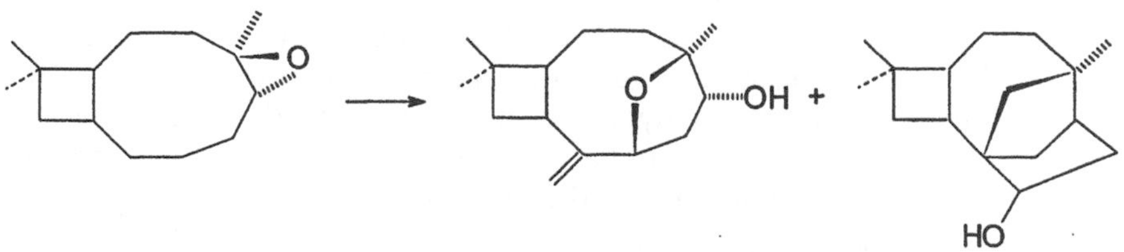

FIGURE 6.20
Transformation of caryophyliene oxide by *Botrytis cinerea*.

The degradation of epoxides is quite complex and several distinct pathways have been observed:

1. Degradation of epichlorohydrin (1-chloro-2,3-epoxypropane) may proceed by hydrolysis of the epoxide to 3-chloro-1,2-propanediol that is then converted successively to 3-hydroxy-1,2-epoxypropane (glycidol) followed by hydrolysis to glycerol before degradation (van den Wijngaard et al. 1989). Epoxide hydrolases have been isolated and characterized from bacteria that are able to use epoxides as growth substrates. A *Corynebacterium* sp. is able to grow with alicyclic epoxides, and the sequence of the hydrolase (Misawa et al. 1998) is similar to the enzyme from *Agrobacterium radiobacter* strain AD1 that used epichlorohydrin (1-chloro-2,3-epoxypropane) as growth substrate (Rink et al.1997). Examination of mutants of this strain prepared by site-directed mutagenesis showed that the mechanism involves nucleophilic attack by Asp107 at the terminal position of the substrate followed by hydrolysis of the resulting ester mediated by His 275. Analogy may be noted with the inversion accompanying hydrolysis of 2-haloacids mediated by L-2-haloacid hydrolase (Section 6.4.2). Limonene-1,2-epoxide hydrolase from *Rhodococcus erythropolis* catalyzes the formation of the *trans*-1,2-diol that is an intermediate in the degradation of limonen (Van der Werf et al. 1999), differs from these groups of enzymes, and does not involve the catalytic function of histidine residues (van der Werf et al. 1998).

2. Hydrolysis to the diol followed by dehydration to the aldehyde and oxidation to the carboxylic acid is used by a propene-utilizing species of *Nocardia* (de Bont et al. 1982). Although an ethene-utilizing strain of *Mycobacterium* sp. strain E44 degrades ethane-1,2-diol by this route, the diol is not an intermediate in the metabolism of the epoxide (Wiegant and de Bont 1980).

3. The aldehyde may be formed directly from the epoxide, and this reaction is involved in the metabolism of ethene by *Mycobacterium* sp. strain E44 (Wiegant and de Bont 1980), of styrene by a strain of *Xanthobacter* sp. strain 124X (Hartmans et al. 1989), and by *Corynebacterium* sp. strain ST-5 and AC-5 (Itoh et al. 1997). The reductase in the coryneforms has a low substrate specificity and is able to reduce acetophenone to 3-phenylethan-2-ol with an enantiomeric excess >96%. In *Rhodococcus rhodochrous*, however, styrene is degraded by ring dioxygenation with the vinyl group intact (Warhurst et al. 1994): 2-vinyl-*cis,cis*-muconate is produced by catechol 1,2-dioxygenase as a terminal metabolite, and complete degradation is carried out by catechol 2,3-dioxygenase activity that is also present.

4. A *Xanthobacter* sp. strain Py2 may be grown with propene or propene oxide. On the basis of amino acid sequences, the monooxygenase that produces the epoxide is related to those that catalyze the

monooxygenation of benzene, and toluene (Zhou et al. 1999). The metabolism of the epoxide takes place by alternative pathways.

a. In the absence of CO_2, by transformation to acetone that is not further degraded; the enzyme responsible is a pyridine nucleotide-disulfide oxidoreductase (Swaving et al. 1996).

b. In the presence of CO_2, 3-ketobutyrate is formed, and this is used partly for cell growth and partly converted into the storage product poly-β-hydroxybutyrate (Small and Ensign 1995). Kinetic and ^{13}C NMR experiments confirm that acetoacetate is the first product from which β-hydroxybutyrate is formed as a secondary metabolite and acetone as the terminal metabolite (Allen and Ensign 1996). The metabolism of acetate is accomplished by an ATP-dependent carboxylase (Sluis et al. 1996). The epoxide carboxylase is a three-component enzyme, all three of which are necessary for activity (Allen and Ensign 1997). Component II is a flavin containing NADPH: disulfide oxidoreductase that is identical to that noted above for degradation of epoxides in the absence of CO_2. The mechanism proposed for epoxypropane degradation involves nucleophilic reaction with the disulfide to produce a β-thioketone that is carboxylated to the β-ketoacid. A comparable mechanism operates in the degradation of epichlorohydrin (1-chloro-2,3-epoxypropane) by the same strain (Small et al. 1995).

5. *Rhodococcus* sp. strain AD45 carried out the transformation of 2-methyl-1,3-butadiene (isoprene), and both *cis*- and *trans*-dichloroethenes to the epoxides (van Hylckama Vlieg et al. 1998). The degradation of the dienes takes place by a pathway involving a glutathione S-transferase that is able to react with the epoxides and a conjugate-specific dehydrogenase that produces 2-glutathionyl-2-methylbut-3-enoate (van Hylckama Vlieg et al. 1999).

Epoxides may be formed from alkenes during degradation by *Pseudomonas oleovorans*, but oct-1,2-epoxide is not further transformed, and degradation of oct-1-ene takes place by ω-oxidation (May and Abbot 1973; Abbott and Hou 1973). The ω-hydroxylase enzyme is able to carry out either hydroxylation or epoxidation (Ruettinger et al. 1977).

Considerable attention has, however, been directed to the epoxidation of alkenes due to industrial interest in these compounds as intermediates. The wide metabolic capability of methane monooxygenase has been noted above and has been applied to the epoxidation of C_2, C_3, and C_4 alkenes (Patel et al. 1982). A large number of propane-utilizing bacteria are also effective in carrying out the epoxidation of alkenes (Hou et al. 1983). Especially valuable is the possibility of using microorganisms for resolving racemic mixtures of epoxides; for example, this has been realized for *cis*- and *trans*-2,3-epoxypentanes using a *Xanthobacter* sp. which is able to degrade only one of the pairs of enantiomers leaving the other intact (Figure 6.21) (Weijers et al. 1988). Bacterial epoxidation of alkenes and fungal enzymatic hydrolysis of epoxides have been reviewed in the context of their application to the synthesis of enantiomerically pure epoxides and their derivatives (Archelas and Furstoss 1997); further comment is given in Section 6.11.2. One of the disadvantages of using bacteria that may carry out undesirable degradation may sometimes be overcome by the use of fungi (Archelas and Furstoss 1992), although the initially formed epoxides are generally hydrolyzed by epoxide hydrolase activity.

Other aspects of epoxide formation and degradation are worth noting, particularly because of their biotechnological relevance.

- *Mycobacterium* sp. strain E3 is able to degrade ethene via the epoxide, but the epoxide-degrading activity is highly specific for epoxyethane, and the higher alkyl epoxides are not degraded and are favored by reductant generated from glycogen or trehalose storage material (de Haan et al. 1993).

- In *Xanthobacter* sp. strain Py2 both the alkene monooxygenase and the epoxidase are induced by C_2, C_3, and C_4 alkenes, and also by chlorinated alkenes including vinyl chloride, *cis*- and *trans*-dichloroethene and 1,3-dichloropropene (Ensign 1996).

6.1.4 Alkynes

The degradation of alkynes has been the subject of sporadic but effective interest for many years so that the pathway has been clearly delineated. It is quite distinct from those used for alkanes and alkenes, and is a reflection of the enhanced nucleophilic character of the alkyne C–C triple bond: the primary step is therefore hydration of the triple bond followed by ketonization of the initially formed enol. This reaction operates during the degradation of

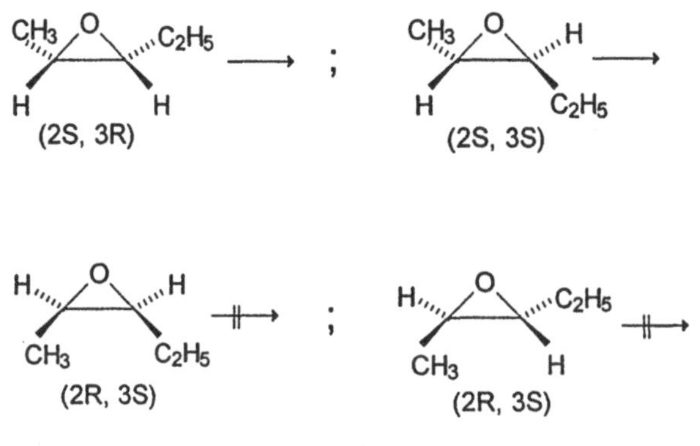

FIGURE 6.21
Biodegradability of enantiomeric of epoxides of *cis*- and *trans*-pent-2-enes.

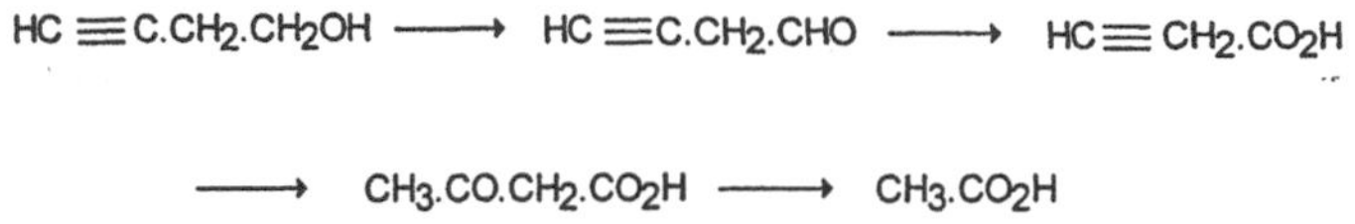

$$HC \equiv C.CH_2.CH_2OH \longrightarrow HC \equiv C.CH_2.CHO \longrightarrow HC \equiv CH_2.CO_2H$$

$$\longrightarrow CH_3.CO.CH_2.CO_2H \longrightarrow CH_3.CO_2H$$

FIGURE 6.22
Aerobic biodegradation of but-3-ynol.

acetylene itself (de Bont and Peck 1980), acetylene carboxylic acids (Yamada and Jakoby 1959), and more complex alkynes (Figure 6.22) (Van den Tweel and de Bont 1985). It is also appropriate to note that the degradation of acetylene by anaerobic bacteria proceeds by the same pathway (Schink 1985b).

6.2 Aerobic Degradation of Aromatic Hydrocarbons and Related Compounds

The degradation of aromatic compounds has attracted interest over many years, for at least four reasons:

1. They are significant components of creosote and tar that have traditionally been used for wood preservation.
2. They are components of unrefined oil and there has been serious concern over the hazard associated with their discharge into the environment after accidents at sea.
3. Many of the polycyclic representatives have been shown to be human carcinogens.
4. There has been increased concern over their presence in the atmosphere as a result of combustion processes and consequent air pollution.

Although growth at the expense of aromatic hydrocarbons has been known for many years (Söhngen 1913; Tausson 1927; Gray and Thornton 1928), it was many years later before details of the ring-cleavage reactions began to emerge. Two converging lines of investigations have examined in detail (1) the degradation of the monocyclic aromatic hydrocarbons benzene and toluene and (2) the degradation of oxygen-substituted compounds such as benzoate, hydroxybenzoates, and phenols. As a result of this activity, the pathways of degradation and their regulation are now known in considerable detail, and ever-increasing attention has been directed to the degradation of polycyclic aromatic hydrocarbons (PAH). Since many of these metabolic sequences recur in the degradation of a wide range of aromatic compounds, a brief sketch of the principal reactions may conveniently be presented here. Extensive reviews that include almost all aspects have been given (Hopper 1978; Cripps and Watkinson 1978; Ribbons and Eaton 1982; Gibson and Subramanian 1984; Neilson and Allard 1998). Developments in regulatory aspects have been presented (Rothmel et al. 1991; van der Meer et al. 1992; Parales and Harwood 1993).

It is important at the outset to appreciate two important facts:

1. For complete degradation of an aromatic hydrocarbon to occur, it is necessary that the products of ring oxidation and cleavage can be further degraded to molecules that enter anabolic and energy-producing reactions.

2. Essentially different mechanisms operate in bacteria and fungi, and these differences have important consequences. In bacteria, the initial reaction is carried out by dioxygenation and results in the synthesis of a *cis*-1,2-dihydro-1,2-diol which is then dehydrogenated to a catechol before ring cleavage by the further action of dioxygenases. In fungi, on the other hand, the first reaction is monooxygenation to an epoxide followed by hydrolysis to a *trans*-1,2-dihydro-1,2-diol and dehydration to a phenol: ring cleavage of PAHs does not generally occur in fungi, so that these reactions are essentially biotransformations. These reactions are schematically illustrated in Figure 6.23. It should be noted, however, that both fungi and yeasts are able to degrade simpler substituted aromatic compounds such as vanillate (Ander et al. 1983) (Figure 6.24) and 3,4-dihydroxybenzoate (Cain et al. 1968). It may also be noted that the degradation of 3,4-dihydroxybenzoate by the yeast *Trichosporon cutaneum* proceeds initially by a pathway different from that used by bacteria and involves hydroxylative decarboxylation to 1,2,4-trihydroxybenzoate prior to ring cleavage (Anderson and Dagley 1980).

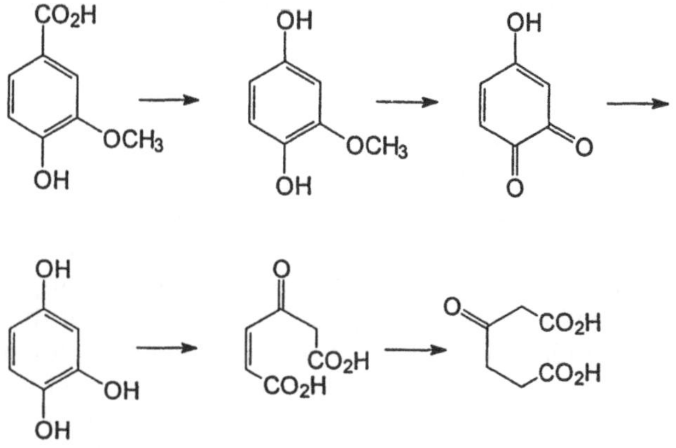

FIGURE 6.24
Biodegradation of vanillic acid by fungi.

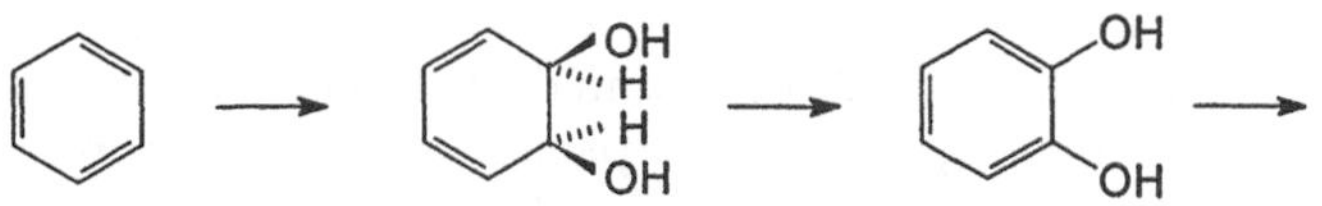

FIGURE 6.25
Biodegradation of benzene by *Pseudomonas putida*.

6.2.1 Bacterial Degradation of Monocyclic Aromatic Compounds

Benzene and alkyl benzenes

Details of the metabolism of benzene and alkylated benzenes have been established as a result of the classic studies of David Gibson and his collaborators (Gibson et al. 1968; 1970). The key intermediate from benzene is catechol that is formed by dioxygenation followed by dehydrogenation (Figure 6.25). Subsequent reactions involve fission of the aromatic ring by two pathways.

1. The reaction in which the bond between the two oxygen-bearing atoms is broken with the formation of a dicarboxylic acid has been termed the *ortho, endo, intra*diol or 1,2-cleavage—of which the last two seem most descriptive and appropriate since the enzyme carrying out this reaction is designated as a catechol-1,2-dioxygenase.

2. Alternatively, the bond between one of the oxygen-bearing atoms and the adjacent unsubstituted atom may be broken with the formation of a monocarboxylic acid and an aldehyde, and by analogy has been designated the *meta, exo, extra*diol or 2,3-cleavage.

The choice between these depends on the organism: for example, whereas in *Pseudomonas putida*, this is mediated by an extradiol 2,3-dioxygenase (Figure 6.26a), an intradiol 1,2-dioxygenase is involved (Figure 6.26b) in a species of *Moraxella* (Högn and Jaenicke 1972). The significance of which of these pathways is followed in halogenated aromatic compounds is discussed in Section 6.5.1.2, since it has particular significance in situations when two different substrates are simultaneously present. In addition, there are differences

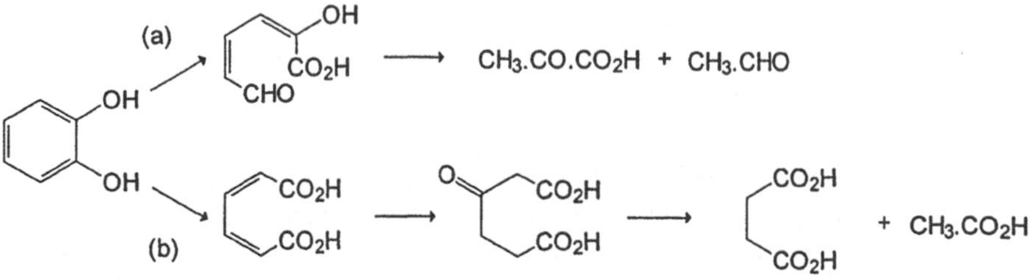

FIGURE 6.26
Metabolism of catechol (a) by catechol 2,3-dioxygenase in *Pseudomonas putida* and (b) by catechol 1,2-dioxygenase in *Moraxella* sp.

in the details of the 1,2-pathway for prokaryotic and eukaryotic cells, and these refinements have been discussed in detail (Cain et al. 1968; Cain 1988).

Although toluene degradation in pseudomonads may be induced by growth with the substrate or closely related aromatic compounds, it may also be induced by exposure to apparently unrelated substrates: (1) by trichloroethene in a strain of *P. putida* (Heald and Jenkins 1994) and (2) in *P. mendocina* strain KR1 by trichloroethene, pentane, and hexane, although not in *Burkholderia* (*Pseudomonas*) *cepacia*, or *P. putida* strain F1 (McClay et al. 1995).

For monoalkylated benzenes, there are two additional factors: (1) the genes may be either chromosomal or carried on plasmids and (2) oxidation may be initiated either on the aromatic ring or at the alkyl substituent. For example, toluene may be degraded by several different pathways.

1. When the catabolic genes are carried on the TOL plasmid, degradation takes place by sequential side-chain oxidation of the methyl group to a carboxylate (Whited et al 1986; Abril et al. 1989), followed by dioxygenation of the resulting benzoate to catechol that is cleaved by 2,3-dioxygenation (Figure 6.27a). The genes in the upper operon of the TOL plasmid of *P. putida* pWW0 occur in the order *xylCMABN* and encode, respectively, benzaldehyde dehydrogenase xylC, the two subunits of xylene oxygenase xylMA, benzyl alcohol dehydrogenase (xylB), and a protein with unknown function (Harayama et al. 1989). The enzymes encoded by this plasmid have a relaxed specificity which is consistent with the ability of organisms carrying the plasmid to degrade other alkyl benzenes such as xylenes and 1,2,4-trimethyl benzene.

2. In the second sequence, degradation is mediated by a 2,3-dioxygenase reaction with the methyl group intact (Figure 6.27b), and this pathway is followed in the metabolism of alkylated benzenes such as ethylbenzene and isopropylbenzene (Eaton and Timmis 1986).

3. Ring monooxygenation of toluene has been found in various taxa.

   a. In *P.* (*Burkholderia*) *cepacia* G4 that produces 2-methylphenol (Shields et al. 1989), in *P. mendocina* KR that produces 4-methylphenol (Whited and Gibson 1991) (Figure 6.27c), and in *P. pickettii* PKO1 that produces 3-methylphenol (Olsen et al. 1994). The last is mediated by a monooxygenase that can be induced by benzene, toluene, and ethylbenzene, and also by xylenes and styrene.

   b. *Pseudomonas* (*Burkholderia*) sp. strain JS150 contains both monooxygenase nooxygenase and dioxygenase activities, and initial products from the metabolism of toluene are therefore 3-methyl catechol produced by 2,3-dioxygenation, 4-methylcatechol by 4-monooxygenation and subsequent *ortho*-monooxygenation, and both 3- and 4-methylcatechols by 2-monooxygenation followed by *ortho*-monooxygenation (Johnson and Olsen 1997).

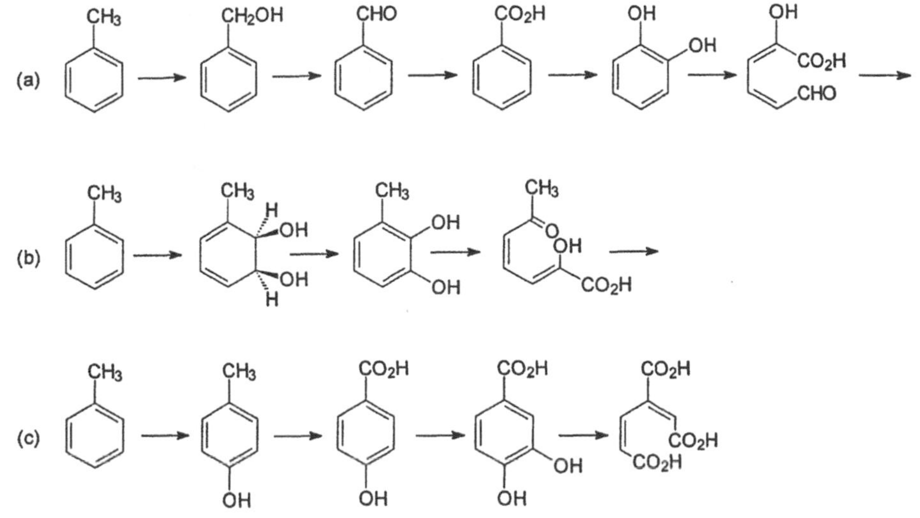

FIGURE 6.27
Biodegradation of toluene (a) by side-chain oxidation, (b) with the methyl group intact, and (c) by hydroxylation.

    c. The alkene monooxygenase in *Xanthobacter* sp. strain Py2 is able to carry out hydroxylation of benzene to phenol, and toluene to a mixture of 2-, 3-, and 4-methylphenol (Zhou et al. 1999).

The degradation of dialkylbenzenes such as the dimethylbenzenes (xylenes) depends critically on the position of the methyl groups (Baggi et al. 1987). Two distinct pathways have been found for the 1,4-isomer (Davey and Gibson 1974; Gibson et al. 1974) and these are illustrated in Figure 6.28.

Phenols and Benzoates

Considerable effort has been devoted to the bacterial metabolism of oxygenated compounds including phenol, catechol, benzoate, and hydroxybenzoates which are much more readily degraded than the parent hydrocarbons, and some of the details have been tacitly assumed in the foregoing discussion. An account of the appropriate oxygenases has been given in Chapter 4, Section 4.4.2, so that only a brief summary of the initial reactions is justified here.

    1. Hydroxylation of phenols by monooxygenases (Nurk et al. 1991);
    2. Decarboxylating dioxygenation of benzoate by a dioxygenases (Neidle et al. 1991);
    3. Hydroxylation and decarboxylation of salicylate by a monooxygenase (You et al. 1990).

From all of these substrates, catechol is formed, and this is metabolized by extradiol (2:3) or intradiol (1:2) ring cleavage mediated by dioxygenases and

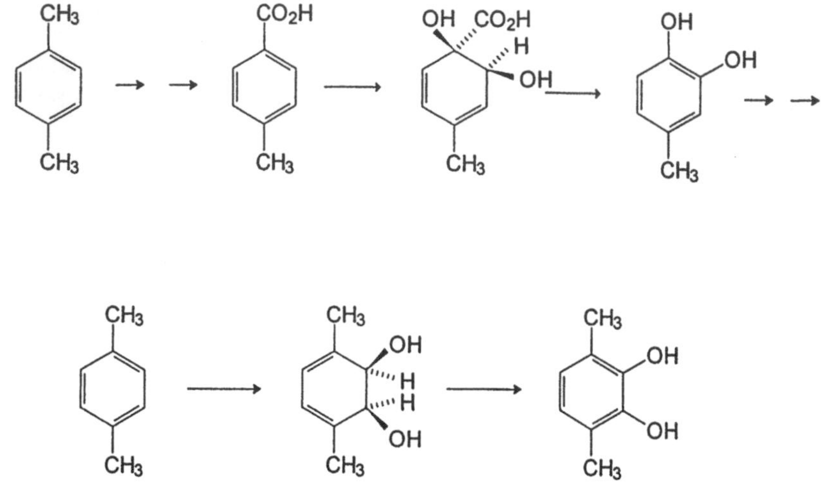

FIGURE 6.28
Biodegradation of 1,4-dimethylbenzene.

involve reactions that are similar to those used for the degradation of aromatic hydrocarbons. The regulation of some of the pathways has been briefly noted in Section 4.8, and experimental aspects have been discussed in Chapter 5, Section 5.5.2. Although the ring cleavage reactions are generally mediated by 1,2- or 2,3-dioxygenases after formation of the 1,2-dihydroxy compounds, there are important variations in the pathways used by various groups of microorganisms.

1. The pathways and their regulation during the degradation of catechol and 3,4-dihydroxybenzoate in *P. putida* have been elucidated in extensive studies (Ornston 1966). In this organism, metabolism proceeds by a 1,2-dioxygenase ring cleavage to produce β-ketoadipate (Figure 6.29). The stereochemistry of the reactions after ring cleavage has been examined in detail (Kozarich 1988), and the regulation and genetics in a range of organisms has been reviewed (Harwood and Parales 1996).

2. Pseudomonads of the acidovorans group, on the other hand, use a 4,5-dioxygenase system to produce pyruvate and formate (Wheelis et al. 1967) (Figure 6.30a).

3. The third alternative for the ring cleavage of 3,4-dihydroxybenzoate is exemplified in *Bacillus macerans* and *B. circulans* that use a 2,3-dioxygenase to accomplish this (Figure 6.30b) (Crawford 1975b; 1976). It may be noted that a 2,3-dioxygenase is elaborated by Gram-negative bacteria for the degradation of 3,4-dihydroxyphenylacetate (Sparnins et al. 1974) and by Gram-positive bacteria for the degradation of L-tyrosine via 3,4-dihydroxyphenylacetate (Sparnins and Chapman 1976).

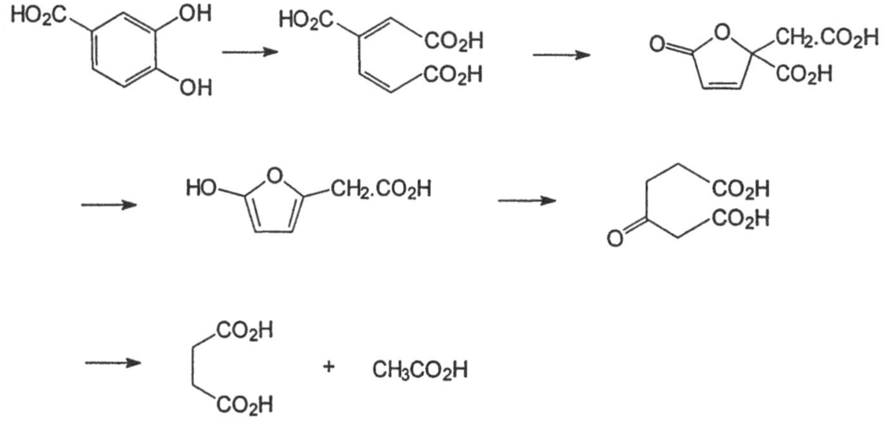

FIGURE 6.29
The β-ketoadipate pathway.

4. The enzymes of both pathways may be induced in a given strain by growth on different substrates: for example, growth of *P. putida* R1 with salicylate induces enzymes of the extradiol cleavage pathway, whereas growth with benzoate induces those of the intradiol pathway (Chakrabarty 1972). As a broad generalization, the extradiol cleavage is used in the degradation of more complex compounds such as toluene, naphthalene, and biphenyl (Furukawa et al. 1983), and is noted again in Sections 6.2.3 and 6.5.1.

The intradiol and extradiol enzymes are entirely specific for their respective substrates, and whereas all of the first group contain Fe^{3+}, those of the latter contain Fe^{2+} (Wolgel et al. 1993).

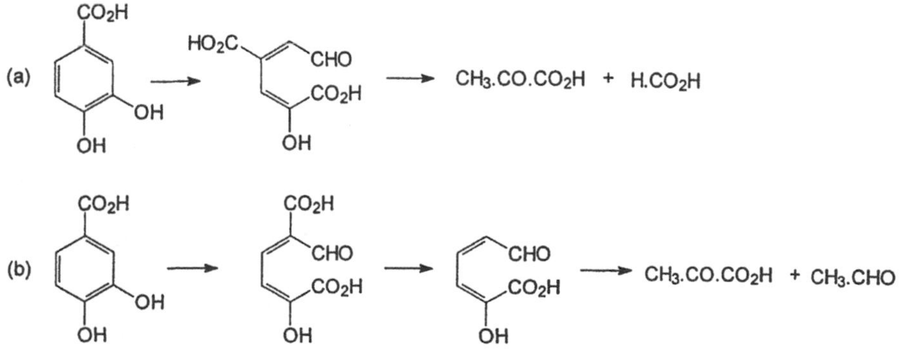

FIGURE 6.30
Biodegradation of 3,4-dihydroxybenzoate mediated by (a) a 4,5-dioxygenase system in *Pseudomonas acidovorans* and (b) by a 2,3-dioxygenase system in *Bacillus macerans*.

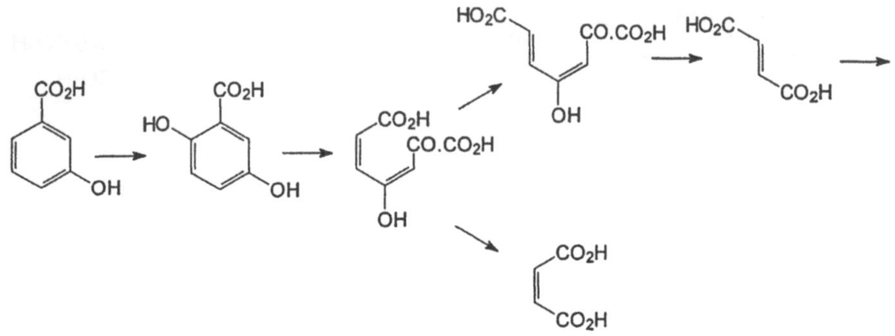

FIGURE 6.31
The gentisate pathway.

In some cases, hydroxylation to 1,4-dihydroxy compounds activates the ring to oxidative cleavage. This alternative pathway is followed during the degradation of 3-methylphenol, 3-hydroxybenzoate, and salicylate by a number of bacteria including species of *Pseudomonas* and *Bacillus*, and involves gentisate (2,5-dihydroxybenzoate) as an intermediate; ring cleavage then produces pyruvate and fumarate or maleate (Crawford 1975a; Poh and Bayly 1980) (Figure 6.31). This pathway may plausibly be involved in the degradation even of benzoate itself by a denitrifying strain of *Pseudomonas* sp. in which the initial reaction is the formation of 3-hydroxybenzoate (Altenschmidt et al. 1993). The gentisate pathway is used for the degradation of salicylate produced from naphthalene by a *Rhodococcus* sp. (Grund et al. 1992) rather than by the more usual sequence involving hydroxylative decarboxylation of salicylate to catechol. Gentisate is also formed in an unusual rearrangement reaction from 4-hydroxybenzoate by a strain of *Bacillus* sp. (Crawford 1976) that is formally analogous to the formation of 2,5-dihydroxyphenylacetate from 4-hydroxyphenylacetate by *P. acidovorans* (Hareland et al. 1975). It may be noted that the formal hydroxylation of 4-methylphenol to 4-hydroxybenzyl alcohol before conversion to 3,4-dihydroxybenzoate and ring cleavage is accomplished by initial dehydrogenation to a quinone methide followed by hydration (Hopper 1988).

Although benzoate is generally metabolized by oxidative decarboxylation to catechol followed by ring cleavage, nonoxidative decarboxylation may also occur: (1) strains of *Bacillus megaterium* transform vanillate to guaiacol by decarboxylation (Crawford and Olson 1978) and (2) a number of decarboxylations of aromatic carboxylic acids by facultatively anaerobic Enterobacteriaceae have been noted in Chapter 4, Section 4.3.2.

Decarboxylation is an integral part of the pathway for degradation of *o*-phthalate—under both aerobic and denitrifying conditions (Section 6.7.3.1). The degradation of *o*-phthalate by *P. fluorescens* PHK takes place by initial dioxygenation and dehydrogenation to 4,5-dihydroxyphthalate followed by decarboxylation to 3,4-dihydroxybenzoate (Pujar and Ribbons

1985), and the degradation of 5-hydroxyisophthalate takes place similarly via 4,5-dihydroxyisophthalate and decarboxylation to 3,4-dihydroxybenzoate (Elmorsi and Hopper 1979). The pathway for the degradation of *o*-phthalate in *Micrococcus* sp. strain 12B, however, involves 3,4-dihydroxybenzoate as intermediate (Eaton and Ribbons 1982). It is worth noting that by contrast, the degradation of 4-methyl-*o*-phthalate by *P. fluorescens* strain JT701 takes place by oxidative decarboxylation analogous to that of benzoate with the formation of 4-methyl-2,3-dihydroxybenzoate (Ribbons et al. 1984).

Anilines

The degradation of anilines is, in principle, straightforward and involves oxidative deamination followed by ring cleavage of the resulting catechols by either intradiol or extradiol ring fission. The deamination to catechol is apparently carried out in a strain of *Nocardia* by a dioxygenase (Bachofer and Lingens 1975), although details of the enzyme are not fully resolved (Fukumori and Saint 1997). The range of substituted anilines that have been examined includes the following: 2-, 3-, and 4-chloroanilines and 4-fluoroaniline by a *Moraxella* sp. strain G (Zeyer et al. 1985), 3-, and 4-methylanilines by *P. putida* mt-2 (McClure and Venables 1986), 2-methylaniline and 4-chloro-2-methylaniline by *R. rhodochrous* strain CTM (Fuchs et al. 1991), 3-chloro-4-methyl aniiline by *P. cepacia* strain CMA1(Stockinger et al. 1992), and aniline-2-carboxylate (anthranilate) (Taniuchi et al. 1964). Several aspects of the regulation of their metabolism are worth noting:

1. A derivative of *P. putida* mt-2 that was able to degrade aniline contained a plasmid pTDN1 that encodes the ability for degradation of aromatic amines (McClure and Venables 1987).

2. Aniline degradation is generally induced by aniline although both 3- and 4-chloroaniline that are poor substrates were able to induce the enzymes for aniline degradation in a strain of *Pseudomonas* sp. that was able to degrade aniline in the presence of readily degradable substrates such as lactate (Konopka et al. 1989).

3. The degradation of 3- and 4-chloroaniline may require the presence of either aniline or glucose (references in Zeyer et al. 1985) while the metabolism of methyl anilines required the addition of ethanol as additional carbon source (Fuchs et al. 1991).

4. The degradation of 3-chloro-4-methlaniline by *P. cepacia* strain CMA1 involved ring fission by an intradiol enzyme (Stockinger et al. 1992).

It is important to note that since anilines may be incorporated into humic material (Section 3.2.4) their fate is not determined solely by biodegradation. Further comments in the context of bioremediation are provided in Chapter 8, Section 8.2.4.4.

Acetophenones and Related Compounds

There are two quite different pathways that may be used for aromatic compounds with a C_2 side chain containing a carbonyl group adjacent to the benzene ring; this includes not only acetophenones but also reduced compounds that may be oxidized to acetophenones.

1. The mandelate pathway in *P. putida* proceeds by successive oxidation to benzoyl formate and benzoate that is further metabolized via catechol and the β-ketoadipate pathway (Figure 6.32a) (Hegeman 1966). Both enantiomers of mandelate were degraded through the activity of a mandelate racemase (Hegeman 1966), and the racemase (mdlA) is encoded in an operon that includes the following two enzymes in the pathway of degradation, *S*-mandelate dehydrogenase (mdlB) and benzoylformate decarboxylase (mdlC) (Tsou et al. 1990).

   A formally comparable pathway is used by a strain of *Alcaligenes* sp. that degrades 4-hydroxyacetophenone via 4-hydroxybenzoyl methanol to 4-hydroxybenzoate; this is further metabolized to β-ketoadipate via 3,4-dihydroxybenzoate (Figure 6.32b) (Hopper et al. 1985).

2. An entirely different sequence is followed during the metabolism of acetophenone by Gram-positive strains of *Arthrobacter* sp. and *Nocardia* sp. (Cripps et al. 1978) and of 4-hydroxyacetophenone and 4-ethylphenol by *P. putida* strain JD1 (Darby et al. 1987). Acetophenone is converted by a Baeyer–Villiger oxidation to phenyl acetate that is hydrolyzed to phenol and then hydroxylated to catechol before ring cleavage (Figure 6.32c). Similarly, 4-hydroxyacetophenone is oxidized to 4-hydroxyphenyl acetate that is hydrolyzed to 1,4-dihydroxybenzene before ring cleavage to β-ketoadipate. A mixed culture of an *Arthrobacter* sp. and a *Micrococcus* sp. was able to degrade 4-chloroacetophenone by an analogous sequence via 4-chlorophenyl acetate, 4-chlorophenol, and 4-chlorocatechol (Havel and Reineke 1993). Formally similar Baeyer–Villiger monooxidation of cycloalkanones has been noted previously in Section 6.1.2.

Although it can be concluded that reactions formally similar to those that have been considered in this section are involved in the aerobic degradation of a wide range of aromatic compounds, it may be convenient to summarize here some of the major exceptions to ring dioxygenation:

1. Some heterocyclic aromatic compounds—particularly those containing nitrogen—are degraded by reactions involving hydroxylation rather than dioxygenation before rupture of the rings (Section 6.3.1.1).

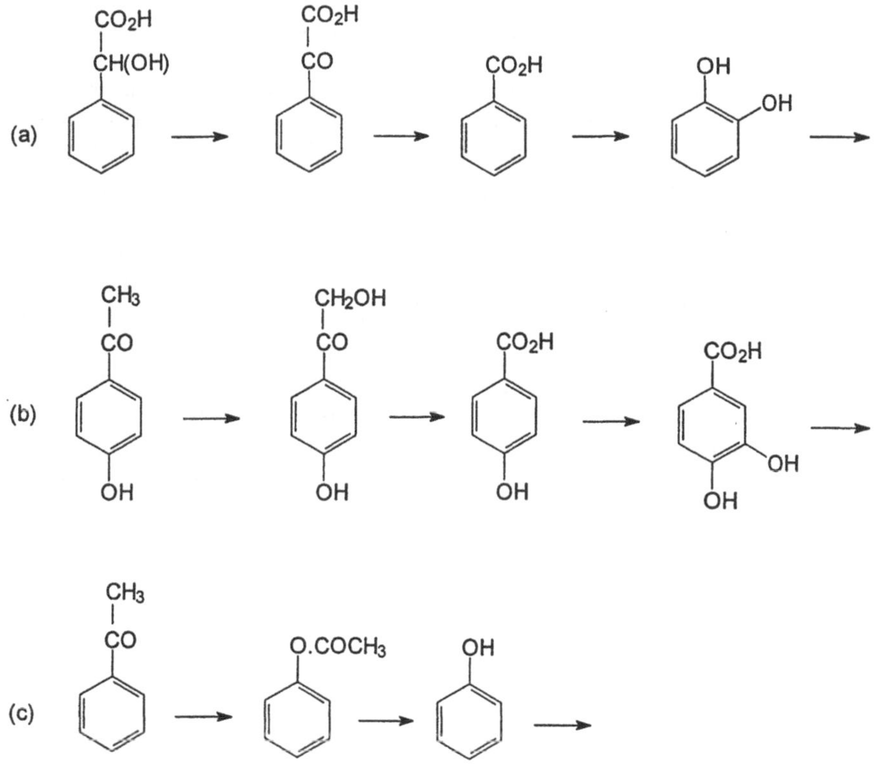

FIGURE 6.32
Degradation of (a) mandelate, (b) 4-hydroxyacetophenone by side-chain oxidation pathways, and (c) acetophenone by Baeyer–Villiger monooxygenation.

2. For halogenated phenols, a number of alternatives to direct hydroxylation (Beadle and Smith 1982) followed by ring cleavage are available (Section 6.5.1.2.).

6.2.2 Metabolism of Polycyclic Aromatic Hydrocarbons by Fungi and Yeasts

Fungal metabolism of PAHs has been studied in different contexts: (1) the analogy between the metabolic pathways used by fungi and by higher organisms (Smith and Rosazza 1983), (2) as a detoxification mechanism, and (3) due to interest in their use in bioremediation programs.

1. Extensive studies have been carried out with *Cuninghamella elegans* and this has undoubtedly been stimulated by concern with PAHs as human carcinogens, and there is no reason to doubt that the reactions carried out by *C. elegans* are representative of those carried out by many other fungi.

There are a number of aspects of the metabolism of PAHs by fungi which are worth noting since they differ from the reactions mediated by bacteria.

a. The phenol which is formed by rearrangement from the initially produced *trans* dihydrodiol may be conjugated to form sulfate esters or glucuronides (Cerniglia et al. 1982a; Golbeck et al. 1983; Cerniglia et al. 1986; Lange et al. 1994), and the less common glucosides have also been identified: 1-phenanthreneglucopyrano-side is produced from phenanthrene by *C. elegans* (Cerniglia et al. 1989) and 3-(8-hydroxyfluoranthene)-glucopyranoside from fluo-ranthene by the same organism (Pothuluri et al. 1990). As a further example of the range of carbohydrates that may be conjugated, the xylosylation of 4-methylguaiacol and vanillin by the basisiomycete *Coriolus versicolor* (Kondo et al. 1993) may be given.

b. The biotransformation of a considerable number of PAHs has been examined using *Cunninghamella elegans*: reactions are generally confined to oxidation of the rings with formation of phenols, cat-echols, and quinones, and ring cleavage does not generally take place. Different rings may be oxygenated, for example, in 7-meth-ylbenz[*a*]anthracene (Cerniglia et al. 1982b) (Figure 6.33), or oxida-tion may take place in several rings, for example, in fluoranthene (Pothuluri et al. 1990) (Figure 6.34).

c. Although the biotransformation of PAHs by fungi bears a rather close resemblance to that carried out by mammalian systems (Smith and Rosazza 1983), there is one very significant differ-ence—and that is the stereochemistry of the products. *trans*-1,2-Dihydroxy-1,2-dihydroanthracene and *trans*-1,2-dihydroxy-1,2-dihydrophenanthrene are formed from the hydrocarbons by

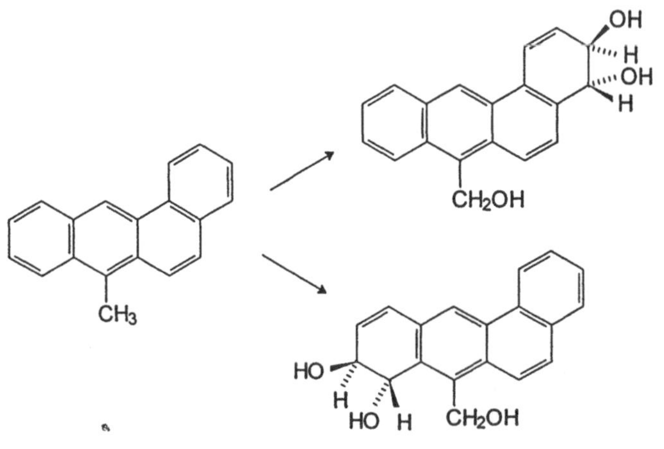

FIGURE 6.33
Biotransformation of 7-methylbenz[a]anthracene by *Cunninghamellaelegans*.

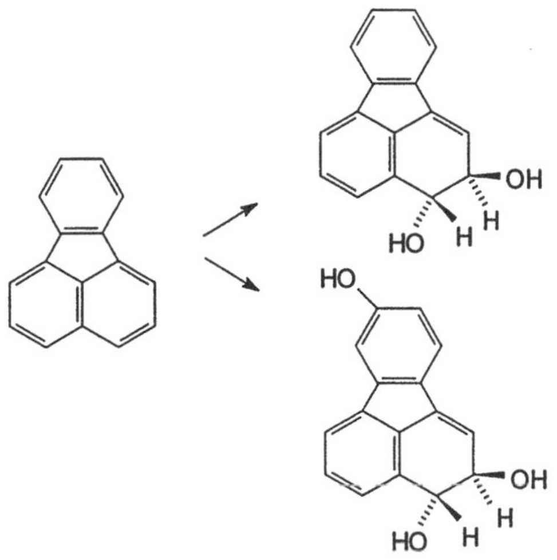

FIGURE 6.34
Alternative pathways for the biotransformation of fluoranthene by *Cunninghamella elegans*.

C. elegans, but these dihydrodiols have the *S,S* configuration in contrast to the *R,R* configuration of the metabolites from rat liver microsomes (Cerniglia and Yang 1984). It has become clear, however, that the situation among a wider range of fungi is much less straightforward. For example, the *trans*-9,10-dihydrodiol produced by *Phanerochaete chrysosporium* was predominantly the 9*S*,10*S* enantiomer whereas those produced by *C. elegans* and by *Syncephalastrum racemosum* were dominated by the 9*R*,10*R* enantiomers (Sutherland et al. 1993). Comparable differences were also observed for the *trans*-1,2-dihydrodiols and *trans*-3,4-dihydrodiols so that generalizations on the stereoselectivity of these reactions are currently unwarranted.

d. In reactions involving monooxygenase systems and formation of intermediate arene oxides, rearrangement of substituents may take place (Figure 6.35a); this is an example of the "NIH shift" which plays an important role in the metabolism of xenobiotics by mammalian systems (Daly et al. 1972), and has been observed in a few fungal systems (Figure 6.35b) (Faulkner and Woodcock 1965; Smith et al. 1981; Cerniglia et al. 1983), and bacterial systems (Figure 6.35c) (Dalton et al. 1981; Cerniglia et al. 1984; Adriaens 1994) including the marine cyanobacterium *Oscillatoria* sp. (Narro et al. 1992).

e. The metabolism of phenanthrene and pyrene by *Aspergillus niger* produces the corresponding 1-methoxy compounds (Sack et al. 1997).

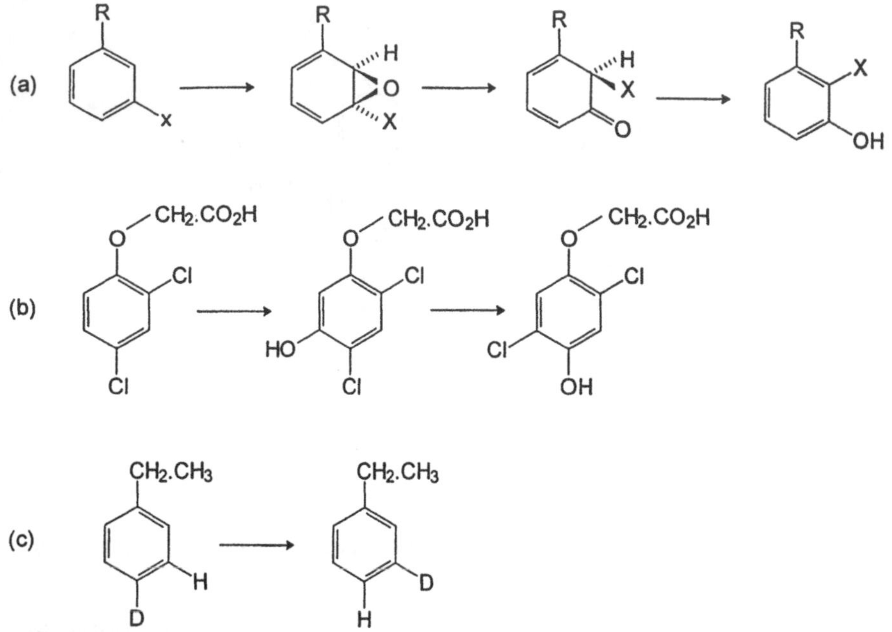

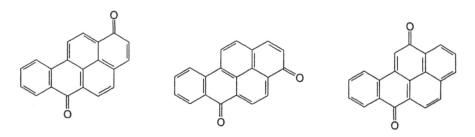

FIGURE 6.35
Examples of the NIH shift.

2. Increasing numbers of studies have been devoted to a wider range of fungi.

> a. White-rot fungi including *Phanerochaete chrysosporium* can partially mineralize a number of PAHs including phenanthrene and pyrene (Bumpus 1989; Hammel et al. 1992; Sack et al. 1997; Hofrichter et al. 1998). Lignin peroxidase from *Phr. chrysosporium* may produce quinones from PAHs. For example, benzo[*a*]pyrene is metabolized to the 1,6-, 3,6-, and 6,12-quinones (Figure 6.36) (Haemmerli et al. 1986), and it is interesting to note that the same quinones are among

FIGURE 6.36
Quinones produced during the metabolism of benzo[*a*]pyrene by *Phanerochaete chrysosporium*.

the metabolites produced by fish from the same substrate (Little et al. 1984). Different products are formed from phenanthrene by the activity of lignin peroxidases and by the monooxygenase systems of *Ph. chrysosporium* that are synthesized under different nitrogen regimes: the peroxidases produce 2,2'-diphenic acid via phenanthrene-9,10-quinone (Hammel et al. 1992), whereas the monooxygenase produces phenanthrene 3,4-oxide and phenanthrene 9,10-oxide that are further transformed into phenanthrene *trans*-dihydrodiols and phenanthrols (Sutherland et al. 1991). *Ph. chrysosporium* is unusual among eukaryotic microorganisms in its ability to carry out ring fission of aromatic hydrocarbons: anthracene is oxidized to anthra-9,10-quinone that is then cleaved to phthalate (Hammel et al. 1991). Comparable reactions may occur in *Ph. laevis* in which PAH quinones were not accumulated significantly (Bogan and Lamar 1996). The metabolism of phenanthrene by the white-rot fungus *Pleurotus ostreatus* has been studied, and proceeds analogously though by cytochrome P-450-mediated epoxidation, hydrolysis to the 9,10,dihydrodiol, and oxidative fission to 2,2'-diphenic acid (Bezalel et al. 1997).

b. The metabolism of phenanthrene and pyrene has been studied in *Nematoloma frowardii* (Sack et al. 1997), that carried out mineralization and transformation to phenanthrene-9,10-dihydrodiol and pyrene-3,4-dihydrodiol, respectively. It has therefore become evident that quite extensive mineralization can be carried out by white-rot fungi, that the extent of this may vary considerably among taxa, and that *Ph. chrysosporium* may indeed be one of the less effective organisms.

c. The basisiomycete *Crinipellis stitpitata* metabolized pyrene to a number of phenolic compounds including 1-hydroxypyrene, *trans*-pyrene-4,5-dihydrodiol, and 1,6- and 1,8-dihydroxypyrenes that are precursors to the corresponding quinones (Lange et al. 1994).

A more extensive discussion of other PAHs is given by Neilson and Allard (1998).

The transformation of a few polycyclic aromatic hydrocarbons has also been investigated in yeasts. The metabolism of naphthalene, biphenyl, and benzo[*a*]pyrene has been examined in a strain of *Debaryomyces hansenii* and in number of strains of *Candida* sp. The results using *C. lipolytica* showed that the transformations were similar to those carried out by fungi: the primary reaction was formation of the epoxides that were then rearranged to phenols (Cerniglia and Crow 1981). Benzo[*a*]pyrene is transformed by *Saccharomyces cerevisiae* to the 3- and 9-hydroxy compounds and the 9,10-dihydrodiol, and the cytochrome P-448 that mediates the monooxygenation has been purified and characterized (King et al. 1984).

6.2.3 Metabolism by Bacteria of PAHs and Related Phenols and Carboxylic Acids

In contrast to the situation with fungi, bacteria may grow at the sole expense of PAHs: ring cleavage takes place after dioxygenation and dehydrogenation. For polycyclic aromatic compounds, successive ring degradation may occur, so that the structure is ultimately degraded to molecules which enter central anabolic pathways. An example of these reactions has already been given for the simplest representative—benzene itself—and reference may be made to a review (Neilson and Allard 1998) that includes further details of the enzymology and specific comments on individual PAHs.

There are a number of general conclusions that can be drawn from the extensive studies that have been carried out on the degradation of a wide range of aromatic hydrocarbons.

Naphthalene

Degradation of naphthalene is readily carried out by many bacteria, and both the details of the initial steps (Jeffrey et al. 1975) and their enzymology have been elucidated. The enzymes for the two key steps—naphthalene dioxygenase (Patel and Barnsley 1980; Ensley and Gibson 1983) and *cis*-naphthalene dihydrodiol dehydrogenase (Patel and Gibson 1974)—have been purified, while further details of the subsequent steps have been added (Eaton and Chapman 1992). The degradation is mediated by (1) naphthalene 1,2-dioxygenase, (2) *cis*-1,2-dihydro-1,2-dihydroxynaphthalene dehydrogenase, (3) 1,2-dihydroxynaphthalene dioxygenase, (4) 2-hydroxychromene-2-carboxylate isomerase, (5) *cis*-2-hydroxybenzylidenepyruvate aldolase, and (6) salicylaldehyde dehydrogenase, and the overall pathway is shown in Figure 6.37. The sequence of the genes encoding the enzymes in the plasmid NAH7 cloned from *Pseudomonas putida* G1064 has been determined as *nahA, nahB, nahF, nahC, nahE, nahD*—naphthalene dioxygenase, naphthalene *cis*-dihydroxydiol dehydrogenase, 1,2-dihydroxynaphthalene dioxygenase, 2-hydroxychromene-2-

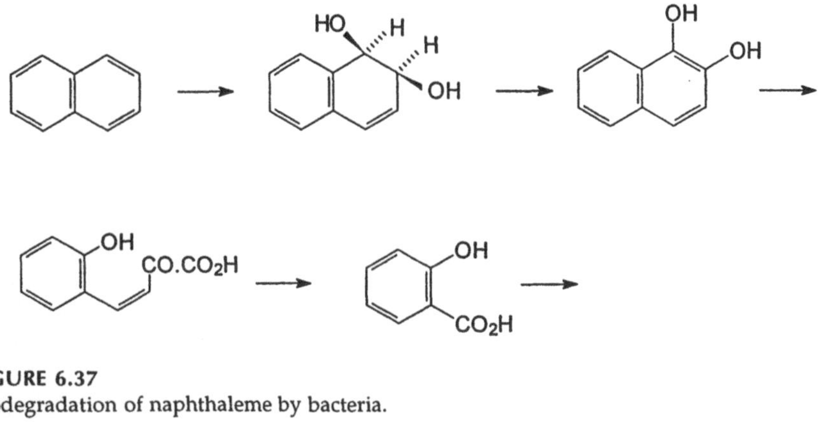

FIGURE 6.37
Biodegradation of naphthaleme by bacteria.

carboxylate isomerase, *cis*-2-hydroxybenzylidenepyruvate aldolase. Various details deserve further comment.

- In some strains of pseudomonads, the degradation of the interme- diate catechol produced by the activity of salicylate hydroxylase may proceed by the alternative intradiol cleavage pathway (Barns- ley 1976). Reference has been made to the alternative gentisate pathway for the degradation of the intermediate salicylate (Grund et al. 1992).

- The enzymes for the complete sequence of enzymes involved in the oxidative degradation of naphthalene to catechol and for the extradiol (*meta*) cleavage pathway for degradation of the catechol are inducible by growth with salicylate in *P. putida* (Austen and Dunn 1980).

- The naphthalene dihydrodiol dehydrogenase NahB from *P. putida* strain G7 has been purified as the his-tagged enzyme, and shown to catalyze also the dehydrogenation of biphenyl-2,3-dihydrodiol. biphenyl-3,4-dihydrodiol, and 2,2′,5,5′-tetrachlorobiphenyl-3,4- dihydrodiol (Barriault et al. 1998). In addition, 1,2-dihydroxynaph- thalene dioxygenase carried out extradiol fission of 3,4-dihydrox- ybiphenyl at both the 2,3- and 4,5-positions.

- It has been noted in Section 4.4.1.1 that naphthalene dioxygenase from a strain of *Pseudomonas* sp. also carries out enantiomeric monooxygenation of indan and dehydrogenation of indene (Gibson et al. 1995), and the stereospecific hydroxylation of (*R*)-1-indanol and (*S*)-1-indanol to *cis*-indan-1,3-diol and *trans* (1*S*,3*S*)-indan-1,3- diol (Lee et al. 1997); the indantriols are also formed by further reactions. Essentially comparable reactions have been observed with *Rhodococcus* sp. strain NCIMB 12038 (Allen et al. 1997).

Alkylated Naphthalenes

Methyl naphthalenes are important components of crude oils, and their deg- radation follows in principle that used in the initial stages for alkylated ben- zenes. The degradation of 2,6-dimethylnaphthalene by flavobacteria involves a pathway analogous to that for dimethylbenzenes—successive oxi- dation of the methyl group to carboxylate, dioxygenation to 1,2-dihydroxy- 6-methylnaphthalene, and ring fission to 5-methylsalicylate, followed by fur- ther degradation by pathways established for naphthalene itself (Barnsley 1988). Degradation of 1-methylnaphthalene took place with formation of 3- methyl catechol, whereas 2-methylnaphthalene produced 4-hydroxymethyl- catechol (Mahajan et al. 1994). Oxidation of 1,5-, 2,6-, 2,7-, and 1,8-dimethyl naphthalenes by a recombinant strain of *P. aeroginosa* PAO1 involved succes- sive oxidation of only a single methyl group to the monocarboxylates, except for 1,8-dimethylnaphthalene in which both methyl groups were oxidized to the dicarboxylate (Selifonov et al. 1996). The oxidation of a wide range of dimethyl naphthalenes has been examined in *Sphingomonas paucimobilis* strain 2322 (Dutta et al 1998). Degradative pathways involved successive oxidation of a methyl group in one of the rings to the corresponding carboxylate; this was then degraded either by (1) decarboxylative dioxygenation to a 1,2-dihydroxynaphthalene followed by 1:1a extradiol fission and formation of salicylate or by (2) hydroxylation to a 1-hydroxynaphthalene-2-carboxylate and 1:2 intradiol ring fission with formation of *o*-phthalate.

Naphthols and Naphthalene Carboxylates

Less attention seems to have been devoted to the degradation of naphthols and naphthalene carboxylates.

1. The degradation of naphth-1-ol has been described (Bollag et al. 1975) and the specificity of oxygen uptake in whole cells—although not in cell-free extracts—has been studied (Larkin 1988). Two strains, *Pseudomonas* sp. strain 12043 and *Rhodococcus* sp. strain NCIB 12038, were less versatile, whereas *Pseudomonas* sp. strain 12042 was more versatile showing oxygen uptake with 1- and 2-naphthols, 1,5- and 2-7-dihydroxynaphthalene, and naphthalene itself. The subsequent degradation of the 1,2-dihydroxynaphthalene formed from the naphthols may plausibly be assumed to proceed by ring fission and subsequent established reactions that produce salicylate and then catechol (strain 12042) or gentisate (strains 12043 and 12038).

2. The degradation of naphthalene-2-carboxylate by *Burkholderia* sp. strain JT 1500 involves formation of naphthalene-1-hydroxy-2-carboxylate rather than initial oxidative decarboxylation. Naphthalene-1,2-dihydrodiol-2-carboxylate is not involved, so that possibly the reaction is carried out by a monooxygenase. Subsequent reactions produce pyruvate and *o*-phthalate that is degraded via 4,5-dihydroxyphthalate (Morawski et al. 1997); this has already been noted. Degradation of naphthalene carboxylates formed by oxidation of methyl groups was noted above.

Hydrocarbons with More Than Two Rings

Degradation becomes increasingly difficult for the more highly condensed compounds and depends not only on the number of rings but on the degree of annelation. Strains that degrade a range of PAHs have been isolated and include the following:

- Naphthalene, phenanthrene, fluoranthene, and 3-methylcholanthrene by a *Mycobacterium* sp. (Heitkamp et al. 1988a);

- Naphthalene, phenanthrene, and anthracene by *Alcaligenes denitrificans* strain WW1 (Weissenfels et al. 1991);
- Phenanthrene, pyrene, and fluoranthene by *Mycobacterium* sp. strain BB1 (Boldrin et al. 1993);
- Naphthalene, phenanthrene, anthracene, and fluorene by *P. cepacia* strain F297 (Grifoll et al. 1995);
- Naphthalene, anthracene, and phenanthrene by *P. putida* strain GZ44 (Goyal and Zylstra 1996);
- Pyrene, benzo[*a*]pyrene, and benz[*a*]anthracene by *Mycobacterium* sp. strain RJGII-135 (Schneider et al. 1996);
- Degradation of benz[*a*]anthracene, chrysene, and benzo[*a*]pyrene by *P. saccharophila* (Aitken et al. 1998);
- Transformation of biphenyl, phenanthrene, anthracene, fluoranthene, benzo[*b*]fluorene, chrysene, and pyrene by *P. paucimobilis* strain EPA505 (Mueller et al. 1990), and degradation of fluoranthene, pyrene, benz[*a*]anthracene, chrysene, benzo[*a*]pyrene, and benzo[*b*]fluoranthene by the same strain (Ye et al. 1996);
- Cell suspensions of a strain of *Burkholderia cepacia* VUN 10,001 were able to metabolize phenanthrene, pyrene, benz[*a*]anthracene, benzo[*a*]pyrene, dibenz[*a*,*h*]anthracene, and coronene (Juhasz et al. 1997).

Phenanthrene is more readily degraded than anthracene, and two different pathways are followed after cleavage of the peripheral ring: in one, the naphthalene pathway via salicylate is used (Evans et al. 1965), whereas in the other *o*-phthalate and 3,4-dihydroxybenzoate are involved (Kiyohara et al. 1976; Kiyohara and Nagao 1978; Barnsley 1983) (Figure 6.38). The enzymes involved, 1-hydroxy-2-naphthoate dioxygenase (Iwabuchi and Harayama

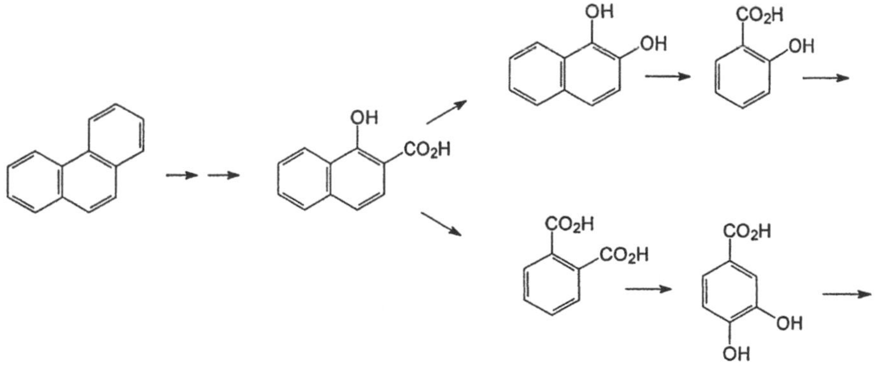

FIGURE 6.38
Alternative pathways for the biodegradation of phenanthrene.

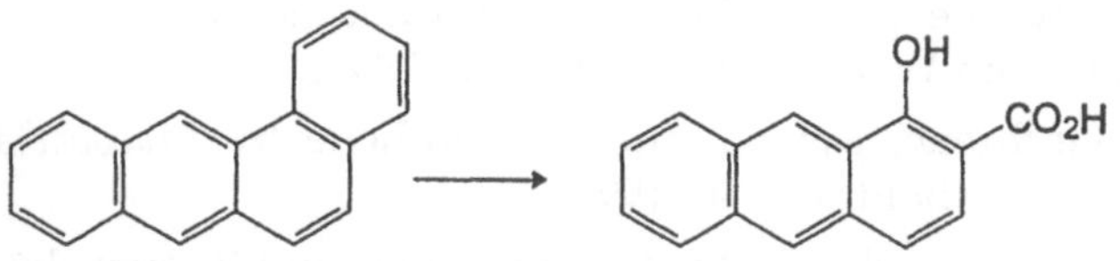

FIGURE 6.39
Biotransformation of benz[*a*]anthracene by *Beijerinckia* sp.

1998a), the *trans*-2'-carboxybenzalpyruvate hydratase-aldolase (Iwabuchi and Harayama 1998b), and the 2-carboxybenzaldehyde dehydrogenase that catalyzes the formation of phthalate (Iwabuchi and Harayama 1997), have been purified and characterized from *Nocardioides* sp. strain KP7. The purified dioxygenase was used to determine further details of the degradation pathways (Adachi et al. 1999): the initial product from ring fission is in equilibrium with the lactone formed by reaction between the carboxyl group and the activated double bond of the benzalpyruvate in a reaction formally comparable with the formation of muconolactones during degradation of monocyclic aromatic compounds. An important exception to the generalization that dioxygenation is used for aromatic hydrocarbons is provided by *Streptomyces flavovirens* that produces (-)*trans*-[9*S*,10*S*]-9,10-dihydrodihydroxyphenanthrene with minor amounts of 9-hydroxyphenanthrene (Sutherland et al. 1990). This is mediated by a cytochrome P-450 system of which further examples have been given in Section 4.4.1.2.

The situation with higher PAHs is more complex, and details of all the pathways have not been conclusively established; it seems, however, that degradation of all rings is not readily accomplished. For example, a strain of *Beijerinckia* sp. although unable to grow at the expense of benz[*a*]anthracene readily formed 1-hydroxy-2-carboxyanthracene which was at least partly further degraded (Figure 6.39) (Mahaffey et al. 1988). For compounds with peri fused structures such as fluoranthrene and pyrene, degradation has also been established. Degradation of pyrene by a *Mycobacterium* sp. has been established and a plausible pathway has been proposed (Heitkamp et al. 1988) (Figure 6.40). One interesting and unusual observation for a bacterial system was the formation of both *cis*- and *trans*-dihydrodiols: whereas the former is to be expected as the product of a dioxygenase system, the latter must be produced by a monooxygenase which might be comparable to that involved in the epoxidation of alkenes to epoxides by a *Mycobacterium* sp. (Hartmans et al. 1991). The strain of *Mycobacterium* sp. that degraded pyrene also degraded fluoranthene with the initial formation of 1-acenaphthenone and of fluoren-9-one that was then further metabolized (Kelley et al. 1993).

The effect of substituents including halogen, sulfonate, and nitro will be discussed in more detail in Sections 6.5.1, 6.8.1, and 6.8.2, but a few general comments may usefully be inserted here. The presence of these electron-attracting substituents generally increases resistance to bacterial degradation, although at least for naphthalenes, the substituted compounds can be

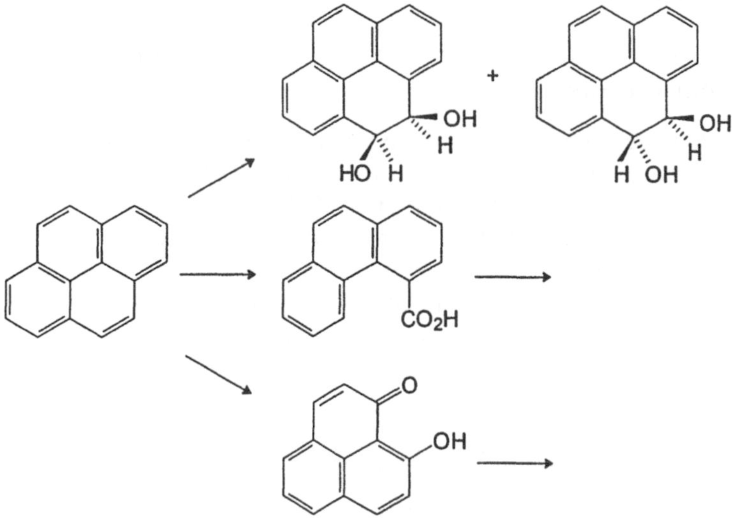

FIGURE 6.40
Biodegradation of pyrene.

degraded. There are also significant differences depending on whether or not mechanisms exist for the early elimination of the substituents. Depending on its position, sulfonate may be eliminated at an early stage and the resulting intermediate is then readily degraded. On the other hand, chlorine is generally retained, and in compounds containing several aromatic rings, the ring with fewest chlorine substituents is degraded first; for example 1,4-dichloronaphthalene is degraded to 3,6-dichlorosalicylate (Figure 6.41) (Durham and Stewart 1987) and 2,4,4′-trichlorobiphenyl is degraded to 2,4-dichlorobenzoate (Furukawa et al. 1979). The degradation of polychlorinated biphenyls is discussed further in Section 6.5.1.1.

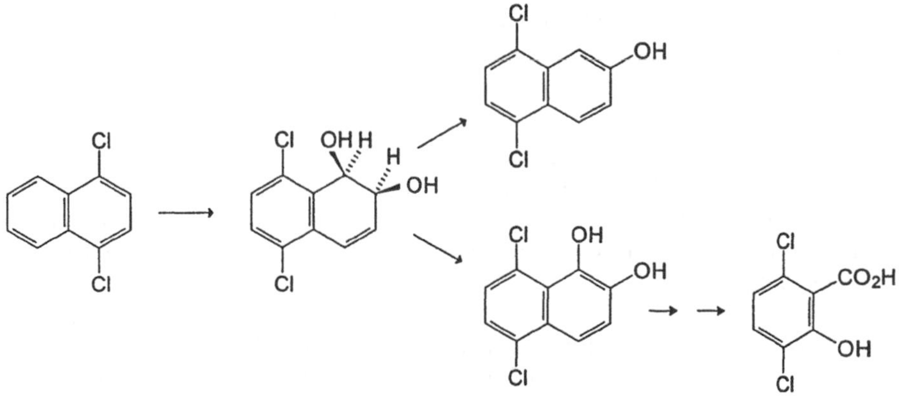

FIGURE 6.41
Biodegradation of 1,4-dichloronaphthalene.

6.3   Aerobic Degradation of Heterocyclic Aromatic Compounds

Quite extensive investigations have been directed to the biodegradation of heterocyclic aromatic compounds since a number of these are constituents of crude oil and creosote (Sundström et al. 1986; Herod 1998), some are used as agrochemicals, and many of them are important chemical intermediates. On the other hand, the polyhalogenated dibenzo-1-4-dioxins and dibenzofurans that have not been deliberately produced may be said to have achieved notoriety. Reference may be made to a more extensive review (Neilson and Allard 1998).

6.3.1   Reactions Mediated by Bacteria

The chemical reactivity of simple heterocyclic aromatic compounds varies widely: in electrophilic substitution reactions, thiophene is similar to benzene and pyridine is less reactive than benzene, while furan and pyrrole are susceptible to polymerization reactions; conversely, pyridine is more readily susceptible than benzene to attack by nucleophilic reagents. These differences are to a considerable extent reflected in the susceptibility of these compounds and their benzo analogues to microbial degradation. In contrast to the almost universal dioxygenation reaction used for the bacterial degradation of aromatic hydrocarbons, two broad mechanisms operate for heterocyclic aromatic compounds:

1. Hydroxylation of the ring by reactions formally comparable to those carried out by salicylate hydroxylase (White-Stevens et al. 1972), 4-hydroxybenzoate hydroxylase (Howell et al. 1972), or 2,4-dichlorophenol hydroxylase (Beadle and Smith 1982; Perkins et al. 1990). These have been discussed in Section 4.4.1.1.

2. Dioxygenation presumably by enzymes formally comparable to those used for the degradationof aromatic hydrocarbons.

These reactions are not mutually exclusive, however, and both may operate in sequence. Indeed, monooxygenation may, in some cases, be the result of dioxygenation followed by elimination. A review of the metabolism of aromatic compounds with N, O, and S hetero atoms has been given (Neilson and Allard 1998).

- Naphthalene 1,2-dioxygenase activity induced in *Pseudomonas* sp. strain NCIB 9816-4 and from *Beijerinckia* sp. strain B8/36 by incubation with carbazole produced 3-hydroxycarbazole (Resnick et al. 1993). Although this product could have resulted from the initial production of *cis*-3,4-dihydro-3,4-dihydroxycarbazole, in light of the demonstration of monoxygenase activity mediated by naphthalene 1,2-dioxygenase (Gibson et al. 1995), direct monooxygenation cannot be excluded.

- The second step in the degradation of quinoline by *P. putida* strain 86 involves introduction of oxygen at the 8-position, and consists of two components, a reductase containing FAD and a [2Fe–2S] cluster, and an oxygenase containing Rieske-type (2Fe–2S) clusters (Rosche et al. 1995). The amino acid sequence of the reductase is similar to class 1B reductases whereas that for the oxygenase suggests an affinity with class 1A oxygenases (Rosche et al. 1997).

It should be noted, however, that for *N*-heterocyclic compounds in which degradation takes place by hydroxylation of the hetero ring, the oxygen atom comes from water and the enzymes are oxidoreductases.

6.3.1.1 Hydroxylation Reactions

The degradation of furan-2-carboxylate (Trudgill 1969), thiophene-2-carboxylate (Cripps 1973), and pyrrrole-2-carboxylate (Hormann and Andreesen 1991) proceeds by initial ring hydroxylation: after ring cleavage, 2-ketoglutarate is produced from all of these compounds (Figure 6.42a to c) and this then enters central anabolic and catabolic pathways. A strain of *Xanthobacter*

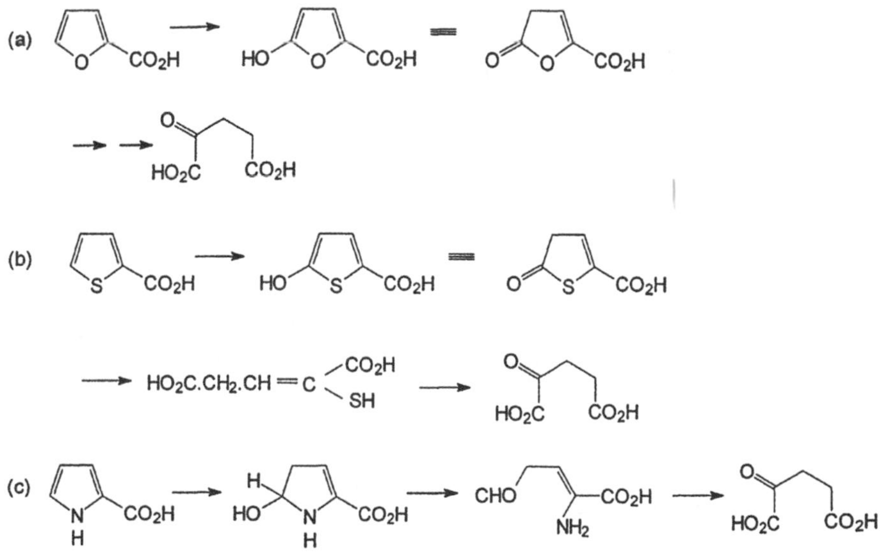

FIGURE 6.42
Biodegradation of (a) furan-2-carboxylate, (b) thiophene-2-carboxylate, and (c) pyrrole-2-carboxylate.

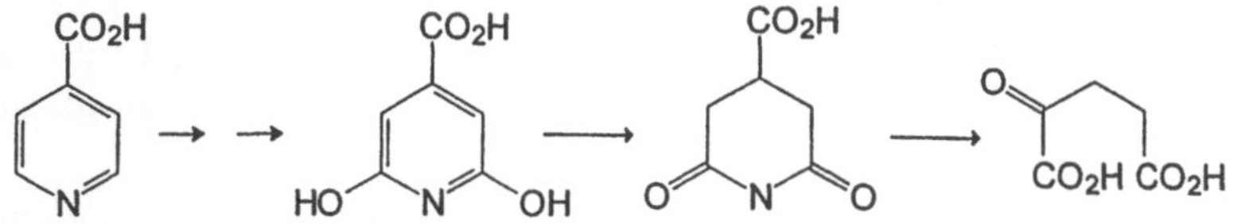

FIGURE 6.43
Biodegradation of pyridine-4-carboxylate.

tagetidis was able to use thiophene 2- and 3-carboxylates, furan-2-carboxylate, and pyrrole-2-carboxylate for growth, and may possibly degrade these by the same mechanism (Padden et al. 1997).

The situation with pyridines is less uniform, and this is consistent with the significant chemical differences between the five- and six-membered N-heterocyclic systems. For pyridines, both hydroxylation and the dioxygenation typical of truly aromatic compounds have been observed, and these are often accompanied by reduction of one or more of the double bonds in the pyridine ring. Some examples are given to illustrate the various metabolic possibilities.

1. The degradation of pyridine-4-carboxylate takes place by successive hydroxylations before reduction and ring cleavage (Kretzer and Andreesen 1991) (Figure 6.43).

2. The degradation of 6-methylpyridine-3-carboxylate takes place by selective hydroxylation at C-2 (Tinschert et al. 1997).

3. Hydroxylation occurs as the initial step during the degradation of 4-hydroxypyridine and is followed by dioxygenation before ring cleavage (Figure 6.44) (Watson et al. 1974).

4. Degradation of pyridine itself by a species of *Nocardia* takes place by hydroxylation after initial reduction of the ring (Watson and Cain 1975).

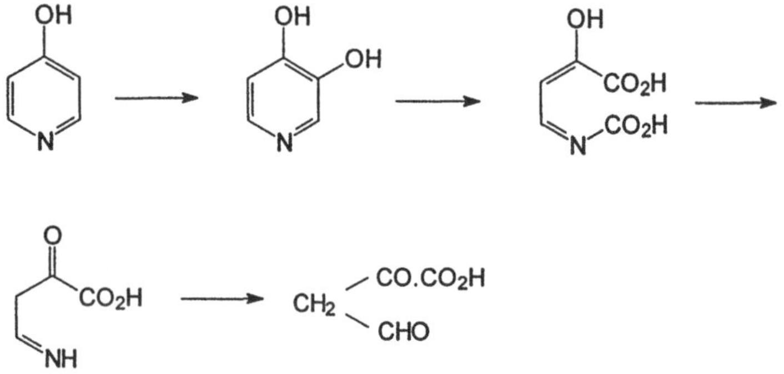

FIGURE 6.44
Biodegradation of 4-hydroxypyridine.

5. Degradation of quinoline by pseudomonads and by a strain of *Rhodococcus* sp. has been investigated (Schwarz et al. 1989): the initial reaction was hydroxylation of the heterocyclic ring for all the organisms examined, although subsequent reactions were different for the Gram-negative and Gram-positive organisms (Figure 6.45a,b). The enzymes from *Comamonas testosteroni* for 2-hydroxylation (quinoline 2-oxidoreductase), and the dioxygenase responsible for the introduction of oxygen in the benzene ring (2-oxox-1,2-dihydroquinoline-5,6-dioxygenase) have been described (Schach et al. 1995). The hydroxylase showed similarities to the enzyme from the other quinoline-degrading organisms *P. putida* strain 86 and *Rhodococcus* sp. strain B1. The initial step in the degradation of isoquinoline by *P. diminuta* strain 7 is also a hydroxylation that leads to the successive formation of 1-hydroxyisoquinoline, *o*-phthalate, and 3,4-dihydroxybenzoate that undergoes extradiol cleavage (Röger et al. 1995). When the position adjacent to the N-atom is substituted by a methyl group, hydroxylation by quinaldine oxidoreductase takes place at the 4-position (De Beyer and Lingens 1993).

It may therefore be concluded that hydroxylation of the pyridine ring is a widely distributed metabolic reaction and this has been confirmed using $H_2^{18}O$ during the hydroxylation of both nicotinate and 6-hydroxynicotinate (Hirschberg and Ensign 1971).

6.3.1.2 Dioxygenation Reactions

As noted above, the degradation of pyridines may involve either hydroxylation or dioxygenation reactions—and frequently both of them. Dioxygenation may be illustrated by three examples:

1. The degradation of pyridine by a species of *Bacillus* (Watson and Cain 1975) produces formate and succinate (Figure 6.46).
2. The degradation of 2,5-dihydroxypyridine has been examined using $^{18}O_2$ and $H_2^{18}O$ (Gauthier and Rittenberg 1971) and shown to involve the incorporation of both atoms of oxygen, one each into formate and maleamate (Figure 6.47).
3. The degradation of quinol-4-one involves initial hydroxylation at C_3 but ring cleavage is then mediated by a dioxygenase with the formation of anthranilate (Block and Lingens 1992).

Whereas degradation of the carboxylates of the monocyclic furan, thiophene, and pyrrole involves hydroxylation, metabolic pathways for their benzo analogues apparently involve dioxygenation. This has been proposed for indole (Figure 6.48) (Fujioka and Wada 1968), carbazole (Sato et al.

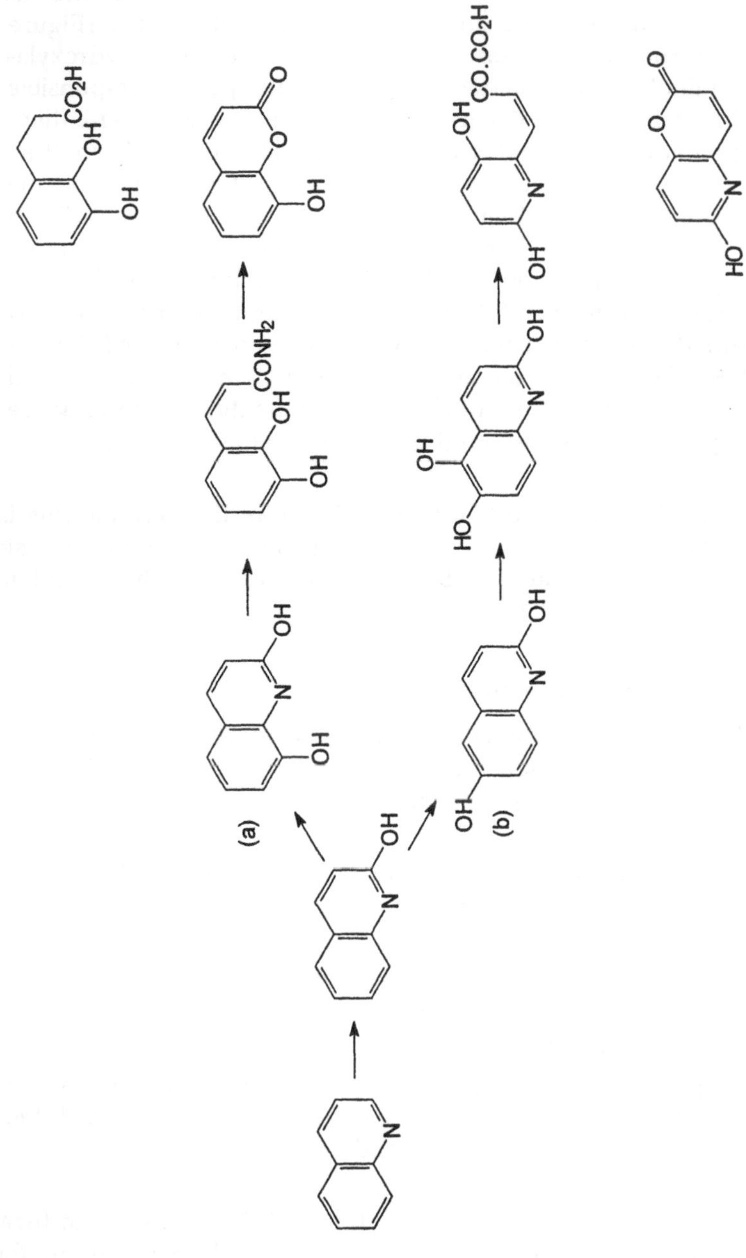

FIGURE 6.45
Alternative pathways for the biodegradation of quinoline.

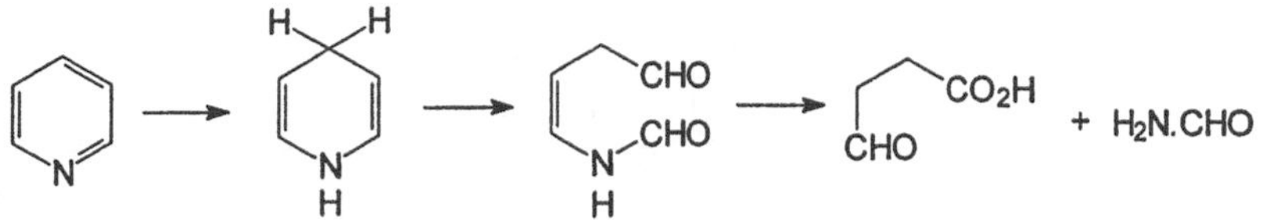

FIGURE 6.46
Biodegradation of pyridine.

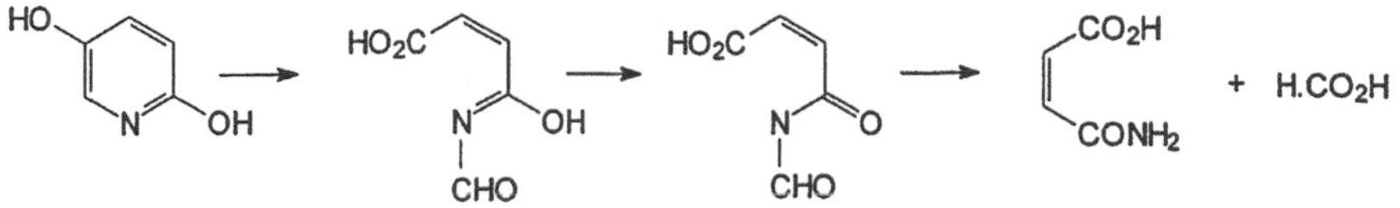

FIGURE 6.47
Biodegradation of 2,5-dihydroxypyridine.

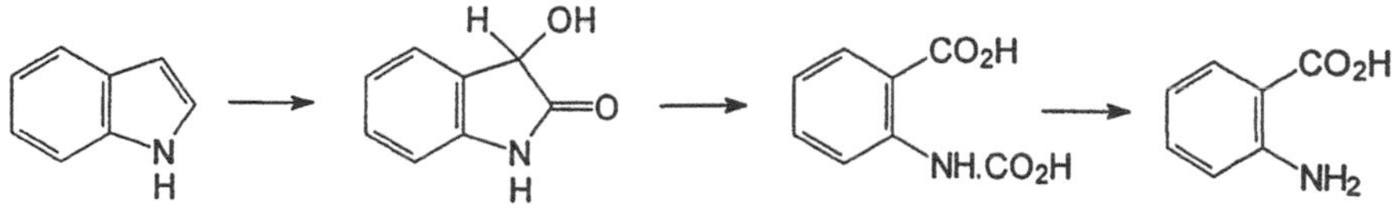

FIGURE 6.48
Biodegradation of indole.

1997a,b), dibenzofuran (Figure 6.49) (Fortnagel et al. 1990) and dibenzo-1,4-dioxin (Figure 6.50) (Wittich et al. 1992).

The degradation by *Pseudomonas* sp. strain LD2, which uses carbazole as sole source of C, N, and energy, revealed a complex set of intermediates (Gieg et al. 1996): anthranilic acid and catechol were intermediates, together with a number of terminal transformation products including indole-3-acetate. These products may plausibly arise from angular 1,9-dioxygenation, and from 1,2-dioxygenation (Figure 6.51). Carbazole 1,9a-dioxygenase that produces 2'-amino-biphenyl-2,3-diol has been examined in *Pseudomonas* sp. strain CA10 and consisted of a terminal dioxygenase, a ferredoxin, and a ferredoxin reductase (Sato et al. 1997b) that were products of the genes *carAa*, *carAc*, and *carAd*. The extradiol fission enzyme and the hydrolase that produced anthranilic acid were products of *carB* and *carC*, respectively (Sato et al. 1997a).

Some details of the enzymology of the dioxygenation reactions in the oxaarenes have been elucidated: for example, the 3,4-dihydroxyxanthone dioxygenase (Chen and Tomasek 1991) used for the degradation of xanthone by a strain of *Arthrobacter* sp. (Figure 6.52) (Tomasek and Crawford 1986), and that involved in the angular dioxygenation of dibenzofuran by a strain of *Sphingomonas* sp. have been characterized (Bünz and Cook 1993). The pathways

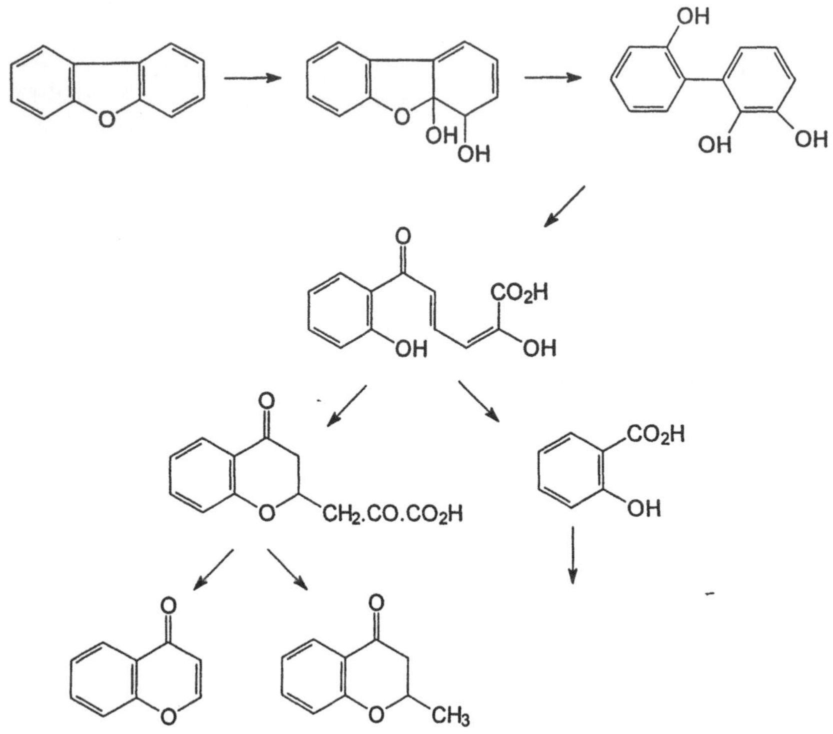

FIGURE 6.49
Biodegradation of dibenzofuran.

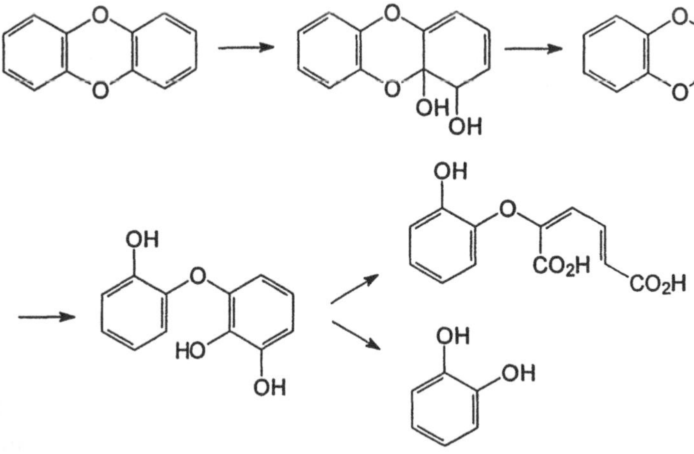

FIGURE 6.50
Biodegradation of dibenzo[1,4]dioxin.

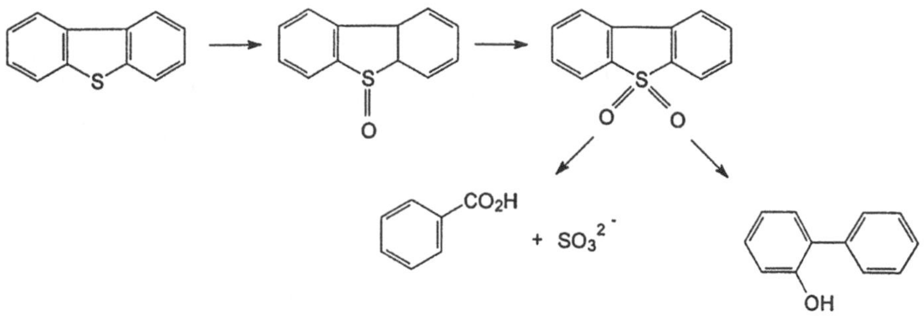

FIGURE 6.51
Degradation of carbazole.

for these oxygen heterocyclic compounds require little further comment except to note the occurrence of the angular dioxygenation reaction in the degradation of dibenzofuran. This is supported by the isolation of the 1,10-dihydrodiol of fluorene-9-one during degradation of fluorene by a dibenzo-furan-degrading strain of *Brevibacterium* sp.(Engesser et al. 1989), and the overall pathway by the identification of a trihydroxybiphenyl from dibenzo-furan before further degradation to salicylate (Strubel et al. 1991). The key enzyme for ring fission of 2,2′,3-trihydroxybiphenyl has been biochemically and genetically analyzed in a strain of *Sphingomonas* sp. (Happe et al. 1993), and three extradiol dioxygenases are involved in the degradation of dibenzo-furan by *Terrabacter* strain DPO360 (*Brevibacter* sp. strain DPO360): two in ring fission of the intermediate 2,2′,3-trihydroxybiphenyl (BphC1 and BphC2) and a catechol 2,3-dioxygenase (Schmid et al. 1997). The oxidation of dibenzo-1,4-dioxin by cultures of a strain of *Sphingomonas* sp. grown with diphenyl ether

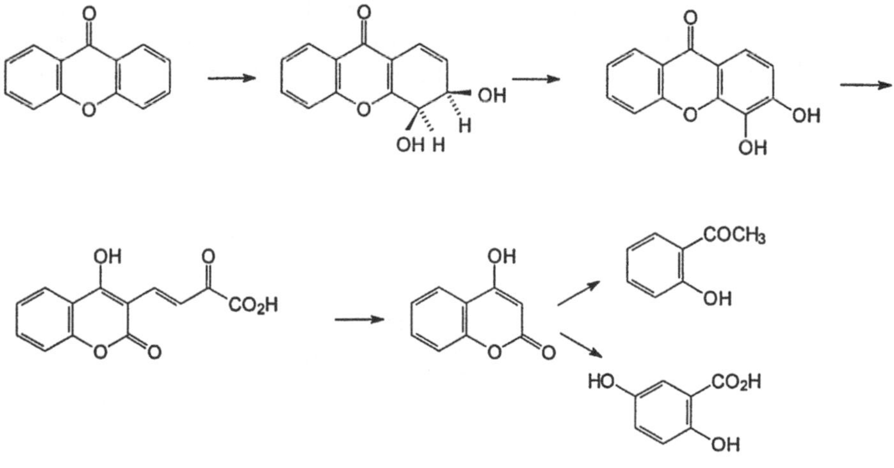

FIGURE 6.52
Biodegradation of xanthone.

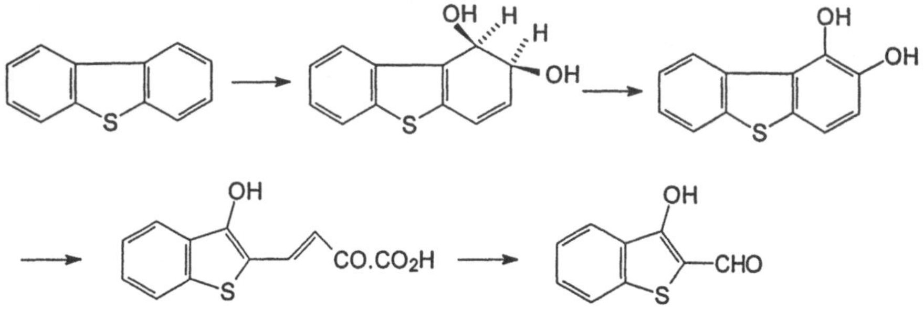

FIGURE 6.53
Alternative pathways for the biodegradation of dibenzothiophene.

is noted in Section 6.9.1, and the versatility of members of the genus has been associated with the scattering of the genes for the component proteins of the dioxygenase system around the genome of the dibenzo[1,4]dioxin-degrading *Sphingomomas* sp. strain RW1 (Armengaud et al. 1998); this paper includes a valuable summary of three-component dioxygenases. A summary of the reactions involved in the biodegradation of dibenzo[1,4]dioxins, dibenzo-furans, diphenyl ethers, and fluoren-9-one has been given (Wittich 1998).

A greater diversity of pathways for the degradation of dibenzothiophene has emerged and the following alternatives are worth specific comment:

1. Dioxygenation of one of the rings occurs, and after dehydrogena-tion to a catechol, ring cleavage takes place (Figure 6.53) (Kodama et al. 1973): this pathway is strictly analogous to that used for the degradation of naphthalene and the isoelectronic anthracene.

2. Successive oxidation at the sulfur atom occurs with formation of the sulfoxide followed by elimination of sulfite to yield either 2-hydroxy-biphenyl (Omori et al. 1992) or benzoate (van Afferden et al. 1990) (Figure 6.54). The pathway for the formation of 2-hydroxybiphenyl

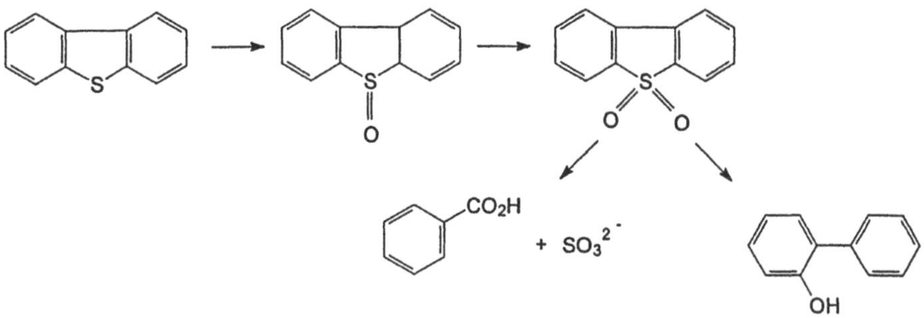

FIGURE 6.54
Alternative pathways for the biodegradation of dibenzothiophene.

in *Rhodococcus* sp. strain IGTS8 (ATCC 53968) has been elucidated. The enzymes are carried on an operon *dsz* which encodes three proteins Dsz C, Dsz A, and DszD which carry out oxidation to the sulfoxide, ring fission to 2′-hydroxybiphenyl-2-sulfinate and reductive scission to 2-hydroxybiphenyl and sulfite (Oldfield et al. 1997). In *Rhodococus* sp. strain ECRD-1, degradation involves successive oxidation to the sulfoxide that is further degraded primarily via 2′-hydroxybiphenyl-2-sulfonate to 2-hydroxybiphenyl and sulfate (Macpherson et al. 1998); the cyclization sultones and sultines were also isolated.

3. The *Corynebacterium* sp. which utilizes dibenzothiophene as a sulfur source produces 2-hydroxybiphenyl and subsequently nitrates this using the nitrate in the growth medium to form two hydroxynitrobiphenyls (Omori et al. 1992) (Figure 6.55a). This reaction is reminiscent of a similar one that takes place during the metabolism of 4-chlorobiphenyl (Figure 6.55b) (Sylvestre et al. 1982). These products are plausibly formed from arene oxide intermediate produced by monooxygenase systems (Section 4.4.1.1).

Attention has been drawn in Section 4.4.2 to the ability of naphthalene dioxygenase from *Pseudomonas* sp. strain 9816-4 to produce dibenzothiophene *cis*-1,2-dihydrodiol (Resnick and Gibson 1996). The transformation of naphtho[2,1-*b*]thiophene and naphtho[2,3-*b*]thiophene by *Pseudomonas* sp. strains grown with 1-methylnaphthalene has been examined (Knopp et al. 1997); as for the corresponding PAHs, the angular substrate (naphtho[2,1-*b*]thiophene) was more readily tranformed with the formation of hydroxybenzothiophene carboxylates by dioxygenation of the outer benzene ring followed by extradiol fission.

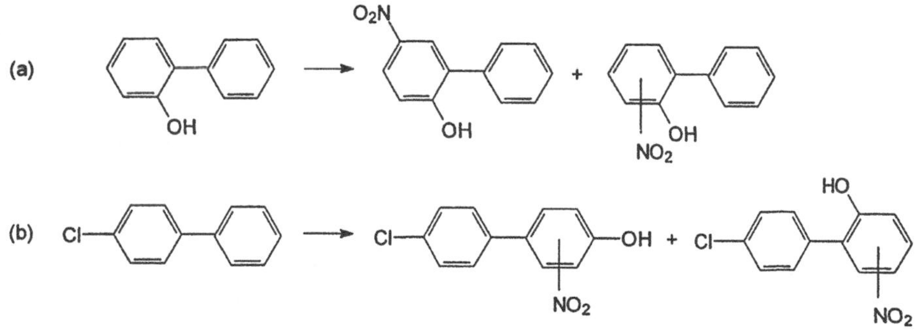

FIGURE 6.55
Formation of nitrohydroxybiphenyls during metabolism of (a) dibenzothiophene and (b) 4-chlorobiphenyl.

6.3.1.3 Reductive Reactions

In many of the degradations of pyridine that have been noted in Sections 6.3.1.1 and 6.3.1.2, successive hydroxylation and reduction reactions are involved. The degradation of pyridine by *Bacillus* sp. to produce succinate has already been noted as an example of dioxygenation and plausibly involves initial 1,4-reduction of the ring followed by dioxygenative fission (Watson and Cain 1975). On the other hand, the degradation by *Nocardia* sp. (Watson and Cain 1975) may reasonably involve hydroxylation of the initial 1,2,3,4-tetrahydropyridine with the formation of glutarate. A strain of *Azoarcus* sp. that can degrade pyridine under both aerobic and denitrifying conditions was presumed to use the latter pathway (Rhee et al. 1997). A purely reductive pathway is used even by some aerobic organisms for the degradation of hydroxylated and aminated pyrimidines. For example, the degradation of uracil and related compounds that are used as N-sources for the growth of *P. stutzeri* (Xu and West 1992) and *Burkholderia* sp. (West 1997) takes place by a reductive pathway with the formation of the 4,5-dihydro compounds before hydrolytic ring cleavage to form *N*-carbamoyl-β-alanine. This pathway is also used during pyrimidine degradation by anaerobic clostridia and is discussed again in Section 6.7.4.

6.3.1.4 Hydrolytic Reactions Resulting in Ring Cleavage

Derivatives of 1,3,5-triazine are important herbicides so that attention has been directed to their persistence particularly in the terrestrial environment. Some examples have been given in Section 5.2.3 of the utilization of amino and thiol substituents of 1,3,5-triazines as sources of nitrogen and sulfur; subsequent degradation of these hydroxylated metabolites is mediated by hydrolytic reactions which result in the cleavage of the ring with the formation of CO_2 and NH_4^+ (Jutzi et al. 1982).

6.3.2 Reactions Mediated by Fungi

By analogy with their ability for biotransformation of PAHs, fungi are capable of carrying out a number of transformations of heterocyclic aromatic compounds though—as for the hydrocarbons—these do not generally result in the total degradation of the substrate.

1. The metabolism of the fluoroquinolone antibiotic enrofloxacin has been examined in the fungus *Gloeophyllum striatum* and produced a complex series of metabolites (Wetzstein et al. 1997): among these, one was tentatively proposed to have been formed by oxidative decarboxylation of the quinolone ring followed by further oxidation to an *N*-cyclopropylisatin and the corresponding anthranilic acid, or by hydrolysis loss of fluoride (Figure 6.56). The metabolism of ciprofloxacin which is identical except for the lack of an *N*-ethyl

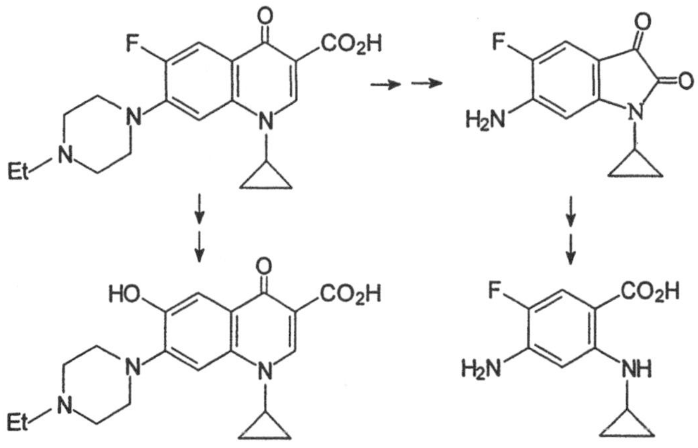

FIGURE 6.56
Degradation of the fluoroquinolone enrofloxacin.

substituent in the piperazine ring is noted in Section 6.10.2. The reactions may be compared with those during the photochemical transformation (Chapter 4, Section 4.1.1).

2. Three reactions of flavanones may be used to represent the broad classes of biotransformation reactions which have been observed in oxaarenes: dehydrogenation, hydroxylation, and ring scission (Figure 6.57) (Ibrahim and Abul-Hajj 1990).

A number of other rather unusual reactions have, however, been observed during the degradation of oxygen heterocyclic compounds and these are worth illustrating:

1. The degradation of rutin by *Aspergillus flavus* proceeds by hydrolysis to the aglycone followed by release of the unusual metabolite carbon monoxide (Figure 6.58) (Simpson et al. 1963; Krishnamurty and Simpson 1970).

2. Degradation of 5-hydroxyisoflavones by strains of *Fusarium* sp. involves initial reduction of the heterocyclic ring before further oxidative fission (Figure 6.59) (Willeke and Barz 1982).

3. Reductive fission of the dihydrofuran ring of 6a-hydroxyinermin is an early step in the total degradation of pisatin by strains of *F. oxysporium* (Figure 6.60) (Fuchs et al. 1980).

In view of current interest in the persistence of recalcitrant organochlorine compounds, it is appropriate to note the degradation of 2,7-dichlorobenzo-1,4-dioxin by *Phanerochaete chrysosporium* (Valli et al. 1992a). Degradation involves dechlorination and ring scission as the initial reactions (Figure 6.61);

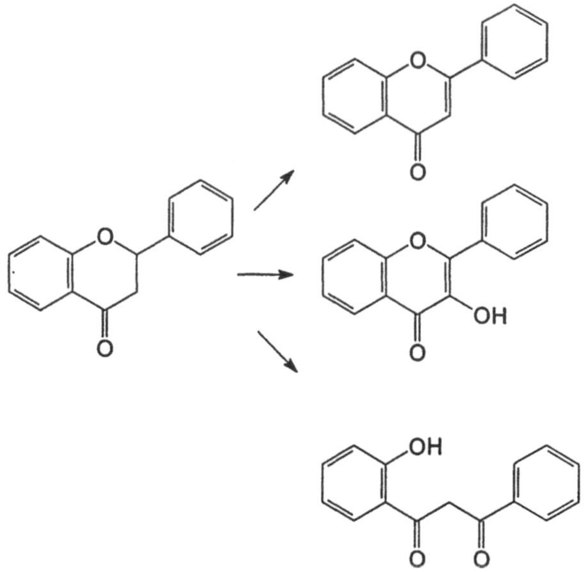

FIGURE 6.57
Alternative pathways for the biodegradation of flavanones.

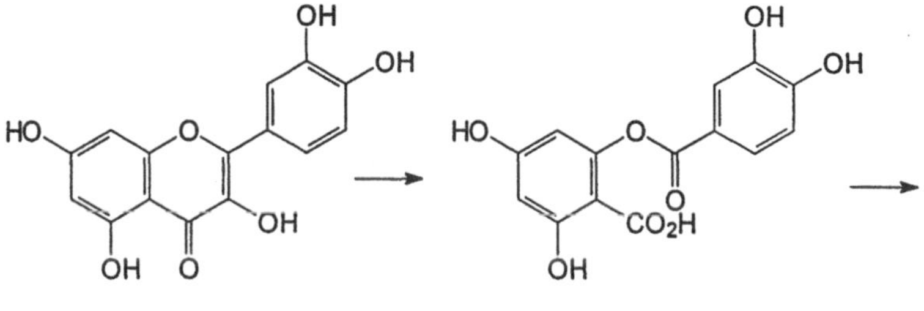

FIGURE 6.58
Biodegradation of rutin by *Aspergillus flavus*.

the products are then degraded by pathways similar to those used for the degradation of 2,4-dichlorophenol by the same organism (Valli and Gold 1991).

6.4   Degradation of Halogenated Alkanes and Alkenes

6.4.1   Elimination Reactions

By analogy with chemical reactions, the biodegradation of halogenated alkanes may take place by elimination, by nucleophilic displacement, or by

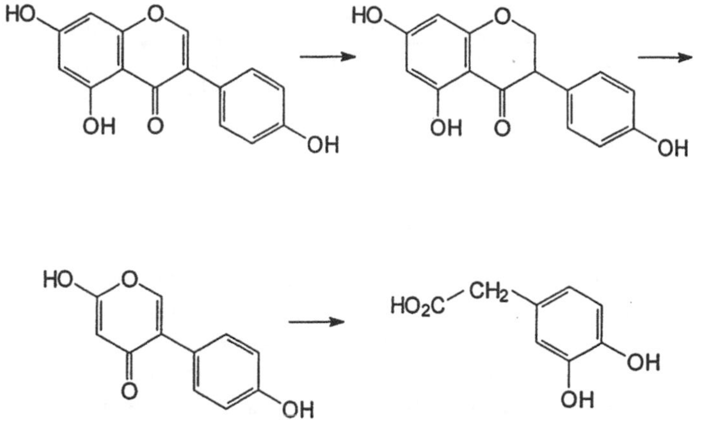

FIGURE 6.59
Biodegradation of a 5-hydroxy*iso*flavone by *Fusarium* sp.

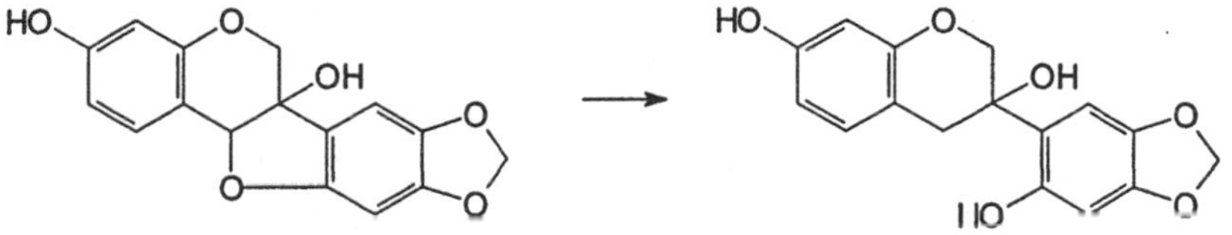

FIGURE 6.60
Biodegradation of 6a-hydroxyinermin by *Fusarium oxysporium*.

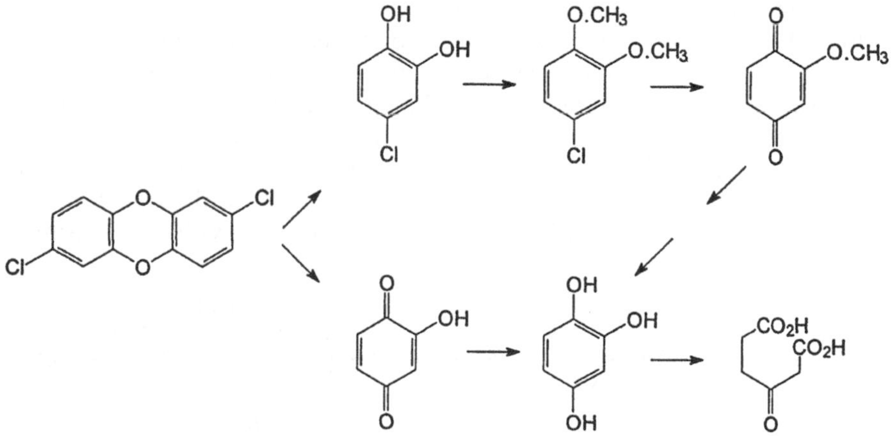

FIGURE 6.61
Biodegradation of 2-7-dichlorobenzo-1,4-dioxin by *Phanerochaete chrysosporium*.

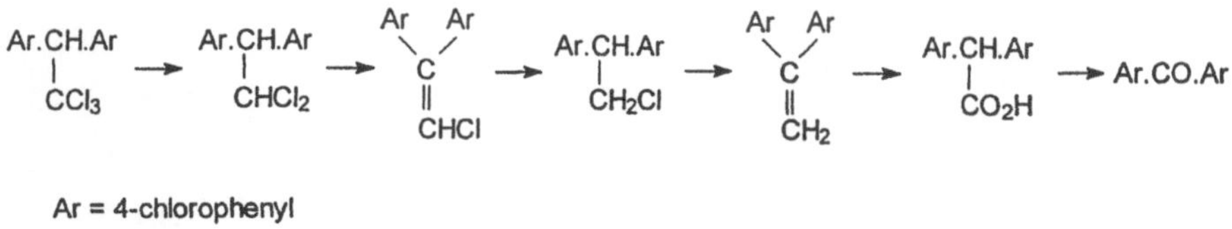

Ar = 4-chlorophenyl

FIGURE 6.62
Biodegradation of DDT by *Aerobacter aerogenes*.

reduction; in contrast to chemical reactions, however, the first two do not occur simultaneously, and indeed elimination is relatively uncommon. It is, however, one of the initial reactions in the degradation of DDT by the facultatively anaerobic bacterium *Aerobacter aerogenes* (Figure 6.62) (Wedemeyer 1967), and the recovery of the elimination product (DDE) from environmental samples long after restriction on the use of DDT suggests the high degree of persistence of that metabolite (Section 4.5.2). Alternative pathways should also be noted:

1. Degradation of DDT may also occur by hydroxylation of the ring and displacement of the aromatic ring chlorine atom by hydroxyl (Figure 6.63) (Massé et al. 1989)

2. Degradation of DDT by *Alcaligenes eutrophus* strain A5 resulted in the formation of both ring-hydroxylated metabolites and 4-chlorobenzoate (Nadeau et al. 1994). It was subsequently shown (Nadeau et al. 1998), however, that the phenolic compounds were artifacts produced by nonenzymatic reactions and do not lie on the pathway of degradation, and that cells of this strain grown with biphenyl metabolized DDT to the *cis*-2,3-dihydrodiol.

An elimination reaction is apparently one of the steps in the degradation of γ-hexachloro[*aaaeee*]cyclohexane from which pentachlorobenzene (Tu 1976) or γ-2,3,4,6-tetrachlorocyclohex-1-ene may be formed (Jagnow et al. 1977) (Figure 6.64). The formation of both 2,5-dichlorophenol and 2,4,5-trichlorophenol during the aerobic degradation of γ-hexachloro[*aaaeee*]cyclohexane by *Pseudomonas paucimobilis* presumably occurs by comparable elimination reactions (Senoo and Wada 1989) and further details of the transformation that produces also 1,2,4-trichlorobenzene have been provided (Nagasawa et

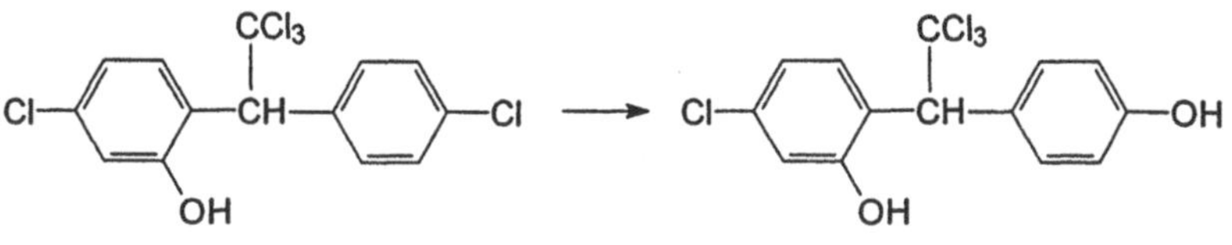

FIGURE 6.63
Alternative pathway for the biodegradation of DDT.

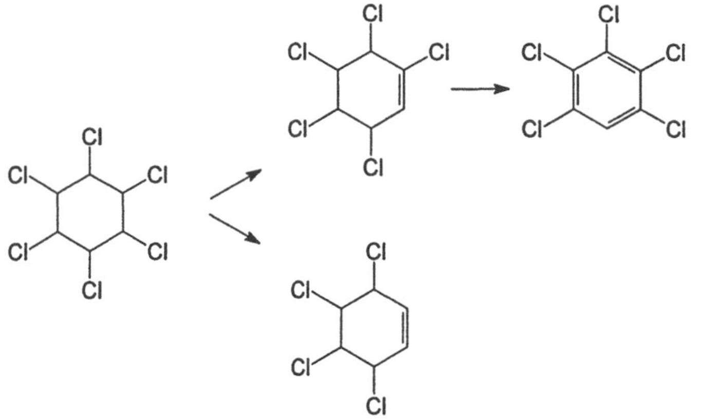

FIGURE 6.64
Pathway for the biotransformation of γ-hexachlorocyclohexane.

al. 1993a). The degradation is, however, more complex since these compounds are terminal metabolites, and degradation occurs by a sequence of reactions involving sequential hydrolytic displacement of chloride (Figure 6.65) (Nagasawa et al. 1993b). The genes leading successively to 2,5-dichlorohydroquinone have been designated *linA*, *linB*, and *linC* (Nagata et al. 1994) and encode enzymes that carry out, respectively, two successive elimination reactions, two successive hydrodehalogenations, and a dehydrogenase (Nagata et al. 1994). The hydrolase (LinB) has been cloned, and in a constructed strain of *Escherichia coli* that overexpresses the dehalogenase gene (linB), a range of chloroalkanes was dechlorinated (Nagata et al. 1993; 1997): these included 1-chloroalkanes (C_3 to C_{10}), α,ω-dichloroalkanes, and some bromoalkanes. Since the elimination reactions that are involved are not themselves dependent on the presence of oxygen, they may occur under anaerobic conditions. For example, pure cultures of strictly anaerobic methanogenic bacteria produce ethene from 1,2-dibromoethane and ethyne from 1,2-dibromoethene (Figure 6.66) (Belay and Daniels 1987).

6.4.2 Hydrolytic Reactions

Displacement of halogen by hydroxyl is a widely distributed reaction in the degradation of haloalkanes and haloalkanoates. An apparently simple pathway involving two displacement steps is illustrated in Figure 6.67, but it should be emphasized that the enzymology of hydrolytic dehalogenation is much more complex than might appear: for example, two different dehalogenases are involved in the dechlorination of 1,2-dichloroethane and of chloroacetate (van den Wijngaard et al. 1992). Indeed, a number of distinct dehalogenases exist, and they may differ significantly in their substrate specificity (Scholtz et al. 1988; Salis et al. 1990) with respect to chain length and the influence of halogen atoms at the ω-position. The crystal structures of the

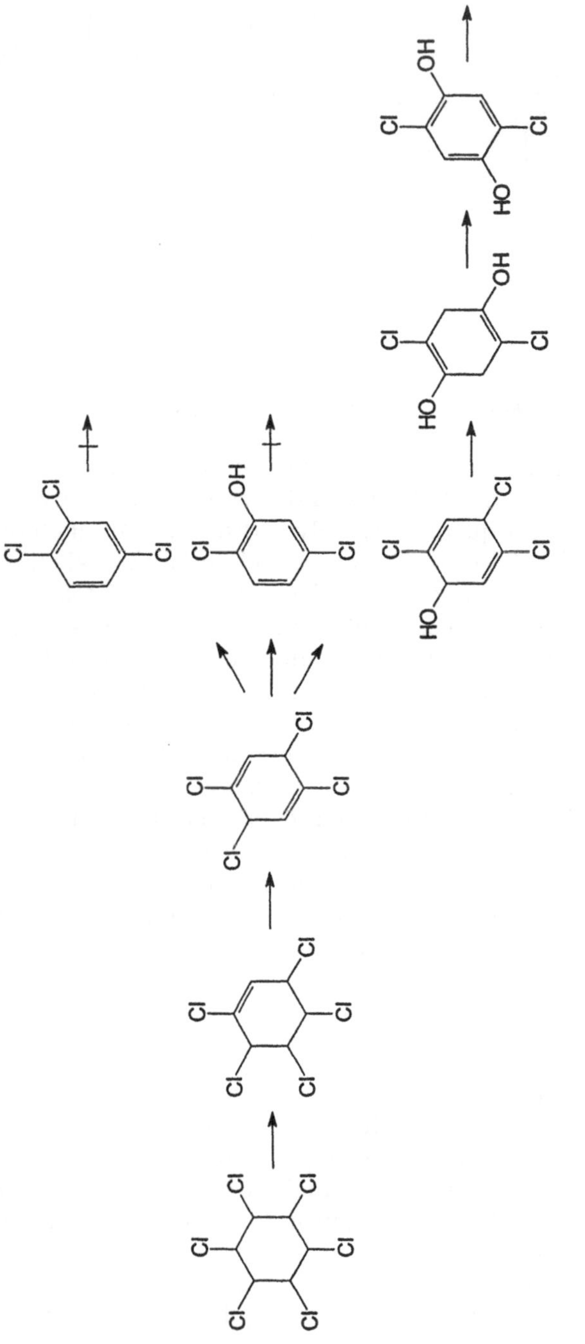

FIGURE 6.65
Pathway for the biodegradation of γ-hexachlorocyclohexane.

$$Br.CH_2.CH_2.Br \longrightarrow CH_2=CH_2 \quad ; \quad Br.CH=CH.Br \longrightarrow CH\equiv CH$$

FIGURE 6.66
Metabolism of 1,2-dibromoethane.

$$Cl.CH_2.CH_2.Cl \longrightarrow Cl.CH_2.CH_2OH \longrightarrow Cl.CH_2CHO \longrightarrow Cl.CH_2.CO_2H \longrightarrow HOCH_2.CO_2H$$

FIGURE 6.67
Biodegradation of 1,2-dichloroethane.

complexes between the L-2-haloacid dehalogenase from *Pseudomonas* sp. strain YL and 2-chloroalkanoates have been determined (Li et al. 1998), and reveal that the hydrolytic inversion involves the Arg41 and Asp10 sites as electrophiles and nucleophiles, respectively, followed by interaction of the Asp10 ester with Ser118.

The metabolism of 1-haloalkanes by *Rhodococcus rhodochrous* NCIM 13064 (Curragh et al. 1994) has revealed a number of pathways:

1. Hydrolysis of 1-chlorobutane to butan-1-ol followed by further metabolism of butyrate;
2. ω-Hydroxylation of 1-chlorohexadecane followed by oxidation to the carboxylic acid, β-oxidation, and nonenzymatic formation of γ-butyrolactone;
3. Incorporation of the ω-chloroalkanoic acid into cellular lipids.

The enantiomeric specific dehalogenase from *Pseudomonas* sp. strain DL-DEX is able to use both enantiomers of 2-haloalkanoic acids as substrates forming the products with inversion of configuration (Nardi-Del et al. 1997). The degradation of 2,2-dichloropropionate involves dehalogenation to pyruvate but even here two different dehalogenases are synthesized (Allison et. al. 1983). A strain of *Pseudomonas* sp. is able to degrade a wide range of 1-bromoalkanes having 6 to 10 carbon atoms (Shochat et al. 1993), although the range of unsubstituted or chloroalkanes that is degraded is quite restricted.

Dichloroacetate and trichloroacetate are degradable, although not apparently by the same groups of organisms. Mono- and dichloroacetate are effectively degraded by a strain of "*Pseudomonas dehalogenans*" that has only limited effect on trichloroacetate, whereas, conversely, strains of *Arthrobacter* sp. readily degraded trichloroacetate—although not monochloroacetate (Jensen 1960). Significant differences between chlorine- and bromine-substituted compounds may exist. For example, although *Xanthobacter autotrophicus* is able to grow with 1,2-dichloroethane and the dehalogenases can debrominate 1,2-dibromoethane and bromoacetate, these substrates are unable to support growth of the organism; several reasons have been suggested including the toxicity of bromoacetaldehyde (van der Ploeg et al. 1995). This is consistent with the facts that, whereas in this strain the initially produced 2-chloroethanol is oxidized to the aldehyde by an alkanol dehydrogenase and then to the carboxylic acid before loss of chloride and mineralization, the corresponding 2-bromoethanol is metabolized by *Mycobacterium sp.* strain GP1 to the epoxide, followed by mineralization without the intermediacy of bromoacetaldehyde (Poelarends et al. 1999).

Although all these reactions are formally nucleophilic displacements of the chlorine atoms by hydroxyl groups, a different mechanism clearly operates in the degradation of dichloromethane by *Hyphomicrobium* sp. since the enzyme is glutathione dependent (Stucki et al. 1981) and the reaction presumably involves at least two steps. The degradation of tetrachloromethane by a strain of *Pseudomonas* sp. presents a number of interesting features: although $^{14}CO_2$ was a major product from the metabolism of $^{14}CCl_4$, a substantial part of the label was retained in nonvolatile water-soluble resides (Lewis and Crawford 1995). The nature of this was revealed by isolation of adducts with cysteine and N,N'-dimethylethylenediamine in which intermediates formally equivalent to $COCl_2$ and $CSCl_2$ were trapped, presumably formed by reaction of the substrate with water and a thiol, respectively.

6.4.3 Monooxygenase Systems

Degradation of halogenated alkenes by direct displacement of halogen is not expected on purely chemical grounds, although this reaction apparently occurs during the degradation of vinyl chloride by a strain of *Pseudomonas* sp. that carries out direct hydrolysis to acetaldehyde followed by mineralization to CO_2 (Castro et al. 1992b). Although the degradation of 3-chlorocrotonate could take place by a comparable reaction, it seems more plausible that degradation involves initial addition of the elements of water to the α,β-unsaturated ketone followed by elimination of chloride (Figure 6.68) (Kohler-Staub and Kohler 1989). The dechlorination of 1,10-dichlorododecane by cells of *Pseudomonas* sp. strain 273 grown with dodecane probably involves monooxygenation (Wischnak et al. 1998) in contrast to the hydrolytic mechanism noted above.

Unactivated halogenated alkenes may be degraded by a completely different pathway that involves a monooxygenase system. Attention has already been drawn to the remarkable spectrum of structures which are amenable to attack by the monooxygenase system of methanotrophic and methylotrophic bacteria; the degradation of trichloroethene provides a good example of the operation of this system (Figure 6.69) (Little et al. 1988). This degradation has

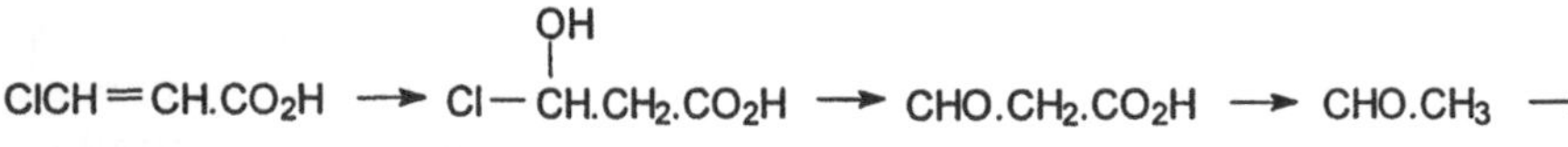

$$ClCH = CH.CO_2H \longrightarrow Cl-\overset{OH}{\underset{|}{CH}}.CH_2.CO_2H \longrightarrow CHO.CH_2.CO_2H \longrightarrow CHO.CH_3 \longrightarrow$$

FIGURE 6.68
Biodegradation of 3-chlorocrotonic acid.

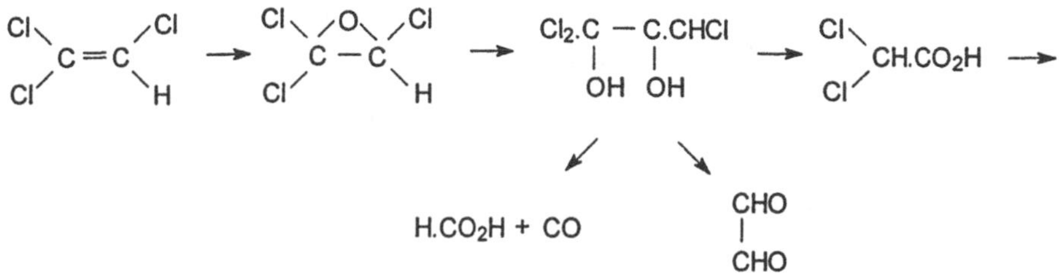

FIGURE 6.69
Biodegradation of trichloroethene.

also been used in Section 4.8.2 to illustrate the problem presented by the synthesis of the toxic metabolite carbon monoxide. In addition, 2,2,2-trichloroacetaldehyde is produced during oxidation of trichloroethene by several methanotrophs and undergoes a dismutation to form trichloroethanol and trichloroacetate (Newman and Wackett 1991); at least formally, the transformation of trichloroethene to 2,2,2-trichloroacetaldehyde is analogous to an NIH shift (Section 6.2.2). As might be expected, a number of haloalkanes including dichloromethane, chloroform, 1,1-dichloroethane, and 1,2-dichloroethane may also be degraded by the soluble methane monooxygenase system of *Methylosinus trichosporium* (Oldenhuis et al. 1989), although *cis*- and *trans*-1,2-dichloroethene formed stable epoxides, and tetrachloroethene was resistant to attack. The aerobic degradation of vinyl chloride by *Mycobacterium aurum* also proceeded by initial formation of an epoxide mediated by an alkene monooxygenase (Hartmans and De Bont 1992), and this reaction has also been demonstrated to occur with *Methylosinus trichosporium* even though subsequent conversion to glycollate involves purely chemical reactions (Castro et al. 1992a).

Although reductive pathways are discussed in Section 6.4.4, examples involving both oxidative and reductive pathways have been observed, for example, in *Pseudomonas putida* strain G-786 in which the synthesis of cytochrome P-450 monooxygenase was induced by growth with camphor.

- Trichloronitromethane, dichloronitromethane, and bromotrichloromethane were reductively dehalogenated (Castro et al. 1985).
- Trichloroethane was degraded by a dominant aerobic pathway to chloroacetate and glyoxylate, and simultaneously by a minor reductive reaction which must also involve an elimination reaction with the formation of vinyl chloride (Figure 6.70) (Castro and Belser 1990).
- Both reductive and oxidative activities have been achieved in a genetically engineered strain of *P. putida* that is discussed further below.

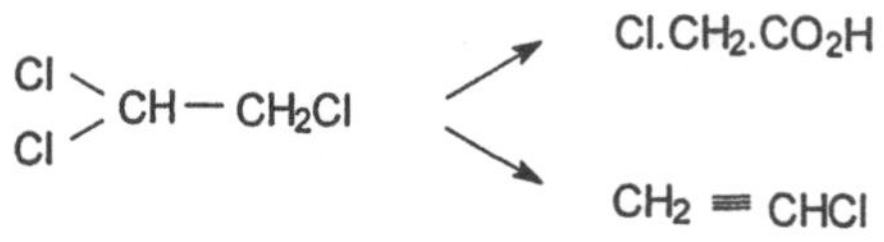

$$Cl\text{-}CH_2\text{-}CO_2H$$

$$\begin{array}{l} Cl \\ \diagdown \\ CH-CH_2Cl \\ \diagup \\ Cl \end{array} \longleftarrow$$

$$CH_2 = CHCl$$

FIGURE 6.70
Biodegradation of trichloroethane by *Pseudomonas putida*.

The metabolism of chloroform has been studied in a group of organisms (Hamamura et al. 1997), and from the inhibitory effect of acetylene, it was concluded that a monooxygenase was involved.

- Cells of *Methylosinus trichosporium* strain OB3b grown with methanol and incubated with formate as electron donor degraded chloroform with the release of 2.1 mol of chloride per mole substrate, and at substrate concentrations up to 38.6 μM.

- Cells of *P. butyrovora* grown with butane and incubated with butyrate as electron donor degraded chloroform with release of 1.7 mol chloride per mole substrate, and degradation was incomplete even at concentrations of 12.9 μM. Butane inhibited the degradation. This organism also partially degraded other chloroalkanes and alkanes including vinyl chloride, 1,2-*trans*-dichloroethene and trichloroethene.

- Butane-grown cells of *Mycobacterium vaccae* strain JOB5 were able to degrade chloroform without addition of an electron donor.

Monooxygenase systems are sometimes involved in the degradation of toluene (Section 6.2.1) and plausibly in the transformation of aromatic compounds to arene oxide intermediates (Chapter 4, Sections 4.2.3 and 4.4.1.1, and in Sections 6.2.2 and 6.3.1.2). There is a complex relation between toluene monooxygenase activity and the degradation of chlorinated hydrocarbons.

Whereas toluene oxidation is accomplished by *P. mendocina* strain KR1, *P. putida* strain F1, *P. picketii* strain PKO1, and *Burkholderia cepacia* strain G4 that possess toluene monooxygenase activities, chloroform, and 1,2-dichloroethane are degraded only by the toluene-4-monooxygenase strains *P. mendocina* KR! and *Pseudomonas* sp. strain ENVPC5 (McClay et al. 1996). In these strains, toluene oxidation was induced by trichloroethene which was subsequently degraded, whereas trichloroethene did not induce toluene oxidation in *B. cepacia* strain G4 or *P. putida* strain F1 (McClay et al. 1995). The degradation of trichloroethene by the three components of toluene 2-monooxygenase of *B. cepacia* involves initial formation of the epoxide followed by spontaneous decomposition to carbon monoxide, formate, and glycollate (Newman and Wackett 1997). In contrast to degradation by methane monooxygenase or cytochrome P-450 monooxygenase, chloral hydrate was not formed.

The induction of trichloroethene degradation has been discussed in Chapter 4, Sections 4.4.1.1 and 4.5.2, and its efficient degradation by constructed strains has been accomplished.

1. In *Pseudomonas pseudoalcaligenes* strain KF707 in which the *bphA1* gene that codes for the large subunit of biphenyl dioxygenase and which determines the specificity of the dioxygenase was replaced with the corresponding gene *todC1* from *P. putida* F1 whose product carries out the ring dioxygenation of toluene (Suyama et al. 1996).

2. Transposon mutagenesis of *Alcaligenes eutrophus* JMP134 that harbors a plasmid with genes for the degradation of 2,4-dichlorophenoxyacetate produced mutants lacking both phenol hydroxylase and catechol 2,3-dioxygenase. From this, a mutant carrying a recombinant plasmid with phenol hydroxylase activity degraded a concentration of 200 μm trichloroethene in 2 days without induction (Kim et al. 1996).

6.4.4 Reductive Dehalogenation Reactions of Halogenated Aliphatic and Alicyclic Compounds

Considerable effort has been devoted to the anaerobic transformation of polychlorinated C_1 alkanes and C_2 alkenes in view of their extensive use as industrial solvents and their identification as widely distributed groundwater contaminants. Early experiments which showed that tetrachloroethene was transformed into vinyl chloride (Vogel and McCarty 1985) (Figure 6.71) aroused concern, although it has now been shown that complete dechlorination can occur under some conditions. A number of investigations have examined the persistence of tetrachloroethene under anaerobic conditions with the result that widely differing pathways for biotransformation have emerged.

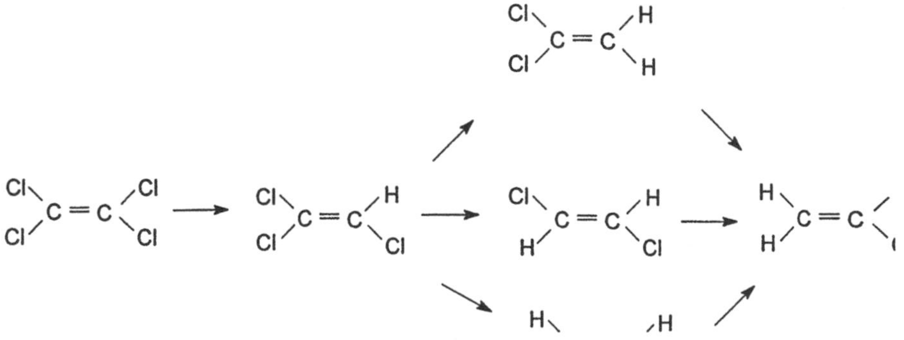

FIGURE 6.71
Anaerobic dechlorination of tetrachloroethene.

- A strain of *Dehalospirillum multivorans* transformed tetrachloroethene to trichloroethene and *cis*-1,2-dichloroethene (Neumann et al. 1994) using pyruvate as electron donor, and some properties of the dehalogenase have been reported (Neumann et al. 1995). A mole of the dehalogenase contained 1 mol of corrinoid, 9.8 mol of Fe, and 8.0 mol of acid-labile sulfur (Neumann et al. 1996), and the genes have been cloned and sequenced (Neumann et al. 1998). Although comparable values have been reported for the enzyme from *Desulfitobacterium* sp. strain PCE-S (Miller et al. 1997), the N-terminal sequence of the enzyme showed little similarity to that from *Dehalospirillum multivorans* (Miller et al. 1998). The important role of a corrinoid was suggested for both organisms (Neumann et al. 1996; Miller et al. 1997). A comparable reductive dehalogenation—although only as far as trichloroethene—is carried out by *Desulfomonile tiedjei* and the reductase is similar to that involved in dehalogenation of 3-chlorobenzoate (Townsend and Suflita 1996).

- Under methanogenic conditions, a strain of *Methanosarcina* sp. transformed tetrachloroethene to trichloroethene (Fathepure and Boyd 1988), and in the presence of suitable electron donors such as methanol, complete reduction of tetrachloroethene to ethene may be achieved in spite of the fact that the dechlorination of vinyl chloride appeared to be the rate-limiting step (Freedman and Gossett 1989).

- A nonfermentative organism putatively assigned to *Desulfuromonas acetexigens* reduced tetrachloroethene to *cis*-dichloroethene using acetate as electron donor (Krumholz et al. 1996), and a similar species *D. chloroethenica* used both tetrachloroethene and trichloroethene as electron acceptors with production of *cis*-dichloroethene and acetate or pyruvate as electron donors (Krumholz 1997).

- Complete dechlorination of high concentrations of tetrachloroethene in the absence of methanogenesis using methanol as electron donor has been achieved (DiStefano et al. 1991), although these conditions will clearly seldom occur in natural environments.

- Reductive transformation of tetrachloroethene to the fully reduced ethane has also been observed in a fixed-bed reactor (De Bruin et al. 1992).

Some of the unresolved issues are illustrated by the different effects of electron donors on the partial dechlorination of tetrachloroethene in anaerobic microcosms (Gibson and Sewell 1992).

An organism that is able to use methyl chloride as energy source and converts this into acetate has been isolated (Traunecker et al. 1991), although it appears that different enzyme systems are involved in the dehalogenation of this substrate and in the O-demethylation of methoxylated aromatic compounds such as vanillate (Meßmer et al. 1993); further comment on these

O-demethylation reactions by acetogenic bacteria is given in Section 6.7.3.1. Cultures of a number of anaerobic bacteria are able to dechlorinate tetrachloromethane, and *Acetobacterium woodii* formed dichloromethane as the final chlorinated metabolite by successive dechlorination, although CO_2 was also produced by an unknown mechanism (Egli et al. 1988). By formally similar dechlorinations, a strain of *Clostridium* sp. transformed 1,1,1-trichloroethane to 1,1-dichloroethane, and tetrachloromethane successively to trichloromethane and dichloromethane (Gälli and McCarty 1989). Partial reductive dechlorination of trichlorofluoromethane to dichlorofluoromethane has been shown in a mixed culture containing sulfate-reducing bacteria (Sonier et al. 1994). Although dichloromethane is a terminal metabolite in many of these transformations, an organism assigned to *Dehalobacterium formicoaceticum* is able to use this as a source of carbon and energy (Mägli et al. 1996; 1998). Dichloromethane is converted to methylene tetrahydrofolate from which formate is produced by oxidation and acetate by incorporation of CO_2 catalyzed by CO dehydrogenase and acetyl coenzyme A synthase.

Reductive dechlorination in combination with the elimination of chloride has been demonstrated in a strain of *Clostridium rectum* (Ohisa et al. 1982): γ-hexachlorocyclohexene formed 1,2,4-trichlorobenzene and γ-1,3,4,5,6-pentachlorocyclohexene formed 1,4-dichlorobenzene (Figure 6.72). It was suggested that this reductive dechlorination is coupled to the synthesis of ATP, and this possibility has been clearly demonstrated during the dehalogenation of 3-chlorobenzoate coupled to the oxidation of formate in *Desulfomonile tiedjei* (Mohn and Tiedje 1991). Combined reduction and elimination has also been demonstrated in methanogenic cultures that transform 1,2-dibromoethane to ethene and 1,2-dibromoethene to ethyne (Belay and Daniels 1987).

Cells of *Pseudomonas putida* G786 containing cytochrome P-450$_{CAM}$ genes on the CAM plasmid carried out the reductive dehalogenation of a number of halogenated alkanes under anaerobic conditions (Hur et al. 1994). The following are illustrative examples: (1) dibromodichloromethane to bromodichloromethane, (2) trichlorofluoromethane to carbon monoxide, (3) hexachloroethane to tetrachloroethene, (4) 1,1,1-trichloro-2,2,2-trifluoroethane to 1,1-dichloro-2,2-difluoroethene (Li and Wackett 1993). When a plasmid containing genes encoding toluene dioxygenase was incorporated into this strain, complete degradations that involved *both* reductive and oxidative steps were accomplished; for example, pentachloroethane was degraded via

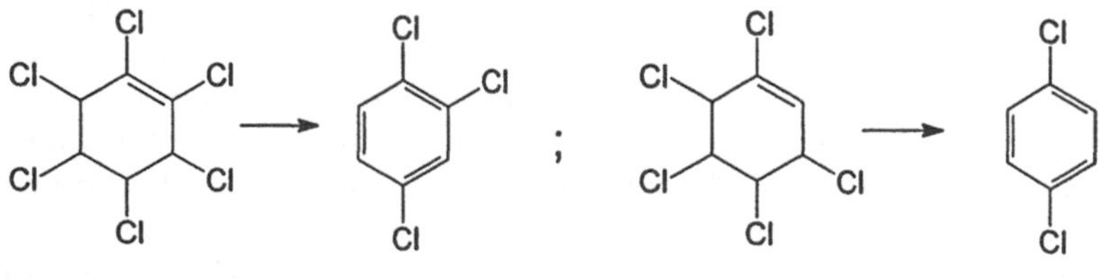

FIGURE 6.72
Biotransformation of γ-hexachlorocyclohexene and γ-1,3,4,5,6-pentachlorocyclohexene.

trichloroethene to glyoxylate and 1,1-dichloro-2,2-difluoroethene to oxalate (Wackett et al. 1994).

From an environmental point of view, it therefore appears that the complete degradation of chlorinated alkenes and alkanes will often require the operation of both anaerobic and aerobic steps: for example, partial or complete dehalogenation may occur under anaerobic conditions, and aerobic degradation of the partially dechlorinated metabolites such as dichloromethane (La Roche and Leisinger 1991) and vinyl chloride (Castro et al. 1992a,b; Hartmans and de Bont 1992) may then subsequently take place. The combination of the two activities in a single strain has been exploited in the genetically constructed strain of *P. putida* and has been discussed above.

A mechanistic basis for such dechlorinations has been formulated using the results of experiments with 1,2-dichloroethane which is transformed to ethene and chloroethane by cell suspensions of methanogenic bacteria. Further investigations showed that cobalamin, or factor F_{430}, or boiled cell extracts from *Methanosarcina barkeri* catalyzed this transformation in the presence of Ti(III) citrate as electron donor (Holliger et al. 1992a). In addition, crude cell extracts from *Methanobacterium thermoautotrophicum* carried out the same transformation using H_2 as electron donor (Holliger et al. 1992b). These observations have been incorporated into a model scheme for the dechlorination reaction, and it is reasonable to assume that these reactions and the purely abiotic reactions outlined in Section 4.1.3.2 have a common mechanistic basis.

6.5 Aerobic Degradation of Halogenated Aromatic Compounds

6.5.1 Bacterial Systems

Because of their potentially adverse environmental consequences including their toxicity and their potential for concentration in biota, enormous activity has been devoted to studies on the biodegradation of halogenated—and particularly chlorinated—aromatic compounds including hydrocarbons, phenols, anilines, and carboxylic acids. Many of these are constituents of valuable commercial products or are important intermediates in the synthesis of agrochemicals; in addition, a range of structurally diverse chlorinated aromatic compounds is produced during production of bleached pulp using conventional technologies with molecular chlorine (references in Neilson et al. 1991). Attention is drawn to reviews by one of the major contributors to this field (Reineke 1984; Reineke et al. 1988).

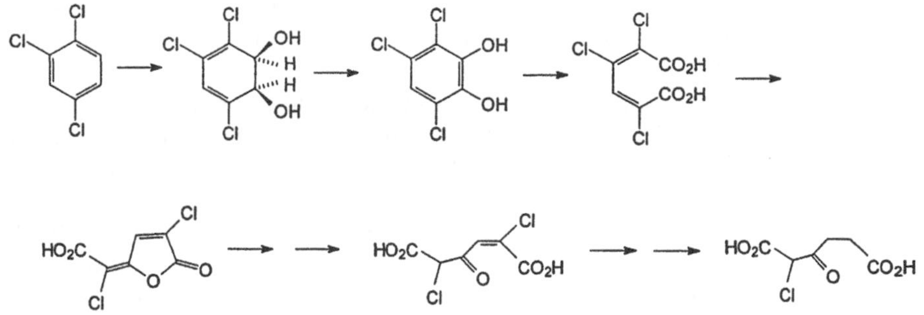

FIGURE 6.73
Biodegradation of 1,2,4-trichlorobenzene.

6.5.1.1 Halogenated Aromatic Hydrocarbons

Ultimately, ring cleavage of halogenated aromatic hydrocarbons must occur if degradation—rather than merely biotransformation—is to be accomplished. The pathways outlined in Section 6.2.3 for unsubstituted hydrocarbons are generally followed: formation of a *cis*-dihydrodiol followed by dehydrogenation to the catechol and ring cleavage with production of muconic acids or hydroxymuconate semialdehydes from monocyclic compounds, or from PCBs, chlorobenzoates. This pathway has already been illustrated with 1,4-dichloronaphthalene.

There are, however, important alternatives: although this dioxygenation pathway is used for the degradation of 1,2,4-trichlorobenzene (Figure 6.73), and 1,2,3,4-tetrachlorobenzene (Potrawfke et al. 1998), a chlorine atom is lost before production of the chlorocatechol during degradation of 1,2,4,5-tetrachlorobenzene (Figure 6.74) (Sander et al. 1991; see Section 4.4.2). The genes encoding the α- and β-subunits of the terminal oxidase, the ferredoxin, and the reductase from this strain now designated *Burkholderia* sp. have been cloned into *Escherichia col;* (Beil et al. 1997). It was confirmed that whereas metabolism of 1,2-, and 1,4-dichlorobenzene, 1,2,4-tri- and 1,2,3,4-tetrachlorobenzene produced stable *cis*-dihydrodiols, dechlorination concomitant with dioxygenation took place with 1,2,4,5-tetrachlorobenzene, and was confirmed using $^{18}O_2$. In addition, dioxygenation was also observed in this constructed

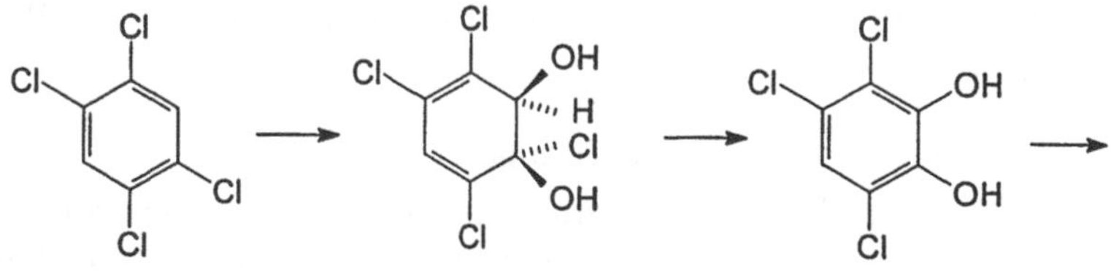

FIGURE 6.74
Biodegradation of 1,2,4,5-tetrachlorobenzene.

strain with biphenyl, dibenzo[1,4]dioxin, dibenzofuran, and naphthalene. An additional example of simultaneous dechlorination and dioxygenation in the degradation of PCB congeners is given below.

The importance of both the number and the position of chlorine substituents is illustrated by the results of a study with *Xanthobacter flavus* strain 14p1: the strain grew *only* with 1,4-dichlorobenzene that was degraded by dioxygenation, dehydrogenation, and intradiol ring fission, whereas 1,3-dichlorobenzene did not induce dioxygenase activity (Sommer and Görisch 1997). Although chlorobenzene induced several of the enzymes essential for degradation, its inability to support growth was assumed to be the result of the accumulation of toxic intermediates.

Some additional features have emerged from studies on the biodegradation of chlorinated and polychlorinated biphenyls (designated here collectively—although not strictly etymologically—as PCBs), and some of these may be summarized briefly. The biodegradation of PCBs has attracted enormous attention, and a number of organisms able to degrade a range of congeners have been isolated generally from biphenyl enrichments. These include both Gram-negative strains—*Alcaligenes eutrophus* strain H850 (Bedard et al. 1987a), *Pseudomonas pseudoalcaligenes* strain KF707 (Furukawa and Miyazaki 1986), *Pseudomonas* sp. strain LB400 (Bopp 1986), *P. testosteroni* (Ahmad et al. 1990)—and Gram-positive strains—*Rhodococcus globerulus* strain P6 (formerly *Acinetobacter* sp. strain P6) (Furukawa et al. 1979; Asturias and Timmis 1993), *R. erythropolis* strain TA421 (Maeda et al. 1995, and *Rhodococcus* sp. strain RHA1 (Seto et al. 1995). Changes in the nomenclature of some of these taxa are given in Section 5.9.

Attention is briefly directed to some important details of PCB degradation that have emerged from studies using these strains.

1. *Enzymes of degradation*

The degradation is carried out by a suite of enzymes comprising biphenyl-2,3-dioxygenase, biphenyl-2,3-dihydrodiol dehydrogenase, 2,3-dihydroxybiphenyl dioxygenase, and the hydrolytic enzymes that produce benzoate encoded by the genes *bphA*, *bphB*, *bphC*, and *bphD* (Furukawa and Miyazaki 1986; Ahmad et al. 1991; Taira et al. 1992). There are several 2,3-dihydroxybiphenyl dioxygenases, and the "cryptic" enzyme BphC2 from a naphthalene sulfonate-degrading strain of *Sphingomonas* displays only low activity toward 3,4-dihydroxybiphenyl, and lower activity for 4-catechol than for 3-catechol (Heiss et al. 1997).

2. *The number and position of substituents*

Attention has already been briefly drawn to the fact that where alternatives exist, the ring with fewer halogen substituents is degraded first. This has already been illustrated for the degradation of 1,4-dichloronaphthalene, and is much more extensively supported by the results from experiments that examined a large number of PCB congeners in which the less highly substituted ring is almost invariably degraded to a carboxylic acid apparently by *extra*diol cleavage of the intermediate catechol: for example, in the 2,4,5-, 2,2′,5′-, and 2,3,4,5-congeners (Figure 6.75). On the other hand, care should be exercised in attempting to promulgate rules of widespread applicability to PCBs since both steric and electronic factors are important determinants of biodegradability: congeners with substituents at the 2- or 6-positions of either or both rings are apparently less susceptible to degradation than other isomers (Furukawa et al. 1979). The analysis of nonplanar PCBs has been discussed in Section 2.4.2. In addition, complexities are introduced by the demonstration that both dehalogenation and dioxygenation may occur, for example, in the degradation of 2,4′-dichloro- and 4,4′-dichlorobiphenyl (Ahmad et al. 1991). This alternative is discussed further in the context of the enzymology of degradation (Section 4.4.2). All these results clearly show that it is not merely the number of substituents, but also their orientation that are determinative factors.

3. Strain differences and dioxygenase systems

There are considerable differences in the oxidative activity of different organisms toward PCB congeners: for example, *Pseudomonas* sp. strain LB400 has a much greater versatility than *P. pseudoalcaligenes* strain KF707 (Gibson et al. 1993) or *P. testosteroni* (Ahmad et al. 1990), and the range of congeners transformed by *Rhodococcus* sp. strain RHA1 differs from that of *Pseudomonas* sp. strain LB400 and *P. paucimobilis* strain KF707 (Seto et al. 1995).

The degradation of PCBs proceeds by dioxygenation to *cis*-2,3-dihydro-2,3-diols, dehydrogenation to the catechol, extradiol ring cleavage between C1 and C2 positions, and hydrolysis of the vinylogous β-keto acid to benzoate. A strain of *Pseudomonas* sp. able to degrade a range of polychlorinated biphenyl congeners had both 2,3-dioxygenase and 3,4-dioxygenase activity, and four of the open reading frames were homologous to components of toluene dioxygenase (Erickson and Mondelo 1992). It was not been resolved whether a single dioxygenase was able to introduce oxygen at the 2,3- or the 3,4-positions or whether there are two different enyzmes. In *R. globerulus* strain P6 (*Acinetobacter* sp. strain P6), there are, however, several nonhomologous 2,3-dihydroxybiphenyl-1,2-dioxygenases with a narrow substrate specificity (Asturias and Timmis 1993). The 2,3-dihydroxybiphenyl dioxygenase has been purified from *P. pseudoalcaligenes* (Furukara and Arimura 1987) and the biphenyl 2,3-dioxygenase from *Pseudomonas* sp. strain LB400 (Haddock et al. 1995). Although whole-cell studies using oxygen uptake had clearly shown the versatility of this organism in the degradation of PCB congeners containing up to four chlorine substituents, the study with purified biphenyl 2,3-dioxygenase supplied additional valuable mechanistic details:

a. The dihydrodiol was produced from all congeners, and a 2-chloro substituent could be eliminated from the 2,3-dihydrodiol to form the catechol.

b. 4,5-Dihydrodiols were produced from 3,3', 2,2',5, and the 2,2',5,5'
   congeners, but these are not substrates for the dihydrodiol
   dehydrogenase.

These results are consistent with the results accumulated during many years
and with the operation of single dioxygenase enzyme. An interesting observation has been made with 2,3-dihydroxybiphenyl dioxygenase in a naphthalene
sulfonate-degrading pseudomonad (strain BN6): the 1,2-dihydroxynaphtha-
lene dioxygenase oxidized both 1,2-dihydroxynaphthalene and 2,3-dihydrox-
ybiphenyl (Kuhm et al. 1991), but the organism also synthesizes a distinct 2,3-
dihydroxybiphenyl dioxygenase that is not involved in the degradation of
naphthalene sulfonates (Heiss et al. 1995). Attention has been drawn in Section
6.2.3 to studies (Barriault et al. 1998) using naphthalene-1,2-dihydrodiol dehydrogenase and 1,2-dihydroxynaphthalene dioxygenase with a view to using
them to complement the enzymes used for biphenyl degradation.

A number of important determinants have emerged from detailed analysis
of the substrate specificity of PCB-degrading organisms. There are two broad
groups of organisms: those such as *Pseudomonas* (*Burkholderia*) *cepacia* strain
LB400 that degrades a range of congeners but does not effectively degrade
4,4'disubstituted congeners, and strains such as *P. pseudoalkaligenes* KF707
with the opposite specificity. The specificity reflects differences in the protein
sequences in two regions (III and IV) of the BphA large α-subunit of the dioxygenase (Kimura et al. 1997; Mondello et al. 1997). This offers the possibility
of expanding the range of degradable PCBs by alterations of specific amino
acids. As noted in Section 4.4.2, however, the substrate specificities of the four
chimeras constructed from the respective α- and β-subunits of the terminal
dioxygenase ISP$_{BPH}$ of *Pseudomonas* sp. strain LB400 and *Comamonas*
(*Pseudomonas*) *testosteroni* strain B-356 were dependent on the presence of
both proteins (Hurtubise et al. 1998). There is therefore a complex dependency on the presence of both subunits.

4. Metabolites

Total degradation of PCBs necessitates degradation of the chlorobenzoates
produced by the foregoing reactions. Metabolism of chlorobenzoates produces chlorocatechols and thence muconic acids, and it has been shown with
P. testosteroni strain B-356 that the metabolites from 3-chlorobenzoate strongly
inhibited the activity of 2,3-dihydroxybiphenyl-1,2-dioxygenase and therefore
the degradation of the original substrates (Sondossi et al. 1992). This inhibition
is reminiscent of the inhibition of catechol-2,3-dioxygenase by 3-chlorocate-
chol (Chapter 4, Section 4.8.2) that is noted in Section 6.5.1.2. It has been suggested that the inability of strains to degrade chlorobenzoates formed from
some PCB congeners may be related to the restricted metabolism of chlorobenzoate fission products (Hernandez et al. 1995). An interesting metabolic
situation involves the formation of protoanemonin (4-methylenebut-2-ene-4-
olide) as an intermediate in the degradation of 4-chlorobenzoate formed by

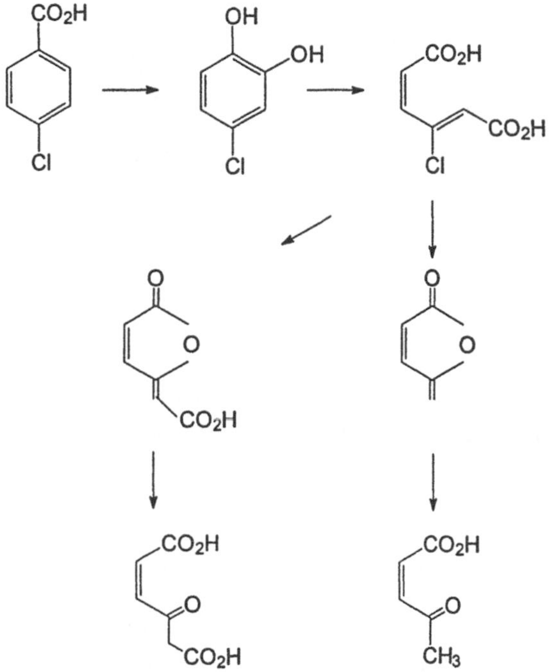

FIGURE 6.76
Formation of protoanemonin from 4-chlorobenzoate.

partial degradation of 4-chlorobiphenyl. This is formed by intradiol fission of 4-chlorocatechol to 3-chloro-*cis*,*cis*-muconate followed by loss of CO_2 and chloride (Blasco et al. 1995) (Figure 6.76). The synthesis of this metabolite adversely affects the survival in soil microcosms of organisms that metabolize 4-chlorobiphenyl, although its formation can be obviated by organisms using a modified pathway that produces the *cis*-dienelactone (Blasco et al. 1997).

Alternative metabolites may also be formed. Although the ultimate products from degradation of PCBs are generally chlorinated muconic acids, the unusual metabolite 2,4,5-trichloroacetophenone has been isolated (Bedard et al. 1987b) from the degradation of 2,4,5,2′,4′,5′-hexachlorobiphenyl (Figure 6.77) by *Alcaligenes eutrophus* H850 which has an unusually wide spectrum of degradative activity for PCB congeners. The essential reaction in its formation is analogous to that whereby acetophenone is produced from cinnamate by a pseudomonad (Hilton and Cain 1990): this presumably involves addition of the elements of water to the α,β-unsaturated C=C bond, followed by oxidation and decarboxylation. The metabolism of chloroacetophenones (Havel and Reineke 1993) has been discussed in Section 6.2.1.

An example of the NIH shift occurs during the transformation of 2-chlorobiphenyl to 2-hydroxy-3-chlorobiphenyl by a methanotroph, and is consistent with the formation of an intermediate arene oxide (Adriaens 1994). The

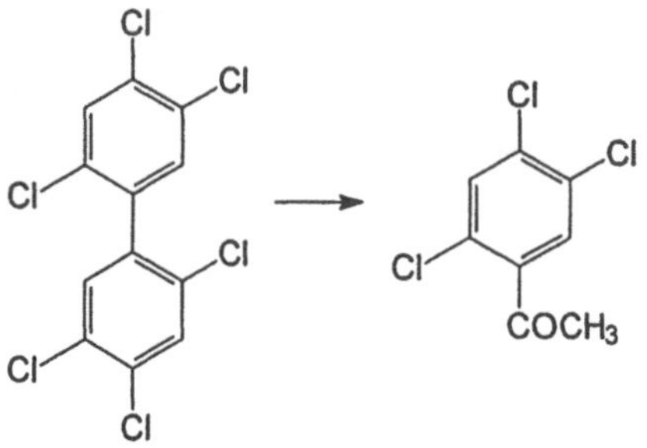

FIGURE 6.77
Biodegradation of 2,4,5,2′,4′,5′-hexachlorobiphenyl.

occurrence of such intermediates offers plausible mechanisms for the formation of nitro-containing metabolites that have been noted in Sections 4.4.1.1 and 6.3.1.2.

5. Induction of PCB Metabolism

A range of related compounds has been examined (Donnelly et al. 1964) for their capacity to support the growth of *A. eutrophus* H850, *P. putida* LB400, and *Corynebacterium* sp. MB1. For strains H850 and MB1, growth with biphenyl was equaled by that using a wide range of substrates including naringin, catechin, and myricitin (Figure 6.78). The pattern of metabolism of PCB congeners by strain H850 was identical after growth with biphenyl and naringin, for strain LB400 using biphenyl, myricitin, and for strain MB1 using biphenyl, coumarin and naringin. These results suggest that natural plant metabolites are able to mediate growth of PCB-degrading organism, and may be relevant in the context of bioremediation noted in Chapter 8, Section 8.2.4.2.

6.5.1.2   Halogenated Aromatic Compounds Carrying Additional Substituents

The degradation of a wide range of halogenated aromatic compounds bearing additional substituents such as hydroxyl or carboxyl groups proceeds by two broadly distinct pathways differing by whether or not halogen is removed from the ring before fission. These possibilities may be defined more specifically:

1. Direct oxidation of the ring may be mediated by dioxygenation or hydroxylation followed by dehydrogenation and disruption of the ring; halogen is subsequently eliminated during further degradation of the acyclic product.

2. Direct elimination of halide from the ring may occur by the hydrolytic, reductive, or decarboxylative displacement of halogen; the products then undergo oxidation followed by disruption of the aromatic ring.

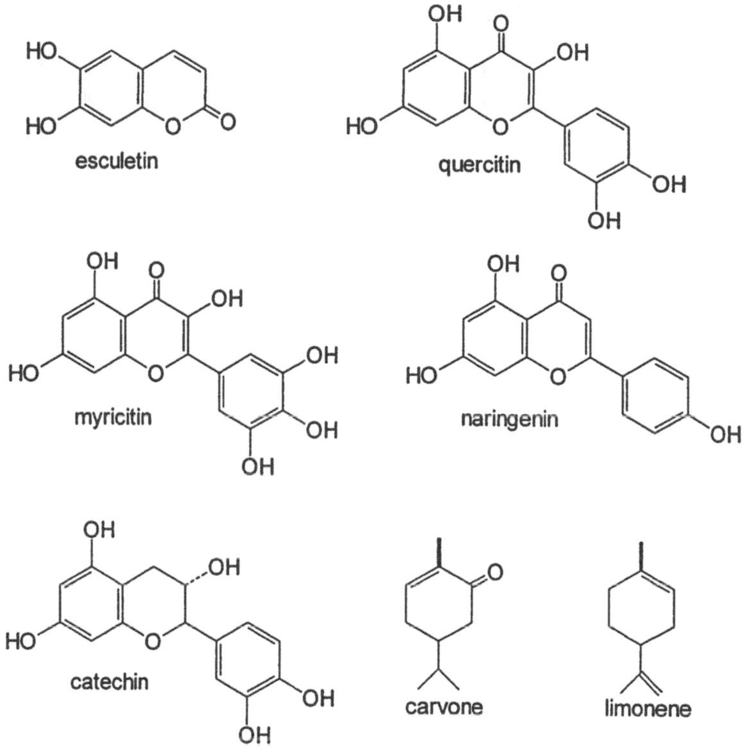

FIGURE 6.78
Naturally occurring products used as substrates or inducers.

Halogen Elimination after Ring Cleavage

There are a number of important details which merit fuller discussion.

1. *The central role of halogenated catechols*—In those cases where the loss of halogen takes place after ring cleavage, halogenated phenols with three or fewer halogen substituents are converted into halogenated catechols by hydroxylase systems (Beadle and Smith 1982; Perkins et al. 1990) that are discussed in Chapter 4, Section 4.4.1, whereas all other aromatic compounds such as halogenated carboxylic acids, halogenated amines, and most halogenated hydrocarbons are converted initially into *cis*-dihydrodiols before dehydrogenation to halogenated catechols which therefore occupy a central position in the metabolism of all these compounds (Figure 6.79). Hydroxylation and dioxygenation are not, however, mutually exclusive since the toluene dioxygenase from *Pseudomonas putida* F1 hydroxylates both phenol and 2,5-dichlorophenol with the introduction of only one atom of oxygen (Spain et al. 1989). In addition, dioxygenation may also result in the simultaneous elimination of halogen and CO_2 that is discussed again below. This has been illustrated for 1,2,4,5-tetrachlorobenzene (see Figure 6.74)(Sander et al. 1991) and 2,2′,5-trichlorobiphenyl (Haddock et al. 1995), and apparently also occurs during the degradation

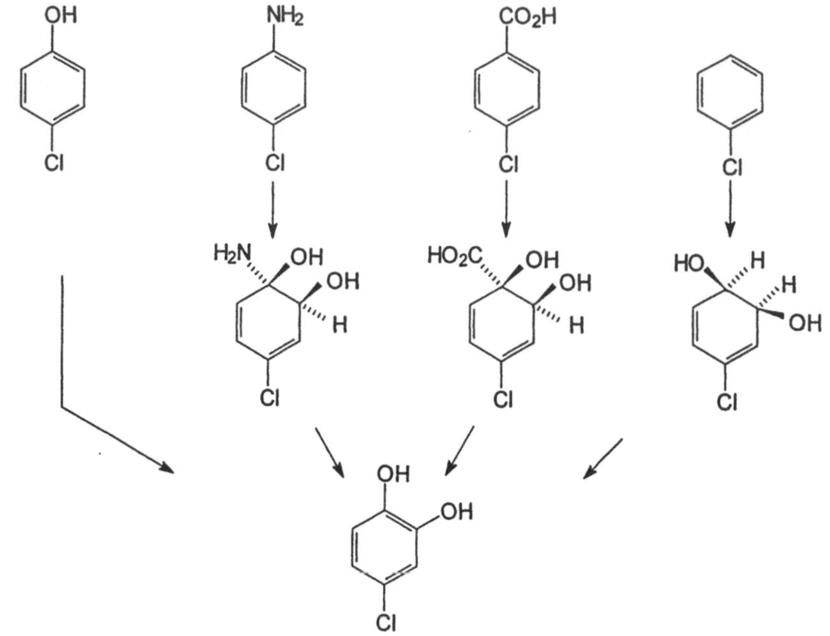

FIGURE 6.79
Outline of the biodegradation of chlorobenzene, 4-chlorophenol, 4-chlorobenzoate, and 4-chloroaniline.

of both 2,3-dichloro- and 2,5-dichlorobenzoate (Hickey and Focht 1990), and a number of other 2-substituted benzoates (Romanov and Hausinger 1994). Thereby, a halogen atom is lost before production of the chlorocatechols (Figure 6.80; see Section 4.4.2). Subsequent degradation follows established pathways in which chloride is eliminated from muconic acids.

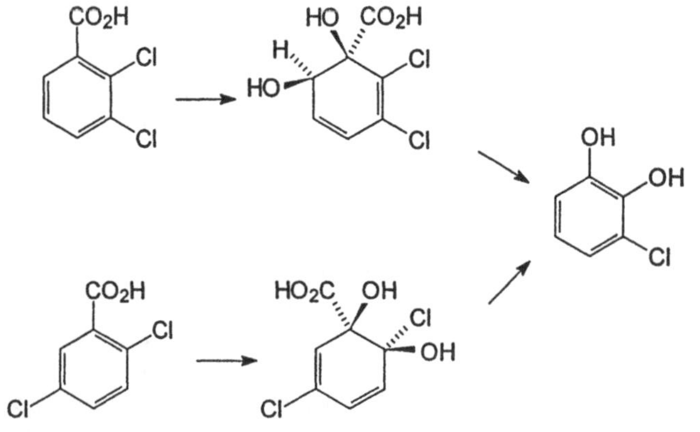

FIGURE 6.80
Biodegradation of dichlorobenzoates.

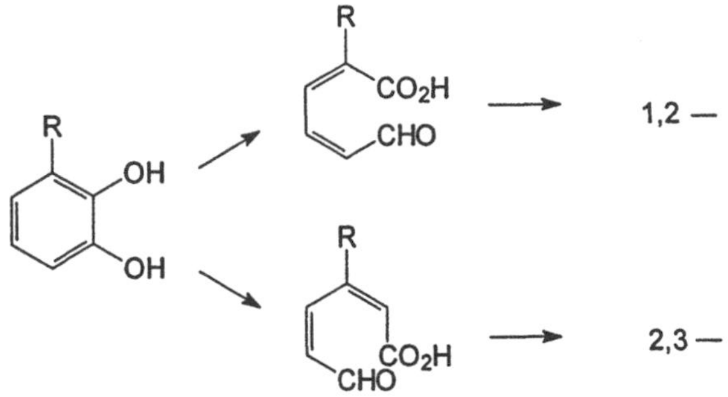

FIGURE 6.81
Ring-cleavage pathways for the biodegradation of substituted catechols.

2. *Mechanisms for the degradation of chlorobenzoates*—Chlorobenzoates may be formed during initial steps in the aerobic degradation of PCBs, and their further metabolism illustrates a number of pathways.

There are two types of initial reaction, dioxygenation, and, for 4-halobenzoates, hydrolytic dehalogenation. For *ortho*-halogenated benzoates degradation may occur by dioxygenation with simultaneous decarboxylation with loss of halogen (Section 4.4.2, Figure 4.48), for example, by a broad-spectrum two-component 1,2-dioxygenase that is used by *P. cepacia* 2CBS for 2-halobenzoates, and a three-component 1,2-dioxygenase for 2-chloro- and 2,4-dichlorobenzoate by *P. aeruginosa* strain 142 (Romanov and Hausinger 1994). The alternative 2,3-dioxygenation to give 2,3-dihydroxybenzoate with loss of chloride has been observed in a *Pseudomonas* sp. (Fetzner et al. 1989). For 3- and 4-halogenated benzoates dioxygenation with decarboxylation is widely used: 3-chlorobenzoate would produce 3-chlorocatechol by 1,2-dioxygenation or 4-chlorocatechol by 1,6-dioxygenation, and 4-chlorobenzoate would produce 4-chlorocatechol. The pathway for the further metabolism of the chlorocatechols may be critical due to the formation of toxic metabolites (Chapter 4, Section 4.8.2), and is discussed below.

Hydrolytic and reductive dehalogenation are discussed in the context of pathways for the loss of halogen before ring fission (see later in this section).

3. *Mechanisms for the ring fission of substituted catechols*—An outline of the ring fission mechanisms has been given in Section 6.2.1, and the two pathways are compared in Figure 6.81: which of them is followed depends both on the organism and on the substrate that is being metabolized.

For 3-chlorocatechol formed from 3-chlorobenzoate by dioxygenation (Figure 6.82), intradiol ring fission with the production of succinate and acetate is preferred to extradiol fission that would result in synthesis of the toxic acyl chloride. On the other hand, *P. putida* strain GJ31 degrades chlorobenzene via 3-chlorocatechol and extradiol fission (Kaschabek et al. 1998). This is accomplished by a chlorocatechol 2,3-dioxygenase that hydrolyses the initially formed *cis,cis*-hydroxymuconacyl chloride to 2-hydroxymuconate: thereby the irreversible reaction of the acid chloride with nucleophiles, or the

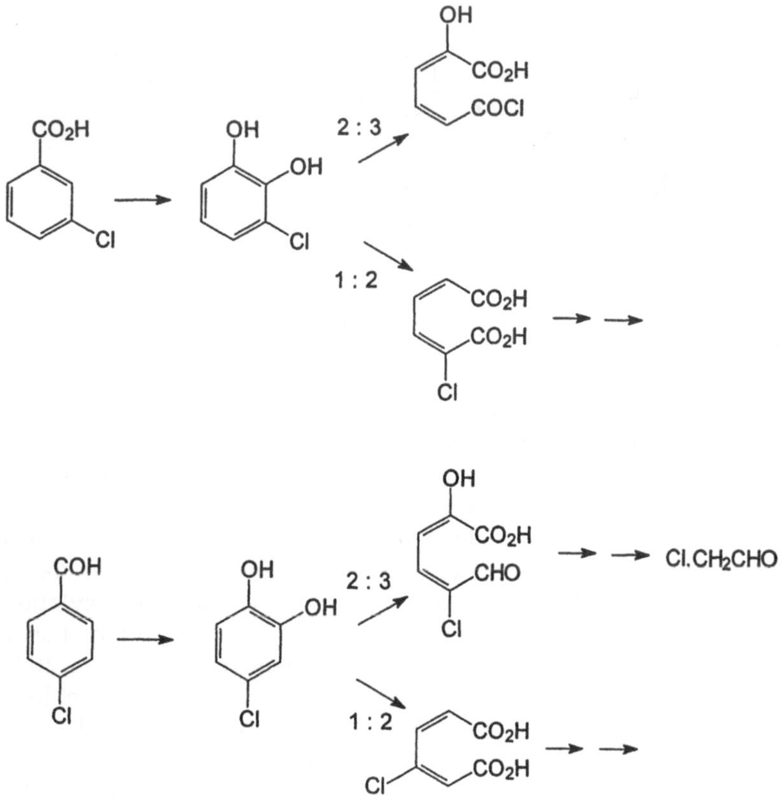

FIGURE 6.82
Ring-cleavage pathways for the biodegradation of chlorocatechols formed from chlorobenzoates.

formation of pyr-2-one-6-carboxylate as a terminal metabolite, is avoided. The alternative extradiol fission between C1 and C6 (distal fission) of 3-chlorocatechol has been observed in *Azotobacter vinelandii* strain 206 (Sala-Trepat and Evans 1971) and in *P. putida* strain GJ31 (Kaschabek et al. 1998).

For 3-chlorobenzoate, an alternative pathway involves 3,4- or 4,5-dioxygenation, that results in catechols that, after ring fission, produce pyruvate and oxalacetate (Nakatsu and Wyndmam 1993).

A comparable problem occurs for the degradation of 4-chlorobenzoate through 4-chlorocatechol by a pathway in which chlorine is retained in the ring before fission: the 2,3-pathway would produce toxic chloroacetaldehyde. For 4-halogenated benzoates, the pathway involving hydrolytic loss of chlorine before ring fission is, however, an important alternative, and the three-component dehalogenase requires initial formation of the CoA ester (Chang et al. 1992). 4-Hydroxybenzoate is readily degraded by hydroxylation to 3,4-dihydroxybenzoate followed by ring fission. An additional issue for 4-chlorobenzoate formed during the degradation of 4-chlorobiphenyl involves the synthesis of protoanemonin (Blasco et al. 1997), and has been noted in Section 6.5.1.1.

In summary, a range of pathways are available for the degradation of halogenated benzoates:

1. Dioxygenation with simultaneous loss of carboxyl and halogen to produce catechol (2-halobenzoates);
2. Dioxygenation without loss of halogen to produce chlorocatechols that may be degraded by several ring-fission pathways (3- and 4-chlorobenzoates);
3. Dioxygenation with loss of halogen only to produce dihydroxy-benzoate (2-chlorobenzoate);
4. Dehalogenation of the CoA ester to hydroxybenzoate (4-haloben-zoates);
5. Both hydrolytic and reductive elimination of halogen (2,4-dichlo-robenzoate) (van der Tweel et al. 1987; Romanov and Hausinger 1996) that are noted below.

There is an additional problem which has important implications for biotechnology: the situation in which two substrates such as a chlorinated and an alkylated aromatic compound are simultaneously present. The 2,3-pathway is generally used for the degradation of alkylbenzenes (Figure 6.83), but this may be incompatible with the degradation of chlorinated aromatic compounds since the 3-chlorocatechol produced inhibits the activity of the catechol-2,3-oxygenase (Klecka and Gibson 1981; Bartels et al. 1984).

This has been overcome in some strains: (1) mutant strains which have successfully reconciled this incompatibility (Taeger et al. 1988; Pettigrew et al. 1991), (2) 2,3-dihydroxybiphenyl-1,2-dioxygenase from the naphthalene sulfonate-degrading *Sphingomonas* sp. strain BN6 that metabolized 3-chlorocatechol by extradiol fission between the 1- and 6-positions (Riegert et al. 1998),

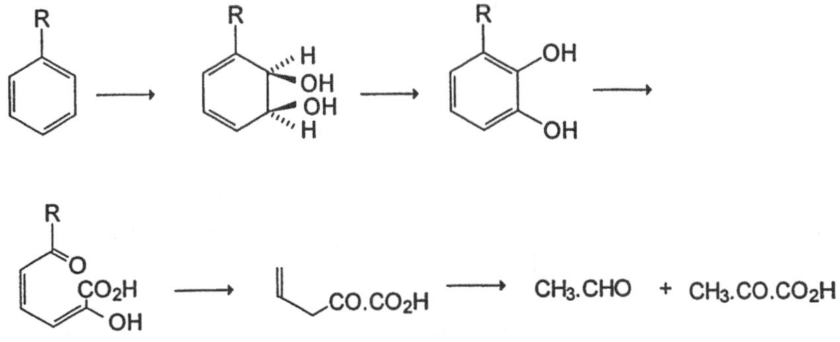

FIGURE 6.83
Biodegradation of alkylbenzenes.

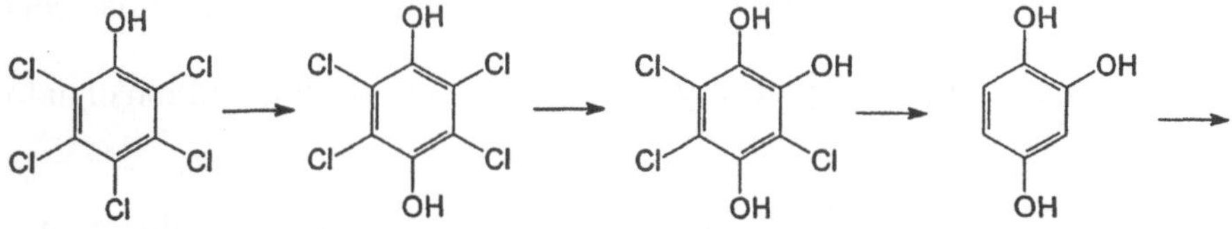

FIGURE 6.84
Biodegradation of pentachlorophenol.

and (3) chlorocatechol 2,3-dioxygenase from *P. putida* GJ31 metabolized 3-chlorocatechol with concominant elimination of chloride to form 2-hydroxy-muconate (Kaschabek et al. 1998), and the catechol 2,3-dioxygenase from this strain encoded by *cbzE* is plasmid borne and is capable metabolizing both 3-chlorocatechol and 3-methylcatechol; it belongs to the 2.C subfamily of type 1 extradiol dioxygenases (Mars et al. 1999). The regulation of the metabolism of chlorinated aromatic compounds has been briefly discussed in Chapter 4, Section 4.8.2.

Loss of Halogen before Ring Cleavage

Some examples of the loss of halogen from halobenzoates prior to ring fission have been noted above. An attempt is made here to summarize and illustrate the various pathways that may be used for the elimination of halogen from phenolic compounds. This may occur by hydrolytic or reductive pathways, or by both.

In contrast to the situation for monochloro- and dichlorophenols, phenols with three or more chlorine substituents undergo degradation by pathways involving displacement of the chlorine atoms by hydroxyl groups before ring cleavage. This reaction has been shown to be the initial step in the degradation of a number of chlorophenols, and a few examples may be given as illustration of this important alternative: for example, in the metabolism of pentachlorophenol (Figure 6.84) by strains of the Gram-positive organisms *Mycobacterium* sp. (Suzuki 1983) and *Rhodococcus chlorophenolicus* (Apajalahti and Salkinoja-Salonen 1987b), and in the Gram-negative organisms *Flavobacterium* sp. (Steiert and Crawford 1986) and *Sphingomonas chlorophenolica* strain RA2 (Miethling and Karlson, 1996). In the application of such strains to bioremediation of contaminated sites, there are a number of important considerations. These are considered in greater detail in Chapter 8, Section 8.2.3, and include both the tolerance of the strains to high concentrations of pentachlorophenol and the longevity of the strains in the environment.

The following examples of loss of halogen before ring fission involve monooxygenases (Chapter 4, Section 4.4.1.1).

1. The initial step in the degradation of pentachlorophenol by strain ATCC 39723 which involves the introduction of oxygen is carried out by an O_2-requiring monooxygenase in both *Flavobacterium* sp.

(Xun et al. 1992a) and *R. chlorophenolicus* (Uotila et al. 1992); in the latter organism, the system converting the initial hydrolysis product, tetrachlorohydroquinone, to 1,2,4-trihydroxybenzene does not, however, require O_2.

2. In the degradation of 2,4,6-trichlorophenol by a strain of *Azotobacter* sp. (Li et al. 1991), the monooxygenase that produces 2,6-dichlorohydroquinone as the initial metabolite has been purified and characterized (Wieser et al. 1997).

3. The degradation of 2,4,5-trichlorophenoxyacetate by *Burkholderia* (*Pseudomonas*) *cepacia* AC1100 (Haugland et al. 1990; Daubaras et al. 1995) proceeds via 2,4,5-trichlorophenol that is degraded by initial oxygenation to 2,5-dichlorohydroquinone. This strain also brings about *para*-hydroxylation of dichlorophenols, whether or not this position carries a chlorine substituent (Tomasi et al. 1995).

Phenols carrying five substituents may be degraded by comparable pathways.

1. The pentachlorophenol monooxygenase from *Flavobacterium* sp. (*Sphingomonas chlophenolica* ATCC 39723) (Xun et al. 1992b) is able to hydroxylate a range of other substituted phenols with elimination of, for example, chloride, bromide, iodide, cyanide, and nitrite from the 4-position.

2. Loss of fluoride from pentafluorophenol and of bromide from pentabromophenol is catalyzed by the monooxygenase from *R. chlorophenolicus* (Uotila et al. 1992).

The metabolism of 1,3,5-triazine herbicides containing chlorine substituents involves hydrolytic displacement of chloride (Cook 1987), although it seems not to have been established whether this is a hydroxylase or a monooxygenase. The halohydrolase from a strain of *Pseudomonas* sp. that uses atrazine as a nitrogen source has been cloned into *Escherichia coli* (de Souza et al. 1995), while the inducible enzyme that has been purified from *R. corallinus* (Mulbry 1994) is also capable of carrying out deamination of amino-substituted 1,3,5-triazines.

These investigations have dealt primarily with chlorine compounds and the conclusions may not necessarily be extrapolated to bromine compounds. For example, whereas a strain of *P. aeruginosa* degrades 2-bromobenzoate by a hydrolytic pathway with the initial formation of salicylate (Higson and Focht 1990), a mutant strain of *P. putida* accumulates 4-bromo-*cis*-2,3-dihydroxy-cyclohexa-4,6-diene-1-carboxylate from 4-bromobenzoate (Taylor et al. 1987). It would therefore seem premature to draw any general conclusion from these results which may depend both on the fact that bromine is involved and on its position on the ring.

In addition to hydrolytic replacement of halogen, reductive displacement has been shown to occur during the degradation of a few aromatic compounds. For example, both hydrolytic and reductive reactions are involved in the degradation of a number of chlorophenols and chlorobenzoates:

1. Degradation of pentachlorophenol by *R. chlorophenolicus* (Apajalahti and Salkinoja-Salonen 1987b) (see Figure 6.84) and 2,4,6-trichlorophenol by a strain of *Azotobacter* sp. (Latus et al. 1995) both involve formation of 1,2,4-trihydroxybenzene as an intermediate.

2. A comparable reductive dechlorination is used by a pentachlorophenol-degrading *Flavobacterium* sp. and the enzyme has been purified: the successive dechlorinations to form 2-chloro-1,4-dihydroxybenzene are probably mediated by a glutathione *S*-transferase system (Xun et al. 1992c). It should be noted that the evidence so far shows that the enzyme involved in this reaction is distinct from that involved in the anaerobic dechlorination of 3-chlorobenzoate by *Desulfomonile tiedjei* DCB-1(Ni et al. 1995) that is discussed further in Section 6.6.

3. 2,6-Dichlorohydroquinone is formed by *Sphingomonas* (*Flavobacterium*) *chlorophenolica* by reductive dechlorination of tetrachlorohydroquinone (Xun et al. 1992c; Chanama and Crawford 1997), and the chlorohydrolase from *Flavobacterium* sp. strain 39723 that converts this to 2-chloro-6-hydroxyhydroquinone has been purified and characterized (Lee and Xun 1997).

4. Degradation of 2,4,5-trichlorophenol by *Burkholderia* (*Pseudomonas*) *cepacia* involves 5-chloro-1,2,4-trihydroxybenzene as intermediate (Daubaras et al. 1995). The conversion is catalyzed by a single FAD-containing monooxygenase (Xun 1996), and the fission of 1,2,4-trihydroxybenzene to maleylacetate is accomplished by an intradiol oxygenase that is highly specific and is unable to use either caetchol or tetrahydroxybenzene (presumably 1,2,4,5-tetrahydroxybenzene) (Daubaras et al. 1996).

5. Both hydrolytic and reductive reactions are involved in the degradation of 2,4-dichlorobenzoate by *Alcaligenes denitrificans* strain NTB-1 (van der Tweel et al. 1987) (Figure 6.85), and via the CoA ester by *Corynebacterium sepedonicum* (Romanov and Hausinger 1996).

In addition to these reactions, loss of both chlorine and CO_2 may take place simultaneously during dioxygenation (Chapter 4, Sections 4.4.2 and 6.5.1.2). It is used by a strain of *P. aeruginosa* for the degradation of a range of benzoates carrying a chlorine substituent at the *ortho*-position, and is mediated by a three-component *ortho*-halobenzoate-1,2-dioxygenase (Romanov and Hausinger 1994). The intermediate *cis* diol formed by the dioxygenase is unstable, but the corresponding product from 2-trifluoromethyl benzoate that is a terminal metabolite has been characterized unambiguously (Selifonov et al. 1995).

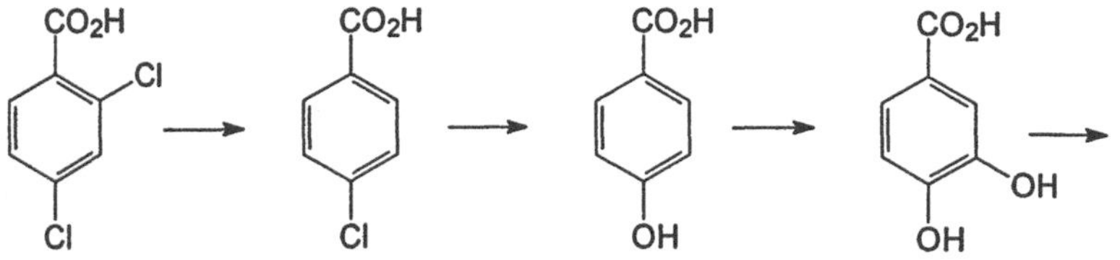

FIGURE 6.85
Reductive steps in the biodegradation of 2,4-dichlorobenzoate.

6.5.2 Fungal Systems

Although these have been less exhaustively investigated than their bacterial counterparts, the results of these investigations have revealed some interesting and significant features.

1. The NIH shift (Daly et al. 1972) with translocation of chlorine has been demonstrated during the biotransformation of 2,4-dichlorophenoxyacetate by *Aspergillus niger* (Figure 6.86) (Faulkner and Woodcock 1965). The NIH shift is not restricted to fungi since it has also been demonstrated with protons—although not apparently with other substituents—in prokaryotic systems (Sections 6.2.2 and 6.5.1.1).

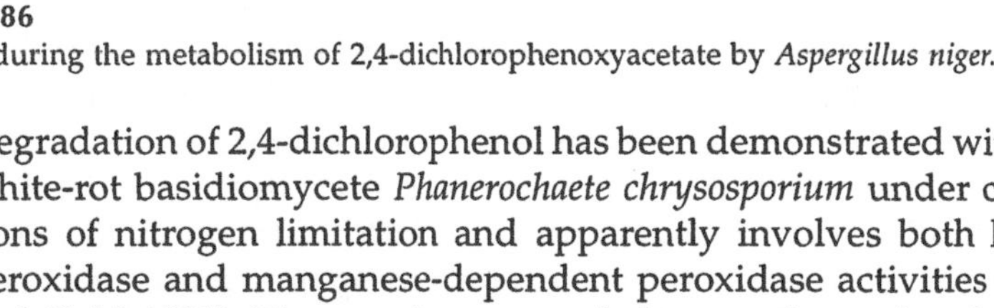

FIGURE 6.86
NIH shift during the metabolism of 2,4-dichlorophenoxyacetate by *Aspergillus niger.*

2. Degradation of 2,4-dichlorophenol has been demonstrated with the white-rot basidiomycete *Phanerochaete chrysosporium* under conditions of nitrogen limitation and apparently involves both lignin peroxidase and manganese-dependent peroxidase activities (Valli and Gold 1991). The reaction proceeds by a pathway involving a series of oxidation and reductions (Figure 6.87) which is entirely different from the sequence that is employed by bacteria, and an essentially similar sequence is involved in the degradation of 2,4,5-trichlorophenol that is accomplished more rapidly, possibly due to less interference from polymerization reactions (Joshi and Gold 1993). In addition, 2-chloro-1,4-dimethoxybenzene is produced from 2,4-dichlorophenol and 2,5-dichloro-1,4-dimethoxybenzene from 2,4,5-trichlorophenol. 2,4,5-Trichlorophenol also gives rise to

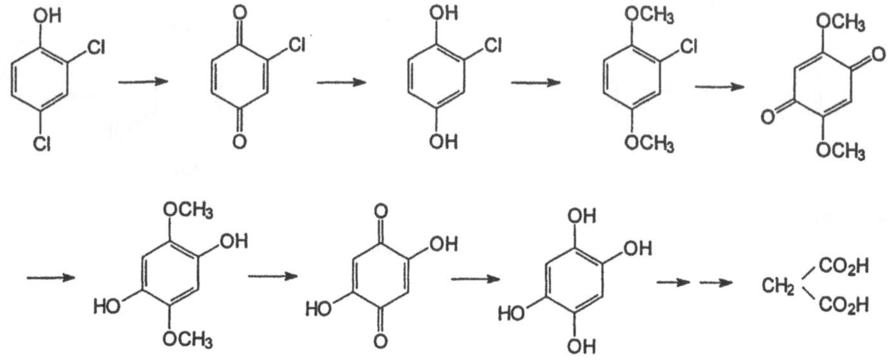

FIGURE 6.87
Biodegradation of 2,4-dichlorophenol by *Phanerochaete chrysosporium*.

small amounts of a dimeric product that was tentatively identified as 2,2'-dihydroxy-3,3',5,5',6,6'-hexachlorobiphenyl. A formally comparable sequence of reactions is used for the degradation of 2,4-dinitrotoluene (Valli et al. 1992b) that is discussed in Section 6.8.2. The degradation of 2,4,6-trichlorophenol by *P. chrysosporium* is comparable to that of 2,4-dichlorophenol and forms 2-chloro-1,4-dihydroxybenzene that is converted by alternative pathways to 1,2,4-trihydroxybenzene before ring cleavage (Reddy et al. 1998).

6.6 Anaerobic Metabolism of Halogenated Aromatic Compounds

There has been substantial interest in the fate of chlorinated aromatic compounds under anaerobic conditions since some of these, such as PCBs, polychlorinated phenols, polychlorinated catechols, and polychlorinated anilines, have been recovered from anaerobic sediment samples. The persistence of these compounds is therefore determined by the activity of anaerobic dehalogenating bacteria and extensive effort has therefore been devoted to isolating the relevant organisms. An increasing number of strains have been obtained in pure culture. They display different specificities, some use the aromatic substrate as an electron acceptor, and the role of reductive dechlorination in energy metabolism has been reviewed (Holliger et al. 1999).

1. *Desulfomonile tiedjei* is a sulfate-reducing bacterium capable of reducing 3-chlorobenzoate to benzoate (DeWeerd et al. 1990) during which ATP is synthesized by coupling proton translocation to dechlorination (Dolfing 1990; Mohn and Tiedje 1991). Cells induced by growth with 3-chlorobenzoate were able to partially dechlorinate

polychlorinated phenols specifically at the 3-position, whereas the monochlorophenols were apparently resistant to dechlorination (Mohn and Kennedy 1992). The membrane-bound reductive dehalogenase from *D. tiedjei* has been solubilized and purified (Ni et al. 1995): it is distinct from the tetrachlorohydroquinone enzyme from a strain of *Flavobacterium* sp. (Xun et al. 1992c), and it plausibly plays a role in the energy transduction of *D. tiedjei*. A membrane-bound cytochrome *c* is coinduced with the activity for reductive dechlorination, and has been purified and shown to be a high-spin diheme cytochrome distinct from previously characterized *c*-type cytochromes (Louie et al. 1997). It has been suggested that a chemiosmotic process may be used to rationalize the coupling of energy production with concomitant dechlorination (Louie and Mohn 1999).

2. A spore-forming organism has been isolated that does not reduce sulfate and generally dechlorinates chlorophenols preferentially at the 2-position (Madsen and Licht 1992).

3. A sulfite-reducing organism *Desulfitobacterium dehalogenans* that is unable to carry out dissimilatory reduction of sulfate dechlorinates 3-chloro-4-hydroxyphenylacetate to 4-hydroxyphenylacetate as a terminal metabolite during growth with pyruvate (Utkin et al. 1994), and the specificity of partial dechlorination of chlorinated phenols has been examined (Utkin et al. 1995). *D. frappieri* strain PCP-1 is able to dechlorinate a range of polychlorinated aromatic substrates including phenols, catechols, anilines, pentachloronitrobenzenes, and pentachloropyridine (Dennie et al. 1998).

4. An organism affiliated to the myxobacteria is capable of dechlorinating 2-chlorophenol to phenol and 2,6-dichlorophenol to 2-chlorophenol and phenol using acetate as electron donor (Cole et al. 1994).

5. A spore-forming strain of *D. chlororespirans* couples the dechlorination of 3-chloro-4-hydroxybenzoate to the oxidation of lactate to acetate, pyruvate, or formate (Sanford et al. 1996). The dehalogenase is membrane bound, was not oxygen sensitive, and was not inhibited by sulfate (Löffler et al. 1996).

There is a significant difference between the design of experiments that have been discussed previously and those that are summarized below. Compared with the large number of pure strains of aerobic bacteria that have been used for studies in biodegradation, relatively fewer strains of anaerobic bacteria that are capable of degrading halogenated aromatic compounds have been isolated in pure culture; metabolic studies have therefore of necessity used mixed cultures. The design of these investigations differs considerably, and it is important to distinguish between the various experimental conditions since these may critically affect the interpretation of the results:

- Unenriched slurries containing both bacteria and significant amounts of sediment or sludge;
- Enrichment cultures obtained after only a small number of successive transfers;
- Stable enrichment cultures which have been obtained after successive transfer over extended periods of time and in which organic matter from the sludge or sediment has been removed by dilution.

Although the use of metabolically stable mixed cultures enables an acceptable degree of repeatability to be attained, the use of suspensions or slurries containing sediment or metabolites from previous additions of enrichment substrates introduces undesirable ambiguity in the interpretaton of analytical results. This has been discussed more fully in Section 5.3.1. Inevitably, however, most examples of dechlorination depend on the results of studies with mixed cultures. There are several important issues which have emerged from these quite extensive investigations and these merit brief discussion.

1. Although complete dechlorination of polyhalogenated compounds under anaerobic conditions has been observed, the most common situation is that in which only partial dehalogenation occurs; all of these reactions are therefore strictly biotransformations. Illustrative examples include pentachlorophenol (Figure 6.88a) (Mikesell and Boyd 1986), hexachlorobenzene (Figure 6.88b) (Fathepure et al. 1988), 3,4,5-trichlorocatechol (Figure 6.88c) (Allard et al. 1991), 2,3,5,6-tetrachlorobiphenyl (Van Dort and Bedard 1991) (Figure 6.88d), chloroanilines (Figure 6.88e) (Kuhn et al. 1990), and 2,4,6-chlorobenzoate (Gerritse et al. 1992) (Figure 6.88f). Partial dechlorination of chlorinated dibenzo[1,4]dioxins has been observed in slurries: the 1,2,3,4-tetrachloro compound produced predominantly the 1,3-dichloro compound (Beurskens et al. 1996; Ballerstedt et al. 1997), and for compounds with 5-7 chlorine substituents, chlorine was removed from both the peri and the lateral positions (Barkowskii and Ariaens 1995). The pattern for removal of chlorine substituents from PCBs depends critically on their position; further comments are given in Chapter 8, Section 8.2.4.2, including the "priming" of dechlorination of highly chlorinated congeners by bromobiphenyls (Bedard et al. 1998).

Collectively, these results show that the less highly substituted congeners are more resistant to dechlorination. In general, this is the reverse of the situation pertaining under aerobic conditions, and this suggests that the complete degradation of these polyhalogenated compounds will probably involve both kinds of reactions; a similar situation has been suggested for halogenated alkanes and alkenes (Section 6.4.4). For polyhalogenated compounds that have been partitioned into the sediment phase, partial anaerobic dechlorination to less highly chlorinated congeners is therefore likely to occur; if these were subsequently released into the water mass aerobic degradation could presumably occur provided that the compounds existed in a state accessible to the relevant microorganisms.

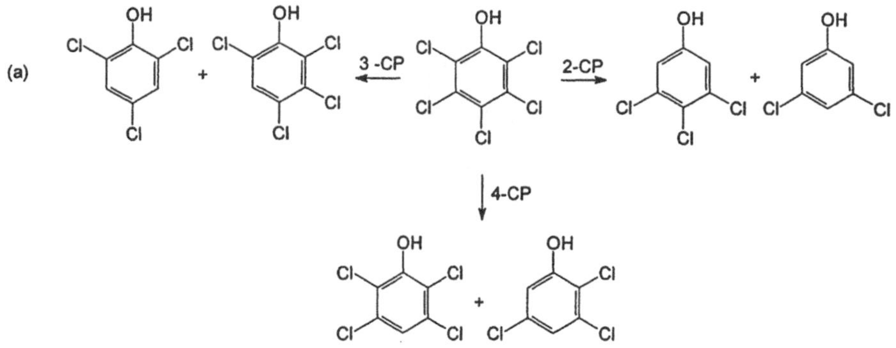

FIGURE 6.88
The partial dechlorination of (a) pentachlorophenol, (b) hexachlorobenzene, (c) 3,4,5-trichloro-catechol, (d) 2,3,5,6-tetrachlorobiphenyl, (e) chloroanilines, and (f) 2,4,6-trichlorobenzoate.

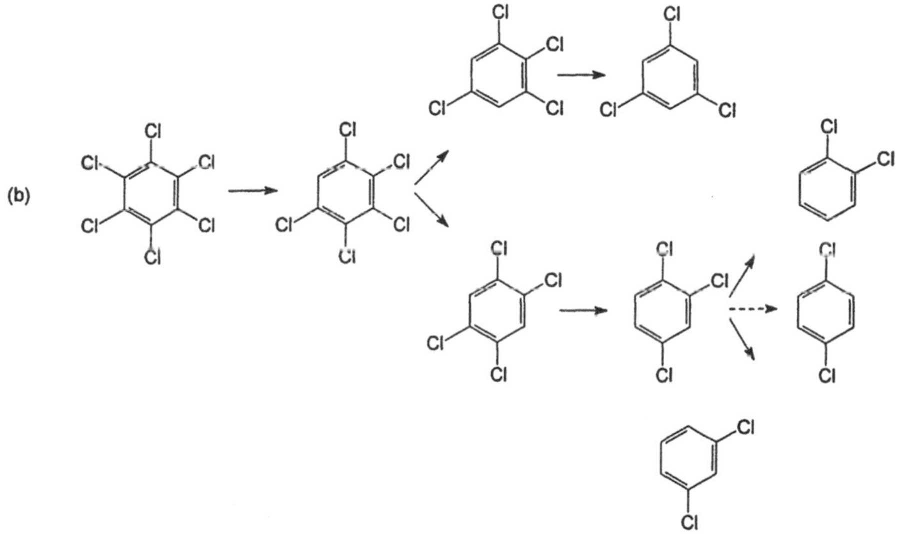

FIGURE 6.88 (b)   (continued)

2. The physiology of a mixed culture is inevitably incompletely defined, and organisms differing as basically as methanogenic and sulfidogenic bacteria may be simultaneously present. Considerable attention has therefore been devoted to the role of sulfate in dechlorination reactions and two essentially different situations have emerged.

   a. Degradation of 4-chlorophenol and 2,4-dichlorophenol has been demonstrated both in methanogenic cultures in the presence of sulfate (Kohring et al. 1989) and in nonmethanogenic cultures reducing sulfate (Häggblom and Young 1990).

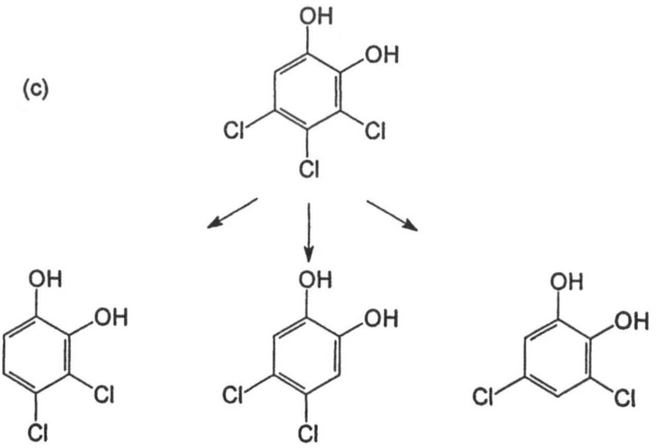

FIGURE 6.88 (c)   (continued)

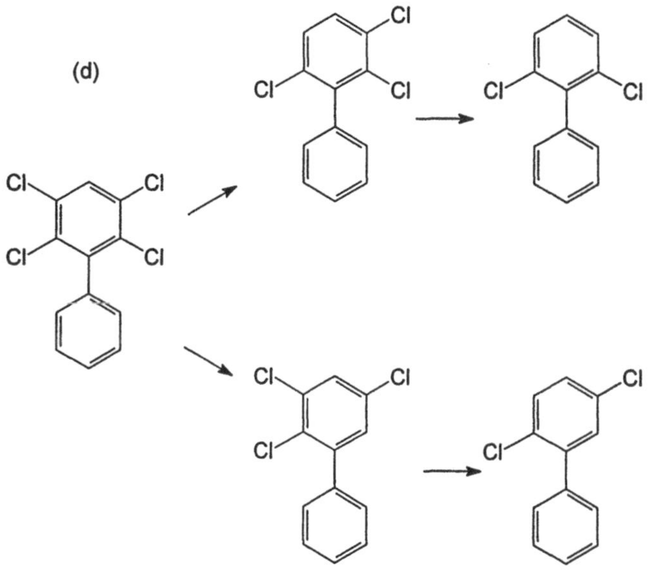

FIGURE 6.88 (d)   (continued)

b. On the other hand, the rate of dechlorination of a series of chloro-anilines by methanogenic groundwater mixed cultures was diminished by the presence of sulfate (Kuhn et al. 1990), and the dechlorination of pentachlorophenol was inhibited by the presence of sulfate (Madsen and Aamand 1991).

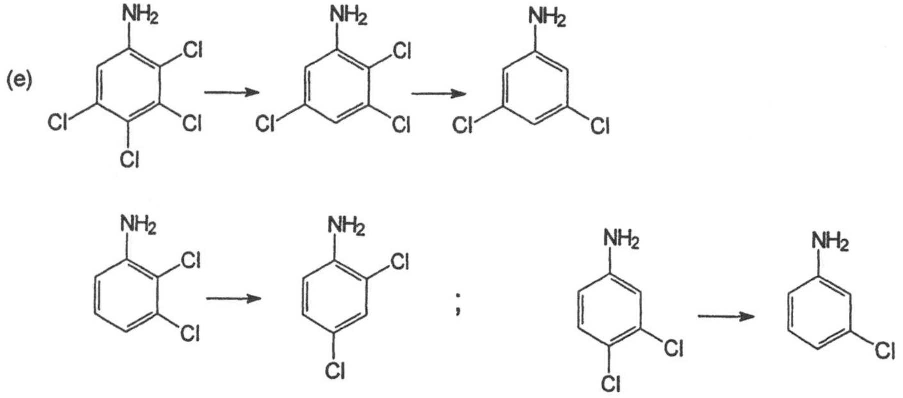

FIGURE 6.88 (e) (continued)

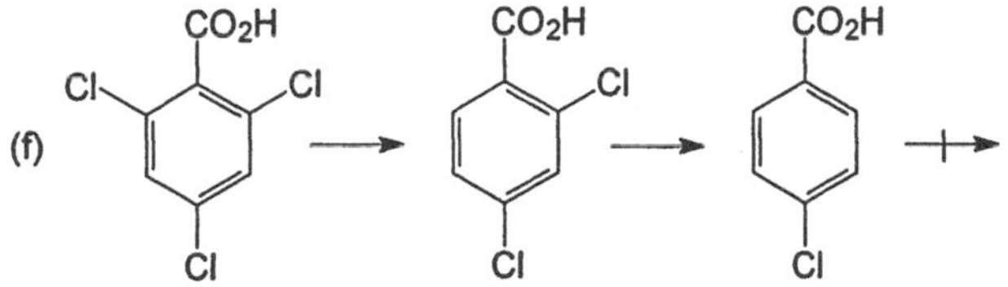

FIGURE 6.88 (f) (continued)

c. The effect of adding sulfate to enrichment cultures has been examined, and has been found to be counterselective to the development of cultures with dechlorination capability (Genther et al. 1989; Allard et al. 1992). Presumably in these cases, sulfate effectively competes for reducing equivalents provided by the organic substrate. On the other hand, addition of sulfate to pure cultures of *Desulfomonile tiedjei* (DeWeerd et al. 1991) or to mixed cultures from several enrichments which possessed dechlorination capability (Allard et al. 1992) did not inhibit dechlorination. These apparently conflicting results may be rationalized on the basis of investigations with *D. tiedjei* (Townsend and Suflita 1997). They showed that the presence of sulfate during growth with 3-chlorobenzoate inhibited dehalogenation activity, although it had no effect on this in cell extracts.

Although considerable effort has been directed to the dechlorination of aromatic compounds, fewer studies have examined debromination. Some examples will be given to illustrate the probably widespread distribution of this reaction:

1. Aquifer slurries under methanogenic conditions debrominated the agrochemical bromacil (5-bromo-3-*sec*-butyl-6-methyluracil) (Adrian and Suflita 1990).

2. Marine sediment slurries debrominated the naturally occurring 2,4-dibromophenol and also 2,4,6-tribromophenol to phenol that was then further degraded (King 1988), and a strain of *Desulfovibrio* sp. is able to dehalogenate 2,4,6-tribromophenol to phenol during growth with lactate (Boyle et al. 1999).

3. Freshwater sediment slurries debrominated 2-bromobenzoate to benzoate that was then degraded by methanogenic organisms (Horowitz et al. 1983).

4. Metabolically stable enrichment cultures reduced the aldehyde group and debrominated 5-bromovanillin to 4-methylcatechol, and debrominated 3-bromo-4-hydroxybenzaldehyde to 4-hydroxybenzoate (Neilson et al. 1988a).

5. The debromination of all isomers of bromophenol and bromobenzoates was examined in sediment slurries under methanogenic, sulfidogenic, and Fe-reducing conditions, although the results were presented only as half-lives (Monserrate and Häggblom 1997).

6. Debromination of polybrominated biphenyls has been observed in cultures obtained from sediments contaminated with the corresponding chlorobiphenyls (Morris et al. 1992). Complete debromination has been observed in sediment slurries supplemented with malate (Bedard and van Dort 1998): biphenyl was the end product from all congeners, and the acclimation time for 2- and 4-monobromo and the 2,4-, 2,5-, and 2,6-dibromo compounds was much less than for the corresponding chloro compounds. All of the intermediates were isolated and this enabled a complete sequence of debromination to be established.

6.7 Reactions Carried Out by Anaerobic Bacteria Other Than Dehalogenation

The comments made at the beginning of Section 6.6 apply equally to some of the reactions described in this section and in particular to the degradation of aromatic compounds that is discussed in Section 6.7.3. In addition, attention should be drawn to the widely differing physiologies of the organisms that include strictly fermentative organisms, phototrophic organisms, and organisms that use nitrate or sulfate or carbonate as electron acceptors. The sole unifying feature of the reactions is that they do not involve incorporation of molecular oxygen.

6.7.1 Aliphatic Compounds

The degradation of aliphatic carboxylic acids is of great ecological importance since compounds such as acetate, propionate, or butyrate may be the terminal fermentation products of organisms degrading more complex compounds including carbohydrates, proteins, and lipids (Zeikus 1980; Mackie et al. 1991). Degradation of aliphatic carboxylic acids by sulfate-reducing bacteria was traditionally restricted to lactate and its near relative, pyruvate, but recent developments have radically altered the situation and increased the spectrum of compounds which can be oxidized to CO_2 at the expense of sulfate reduction. In the following paragraphs, an attempt will be made to present a brief summary of the anaerobic degradation of the main groups of aliphatic compounds. Greatest attention has been directed to compounds with functional groups and in particular alkanoic acids and amino acids.

Alkanes

Complete oxidation of hexadecane by a sulfate-reducing bacterium has been reported (Aeckersberg et al. 1991), and presents a particularly interesting addition to the range of highly reduced compounds which can be oxidized under anaerobic conditions. Further studies with additional strains indicate the complexity of the reaction and differences between the strains, but are not supportive of an initial desaturation to the alk-1-en (Aeckersberg et al. 1998). Whether or not degradation takes place by an initial addition reaction comparable with that involved in the anaerobic degradation of toluene is unresolved. In experiments with sediment slurries from contaminated marine areas, $^{14}CO_2$ was recovered from ^{14}C-hexadecane (Coates et al. 1997) and was inhibited by molybdate that suggested the involvement of sulfate reduction, and $^{14}CO_2$ was produced from $^{14}C[14,15]$octacosane ($C_{28}H_{58}$) under sulfate-reducing conditions (Caldwell et al. 1998).

Alkenes

Degradation of hex-1-ene has been observed in a methanogenic consortium (Schink 1985a) that converted the substrate to methane, and a plausible pathway involving hydration and oxidation was suggested.

Alkynes

Acetylene supports the growth of *Pelobacter acetylenicus* (Schink 1985b) and in mixed cultures of organisms, undergoes initial hydration to acetaldehyde followed by dismutation into acetate and ethanol. The hydratase contains tungsten, and although the enzyme is stable in air, it requires a strong reductant such as Ti(III) or dithionite for its activity (Rosner and Schink 1995).

Alkanols

Oxidation of methanol—although not incorporated into cellular material—has been observed in a sulfate-reducing bacterium (Braun and Stolp 1985), while ethanol may be converted into methane by *Methanogenium organophilum* (Widdel 1986; Frimmer and Widdel 1989), and oxidation of primary alkanols has been demonstrated in *Acetobacterium carbinolicum* (Eichler and Schink 1984). Secondary alcohols such as propan-2-ol and butan-2-ol may be used as hydrogen donors for methanogenesis with concomitant oxidation to the corresponding ketones (Widdel et al. 1988). An NAD-dependent alcohol dehydrogenase has been purified from *Desulfovibrio gigas* that can oxidize ethanol, and it has been shown that the enzyme does not bear any relation to classical alcohol dehydrogenases (Hensgens et al. 1993). The metabolism of 1,2-diols has attracted considerable attention, and in particular that of glycerol in view of its ubiquity as a component of lipids. Widely different pathways have been found of which two are given as illustration:

1. *Anaerovibrio glycerini* ferments glycerol to propionate (Schauder and Schink 1989) and *D. carbinolicus* to 3-hydroxypropionate (Nanninga and Gottschal 1987).

2. *D. alcoholovorans* converts glycerol to acetate, and 1,2-propandiol to acetate and propionate (Qatibi et al. 1991).

Alkanones

The degradation of acetone and some higher ketones has been studied, and involves initial carboxylation to acetoacetate although this reaction has not been demonstrated *in vitro*. There are two subsequent pathways: (1) oxidation to CO_2 in denitrifying and sulfate-reducing bacteria by the acetyl-CoA-CO dehydrogenase pathway, and (2) an anabolic pathway by a modified tricarboxylate-glyoxylate cycle (Janssen and Schink 1995a).

Alkylamines

Few investigations have been devoted to the degradation of alkylamines although there has been considerable interest in the metabolism of trimethylamine, choline, and glycine betaine as a source of methane in marine sediments (King 1984). Two aspects of the metabolism of N-alkyl compounds may be used as illustration.

1. Methane formation from trimethylamine by *Methanosarcina barkeri* has been demonstrated (Hippe et al. 1979; Patterson and Hespell 1979), and the metabolically versatile organisms *Methanococcoides methylutens* (Sowers and Ferry 1983) and *Methanolobus tindarius* (Konig and Stetter 1982) that use methylamines and methanol for methane formation have been described.

2. The metabolism of betaine that is an important osmoregulatory solute in many organisms has been studied and different metabolic pathways have been revealed.

- Betaine is demethylated to dimethylglycine by *Eubacterium limosum* (Müller et al. 1981) and by strains of *Desulfobacterium* sp. (Heijthuijsen and Hansen 1989a).
- Betaine is a fermented strain of *Desulfuromonas acetoxidans* with the production of trimethylamine and acetate (Heijthuijsen and Hansen 1989b), and the same products are formed by *Clostridium sporogenes* in a Stickland reaction with alanine, valine, leucine, or isoleucine (Naumann et al. 1983).

Alkanoic Acids

For a number of reasons, a great deal of effort has been directed to the degradation of alkanoic acids: acetate, propionate, and butyrate are fermentation products of carbohydrates and are metabolites of the aerobic degradation of alklanes and related compounds, while long-chain acids are produced by the hydrolysis of lipids. Studies on the degradation of alkanoic acids have been carried out using both pure cultures and syntrophic associations that have been discussed in Section 4.5.1.

For the degradation of acetate, two different reactions may take place: oxidation to CO_2 and dismutation to methane and CO_2. To some extent, as will emerge, segments of both pathways are at least formally similar although the mechanisms for anaerobic degradation of these apparently simple compounds are quite subtle. Degradation of the short-chain carboxylic acids acetate and butyrate can be accomplished by *Desulfotomaculum acetoxidans* (Widdel and Pfennig 1981), and of acetate by *Desulfuromonas acetoxidans* (Pfennig and Biebl 1976) and species of *Desulfobacter* (Widdel 1987), while propionate is degraded by species of *Desulfobulbus* (Widdel and Pfennig 1982; Samain et al. 1984). A sulfur-reducing organism with a much wider degradative capability than *Desulfuromonas acetoxidans* has been isolated (Finster and Bak 1993) and this organism is capable of accomplishing the complete oxidation of, for example, propionate, valerate, and succinate.

The oxidation of acetate under anaerobic conditions can occur by two completely different pathways, both of which have been investigated in detail and the enzymology has been delineated (Thauer et al. 1989).

1. Oxidation may take place by a modified tricarboxylic acid cycle in which the production of CO_2 is coupled to the synthesis of NADPH and reduced ferredoxin, and the dehydrogenation of succinate to fumarate is coupled to the synthesis of reduced menaquinone. This pathway is used for example by *D. acetoxidans* and in modified form by *Desulfobacter postgatei*.

2. On the other hand, dissimilation of acetate may take place by reversal of the pathway used by organisms such as *Clostridium thermoaceticum* for the synthesis of acetate from CO_2. In the degradation of acetate, the pathway involves a dismutation in which the methyl group is successively oxidized via methyl tetrahydrofolate to CO_2 while the carbonyl group is oxidized via bound carbon monoxide. Such THF-mediated reactions are of great importance in the anaerobic degradation of purines that will be discussed in Section 6.7.4.1.

Acetate may also be converted into methane by a few methanogens belonging to the genus *Methanosarcina*. The methyl group is initially converted into methyltetrahydromethanopterin (corresponding to methyltetrahydrofolate in the acetate oxidations discussed above) before reduction to methane via methyl-coenzyme M; the carbonyl group of acetate is oxidized via bound CO to CO_2.

Both the synthesis of propionate and its metabolism may take place under anaerobic conditions, and in *Desulfobulbus propionicum*, the degradation plausibly takes place by reversal of the steps used for its synthesis from acetate (Stams et al. 1984): carboxylation of propionate to methylmalonate followed by coenzyme B_{12}-mediated rearrangement to succinate which then enters the tricarboxylic acid cycle. Growth of syntrophic propionate-oxidizing bacteria in the absence of methanogens has been accomplished using fumarate as the sole substrate (Plugge et al. 1993). Fumarate plays a central role in the metabolism of this organism since it is produced from propionate via methylmalonate and succinate, and fumarate itself is metabolized by the acetyl-CoA cleavage pathway via malate, oxalacetate, and pyruvate. This has already been noted in Chapter 4, Section 4.5.1.

The degradation of long-chain carboxylic acids is important in the anaerobic metabolism of lipids and an extensive compilation of the organisms that can accomplish this has been given (Mackie et al. 1991). This capability has been demonstrated in syntrophic bacteria in the presence of hydrogen-utilizing bacteria: for example, β-oxidation of C_4 to C_8, and C_5 and C_7 carboxylic acids was carried out by the *Syntrophomonas wolfei* association (McInerney et al. 1981), and of C_4 to C_{10}, and C_5 to C_{11} by the *Clostridium bryantii* syntroph (Stieb and Schink 1985). Acetate and propionate were the respective terminal products from the even- and odd-numbered acids. Single cultures of many sulfate-reducing bacteria are also able to carry out comparable reactions (Mackie et al. 1991): for example, *Desulfobacterium cetonicum* degrades butyrate to acetate by a typical β-oxidation pathway (Janssen and Schink 1995b).

Brief attention has been directed in Chapter 4, Section 4.3.4 to anaerobic bacteria that are important in the decarboxylation of dicarboxylic acids including oxalate and malonate—and succinate that is produced as a fermentation product of carbohydrates, and in anaerobic respirations involving fumarate. The various anaerobic organisms, and their metabolic capabilities are briefly summarized here:

1. Oxalate—*Oxalobacter vibrioformis* uses this as sole energy source with acetate as principal carbon source (Dehning and Schink 1989a). In *O. formigenes*, oxalate and its decarboxylation product formate form a one-to-one antiport system which involves the consumption of an internal proton during decarboxylation and serves as a proton pump to generate ATP by decarboxylative phosphorylation (Anantharam et al. 1989).

2. Malonate—*Malonomonas rubra* is a microaerotolerant fermenting organism that decarboxylates this to acetate (Dehning and Schink 1989b). The organism contains high concentrations of c-type cytochromes that are not involved in the metabolism of the substrate, and are presumably remnants of the sulfur-reducing relatives of the organism (Kolb et al. 1998). Biotin-dependent carboxylases couple the decarboxylation to the transport of Na^+ across the cytoplasmic membrane and use the electrochemical potential $\Delta\mu$ Na^+ to mediate the synthesis of ATP (Dimroth and Hilbi 1997).

3. Succinate—*Propionigenium modestum* (Schink and Pfennig 1982a), *Sporomusa malonica* (Dehning et al. 1989), *Selenomonas acidaminovorans* (Guangsheng et al. 1992), and *P. maris* (Janssen and Liesack 1995) decarboxylate succinate to propionate.

Decarboxylation is a key step in the degradation of glutarate to butyrate and isobutyrate by dehydrogenation to glutaconate followed by decarboxylation to crotonate (Matties and Schink 1992). This pathway has been proposed for the degradation of pimelate that is a possible intermediate in the anaerobic degradation of benzoate (Härtel et al. 1993; Gallus and Schink 1994).

In summary, the anaerobic degradation of alkanoic acids may truly be described as ubiquitous and is carried out by organisms with widely different taxonomic affinity both in pure culture and in syntrophic associations.

Amino Acids

The degradation pathways of amino acids have been examined in detail particularly in clostridia (Barker 1981), and these investigations have revealed a number of important reactions not encountered in other degradations.

1. Coenzyme B_{12}-mediated rearrangements have been elucidated as an important reaction in the degradation of glutamate and ornithine, and subsequently led to a detailed investigation into the role of this rearrangement in other reactions (Barker 1972). An outline of the β-methylaspartate pathway for the degradation of glutamate by *C. tetanomorphum* is given in Figure 6.89. The involvement of radicals in a number of rearrangements carried

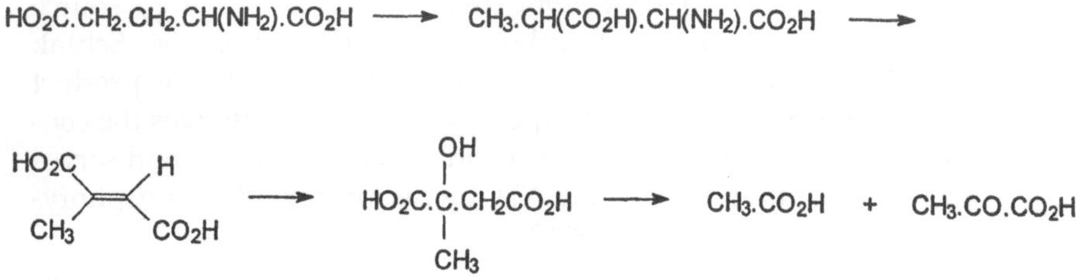

$$HO_2C.CH_2.CH_2.CH(NH_2).CO_2H \longrightarrow CH_3.CH(CO_2H).CH(NH_2).CO_2H \longrightarrow$$

$$\underset{\substack{CH_3 \\ }}{\overset{\substack{HO_2C \qquad H}}{\diagdown}} \underset{CO_2H}{\diagup} \longrightarrow HO_2C.\underset{\substack{| \\ CH_3}}{\overset{\substack{OH \\ |}}{C}}.CH_2CO_2H \longrightarrow CH_3.CO_2H + CH_3.CO.CO_2H$$

FIGURE 6.89
Biodegradation of glutamate by *Clostridium tetanomorphum.*

    out by clostridia during fermentation of amino acids is discussed in a review (Buckel and Golding 1999). These include *S*-lysine to *S*-β-lysine (lysine 2,3-aminomutase), *S*-glutamate to (2*S*,3*S*)-3-methylaspartate (glutamate mutase), and 2-methyleneglutarate — produced during the fermentation of nicotinate by *C. barkeri* — to (*R*)-3-methylitaconate (2-methyleneglutarate mutase).

2. The transformation of aromatic amino acids has been examined, and it has been shown (Elsden et al. 1976; D'Ari and Barker 1985) that in *Cl. difficile* these compounds are possible biogenic sources of 4-methyl phenol, and that in an organism designated "*Cl. aerofoetidum*" toluene may be produced (Pons et al. 1984): the postulated reactions are schematically shown in Figure 6.90.

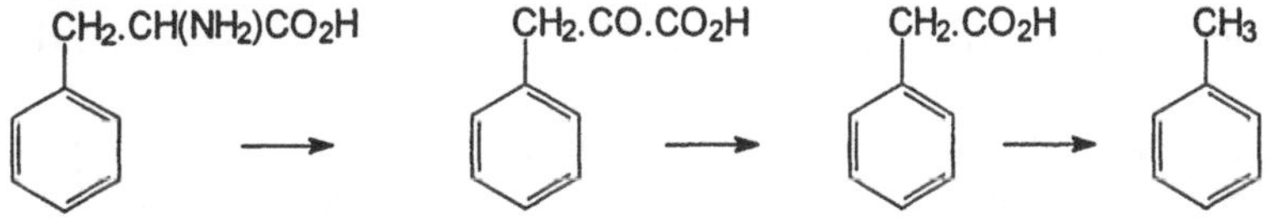

FIGURE 6.90
Metabolism of phenylalanine.

3. An unusual dismutation reaction involving pairs of amino acids (Strickland reaction) has been studied extensively in *Cl. sporogenes* (Stickland 1934; 1935a,b). The reaction can be carried out by many other clostridia and has been summarized (Barker 1961). The products from proline and alanine (Stickland 1935a) illustrate the reaction and are shown in Figure 6.91, and the mechanism whereby glycine is reduced to acetate by glycine reductase from *Cl. sticklandii* has been elucidated (Arkowitz and Abeles 1989). The involvement of betaine in the reaction has already been noted. Formally similar dismutations have been suggested for the biotransformation of γ-hexachloro[*aaaeee*]cyclohexane (Ohisa et al. 1980) and of 5-chlorovanillin (Neilson et al. 1988a).

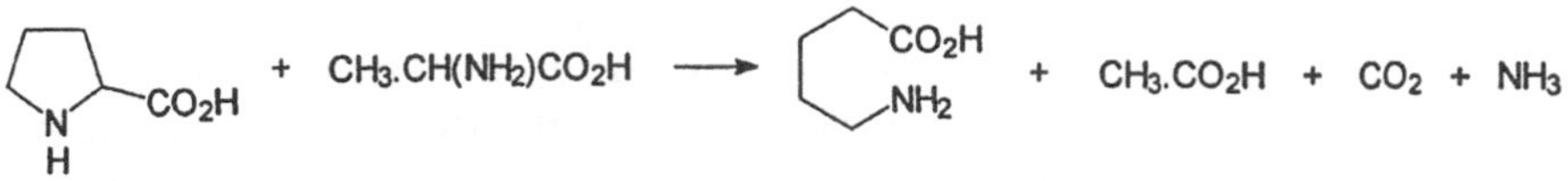

FIGURE 6.91
The Stickland reaction between proline and alanine.

In summary, therefore, pathways exist for the metabolism of a wide range of aliphatic compounds, and a net of metabolic interactions operate for their complete degradation to CO_2 and cell material.

6.7.2 Biotransformation of Alicyclic Compounds Containing Several Rings

The transformation of sterols and bile acids has been examined quite thoroughly in the context of their intestinal metabolism, and a number of reactions that are otherwise quite unusual have been observed, most frequently in organisms belonging to the genus *Eubacterium*. These reactions may be very briefly summarized: (1) reduction of the $\Delta^{4,5}$ bond with production of 5β-reduced compounds (Mott et al. 1980) and (2) reductive dehydroxylation of 7α-hydroxy bile acids (Masuda et al. 1984) and 16α-hydroxy and 21-hydroxy corticosterols (Bokkenheuser et al. 1980).

6.7.3 Aromatic Compounds

Considerable effort has been devoted to the anaerobic degradation of aromatic compounds, in particular to their conversion to methane. In spite of this, however, pure cultures of organisms have not always been achieved so that many investigations have used mixed cultures with the result that the proposed pathways are inevitably more speculative. It is important to note that several distinct groups of organisms are involved: (1) strictly anaerobic fermentative bacteria, (2) strictly anaerobic photoheterotrophic bacteria, (3) anaerobic sulfate-reducing bacteria, and (4) organisms using nitrate as electron acceptor under anaerobic conditions. This section will summarize first reactions carried out by pure cultures—and then those accomplished by various types of consortia.

6.7.3.1 Pure Cultures

An attempt is made to provide a spectrum of the reactions carried out by pure cultures of a wide range of bacteria under anaerobic conditions.

Anaerobic Bacteria

1. Aromatic methoxy groups may be used to support the growth of methoxybenzoates (Bache and Pfennig 1981): acetogenic organisms such as

Acetobacterium woodii use the methyl group absence of sulfate (Tasaki et al. 1992; Hanselmann et al. 1995). The methyl group may be used for growth, and sulfate-reducing organisms can carry out the demethylation reaction with subsequent conversion to butyrate by *C. pfennigii* (Krumholz and Bryant 1985) or into methanethiol and dimethyl sulfide by inorganic sulfide in the growth medium (Bak et al. 1992). The details of this complex demethylation reaction have been examined in a strain of a sulfide-methylating homoaceto-genic bacterium that degraded gallate by the same pathway as that employed by *Pelobacter acidgallici* (Kreft and Schink 1993). The metabolism of 4-hydroxy-3,5-dimethoxybenzoate by an organism that has an obligate dependence on H_2 (Liu and Suflita 1993) indicates additional complexities of these apparently straightforward O-demethylation reactions.

2. Considerable attention has been directed to the degradation of phenolic compounds.

a. Catechol and aniline are degraded to CO_2 by *Desulfobacterium anilini* (Schnell et al. 1989), and catechol is slowly degraded by *D. catecholicum* (Szewzyk and Pfennig 1987) and by a strain of *Desulfotomaculum* sp. (Kuever et al. 1993). Aniline is carboxylated to 4-aminobenzoate which is then reductively deaminated to benzoate (Schnell and Schink 1991). By analogy with phenol itself (Section 6.7.3.1), and consistent with the CO_2-dependence of the catechol-degrading *Desulfotomaculum* sp. (Kuever et al. 1992), catechol is carboxylated to 3,4-dihydroxybenzoate by a strain of *Desulfobacterium* sp. (Gorny and Schink 1994b): this is then dehydroxylated to benzoate before further degradation. The reversible decarboxylation of 3,4-dihydroxybenzoate has been described (He and Wiegel 1996) in *C. hydroxybenzoicum*, and the enzyme has been purified.

b. The degradation of hydroquinone takes place by carboxylation followed by dehydroxylation to benzoate that was further degraded by reduction of the aromatic ring (Gorny and Schink 1994a).

c. 3,4,5-Trihydroxybenzoate is degraded by *P. acidigallici* to acetate and CO_2 (Schink and Pfennig 1982b) and by *Eubacterium oxidoreducens* in the presence of exogenous H_2 or formate to acetate, butyrate, and CO_2 (Krumholz et al. 1987). The degradation has been studied in detail, and takes place by the unusual pathway (Figure 6.92) (Krumholtz and Bryant 1988; Brune and Schink 1992). The formation of phloroglucinol in *P. acidigallici* involves a series of intramolecular hydroxyl transfer reactions (Brune and Schink 1990) (Figure 6.93), and the reaction in *E. oxidoreducens* also involves 1,2,3,5-tetrahydroxybenzene although details of the pathway may be different (Haddock and Ferry 1993). Subsequent reduction and ring cleavage follow a pathway comparable to that involved in the degradation of resorcinol by species of clostridia (Tschech and Schink 1985), although an alternative pathway involving direct hydrolysis has been observed in denitrifying bacteria (Gorny et al. 1992). *P. massiliensis* is able to degrade all three trihydroxybenzenes (1,2,3-, 1,3,5, 1,2,4-) by the transhydroxylation pathway (Brune et al. 1992), whereas *Desulfovibrio inopinatus* degrades 1,2,4-trihydroxybenzene by a different pathway (Reichenbecher and Schink 1997).

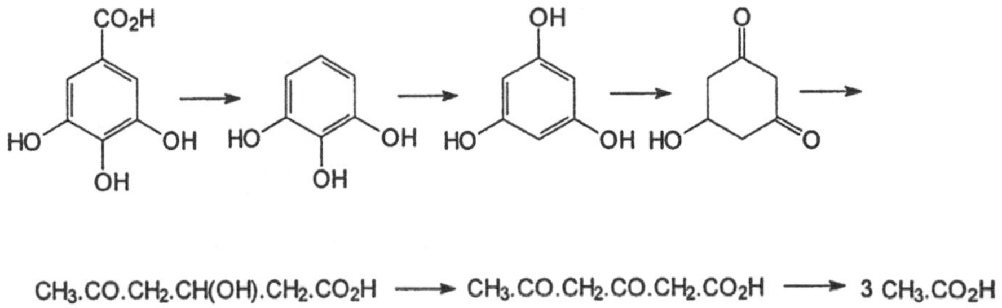

$$CH_3.CO.CH_2.CH(OH).CH_2.CO_2H \longrightarrow CH_3.CO.CH_2.CO.CH_2.CO_2H \longrightarrow 3\ CH_3.CO_2H$$

FIGURE 6.92
Pathway for the biodegradation of 3,4,5-trihydroxybenzoate.

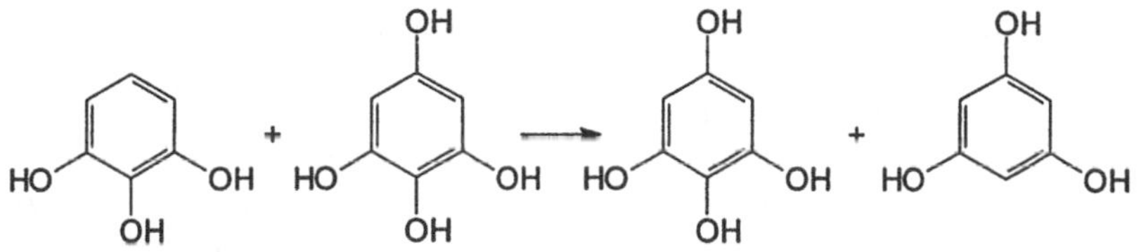

FIGURE 6.93
Intramolecular hydroxyl group transfers in the biodegradation of 1,2,3-trihydroxybenzoate.

3. Vanillate is transformed by *C. thermoaceticum* to catechol and phenol and the CO_2 produced relieves the requirement for supplemental CO_2 (Hsu et al. 1990): the metabolically produced CO_2 is then able to enter a sequence of reactions which results in acetate synthesis. Aromatic aldehydes may be involved in both reductive and oxidative reactions under anaerobic conditions, and in some cases the carboxylic acid is further decarboxylated.

a. Oxidation of substituted benzaldehydes to benzoates at the expense of sulfate reduction has been demonstrated in strains of *Desulfovibrio* sp., although the carboxylic acids produced were apparently stable to further degradation (Zellner et al. 1990). Vanillin was, however, used as a substrate for growth by a strain of *Desulfotomaculum* sp. and was metabolized via vanillate and catechol (Kuever et al. 1993). Vanillin is transformed by *C. formicoaceticum* sequentially to vanillate and 3,4-dihydroxybenzoate, while the methyl group is converted into acetate via acetyl CoA (Göbner et al. 1994). Different aldehyde oxidoreductases have been isolated from *C. thermoaceticum* and *Cl. formicoaceticum* (White et al. 1993): whereas, however, both cinnamate and cinnamaldehyde were good substrates for the Mo-containing enzyme from the latter, benzoate was an extremely poor substrate. The W-containing enzyme from *Cl. formicoaceticum*, however, displays high activity toward a greater range of substituted benzoates (White et al. 1991).

b. On the other hand, cell extracts of *Cl. formicoaceticum* reduce benzoate to benzyl alcohol at the expense of carbon monoxide (Fraisse and Simon 1988), and *Desulfomicrobium escambiense* reduced benzoate, 3-chlorobenzoate, and 3-bromobenzoate to the corresponding benzyl alcohols at the expense of pyruvate that was oxidized to acetate, lactate, and succinate (Sharak Genther et al. 1997).

In addition, some important transformations are carried out by pure strains of anaerobic bacteria.

a. Utilization of toluene-4-sulfonate as a sulfur source following desulfonation by a *Clostridium* sp. (Denger et al. 1996).

b. Reduction of the nitro group of nitrodiphenylamines was accomplished by strains of *Desulfovibrio* sp., *Desulfococcus* sp., and *Desulfomicrobium* sp., during growth with the appropriate carbon sources (lactate or benzoate). The condensation products phenazine and 4-aminoacridine were formed during incubations with 2-aminodiphenylamine (Drzyzga et al. 1996), although the mechanism whereby the C-10 of 4-aminoacridine is formed from the growth substrate lactate was not resolved.

Bacteria Using Nitrate Electron Acceptor under Anaerobic Conditions and Anaerobic Phototrophs

It seems clear that benzoate occupies a central position in the anaerobic degradation of both phenols and alkaryl hydrocarbons, and that carboxylation, hydroxylation, and reductive dehydroxylation are important—and less expected—reactions.

1. By analogy with the initial steps for the degradation of aniline, 2-aminobenzoate is degraded under denitrifying conditions by a *Pseudomonas* sp. to benzoate which is then reduced to cyclohexene-1-carboxylate (Lochmeyer et al. 1992).

2. The metabolism of cinnamate and ω-phenylalkane carboxylates has been studied in *Rhodopseudomonas palustris* (Elder et al. 1992) and for growth with the higher homologues additional CO_2 was necessary. The key degradative reaction was β-oxidation: for compounds with chain lengths of 3-, 5-, and 7-carbon atoms, benzoate

was formed and further metabolized, but for the even-numbered compounds with 4-,6-, and 8-carbon atoms phenylacetate was a terminal metabolite.

3. The anaerobic metabolism of L-phenylalanine by *Thauera aromatica* under denitrifying conditions involves a number of steps that result in the formation of benzoyl-CoA: transamination to phenylacetyl-CoA, oxidation to phenylglyoxalate, and decarboxylation to benzoyl-CoA (Schneider et al. 1997). The membrane-bound phenylacetyl-CoA:acceptor oxidoreductase is induced under denitrifying conditions during growth with phenylalanine or phenylacetate and has been purified (Schneider et al. 1998); the enzyme level was low in cells grown with phenylglyoxalate, was insensitive to oxygen, and was absent in cells grown aerobically with phenylacetate.

4. Phenylacetate and 4-hydroxyphenylacetate are oxidized sequentially under anaerobic conditions by a denitrifying strain of *Pseudomonas* sp. to the phenylglyoxylates and benzoate (Mohamed et al. 1993).

Degradation of Benzoates—Benzoate occupies a central role in the metabolism of a range of aromatic compounds, and the pathway for its anaerobic degradation under denitrifying or phototrophic conditions involves three cardinal reactions (further references in Harwood et al. 1999) (Figure 6.94a,b): the designation of the enzymes corresponds to those for the genes discussed below.

1. Formation of the CoA thioester mediated by a ligase which has been demonstrated in a number of organisms including *Pseudomonas* sp. strain K172 (*Thauera* sp. strain K172) (Dangel et al. 1991) and *Rhodopseudomonas palustris* (Egland et al. 1995).

2. A benzoyl-CoA reductase reduces one or more of the double bonds in the aromatic ring (Gibson and Gibson 1992; Koch and Fuchs 1992). The pathways in *T. aromatica* and *Rh. palustris* differ slightly in detail.

   a. In *T. aromatica* strain K172, the gene sequence for the enzymes has been determined: *bcrCBAD* for the benzoyl-CoA reductase, *dch* for the dienyl-CoA hydratase, *had* for 6-hydroxycyclohex-1-ene-1-carboxy-CoA dehydrogenase, and they occur in the order *had*, *dch*, *bcrCBAD* (Breese et al. 1998).

   b. In *Rh. palustris*, the genes are *badDEFG* for the benzoyl-CoA reductase, *badK* for the cyclohex-1ene-1-carboxyl-CoA hydratase, *badH* for the 2-hydroxycyclohexane-1-carboxyl-CoA dehydrogenase, and *badI* for the ketohydrolase, and they occur in the order *badK, badI, badH, badD, badE, badF, badG*, the first three transcribing in the opposite direction to the others (Harwood et al. 1999; Egland and Harwood 1999).

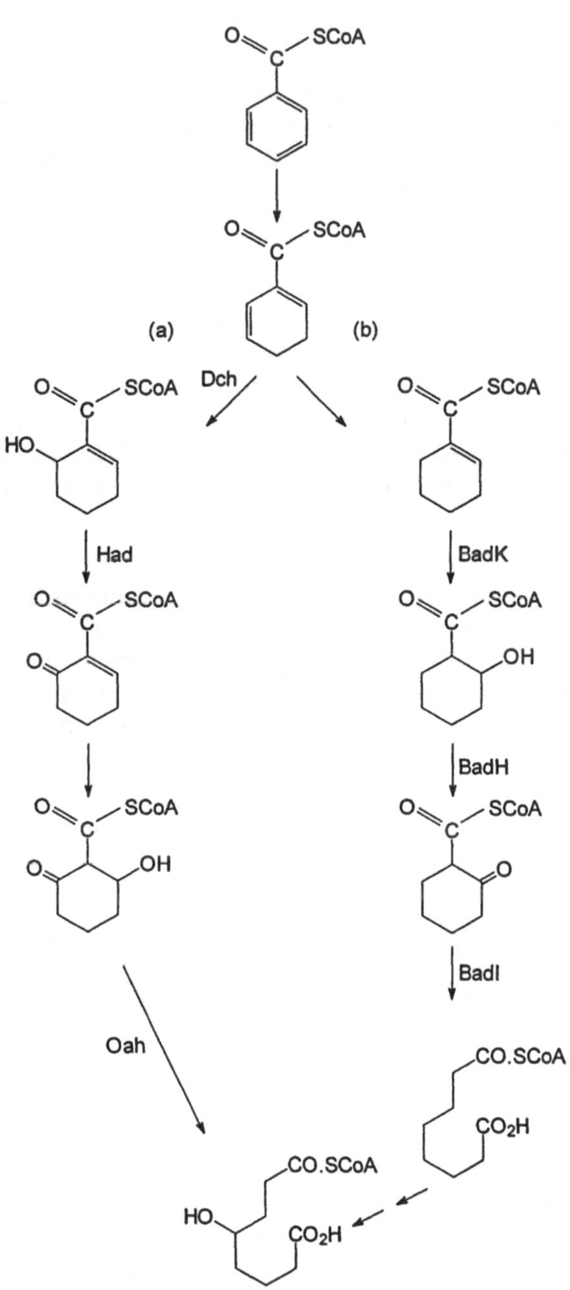

Organic Chemicals: An Environmental Perspective

FIGURE 6.94
Anaerobic degradation of benzoate.

3. Fission of the ring—In *T. aromatica*, the initial ring fission product is hepta-3-ene-1,7-dicarboxylyl-CoA formed by 3-ketoacyl-CoA hydrolase encoded by *oah* that lies immediately to the right of *had*. The C_7 dicarboxylate 3-hydroxpimelyl-CoA is then formed degraded by β-oxidation to glutaryl-CoA. With the minor alteration noted above, the same pathway has been demonstrated in *R. palustris* and the ring fission enzyme 2-ketocyclohexanecarboxyl-CoA hydrolase which produces pimelyl-CoA has been purified (Pelletier and Harwood 1998). The gene for the ring-fission enzyme *badI* lies to the left of *badH*.

It was shown that in *Rh. palustris*, the suite of enzymes in the sequence from cyclohexen-1-carboxylate onward was induced during growth on benzoate but not on succinate (Perrotta and Harwood 1994). The further degradation of glutarate takes place by a pathway involving dehydrogenation and decarboxylation to crotonyl-CoA and subsequent formation of acetate (Härtel et al. 1993). After growth of a denitrifying organism on either benzoate or pimelate, both glutaryl-CoA dehydrogenase and glutaconyl-CoA decarboxylase activities were induced, although not the enzymes leading from pimelyl-CoA to glutaryl-CoA; this is consistent with the involvement of 3-hydroxypimelyl-CoA in the degradation of benzoyl-CoA (Gallus and Schink 1994). The initial steps in the degradation of glutarate involving dehydrogenation, decarboxylation, and the β-oxidation pathway have also been demonstrated in other organisms that produce butyrate and *iso*butyrate (Matties and Schink 1992). Details of the various pathways for the anaerobic metabolism of acetate have been reviewed (Thauer et al. 1989).

The degradation of *o*-, *m*-, and *p*-phthalates under denitrifying conditions has been examined (Nozawa and Maruyama 1988), and after an initial decarboxylation to benzoate followed the pathway for the anaerobic degradation of benzoate that has been noted above: formation of the CoA ester, reduction to cyclohex-1-ene carboxylate, hydroxylation to 2-hydroxycyclohexane carboxylate, and ring fission to pimelic acid that was further degraded by β-oxidation.

Degradation of Toluene and Alkyl Benzenes—The anaerobic degradation of toluene under denitrifying conditions has been extensively investigated and different pathways have emerged.

1. *Dehydrogenation of the methyl group*—Strains of denitrifying bacteria have been shown to degrade toluene in the absence of oxygen using N_2O as electron acceptor (Schocher et al. 1991), and the data are consistent with a pathway involving successive dehydrogenation of the ring methyl group with the formation of benzoate. The details of this pathway involving benzyl alcohol and benzaldehyde have been clearly demonstrated with a strain of *Thauera* (*Pseudomonas*) sp. under denitrifying conditions (Altenschmidt and Fuchs 1992a), and are supported by the demonstration of benzyl alcohol dehydrogenase, benzaldehyde dehydrogenase, benzoyl-CoA ligase, and benzoyl-CoA

reductase activities in cell extracts (Biegert and Fuchs 1995). The benzyl alcohol dehydrogenase from benzyl-alcohol-grown cells was similar in many of its properties to those from the aerobic bacteria *Acinetobacter calcoaceticus* and *P. putida* (Biegert et al. 1995). A number of organisms are able to degrade alkyl benzenes under anaerobic conditions in the presence of nitrate (Rabus and Widdel 1995). They exhibit different specificities for toluene, ethylbenzene, and *n*-propylbenzene, but all of them belong to the phylogenetic groups that include *T. selenatis* and *Azoarcus indigens*. Brief reference has been made to these genera in Section 5.9.

2. *Condensation reactions*—It has been suggested that the degradation of toluene could proceed by condensation with acetate to form phenylpropionate and benzoate before ring cleavage (Evans et al. 1992): the same terminal metabolites are also produced during anaerobic degradation of toluene by sulfate-reducing enrichment cultures (Beller et al. 1992). A mechanism for the oxidation of the methyl group has been proposed for *A. tolulyticus*, and involves a condensation reaction with acetyl-CoA followed by dehydrogenation to cinnamoyl-CoA (Migaud et al. 1996). This is then either transformed into benzylsuccinate and benzylfumarate that are apparently terminal metabolites that have also been isolated from a denitrifying organism (Evans et al. 1992), or by further degradation to benzoyl-CoA that is the substrate for the ring reductase (Figure 6.95).

3. *Addition reactions*—An alternative sequence involves the direct formation of benzylsuccinate by reaction of toluene with fumarate in *T. aromatica* (Biegert et al. 1996), and in a strain designated *Pseudomonas* sp. strain T2 (Beller and Spormann 1997). It has been shown in the denitrifying strain T2 showed that the reaction between toluene and fumarate is stereospecific yielding (+)-benzylsuccinate, and that the proton abstracted from toluene is incorporated into the benzylsuccinate (Beller et al. 1998). The synthesis of benzylsuccinate during the anaerobic degradation of toluene under denitrifying conditions involves activation of toluene to a benzyl radical by a mechanism involving glycine-radical catalysis (Leuthner et al. 1998). The amino acid sequence of the large subunit of the purified enzyme from *T. aromatica* revealed a high homology to glycine radical enzymes and particularly to pyruvate formate lyase. It has been shown (Coschigano et al. 1998) in two mutants of strain T1 that the genes *tutD* and *tutE* involved in the anaerobic degradation of toluene

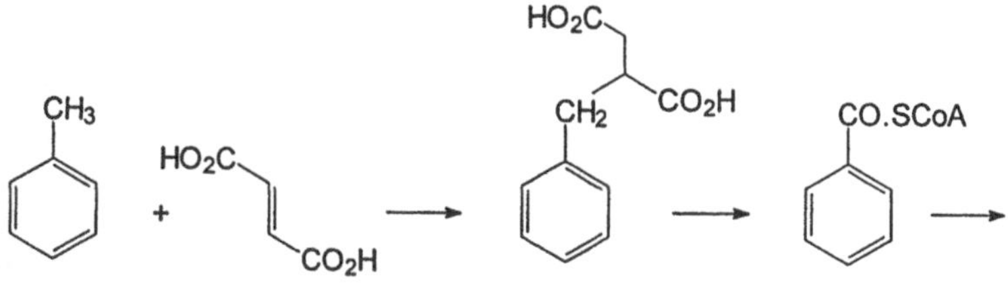

FIGURE 6.95
Anaerobic degradation of toluene via benzylsuccinate.

encode proteins with molecular masses of 97.6 and 41.3 kDa that possess homologies to pyruvate formate lyase and its activating enzyme, respectively. A free radical at glycine-734 is involved in the operation of pyruvate formate lyase (Wagner et al. 1992) so that there is therefore a formal similarity in at least one step in the two reaction pathways. The resulting benzylsuccinate is then metabolized to benzoyl-CoA followed by its metabolism that has already been noted.

A sulfate-reducing bacterium strain PRTOL1 also catalyses the reaction between toluene and fumarate with the formation of benzylsuccinate, so that this is apparently a widespread reaction (Beller and Spormann 1997) that is used also by *Desulfobacula toluolytica* (Rabus and Heider 1998). The benzyl-succinate pathway is also used in a strain of a sufate-reducing bacterium that mineralizes 80% of the substrate to CO_2: benzylsuccinate from toluene, 2-methylbenzylsuccinate from 1,2-dimethylbenzene, and 4-methylbenzoate from 1,4-dimethylbenzene were formed as terminal metabolites (Beller et al. 1996). The analogous succinates and fumarates that would be produced from dimethyl benzenes have been used as markers for the anaerobic degradation of these substrates in an aquifer (Beller et al. 1995).

4. Other reactions—Alternatives may be envisaged based on the anaerobic hydroxylation of toluene to 4-hydroxytoluene followed by oxidation of the methyl group and dehydroxylation to benzoate (Rudolphi et al. 1991). Although the role of this in the degradation has not been clearly established, two possibilities can be suggested: (a) carboxylation and dehydroxylation to 3-methylbenzoate that is produced from 1,3-dimethylbenzene by *Thauera* sp. strain K172 under denitrifying conditions (Biegert and Fuchs 1995), (b) oxidation to 4-hydroxybenzoate and further degradation by the pathway noted above (Rudolphi et al. 1991), although the occurrence of hydroxylation and dehydroxylation at different stages of the pathway has no obvious rationale.

An alternative pathway includes carboxylation to phenyl acetate and then oxidation via phenylglyoxylate to benzoate which is an established pathway for phenyl acetate itself (Dangel et al. 1991). The degradation of ethylbenzene by the denitrifying strain EbN1 involved successive oxidation to 1-phenyle-thanol and acetophenone, followed by carboxylation and conversion to ben-zoyl-CoA and acetyl-CoA (Rabus and Heider 1998).

Attention is drawn to further examples of the metabolic diversity of anaer-obic sulfate-reducing bacteria (see Section 6.7.1). Two anaerobes affiliated with known sulfate-reducing bacteria isolated from enrichments with crude oil were able to grow at the expense of a number of alkylated ben-zenes—strain oXyS1 with toluene, *o*-xylene, and *o*-ethyltoluene, and strain mXyS1 with toluene, *m*-xylene, and *m*-ethyltoluene (Harms et al. 1999). Although no pathway has been established, it is worth noting the degrada-tion assessed from the diminishing substrate concentration of anthracene, phenanthrene, and pyrene by denitrifying pseudomonads isolated from diverse environments under both aerobic and anaerobic denitrifying condi-tions (McNally et al. 1998).

Degradation of Phenols—Phenol is transformed into benzoyl CoA by a denitrifying strain of *Pseudomonas* sp., and it has been suggested that phenyl phosphate may be the actual substrate for carboxylation (Lack and Fuchs 1992). The central role of 4-hydroxybenzoyl-CoA in the metabolism of phenol, 4-methylphenol, 4-hydroxyphenylacetate, and related aromatic compounds by a denitrifying strain of *Pseudomonas* sp. is supported by the isolation and purification of the dehydroxylating enzyme (Brackmann and Fuchs 1993), and some plausible mechanistic arguments have been put forward. A similar dehydroxylase is also involved in the metabolism of 4-hydoxybenzoate by *Rhodopseudomonas palustris* (Gibson et al. 1997), and is a molybdenum-containing enzyme. Degradation of 3-methylphenol by oxidation to 3-hydroxybenzoate followed by analogous dehydroxylation has been observed in a "Pseudomonas-like" organism under denitrifying conditions (Bonting et al. 1995), and in *Desulfotomaculum* sp. strain Gross (Londry et al. 1997). The oxidation is formally comparable to the reactions in the metabolism of toluene that have been described above, and a carboxylation pathway is apparently not involved.

6.7.3.2 Mixed Cultures

There is a substantial literature dealing with the anaerobic degradation of compounds such as benzoate and phenol and, more recently, of benzene, toluene, and the xylenes by mixed cultures. There can be no doubt that such compounds can be degraded to methane in some cases, or to aliphatic carboxylic acids and CO_2 in others. It is virtually impossible, however, to present a detailed account of the reactions involved since no pure cultures have been investigated, and both methanogenic and nonmethanogenic cultures have been examined. Instead a summary of the individual reactions which have been demonstrated is presented but an attempt to synthesize these to suggest complete degradative pathways is probably not currently justified.

1. Benzene and toluene are anaerobically hydroxylated to phenol and 4-hydroxytoluene by mixed methanogenic cultures, and the oxygen atom comes from water (Vogel and Grbic-Galic 1986), and sulfidogenic mixed cultures degraded benzene (Weiner and Lovley 1998), and toluene and xylene to CO_2 (Edwards et al. 1992). An enrichment culture was able to grow at the expense of 1,3,5-trimethyl- and 1,2,4-trimethylbenzene using N_2O as electron acceptor (Häner et al. 1997), and the mechanisms for the degradation may resemble, or be identical to those used by pure cultures of denitrifying bacteria that were discussed above. Benzene is degradable under methanogenic, sulfate-reducing and Fe(III)-reducing (Kazumi et al. 1997), and nitrate-reducing (Burland and Edwards 1999) conditions.

2. Carboxylation of phenols has been demonstrated under a number of conditions and this reaction is presumably analogous to that demonstrated in pure cultures of *Desulfobacterium anilini* during metabolism of aniline and in denitrifying bacteria during metabolism of phenol (Section 6.7.3.1). The following examples are used to illustrate some important details of these

reactions; it is worth noting that metabolism of the methylphenols by mixed cultures proceeds by ring carboxylation in contrast to the oxidation pathway that has been demonstrated in pure cultures under denitrifying conditions.

a. Phenol is carboxylated by a defined obligate syntrophic consortium to benzoate which is then degraded to acetate, methane, and CO_2 (Knoll and Winter 1989).

b. 2-Methylphenol is carboxylated by a methanogenic consortium to 4-hydroxy-3-methylbenzoate which was dehydroxylated to 3-methylbenzoate which was stable to further transformation (Bisaillon et al. 1991).

c. 3-Methylphenol is carboxylated to 2-methyl-4-hydroxybenzoate by a methanogenic enrichment culture before degradation to acetate (Figure 6.96a) (Roberts et al. 1990): ^{14}C-labeled bicarbonate produced carboxyl-labeled acetate, while ^{14}C-methyl-labeled 3-methylphenol yielded methyl-labeled acetate. On the other hand, 2-methylbenzoate formed by dehydroxylation of 2-methyl-4-hydroxybenzoate was not further metabolized (Figure 6.96b). A similar reaction occurs with a sulfate-reducing mixed culture (Ramanand and Suflita 1991).

d. An unusual reaction occurred during the degradation of 3-methylphenol by a methanogenic consortium (Londry and Fedorak 1993): although carboxylation to 2-methyl-4-hydroxybenzoate took place as in the preceding example, further metabolism involved loss of the methyl group with the formation of methane before dehydroxylation to benzoate (Figure 6.96c).

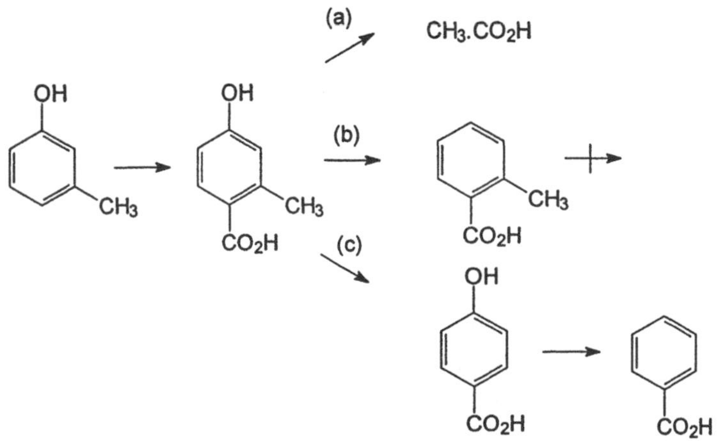

FIGURE 6.96
Anaerobic biodegradation of 3-methylphenol.

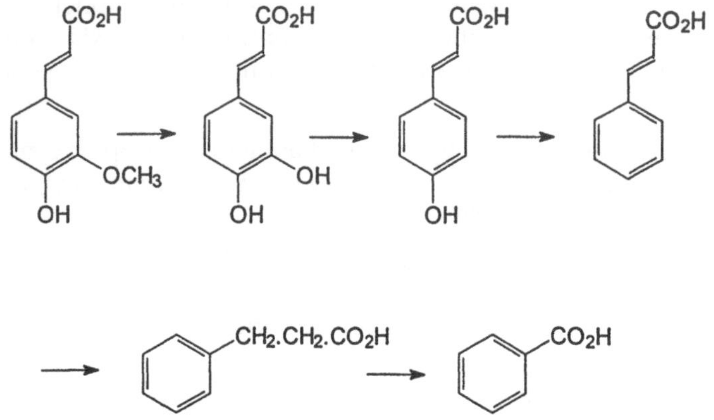

FIGURE 6.97
Anaerobic biodegradation of ferulate.

It may be concluded from all these observations that whereas benzoate produced by the carboxylation of phenols can be degraded, dehydroxylation with the formation of substituted benzoates may produce stable terminal metabolites.

3. Benzoate has been identified as a transient metabolite during methanogenic degradation of ferulate (Figure 6.97) (Grbic-Galic and Young 1985), toluene (Grbic-Galic and Vogel 1987), and styrene (Figure 6.98) (Grbic-Galic et al. 1990), and there seems little doubt that benzoate can be degraded anaerobically although details of the mechanism of methanogenesis have not been finally established.

4. Metabolically stable enrichment cultures of anaerobic bacteria carried out a reaction in which aromatic hydroxyaldehydes were concomitantly transformed into the corresponding carboxylic acid and the benzyl alcohol (Neilson et al. 1988a); in some cases the carboxylic acid was decarboxylated to produce phenols or catechols (Figure 6.99). Decarboxylation accompanied by dechlorination in freshwater sediments has been demonstrated with 3-chloro-4-hydroxybenzoate that produces phenol before carboxylation to benzoate and further degradation (Zhang and Wiegel 1992). The fungal metabolite 3,5-dichloro-4-methoxylbenzyl alcohol was transformed under methanogenic conditions to 2,6-dichlorophenol, and a dimeric product was formed by an abiotic reaction (Verhagen et al. 1998; see Section 4.2.3) (Figure 6.100): the initial dehydration is formally comparable to the dehydrogenation involved in the aerobic degradation of 4-methylphenol (Section 4.4.1.1).

5. The anaerobic degradation of a few PAHs has been investigated with apparently conflicting results that are not, however, irreconcilable with the use of mixed cultures from widely different habitats. Under sulfate-reducing conditions, [1-^{14}C]naphthalene and [9-^{14}C]phenanthrene were oxidized to $^{14}CO_2$ (Coates et al. 1996; 1997), and under anaerobic nitrate-reducing conditions, both [ring ^{14}C]toluene and [1-^{14}C]naphthalene were oxidized to $^{14}CO_2$

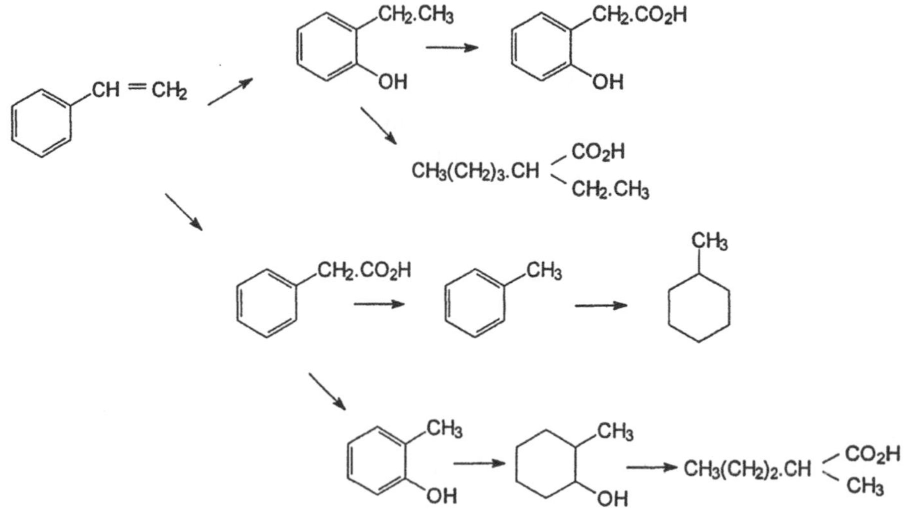

FIGURE 6.98
Anaerobic biodegradation of styrene.

(Bregnard et al. 1996). Under sulfidogenic conditions, naphthalene and phenanthrene were mineralized and their metabolism involved carboxylation to the corresponding carboxylic acids (Zhang and Young 1997). The pathways by which these reactions are accomplished have not been established but may, by analogy with the degradation of benzoate, involve carboxylation, reduction of the ring, hydroxylation, oxidation, and ring fission.

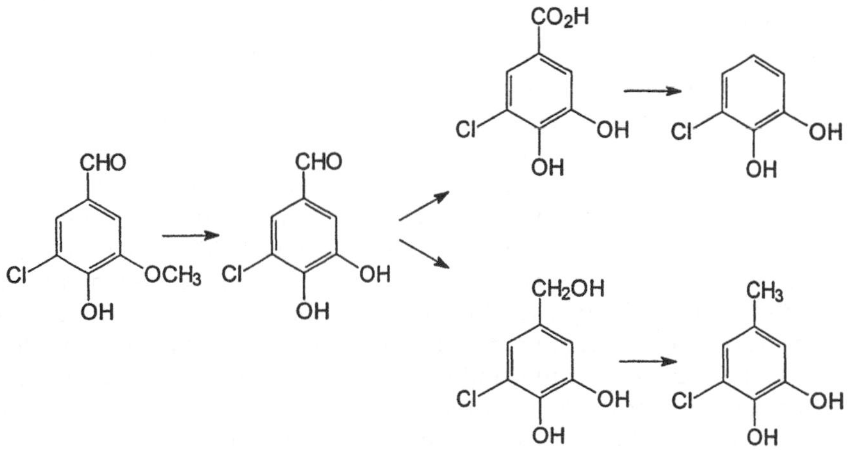

FIGURE 6.99
Biotransformation of 5-chlorovanillin.

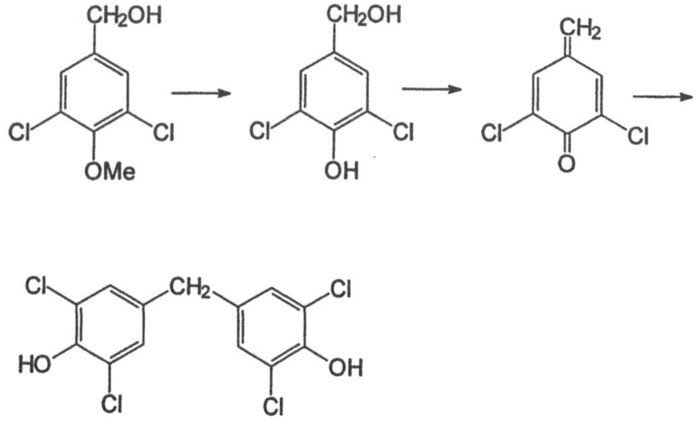

FIGURE 6.100
Transformation of 3,5-dichloro-4-methoxybenzyl alcohol.

6.7.4 Heterocyclic Aromatic Compounds

6.7.4.1 Nitrogen-Containing Heterocyclic Compounds

Whereas the metabolism of a structurally diverse range of heterocyclic aromatic compounds has been examined under aerobic conditions, a much more limited range has been investigated under anaerobic conditions, and these investigations have apparently centered on the N-heterocyclic compounds. Demonstration of the possibility of anaerobic degradation of purines belongs to the golden age of microbiology and appropriately was discovered in Beijerinck's laboratory in Delft. Liebert (1909) obtained a pure culture of an organism that was able to grow anaerobically with 2,6,8-trihydroxypurine (uric acid) and which he named *Bacillus acidi urici* (Liebert 1909), although the organism is now known as *Clostridium acidurici*. Subsequently, several other purinolytic clostridia have been isolated and the details of their metabolism of purines has been extensively delineated by Barker and his collaborators.

Detailed investigations using pure cultures have elucidated the degradation pathways, and their enzymology is known in considerable detail. In the following paragraphs, the pathways used for the degradation of pyridines, pyrimidines, and purines by anaerobic bacteria will be briefly summarized. Complete mineralization may not always be achieved although this is clearly the case for nicotinate that is degraded by *Desulfococcus niacinii* and for the degradation of indoles by *Desulfobacterium indolicum* (Bak and Widdel 1986). For the sake of completeness, the degradation of quinoline and indole by undefined cultures occurred under nitrate-reducing, sulfate-reducing, or methanogenic conditions is noted (Licht et al. 1996): by contrast, benzothiophene and benzofuran were recalcitrant during incubation for 100 days.

Broadly, four different types of reactions have been found.

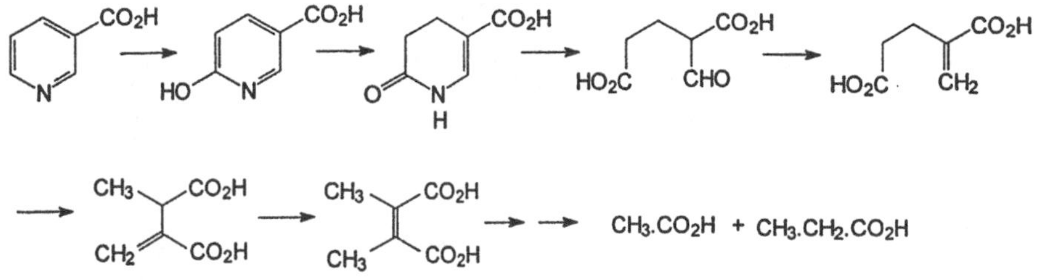

FIGURE 6.101
Anaerobic biodegradation of nicotinate.

1. Hydroxylation of the heterocyclic ring—The metabolism of nicotinate has been extensively studied in clostridia, and the details of the pathway (Figure 6.101) have been delineated in a series of studies (Kung and Tsai 1971). The degradation is initiated by hydroxylation of the ring and the level of nicotinic acid hydroxylase is substantially increased by the addition of selenite to the medium (Imhoff and Andreesen 1979). The most remarkable feature of the pathway is the mechanism whereby 2-methylene-glutarate is converted into methylitaconate by a coenzyme B_{12}-mediated reaction (Kung and Stadtman 1971). This is representative of a group of related reactions which have been discussed in a classic review (Barker 1972). Whereas in *C. barkeri*, the end products are carboxylic acids, CO_2, and ammonium, the anaerobic sulfate-reducing *Desulfococcus niacinii* degrades nicotinate completely to CO_2 (Imhoff-Stuckle and Pfennig 1983) although the details of the pathway remain unresolved.

2. Ring-cleavage reactions : the degradation of substituted purines—Very substantial effort has been devoted to elucidating the details of the anaerobic degradation of purines containing hydroxyl and amino groups. The most-studied group of organisms are the clostridia including *Cl. acidurici*, *Cl. cylindrospermum*, and *Cl. purinilyticum* (Schiefer-Ullrich et al. 1984), but attention should also be drawn to the nonsporeforming *Eubacterium angustum* (Beuscher and Andreesen 1984) and *Peptostreptococcus barnesae* (Schieffer-Ullrich and Andreesen 1985). Many of the basic investigations on the mechanisms of purine degradation by clostridia were carried out by Barker and his colleagues (Barker 1961) and more recent developments have been presented by Dürre and Andreesen 1983. One of the significant findings was that of selenium dependency due to its requirement for the synthesis of several critical enzymes—xanthine dehydrogenase, formate dehydrogenase, and glycine reductase. The pathways for the degradation of purines containing amino and/or hydroxyl groups converge on the synthesis of xanthine (2,6-dihydroxypurine) and are followed by its

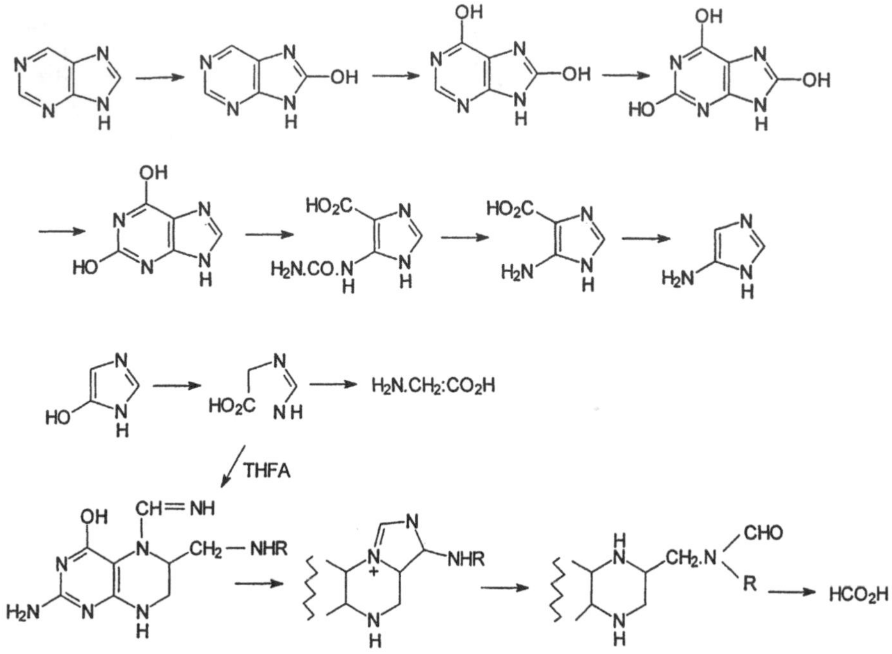

FIGURE 6.102
Anaerobic biodegradation of purine.

degradation to formiminoglycine (Figure 6.102). This compound is then used for the synthesis of glycine and 5-formiminotetrahydro-folate whose further metabolism to formate results in the synthesis of ATP. The energy requirements of the cell are supplemented by the contribution of ATP produced during the reduction of glycine to acetate in an unusual reaction catalyzed by glycine reductase (Arkowitz and Abeles 1989). An essentially similar pathway is used by *Methanococcus vannielii* that uses a number of purines as nitrogen sources (DeMoll and Auffenberg 1993).

3. Reduction of the heterocyclic ring—Reduction of the pyrimidine ring has been shown to be the first step in the degradation of orotic acid by *Cl.* (*Zymobacterium*) *oroticum* (Lieberman and Kornberg 1955) (Figure 6.103a), and of uracil by *Cl. uracilicum* (Campbell 1960) (Figure 6.103b). Comparable pathways used for the degrada-tion of substituted pyrimidines by aerobic bacteria have already been noted in Section 6.3.1.3.

4. Cleavage of the iminazole ring of purines—Whereas the degrada-tion of substituted purines generally takes place by initial cleavage of the pyrimidine ring with the formation of formiminoglycine as a key metabolite, an alternative pathway exists for *Cl. purinilyticum*. This organism degrades a wide range of purines and was shown

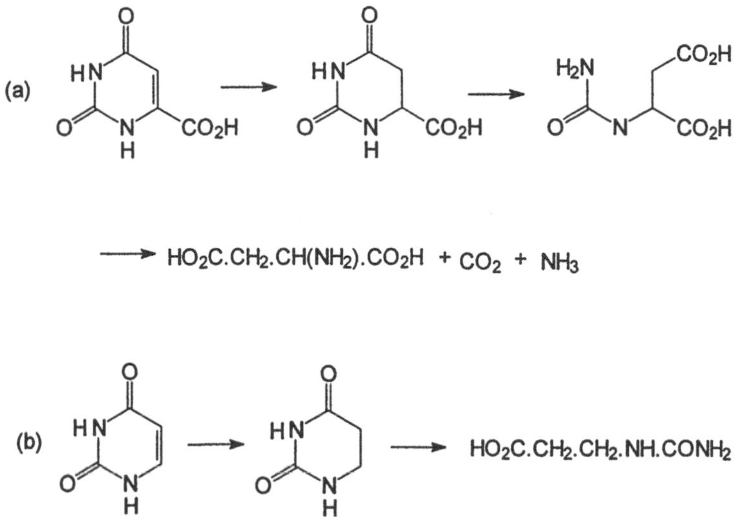

$$\longrightarrow HO_2C.CH_2.CH(NH_2).CO_2H + CO_2 + NH_3$$

$$\longrightarrow HO_2C.CH_2.CH_2.NH.CONH_2$$

FIGURE 6.103
Anaerobic biodegradation of (a) orotic acid and (b) uracil.

to be obligately dependent on selenium for growth (Dürre et al. 1981): under conditions of selenium starvation, this organism degrades uric acid by cleavage of the iminazole ring to produce 5,6-diaminouracil which is then degraded to formate, acetate, glycine, and CO_2 (Dürre and Andreesen 1982).

6.7.4.2 Oxygen-Containing Heterocyclic Compounds

Probably all higher plants synthesize flavanoids of widely varying structure and these compounds make a significant contribution to the food intake of both herbivores and humans. Their metabolism has therefore been studied under anaerobic conditions which prevail in their digestive tracts. Many of these compounds exist naturally as glycosides and these are readily hydrolyzed to the aglycones. The primary degradative reaction is generally reductive cleavage of the heterocyclic ring of the aglycones with the formation of aromatic carboxylic acids and phenolic compounds which are then further metabolized. This reduction has been observed in strains of a *Clostridium* sp. (Winter et al. 1989) and a strain of *Butyrivibrio* sp. (Krishnamurty et al. 1970; Cheng et al. 1971), and an illustrative example is given in Figure 6.104. It should be noted that such reactions are not carried out exclusively by anaerobic bacteria since they have been observed also during the aerobic metabolism of the pentahydroxyflavone quercitin by strains of *Rhizobium loti* and *Bradyrhizobium* sp. (Rao et al. 1991), and C-ring fission has been observed in the metabolism of a number of flavonoids by rhizobia (Rao and Cooper 1994). Initial products from flavanoids were formed by reductive fission of the O–C_2

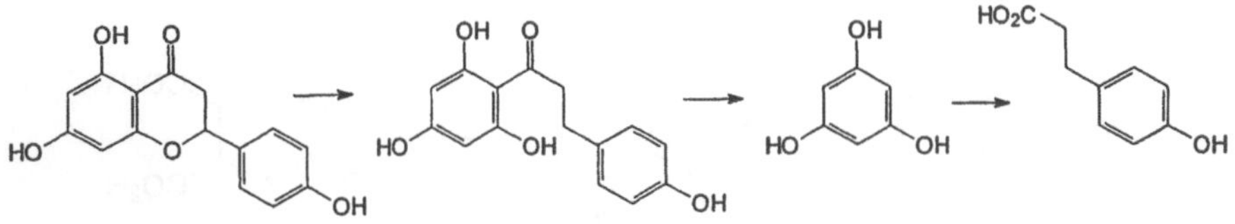

FIGURE 6.104
Anaerobic biodegradation of flavanoids.

bond followed by release of products with the A or B rings intact, whereas in isoflavones this fission is followed by rearrangement of the C_3 phenyl residue to C_2 (Rao and Cooper 1994).

6.7.4.3 Sulfur-Containing Heterocyclic Compounds

Sulfate-reducing anaerobic bacteria are able to use dibenzothiophene as sole source of sulfur (Lizama et al. 1995), and the product of desulfurization by a number of strains has been shown to be biphenyl that is formed in yields from 0.22 to 1.14% (Armstrong et al. 1995). This is formally similar to the reduction of dibenzyl disulfide to toluene (Section 6.9.3).

6.8 Aerobic Degradation of Aromatic Compounds Containing Nitro or Sulfonate Groups

Aromatic sulfonates are important structural elements of many industrially important dyestuffs, while aromatic nitro compounds are industrially important both as intermediates for the synthesis of a wide range of compounds including dyestuffs and pharmaceutical products and as explosives. Increased attention has therefore been directed to their biodegradation particularly in biological treatment systems.

6.8.1 Aromatic Sulfonates

Whereas chlorine is only exceptionally removed from aromatic rings during dioxygenation, this is more generally the case for carboxyl and sulfonate groups. The pathway for the degradation of aromatic sulfonates has been elucidated in a detailed study (Cain and Farr 1968) and is illustrated for 4-toluenesulfonate in Figure 6.105. The basic reaction is a dioxygenation with subsequent elimination of sulfite is illustrated for naphthalene-1-sulfonate (Figure 6.106), and the dioxygenase has been purified from a naphthalene-2-sulfonate-degrading pseudomonad (Kuhm et al. 1991). Three variants of this pathway have emerged.

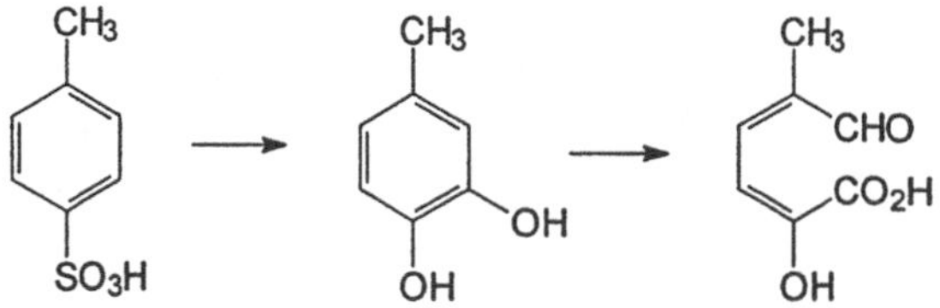

FIGURE 6.105
Biodegradation of 4-toluenesulfonate.

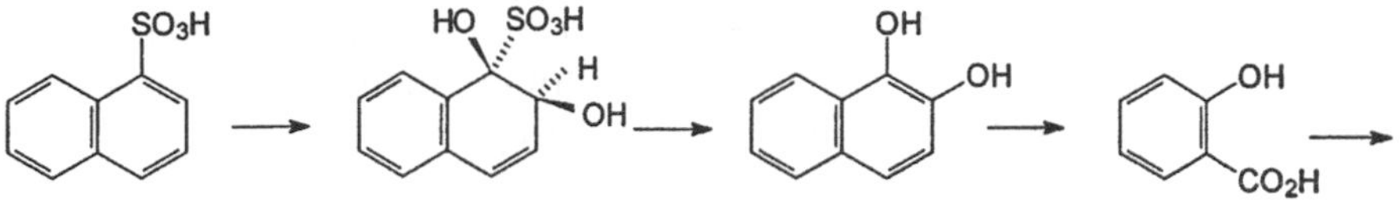

FIGURE 6.106
Biodegradation of naphthalene-1-sulfonate.

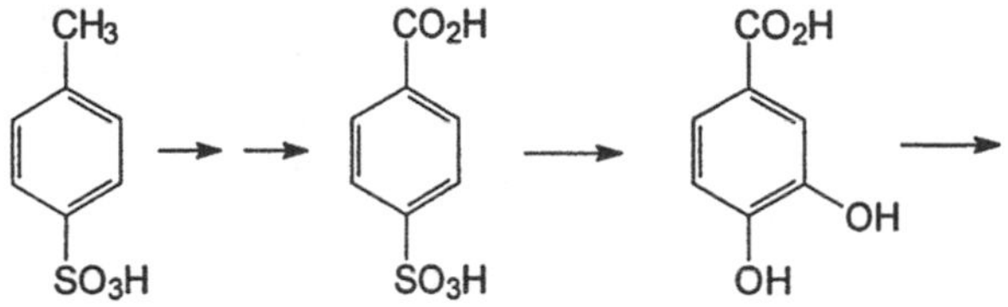

FIGURE 6.107
Alternative pathway for the biodegradation of 4-toluenesulfonate.

1. Oxidation of the methyl group in 4-toluenesulfonate to a carboxy group may precede elimination of sulfite (Locher et al. 1991) (Figure 6.107).

2. Considerable attention has been directed to the degradation of naphthalenesulfonates, and one additional reaction has emerged for the elimination of sulfite. The pathway for the degradation of naphthalene 1,6- and 2,6-disulfonates involves the expected elimination of the 1- or 2-sulfonate groups with the formation of 5-sulfosalicylate (Chapter 4, Section 4.4.2); this is then, however, converted into 2,5-dihydroxybenzoate by what formally may be represented as a hydroxylation with elimination of sulfite (Figure 6.108) (Wittich et al. 1988). The 1,2-dihydroxynaphthalene dioxygenase from *Pseudomonas* strain (BN6) that degrades naphthalene sulfonates oxidized 1,2-dihydroxynaphthalene and 2,3-dihydroxybiphenyl (Kuhm et al. 1991), and the same organism also synthesizes a distinct 2,3-dihydroxybiphenyl dioxygenase that is not involved in the degradation of naphthalene sulfonates (Heiss et al. 1995).

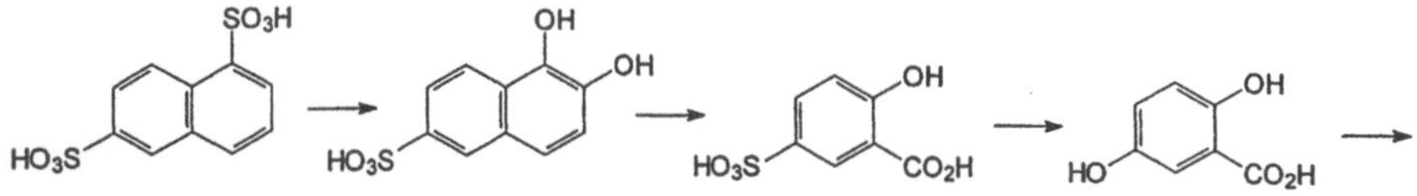

FIGURE 6.108
Biodegradation of naphthalene-1,6-disulfonate.

3. Sulphite may not necessarily be elimination before ring fission, and the degradation of sulfanilate by a mixed culture of *Hydrogenophaga palleroni* and *Agrobacterium radiobacter* proceeds via catechol-4-sulfonate and intradiol ring fission: sulfite is then elimination from the muconolactone (Hammer et al. 1996).

6.8.2 Aromatic Nitro Compounds

A number of different reactions are carried out by microorganisms during the degradation and transformation of aromatic nitro compounds. Fungal transformation of arenes has been discussed in Section 6.2.2, and comparable reactions have been observed with nitroarenes. For example, *Cunninghamella elegans* is able to transform 2- and 3-nitrofluoranthenes to the 8- and 9-sulfates, and 6-nitrochrysene to the 1- and 2-sulfates (references in Pothuluri et al. 1998). In this illustration, the nitro group is left intact whereas in bacteria this undergoes a number of reactions that include:

- Elimination of nitrite followed by ring cleavage;
- Reduction to the amine followed in some cases by acetylation;
- Partial reduction to phenylhydroxylamines followed by rearrangement to aminophenols;
- Reactions involving dimerization of partially reduced metabolites and intramolecular cyclization.

It will become apparent that, in some cases, these reactions are merely biotransformations and that they are therefore tangential to degradation, and as for halogen substituents, the position of the nitro groups may be critical in determining the outcome of the microbial reactions. An attempt will be made to illustrate the various possibilities, and examples will be drawn from the metabolism of aromatic nitro compounds by both bacteria and fungi. A review (Spain 1995a) and a book (Spain 1995b) include many illustrative examples of the various reactions.

Elimination of Nitrite

This is a rather widely distributed reaction among bacteria, and may be mediated by either monooxygenation (hydroxylation) or dioxygenation reactions.

1. Monooxygenation is involved in the degradation of 4-nitrophenol by a strain of *Moraxella* sp. (Spain and Gibson 1991), and the enzyme appears to be a particulate flavoprotein in contrast to the dioxygenase-mediated reaction in *Pseudomonas putida* B2 (Zeyer and Kocher 1988). In the Gram-positive organisms *Arthrobacter* sp. (Jain et al. 1994) and *Bacillus sphaericus* strain JS905 (Kadiyala and Spain 1998), 4-nitrophenol degradation involves initial hydroxyla-tion to 4-nitrocatechol followed by elimination of nitrite with the formation of benzene-1,2,4-triol before ring fission to β-ketoadipate. In *B. sphaericus*, the first two reactions are carried out by a single two-component enzyme composed of a flavoprotein reductase and an oxygenase (Figure 6.109).

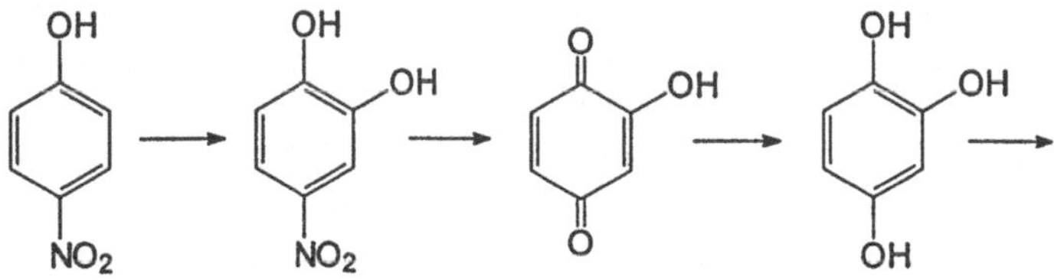

FIGURE 6.109
Degradation of 4-nitrophenol.

2. Dioxygenation may result in the elimination of nitrite and com-pounds and is analogous to the elimination of sulfite from aromatic sulfonates, and halogen from halobenzoates (Chapter 4, Section 4.4.2). The degradation of both nitrobenzene (Nishino and Spain 1995) by *Comamonas* sp. strain JS765, and of 2-nitrotoluene (Haigler et al. 1994) by *Pseudomonas* sp. strain JS42 is mediated by dioxy-genation with the formation of catechol and 3-methylcatechol, respectively. The enzyme involved in the degradation of 2-nitro-toluene by this strain has been purified and shown to consist of three components: an Fe-S protein ISP_{2NT} that serves as the terminal oxygenase, a reductase$_{2NT}$ that may be a flavoprotein, and an elec-tron-transfer protein ferredoxin$_{2NT}$ similar to Rieske-type ferredox-ins (An et al. 1994).

Nitrocatechols may be initially formed from 1,3-dinitroben-zenes, and nitrite is then eliminated in a second step (Figure 6.110) (Dickel and Knackmuss 1991; Spanggord et al. 1991). The dioxyge-nase involved in the degradation of 2,4-dinitrotoluene by *Burkhold-eria* (*Pseudomonas*) sp. strain DNT consists of three components—a terminal oxygenase, an Fe-S ferredoxin, and a reductase—and is broadly similar to naphthalene dioxygenase (Suen et al. 1996). Nonetheless, although the 2-nitrotoluene dioxygenase from *Pseudomonas* sp. strain JS42, the 2,4-dinitrotoluene dioxygenase from *B. cepacia*, and naphthalene dioxygenase from *Pseudomonas*

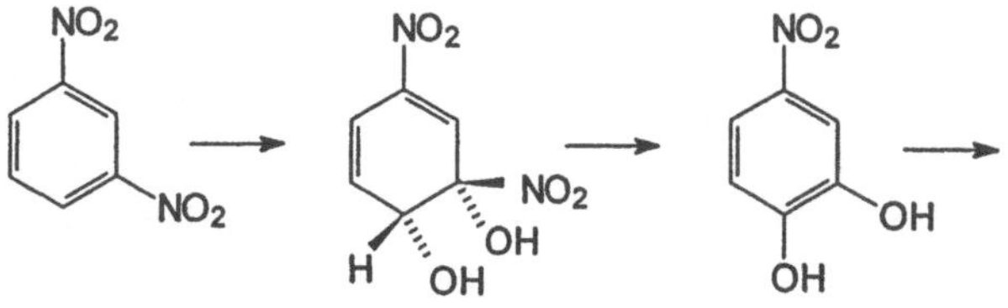

FIGURE 6.110
Biodegradation of 1,3-dinitrobenzene.

sp. strain 9816-4 have comparable dioxygenase activities toward naphthalene, they have widely different specificities to the isomeric mononitrotoluenes and to 2,4-dinitrotoluene (Parales et al. 1998). The degradation of 2,4-dinitrotoluene by *Burckholderia* sp. strain DNT is accomplished by enzymes for dioxygenation to 4-methyl-5-nitrocatechol, monooxygenation to 2-hydroxy-5-methylbenzo-quinone, dehydrogenation to 4-methyl-5-hydroxycatechol, and extradiol dioxygenation of 2,4,5-trihydroxytoluene encoded by genes designated *dntA*, *dntB*, *dntC*, and *dntD* (Haigler et al. 1999).

The bacterial degradation of 2-nitrophenol by *P. putida* B2 is mediated by a dioxygenase (Zeyer and Kocher 1988), and results in the production of catechol possibly via a quinone analogous to the monooxygenase-catalyzed loss of nitrite from 4-nitrophenol (Spain and Gibson 1991). The pathway for the degradation of 3-nitrophenol is apparently different (Zeyer and Kearney 1984) and, as noted below, may involve rearrangement of the initially formed hydroxylamine (Meulenberg et al. 1996).

Whereas the degradation of 3-nitrobenzoate by *Pseudomonas* sp. strain JS51 and *Comamonas* sp. strain JS46 is mediated by dioxy-genation to 3,4-dihydroxybenzoate (Nadeau and Spain 1995), the degradation of 4-nitrobenzoate that is noted below proceeds by a different pathway.

An unusual reaction involving reductive elimination of nitrite has been observed in cultures of a *Pseudomonas* sp. that can use 2,4,6-trinitrotoluene as N-source (Duque et al. 1993). The substrate is transformed by successive reductive loss of nitro groups with the formation of toluene: although this product cannot be metabolized by this strain, it can be degraded by a transconjugant containing the TOL plasmid from *P. putida*.

Reduction of Nitro Groups

Reduction of nitro groups may occur under either aerobic or anaerobic con-ditions, and the complete sequence of reduction products is produced from 2,6-dinitrotoluene by *Salmonella typhimurium* strain TA 98 (Sayama et al.

1992): 2-nitroso-6-nitrotoluene, 2-hydroxylamino-6-nitrotoluene, and 2-amino-6-nitrotoluene. The degradation of 2,4-dinitrotoluene by *Phanerochaete chrysosporium* has been investigated (Valli et al. 1992b), and involves operation of both the manganese-dependent and the lignin peroxidase systems. The pathway is shown in Figure 6.111, and is reminiscent of that used by the same organism for the degradation of 2,4-dichlorophenol (Valli and Gold 1991). The second step is catalyzed by the manganese peroxidase system with elimination of methanol, and the subsequent loss of nitrite with partial de-*O*-methylation is carried out by the lipid peroxidase system.

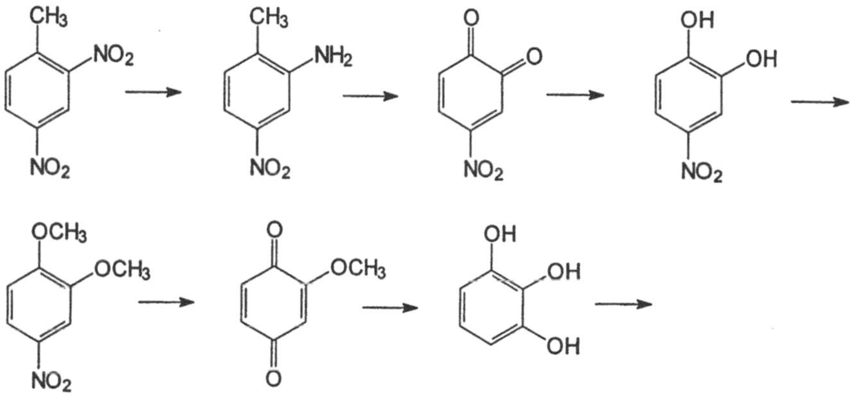

FIGURE 6.111
Biodegradation of 2,4-dinitrotoluene by *Phanerochaete chrysosporium*.

Complete reduction of the nitro group to an amine may not necessarily occur and partial reduction of the nitro group has been demonstrated: (1) during the metabolism of 4-chloronitrobenzene by a strain of the yeast *Rhodosporodinium* sp., the hydroxylamine that is formed may undergo rearrangement (Corbett and Corbett 1981) and (2) hydroxylamines may also be intermediates during the degradation of a number of aromatic nitro compounds by bacteria. The hydroxylamines then rearrange to amino phenols that are further degraded by different pathways involving loss of NH_4^+: (1) before ring fission with the formation of catechols or (2) after ring fission from 2-aminomuconaldehyde. Further details are given in the following illustrations.

1. During the degradation of 4-nitrobenzoate by *C. acidovorans*, 4-nitroso and 4-hydroxylaminobenzoate were formed successively, and the latter was then metabolized to 3,4-dihydroxybenzoate with the elimination of NH_4^+ (Groenewegen et al. 1992). It should be noted that these cells were not adapted to growth with either 4-aminobenzoate or 4-hydroxybenzoate that are alternative plausible intermediates. A comparable pathway is used by strains of *Pseudomonas* sp. for the degradation of 4-nitrotoluene that is

initially oxidized to 4-nitrobenzoate via 4-nitrobenzyl alcohol and 4-nitrobenzaldehyde (Haigler and Spain 1993; Rhys-Williams et al. 1993; James and Williams 1998). The degradation of 3-nitrophenol by *P. putida* strain B2 proceeds by comparable reactions involving successive anaerobic reduction to 3-nitroso- and 3-hydroxylaminophenol with subsequent loss of NH_4^+ to produce benzene-1,2,4-triol that is degraded by oxygenation and ring fission (Meulenberg et al. 1996). The hydroxylamine lyases that carry out formally Bamberger rearrangements of aryl hydroxylamines are involved in the degradation of both 4-nitrobenzoate and 3-nitrophenol, and appear to have very narrow substrate specificity. A plausibly comparable sequence is probably involved in the aerobic degradation of 3-nitrophenol by *Ralstonia* (*Alcaligenes*) *eutropha* strain JMP 134 (Schenzle et al. 1999).

2. The biodegradation of nitrobenzene by a strain of *P. pseudoalcaligenes* takes place by an enzymatically mediated rearrangement of the initially formed phenylhydroxylamine to 2-aminophenol (Bamberger rearrangement) that is then degraded by extradiol ring cleavage (Nishino and Spain 1993; Lendenmann and Spain 1996) to 2-aminomuconic semialdehyde that is further degraded by dehydrogenation and deamination to 2-hydroxymuconate (4-oxalocrotonate) (He and Spain 1997; 1998). Although nitrosobenzene was not detected as an intermediate, it is a substrate for the nitroreductase, whereas hydroxylaminobenzene is not further reduced by the enzyme (Somerville et al. 1995). This pathway is also used by a strain of *Mycobacterium* sp. for the degradation of 4-nitrotoluene (Spiess et al. 1998) with the formation of 3-hydroxy-4-aminotoluene as an intermediate before further degradation.

3. The 3-hydroxylaminophenol mutase from cells of *R. eutropha* JMP134 grown with 3-nitrophenol as N-source has been purified (Schenzle et al. 1999), and is able to catalyze the rearrangement of a number of substituted aromatic hydroxylamines, and the formation of both 2- and 4-hydroxyphenol from hydroxylaminobenzene that is formally comparable to the classic Bamberger rearrangement.

4. Anaerobic extracts of *Clostridium acetobutylicum* reduced 2,4,6-trinitrotoluene to 2,4-dihydroxylamino-6-nitrobenzene that underwent an enzymatic Bamberger-type rearrangement to 2-amino-4-hydroxylamino-5-hydroxy-6-nitrotoluene (Hughes et al. 1998). This is especially remarkable since the enzymatic activity was not dependent on the presence of nitroaromatic compounds as growth substrates.

It is convenient to summarize a few of the metabolic complexities for this straightforward reaction.

1. The amines may not be further metabolized by the organism carrying out the reduction so that their formation is a biotransformation;

this is noted again later. The amines may not even lie on the degradative pathway: this was clearly the case for the aerobic metabolism of nitrobenzoates (Cartwright and Cain 1959), for some dinitro compounds (Schackmann and Müller 1991), and for 4-nitrobenzoate (Groenewegen et al. 1992).

2. The established carcinogenicity of many aromatic amines has led to concern over the health risk associated with exposure to nitrated aromatic compounds, and this has motivated studies on their reductive metabolism by intestinal and other anaerobic bacteria: a range of pure cultures has been examined including *Veilonella alkalescens* (McCormick et al. 1976), strains of clostridia and a *Eubacterium* sp. (Rafii et al. 1991), and a range of strains of methanogenic and sulfidogenic bacteria and clostridia (Gorontzy et al. 1993).

3. In some aerobic bacteria, the amines are acetylated presumably to less toxic metabolites (Van Alfen and Kosuge 1974; Beunink and Rehm 1990; Gilcrease and Murphy 1995).

4. As noted above for 2,6-dinitrotoluene, reduction to the amine presumably proceeds in stages, and the intermediate hydroxylamines may undergo chemical rearrangement: the metabolism of 4-chloronitrobenzene by a strain of the yeast *Rhodosporodinium* sp. illustrates this possibility as well as the complete reduction to the amine followed by acetylation (Figure 6.112) (Corbett and Corbett 1981).

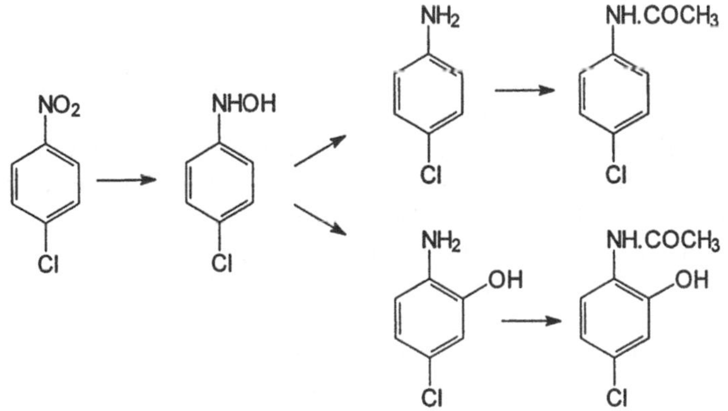

FIGURE 6.112
Metabolism of 4-chloronitrobenzene by *Rhodosporodinium* sp.

Biotransformation Reactions with Retention of the Nitro Substituents

A number of biotransformation reactions in which reduction of the nitro group occurs have already been noted, but two examples are given in which the nitro group is retained.

1. The biotransformation of nitrotoluenes has been examined in toluene-grown cells of pseudomonads which contain a toluene dioxygenase. The products depended on the relative position of the substituents and clearly illustrated the various metabolic reactions that may be mediated by this dioxygenase: 2-nitro- and 3-nitrotoluene were converted into the corresponding benzyl alcohols, whereas 4-nitrotoluene produced 3-methyl-6-nitrocatechol and 2-methyl-5-nitrophenol (Figure 6.113) (Robertson et al. 1992).

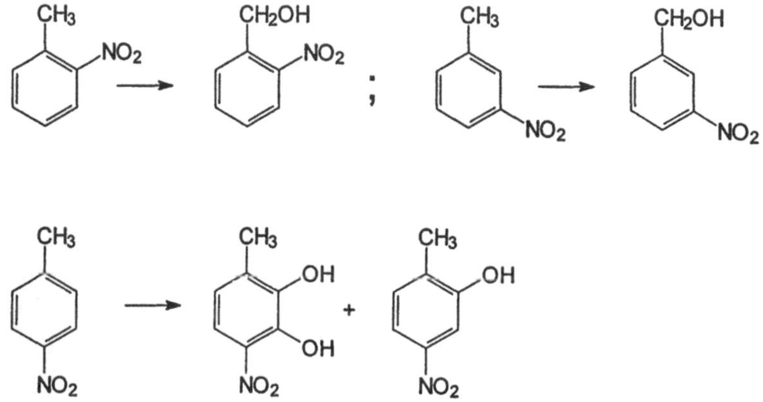

FIGURE 6.113
Biotransformation of nitrotoluenes.

2. The degradation of phenols containing several nitro groups has revealed an unusual reaction. Although this may involve the elimination of nitrite, it does not apparently involve any of the pathways illustrated above, and terminal metabolites may be formed by reduction of the aromatic ring. For example, 4,6-dinitrohexanoate is produced from 2,4-dinitrophenol (Lenke et al. 1992), 2,4,6-trinitrocyclohexanone from 2,4,6-trinitrophenol (Lenke and Knackmuss 1992), and 3-nitroadipate is produced during degradation of 2,4-dinitrophenol by *Rhodococcus* sp. strain RB1 and is further degraded with loss of nitrite (Blasco et al. 1999). The reductive pathways may plausibly be attributed to the presence of the strongly electron-withdrawing nitro groups which facilitate hydride transfer. In *Nocardioides simplex* strain FJ2-1A that degrades both 2,4-dinitro- and 2,4,6-trinitrophenol, this is accomplished by a two-component enzyme system. One component functions as an NADPH-dependent reductase of coenzyme F_{420}, and the other as a transferase from the reduced coenzyme to the nitrophenol (Ebert et al. 1999). The formation of 2,4-dinitrophenol from 2,4,6-trinitrophenol by *Rh erythropolis* (Lenke and Knackmuss 1992; 1996; Rieger

et al. 1999) and by *Nocardiodides* sp. strain CB 22-2 (Behrend and Heesche-Wagner 1999) is consistent with a pathway involving loss of nitrite from a Meisenheimer-like hydride complex: an analogous complex has been identified from a strain of *Mycobacterium* sp. during metabolism of 2,4,6-trinitrotoluene (Vorbeck et al. 1994; 1998). The general role of these complexes is supported by the fact that purified pentaerythritol tetranitrate reductase from *Enterobacter cloacae* reduced 2,4,6-trinitrotoluene successively to the mono- and dihydride Meisenheimer complexes (French et al. 1998).

Interactions between Reduction Products at Different Oxidation Levels

The formation of coupling products by interactions between metabolites at different levels of oxidation may reasonably be presumed to account for the identification of azoxy compounds as biotransformation products of 2,4-dinitrotoluene by the fungus *Mucrosporium* sp. (Figure 6.114) (McCormick et al. 1978), and by a *Pseudomonas* sp. that uses 2,4,6-trinitrotoluene as a N-source (Duque et al. 1993).

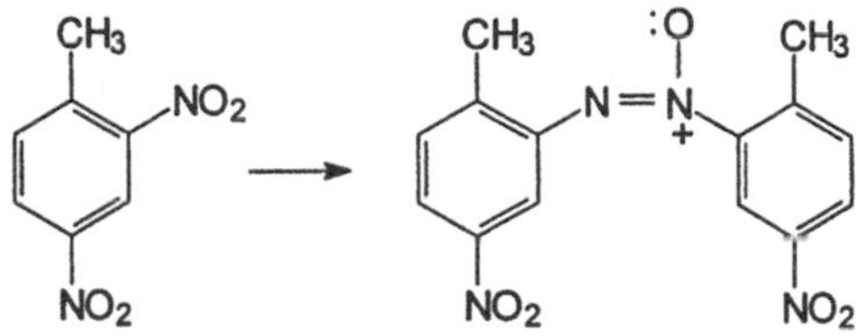

FIGURE 6.114
Metabolism of 2,4-dinitrotoluene by *Mucrosporium* sp.

Intramolecular Reactions

Cyclization to form benziminazoles may take place in compounds in which the nitro group is vicinal to an amino or substituted amine group. This has been demonstrated as one of the pathways during the fungal metabolism of dinitramine (Figure 6.115) (Laanio et al. 1973), and in a mixed bacterial culture which metabolized 2-nitroaniline to 2-methylbenziminazole (Hallas and Alexander 1983).

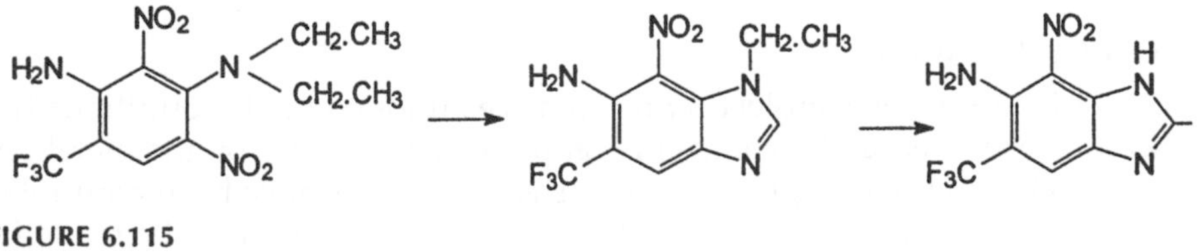

FIGURE 6.115
Biodegradation of dinitramine.

6.8.3   Aromatic Azo Compounds

Aromatic azo compounds are components of many commercially important dyes and pigments, so that attention has been directed to their degradation and transformation. Possibly the most significant discovery on the metabolism of aromatic azo compounds that had implications which heralded the age of modern chemotherapy concerned the bactericidal effect of the azo dye Prontosil. It was shown that the effect of Prontosil *in vivo* was in fact due to the action of its transformation product, sulfanilamide, which is an antagonist of 4-aminobenzoate that is required for the synthesis of the vitamin folic acid. Indeed, this reduction is the typical reaction involved in the first stage of the biodegradation of aromatic azo compounds. The reaction is readily accomplished under anaerobic conditions (Haug et al. 1991), and the azoreductase and nitroreductase from *Cl. perfringens* apparently involve the same protein (Rafii and Cerniglia 1993). Bacterial azoreductases have also been purified from aerobic organisms including strains of *Pseudomonas* sp. adapted to grow at the expense of azo dyes (Zimmermann et al. 1984). The amines resulting from all these reductive transformations then enter well-established metabolic pathways for the degradation of anilines by oxidative deamination to catechol followed by ring fission (McClure and Venables 1988; Fuchs et al. 1991). It should, however, be noted that the sulfanilic acid formed from sulfonated azo dyes may be excreted into the medium or, after further oxygenation, polymerized to terminal products (Kulla et al. 1983). Sulfanilate is, however, degraded by a mixed culture of *Hydrogenophaga palleroni* and *Agrobacterium radiobacter*, and proceeds via catechol-4-sulfonate followed by intradiol ring fission: sulfite is then elimination from the muconolactone (Hammer et al. 1996). This pathway has been confirmed using an adapted strain of *H. palleroni* strain S1(Blümel et al. 1998).

6.9   Aliphatic Compounds Containing Oxygen, Nitrogen, Sulfur, and Phosphorus

6.9.1   Ethers

The degradation of a structurally wide range of ethers under both aerobic and anaerobic conditions is covered in a review (White et al. 1996a) that should be consulted for details particularly of polyethers.

Aerobic Pathways

Degradation of the simplest representatives under aerobic conditions has been investigated in the course of comprehensive studies on the metabolic potential of methylotrophic bacteria. These reactions have been admirably summarized in a book (Anthony 1982) to which reference may be made for details. Only two simple examples will therefore be used for illustration:

- Diethyl ether is metabolized by monooxidation at C_1 to ethanol and acetaldehyde.

- By successive formation of formaldehyde, tetramethylammonium salts are oxidized either directly to formaldehyde or indirectly via *N*-methylglutamate: the C_1 metabolites then enter the conventional pathways for their further degradation to CO_2.

Although both reactions are typically carried out by the monooxygenases of methylotrophs (references in Anthony 1982), strain G4/PR1 of *Burkholderia cepacia* that is constitutive for the synthesis of toluene-2-monooxygenase is able to degrade a number of ethers including diethyl ether and *n*-butyl methyl ether, but not *t*-butyl methyl ether (Hur et al. 1997). Degradation of *t*-butyl methyl ether with the intermediate formation of *t*-butanol has been accomplished with a mixed culture (Salanitro et al. 1994). Propane-grown cells of *Mycobacterium vaccae* to JOB5 and a strain ENV425 obtained by pro-pane enrichment degraded *t*-butyl methyl ether to *t*-butanol, and by hydrox-ylation of the quaternary methyl group to 2-hydroxyisobutyric acid that was not, however, used as a growth substrate for the organisms (Steffan et al. 1997). The hydroxylation reaction may plausibly involve a cytochrome P-450 system (Chapter 4, Section 4.4.1.2). The metabolism of diethyl ether has been studied in the fungus *Graphium* sp. strain ATCC 58400 that was able to use this as sole source of carbon and energy. When grown with *n*-butane, the fun-gus was able to transform, but not to degrade *t*-butyl methyl ether to *t*-butanol and *t*-butyl formate (Hardison et al. 1997).

Degradation of symmetric long-chain dialkyl ethers may be used to illus-trate an entirely different metabolic pathway. The di-*n*-heptyl-, di-*n*-octyl-, di-*n*-nonyl-, and di-*n*-decyl ethers are degraded by a strain of *Acinetobacter* sp. to two different groups of metabolites (Figure 6.116):

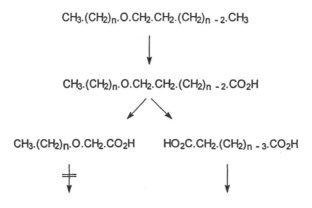

FIGURE 6.116
Biodegradation of di-*n*-heptyl ether by *Acinetobacter* sp.

1. To *n*-heptan-, *n*-octan-, *n*-nonan-, and *n*-decanol-1-acetic acids which are not metabolized further;

2. To glutaric, adipic, pimelic, and suberic acids which serve as sources of carbon and energy: these compounds are formed by terminal oxidation followed by an unusual oxidation at the carbon atom β to the ether bond (Modrzakowski and Finnerty 1980).

A less usual reaction is involved in the degradation of succinyloxyacetate by *Zoogloea* sp. (Peterson and Llaneza 1974). This is accomplished by a lyase that resulted in the production of fumarate and glycollic acid by an elimination reaction.

The possible persistence in the environment of alkylethoxy sulfates has led to extensive investigations of their degradability. Although the role of alkyl sulfatase activity is noted in Section 6.11.1, ether-cleavage has been shown to be the major pathway for the degradation of dodecyltriethoxy sulfate (Hales et al. 1986).

Three groups of xenobiotics containing aryl ether groups have been extensively used and have given rise to environmental concern.

1. The considerable interest in the persistence of chlorinated phenoxy-alkanoates—and of phenoxyacetates in particular that have been used as agrochemicals—has stimulated studies on the degradation of aromatic ethers. It should, however, be pointed out that considerably greater attention has been devoted to elucidating the subsequent steps that culminate in the fission of the aromatic ring that have been discussed in Section 6.5.1.2. The first step in the degradation of phenoxyalkanoates is dealkylation to the corresponding phenol with the formation of glyoxylate (Evans et al. 1971; Pieper et al. 1988). Although this reaction involves monooxygenation of the aromatic ring (Gamar and Gaunt 1971), the enzyme is in fact a 2-ketoglutarate-dependent dioxygenase (Fukumori and Hausinger 1993). The reaction has been studied in cell extracts of *Sphingomonas herbicidovorans* and, by using $^{18}O_2$, it was shown that one atom of this is incorporated into each of pyruvate and succinate (Nickel et al. 1997). The enantiomeric specificity is discussed in Chapter 4, Section 4.2.2.

2. Concern has also been expressed over the extensive use of alkylphenol polyethoxylates that are used as detergents, since some of the metabolites are apparently persistent compounds: although the alkyl phenols may be formed as a result of complete oxidation of the polyethoxylate side chains, partially degraded metabolites are in some cases apparently resistant to further degradation (Ball et al. 1989). The degradation of a highly branched nonylphenol polyethoxylate by *Pseudomonads putida* isolated from activated sludge proceeded with loss of single ethoxylate groups as acetaldehyde until two ethoxylate residues remained (John and White 1998). The

mechanism is reminiscent of that involved in the anaerobic degradation of polyethylene glycols that is noted below.

3. Halogenated derivatives of diphenyl ether have been used as herbicides (Scalla et al. 1990) and flame retardants (References in Sellström et al. 1998), and occur naturally (Voinov et al. 1991). Attention has therefore been directed to this class of compounds that formally includes dibenzofurans and dibenzo[1,4]dioxins that have already been discussed in Section 6.3.1.2. The degradation of diphenyl ether itself by *P.cepacia* has been examined (Pfeifer et al. 1989; 1993) and yields 2-pyrone-6-carboxylate as a stable end product: it may be formed from the initially produced 2,3-dihydroxydiphenyl ether in a reaction formally analogous (Figure 6.117) to that whereby 3-*O*-methylgallate is converted to 2-pyrone-4,6-dicarboxylate by 3,4-dihydroxybenzoate-4,5-dioxygenase in pseudomonads (Kersten et al. 1982). The degradation of diphenyl ether by a strain of *Sphingomonas* sp. proceeded with degradation of both rings (Schmidt et al. 1992) and cells grown with diphenyl ether were able to oxidize dibenzo[1,4]dioxin to 2-(2-hydroxyphenoxy)-*cis,cis*-muconate. After adaptation to growth with 4,4'-difluorodiphenyl ether, the organism grew with the corresponding chloro, but not the bromo, compound (Schmidt et al. 1993). In general, however, some degree of recalcitrance seems to be associated with substituted diaryl ethers.

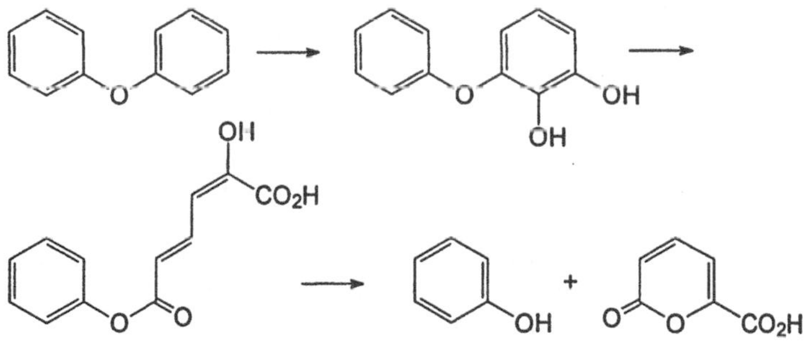

FIGURE 6.117
Biodegradation of diphenyl ether.

Anaerobic Pathways

The anaerobic degradation of polyethylene glycol has been investigated in a variety of organisms including *Pelobacter venetianus* (Schink and Stieb 1983), an *Acetobacterium* sp. (Schramm and Schink 1991), *Desulfovibrio desulfuricans*, and *Bacteroides* sp. (Dwyer and Tiedje 1986). The initial product is acetaldehyde, which is formed in two stages by the action of a diol dehydratase and a polyethylene glycol acetaldehyde lyase (Frings et al. 1992) (Figure 6.118) which is apparently found in all PGE-degrading anaerobic bacteria.

$$HO.(CH_2.CH_2.O)_n.CH_2.CH_2.OH \longrightarrow HO.(CH_2.CH_2.O)_n.CH(OH).CH_3$$

$$\longrightarrow HO.(CH_2.CH_2.O)_{n-1}.CH_2.CH_2OH + CH_3.CHO$$

FIGURE 6.118
Anaerobic biodegradation of polyethylene glycol.

6.9.2 Aliphatic Amines

The two mechanisms involved in the degradation of primary aliphatic amines by the action of dehydrogenases or oxidases have been noted in Chapter 4, Section 4.4.4 and only brief comments are given here. An inducible primary amine dehydrogenase in a strain of *Mycobacterium convolutum* with diverse degradative capability had a broad specificity, and was involved in the degradation of 1- and 2-aminopropane and 1,3-diaminopropane: the products were assimilated by the methylmalonate pathway or by formation of $C_2 + C_1$ fragments (Cerniglia and Perry 1975). On the other hand, the degradation of secondary or tertiary amines involves monooxygenation. Species of *Paracoccus* able to degrade tetramethylammonium hydroxide and *N,N'*-dimethylformamide have been assigned to the species *Pa. kocurii* (Ohara et al. 1990) and *Pa. aminophilus*, and *Pa. aminovorans*, respectively (Urakami et al. 1990). Although *Aminobacter aminovorans* (*Pseudomonas aminovorans*) is able to utilize methylamine and trimethylamine, it is unable to use methane, methanol, or dimethylamine (Urakami et al. 1992). The enzymological complexity of some of these apparently straightforward monooxygenation reactions is revealed by the fact that although the degradation of nitrilotriacetate takes place by successive loss of glyoxylate (Cripps and Noble 1973; Firestone and Tiedje 1978), the monooxygenase system consists of two components both of which are necessary for hydroxylation (Uetz et al. 1992). The degradation of EDTA involves formation of glyoxal and ethylene diamine, and is mediated by a flavin-dependent monooxygenase which has has been purified (Witschel et al. 1997; Payne et al. 1998).

6.9.3 Aliphatic Nitro Compounds

In contrast to aromatic nitro compounds, less attention appears to have been directed to their aliphatic counterparts. Two examples may be used to illustrate aerobic and anaerobic reactions.

1. 2-Nitropropane is oxidized by *Hansenula mrakii* and the enzyme has been purified: this is a flavodioxygenase that converts two molecules of the substrate to two molecules of acetone with the elimination of nitrite (Kido and Soda 1976) (Figure 6.119), and a similar enzyme has been isolated frm *Neurospora crassa* (Gorlatova

et al. 1998). On the other hand, the enzyme from *Fusarium oxysporium* is an oxidase (Kido et al. 1978a). Both enzymes have been noted in Sections 4.4.3 and 4.4.4.

2. The plant toxins 3-nitropropionic acid and 3-nitropropanol are converted into the corresponding amines by mixed cultures of ruminant microorganisms and the β-alanine resulting from the former is then further degraded (Anderson et al. 1993).

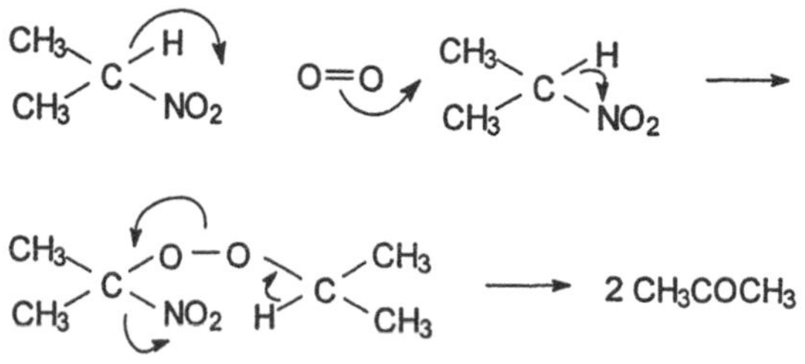

FIGURE 6.119
Metabolism of 2-nitropropane.

6.9.4 Sulfides, Disulfides, and Related Compounds

Interest in the possible persistence of aliphatic sulfides has arisen from the fact that they are produced in marine anaerobic sediments and from an appreciation of the possible implication of dimethylsulfide in climate alteration (Charlson et al. 1987) that has been noted briefly in Chapter 4, Section 4.1.2.

Sulfides and related compounds may be degraded by a number of quite different pathways which may be illustrated by the following examples.

1. Anaerobic reduction of dimethyl sulfide with production of methane (Oremland et al. 1989), and of dibenzyl disulfide to toluenethiol and finally toluene (Miller 1992). On the other hand, dimethyl sulfide that is used as an electron donor for the growth of *Rhodobacter sulfidophilus* is converted into dimethyl sulfoxide (Hanlon et al. 1994).

2. Elimination reactions have been implicated in the degradation of a variety of simple sulfur compounds:

   • The formation of pyruvate from cysteine is mediated by a desulfhydrase (Kredich et al. 1973) (Figure 6.120a) and pyruvate is also formed by a similar reaction from aminoethylcysteine (Rossol and Pühler 1992).

   • The formation of ethene from methionine by *Escherichia coli* takes place by an elimination reaction (Ince and Knowles 1986) (Figure 6.120b), though considerable complexities in the control and regulation of this reaction have emerged (Mansouri and Bunch 1989). It should be noted that the synthesis of ethene in plants proceeds by an entirely different reaction via *S*-adenosyl methionine and 1-aminocyclopropane-1-carboxylate (Kende 1989).

- Elimination is one of the pathways used for the degradation of dimethylsulfoniopropionate, and has been demonstrated in a strain of *Clostridium* sp. (Wagner and Stadtman 1962) (Figure 6.120c), and the enzyme has been purified from a strain of an *Alcaligenes*-like organism (de Souza and Yoch 1995).

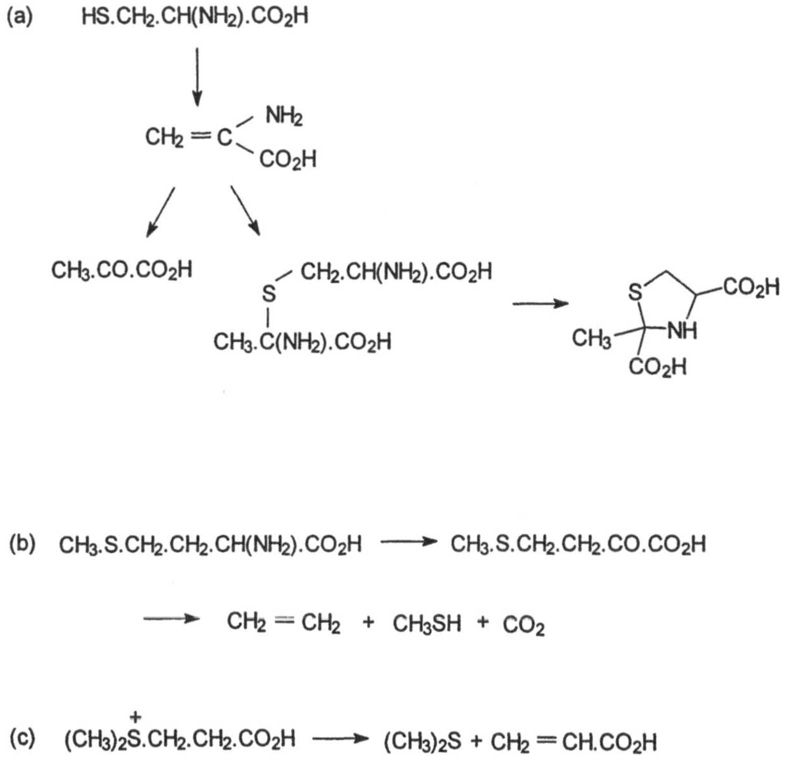

(a) $HS.CH_2.CH(NH_2).CO_2H$

$$CH_2 = C \begin{array}{c} NH_2 \\ CO_2H \end{array}$$

$CH_3.CO.CO_2H$

$CH_2.CH(NH_2).CO_2H$
S
$CH_3.C(NH_2).CO_2H$

$\longrightarrow$

$CH_3 \overset{S}{\underset{NH}{\diagup}} CO_2H$ CO_2H

(b) $CH_3.S.CH_2.CH_2.CH(NH_2).CO_2H \longrightarrow CH_3.S.CH_2.CH_2.CO.CO_2H$

$\longrightarrow CH_2 = CH_2 + CH_3SH + CO_2$

(c) $(CH_3)_2\overset{+}{S}.CH_2.CH_2.CO_2H \longrightarrow (CH_3)_2S + CH_2 = CH.CO_2H$

FIGURE 6.120
Elimination reactions during metabolism of aliphatic sulfur compounds (a) cysteine, (b) methionine, (c) dimethylsulfoniopropionate.

3. Dimethyldisulfide is degraded by autotrophic sulfur bacteria with the formation of sulfate and CO_2 which then enters the Benson–Calvin cycle (Smith and Kelly 1988). On the other hand, dimethyl sulfide and dimethyl sulfoxide are degraded by a strain of *Hyphomicrobium* sp. by pathways involving the formation from both carbon atoms of formaldehyde which subsequently enters the serine pathway (Suylen et al. 1986) (Figure 6.121). The key enzyme is methyl mercaptan oxidase which converts methyl sulfide into formaldehyde, sulfide, and peroxide (Suylen et al. 1987). A strain

of *Thiobacillus* sp. metabolizes dimethyl sulfide by an alternative pathway involving transfer of the methyl group probably to tetrahydrofolate by a cobalamin carrier (Visscher and Taylor 1993). Oxygen is not involved in the removal of the methyl groups so that the reaction may proceed anaerobically.

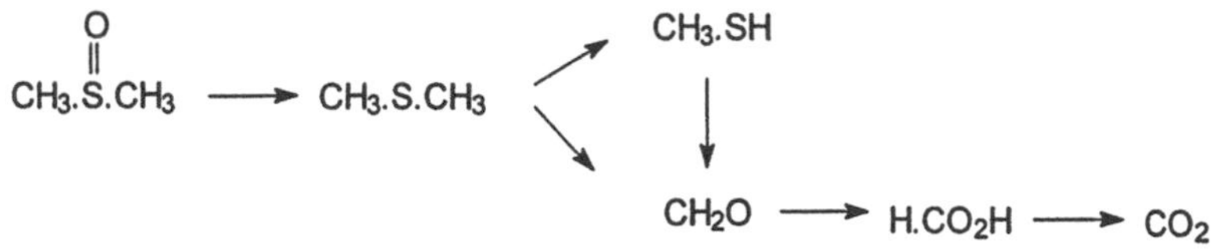

$$CH_3.S.CH_3 \longrightarrow \overset{\overset{\displaystyle O}{\|}}{CH_3.S.CH_3} \quad \begin{array}{l} CH_3.SH \\ \downarrow \\ CH_2O \longrightarrow H.CO_2H \longrightarrow CO_2 \end{array}$$

FIGURE 6.121
Biodegradation of dimethyl sulfide by *Hyphomicrobium* sp.

4. Strains of methylotrophic bacteria degrade methane sulfonate by initial oxidation to formaldehyde and sulfite (Kelly et al. 1994; Thompson et al. 1995), and do not involve hydrolytic steps. A formally comparable reaction is used in the biodegradation of sulfosuccinate to oxalacetate (Quick et al. 1994). The aerobic strains capable of degrading methanesulfonate have been assigned to the new genera *Methylosulfonomonas* and *Marinosulfonomonas* (Holmes et al. 1997). Degradation by *Methylosulfonomonas methylovora* strain M2 is initiated by monooxygenation (hydroxylation) by a multicomponent enzyme. One component of this is an electron transfer protein of which the larger subunit contains a Rieske [2Fe–2S] center, and both this and the small subunit show a high degree of homology with dioxygenase enzymes (De Marco et al. 1999).

5. Dimethylsulfoniopropionate is S-demethylated to 3-methylthiopropionate by a strain of *Desulfobacterium* sp. (Van der Maarel 1993) that also degrades betaine to dimethylglycine by a formally analogous reaction, and both 3-methylthiopropionate and methyl sulfide are produced by anoxic marine sediments (Kiene and Taylor 1988). An alternative degradation pathway has been demonstrated in *Desulfovibrio acrylicus*, which involves elimination with formation of acrylate that is reduced to propionate (van der Maarel et al. 1996). This organism can use both sulfate and acrylate as electron acceptors and a range of electron donors including lactate, succinate, ethanol, propanol, glycerol, glycine, and alanine.

6. Strains of facultatively heterotrophic and methylotrophic bacteria can use CS_2 as sole energy source, and under aerobic conditions also COS, dimethyl sulfide, dimethyl disulfide, and thioacetate (Jordan et al. 1995). It was proposed that the strains belonged to the genus *Thiobacillus* although they are clearly distinct from previously described species. These strains have now been assigned to *Paracoccus denitrificans* (Jordan et al. 1997).

6.9.5 Phosphonates

Considerable interest has been directed to the biodegradation of phosphonates which contain the rather unusual C–P bond since a number of them are
naturally occurring compounds, and one of them—glyphosate—has been
extensively used as a herbicide.

Many bacteria are able to use phosphonates as a source of phosphorus
(Cook et al. 1978a; Schowanek and Verstraete 1990) and this involves cleavage of the C–P bond with the formation of inorganic phosphate. This is
accomplished by a carbon–phosphorus lyase although the enzymology of
this, and its regulation is extremely complex (Chen et al. 1990). The gene cluster required for the utilization of phosphonates is induced in *E. coli* by phosphate limitation, and genetic evidence suggests a connection between the
metabolism of phosphonates and phosphites. On the basis of this, the interesting suggestion has been made that there may exist a phosphorus redox
cycle, and that phosphorus is involved not only at the +5 oxidation level but
at lower oxidation levels (Metcalf and Wanner 1991).

Some illustrative examples of the metabolism of phosphonates are noted
briefly.

1. In *Bacillus cereus*, 2-aminoethylphosphonate is initially oxidized to
   2-phosphonoacetaldehyde (La Nauze and Rosenberg 1968) before
   cleavage of the C–P bond (La Nauze et al. 1970) (Figure 6.122).
   The degradation of 2-aminoethylphosphonate by *Pseudomonas
   putida* strain NG2 is carried out even in the presence of phosphate,
   and is mediated by pyruvate aminotransferase and phosphonoac-
   etaldehyde hydrolase activities that are induced by 2-aminoeth-
   ylphosphonate (Ternan and Quinn 1998). On the other hand, in
   Enterobacter aerogenes strain IFO 12010, these activities are induced
   only under conditions of phosphate limitation.

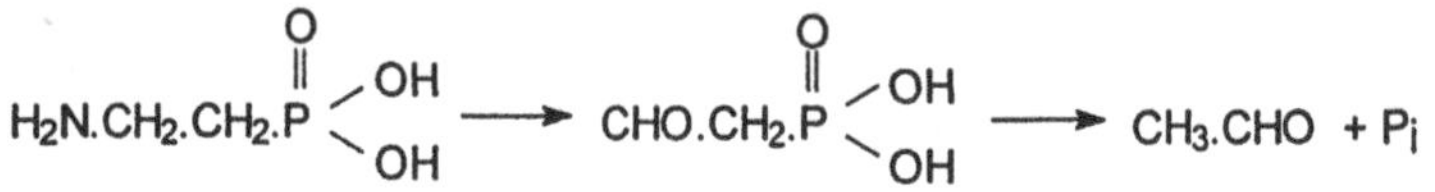

FIGURE 6.122
Biodegradation of 2-aminomethylphosphonate.

2. Alkyl phosphonates and phosphites are degraded by a number of
   bacteria including species of *Klyvera* and *Klebsiella* (Wackett et al.
   1987a) (Figure 6.123), and by *E. coli* (Wackett et al. 1987b) by a
   pathway in which the alkyl groups are reduced to alkanes. The
   metabolism of methylphenylphosphonate to benzene (Cook et al.
   1979) by *Klebsiella pneumoniae* has been noted in Section 4.3.2.

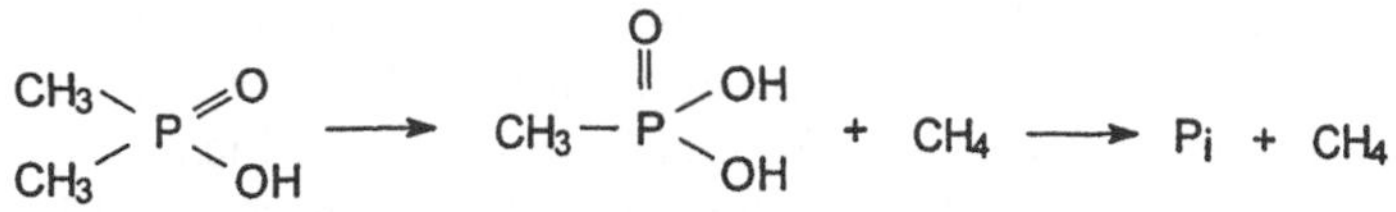

FIGURE 6.123
Reductive biodegradation of alkyl phosphonates and phosphites.

3. In view of its importance as a herbicide, the degradation of glyphosate has been investigated in a number of organisms, and two pathways have been elucidated, differing in the stage at which the C–P bond is cleaved:

   a. Loss of a C_2 fragment—formally glyoxylate—with the formation of aminomethylphosphonate (Pipke and Amrhein 1988) which may be further degraded by cleavage of the C–P bond to methylamine and phosphate (Jacob et al. 1988) (Figure 6.124a).

   b. Initial cleavage of the C–P bond with the formation of sarcosine which is then metabolized to glycine (Pipke et al. 1987; Liu et al. 1991) (Figure 6.124b).

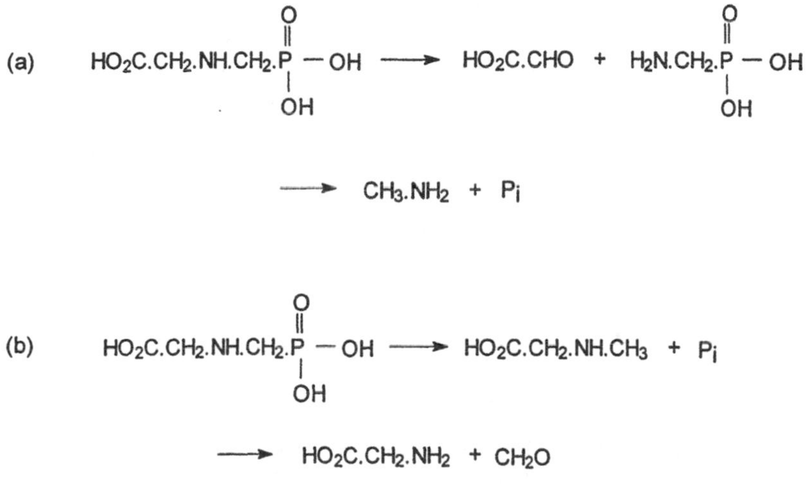

FIGURE 6.124
Alternative pathways for the biodegradation of glyphosate.

6.9.6 Organometallic and Related Compounds

Metal–carbon bonds that are stable under environmental conditions are known from elements ranging from Hg to Sn, Pb, and As: the simple alkylated compounds are nonpolar with low solubilities in water and are strongly lipophilic. On account of the extreme toxicity of methyl mercury, considerable attention has been directed to the methylation and demethylation of mercury, and to the degradation (Tezuka and Tonomura 1978) of phenyl mercuric acetate that was formerly used as a biocide. Arsenical drugs including derivatives of phenylarsonic acids and arsenobenzene were formerly used in medicine, and a cycle linking inorganic and organic As compounds including the naturally occurring arsenobetaine and arsenocarbohydrates has been put forward (Blanck et al. 1989). An unidentified Gram-negative organism was able to cleave the C–As bond a variety of organoarsenic compounds including arsonoacetic acid, methyl arsonic acid, and 2-aminoethylarsonic acid (Quinn and McMullan 1995). Various alkylated compounds of lead, generally tetraethyl lead, were used for many years as components of automobile gasoline, and attention has been given to its persistence both in view of its toxicity and its presence in areas formerly occupied by gasoline filling stations. It has been shown using (1-^{14}C-ethyl) tetraethyl lead that this was biodegradable, and that the rate of mineralization was adversely affected by the presence of hydrocarbons that are generally simultaneous contaminants (Mulroy and Ou 1998).

Organosilicon compounds are components of silicones and of tetraalkoxysilanes. The latter are used for a number of purposes including heat-exchange fluids, and an undefined enrichment culture was able to hydrolyze the tetrabutoxy- and tetra(2-ethylbutoxy)silanes and degrade the alkyl groups (Vancheeswaran et al. 1999). Polydimethylsiloxanes are at least partially degraded in soil to dimethylsilanediol (Carpenter et al. 1995; Fendinger et al. 1997), and the mineralization in soil of the monomer—dimethylsilanediol—has been shown (Sabourin et al. 1996). The interesting observations were made that in liquid cultures degradation could be obtained with *Fusarium oxysporium* growing concurrently with propan-2-ol or acetone, and by a strain of *Arthrobacter* sp. growing concurrently with dimethylsulfone.

6.10   Organofluorine Compounds

These deserve a section to themselves since fluorine differs so much from the other halogens, even from that of its nearest relative, chlorine. There are three outstanding features of fluorine: (1) its high electron-attracting power or electronegativity 3.98 on the Pauling scale, (2) its small atomic radius so that the length of the C–F bond is 1.36 Å compared with 1.755 Å for the C–Cl bond so that it is therefore closer to that of a normal saturated C–H bond with a length of 1.09 Å., and (3) the bond enthalpy of the C–F bond with a value of 448 kJ.mol^{-1}.

Organofluorine compounds have achieved enormous industrial significance in view of their generally great chemical stability (e.g., polytetrafluoroethylene) or their volatility (e.g., fluoroalkanes used as refrigerants), while some of them are valuable pharmaceutical products. Indeed, there has been increasing interest in perfluorinated aliphatic compounds as surfactants

(Dams 1993) and in perfluoropolyethers (Gurarini et al. 1993). Extensive discussion of the industrial use of organofluorine compounds is given in a book (Banks et al. 1994). In addition, some organofluorine compounds are toxic, including simple naturally occurring compounds such as fluoroacetate, whereas others such as the phosphorofluoridates have been deliberately synthesized as nerve poisons with potential military application. Systematic investigation of the biodegradability of organofluorine compounds appears not to have been made, but an attempt will be made to summarize the principles which have hitherto emerged.

6.10.1   Aliphatic Fluoro Compounds

Only relatively few details are available on the biodegradation of aliphatic fluoro compounds. Lack of definitive information on a wider range of compounds, and the fairly limited extent of the effort hitherto directed to studying their degradation suggests caution in determining whether or not they are recalcitrant: chemical evidence would, however, certainly suggest their resistance to biodegradation. The current status is illustrated by the following examples.

1. A pseudomonad has been shown to degrade fluoroacetate (Meyer et al. 1990), and the enzymology has been examined (Goldman 1965).

2. *Nitrosomonas europaea* is capable of oxidizing methyl fluoride to formaldehyde (Hyman et al. 1994).

3. A recombinant strain of *Pseudomonas* sp. containing the camphor plasmid (CAM) and toluene dioxygenase genes reduced 1,1,1-tetrachloro-2,2-difluoroethane to 1,1-dichloro-2,2-difluoroethene using cytochrome P-450 $_{cam}$: toluene dioxygenase may then oxidize this to oxalate (Hur et al. 1994; Wackett et al. 1994).

4. The same strain lacking toluene dioxygenase was able to convert trichlorofluoromethane to carbon monoxide and 1,1,1-trichloro-2,2,2-trifluoroethane to 1,1-dichloro-2,2-difluoroethene (Li and Wackett 1993).

5. *Pseudomonas* sp. strain D2 was able to use difluoromethane sulfonate as sulfur—but not carbon—source with the release of fluoride, but although it could use trifluoromethane and 1*H*,1*H*,2*H*,2*H*-tetradecafluorooctane sulfonate as sulfur sources, fluoride was not eliminated (Key et al. 1998).

6.10.2   Aromatic Fluoro Compounds

Substantial activity has been directed to the degradation particularly of fluorobenzoates under aerobic conditions. It should be noted that the ability to use these as sole source of carbon and energy is more restricted than for the corresponding chlorobenzoates, although they may be degraded under conditions of concurrent metabolism in the presence of suitable growth substrates. Pathways for the degradation of the chlorinated analogues may not be viable since, for example, 3-fluorocatechol is not generally a good substrate for the catechol-cleaving enzymes. There are also different mechanisms whereby loss of fluoride is accomplished. There are therefore alternative strategies that are used for the degradation of fluorobenzoates.

1. Elimination of fluoride from 2-fluorobenzoate may take place concomitantly with dioxygenation and carboxylation (Engesser et al. 1980) (Figure 6.125a) (Section 4.4.2), and in the metabolism of 2,5-difluorobenzoate that is metabolized via 4-fluorocatechol (Cass et al. 1987).

2. Degradation of 3-fluoro- and 4-fluorobenzote may proceed by dioxygenation with retention of fluorine which is subsequently eliminated from 3-fluoromuconate (Engesser et al. 1990a) (Figure 6.125b). Several strains of *Alcaligenes eutrophus* are able to grow with 4-fluorobenzoate, but not with the corresponding 4-chlorobenzoate (Schlömann et al. 1990a). 4-Fluorocatechol is formed and ring fission is achieved by a catechol-1,2-dioxygenase: loss of fluoride occurs during hydrolysis of the lactone produced by muconate cycloisomerse and not, as generally with chloro analogues, during lactone formation (Schlömann et al. 1990b). It was suggested that this might be a general pathway for other compounds metabolized via 4-fluorocatechol such as 2,5-difluorobenzoate (Cass et al. 1987) and 4-fluoroaniline (Zeyer et al. 1985).

3. A strain of *Aureobacterium* sp. degrades 4-fluorobenzoate by hydrolytic elimination of fluoride and initial formation of 4-hydroxybenzoate (Oltmanns et al. 1989) (Figure 6.125c).

4. 4-Hydroxybenzoate hydroxylase is able to catalyze loss of fluoride from 3,5-difluoro-4-hydroxybenzoate and 2,3,5,6-tetrafluoro-4-hydroxybenzoate with the formation of catechols (Husain et al. 1980), and in an apparently analogous reaction pentachlorophenol hydroxylase converts pentafluorophenol into 2,3,5,6-tetrafluoro-1,4-dihydroxybenzene (Uotila et al. 1992).

5. The metabolism of enrofloxacin (Wetzstein et al. 1997) has been discussed in Section 6.3.2 in the context of the degradation of the heterocyclic ring by *Gleophyllum striatum*.The metabolism of the related ciprofloxacid by the same organism is comparable in many respects: the main reactions are complete or partial degradation of the piperazine ring, and displacement of the fluorine atom by a hydroxyl group (Wetzstein et al. 1999). The reactions may be compared with those during the photochemical transformation (Chapter 4, Section 4.1.1).

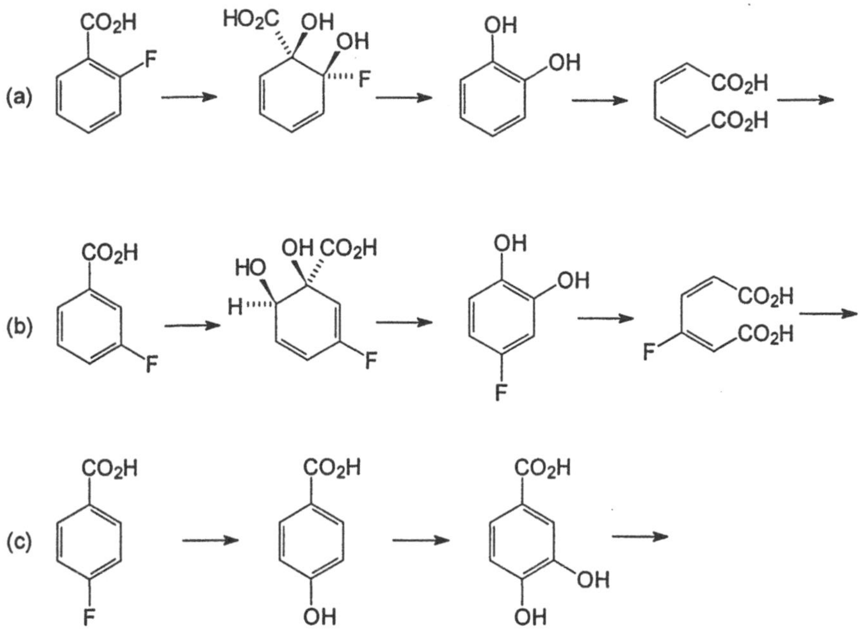

FIGURE 6.125
Pathways for the biodegradation of (a) 2-fluorobenzoate, (b) 3-fluorobenzoate, (c) 4-fluoroben-
zoate.

6. The metabolism of 2,3-, 2,4-, and 2,5-difluoro- and 2,3,4-, 2,3,5-, and
   2,4,5-trifluorophenol was studied in several species of *Rhodococcus*
   using ^{19}F NMR. Catechols were produced both by hydroxylation
   and by dioxygenation with elimination of fluoride, and 2,3-diflu-
   oromuconate was formed as the ring fission product from 2,3-
   difluorophenol before further transformation (Bondar et al. 1998).

7. A range of 3-fluorosubstituted benzenes has been examined using
   a toluene-degrading strain of *Pseudomonas* sp. (Renganathan 1989)
   and two reactions were observed: one in which the fluorine was
   eliminated during dioxygenation and the other in which it was
   retained (Figure 6.126).

The degradation of 5-fluorouracil by the fungus *Nectria haematococca* proceeds by the established pathway for the nonfluorinated compound with the
formation of the stable α-fluoro-β-alanine (Parisot et al. 1991).

The degradation of aromatic fluoro compounds under anaerobic conditions appears to have attracted less attention but a single example will be
used as illustration: the pathway for the nonfluorinated analogue 3-meth-
ylphenol has already been noted in Section 6.7.3.2, and the principles used for
delineating the pathway in Chapter 5, Section 5.5.3. Under methanogenic
conditions, 6-fluoro-3-methylphenol was degraded to 3-fluorobenzoate via 5-

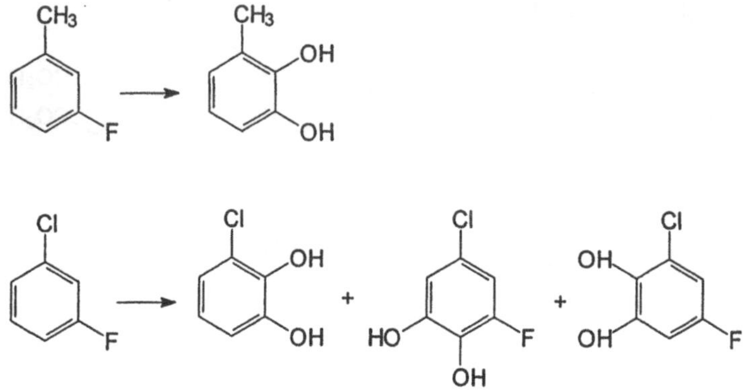

FIGURE 6.126
Biodegradation of 3-fluorotoluene and 3-chlorofluorobenzene.

fluoro-4-hydroxy-2-methylbenzoate and 3-fluoro-4-hydroxybenzoate (Figure 6.127) (Londry and Fedorak 1993). Comparable carboxylation and dehydroxylation reactions have also been demonstrated with 2- and 3-fluorophenol that form 2- and 3-fluorobenzoate (Genther et al. 1990). It should be noted that in neither case was the fluorine atom eliminated even under these highly reducing conditions so that the persistence of aromatic fluoro compounds under anaerobic conditions may, at least provisionally, be accepted.

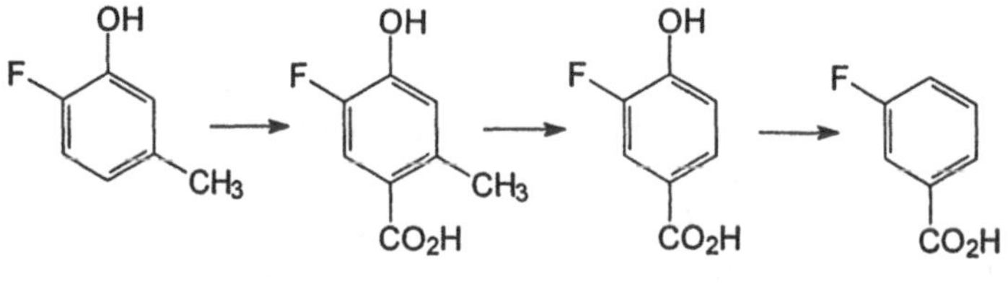

FIGURE 6.127
Biotransformation of 6-fluoro-3-methylphenol.

6.10.3 Trifluoromethyl Compounds

The extraordinary chemical stability of the trifluoromethyl group is illustrated by the original synthesis of trifluoracetic acid by chromic acid oxidation of 3-trifluoromethylaniline (Swarts 1922); this recalcitrance is also revealed by the results of studies on the biodegradation of compounds containing the CF_3 group. Trifluoromethylbenzoates cannot support growth of aerobic bacteria, even though they are apparently effective substrates for the enzymes used for the metabolism of the methylbenzoates. Ring dioxygenation and dehydrogenation take place with the formation of trifluoromethyl catechols which may be further degraded with ring fission to form terminal metabolites in which the trifluoromethyl group remains intact (Engesser et al.

1988 a,b) (Figure 6.128): this has been confirmed in a more extensive study (Engesser et al. 1990b). Attention has already been directed in Chapter 4, Section 4.1.1 to the degradability of 3-trifluoromethyl-4-nitrophenol photochemically (Carey and Cox 1981), and of 3- and 4-trifluoromethylbenzoates by sequential microbial and photochemical treatment (Taylor et al. 1993). A comparable situation exists for 2-trifluoromethylphenol that is sequentially hydroxylated and degraded by extradiol ring fission to 2-hydroxy-6-keto-7,7,7-trifluoro-2,4-dienoate that undergoes photochemical loss of the CF_3 group (Reinscheid et al. 1998).

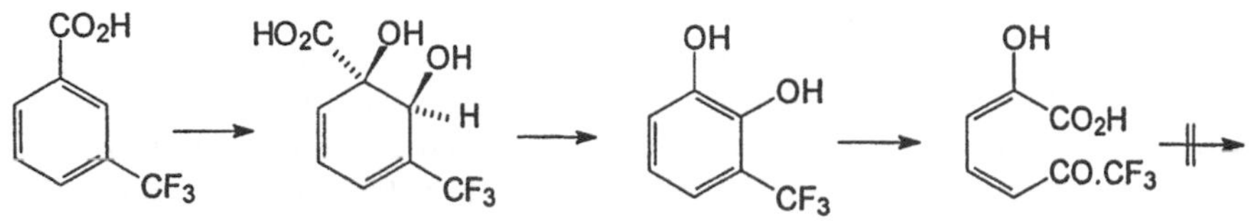

FIGURE 6.128
Metabolism of 3-trifluoromethylbenzoate.

6.11 Biotransformations

In the course of previous discussions, a number of examples of biotransformations have been illustrated, but it is convenient to summarize briefly some others that have not been discussed and that involve structural groups that are represented in diverse xenobiotics. Biotransformations have attracted interest for at least two reasons: first from their biotechnological importance in the production of valuable organic compounds including pharmaceuticals, and second due to their environmental significance. In addition, some biotransformation reactions are a response elicited from an organism by exposure to a toxicant and therefore represent a detoxification mechanism for the organism. Four sections are devoted to selected illustrations of biotransformation reactions, while the fifth attempts to illustrate the possible adverse environmental significance of some of the reactions. Reference should be made to the book by Faber (1997) for an extended discussion of a much wider range of biochemical transformations than is possible here, and to the book by Hudlicky (1990) that is devoted to oxidation and contains many examples of microbiological transformations.

6.11.1 Hydrolysis of Esters, Amides, and Nitriles

Hydrolytic activity toward a range of esters and related compounds is widely distributed among microorganisms and reference has already been made to some of them. A somewhat brief summary may therefore suffice.

Carboxylic Acid Esters

Two examples may be given merely as illustration of the diversity of potential substrates: the hydrolysis of dibutyl phthalate (Eaton and Ribbons 1982), and of cocaine to benzoate and ecgonine methyl ester (Britt et al. 1992).

Sulfate Esters

Alkyl sulfates and alkylethoxy sulfates have been extensively used as detergents so that concern has been expressed on their biodegradability: a review (Cain 1981) covers the degradation of a wide range of surfactants including both of these groups, and one by White and Russell (1994) discusses in detail the biodegradation of alkyl sulfates including the enzymology and regulation. It should also be appreciated that the hydrolysis of a range of naturally occurring sulfate esters may make a contribution to the sulfate present in aerobic soils (Fitzgerald 1976) quite apart from that contributed from the anthropogenic input of SO_x.

Long-chain unbranched aliphatic sulfate esters are generally degraded by initial hydrolysis to sulfate and the alkanol which is then degraded by conventional pathways: the alkylsulfatases show a diversity of specificity (Dodson and White 1983) generally for compounds with chain lengths >5, although organisms have been isolated that degrade short-chain (C_1–C_4) primary alkyl sulfates (White et al. 1987). *Pseudomonas* sp. strain C12B produces two primary (P1 and P2) and three secondary (S1, S2, and S3) alkylsulfatases, and hydrolysis by the P1 enzyme takes place by O–S bond fission. The S1 and S2 enzymes are constitutive, and for chiral compounds, such as octan-2-yl sulfate, hydrolysis proceeds with inversion of configuration by cleavage of the alkyl–oxygen bond (Bartholomew et al. 1977). As an alternative pathway for degradation, oxidation may precede elimination of sulfate: examples are found in the degradation of propan-2-yl sulfate (Crescenzi et al. 1985) and of monomethyl sulfate (Davies et al. 1990; Higgins et al. 1993) (Figure 6.129). For alkylethoxy sulfates, a greater range of possibilities exist including ether-cleavage reactions that have been noted in Section 6.9.1, and direct removal of sulfate may be of less significance (Hales et al. 1986). It should be noted, however, that an unusual reaction may occur simultaneously: chain elongation of the carboxylic acid. For example, during degradation of dodecyl sulfate, lipids containing 14, 16, and 18 carbon atoms were synthesized (Thomas and White 1989).

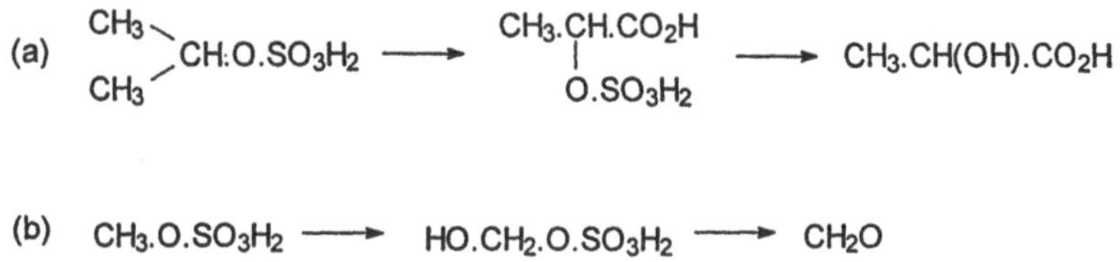

(a) CH_3
     $\underset{CH_3}{\overset{CH_3}{>}}CH{:}O.SO_3H_2 \longrightarrow \underset{O.SO_3H_2}{\overset{CH_3.CH.CO_2H}{|}} \longrightarrow CH_3.CH(OH).CO_2H$

(b) $CH_3.O.SO_3H_2 \longrightarrow HO.CH_2.O.SO_3H_2 \longrightarrow CH_2O$

FIGURE 6.129
Degradation of (a) propan-2-yl sulfate and (b) methyl sulfate.

Aryl sulfates are widely synthesized from phenolic substrates and serve as a detoxification mechanism both for microorganisms (Sections 6.2.2 and 6.11.5) and for fish (Chapter 7, Section 7.5.1). The hydrolysis of aryl sulfates has traditionally been a useful taxonomic character in the genus *Mycobacterium* (Wayne 1961). A test for the hydrolysis of phenolphthalein sulfate has been routinely incorporated into mycobacterial taxonomy, and a positive result in the 3-day aryl sulfatase test has been particularly valuable for distinguishing members of the rapidly-growing *M. fortuitum* group which are potentially pathogenic to humans: slow-growing *M. tuberculosis* and *M. bovis* are negative even after 10 days. Regulation of the synthesis of tyrosine sulfate sulfohydrolase has been examined in a strain of *Comamonas terrigena* (Fitzgerald et al. 1979), and both inducible and constitutive forms of the enzyme exist. These are, however, apparently distinct from the aryl sulfate sulfohydrolase which has found application in taxonomic classification that has been noted above.

Phosphate Esters and Related Compounds

Concern over the persistence and the biodegradability of organophosphate and organophosphorothioates that are used as agrochemicals has stimulated studies into their degradation. Investigations have also been directed to the use of their degradation products as a source of phosphate for the growth of bacteria: a wide range of phosphates, phosphorothioates, and phosphonates has therefore been examined as suitable P-sources (Cook et al. 1978b). The first step in their degradation of all these phosphate and phosphorothioate esters is hydrolysis, and substantial effort has been directed to all of these groups: a summary of investigations in this important area has been given (Munnecke et al. 1982), and it is therefore sufficient merely to provide a typical illustration (Figure 6.130a). In addition, it is worth noting that the initial metabolites after hydrolysis such as the nitrophenols may be both toxic and sometimes resistant to further degradation. The monoxygenation of nitrophenols has been briefly discussed in Chapter 4, Section 4.4.11, the degradation of nitrophenols in Section 6.8.2, and the issue of toxic substrates in Chapter 5, Section 5.3.1.

Nitrate Esters

Compared with the fairly numerous investigations on the microbial hydrolysis of carboxylic acid, sulfate, and phosphate esters, data on the hydrolysis of nitrate esters is much more fragmentary. This has been clearly revealed in a review (White and Snape 1993) that summarizes existing knowledge on the microbial degradation of nitrate esters including glycerol trinitrate and its close relatives, and the pharmaceutical products pentaerythritol tetranitrate and isosorbide 2,5-dinitrate. As an alternative to purely hydrolytic reactions, both reductive reactions and reactions involving glutathione transferases seem to be important in eukaryotic microorganisms (White et al. 1996b). The biotransformation of gylcerol trinitrate by strains of *Bacillus thuringiensis/ cereus* or *Enterobacter agglomerans* (Meng et al. 1995) and by strains of

Pseudomonas sp., and some Enterobacteriaceae (Blehert et al. 1997) involves the expected successive loss of nitrite with formation of glycerol. The biotransformation of pentaerythritol tetranitrate by *Enterobacter cloacae* proceeds comparably with metabolism of two hydroxymethyl groups produced by loss of nitrite to the aldehyde (Binks et al. 1996). The enzymes that produce nitrite from the nitrate esters are apparently reductases, and the enzyme from E. *cloacae* is strongly inhibited by steroids and is capable of the reduction of cyclohex-2-ene-1-one (French et al. 1996). In a medium containing glucose and ammonium nitrate, glyceryl trinitrate is degraded by *Penicillium corylophilum* to the dinitrate and the mononitrate before complete degradation. On the other hand, the metabolism of glyceryl trinitrate by *Phanerochaete chrysosporium* involves the production of nitric oxide (Servent et al. 1991). The formation of nitric oxide during conversion of L-arginine to L-citrulline by a strain of *Nocardia* sp. (Chen and Rosazza 1995) has been briefly noted in Chapter 4, Section 4.3.3.

Amides and Related Compounds

A number of important agrochemicals are aromatic amides, carbamates, and ureas, and the the first step in their biodegradation is mediated by the appropriate amidases, ureases, and carbamylases (Figure 6.130b, c, and d). It should, however, be pointed out that, for example, the chloroanilines that are formed from many of them as initial products may be substantially more resistant to further degradation. Application of tests for amidase activity, particularly using pyrazinamidase, has also been widely used in the classification of mycobacteria (Wayne et al. 1991). Sequential hydrolysis of nitriles to amides and carboxylic acids is well established both in aliphatic (Miller and Gray 1982; Nawaz et al. 1992) and in aromatic compounds (Harper 1977; McBride et al. 1986), although it should be noted that the herbicide bromoxynil may also be degraded by the elimination of cyanide from the ring with the initial formation of 2,6-dibromohydroquinone (Topp et al. 1992) (Figure 6.131). There may be a high degree of specificity in the action of these nitrilases and this may have considerable interest in biotechnology. Some examples may be given as illustration.

1. Racemic 2-(4'-isobutylphenyl)propionitrile is converted by a strain of *Acinetobacter* sp. to S-(+)-2-(4'-isobutylphenyl)propionic acid with an optical purity >95% (Yamamoto et al. 1990).

2. The nitrilase from a number of strains of *Pseudomonas* sp. mediated an enantiomerically selective hydrolysis of racemic O-acetylmandelonitrile to D-acetylmandelic acid [(R (-)-acetylmandelic acid] (Layh et al. 1992).

3. The amidase from *Rhodococcus erythopolis* strain MP50 was used to convert selectively racemic 2-phenylpropionamide to S-2-phenyl-propiohydroxamate. This was converted to the isocyanate by Lossen rearrangement and then by hydrolysis to S(-)-phenyl-ethylamine (Hirrlinger and Stolz 1997).

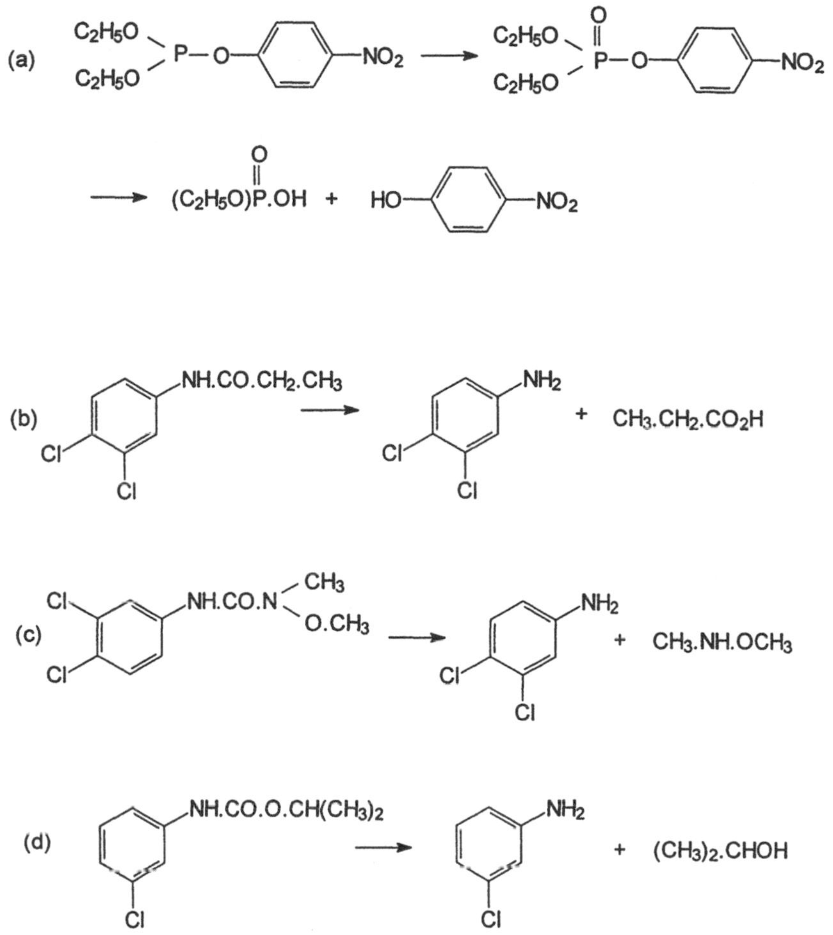

FIGURE 6.130
Hydrolysis of (a) a phosphate triester, (b) acetanilides, (c) phenylureas, and (d) carbamates.

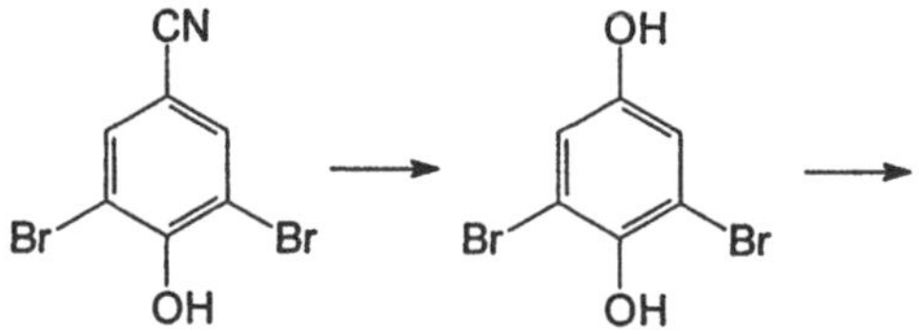

FIGURE 6.131
Metabolism of bromoxynil.

Although the degradation of thiocyanate is accomplished by a few bacteria, the chemolithoautrophic *Thiobacillus thioparus* strain THI 115 is unusual in being able to use this as an energy source. The hydrolase that converts the substrate to NH_4^+ and COS has been purified and consists of three subunits (Katayama et al. 1998): the hydrolase showed high homology to nitrile hydrolase and this suggests a close similarity in the pathways for their degradation: hydrolysis to the amines followed by deamination.

6.11.2 Hydroxylations, Oxidations, Dehydrogenations, and Reductions

Single-step reactions involving hydroxylation, oxidation, dehydrogenation, or reduction in which the microorganisms function as biocatalysts have been extensively exploited in the synthesis of commercially valuable compounds particularly sterols. An attempt will be made here to provide a wider perspective and to summarize briefly some other examples of microbial biotransformations which have the potential for application in biotechnology. A valuable compilation of a wide range of oxidations carried out by diverse microorganisms including bacteria, fungi, and yeasts has been provided (Hudlický 1990) and should be consulted for details and further illustrations. Only a few examples which have rather different motivation will be used to illustrate the extremely wide possibilities. It is worth noting the versatility of enzymes from different organisms to effect transformations of a given substrate: an illustrative example (Lee and Gibson 1996) is provided by the transformation of styrene carried out by (1) naphthalene dioxygenase, (2) toluene and styrene 2,3-dioxygenases, and (3) monooxygenation by cytochrome P-450 (Figure 6.132).

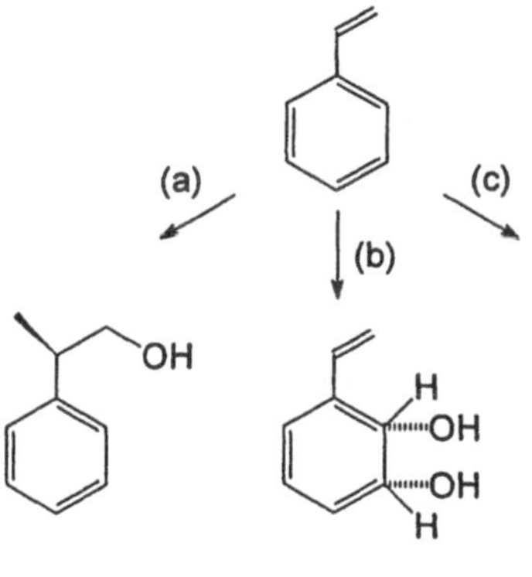

FIGURE 6.132
Transformations of styrene.

6.11.2.1 *Hydroxylation*

Interest in fungal metabolism as a model for that of higher organisms has been noted briefly in Chapter 4, Section 4.3.6, and reactions catalyzed by cytochrome P-450 mediated in both prokaryotes and eukaryotes have been discussed in Chapter 4, Section 4.4.1.2. The use of microorganisms, particularly fungi, to produce hydroxylated sterols of therapeutic interest has been extensively explored, and greatest attention has been directed to fungal hydroxylation of steroids at the 11α-, 12β-, 15-, 17-, 19-, and 21-positions (references in Hudlický 1990), although different fungi may produce different products from the same substrate:

a. *Calonectria decora, Rhizopus nigricans,* and *Aspergillus ochraceus* produce respectively the 12β, 15α-, 11α,16β-, and 6β,11α-diols from 3-keto-5α-androstane (Bird et al. 1980).

b. The biotransformation of pregna-4,17(20)-*cis*-diene-3,16-dione by *A. niger* produced metabolites with hydroxyl groups at the 7β-, or 7β- and 15β-positions whereas *Cephalosporium aphidicola* produces metabolites with hydroxyl groups at the 11α-, or 11α- and 15β-positions (Atta-ur-Rahman et al. 1998).

A recent departure has been motivated by the need for unusual products which may be sterol metabolites. Studies have been directed to the biosynthesis, for example, of the otherwise rare and inaccessible derivatives of progesterone hydroxylated at the 6-, 9-, 14-, or 15-positions that could be accomplished by incubating progesterone in a complex medium with the fungus *Apiocrea chrysosperma* (Smith et al. 1988). The hydroxylation of progesterone and closely related compounds at the 15β-position by cell extracts of *Bacillus megaterium* has been noted in Section 4.4.1.2.

Hydroxylation of a diverse group of other natural products has also been demonstrated:

• A synthetic diterpenoid butenolide by *Cunninghamella elegans* at the 5α- and both 7α- and 7β-positions as well as on the isopropyl side chain (Milanova et al. 1994).

• Sclareol by *C. elegans* at the 2α-, 3β-. 18-, and 19-positions (Abraham 1994).

• Stemodin by *Cephalosporium aphidicola* at the 7α-, 7β-, 8β-, 18-, and 19-positions (Hanson et al. 1994).

The spectrum of substrates has been extended to xenobiotics, and a single example may suffice: the fungus *Beauveria bassiana* metabolized the hydroxycoumarin rodenticide warfarin not only to the hydroxylated and oxidized metabolites that have already been established in mammalian systems or mediated by fungal cytochrome P-450 monooxygenase systems, but also to novel products resulting from the reduction of the keto group originating from the 4-hydroxyl group (Griffiths et al. 1992). Bacterial hydroxylation by cytochrome P-450 systems is well established in actinomyces (O'Keefe and Harder 1991) and some examples have already been given in Chapter 4, Section 4.4.1.2. Stereospecific hydroxylation of α-ionone—although not β-ionone—has been observed with strains of *Streptomyces* sp. (Lutz-Wahl et al. 1998), and the racemic substrate is hydroxylated to the (3R,6R)- and (3R,6S)-hydroxy-α-ionones.

6.11.2.2 Oxidases

Microorganisms including yeasts and bacteria are able to produce D-amino acid oxidases, and interest has centered particularly on the use of *n*-alkane as growth substrates (Kawamoto et al. 1977). The enzyme has achieved importance for its role in carrying out the first step in the transformation of cephalosporin C to 7-aminocephalosporanic acid that is an intermediate for the synthesis of semisynthetic cephalosporins. The nucleotide sequence of the enzyme from *Rhodotorula gracilis* ATCC 26217 has been established and the gene could be overexpressed in *Escherichia coli* (Alonso et al. 1998).

6.11.2.3 Halogenation

Haloperoxidase from the fungus *Caldariomyces fumago* have been used to accomplish a number of potentially valuable synthetic reactions including the following:

- α,β-Halohydrins from ethene and propene (Geigert et al. 1983a);
- Vicinal dihalides from alkenes and alkynes (Geigert et al. 1983b);
- α-Halogenated ketones from alkynes and 3-hydroxyhalides from cyclopropanes (Geigert et al. 1983c).

In all cases, reactive hypohalous acid is probably the active reagent.

6.11.2.4 Chiral Synthesis

Increasing interest has been directed to the microbial synthesis of chiral compounds that would not be readily accessible by chemical synthesis, and a few examples are given to illustrate the diversity of reactions that have been examined: the stereoselective hydrolysis of nitriles to carboxylic acids has already been noted. Both by microbiological reactions and by combinations of microbiological transformation of aromatic compounds followed by designed chemical transformation (see Chapter 4, Section 4.2.3) have been used. Illustrations of both will be given.

1. Direct Microbiological Synthesis

These include a variety of reactions including epoxidation, hydroxylation, and stereospecific synthesis either directly or by degradation of the alternative enantiomer.

a. Attention has already been directed (Section 6.1.3) to the use of a *Xanthobacter* sp. for resolution of 2,3-epoxyalkanes by selective metabolism of one of the enantiomers (Weijers et al. 1988). A review of the application of these enantiomerically specific reactions has been given (Archelas and Furstoss 1997), and two examples may suffice as illustration: (i) the synthesis of 2*R*,3*S*-pityol was

accomplished using two stereospecific microbiological transformations: reduction of the ketone with *Saccharomyces cerevisiae* and epoxidation of the alkene with *Aspergillus niger* which, in this case, did
not produce the *trans* diol by the further activity of an epoxide
hydrolase (Archelas and Furstoss 1992) (Figure 6.133), and (ii) the
epoxidization of *cis*-prop-1-enylphosphonic acid to the epoxide fosfomycin by a strain of *Penicillium decumbans* (Watanabe et al. 1999):
expression of the *bar* reporter gene was induced by the substrate and
its product, and by phosphite but not phosphate (*bar* is the bialaphos
resistance gene encoding phosphinothricin acetyltransferase).

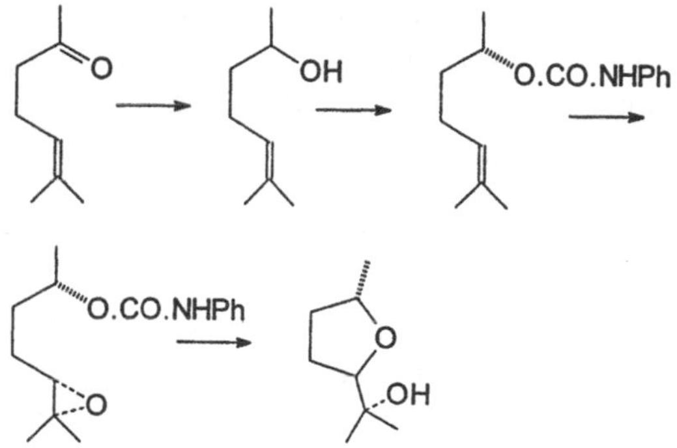

FIGURE 6.133
Synthesis of optically pure 2R,3S-pityol.

b. Hydroxylation of quaternary methyl groups by cytochrome P-
450$_{CAM}$, and an enantiomerically specific oxidation of a cyclic
ketone to a lactone have been noted in Section 6.1.2.

c. A number of compounds which would be difficult to synthesize
chemically have been produced as pure enantiomers by the oxidation of aromatic compounds by mutant strains of bacteria whereby
catechols or dihydrodiols may be produced (Ribbons et al. 1990).

d. A strain of *Rhodococcus* sp. carries out allylic oxidation of α-cedrene
to (R)-10-hydroxycedrene that undergoes rearrangement to α-cur-
cumene (Figure 6.134) (Takigawa et al. 1993).

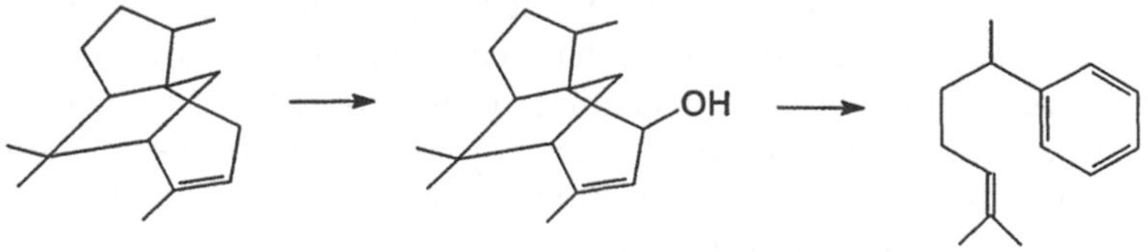

FIGURE 6.134
Biotransformation of α-cedrene by *Rhodococcus* sp.

e. The unusual reduction of the α,β-double bond in 2-nitro-1-phenyl-1-propene by strains of *Rhodococcus rhodochrous* and several species of *Nocardia* results in a preponderance of one enantiomer (Sakai et al. 1985), and this is of interest in the synthesis of physiologically active norephedrin-type compounds. Enone reductases from *Saccharomyces cerevisiae* have been purified and characterized, and are able to carry out reduction of the C=C bonds in aliphatic aldehydes and ketones, ring double bonds in cyclohexenones, and both *cis-* and *trans*-2-phenylbut-2-enal to (*R*)-2-phenylbutanal (Wanner and Tressel 1998).

f. Reduction of perfluoroalkylated ketones and of the double bond in perfluoroalkenoic acid esters has been accomplished with baker's yeast (*S. cerevisiae*), and the products had a high optical purity ranging from 67 to 96% (Kitazume and Ishikawa 1983).

g. The hydration of the double bond in octadec-9-enoate to 10-hydroxyoctadecanoate may be carried out by a strain of *Pseudomonas* sp. with exclusive formation of the 10(*R*) compound (Yang et al. 1993).

h. At low concentrations of 1,3-dichloro-propan-2-ol, cells of a strain of *Corynebacterium* sp. induced with glycerol or a number of chlorinated alcohols were able to convert the substrate into (*R*)-3-chloropropan-1,2-diol in a yield of 97% and with an optical purity exceeding 80% (Nakamura et al. 1993).

i. Enantiomerically pure alkyl aryl sulfoxides have been obtained by microbial oxidation of the corresponding sulfides (Holland 1988): both *Corynebacterium equi* and fungi including *Aspergillus niger*, species of *Helminthosporium*, and *Mortierella isabellina* were effective although the same fungi were not able to carry out enantiomeric-selective oxidation of ethylmethylphenyl phosphine due apparently to the intrusion of nonselective chemical autoxidation (Holland et al. 1993).

2. *Microbial Oxidation of Aromatic Compounds Followed by Chemical Transformation*

It has already been shown that biodegradation of many aromatic compounds proceeds by initial dioxygenation to *cis*-dihydrodiols, followed by dehydrogenation and ring fission: the high enantiomeric purity of *cis*-dihydrodiols produced by bacterial dioxygenases has been emphasized. Mutant strains in which dehydrogenase activity is lacking produce only the dihydrodiols, and there has been increasing interest in developing the use of these as synthons for the production of novel compounds that would not be readily available by conventional chemical synthesis.

For dihydrodiols derived from substituted benzenes, the key to their significance lies in the availability of two adjacent chiral centers with an established absolute stereochemistry. The dihydrodiol from benzene is, of course, the meso compound, but enantiomers produced by subsequent reaction with a chiral reagent are readily separated. Useful reviews containing many applications have been given (Carless et al. 1989; 1992; Ribbons et al. 1989) many of which involve, in addition, the use of *cis*-fluoro, *cis*-chloro, or *cis*-bromobenzene-2,3-dihydrodiols. Several groups of reactions have been most extensively exploited:

- Formation of *cis*-1,2-epoxides and conversion into *trans*-1,2-diols;
- Osmylation to provide *cis*-1,2-diols;
- 1:4 addition to the cyclohexadiene using nitroso compounds, singlet oxygen, or Diels–Alder reactions;
- Fission of the double bonds in the cyclohexadiene ring with ozone.

Although the product from the transformation of toluene by mutants of *Pseudomonas putida* lacking dehydrogenase activity is the *cis*-2R,3S dihydrodiol, the *cis*-2S,3R dihydrodiol has been synthesized from 4-iodotoluene by a combination of microbiological and chemical reactions. *P. putida* strain UV4 was used to prepare both enantiomers of the *cis*-dihydrodiol, and iodine chemically removed with H_2-Pd/C. Incubation of the mixture of enantiomers with *P. putida* NCIMB 8859 selectively degraded the 2R,3S compound to produce toluene *cis*-2S,3R dihydrodiol (Allen et al. 1995). A few illustrative syntheses using benzene and toluene *cis*-dihydrodiols are given.

- Racemic pinitol from benzene *cis*-dihydrodiol benzoate by successive epoxidation and osmylation (Figure 6.135a) (Ley at al. 1987);
- Conduramine A1 tetraacetate using an activated nitroso-mannose derivative (Figure 6.135b) (Felber et al. 1986; Werbitzky et al. 1990);
- (-)-Laminitol from toluene *cis*-dihydrodiol by successive epoxidations (Figure 6.135c) (Carless et al. 1991);
- Analogues of conduritols from toluene *cis*-dihydrodiol by reaction with singlet oxygen followed by fission with thiourea (Figure 6.135d) (Carless et al. 1989);
- *cis*-1,2-Dihydroxycyclopent-3-ene-5-one by successive ozonation of the cyclohexadiene ring (Figure 6.135e) (Hudlicky et al. 1988), and the corresponding 5-isopropylidene compound by dioxygenation of 6,6-dimethylfulvene with a mutant of *P. putida* RE204 that produces isopropylbenzene-2,3-dioxygenase (Figure 6.135f) (Eaton and Selifonov 1996).

There is clearly enormous potential using other *cis*-dihydrodiols produced from benzocycloalkenes (Allen et al. 1995) or from the numerous dihydrodiols produced from polycyclic carbocyclic and heterocyclic substrates.

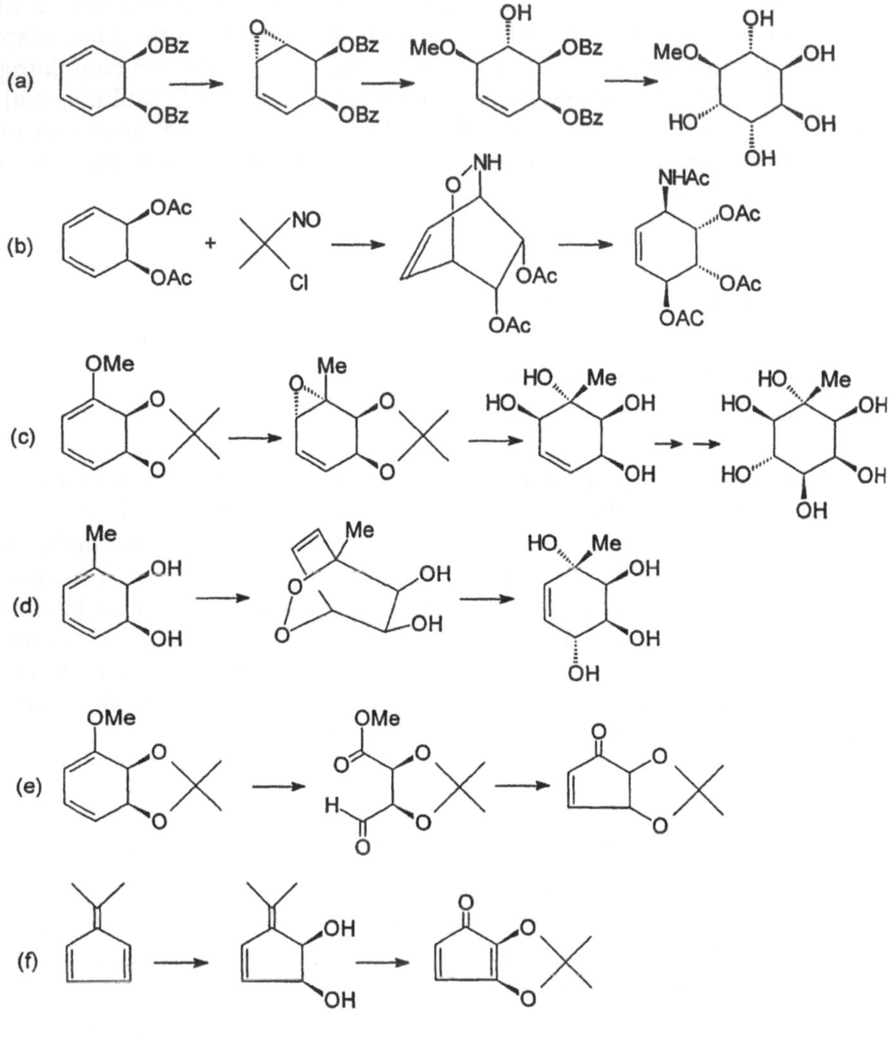

FIGURE 6.135
Chemical syntheses based on cyclohexadiene *cis*-dihydrodiols.

6.11.3 The Formation of Dimeric Products from Aromatic Amines

It has been established that aromatic amines may be degraded by bacteria (McClure and Venables 1986; Fuchs et al. 1991) and it has already been noted that amines may be formed by reduction of the corresponding nitro and azo compounds (Sections 6.8.2 and 6.8.3). On the other hand, fungi may oxidize aromatic amines. For example, *Fusarium oxysporum* oxidizes 4-chloroaniline to 4-chlorophenylhydroxylamine, 4-chloronitrosobenzene, and 4-chloronitrobenzene. Condensation of suitable pairs of these intermediates would

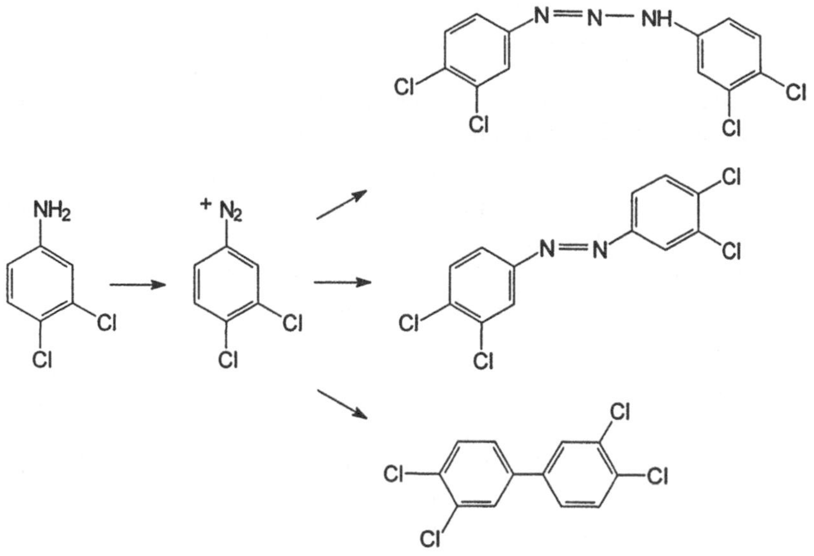

FIGURE 6.136
Condensation products from 3,4-dichloroaniline via diazo compounds.

then reasonably account for the formation of 4,4'-dichloroazoxybenzene and 4,4'-dichloroazobenzene by this organism (Kaufman et al. 1973). The same organism produced 3,3',4,4'-tetrachloroazoxybenzene from 3,4-dichloroaniline (Kaufman et al. 1972). A number of bacteria growing in the presence of 4-chloroaniline and nitrate produce 4,4'-dichloroazobenzene, and more-detailed investigations with *Escherichia coli* and 3,4-dichloroaniline revealed that in addition to the corresponding azo compound and the triazine, the unexpected 3,3',4,4'-tetrachlorobiphenyl was also formed (Figure 6.136) (Corke et al. 1979). These reactions mediated by bacteria are quite different from the fungal oxidations, and plausibly involve the diazonium compound that is produced chemically from 4-chloroaniline and nitrite that is microbiologically formed from nitrate. Although the reactions mediated by fungi may not have a direct major impact on the aquatic environment, their occurrence in landfills with subsequent leaching of the metabolites into watercourses cannot be discounted.

6.11.4 Methylation Reactions

The O-methylation of halogenated phenols has been briefly noted in Chapter 4 as an example of a biotransformation reaction, and they have attracted attention in view of the unacceptable flavor that chloroanisoles impart to broiler chickens (Gee and Peel 1974), freshwater fish (Paasivirta et al. 1987), and wine corks (Buser et al. 1982). A few additional comments may usefully be added to place these observations in a wider perspective.

1. The *O*-methylation of halogenated phenolic compounds by both Gram-positive and Gram-negative bacteria has been demonstrated (Allard et al. 1987), and in cell extracts of representatives of these bacteria, it appears that the methyl group is provided as expected by *S*-adenosyl methionine (Neilson et al. 1988b).

2. Many fungi of the family *Hymenochaetaceae* produce chloromethane in substantial quantities by methylation of chloride (Harper and Hamilton 1988), and it has been suggested that this could result in a significant biotic input into the environment, independently of anthropogenic discharge. In addition, fungi are able to carry out a transmethylation reaction between methyl chloride and compounds with reactive hydroxyl groups including phenols and both aliphatic and aromatic carboxylic acids (Harper et al. 1989; McNally et al. 1990). In *Phanerochaete chrysosporium*, chloromethane is able to function as a methyl donor in the biosynthesis of veratryl alcohol that fulfills an important function in the regulation of lignolytic systems (Harper et al. 1990). In addition, it has been shown that, in whole cells of *Ph. chrysosporium O*-methylation of acetovanillone may be accomplished both by transmethylation from chloromethane and from *S*-adenosylmethionine (Coulter et al. 1993). Care should therefore be exercised in assigning the ultimate source of halogenated anisoles which are widely distributed in the environment (Wittlinger and Ballschmitter 1990), and which have been identified in environmental samples including the marine atmosphere (Atlas et al. 1986), sediment samples (Tolosa et al. 1992), and biota (Watanabe et al. 1983): they may possibly originate from halogenated phenols that have been identified in automobile exhaust (Müller and Büser 1986), or from brominated phenols that have been used as flame retardants. The origin of some of these anisoles such as those that have been established as metabolites of agrochemicals incuding 2,4-dichloro- and 2,4,5-trichlorophenoxy-acetic acids (McCall et al. 1981; Smith 1985) and 2,4,5-trichloro-2-pyridinyloxyacetate (Lee et al. 1986) is, of course, unequivocal.

3. Cell extracts of the protozoan *Tetrahymena thermophila* carry out the methylation of sulfide and selenide to methyl sulfide and methyl selenide (Drotar et al. 1987a), and the importance of microbial reactions in its synthesis has been examined (Drotar et al. 1987b). The environmental significance of the production of methyl sulfide has already been noted in Section 4.1.2. Aromatic thiols are also *S*-methylated by *T. thermophila* (Drotar and Fall 1986) as well as by *Euglena gracilis* (Drotar and Fall 1985).

4. An apparently analogous reaction is the formation of isobutyral-dehyde oxime during the metabolism of valine by a number of Gram-negative bacteria (Harper and Nelson 1982) and the enzyme

has been purified from a strain of *Pseudomonas* sp. (Harper and Kennedy 1985).

6.11.5 Environmental Consequences of Biotransformation

Throughout Chapters 4 and 6, many examples have been given of the biotransformation of xenobiotics that results in the formation of stable metabolites rather than mineralization. It is therefore appropriate to the environmental importance of these metabolites in a wider context. There are two opposing aspects of their significance: (1) their function as a detoxification mechanism toward a potential toxicant and (2) the adverse effect of metabolites on organisms other than those producing them. In addition, these transformation products may have physicochemical properties very different from those of their precursors: they may therefore be accumulated in higher organisms even if they cause no palpably adverse effect, or disseminated within the aquatic environment (Section 3.5.2).

Biotransformation as an Effective Mechanism of Detoxification

This is discussed more fully in Section 7.5 in the wider context of toxicology. Only brief attention will therefore be directed here to the less fully documented role of analogous reactions in microorganisms.

1. One of the most significant illustrations of detoxification is, of course, provided by the mechanisms whereby bacteria develop resistance to antibiotics, and a full discussion of this is available (Franklin and Snow 1981). A variety of different reactions may occur, and whereas all of them leave the structure of the antibiotics largely intact, they effectively destroy the biological activity (toxicity) of the drug. A few of these transformations will be used as illustration.

   - Hydrolysis of the β-lactam ring in penicillins with the formation of inactive penicilloic acid is a serious problem in the development of resistance to penicillin, although this may be at least partially overcome in the semisynthetic penicillins.
   - Resistance to chloramphenicol is mediated by the synthesis of the 1-acetate and 1,3-diacetate while the corresponding resistance to streptomycin involves phosphorylation.

2. The reductive cleavage of methylmercury to methane (Spangler et al. 1973) and of phenylmercuric acetate to benzene (Furukawa and Tonomura 1971; Nelson et al. 1973) may be considered as mechanisms of detoxification in a wider context since these organomercury compounds are extremely toxic to higher organisms. There are apparently two different enzymes in *Pseudomonas* sp. K-62 that bring about splitting of the phenyl–Hg and methyl–Hg bonds: the S-1

enzymes accept both substrates whereas the S-2 enzyme is specific to substrates with phenyl–Hg bonds (Tezuka and Tonomura 1978).

3. Attention has already been made to the formation of glycoside and sulfate conjugates of phenolic metabolites during the biotransformation of polycyclic aromatic hydrocarbons by fungi (Section 6.2.2). Formation of sulfate esters is not, however, limited to fungal systems since this has been observed, for example, following 4′-hydroxylation of 5-hydroxyflavone in *Streptomyces fulvissimus* (Ibrahim and Abul-Hajj 1989), and the enzyme has been partly purified from a *Eubacterium* sp. (Koizumi et al. 1990).

4. Although the biodegradation of phenols and anilines is well established, these are relatively toxic compounds and some microorganisms detoxify them by acylation as an alternative to biodegradation. For example, acetylation of pentachlorophenol has been observed (Rottt et al. 1979), and acetylation of substituted anilines has been established: 4-chloroaniline is converted into 4-chloroacetanilide by *Fusarium oxysporum* (Kaufman et al. 1973) and 2-nitro-4-aminotoluene that is produced from 2,4-dinitrotoluene by a species of *Mucrosporium* sp. (McCormick et al. 1978) is converted into 3-nitro-4-methylacetanilide. All these neutral compounds may plausibly be assumed to be less toxic than their precursors.

Potentially Adverse Ecological Consequences of Biotransformation

A number of different metabolic possibilities that may have extensive—and possibly adverse—environmental repercussions may be recognized: these involve the synthesis of metabolites that are (1) lipophilic and less polar than their precursors, (2) highly water-soluble, or (3) toxic to other biota in the ecosystem. A number of diverse examples will be used to illustrate what is probably a widespread phenomenon.

1. The synthesis of lipophilic metabolites has been demonstrated in a number of studies of the transformation of xenobiotics, and the following are given as illustration.

   a. The classic case is the biomethylation of mercury (Jensen and Jernelöv 1969) whereby cationic inorganic Hg is transformed into an organic form that is both lipophilic and more toxic to higher organisms. Both abiotic and biotic processes may be involved, and microbial methylation has been confirmed both in the sediment phase (Furutani and Rudd 1980) and probably even in the intestines of fish (Rudd et al. 1980). A range of microorganisms that are capable of methylating Hg^{2+} has been isolated, and these include aerobic bacteria and fungi (Vonk and Sijpesteijn 1973), and facultatively anaerobic (Hamdy and Noyes 1975) and anaerobic bacteria (Yamada and Tonomura 1972).

Methylation of other inorganic compounds has also been observed, although these have been less extensively investigated, and includes compounds of Sn (Jackson et al. 1982), as in which reduction of As(III or IV) is also involved (Cullen et al. 1984), and S and Se; methylation of the last two has been discussed briefly in Section 6.11.4. In all of these cases, the organic forms of these elements should therefore be taken into consideration in constructing global cycles.

b. The *O*-methylation of halogenated phenolic compounds has been systematically investigated (Allard et al. 1987; Neilson et al. 1988b): these metabolites have a high bioconcentration potential and therefore present a potential environmental hazard—quite apart from their demonstrated toxicity to zebra fish larvae (Neilson et al. 1984). These and some related reactions have been discussed in Section 6.11.4. The occurrence of such reactions provides a plausible reason for the occurrence of, for example, the *O*-methyl ether of the diphenyl ether antibacterial agent triclosan in aquatic biota (Miyazaki et al. 1984).

c. It has already been noted that some microorganisms do not degrade long-chain alkanols completely but synthesize esters by reaction between the carboxylic acid produced and the alkanol (Hankin and Kolattukudy 1968): these metabolites are highly lipophilic, although any adverse effect may be restricted through their subsequent hydrolysis after uptake by higher organisms.

d. Attention has been drawn in Section 4.3.2 to a number of decarboxylation reactions carried out by Enterobacteriaceae: the products are much less polar than their precursors and may therefore exhibit the potential for concentration in biota. The significance of this reaction is seldom taken into account but merits serious consideration.

2. The synthesis of water-soluble conjugates of metabolites may bring about the widespread dissemination of the metabolites and this has been briefly noted in Section 3.5.2. For example, many xenobiotics are metabolized both by fungi and higher organisms to hydroxylated metabolites which are then conjugated to form sulfate esters, glycosides, or glucuronides. Some examples of the reactions mediated by microorganisms have been discussed in this chapter and those carried out by higher biota will be discussed in Chapter 7. The significance of these metabolites in a wider context lies in the possibility for the widespread dispersion of these water-soluble compounds in water masses, and their subsequent hydrolysis by bacteria to the original phenolic compounds. For example, it is well established that *Escherichia coli* readily hydrolyses a number of phenolic glucuronides.

3. The synthesis of stable microbial (and other) transformation products should be assessed in the wider context of the possible adverse effect of these metabolites on other components of the ecosystem. One of the earliest examples that demonstrated the toxicological significance of metabolites is provided by the drug Prontosil that is noted in Section 7.5, the toxicity of methyl mercury to humans, and of chlorinated veratroles to zebra fish larvae has been noted above. Another example is the formation of established carcinogenic nitrosamines that may be mediated by microbiologically mediated reactions that are discussed in Chapter 4, Section 4.2.3. Whereas PAHs themselves are generally relatively nontoxic, they can be activated by various mechanisms to compounds that are well established as inducers of tumors in mammals (Cavalieri and Rogan 1998). In addition, phenolic metabolites may be toxic to other biota in natural ecosystems. For example, pyrene is transformed by monooxygenation to 1-hydroxypyrene in a number of microorganisms including the fungi *Cunninghamella elegans* (Cerniglia et al. 1986), *Penicillium janthinellum* and *Syncephalastrum racemosum* (Launen et al. 1995), and *Crinipellis stipitata* (Lambert et al. 1994; Lange et al. 1994), and possibly in some bacteria. 1-Hydroxypyrene has been shown to be highly toxic to the terrestrial nematode *Chaenorhabditis elegans* (Lambert et al. 1995). In a wider context, it is noted that this metabolite is also formed from pyrene by the isopod *Porcellio scaber* and in trees; these are noted in Section 7.3.6.

Clearly, therefore, microbial metabolites may have potentially widespread adverse environmental effects on higher biota including humans.

6.12 Summary of Basic Microbial Reactions

In the preceding sections, an account has been given of the pathways followed during biodegradation and biotransformation of a range of organic compounds. It may be useful to present this information in an alternative way: in terms of the types of reaction which may be carried out. No references to the literature are given since those that are relevant may be found in the earlier sections. Some examples of typical reactions are given by way of illustration together with references to the appropriate section in which they have already been discussed.

Details of oxygenation reactions have been discussed in detail in Section 4.4.

1. Hydrolysis—esters, amides, and related compounds, and nitriles (Section 6.11.1)

2. Nucleophilic substitution—hydrolysis of organohalogen compounds (Sections 6.4.2 and 6.5.1)

3. Addition of water—alkynes (Sections 6.1.4 and 6.7.1); azaarenes (Section 6.7.4.1)

4. Elimination—halogen (Section 6.4.1); aliphatic sulfides (Section 6.9.3)

5. Carboxylation
   — Aerobic degradation of branched alkanes (Section 6.1.1)
   — Anaerobic degradation of phenols (Section 6.7.3)

6. Decarboxylation (Section 6.2.1 and Chapter 4, Section 4.3.1); with elimination of halogen (Section 6.5.1.2)

7. Introduction of oxygen: hydroxylases, monooxygenases, and dioxygenases
   — Aliphatic and alicyclic compounds: hydroxylation (Sections 6.1.1, 6.1.2, and 6.11.2)
   — Alkenes: epoxidation (Sections 6.1.3 and 6.11.2)
   — Oxygenation of cycloalkanones (Section 6.1.2)
   — Aromatic compounds: monooxygenation and dioxygenation (Sections 6.2 and 6.3.1)
   — Halogenated aromatic compounds: hydroxylation (Section 6.5.1.2)
   — Aromatic nitro compounds and sulfonates: elimination of nitrite or sulfite by dioxygenation (Sections 6.8.1 and 6.8.2)

8. Oxidation—alkanols to aldehydes or ketones (Sections 6.1.1 and 6.1.2)

9. Dehydrogenation of alicyclic rings (Section 6.1.2)

10. Reduction
    — Anaerobic dehydroxylation (Sections 6.7.2 and 6.7.3)
    — Nitro and azo compounds (Sections 6.8.2 and Section 6.8.3)
    — Aromatic aldehydes (Section 6.7.3)
    — Dehalogenation (Sections 6.4.4, 6.5.1.2, and 6.6)

11. Reductive cleavage
    — The C–P bond in organophosphonates (Section 6.9.4)
    — The C–Hg bond in organomercurials (Section 6.11.5)

12. Conjugation—sulfates and glucuronides of phenols: excretion of water-soluble metabolites (Sections 6.2.2 and 6.11.5)

13. Rearrangement
    — Coenzyme B_{12}-mediated reactions
      • Anaerobic degradation of amino acids (Section 6.7.1)
      • Anaerobic degradation of polyethylene glycols (Section 6.9.1)
    — The NIH shift (Sections 6.2.2 and 6.5.2)

— The Bamberger rearrangement of aryl hydroxylamines

— Terpenes (Sections 6.1.2 and 6.11.2)

14. Dismutation—the Stickland reaction between pairs of amino acids (Section 6.7.1)

15. Methylation reactions (Section 6.11.4)

References

Abbott, B.J. and C.T. Hou. 1973. Oxidation of 1-alkenes to 1,2-epoxides by *Pseudomonas oleovorans*. *Appl. Microbiol.* 26: 86–91.

Abraham, W.-R. 1994. Microbial hydroxylation of sclareol. *Phytochemistry* 36: 1421–1424.

Abril, M.-A., C. Michan, K.N. Timmis, and J.L. Ramos. 1989. Regulator and enzyme specificities of the TOL plasmid-encoded upper pathway for degradation of aromatic hydrocarbons and expansion of the substrate range of the pathway. *J. Bacteriol.* 171: 6782–6790

Adachi, K., T. Iwabuchi, H. Sano, and S. Harayama. 1999. Structure of the ring cleavage product of 1-hydroxy-2-naphthoate, an intermediate of the phenanthrene-degradative pathway of *Nocardioides* sp. strain KP7. *J. Bacteriol.* 181: 757–763.

Adriaens, P. 1994. Evidence for chlorine migration during oxidation of 2-chlorobiphenyl by a type II methanotroph. *Appl. Environ. Microbiol.* 60: 1658–1662.

Adrian, N.R. and J.M. Suflita. 1990. Reductive dehalogenation of a nitrogen heterocyclic herbicide in anoxic aquifer slurries. *Appl. Environ. Microbiol.* 56: 292–294.

Aeckersberg, F., F. Bak, and F. Widdel. 1991. Anaerobic oxidation of saturated hydrocarbons to CO_2 by a new type of sulfate-reducing bacterium. *Arch. Microbiol.* 156: 5–14.

Aeckersberg, F., F.A. Rainey, and F. Widdel. 1998. Growth, natural relationships, cellular fatty acids and metabolic adaptation of sulfate-reducing bacteria that utilize long-chain alkanes under anoxic conditions. *Arch. Microbiol.* 170: 361–369.

Ahmad, D., R. Massé, and M. Sylvestre. 1990. Cloning, physical mapping and expression in *Pseudomonas putida* of 4-chlorobiphenyl transformation genes from *Pseudomonas testosteroni* strain B-356 and their homology to the genomic DNA from other PCB-degrading bacteria. *Gene* 86: 53–61.

Ahmad, D., M. Sylvestre, and M. Sondossi. 1991. Subcloning of *bph* genes from *Pseudomonas testosteroni* B-356 in *Pseudomonas putida* and *Escherichi coli*: evidence for dehalogenation during initial attack on chlorobiphenyls. *Appl. Environ. Microbiol.* 57: 2880–2887.

Aitken, M.D., W.T. Stringfellow, R.D. Nagel, C. Kazuga, and S.-H. Chen. 1998. Characteristics of phenanthrene-degrading bacteria isolated from soils contaminated with polycyclic aromatic hydrocarbons. *Can. J. Microbiol.* 44: 743–752.

Aleu, J., J.R. Hanson, R.H. Galán, and I.G. Collado. 1999. Biotransformation of the fungistatic sesquiterpenoid patchoulol by *Botrytis cinerea*. *J. Nat. Prod.* 62: 437–440.

Alexander, M. 1981. Biodegradation of chemicals of environmental concern. *Science* 211: 132–138.

Allard, A.-S., M. Remberger, and A.H. Neilson. 1987. Bacterial O-methylation of halogen-substituted phenols. *Appl. Environ. Microbiol.* 53: 839–845.

Allard, A.-S., C. Lindgren, P.-Å. Hynning, M. Remberger, and A.H. Neilson. 1991. Dechlorination of chlorocatechols by stable enrichment cultures of anaerobic bacteria. *Appl. Environ. Microbiol.* 57: 77–84.

Allard, A.-S., P.-Å. Hynning, M. Remberger, and A.H. Neilson.1992. Role of sulfate concentration in dechlorination of 3,4,5-trichlorocatechol by stable enrichment cultures grown with coumarin and flavanone glycones and aglycones. *Appl. Environ. Microbiol.* 58: 961–968.

Allen, C.C.R., D.R. Boyd, H. Dalton, N.D. Sharma, I. Brannigan, N.A. Kerley, G.N. Sheldrake, and S.C. Taylor. 1995. Enantioselective bacterial biotransformation routes to *cis*-diol metabolites of monosubstituted benzenes, naphthalene and benzocycloalkenes of either absolute configuration. *J. Chem. Soc. Chem. Commun.* 117–118.

Allen, C.C., D.R. Boyd, M.J. Larkin, K.A. Reid, N.D. Sharma, and K. Wilson. 1997. Metabolism of naphthalene, 1-naphthol. indene, and indole by *Rhodococcus* sp. strain NCIMB 12038. *Appl. Environ. Microbiol.* 63: 151–155.

Allen, J.R. and S.A. Ensign. 1996. Carboxylation of epoxides to β-keto acids in cell extracts of *Xanthobacter* strain Py2. *J. Bacteriol.* 178: 1469–1472.

Allen, J.R. and S.A. Ensign. 1997. Characterization of three protein components required for functional reconstitution of the epoxide carboxylase multienzyme complex from *Xanthobacter* strain Py2. *J. Bacteriol.* 179: 3110–3115.

Allison, N., A.J. Skinner, and R.A. Cooper. 1983. The dehalogenases of a 2,2-dichloropropionate-degrading bacterium. *J. Gen. Microbiol.* 129: 1283–1293.

Alonso, J. J.L. Barredo, B. Díez, E. Mellado, F. Salto, J.L. García, and E. Cortés. 1998. D-Amino-acid oxidase gene from *Rhodotorula gracilis* (*Rhodosporidium toruloides*) ATCC 26217. *Microbiology* (U.K.) 144: 1095–1101.

Altenschmidt, U. and G. Fuchs. 1992a. Anaerobic toluene oxidation to benzyl alcohol and benzaldehyde in a denitrifying *Pseudomonas* strain. *J. Bacteriol.* 174: 4860–4862.

Altenschmidt, U. and G. Fuchs. 1992b. Novel aerobic 2-aminobenzoate metabolism. *Eur. J. Biochem.* 205: 721–727.

Altenschmidt, U., B. Oswald, E. Steiner, H. Herrmann, and G. Fuchs. 1993. New aerobic benzoate oxidation pathway via benzoyl-coenzyme A and 3-hydroxybenzoyl-coenzyme A in a denitrifying *Pseudomonas* sp. *J. Bacteriol.* 175: 4851–4858.

An, D., D.T. Gibson, and J.C. Spain. 1994. Oxidative release of nitrite from 2-nitrotoluene by a three component enzyme system from *Pseudomomas* sp. strain JS42. *J. Bacteriol.* 176: 7462–7467.

Anantharam, V., M.J. Allison, and P.C. Maloney. 1989. Oxalate: formate exchange. *J. Biol. Chem.* 264: 7244–7250.

Ander, P., K.-E. Eriksson, and H.-S. Yu. 1983. Vanillic acid metabolism by *Sporotrichium pulverulentum*: evidence for demethoxylation before ring-cleavage. *Arch. Microbiol.* 136: 1–6.

Anderson, J.J. and S. Dagley. 1980. Catabolism of aromatic acids in *Trichosporon cutaneum*. *J. Bacteriol.* 141: 534–543.

Anderson, R.C., M.A. Rasmussen, and M.J. Allison. 1993. Metabolism of plant toxins nitropropionic acid and nitropropanol by ruminal microorganisms. *Appl. Environ. Microbiol.* 59: 3056–3061.

Anthony, C. 1982. *The Biochemistry of Methylotrophs*. Academic Press, London.

Anthony, C. 1992. The structure of bacterial quinoprotein dehydrogenases. *Int. J. Biochem.* 24: 29–39.

Apajalahti, J.H.A. and M.S. Salkinoja-Salonen. 1987a. Dechlorination and para-hydroxylation of polychlorinated phenols by *Rhodococcus chlorophenolicus. J. Bacteriol.* 169: 675–681.

Apajalahti, J.H.A. and M.S. Salkinoja-Salonen. 1987b. Complete dechlorination of tetrachlorohydroquinone by cell extracts of pentachlorophenol-induced *Rhodococcus chlorophenolicus. J. Bacteriol.* 169: 5125–5130.

Archelas, A. and Furstoss. 1992. Synthesis of optically pure pityol—a pheromone of the bark beetle *Pityophthorus pityographus*—using a chemoenzymatic route. *Tetrahedron Lett.* 33: 5241–5242.

Archelas, A. and Furstoss. 1997. Synthesis of enantiopure epoxides through biocatalytic approaches. *Annu. Rev. Microbiol.* 51: 491–525.

Archelas, A., J.-D. Fourneron, R. Furstoss, M. Cesario, and C. Pascard. 1988. Microbial transformations. 8. First example of a highly enentioselective microbiological hydroxylation process. *J. Am. Chem. Soc.* 53: 1797–1799.

Arkowitz, R.A. and R.H. Abeles. 1989. Identification of acetyl phosphate as the product of clostridial glycine reductase: evidence for an acyl enzyme intermediate. *Biochemistry* 28: 4639–4644.

Armengaud, J., B. Happe, and K.N. Timmis. 1998. Genetic analysis of dioxin dioxygenase of *Sphingomonas* sp. strain RW1: catabolic genes dispersed on the genome. *J. Bacteriol.* 180: 3954–3966.

Armstrong, S.M., B.M. Sankey, and G. Voodouw. 1995. Conversion of dibenzothiophene to biphenyl by sulfate-reducing bacteria isolated from oil field production facilities. *Biotechnol. Lett.* 17: 1133–1136.

Asturias, J.A. and K.N. Timmis. 1993. Three different 2,3-dihydroxybiphenyl-1,2-dioxygenase genes in the Gram-positive polychlorobiphenyl-degrading bacterium *Rhodococcus globerulus* P6. *J. Bacteriol.* 175: 4631–4640.

Atlas, E., K. Sullivan, and C.S. Giam. 1986. Widespread occurrence of polyhalogenated aromatic ethers in the marine atmosphere. *Atmos. Environ.* 20: 1217–1220.

Atta-ur-Rahman, M.I. Choudhary, F. Shaheen, M. Ashraf, and S. Jahan. 1998. Microbial transformation of hypolipemic E-guggulsterone. *J. Nat. Prod.* 61: 428–431.

Austen, R.A. and N.W. Dunn. 1980. Regulation of the plasmid-specified naphthalene catabolic pathway of *Pseudomonas putida. J. Gen. Microbiol.* 117: 521–528.

Bache, R. and N. Pfennig. 1981. Selective isolation of *Acetobacterium woodii* on methoxylated aromatic acids and determination of growth yields. *Arch. Microbiol.* 130: 255–261.

Bachofer, R. and F. Lingens. 1975. Conversion of aniline into pyrocatechol by a *Nocardia* sp.: incorporation of oxygen-18. *FEBS Lett.* 50: 288–290.

Baggi, G., P. Barbieri, E. Galli, and S. Tollari. 1987. Isolation of a *Pseudomonas stutzeri* strain that degrades o-xylene. *Appl. Environ. Microbiol.* 53: 2129–2132.

Bak, F. and E. Widdel. 1986. Anaerobic degradation of indolic compounds by sulfate-reducing enrichment cultures, and description of *Desulfobacterium indolicum* gen. nov., sp. nov. *Arch. Microbiol.* 146: 170–176

Bak, F., K. Finster, and F. Rothfub. 1992. Formation of dimethylsulfide and methanthiol from methoxylated aromatic compounds and inorganic sulfide by newly isolated anaerobic bacteria. *Arch. Microbiol.* 157: 529–534.

Ball, H.A., M. Reinhard, and P.L. McCarty. 1989. Biotransformation of halogenated and nonhalogenated octylphenol polyethoxylate residues under aerobic and anaerobic conditions. *Environ. Sci. Technol.* 23: 951–961.

Ballerstedt, H., A. Kraus, and U. Lechner. 1997. Reductive dechlorination of 1,2,3,4-tetrachlorodibenzo-*p*-dioxin and its products by anaerobic mixed cultures from Saale River sediment. *Environ. Sci. Technol.* 31: 1749–1753.

Banks, R.E., B.E. Smart, and J.C. Tatlow (Eds.). 1994. *Organofluorine Chemistry. Principles and Commerial Applications.* Plenum Press, New York.

Barker, H.A. 1961. Fermentations of nitrogenous organic compounds pp.151–207. In *The Bacteria*, Vol. 2 (Eds. I.C. Gunsalus and R.Y. Stanier). Academic Press, New York.

Barker, H.A. 1972. Corrinoid-dependent enzymatic reactions. *Annu. Rev. Biochem.* 41: 55–90.

Barker, H.A. 1981. Amino acid degradation by anaerobic bacteria. *Annu. Rev. Biochem.* 50: 23–40.

Barnsley, E.A. 1976. Role and regulation of the *ortho* and *meta* pathways of catechol metabolism in pseudomonads metabolizing naphthalene and salicylate. *J. Bacteriol.* 125: 404–408.

Barnsley, E.A. 1983. Phthalate pathway of phenanthrene metabolism: formation of 2'-carboxybenzalpyruvate. *J. Bacteriol.* 154: 113–117.

Barriault, D., J. Durand, H. Maaroufi, L.D. Eltis, and M. Sylvestre. 1998. Degradation of polychlorinaterd biphenyl metabolites by naphthalene-catabolizing enzymes. *Appl. Environ. Microbiol.* 64: 4637–4642.

Bartels, I., H.-J. Knackmuss, and W. Reineke. 1984. Suicide inactivation of catechol 2,3-dioxygenase from *Pseudomonas putida* mt-2 by 3-halocatechols. *Appl. Environ. Microbiol.* 47: 500–505.

Bartholomew, B., K.S. Dodgson, G.W.J. Matcham, D.J. Shaw, and G.F. White. 1977. A novel mechanism of enzymatic hydrolysis. Inversion of configuration and carbon-oxygen bond cleavage by secondary alkylsulphohydrolases from detergent-degrading micro-organisms. *Biochem. J.* 165: 575–580.

Bartholomew, B.A., M.J. Smith, P.W. Trudgill, and D.J. Hopper. 1996. Atropine metabolism by *Pseudomonas* sp. strain AT3: evidence for nortropine as an intermediate in atropine breakdown and reactions leading to succinate. *Appl. Environ. Microbiol.* 62: 3245–3250.

Beadle, C.A. and A.R.W. Smith. 1982. The purification and properties of 2,4-dichlorophenol hydroxylase from a strain of *Acinetobacter* sp. *Eur. J. Biochem.* 123: 323–332.

Bedard, D.L. and H. M. van Dort. 1998. Complete reductive dehalogenation of brominated biphenyls by anaerobic microorganisms in sediment. *Appl. Environ. Microbiol.* 64: 940–947.

Bedard, D.L., M.L. Haberl, R.J. May, and M.J. Brennan. 1987a. Evidence for novel mechanisms of polychlorinated biphenyl metabolism in *Alcaligenes eutrophus* H 850. *Appl. Environ. Microbiol.* 53: 1103–1112.

Bedard, D.L., R.E. Wagner, M.J. Brennan, M.L. Haberl, R.J. May, and J.F.Brown, Jr. 1987b. Extensive degradation of Arochlors and environmentally transformed polychlorinated biphenyls by *Alcaligenes eutrophus* H 850. *Appl. Environ. Microbiol.* 53: 1094–1102.

Bedard, D.L., H. van Dort, and K.A. Deweerd. 1998. Brominated biphenyls prime extensive microbial reductive dehalogenation of Arochlor 1260 in Housatonic River sediment. *Appl. Environ. Microbiol.* 64: 1786–1795.

Behrend, C. and K. Heesche-Wagner. 1999. Formation of hydride–Meisenheimer complexes of picric acid (2,4,6-trinitrophenol) and 2,4-dinitrophenol during mineralization of picric acid by *Nocardiodes* sp. strain CB 22-2. *Appl. Environ. Microbiol.* 65: 1372–1377.

Beil, S., B. Happe, K.N. Timmis, and D.H. Pieper. 1997. Genetic and biochemical characterization of the broad spectrum chlorobenene dioxygenase from *Burkholderia* sp. strain PS12. Dechlorination of 1,2,4,5-tetrachlorobenzene. *Eur. J. Biochem.* 247: 190–199.

Belay, N. and L. Daniels. 1987. Production of ethane, ethylene, and acetylene from halogenated hydrocarbons by methanogenic bacteria. *Appl. Environ. Microbiol.* 53: 1604–1610.

Beller, H.R. and A.M. Spormann. 1997a. Anaerobic activation of toluene and *o*-xylene by addition to fumarate in denitrifying strain T. *J. Bacteriol.* 179: 670–676.

Beller, H.R. and A.M. Spormann. 1997b. Benzylsucccinate formation as a means of anaerobic toluene activation by sulfate-reducing strain PRTOL1. *Appl. Environ. Microbiol.* 63: 3729–3731.

Beller, H.R. and A.M. Spormann. 1998. Analysis of the novel benzylsuccinate synthase reaction for anaerobic toluene activation based on structural studies of the product. *J. Bacteriol.* 180: 5454–5457.

Beller, H.R., M. Reinhard, and D. Grbic-Galic. 1992. Metabolic by-products of anaerobic toluene degradation by sulfate-reducing enrichment cultures. *Appl. Environ. Microbiol.* 58: 3192–3195.

Beller, H.R., W.-H. Ding, and M. Reinhard. 1995. By-products of anaerobic alkylbenzene metabolism useful as indicators of *in situ* bioremediation. *Environ. Sci. Technol.* 29: 2864–2870.

Beller, H.R., A.M. Spormann, P.K. Sharma, J.R. Cole, and M. Reinhard. 1996. Isolation and characterization of a novel toluene-degrading sulfate-reducing bacterium. *Appl. Environ. Microbiol.* 62: 1188–1196

Best, D.J., N.C. Floyd, A. Magalhaes, A. Burfield, and P.M. Rhodes. 1987. Initial steps in the degradation of *alpha*-pinene by *Pseudomonas fluorescens* NCIMB 11671. *Biocatalysis:* 147–159.

Beunink, J. and H.-J. Rehm. 1990. Coupled reductive and oxidative degradation of 4-chloro-2-nitrophenol by a co-immobilized mixed culture System. *Appl. Microbiol. Biotechnol.* 34: 108–115.

Beurskens, J.E.M., M. Toussaint, J. de Wolf, J.M.D. van der Steen, P.C. Slot, L.C.M. Commandeur, and J.R. Parsons. 1995. Dehalogenation of chlorinated dioxins by an anaerobic consortium from sediment. *Environ. Toxicol. Chem.* 14: 939–943.

Beuscher, H.U. and J.R. Andreesen. 1984. *Eubacterium angustum* sp. nov., a Grampositive anaerobic, non-sporeforming, obligate purine fermenting organism. *Arch. Microbiol.* 140: 2–8.

Bezalel, L., Y. Hadar, and C.E. Cerniglia. 1997. Enzymatic mechanisms involved in phenanthrene degradation by the white-rot fungus *Pleurotus ostreatus*. *Appl. Environ. Microbiol.* 63: 2495–2501.

Biegert, T. and G. Fuchs. 1995. Anaerobic oxidation of toluene (analogues) to benzoate (analogues) by whole cells and by cell extracts of a denitrifying *Thauera* sp. *Arch. Microbiol.* 163: 407–417.

Biegert, T., U. Altenschmidt, C. Eckerskorn, and G. Fuchs. 1995. Purification and properties of benzyl alcohol dehydrogenase from a denitrifying *Thauera* sp. *Arch. Microbiol.* 163: 418–423.

Biegert, T., G. Fuchs, and J. Heider. 1996. Evidence that anaerobic oxidation of toluene in the denitrifying bacterium *Thauera aromatica* is initiated by formation of benzylsuccinate from toluene and fumarate. *Eur. J. Biochem.* 238: 661–668.

Binks, P.R., C.E. French, S. Nicklin, and N.C. Bruce. 1996. Degradation of pentaerythritol tetranitrate by *Enterobacter cloacae* PB2. *Appl. Environ. Microbiol.* 62: 1214–1219.

Bird, T.G., P.M. Fredricks, E.R.H. Jones, and G.D. Meakins. 1980. Microbiological hydroxylations. Part 23. Hydroxylations of fluoro-5α-androstanones by the fungi *Calonectria decora*, *Rhizopus nigricans*, and *Aspergillus ohraceus*. *J. Chem. Soc. Perkin* I: 750–755.

Bisaillon, J.-G., F. Lépine, R. Beaudet, and M. Sylvestre. 1991. Carboxylation of o-cresol by an anaerobic consortium under methanogenic conditions. *Appl. Environ. Microbiol.* 57: 2131–2134.

Blakley, E.R. 1974. The microbial degradation of cyclohexanecarboxylic acid: a pathway involving aromatization to form p-hydroxybenzoic acid. *Can. J. Microbiol.* 20: 1297–1306.

Blanck, H., K. Holmgren, L. Landner, H. Norin, M. Notini, A. Rosmarin, and B. Sundelin. 1989. Advanced hazard assessment of arsenic in the Swedish environment pp. 256–328. In *Chemicals in the Aquatic Environment. Advanced Hazard Assessment* (Ed. L. Landner). Springer-Verlag, Berlin.

Blasco, R., R.-M. Wittich, M. Mallavarapu, K.N. Timmis, and D.H. Pieper. 1995. From xenobiotic to antibiotic, formation of protoanemonin from 4-chlorocatechol by enzymes of the 3-oxoadipate pathway. *J. Biol. Chem.* 270: 29229–29235.

Blasco, R., E. Moore, V. Wray, D. Pieper, K. Timmis, and F. Castillo. 1999. 3-Nitroadipoate, a metabolic intermediate for mineralization of 2,4-dinitrophenol by a new strain of a *Rhodococcus* species. *J. Bacteriol.* 181: 149–152.

Blasco, R., M. Mallavarapu, R.-M. Wittich, K.N. Timmis, and D.H. Pieper. 1997. Evidence that formation of protoanemonin from metabolites of 4-chlorobiphenyl degradation negatively affects the survival of 4-chlorobiphenyl-cometabolizing microorganisms. *Appl. Environ. Microbiol.* 63: 427–434.

Blehert, D.S., K.L. Knoke, B.G. Fox, and G.H. Chambliss. 1997. Regioselectivity of nitroglycerine denitration by flavoprotein nitroester reductases purified from two *Pseudomonas* species. *J. Bacteriol.* 179: 6912–6920.

Block, D.W. and F. Lingens. 1992. Microbial metabolism of quinoline and related compounds XIV. Purification and properties of 1*H*-3-hydroxy-4-oxoquinoline oxygenase, a new extradiol cleavage enzyme from *Pseudomonas putida* strain 33/1. *Biol. Chem. Hoppe-Seyler* 373: 343–349.

Blümel, S., M. Contzen, M. Lutz, A. Stolz, and H.-J. Knackmnuss. 1998. Isolation of a bacterial strain with the ability to utilize the sulfonated azo compound 4-carboxy-4′-sulfoazobenzene as the sole source of carbon and energy. *Appl. Environ. Microbiol.* 64: 2315–2317.

Boethling, R.S., B. Gregg, R. Frederick, N.W. Gabel, S.E. Campbell, and A. Sabljic. 1989. Expert systems survey on biodegradation of xenobiotic chemicals. *Ecotoxicol. Environ. Saf.* 18: 252–267.

Bogan, B.W. and R.T. Lamar. 1996. Polycyclic aromatic hydrocarbon-degrading capabilities of *Phanerochaete laevis* HHB-1625 and its extracellular lignolytic enzymes. *Appl. Environ. Microbiol.* 62: 1597–1603.

Bokkenheuser, V.D., J. Winter, S. O'Rourke, and A.E. Ritchie. 1980. Isolation and characterization of fecal bacteria capable of 16α-dehydroxylating corticoids. *Appl. Environ. Microbiol.* 40: 803–808.

Boldrin, B., A. Tiehm, and C. Fritzsche. 1993. Degradation of phenanthrene, fluorene, fluoranthene, and pyrene by a *Mycobacterium* sp. *Appl. Environ. Microbiol.* 59: 1927–1930

Bollag, J.-M., E.J. Czaplicki, and R.D. Minard. 1975. Bacterial metabolism of 1-naphthol. *J. Agric. Food Chem.* 23: 85–90.

Bonting, C.F.C., S. Schneider, G. Schmidtberg, and G. Fuchs. 1995. Anaerobic degradation of *m*-cresol via methyl oxidation to 3-hydroxybenzoate by a denitrifying bacterium. *Arch. Microbiol.* 164: 63–69.

Bopp, L.H. 1986. Degradation of highly chlorinated PCBs by *Pseudomonas* strain LB400. *J. Indust. Microbiol.* 1: 23–29.

Bouquard, C., J. Ouazzani, J.-C. Promé, Y. Michel-Briand, and P. Plésiat. 1997. Dechlorination of atrazine by a *Rhizobium* sp. isolate. *Appl. Environ. Microbiol.* 63: 762–866.

Boyle, A.W., C.D. Phelps, and L.Y. Young. 1999. Isolation from estuarine sediments of a *Desulfovibrio* strain which can grow on lactate coupled to the reductive dehalogenation of 2,4,6-tribromophenol. *Appl. Environ. Microbiol.* 65: 1133–1140.

Brackmann, R. and G. Fuchs. 1993. Enzymes of anaerobic metabolism of phenolic compounds. 4-Hydroxybenzoyl-CoA reductase (dehydroxylating) from a denitrifying *Pseudomonas* sp. *Eur. J. Biochem.* 213: 563–571.

Branchaud, B.P. and C.T. Walsh. 1985. Functional group diversity in enzymatic oxygenation reactions catalyzed by bacterial flavin-containing cyclohexanone oxygenase. *J. Am. Chem. Soc.* 107: 2153–2161.

Braun, M. and H. Stolp. 1985. Degradation of methanol by a sulfate reducing bacterium. *Arch. Microbiol.* 142: 77–80.

Breese, K., M. Boll, J. Alt-Mörbe, H. Schäggrer, and G. Fuchs. 1998. Genes encoding the benzoyl-CoA pathways of anaerobic aromatic metabolism in the bacterium *Thauera aromatica. Eur. J. Biochem.* 256: 148–154.

Bregnard, T.P.-A., P. Höhener, A. Häner, and J. Zeyer. 1997. Degradation of weathered diesel fuel by microorganisms from a contaminated aquifer in aerobic and anaerobic conditions. *Environ. Toxicol. Chem.* 15: 299–307.

Britt, A.J., N.C. Bruce, and C.R Lowe. 1992. Identification of a cocaine esterase in a strain of *Pseudomonas maltophilia. J. Bacteriol.* 174: 2087–2094.

Britton, L.N. 1984. Microbial degradation of aliphatic hydrocarbons pp. 89–130. In *Microbial Degradation of Organic Compounds* (Ed. D.T. Gibson). Marcel Dekker, New York.

Broadway, N.M., F.M. Dickinson, and C. Ratledge. 1993. The enzymology of dicarboxylic acid formation by *Corynebacterium* sp. strain 7E1C grown on *n*-alkanes. *J. Gen. Microbiol.* 139: 1337–1344.

Brune, A. and B. Schink. 1990. Pyrogallol-to-phloroglucinal conversion and other hydroxyl-transfer reactions catalyzed by cell extracts of *Pelobacter acidigallici. J. Bacteriol.* 172: 1070–1076.

Brune, A. and B. Schink. 1992. Phloroglucinol pathway in the strictly anaerobic *Pelobacter acidigallici* fermentation of trihydroxybenzenes to acetate via triacetic acid. *Arch. Microbiol.* 157: 417–424.

Brune, A., S. Schnell, and B. Schink. 1992. Sequential transhydroxylations converting hydroxyhydroquinone to phloroglucinol in the strictly anaerobic, fermentative bcterium. *Pelobacter massiliensis. Appl. Environ. Microbiol.* 58: 1861–1868.

Buckel, W. and B.T. Golding. 1999. Radical species in the catalytic pathways of enzymes from anaerobes. *FEMS Microbiol. Rev.* 22: 523–541.

Bumpus, J.A. 1989. Biodegradation of polycyclic aromatic hydrocarbons by *Phanerochaete chryspsporium. Appl. Environ. Microbiol.* 55: 154–158.

Bumpus, J.A. and S.A. Aust. 1987. Biodegradation of DDT [1,1,1-trichloro-2,3-bis(4-chlorophenyl)ethane] by the white rot fungus *Phanerochaete chrysosporium. Appl. Environ. Microbiol.* 53: 2002–2008.

Bünz, P.V. and A.M. Cook. 1993. Dibenzofuran 4,4a-dioxygenase from *Sphingomonas* sp. strain RW1: angular dioxygenation by a three-component system. *J. Bacteriol.* 175: 6467–6475.

Burland, S.M. and E.A. Edwards. 1999. Anaerobic benzene biodegradation linked to nitrate reduction. *Appl. Environ. Microbiol.* 65: 529–533.

Buser, H.-R., C. Zanier, and H. Tanner. 1982. Identification of 2,4,6-trichloroanisole as a potent compound causing cork taint in wine. *J. Agric. Food Chem.* 30: 359–362.

Cain, R.B. 1981. Microbial degradation of surfactants and "builder" components pp. 325–370. In *Microbial Degradation of Xenobiotics and Recalcitrant Compounds* (Eds. T. Leisinger, A.M. Cook, R. Hütter, and J. Nüesch. Academic Press, London.

Cain, R.B. 1988. Aromatic metabolism by mycelial organisms: actinomycete and fungal strategies pp. 101–144. In *Microbial Metabolism and the Carbon Cycle* (Eds. S.R. Hagedorn, R.S. Hanson, and D.A. Kunz). Harwood Academic Publishers, Chur, Switzerland.

Cain, R.B. and D.R. Farr. 1968. Metabolism of arylsulphonates by micro-organisms. *Biochem. J.* 106: 859–877.

Cain, R.B., R.F. Bilton, and J.A. Darrah. 1968. The metabolism of aromatic acids by micro-organisms. Metabolic pathways in the fungi. *Biochem. J.* 108: 797–828.

Caldwell, M.E., R.M. Garrett, R.C. Prince, and J.M. Suflita. 1998. Anaerobic biodegradation of long-chain *n*-alkanes under sulfate-reducing conditions. *Environ. Sci. Technol.* 32: 2191–2195.

Campbell, L.L. 1960. Reductive degradation of pyrimidines. V. Enzymatic conversion of N-carbamyl-beta-alanine to beta-alanine, carbon dioxide, and ammonia. *J. Biol. Chem.* 235: 2375–2378.

Carey, J.H. and M.E. Cox. 1981. Photodegradation of the lampricide 3-trifluoromethyl-4-nitrophenol (TFM) 1. Pathway of the direct photolysis in solution. *J. Great Lakes Res.* 7: 234–241.

Carless, H.A.J. 1992. The use of cyclohexa-3,5-diene-1,2-diols in enantiospecific synthesis. *Tetrahedron Asymmetry* 3: 795–826

Carless, H.A.J., J.R. Billinge, and O.Z. Oak. 1989. Photochemical routes from arenes to inositol intermediates: the photo-oxidation of substituted *cis*-cyclohexane,3,5-diene-1,2-diols. *Tetrahedron Lett.* 30: 3113–3116.

Carpenter, J.C., J.A. Cella, and S.B. Dorn, 1995. Study of the degradation of polydimethylsiloxanes on soil. *Environ. Sci. Technol.* 29: 864–868.

Carter, S.F. and D.J. Leak. 1995. The isolation and characterisation of a carbocyclic epoxide-degrading *Corynebacterium* sp. *Biocatal. Biotrans.* 13: 111–129.

Cartwright, N.J. and R.B. Cain. 1959. Bacterial degradation of the nitrobenzoic acids 2. Reduction of the nitro group. *Biochem J.* 73: 305–314.

Cass, A.E.G., D.W. Ribbons, J.T. Rossiter, and S.R. Williams. 1987. Biotransformation of aromatic compounds. Monitoring fluorinated analogues by NMR. *FEBS Lett.* 220: 353–347.

Castro, C.E. and N.O. Belser. 1990. Biodehalogenation: oxidative and reductive metabolism of 1,1,2-trichloroethane by *Pseudomonas putida*—biogeneration of vinyl chloride. *Environ. Toxicol. Chem.* 9: 707–714.

Castro, C.E., D.M. Riebeth, and N.O. Belser. 1992a. Biodehalogenation: the metabolism of vinyl chloride by *Methylosinus trichosporium* OB-3b. A sequential oxidative and reductive pathway through chloroethylene oxide. *Environ. Toxicol. Chem.* 11: 749–755.

Castro, C.E., R.S. Wade, D.M. Riebeth, E.W. Bartnicki, and N.O. Belser. 1992b. Biode-halogenation: rapid metabolism of vinyl chloride by a soil *Pseudmonas* sp. Direct hydrolysis of a vinyl C–Cl bond. *Environ. Toxicol. Chem.* 11: 757–764

Cavalieri, E. and E. Rogan. 1998. Mechanisms of tumor initiation by polycycli aromatic hydrocarbons in mammals pp. 81–117. In *PAHs and Related Compounds.* Vol. 3.J (Ed. A.H. Neilson), Springer-Verlag, Berlin.

Cerniglia, C.E. and S.A. Crow. 1981. Metabolism of aromatic hydrocarbons by yeasts. *Arch. Microbiol.* 129: 9–13.

Cerniglia, C.E. and J.J. Perry. 1975. Metabolism of *n*-propylamine, isopropylamine, and 1,3-propane diamine by *Mycobacterium convolutum. J. Bacteriol.* 124: 285–289.

Cerniglia, C.E. and S.K. Yang. 1984. Stereoselective metabolism of anthracene and phenanthrene by the fungus *Cunninghamella elegans. Appl. Environ. Microbiol.* 47: 119–124.

Cerniglia, C.E., J.P. Freeman, and R.K. Mitchum. 1982a. Glucuronide and sulfate conjugation in the fungal metabolism of aromatic hydrocarbons. *Appl. Environ. Microbiol.* 43: 1070–1075.

Cerniglia, C.E., P.P. Fu, and S.K. Yang. 1982b. Metabolism of 7-methylbenz[*a*]anthracene and 7-hydroxymethylbenz[*a*]anthracene by *Cunninghamella elegans. Appl. Environ. Microbiol.* 44: 682–689.

Cerniglia, C.E., J.R. Althaus, F.E. Evans, J.P. Freeman, R.K. Mitchum, and S.K. Yang. 1983. Stereochemistry and evidence for an arene oxide-NIH shift pathway in the fungal metabolism of naphthalene. *Chem. Biol. Interactions* 44: 119–132.

Cerniglia, C.E., J.P. Freeman, and F.E. Evans. 1984. Evidence for an arene oxide-NIH shift pathway in the transformation of naphthalene to 1-naphthol by *Bacillus cereus. Arch. Microbiol.* 138: 283–286.

Cerniglia, C.E., D.W. Kelly, J.P. Freman, and D.W. Miller. 1986. Microbial metabolism of pyrene. *Chem. Biol. Interactions* 57: 203–216.

Cerniglia, C.E., W.L. Campbell, J.P Freeman, and F.E. Evans 1989. Identification of a novel metabolite in phenanthrene metabolism by the fungus *Cunninghamella elegans. Appl. Environ. Microbiol.* 55: 2275–2279.

Chakrabarty, A.M. 1972. Genetic basis of the biodegradation of salicylate in *Pseudomonas. J. Bacteriol.* 112: 815–823.

Chanana, S. and R.L. Crawford. 1997. Mutational analysis of *pcpA* and its role in pentachlorophenol degradation by *Sphingomonas (Flavobacterium) chlorophenolica* ATCC39723. *Appl. Environ. Microbiol.* 63: 4833–4838.

Chang, K.H., P.-H. Liang, W. Beck, J.D. Scholten, and D. Dunaway-Mariano. 1992. Isolation and characterization of the three polypeptide components of 4-chlorobenzoate dehalogenase from *Pseudomonas* sp. strain CBS-3. *Biochemistry* 31: 5605–5610.

Chapman, P.J., G. Meerman, J.C. Gunsalus, R. Srinivasan, and K.L. Rinehart. 1966. A new acyclic metabolite in camphor oxidation. *J. Am. Chem. Soc.* 88: 618–619.

Charlson, R.J., J.E. Lovelock, M.O. Andreae, and S.G. Warren. 1987. Oceanic phytoplankton, atmospheric sulphur, cloud albedo and climate. *Nature* 326: 655–661.

Chen, C.-M. and P.H. Tomasek. 1991. 3,4-Dihydroxyxanthone dioxygenase from *Arthrobacter* sp. strain GFB 100. *Appl. Environ. Microbiol.* 57: 2217–2222.

Chen, C.-M., Q.-Z. Zhuang, Z. Zhu, B.L. Wanner, and C.T. Walsh. 1990. Molecular biology of carbon–phosphorus bond cleavage. Cloning and sequencing of the *phn* (*psiD*) genes involved in alkylphosphonate uptake and C-P lyase activity in *Escherichia coli. J. Biol. Chem.* 265: 4461–4471.

Chen, Y. and J.P.N. Rosazza. 1995. Purification and characterization of nitric oxide synthase NOS$_{Noc}$ from a *Nocardia* sp. *J. Bacteriol.* 177: 5122–5128.

Cheng, K.-J., H.G. Krishnamurty, G.A. Jones, and F.J. Simpson. 1971. Identification of products produced by the anaerobic degradation of naringin by *Butyrivibrio* sp. C$_3$. *Can. J. Microbiol.* 17: 129–131.

Coates, J.D., R.T. Anderson, and D.R. Lovley. 1996. Oxidation of polycyclic aromatic hydrocarbons under sulfate-reducing conditions. *Appl. Environ. Microbiol.* 62: 1099–1101.

Coates, J.D., J. Woodward, J. Allen, P. Philip, and D.R. Lovley. 1997. Anaerobic degradation of polycyclic aromatic hydrocarbons and alkanes in petroleum-contaminated marine harbor sediments. *Appl. Environ. Microbiol.* 63: 3589–3593.

Cole, J.R., A.L. Cascarelli, W.W. Mohn, and J.M. Tiedje. 1994. Isolation and characterization of a novel bacterium growing via reductive dehalogenation of 2-chlorophenol. *Appl. Environ. Microbiol.* 60: 3536–3542.

Cook, A.M. 1987. Biodegradation of s-triazine xenobiotics. *FEMS Microbiol. Rev.* 46: 93–116.

Cook, A.M., C.G. Daughton, and M. Alexander. 1978a. Phosphonate utilization by bacteria. *J. Bacteriol.* 133: 85–90.

Cook, A.M., C.G. Daughton, and M. Alexander. 1978b. Phosphorus-containing pesticide breakdown products: quantitative utilization as phosphorus sources by bacteria. *Appl. Environ. Microbiol.* 36: 668–672.

Cook, A.M., C.G. Daughton, and M. Alexander. 1979. Benzene from bacterial cleavage of the carbon-phosphorus bond of phenylphosphonates. *Biochem. J.* 184: 453–455.

Corbett, M.D. and B.R. Corbett. 1981. Metabolism of 4-chloronitrobenzene by the yeast *Rhodosporidium* sp. *Appl. Environ. Microbiol.* 41: 942–949.

Corke, C.T., N.J. Bunce, A.-L. Beaumont, and R.L. Merrick. 1979. Diazonium cations as intermediates in the microbial transformations of chloroanilines to chlorinated biphenyls, azo compounds and triazenes. *J. Agric. Food Chem.* 27: 644–646.

Coschigano, P.W., T.S. Wehrman, and L.Y. Young. 1998. Identification and analysis of genes involved in anaerobic toluene metabolism by strain T1: putative role of a glycine free radical. *Appl. Environ. Microbiol.* 64: 1650–1656.

Coulter, C., J.T.G. Hamilton, and D.B. Harper. 1993. Evidence for the existence of independent chloromethane- and S-adenosylmethionine-utilizing systems for methylation in *Phanerochaete chrysosporium*. *Appl. Environ. Microbiol.* 59: 1461–1466.

Crawford, R.L. 1975a. Degradation of 3-hydroxybenzoate by bacteria of the genus *Bacillus*. *Appl. Microbiol.* 30: 439–444.

Crawford, R.L. 1975b. Novel pathways for degradation of protocatechuic acid in *Bacillus* species. *J. Bacteriol.* 121: 531–536.

Crawford, R.L. 1976. Pathways of 4-hydroxybenzoate degradation among species of *Bacillus*. *J. Bacteriol.* 127: 204–290.

Crawford, R.L. and P.P. Olson. 1978. Microbial catabolism of vanillate: decarboxylation to guaiacol. *Appl. Environ. Microbiol.* 36: 539–543.

Crescenzi, A.M.V., K.S. Dodgson, G.F. White, and W.J. Payne. 1985. Initial oxidation and subsequent desulphation of propan-2-yl sulphate by *Pseudomonas syringae* strain GG. *J. Gen. Microbiol.* 131: 469–477.

Cripps, R.E. 1973. The microbial metabolism of thiophen-2-carboxylate. *Biochem. J.* 134: 353–366.

Cripps, R.E. and A.S. Noble. 1973. The metabolism of nitrilotriacetate by a pseudomonad. *Biochem. J.* 136: 1059–1068.

Cripps, R.E. and R.J. Watkinson. 1978. Polycyclic hydrocarbons: metabolism and environmental aspects pp. 113–134. In *Developments in Biodegradation of Hydrocarbons*—1 (Ed. R.J. Watkinson). Applied Science Publishers Ltd., London.

Cripps, R.E., P.W. Trudgill, and J.G. Whateley. 1978. The metabolism of 1-phenylethanol and acetophenone by *Nocardia* T5 and an *Arthrobacter* species. *Eur. J. Biochem.* 86: 175–186.

Cullen, W.R., B.C. McBride, and J. Reglinski. 1984. The reduction of trimethylarsine oxide to trimethylarsine by thiols: a mechanistic model for the biological reduction of arsenicals. *J. Inorg. Biochem.* 21: 45–60.

Curragh, H., O. Flynn, M.J. Larkin, T.M. Stafford, J.T.G. Hamilton, and D.B. Harper. 1994. Haloalkane degradation and assimilation by *Rhodococcus rhodochrous* NCIMB 13064. *Microbiology* (U.K.) 140: 1433–1442.

D'Ari, L. and W.A. Barker. 1985. *p*-Cresol formation by cell-free extracts of *Clostridium difficile*. *Arch. Microbiol.* 143: 311–312.

Dalton, H., B.T. Golding, B.W. Waters, R. Higgins, and J.A. Taylor. 1981. Oxidation of cyclopropane, methylcyclopropane, and arenes with the mono-oxygenase systems from *Methylococcus capsulatus*. *J. Chem. Soc. Chem. Commun.* 482–483.

Daly, J. W., D.M. Jerina, and B. Witkop. 1972. Arene oxides and the NIH shift: the metabolism, toxicity and carcinogenicity of aromatic compounds. *Experientia* 28: 1129–1149.

Dams, R. 1993. Fluorochemical surfactants for hostile environments. *Speciality Chemicals* 13: 4–6.

Dangel, W., R. Brackmann, A. Lack, M. Mohamed, J. Koch, B. Oswald, B. Seyfried, A. Tschech, and G. Fuchs. 1991. Differential expression of enzyme activities initiating anoxic metabolism of various aromatic compounds via benzoyl-CoA. *Arch. Microbiol.* 155: 256–262.

Darby, J.M., D.G. Taylor, and D.J. Hopper. 1987. Hydroquinone as the ring-fission substrate in the catabolism of 4-ethylphenol and 4-hydroxyacetophenone by *Pseudomonas putida* D1. *J. Gen. Microbiol.* 133: 2137–2146.

Daubaras, D.L., C.D. Hershberger, K. Kitano, and A.M. Chakrabarty. 1995. Sequence analysis of a gene cluster involved in metabolism of 2,4,5-trichlorophenoxyacetic acid by *Burkholderia cepacia* AC1100. *Appl. Environ. Microbiol.* 61: 1279–1289.

Daubaras, D.L., K. Saido, and A.M. Chakrabarty. 1996. Purification of hydroxyquinol 1,2-dioxygenase and maleylacetate reductase: the lower pathway of 2,4,5-trichlorophenoxyacetic acid metabolism by *Burkholderia cepacia* AC1100. *Appl. Environ. Microbiol.* 62: 4276–4279.

Davey, J.F. and D.T. Gibson. 1974. Bacterial metabolism of para- and meta-xylene: oxidation of a methyl substituent. *J. Bacteriol.* 119: 923–929.

Davies, I., G.F. White, and W.J. Payne. 1990. Oxygen-dependent desulphation of monomethyl sulphate by *Agrobacterium* sp. M3C. *Biodegradation* 1: 229–241

de Beyer, A., and F. Lingens. 1993. Microbial metabolism of quinoline and related compounds XVI. Quinaldine oxidoreductase from *Arthrobacter* spec. Rü 61a: a molybdenum-containing enzyme catalysing hydroxylation at C-4 of the heterocycle. *Biol. Chem. Hoppe-Seyler* 374: 101–110.

de Bont, J.A.M. and M.W. Peck. 1980. Metabolism of acetylene by *Rhodococcus* A1. *Arch. Microbiol.* 127: 99–104.

de Bont, J.A.M., J.P. van Dijken, and C.G. van Ginkel. 1982. The metabolism of 1,2-propanediol by the propylene oxide utilizing bacterium *Nocardia* A60. *Biochim. Biophys. Acta* 714: 465–470.

de Bruin, W.P., M.J.J. Kotterman, M.A. Posthumus, G. Schraa, and A.J.B. Zehnder. 1992. Complete biological reductive transformation of tetrachloroethene to ethane. *Appl. Environ. Microbiol.* 58: 1996–2000.

de Haan, A., M.R. Smith, W.G.B. Voorhorst, and J.A.M. de Bont. 1993. Co-factor regeneration in the production of 1,2-epoxypropane by *Mycobacterium* strain E3: the role of storage material. *J. Gen. Microbiol.* 139: 3017–3022.

De Marco, P., P. Moradas-Ferreira, T.P. Higgins, I. McDonald, E.M. Kenna, and J.C. Murrell. 1999. Molecular analysis of a novel methanesulfonic acid monooxygenase from the methylotroph *Methylosulfonomonas methylovora. J. Bacteriol.* 181: 2244–2251.

de Souza, M.P. and D.C. Yoch. 1995. Purification and characterization of dimethyl-sulfoniopropionate lyase from an *Alcaligenes*-like dimethyl sulfide-producing marine isolate. *Appl. Environ. Microbiol.* 61: 21–26.

de Souza, M.L., L.P. Wackett, K.L. Boundy-Mills, R.T. Mandelbaum, and M.J. Sadowsky. 1995. Cloning, characterization, and expression of a gene region from *Pseudomonas* sp. strain ADP involved in the dechlorination of atrazine. *Appl. Environ. Microbiol.* 61: 3373–3378.

Dehning, I. and B. Schink. 1989a. Two new species of anaerobic oxalate-fermenting bacteria, *Oxalobacter vibrioformis* sp. nov. and *Clostridium oxalicum* sp. nov. from sediment samples. *Arch. Microbiol.* 153: 79–84.

Dehning, I. and B. Schink. 1989b. *Malonomonas rubra* gen. nov. sp. nov., a microaerotolerant anaerobic bacterium growing by decarboxylation of malonate. *Arch. Microbiol.* 151: 427–433.

Dehning, I., M. Stieb, and B. Schink.1989. *Sporomusa malonica* sp. nov., a homoacetogenic bacterium growing by decarboxylation of malonate or succinate. *Arch. Microbiol.* 151: 421–426.

DeMoll, E. and T. Auffenberg. 1993. Purine metabolism in *Methanococcus vannielii. J. Bacteriol.* 175: 5754–5761.

Denger, K., M.A. Kertesz, E.M. Vock, R. Schön, A. Mägli, and A.M. Cook. 1996. Anaerobic desulfonation of 4-tolylsulfonate and 2-4-sulfophenylbutyrate by a *Clostridium* sp. *Appl. Environ. Microbiol.* 62: 1526–1530.

Dennie, D., Gladu, D.D.I., F. Lépine, R. Villemur, J.-G. Bisaillon, and R. Beaudet. 1998. Spectrum of the reductive dehalogenation activity of *Desulfitobacterium frappieri* PCP-1. *Appl. Environ. Microbiol.* 64: 4603–4606.

DeWeerd, K.A., L. Mandelco, R.S. Tanner, C.R. Woese, and J.M. Suflita. 1990. *Desulfomonile tiedjei* gen. nov. and sp. nov., a novel anaerobic, dehalogenating, sulfate-reducing bacterium. *Arch. Microbiol.* 154: 23–30.

DeWeerd, K.A., F. Concannon, and J.M. Suflita. 1991. Relationship between hydrogen consumption, dehalogenation, and the reduction of sulfur oxyanions by *Desulfomonile tiedjei. Appl. Environ. Microbiol.* 57: 1929–1934.

Dickel, O.D. and H.-J. Knackmuss. 1991. Catabolism of 1,3-dinitrobenzene by *Rhodococcus* sp. QT-1. *Arch. Microbiol.* 157: 76–79.

Dimroth, P. and H. Hilbi. 1997. Enzymatic and genetic basis for bacterial growth on malonate. *Mol. Microbiol.* 25: 3–10.

DiStefano, T.D., J.M. Gossett, and S.H. Zinder. 1991. Reductive dechlorination of high concentrations of tetrachloroethene to ethene by an anaerobic enrichment culture in the absence of methanogenesis. *Appl. Environ. Microbiol.* 57: 2287–2292.

Dodson, K.S. and G.F. White. 1983. Some microbial enzymes involved in the biodegradation of sulfated surfactants pp. 90–155. In *Topics in Enzyme and Fermentation Technology*, Vol. 7 (Ed. A. Wiseman). Ellis-Horwood, Chichester.

Dolfing, J. 1990. Reductive dechlorination of 3-chlorobenzoate is coupled to ATP production and growth in an anaerobic bacterium, strain DCB-1. *Arch. Microbiol.* 153: 264–266.

Donnelly, P.K., R.S. Hegde, and J.S. Fletcher. 1994. Growth of PCB-degrading bacteria on compounds from photosynthetic plants. *Chemosphere* 28: 981–988.

Drotar, A.-M. and R. Fall. 1985. Methylation of xenobiotic thiols by *Euglena gracilis*: characterization of a cytoplasmic thiol methyltransferase. *Plant Cell Physiol.* 26: 847–854.

Drotar, A.-M. and R. Fall. 1986. Characterization of a xenobiotic thiol methyltransferase and its role in detoxication in *Tetrahymena thermophila*. *Pestic. Biochem. Physiol.* 25: 396–406.

Drotar, A.-M., L.R. Fall, E.A. Mishalanie, J.E. Tavernier, and R. Fall. 1987a. Enzymatic methylation of sulfide, selenide, and organic thiols by *Tetrahymena thermophila*. *Appl. Environ. Microbiol.* 53: 2111–2118.

Drotar, A.-M., G.A. Burton, J.E. Tavernier, and R. Fall. 1987b. Widespread occurrence of bacterial thiol methyltransferases and the biogenic emission of methylated sulfur gases. *Appl. Environ. Microbiol.* 53: 1626–1631.

Drzyzga, O., A. Schmidt, and K.-H. Blotevogel. 1996. Cometabolic transformation and cleavage of nitrodiphenylamines by three newly isolated sulfate-reducing bacterial strains. *Appl. Environ. Microbiol.* 62: 1710–1716.

Duque, E., A. Haidour, F. Godoy, and J.L. Ramos. 1993. Construction of a *Pseudomonas* hybrid strain that mineralizes 2,4,6-trinitrotoluene. *J. Bacteriol.* 175: 2278–2283.

Duran, R., E. Corrales, R. Hernández-Galán, and I.G. Collado. 1999. Biotransformation of caryophyllene oxide by *Botrytis cinerea*. *J. Nat. Prod.* 62: 41–33.

Durham, D.R. and D.B. Stewart. 1987. Recruitment of naphthalene dissimilatory enzymes for the oxidation of 1,4-dichloronapththalene to 3,6-dichlorosalicylate, a precursor for the herbicide dicamba. *J. Bacteriol.* 169: 2889–2892.

Dürre, P. and J.R. Andreesen. 1982. Anaerobic degradation of uric acid via pyrimidine derivatives by selenium-starved cells of *Clostridium purinolyticum*. *Arch. Microbiol.* 131: 255–260.

Dürre, P. and J.R. Andreesen. 1983. Purine and glycine metabolism by purinolytic clostridia. *J. Bacteriol.* 154: 192–199.

Dürre, P., W. Andersch, and J.R. Andreesen. 1981. Isolation and characterization of an adenine-utlizing, anaerobic sporeformer, *Clostridium purinolyticum* sp. nov. *Int. J. Syst. Bacteriol.* 31: 184–194.

Dutta, T.K., S.A. Selifonov, and I.C. Gunsalus. 1998. Oxidation of methyl-substituted naphthalenes: pathways in a versatile *Sphingomomas paucimobilis* strain. *Appl. Environ. Microbiol.* 64: 1884–1889.

Dwyer, D.F. and J.M. Tiedje. 1986. Metabolism of polyethylene glycol by two anaerobic bacteria, *Desulfovibrio desulfuricans* and a *Bacteroides* sp. *Appl. Environ. Microbiol.* 52: 852–856.

Eaton, R.W. and P.J. Chapman. 1992. Bacterial metabolism of naphthalene: construction and use of recombinant bacteria to study ring cleavage of 1,2-dihydroxynaphthalene and subsequent reactions. *J. Bacteriol.* 174: 7542–7554.

Eaton, R.W. and D.W. Ribbons. 1982. Metabolism of dibutylphthalate and phthalate by *Micrococcus* sp. strain 12B. *J. Bacteriol.* 151: 48–57.

Eaton, R.W. and S.A. Selifonov. 1996. Biotransfomation of 6,6-dimethylfulvene by *Pseudomonas putida* RE 213. *Appl. Environ. Microbiol.* 62: 756–760.

Eaton, R.W. and K.N. Timmis. 1986. Characterization of a plasmid-specified pathway for catabolism of isopropylbenzene in *Pseudomonas putida* RE204. *J. Bacteriol.* 168: 123–131.

Eaton, S.L., S.M. Resnick, and D.T. Gibson. 1996. Initial reactions in the oxidation of 1,2-dihydronaphthalene by *Sphingomonas yanoikuyae* strains. *Appl. Environ. Microbiol.* 62: 4388–4394.

Edwards, E.A., L.E. Wills, M. Reinhard, and D. Grbic-Galic. 1992. Anaerobic degradation of toluene and xylene by aquifer microorganisms under sulfate-reducing conditions. *Appl. Environ. Microbiol.* 58: 794–800.

Egland, P.G. and C.S. Harwood. 1999. BadR, a new MarR family member regulates anaerobic benzoate degradation by *Rhodopseudomonas palustris* in concert with AadR, an Fnr family member. *J. Bacteriol.* 181: 2102–2109.

Egland, P.G., Gibson, J., and Harwood, C.S. 1995. Benzoate-coenzyme A ligase, encoded *badA*, is one of three ligases to catalyze benzoyl-coenzyme A formation during anaerobic growth of *Rhodopseudomonas palustris* on benzoate. *J. Bacteriol.* 177: 6545–6551

Egli, C., T. Tschan, R. Scholtz, A.M. Cook, and T. Leisinger. 1988. Transformation of tetrachloromethane to dichloromethane and carbon dioxide by *Acetobacterium woodii*. *Appl. Environ. Microbiol.* 54: 2819–2824.

Eichler, B. and B. Schink. 1984. Oxidation of primary aliphatic alcohols by *Acetobacterium carbinolicum* sp. nov., a homoacetogenic anaerobe. *Arch. Microbiol.* 140: 147–152.

Elder, D.J.E., P. Morgan, and D.J. Kelly 1992. Anaerobic degradation of *trans*-cinnamate and ω-phenylalkane carboxylic acids by the photosynthetic bacterium *Rhodopseudomonas palustris*: evidence for a beta-oxidation mechanism. *Arch. Microbiol.* 157: 148–154.

Elmorsi, E.A. and D.J. Hopper. 1979. The catabolism of 5-hydroxyisophthalate by a soil bacterium. *J. Gen. Microbiol.* 111: 145–152.

Elsden, S.R., M.G. Hilton, and D.A. Hopwood. 1976. The end products of the metabolism of aromatic acids by clostridia. *Arch. Microbiol.* 107: 283–288.

Engesser, K.H., E. Schmidt, and H.-J. Knackmuss. 1980. Adaptation of *Alcaligenes eutrophus* B9 and *Pseudomonas* sp. B13 to 2-fluorobenzoate as growth substrate. *Appl. Environ. Microbiol.* 39: 68–73.

Engesser, K.H., M.A. Rubio, and D.W. Ribbons. 1988a. Bacterial metabolism of side chain fluorinated aromatics: cometabolism of 4-trifluoromethyl (TFM)-benzoate by 4-isopropylbenzoate grown *Pseudomonas putida* JT strains. *Arch. Microbiol.* 149: 198–206.

Engesser, K.H., R.B. Cain, and H. J. Knackmuss. 1988b. Bacterial metabolism of side chain fluorinated aromatics: cometabolism of 3-trifluoromethyl (TFM)-benzoate by *Pseudomonas putida* (*arvilla*) mt-2 and *Rhodococcus rubropertinctus* N657. *Arch. Microbiol.* 149: 188–197.

Engesser, K.H., V. Strubel, K. Christoglou, P. Fischer, and H.G. Rast. 1989. Dioxygenolytic cleavage of aryl ether bonds: 1,10-dihydro-1,10-dihydroxyfluorene-9-one, a novel arene dihydrodiol as evidence for angular dioxygenation of dibenzofuran. *FEMS Microbiol. Lett.* 65: 205–210.

Engesser, K.H., G. Auling, J. Busse, and H.-J. Knackmuss. 1990a. 3-Fluorobenzoate enriched bacterial strain FLB 300 degrades benzoate and all three isomeric monofluorobenzoates. *Arch. Microbiol.* 153: 193–199.

Engesser, K.H., M.A. Rubio, and H.-J. Knackmuss. 1990b. Bacterial metabolism of side-chain-fluorinated aromatics: unproductive meta cleavage of 3-trifluorome-thylcatechol. *Appl. Microbiol. Biotechnol.* 32: 600–608.

Ensign, S.A. 1996. Aliphatic and chlorinated alkenes and epoxides as inducers of alkene monooxyganase and epoxidase activities in *Xanthobacter* strain Py2. *Appl. Environ. Microbiol.* 62: 61–66.

Ensley, B.D. and D.T. Gibson. 1983. Naphthalene dioxygenase: purification and prop-erties of a terminal oxygenase component. *J. Bacteriol.* 155: 505–511.

Erickson, B.D. and F.J. Mondelo. 1992. Nucleotide sequencing and transcriptional mapping of the genes encoding biphenyl dioxygenase, a multicomponent poly-chlorinated-biphenyl-degrading enzyme in *Pseudomonas* strain LB 400. *J. Bacte-riol.* 174: 2903–2912.

Evans, P.J., W. Ling, B. Goldschmidt, E.R. Ritter, and L.Y. Young. 1992. Metabolites formed during anaerobic transformation of toluene and *o*-xylene and their pro-posed relationship to the initial steps of toluene mineralization. *Appl. Environ. Microbiol.* 58: 496–501.

Evans, W.C. and G. Fuchs. 1988. Anaerobic degradation of aromatic compounds. *Annu. Rev. Microbiol.* 42: 289–317.

Evans, W.C., H.N. Fernley, and E. Griffiths. 1965. Oxidative metabolism of phenan-threne and anthracene by soil pseudomonads. The ring-fission mechanism. *Bio-chem. J.* 95: 819–831.

Evans, W.C., B.S.W. Smith, H.N. Zernley, and J.I. Davies. 1971. Bacterial metabolism of 2,4-dichlorophenoxyacetate. *Biochem. J.* 122: 543–551.

Faber, K. 1997. *Biotransformations in Organic Chemistry,* 3rd ed., Springer-Verlag, Berlin.

Fall, R.R., J.I. Brown, and T.L. Schaeffer. 1979. Enzyme recruitment allows the bio-degradation of recalcitrant branched hydrocarbons by *Pseudomonas citronellolis.* *Appl. Environ. Microbiol.* 38: 715–722.

Fathepure, B.Z. and S. A. Boyd. 1988. Dependence of tetrachloroethylene dechlori-nation on methanogenic substrate consumption by *Methanosarcina* sp. strain DCM. *Appl. Environ. Microbiol.* 54: 2976–2980.

Fathepure, B.Z., J.M. Tiedje, and S. A. Boyd. 1988. Reductive dechlorination of hexachlorobenzene to tri- and dichlorobenzenes in anaerobic sewage sludge. *Appl. Environ. Microbiol.* 54: 327–330.

Faulkner, J.K. and D. Woodcock. 1965. Fungal detoxication. Part VII. Metabolism of 2,4-dichlorophenoxyacetic and 4-chloro-2-methylphenoxyacetic acids by *As-pergillus niger. J. Chem. Soc.* 1187–1191.

Felber, H., G. Ktresze, R. Prewo, and A. Vasella. 1986. Diastereoselectivity and reac-tivity in the *Diels–Alder* reactions of α-chloronitroso ethers. *Helv. Chim. Acta* 69: 1137–1146.

Fendinger, N.J., D.C. Mcavoy, W.S. Eckhoff, and B.B. Price. 1997. Environmental occurrence of polymethylsiloxane. *Environ. Sci. Technol.* 31: 1555–1563.

Fennewald, M., S. Benson, M. Oppici, and J. Shapiro. 1979. Insertion element analysis and mapping of the *Pseudomonas* plasmid *alk* regulon. *J. Bacteriol.* 139: 940–952.

Fetzner, S., R. Müller, and F. Lingens. 1989. A novel metabolite in the degradation of 2-chlorobenzoate. *Biochem. Biophys. Res. Commun.* 161: 700–705.

Fetzner, S., R. Müller, and F. Lingens. 1992. Purification and some properties of 2-halobenzoate 1,2-dioxygenase, a two-component enzyme system from *Pseudomonas cepacia* 2CBS. *J. Bacteriol.* 174: 279–290.

Finster, K. and F. Bak. 1993. Complete oxidation of propionate, valerate, succinate, and other organic compounds by newly isolated types of marine, anaerobic, mesophilic, Gram-negative sulfur-reducing eubacteria. *Appl. Environ. Microbiol.* 59: 1452–1460.

Firestone, M.K. and J.M. Tiedje. 1978. Pathway of degradation of nitrilotriacetate by a *Pseudomonas* sp. *Appl. Environ. Microbiol.* 35: 955–961.

Fitzgerald, J.W. 1976. Sulfate ester formation and hydrolysis: a potentially important yet often ignored aspect of the sulfur cycle of aerobic soils. *Bacteriol. Rev.* 40: 698–721.

Fitzgerald, J.W., H.W. Maca, and F.A. Rose. 1979. Physiological factors regulating tyrosine-sulphate sulphohydrolase activity in *Comamonas terrigena*: occurrence of constitutive and inducible enzymes. *J. Gen. Microbiol.* 111: 407–415.

Fortnagel, P., H. Harms, R.-M. Wittich, S. Krohn, H. Meter, V. Sinnwell, H. Wilkes, and W. Francke. 1990. Metabolism of dibenzofuran by *Pseudomonas* sp. strain HH 69 and the mixed culture HH27. *Appl. Environ. Microbiol.* 56: 1148–1156.

Foss, S. and J. Harder. 1998. *Thauera linaloolentis* sp. nov. and *Thauera terpenica* sp. nov., isolated on oxygen-containing monoterpenes (linalool, menthol, and eucalyptol) and nitrate. *Syst. Appl. Microbiol.* 21: 365–373.

Fox, B.G., W.A. Froland, J.E. Dege, and J.D. Lipscomb. 1989. Methane monooxygenase from *Methylosinus trichosporium* OB3b. Purification and properties of a three-component system with a high specific activity from a type II methanotroph. *J. Biol. Chem.* 264: 10023–10033.

Fraisse, L. and H. Simon. 1988. Observations on the reduction of non-activated carboxylates by *Clostridium formicoaceticum* with carbon monoxide or formate and the influence of various viologens. *Arch. Microbiol.* 150: 381–386.

Franklin, T.J. and G.A. Snow. 1981. *Biochemistry of Antimicrobial Action*. Chapman and Hall, London.

Freedman, D.L. and J.M. Gossett. 1989. Biological reductive dechlorination of tetrachloroethylene and trichloroethylene under methanogenic conditions. *Appl. Environ. Microbiol.* 55: 2144–2151.

French, C.E., Nicklin, S., and Bruce, N.C. 1996. Sequence and properties of pentaerythritol tetranitrate reductase from *Enterobacter cloacae* PB2. *J. Bacteriol.* 178: 6623–6627.

French, C.E., S. Nicklin, and N.C. Bruce. 1998. Aerobic degradation of 2,4,6-trinitrotoluene by *Enterobacter cloacae* PB2 and by pentaerythritol tetranitrate reductase. *Appl. Environ. Microbiol.* 64: 2864–2868.

Frimmer, U. and F. Widdel. 1989. Oxidation of ethanol by methanogenic bacteria. Growth experiments and enzymatic studies. *Arch. Microbiol.* 152: 479–483.

Frings, J., E. Schramm, and B. Schink. 1992. Enzymes involved in anaerobic polyethylene glycol degradation by *Pelobacter venetianus* and *Bacteroides* strain PG1. *Appl. Environ. Microbiol.* 58: 2164–2167.

Fuchs, A., W. de Vries, and M.P. Sanz. 1980. The mechanism of pisatin degradation by *Fusarium oxysporum* f. sp. *pisi*. *Physiol. Plant Pathol.* 16: 119–133.

Fuchs, K., A. Schreiner, and F. Lingens. 1991. Degradation of 2-methylaniline and chlorinated isomers of 2-methylaniline by *Rhodococcus rhodochrous* strain CTM. *J. Gen. Microbiol.* 137: 2033–2039.

Fujioka, M. and H. Wada. 1968. The bacterial oxidation of indole. *Biochim. Biophys. Acta* 158: 70–78.

Fukumori, F. and C.P. Saint. 1997. Nucleotide sequences and regulational analysis of genes involved in conversion of aniline to catechol in *Pseudomonas putida* UCC22 (pTDN1). *J. Bacteriol.* 179: 399–408.

Fukumori, F. and R.P. Hausinger. 1993. *Alcaligenes eutrophus* JMP 134 "2,4-dichlorophenoxyacetate monooxygenase" is an a-ketoglutarate-dependent dioxygenase. *J. Bacteriol.* 175: 2083–2086.

Furukawa, K. and N. Arimura. 1987. Purification and properties of 2,3-dihydroxybiphenyl dioxygenase from polychlorinated biphenyl-degrading *Pseudomonas pseudoalcaligenes* and *Pseudomonas aeruginosa* carrying the cloned *bphC* gene. *J. Bacteriol.* 169: 924–927.

Furukawa, K. and T. Miyazaki. 1986. Cloning of a gene cluster encoding biphenyl and chlorobiphenyl degradation in *Pseudomonas pseudoalcaligenes*. *J. Bacteriol.* 166: 392–398.

Furukawa, K., F. Matsumara, and K. Tonomura. 1978. *Alcaligenes* and *Acinetobacter* strains capable of degrading polychlorinated biophenyls. *Agric. Biol. Chem.* 42: 543–548.

Furukawa, K., N. Tomizuka, and A. Kamibayashi. 1979. Effect of chlorine substitution on the bacterial metabolism of various polychlorinated biphenyls. *Appl. Environ. Microbiol.* 38: 301–310.

Furukawa, K., J.R. Simon, and A.M. Chakrabarty. 1983. Common induction and regulation of biphenyl. xylene/toluene, and salicylate catabolism in *Pseudomonas paucimobilis*. *J. Bacteriol.* 154: 1356–1362.

Furukawa, N. and K. Tonomura. 1971. Enzyme system involved in the decomposition of phenyl mercuric acetate by mercury-resistant *Pseudomonas*. *Agric. Biol. Chem.* 35: 604–610.

Furutani, A. and J.W.M. Rudd. 1980. Measurement of mercury methylation in lake water and sediment samples. *Appl. Environ. Microbiol.* 40: 770–776.

Gälli, R. and P.L. McCarthy. 1989. Biotransformation of 1,1,1-trichloroethane, trichloromethane, and tetrachloromethane by a *Clostridium* sp. *Appl. Environ. Microbiol.* 55: 837–844.

Gallus, C. and B. Schink. 1994. Anaerobic degradation of pimelate by newly isolated denitrifying bacteria. *Microbiology* (U.K.) 140: 409–416.

Gamar, Y. and J.K. Gaunt. 1971. Bacterial metabolism of 4-chloro-2-methylphenoxyacetate (MCPA): formation of glyoxylate by side-chain cleavage. *Biochem. J.* 122: 527–531.

Gauthier, J.J. and S.C. Rittenberg. 1971. The metabolism of nicotinic acid. II. 2,5-Dihydroxypyridine oxidation, product formation, and oxygen-18 incorporation. *J. Biol. Chem.* 246: 3743–3748.

Gee, J.M. and J.L. Peel. 1974. Metabolism of 2,3,4,6-tetrachlorophenol by micro-organisms from broiler house litter. *J. Gen. Microbiol.* 85: 237–243.

Geigert, J., S.L. Neidleman, D.J. Dalietos, and S.K. DeWitt. 1983a. Haloperoxidases: enzymatic synthesis of α,β-halohydríns from gaseous alkenes. *Appl. Environ. Microbiol.* 45: 366–374.

Geigert, J., S.L. Neidleman, D.J. Dalietos, and S.K. DeWitt. 1983b. Novel haloperoxidase reaction: synthesis of dihalogenated products. *Appl. Environ. Microbiol.* 45: 1575–1578.

Geigert, J., S.L. Neidleman, and D.J. Dalietos. 1983c. Novel haloperoxidase substrates. Alkynes and cyclopropanes. *J. Biol. Chem.* 258: 2273–2277.

Geissler, J.F., C.S. Harwood, and J. Gibson. 1988. Purification and properties of benzoate-coenzyme A ligase, a *Rhodopseudomonas palustris* enzyme involved in the anaerobic degradation of benzoate. *J. Bacteriol.* 170: 1709–1714.

Genther, B.R.S., W.A. Price, and P.H. Pritchard. 1989. Anaerobic degradation of chloroaromatic compounds in aquatic sediments under a variety of enrichment conditions. *Appl. Environ. Microbiol.* 55: 1466–1471.

Genther, B.R.S., G.T. Townsend, and P.J. Chapman. 1990. Effect of fluorinated analogues of phenol and hydroxybenzoates on the anaerobic transformation of phenol to benzoate. *Biodegradation* 1: 65–74.

Gerritse, J., B.J. van der Woude, and J.C. Gottschal. 1992. Specific removal of chlorine from the *ortho*-position of halogenated benzoic acids by reductive dechlorination in anaerobic enrichment cultures. *FEMS Microbiol Lett.* 100: 273–280.

Gibson, D.T. and V. Subramanian. 1984. Microbial degradation of aromatic hydrocarbons pp. 181–252. In *Microbial Degradation of Organic Compounds* (Ed. D.T. Gibson). Marcel Dekker, New York.

Gibson, D.T., J.R. Koch, and R.E. Kallio. 1968. Oxidative degradation of aromatic hydrocarbons. I. Enzymatic formation of catechol from benzene. *Biochemistry* 9: 2653–2662.

Gibson, D.T., G.E. Cardini, F.C. Masales, and R.E. Kallio. 1970. Incorporation of oxygen-18 into benzene by *Pseudomonas putida*. *Biochemistry* 9: 1631–1635.

Gibson, D.T., V. Mahadevan, and J.F. Davey. 1974. Bacterial metabolism of para- and meta-xylene: oxidation of the aromatic ring. *J. Bacteriol.* 119: 930–936.

Gibson, D.T., D.L. Cruden, J.D. Haddock, G.J. Zylstra, and J.M. Brande. 1993. Oxidation of polychlorinated biphenyls by *Pseudomonas* sp. strain LB400 and *Pseudomonas pseudoalcaligenes* KF707. *J. Bacteriol.* 175: 4561–4564.

Gibson, D.T., S.M. Resnick, K. Lee, J.M. Brand, D.S. Torok, L.P. Wackett, M.J. Schocken, and B.E. Haigler. 1995. Desaturation, dioxygenation, and monooxygenation reactions catalyzed by naphthalene dioxygenase from *Pseudomonas* sp. strain 9816-4. *J. Bacteriol.* 177: 2615–2621.

Gibson, J., M. Dispensa, and C. S. Harwood. 1997. 4-Hydroxybenzoyl coenzyme A reductase dehydroxylating is required for anaerobic degradation of 4-hydrozybenzoate by *Rhodopseudomonas palustris* and shares features with molybdenum-containing hydroxylases. *J. Bacteriol.* 179: 634–642.

Gibson, K.J. and J. Gibson. 1992. Potential early intermediates in anaerobic benzoate degradation by *Rhodopseudomonas palustris*. *Appl. Environ. Microbiol.* 58: 696–698.

Gibson, S.A. and G.W. Sewell. 1992. Stimulation of reductive dechlorination of tetrachloroethene in anaerobic aquifer microcosms by addition of short-chain acids or alcohols. *Appl. Environ. Microbiol.* 58: 1392–1393.

Gieg, L.M., A. Otter, and P.M. Fedorak. 1996. Carbazole degradation by *Pseudomonas* sp. LD2: metabolic characteristics and identification of some metabolites. *Environ. Sci. Technol.* 30: 575–585.

Gilcrease, P.C. and V.G. Murphy. 1995. Bioconversion of 2,4-diamino-6-nitrotoluene to a novel metabolite under anoxic and aerobic conditions. *Appl. Environ. Microbiol.* 61: 4209–4214.

Golbeck, J.H., S.A. Albaugh, and R. Radmer. 1983. Metabolism of biphenyl by *Aspergillus toxicarius*: induction of hydroxylating activity and accumulation of water-soluble conjugates. *J. Bacteriol.* 156: 49–57.

Goldman, P. 1965. The enzymatic cleavage of the carbon-fluorine bond in fluoroacetate. *J. Biol. Chem.* 240: 3434–3438.

Gorlatova, N., M. Tchorzewski, T. Kurihara, K. Soda, and N. Esaki. 1998. Purification, characterization, and mechanism of a flavin mononucleotide-dependent 2-nitropropane dioxygenase from *Neurospora crassa*. *Appl. Environ. Microbiol.* 64: 1029–1033.

Gorny, N. and B. Schink. 1994a. Hydroquinone degradation via reductive dehydroxylation of gentisyl-CoA by a strictly anaerobic fermenting bacterium. *Arch. Microbiol.* 161: 25–32.

Gorny, N. and B. Schink. 1994b. Anaerobic degradation of catechol by *Desulfobacterium* sp. strain cat2 proceeds via carboxylation to protocatechuate. *Appl. Environ. Microbiol.* 60: 3396–3400.

Gorny, N., G. Wahl, A. Brune, and B. Schink. 1992. A strictly anaerobic nitrate-reducing bacterium growing with resorcinol and other aromatic compounds. *Arch. Microbiol.* 158: 48–53.

Gorontzy, T., J. Küver, and K.-H. Blotevogel. 1993. Microbial reduction of nitroaromatic compounds under anaerobic conditions. *J. Gen. Microbiol.* 139: 1331–1336.

Gößner, A., S.L. Daniel, and H.L. Drake. 1994. Acetogenesis coupled to the oxidation of aromatic aldehyde groups. *Arch. Microbiol.* 161: 126–131.

Goyal A.K. and G.J. Zylstra. 1996. Molecular cloning of novel genes for polycyclic aromatic hydrocarbon degradation from *Comamonas testosteroni*. *Appl. Environ. Microbiol.* 62: 230–236

Gray, P.H.H. and H.G. Thornton. 1928. Soil bacteria that decompose certain aromatic compounds. *Zentralbl. Bakteriol. Parasitenkd. Infektionskr.* (2 Abt.) 73: 74–96.

Grbic-Galic, D. and T.M. Vogel. 1987. Transformation of toluene and benzene by mixed methanogenic cultures. *Appl. Environ. Microbiol.* 53: 254–260.

Grbic-Galic, D. and L.Y. Young. 1985. Methane fermentation of ferulate and benzoate: anaerobic degradation pathways. *Appl. Environ. Microbiol.* 50: 292–297.

Grbic-Galic, D., N. Churchman-Eisel, and I. Mrakovic. 1990. Microbial transformation of styrene by anaerobic consortia. *J. Appl. Bacteriol.* 69: 247–260.

Griffin, M. and P.W. Trudgill. 1972. The metabolism of cyclopentanol by *Pseudomonas* N.C.I.B. 9872. *Biochem. J.* 129: 595–603.

Griffiths, D.A., D.E. Brown, and S.G. Jezequel. 1992. Biotransformation of warfarin by the fungus *Beauveria bassiana*. *Appl. Microbiol. Biotechnol.* 37: 169–175.

Griffiths, E.T., S.M. Bociek, P.C. Harries, R. Jeffcoat, D.J. Sissons, and P.W. Trudgill. 1987. Bacterial metabolism of α-pinene: pathway from α-pinene oxide to acyclic metabolites in *Nocardia* sp. strain P18.3. *J. Bacteriol.* 169: 4972–4979.

Grifoll, M., S.A. Selifonov, C.V. Gatlin, and P.J. Chapman. 1995. Actions of a versatile fluorene-degrading bacterial isolate on polycyclic aromatic compounds. *Appl. Environ. Microbiol.* 61: 3711–3723

Groenewegen, P.E.G., P. Breeuwer, J.M.L.M. van Helvoort, A.A.M. Langenhoff, F.P. de Vries, and J.A.M. de Bont. 1992. Novel degradative pathway of 4-nitrobenzoate in *Comamonas acidovorans* NBA-10. *J. Gen. Microbiol.* 138: 1599–1605.

Grund, A., J. Shapiro, M. Fennewald, P. Bacha, J. Leahy, K. Markbreiter, M. Nieder, and M. Toepfer. 1975. Regulation of alkane oxidation in *Pseudomonas putida*. *J. Bacteriol.* 123: 546–556.

Grund, E., B. Denecke, and R. Eichenlaub. 1992. Naphthalene degradation via salicylate and gentisate by *Rhodococcus* sp. strain B4. *Appl. Environ. Microbiol.* 58: 1874–1877.

Guangsheng, C., C.M. Plugge, W. Roelofsen, F.P. Houwen, and A.J.M. Stams. 1992. *Selenomonas acidaminovorans* sp. nov., a versatile thermophilic proton-reducing anaerobe able to grow by decarboxylation of succinate to propionate. *Arch. Microbiol.* 157: 169–175.

Gurarini, A., G. Guglielmetti, M. Vincenti, P. Guarda, and G. Marchionni. 1993. Characterization of perfluoropolyethers by desorption chemical ionization and tandem mass spectrometry. *Anal. Chem.* 65: 970–975.

Hacisalihoglu, A., J.A. Jongejan, and J.A. Duine. 1997. Distribution of amine oxidases and amine dehydrogenases in bacteria grown on primary amimes and characterization of the amine oxidase from *Klebsiella oxytoca*. *Microbiology* (U.K.) 143: 505–512.

Haddock, J.D. and J.G. Ferry. 1993. Initial steps in the anaerobic degradation of 3,4,5-trihydroxybenzoate by *Eubacterium oxidoreducens*: characterization of mutamts and role of 1,2,3,5-tetrahydroxybenzene. *J. Bacteriol.* 175: 669–673.

Haddock, J.D. and D.T.Gibson. 1995. Purification and characterization of the oxygenase component of biphenyl 2,3-dioxygenase from *Pseudomonas* sp. strain LB400. *J. Bacteriol.* 177: 5834–5839.

Haddock, J.D., J.R. Horton, and D.T. Gibson. 1995. Dihydroxylation and dechlorination of chlorinated biphenyls by purified biphenyl 2,3-dioxygenase from *Pseudomonas* sp. strain LB400. *J. Bacteriol.* 177: 20–26.

Haemmerli, S.D., M.S.A. Leisola, D. Sanglard, and A. Fiechter. 1986. Oxidation of benzo[a]pyrene by extracellular ligninases of *Phanerochaete chrysosporium*. *J. Biol. Chem.* 261: 6900–6903.

Hagedorn, S.R., R.S. Hanson, and D.A. Kunz (Eds.).1988. *Microbial Metabolism and the Carbon Cycle*. Harwood Academic Publishers, Chur, Switzerland.

Häggblom, M.M. and L.Y. Young. 1990. Chlorophenol degradation coupled to sulfate reduction. *Appl. Environ. Microbiol.* 56: 3255–3260.

Haigler, B.E. and J.C. Spain. 1993. Biodegradation of 4-nitrotoluene by *Pseudomonas* sp. strain 4NT. *Appl. Environ. Microbiol.* 59: 2239–2243.

Haigler, B.E., W.H. Wallace, and J.C. Spain. 1994. Biodegradation of 2-nitrotoluene by *Pseudomonas* sp. strain JS 42. *Appl. Environ. Microbiol.* 60: 3466–3469.

Hales, S.G., G.K. Watson, K.S. Dodson, and G.F. White. 1986. A comparative study of the biodegradation of the surfactant sodium dodecyltriethoxy sulphate by four detergent-degrading bacteria. *J. Gen. Microbiol.* 132: 953–961.

Hallas, L.E. and M. Alexander. 1983. Microbial transformation of nitroaromatic compounds in sewage effluent. *Appl. Environ. Microbiol.* 45: 1234–1241.

Hamamura, N., C. Page, T. Long, L. Semprini, and D.H. Arp. 1997. Chloroform cometabolism by butane-grown CF8, *Pseudomonas butanovora*, and *Mycobacterium vaccae* JOB5 and methane-grown *Methylosinus trichosporium* OB3b. *Appl. Environ. Microbiol.* 63: 3607–3613.

Hamdy, M.K. and O.R. Noyes. 1975. Formation of methyl mercury by bacteria. *Appl. Microbiol.* 30: 424–432.

Hammel, K.E., B. Green, and W.Z. Gai. 1991. Ring fission of anthracene by a eukaryote. *Proc. Natl. Acad. Sci. U.S.A.* 88: 10605–10608.

Hammel, K.E., W.Z. Gai, B. Green, and M.A. Moen. 1992. Oxidative degradation of phenanthrene by the lignolytic fungus *Phanerochaete chrysosporium*. *Appl. Environ. Microbiol.* 58: 1832–1838.

Hammer, A., A. Stolz, and H.-J. Knackmuss. 1996. Purification and characterization of a novel type of protocatechuate 3,4-dioxygenase with the ability to oxidize 4-sulfocatechol. *Arch. Microbiol.* 166: 92–100.

Häner, A., P. Höhener, and J. Zeyer. 1997. Degradation of trimethylbenzene isomers by an enrichment culture under N_2O-reducing conditions. *Appl. Environ. Microbiol.* 63: 1171–1174.

Hankin, L. and P.E. Kolattukudy. 1968. Metabolism of a plant wax paraffin (*n*-nonacosane) by a soil bacterium (*Micrococcus cerificans*). *J. Gen. Microbiol.* 51: 457–463.

Hanlon, S.P., R.A. Holt, G.R. Moore, and A.G. McEwan. 1994. Isolation and characterization of a strain of *Rhodobacter sulfidophilus*: a bacterium which grows autotrophically with dimethylsulphide as electron donor. *Microbiology* (U.K.) 140: 1953–1958.

Hanselmann, K.W., J.P. Kaiser, M. Wenk, R. Schön, and R. Bachofen. 1995. Growth on methanol and conversion of methoxylated aromatic substrates by *Desulfomaculum orientis* in the presence and absence of sulfate. *Microbiol. Res.* 150: 387–401.

Hanson, J.R., P.B. Reese, J.A. Takahashi, and M.R. Wilson, 1994. Biotransformation of some stemodane diterpenoids by *Cephalosporium aphidicola*. *Phytochemistry* 36: 1391–1393.

Hanson, R.S. and T.E. Hanson. 1996. Methanotrophic bacteria. *Microbiol. Rev.* 60: 439–471.

Harayama, S., M. Rekik, M. Wubbolts, K. Rose, R.A. Leppik, and K.N. Timmis. 1989. Characterization of five genes in the upper-pathway operon of TOL plasmid pPWW0 from *Pseudomonas putida* and identification of the gene products. *J. Bacteriol.* 171: 5048–5055.

Hardison, L., S.S. Curie, L.M. Ciuffeti, and M.R. Hyman. 1997. Metabolism of diethyl ether and cometabolism of methyl *tert*-butyl ether by a filamentous fungus, a *Graphium* sp. *Appl. Environ. Microbiol.* 63: 3059–3167.

Hareland, W.A., R.L. Crawford, P.J. Chapman, and S. Dagley. 1975. Metabolic function and properties of 4-hydroxyphenylacetic acid 1-hydrolase from *Pseudomonas acidovorans*. *J. Bacteriol.* 121: 272–285.

Harms, G., K. Zengle, R. Rabus, F. Aeckersberg, D. Minz, R. Rosselló-Mora, and F. Widdel. 1999. Anaerobic oxidation of *o*-xylene, *m*-xylene, and homologous alkylbenzenes by new types of sulfate-reducing bacteria. *Appl. Environ. Microbiol.* 65: 999–1004.

Harper, D.B. 1977. Microbial metabolism of aromatic nitriles. Enzymology of C-N cleavage by *Nocardia* sp. (Rhodochrous group) N.C.I.B. 11216. *Biochem. J.* 165: 309–319.

Harper, D.B. and J.T.G. Hamilton. 1988. Biosynthesis of chloromethane in *Phellinus pomaceus*. *J. Gen. Microbiol.* 13 4: 2831–2839.

Harper, D.B. and J.T. Kennedy. 1985. Purification and properties of *S*-adenosylmethionine: aldoxime *O*-methyltransferase from *Pseudomonas* sp. N.C.I.B. 11652. *Biochem. J.* 226: 147–153.

Harper, D.B. and J. Nelson. 1982. The bacterial biogenesis of isobutyraldoxine *O*-methyl ether, a novel volatile secondary metabolite. *J. Gen. Microbiol.* 128: 1667–1678.

Harper, D.B., J.T.G. Hamilton, J.T. Kennedy, and K.J. McNally. 1989. Chloromethane, a novel methyl donor for biosynthesis of esters and anisoles in *Phellinus pomaceus*. *Appl. Environ. Microbiol.* 55: 1981–1989.

Harper, D.B., J.A. Buswell, J.T. Kennedy, and J.T.J. Hamilton. 1990. Chloromethane, methyl donor in veratryl alcohols biosynthesis in *Phanerochaete chrysosporium* and other lignin degrading fungi. *Appl. Environ. Microbiol.* 56: 3540–3547.

Härtel, U., E. Eckel, J. Koch, G. Fuchs, D. Linder, and W. Buckel. 1993. Purification of glutaryl-CoA dehydrogenase from *Pseudomonas* sp., an enzyme involved in the anaerobic degradation of benzoate. *Arch. Microbiol.* 159: 174–181.

Hartmans, S. and J.A.M. de Bont. 1992. Aerobic vinyl chloride metabolism in *Mycobacterium aurum* L1. *Appl. Environ. Microbiol.* 58: 1220–1226.

Hartmans, S., J.P. Smits, M.J. van der Werf, F. Volkering, and J.A.M. de Bont. 1989. Metabolism of styrene oxide and 2-phenylerthanol in the styrene-degrading *Xanthobacter* strain 124X. *Appl. Environ. Microbiol.* 55: 2850–2855.

Hartmans, S., F.J. Weber, B.P.M. Somhorst, and J.A.M. de Bont. 1991. Alkene monooxygenase from *Mycobacterium* E3: a multicomponent enzyme. *J. Gen. Microbiol.* 137: 2555–2560.

Harwood, C.S. and Parales, R.E. 1996. The β-ketoadipate pathway and the biology of self-identity. *Annu. Rev. Microbiol.* 50: 553–590.

Harwood, C.S., G. Burchardt, H. Herrmann, and G. Fuchs. 1999. Anaerobic metabolism of aromatic compounds via the benzoyl-CoA pathway. *FEMS Microbiol. Rev.* 22: 439–458.

Haug, W., A. Schmidt, B. Nörtemann, D.C. Hempel, A. Stolz, and H.-J. Knackmuss. 1991. Mineralization of the sulfonated azo dye mordant Yellow 3 by a 6-aminonaphthalene-2-sulfonate-degrading bacterial consortium. *Appl. Environ. Microbiol.* 57: 3144–3149.

Haugland, R.A., D.J. Schlemm, R.P. Lyons, P.R. Sferra, and A. M. Chakrabarty. 1990. Degradation of the chlorinated phenoxyacetate herbicides 2,4-dichlorophenoxyacetic acid and 2,4,5-trichlorophenoxyacetic acid by pure and mixed bacterial cultures. *Appl. Environ. Microbiol.* 56: 1357–1362.

Havel, J. and W. Reineke. 1993. Microbial degradation of chlorinated acetophenones. *Appl. Environ. Microbiol.* 59: 2706–2712.

He, Z. and J.C. Spain. 1997. Studies of the catabolic pathway of degradation of nitrobenzene by *Pseudomonas pseudoalcaligenes* JS45: removal of the amino group from 2-aminomuconic semialdehyde. *Appl. Environ. Microbiol.* 63: 4839–4843.

He, Z. and J.C. Spain. 1998. A novel 2-aminomuconate deaminase in the nitrobenzene degradation pathway of *Pseudomonas pseudoalcaligenes* JS45. *J. Bacteriol.* 180: 2502–2506.

He, Z. and J. Wiegel. 1996. Purification and characterization of an oxygen-sensitive, reversible 3,4-dihydroxybenzoate decarboxylase from *Clostridium hydroxybenzoicum. J. Bacteriol.* 178: 3539–3543.

Heald, S. and R.O. Jenkins. 1994. Trichloroethylene removal and oxidation toxicity mediated by toluene dioxygenase of *Pseudomonas putida. Appl. Environ. Microbiol.* 60: 4634–4637.

Hegeman, G.D. 1966. Synthesis of the enzymes of the mandelate pathway by *Pseudomonas putida.* I. Synthesis of enzymes of the wild type. *J. Bacteriol.* 91: 1140–1154.

Heijthuijsen, J.H.F.G. and T.A. Hansen. 1989a. Anaerobic degradation of betaine by marine *Desulfobacterium* strains. *Arch. Microbiol.* 152: 393–396.

Heijthuijsen, J.H.F.G. and T.A. Hansen. 1989b. Betaine fermentation and oxidation by marine *Desulfuromonas* strains. *Appl. Environ. Microbiol.* 55: 965–969.

Heiss, G., A. Stolz, A.E. Kuhm, C. Müller, J. Klein, J. Altenbuchner, and H.-J. Knackmuss. 1995. Characterization of a 2,3-dihydroxybiphenyl dioxygenase from the naphthalenesulfonate-degrading bacterium strain BN6. *J. Bacteriol.* 177: 5865–5871.

Heiss, G., C. Müller, J. Altenbuchner, and A. Stolz. 1997. Analysis of a new dimeric extradiol dioxygenase from a naphthalenesulfonate-degrading sphingomonad. *Microbiology* (U.K.) 143: 1691–1699.

Heitkamp, M.A., J.P. Freeman, D.C. McMillan, and C.E.Cerniglia. 1985. Fungal metabolism of *tert*-butylphenyl diphenyl phosphate. *Appl. Environ. Microbiol.* 50: 265–273.

Heitkamp, M.A., W. Franklin, and C.E. Cerniglia. 1988a. Microbial metabolism of polycyclic aromatic hydrocarbons: isolation and characterization of a pyrene-degrading bacterium. *Appl. Environ. Microbiol.* 54: 2549–2555

Heitkamp, M.A., J.P. Freeman, D.W. Miller, and C.E. Cerniglia. 1988b. Pyrene degradation by a *Mycobacterium* sp.: identification of ring oxidation and ring fission products. *Appl. Environ. Microbiol.* 54: 2556–2565.

Hensgens, C.M.H., J. Vonck, J. Van Beeumen, E.F.J. van Bruggen, and T.A. Hansen. 1993. Purification and characterization of an oxygen-labile, NAD-dependent alcohol dehydrogenase from *Desulfovibrio gigas*. *J. Bacteriol.* 175: 2859–2863.

Hernandez, B.S., J.J. Arensdorf, and D.D. Focht. 1995. Catabolic characteristic of biphenyl-utilizing isolates which cometabolize PCBs. *Biodegradation* 6: 75–82.

Herod, A.A. 1998. Azaarenes and thiaarenes, pp. 271–323. In *Handbook of Environmental Chemistry* (Ed. A.H. Neilson) Vol. 3, Part I. Springer-Verlag, Berlin.

Hickey, W.J. and D.D. Focht. 1990. Degradation of mono-, di-, and trihalogenated benzoic acids by *Pseudomonas aeruginosa* JB2. *Appl. Environ. Microbiol.* 56: 3842–3850.

Higgins, T.P., J.R. Snape, and G.F. White. 1993. Comparison of pathways for biodegradation of monomethyl sulphate in *Agrobacterium* and *Hyphomicrobium* species. *J. Gen. Microbiol.* 139: 2915–2920.

Higson, F.K. and D.D. Focht. 1990. Degradation of 2-bromobenzoate by a strain of *Pseudomonas aeruginosa*. *Appl. Environ. Microbiol.* 56: 1615–1619.

Hilton, M.D. and W.J. Cain. 1990. Bioconversion of cinnamic acid to acetophenone by a pseudomonad: microbial production of a natural flavor compound. *Appl. Environ. Microbiol.* 56: 623–627.

Hippe, H., D. Caspari, K. Fiebig, and G. Gottschalk. 1979. Utilization of trimethylamine and other N-methyl compounds for growth and methane formation by *Methanosarcina barkeri*. *Proc. Natl. Acad. Sci. U.S.A.* 76: 494–498.

Hirrlinger, B. and A. Stolz. 1997. Formation of a chiral hydroxamic aid with an amidase from *Rhodococcus erythropolis* MP50 and subsequent chemical Lossen rearrangement to a chiral amine. *Appl. Environ. Microbiol.* 63: 3390–3393.

Hirschberg, R. and J.C. Ensign. 1971. Oxidation of nicotinic acid by a *Bacillus* species: source of oxygen atoms for the hydroxylation of nicotinic acid and 6-hydroxynicotinic acid. *J. Bacteriol.* 108: 757–759.

Hofrichter, M., K. Scheibner, I. Schneegaß, and W. Fritsche. Enzymatic combusion of aromatic and aliphatic compounds by manganese peroxidase from *Nematoloma frowardii*. *Appl. Environ. Microbiol.* 64: 399–404.

Högn, T. and L. Jaenicke. 1972. Benzene metabolism of *Moraxella* species. *Eur. J. Biochem.* 30: 369–375.

Holland, H.L. 1988. Chiral sulfoxidation by biotransformation of organic sulfides. *Chem. Rev.* 88: 473–485.

Holland, H.L., M. Carey, and S. Kumaresan. 1993. Fungal biotransformation of organophosphines. *Xenobiotica* 23: 519–524.

Holliger, C., G. Schraa, E. Stuperich, A.J.M. Stams, and A.J.B. Zehnder. 1992a. Evidence for the involvement of corrinoids and factor F_{430} in the reductive dechlorination of 1,2-dichloroethane by *Methanosarcina barkeri*. *J. Bacteriol.* 174: 4427–4434.

Holliger, C., S.W.M. Kengen, G. Schraa, A.J.M. Stams, and AJ.B. Zehnder. 1992b. Methyl-coenzyme M reductase of *Methanobacterium thermoautrotrophicum* delta H catalyzes the reductive dechlorination of 1,2-dichloroethane to ethylene and chloroethane. *J. Bacteriol.* 174: 4435–4443.

Holliger, C., G. Wohlfarth, and G. Diekert. 1999. Reductive dechlorination in the energy metabolism of anaerobic bacteria. *FEMS Microbiol. Rev.* 22: 383–398.

Holmes, A.J., D.P. Kelly, S.C. Baker, A.S. Thompson, P. de Marco, E.M. Kenna, and J.C. Murrell. 1997. *Methylosulfonomonas methylovora* gen. nov., sp. nov., and *Marinosulfonomonas methylotropha* gen. nov., so. nov.: novel methylotrophs able to grow on methansulfonic acid. *Arch. Microbiol.* 167: 46–53.

Hopper, D.J. 1978. Microbial degradation of aromatic hydrocarbons pp. 85–112. In *Developments in Biodegradation of Hydrocarbons*—1 (Ed. R.J. Watkinson). Applied Science Publishers Ltd., London.

Hopper, D.J. 1988. Properties of *p*-cresol methylhydroxylases. In *Microbial Metabolism and the Carbon Cycle* pp. 247–258 (Eds. S.R. Hagedorn, R.S. Hanson, and D.A. Kunz). Harwood Academic Publishers, Chur, Switzerland.

Hopper, D.J., H.G. Jones, E.A. Elmorisi, and M.E. Rhodes-Roberts. 1985. The catabolism of 4-hydroxyacetophenone by an *Alcaligenes* sp. *J. Gen. Microbiol.* 131: 1807–1814.

Hormann, K. and J.R. Andreesen. 1991. A flavin-dependent oxygenase reaction initiates the degradation of pyrrole-2-carboxylate in *Arthrobacter* strain Py1 (DSM 6386). *Arch. Microbiol.* 157: 43–48.

Horowitz, A., J.M. Suflita, and J.M. Tiedje. 1983. Reductive dehalogenations of halobenzoates by anaerobic lake sediment microorganisms. *Appl. Environ. Microbiol.* 45: 1459–1465.

Hou, C.T., R. Patel, A.I. Laskin, N. Barnabe, and I. Barist. 1983. Epoxidation of short-chain alkenes by resting-cell suspensions of propane-grown bacteria. *Appl. Environ. Microbiol.* 46: 171–177.

Howell, L.G., T. Spector, and V. Massey. 1972. Purification and properties of *p*-hydroxybenzoate hydroxylase from *Pseudomonas fluorescens*. *J. Biol. Chem.* 247: 4340–4350.

Hsu, T., S.L. Daniel, M.F. Lux, and H.L. Drake. 1990. Biotransformation of carboxylated aromatic compounds by the acetogen *Clostridium thermoaceticum*: generation of growth-supportive CO_2 equivalents under CO_2-limited conditions. *J. Bacteriol.* 172: 212–217.

Hudlický, M. 1990. *Oxidations in Organic Chemistry.* ACS Monograph 186. American Chemical Society, Washington, D.C.

Hudlicky, T., H. Luna, G. Barbieri, and L.D. Kwart. 1988. Enantioselective synthesis through microbial oxidation of arenes. 1. Efficient preparation of terpene and prostanoid synthons. *J. Am. Chem. Soc.* 110: 4735–4741

Hughes, J.B., C. Wang, K. Yesland, A. Richardson, R. Bhadra, G. Bennett, and F. Rudolph. 1998. Bamberger rearrangement during TNT metabolism by *Clostridium acetobutylicum*. *Environ. Sci. Technol.* 32: 494–500.

Hur, H.-G., M. Sadowsky, and L.P. Wackett. 1994. Metabolism of chlorofluorocarbons and polybrominated compounds by *Pseudomonas putida* G786pHG-2 via an engineered metabolic pathway. *Appl. Environ. Microbiol.* 60: 4148–4154.

Hur, H.-G., L.M. Newman, L.P. Wackett, and M.J. Sadowsky. 1997. Toluene 2-monooxygenase-dependent growth of *Burkholderia cepacia* G4/PR1 on diethyl ether. *Appl. Environ. Microbiol.* 63: 1606–1609.

Hurtubise, Y., D. Barriault, and M. Sylvestre. 1998. Involvement of the terminal oxygenase β subunit in the biphenyl dioxygenase reactivity pattern towards chlorobiphenyls. *J. Bacteriol.* 180: 5828–5835.

Husain, M., B. Entsch, D.P. Ballou, V. Massey, and P.J Chapman. 1980. Fluoride elimination from substrates in hydroxylation reactions catalyzed by *p*-hydroxybenzoate hydroxylase. *J. Biol. Chem.* 255: 4189–4197.

Hyman, M.R., C.L. Page, and D.J. Arp. 1994. Oxidation of methyl fluoride and dimethyl ether by ammonia monooxygenase in *Nitrosomonas europaea*. *Appl. Environ. Microbiol.* 60: 3033–3035.

Ibrahim, A.-R. and Y.J. Abul-Hajj. 1989. Aromatic hydroxylation and sulfation of 5-hydroxyflavone by *Streptomyces fulvissimus*. *Appl. Environ. Microbiol.* 55: 3140–3142.

Ibrahim, A.-R.S. and Y.J. Abul-Hajj. 1990. Microbiological transformation of flavone and isoflavone. *Xenobiotica* 20: 363–373.

Imhoff, D. and J.R. Andreesen. 1979. Nicotinic acid hydroxylase from *Clostridium barkeri*: selenium-dependent formation of active enzyme. *FEMS Microbiol. Lett.* 5: 155–158.

Imhoff-Stuckle, D. and N. Pfennig. 1983. Isolation and characterization of a nicotinic acid-degrading sulfate-reducing bacterium, *Desulfococcus niacini* sp. nov. *Arch. Microbiol.* 136: 194–198.

Ince, J.E. and C.J. Knowles. 1986. Ethylene formation by cell-free extracts of *Escherichia coli*. *Arch. Microbiol.* 146: 151–158

Ingledew, W.M., M.E.F. Tresguerres, and J.L. Cánovas. 1971. Regulation of the enzymes of the hydroaromatic pathway in *Acinetobacter calco-aceticus*. *J. Gen. Microbiol.* 68: 273–282.

Itoh, N., R. Morihama, J. Wang, K. Okada, and N. Mizuguchi. 1997. Purification and characterization of phenylacetaldehyde reductase from a styrene-assimilating *Corynebacterium* strain ST-10. *Appl. Environ. Microbiol.* 63: 3783–3788.

Iwabuchi, T. and S. Harayama. 1997. Biochemical and genetic characterization of 2-carboxybenzaldehyde dehydrogenase, an enzyme involved in phenanthrene degradation by *Nocardioides* sp. strain KP7. *J. Bacteriol.* 179: 6488–6494.

Iwabuchi, T. and S. Harayama. 1998a. Biochemical and molecular characterization of 1-hydroxy-2-naphthoate dioxygenase from *Nocardioides* sp. KP7. *J. Biol. Chem.* 273: 8332–8336.

Iwabuchi, T. and S. Harayama. 1998b. Biochemical and genetic characterization of *trans*-2'-carboxybenzalpyruvate hydratase-aldolase from a phenanthrene-degrading *Nocardioides* strain. *J. Bacteriol.* 180: 945–949.

Jackson, J.-A., W.R. Blair, F.E. Brinckman, and W.P. Iverson. 1982. Gas-chromatographic speciation of methylstannanes in the Chesapeake Bay using purge and trap sampling with a tin-selective detector. *Environ. Sci. Technol.* 16: 110–119.

Jacob, G.S., J.R. Garbow, L.E. Hallas, N.M. Kimack, G.N. Kishore, and J. Schaefer. 1988. Metabolism of glyphosate in *Pseudomonas* sp. strain LBr. *Appl. Environ. Microbiol.* 54: 2953–2958.

Jagnow, G., K. Haider, and P.-C. Ellwardt. 1977. Anaerobic dechlorination and degradation of hexachlorocyclohexane by anaerobic and facultatively anaerobic bacteria. *Arch. Microbiol.* 115: 285–292.

Jain, R.K., J.H. Dreisbach, and J.C. Spain. 1994. Biodegradation of *p*-nitrophenol via 1,2,4-benzenetriol by an *Arthrobacter* sp. *Appl. Environ. Microbiol.* 60: 3030–3032.

James, K.D. and P.A. Williams. 1998. *ntn* genes determining the early steps in the divergent catabolism of 4-nitrotoluene and toluene in *Pseudomonas* sp. strain TW3. *J. Bacteriol.* 180: 2043–2049.

Janssen, D.B., F. Pries, and J.R. van der Ploeg. 1994. Genetics and biochemistry of dehalogenating enzymes. *Annu. Rev. Microbiol.* 48: 163–191.

Janssen, P.H. and W. Liesack. 1995. Succinate decarboxylation by *Propionigenium maris* sp. nov., a new anaerobic bacterium from an estuarine sediment. *Arch. Microbiol.* 164: 29–35.

Janssen, P.H. and B. Schink. 1995a. Metabolic pathways and energetics of the acetone-oxidizing sulfate-reducing bacterium *Desulfobacterium cetonicum*. *Arch. Microbiol.* 163: 188–194.

Janssen, P.H. and B. Schink. 1995b. Pathway of butyrate catabolism by *Desulfobacterium cetonicum*. *J. Bacteriol.* 177: 3870–3872.

Jeffrey, A.M., H.J.C. Yeh, D.M. Jerina, T.R. Patel, J.F. Davey, and D.T. Gibson. 1975. Initial reactions in the oxidation of naphthalene by *Pseudomonas putida*. *Biochemistry* 14: 575–584.

Jensen, H.L. 1960. Decomposition of chloroacetates and chloropropionates by bacteria. *Acta Agric. Scand.* 10: 83–103.

Jensen, S. and A. Jernelöv. 1969. Biological methylation of mercury in aquatic organisms. *Nature* (London) 233: 753–754.

John, D.M. and G.F. White. 1998. Mechanism for biotransformation of nonylphenol polyethoxylates to xenoestrogens in *Pseudomonas putida*. *J. Bacteriol.* 180: 4332–4338.

Johnson, G.R. and R.H. Olsen. 1995. Nucleotide sequence analysis of genes encoding a toluene/benzene-2-monooxygenase from *Pseudomonas* sp. strain JS150. *Appl. Environ. Microbiol.* 61: 3336–3346.

Johnson, G.R. and R.H. Olsen. 1997. Multiple pathways for toluene degradation in *Burkholderia* sp. strain JS150. *Appl. Environ. Microbiol.* 63: 4047–4052.

Jones, K.H., R.T. Smith, and P.W. Trudgill. 1993. Diketocamphane enantiomer-specific 'Bayer-Villiger' monooxygenases from camphor-grown *Pseudomonas putida* ATCC 17453. *J. Gen. Microbiol.* 139: 797–805.

Jordan, S.L., A.J. Kraczkiewicz-Dowjat, D.P. Kelly, and A.P. Wood. 1995. Novel eubacteria able to grow on carbon disulfide. *Arch. Microbiol.* 163: 131–137.

Jordan, S.L., I.R. McDonald, A.J. Kraczkiewicz-Dowjat, D.P. Kelly, F.A. Rainey, J.-C. Murrell, and A.P. Wood. 1997. Autotrophic growth on carbon disulfide is a property of novel strains of *Paracoccus denitrificans*. *Arch. Microbiol.* 168: 225–236.

Joshi, D.K. and M.H. Gold. 1993. Degradation of 2,4,5-trichlorophenol by the lignin-degrading basidiomycete *Phanerochaete chrysosporium*. *Appl. Environ. Microbiol.* 59: 1779–1785.

Juhasz, A.L., M.L. Britz, and G.A. Stanley. 1997a. Degradation of fluoranthene, pyrene, benz(*a*)anthracene and dibenz(*a,h*)anthracene by *Burkholderia cepacia*. *J. Appl. Microbiol.* 83: 189–198.

Juhasz, A.L., M.L. Britz, and G.A. Stanley. 1997b. Degradation of benzo[*a*]pyrene, dibenz[*a,h*]anthracene and coronene by *Burkholderia cepacia*. *Water Sci. Technol.* 36: 45–51.

Jutzi, K., A.M. Cook, and R. Hütter. 1982. The degradative pathway of the *s*-triazine melamine. *Biochem. J.* 208: 679–684.

Kadiyala, V. and J.C. Spain. 1998. A two-component monoxygenase catalyzes both the hydroxylation of *p*-nitrophenol and the oxidative release of nitrite from 4-nitrocatechol in *Bacillus sphaericus* JS905. *Appl. Environ. Microbiol.* 64: 2479–2484.

Käppeli, O. 1986. Cytochromes P-450 of yeasts. *Microbiol. Rev.* 50: 244–258.

Kaschabek, S.R., T. Kasberg, D. Müller, A.E. Mars, D.B. Janssen, and W. Reineke. 1998. Degradation of chloroaromatics: purification and characterization of a novel type of chlorocatechol 2,3-dioxygenase of *Pseudomonas putida* GJ31. *J. Bacteriol.* 180: 296–302.

Katayama, Y., Y. Matsushita, M. Kaneko, M. Kondo, T. Mizuno, and H. Nyunoya. 1998. Cloning of genes coding for the three subunits of thiocyanate hydrolase of *Thiobacillus thioparus* THI 115 and their evolutionary relationships to nitrile hydratase. *J. Bacteriol.* 180: 2583–2589.

Kaufman, D.D., J. Plimmer, J. Iwan, and U.I. Klingebiel. 1972. 3,3′,4,4′-Tetrachloroazoxybenzene from 3,4-dichloroaniline. *J. Agric. Food Chem.* 20: 916–919.

Kaufman, D.D., J. Plimmer, and U.I. Klingebiel. 1973. Microbial oxidation of 4-chloroaniline. *J. Agric. Food Chem.* 21: 127–132.

Kawamoto, S., M. Kobayashi, A. Tanaka, and S. Fukui. 1977. Production of D-amino acid oxidase by *Candida tropicalis*. *J. Ferment. Technol.* 55: 13–18.

Kazumi, J., M.E. Caldwell, J.M. Suflita, D.R. Lovley, and L.Y. Young. 1997. Anaerobic degradation of benzene in diverse anoxic environments. *Environ. Sci. Technol.* 31: 813–818.

Kelley, I., J.P. Freeman, F.E. Evans, and C.E. Cerniglia. 1993. Identification of metabolites from the degradation of fluoranthene by *Mycobacterium* sp. strain PYR-1. *Appl. Environ. Microbiol.* 59: 800–806.

Kelly, D.P., S.C. Baker, J. Trickett, M. Davey, and J.C. Murrell. 1994. Methanesulphonate utilization by a novel methylotrophic bacterium involves an unusual monooxygenase. *Microbiology* (U.K.) 140: 1419–1426.

Kende, H. 1989. Enzymes of ethylene biosynthesis. *Plant Physiol.* 91: 1–4.

Kersten, P.J., S. Dagley, J.W. Whittaker, D.M. Arciero, and J.D. Lipscomb. 1982. 2-Pyrone-4,6-dicarboxylic acid, a catabolite of gallic acids in *Pseudomonas* species. *J. Bacteriol.* 152: 1154–1162.

Key, B.D., R.D. Howell, and C.S. Criddle. 1998. Defluorination of organofluorine sulfur compounds by *Pseudomonas* sp. strain D2. *Environ. Sci. Technol.* 32: 2283–2287.

Kido, T. and K. Soda. 1976. A new oxygenase, 2-nitropropane dioxygenase of *Hansenula mrakii*. Enzymologic and spectrophotometric properties. *J. Biol. Chem.* 251: 6994–7000.

Kiene, R.P. and B.F. Taylor. 1988. Demethylation of dimethylsulfoniopropionate and production of thiols in anoxic marine sediments. *Appl. Environ. Microbiol.* 54: 2208–2212.

Kim, Y., P. Ayoubi, and A.R. Harker. 1996. Constitutive expression of the cloned phenol hydroxylase gene(s) from *Alcaligenes eutrophus* JMP134 and concomitant trichloroethene oxidation. *Appl. Environ. Microbiol.* 62: 3227–3233.

Kimura, N., A. Nishi, M. Goto, and K. Furukawa. 1997. Functional analysis of a variety of chimeric dioxygenases constructed from two biphenyl dioxygenases that are similar structurally but different functionally. *J. Bacteriol.* 179: 3936–3943

King, D.J., M.R. Azari, and A. Wiseman. 1984. Studies on the properties of highly purified cytochrome P-448 and its dependent activity benzo[*a*]pyrene hydroxylase, from *Saccharomyces cerevisiae*. *Xenobiotica* 14: 187–206.

King, G.M. 1984. Metabolism of trimethylamine, choline and glycine betaine by sulfate-reducing and methanogenic bacteria in marine sediments. *Appl. Environ. Microbiol.* 48: 719–725.

King, G.M. 1988. Dehalogenation in marine sediments containing natural sources of halophenols. *Appl. Environ. Microbiol.* 54: 3079–3085.

Kitazume, T. and N. Ishikawa. 1983. Asymmetrical reduction of perfluoroalkylated ketones, ketoesters and vinyl compounds with baker's yeast. *Chem. Lett.* 237–238.

Kiyohara, H. and K. Nagao. 1978. The catabolism of phenanthrene and naphthalene by bacteria. *J. Gen. Microbiol.* 105: 69–75

Kiyohara, H., K. Nagao, and R. Nomi. 1976. Degradation of phenanthrene through *o*-phthalate by an *Aeromonas* sp. *Agric. Biol. Chem.* 40: 1075–1082.

Klecka, G.M. and D.T. Gibson. 1981. Inhibition of catechol 2,3-dioxygenase from *Pseudomonas putida* by 3-chlorocatechol. *Appl. Environ. Microbiol.* 41: 1159–1165.

Knoll, G. and J. Winter. 1989. Degradation of phenol via carboxylation to benzoate by a defined, obligate syntrophic consortium of anaerobic bacteria. *Appl. Microbiol. Biotechnol.* 30: 318–324.

Koch, J. and G. Fuchs. 1992. Enzymatic reduction of benzoyl-CoA to alicyclic compounds, a key reaction in anaerobic aromatic metabolism. *Eur. J. Biochem.* 205: 195–202

Kodama, K., K. Umehara, K. Shikmizu, S. Nakatani, Y. Minoda, and K. Yamada. 1973. Identification of microbial products from dibenzothiophene and its proposed oxidation pathway. *Agric. Biol. Chem.* 37: 45–50.

Kohler-Staub, D. and H.-P.E. Kohler. 1989. Microbial degradation of beta-chlorinated four-carbon aliphatic acids. *J. Bacteriol.* 171: 1428–1434.

Kohring, G.-W., X. Zhang, and J. Wiegel. 1989. Anaerobic dechlorination of 2,4-dichlorophenol in freshwater sediments in the presence of sulfate. *Appl. Environ. Microbiol.* 55: 2735–2737.

Koizumi, M., M. Shimizu, and K. Kobashi. 1990. Enzymatic sulfation of quercitin by arylsulfotransferase from a human intestinal bacterium. *Chem. Pharm. Bull.* 38: 794–796.

Kok, M., R. Oldenius, M.P.G. van der Linden, C.H.C. Meulenberg, J. Kingma, and B. Witholt. 1989a. The *Pseudomonas oleovorans alkBAC* operon encodes two structurally related rubredoxins and an aldehyde dehydrogenase. *J. Biol. Chem.* 264: 5442–5451.

Kok, M., R. Oldenius, M.P.G. van der Linden, P. Raatjes, J. Kingma, P.H. van Lelyveld, and B. Witholt. 1989b. The *Pseudomonas oleovorans* alkane hydrolysis gene. Sequence and expression. *J. Biol. Chem.* 264: 5436–5442.

Kolattukudy, P.E. and L. Hankin. 1968. Production of ω-haloesters from alkyl halides by *Micrococcus cerificans*. *J. Gen. Microbiol.* 54: 333–336.

Kolb, S., S. Seeliger, N. Springer, W. Ludwig, and B. Schink. 1998. The fermenting bacterium *Malonomonas rubra* is phylogenetically related to sulfur-reducing bacteria and contains a *c*-type cytochrome similar to those of sulfur and sulfate reducers. *Syst. Appl. Microbiol.* 21: 340–345.

Kondo, R., H. Yamagami, and K. Sakai. 1993. Xylosation of phenolic hydroxyl groups of the monomeric lignin model compounds 4-methylguaiacol and vanillyl alcohol by *Coriolus versicolor*. *Appl. Environ. Microbiol.* 59: 438–441.

Konig, H. and K.O. Stetter. 1982. Isolation and characterization of *Methanolobus tindarius*, sp. nov., a coccoid methanogen growing only on methanol and methylamines. *Zentralbl. Bakteriol. Parasitenkd. Infektionskr. Hyg. Abt 1* C3: 478–490.

Konopka, A., D. Knoght, and R.F. Turco. 1989. Characterization of a *Pseudomonas* sp. capable of aniline degradation in the presence of secondary carbon sources. *Appl. Environ. Microbiol.* 55: 385–389.

Kozarich, J.W. 1988. Enzyme chemistry and evolution in the β-ketoadipate pathway. In *Microbial Metabolism and the Carbon Cycle* pp. 283–302. (Eds. S.R. Hagedorn, R.S. Hanson, and D.A. Kunz). Harwood Academic Publishers, Chur, Switzerland.

Kredich, N.M., L.J. Foote, and B.S. Keenan. 1973. The stoichiometry and kinetics of the inducible cysteine desulfhydrase from *Salmonella typhimurium*. *J. Biol. Chem.* 248: 6187–6196.

Kreft, J.-U. and B. Schink. 1993. Demethyation and degradation of phenylmethylethers by the sulfide-methylating homoacetogenic bacterium strain TMBS 4. *Arch. Microbiol.* 159: 308–315.

Kretzer, A. and J.R. Andreesen. 1991. A new pathway for isonicotinate degradation by *Mycobacterium* sp. INA1. *J. Gen. Microbiol.* 137: 1073–1080.

Krishnamurty, H.G. and F.J. Simpson. 1970. Degradation of rutin by *Aspergillus flavus*. Studies with oxygen 18 on the action of a dioxygenase on quercitin. *J. Biol. Chem.* 245: 1467–1471.

Krishnamurty, H.G., K.-J. Cheng, G.A. Jones, F.J. Simpson, and J.E. Watkin. 1970. Identification of products produced by the anaerobic degradation of rutin and related flavonoids by *Butyrivibrio* sp. C_3. *Can. J. Microbiol.* 16: 759–767.

Kropp, K.G., J.T. Andersson, and P.M. Fedorak. 1997. Bacterial transformations of naphthothiophenes. *Appl. Environ. Microbiol.* 63: 3463–3473.

Krumholz, L.R. 1997. *Desulfuromonas chloroethenica* sp. nov. uses tetrachloroethylene and trichloroethylene as electrom donors. *Int. J. Syst. Bacteriol.* 47: 1262–1263.

Krumholz, L.R. and M.P. Bryant. 1985. *Clostridium pfennigi* sp. nov. uses methoxyl groups of monobenzenoids and produces butyrate. *Appl. Environ. Microbiol.* 35: 454–456.

Krumholz, L.R. and M.P. Bryant. 1988. Characterization of the pyrogallol-phloroglucinol isomerase of *Eubacterium oxidoreducens*. *J. Bacteriol.* 170: 2472–2479.

Krumholz, L.R., R.L. Crawford, M.E. Hemling, and M.P. Bryant. 1987. Metabolism of gallate and phloroglucinol in *Eubacterium oxidoreducens* via 3-hydroxy-5-oxohexanoate. *J. Bacteriol.* 169: 1886–1890.

Krumholz, L.R., R. Sharp, and S.S. Fishbain. 1996. A freshwater anaerobe coupling acetate oxidation to tetrachloroethylene dehalogenation. *Appl. Environ. Microbiol.* 62: 4108–4113.

Kuever, J., J. Kulmer, S. Janssen, U. Fischer, and K.-H. Blotevogel. 1993. Isolation and characterization of a new spore-forming sulfate-reducing bacterium growing by complete oxidation of catechol. *Arch. Microbiol.*159: 282–288.

Kuhm, A.E., A. Stolz, K.-L. Ngai, and H.-J Knackmuss. 1991. Purification and characterization of a 1,2-dihydroxynaphthalene dioxygenase from a bacterium that degrades naphthalenesulfonic acids. *J. Bacteriol.* 173: 3795–3802.

Kuhn, E.P., G.T. Townsend, and J.M. Suflita. 1990. Effect of sulfate and organic carbon supplements on reductive dehalogenation of chloroanilines in anaerobic aquifer slurries. *Appl. Environ. Microbiol.* 56: 2630–2637.

Kulla, H.G., F. Klausener, U. Meyer, B. Lüdeke, and T. Leisinger. 1983. Interference of aromatic sulfo groups in the microbial degradation of the aza dyes Orange I and Orange II. *Arch. Microbiol.* 135: 1–7.

Kung, H.-F. and T.C. Stadtman. 1971. Nicotinic acid metabolism VI. Purification and properties of alpha-methyleneglutarate mutase (B_{12}-dependent) and methylitaconate isomerase. *J. Biol. Chem.* 246: 3378–3388.

Kung, H.-F. and L. Tsai. 1971. Nicotinic acid metabolism VII. Mechanism of action of clostridial alpha-methyleneglutarate mutase (B_{12}-dependent) and methylitaconate isomerase. *J. Biol. Chem.* 246: 6436–6443.

Kunz, D.A. and P.J. Weimer. 1983. Bacterial formation and metabolism of 6-hydroxyhexanoate: evidence of a potential role for ω-oxidation. *J. Bacteriol.* 156: 567–575.

La Nauze, J.M. and H. Rosenberg. 1968. The identification of 2-phosphonoacetaldehyde as an intermediate in the degradation of 2-aminoethylphosphonate by *Bacillus cereus*. *Biochim. Biophys. Acta* 165: 438–447.

La Nauze, J.M., H. Rosenberg, and D.C. Shaw. 1970. The enzymatic cleavage of the carbon–phosphorus bond: purification and properties of phosphonatase. *Biochim. Biophys. Acta* 212: 332–350.

La Roche, S.D. and T. Leisinger. 1991. Identification of *dcmR*, the regulatory gene governing expression of dichloromethane dehalogenase in *Methylobacterium* sp. strain DM4. *J. Bacteriol.* 173: 6714–6721.

Laanio, T.L, P.C. Kearney, and D.D. Kaufman. 1973. Microbial metabolism of dinitramine. *Pestic. Biochem. Physiol.* 3: 271–277.

Lack, A. and G. Fuchs. 1992. Carboxylation of phenylphosphate by phenol carboxylase, an enzyme system of anaerobic phenol metabolism. *J. Bacteriol.* 174: 3629–3636.

Lambert, M., S. Kremer, O. Sterner, and A. Anke. 1994. Metabolism of pyrene by the basidiomycete *Crinipellis stipitaria* and identification of pyrenequinones and their hydroxykated precursors in strain JK375. *Appl. Environ. Microbiol.* 60: 3597–3601.

Lange, B., S. Kremer, O. Sterner, and A. Anke. 1994. Pyrene metabolism in *Crinipellis stipitaria*: identification of *trans*-4,5-dihydro-4,5-dihydroxypyrene and 1-pyrenylsulfate in strain KJ364. *Appl. Environ. Microbiol.* 60: 3602–3607.

Larkin, M.J. 1988. The specificity of 1-naphthol oxygenases from three bacterial isolates, *Pseudomonas* spp. (NCIB 12042 and 12043) and *Rhodococcus* sp. (NCIB 12038) isolated from garden soil. *FEMS Microbiol. Lett.* 52: 173–176.

Latus, M., H.-J. Seitz, J. Eberspächer, and F.Lingens. 1995. Purification and characterization of hydroxyquinol 1,2-dioxygenase from *Azotobacter* sp. strain GP1. *Appl. Environ. Microbiol.* 61: 2453–2460.

Launen, L., L. Pinto, C. Wiebe, E. Kiehlmann, and M. Moore. 1995. The oxidation of pyrene and benzo(*a*)pyrene by nonbasidiomycete soil fungi. *Can. J. Microbiol.* 41: 477–488.

Layh, N., A. Stolz, S. Förster, F. Effenberger, and H.-J. Knackmuss. 1992. Enantioselective hydrolysis of *O*-acetylmandelonitrile to *O*-acetylmandelic acid by bacterial nitrilases. *Arch. Microbiol.* 158: 405–411.

Lee, C.H., P.C. Oloffs, and S.Y. Szeto. 1986. Persistence, degradation, and movement of triclopyr and its ethylene glycol butyl ether ester in a forest soil. *J. Agric. Food Chem.* 34: 1075–1079.

Lee, J.-Y. and L. Xun. 1997. Purification and characterization of 2,6-dichloro-*p*-hydroquinone chlorohydrolase from *Flavobacterium* sp. strain ATCC 39723. *J. Bacteriol.* 179: 1521–1524.

Lee, K. and D.T. Gibson. 1996. Stereospecific dihydroxylation of the styrene vinyl group by purified naphthalene dioxygenase from *Pseudomonas* sp. strain NCIB 9816-4. *J. Bacteriol.* 178: 3353–3356.

Lee, K., S.M. Resnick, and D.T. Gibson. 1997. Stereospecific oxidation of (*R*)- and (*S*)-1-indanol by naphthalene dioxygenase from *Pseudomonas* sp. strain NCIB 9816-4. *Appl. Environ. Microbiol.* 63: 2067–2070.

Lendenmann, U. and J.C. Spain. 1996. 2-Aminophenol 1,6-dioxygenase: a novel aromatic ring cleavage enzyme purified from *Pseudomonas pseudoalcaligenes* JS 45. *J. Bacteriol.* 178: 6227–6232.

Lenke, H. and H.-J. Knackmuss. 1992. Initial hydrogenation during catabolism of picric acid by *Rhodococcus erythropolis* HL 24-2. *Appl. Environ. Microbiol.* 58: 2933–2937.

Lenke H. and H.-J. Knackmuss. 1996. Initial hydrogenation and extensive reduction of substituted 2,4-dinitrophenols. *Appl. Environ. Microbiol.* 62: 784–790.

Lenke, H., D.H. Pieper, C. Bruhn, and H.-J. Knackmuss. 1992. Degradation of 2,4-dinitrophenol by two *Rhodococcus erythropolis* strains, HL 24-1 and HL 24-2. *Appl. Environ. Microbiol.* 58: 2928–2932.

Leppik, R.A. 1989. Steroid catechol degradation: disecoandrostane intermediates accumulated by *Pseudomonas* transposon mutant strains. *J. Gen. Microbiol.* 135: 1979–1988.

Leuthner, B., C. Leutwein, H. Schulz, P. Hörth, W. Haehnel, E. Schiltz, H. Schägger, and J. Heider. 1998. Biochemical and genetic characterization of benzylsuccinate synthase from *Thauera aromatica*: a new glycyl-radical catalysing the first step in anaerobic toluene degradation. *Mol. Microbiol.* 28: 515–628.

Levitt, M.S., R.F. Newton, S.M. Roberts, and A.J. Willetts. 1990. Preparation of optically active 6'-fluorocarbocyclic nucleosides utilizing an enantiospecific enzyme-catalysed Baeyer-Villiger type oxidation. *J. Chem. Soc. Chem. Commun.* 619–620.

Lewis, T.A. and R.L. Crawford. 1995. Transformation of carbon tetrachloride via sulfur and oxygen substitution by *Pseudomonas* sp. strain KC. *J. Bacteriol.* 177: 2204–2208.

Ley, S.V., F. Sternfield, and S.Taylor. 1987. Microbial oxidation in synthesis: a six step preparation of (+/–)-pinitol from benzene. *Tetrahedron Lett.* 28: 225–226.

Li, D.-Y., J. Eberspächer, B. Wagner, J. Kuntzer, and F. Lingens. 1992. Degradation of 2,4,6-trichlorophenol by *Azotobacter* sp. strain GP1. *Appl. Environ. Microbiol.* 57: 1920–1928.

Li, S. and L.P. Wackett. 1993. Reductive dehalogenation by cytochrome $P450_{CAM}$: substrate binding and catalysis. *Biochemistry* 32: 9355–9361.

Li, Y.-F., Y. Hata, T. Fujii, T. Hisano, M. Nishihara, T. Kurihara, and N. Esaki. 1998. Crystal structure of reaction intermediates of L-2-haloacid dehalogenase and implications for the reaction mechanism. *J. Biol. Chem.* 273: 15035–15044.

Licht, D., B.K. Ahring, and E. Arvin. 1996. Effects of electron acceptors, reducing agents, and toxic metabolites on anaerobic degradation of heterocyclic compounds. *Biodegradation* 7: 83–90.

Lieberman, I. and A. Kornberg. 1955. Enzymatic synthesis and breakdown of a pyrimidine, orotic acid III. Ureidosuccinase. *J. Biol. Chem.* 212: 909–920.

Liebert, F. 1909. The decomposition of uric acid by bacteria. *Proc. K. Acad. Ned. Wet.* 12: 54–64.

Little, C.D, A.V. Palumbo, S.E. Herbes, M.E. Lidstrom, R.L. Tyndall, and P.J. Gilmer. 1988. Trichloroethylene biodegradation by a methane-oxidizing bacterium. *Appl. Environ. Microbiol.* 54: 951–956.

Little, P.J., M.O. James, J.B. Pritchard, and J.R. Bend. 1984. Benzo(*a*)pyrene metabolism in hepatic microsomes from feral and 3-methylcholanthrene-treated southern flounder, *Paralichthys lethostigma*. *J. Environ. Pathol. Toxicol. Oncol.* 5: 309–320.

Liu, C.-M., P.A. McLean, C.C. Sookdeo, and F.C. Cannon. 1991. Degradation of the herbicide glyphosate by members of the family Rhizobiaceae. *Appl. Environ. Microbiol.* 57: 1799–1804.

Liu, S. and J.M. Suflita. 1993. H_2-CO_2-dependent anaerobic O-demethylation activity in subsurface sediments and by an isolated bacterium. *Appl. Environ. Microbiol.* 59: 1325–1331.

Lizama, H.M., L.A. Wilkins, and T.C. Scott. 1995. Dibenzothiophene sulfur can serve as the sole electron acceptor during growth by sulfate-reducing bacteria. *Biotechnol. Lett.* 17: 113–116.

Lobos, J.H., T.K. Leib, and T.-M. Su. 1992. Biodegradation of bisphenol A and other bisphenols by a Gram-negative aerobic bacterium. *Appl. Environ. Microbiol.* 58: 1823–1831.

Locher, H.H., T. Leisinger, and A.M. Cook. 1991. 4-Toluene sulfonate methyl-monooxygenase from *Comamonas testosteroni*: purification and some properties of the oxygenase component. *J. Bacteriol.* 173: 3741–3748.

Lochmeyer, C., J. Koch, and G. Fuchs. 1992. Anaerobic degradation of 2-aminobenzoic acid (anthranilic acid) via benzoyl-coenzyme A (CoA) and cyclohex-1-enecarboxyl-CoA in a denitrifying bacterium. *J. Bacteriol.* 174: 3621–3628.

Löffler F.E., R.A. Sanford, and J.M. Tiedje. 1996. Initial characterization of a reductive dehalogenase from *Desulfitobacterium chlororespirans* Co23. *Appl. Environ. Microbiol.* 62: 3809–3813.

Londry, K.L. and P.M. Fedorak. 1993. Use of fluorinated compounds to detect aromatic metabolites from *m*-cresol in a methanogenic consortium: evidence for a demethylation reaction. *Appl. Environ. Microbiol.* 59: 2229–2238.

Londry, K.L., P.M. Fedorak, and J.M. Suflita. 1997. Anaerobic degradation of *m*-cresol by a sulfate-reducing bacterium. *Appl. Environ. Microbiol.* 63: 3170–3175.

Lontoh, S. and J.D. Semrau. 1998. Methane and trichloroethylene degradation by *Methylosinus trichosporium* OB3b expressing particulate methane monooxygenase. *Appl. Environ. Microbiol.* 64: 1106–1114.

Louie, T.M. and W.W. Mohn. 1999. Evidence for a chemiosmotic model of dehalorespiration in *Desulfomonile tiedjei* DCB-1. *J. Bacteriol.* 181: 40–46.

Louie, T.M., S. Ni., L. Xun, and W.W. Mohn. 1997. Purification, characterization and gene sequence analysis of a novel cytochrome *c* coinduced with reductive dechlorination activity in *Desulfomonile tiedjei* DCB-1. *Arch. Microbiol.* 168: 520–527.

Lutz-Wahl, S., P. Fischer, C. Schmidt-Dannert, W. Wohlleben, B. Hauer, and R.D. Schmid. 1998. Stereo- and regioselective hydroxylation of α-ionone by *Streptomyces* strains. *Appl. Environ. Microbiol.* 64: 3878–3881.

Mackie, R.I., B.A. White, and M.P. Bryant. 1991. Lipid metabolism in anaerobic ecosystems. *Crit. Rev. Microbiol.* 17: 449–479.

Macpherson, T., C.W. Greer, E. Zhou, A.M. Jones, G. Wisse, P.C.K. Lau, B. Sankey, M.J. Grossman, and J. Hawari. 1998. Application of SPME/GC-MS to characterize metabolites in the biodesulfurization of organosulfur model compounds in bitumen. *Environ. Sci. Technol.* 32: 421–426.

Madsen, T. and J. Aamand. 1991. Effects of sulfuroxy anions on degradation of pentachlorophenol by a methanogenic enrichment culture. *Appl. Environ. Microbiol.* 57: 2453–2458.

Madsen, T. and D. Licht. 1992. Isolation and characterization of an anaerobic chlorophenol-transforming bacterium. *Appl. Environ. Microbiol.* 58: 2874–2878.

Maeda, M., S.-Y. Chung, E. Song, and T. Kudo. 1995. Multiple genes encoding 2,3-dihydroxybiphenyl 1,2-dioxygenase in the Gram-positive polychlorinated biphenyl-degrading bacterium *Rhodococcus erythropolis* TA421, isolated from a termite ecosystem. *Appl. Environ. Microbiol.* 61: 549–555.

Mägli, A., M. Wendt, and T. Leisinger. 1996. Isolation and characterization of *Dehalobacterium formicoaceticum* gen. nov., sp. nov., a strictly anaerobic bacterium utilizing dichloromethane as source of carbon and energy. *Arch. Microbiol.* 166: 101–108.

Mägli, A., M. Messmer, and T. Leisinger. 1998. Metabolism of dichloromethane by the strict anaerobe *Dehalobacterium formicoaceticum. Appl. Environ. Microbiol.* 64: 646–650.

Mahaffey, W.R., D.T. Gibson, and C.E. Cerniglia. 1988. Bacterial oxidation of chemical carcinogens: formation of polycyclic aromatic acids from benz[a]anthracene. *Appl. Environ. Microbiol.* 54: 2415–2423.

Mandelstam, J., K. McQuillen, and I. Dawes.1982. *Biochemistry of Bacterial Growth.* Blackwell Scientific Publications, Oxford.

Mansouri, S., and A.W. Bunch. 1989. Bacterial synthesis from 2-oxo-4-thiobutyric acid and from methionine. *J. Gen. Microbiol.* 135: 2819–2827.

Mars, A.E., J. Kingma, S.R. Kaschabek, W. Reinke, and D.B. Janssen. 1999. Conversion of 3-chlorocatechol by various catechol 2,3-dioxygenases and sequence analysis of the chlorocatechol dioxygense region of *Pseudomonas putida* GJ31. *J. Bacteriol.* 181: 1309–1318.

Massé, R., D. Lalanne, F. Mssier, and M. Sylvestre. 1989. Characterization of new bacterial transformation products of 1,1,1-trichloro-2,2-bis-(4-chlorophenyl)ethane (DDT) by gas chromatography/mass spectrometry. *Biomed. Environ. Mass Spectrom.* 18: 741–752.

Masuda, N., H. Oda, S. Hirano, M. Masuda, and H. Tanaka. 1984. 7α-Dehydroxylation of bile acids by resting cells of a *Eubacterium lentum*-like intestinal anaerobe, strain c-25. *Appl. Environ. Microbiol.* 47: 735–739.

Matties, C. and B. Schink. 1992. Fermentative degradation of glutarate via decarboxylation by newly isolated strictly anaerobic bacteria. *Arch. Microbiol.* 157: 290–296.

May, S.W. and B.J. Abbott. 1973. Enzymatic epoxidation. II. Comparison between the epoxidation and hydroxylation reactions catalyzed by the omega-hydroxylation system of *Pseudomonas oleovorans. J. Biol. Chem.* 248: 1725–1730.

McBride, K.E., J.W. Kenny, and D.M. Stalker. 1986. Metabolism of the herbicide bromoxynil by *Klebsiella pneumoniae* subspecies *ozaenae. Appl. Environ. Microbiol.* 52: 325–330.

McCall, P.J., S.A. Vrona, and S.S. Kelly. 1981. Fate of uniformly carbon-14 ring labelled 2,4,5-trichlorophenoxyacetic acid and 2,4-dichlorophenoxyacetic acid. *J. Agric. Food Chem.* 29: 100–107.

McClay, K., S.H. Streger, and R.J. Steffan. 1995. Induction of toluene oxidation in *Pseudomonas mendocina* KR1 and *Pseudomonas* sp. strain ENVPC5 by chlorinated solvents and alkanes. *Appl. Environ. Microbiol.* 61: 3479–3481.

McClay, K., B.G. Fox, and R.J. Steffan. 1996. Chloroform mineralization by tolueneoxidizing bacteria. *Appl. Environ. Microbiol.* 62: 2716–2732.

McClure, N.C. and W.A. Venables. 1986. Adaptation of *Pseudomonas putida* mt-2 to growth on aromatic amines. *J. Gen. Microbiol.* 132: 2209–2218.

McClure, N.C. and W.A. Venables. 1987. pTDN1, a catabolic plasmid involved in aromatic amine catabolism in *Pseudomonas putida* mt-2. *J. Gen. Microbiol.* 133: 2073–2077.

McCormick, N.G., F.E. Feeherry, and H.S. Levinson. 1976. Microbial transformation of 2,4,6-trinitrotoluene and other nitroaromatic compounds. *Appl. Environ. Microbiol.* 31: 949–958.

McCormick, N.G., J.H. Cornell, and A.M. Kaplan. 1978. Identification of biotransformation products from 2,4-dinitrotoluene. *Appl. Environ. Microbiol.* 35: 945–948.

McInerney, M.J., M.P. Bryant, R.B. Hespell, and J.W. Costerton. 1981. *Syntrophomonas wolfei* gen. nov. sp. nov., an anaerobic, syntrophic, fatty acid-oxidizing bacterium. *Appl. Environ. Microbiol.* 41: 1029–1039.

McKenna, E.J. and R.E. Kallio. 1971. Microbial metabolism of the isoprenoid alkane pristane. *Proc. Natl. Acad. Sci. U.S.A.* 68: 1552–1554.

McNally, D.L., J.R. Mihelcic, and D.R. Lueking. 1998. Biodegradation of three- and four-ring polycyclic aromatic hydrocarbons under aerobic and denitrifying conditions. *Environ. Sci. Technol.* 32: 2633–2639.

McNally, K.J., J.T.G. Hamilton, and D.B. Harper. 1990. The methylation of benzoic and *n*-butyric acids by chloromethane in *Phellinus pomaceus. J. Gen. Microbiol.* 136: 1509–1515.

Meng, M., W.-Q. Sun, L.A. Geelhaar, G. Kumar, A.R. Patel, G.F. Payne, M.K. Speedie, and J.R. Stacy. 1995. Denitration of glycerol trinitrate by resting cells and cell extracts of *Bacillus thuringiensis/cereus* and *Enterobacter agglomerans. Appl. Environ. Microbiol.* 61: 2548–2553.

Meßmer, M., G. Wohlfarth, and G. Diekert. 1993. Methyl chloride metabolism of the strictly anaerobic, methyl chloride-utilizing homoacetogen strain MC. *Arch. Microbiol.* 160: 383–387.

Metcalf, W.W. and B.L. Wanner. 1991. Involvement of the *Escherichia coli phn* (*psiD*) gene cluster in assimilation of phosphorus in the form of phosphonates, phosphite, P_i esters, and P_i. *J. Bacteriol.* 173: 587–600.

Meulenberg, R., M. Pepi, and J.A.M. de Bont. 1996. Degradation of 3-nitrophenol by *Pseudomonas putida* B2 occurs via 1,2,4-benzenetriol. *Biodegradation* 7: 303–311.

Meyer, J.J.M., N. Grobbelaar, and P.L. Steyn. 1990. Fluoroacetate-metabolizing pseudomonad Isolated from *Dichapetalum cymosum. Appl. Environ. Microbiol.* 56: 2152–2155.

Michaelsen, M., R. Hulsch, T. Höpner, and L. Berthe-Corti. 1992. Hexadecane mineralization in oxygen-controlled sediment-seawater cultivations with autochthonous microorganisms. *Appl. Environ. Microbiol.* 58: 3072–3077.

Migaud, M.E., J.C. Chee-Sandford, J.M. Tiedje, and J.W. Frost. 1996. Benzylfumaric, benzylmaleic, and Z- and E-phenylitaconic acids: synthesis, characterization and correlation with a metabolite generated by *Azoarcus tolulyticus* Tol-4 during anaerobic toluene degradation. *Appl. Environ. Microbiol.* 62: 974–978.

Mikesell, M.D. and S.A. Boyd. 1986. Complete reductive dechlorination and mineralization of pentachlorophenol by anaerobic microorganisms. *Appl. Environ. Microbiol.* 52: 861–865.

Milanova, R., Moore, M., and Hirai, Y. 1994. Hydroxylation of synthetic abietane diterpenes by *Aspergillus* and *Cunninghamella* species: novel route to the family of diterpenes isolated from *Tripterygium wilfordii. J. Nat. Prod.* 57: 882–889.

Miller, E., G. Wohlfarth, and G. Diekert. 1997. Comparative studies on tetrachloethene reductive dechlorination mediated by *Desulfitobacterium* sp. strain PCE-S. *Arch. Microbiol.* 168: 513–519.

Miller, E., G. Wohlfarth, and G. Dielkert. 1998. Purification and characterization of the tetrachloroethene reductive dehalogenase of strain PCE-S. *Arch. Microbiol.* 169: 497–502.

Miller, J.M. and D.O. Gray. 1982. The utilization of nitriles and amides by a *Rhodococcus* species. *J. Gen. Microbiol.* 128: 1803–1809.

Miller, K.W. 1992. Reductive desulfurization of dibenzyldisulfide. *Appl. Environ. Microbiol.* 58: 2176–2179.

Misawa, E., C.K.C.C.K. Chion, I.V. Archer, M.P. Woodland, N.-Y. Zhou, S.F. Carter, D.A. Widdowson, and D.J. Leak. 1998. Characterization of a catabolic epoxide hydrolase from a *Corynebacterium* sp. *Eur. J. Biochem.* 253: 173–183.

Miura, Y. and A.J. Fulco. 1975. ω-1, ω-2, and ω-3 hydroxylation of long-chain fatty acids, amides and alcohols by a soluble enzyme system from *Bacillus megaterium*. *Biochim. Biophys. Acta* 388: 305–317.

Miyazaki, T., T. Yamagishi, and M. Matsumoto. 1984. Residues of 4-chloro-1-(2,4-dichlorophenoxy)-2-methoxybenzene (triclosan methyl) in aquatic biota. *Bull. Environ. Contam. Toxicol.* 32: 227–232.

Modrzakowski, M.C. and W.R. Finnerty. 1980. Metabolism of symmetrical dialkyl ethers by *Acinetobacter* sp. HO1-N. *Arch. Microbiol.* 126: 285–290.

Mohamed, M.E.-S., B. Seyfried, A. Tschech, and G. Fuchs. 1993. Anaerobic oxidation of phenylacetate and 4-hydroxyphenylacetate to benzoyl-coenzyme A and CO_2 in denitrifying *Pseudomonas* sp. *Arch. Microbiol.* 159: 563–573.

Mohn, W.W. and K.J. Kennedy. 1992. Reductive dehalogenation of chlorophenols by *Desulfomonile tiedjei* DCB-1. *Appl. Environ. Microbiol.* 58: 1367–1370.

Mohn, W.W. and J.M. Tiedje. 1991. Evidence for chemiosmotic coupling of reductive dechlorination and ATP synthesis in *Desulfomonile tiedjei*. *Arch. Microbiol.* 157: 1–6.

Mondello, F.J., M.P. Turcich, J.H. Lobos, and B.D. Erickson. 1997. Identification and modification of biphenyl dioxygenase sequences that determine the specificity of polychlorinated biphenyl degradation. *Appl. Environ. Microbiol.* 63: 3096–3103.

Montserrate, E. and M.M. Häggblom. 1997. Dehalogenation and biodegradation of brominated phenols and benzoic acids under iron-reducing, sulfidogenic, and methanogenic conditions. *Appl. Environ. Microbiol.* 63: 3911–3915.

Morawski, B., R.W. Eaton, J.T. Rossiter, S. Guoping, H. Griengl, and D.W. Ribbons. 1997. 2-Naphthoate catabolic pathway in *Burkholderia* strain JT 1500. *J. Bacteriol.* 179: 115–121.

Morris, P.J., J.F. Quensen III, J.M. Tiedje, and S.A. Boyd. 1992. Reductive debromination of the commercial polybrominated biphenyl mixture Firemaster BP6 by anaerobic microorganisms from sediments. *Appl. Environ. Microbiol.* 58: 3249–3256.

Mott, G.E., A.W. Brinkley, and C.L. Mersinger. 1980. Biochemical characterization of cholesterol-reducing *Eubacterium*. *Appl. Environ. Microbiol.* 40: 1017–1022.

Mountfort, D.O. 1987. The rumen anaerobic fungi. *FEMS Microbiol. Rev.* 46: 401–408.

Mueller J.G., P.J. Chapman, B.O. Blattman, and P.H. Pritchard. 1990. Isolation and characterization of a fluoranthene-utilizing strain of *Pseudomonas paucimobilis*. *Appl. Environ. Microbiol.* 56: 1079–1086.

Mulbry, W.W. 1994. Purification and characterization of an inducible *s*-triazine hydrolase from *Rhodococcus corallinus* NRRL B-15444R. *Appl. Environ. Microbiol.* 60: 613–618.

Müller, E., K. Fahlbusch, R. Walther, and G. Gottschalk. 1981. Formation of *N,N*-dimethylglycine, acetic acid and butyric acid from betaine by *Eubacterium limosum*. *Appl. Environ. Microbiol.* 42: 439–445.

Müller, M.D. and H.-R. Büser. 1986. Halogenated aromatic compounds in autmotive emissions from leaded gasoline additives. *Environ. Sci. Technol.* 20: 1151–1157.

Mulroy, P.T. and L.-T. Ou. 1998. Degradation of tetraethyllead during the degradation of leaded gasoline hydrocarbons in soil. *Environ. Toxicol. Chem.* 17: 777–782.

Munnecke, D.M., L.M. Johnson, H.W. Talbot, and S. Barik. 1982. Microbial metabolism and enzymology of selected pesticides pp.1–32. In *Biodegradation and Detoxification of Environmental Pollutants* (Ed. A.M. Chrakrabarty). CRC Press, Boca Raton, FL.

Nadeau, I.J. and J.C. Spain. 1995. The bacterial degradation of *m*-nitrobenzoic acid. *Appl. Environ. Microbiol.* 61: 840–843.

Nadeau, L.F., G.S. Sayler, and J.C. Spain. 1998. Oxidation of 1,1,1-trichloro-2,2-bis(4-chlorophenyl)ethane (DDT) by *Alcaligenes eutrophus* A5. *Arch. Microbiol.* 171: 44–49.

Nadeau, L.J., F.-M. Menn, A. Breen, and G.S. Sayler. 1994. Aerobic degradation of 1,1,1-trichloro-2,2-bis(4-chlorophenyl)ethane (DDT) by *Alcaligenes eutrophus* A5. *Appl. Environ. Microbiol.* 60: 51–55.

Nagasawa, S., R. Kikuchi, Y. Nagata, M. Takagi, and M. Matsuo. 1993a. Stereochemical analysis of γ-HCH degradation by *Pseudomonas paucimobilis* UT26. *Chemosphere* 26: 1187–1201.

Nagasawa, S., R. Kikuchi, Y. Nagata, M. Takagi, and M. Matsuo. 1993b. Aerobic mineralization of γ-HCH by *Pseudomonas paucimobilis* UT26. *Chemosphere* 26: 1719–1728.

Nagata, Y., T. Nariya, R. Ohtomo, M. Fukuda, K. Yano, and M. Takagi. 1993. Cloning and sequencing of a dehalogenase gene encoding an enzyme with hydrolase activity involved in the degradation of γ-hexachlorocyclohexane in *Pseudomonas paucimobilis*. *J. Bacteriol.* 175: 6403–6410.

Nagata, Y., K. Miyauchi, J. Damborsky, K. Manova, A. Ansorgova, and M. Takagi. 1997. Purification and characterization of a haloalkane dehalogenase of a new substrate class from a γ-hexachlorocyclohexane-degrading bacterium, *Sphingomonas paucimobilis* UT26. *Appl. Environ. Microbiol.* 63: 3703–3710.

Nagata, Y., R. Ohtomo, K. Miyauchi, M. Fukuda, K. Yano, and M. Takagi. 1994. Cloning and sequencing of a 2,5-dichloro-2,5-cyclohexadiene-1,4-diol dehydrogenase involved in the degradation of γ hexachlorocyclohexane in *Pseudomonas paucimobilis*. *J. Bacteriol.* 176: 3117–3125.

Nakamura, T., F. Yu, W. Mizunashi, and I. Watanabe. 1993. Production of (R)-3-chloro-1,2-propandiol from prochiral 1,3-dichloro-2-propanol by *Corynebacterium* sp. strain N-1074. *Appl. Environ. Microbiol.* 59: 227–230.

Nakatsu, C.H. and R.C. Wyndham. 1993. Cloning and expression of the transposable chlorobenzoate-3,4-dioxygenase genes of *Alcaligenes* sp. strain BR60. *Appl. Environ. Microbiol.* 59: 3625–3633.

Nanninga, H.J. and J.C. Gottschal. 1987. Properties of *Desulfovibrio carbinolicus* sp. nov. and other sulfate-reducing bacteria isolated from an anaerobic-purification plant. *Appl. Environ. Microbiol.* 53: 802–809.

Nardi-Del, V.T., C. Kutihara, C. Park, N. Esaki, and K. Soda. 1997. Bacterial DL-2-haloacid dehalogenase from *Pseudomonas* sp. strain 113: gene cloning and structural comparison with D- and L-2-haloacid dehalogenases. *J. Bacteriol.* 179: 4232–4238.

Narhi, L.O. and A.J. Fulco. 1987. Identification and characterization of two functional domains in cytochrome P-450$_{BM-3}$, a catalytically self-sufficient monooxygenase induced by barbiturates in *Bacillus megaterium*. *J. Biol. Chem.* 262: 6683–6690.

Narro, M.L., C.E. Cerniglia, C. Van Baalen, and D.T. Gibson. 1992. Evidence for an NIH shift in oxidation of naphthalene by the marine cyanobacterium *Oscillatoria* sp. strain JCM. *Appl. Environ. Microbiol.* 58: 1360–1363.

Naumann, E., H. Hippe, and G. Gottschalk. 1983. Betaine: new oxidant in the Stickland reaction and methanogenesis from betaine and L-alanine by a *Clostridium sporogenes–Methanosarcina barkeri* coculture. *Appl. Environ. Microbiol.* 45: 474–483.

Nawaz, M.S., T.M. Heinze, and C.E. Cerniglia. 1992. Metabolism of benzonitrile and butyronitrile by *Klebsiella pneumoniae*. *Appl. Environ. Microbiol.* 58: 27–31.

Neidle, E.L., C. Harnett, L.N. Ornston, A. Bairoch, M. Rekik et al. 1991. Nucleotide sequences of the *Acinetobacter calcoaceticus benABC* genes for benzoate 1,2-dioxygenase reveal evolutionary relationships among multi-component enzymes. *J. Bacteriol.* 173: 5385–5395.

Neilson, A.H. and A.-S. Allard. 1998. Microbial metabolism of PAHs and heteroarenes, pp. 1–80. In *Handbook of Environmental Chemistry* (Ed. A.H. Neilson). Vol. 3, Part J, Springer-Verlag, Berlin.

Neilson, A.H., A.-S. Allard, P.-Å. Hynning, and M. Remberger. 1988a. Transformations of halogenated aromatic aldehydes by metabolically stable anaerobic enrichment cultures. *Appl. Environ. Microbiol.* 54: 2226–2236.

Neilson, A. H., C. Lindgren, P.-Å. Hynning, and M. Remberger. 1988b. Methylation of halogenated phenols and thiophenols by cell extracts of Gram-positive and Gram-negative bacteria. *Appl. Environ. Microbiol.* 54: 524–530.

Neilson, A.H., A.-S. Allard, P.-Å. Hynning, and M. Remberger. 1991. Distribution, fate and persistence of organochlorine compounds formed during production of bleached pulp. *Toxicol. Environ. Chem.* 30: 3–41.

Nelson, J.D., W. Blair, F.E. Brinckman, R.R. Colwell, and W.P. Iverson. 1973. Biodegradation of phenylmercuric acetate by mercury-resistant bacteria. *Appl. Microbiol.* 26: 321–326.

Neumann, A., H. Scholz-Muramatsu, and G. Diekert. 1994. Tetrachloroethene metabolism of *Dehalospirillum multivorans*. *Arch. Microbiol.* 162: 295–301.

Neumann, A., G. Wohlfart, and G. Diekert. 1995. Properties of tetrachloroethene and trichloroethene dehalogenase of *Dehalospirillum multivorans*. *Arch. Microbiol.* 163: 276–281.

Neumann, A., G. Wohlfarth, and G. Diekert. 1996. Purification and characterization of tetrachloroethene dehalogenase from *Dehalospirillum multivorans*. *J. Biol. Chem.* 271: 16515–16519.

Neumann, A., G. Wohlfarth, and G. Diekert. 1998. Tetrachloroethene dehalogenase from *Dehalospirillum multivorans*: cloning, sequencing of the encoding genes, and expression of the *pceA* gene in *Escherichia coli*. *J. Bacteriol.* 180: 4140–4145.

Newman, L.M. and L.P. Wackett. 1991. Fate of 2,2,2-trichloroacetaldehyde (chloral hydrate) produced during trichloroethylene oxidation by methanotrophs. *Appl. Environ. Microbiol.* 57: 2399–2402.

Newman, L.M. and L.P. Wackett. 1997. Trichloroethylene by purified toluene 2-monooxygenase: products, kinetics, and turnover-dependent inactivation. *J. Bacteriol.* 179: 90–96.

Ni, S., J.K. Fredrickson, and L. Xun. 1995. Purification and characterization of a novel 3-chlorobenzoate-reductive dehalogenase from the cytoplasmic membrane of *Desulfomonile tiedje* DVCB-1. *J. Bacteriol.* 177: 5135–5139.

Nishino, S.F. and J.C. Spain. 1993. Degradation of nitrobenzene by a *Pseudomonas pseudoalcaligenes*. *Appl. Environ. Microbiol.* 59: 2520–2525.

Nishino, S.F. and J.C. Spain. 1995. Oxidative pathways for the degradation of nitrobenzene by *Comamonas* sp. strain JS 765. *Appl. Environ. Microbiol.* 61: 2308–2313.

Nozawa, T. and Y. Maruyama. 1988. Anaerobic metabolism of phthalate and other aromatic compounds by a denitrifying bacterium. *J. Bacteriol.* 170: 5778–5784.

Nurk, A., L. Kasak, and M. Kivisaar. 1991. Sequence of the gene (*pheA*) encoding phenol monooxygenase from *Paseudomonas* sp. EST1001: expression in *Escherichi coli* and *Pseudomonas putida*. *Gene* 102: 13–18.

O'Keefe, D.P. and P.A. Harder. 1991. Occurrence and biological function of cytochrome P-450 monoxygenase in the actinomycetes. *Mol. Microbiol.* 5: 2099–2105.

Ohara, M., Y. Katayama, M. Tsuzaki, S. Nakamoto, and H. Kuraishi. 1990. *Paracoccus kocurii* sp. nov., a tetramethylammonoum-assimilating bacterium. *Int. J. Syst. Bacteriol.* 40: 292–296.

Ohisa, N., M. Yamaguchi, and N. Kurihara. 1980. Lindane degradation by cell-free extracts of *Clostridium rectum*. *Arch. Microbiol.* 125: 221–225.

Ohisa, N., N. Kurihara, and M. Nakajima. 1982. ATP synthesis associated with the conversion of hexachlorocyclohexane related compounds. *Arch. Microbiol.* 131: 330–333.

Oldenhuis, R., R.L.J.M. Vink, D.B. Janssen, and B. Witholt. 1989. Degradation of chlorinated aliphatic hydrocarbons by *Methylosinus trichosporium* OB3b expressing soluble methane monooxygenase. *Appl. Environ. Microbiol.* 55: 2819–2816.

Oldfield, C., O. Pogrebinsky, J. Simmonds, E.S. Olson, and C.F. Kulpa. 1997. Elucidation of the metabolic pathway for dibenzothiophene desulphurization by *Rhodococcus* sp. strain IGTS8 (ATCC 53968). *Microbiology* (U.K.) 143: 2961–2973.

Olsen, R.H., J.J. Kukor, and B. Kaphammer. 1994. A novel toluene-3-monooxygenase pathway cloned from *Pseudomonas pickettii* PKO1. *J. Bacteriol.* 176: 3749–3756.

Oltmanns, R.H., R. Müller, M.K. Otto, and F. Lingens. 1989. Evidence for a new pathway in the bacterial degradation of 4-fluorobenzoate. *Appl. Environ. Microbiol.* 55: 2499–2504.

Omori, T., L. Monna, Y. Saiki, and T. Kodama. 1992. Desulfurization of dibenzothiophene by *Corynebacterium* sp. strain SY1. *Appl. Environ. Microbiol.* 58: 911–915.

Oremland, R.S., R.S. Kiene, I. Mathrani, M.J. Whiticar, and D.R. Boone. 1989. Description of an estuarine methylotrophic methanogen which grows on dimethyl sulfide. *Appl. Environ. Microbiol.* 55: 994–1002.

Ornston, L.N. 1966. The conversion of catechol and protocatechuate to beta-ketoadipate by *Pseudomonas putida*. IV. Regulation. *J. Biol. Chem.* 241: 3800–3810.

Ougham, H.J. and P.W. Trudgill. 1982. Metabolism of cyclohexaneacetic acid and cyclohexanebutyric acid by *Arthrobacter* sp. strain CA1. *J. Bacteriol.* 150: 1172–1182.

Ougham, H.J., D.G. Taylor, and P.W. Trudgill. 1983. Camphor revisited: involvement of a unique monooxygenase in metabolism of 2-oxo-Δ^3-4,5,5-trimethylcyclopentenylacetic acid by *Pseudomonas putida*. *J. Bacteriol.* 153: 140–152.

Paasivirta, J., P. Klein, M. Knuutila, J. Knuutinen, M. Lahtiperä, R. Paukku, A. Veijanen, L. Welling, M. Vuorinen, and P.J. Vuorinen. 1987. Chlorinated anisoles and veratroles in fish. Model compounds. Instrumental and sensory determinations. *Chemosphere* 16: 1231–1241.

Padden, A.N., F.A. Rainey, D.P. Kelly, and A.P. Wood. 1997. *Xanthobacter tagetidis* sp. nov., an organism associated with *Tagetes* species and able to grow on substituted thiophenes. *Int. J. Syst. Bacteriol.* 47: 394–401.

Parales, R.E. and C.S. Harwood. 1993. Regulation of the *pcaIJ* genes for aromatic acid degradation in *Pseudomonas putida*. *J. Bacteriol.* 175: 5829–5838.

Parales, J.V., R.E. Parales, S.M. Resnick, and D.T. Gibson. 1998. Enzyme specificity of 2-nitrotoluene 2,3-dioxygenase from *Pseudomonas* sp. strain JS42 is determined by the C-terminal region of the α subunit of the oxygenase component. *J. Bacteriol.* 180: 1194–1199.

Parekh, V.R., R.W. Traxler, and J.M. Sobek. 1977. *n*-Alkane oxidation enzymes of a pseudomonad. *Appl. Environ. Microbiol.* 33: 881–884.

Parisot, D., M.C. Malet-Martino, R. Martino, and P. Crasnier. 1991. ^{19}F nuclear magnetic resonance analysis of 5-fluorouracil metabolism in four differently pigmented strains of *Nectria haematococca*. *Appl. Environ. Microbiol.* 57: 3605–3612.

Patel, R.N., C.T. Hou, A.I. Laskin, and A. Felix. 1982. Microbial oxidation of hydrocarbons: properties of a soluble methane monooxygenase from a facultative methane-utilizing organism, *Methylobacterium* sp. strain CRL-26. *Appl. Environ. Microbiol.* 44: 1130–1137.

Patel, T.R. and E.A. Barnsley. 1980. Naphthalene metabolism by pseudomonads: purification and properties of 1,2-dihydroxynaphthalene oxygenase. *J. Bacteriol.* 143: 668–673.

Patel, T.R. and D.T. Gibson. 1974. Purification and properties of (+)-*cis*-naphthalene dihydrodiol dehydrogenase of *Pseudomonas putida*. *J. Bacteriol.* 119: 879–888.

Patterson, J.A. and R.B. Hespell. 1979. Trimethylamine and methylamine as growth substrates for rumen bacteria and *Methanosarcina barkeri*. *Curr. Microbiol.* 3: 79–83.

Payne, J.W., H. Bolton, J.A. Campbell, and L. Xun. 1998. Purification and characterization of EDTA monooxygenase from the EDTA-degrading bacterium BNC1. *J. Bacteriol.* 180: 3823–3827.

Peelen, S., I.M.C.M. Rietjens, W.J.H. van Berkel, W.A.T. van Workum, and J. Vervoort. 1993. ^{19}F-NMR study on the pH-dependent regioselectivity and rate of the *ortho*-hydroxylation of 3-fluorophenol by phenol hydroxylase from *Trichosporon cutaneum*. *Eur. J. Biochem.* 218: 345–353.

Pelletier, D.A. and C.S. Harwood. 1998. 2-Ketohexanecarboxyl coenzyme A hydrolase, the ring cleavage enzyme required for anaerobic benzoate degradation by *Rhodopseudomonas palustris*. *J. Bacteriol.* 180: 2330–2336.

Perkins, E.J., M.P. Gordon, O. Caceres, and P.F. Lurquin. 1990. Organization and sequence analysis of the 2,4-dichlorophenol hydroxylase and dichlocatechol oxidative operons of plasmid pJP4. *J. Bacteriol.* 172: 2351–2359.

Perrotta, J.A. and C.S. Harwood. 1994. Anaerobic metabolism of cyclohex-1-ene-1-carboxylate, a proposed intermediate of benzoate degradation by *Rhodopseudomonas palustris*. *Appl. Environ. Microbiol.* 60: 1775–1782

Peterson, D. and J. Llaneza. 1974. Identification of a carbon-oxygen lyase activity cleaving the ether linkage in carboxymethyloxysuccinic acid. *Arch. Biochem. Biophys.* 162: 135–146.

Pettigrew, C.A., B.E. Haigler, and J.C. Spain. 1991. Simultaneous biodegradation of chlorobenzene and toluene by a *Pseudomonas* strain. *Appl. Environ. Microbiol.* 57: 157–162.

Pfeifer, F., S. Schacht, J. Klein, and H.G. Trüper. 1989. Degradation of diphenyl ether by *Pseudomonas cepacia*. *Arch. Microbiol.* 152: 515–519.

Pfeifer, F., H.G. Trüper, J. Klein, and S. Schacht. 1993. Degradation of diphenylether by *Pseudomonas cepacia* Et4: enzymatic release of phenol from 2,3-dihydroxydiphenylether. *Arch. Microbiol.* 159: 323–329.

Pieper, D.H., W. Reineke, K.-H. Engesser, and H.-J. Knackmuss. 1988. Metabolism of 2,4-dichlorophenoxyacetic acid, 4-chloro-2-methylphenoxyacetic acid and 2-methylphenoxyacetic acid by *Alcaliges eutrophus* JMP 134. *Arch. Microbiol.* 150: 95–102.

Pipke, R. and N. Amrhein. 1988. Degradation of the phosphonate herbicide glyphosate by *Arthrobacter atrocyaneus* ATCC 13752. *Appl. Environ. Microbiol.* 54: 1293–1296.

Pipke, R., N. Amrhein, G.S. Jacob, J. Schaefer, and G.M. Kishore. 1987. Metabolism of glyphosate in an *Arthrobacter* sp. GLP-1. *Eur. J. Biochem.* 165: 267–273.

Pirnik, M.P., R.M. Atlas, and R. Bartha. 1974. Hydrocarbon metabolism by *Brevibacterium erythrogenes:* normal and branched alkanes. *J. Bacteriol.* 119: 868–878.

Plugge, C.M., C. Dijkema, and A.J.M. Stams. 1993. Acetyl-CoA cleavage pathways in a syntrophic propionate oxidizing bacterium growing on fumarate in the absence of methanogens. *FEMS Microbiol. Lett.* 110: 71–76.

Poelarends, G.J., J.E.T. van Hylckama Vlieg, J.R. Marchesi, L.M. Freitas dos Santos, and D.B. Janssen. 1999. Degradation of 1,2-dibromoethane by *Mycobacterium* sp. strain GP1. *J. Bacteriol.* 181: 2050–2058.

Poh, C.L. and R.C. Bayly. 1980. Evidence for isofunctional enzymes used in *meta*-cresol and 2,4-xylenol degradation via the gentisate pathway in *Pseudomonas alcaligenes. J. Bacteriol.* 143: 59–69.

Pons, J.-L., A. Rimbault, J.C. Darbord, and G. Leluan. 1984. Biosynthèse de toluène chez *Clostridium aerofoetidum* souche WS. *Ann. Microbiol. (Inst. Pasteur)* 135B: 219–222.

Poth, M. and D.D. Focht. 1985. [15]N kinetic analysis of N_2O production by *Nitrosomonas europaea:* an examination of nitrifier denitrification. *Appl. Environ. Microbiol.* 49: 1134–1141.

Pothuluri, J.V., J.B. Sutherland, J.P. Freeman, and C.E. Cerniglia. 1998. Fungal biotransformation of 6-nitrochrysene. *Appl. Environ. Microbiol.* 64: 3106–3109.

Pothuluri, J.V., J.P. Freeman, F.E. Evans, and C.E. Cerniglia. 1990. Fungal transformation of fluoranthene. *Appl. Environ. Microbiol.* 56: 2974–2983.

Potrawfke, T., K.N. Timmis, and R.-M. Wittich. 1998. Degradation of 1,2,3,4-tetrachlorobenzene by *Pseudomonas chlororaphis* RW71. *Appl. Environ. Microbiol.* 64: 3798–3806.

Prairie, R.L. and P. Talalay. 1963. Enzymatic formation of testololactone. *Biochemistry* 2: 203–208.

Pujar, B.G. and D.W. Ribbons. 1985. Phthalate metabolism in *Pseudomonas florescens* PHK: purification and properties of 4,5-dihydroxyphthalate decarboxylase. *Appl. Environ. Microbiol.* 49: 374–376.

Qatibi, A.I., V. Niviére, and J.L. Garcia. 1991. *Desulfovibrio alcoholovorans* sp. nov., a sulfate-reducing bacterium able to grow on glycerol, 1,2- and 1,3-propanediol. *Arch. Microbiol.* 155: 143–148.

Quick, A., N.J. Russell, S.G. Hales, and G.F. White. 1994. Biodegradation of sulphosuccinate: direct desulphonation of a secondary sulphonate. *Microbiology* (U.K.) 140: 2991–2998.

Quinn, J.P. and G. McMullan. 1995. Carbon-arsenic bond cleavage by a newly isolated Gram-negative bacterium, strain ASV2. *Microbiology* (U.K.) 141: 721–727.

Rabus, R. and J. Heider. 1998. Initial reactions of anaerobic metabolism of alkylbenzenes in denitrifying and sulfate-reducing bacteria. *Arch. Microbiol.* 170: 377–384.

Rabus, R. and F. Widdel. 1995. Anaerobic degradation of ethylbenzene and other aromatic hydrocarbons by new denitrifying bacteria. *Arch. Microbiol.* 163: 96–103.

Rabus, R., R. Nordhaus, W. Ludwig, and F. Widdel. 1993. Complete oxidation of toluene under strictly anoxic conditions by a new sulfate-reducing bacterium. *Appl. Environ. Microbiol.* 59: 1444–1451.

Rafii, F. and C.E. Cerniglia. 1993. Comparison of the azoreductase and nitroreductase from *Clostridium perfringens. Appl. Environ. Microbiol.* 59: 1731–1734.

Rafii, F., W. Franklin, R.H. Heflich, and C.E. Cerniglia. 1991. Reduction of nitroaromatic compounds by anaerobic bacteria isolated from the human gastrointestinal tract. *Appl. Environ. Microbiol.* 57: 962–968.

Ramamoorthi, R. and M.E. Lidstrom. 1995. Transcriptional analysis of *pqqD* and study of the regulation of pyrroloquinoline quinone biosynthesis in *Methylobacterium extorquens* AM1. *J. Bacteriol.* 177: 206–211.

Ramanand, K. and J.M. Suflita. 1991. Anaerobic degradation of *m*-cresol in anoxic aquifer slurries: carboxylation reactions in a sulfate-reducing bacterial enrichment. *Appl. Environ. Microbiol.* 57: 1689–1695.

Rao, J.R., and J.E. Cooper 1994. Rhizobia catabolize *nod* gene-inducing flavonoids via C-ring fission mechanisms. *J. Bacteriol.* 176: 5409–5413.

Rao, J.R., N.D. Sharma, J.T.G. Hamilton, D.R. Boyd, and J.E. Cooper. 1991. Biotransformation of the pentahydroxy flavone quercitin by *Rhizobium loti* and *Bradyrhizobium* strains (Lotus). *Appl. Environ Microbiol.* 57: 1563–1565.

Ratledge, C. 1978. Degradation of aliphatic hydrocarbons pp. 1-46. In *Developments in Biodegradation of Hydrocarbons—1* (Ed. R.J. Watkinson), Applied Science Publishers, London.

Reddy, G.V.B., M.D.S. Gelpke, and M.H. Gold. 1998. Degradation of 2,4,6-trichlorophenol by *Phanerochaete chrysosporium*: involvement of reductive dechlorination. *J. Bacteriol.* 180: 5159–5164.

Reichenbecher, W.W. and B. Schink. 1997. *Desulfovibrio inopinatus* sp. nov., a new sulfate-reducing bacterium that degrades hydroxyhydroquinone (1,2,4-trihydroxybenzene). *Arch. Microbiol.* 168: 338–344.

Reineke, W. 1984. Microbial degradation of halogenated aromatic compounds pp. 319–360. In *Microbial Degradation of Organic Compounds* (Ed. D.T. Gibson). Marcel Dekker, New York.

Reineke, W. and H.-J. Knackmuss. 1988. Microbial degradation of haloaromatics. *Annu. Rev. Microbiol.* 42: 263–287.

Renganathan, V. 1989. Possible involvement of toluene-2,3-dioxygenase in defluorination of 3-fluoro-substituted benzenes by toluene-degrading *Pseudomonas* sp. strain T-12. *Appl. Environ. Microbiol.* 55: 330–334.

Resnick, S.M. and D.T. Gibson. 1996. Regio- and stereospecific oxidation of fluorene, dibenzofuran, and dibenzothiophene by naphthalene dioxygenase from *Pseudomonas* sp. strain NCIB-4. *Appl. Environ. Microbiol.* 62: 4073–4080.

Resnick, S.M., D.S. Torok, and D.T. Gibson. 1993. Oxidation of carbazole to 3-hydroxycarbazole by naphthalene 1,2-dioxygenase and biphenyl 2,3-dioxygenase. *FEMS Microbiol. Lett.* 113: 297–302.

Rhee, S.-K., G.M. Lee, J.-H. Yoon, Y.-H. Park, H.-S. Bae, and S.-T. Lee. 1997. Anaerobic and aerobic degradation of pyridine by a newly isolated denitrifying bacterium. *Appl. Environ. Microbiol.* 63: 2578–2585.

Rhys-Williams, W., S.C. Taylor, and P.A. Williams. 1993. A novel pathway for the catabolism of 4-nitrotoluene by *Pseudomonas. J. Gen. Microbiol.* 139: 1967–1972.

Ribbons, D.W. and R.W. Eaton. 1982. Chemical transformations of aromatic hydrocarbons that support the growth of microorganisms pp. 59-84. In *Biodegradation and Detoxification of Environmental Pollutants* (Ed. A.M. Chakrabarty). CRC Press, Boca Raton, FL.

Ribbons, D.W., P. Keyser, D.A. Kunz, B.F. Taylor, R.W. Eaton, and B.N. Anderson. 1984. Microbial degradation of phthalates pp. 371–395. In *Microbial Degradation of Organic Compounds* (D.T. Gibson Ed.). Marcel Dekker, New York.

Ribbons, D.W., S.J.C. Taylor, C.T. Evans, S.T. Thomas, J.T. Rossiter, D.A. Widdowson, and D.J. Williams. 1990. Biodegradations yield novel intermediates for chemical synthesis pp. 213–245. In *Biotechnology and Biodegradation* (Eds. D. Kamely, A. Chakrabarty, and G.S. Omenn). Gulf Publishing Company, Houston, TX.

Rieger, P.-G., V. Sinnwell, A. Preuß, W. Francke, and H.-J. Knackmuss. 1999. Hydride-Meisenheimer complex formation and protonation as key reactions of 2,4,6-trinitrophenol biodegradation by *Rhodococcus erythropolis*. *J. Bacteriol.* 181: 1189–1195.

Riegert, U., G. Heiss, P. Fischer, and A. Stolz. 1998. Distal cleavage of 3-chlorocatechol by an extradiol dioxygenase to 3-chloro-2-hydroxymuconic semialdehyde. *J. Bacteriol.* 180: 2849–2853.

Rink, R., M. Fennema, M. Smids, U. Dehmel, and D.B. Janssen. 1997. Primary structure and catalytic mechanism of the epoxide hydrolase from *Agrobacterium radiobacter* AD1. *J. Biol. Chem.* 272: 14650–14657.

Roberts, D.J., P.M. Fedorak, and S.E. Hrudey. 1990. CO_2 incorporation and 4-hydroxy-2-methylbenzoic acid formation during anaerobic metabolism of *m*-cresol by a methanogenic consortium. *Appl. Environ. Microbiol.* 56: 472–478.

Robertson, J.B., J.C. Spain, J.D. Haddock, and D.T. Gibson. 1992. Oxidation of nitrotoluenes by toluene dioxygenase: evidence for a monooxygenase reaction. *Appl. Environ. Microbiol.* 58: 2643–2648.

Röger, P., G. Bär, and F. Lingens. 1995. Two novel metabolites in the degradation pathway of isoquinoline by *Pseudomonas putida*. *FEMS Microbiol. Lett.* 129: 281–286.

Romanov, V. and R.P. Hausinger. 1994. *Pseudomonas aeruginosa* 142 uses a three-component *ortho*-halobenzoate 1,2-dioxygenase for metabolism of 2,4-dichloro-and 2-chlorobenzoate. *J. Bacteriol.* 176: 3368–3374.

Romanov, V. and R.P. Hausinger. 1996. NADPH-dependent reductive *ortho* dehalogenation of 2,4-dichlorobenzoic acid in *Corynebacterium sepedonicum* KZ-4 and coryneform bacterium strain NTB-1 via 2,4-dichlorobenzoyl coenzyme A. *J. Bacteriol.* 178: 2656–2661.

Rontani, J.-F., M.J. Gilewicz, V.D. Micgotey, T.L. Zheng, P.C. Bonin, and J.-C. Bertrand. 1997. Aerobic and anaerobic metabolism of 6,10,14-trimethylpentadecan-2-one by a denitrifying bacterium isolated from marine sediments. *Appl. Environ. Microbiol.* 63: 636–643.

Rosche. B., B. Tshisuaka, S. Fetzner, and F. Lingens. 1995. 2-Oxo-1,2-dihydroquinoline 8-monooxygenase, a two-component enzyme system from *Pseudomonas putida* 86. *J. Biol. Chem.* 270: 17836–17842.

Rosche, B., B. Tshisuaka, B. Hauer, F. Lingens, and S. Fetzner. 1997. 2-Oxo-1,2-dihydroquionoline 8-monooxygenase: phylogenetic relationship to other multicomponent nonheme iron oxygenases. *J. Bacteriol.* 179: 3459–3554.

Rosner, B.M. and B. Schink. 1995. Purification and characterization of acetylene hydratase of *Pelobacter acetylenicus*, a tungsten iron-sulfur protein. *J. Bacteriol.* 177: 5767–5772.

Rossol, I. and A. Pühler. 1992. The *Corynebacterium glutamicum aecD* gene encodes a C-S lyase with alpha-beta-elimination activity that degrades aminoethylcysteine. *J. Bacteriol.* 174: 2968–2977.

Rothmel, R.K., D.L. Shinbarger, M.R. Parsek, T.L. Aldrich, and A.M. Chakrabarty. 1991. Functional analysis of the *Pseudomonas putida* regulatory protein CatR: transcriptional studies and determination of the CatR DNA-binding site by hydroxyl-radical footprinting. *J. Bacteriol.* 173: 4717–4724.

Rott, B., S. Nitz, and F. Korte. 1979. Microbial decomposition of pentachlorophenolate. *J. Agric. Food Chem.* 27: 306–310.

Rudd, J.W.M., A. Furutani, and M.A. Turner. 1980. Mercury methylation by fish intestinal contents. *Appl. Environ. Microbiol.* 40: 777–782.

Rudolphi, A., A. Tschech, and G. Fuchs. 1991. Anaerobic degradation of cresols by denitrifying bacteria. *Arch. Microbiol.* 155: 238–248.

Ruettinger, R.T., S.T. Olson, R.F. Boyer, and M.J. Coon. 1974. Identification of the ω-hydroxylase of *Pseudomonas oleovorans* as a non-heme iron protein requiring phospholipid for catalytic activity. *Biochem. Biophys. Res. Commun.* 57: 1011–1017.

Ruettinger, R.T., G.R. Griffith, and M.J. Coon. 1977. Characteristics of the ω-hydroxylase of *Pseudomonas oleovorans* as a non-heme iron protein. *Arch. Biochem. Biophys.* 183: 528–537.

Sabourin, C.L., Carpenter, J.C., Leib, T.K., and Spivack, J.L. 1996. Biodegradation of dimethylsilanediol in soils. *Appl. Environ. Microbiol.* 62: 4352–4360.

Sack, U., M. Hofrichter, and W. Fritsche. 1997. Degradation of phenanthrene and pyrene by *Nematoloma frowardii*. *J. Basic Microbiol.* 4: 287–293.

Sack, U., T.M. Heinze, J. Deck, C.E. Cerniglia, M.C. Cazau, and W. Fritsche. 1997. Novel metabolites in phenanthrene and pyrene transformations by *Aspergillus niger*. *Appl. Environ. Microbiol.* 63: 2906–2909.

Sakai, K., A. Nazakawa, K. Kondo, and H. Ohta. 1985. Microbial reduction of nitroolefins. *Agric. Biol. Chem.* 49: 2331–2335.

Salanitro, J.P., L.A. Diaz, M.P. Williams, and H.L. Wisniewski. 1994. Isolation of a bacterial culture that degrades methyl *t*-butyl ether. *Appl. Environ. Microbiol.* 60: 2593–2596.

Sala-Trepat, J.M. and W.C. Evans. 1971. The meta cleavage of catechol by *Azotobacter* species: 4-oxalocrotonatev pathway. *Eur. J. Biochem.* 20: 400–413.

Sallis, P.J., S.J. Armfield, A.T. Bull, and D.J. Hardman. 1990. Isolation and characterization of a haloalkane halidohydrolase from *Rhodococcus erythropolis* Y2. *J. Gen. Microbiol.* 136: 115–120.

Samain, E., H.C. Dubourguier, and G. Albagnac. 1984. Isolation and characterization of *Desulfobulbus elongatus* sp. nov. from a mesophilic industrial digester. *Syst. Appl. Microbiol.* 5: 391–401.

Sander, P., R.-M. Wittich, P. Fortnagel, H. Wilkes, and W. Francke. 1991. Degradation of 1,2,4-trichloro- and 1,2,4,5-tetrachlorobenzene by *Pseudomonas* strains. *Appl. Environ. Microbiol.* 57: 1430–1440.

Sanford, R.A., J.R. Cole, F.E. Löffler, and J.M. Tiedje. 1996. Characterization of *Desulfitobacterium chlororespirans*. sp. nov., which grows by coupling the oxidation of lactate to the reductive dechlorination of 3-chloro-4-hydroxybenzoate. *Appl. Environ. Microbiol.* 62: 3800–3808.

Sato, S.-I., N. Ouchiyama, T. Kimura, H. Nojiri, H. Yamane, and T. Omori. 1997a. Cloning of genes involved in carbazole degradation of *Pseudomonas* sp. strain CA10: nucleotide sequences of genes and characterization of *meta* cleavage enzymes and hydrolase. *J. Bacteriol.* 179: 4841–4849.

Sato, S.-I., J.-W. Nam, K. Kasuga, H. Nojiri, H. Yamane, and T. Omori. 1997b. Identification and characterization of genes encoding carbazole 1,9a-dioxygenase in *Pseudomonas* sp. strain CA10. *J. Bacteriol.* 179: 4850–4858.

Sayama, M., M. Inoue, M.-A. Mori, Y. Maruyama, and H. Kozuka. 1992. Bacterial metabolism of 2,6-dinitrotoluene with *Salmonella typhimurium* and mutagenicity of the metabolites of 2,6-dinitrotoluene and related compounds. *Xenobiotica* 22: 633–640.

Scalla, R., M. Matringe, J.-M. Camadroo, and P. Labbe. 1990. Recent advances in the mode of action of diphenyl ethers and related herbicides. *Z. Naturforsch.* 45c: 503–511.

Schach, S., B. Tshisuaka, S. Fetzner, and F. Lingens. 1995. Quinoline 2-oxidoreductase and 2-oxo-1,2-dihydroquinoline-5,6-dioxygenase from *Comamonas testeteroni* 63. the first two enzymes in quinoline and 3-methylquinoline degradation. *Eur. J. Biochem.* 232: 536–544.

Schackmann, A. and R. Müller. 1991. Reduction of nitroaromatic compounds by different *Pseudomonas* species under aerobic conditions. *Appl. Microbiol. Biotechnol.* 34: 809–813.

Schauder, R. and B. Schink. 1989. *Anaerovibrio glycerini* sp. nov., an anaerobic bacterium fermenting glycerol to propionate, cell matter and hydrogen. *Arch. Microbiol.* 152: 473–478.

Schenzle, A., H. Lenke, P. Fischer, P.A. Williams, and H.-J. Knackmuss. 1997. Catabolism of 3-nitrophenol by *Ralstonia eutropha* JMP 134. *Appl. Environ. Microbiol.* 63: 1421–1427.

Schenzle, A., H. Lenke, J.V. Spain, and H.-J. Knackmuss. 1999. 3-Hydroxylaminophenol mutase from *Ralstonia eutropha* JMP134 catalyzes a Bamberger rearrangement. *J. Bacteriol.* 181: 1444–1450.

Schieffer-Ullrich, H. and J.R. Andreesen. 1985. *Peptostreptoccus barnesae* sp. nov., a Gram-positive, anaerobic, obligately purine utilizing coccus from chicken feces. *Arch. Microbiol.* 143: 26–31.

Schieffer-Ullrich, H., R. Wagner, P. Dürre, and J.R. Andreesen. 1984. Comparative studies on physiology and taxonomy of obligately purinolytic clostridia. *Arch. Microbiol.* 138: 345–353.

Schink, B. 1985a. Degradation of unsaturated hydrocarbons by methanogenic enrichments cultures. *FEMS Microbiol. Ecol.* 31: 69–77.

Schink, B. 1985b. Fermentation of acetylene by an obligate anaerobe, *Pelobacter acetylenicus* sp. nov. *Arch. Microbiol.* 142: 295–301.

Schink, B. and N. Pfennig. 1982a. Fermentation of trihydroxybenzenes by *Pelobacter acidigallici* gen. nov. sp. nov., a new strictly anaerobic, non-sporeforming bacterium. *Arch. Microbiol.* 133: 195–201.

Schink, B. and N. Pfennig. 1982b. *Propionigenium modestum* gen. nov. sp. nov. a new strictly anaerobic, nonsporing bacterium growing on succinate. *Arch. Microbiol.* 133: 209–216.

Schink, B. and M. Stieb. 1983. Fermentative degradation of polyethylene glycol by a strictly anaerobic, Gram-negative, nonsporeforming bacterium, *Pelobacter venetianus*. sp. nov. *Appl. Environ. Microbiol.* 45: 1905–1913.

Schlömann, M., E. Schmidt, and H.-J. Knackmuss. 1990a. Different types of dienelactone hydrolase in 4-fluorobenzoate-utilizing bacteria. *J. Bacteriol.* 172: 5112–5118.

Schlömann, M., P. Fischer, E. Schmidt, and H.-J. Knackmuss. 1990b. Enzymatic formation, stability, and spontaneous reactions of 4-fluoromuconolactone, a metabolite of the bacterial degradation of 4-fluorobenzoate. *J. Bacteriol.* 172: 5119–5129.

Schmid, A.S., B. Rothe, J. Altenbuchnerm, W. Ludwig, and K.-H. Engesser. 1997. *Terrabacter* sp. strain DPO 360. *J. Bacteriol.* 179: 53–62.

Schmidt, S., R.-M. Wittich, D. Erdmann, H. Wilkes, W. Francke, and P. Fortnagel. 1992. Biodegradation of diphenyl ether and its monohalogenated derivatives by *Sphingomonas* sp. strain SS3. *Appl. Environ. Microbiol.* 58: 2744–2750.

Schneider, J., R. Grosser, K. Jayasimhulu, W. Xue, and D. Warshawsky. 1996. Degradation of pyrene, benz[*a*]anthracene, and benzo[*a*]pyrene by *Mycobacterium* sp. strain RGHII-135, isolated from a former coal gasification site. *Appl. Environ. Microbiol.* 62: 13–19

Schneider, S., and G. Fuchs. 1998. Phenylacetyl-CoA: acceptor oxidoreductase, a new a-oxidizing enzyme that produces phenylglyoxylate. Assay, membrane localization, and differential production in *Thauera aromatica*. *Arch. Microbiol.* 169: 509–516.

Schneider, S., M. El-Said Mohamed, and G. Fuchs. 1997. Anaerobic metabolism of L-phenylalanine via benzoyl-CoA in the denitrifying bacterium *Thauera aromatica*. *Arch. Microbiol.* 168: 310–320.

Schnell, S. and B. Schink. 1991. Anaerobic aniline degradation via reductive deamination of 4-aminobenzoyl-CoA in *Desulfobacterium anilini*. *Arch. Microbiol.* 155: 183–190.

Schnell, S., F. Bak, and N. Pfennig. 1989. Anaerobic degradation of aniline and dihydroxybenzenes by newly isolated sulfate-reducing bacteria and description of *Desulfobacterium anilini*. *Arch. Microbiol.* 152: 556–563.

Schocher, R.J., B. Seyfried, F. Vazquez, and J. Zeyer. 1991. Anaerobic degradation of toluene by pure cultures of denitrifying bacteria. *Arch. Microbiol.* 157: 7–12.

Scholtz, R., F. Messi, T. Leisinger, and A.M. Cook. 1988. Three dehalogenases and physiological restraints in the biodegradation of haloalkanes by *Arthrobacter* sp. strain HA1. *Appl. Environ. Microbiol.* 54: 3034–3038.

Schowanek, D. and W. Verstraete. 1990. Phosphonate utilization by bacterial cultures and enrichments from environmental samples. *Appl. Environ. Microbiol.* 56: 895–903.

Schramm, E. and B. Schink. 1991. Ether-cleaving enzyme and diol dehydratase involved in anaerobic polyethylene glycol degradation by a new *Acetobacterium* sp. *Biodegradation* 2: 71–79.

Schwarz, G., R. Bauder, M. Speer, T.O. Rommel, and F. Lingens. 1989. Microbial metabolism of quinoline and related compounds II. Degradation of quinoline by *Pseudomonas fluorescens* 3, *Pseudomonas putida* 86 and *Rhodococcus* sp. B1. *Biol. Chem. Hoppe-Seyler* 370: 1183–1189.

Selifonov, S.A., J.E. Gurst, and L.P. Wackett. 1995. Regioselective dioxygenation of *ortho*-trifluoromethylbenzoate by *Pseudomonas aeruginosa* 142: evidence for 1,2-dioxygenation as a mechanism in *ortho*-halobenzoate dehalogenation. *Biochem. Biophys. Res. Commun.* 213: 759–767.

Sellström, U., A. Kierkegaard, C. de Witt, and B. Jansson. 1998. Polybrominated diphenyl ethers and hexabromocyclododecane in sediment and fish from a Swedish river. *Environ. Toxicol. Chem.* 17: 1065–1072.

Sendoo, K. and H. Wada. 1989. Isolation and identification of an aerobic γ-HCH-decomposing bacterium from soil. *Soil Sci. Plant Nutr.* 35: 79–87.

Servent, D., C. Ducrorq, Y. Henry, A. Guissani, and M. Lenfant. 1991. Nitroglycerin metabolism by *Phanerochaete chrysosporium*: evidence for nitric oxide and nitrite formation. *Biochim. Biophys. Acta* 1074: 320–325.

Seto, M., E. Masai, M. Ida, T. Hatta, K. Kimbara, M. Fukuda, and K. Yano. 1995. Multiple polychlorinated biphenyl transformation systems in the Gram-positive bacterium *Rhodococcus* sp. strain RHA1. *Appl. Environ. Microbiol.* 61: 4510–4513.

Sharak Genther, B.R., G.T. Townsend, and B.O. Blattmann. 1997. Reduction of 3-chlorobenzoate, 3-bromobenzoate, and benzoate to corresponding alcohols by *Desulfomicrobium escambiense*, isolated from a 3-chlorobenzoate-dechlorinating coculture. *Appl. Environ. Microbiol.* 63: 4698–4703.

Shields, M.S., S.O. Montgomery, P.J. Chapman, S.M. Cuskey, and P.H. Pritchard. 1989. Novel pathway of toluene catabolism in the trichloroethylene-degrading bacterium G4. *Appl. Environ. Microbiol.* 55: 1624–1629.

Shimizu, S., S. Jareonkitmongkol, H. Kawashima, K. Akimoto, and H. Yamada. 1991. Production of a novel ω1-eicosapentaenoic acid by *Mortierella alpina* 1S-4 grown on 1-hexadecene. *Arch. Microbiol.* 156: 163–166.

Shochat, E., I. Hermoni, Z. Cohen, A. Abeliovich, and S. Belkin. 1993. Bromoalkane-degrading *Pseudomonas* strains. *Appl. Environ. Microbiol.* 59: 1403–1409.

Simpson, F.J., N. Narasimhachari, and D.W.S. Westlake. 1963. Degradation of rutin by *Aspergillus flavus*. The carbon monoxide producing system. *Can. J. Microbiol.* 9: 15–25.

Singer, M.E. and W.R. Finnerty. 1985a. Fatty aldehyde dehydrogenases in *Acinetobacter* sp. strain HO1-N: role in hexadecane and hexadecanol metabolism. *J. Bacteriol.* 164: 1011–1016.

Singer, M.E. and W.R. Finnerty. 1985b. Alcohol dehydrogenases in *Acinetobacter* sp. strain HO1-N: role in hexadecane and hexadecanol metabolism. *J. Bacteriol.* 164: 1017–1024.

Sluis, M.K., F.J. Small, J.R. Allen, and S.A. Ensign. 1996. Involvement of an ATP-dependent carboxylase in a CO_2-dependent pathway of acetone metabolism by *Xanthobacter* strain Py2. *J. Bacteriol.* 178: 4020–4026.

Small, F.J. and S.A. Ensign. 1995. Carbon dioxide fixation in the metabolism of propylene and propylene oxide by *Xanthobacter* strain Py2. *J. Bacteriol.* 177: 6170–6175.

Small, F.J., J.K. Tilley, and S.A. Ensign. 1995. Characterization of a new pathway for epichlorohydrin degradation by whole cells of *Xanthobacter* strain Py2. *Appl. Environ. Microbiol.* 61: 1507–1513.

Smith, A.E. 1985. Identification of 2,4-dichloroanisole and 2,4-dichlorophenol as soil degradation products of ring-labelled [^{14}C] 2,4-D. *Bull. Environ. Contam. Toxicol.* 34: 150–157.

Smith, K.E., S. Latif, D.N. Kirk, and K.A. White. 1988. Microbial transformations of steroids—I. Rare transformations of progesterone by *Apiocrea chrysosperma*. *J. Steroid Biochem.* 31: 83–89.

Smith, N.A. and D.P. Kelly. 1988. Isolation and physiological characterization of autotrophic sulphur bacteria oxidizing dimethyl disulphide as sole source of energy. *J. Gen. Microbiol.* 134: 1407–1417.

Smith, R.V. and J.P. Rosazza. 1983. Microbial models of mammalian metabolism. *J. Nat. Prod.* 46: 79–91.

Smith, R.V., P.J. Davies, A.M. Clark, and S.K. Prasatik. 1981. Mechanism of hydroxylation of biphenyl by *Cunninghamella echinulata*. *Biochem. J.* 196: 369–371.

Söhngen, N.L. 1913. Benzin, Petroleum, Paraffinöl und Paraffin als Kohlenstoff- und Energiequelle für Mikroben. *Zentralbl. Bakteriol. Parasitenkd. Infektionskr. (2 Abt.)* 37: 595–609.

Somerville, C.C., S.F. Nishino, and J.C. Spain. 1995. Purification and characterization of nitrobenzene nitroreductase from *Pseudomonas pseudoalcaligenes* JS45. *J. Bacteriol.* 177: 3837–3842.

Sommer, C. and H. Görisch. 1997. Enzymology of the degradation of (di)chlorobenzenes by *Xanthobacter flavus* 14p1. *Arch. Microbiol.* 167: 384–391.

Sondossi, M., M. Sylvestre, and D. Ahmad. 1992. Effects of chlorobenzoate transformation on the *Pseudomonas testosteroni* biphenyl and chlorobiphenyl degradation pathway. *Appl. Environ. Microbiol.* 58: 485–495.

Sonier, D.N., N.L. Duran, and G.B. Smith. 1994. Dechlorination of trichlorofluoromethane CFC-11 by sulfate-reducing bacteria from an aquifer contaminated with halogenated aliphatic compounds. *Appl. Environ. Microbiol.* 60: 4567–4572.

Sowers, K.R. and J.G. Ferry. 1983. Isolation and characterization of a methylotrophic marine methanogen, *Methanococcoides methylutens* gen. nov., sp. nov. *Appl. Environ. Microbiol.* 45: 684–690.

Spain, J.C. 1995a. Biodegradation of nitroaromatic compounds. *Annu. Rev. Microbiol.* 49: 523–555.

Spain, J.C. Ed. 1995b. *Biodegradation of Nitroaromatic Compounds*. Plenum Press, New York.

Spain, J.C. and D.T. Gibson. 1991. Pathway for biodegradation of *p*-nitrophenol in a *Moraxella* sp. *Appl. Environ. Microbiol.* 57: 812–819.

Spain, J.C., G.J. Zylstra, C.K. Blake, and D.T. Gibson. 1989. Monohydroxylation of phenol and 2,5-dichlorophenol by toluene dioxygenase in *Pseudomonas putida* F1. *Appl. Environ. Microbiol.* 55: 2648–2652.

Spanggord, R.J., J.C. Spain, S.F. Nishino, and K.E. Mortelmans. 1991. Biodegradation of 2,4-dinitrotoluene by a *Pseudomonas* sp. *Appl. Environ. Microbiol.* 57: 3200–3205.

Spangler, W.J., J.L. Spigarelli, J.M. Rose, R.S. Flipin, and H.H. Miller. 1973. Degradation of methylmercury by bacteria isolated from environmental samples. *Appl. Microbiol.* 25: 488–493.

Sparnins, V.L., and P.J. Chapman. 1976. Catabolism of L-tyrosine by the homoprotocatechuate pathway in Gram-positive bacteria. *J. Bacteriol.* 127: 363–366.

Sparnins, V.L., P.J. Chapman, and S. Dagley. 1974. Bacterial degradation of 4-hydroxyphenylacetic acid and homoprotocatechuic acid. *J. Bacteriol.* 120: 159–167.

Spiess, T., F. Desiere, P. Fischer, J.C. Spain, H.-J. Knackmuss, and H. Lenke. 1998. A new 4-nitrotoluene degradation pathway in a *Mycobacterium* strain. *Appl. Environ. Microbiol.* 64: 446–452.

Stams, A.J.M., D.R. Kremer, K. Nicolay, G.H. Weenk, and T.A. Hansen. 1984. Pathway of propionate formation in *Desulfobulbus propionicus*. *Arch. Microbiol.* 139: 167–173.

Stanier, R.Y. 1968. Biochemical and immunological studies on the evolution of a metabolic pathway in bacteria pp. 201–225. In *Chemotaxonomy and Serotaxonomy* (Ed. J.G. Hawkes). Systematics Association Special Vol. 2. Academic Press, London.

Steffan, R.J., K. McClay, S. Vainberg, C.W. Condee, and D. Zhang. 1997. Biodegradation of the gasoline oxygenates methyl *tert*-butyl ether, ethyl *tert*-butyl ether, and amyl *tert*-butyl ether by propane-oxidizing bacteria. *Appl. Environ. Microbiol.* 63: 4216–4222.

Steiert, J.G. and R.L. Crawford. 1986. Catabolism of pentachlorophenol by a *Flavobacterium* sp. *Biochem. Biophys. Res. Commun.* 141: 1421–1427.

Stickland, L.H. 1934. Studies in the metabolism of the strict anaerobes (genus *Clostridium*). I. The chemical reactions by which *Cl. sporogenes* obtains its energy. *Biochem. J.* 28: 1746–1759.

Stickland, L.H. 1935a. Studies in the metabolism of the strict anaerobes (genus *Clostridium*). II. The reduction of proline by *Cl. sporogenes*. *Biochem. J.* 29: 288–290.

Stickland, L.H. 1935b. Studies in the metabolism of the strict anaerobes (genus *Clostridium*). III. The oxidation of alanine by *Cl. sporogenes*. IV. The reduction of glycine by *Cl. sporogenes*. *Biochem. J.* 29: 898.

Stieb, M. and B. Schink. 1985. Anaerobic oxidation of fatty acids by *Clostridium bryantii* sp. nov., a sporeforming, obligately syntrophic bacterium. *Arch. Microbiol.* 140: 387–390.

Stirling, L.A., R.J. Watkinson, and I.J. Higgins. 1977. Microbial metabolism of alicyclic hydrocarbons: isolation and properties of a cyclohexane-degrading bacterium. *J. Gen. Microbiol.* 99: 119–125.

Stockinger, J., C. Hinteregger, M. Loidl, A. Ferschl, and F. Streichsbier. 1992. Mineralizarion of 3-chloro-4-methylamiline via an *ortho*-cleavage pathway by *Pseudomonas cepacia* strain CMA1. *Appl. Microbiol. Biotechnol.* 38: 421–428.

Strubel, V., K.-H. Engesser, P. Fischer, and H.-J. Knackmuss. 1991. 3-(2-Hydroxyphenyl)catechol as substrate for proximal *meta* ring cleavage in dibenzofuran degradation by *Brevibacterium* sp. strain DPO 1361. *J. Bacteriol.* 173: 1932–1937.

Stubbe, J. and W.A. van der Donk. 1998. Protein radicals in enzyme catalysis. *Chem. Rev.* 98: 705–762.

Stucki, G., R. Gälli, H.R. Ebersold, and T. Leisinger. 1981. Dehalogenation of dichloromethane by cell extracts of *Hyphomicrobium* DM2. *Arch. Microbiol.* 130: 366–371.

Suen, W.-C., B.E. Haigler, and J.C. Spain 1996. 2,4-Dinitrotoluene dioxygenase from *Burkholderia* sp. strain DNT: similarity to naphthalene dioxygenase. *J. Bacteriol.* 178: 4926–4934.

Sundström, G., Å. Larsson, and M. Tarkpea. 1986. Creosote pp. 159–205. In *Handbook of Environmental Chemistry* (Ed. O. Hutzinger) Vol. 3/Part D. Springer-Verlag, Berlin.

Sutherland, J.B., J.P. Freeman, A.L. Selby, P.P. Fu, D.W. Miller, and C.E. Cerniglia. 1990. Stereoselective formation of a K-region dihydrodiol from phenanthrene by *Streptomyces flavovirens*. *Arch. Microbiol.* 154: 260–266.

Sutherland, J.B., A.L. Selby, J.P. Freeman, F.E. Evans, and C.E. Cerniglia. 1991. Metabolism of phenanthrene by *Phanerochaete chrysosporium*. *Appl. Environ. Microbiol.* 57: 3310–3316.

Sutherland, J.B., P.P. Fu, S.K. Yang, L.S. von Tungelnm,, R.P. Casillas, S.A. Crow, and C.E. Cerniglia. 1993. Enantiomeric composition of the *trans*-dihydrodiols produced from phenanthrene by fungi. *Appl. Environ. Microbiol.* 59: 2145–2149.

Suyama, A., R. Iwakiri, N. Kimura, A. Nishi, K. Nakamura, and K. Furukawa. 1996. Engineering hybrid pseudomonads capable of utilizing a wide range of aromatic hydrocarbons and of efficient degradation of trichloroethylene. *J. Bacteriol.* 178: 4039–4046.

Suylen, G.M.H., G.C. Stefess, and J.G. Kuenen. 1986. Chemolithotrophic potential of a *Hyphomicrobium* species, capable of growth on methylated sulphur compounds. *Arch. Microbiol.* 146: 192–198.

Suylen, G.M.H., P.J. Large, J.P. van Dijken, and J.G. Kuenen. 1987. Methyl mercaptan oxidase, a key enzyme in the metabolism of methylated sulphur compounds by *Hyphomicrobium* EG. *J. Gen. Microbiol.* 133: 2989–2997.

Suzuki, M., T. Hayakawa, J.P. Shaw, M. Rekik, and S. Harayama. 1991. Primary structure of xylene monooxygenase: similarities to and differences from the alkane hydroxylation system. *J. Bacteriol.* 173: 1690–1695.

Suzuki, T. 1983. Methylation and hydroxylation of pentachlorophenol by *Mycobacterium* sp. isolated from soil. *J. Pestic. Sci.* 8: 419–428.

Swarts, F. 1922. Sur l'acide trifluoracétique. *Bull. Acad. R. Belg.* 8: 343–370.

Swaving, J., J.A.M. de Bont, A. Westphal, and A. de Kok. 1996. A novel type of pyridine nucleotide-disulfide oxidoreductase is essential for NAD+- and NADPH-dependent degradation of epoxyalkanes by *Xanthobacter* strain Py2. *J. Bacteriol.* 178: 6644–6646.

Sylvestre, M., R. Massé, F. Messier, J. Fauteux, J.-G. Bisaillon, and R. Beaudet. 1982. Bacterial nitration of 4-chlorobiphenyl. *Appl. Environ. Microbiol.* 44: 871–877.

Szewzyk, R. and N. Pfennig. 1987. Complete oxidation of catechol by the strictly anaerobic sulfate-reducing *Desulfobacterium catecholicum* sp. nov. *Arch. Microbiol.* 147: 163–168.

Taeger, K., H.-J. Knackmuss, and E. Schmidt. 1988. Biodegradability of mixtures of chloro- and methylsubstituted aromatics: simultaneous degradation of 3-chlorobenzoate and 3-methylbenzoate. *Appl. Microbiol. Biotechnol.* 28: 603–608.

Taira, K., J. Hirose, S. Hayashida, and K. Furukawa. 1992. Analysis of *bph* operon from the polychlorinated biphenyl-degrading strain of *Pseudomonas pseudoalcaligenes* KF707. *J. Biol. Chem.* 267: 4844–4853.

Takigawa, H., H. Kubota, H. Sonohara, M. Okuda, S. Tanaka, Y. Fujikura, and S. Ito. 1993. Novel allylic oxidation of α-cedrene to *sec*-cedrenol by a *Rhodococcus* strain. *Appl. Environ. Microbiol.* 59: 1336–1341.

Tanaka, A. and M. Ueda. 1993. Assimilation of alkanes by yeasts: functions and biogenesis of peroxisomes. *Mycol. Res.* 98: 1025–1044.

Taniuchi, H., M. Hatanaka, S. Kuno, O. Hayashi, M. Nakazima, and N. Kurihara. 1964. Enzymatic formation of catechol from anthranilic acid. *J. Biol. Chem.* 239: 2204–2211.

Tasaki, M., Y. Kamagata, K. Nakamura, and E. Mikami. 1992. Utilization of methoxylated benzoates and formation of intermediates by *Desulfotomaculum thermobenzoicum* in the presence or absence of sulfate. *Arch. Microbiol.* 157: 209–212.

Tausson, W.O. 1927. Naphthalin als Kohlenstoffquelle für Bakterien. *Planta* 4: 214–256.

Taylor, B.F., J.A. Amador, and H.S. Levinson. 1993. Degradation of *meta*-trifluoromethylbenzoate by sequential microbial and photochemical treatments. *FEMS Microbiol. Lett.* 110: 213–216.

Taylor, D.G. and P.W. Trudgill. 1978. Metabolism of cyclohexane carboxylic acid by *Alcaligenes* strain W1. *J. Bacteriol.* 134: 401–411.

Taylor, S.J.C., D.W. Ribbons, A.M.Z. Slawin, D.A. Widdowson, and D.J. Williams. 1987. Biochemically generated chiral intermediates for organic synthesis: the absolute stereochemistry of 4-bromo-*cis*-2,3-dihydroxycyclohexa-4,6-diene-1-carboxylic acid formed from 4-bromobenzoic acid by a mutant of *Pseudomonas putida*. *Tetrahedron Lett.* 28: 6391–6392.

Ternan, N.G. and J.P. Quinn. 1998. Phosphate starvation-independent 2-aminoethylphosphonic acid biodegradation in a newly isolated strain of *Pseudomonas putida*, NG2. *Syst. Appl. Microbiol.* 21: 346–352.

Textor, S., V.F. Wendich, A.A. De Graaf, U. Müller, M.I. Linder, D. Linder, and W. Buckel. 1997. Propionate oxidation in *Escherichia coli*: evidence for operation of a methylcitrate cycle in bacteria. *Arch. Microbiol.* 168: 428–436.

Tezuka, T. and K. Tonomura. 1978. Purification and properties of a second enzyme catalyzing the splitting of carbon–mercury linkages from mercury-resistant *Pseudomonas* K-62. *J. Bacteriol.* 135: 138–143.

Thauer, R.K., D. Möller-Zinjhan, and A.M. Spormann. 1989. Biochemistry of acetate catabolism in anaerobic chemotrophic bacteria. *Annu. Rev. Microbiol.* 43: 43–67

Thomas, O.R.T. and G.F. White. 1989. Metabolic pathway for the biodegradation of sodium dodecyl sulfate by *Pseudomonas* sp. C12B. *Biotechnol. Appl. Biochem.* 11: 318–327.

Thompson, A.S., N.J.P. Owens, and J.C. Murrell. 1995. Isolation and characterization of methansulfonic acid-degrading bacteria from the marine environment. *Appl. Environ. Microbiol.* 61: 2388–2393.

Tinschert, A., A. Kiener, K. Heinzmann, and A. Tschech. 1997. Isolation of new 6-methylnicotinic-acid-degrading bacteria, one of which catalyses the regioselective hydroxylation of nicotinic acid at position C2. *Arch. Microbiol.* 168: 355–361.

Tipton, C.L. and N.M. Al-Shathir. 1974. The metabolism of cyclopropane fatty acids by *Tetrahymena pyriformis*. *J. Biol. Chem.* 249: 886–889.

Tolosa, I., J.M. Bayona, and J. Albaigés. 1992. Identification and occurrence of brominated and nitrated phenols in estuarine sediments. *Mar. Pollut. Bull.* 22: 603–607.

Tomasek, P.H. and R.L. Crawford. 1986. Initial reactions of xanthone biodegradation by an *Arthrobacter* sp. *J. Bacteriol.* 167: 818–827.

Tomasi, I., I. Artaud, Y. Bertheau, and D. Mansuy. 1995. Metabolism of polychlorinated phenols by *Pseudomonas cepacia* AC1100: determination of the first two steps and specific inhibitory effect of methimazole. *J. Bacteriol.* 177: 307–311.

Topp, E., L. Xun, and C.S. Orser. 1992. Biodegradation of the herbicide bromoxynil (3,5-dibromo-4-hydroxybenzonitrile) by purified pentachlorophenol hydroxylase and whole cells of *Flavobacterium* sp. strain ATCC 39723 is accompanied by cyanogenesis. *Appl. Environ. Microbiol.* 58: 502–506.

Townsend, G.T. and J.M. Suflita. 1996. Characterization of chloroethylene dehalogenation by cell extracts of *Desulfomonile tiedjei* and its relationship to chlorobenzoate dehalogenation. *Appl. Environ. Microbiol.* 62: 2850–2853.

Townsend, G.T. and J.M. Suflita. 1997. Influence of sulfur oxyanions on reductive dehalogenation activities in *Desulfomonile tiedjei*. *Appl. Environ. Microbiol.* 63: 3594–3599.

Traunecker, J., A. Preub, and G. Diekert. 1991. Isolation and characterization of a methyl chloride utilizing strictly anaerobic bacterium. *Arch. Microbiol.* 156: 416–421.

Trower, M.K., R.M. Buckland, R. Higgins, and M. Griffin. 1985. Isolation and characterization of a cyclohexane-metabolizing *Xanthobacter* sp. *Appl. Environ. Microbiol.* 49: 1282–1289.

Trudgill, P.W. 1969. The metabolism of 2-furoic acid by *Pseudomonas* F2. *Biochem. J.* 113: 577–587.

Trudgill, P.W. 1978. Microbial degradation of alicyclic hydrocarbons pp. 47–84. In *Developments in Biodegradation of Hydrocarbons—1* (Ed. R.J. Watkinson). Applied Science Publishers, London.

Trudgill, P.W. 1984. Microbial degradation of the alicyclic ring: structural relationships and metabolic pathways pp. 131–180. In *Microbial Degradation of Organic Compounds* (Ed. D.T. Gibson). Marcel Dekker, New York.

Trudgill, P.W. 1994. Microbial metabolism and transformation of selected monoterpenes pp.33–61. In *Biochemistry of Microbial Degradation* (Ed. C. Ratledge). Kluwer Academic Publishers, Dordrecht, the Netherlands.

Tschech, A. and B. Schink. 1985. Fermentative degradation of resorcinol and resorcylic acids. *Arch. Microbiol.* 143: 52–59.

Tsou, A.Y., S.C. Ransom, J.A. Gerlt, D.D. Buechter, P.C. Babbitt, and G.L. Kenyon. 1990. Mandelate pathway of *Pseudomonas putida*: sequence relationships involving mandelate racemase, (S)-mandelate dehydrogenase, and benzoylformate decarboxylase and expression of benzoylformate decarboxylase in *Escherichia coli*. *Biochemistry* 29: 9856–9862.

Tu, C.M. 1976. Utilization and degradation of lindane by soil microorganisms. *Arch. Microbiol.* 108: 259–263.

Uchiyama, H. and T. Tabuchi. 1976. Properties of methylcitrate synthase from *Candida lipolytica*. *Agric. Biol. Chem.* 40: 1411–1418.

Uetz, T., R. Schneider, M. Snozzi, and T. Egli. 1992. Purification and characterization of a two-component monooxygenase that hydroxylates nitrilotriacetate from "*Chelatobacter*" strain ATCC 29600. *J. Bacteriol.* 174: 1179–1188.

Uotila, J.S., V.H. Kitunen, T. Saastamoinen, T. Coote, M.M. Häggblom, and M.S. Salkinoja-Salonen. 1992. Characterization of aromatic dehalogenases of *Mycobacterium fortuitum* CG-2. *J. Bacteriol.* 174: 5669–5675.

Urakami, T., H. Araki, H. Oyanagi, K.-I. Suzuki, and K. Komagata. 1992. Transfer of *Pseudomonas aminovorans* (den Dooren de Jong 1926) to *Aminobacter* gen. nov. as *Aminobacter aminovorans* comb. nov. and description of *Aminobacter aganoensis* sp. nov. and *Aminobacter niigataensis* sp. nov. *Int. J. Syst. Bacteriol.* 42: 84–92.

Urakami, T., H. Araki, H. Oyanagi, K.-I. Suzuli, and K. Komagata. 1990. *Paracoccus aminophilus* sp. nov. and *Paracoccus aminovorans* sp. nov., which utilize N,N'-dimethylformamide. *Int. J. Syst. Bacteriol.* 40: 287–291.

Utkin, I., C. Woese, and J. Wiegel. 1994. Isolation and characterization of *Desulfitobacterium dehalogenans* gen. nov., sp.nov., an anaerobic bacterium which reductively dechlorinates chlorophenolic compounds. *Int. J. Syst. Bacteriol.* 44: 612–619.

Utkin, I., D.D. Dalton, and J. Wiegel. 1995. Specificity of reductive dehalogenation of substituted *ortho*-chlorophenols by *Desulfitobacterium dehalogenans* JW/IU-DC1. *Appl. Environ. Microbiol.* 61: 346–351.

Valli, K. and M.H. Gold. 1991. Degradation of 2,4-dichlorophenol by the lignin-degrading fungus *Phanerochaete chrysosporium*. *J. Bacteriol.* 173: 345–352.

Valli, K., H. Wariishi, and M.H. Gold. 1992a. Degradation of 2,7-dichlorodibenzo-p-dioxin by the lignin-degrading basidiomycete *Phanerochaete chrysosporium*. *J. Bacteriol.* 174: 2131–2137.

Valli, K., B.J. Brock, D.K. Joshi, and M.H. Gold. 1992b. Degradation of 2,4-dinitrotoluene by the lignin-degrading fungus *Phanerochaete chrysosporium*. *Appl. Environ. Microbiol.* 58: 221–228.

Van Afferden, M., S. Schacht, J. Klein, and H.G. Trüper. 1990. Degradation of dibenzothiophene by *Brevibacterium* sp. DO. *Arch. Microbiol.* 153: 324–328.

Van Alfen, N.K. and T. Kosuge. 1974. Microbial metabolism of the fungicide 2,6-dichloro-4-nitroaniline. *J. Agric. Food Chem.* 22: 221–224.

Van den Tweel, W.J.J. and J.A.M. de Bont. 1985. Metabolism of 3-butyl-1-ol by *Pseudomonas* BB1. *J. Gen. Microbiol.* 131: 3155–3162.

Van den Tweel, W.J.J., J.B. Kok, and J.A.M. de Bont. 1987. Reductive dechlorination of 2,4-dichlorobenzoate to 4-chlorobenzoate and hydrolytic dehalogenation of 4-chloro, 4-bromo-, and 4-iodobenzoate by *Alcaligenes denitrificans* NTB-1. *Appl. Environ. Microbiol.* 53: 810–815.

Van den Wijngaard, A.J., D.B. Janssen, and B. Withold. 1989. Degradation of epichlorohydrin and halohydrins by bacterial cultures isolated from freshwater sediment. *J. Gen. Microbiol.* 135: 2199–2208.

Van den Wijngaard, A.J., K.W.H.J. van der Kamp, J. van der Ploeg, F. Pries, B. Kazemier, and D.B. Janssen. 1992. Degradation of 1,2-dichloroethane by *Ancyclobacter aquaticus* and other facultative methylotrophs. *Appl. Environ. Microbiol.* 58: 976–983.

Van der Maarel, M.J.E.C., P. Quist, L. Dijkhuizen, and T.A. Hansen. 1993. Anaerobic degradation of dimethylsulfoniopropionate to 3-S-methylmercaptopropionate by a marine *Desulfobacterium* strain. *Arch. Microbiol.* 166: 411–412.

Van der Maarel, M.J.E.C., S. van Bergeijk, A.F. van Werkhoven, A.M. Laverman, W.G. Maijer, W.T. Stam, and T.A. Hansen. 1996. Cleavage of dimethylsulfoniopropionate and reduction of acrylate by *Desulfovibrio acrylicus* sp. nov. *Arch. Microbiol.* 166: 109–115.

Van der Meer, J.R., W.M. de Vos, S. Harayama, and A.J.B. Zehnder. 1992. Molecular mechanisms of genetic adaptation to xenobiotic compounds. *Microbiol. Rev.* 56: 677–694.

van der Ploeg, J., M. Willemsen, G. van Hall, and D.B. Janssen. 1995. Adaptation of *Xanthobacter autotrophicus* GJ10 to bromoacetate due to activation and mobilization of the haloacetate dehalogenase gene by insertion element *IS1247*. *J. Bacteriol.* 177: 1348–1356.

van der Werf, M.J., K.M. Overkamp, and J.A.M. de Bont. 1998. Limonene-1,2-epoxide hydrolase from *Rhodococcus erythropolis* DCL14 belongs to a novel class of epoxide hydrolases. *J. Bacteriol.* 180: 5052–5057.

Van Dort, H.M. and D.L. Bedard. 1991. Reductive *ortho* and *meta* dechlorination of a polychlorinated biphenyl congener by anaerobic microorganisms. *Appl. Environ. Microbiol.* 57: 1576–1578.

Van Ginkel, C.G. and J.A.M. de Bont. 1986. Isolation and characterization of alkene-utilizing *Xanthobacter* spp. *Arch. Microbiol.* 145: 403–407.

van Hylckama Vlieg, J.E.T., J. Kingma, A.J. van den Wijngaard, and D.B. Janssen. 1998. A gluathione S-transferase with activity towards *cis*-1,2-dichloroepoxyethane is involved in isoprene utilization by *Rhodococcus* strain AD 45. *Appl. Environ. Microbiol.* 64: 2800–2805.

van Hylckama Vlieg, J.E.T., J. Kingma, W. Kruizinga, and D.B. Janssen. 1999. Purification of a glutathione S-transferase and a glutathione conjugate-specific dehydrogenase is involved in isoprene metabolism by *Rhodococcus* sp. strain AD 45. *J. Bacteriol.* 181: 2094–2101.

Vancheesewaran, S., R.U. Halden, K.J. Williamson, J.D. Ingle, and L. Semprini. 1999. Abiotic and biological transformations of tetraalkoxysilanes and trichloroethene/*cis*-1,2-dichloroethene cometabolism driven by tetrabutoxylsilane-degrading microorganisms. *Environ. Sci. Technol.* 33: 1077–1085.

Verhagen, F.J.M., H.J. Swarts, J.B.P.A. Wijnberg, and J.A. Field. 1998. Biotransformation of the major fungal metabolite 3,5-dichloro-*p*-anisyl alcohol under anaerobic conditions and its role in formation of bis(3,5-dichloro-4-hydroxyphenyl)methane. *Appl. Environ. Microbiol.* 64: 3225–3231.

Vogel, T.M. and D. Grbic-Galic. 1986. Incorporation of oxygen from water into toluene and benzene during anaerobic fermentative transformation. *Appl. Environ. Microbiol.* 52: 200–202.

Vogel, T.M. and P.L. McCarty. 1985. Biotransformation of tetrachloroethylene to trichloroethylene, dichloroethylene, vinyl chloride, and carbon dioxide under methanogenic conditions. *Appl. Environ. Microbiol.* 49: 1080–1083.

Voinov, V.G., Yu.N. El'kin, T.A. Kuznetsova, I.I. Mal'tsev, V.V. Mikhailov, and V.A. Sasunkevich. 1991. Use of mass spectrometry for the detection and identification of bromine-containing diphenyl ethers. *J. Chromatogr.* 586: 360–362.

Vonk, J.W. and A.K. Sijpesteijn. 1973. Studies on the methylation of mercuric chloride by pure cultures of bacteria and fungi. *Antonie van Leeuwenhoek* 39: 505–513.

Vorbeck, C., H. Lenke, P. Fischer, and H.-J. Knackmuss. 1994. Identification of a hydride-Meisenheimer complex as a metabolite of 2,4,6-trinitrotoluene by a *Mycobacterium* strain. *J. Bacteriol.* 176: 932–934.

Vorbeck, C., H. Lenke, P. Fischer, J.C. Spain, and H.-J. Knackmuss. 1998. Initial reductive reactions in aerobic microbial metabolism of 2,4,6-trinitrotoluene. *Appl. Environ. Microbiol.* 64: 246–252.

Wackett, L.P., S.L. Shames, C.P. Venditti, and C.T. Walsh. 1987a. Bacterial carbon-phosphorus lyase: products, rates and regulation of phosphonic and phosphinic acid metabolism. *J. Bacteriol.* 169: 710–717.

Wackett, L.P., B.L. Wanner, C.P. Venditti, and C.T. Walsh. 1987b. Involvement of the phosphate regulon and the *psiD* locus in carbon-phosphorus lyase activity of *Escherichi coli* K-12. *J. Bacteriol.* 169: 1753–1756.

Wackett, L.P., M.J. Sadowsky, L.M. Newman, H.-G. Hur, and S. Li. 1994. Metabolism of polyhalogenated compounds by a genetically engineered bacterium. *Nature* 368: 627–629.

Wagner, A.F.V., M. Frey, F.A. Neugebauer, W. Schäfer, and J. Knappe. 1992. The free radical in pyruvate formate-lyase is located on glycine-734. *Proc. Natl. Acad. Sci. U.S.A.* 89: 996–1000.

Wagner, C. and E.R. Stadtman. 1962. Bacterial fermentation of dimethyl-β-propiothetin. *Arch. Biochem. Biophys.* 98: 331–336.

Wanner, P. and R. Tressel. 1998. Purification and characterization of two enone reductases from *Saccharomyces cerevisiae*. *Eur. J. Biochem.* 255: 271–278.

Warhurst, A.M., K.F. Clarke, R.A. Hill, R.A. Holt, and C.A. Fewson. 1994. Metabolism of styrene by *Rhodococcus rhodochrous* NCIMB 13259. *Appl. Environ. Microbiol.* 60: 1137–1145.

Watanabe, I., T. Kashimoto, and R. Tatsukawa. 1983. Polybromianted anisoles in marine fish, shellfish, and sediments in Japan. *Arch. Environ. Contam. Toxicol.* 12: 615–620.

Watanabe, M., N. Sumida, S. Murakami, H. Anzai, C.J. Thompsom, Y. Tateno, and T. Murakami. 1999. A phosphonate-induced gene which promotes *Penicillium*-mediated bioconversion of *cis*-propenylphosphonic acid to fosfomycin. *Appl. Environ. Microbiol.* 65: 1036–1044.

Watson, G.K. and R.B. Cain. 1975. Microbial metabolism of the pyridine ring. Metabolic pathways of pyridine biodegradation by soil bacteria. *Biochem. J.* 146: 157–172.

Watson, G.K., C. Houghton, and R.B. Cain. 1974. Microbial metabolism of the pyridine ring. The metabolism of pyridine-3,4-diol (3,4-dihydroxypyridine) by *Agrobacterium* sp. *Biochem. J.* 140: 277–292.

Wayne, L.G. 1961. Recognition of *Mycobacterium fortuitum* by means of the 3-day phenolphthalein sulfatase test. *Am. J. Clin. Pathol.* 36: 185–187.

Wayne, L.G. and 17 coauthors. 1991. Fourth report of the cooperative, open-ended study of slowly growing mycobacteria by the International Working Group on Mycobacterial Taxonomy. *Int. J. Syst. Bacteriol.* 41: 463–472.

Wedemeyer, G. 1967. Dechlorination of 1,1,1-trichloro-2,2-bis[*p*-chlorophenyl]ethane by *Aerobacter aerogenes. Appl. Microbiol.* 15: 569–574.

Wegener, W.S., H.C. Reeves, R. Rabin, and S.J. Ajl. 1968. Alternate pathways of metabolism of short-chain fatty acids. *Bacteriol. Rev.* 32: 1–26.

Weijers, C.A.G.M., A. de Haan, and J.A.M. de Bont. 1988. Chiral resolution of 2,3-epoxyalkanes by *Xanthobacter* Py2. *Appl. Microbiol. Biotechnol.* 27: 337–340.

Weiner, J.M. and D.R. Lovley. 1998. Anaerobic benzene degradation in petroleum-contaminated aquifer sediments after inoculation with a benzene-oxidizing enrichment. *Appl. Environ. Microbiol.* 64: 775–778.

Weissenfels, W.D., M. Beyer, J. Klein, and H.J. Rehm. 1991. Microbial metabolism of fluoranthene: isolation and identification of fission products. *Appl. Microbiol. Biotechnol.* 34: 528–535.

Werbitzky, O., K. Klier, and H. Felber. 1990. Asymmetric induction of four chiral centers by hetero Diels-Alder reaction of a chiral nitrosodienmophile. *Liebigs Ann. Chem.* 267–270.

West, T.P. 1997. Reductive catabolism of uracil and thymine by *Burkholderia cepacia. Arch. Microbiol.* 168: 237–239.

Wetzstein, H.-G., N. Schmeer, and W. Karl. 1997. Degradation of the fluoroquinolone enrofloxacin by the brown-rot fungus *Gloeophyllum striatum*: identification of metabolites. *Appl. Environ. Microbiol.* 63: 4272–4281.

Wetzstein, H.-G., M. Stadler, H.-V. Tichy, A. Dalhoff, and W. Karl. 1999. Degradation of ciprofloxacin by basidiomycetes and identification of metabolites generated by the brown rot fungus *Gloeophyllum striatum. Appl. Environ. Microbiol.* 65: 1556–1563.

Wheelis, M., N.J. Palleroni, and R.Y. Stanier. 1967. The metabolism of aromatic acids by *Pseudomonas testosteroni* and *P. acidovorans. Arch. Mikrobiol.* 59: 302–314.

White, G.F. and N.J. Russell. 1994. Biodegradation of anionic surfactants and related molecules pp. 143–177. In *Biochemistry of Microbial Degradation* (Ed. C. Ratledge). Kluwer Academic Publishers, Dordrecht, the Netherlands.

White, G.F. and J.R. Snape. 1993. Microbial cleavage of nitrate esters: defusing the environment. *J. Gen. Microbiol.* 139: 1947–1957.

White, G.F., K.S. Dodson, I. Davies, P.J. Matts, J.P. Shapleigh, and W.J. Payne. 1987. Bacterial utilization of short-chain primary alkyl sulphate esters. *FEMS Microbiol. Lett.* 40: 173–177.

White, G.F., N.J. Russell, and E.C. Tidswell. 1996a. Bacterial scission of ether bonds. *Microbiol. Rev.* 60: 216–232.

White, G.F., J.R. Snape, and S. Nicklin. 1996b. Bacterial biodegradation of glycerol trinitrate. *Int. Biodeterior. Biodeg.* 38: 77–82.

White, H., R. Feicht, C. Huber, F. Lottspeich, and H. Simon. 1991. Purification and some properties of the tungsten-containing carboxylic acid reductase from *Clostridium formicoaceticum. Biol. Chem. Hoppe-Seyler* 372: 999–1005.

White, H., C. Huber, R. Feicht, and H. Simon. 1993. On a reversible molybdenum-containing aldehyde oxidoreductase from *Clostridium formicoaceticum. Arch. Microbiol.* 159: 244–249.

White, R.E., M.B. McCarthy, K.D. Egeberg, and S.G. Sligar. 1984. Regioselectivity in the cytochromes P-450: control by protein constraints and by chemical reactivities. *Arch. Biochem. Biophys.* 228: 493–502.

Whited, G.M. and D.T. Gibson. 1991. Separation and partial characterization of the enzymes of the toluene-4-monooxygenase catabolic pathway in *Pseudomonas mendocina* KR1. *J. Bacteriol.* 173: 3017–3020.

Whited, G.M., W.R. McCombie, L.D. Kwart, and D.T. Gibson. 1986. Identificaton of cis-diols as intermediates in the oxidation of aromatic acids by a strain of *Pseudomonas putida* that contains a TOL plasmid. *J. Bacteriol.* 166: 1028–1039.

White-Stevens, R.H., H. Kamin, and Q.H. Gibson. 1972. Studies of a flavoprotein, salicylate hydroxylase. II. Enzyme mechanism. *J. Biol. Chem.* 247: 2371–2381.

Widdel, F. 1986. Growth of methanogenic bacteria in pure culture with 2-propanol and other alcohols as hydrogen donors. *Appl. Environ. Microbiol.* 51: 1056–1062.

Widdel, F. 1987. New types of acetate-oxidizing, sulfate-reducing *Desulfobacter* species, *D. hydrogenophilus* sp. nov., *D. latus* sp. nov., and *D. curvatus* sp. nov. *Arch. Microbiol.* 148: 286–291.

Widdel, F. and N. Pfennig. 1981. Sporulation and further nutritional characteristics of *Desulfotomaculum acetoxidans* (emend.). *Arch. Microbiol.* 112: 119–122.

Widdel, F. and N. Pfennig. 1982. Studies on dissimilatory sulfate-reducing bacteria that decompose fatty acids. II. Incomplete opxidation of propionate by *Desulfobulbus propionicus* gen. no., sp. nov. *Arch. Microbiol.* 131: 360–365.

Widdel, F., P.E. Rouvière, and R.S. Wolfe. 1988. Classification of secondary alcohol-utilizing methanogens including a new thermophilic isolate. *Arch. Microbiol.* 150: 477–481.

Wiegant, W.M. and J.A.M. de Bont. 1980. A new route for ethylene glycol metabolism in *Mycobacterium* E44. *J. Gen. Microbiol.* 120: 325–331.

Wieser, M., B. Wagner, J. Eberspächer, and F. Lingens. 1997. Purification and characterization of 2,4,6-trichlorophenol-4-monooxygenase, a dehalogenating enzyme from *Azotobacter* sp. strain GP1. *J. Bacteriol.* 179: 202–208.

Willeke, U. and W. Barz. 1982. Catabolism of 5-hydroxyisoflavones by fungi of the genus *Fusarium*. *Arch. Microbiol.* 132: 266–269.

Winter, J., L.H. Moore, V.R. Dowell, and V.D. Bokkenheuser. 1989. C-ring cleavage of flavonoids by human intestinal bacteria. *Appl. Environ. Microbiol.* 55: 1203–1208.

Wischnak, C., F.E. Löffler, J. Li, J.W. Urbance, and R. Müller. 1998. *Pseudomonas* sp. strain 273, an aerobic α,ω-dichloroalkane-degrading bacterium. *Appl. Environ. Microbiol.* 64: 3507–3511.

Wittich, R.-M. 1998. Degradation of dioxin-like compounds by microorganisms. *Appl. Microbiol. Biotechnol.* 49: 489–499.

Wittich, R.-M., H.G. Rast, and H.-J. Knackmuss. 1988. Degradation of naphthalene-2,6- and naphthalene-1,6-disulfonic acid by a *Moraxella* sp. *Appl. Environ. Microbiol.* 54: 1842–1847.

Wittich, R.-M., H. Wilkes, V. Sinnwell, W. Francke, and P. Fortnagel. 1992. Metabolism of dibenzo-p-dioxin by *Sphingomonas* sp. strain RW1. *Appl. Environ. Microbiol.* 58: 1005–1010.

Wittlinger, R. and K. Ballschmiter 1990. Studies of the global baseline pollution XIII. C6-C14 organohalogens (α and γ-HCH, HCB, PCB, 4,4'-DDT, 4,4-DDE, cis- and trans-chlordane, trans-nonachlor, anisoles) in the lower troposphere of the southern Indian Ocean. *Fresenius Jz. Anal. Chem.* 336: 193–200.

Wolgel, S.A., J.E. Dege, P.E. Perkins-Olson, C.H. Juarez-Garcia, R.L. Crawford, E. Münck, and J.D. Lipscomb. 1993. Purification and characterization of protocatechuate 2,3-dioxygenase from *Bacillus macerans*: a new extradiol catecholic dioxygenase. *J. Bacteriol.* 175: 4414–4426.

Xu, G. and T.P. West. 1992. Reductive catabolism of pyrimidine bases by *Pseudomonas stutzeri*. *J. Gen. Microbiol.* 138: 2459–2463.

Xun, L. 1996. Purification and characterization of chlorophenol 4-monooxygenase from *Burckholderia cepacia* AC1100. *J. Bacteriol.* 178: 2645–2649.

Xun, L., E. Topp, and C.S. Orser. 1992a. Confirmation of oxidative dehalogenation of pentachlorophenol by a *Flavobacterium* pentachlorophenol hydroxylase. *J. Bacteriol.* 174: 5745–5747.

Xun, L., E. Topp, and C.S. Orser. 1992b. Diverse substrate range of a *Flavobacterium* pentachlorophenol hydroxylase and reaction stoichiometries. *J. Bacteriol.* 174: 2898–2902.

Xun, L., E. Topp, and C.S. Orser. 1992c. Purification and characterization of a tetra-chloro-*p*-hydroquinone reductive dehalogenase from a *Flavobacterium* sp. *J. Bacteriol.* 174: 8003–8007.

Yamada, E.W., and W.B. Jakoby. 1959. Enzymatic utilization of acetylenic compounds II. Acetylenemonocarboxylic acid hydrase. *J. Biol. Chem.* 234: 941–945.

Yamada, M. and K. Tonomura. 1972. Further study of formation of methylmercury from inorganic mercury by *Clostridium cochlearium* T-2. *J. Ferment. Technol.* 50: 893–900.

Yamamoto, K., Y. Ueno, K. Otsubo, K. Kawakami, and K.-I. Komatsu. 1990. Production of *S*-(+)-ibuprofen from a nitrile compound by *Acinetobacter* sp. strain AK 226. *Appl. Environ. Microbiol.* 56: 3125–3129.

Yang, W., L. Dostal, and J.P.N. Rosazza. 1993. Stereospecificity of microbial hydrations of oleic acid to 10-hydroxystearic acid. *Appl. Environ. Microbiol.* 59: 281–284.

Ye, D., M.A. Siddiqui, A.E. Maccubbin, S. Kumar, and H.C. Sikka. 1996. Degradation of polynuclear aromatic hydrocarbons by *Sphingomonas paucimobilis*. *Environ. Sci. Technol.* 30: 136–142.

Yoon, J.-H., S.-K. Rhee, J.-S. Lee, Y.-H. Park, and S.T. Lee. 1997. *Nocardiodes pyridinolyticus* sp. nov., a pyridine-degrading bacterium isolated froim the oxic zone of an oil shale column. *Int. J. Syst. Bacteriol.* 47: 933–938.

You, I.-S., R.I. Murray, D. Jollie, and I.C. Gunsalus. 1990. Purification and characterization of salicylate hydroxylase from *Pseudomonas putida* PpG7. *Biochem. Biophys. Res. Commun.* 169: 1049–1054.

Zeikus, J.G. 1980. Chemical and fuel production by anaerobic bacteria. *Annu. Rev. Microbiol.* 34: 423–464.

Zellner, G., H. Kneifel, and J. Winter. 1990. Oxidation of benzaldehydes to benzoic acid derivatives by three *Desulfovibrio* strains. *Appl. Environ. Microbiol.* 56: 2228–2233.

Zeyer, J. and P.C. Kearney. 1984. Degradation of *o*-nitrophenol and *m*-nitrophenol by a *Pseudomonas putida*. *J. Agric. Food Chem.* 32: 238–242.

Zeyer, J. and H.P. Kocher. 1988. Purification and characterization of a bacterial nitro-phenol oxygenase which converts *ortho*-nitrophenol to catechol and nitrite. *J. Bacteriol.* 170: 1789–1794.

Zeyer, J., A. Wasserfallen, and K.N. Timmis. 1985. Microbial mineralization of ring-substituted anilines though an *ortho*-cleavage pathway. *Appl. Environ. Microbiol.* 50: 447–453.

Zhang, X. and J. Wiegel. 1992. The anaerobic degradation of 3-chloro-4-hydroxybenzoate in freshwater sediment proceeds via either chlorophenol or hydroxybenzoate to phenol and subsequently to benzoate. *Appl. Environ. Microbiol.* 58: 3580–3585.

Zhang, X. and L.Y. Young. 1997. Carboxylation as an initial reaction in the anaerobic metabolism of naphthalene and phenanthrene by sulfidogenic consortia. *Appl. Environ. Microbiol.* 63: 4759–4764.

Zhang, Y.Z., S.T. Sundaram, A. Sharma, and B.W. Brodman. 1997. Biodegradation of glyceryl trinitrate by *Penicillium corylophilum* Dierckx. *Appl. Environ. Microbiol.* 63: 1712–1714.

Zhou, N.-Y., A. Jenkins, C.K.N. Chan kwo Chion, and D.J. Leak. 1999. The alkene monooxygenase from *Xanthobacter* strain Py2 is closely related to aromatic monooxygenases and catalyzes aromatic monooxygenation of benzene, toluene, and phenol. *Appl. Environ. Microbiol.* 65: 1589–1595.

Zimmermann, T., F. Gasser, H.G. Kulla, and T. Leisinger. 1984. Comparison of two bacterial azoreductases acquired during adaptation to growth on azo dyes. *Arch. Microbiol.* 138: 37–43.

7

Ecotoxicology

SYNOPSIS   The basic input required for assessing the toxicity of xenobiotics is summarized and includes data both on the exposure to the toxicant and an evaluation of its biological effect in terms of numerically determined end points. A brief discussion is directed to the choice of test species, to the range of acceptable end points, and to the choice of media for laboratory tests. Some comments are provided on the commonly used tests using single organisms that include representatives of algae, crustaceans, and fish with emphasis on assays for sublethal effects. Brief descriptions are given of less widely used assays using rotifers, mayflies, tadpoles, and sea urchins. Assays for evaluating toxicity in sediments and toward terrestrial organisms are discussed, and procedures for detecting genotoxic and estrogenic effects are noted. A discussion is presented on multicomponent test systems including different types of mesocosms. Attention is directed to the important question of metabolism by the test organisms with emphasis on fish. The use of biomarkers is briefly discussed and includes application of both biochemical and physiological parameters. Some cautionary comments are given on the application of these to feral fish. Attention is briefly drawn to the application of toxicity-equivalent factors and the role of epidemiology. It is suggested that a hierarchical approach to evaluating toxicity could be used, and that assays at the higher levels are justified if no effects are observed at the lower ones. Any system should be flexible and should be able to incorporate studies of partition and additional factors relevant to a particular environment.

Introduction

Previous chapters have been devoted to the distribution and persistence of xenobiotics after discharge into the aquatic environment. This chapter is devoted to the effect of xenobiotics on aquatic organisms. Its depth and orientation should be clearly recognized: like Chapter 2 on analytical procedures, this chapter is not directed to the professional ecotoxicologist. The aim has been to provide an overview of the kinds of bioassays that are being used in environmental research and to indicate a few of the areas to which further attention might profitably be directed. No attempt has been made to provide protocols for standardized procedures, nor to indicate which of the many possible assay procedures are acceptable to the administrative authorities that issue discharge permits. Attention is directed to detailed presentation of procedures aimed at determining the ecological effects of xenobiotics on the structure of populations and communities (Petersen and Petersen 1989) and procedures for community testing (Landner et al. 1989).

Toxicology may be defined as the science of poisons, and has traditionally been devoted to the effect of poisons on higher organisms including humans and domestic animals. The cardinal concepts are those of dose, which was introduced by Paracelsus in the 16th century, and the correlation of effect with the chemical nature of the toxicant promulgated by Mattieu Orfila in the 19th century. Toxicology is concerned with four interacting elements—the cause, the organism, the effect, and the consequence. The term *ecotoxicology* has been coined to include the effect of toxicants on biota both in natural eco-systems and in laboratory test systems: these may evaluate the effect on a wide spectrum of organisms ranging from bacteria through algae and crusta-ceans to vertebrates. In addition, attention has increasingly been directed to the application of a number of parameters in routine clinical use, and to their adaptation for use both in laboratory experiments and in evaluations using feral fish; for example, the levels of specific enzymes, morphological changes in organs such as liver, and blood parameters have been successfully used. The two approaches are not, of course, mutually exclusive.

Humans are a predator of organisms at higher trophic levels so that, for example, the consumption of fish provides a mechanism whereby humans may be indirectly exposed to xenobiotics. In this case, the critical question therefore is the degree of contamination of fish by toxicants; the mechanisms whereby xenobiotics may be accumulated in biota have been discussed in Section 3.1, and the metabolism of these by higher aquatic biota will be reviewed briefly later in Section 7.5. There is an enormous literature on human toxicology from which many useful ideas applicable to ecotoxicology may be gleaned, and these can profitably be adapted with only minor modi-fication. In human toxicology, a number of basic problems have been exten-sively explored and these include:

1. The fundamental issue of the relation between the dose of a toxicant and the response elicited;
2. The vexatious question of the existence or otherwise of threshold concentrations below which toxicity is not displayed;
3. The statistical design of experiments in toxicology.

It is appropriate therefore to make brief reference to some essentially popular accounts in which these are illustrated by readily understood examples

(Ottoboni 1984; Rodricks 1992) in addition to the substantial discussions presented in the classic text by Casarett and Doull (Klaasen et al. 1986).

Most of the present discussion will be illustrated by examples from experiments with pure organic compounds, but it should be appreciated that the discharge of single compounds into the aquatic environment is exceptional: almost invariably the effluents consist of a complex mixture of compounds and these are generally evaluated as nonfractionated effluents or occasionally on the basis of the effect of their major components. The question of synergistic or antagonistic effects therefore remains essentially unresolved. Toxicity equivalent factors (TEFs) have been used to provide an overall estimate of toxicity in situations where mixtures of compounds are present, and this is discussed again in Section 7.7.1. Briefly, the toxicity of each component is evaluated using a given test system and this value is multiplied by the concentration of that component in the mixture; these values are then summed to provide an estimate of the toxicity of the mixture.

The problem of assessing effects on natural ecosystems is so complex that it is generally simplified to a greater or lesser extent: experiments may be conducted in the laboratory or outdoors in model ecosystems, and a plethora of single species have been used for assaying biological effects. These assays attempt to encompass various trophic levels, and differencies in physiology and metabolism. For example, representatives of algae are generally used to assess effects on photosynthesis and on primary production, crustaceans and fish may be used to evaluate effects on secondary producers, while communities may be used to explore interactions among components of natural ecosystems. In the final analysis, however, there are three well-defined stages in all test systems:

1. Exposure of the test organism(s) to the toxicant;
2. Evaluation of the effect(s) in terms of numerically accessible end points;
3. Analysis of the data to provide a single value representing the biological effect (toxicity).

In the past, many industrial effluents were significantly toxic so that tests relying on acute toxicity to fish were routinely used: these usually involved exposure to the toxicant for a maximum of 96 h. Rainbow trout were traditionally used and the results were reported as LD_{50} values. With increased demand for less toxic effluents before discharge into aquatic systems and increased appreciation of the complexity of ecosystem effects (Rosenthal and Alderdice 1976), assays for acute toxicity have gradually been replaced by considerably more-sophisticated test systems. A review has been given that discusses not only the broad mechanisms whereby PAHs exert their toxicity on aquatic organisms, but the cardinal issue of bioavailability (van Brummelen et al. 1998).

It is desirable to distinguish clearly and appreciate the differences between the various terms; in this account, the following usage has been adopted:

Acute implies that the organism does not survive the exposure and often—although not necessarily—implies a short-term exposure. It should be appreciated that organisms at various stages of development may be used, and that earlier stages will generally display greater sensitivity to the xenobiotic.

Subacute or *sublethal* implies that the test organism survives exposure, but is nonetheless impaired in some specific way: a test may, for example, examine the effect of a toxicant on growth or reproduction. An old though stimulating review with valuable references to the basic literature is available (Sprague 1971).

Chronic tests aim at examining the effect of prolonged exposure and will therefore of necessity examine subacute effects. Considerable ambiguity surrounds the application of the term *chronic*, and this has been carefully analyzed by Suter et al. (1987); the length of the exposure is not rigorously defined, but should probably be related to both the growth rate and the life expectation of the test organism. Life-cycle tests using, for example, fish, clearly represent a truly chronic assay. The introduction of the term *subchronic* and its use in the context of relatively short term tests lasting 7 days (Norberg and Mount 1985; Norberg-King 1989) seems therefore regrettable.

Reproduction tests may be directed to a single generation or, especially for fish, to a complete life cycle: growth of fish from spawn to maturity followed by growth of the next generation. The term *fecundity* has been used for tests that examine early stages in the development of the test organism.

There has been increased interest in the pathology of organisms exposed to xenobiotics, and tumor induction in fish is noted in Sections 7.3.3 and 7.5.1. Particular concern has been expressed over the effect of exposure to xenobiotics on the early life stages of fish, and this may be illustrated by investigations with 2,3,7,8-tetrachlorodibenzo[1,4]dioxin. Fertilized eggs of lake trout (*Salvelinus namaycush*) were exposed for 48 h to the toxicant and then maintained in flowing water; subcutaneous hemorrhages developed in embryos and fry, and survivors displayed severe edemas and necroses (Spitsbergen et al. 1991). Broadly similar observations were made (Walker et al. 1992) when the toxicant was injected into the fertilized eggs of rainbow trout (*Oncorhynchus mykiss*). Injection of 4-chloroaniline into the fertilized eggs of zebra fish (*Brachydanio rerio*) resulted in delayed hatching and a large number of dose-related cytological alterations in the proximal renal tubule of 4- and 6-day old fish (Oulmi and Braunbeck 1996).

The exclusive use of single-species test systems has been the subject of justified criticism (Cairns 1984), and it should be pointed out that fundamental objections have been raised to the application of conventional bioassays to predictive assessment. A valuable critique has been provided (Maltby and Calow 1989) in which the limitations of conventional approaches were carefully delineated: it was suggested that the intrinsic limitations of inductive procedures make these of low predictive value except in restricted circumstances.

Attention is directed to some general issues.

1. The biological level of the system used for assay may range from the subcellular to the individual to the population, and the interpretation of the results should take this into account. For example, tests at the subcellular level cannot consider the important issue of transport into the cell, and assays with individual organisms cannot validly be extrapolated to effects on populations.

2. The effect of a potential toxicant is a function of a number of factors: (a) the organism that is exposed and its trophic level, (b) its physiology and biochemistry including the capacity for repair or excretion, and (c) the stage in its life cycle.

7.1 Choice of Test Species in Laboratory Tests

Broadly, three different philosophies may be adopted although these should not be regarded as mutually exclusive.

1. Internationally recognized organisms using standardized protocols may be used, and for judicial purposes may be obligatory. These procedures have the advantage of enabling comparison with extensive published data, although their relevance to a specific ecosystem may be restricted.

2. Alternatively, use may be made of indigenous species, in which case there can be no doubt of their relevance to the ecosystem from which they were isolated: standardized protocols must exist for these organisms also, and cloned cultures of taxonomically defined individuals should be used. An interesting example is afforded by the widespread European use of zebra fish (*Brachidanio rerio*) as a test organism. This is, of course, a tropical fish and it may reasonably be questioned whether this is appropriate for application to the cold-water environment of northern Europe. A study using 3,4-dichloroaniline has revealed, however, that in early life-stage studies there was no significant difference between the results from tests using zebra fish and those using perch (*Perca fluviatilis*) that is widely distributed in Europe (Schäfers and Nagel 1993). This investigation alone clearly does not provide a *nihil obstat* to the use of zebra fish but provides valuable support for its widespread application in ecotoxicology.

3. Feral organisms have sometimes been used for each test series and stocks of these have not been maintained in the laboratory. This is, however, a questionable procedure for at least three reasons:

- Variations in the natural population may remain unnoticed.

- Experiments may be restricted to particular seasons of the year.

- In view of the virtual global dissemination of xenobiotics including potentially toxic organochlorines and PAHs, the test organisms may already have been exposed to background levels of such toxicants.

It should be emphasized that the choice of test organism is not primarily determined by the requirement for maximum sensitivity to toxicants: indeed, extreme sensitivity may be a disadvantage if it results in problems of repeatability or reproducibility. On the other hand, it should be clearly appreciated that some groups of organisms may be significantly more sensitive than others to a given class of toxicant—that may have a common mode of action. For example, the substituted diphenyl ether pyrethroids permethrin, fenvalerate, and cypermethrin were generally more toxic to marine invertebrates than to marine fish, and among the former the mysid *Mysidopsis bahia* was the most sensitive (Clark et al. 1989).

There exist also technical issues of considerable importance.

1. Some organisms such as strains of algae or cultures of crustaceans may be maintained in the laboratory as stocks: this has the advantage that putatively unaltered test strains are always available, but it is important that additional stress or selection is not imposed during maintenance.

2. Many species of fish may be available from commercial breeders, and this avoids extensive labor in keeping such stocks. On the other hand, absolute uniformity cannot be guaranteed and genetic variations are clearly possible. One possible danger results from the use of antibiotics by fish breeders to maintain healthy stocks free from microbial infection.

3. Relatively little attention has been directed to genetic variation within specific taxa. Examples of the care which should be exercised are provided by studies with the midge *Chironomus tentans*, the water flea *Daphnia magna*, and the viviparous fish *Poeciliopsis monacha-lucida*.

   a. Analysis of gel electrophoresis patterns of a number of glycolytic enzymes in different strains of *C. tentans* was used to assess a number of relevant genetic parameters including heterozygosity, percent polymorphic loci, and genetic distance within the populations (Woods et al. 1989). The observed variations between strains of the same organism from different sources were considerable, and the results strongly underscore the critical importance of taking this into consideration in assessing the effects of toxicants on different populations of the same organism.

b. Clonal variations in various strains of *D. magna* to cadmium chloride and 3,4-dichloroaniline were examined (Baird et al. 1990) in both acute and in 21-day life-cycle (chronic) tests. There were wide differences in results from the acute tests, particularly for cadmium chloride, although these were relatively small for the life-cycle tests; it was therefore concluded that different mechanisms of toxicity operate in the two test systems and this illustrates, in addition, the advantage of including both kinds of assay.

c. Cytochrome P-450 CYP2E1 mediates dealkylaton of nitrosodialkylamines, and its activity among *P. lucida-monacha* hybrids varied markedly: values in liver microsomes ranging from 0.4 to 3.6 μg 6-hydroxychlorzoxazone/min/mg protein (Crivello and Schultz 1995). It should be noted that fish in the genus *Poeciliopsis* are unisexual, and that as many as 20 different "hemiclones" have been identified in river systems.

All these results not only reinforce the conclusions on the possible significance of genetic variations in test species within the same taxon, but also indicate the subtleties in expression of toxicity that may be revealed under different exposure conditions. A commendable development is therefore the assessment of the genetic structure in the mayfly *Cloeon triangulifer* that has been proposed as a suitable assay organism (Sweeney et al. 1993). Since the analysis of alloenzyme composition for enzymes representing different genetic foci is well developed and generally straightforward, more widespread application of this technique could profitably be made to other populations of organisms already used for bioassay.

7.2 Experimental Determinants

There are a number of important experimental considerations which affect both the design of the experiments and the interpretation of the data, and some of the most important will therefore be briefly summarized. It should be noted that in order to display its effect the potential toxicant must be transported into the cells and that the compound may then be metabolized. In assays for toxicity, no distinction can therefore be made between transport into the cell, toxicity of the compound supplied, or that of potential metabolites: these effects are assayed collectively and indiscriminately. The situation has therefore certain features in common with that of bioconcentration, which are discussed in Section 3.1. Indeed, because of their close relation, it may be expedient to examine the potential for bioconcentration and bioaccumulation, metabolism and excretion, and toxicity in the same experiment.

7.2.1 Exposure

Exposure conditions should simulate as far as possible those to which the organisms will be exposed in the environment that is being evaluated.

Aquatic Organisms

The most widely used exposure of the test organisms is to aqueous solutions of the toxicant prepared in media that are suitable for their growth and reproduction; this represents the situation for many potential toxicants, although attention is briefly drawn later to alternative procedures that have been applied to compounds with poor water solubility. Defined synthetic mineral media in deionized organic-free water are generally used for the sake of reproducibility and since this minimizes the possible ameliorating effect of organic components in natural waters. An interesting departure from the conventional practice of reporting toxicant concentrations in the test medium is provided by a study with sea urchin genetes and embryos in which estimates of toxicity were based on concentrations of the toxicant in the embryos (Anderson et al. 1994). This has also been used for terrestrial organisms that are discussed in Section 7.3.6.

Different exposure protocols have been used: (1) static systems without renewal of the toxicant during exposure, (2) semistatic systems in which the medium containing the toxicant is renewed periodically during the test, and (3) continuous flow-through systems in which the concentration of the toxicant is essentially constant. In the first procedure—and to a lesser extent in the second—toxicant concentrations will not remain constant during exposure, and the range of exposure concentrations during the test could advantageously be reported. In any case, analytically controlled concentrations of pure compounds should be provided since some of them may be unstable under the conditions used for testing. Whereas exposure for a predetermined time is acceptable for experiments using crustaceans and fish, this cannot be done for growth tests using algae since an unpredictable lag phase may exist before growth commences. The length of the lag phase may provide useful information on adaptation to the toxicant, but growth must be measured during the whole of the test and into the stationary phase.

A particularly troublesome problem arises with compounds having only low solubility in water, and this is particularly acute if the compound is only slightly toxic since high concentrations are then necessary to elicit a response from the test organism. Solutions of such compounds have been prepared in water-miscible organic solvents such as acetone, ethanol, dimethyl sulfoxide, or dimethylformamide, which are added to the test medium. Although the toxicity of these solvents can readily be evaluated, a much greater uncertainty surrounds the true state of the toxicant: Is a true solution attained or merely a suspension of finely divided particles? It has also been shown that the acute toxicity of three xenobiotics toward a number of crustacean and rotifers was influenced by the organic solvent independently of the possible effect of the solvent alone (Calleja and Persoone 1993). It is therefore preferable to use saturated solutions of the toxicant in the test medium: these may conveniently be prepared by passing the test medium through a column containing glass beads or other sorbants coated with the toxicant (Veith and Comstock 1975; Billington et al. 1988). Although this is the method of choice, preparation of large volumes for testing may be impractical although a design incorporating continuous flow has been developed and is clearly attractive (Veith and Comstock 1975). There is an additional problem that may be encountered with compounds that have extremely low water solubility: in an investigation on the uptake of highly chlorinated dibenzo[1,4]dioxins by fish, it was found that the concentrations in saturated solutions prepared by this procedure exceeded the established water solubility of the octachloro congener (Muir et al. 1986). It was hypothesized that the compound was associated with low concentrations of dissolved organic carbon in the water, and that the very low BCF values that were measured could be the result of the poor bioavailability of the compound; this is equally relevant to the toxicity and is an issue to which further attention should be devoted. The organic C content of the water, although low, may be sufficient to mediate associations with low concentrations of toxicants.

There are, however, situations in which organisms in natural systems may not be exposed to the essentially constant concentrations of the toxicant used in flow-through laboratory systems. This may be the case in quite different field situations:

1. Accidental spills that result in sudden and temporarily high concentrations of the toxicant;
2. Anadromous and catadromous fish that may be exposed temporarily to a plume of toxicant on their way to the spawning ground;
3. Nonstationary fish that are therefore exposed to varying concentrations of a toxicant.

The question then arises of the extent to which any nonlethal effect is reversible after removal of the toxicant. The answer seems to be that in the few cases which have been examined, this may indeed be the case: all of them have examined phenolic compounds, one (McCahon et al. 1990) using the crustacean *Asellus aquaticus*, one assessing respiratory/cardiovascular effects on rainbow trout (Bradbury et al. 1989), and the third (Neilson et al. 1990) using the embryo/larvae assay with zebra fish (*Brachydanio rerio*). The last of these has been developed into a protocol that is modeled on the standard bioconcentration procedure in which a period of depuration is included after exposure to the toxicant.

For compounds with extremely low water solubility such as 2,3,7,8-tetra-chlorodibenzo[1,4]dioxin, two other exposure regimes have been used: (1) egg injection using a single dose for rainbow trout (Walker and Peterson 1991; Walker et al. 1992) or bird eggs (Powell et al. 1998) and (2) intraperitoneal injection into fingerling (Spitsbergen et al. 1988) or juvenile (van der Welden et al. 1992) rainbow trout. Although these provide valuable data and provide a solution to a technical problem, the extent to which they simulate environmental exposure appears questionable in most circumstances.

Soils and sediments

It is difficult to obtain a truly homogeneous distribution of toxicants spiked into soils and sediments, and two methods have been most extensively used.

1. Dilute solutions in a volatile organic solvent are thoroughly mixed with the sample, and the solvent removed under vacuum at low temperature. Alternatively, the solution may be added to a suitable inert solid sorbent and after removal of solvent, mixed with the soil or sediment.

2. If the toxicant is insoluble in organic solvents, there is little alternative to mixing the finely powdered toxicant with the soil or sediment followed by thorough mixing, for example, by tumbling.

3. For terrestrial organisms, there remains the choice of natural or artificial soil. The former is clearly more directly relevant, but suffers from the lack of repeatability. The problem with artificial soils is the range of types that may be required, varying critically in their organic content that is a primary determinant of bioavailability and toxicity.

4. Other procedures using, for example, exposure on filter paper impregnated with the toxicant for earthworms clearly presents an ideal situation that does not appear to be environmentally realistic. Comparison of different routes of exposure to the toxicant has, however, been made with *Eisenia andrei* (Belfroid et al. 1993), and the acute toxicity was given as lethal body burden (LBB) after exposure to 1,2,3-trichloro- and pentachlorobenzene. Exposure was carried out in water, in soil via food, and on filter paper, and when LBBs were normalized to lipid content the values for pentachlorobenzene were comparable, although the value for 1,2,3-trichlorobenzene for exposure on filter paper was higher than for the other routes of exposure.

7.2.2   End Points

Any parameter that can be assessed numerically may be used, and these are generally adapted to the test organism and to experimental accessibility. An interesting survey (Maltby and Calow 1989) of papers during the periods pre-1979 and 1979 to 1987 using single-species laboratory tests revealed that survival was by far the most common end point, followed by growth and reproduction; physiological/biochemical, behavioral, and morphological end points were much less common. The chosen end points differ widely. For example, the growth of algae may be estimated using cell number or turbidity or chlorophyll concentration—although care must be exercised in using the last since the toxicant may affect chlorophyll levels without significantly affecting growth. Although growth has been proposed (Norberg-King 1989) as a simple end point in tests using fathead minnows (*Pimephales promelas*), growth has not generally been used for larger fish since it is a relatively insensitive parameter. For reproduction tests using crustaceans, the number of offspring is unambiguous, although for life-cycle tests using fish, a number of different end points have been suggested including the number of eggs produced after spawning. The design of the test may also determine the end point: for example, in one version of the zebra fish embryo/larvae test, food is not provided during exposure and the test uses the median survival time of the larvae which ingest the egg yolk until starvation (Landner et al. 1985). Tests using physiological or biochemical or histological parameters have access to a much wider array of end points, which are discussed in more detail in Section 7.6.

Some examples are given as illustration of other less common parameters that have been used as end points.

1. The scope for growth (SfG) which is the difference between the energy absorbed and that metabolized indicates how much energy is available for growth and reproduction. This has been used with marine invertebrates particularly the mussel *Mytilus edulis*, and has also been examined (Maltby et al. 1990a) in the amphipod *Gammarus pulex*. Concentrations of 0.5 mg/l 3,4-dichloroaniline significantly reduced the SfG and this concentration may be compared with the LC_{50} (48 h) value of 7.9 mg/l. The test has also been evaluated in a field bioassay (Maltby et al. 1990b) and in mesocosms (Maltby 1992). This concept has been extended to the scope for change in ascendency (SfCA): although this is a comparative index that has been illustrated with atrazine in a microcosm system (Genoni 1992), it appears to be capable of further refinement as an index for chronic effects.

2. Behavioral responses have been used since these may be important determinants of the ability for a species to survive and reproduce under natural conditions. This has been examined in the amphipod *Pontoporeia affinis*, which is an important food for fish in the Baltic Sea. In organisms exposed to sublethal concentrations of phenols and styrene in a flow-through system (Lindström and Lindström 1980), there was an initial stimulation of motility, although the

long-term effect was an impairment of swimming at toxicant concentrations that had no effect on mortality. The response of fish to toxicants has been examined in a rotary-flow system (Lindahl et al. 1976), and this has been incorporated into chronic exposure tests (Bengtsson 1980). The impairment of swimming could be highly significant in determining the survival and reproduction of feral fish exposed to toxicants in natural ecosystems. The effect of the carbamate biocide carbaryl on the swimming behavior of *Daphnia pulex* has been examined (Dodson et al.1995), and nine different end points evaluated: of these, six were the most useful—velocity, average turning angle, upward hopping angle, downward sinking angle, variance in vertical position, and hopping rate. In a wider context, the significance of this behavioral alteration was illustrated by the results of an experiment in which *D. pulex* previously exposed to carbaryl were preferentially predated upon by bluegill sunfish (*Lepomis macrochirus*).

3. A single example is given to illustrate the fact that virtually any quantifiable effect may be used as an end point. The net spinning behavior of the caddis fly *Hydropsyche angustipennis* has been used to evaluate the aquatic toxicity of 4,5,6-trichloroguaiacol: both the increased frequency of different types of net distortions and the time necessary for pupilation were examined (Petersen and Petersen 1984). Since these insects may be common in ecosystems and are themselves an important source of food for fish, the effect of toxicants on them may have widespread repercussions.

The results from such experiments will provide a relation between the exposure concentration of the toxicant and the observed biological effect—a dose–response relationship. From this data, assessments of toxicity can be calculated in terms of the relative effect, for example, EC_{50}, EC_{20}, or EC_{10}—the concentrations eliciting 50, 20, or 10% of the maximally observed effect. The use of lethal body burden for assessing the toxicity toward earthworms has been described in Section 7.2.1. It is valuable to calculate threshold toxic concentrations—the lowest concentrations causing an observed effect (LOEC) that are significant compared with those causing no observed effect (NOEC). It should be carefully noted that the term *observed* is used; use of *observable* is clearly misleading since effects might be registered under other more sensitive test conditions. The existence of threshold concentrations below which biological effects are not manifested has been extensively discussed in connection with exposure of humans to carcinogens, but has been less fully explored in the context of ecotoxicology (Cairns 1992). The existence or otherwise of such thresholds is extremely important and seems not unreasonable in view of the concentration dependence of mechanisms whereby toxicants are transported into cells. It should be appreciated that the range of concentrations of toxicants that are examined determines the accuracy of the value assigned to LOEC; there may therefore be a considerable gap between

the values of NOEC and LOEC, and this can only be diminished at considerable expense by repeating the experiment with a more appropriate range of
concentrations. Indeed, as noted in Chapter 2, the range may be a more realistic value (Miskiewicz and Gibbs 1992).

The assessment of effects on community structure which are applicable to
microcosm and mesocosm experiments is very much more complex and will
generally necessitate careful statistical analysis. A summary of the most commonly used indexes is given by Maltby and Calow (1989).

7.2.3 Test Conditions

One basic consideration is the choice of the test medium—which may conveniently be termed the dilution medium for the toxicant in aquatic systems and
the matrix for soils and sediments. It is desirable that this medium is defined
as closely as possible, and in an ideal situation a completely defined medium
using laboratory distilled water supplemented with inorganic nutrients and
adjusted to a suitable degree of hardness and salinity would be used. In view
of the possible complicating role of organic C in the water (Section 7.2.1), use
of commercial water purification systems that incorporate reverse osmosis
and a carbon filter are recommended. This method is possible for many algae
that are used as test organisms since these organisms may generally be grown
in defined mineral media, supplemented if necessary with trace elements and
vitamins. Even here, however, problems may arise. A good illustration is provided by the variation in the sensitivity of a range of marine algae toward tet-
rabromobisphenol-A which depends critically on the test medium, and that
could not be rationalized on a simple basis (Walsh et al. 1987).

This apparently ideal situation may prove, however, to be impossible for
many other groups of test organisms: on the one hand, some are highly sensitive to the impurities unavoidably introduced into such media, while on the
other hand many have obligate nutritional requirements that are accessible at
low concentrations in natural waters but are not provided by synthetic media.
If the laboratory is situated close to a clean and unpolluted site, suitable natural
water may readily be obtained, but unfortunately those who decide laboratory
locations do not always consider such important matters. Most higher organisms have complex nutritional requirements which have seldom been defined,
so that food is supplied, for example, to fish as commercial preparations, or to
crustaceans in cookbook-type preparations. Addition of food during the test
period also inevitably introduces several complications through sorption of
the toxicant so that a multiple exposure route may exist; this has already been
discussed in Sections 3.1.1 and 3.1.2 in the particular context of hydrophobic
compounds. Also, the organically rich test medium may favor the development of microbial populations of bacteria or fungi which may be pathogenic to
the test organism. This problem can be especially severe when industrial effluents which have been subjected to biological treatment are tested.

Test media are generally chosen to be as nearly as possible optimal for
growth and reproduction of the test organism which should not be subjected

to any stress additional to that imposed by the presence of the toxicant. This means, however, that dilution media for different organisms may differ appreciably in pH, salinity, or degree of hardness. Two examples may be used to illustrate these problems.

1. For ionizable organic compounds, such as carboxylic acids and phenols, or amines it is generally assumed that it is the free acid (or free base) that is transported into the cells and which mediates the toxicity. There is, however, evidence that both free and ionized fractions may be transported into cells (Pärt 1990; Escher and Schwartzenbach 1996). The relative proportion of the free acid is a function of the degree of its dissociation and of the pH of the medium. If test media have different pH, the effect of the toxicant may be compounded by the pH of the medium as well as by the sensitivity of the test organism. The problem probably becomes significant for compounds with pK_a values below ~6 and for organisms grown in media with pH > 7.5 which is the case for most fish and crustaceans. Most test media are only weakly buffered, generally with bicarbonate, and only for algal systems and the MICRO-TOX system has the use of buffered media been consistently investigated (Neilson and Larsson 1980; Neilson et al. 1990). Even for neutral compounds, the pH of the test medium may significantly affect toxicity. For example, whereas assays for the acute toxicity of 4-nitrophenol and 2,4-dinitrophenol using rainbow trout and the amphipod *Gammarus pseudolimnaeus* showed the expected decrease in toxicity between pH 6.5 and 9.5, the toxicity of the neutral phosphate triester trichlorophon increased with increasing pH in the same range (Howe et al. 1994).

2. The presence of Ca^{2+} or Mg^{2+} cations in test media of appreciable hardness may introduce complications due to complex formation with the toxicant or, in extreme cases, formation of precipitates which make it impossible to carry out the test. One example of this is provided by 2,5-dichloro-3,6-dihydroxybenzo-1,4-quinone which could be examined in the zebra fish system although not in the *Ceriodaphnia dubia* system using dilution media of greater hardness (Remberger et al. 1991). The additional sigificance of ionic strength is noted briefly in Section 7.3.1.4.

Other metal cations that may form complexes with the toxicant may have a profound effect on toxicity. The toxicity of ethylenediamine tetraacetic acid (EDTA) and diethylenetriamine pentaacetic acid (DTPA) and their metal complexes has been examined in daphnids. The results are relevant both for the effect on the toxicity of metal cations and for the influence of complexation on these toxicities. For the acute toxicity (EC_{50}, 24 h) to *Daphnia magna*, the following results (mg/l) were obtained (Sorvari and Silanpää 1996):

(1) the toxicity of ETDA was less than that of the Fe, Cu, and Zn complexes and (2) the toxicity of the EDTA complexes of Mn,Cu, Zn, and Cd was less than that of the free cations. The acute and chronic toxicity to *D. carinata* has been examined for DPTA and for its Fe complex (van Dam et al. 1996). Details of this complex experiment were provided but only the conclusions are briefly summarized: (1) the acute toxicities (LC_{50}, 48 h) were 245 mg/l for DTPA, whereas that for the Fe complex exceeded 1000 mg/l and (2) the values of NOEC were 1 mg/l for DTPA and 67 mg/l for Fe-DTPA and of LOEC were 10 and 134 mg/l.

From these results, it may be concluded that free EDTA and DTPA display low toxicity, that this is substantially reduced by complexation, and that the toxicity of the metals in the complexes may also be diminished by chelation.

The same principles apply equally to solid matrices, both natural and artificial, and the pH both of the matrix and the solution of the toxicant should be carefully taken into consideration. In general, pH control of the matrices is difficult to maintain and reliance must be made on the natural buffering capacity of soils containing humic material.

7.2.4 Evaluation of Variability in the Sensitivity of Test Organisms

It is obligatory to assess periodically the sensitivity of the organisms used in laboratory tests, and there are some basic requirements that have been generally accepted:

- That the standard compound is available in high purity and is readily soluble in the test medium;
- That it is stable under the test conditions;
- That it is sufficiently toxic to provide a reasonable response in the test organism.

It is also desirable that the test compound can be analyzed to provide data on the true exposure concentration. The choice of the compounds fulfilling these specifications is less easy. It seems clearly unsatisfactory to use a compound such as potassium dichromate to control the sensitivity of organisms used for evaluating organic toxicants. The problem with organic compounds is that of choosing one or a few from the huge structural range that is available. As a general rule, it might be stated that the compound should ideally bear some structural resemblance to the compounds which are to be evaluated and, if possible, related to the mechanism of their toxicity: it should be a surrogate. Halogenated phenols have been quite widely used for one important class of toxicants, but for neutral compounds, the choice is more difficult primarily due to volatility on the one hand (e.g., naphthalene) or poor water solubility (e.g., anthracene and PCBs) on the other. In the last analysis, probably an unsatisfactory compromise must be accepted, and advantage be taken of the established sensitivity of the organism to as wide a range of toxicants as is realistic. The systematic collection and availability of these basic data for all test organisms would be extremely valuable.

7.3   Test Systems: Single Organisms

7.3.1   Aquatic Organisms

A wide range of organisms is available, and many of them have been extensively evaluated with structurally diverse organic compounds. The choice of organism and the design of the study will depend critically on its objective. If, for example, the aim is the elucidation of the mode of action of a toxicant, detailed investigations may be devoted to one or only a few taxa. If, on the other hand, the aim is to evaluate the effect of a toxicant on an ecosystem, several taxonomically and physiologically diverse taxa should be included. The data from these may then be incorporated into a hierarchical system for environmental impact assessment of the toxicant (Section 7.8). In the following paragraphs, an attempt is made to describe briefly the major groups of organisms which have been most widely used in ecotoxicological studies.

It is important to take into account the different modes of exposure to the toxicant: this may take place directly from the aqueous phase, from interstitial water, or via food to which the toxicant is bound. The whole issue of partition must, therefore, be taken into account. In addition, the issue of true tissue levels and the role of metabolism and elimination, for example, by fish must be assessed. This has already been discussed in Chapter 3, Section 3.1.5.

7.3.1.1   The MICROTOX System

This is a surrogate test for toxicity and uses a luminescent strain of a marine bacterium: the end point is the inhibition of luminescence measured after 5 or 15 min exposure to the toxicant. The rapidity and low cost of this assay make this an attractive screening system, although care should be exercised in drawing conclusions of environmental relevance from the data. Many correlations between the acute toxicity measured in this system with values using higher organisms such as crustaceans and fish have been made, but it should be noted that these are made on a log–log basis so that agreement may really be no better than within a power of 10. Attempts have been made to use buffered media, but even in this case, correlation with the acute toxicity of a range of phenolic compounds to other organisms was singularly unimpressive (Neilson et al. 1990), although within a structural class of compounds this may be a valuable test system.

The general issue of multicomponent microbial test systems has been discussed (Cairns et al. 1992), although it should be clearly appreciated that, from a biochemical and evolutionary point of view, there are very substantial and important differences between bacteria and higher organisms.

7.3.1.2 Algae

Algae have been used for toxic assessment over many years since, as primary producers, they represent an important element in aquatic food chains. In addition, unicellular strains may generally be isolated, maintained, and grown using standard bacteriological techniques. In spite of this, however, there are a number of important issues to which attention should be directed since these influence the conclusions that may be drawn from the results of a given test.

1. Use of axenic cultures is attractive in view of the high degree of reproducibility, and avoids potentially serious problems due to the possible occurrence of bacterial transformation of the toxicant during tests with unialgal cultures containing bacteria.

2. The algal kingdom includes a very wide range of biochemically, physiologically, and morphologically distinct organisms of which only a very few have been used in ecotoxicology. The choice is naturally dictated largely by their ease of cultivation, and representatives of Chlorophyceae, Xanthophyceae, and Bacillariophyceae have been used: some Cyanobacteria (Cyanophyta) have been used less frequently. Widely differing sensitivities of representatives of these groups toward a range of toxicants has been observed (Blanck et al. 1984; Neilson et al. 1990), so that no single organism can reflect the spectrum of responses observed. Use of a group of taxonomically diverse strains is therefore to be preferred to the use of a single standardized species such as *Selenastrum capricornutum*.

3. Two essentially different end points have been widely used in experiments with algae—growth rate or biomass—and estimates of toxicity using these may not be comparable (Nyholm 1985). Measurements of $H^{14}CO_3$ uptake include both the increase in the biomass and growth rate, and have generally been applied to undefined populations of algae taken from receiving waters. Clearly, comparison between the results of such experiments carried out at different times is not possible, although the results are valuable in assessing the effects of a toxicant on organisms in a given ecosystem. Attention has been drawn to possible ambiguities resulting from measurements of chlorophyll concentrations for growth measurements (Neilson and Larsson 1980).

4. The laminarian marine alga *Macrocystis pyrifera* has been used in a laboratory assay (Anderson and Hunt 1988) and takes advantage of the different stages in the life cycle of these algae: suitable end points are the germination and growth of the spores which have been incorporated into a short-term (48 h) assay, or the production of sporophytes by fertilization of female gametophytes that may be used in a 16-day assay. The possibility of using this or other algae with a comparable life cycle clearly merits further attention.

5. An interesting assay has been developed that uses communities of naturally occurring algae. The assay is based on the concept of pollution-induced community tolerance (PICT) (Blanck et al. 1988) whereby exposure to a toxicant results in the elimination of sensitive species and the dominance of tolerant ones: quantification is achieved by using short-term assays for the inhibition of photosynthesis under laboratory conditions (Blanck and Wängberg 1988). One of the attractive features of the system is that it can be used in widely different situations ranging from microcosms and mesocosms to natural ecosystems.

7.3.1.3 Higher Plants

Growth of higher plants has been used for assessing toxicity in aquatic systems, although these assays have not apparently been widely adopted. A general overview has been given (Kristen 1997), and their use in terrestrial systems is discussed in Section 7.3.6. The use of higher plants may be particularly advantageous when discharge of toxicants is made to specific ecosystems where reliance solely on evaluating the effects on algae may be considered insufficient or where water is used for irrigation of agricultural crops. The possible bioconcentration of the xenobiotic (Section 3.1.3) may also be conveniently evaluated. A range of different plants including both monocotyledons, such as oats and wheat, and dicotyledons, such as beans, carrots, and cucumbers, has been used. Suitable end points are the frequency of seed germination and the extent of root elongation, and the tests are readily carried out even though systematic evaluation has apparently been restricted to only a few species (Wang 1991): inhibition of root development has been used to assay a few phenols (Wang 1987), and millet (*Panicum miliaceum*) was the most sensitive compared with cucumber (*Cucumis sativus*) and lettuce (*Lactuca sativa*). The toxicity of a range of compounds including substituted phenols, anilines, chlorobenzenes, aromatic hydrocarbons, and pesticides to lettuce (*L. sativa*) has been examined both in a semistatic assay in nutrient solutions and in soil (Hulzebos et al. 1993), and the toxicity correlated with values of P_{ow}.

Pollen germination and pollen tube growth have been used as end points. Pollen grains from a number of plants have been used, and an assay for growth of pollen from *Nicotiana sylvestris* has been developed (Kristen and Kappler 1995).

The duckweed *Lemna minor* has attracted considerable attention and has been proposed as a suitable representative assay organism (Taraldsen and Norberg-King 1990): after 96 h exposure to the toxicant, the biomass is assessed from the number of fronds or from the chlorophyll concentration. Presumably these experiments could also be adapted to measure growth in which, for example, root length is measured. In many ways, the experimental conditions are similar to those used for algae with the added advantage that the medium can be renewed during the test. In experiments using a few substituted phenols, *L. minor* was comparable in sensitivity with the algae tested (Neilson et al. 1990), although this should not be regarded as the determining advantage of the test.

Higher plants have been used in tests for genotoxicity that are discussed in more detail in Section 7.3.3. For example, the onion *Allium cepa* has been used not only in root elongation assays for toxicity but to detect mitotic, cytokinetic, and chromosomal aberration in the root tips (Fiskesjö 1995). A micronucleus assay that detects chromosome breaks in *Tradescantia* sp. planted in areas suspected of pollution has been used (Sandhu et al. 1991) and suggests the possible value of such assays in the terrestrial environment or as an indicator of atmospheric pollution.

7.3.1.4 Crustaceans

Several freshwater daphnids have been extensively used, including *Daphnia magna*, *D. pulex*, and *Ceriodaphnia dubia*, and standardized protocols have been developed. It has already been noted that only a single stress induced by the toxicant should be imposed during exposure. The importance of using the same media for the cultivation of *D. magna* and in toxicity tests has emerged together with the possible influence not only of the hardness of the water but of its specific ionic composition (Buhl et al. 1993). All of these organisms have been used both in acute tests and in reproduction tests, and a life-cycle test for *D. magna* has been proposed (Meyerhoff et al. 1985) although it has been less extensively applied. There has been an increased tendency to use *C. dubia* which is apparently less susceptible to some of the cultivation problems associated especially with *D. magna*, and it seems (Winner 1988) that *C. dubia* is at least as—or even more sensitive than—*D. magna*. Quite extensive effort has been devoted to development of the reproduction test using *C. dubia*, although some of the experimental problems appear to remain incompletely resolved. For the sake of completeness, it may be noted that a cloned strain of *D. pulicaria* from Lake Erie has been used to assess the toxicity of 2,2'-dichlorobiphenyl in full life-cycle tests, and that this experiment demonstrated effects at the extremely low concentrations in the range of 50 to 100 ng/l (Bridgham 1988).

The harpacticoid copepod *Nitocra spinipes* which is common in the brackish waters of the Baltic Sea has been cloned and used extensively for toxic evaluation in Sweden (Bengtsson 1978), and a reproduction test in a flow-through system has been developed (Bengtsson and Bergström 1987). This is a valuable organism since it has an appreciable tolerance of the varying salinities which exist in the Gulf of Bothnia and the Baltic Sea, although it has possibly a more-restricted geographic application than other test organisms.

7.3.1.5 Fish

Fish have traditionally been used for assessing the toxicity of effluents and the 96-h acute test with rainbow trout belongs to the classic era of ecotoxicology. There have been at least three significant developments over the intervening years: (1) use of a wider range of test fish, (2) development of tests aimed at evaluating reproduction efficiency and long-term (chronic) exposure, and (3) the use of biochemical and physiological parameters to assess the effect of toxicants.

Probably most investigations have used one or more of the following fish: rainbow trout (*Oncorhynchus mykiss*, syn. *Salmo gairdneri*), guppy (*Poecilia reticulata*), fathead minnow (*Pimephales promelas*), sheepshead minnow (*Cyprinidon variegatus*), channel catfish (*Ictalurus punctatus*), bluegill sunfish (*Lepomis macrochirus*), zebra fish (*Brachydanio rerio*), and flagfish (*Jordanella floridae*). Efforts have been devoted to developing standardized protocols for all of these, and for many of them accessible tests for assessing sublethal or chronic effects have been developed. Less attention has been directed to the use of marine fish although limited application has been made (Shenker and Cherr 1990) of assays using larvae of English sole (*Parophrys vetulus*) and top-smelt (*Atherinops affinis*).

The Sensitivity of the Test Species

It may be valuable to attempt an assessment of the relative sensitivities of different fish toward organic toxicants. Short-term LC_{50} values for several compounds including pentachlorophenol and picloram (4-amino-3,4,5-trichloropicolinic acid) have been compared using rainbow trout, zebra fish, and flagfish. The results showed considerable variation in sensitivity among the test fish, although these were not judged to be significantly greater than those encountered using the same fish in the same laboratory during repeated testing (Fogels and Sprague 1977). In a similar way, comparison of the data for the sensitivity to pentachlorophenol of zebra fish, rainbow trout, and fathead minnows in early life-stage tests showed that these fish were quite similar (Neilson et al. 1990). A much more extensive correlation of acute toxicity has been carried out (Doherty 1983) using data for rainbow trout, bluegill sunfish, and fathead minnow. Although reasonable correlation was obtained between pairs of data, it seems perilous to evaluate applications for a discharge permit on the basis of data for a single organism instead of relying on the established practice of using the results for one cold-water fish, one warm-water fish, and an invertebrate.

It is critically important to assess the significance of the differences in the response of various fish to the same toxicant: for example, how significant is the difference between the 96-h acute toxicity of chlorothalonil (2,4,5,6-tetra-chloro-1,3-dicyanobenzene) to channel catfish (52 µg/l) and to rainbow trout (18 µg/l) (Gallagher et al. 1992). Such differences would probably not emerge as significant in a correlation study that plots data on a log–log basis—although it could be highly relevant for a given receiving water. In addition, the conclusion that the use of a single test organism is acceptable would not be supported by either the increased appreciation of the limitations of using only data for acute toxicity in environmental hazard assessments, or the philosophy of implementing a hierarchical evaluation system (Svanberg and Renberg 1988).

Experimental Determinants

In attempting to provide an overview of the effect of toxicants on fish, a number of significant determinative parameters should be taken into consideration.

1. Marked differences in the effect of temperature on the toxicity of pentachlorophenol to rainbow trout have been observed (Hodson and Blunt 1981), and early life-cycle stages were more adversely affected in fish exposed to a cold-water regime (6°C) than with those exposed to a warm-water regime (10°C). These results could have serious implications for natural populations exposed to pentachlorophenol during low temperatures when spring egg development occurs.

2. In most laboratory tests, fish are exposed to an essentially continuous concentration of the toxicant during the exposure regime. It has been shown, however, that even brief exposure of fathead minnows to pesticides such as chloropyrifos, endrin, or fenvalerate at high concentration may induce chronic effects including deformation and reduced growth (Jarvinen et al. 1988). A test in which adult zebra fish are exposed to a toxicant before examining the sensitivity of the offspring to the same toxicant has been used to simulate, in principle, the exposure of anadromous or nonstationary fish to toxicants in natural ecosystems; increased sensitivity by a factor of ~5 was observed with pure test compounds, but this ratio appears to be related to the magnitude of the toxic effect, i.e., it decreases with decreasing toxicity (Landner et al. 1985; Neilson et al. 1990). These represent ecologically important considerations that could readily be incorporated into standardized protocols.

3. It has already been briefly pointed out in Section 3.1.5 that probably most xenobiotics can be metabolized by the test organisms, albeit to varying extents, and that this may compromise estimates of bioconcentration potential. The same is also true for toxicity, since

whereas metabolism may in many cases serve as a detoxification mechanism, it may also result in the synthesis of toxic metabolites (Section 7.5); the question of enzyme induction and its implication for toxicity has been discussed in detail (Kleinow et al. 1987), and some examples of the metabolism of xenobiotics by fish are given in Section 7.5.1.

Subacute and Chronic Tests

The classic experiments of McKim et al. (1976) on the chronic effects of toxicants in which three generations of brook trout (*Salvelinus fontinalis*) were exposed to methyl mercuric chloride led to a scientifically based appreciation of the fact that the early life stages could be used for assessing chronic toxicity (McKim 1977). Due to the cost of complete life-cycle tests and the time required for their completion, increasing attention has been directed to truncated tests for chronic toxicity, and to the effects of toxicants on sensitive life stages. Nonetheless, there is substantial evidence that clearly exposes the limitations in the conclusions which can be drawn if *only* early life-cycle stages are examined.

Attention is directed both to two end points that have been examined and to the different concentrations of the toxicant at which the same end point is affected in successive generations. There is clear evidence for a number of fish that toxicants may have significant effects on their fecundity even though the effects on early life stages are marginal (Suter et al. 1987). A dramatic example is provided by the differences in the effect of chloropyrifos that were observed during a chronic test using fathead minnows lasting 200 days (Jarvinen et al. 1983): growth of first-generation fish was reduced at a concentration of 2.68 µg/l within 30 days, whereas the comparable value for second-generation fish was 0.12 µg/l. The results of such studies have revealed their value and justifiably led to a revival of interest in full life-cycle tests.

The use of the term *subchronic* for a test with fathead minnows lasting 7 days seems regrettable, and the reporting of the results in terms of a chronic value which is the geometric mean of the LOEC and NOEC values (Norberg-King 1989) appears potentially misleading: such data clearly do not represent the effect of long-term exposure, and cannot therefore be considered chronic in any etymologically acceptable sense of the word. This does not mean, however, that the results of such evaluations are not valuable as a basis for further examination, and some descriptive term such as *early life-cycle test* would seem more appropriate for such experiments.

It is important to appreciate methodological differences in the procedures by which tests for subacute toxicity are carried out, and differences in the results that may be obtained with fish having different reproduction strategies. The two test protocols for zebra fish may be used as illustration of the first, and the divergent results for a single substance using zebra fish and guppy used as illustration of the second.

In one protocol using zebra fish, the spawn from unexposed adults are collected, the fertilization rate determined after 24 h and the test continued with fertilized ova; the survival rate and body length of the larvae are then determined after 6 weeks during which time the larvae are exposed to the toxicant and provided with food (Nagel et al. 1991).

In an alternative procedure also using zebra fish, the end point is the median survival time of the larvae which are exposed to the toxicant but are not provided with food; the termination of the test is determined by the time of survival that may be achieved by ingestion of the yolk sac (Landner et al. 1985). In these experiments, food is withheld for two specific reasons: to inhibit growth of microorganisms particularly in industrial effluents which have been treated biologically and to circumvent the possibility of sorption of the toxicant to the food.

Guppy and zebra fish are both warm-water fish with short generation times, and both of them reproduce throughout the year; the guppy is, however, viviparous with a high energy cost per larva (K-strategy) compared with the low energy cost per larva (r-strategy) of the zebra fish. The acute toxicity of 3,4-dichlorophenol is similar in both fish, with values of 8.4 mg/l (zebra fish) and 8.7 to 9.0 mg/l (guppy). For zebra fish, however, survival of the larvae is the most sensitive parameter, whereas reproduction is the most sensitive for guppy. Reproduction of guppy was reduced by 35% at a concentration of 2 µg/l which does not affect zebra fish. On the other hand, at a concentration of 200 µg/l, zebra fish populations will be eliminated, whereas in guppies this results only in a reduction of 40% in the number of offspring (Schäfers and Nagel 1991). Whether this conclusion may be extended to cold-water fish with an r-strategy—which probably represents most European fish—is at present unknown.

By way of offering a wider perspective on the range of compounds that have been evaluated and the different fish that have been used in early life-stage evaluations, it may be mentioned that the Californian grunion (*Leuresthes tenuis*) has been employed for assessing the toxicity of chlorpyrifos (*O,O*-diethyl-*O*-(3,5,6-trichloro-2-pyridyl) phosphorothioate (Goodman et al. 1985) and pike (*Esox lucius*) for 2,3,7,8-tetrachlorodibenzo[1,4]dioxin (Helder 1980).

A truly chronic test using zebra fish and extending over three generations has been described and evaluated using 4-chloroaniline (Bresch et al. 1990). Reduction in egg release was the most sensitive parameter, and this was affected at a concentration of 40 µg/l which is some 10 times lower than the threshold toxic concentration for growth. These results illustrate the valuable information that may be gained from such long-term toxicity tests; their only serious limitations are the time required for carrying them out and the expense involved. Possibly some compromise such as that outlined above employing preexposure of adults before assessing the sensitivity of their offspring to the toxicant might be acceptable, although there clearly remain serious difficulties in estimating concentrations which are environmentally innocuous—if indeed this value is scientifically accessible.

Integration with Other Criteria

In all these experiments using relatively long-term exposure, advantage may usefully be taken of the opportunity to examine behavioral, morphological, metabolic, and biochemical effects. Some of these are discussed in greater detail in Sections 7.5 and 7.6 so that only three examples will be used to illustrate the possibilities.

1. During experiments investigating the toxicity and bioconcentration of chlorinated veratroles in zebra fish, the metabolism of these compounds was also examined. Successive O-demethylation of 3,4,5-trichloroveratrole to 3,4,5- and 4,5,6-trichloroguaiacol and 3,4,5-trichlorocatechol occurred, and these were then further metabolized to form sulfate and glucuronic acid conjugates (Neilson et al. 1989).

2. Exposure of zebra fish or rainbow trout to 4-chloroaniline resulted in numerous morphological alterations in the ultrastructure of the liver (Braunbeck et al. 1990a). In zebra fish, the effects were observed at concentrations as low as 40 µg/l and included hepatic compartmentation, invasion of macrophages, effects on the rough endoplasmic reticulum, and increase in the number of lysosomes, of autophagosomes, and of myelinated bodies. These changes apparently indicate response to stress and induction of biotransformation (detoxification) processes. Morphological changes including deformation of the spine have also been observed in the F-1 generation after 8 months exposure to 4-chloroaniline (Bresch et al. 1990). Long-term exposure of zebra fish to γ-hexachloro[*aaaeee*]cyclohexane (lindane) during a full life-cycle test resulted in a number of pathological effects including hepatic steatosis, glycogen depletion, and occurrence of deformed mitochondria (Braunbeck et al. 1990b). Exposure of zebra fish larvae to chloroveratroles resulted in curvature of the larvae and deformation of the notochord (Neilson et al. 1984) and this is at least formally comparable to the serious skeletal deformations that have been observed after exposure to a number of other neutral xenobiotics (references in Van den Avyle et al. 1989). The observation of such pathologies is an important supplement to the diagnosis of adverse effects in feral fish.

3. The behavior of fish in a rotary-flow system (Lindahl et al. 1976) has been incorporated into chronic exposure tests (Bengtsson 1980), and such behavioral alterations could be particularly important to fish which are exposed to toxicants in natural habitats.

7.3.1.6 *Other Aquatic Organisms*

The preceding organisms represent groups of organisms that have been most widely used in ecotoxicology. Increased appreciation of the need for evaluating the effect of a toxicant on a spectrum of organisms has directed attention to other groups of organisms, and these developments have encompassed important biotic components of both freshwater and marine ecosystems. A brief discussion of some of these seems therefore appropriate.

1. Rotifers are major components of the zooplankton, and may graze on phytoplankton and themselves serve as food for larval fish. A 2-day life-cycle test using *Brachionus calyciflorus* has been developed (Snell and Moffat 1992), and is unusual in several respects: the rate of *population* growth is used as the end point, and all life stages are present during the test including embyros, juveniles, and adults. For the relatively few organic compounds which have so far been examined, the sensitivity appears to be comparable to that of *Ceriodaphnia dubia*.

2. Aquatic insects are an important component of stream and river ecosystems and may be particularly vulnerable to the adverse effect of xenobiotics. An assay has been developed using the mayfly *Cloeon triangulifer* that exhibits many features that distinguish it from other insects and make it accessible to laboratory bioassays: it has a relatively short life cycle, egg and larval stages can be reared under laboratory conditions, and important parameters in the life cycle can be quantified. This organism has been used to evaluate the effect of a commercial mixture of chlorinated camphenes on all stages of the life cycle (Sweeney et al. 1993), although certain critical aspects of the design including the exposure regime merit further development.

3. Evaluation of toxicants toward amphibians has been examined (references in Berrill et al. 1998), and a single example is given as illustration. Embryos and tadpoles of three species of frogs *(Rana sylvatica, Bufo americanus,* and *R. clamitans)* were used to examine the toxicity of endosulfan in a 96-h assay (Berrill et al. 1998). Tadpoles showed abnormal behavior including paralysis during exposure. Postexposure mortality was greater for 2-week-old than for newly hatched tadpoles, and occurred at the lowest concentrations evaluated—68, 41, and 53 µg/l, respectively, for the three taxa.

4. In general, less attention has been given to the development of test systems using marine organisms but two of these are briefly noted even though they do not appear to have been extensively evaluated with organic compounds, and in the first two examples cited, azide has been used as a reference toxicant.

   a. A study based on assessing fertilization and embryo/larval development in the mussel *Mytilus californianus* (Cherr et al. 1990) has addressed important factors in the design of the test protocol including the use of the alga *Isochrysis galbana* to induce spawning, determination of the optimal sperm/egg ratio, and the use of small chambers to avoid selective adherence of different development stages to the walls of the containing vessel. The end point was the frequency of development of veliger larvae with a complete shell and this was assessed using polarization microscopy.

b. Sea urchin (*Strongylocentrotus purpuratus*) sperm has been used to assess the toxicity of some components of bleachery effluents (Cherr et al. 1987): the sperm is preincubated with the toxicant, and then both the sperm and eggs are exposed to the toxicant. The end points are the inhibition of fertilization and the effect on the motility of the sperm: as in the assay using *M. californicus*, the sperm/egg ratio is critical. One of the important results from this study was the difference in the toxicity of some of the compounds compared with that which had been estimated from the effects on juvenile salmonid fish. Further studies with the high-molecular-weight fraction (>10 kDa) showed that this inhibited fertilization of the eggs and the acrosome reaction in response to egg jelly (Pillai et al. 1997). The mechanism of inhibition was postulated to involve blocking egg jelly interaction with the sperm surface, and in view of the observation that the calcium ionophore A23187 and ionomycin were able to overcome this inhibition, it seems that ionic events including increasing intracellular calcium are involved. In a more general context, this study draws attention to the possibility of biological effects of sulfated macromolecules at the cellular level.

c. Sea urchin (*S. purpuratus*) gametes and embryos were used to assess the toxicity of phenol, benzidine, and pentachlorophenol, and an interesting departure from convention was the evaluation of toxicity in terms of tissue concentration rather than concentration in the exposure medium (Anderson et al. 1994).

7.3.2 Organisms for Evaluating the Toxicity of Sediments

The association between xenobiotics and sediments has been discussed in Section 3.2, and it is now clearly established that many xenobiotics can be recovered from contaminated sediments which may act as a "sink" for these compounds. Whereas it is certainly true that during aging in both soil and in sediments, these compounds become less accessible to biota, it seems rather dangerous to postulate that their hazard also diminishes with time (Alexander 1995). Desorption may occur slowly over long periods of time (ten Hulscher et al. 1999), and levels after bioconcentration may reach unacceptable levels and, in extreme situations, pose a toxic hazard to humans through consumption of fish. In a wider context, the toxicity of xenobiotics to terrestrial organisms is addressed in Section 7.3.6.

Attention has been increasingly directed to assessing the toxicity of contaminated sediments (Burton and Scott 1992), and assays have used any of three different exposure strategies: (1) to whole sediments, (2) to elutriates or (3) to interstitial water. Indigenous organisms have been extensively used and bred in the laboratory; in general, however, these organisms have complex nutritional requirements, and the degree and route of exposure to the toxicant are significantly determined by the affinity of the toxicant for the food supplied as well as by the organic carbon present in the system. The results of sediment assays using elutriates and typical water-dwelling organisms such as daphnids should therefore be interpreted with caution unless there is free desorption of the xenobiotics from the sediments: aspects of this have been discussed in Sections 3.2.2 and 3.3.2. A number of examples of these assays is given, although since the tests have not always been standardized, it is not generally justifiable to make comparisons of their relative sensitivity.

Amphipods

Numerous studies have used freshwater amphipods. For example, *Diporeia* sp. (syn. *Pontoporeia hoyi*)—which is one of the important benthic invertebrates in the Great Lakes—has been used to study basic determinants of both toxicity and bioavailability (Landrum et al. 1991), while *Hyalella azteca* has been quite extensively used for assessing toxicity (Nebeker et al. 1989; Ankley et al. 1991), and *Gammarus pulex* has been proposed as a suitable organism for routine toxicity evaluation (McCahon and Pascoe 1988; Taylor et al. 1991) although the nutritional demands of the organism must be taken into account. The marine amphipod *Rhepoxynius abronius* collected from an unpolluted site in Yaquina Bay (Oregon) has been used to assess the toxicity of creosote-contaminated sediments in Eagle Harbor, Washington (Swartz et al. 1989) and the estuarine amphipod *Leptocheirus plumulosus* has also been examined (McGee et al. 1993). A comparison has been made using a wide range of marine sediments between results from the 10-day survival assays using *R. abronius* and from the 20-day assays using *Nereis arenaceodentata* for both of which standard protocols have been developed (references in Anderson et al. 1998). The *R. abronius* assay displayed good sensitivity toward a range of toxicants and a good correlation with concentrations of metal cations, PAHs, PCBs, and some agrochemicals (Anderson et al. 1998).

Polychaetes

A life-cycle test using interstitial water and the polychaete *Dinophilus gyrociliatus* has been proposed for incorporation into a national screening program (Carr et al. 1989), and a whole-sediment assay using the marine polychaete *N. arenaceodentata* has been developed, although important issues including the feeding regime should be critically evaluated (Dillon et al. 1993). The comparison of results using the *R. abronius* and the *N. arenaceodentata* assays suggested that additional effort be made to improve the statistical power of the latter assay (Anderson et al. 1998).

Oligochaetes

An extensive study has been devoted to examining the use of freshwater oligochaetes as test organisms for whole sediments (Wiederholm et al. 1987): of those tested including *Tubifex tubifex*, three species of *Limnodrilus*, and *Potamothrix hammoniensis*, the first was the preferred organism due to the ease of manipulation. Possible complications arising from the provision of food during the test, and from differences in the nutritional status of sediments from different lake environments were addressed, and the importance of the relevant controls underscored. On the basis of further evidence and development, this organism has been recommended as one component in a group of tests for assessing sediment toxicity (Reynoldson 1991). Protocols have also been developed for bioassays and estimation of bioaccumulation using *Lumbriculus variegatus* (Phipps et al. 1993).

Midges

Midge larvae of several species of *Chironomus* including *C. tentans*, *C. decorus*, and *C. riparius* have been quite extensively used (Nebeker et al. 1984), and these organisms have been incorporated into microcosms (Fisher and Lohner 1987). Two end points have been effectively used: adult emergence and growth of larvae (Nebeker et al. 1988) using *C. tentans*. Both acute toxicity and effects on reproduction (Kosalwat and Knight 1987a,b) have been examined with cupric sulfate using *C. decorus*, and, as expected, larval stages were the more sensitive. It is clearly important to define the development stage and, for practical reasons, the second larval instar of *C. riparius* has been used (Taylor et al. 1991): it was also shown that the lethal toxicity varied during exposure during 240 h, and comparable test protocols could be more generally exploited with advantage. The effect of γ-hexachloro[*aaaeee*]cyclohexane (lindane) on *C. riparius* has also been examined during the complete life cyle (Taylor et al. 1993): the LOEC of 9.9 µg/l was less than that found in assays that focused on growth or emergence although the authors point out that this could also be the result of experimental factors in the life-cycle assay. The possibility of examining larval deformation has also been examined using mentum deformities along a gradient of contamination (van Urk et al. 1992), though further laboratory investigations would be needed to reveal the causative agent. The important issue of the genetic variation in laboratory populations of *C. tentans* (Woods et al. 1989) has already been noted in Section 7.1.

7.3.3 Assays for Genotoxic Effects

There are a large number of compounds which may induce carcinomas in humans. Very substantial effort has therefore been devoted to detecting these compounds so as to eliminate them—or at least to reduce human exposure to them as far as possible. Tests for induction of cancer in laboratory animals are extremely time-consuming and costly, and require strict attention to the design of the experimental protocols. Attention has therefore been directed to developing screening tests which are rapid and reproducible, and whose results display a satisfying correlation with cancer induction. Two widely different organisms have been used for such assays—microorganisms and fish. Attention should, however, also be directed to the use of morphological alterations of the polytene chromosomes in the salivary glands of larval *C. tentans* exposed to established carcinogens (Bentivegna and Cooper 1993): attractive features are the use of whole organisms and the possible direct application to field samples. The use of higher plants for assessing toxicity including chromosomal aberrations has been noted in Section 7.3.1.3.

Microorganisms

Numerous tests for mutagenesis using microorganisms have been developed and, of these, the Ames test using mutant strains of *Salmonella typhimurium* has been widely used. Care should be exercised in correlating mutagenicity with carcinogenicity and this has been consistently pointed out by Ames. The Ames test has been extensively evaluated, and has therefore become the most commonly applied routine test. It is normally combined with a microsomal activation system prepared from rat liver, but two interesting developments are worth noting. One of these uses a preparation from the liver of rainbow trout for activation of promutagens: an important modification of the test protocol is necessitated by the temperature sensitivity of the metabolizing system which is optimal at 10 to 15°C (Johnson 1992). Although the second has probably the greatest relevance to the terrestrial environment, it is clearly applicable to aquatic plants that have generally received somewhat scant attention. Activation of aromatic amines by plant-cell cultures has been examined, and a provisional model incorporating the formation of mutagenic complexes with macromolecules has been proposed (Plewa et al. 1993). Although activation of procarcinogens by cytochrome P-450 is important, some aromatic amines and hydroxymethyl derivatives of PAHs are activated by sulfation (Glatt 1997).

A revised protocol for the Ames test with experimental details and including a wider range of tester strains has been issued (Maron and Ames 1983), and a modification has been proposed that uses higher cell densities and diminishes the addition of microsomes to increase the sensitivity (Kado et al. 1986). A good account of the results using a wide variety of test organisms—including conventional assays using *S. typhimurium*—and their relative response to a selection of compounds has been given (De Serres and Ashby 1981).

The Ames test uses a set of histidine-requiring mutants of *S. typhimurium* and assesses the number of revertants after exposure to various concentrations of the test compound. The test can readily be carried out by an experienced bacteriologist and offers the following advantages:

1. The test strains include those that are able to detect both frameshift and base-pair mutations.

2. A metabolizing preparation of microsomes can be included to evaluate the synthesis of mutagenic metabolites from the test compound.

3. Solutions in solvents such as dimethyl sulfoxide can be used so that problems of poor water solubility are minimized.

4. The test can be carried out within 2 days and the results are generally readily interpreted: since a large number of cells are used, the test is extremely sensitive.

5. The test may be carried out either on plates of agar medium, or as a fluctuation test in liquid medium: the latter is probably preferable in terms of sensitivity.

6. There is a reasonably good correlation between mutagenic effects revealed by the test and the development of cancer in mammals.

Naturally, care should be exercised in using the results, and they should be complemented at least by a cell transformation test using a suitable cell line. Difficulties may emerge in the form of complex or nonlinear dose–response curves, but in any case, quantitative comparison of numerical values should not normally be attempted. The greatest ambiguity probably arises from the microsomal metabolizing system. Most often, a crude preparation termed S-9 prepared from rats after induction with a PCB mixture is used. It must, however, be appreciated that other animals may have widely different metabolizing systems so that both care and judgment should be exercised in interpreting the results. The attractive alternative using microsomes from fish has already been noted.

Another test has been developed which employs a strain of *Escherichia coli* in which an operon fusion brings the *lacZ* gene under control of the *sfiA* gene, an SOS function involved in the inhibition of cell division (Quillardet and Hoffnung 1985). A number of other functions *uvrA* and *rfa* mutations are incorporated that are comparable to those in the Ames test. The assay is a straightforward one for the induction of β-galactosidase activity, and is rapid and simple to carry out. It has been extensively validated by comparison with the Ames set of tester strains, and a high degree of concordance was demonstrated (Quillardet and Hofnung 1993; Mersch-Sundermann et al. 1994). As stated by the inventors of the SOS Chromotest, this assay should be regarded as a complement to the Ames test, and might allow the detection of genotoxic compounds that are inactive in the Ames test (e.g., estradiol). The agreement between results of a comparison between SOS induction and the Ames *S. typhimurium* assay for 330 structurally diverse compounds was interpreted as indicating a common mechanism of action (Mersch-Sundermann et al. 1994).

A number of other microbial systems for evaluating genotoxic effects have been developed although they have been less widely used. These include a forward-mutation assay using *S. typhimurium*, and assays using the yeast *Saccharomyces cerevisiae* or *Bacillus subtilis*, although these have not been as extensively used as the Ames test. An ingenious assay is based on induction of the prophase λ in *E. coli* (Moreau et al. 1976) is as readily carried out as the Ames test, and has the advantage of greater sensitivity toward specific groups of compounds such as chlorinated pesticides (Houk and DeMarini 1987) and chlorophenols (DeMarini et al. 1990). There is, however, no question of the superiority of one of these systems over the other: they are complementary, and both may advantageously be employed.

Probably any compound suspected of entering higher trophic levels and which may therefore be consumed by humans is worth evaluating for its mutagenic potential before carrying out assessment of its carcinogenicity in mammals.

Fish and Fish Embryos

The preceding tests have been directed primarily to the possibility of inducing cancer in humans. It has, however, become increasingly clear that many biota in aquatic systems are exposed to potentially carcinogenic compounds and that feral fish may display pathological evidence of disease in certain organs (Malins et al. 1985; 1987; de Maagd et al. 1998). Both of these objectives may be combined in tests in which fish embryos are injected with the toxicant dissolved in dimethyl sulfoxide: the embryos are allowed to develop into fry and, 8 to 9 months after exposure, samples are examined for the development of neoplasms (Black et al. 1985). Both rainbow trout (*Oncorhynchus mykiss*) and coho salmon (*O. kisutch*) were examined, and although a comprehensive evaluation of the test does not appear to be available, this is an attractive test which is suitable for screening large numbers of samples. Initiatory investigations (Hawkins et al. 1988) have been directed to the use of small laboratory fish such as sheepshead minnow and fathead minnow to detect carcinogenic effects. From these investigations, it is clear that the susceptibility of fish to induction of cancer varies widely among the species evaluated and that the influence of, for example, nutrition should be evaluated before a standardized protocol can be presented. Subsequent investigations (Hawkins et al. 1990) have demonstrated the potential value of both Japanese medaka (*Oryzias latipes*) and guppy (*Poecilia reticulata*), and the differential sensitivity of these to the known human carcinogens benzo[a]pyrene and 7,12-dimethylbenz[a]anthracene. A micronucleus assay to assess genotoxic effects in fish has been developed, and is applicable in principle to samples of feral fish. Its evaluation using white croaker (*Cenyonemus lineatus*) revealed, however, a low sensitivity, variability in the end poimt, and poor correlation between the observed effects and levels of established contaminants (Carrasco et al. 1990).

7.3.4 Assays for Deformation and Teratogenicity

As a result of the widespread concern over the observation of deformed anurans in ponds and lakes in North America (references in La Clair et al. 1998), an assay for detecting chemicals that induce such deformation has been developed. This is the Frog Embryo Teratogenic Assay—*Xenopus* (FETAX) that uses the South African frog *X. laevis* as assay organism. The test has been developed and undergone intralaboratory and interlaboratory validation using both negative (saccharin and cyclamate) and positive (caffeine and 5-fluorouracil) controls (Bantle et al. 1994), and for a few other compounds including β-aminopropionitrile that was positive in mammals (Bantle et. al. 1996). The results showed that the assay was both repeatable and reliable, although in the latter evaluation, interlaboratory values for the coefficients of variation for LC_{50} and EC_{50} (malformation), and MCIG (minimum concentrations that inhibit growth) were rather large for reasons that were not completely resolved. There may be a degree of subjectivity in assessing malformation, and an atlas of abnormalities for this organism has been prepared (Bantle et al. 1991). Further effort has been directed to the potential role of metabolites. These studies used Arochlor 1254–induced rat liver microsomes for activation and analyzed data from an interlaboratory evaluation using cyclophosphamide and caffeine: results were reported as LC_{50}, EC_{50} (malformation), the minimum concentration for inhibition of growth (MCIG), the mortality/malformation index (MMI) defined as LC_{50} / EC_{50}, and the teratogenic index (TI) (Fort et al. 1998). For cyclophosphamide that is metabolized by cytochrome P-450 to the 4-hydroxy compound and phosphoramide mustard, values of the first three were lowered by inclusion of the microsomes, and the MMI increased from 1.54 to a mean value of 3.62. On the other hand, for caffeine that is metabolized to other purines, values of the first two were comparable, whereas values for LC_{50} and MMI decreased. These results clearly reveal the potentially important role of metabolism.

An interesting application of this assay revealed that although the insect growth regulator *S*-methoprene had relatively high values for 96-h LC_{50} and 96-h EC_{50}, the values for metabolites produced by photolysis and hydrolysis were lower (La Clair et al. 1998): in particular those in which hydrolysis of the ester had occurred displayed serious eye deformations.

7.3.5 Assays for Estrogenic Activity

There has been increased concern over the release of compounds that may potentially display adverse hormonal effects (estrogenic and androgenic). A number of compounds have been implicated including 4-alkyl phenols whose activity was observed in 1978 (references in White et al. 1994). As for mutagenic and carcinogenic activity, the ultimate target organism is humans, although concern has been expressed about adverse effects on wildlife, especially fish: in both cases, the significance of this activity depends critically on the degree of exposure, and the concern with fish is reinforced by the identification of nonylphenol, octylphenol, and nonylphenolmonoethoxylate in flounder (*Platichthys flesus*) and in sediments in the United Kingdom (Lye et al. 1999).

The numbers and structures of potential candidates are enormous, so that surrogate assays have been developed. The *in vitro* assays that have been considered and evaluated have been reviewed (Zacharewski 1997; Anderson et al. 1999), and in environmental studies several assays have been most generally used.

1. A strain of *Saccharomyces cerevisiae* was developed (Purvis et al. 1991) in which the DNA sequence of the human estrogen acceptor was stably integrated into the yeast chromosome together with the *lacZ* gene whose activity is used to measure activity. A few illustrative examples of its application are given.

   a. A range of anionic, cationic, nonionic, and amphoteric surfactants and some of their established metabolites has been examined (Routledge and Sumpter 1996). Although none of the original compounds displayed activity, weak activity was established in the degradation products of alkylphenol ethoxylates.

   b. The estrogenic activity in fractionated sewage treatment plant effluents (Section 2.2.3) has been examined (Desbrow et al. 1998). The results showed that the activity of 4-*tert*-octyl phenol was only from 1/100 to 1/1000 that of 17β-estradiol.

   c. A range of flavanoids was examined and their estrogenic potency was between 4000 and 4 million times lower than that of 17β-estradiol (Breinholt and Larsen 1998).

2. An estrogen-responsive human breast cancer cell line MCF-7 transiently cotransfected with the Gal4-human ER anf and a Gal4-responsive luciferase reporter (17m5-G-Luc) was used to analyze activities relative to 17β-estradiol in fractionated samples of urban air particulates. Although activity was shown by benzo[*a*]pyrene, benzo[*a*]anthracene, and chrysene, this was only 1/1000 to 1/10,000 that of 17β-estradiol (Clemons et al. 1998). This assay has also been used to explore further the estrogenic activity of the flavanoids noted above (Breinholt and Larsen 1998). Another assay using MCF-7 cells has been developed in which trypsinized cells were exposed to the test media, and after lysis, the nuclei were counted (Soto et al. 1995). Effects were normalized to those induced by 17β-estradiol and given as the ratio of the doses required to produce maximal cell yields ×100. For 4-alkylphenols at concentrations of 10 μM, values were 0.0003, except for 4-octylphenol, and were 0.0001 for all the PCB congeners at concentrations of 10 μM. In this study, attention was drawn to the potentially important issue of metabolism comparable to its significance in determining mutagenic activity.

3. A strain of *S. cerevisiae* in which the estrogen receptor from rainbow trout has been developed (Petit et al. 1995), and might be more relevant for ecotoxicological investigations.

This assay was used to examine a wide range of xenobiotics including herbicides, fungicides, insecticides, PCBs, and detergent-related compounds (Petit et al. 1998). This was supplemented by the more elaborate assay using trout hepatocyte aggregate cultures. Although 70% of the estrogenic compounds were positive in both assays, 30% exhibited activity in only one of them. It was therefore proposed that both *in vivo* and *in vitro* assays should be made, and that the possible significance of metabolism to active compounds should be considered.

In summary, as with all surrogate assays, care must be exercised by taking into account the sensitivity of different assays, their intrinsic variability, and the possibly cardinal issue of metabolism.

At the whole organism level, it is worth noting the results of experiments in which *Daphnia magna* was exposed to ethanolic solutions of the established estrogen diethylstilbestrol (Baldwin et al. 1995): concentrations up to 0.5 mg/l were tested. In summary, the results showed rather weak effects on the number of preadult molts during exposure of first-generation *D. magna*, and on the ratio of cumulative offspring per female during second-generation exposure (Baldwin et al. 1995).

Reference should be made to the vitellogenin assay that has been used as a biomarker for estrogenic effects in fish (Section 7.6.1).

7.3.6 Terrestrial Organisms

There is a long history of the evaluation of toxicity toward terrestrial organisms. This has been largely motivated by evaluating the specificity of agrochemicals, and ensuring the absence of adverse effects on nontarget organisms. Test organisms have been chosen to represent important chains in the ecosystem and include insects, collembolans, oligochaetes, isopods, and higher plants. The field has, however, taken a much wider perspective in light of utilizing former industrial sites for residential and office building, and in efforts to remove undesirable contaminants from such sites (Chapter 8). This has encouraged developments in the use of a range of terrestrial organisms to assay toxicity. Microbiological aspects of bioremediation are discussed in Chapter 8, but some comments on assays directed at determining both the need for, and the effectiveness of, remediation are discussed briefly here. All these assays are subject to the limitations noted for aquatic organisms in Section 7.3.1, and, as for assays for sediment and soil toxicity (Section 7.3.2), the importance of association with the organic matrix and bioavailability plays a cardinal role in the interpretation and evaluation of the results. The important issues of bioavailability, association with organic matter in soil, aging, and the mechanisms for the interaction between xenobiotics and organic soil components have been discussed in greater detail in

Chapter 3, Sections 3.2.2 and 3.2.4. One important difference from the situation of biota in most aquatic systems is that terrestrial organisms may be able simply to avoid exposure by physically escaping from the adverse effect of toxicants. Analytical issues have been taken up in Chapter 2, Section 2.2.5, and it should be noted once again there may or may not (Robertson and Alexander 1998) be a correlation between chemical extractability and bioavailability.

By analogy with aquatic assays, experiments may be carried out using either a synthetic or a natural matrix, but the nutritional demands of the test organisms may often necessitate the use of complex, ill-defined test media. The experiments involve spiking the matrix with various concentrations of the toxicant, and this may be carried out by simple mixing with solids or of solutions in organic solvents followed by evaporation of the solvent under reduced pressure: this may not, however, achieve a homogeneous distribution of the toxicant. As for fish, experiments may involve exposure either to the compound in the matrix or the food, although in this case the difference may not be clear-cut since components of the matrix may serve as natural food sources. Tests may use end points for both acute toxicity and for effects on reproduction.

Probably most animal studies have used either oligochaetes or terrestrial isopods. It is important to appreciate the large number of genera in these groups so that considerable differences in toxicity are to be expected when different taxa are used for evaluation.

Bacteria

The principle of PICT that has been noted in Section 7.3.1.2 for algae has been applied to bacteria in a terrestrial system (Rutgers et al. 1998). Determination of the community structure based on the taxonomy of the components is based on their nutritional demands. Metabolic activity was assessed using 95 substrates, which do not have any structural relation either to the organic components of the soil or to the toxicant that is being examined.

Oligochaetes

Terrestrial worms of the genera *Lumbriculus*, *Eisenia*, and *Allolobophora* have been used using standardized protocols, and both acute and chronic toxicity have been examined.

1. Exposure has been carried out on filter paper impregnated with the toxicant prepared in an organic solvent, and has been used for pentachlorophenol (Giggleman et al. 1998) and Arochlor 1254 (Fitzpatrick et al. 1992). The results are reported as lethal body burden (LBB) to take into account the amount of the toxicant that has been accumulated. Although this procedure facilitates comparison among different test organisms including *L. terrestris* and *E. fetida*

(Fitzpatrick et al. 1992), the results cannot be extrapolated to natural soils in view of the importance of organic carbon in determining the bioavailability of the toxicant (Sections 3.2.2 and 3.2.3).

2. Values of LC_{50} for a range of chlorophenols were generally greater for *L. fetida* than for *E. andrei*, and differences observed between two different types of soil could be correlated with the solution concentrations of the toxicants (van Gestel and Ma 1988). For 3-chloroaniline, 2,4,5-trichloroaniline, and 2,3,5,6-tetrachloroaniline, however, LC_{50} values for both taxa based on predicted concentrations in the water phase were quite similar (van Gestel and Ma 1993).

3. A valuable comparison of results has been carried out (van Gestel 1992) using, in laboratory studies, the fungicide benomyl (methyl-1-(butylcarbamoyl)benzimidazol-2-ylcarbamate), and the insecticide carbaryl (1-naphthyl methylcarbamate), and in field studies using benomyl, carbendazim, carbofuran (2,3-dihydro-2,2-dimethyl-7-benzofuranyl-N-methylcarbamate), and carbaryl. The results of this comparison are important and deserve brief comment. Exposure conditions differed in the various laboratory experiments, and there is an intrinsic difficulty in determining real measures of exposure after field application.

   a. In laboratory experiments, there was considerable variation among results for acute toxicity of benomyl toward *E. andrei*, *E. fetida*, and *L. terrestris*: results for chronic toxicity toward *E. andrei*, *L. terrestris*, and *A. caliginosa* used different end points and, on the basis of values for NOEC, were indecisive numerically. Comparable differences were reported for acute toxicity toward carbofuran and carbaryl.

   b. In field studies, exposure to the toxicants is dependent both on the heterogeneity of the soil after application and on the vertical distribution of the worms. Broadly, however, there was agreement between the results of laboratory and field experiments.

A study using *Aporrectodea tuberculata* revealed some important issues of general significance (Haimi and Paavola 1998): (1) the toxicity of pentachlorphenol from contaminated and uncontaminated sites was not significantly different, so that preexposure had no effect on the response to the toxicant and (2) these observations could most readily be accommodated by taking into account the fact that earthworms can diminish their susceptibility to toxicants by physical avoidance.

Isopods

PAHs have attracted widespread attention in view of the substantial quantities that are discharged into the atmosphere and subsequently deposited in the form of particulate matter and precipitation. The toxicity of benzo[*a*]pyrene in food toward the terrestrial isopods *Oniscus ascellus* and *Porcellio scaber* has been examined (van Brummelen and Stuijfzand 1993): values of NOEC were 31.6 µg/g, and for both species growth was affected at concentrations of 100 µg/g, and in *P. scaber* the protein level was significantly reduced at 316 µg/g. In a later study (van Brummelen et al. 1996), a wider range of PAHs was studied using the same organisms: the long-term toxicity was generally low, and when positive only marginal. The concentrations that were evaluated were relatively high in the terrestrial environment, but attention should be drawn to the metabolic activity of these organisms, and the adverse effect of metabolites on other organisms. For example, pyrene is metabolized by *P. scaber* with the formation of 1-hydroxypyrene (Stroomberg et al. 1996), and this is toxic to the nematode *Caenorhabditis elegans* (Lambert et al. 1995). This has been noted in Section 3.1.2 in the context of bioaccumulation.

Collembola

The insect *Folsomia candida* has been used in laboratory experiments to determine the toxicity of a range of insecticides (Thompson and Gore 1972), and it was established that temperature played a significant role as a determinant of toxicity.

Nontarget Higher Plants

In this case, there are several issues that should be taken into account.

1. Uptake of nonvolatile substrates occurs through the roots, and its extent is determined by the network of partition among the soil, the water, and the plant root phases, while for atmospheric contaminants in the form of aerosols, the leaves may provide an additional route of exposure. The uptake of a range of agricultural plants including barley, lettuce, carrots, and radish illustrated the different modes of uptake and translocation (Schroll et al. 1994). The modes of exposure and transport are of increasing concern and have been discussed in Section 3.5.3. Aerial exposure is important in view of the established long-distance atmospheric transport of xenobiotics, and is consistent with the use of, for example, pine needles in environmental monitoring of PCBs (Safe et al. 1992), a range of halogenated aromatic compounds (Jensen et al. 1992), and of lichens from the Great Lakes region (Muir et al. 1993). In the last, the pattern of compounds that were found differed from those found in atmospheric or rainfall samples, and illustrated clearly the specific influence of both the chemical structure of plant surfaces and of their surface area (Section 3.5.3).

2. There is accumulating evidence for the capacity of plants to metabolize xenobiotics: for example, phoxim is metabolized by plant organs and cell suspension of soybean (*Glycine max*) (Höhl and

Barz 1995; see Chapter 3, Section 3.1.3) and benzene in leaves of spinach (*Spinacia oleracea*) (Ugrekhelidze et al. 1997; see Chapter 3, Section 3.1.3).

All the plants that have been used for evaluating the effect of toxicants in the aquatic environment (Section 7.3.1.3) are possible candidates—bearing in mind the additional routes of exposure that may be important in the natural environment. Experiments on the toxicity of a range of chlorophenols and chloroanilines to lettuce (*Lactuca sativis*) revealed a number of important determinants (van Gestel et al. 1996).

- There could be considerable differences in the acute toxicity (EC_{50}) determined after 7 or 14 days.
- Values of EC_{50} were substantially higher using an artificial soil than in using a natural loam.
- A comparison of EC_{50} using concentrations in pore water showed that values for the artificial soil were highest followed by values for the nutrient solution.

It was concluded that, on the basis of the desirability for comparison and standardization, the artificial soil recommended by OECD was preferable.

Fluorobenzoates have been used as tracers for the movement of water in soil and groundwater so that their potential toxicity to plants is important. A study using 11 crop species and pentafluorobenzoate and 2,6-difluoro- and 3,4-difluorobenzoate showed the generally low toxicity of these compounds (Bowman et al. 1997). More extensive studies were carried out with alfalfa (*Medico sativa*), barley (*Hordeum vulgare*), and canola (*Brassica napus*):

- The compounds exerted essentially no significant effect on germination.
- Soil solutions at concentrations >125 mg/l caused inhibition of growth.
- Variable concentrations were found in the plants, and were generally greater for canola and for 2,6-difluorobenzoate. In the light of the possible role of lipid content (Section 3.1.3), it would be valuable to determine the lipid content of this oil-producing plant.

The influence of soil heavily contaminated with PAHs of which phenanthrene and fluoranthene were in highest concentration, on the mycorrhizal colonization by *Glomus mosseae* pP2 of leek (*Allium porrum*), maize (*Zea mays*), ryegrass (*Lolium perenne*), and clover (*Trifolium subterraneum*) has been examined (Leyval and Binet 1998). Addition of contaminated soil resulted in decreased colonization of leek and clover, but not of maize and ryegrass. For ryegrass, concentrations of Σ PAH of 1 g/ kg or more reduced growth of both shoots and roots than for roots, established the positive influence of mycorrhizal fungi. These results imply the possible importance of mycorrhizal fungi in the establishment and maintenance of plants in PAH-contaminated soils.

7.4 Test Systems: Several Organisms

7.4.1 Introduction

The relevance of tests using single species for evaluating the effects of xenobiotics on natural ecosystems has been the subject of considerable discussion (Cairns 1984), and no final resolution of the conflicting views is to be expected. On the one hand, it is obvious that few natural ecosystems consist of only one taxon and that the effects of a toxicant should be evaluated in the context of communities. Attention is drawn to a review of procedures aimed at determining the ecological effects of xenobiotics on the structure of populations and communities (Petersen and Petersen 1989). On the other hand, it is clear that the effect will ultimately be displayed in individual organisms as a result of reactions at the molecular level, and it is possible that effects observed in laboratory assays have little relevance in natural ecosystems. Neither approach is therefore exclusive, and evaluations of the biological effect of xenobiotics may profitably be carried out at three levels: (1) the individual, (2) the population, and (3) the community. Detailed discussions of all three approaches have been given (Landner et al. 1989).

The design of the experiments to address these questions is necessarily complex, and complementary monitoring to reveal adverse effects of xenobiotics on wild populations is probably necessary. It should, however, be emphasized that extreme care must be exercised in attributing the observed effect to the exposure of biota to a given toxicant, and this problem is discussed again in Section 7.7.2. There is therefore a clear need for test systems that lie between the two extremes of single species and natural ecosystems; these test systems should attempt to combine the reproducibility of single-species tests with the ecological relevance of field experiments. It is hardly realistic to expect a non-specialist like the author to evaluate cardinal issues such as the design of the systems, the end points used, or the relevance of the results to natural ecosystems. With inevitable shortcomings, some attempt is, however, made to provide at least an overview of what appear to be the critical issues. It should also be appreciated that it may not be possible to present the data collected from such experiments in formats which are as accessible as those from single-species testing: this fact reflects in essence the complexity of the natural systems whose simulation is the primary objective of the experiments.

Two broad types of multispecies test systems have been developed, differing essentially in their dimensions: laboratory microcosms having a volume of a few liters and mesocosm systems with volumes of cubic meters. There is, of course, no absolute distinction between these, and some systems have volumes between the two extreme dimensions.

7.4.2 Microcosms

The application of these two studies of biodegradation has been given in Chapter 5, Section 5.3.2. The present discussion is directed to evaluation of toxicity. The communities that are used generally consist of organisms at different trophic levels, and two rather different approaches have been used.

1. Taube microcosms (Larsen et al. 1986) may include representatives of the following groups of organisms which are added to a synthetic medium: heterotrophic bacteria and fungi, algae, protozoans, amphipods, and ostracods. A sediment component is used generally supplemented with insoluble organic material such as cellulose or chitin.

2. Alternatively, organisms such as protozoans may be collected on polyurethane foam blocks suspended in a stream which the system is designed to simulate and macroinvertebrate communities assembled on rocks held in plastic containers (Pontasch et al. 1989).

In these experiments, a range of end points has been used such as the population densities of individual taxa (Pontasch et al. 1989) or appropriate measures of primary and secondary production including $^{14}CO_2$ uptake, the concentration of chlorophyll, the oxygen concentration, and the rate of respiration (Larsen et al. 1986; Stay et al. 1988). Use of the rate of respiration of glucose to measure of heterotrophic activity (Stay et al. 1988) seems, however, unduly restrictive since many bacteria and even heterotrophic algae are unable to use glucose as a substrate even though they may be able to incorporate it into cell material.

These studies which have examined single compounds (atrazine or fluorene), or complex effluents, were designed to assess the extent to which the results of these test systems could be used to predict effects in specific ecosystems. The overall agreement between the two was good, and the results complemented those using single-species assays. The results also clearly supported the view that the ecological effects of a toxicant should be assessed on the basis of the results from a range of organisms, even within the same taxonomic group. On the other hand, they indicated that significant differences could occur between the effects at population and at ecosystem levels, and between results obtained from microcosms using biota from different environments (Stay et al. 1988). The suggestion (Niederlehner et al. 1990) that such disagreement with predictions implies the need for a review both of the design of the experiments and their objective is highly commendable. It therefore seems unlikely that a single test system will adequately represent a range of natural ecosystems, and the assay system that is used should reflect as far as possible the structure of specific ecosystems. In addition, it is entirely conceivable that some significant variables in all ecosystems remain to be evaluated.

Systems using organisms at other trophic levels have also been used, and two may serve to illustrate.

1. The use of indigenous microbial communities has been suggested, not as surrogates for macroscopic communities, but because they provide communities of comparable complexity and offer the possibility of studying the dynamics of their susceptibility to the effect of xenobiotics. It should be noted that the term *microbial* is generally used in a restricted sense to include only protozoa and algae, and that emphasis has hitherto been placed on the use of the distribution of taxa in the system as an end point (Cairns et al. 1992).

2. An assay system has been developed that used sediment containing micro-, meio-, and macrofauna. The sediment was sieved and the amphipod *Pontoporeia affinis* and the mussel *Macoma baltica* were used as test organisms in a flow-through system (Landner et al. 1989). A number of end points are possible including the survival and reproduction of *P. affinis* during chronic exposure (Sundelin 1983), and population changes in various groups of meiofauna. Apparently, this system has not, however, been systematically evaluated with a range of compounds so that its potential remains unknown.

7.4.3 Mesocosms

A valuable review of the widely different experimental systems which have been used has been provided (Lundgren 1985). Model systems have been developed to simulate at least four different types of natural ecosystems: (1) littoral zones using pools or ponds, (2) benthic zones using tanks, (3) riverine systems using model streams, (4) lake systems. These have already been discussed in Section 5.3.3 in the context of biodegradation and biotransformation.

The Objective and Design of Model Ecosystems

It is valuable at the start to summarize the diverse issues to which mesocosm ecosystems have been applied.

1. The effect on biota of chronic exposure to low concentrations of a toxicant;
2. The relative significance of various routes of exposure of biota to the toxicant;
3. The fate of toxicants and their distribution among the environmental compartments.

One of the very substantial attractions of model ecosystems is the possibility of attaining all three objectives in a single set of experiments. On the other hand, one of the inherent limitations is the virtual impossibility of accurately reproducing the same conditions, so that particular care in the design of experiments should be directed to providing a sufficient number of replicates and an unequivocal control. Climatic variations during the year may be an advantage in providing a degree of realism not readily accomplished in laboratory experiments, although in more northern latitudes, ice formation may exclude the possibility of continuing the experiments during the winter months.

Attention should also be drawn to some fairly obvious experimental issues. Most of these apply equally to tests using single species but they may be particularly important in the more complex mesocosm systems.

1. Particular care is needed in testing compounds with poor water solubility since concomitant application of dispersion agents or organic solvents may lead to overgrowth of the systems with bacteria or algae.

2. In systems with fish belonging to different taxa, care should be exercised in their selection, e.g., use of both largemouth bass (*Micropterus salmoides*) and bluegills (*Lepomis macrochirus*), to maintain both of them in good condition (Deutsch et al. 1992).

3. If physiological or biochemical parameters on fish are assessed, particular care should be taken in the handling of the fish to avoid stress and its interference with the measurements (Munkittrick et al. 1991; Hontela et al. 1992).

4. Analytical control of toxicant concentration in the system should be carried out since some compounds may partition rapidly into the sediment phase: although this partition reflects an important aspect of natural ecosystems, it may clearly influence the exposure concentration to aquatic biota.

A large number of different end points may be used in experiments using model ecosystems, and will generally include those used for single species, e.g., growth and reproduction of fish or planktonic algae. In addition, assessment of the composition of populations of invertebrates in sediments, or of algae attached to surfaces (periphyton) may be made, as well as measurements of any of the physiological and biochemical parameters in fish which are discussed in more detail in Sections 7.6.1 and 7.6.2. An interesting development is the use of physiological energetics in the amphipod *Gammarus pulex* (Maltby 1992) that has the attractive advantage that this may also be applied to laboratory experiments.

A striking example of the toxic effect of a single compound which was initially revealed in experiments with model mecosystems is that of chlorate toward bladder wrack (*Fucus vesiculosus*) which is a component of the Baltic

Sea littoral zone (Rosemarin et al. 1986): in this case, it emerged that, compared with unicellular algae that are generally used for toxicity testing, *F. vesiculosus* was exquisitely sensitive to chlorate. In general, however, model ecosystems provide the possibility of revealing interactions between the various components and they are less fully exploited if they merely measure isolated effects on single species. In a strongly interacting ecosystem, the total effect of a toxicant may be greater than the sum of these single-species effects.

Illustrative Examples

It has been noted that mesocosm experiments have the potential for revealing both the toxicity and the fate of a xenobiotic: results from investigations that illustrate some aspects of these are provided in the following examples.

Fate and Distribution

1. The fate of 2,3,7,8-tetrachlorobenzo[1,4]dioxin (TCDD) was examined in freshwater ponds using ^{14}C-labeled substrate over a period of up to 2 years. Equilibrium concentrations in pondweeds (*Elodea nuttali* and *Cerotophyllon demersum*), in fathead minnows (*Pimephales promelas*) and in sediment were attained after 1, 2, and 6 months, respectively. After 1 year, most of the remaining TCDD was found in the pondweed, and after 2 years, through death of the plants, almost all the TCDD was found in the sediment phase that included plant detritus. Unidentified metabolites were confined to the aqueous phase and the plants (Tsushimoto et al. 1982).

2. The fate and distribution of the pyrethroid insecticide deltamethrin was studied in small outdoor ponds over a period of 306 days (Muir et al. 1985). The compound rapidly partitioned into suspended solids, plants, sediment, and the atmosphere, and at the termination of the experiment, the major sink for the intact compound was the sediment phase. Although fathead minnows (*P. promelas*) concentrated the compound from the aquatic phase, no mortality was observed.

3. The persistence of benzo[*a*]pyrene was investigated in a marine model ecosystem with planktonic primary producers and a heterotrophic benthos. The experiment was conducted over 202 days, and showed that the initial substrate was substantially degraded into polar products, but that after a period of ~2 months, both the remaining substrate and its metabolites were contained in the sediment and were apparently then protected from further degradation or metabolism (Hinga and Pilson 1987).

4. Brackish-water systems simulating the littoral zone of the Baltic Sea were used to examine the fate and distribution of 4,5,6-trichloroguaiacol and its metabolites. The known metabolites including

3,4,5-trichloroveratrole, 3,4,5-trichlorocatechol, and dechlorination products were widely distributed in the biota, and a mass balance showed that the bulk of them was partitioned into the sediment phase. The results effectively confirmed those from previous laboratory experiments (Neilson et al. 1989).

5. A study using mesocosms installed in a lake in Finland was designed to study the long-term fate of bleachery effluent without the complication of processes involving the lake sediment (references in Saski et al. 1997). Although compound-specific details could not be established since sum parameters (EOCl and AOX) were used, these were supplemented by data on the molecular weight distribution of sediment extracts. The results were clear in showing (a) a diminution in levels of EOCl in the water mass, (b) the role of photochemical processes, and (c) alteration in the composition of *de novo* sediment that resulted in the production of compounds that were both more hydrophobic than their putative precursors and were also of higher molecular mass.

The results of all these controlled experiments are consistent with the principles of partitioning that have been discussed in Section 3.2 and with the established recovery of many xenobiotics from sediments.

Effects on Biota

Two kinds of experiments have been carried out. In one, only a single organism such as the water flea *Daphnia longispina* (Crossland and Hillaby 1985) has been the primary organism of interest, but attention should be drawn to the much more sophisticated full life-cycle tests with rainbow trout carried out in model stream systems subjected to a constant input of diluted bleached kraft mill effluent during 3.5 years (Hall et al. 1991). In the second type of mesocosm experiment, advantage is taken of the possibility of examining the population changes in the major components of the biota. In this respect, the procedures are similar to those used in microcosm studies.

1. A study of the effect of 4-nitrophenol used a divided pond and contained algae, macrophytes, and a range of aquatic fauna. There was an alteration in the structure of the algal population in which cryptophytes replaced chlorophytes, and there were reduced populations of the aquatic fauna particularly the crustaceans belonging to the orders Cladocera and Cyclopida. With the exception of the alga *Chara hispida*, the population of macrophytes was eliminated 1 year after application of the toxicant (Zieris et al. 1988). It is important to point out that the results of this study differed significantly from those obtained using single-species tests: the effect on algae was less, and on Cladocera greater. In this case, false predictions

would therefore have resulted from extrapolation of the results of assays using single species.

2. The effect of single applications of esfenvalerate [S-alpha-cyano-3-phenoxybenzyl (S)-2-(4-chlorophenyl)-3-methylbutyrate] was examined (Lozano et al 1992) in a pond system containing plant and animal communities including the following: macrophytes and phytoplankton, micro- and macroinvertebrates and fish (fathead minnows, bluegills, and northern redbelly dace (*Phoxinus eos*)) which were placed in enclosures. A range of toxicant exposure concentrations was employed, but at the lowest that were used, the concentrations of the toxicant were unmeasurable after 24 h. At the higher concentrations tested (1 and 5 µg/l), drastic reduction or elimination of populations was observed, although some invertebrate communities recovered at the lower concentrations. It should be noted that this experiment was designed to assess the effect of low concentrations, and not chronic exposure which is the more common objective. The same compound was examined during application every 2 weeks to a mesocosm system, and a number of dynamic parameters were evaluated, including growth of bluegill, primary production, community respiration, and enumeration of benthic invertebrates and zooplankton (Fairchild et al. 1992). Again, the important conclusion was drawn that although effects on individual organisms such as fish or crustaceans may be assessed from single-species tests, alterations in the composition of the ecosystem as a whole may prove both more subtle and, in the long run, more significant.

These results clearly illustrate the limitations in the conclusions that may be drawn from conventional single-species assays, and support the views of Maltby and Calow (1989).

7.5   Metabolism of Xenobiotics by Higher Organisms

It will have become clear from arguments presented in previous chapters that very few xenobiotics remain unaltered in the environment for any length of time after their release. Although metabolism primarily by microorganisms has been discussed in detail in Chapter 6, some brief comments on metabolism—particularly by fish—are briefly summarized here. The biotransformation of xenobiotics in many higher organisms is mediated by the cytochrome P-450 monooxygenase system and the complexity of factors that regulate the synthesis of this in fish has been reviewed (Andersson and Förlin 1992).

Possibly the motivation for discussing metabolism in a chapter dealing with toxicology should be briefly addressed.

1. The impact of metabolism on experiments on bioconcentration has been outlined briefly in Chapter 3, Section 3.1.5, and similar principles apply equally to toxicity: the kinetics and products of metabolism critically influence the nature and the concentrations of the xenobiotic and its transformation products to which the cells are exposed. Increasing evidence from different sources has shown that the effective toxicant may indeed be a metabolite synthesized from the compound originally supplied and not the xenobiotic itself. It is important to appreciate that—as with toxicity—the extent of metabolism will generally depend on the nature and position of substituents on aromatic rings as well as on their number. For example, although 2,3,4- and 3,4,5-trichloroaniline were N-acetylated in guppy, this did not occur with 2,4,5-trichloroaniline (de Wolf et al. 1993). Metabolism may also be an important determinant of genotoxic effects (Section 7.3.3), estrogenic activity (Section 7.3.5), and teratogenicity (Section 7.3.4).

2. Metabolism may also serve as a detoxification mechanism since initially formed metabolites may be further converted to water-soluble conjugates that are excreted.

A few diverse examples may be given to illustrate the important toxicological consequences of metabolism.

1. The classic case is that of Prontosil (Figure 7.1) in which the compound is active against bacterial infection in animals although inactive against the bacteria in pure culture. The toxicity in animals is the result of reduction to the sulfanilamide (4-aminobenzenesulfonamide) that competitively blocks the incorporation of 4-aminobenzoate into the vitamin folic acid.

2. The case of fluoroacetate in which the toxic substance is fluorocitrate synthesized by incomplete operation of the tricarboxylic acid cycle (Peters 1952) has already been noted in Section 4.8.2.

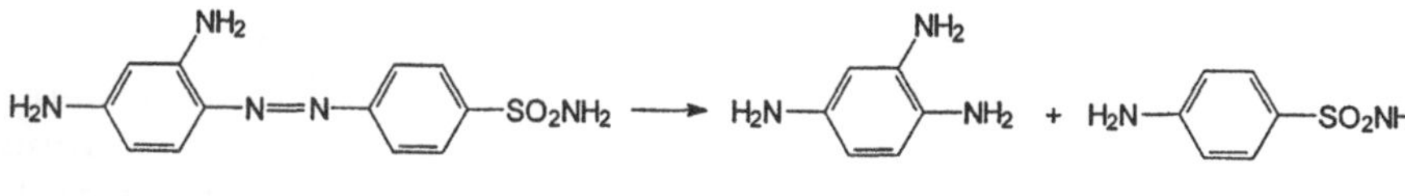

FIGURE 7.1
Metabolism of Prontosil in mammals.

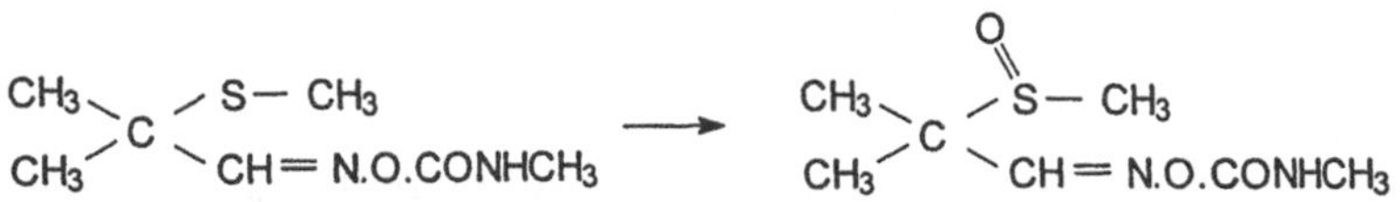

$$CH_3 - C(CH_3) - S - CH_3, CH = N.O.CONHCH_3 \longrightarrow CH_3 - C(CH_3) - S(O) - CH_3, CH = N.O.CONHCH_3$$

FIGURE 7.2
Metabolism of aldicarb by rainbow trout.

3. Considerable attention has been directed to the synthesis of the epoxides and dihydrodiol epoxides of PAHs mediated by the action of cytochrome P-450 systems, and their role in inducing carcinogenesis in fish (Varanasi et al. 1987; de Maagd and Vethaak 1998). Tumors observed in feral fish exposed to PAHs may plausibly—although not necessarily—be the result of this transformation. Even though an apparently causal relationship between exposure of fish to PAHs and disease may be established (Malins et al. 1985; Harshbarger and Mark 1990), caution should be exercised due to the possibility that other—and unknown—substances may have induced carcinogenesis. Further discussion of this is presented in Section 7.7.2. It is also important to appreciate that induction of the metabolic system for PAHs may be induced by other compounds. For example, exposure of rainbow trout to PCBs increases the effectiveness of liver enzymes to transform benzo[a]pyrene to carcinogenic intermediates (Egaas and Varanasi 1982).

4. The carbamate insecticide aldicarb (Figure 7.2) that exerts its effect by inactivating acetylcholinesterase is metabolized by a flavin monooxygenase from rainbow trout to the sulfoxide which is a more effective inhibitor (Schlenk and Buhler 1991).

5. Preexposure to the organophosphate diazinon at exposures half the LC_{50} values increased the LC_{50} value by a factor of about five for guppy (*Poecilia reticulata*), but had no effect on the value for zebra fish (*Brachydanio rerio*). This was consistent with the observation that during preexposure of guppy, there was a marked inhibition in the synthesis of the toxic metabolites diazoxon and pyrimidinol, whereas this did not occur with zebra fish in which the toxicity was mediated primarily by the parent compound (Keizer et al. 1993).

6. Pyrene is metabolized by the fungus *Crinipellis stipitaria* to 1-hydroxypyrene, and this has a spectrum of toxic effects substantially greater than those of pyrene: these include cytotoxicity to HeLa S3 cells, toxicity to a number of bacteria and to the nematode *Caenorhabditis elegans* (Lambert et al. 1995).

At the other extreme, if metabolism of the xenobiotic by the organism does not occur at all—or at insignificant rates—after exposure, the compound will be persistent in the organism, and may therefore be consumed by predators. This has been briefly discussed in the context of biomagnification in Section 3.5.4.

Most of the reactions carried out by fish and by higher aquatic organisms are relatively limited transformation reactions in which the skeletal structure of the compounds remains intact. Three widely distributed reactions are probably of greatest significance:

1. Cytochrome P-450-type monooxygenase systems which have a generally low substrate specificity are widely distributed in the species of fish used for toxicity testing (Funari et al. 1987).

2. Glutathione S-transferases (Nimmo 1987; Donnarumma et al. 1988) which are important in the metabolism of highly reactive compounds containing electrophilic groups, such as epoxides, and aromatic rings with several strongly electron-attracting substituents such as halogen, cyano, or nitro groups.

3. Conjugation of polar groups such as amines, carboxylic acids, and phenolic hydroxyl groups produce water-soluble compounds that are excreted and these reactions therefore function as a detoxification mechanism.

7.5.1 Metabolism by Fish

The metabolic potential of fish may appear restricted compared with that of microorganisms, but it may have been considerably underestimated. For example, metolachlor (2-chloro-*N*-[2-ethyl-6-methylphenyl]-*N*-[2-methoxy-1-methylethyl]acetamide) is metabolized by bluegill sunfish (*Lepomis macrochirus*) by reactions involving initially *O*-demethylation and hydroxylation (Cruz et al. 1993) (Figure 7.3) that are comparable with those carried out by an actinomycete (Krause et al. 1985), while the benzylic hydroxylation is analogous to that involved in the biotransformation of the structurally similar alachlor by the fungus *Cunninghamella elegans* to which reference has been made in Section 4.3.6. The metabolism of PAHs that has attracted considerable attention both in the context of assays for genotoxic effects (Section 7.3.3) and because of the incidence of tumors in natural fish populations exposed

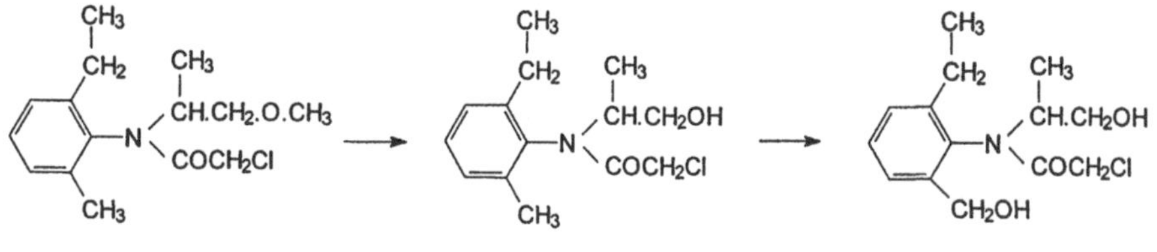

FIGURE 7.3
Metabolism of metolachlor by bluegill sunfish.

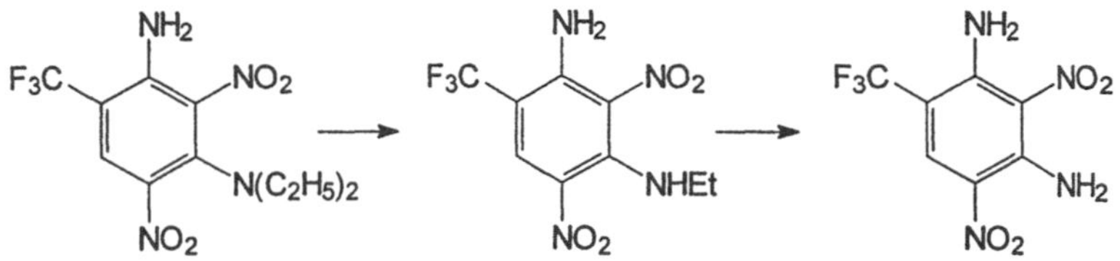

FIGURE 7.4
Metabolism of dinitramine by carp.

to PAHs that has been noted in the introduction to Section 7.5 and briefly in Section 7.7.2). An extensive compilation of the transformation of xenobiotics by fish has been given (Sijm and Oppenhuizen 1989) and only a few examples of these reactions are summarized here by way of illustration

1. *N*-Dealkylation of dinitramine to 1,3-diamino-2,4-dinitro-6-trifluoro-methylbenzene (Olson et al. 1977) by carp (*Cyprinus carpio*) (Figure 7.4).
2. *O*-Demethylation of pentachloroanisole in rainbow trout (Glickman et al. 1977), and of chlorinated veratroles by zebra fish (Neilson et al. 1989).
3. Acetylation of 3-amino ethylbenzoate in rainbow trout (Hunn et al. 1968)
4. Displacement of the nitro group in pentachloronitrobenzene by hydroxyl and thiol groups (Bahig et al. 1981) (Figure 7.5) in golden orfs (*Idus idus*).

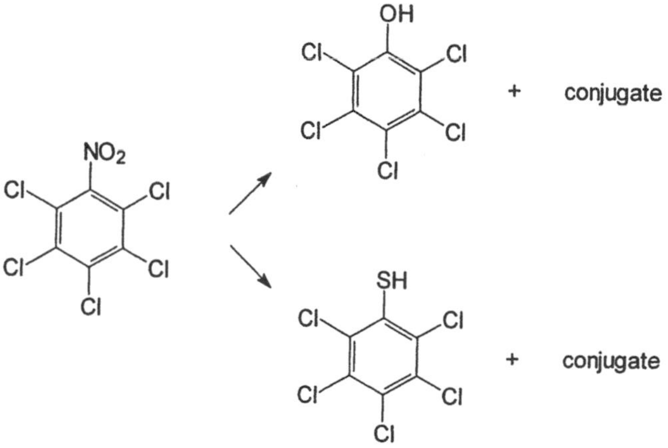

FIGURE 7.5
Metabolism of pentachloronitrobenzene by golden orfs.

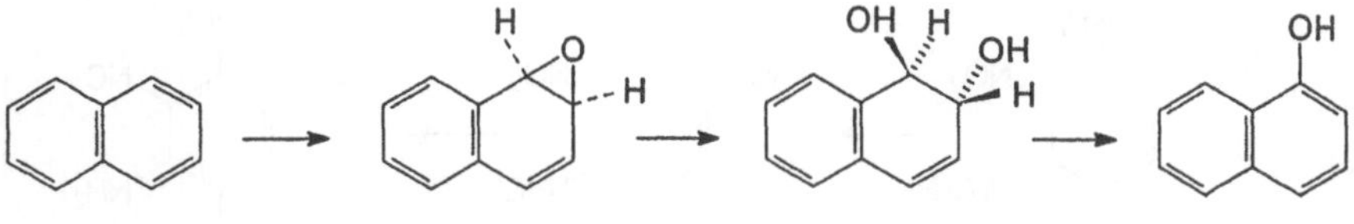

FIGURE 7.6
Metabolism of naphthalene by coho salmon.

5. Oxidation of a number of PAHs has been demonstrated in a variety of fish. A review directed to metabolism and the role of PAH metabolites in inducing tumorogenesis has been given (de Maagd and Vethaak 1998), and only two examples are given here.

    a. Coho salmon (*Oncorhynchus kisutch*) metabolized naphthalene to a number of compounds consistent with oxidation to the epoxide, hydrolysis to the dihydrodiol, and dehydration of the *trans*-dihydrodiol to naphth-1-ol (Figure 7.6) (Collier et al. 1978).

    b. For the carcinogen benzo[*a*]pyrene, a much wider range of metabolites has been identified in southern flounder (*Paralichthys lethostigma*) including the 4,5-, 7,8-, and 9,10-diols, the 1,6-, 3,6-, and 6,12-quinones, as well as the 1-, 3-, and 9-benzopyreneols (Figure 7.7) (Little et al. 1984).

6. *N*-Hydroxylation of aniline and 4-chloroaniline by rainbow trout to hydroxylamines that could plausibly account for the subchronic toxicity of the original compounds (Dady et al. 1991).

Initially formed polar metabolites such as phenols and amines may then be conjugated to terminal metabolites that are excreted into the medium. For

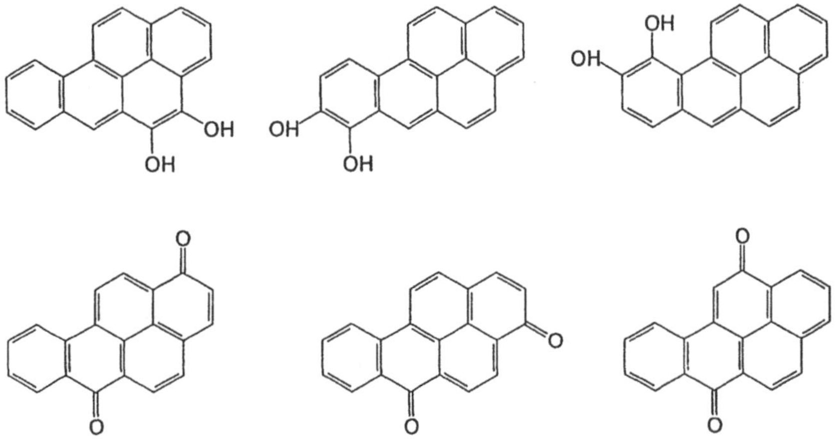

FIGURE 7.7
Metabolism of benzo[*a*]pyrene by southern flounder.

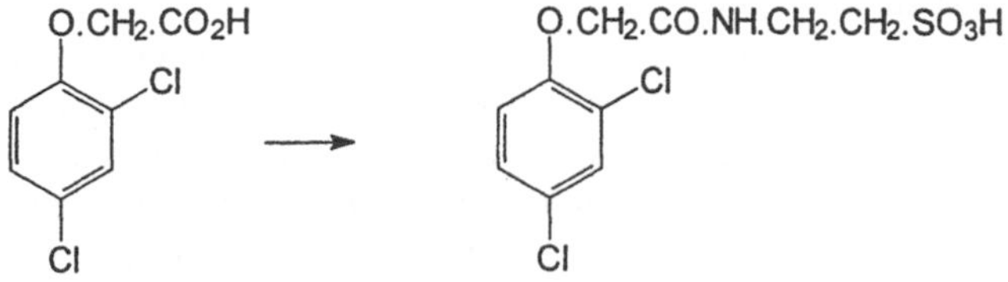

O.CH$_2$.CO$_2$H O.CH$_2$.CO.NH.CH$_2$.CH$_2$.SO$_3$H

$\longrightarrow$

FIGURE 7.8
Conjugation of carboxylic acids with amino acids.

example, pentachlorophenol and pentachlorothiophenol produced from pentachloronitrobenzene were conjugated to yield the major metabolites; although the naphthalene dihydrodiol was the major metabolite produced from naphthalene, the further transformation product naphth-1-ol was also isolated as the sulfate, glucuronic acid, and glucose conjugates. Conjugation of xenobiotics is extremely important since this reaction results in the production of highly water-soluble products that are then excreted from the fish: this reaction therefore functions as an effective detoxification mechanism. Diverse conjugation reactions have been described including the following:

1. Conjugation of phenolic compounds with formation of glucuronides, sulfates, or glucosides as noted above.

2. Reaction of carboxylic acids with the amino groups of glycine (Huang and Collins 1962) or taurine (Figure 7.8) (James and Bend 1976) to form the amides.

3. Reaction between glutathione and reactive chloro compounds such as 1-chloro-2,4-dinitrobenzene (Niimi et al. 1989) or the chloroacetamide group in demethylated metolachlor (Cruz et al. 1993).

One additional consequence of such reactions is that experiments designed to measure bioconcentration may be seriously compromised and this will be the case particularly with compounds which are metabolized at rates comparable with the rate of uptake from the aqueous phase. This has already been discussed in Section 3.1.5, but three examples will be used here to illustrate the care which should be exercised even with groups of apparently recalcitrant compounds.

1. The lower chlorinated dibenzo[1,4]dioxins, and even 2,3,7,8-tetrachlorodibenzofuran (Opperhuizen and Sijm 1990) were apparently metabolized by aquatic organisms so that values of their BCFs could not be predicted on the basis of their P_{ow} values.

2. The discrepancy between P_{ow} and observed BCF values for a series of chloronitrobenzenes in rainbow trout (Niimi et al. 1989) would be consistent with their metabolism, for example, by glutathione conjugation, since the discrepancy was greatest for the more highly chlorinated and, therefore, more reactive congeners.

3. The results of experiments on the bioconcentration of azaarenes in fathead minnows showed significant differences between the observed bioconcentration factors for some azaarenes and the values predicted from their P_{ow} values: this was attributed to metabolism of the test substances that was significant for benz[*a*]acridine and dibenz[*a*,*h*]acridine but not for acridine itself, or for quinoline (Southworth et al. 1980). This could plausibly be correlated with the higher π-electron density of the rings in the benzo compounds.

The metabolic capability of fish has found two complementary applications in monitoring studies.

1. Analysis of xenobiotic conjugates in fish bile has been used to demonstrate the exposure of fish to different classes of xenobiotics;

   a. To conjugates of chlorophenolic compounds—although not for their quantification (Oikari and Kunnamo-Ojala 1987; Wachtmeister et al. 1991; Brumley et al. 1996);

   b. To conjugates of metabolites of aromatic components of oil after a spill to identify the origin of the oil (Krahn et al. 1992): specific use of the metabolites of alkylated dibenzothiophenes (Figure 7.9) was suggested as a valuable indicator of the source of the oil.

2. Biomarkers including the level of monooxygenase enzymes involved in metabolizing xenobiotics have been used to demonstrate the exposure of biota to xenobiotics and this application is discussed in greater detail in Section 7.6.1.

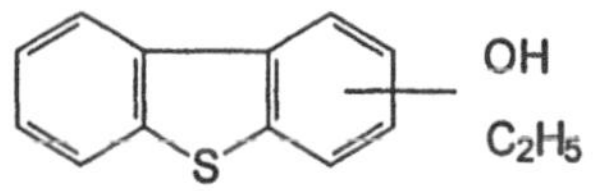

FIGURE 7.9
Alkylated dibenzothiophenes suggested as markers.

7.5.2 Metabolism by Other Higher Aquatic Organisms

1. It has been generally assumed that mussels do not carry out more than the limited reactions of oxidation and conjugation, although variations between summer and winter levels for both cytochrome P-450 and NADPH-independent 7-ethoxycoumarin *O*-deethylase have been found in the common mussel *Mytilus edulis* (Kirchin et al. 1992). Levels of cytochrome P-450 and the rates of metabolism of PAHs are apparently low compared with those found in fish (Livingstone and Farrar 1982), and an investigation using subcellular extracts of the digestive glands from the mussel *M. galloprovincialis*

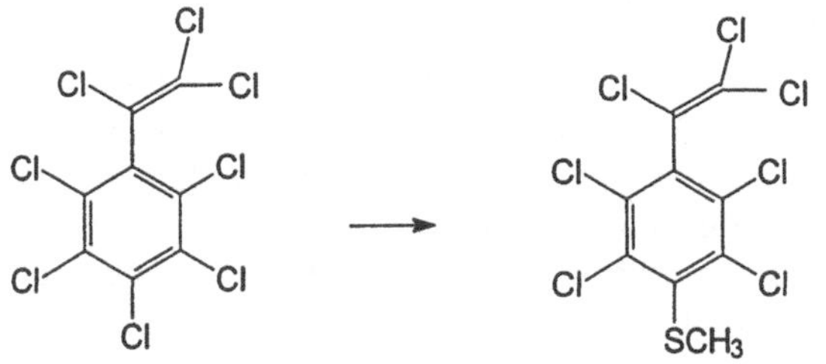

FIGURE 7.10
Metabolism of octachlorostyrene by the blue mussel.

showed that although the formation of diols and phenols from benzo[*a*]pyrene was dependent on NADPH, the quinones that were the major metabolites were produced in the absence of NADPH apparently by radical-mediated reactions involving lipid peroxidase systems (Michel et al. 1992).

2. A reaction presumably mediated by glutathione *S*-transferase is the replacement of the 4-chloro substitutent in octachlorostyrene in the blue mussel (*M. edulis*) by a thiomethyl group (Figure 7.10) (Bauer et al. 1989). A similar reaction of glutathione with arene oxides produced by aquatic mammals from PCBs and DDE results ultimately in the production of methyl sulfones (Bergman et al. 1994; Janák et al. 1998). In the arctic food chain, arctic cod (*Boreogadus saida*)–ringed seal (*Phoca hispidus*)–polar bear (*Ursus maritima*), it has been shown that levels of the dimethyl sulfones of DDE and PCB were low in cod, and that levels in polar bear were the combined result of bioaccumulation from seals and endogenous metabolism (Letcher et al. 1998).

3. The herbicide alachlor is transformed by chironomid larvae and proceeds by *O*-demethylation followed by loss of the chloroacetyl group to produce ultimately 2,6-diethylaniline (Figure 7.11) (Wei and Vossbrinck 1992).

The results of these investigations suggest that caution should be exercised in interpreting not only the results of toxicity assays in which such organism are employed but also data accumulated in monitoring studies that may not have taken into account the existence of metabolites.

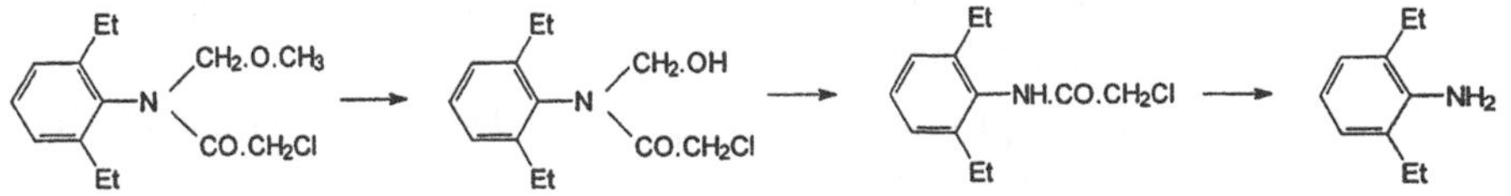

FIGURE 7.11
Metabolism of alachlor by chironomid larvae.

7.5.3   Metabolism and Dissemination by Birds

It seems appropriate to add some brief comments on birds since it was the attention directed to the effect of agrochemicals on these which effectively mobilized public awareness of the environmental hazard of chemicals. Apart from the adverse effects of toxicants on birds, they may also play an important role in the dissemination of xenobiotics and their metabolites produced by endogenous metabolism. Although birds may possess lower levels of microsomal monooxygenase enzymes which are involved in the metabolism of xenobiotics than fish (Walker 1983; Ade et al. 1984), like mammals they seem to have phenobarbital-inducible systems that mediate the metabolism of important PCB congeners (Buhler and Williams 1988).

1. In some aquatic ecosystems, fish-eating birds may be the top predator and may thereby be exposed to a variety of xenobiotics. For example, there is continued concern that the existence of a variety of deformities among fish-eating water birds in the Great Lakes is causally associated with exposure to xenobiotics (Giesy et al. 1994).

2. In terrestrial ecosystems, there is extensive data suggesting that some agrochemicals are more toxic to birds than to mammals. In addition, terrestrial worms may accumulate xenobiotics or their metabolites and thereby transfer these compounds to birds. For example, the concentration in earthworms has been demonstrated both of chlorophenols (van Gestel and Ma 1988) and of their microbial transformation products pentachloroanisole and 2,3,4,6-tetra-chloroanisole (Palm et al. 1991). The consumption of these compounds by predators such as birds is an obvious possibility for their further dissemination.

Attention is briefly be directed to the results of different kinds of investigation.

1. The existence of residues of PCBs in cormorants (*Phalacrocorax carbo sinensis*) captured on the German North Sea coast (Scharenberg 1991), of PCBs and DDT in samples of muscle and in eggs of various birds from Sweden (Andersson et al. 1988), and of chlorinated benzenes in herring gulls from the Great Lakes (Hallett et al. 1982) supports the widespread dissemination of these compounds to which the birds must have been exposed through consumption of contaminated food. This has been discussed in the context of biomagnification in Chapter 3, Section 3.5.4.

2. Evidence of porphyria in herring gulls (*Larus argentatus*) in the Great Lakes (Fox et al. 1988) has been associated with exposure to organochlorine compounds that may plausibly be assumed to have originated in fish.

3. By comparing rates of elimination of 2,3,7,8-tetrachloro-dibenzo[1,4]dioxin in egg-laying and nonlaying hens of ring-necked pheasants (*Phasianus colchicus*), it was shown that egg laying was an important route of elimination of the toxicant, and could contribute up to ~35% of the total amount of the toxicant that was administered. There was no evidence of metabolism during the experiment, and the toxicant was found only in the yolk of the eggs (Nosek et al. 1992). The consumption of eggs by predators therefore would expose them to the xenobiotic.

In addition, migratory birds may be exposed via their food to diverse xenobiotics from widely separated geographic areas. Although numerous questions therefore remain to be answered, sufficient evidence for concern clearly remains.

7.5.4 Metabolism by Invertebrates

The metabolism of xenobiotics by both terrestrial and sediment-dwelling biota has been studied, and provides illustrations of the importance of uptake by food or by sorbed sediment. Some examples of metabolism by terrestrial biota are briefly noted here.

Pyrene is metabolized by the terrestrial isopod *Porcellio scaber* with the formation of 1-hydroxypyrene (Stroomberg et al. 1996), and some sediment-dwelling polychaetes have the potential for metabolizing xenobiotics. For example, *Nereis virens* is able to metabolize PCBs (McElroy and Means 1988) and a number of PAHs (McElroy 1990), while *N. diversicolor* and *Scolecolepides viridis* are able to metabolize benzo[a]pyrene (Driscoll and McElroy 1996). It is worth noting that apart from excretion of the toxicant, polar and much more water-soluble metabolites such as the glycosides formed from pyrene by *Porcellio* sp. (Larsen et al. 1998) may be mobile in the interstitial water of the sediment phase.

7.5.5 Metabolism by Higher Plants

In earlier discussions of the uptake (Chapter 3, Section 3.1.3) and toxicity (Sections 7.3.1.3 and 7.3.6) of xenobiotics to higher plants, it has been pointed out that attention should be given to their metabolic potential (Section 4.3.7). Although the metabolism of xenobiotics by higher plants that are used for toxicity assays in the terrestrial environment seems to have been examined less frequently, attention is drawn to their metabolism for several reasons: (1) the metabolites may be toxic to biota at higher trophic levels and (2) the metabolites may be translocated into the root system, and thence after partition into interstitial water in the soil exert a toxic effect on other terrestrial organisms. For example, the leaves of trees along a heavily trafficed road in Japan contained not only pyrene and 1-hydroxypyrene, but also the β-O-glucoside and β-*O*-glucuronide conjugates (Nakajima et al. 1996), and it has been shown that 1-hydroxypyrene is highly toxic to the nematode *Caenorhabditis elegans* (Lambert et al. 1995). Both these important issues should be considered particularly in the application of higher plants to bioremediation (Section 8.1.1).

7.6 Biomarkers: Biochemical and Physiological End Points

Concomitant with an appreciation of the importance of sublethal effects of xenobiotics on natural populations, there has been increased interest in alternative procedures for assessing these effects. Considerable interest has centered on the application to fish of procedures developed in clinical medicine: these have used biochemical assays for specific enzyme activities and physiological parameters in blood and serum samples. The term *biomarker* has been used, but it should be pointed out that this has also been applied in a completely different context to compounds isolated from samples of sediment, coal, and oil and which plausibly have a biological origin (Simoneit 1998). There seems, however, little reason for confusion in the application of the same term in these widely different contexts.

7.6.1 Biochemical Parameters

Monooxygenase Activity

The previous discussion has illustrated the role of monooxygenase systems in fish for the metabolism of xenobiotics. Measurements of this activity have therefore been used as a measure of the extent to which fish have been exposed to xenobiotics: at the same time, of course, increased levels enable the fish to metabolize xenobiotics effectively (Kleinow et al. 1987). Although specific assays of cytochrome P-450 activity may be made by immunoblot methods (Monosson and Stegeman 1991), it may be expedient to measure specific enzyme activity using defined substrates. Two assays have been widely used: (1) aryl hydrocarbon hydroxylase activity that may be assayed using benzo[*a*]pyrene as substrate, although this substrate has been replaced recently by the less hazardous 2,5-diphenyloxazole, and (2) activity for *O*-deethylation of 7-ethoxyresorufin (EROD) has been extensively used and is a simple and convenient assay.

In interpreting the results of such assays, however, a number of important factors should be considered.

- Induction by exposure to a given xenobiotic may be strongly influenced by previous exposure to other environmental toxicants (Monosson and Stegeman 1991).

- There are several isoenzymes of cytochrome P-450 each of which is specific for a given class of substrate.
- Environmental factors such as the temperature and the feeding regime may influence EROD activity (Jimenez et al. 1988), and both the rate of induction and the levels of activity that are attained may be influenced by the temperature at which fish are maintained (Andersson and Koivusaari 1985). This is consistent with the role of temperature on toxicity that has already been noted (Hodson and Blunt 1981).

Conjugating Enzymes

Other metabolic enzymes have also been used in assessing exposure to toxicants. These include conjugation enzymes that are important in the metabolism and depuration of xenobiotics: for example, an assay for uridine diphosphoglucuronosyl transferase (Andersson et al. 1985) has been used quite extensively in Sweden in conjunction with assays for EROD activity. Two important factors in interpreting the results of such assays are worth pointing out.

1. The pattern of induction of metabolizing and of conjugating enzymes may be quite different, even though, for example, cytochrome P-450 inducers are capable of inducing both glutathione transferase and UDP glucuronosyl transferase activities (Andersson et al. 1985).
2. Levels of UDP glucuronosyl transferase activity after induction with phenolic compounds differed considerably among various fish (Förlin et al. 1989).

Vitellogenin

This is a female egg yolk protein, and its radioimmunoassay has been used to determine possible estrogenic effects in fish exposed to compounds that were isolated and quantified from sewage treatment systems (Section 2.2.3). Rainbow trout (*Oncorhynchus mykiss*) and roach (*Rutilis rutilis*) were exposed during 3 weeks to 17β-estradiol, estrone, and 4-*t*-octylphenol. Effects in trout were observed at concentrations of 100 ng/l of 17β-estradiol and estrone, and in male roach at 100 ng/l of estrone (Routledge et al. 1998). It should be emphasized that values for 4-*t*-octylphenol were from 1/100 to 1/1000 those found for 17β-estradiol. Concentrations in the effluents ranged from 2.7 to 48 ± 6 ng/l for 17β-estradiol, and 1.4 to 76 ± 10 ng/l for estrone, so that the observed concentrations are capable of inducing synthesis of vitellogenin that has been shown to be a sensitive biomarker for the weak estrogenic activity of alkylated phenols, Arochlor 1221, bisphenol A, and 2,4'-DDT (Sumpter and Jobling 1995). An *in vivo* assay has been developed (Tyler et al. 1999) that used fathead minnow (*Pimephales promelas*) and an ELISA system for analysis based on carp (*Cyprinus carpio*). Levels of vitellogenin were significantly increased after exposure to 50 ng/l 17-β-estradiol from 24 h post fertilization to 30 day posthatching or to 100 ng/l from 24 h postfertilization to 10 day posthatching.

Other Enzymatic Assays

1. An assay for plasma leucine aminonaphthylamidase has been examined as a measure of stress in rainbow trout, but variability in the levels due, for example, to diet combined with the rather low sensitivity of the assay do not seem encouraging for its widespread application especially to field samples (Dixon et al. 1985).

2. Metallothionein is a metal-binding protein whose level generally increases in response to exposure to metals (Hennig 1986), although the demonstration that its level is also increased by exposure to hydrocarbon fuel oil even in the absence of metals illustrates the care that must continuously be exercised in interpreting data on enzyme levels (Steadman et al. 1991).

A study in which biochemical indicators were used to assess the effect of hydrocarbon fuel oil on rainbow trout clearly revealed a number of factors which could compromise interpretation of the results (Steadman et al. 1991) and these seem of sufficient general importance to merit brief summary.

1. As a result of the different toxicities of components contained in the mixture of toxicants, enzyme levels may not exhibit a dose–response.

2. Enzyme levels may be affected by the exposure route and by chronic exposure to toxicants, and these possibilities have not generally been systematically examined.

3. At high levels of exposure to a toxicant, fish may no longer synthesize increased levels of detoxification enzymes, but respond by displaying liver hypertrophy and a reduction in spleen size.

4. The increased levels of metallothein in the putative absence of metals to which reference has already been made.

7.6.2 Physiological Parameters

These have been chosen to reflect disturbances in function, and some have been based on those which are routinely used in clinical medicine. Some examples of those that have been most widely used include the following.

1. Effect on liver function: liver somatic index, level of ascorbic acid, together with enzyme levels, e.g., EROD noted above;

2. Gonad growth: gonad somatic index;

3. Carbohydrate metabolism: glycogen content of liver and muscle, glucose and lactate levels in blood;

4. Osmotic and ion regulation: concentrations of Na^+, K^+, Ca^{2+}, Mg^{2+}, and Cl^- in blood plasma;

5. Status of blood cells: red and white blood cell count, hematocrit, methemoglobin.

Many of these parameters have been used to assess the effect of industrial effluents on both laboratory-reared and feral fish (Andersson et al. 1987; Härdig et al. 1988; Larsson et al. 1988). It should, however, be appreciated that the statistical evaluation used in clinical trials to establish cause and effect cannot generally be incorporated into the design of these experiments. Attention is drawn to the established placebo effect in clinical trials and the sometimes counterintuitive significance of randomness (Bennett 1998)

Respiratory-cardiovascular responses in rainbow trout have been used to assess the effects of toxicants, and the results with two phenols and three anilines indicated different narcosis syndromes which in the case of phenol could be reversed by exposure of the fish to uncontaminated medium (Bradbury et al. 1989).

In addition, analysis of hormone levels in serum samples of white sucker (*Catostomus commersoni*) in Lake Superior have been used to complement measurements of physiological response: significant differences in the levels of testosterone and estradiol in females, and of testosterone levels in males have been associated with pollution by bleachery effluents (Munkittrick et al. 1991). It has been observed that feral fish captured from areas exposed to established contamination by PAHs, PCBs, and mercury did not exhibit the increased levels of hydrocortisone normally resulting from capture (Hontela et al. 1992). These results were interpreted as showing the adverse effect on sterol metabolism in fish chronically exposed to such pollutants. Collectively, such data draw attention to the subtler effects of exposure to xenobiotics.

A word of caution should be inserted on the difficulties of drawing unequivocal correlations between effects observed in feral fish and the levels of their exposure to putative toxicants. The data are often interpreted in terms of distance from an established point discharge—sometimes supplemented with analyses for specific toxicants (Goksöyr et al. 1991). Feral fish may, however, be exposed—simultaneously or consecutively—to a number of potential stresses which may be reflected in their physiological and biochemical response; the significance of these different factors may not always be sufficiently carefully evaluated. A good illustration of the difficulties is afforded by observations on the occurrence of skeletal deformities in smallmouth bass

(*Micropterus dolomieui*) which could not be correlated with exposure to any of the toxicants known to induce such symptoms (Van den Avyle et al. 1989): the cause of the effect was not established, and the results clearly underscore the need for a deeper understanding of fish physiology. The whole issue of associating observed biological effects with presumptive exposure to toxicants is discussed more fully in Section 7.7.2.

7.6.3 Chromosomal Alterations

Attention has already been briefly drawn to the use of morphological alterations of the polytene chromosomes in the salivary glands of larval *Chironomus tentans* (Bentivegna and Cooper 1993) as an assay for genotoxicity. This could be applied equally to field samples although care would have to be exercised concerning the variability in natural populations which may contain several species of *Chironomus*.

7.7 A Wider Perspective

7.7.1 Toxic Equivalent Factors (TEF)

Industrial effluents generally contain a range of structurally diverse components, while commercial products such as PCBs, polychlorinated bornanes, chlordane, and polychlorinated alkanes comprise a range of congeners. These effluents contain components which have widely differing toxicities so that the effective toxicity of the effluent or a commercial product is a function both of the relative concentrations of the components and their individual toxicities. To meet such situations, the concept of TEFs has been introduced, and has been developed for PCBs (Safe 1987), PCDDs and PCDFs (Kutz et al. 1990), and PAHs (Delistraty 1998). Briefly, each component is assigned a relative toxicity using a given organism or test system and a reference compound: the concentrations of the components in the mixture or effluent are then multiplied by this factor, and the sum of the products for all the components is used to evaluate the toxicity of the product or effluent. There is, however, at least one important issue that should be clearly appreciated: the numerical values assigned are valid only for the organism used for assessment. For example, if the assessment of toxicity to humans is the objective, toxicity factors should be obtained from values for human toxicity: in this case, of course, the use of surrogate organisms is acceptable provided that the inherent limitations are clearly appreciated. As was wisely stated (Pochin 1983) in the context of the risk to humans of exposure to radionuclides: "I don't know whether we are closer to the dog, the mouse, or the rat in terms of lymph node behavior." The same limitation surrounds the use of such equivalency factors in environmental hazard assessment, and attention should be directed to the differing morphologies, physiologies, and biochemistry of the test organisms: at the very least, TEFs in environmental studies should be assigned on the basis of toxicity results at least for algae, crustaceans and fish, and take into account possible differences in the sensitivity toward individual representatives of these groups. Two different examples are given as illustration.

1. Carcinogenic Effects of PAHs in Mammals

This has been extensively reviewed by Delistraty (1998) who provides discussion of many salient aspects including the uncertainty and validation of the values. Only some brief comments on cardinal issues are given here: (a) the choice of end point and the dose–response model; (b) the fact that PAHs exert their capacity for tumor induction only after metabolism; (c) the dependency both on the animal used for assay and on the tissue that is examined, for example, lung or skin; and (d) the importance of the exposure route, oral, dermal, intraperitoneal, or intrapulmonary.

There have been numerous listings of TEFs for PAHs, and only one of these is used here for illustration. The values of the TEFs in Table 6 of Delistraty (1998) are compared with the concentrations of the components in a liquid coal tar in Table 3 of Howson and Jones (1998). Even if too great an emphasis is not placed on the numerical values, it is clear that there is no relation between the TEF values and the concentrations of PAHs in a tar sample. This is particularly relevant to the question of remediation of terrestrial sites contaminated with PAHs, and since, for example, the hexacyclic dibenzopyrenes that may be present in only low concentration are highly carcinogenic.

PAH	TEF	Concentration in a Liquid Coal Tar (g/kg)
Benzo[a]pyrene	1.0	3.6
Benz[a]anthracene	0.1	4.5
Chrysene	0.01	4.3
Pyrene	0.001	8.5
Fluoranthene	0.001	11.4

2. Toxicity of Chlorophenolic Compounds in Aquatic Biota

A commendable study (Kovacs et al. 1993) has addressed the application of the concept to a number of chlorophenolic compounds and has taken advantage of subacute data from various organisms, including different species of fish, and from the sea urchin sperm assay. The authors carefully point out that the values proposed should be regarded as provisional and that further refinements should take into account the position of substituents and of possible differences in the sensitivity of the test organisms.

There are both merits and demerits in the application of this concept: on the positive side, undeserved attention would not be directed to components existing in low concentration unless they were highly toxic, whereas on the negative side, components of low toxicity—although present in relatively high concentration—might not receive the attention they rightly deserved. In addition, an apparent contradiction may result from the use of different determinants of environmental impact. A good illustration is provided by the more highly chlorinated (six or more chlorine substituents) PCBs and PCDDs which are present in high concentration in some kinds of environmental samples: although these are relatively less toxic than congeners with lower degrees of chlorination, they are apparently highly persistent to microbial attack and to metabolism in higher biota.

Although this procedure may be the only rational approach for assessing the toxicity of complex effluents, it is imperative to be aware of the fact that TEFs are critical functions of the organisms used for their assessments, and that the values obtained using one organism will not necessarily—or even can reasonably—be applicable to any other. Compared even with organisms within the same group, a single taxon may be highly sensitive to a particular toxicant.

7.7.2   Effect and Cause: Ecoepidemiology

It is highly desirable to recognize in advance a potential hazard and thereby prevent the occurrence of seriously adverse effects on ecosystems, and the previous discussions have been directed to procedures that attempt to predict adverse effects of a compound prior to its release into the environment. To provide a scientific basis for restricting the use of a compound, it may, however, be necessary to ascertain the probable cause or causes of environmental perturbations that have already been observed even though they have not been attributed to a specific cause. The complexities involved in determining causative factors for the declining population of amphibians and the relative role of natural fluctuations and human impacts have been critically reviewed (Pechmann and Wilbur 1994), and the analysis merits extension to other biota. The biological effects may indeed be highly visible and quite dramatic: the following examples illustrate the range of biota that have been affected.

1. The extensive deaths of sea mammals including bottlenose dolphins (Sarokin and Schulkin 1992) and several species of seals (Visser et al. 1990);

2. The decline in the population of rockhopper penguins (*Eudyptes chrysocome*) on Campbell Island, New Zealand (Moors 1986);

3. The incidence of tumors in feral fish populations putatively exposed to PAHs (Harshbarger and Clark 1990; de Maagd and Vethaak 1998);

4. The blooming of algae which may produce toxins (Underdal et al. 1989);

5. The incidence of porphyria in fish-eating herring gulls (*Larus argentatus*) in the Great Lakes (Fox et al. 1988);

6. The incidence of abnormalities in amphibians in many parts of the world including pristine areas (La Clair et al. 1998).

Some examples of the difficulties of correlating effects with exposure to specific toxicants have already been given in Sections 7.5 and 7.6, and attention is directed particularly to a careful review that examines the evidence for the adverse effect of exposure to organochlorine compounds on the reproductive success of various populations of different marine mammals (Addison 1989). This activity may conveniently be termed *ecoepidemiology* by analogy with the procedures used in human disease control and prevention. It should be clearly appreciated that there are inherent difficulties in carrying out such an analysis, with the result that it may be impossible to establish an absolute correlation between the observed effect and the putative cause. This is clearly illustrated by investigations into the possible adverse effect on humans who consume fish from the Great Lakes that are contaminated with organochlorine compounds (Swain 1991). Although levels of Pb, Cd, PCB, and DDT in humans are higher in fish-consuming groups, the increased levels of Pb and Cd were apparently influenced primarily by differences in lifestyle rather than by fish consumption (Hovinga et al. 1993). Some of these complicating issues are explored in the next few paragraphs.

Epidemiology is concerned with the health of human populations and may be defined (Last 1983) as "the study of the distribution and determinants of disease and health-related status in populations and related phenomena in human population groups." This is readily translated into the corresponding activity in ecosystems and is then designated ecoepidemiology. It is, however, valuable to make a few brief comments on epidemiology itself since this is a highly developed discipline that has been conveniently formalized (Susser 1986).

Provided that the correlation between the exposure and the observed effect is sufficiently strong, valuable—although temporary—measures to alleviate a problem may be possible without a mechanistic or rational understanding of the cause. A good example quoted by Fox (1991) was the recognition by Snow that the outbreak of cholera in London during 1848 to 1854 was the result of drinking sewage-polluted water, even though the organism responsible was not isolated until many years later by Koch. Similarly, although the investigations of Potts in 1775 demonstrated the association between the frequency of scrotal cancer in chimney sweeps and their occupation, the underlying association with exposure to PAHs was not revealed until the early years of this century. In both of these cases, however, the exposed population was either confined to a restricted geographic locality or to a specific occupation. This is, however, seldom the case with environmental pollutants, many of which now have a global distribution: the difficulty of establishing a correlation between exposure and observed effect is therefore greatly exacerbated.

The ecoepidemiological approach does in fact accept that it may be impossible to establish a strictly causal relationship between the observed effect and the compound putatively responsible. The objective is to establish how persuasive the correlation is, and the degree to which it is legitimate to exclude alternative explanations. A few specific examples of the difficulties may be used as illustration at this stage: these are taken from a review by Sarokin and Schulkin (1992) which includes other interesting examples.

At least three situations may be distinguished.

1. A causal relation between field observations and exposure to a xenobiotic may be supported by laboratory studies that show the induction of the symptoms on exposure of the appropriate organisms to the supposed toxicant. An example of this is provided by the occurrence of liver tumors in feral fish (Harshbarger et al. 1990) which has been associated with exposure to PAHs (Malins et al. 1985; Myers et al. 1991; de Maagd and Vethaak 1998). This correlation is strongly supported by extensive information on the mechanism of carcinogenesis induced by these compounds both in fish (Hawkins et al. 1990) and in other organisms as well as by the specific nature of the tumors. On the other hand, an extensive statistical study (Johnson et al. 1993) of hepatic lesions in winter flounder (*Pleuronectes americanus*) captured from 22 sites on the northeast coast of the United States illustrates the complexities that may be encountered. The single most important biological factor was age, while PAHs, DDT, and chlordanes—but not PCBs—emerged as important risk factors, although it was not possible to evaluate the quantitative contribution of each of them since they were generally present simultaneously. The practical limitation to the further application of this approach is that whereas it is readily carried out using, for example, fish, it cannot be extended to the large sea mammals that have understandably attracted so much concern.

2. The immediate causative agent may have been identified, but the effect may be suspected of being exacerbated by other simultaneous exposures. A good illustration of this is provided by the extensive death of harbor seals (*Phoca vitulina*) in the North Sea. This has been attributed to infection by a phocine distemper virus belonging to a new member of the genus *Morbillivirus* (Visser et al. 1990), although susceptibility to infection may in fact be the ultimate link in a chain of events triggered by exposure of the seals to pollutants such as PCBs. The possible significance of this factor is supported

by the observations that seals fed with PCB-contaminated fish had decreased levels of plasma retinol and thyroxins compared with seals fed with fish containing only low levels of PCB (Brouwer et al. 1989). By analogy with the pathogenic microorganisms that have serious consequences for compromised human beings, these organisms may be termed *opportunistic pathogens*. Attention has already been drawn in Section 7.5.1 to the increase in the level of metabolites of benzo[*a*]pyrene in the liver of rainbow trout exposed to PCBs. A critical review of these and other associations has been given (Addison 1989).

3. Natural ecosystems are complex with regulatory controls that are frequently only partly understood, so that it may be impossible to establish a correlation between the observed effect and a single cause; there may be several interacting factors none of which alone is sufficient. Three examples are given to illustrate a situation that is probably widespread as a result of the innate complexity of biological control mechanisms and of the ubiquity of many xenobiotics in several environmental compartments and in geographically separated areas.

   a. The occurrence of toxic blooms of algae may be due to a number of interacting factors including, for example, pollution, input of humus, acid rain, or temperature, and it is very difficult to establish a correlation with any single one of these to the exclusion of the others: introduction of preventive measures to prevent their occurrence is therefore extremely difficult.

   b. As a result of the investigations into the cause of the massive deaths of bottlenose dolphins (Sarokin and Schulkin 1992), at least three possible—and widely different—explanations have been advanced: (i) massive systemic bacterial infection by opportunistic pathogens, (ii) ingestion of the toxin produced by the alga *Ptychodis brevis*, or (iii) impaired function of immune systems following exposure to toxicants and to PCBs in particular.

   c. The decline in populations and the occurrence of deformities in frogs from various parts of the world have been attributed to a number of factors. These include (1) protozoan infections or increased UV radiation (La Clair et al. 1998), (2) deformities in the toad *Xenopus* sp. that have been attributed to phototransformation products of the insect growth regulator *S*-methoprene (La Clair et al. 1998) (3) deformities in Pacific tree frogs (*Hyla regilla*) as a result of infection with the trematode *Ribeiroia* sp. (Sessions et al. 1999; Johnson et al. 1999) (4) death of frogs in Australia and Panama, (Berger et al. 1998) from infection by the chytrid fungus *Batrachochytrium dendrobatidis* (Longmore et al. 1999). The possibility remains, however, that the resistance of the frogs was, in some way compromised by environmental factors.

d. Changes in natural populations may be the result of chain of circumstances that are not necessarily directly associated with the adverse effects of xenobiotics. As an example, the intrusion of killer whales as predators of sea otters in western Alaska with resulting increase of sea urchins and diminishing kelp beds may be the result of a long chain of effects beginning with decreased abundance of prey for pinnipeds, the reasons for which are unknown (Estes et al. 1998).

This indeterminacy will probably have to be accepted in most cases since there are too many interacting factors, and the determination of a single factor may simply lie beyond the resolution of current experimental science: for example, the lack of agreement over the effect of greenhouse gases, the uncertainties in estimating the global carbon balance, and the existence of large carbon sinks illustrate our current lack of quantitative understanding of the basic processes. All ecosystems are dynamic, and it is therefore difficult to assess over a long period of time the natural perturbations and oscillations that occur.

An additional complicating factor arises from the background level of some apparently ubiquitous contaminants whose partitioning and dissemination have been discussed in Chapter 3, Section 3.5. For example, there is a global—although by no means uniform—distribution of a range of organochlorine compounds including hexachlorocyclohexanes, chlordanes, PCBs, and DDT (Iwata et al. 1993) and the same is also true for PAHs. Whereas the ultimate source of many toxicants can be established with a high degree of certainty, this may not always be the case: an example of two groups of compounds that are apparently ubiquitous in the environment may be used as illustration.

1. Analysis of sediment cores from Japan that have been dated to ~8000 years revealed the presence of the more persistent heptachloro and octachloro congeners of dibenzo[1,4]dioxins (Hashimoto et al. 1990), and at present there appears to be a background level of chlorinated dibenzo[1,4]dioxins and dibenzofurans that is not attributable to the production of pulp using molecular chlorine (Berry et al. 1993). Their production may possibly be associated with incineration processes or with industrial activity such as the manufacture of pentachlorophenol. Extensive data suggest that it is only the most highly chlorinated congeners that are persistent in the atmosphere (Hites 1990), and these are therefore the most likely to be recovered from environmental samples except those of recent origin. This study also showed that, in apparent contradiction of the Japanese study, the concentrations of octachlorodibenzo[1,4]dioxin in samples from a remote area (Siskiwit Lake, Isle Royale, Lake Superior) were extremely low before 1930. Studies

of soil samples from Rothamstead, England have been discussed in Chapter 3, Section 3.6.2 and showed that polychlorinated dibenzo[1,4]dioxins and furans existed in all samples from 1881 onward (Alcock et al. 1998). In England, the source of these compounds cannot, however, be determined since substantial volumes of wood and coal for iron smelting and glass production had been used for many years, and chlorine for bleaching had been introduced during the late 18th century.

2. Although PAHs have been recovered from a plethora of samples of sediment, soil, and biota (Howson and Jones 1998), care should be to establish their structure since a wide range of substituted PAHs are not the products of incineration but are of petrogenic and biogenic origin (Simoneit 1998). Analysis of sediment samples from the Mackenzie River delta and the Beaufort Sea illustrated that in a pristine area even though combustion-related PAHs were present, their concentrations were exceeded by those of petrogenic and biogenic origin (Yunker and Macdonald 1995).

Care should therefore be exercised in drawing conclusions on putative causes of biological effects against the background levels of these virtually ubiquitous compounds.

Certain basic criteria in epidemiology (Susser 1986) have been widely accepted and appear to be equally applicable to ecoepidemiology (Fox 1991). It should, however, be clearly appreciated that these techniques are not designed to establish unambiguous causal relationships, but rather to indicate which criteria may most usefully be used to provide a balanced evaluation in support—or in contradiction—of a given hypothesis: it should be noted that the relatively loose term *association* is consistently used. These five criteria are (1) consistency, (2) strength, (3) specificity, (4) temporal relationship, and (5) coherence.

The application of these principles has been explored more extensively by Fox (1991) in his review, and in specific application to the reproductive success of lake trout (*Salvelinus namaycush*) in the Great Lakes (Mac and Edsall 1991). In that investigation, although the evidence suggested that the survival of fry was a significant factor in reproductive impairment, no specific suggestion for the involvement of a significant pollutant was put forward. Even when the *source* of the perturbation has been plausibly identified, the problem of ameliorative measures is by no means simple. For example, there is extensive and fairly compulsive evidence that impaired reproduction in white sucker (*Catostomus commersoni*) may be associated with the exposure of adults to bleachery effluents (McMaster et al. 1992; Van der Kraak et al. 1992); until, however, the responsible component(s) have been identified in the effluents, it is difficult to determine the causative agent and to implement preventive measures. This example provides incidentally a good illustration of the central role of chemical analysis in ecoepidemiology.

The examples noted at the beginning of this section on highly visible perturbations of natural populations and the lack of consensus on their underlying causes illustrate the tentative nature of ecoepidemiology. In spite of this apparent limitation, there are some very attractive features of ecoepidemiology which may be summarized:

1. It will bring to light often diverging—and even contradictory—views and this is entirely consistent with the inherent complexity of almost all natural situations.

2. It will encourage the careful exploration of different aspects of the problem and reveal the possibility that several factors may be complementary rather than mutually exclusive.

3. It will focus attention on the mechanisms that regulate natural populations, and the interactions whereby populations react to alterations in their environment—and it is hoped will attract research effort to this important area.

Examination of all these aspects has been profitable and—possibly even more valuable—may suggest strategies relevant to the resolution of other unresolved environmental issues.

7.7.3 The Structure of the Toxicant and Its Biological Effect

There has been substantial interest in correlating the toxicity of xenobiotics with their chemical structure. Whereas it is unrealistic to expect that widely applicable principles can be formulated, it is has been convincingly demonstrated that within groups of structurally related compounds, valuable correlations exist between their structure and their biological activity. This approach was pioneered by Hansch and his co-workers (Hansch and Leo 1979), and attention is drawn to its extensive application to ecotoxicology (Kaiser 1987; Hermens 1989) while brief mention has been made in Chapter 4, Section 4.7.1 of its application to problems in biodegradability and biotransformation. A brief summary of salient features of the review by Hermens (1989) is given here.

A number of numerical analytical procedures have been used, and advantage taken of a number of determinants. Advantage may be taken not only of electronic and steric factors, but also of physicochemical parameters that may be considered significant in determining both the bioavailability of the toxicant and its mode of action. These include the following: (a) values of P_{ow} as a measure of hydrophobicity, (b) values of the redox potential when oxidoreductases are involved, (c) use of the Hammett σ constant in aromatic compounds which do not contain interacting groups such as OH, NH_2, NO_2, SO_3H, (d) descriptors such as the Taft constant that take into account steric effects, and (e) factors that take into account the overall topology of the molecule.

The procedure is especially effective when data for large numbers of compounds differing only in minor structural features are available, although it is clearly more limited in its application to completely novel structures. Only brief note will be made here of a few salient issues on the relation between the chemical structure of a compound and its toxicity.

1. Attention has already been drawn (Section 7.5) to the fact that a number of compounds exert their biological effect through metabolites that may be considerably more toxic than their precursors. This clearly cannot be taken into account by such procedures.

2. The toxicity may depend critically not only on the number of substituents on aromatic rings but also on their orientation. This is illustrated by data for mammalian toxicity of chlorinated dibenzo[1,4]dioxins and dibenzofurans (Kutz et al. 1990) and for PCBs (Safe 1987). Although this aspect of toxicity has been much less extensively explored for aquatic organisms, there is certainly no reason to doubt its general relevance.

7.8   A Hierarchical System for Evaluating the Biological Effects of Toxicants

It will have become clear that attempts to assess the environmental impact of a toxicant are fraught with difficulties. No single system is sufficiently embracing, no single organism is uniquely sensitive, life-cycle tests with fish and multispecies systems are complex, time-consuming, and expensive, and inherent ambiguities exist in interpreting measurements of biochemical and physiological parameters. The simplest solution is to cut the Gordian knot and take maximum advantage of each system while recognizing its limitations. The various assay procedures could be built into a hierarchical system such as the following:

Stage 1: Acute tests using for example, algae, crustaceans, and fish

Stage 2: Subacute tests including, for example:
  • Reproduction tests using crustaceans and fish
  • Physiological/biochemical indexes in fish
  • Community-induced tolerance in algae

Stage 3: Life-cycle tests with fish including examination of pathologies

Stage 4: Multicomponent systems
  • Microcosms
  • Mesocosms

The four stages could be regarded as constituting an algorithm: for screening purposes, if a toxicant fails at any level, testing at a higher level is not justified. Conversely, if no effect is observed at a lower level, testing should be continued to the next higher level. Since, however, the cost of preexposure tests and multicomponent systems is relatively high compared with those for other tests, at least some of these other single-species tests should be carried out simultaneously as a routine complement. For completeness, assays for carcinogenicity using either fish (Black et al., 1985) or surrogate assays using microorganisms (De Serres and Ashby, 1981) may be included.

It should also be emphasized that difficulties may remain in correlating results from tests at the different stages (Stay et al. 1988). In addition, the boundaries between them should not be sharply drawn so that at stage 2 for example, a high degree of flexibility should exist, and the various alternatives regarded as complementary rather than mutually exclusive: application of sophisticated biochemical/physiological indexes in fish, for example, might justifiably be considered an element in stage 3 systems. It should also be pointed out that a number of additional features might be incorporated in order to provide a more comprehensive environmental hazard assessment in a given situation. Among these features, the following might, for example, be considered for inclusion:

1. Use of species relevant to a specific ecosystem;
2. Use of media with varying salinity, hardness, or pH simulating those of the receiving system;
3. Assessment of bioconcentration potential either directly or by use of surrogate systems which have been discussed in Section 3.1.

7.9   Conclusions

Opinions differ widely on matters of detail, but it may be valuable to provide some general conclusions in an attempt to provide a wider perspective on this complex area.

1. Increasing attention is being directed to the sublethal and chronic effects of xenobiotics, and this means that assay systems are becoming increasingly sophisticated.
2. A wide range of end points is available and attempts should be made to extract the maximal information from the results of a toxicological assay.
3. No single assay will provide data that make it possible to predict the effect of a xenobiotic on a complex ecosystem, and serious attempts should be made to design assays that will reflect important effects of xenobiotics at the molecular level.

4. It is not practical to examine the toxic effects of all compounds in highly sophisticated assays so that a hierarchical approach may be used: the results of the simpler tests together with an assessment of the magnitude of the problem will then determine the extent to which the application of more complex assay systems is justified.

5. The toxicity of a xenobiotic cannot be dissociated from its transport into the organism and its subsequent metabolism: it is experimentally attractive to combine assays of toxicity with assessments of bioconcentration potential and metabolism.

6. The application of biochemical and physiological assays to feral fish is a valuable complement to laboratory tests for toxicity, although care should be exercised in making correlations between the observed responses and exposure to toxicants: a number of compromising factors may be involved.

References

Addison, R.F. 1989. Organochlorines and marine mammal reproduction. *Can. J. Fish. Aquat. Sci.* 46: 360–368.

Ade, P. et al. 1984. Biochemical and morphological comparison of microsomal preparations from rat, quail, trout, mussel, and water flea. *Ecotoxicol. Environ. Saf.* 8: 423–446.

Alcock, R.E., M.S. Mclachlan, A.E. Johnston, and K.C. Jones. 1998. Evidence for the presence of PCDD/Fs in the environment prior to 1900 and further studies on their temporal trends. *Environ. Sci. Technol.* 32: 1580–1587.

Alexander, M. 1995. How toxic are toxic chemicals in soil? *Environ. Sci. Technol.* 29: 2713–2717.

Anderson, B.S. and J.W. Hunt. 1988. Bioassay methods for evaluating the toxicity of heavy metals, biocides and sewage effluents using microscopic stages of giant kelp *Macrocystis pyrifera* (Agardh): a preliminary report. *Mar. Environ. Res.* 26: 113–134.

Anderson, B.S. and nine coauthors. 1998. Comparison of marine sediment toxicity test protocols for the amphopod *Rhepoxynius abronius* and the polychaete worm *Nereis (Neanthes) arenaceodentata*. *Environ. Toxicol. Chem.* 17: 859–866.

Anderson, H.L. and 27 coauthors. 1999. Comparison of short-term estrogenicity tests for femnification of hormone-disrupting chemicals. *Environ. Health Perspect.* 107, Suppl. 1: 89–108.

Anderson, S.L., J.E. Hose, and J.P. Knezovich. 1994. Genotoxic and developmental effects in sea urchins are sensitive indicators of effects of genotoxic chemicals. *Environ. Toxicol. Chem.* 13: 1033–1041.

Andersson, T. and L. Förlin. 1992. Regulation of the cytochrome P450 enzyme system in fish. *Aquat. Toxicol.* 24: 1–20.

Andersson, T. and U. Koivusaari. 1985. Influence of environmental temperature on the induction of xenobiotic metabolism by β-naphthoflavone in rainbow trout, *Salmo gairdneri*. *Toxicol. Appl. Pharmacol.* 80: 43–50.

Andersson, T., M. Pesonen, and C. Johansson. 1985. Differential induction of cyto-chrome P-450-dependent monoxygenase, epoxide hydrolase, glutathione trans-ferase and UDP glucuronyltransferase activities in the liver of the rainbow trout by β-naphthoflavone or Clophen A50. *Biochem. Pharmacol.* 34: 3309–3314.

Andersson, T., B.-E. Bengtsson, L. Förlin, J. Härdig, and Å. Larsson. 1987. Long-term effects of bleached kraft mill effluents on carbohydrate metabolism and hepatic xenobiotic biotransformation enzymes in fish. *Ecotoxicol. Environ. Saf.* 13: 53–60.

Andersson, Ö., C.-E. Linder, M. Olsson, L. Reutergård, U.-B. Uvemo, and U. Wide-qvist. 1988. Spatial differences and temporal trends of organochlorine com-pounds in biota from the northwestern hemisphere. *Arch Environ. Contam. Toxicol.* 17: 755–765.

Ankley, G.T., M.K. Schubauer-Berigan, and J.R. Dierkes. 1991. Predicting the toxicity of bulk sediments to aquatic organisms with aqueous test fractions: pore water vs. elutriate. *Environ. Toxicol. Chem.* 10: 1359–1366.

Bahig, M.E., A. Kraus, W. Klein, and F. Korte. 1981. Metabolism of pentachloroni-trobenzene-^{14}C quintozene in fish. *Chemosphere* 10: 319–322.

Baird, D.J., I. Barber, and P. Calow. 1990. Clonal variation in general responses of *Daphnia magna* Straus to toxic stress. I. Chronic life-history effects. *Funct. Ecol.* 4: 399–407.

Baldwin, W.S., D.L. Milan, and G.A. LeBlanc. 1995. Physiological and biochemical perturbations in *Daphnia magna* following exposure to the model environmental estrogen diethylstilbestrol. *Environ. Toxicol. Chem.* 14: 945–952.

Bantle, J.A. et al. 1994. FETAX interlaboratory validation study: phase II testing. *Environ. Toxicol. Chem.* 13: 1629–1637.

Bantle, J.A. et al. 1996. FETAX interlaboratory validation study: phase III—Part 1 testing. *J. Appl. Toxicol.* 16: 517–528.

Bauer, I., K. Weber, and W. Ernst. 1989. Metabolism of octachlorostyrene in the blue mussel (*Mytilus edulis*). *Chemosphere* 18: 1573–1579.

Belfroid, A., W. Seinen, K. van Gestel, and J. Hermens. 1993. The acute toxicity of chlorobenzenes for earthworms (*Eisenia andrei*) in different exposure systems. *Chemosphere* 26: 2265–2277.

Bengtsson, B.-E. 1978. Use of a harpacticoid copepod in toxicity tests. *Mar. Pollut. Bull.* 9: 238–241.

Bengtsson, B.-E. 1980. Long-term effects of PCB (Clophen A 50) on growth, repro-duction and swimming performance in the minnow *Phoxinus phoxinus*. *Water Res.* 14: 681–687.

Bengtsson, B.-E. and B. Bergström. 1987. A flowthrough fecundity test with *Nitocra spinipes* (Harpacticoidea Crustaceae) for aquatic toxicity. *Ecotoxicol. Environ. Saf.* 14: 260–268.

Bennett, D.J. 1998. *Randomness*. Harvard University Press, Cambridge, MA.

Bentivegna, C.S. and K.R. Cooper. 1993. Reduced chromosomal puffing in *Chironomus tentans* as a biomarker for potentially genotoxic substances. *Environ. Toxicol. Chem.* 12: 1001–1011.

Berger, L. et al. 1998. Chytridiomycosis causes amphibian mortality associated with population declines in the rain forests of Australia and Central America. *Proc. Natl. Acad. Sci. USA* 85: 9031–9036.

Bergman, Å., R.J. Norstrom, K. Haraguchi, H. Kuroki, and P. Béland. 1994. PCB and DDE methyl sulfones in mammals from Canada and Sweden. *Environ. Toxicol. Chem.* 13: 121–128.

Berrill, M., D. Coulson, L. McGillivray, and B. Pauli. 1998. Toxicity of endosulfan to aquatic stages of anuran amphibians. *Environ. Toxicol. Chem.* 17: 1738–1744.

Berry, R.M., C.E. Luthe, and R.H. Voss. 1993. Ubiquitous nature of dioxins: a comparison of the dioxins content of common everyday materials with that of pulps and papers. *Environ. Sci. Technol.* 27: 1164–1168.

Billington, J.W., G.-L. Huang, F. Szeto, W.Y. Shiu, and D. Mackay. 1988. Preparation of aqueous solutions of sparingly soluble organic substances: I. Single component systems. *Environ. Toxicol. Chem.* 7: 117–124.

Black, J.J., A.E. Maccubbin, and M. Schiffert. 1985. A reliable, efficient, microinjection apparatus and methodology for the *in vivo* exposure of rainbow trout and salmon embryos to chemical carcinogens. *J. Natl. Cancer Inst.* 75: 1123–1128.

Blanck, H. and S.Å. Wänberg. 1988. The validity of an ecotoxicological test system. Short-term and long-term effects of arsenate on marine phytoplankton communities in laboratory systems. *Can. J. Fish. Aquat. Sci.* 45: 1807–1815.

Blanck, H., G. Wallin, and S.-Å. Wänberg. 1984. Species-dependent variation in algal sensitivity to chemical compounds. *Ecotoxicol. Environ. Saf.* 8: 339–351.

Blanck, H., S.-Å. Wänberg, and S. Molander. 1988. Pollution-induced community tolerance—a new ecotoxicological tool. In *Functional Testing of Aquatic Biota for Estimating Hazards of Chemicals* pp. 219–230 (Eds. J. Cairns and J.P. Pratt). ASTM STP 988. American Society for Testing and Materials, Philadelphia.

Bowman, R.S., J. Schroeder, R. Buluau, M. Remmenga, and R. Heightman. 1997. Plant toxicity and plant uptake of fluorobenzoate and bromide water tracers. *J. Environ. Qual.* 26: 1292–1299.

Bradbury, S.P., T.R. Henry, G.J. Niemi, R.W. Carlson, and V.M. Snarski. 1989. Use of respiratory-cardiovascular responses of rainbow trout (*Salmo gairdneri*) in identifying acute toxicity syndromes in fish: Part 3. Polar narcotics. *Environ. Toxicol. Chem.* 8: 247–261.

Braunbeck, T., V. Storch, and H. Bresch. 1990a. Species-specific reaction of liver ultrastructure in zebra fish (*Brachydanio rerio*) and trout (*Salmo gairdneri*) after prolonged exposure to 4-chloroaniline. *Arch. Environ. Contam. Toxicol.* 19: 405–418.

Braunbeck, T., G. Görge, V. Storch, and R. Nagel. 1990b. Hepatoc steatosis in zebra fish (*Brachydanio rerio*) induced by long-term exposure to gamma-hexchlorocyclohexane. *Environ. Toxicol. Chem.* 19: 355–374.

Breinholt, V. and J.C. Larsen. 1998. Detection of weak estrogenic flavanoids using a recombinant yeast strain and a modified MCF7 cell proliferation assay. *Chem. Res. Toxicol.* 11: 622–629.

Bresch, H., H. Beck, D. Ehlermann, H. Schlaszus, and M. Urbanek. 1990. A long-term toxicity test comprising reproduction and growth of zebra fish with 4-chloroaniline. *Arch. Environ. Contam. Toxicol.* 19: 419–427.

Bridgham, S.D. 1988. Chronic effects of 2,2'-dichlorobiphenyl on reproduction, mortality, growth, and respiration of *Daphnia pulicaria*. *Arch. Environ. Contam. Toxicol.* 17: 731–740.

Brouwer, A., P.J.H. Reijnders, and J.H. Koeman. 1989. Polychlorinated biphenyl (PCB)-contaminated fish induces vitamin A and thyroid hormone deficiency in the common seal (*Phoca vitulina*). *Aquat. Toxicol.* 15: 99–106.

Brumley, C., V.S. Haritos, J.T. Ahokas, and D.A. Holdway. 1996. Metabolites of chlorinated syringols in fish bile as biomarkers of exposure to bleached eucalypt pulp effluents. *Ecotoxicol. Environ. Saf.* 33: 253–260.

Buhl, K.J., S.J. Hamilton, and J.C. Schmulbach. 1993. Acute toxicity of the herbicide bromoxynil to *Daphnia magna*. *Environ. Toxicol. Chem.* 12: 1455–1468.

Buhler, D.R. and D.E. Williams. 1988. The role of biotransformation in the toxicity of chemicals. *Aquat. Toxicol.* 11: 19–28.

Burton, G.A. and K. J. Scott. 1992. Sediment toxicity evaluations. Their niche in ecological assessments. *Environ. Sci. Technol.* 26: 2068–2075.

Cairns, J. 1984. Are single species toxicity tests alone adequate for estimating environmental hazard? *Environ. Monitor. Assess.* 4: 259–273.

Cairns, J. 1992. The threshold problem in ecotoxicology. *Ecotoxicology* 1: 3–16.

Cairns, J., P.V. McCormick, and B.R. Niederlehner. 1992. Estimating ecotoxicological risk and impact using indigenous aquatic microbial communities. *Hydrobiologia* 237: 131–145.

Calleja, M.C. and G. Persoone. 1993. The influence of solvents on the acute toxicity of some lipophilic chemicals to aquatic invertebrates. *Chemosphere* 26: 2007–2022.

Carr, R.S., J.W. Williams, and C.T.B. Fragata. 1989. Development and evaluation of a novel marine sediment pore water toxicity test with the polychaete *Dinophilus gyrociliatus*. *Environ. Toxicol. Chem.* 8: 533–543.

Carrasco, K.R., K.L. Tilbury, and M.S. Myers. 1990. Assessment of the piscine micronucleus test as an *in situ* biological indicator of chemical contaminant effects. *Can. J. Fish. Aquat. Sci.* 47: 2123–2136.

Cherr, G.N., J.M. Shenker, C. Lundmark, and K.O. Turner. 1987. Toxic effects of selected bleached kraft mill effluent constituents on sea urchin sperm cell. *Environ. Toxicol. Chem.* 6: 561–569.

Cherr, G.N., J. Shoffner-McGee, and J.M. Shenker. 1990. Method for assessing fertilization and embryonic/larval development in toxicity tests using the California mussel (*Mytilus californianus*). *Environ. Toxicol. Chem.* 9: 1137–1145.

Clark, J.R., L.R. Goodman, P.W. Borthwick, J.M. Patrick, G.M. Cripe, P.M. Moody, J.C. Moore, and M. Lores. 1989. Toxicity of pyrethroids to marine invertebrates and fish: a literature review and test results with sediment-sorbed chemicals. *Environ. Toxicol. Chem.* 8: 393–401.

Clemons, J.H., L.M. Allan, C.H. Marvin, Z. Wu, B.E. Mccarry, D.W. Bryant, and T.R. Zacherewski. 1998. Evidence of estrogen- and TCDD-like activities in crude and fractionated extracts of PM_{10} air particulate material using *in vitro* gene expression assays. *Environ. Sci. Technol.* 32: 1853–1860.

Collier, T.K., L.C. Thomas, and D.C. Malins. 1978. Influence of environmental temperature on disposition of dietary naphthalene in coho salmon (*Oncorhynchus kisutch*): isolation and identification of individual metabolites. *Comp. Biochem. Physiol.* 61C: 23–28.

Crivello, J.F. and R.J. Schultz. 1995. Genetic variation in the temperature dependence of liver microsomal CYP2E1 activity, within and between species of the viviparous fish *Poeciliopis*. *Environ. Toxicol. Chem.* 14: 1–8.

Crossland, N.O. and J.M. Hillaby. 1985. Fate and effects of 3,4-dichloroaniline in the laboratory and in outdoor ponds: II. Chronic toxicity to *Daphnia* spp. and other invertebrates. *Environ. Toxicol. Chem.* 4: 489–499.

Cruz, S.M., M.N. Scott, and A.K. Merritt. 1993. Metabolism of [^{14}C]metolachlor in bluegill sunfish. *J. Agric. Food Chem.* 41: 662–668.

Dady, J.M., S.P. Bradbury, A.D. Hoffman, M.M. Voit, and D.L. Olson. 1991. Hepatic microsomal N-hydroxylation of aniline and 4-chloroaniline by rainbow trout (*Oncorhyncus mykiss*). *Xenobiotica* 21: 1605–1620.

De Maagd, P.G.-J. and A.D. Vethaak. 1998. Biotransformation of PAHs and their carcinogenic effects in fish, pp.265–309. In *PAHs and Related Compounds* Vol. 3.J (A.H. Neilson Ed.) Springer-Verlag, Berlin.

De Serres, F.J. and J. Ashby, Eds. 1981. Evaluation of short-term tests for carcinogenesis. Elsevier/North Holland, New York.

De Wolf, W., W. Seinen, and J.L.M. Hermens. 1993. Biotransformation and toxicokinetics of trichloroanilines in fish in relation to their hydrophobicity. *Arch. Environ. Contam. Toxicol.* 25: 110–117.

Delistraty, D. 1998. A critical revew of the application of toxic equivalency factors to carcinogenic effects of polycyclic aromatic hydrocarbons in mammals pp. 312–359. In *PAHs and Related Compounds,* Vol. 3.J (A.H. Neilson ed.) Springer-Verlag, Berlin.

DeMarini, D.M., H.G. Brooks, and D.G. Parkes. 1990. Induction of prophage lambda by chlorophenols. *Environ. Mol. Mutagen.* 15: 1–9.

Desbrow, C., E.J. Routledge, G.C. Brighty, J.P. Sumpter, and M. Waldock. 1998. Identification of estrogenic chemicals in STW effluent. 1. Chemical fractionation and *in vitro* biological screening. *Environ. Sci. Technol.* 32: 1549–1558.

Deutsch, W.G., E.C. Webber, D.R. Bayne, and C.W. Reed. 1992. Effects of largemouth bass stocking rate on fish populations in aquatic mesocosms used for pesticide research. *Environ. Toxicol. Chem.* 11: 5–10.

Dillon, T.M., D.W. Moore, and A.B. Gibson. 1993. Development of a chronic sublethal bioassay for evaluating contaminated sediment with the marine polychaete worm *Nereis* (*Neanthes*) *arenaceodentata. Environ. Toxicol. Chem.* 12: 589–605.

Dixon, D.G., C.E.A. Hill, P.V. Hodson, E.J. Kempe, and K.L.E. Kaiser. 1985. Plasma leucine aminonaphthylamidase as an indicator of acute sublethal toxicant stress in rainbow trout. *Environ. Toxicol. Chem.* 4: 789–796.

Dodson, S.I., T. Hanazato, and P.R. Gorski. 1995. Behavioural responses of *Daphnia pulex* exposed to carbaryl and *Chaoborus kairomone. Environ. Toxicol. Chem.* 14: 43–50.

Doherty, F.G. 1983. Interspecies correlations of acute aquatic median lethal concentration for four standard testing species. *Environ. Sci. Technol.* 17: 661–665.

Donnarumma, L., G. de Angelis, F. Gramenzi, and L. Vittozzi. 1988. Xenobiotic metabolizing enzyme systems in test fish. III. Comparative studies of liver cytosolic glutathione S-transferases. *Ecotoxicol. Environ. Saf.* 16: 180–186.

Driscoll, S.K. and A.E. McElroy. 1996. Bioaccumulation and metabolism of benzo[a]pyrene in three species of polychaete worms. *Environ. Toxicol. Chem.* 15: 1401–1410.

Egaas, E. and U. Varanasi. 1982. Effects of polychlorinated biphenyls and environmental temperature on *in vitro* formation of benzo[a]pyrene metabolites by liver of trout (*Salmo gairdneri*). *Biochem. Pharmacol.* 31: 561–566.

Estes, J.A., M.T. Tinker, T.M. Williams, and D.F. Doak. 1998. Killer whale predation on sea otters linking oceanic and nearshore ecosystems. *Science* 282: 473–476.

Fairchild, J.F., T.W. La Point, J.L. Zajicek, M.K. Nelson, F.J. Dwyer, and P.A. Lovely. 1992. Population-, community- and ecosystem-level responses of aquatic mesocosms to pulsed doses of a pyrethroid insecticide. *Environ. Toxicol. Chem.* 11: 115–129.

Fisher, S.W. and T.W. Lohner. 1987. Changes in the aqueous behaviour of parathion under varying conditions of pH. *Arch. Environ. Contam. Toxicol.* 16: 79–84.

Fiskesjö, G. 1993. The Allium test in wastewater monitoring. *Environ. Toxicol. Water Qual.* 8: 291–298.

Fiskesjö, G. 1995. Allium test. pp 119–127. In *In vito Testing Protocols. Methods in Molecular Biology* (Ed. S.O'Hare and C.K. Atterwill) Vol. 43. Humana Press, Totowa, NJ.

Fitzpatrick, L.C., R. Sassani, B.J. Venables, and A.J. Goven. 1992. Comparative toxicity of polychlorinated biphenyls to the earthworms *Eisenia foetida* and *Lumbriculus terrestris. Environ. Pollut.* 77: 65–69.

Fogels, A. and J.B. Sprague. 1977. Comparative short-term tolerance of zebra fish, flagfish, and rainbow trout to five poisons including potential reference toxicants. *Water Res.*11: 811–817.

Förlin, L., T. Andersson, and C.A. Wachtmeister. 1989. Hepatic microsomal 4,5,6-trichloroguaiacol glucuronidation in five species of fish. *Comp. Biochem. Physiol.* 93B: 653–656.

Fox, D.J. et al. 1998. Phase III interlaboratory study of FETAX, Part 2. Interlaboratory validation of exogenous metabolic activation system for frog embryo teratogenesis assay—Xenopus (FETAX). *Drug Chem. Toxicol.* 21: 1–14.

Fox, G.A. 1991. Practical causal inference for ecoepidemiologists. *J. Toxicol. Environ. Health* 33: 359–373.

Fox, G.A., S.W. Kennedy, R.J. Norstreom, and D.C. Wigfield. 1988. Porphyria is herring gulls: a biochemical response to chemical contamination of Great Lakes food chains. *Environ. Toxicol. Chem.* 7: 831–839

Funari, E., A. Zoppinki, A. Verdina, G. de Angelis, and L. Vittozzi. 1987. Xenobiotic metabolizing enzyme systems in test fish. I. Comparative studies of liver microsomal monooxygenases. *Ecotoxicol. Environ. Saf.* 13: 24–31.

Gallagher, E.P., G.L. Kedders, and R.T. di Giulio. 1991. Glutathione S-transferase-mediated chlorothalonil metabolism in liver and gill subcellular fractions of channel catfish. *Biochem. Pharmacol.* 42: 139–145.

Gallagher, E.P., R.C. Cattley, and R.T. di Giulio. 1992. The acute toxicity and sublethal effects of chlorothalonil in channel catfish (*Ictalurus punctatus*). *Chemosphere* 24: 3–10.

Genoni, G.P. 1992. Short-term effect of a toxicant on scope for change in ascendency in a microcosm community. *Ecotoxicol. Environ. Saf.* 24: 179–191.

Giesy, J.P., J.P. Ludwig, and D.E. Tillitt. 1994. Deformities in birds of the Great Lakes region. Assigning causality. *Environ. Sci. Technol.* 28: 128A–135A.

Giggleman, M.A., L.C. Fitzpatrick, A.J. Goven, and B.J. Venables. 1998. Effects of pentachlorophenol on survival of earthworms (*Lumbricus terrestris*) and phagocytosis by their immunoactive coelomocytes. *Environ. Toxicol. Chem.* 17: 2391–2394.

Glatt, H. 1997. Sulfation and sulfotransferases 4. Bioactivation of mutagens by sulfation. *FASEB J.* 11: 314–321.

Glickman, A.H., C.N. Statham, A. Wu, and J.J. Lech. 1977. Studies on the uptake, metabolism, and disposition of pentachlorophenol and pentachloroanisole in rainbow trout. *Toxicol. Appl. Pharmacol.* 41: 649–658.

Goksöyr, A.-M. Husöy, H.E. Larsen, J. Klungsöyr, S. Wilhelmsen, A. Maage, E.M. Brevik, T. Andersson, M. Celander, M. Pesonen, and L. Förlin. 1991. Environmental contaminants and biochemical responses in flatfish from the Hvaler archipelago in Norway. *Arch. Environ. Contam. Toxicol.* 21: 486–496.

Goodman, L.R., D.J. Hansen, G.M. Cripe, D.P. Middaugh, and J.C. Moore. 1985. A new early life-stage toxicity test using the Californian grunion (*Leuresthes tenuis*) and results with chlorpyrifos. *Ecotoxicol. Environ. Saf.* 10: 12–21.

Haimi, J. and S. Paavola. 1998. Responses of two earthworm populations with different exposure histories to chlorophenol contamination. *Environ. Toxicol. Chem.* 17: 1114–1117.

Hall, T.J., R.K. Haley, and L.E. LaFleur. 1991. Effects of biologically treated bleached kraft mill effluent on cold water stream productivity in experimental stream channels. *Environ. Toxicol. Chem.* 10: 1051–1060.

Hallett, D.J., R.J. Norstrom, F.I. Onuska, and M.E. Comba. 1982. Chlorinated benzenes in Great Lakes herring gulls. *Chemosphere* 11: 277–285.

Hansch, C. and A. Leo. 1979. *Substituent Constants of Correlation Analysis in Chemistry and Biology.* John Wiley & Sons, New York.

Härdig J., T. Andersson, B.-E. Bengtsson, L. Förlin, and Å. Larsson.1988. Long-term effects of in fish. *Ecotoxicol. Environ. Saf.* 15: 96–106.

Harshbarger, J.C. and J.B. Clark. 1990. Epizootiology of neoplasms in bony fish of North America. *Sci. Total Environ.* 94: 1–32.

Hashimoto, S., T. Wakimoto, and R. Tatsukawa. 1990. PCDDs in the sediments accumulated about 8120 years ago from Japanese coastal areas. *Chemosphere* 21: 825–835.

Hawkins, W.E., R.M. Overstreet, and W.W. Walker. 1988. Carcinogenicity tests with small fish species. *Aquat. Toxicol.* 11: 113–128.

Hawkins, W.E., W.W. Walker, R.M. Overstreet, J.S. Lytle, and T.F. Lytle. 1990. Carcinogenic effects of some polycyclic aromatic hydrocarbons on the Japaneses medaka and guppy in waterborne exposures. *Sci. Total Environ.* 94: 155–167.

Helder, T. 1980. Effects of 2,3,7,8-tetrachlorodibenzo-*p*-dioxin (TCDD) on early life stages of the pike (*Esox lucius* L.). *Sci. Total Environ.* 14: 255–264.

Hennig, H.F.-K.O. 1986. Metal-binding proteins as metal pollution indicators. *Environ. Health Perspect.* 65: 175–187.

Hermens, J.L.M. 1989. Quantitative structure–activity relationships of environmental pollutants. pp. 111-162. In *Handbook of Environmental Chemistry* (Ed. O. Hutzinger) Vol. 2E. Springer-Verlag, Berlin.

Hinga, K.R. and M.E.Q. Pilson. 1987. Persistence of benz[*a*]anthracene degradation products in an enclosed marine ecosystem. *Environ. Sci. Technol.* 21: 648–653.

Hites, R.A. 1990. Environmental behaviour of chlorinated dioxins and furans. *Acc. Chem. Res.* 23: 194–201.

Hodson, P.V. and B.R. Blunt. 1981. Temperature-induced changes in pentachlorophenol chronic toxicity to early life stages of rainbow trout. *Aquat. Toxicol.* 1: 113–127.

Höhl, H.-U. and W. Barz. 1995. Metabolism of the insecticide phoxim in plants and cell suspension cultures of soybean. *J. Agric. Food Chem.* 43: 1052–1056.

Hontela, A., J.B. Rasmussen, C. Audet, and G. Chevalier. 1992. Impaired cortisol stress response in fish from environments polluted by PAHs, PCBs, and mercury. *Arch. Environ. Contam. Toxicol.* 22: 278–283.

Houk, V.S. and D.M. DeMarini. 1987. Induction of prophage lambda by chlorinated pesticides. *Mutat. Res.* 182: 193–201.

Howe, G.E., L.L. Marking, T.D. Bills, J.J. Rach, and F.L. Mayer. 1994. Effects of water temperature and pH on toxicity of terbufos, trichlorfon, 4-nitrophenol and 2,4-dinitrophenol to the amphipod *Gammarus pseudolimnaeus* and rainbow trout (*Oncorhynchus mykiss*). *Environ. Toxicol. Chem.* 13: 51–66.

Howson, M. and K.C. Jones. 1998. Sources of PAHs in the environment pp. 138–174. In *PAHs and Related Compounds* Vol. 3 I (A.H. Neilson ed.) Springer-Verlag, Berlin.

Huang, K.C. and S.F. Collins. 1962. Conjugation and excretion of aminobenzoic acid isomers in marine fishes. *J. Cell. Comp. Physiol.* 60: 49–52.

Hulzebos, E.M., D.M.M. Adema, E.M. Dirven-van Breemen, L. Henzen, W.A. van Dis, H.A. Herbold, J.A. Hoekstra, R. Baerselman, and C.A.M. van Gestel. 1993. Phytotoxicity studies with *Lactuca sativa* in soil and nutrient solution. *Environ. Toxicol. Chem.* 12: 1079–1094.

Hunn, J.B., R.A. Schoettger, and W.A. Willford. 1968. Turnover and urinary excretion of free and acetylated M.S. 222 by rainbow trout, *Salmo gairdneri*. *J. Fish. Res. Bd. Can.* 25: 215–231.

Iwata, H., S. Tanabe, N. Sakai, and R. Tatsukawa. 1993. Distibution of persistent organochlorines in the oceanic air and surface seawater and the role of ocean on their global transport and fate. *Environ. Sci. Technol.* 27: 1080–1098.

James, M.O. and J.R. Bend. 1976. Taurine conjugation of 2,4-dichlorophenoxyacetic acid and phenylacetic acid in two marine species. *Xenobiotica* 6: 393–398.

Jarvinen, A.W., B.R. Nordling, and M.E. Henry. 1983. Chronic toxicity of Dursban (Chloropyrifos) to the fathead minnow (*Pimephales promelas*) and the resultant acetylcholinesterase inhibition. *Ecotoxicol. Environ. Saf.* 7: 423–434.

Jarvinen, A.W, D.K. Tanner, and E.R. Kline. 1988. Toxicity of chlorpyrifos, endrin, or fenvalerate to fathead minnows following episodic or continuous exposure. *Ecotoxicol. Environ. Saf.* 15: 78–95.

Jensen, S., G. Eriksson, H. Kylin, and W.M.J. Strachan. 1992. Atmospheric pollution by persistent organic compounds: monitoring with pine needles. *Chemosphere* 24: 229–245.

Jimenez, B.D., L.S. Burtis, G.H. Ezell, B.Z. Egan, N.E. Lee, J.J. Beauchamp, and J.F. McCarthy. 1988. The mixed function oxidase system of bluegill sunfish, *Lepomis macrochirus*: correlation of activities in experimental and wild fish. *Environ. Toxicol. Chem.* 7: 623–634.

Johnson, B.T. 1992. Rainbow trout liver activation systems with the Ames mutagenicity test. *Environ. Toxicol. Chem.* 9: 1183–1192.

Johnson, L.L., C.M. Stehr, O.P. Olson, M.S. Myers, S.M. Pierce, C.A. Wigren, B.B. McCain, and U. Varanasi. 1993. Chemical contaminants and hepatic lesions in winter flounder (*Pleuronectes americanus*) from the northeastern coast of the United States. *Environ. Sci. Technol.* 27: 2759–2771.

Johnson, P.T.J., K.B. Lunde, E.G. Ritchie, and A.E. Launer. 1999. The effect of trematode infection on amphibian limb development and survivorship. *Science* 284: 802–804.

Kado, N.Y., G.N. Guirguis, C.P. Flessel, R.C. Chan, K.-I. Chang, and J.J. Wesolowski. 1986. Mutagenicity of fine (< 2.4 µm) airborne particles: diurnal variation in community air determined by a Salmonella micro preincubation (microsuspension) procedure. *Environ. Mutagen.* 8: 53–66.

Kaiser, K.L. (Ed.). 1987. QSAR in Environmental Toxicology. Reidel Publishers, Dordrecht.

Keizer, J., G. d'Agostino, R. Nagel, F. Gramenzi, and L. Vittozzi. 1993. Comparative diazinon toxicity in guppy and zebra fish: different role of oxidative metabolism. *Environ. Toxicol. Chem.* 12: 1243–1250.

Kirchin, M.A., A. Wiseman, and D.R. Livingstone. 1992. Seasonal and sex variation in the mixed-function oxygenase system of digestive gland microsomes of the common mussel, *Mytilus edulis* L. *Comp. Biochem. Physiol.* 101C: 81–91.

Klaasen, C.D., M.O. Amdur, and J. Doull. 1986. *Casarett and Doull's Toxicology: The Basic Science of Poisons*. Macmillan, New York.

Kleinow, K.M., M.J. Melancon, and J.J. Lech. 1987. Biotransformation and induction: implications for toxicity, bioaccumulation and monitoring of environmental xenobiotics in fish. *Environ. Health Perspect.* 71: 105–119.

Kosalwat, P. and A.W. Knight. 1986a. Acute toxicity of aqueous and substrate-bound copper to the midge, *Chironomus decorus*. *Arch. Environ. Contam. Toxicol.* 16: 275–282.

Kosalwat, P. and A.W. Knight. 1986b. Chronic toxicity of copper to a partial life stage of the midge, *Chironomus decorus*. *Arch. Environ. Contam. Toxicol.* 16: 283–290.

Kovacs, T.G., P.H. Martel, R.H. Voss, P.E. Wrist, and R.F. Willes. 1993. Aquatic toxicity equivalency factors for chlorinated phenolic compounds present in pulp mill effluents. *Environ. Toxicol. Chem.* 12: 281–289.

Krahn, M.M., D.G. Burrows, G.M. Ylitalo, D.W. Brown, C.A. Wigren, T.K. Collier, S.-L. Chan, and U. Varanasi. 1992. Mass spectrometric analysis for aromatic compounds in bile of fish sampled after the Exxon Valdez oil spill. *Environ. Sci. Technol.* 26: 116–126.

Krause, A., W.G. Hancock, R.D. Minard, A.J. Freyer, R.C. Honeycutt, H.M. LeBaron, D.L. Paulson, S.Y. Liu, and J.M. Bollag. 1985. Microbial transformation of the herbicide metolachlor by a soil actinomycete. *J. Agric. Food Chem.* 33: 584–589.

Kristen, U. 1997. Use of higher plants as screens for toxicity assessment. *Toxicol. in Vitro.* 11: 181–191.

Kristen, U. and R. Kappler. 1995. The pollen tube growth test pp.189–198. In *In Vitro Testing Protocols. Methods in Molecular Biology* (Ed. S. O'Hare and C.K. Atterwill) Vol. 43. Humana Press, Totowa, NJ.

Kutz, F.W., D.G. Barnes, E.W. Bretthauer, D.P. Bottimore, and H. Greim. 1990. The international toxicity equivalency factor (I-TEF) method for estimating risks associated with exposures to complex mixtures of dioxins and related compounds. *Toxicol. Environ. Chem.* 26: 99–109.

La Clair, J.J., J.A. Bantle, and J. Dumont. 1998. Photoproducts and metabolites of a common insect growth regulator produce development deformities in *Xenopus*. *Environ. Sci. Technol.* 32: 1453–1461.

Lambert, M., S. Kremer, and H. Anke. 1995. Antimicrobial, phytotoxic, nematicidal, cytotoxic, and mutagenic activities of 1-hydroxypyrene, the initial metabolite in pyrene metabolism by the basidiomycete *Crinipellis stipitaria*. *Bull. Environ. Contam. Toxicol.* 55: 251–257.

Landner, L., A.H. Neilson, L. Sörensen, A. Tärnholm, and T. Viktor. 1985. Short-term test for predicting the potential of xenobiotics to impair reproductive success in fish. *Ecotoxicol. Environ. Saf.* 9: 282–293.

Landner, L., H. Blanck, U. Heyman, A. Lundgren, M. Notini, A. Rosemarin, and B. Sundelin. 1989. Community testing, microcosm and mesocosm experiments: ecotoxicological tools with high ecological realism pp. 216-254. In *Chemicals in the Aquatic Environment* (Ed. L. Landner). Springer-Verlag, Berlin.

Landrum, P.F., B.J. Eadie, and W.R. Faust. 1991. Toxicokinetics and toxicity of a mixture of sediment-associated polycyclic aromatic hydrocarbons to the amphipod *Diporeia* sp. *Environ. Toxicol. Chem.* 10: 35–46.

Larsen, D.P., F. deNoyelles, F. Stay, and T. Shiroyama. 1986. Comparisons of single-species, microcosm and experimental pond responses to atrzine exposure. *Environ. Toxicol. Chem.* 5: 179–190.

Larsen, O.F.A., I.S. Kozin, A.M. Rija, G.J. Stroomberg, J.A. de Knecht, N.H. Velthorst, and C. Gooijer. 1998. Direct identification of pyrene metabolites in organs of the isopod *Porcello scaber* by fluorescence line narrowing spectroscopy. *Anal Chem.* 70: 1182–1185.

Larsson, Å., T. Andersson, L. Förlin, and J. Härdig. 1988. Physiological disturbances in fish exposed to bleached kraft mill effluents. *Water Sci. Technol.* 20 (2): 67–76.

Last, J.M., Ed. 1983. *A Dictionary of Epidemiology.* Oxford University Press, New York.

Letcher, R.J., R.J. Norstrom, and D.C.G. Muir. 1998. Biotransformation versus bioaccumulation: sources of methyl sulfone PCB and 4,4'-DDE metabolites in polar bear food chain. *Environ. Sci. Technol.* 32: 1656–1661.

Leyal, C. and P. Binet. 1998. Effect of polyaromatic hydrocarbons in soil on arbuscular mycorrhizal plants. *J. Environ. Qual.* 27: 402–407.

Lindahl, P.E., S. Olofsson, and E. Schwanbom. 1976. Improved rotary-flow technique applied to cod (*Gadus morrhua* L). *Water Res.* 10: 833–845.

Lindström, M. and A. Lindström. 1980. Changes in the swimming activity of *Pontoporeia affinis* (Crustacea, Amphipoda) after exposure to sublethal concentrations of phenol, chlorophenol and styrene. *Ann. Zool. Fenn.* 17: 221–231.

Little, P.J., M.O. James, J.B. Pritchard, and J.R. Bend. 1984. Benzo(*a*)pyrene metabolism in hepatic microsomes from feral and 3-methylcholanthrene-treated southern flounder, *Paralichthys lethostigma. J. Environ. Pathol. Toxicol. Oncol.* 5: 309–320.

Livingstone, D.R. and S.V. Farrar. 1984. Tissue and subcellular distribution of enzyme activities of mixed-function oxygenase and benzo[*a*]pyrene metabolism in the common mussel *Mytilis edulis* L. *Sci. Total Environ.* 39: 209–235.

Longcore, J.E., A.P. Pessier, and D.K. Nicols. 1999. *Batrachochytrium dendrobatidis* gen. et sp. nov., a chytrid pathogenic to amphibians. *Mycologia* 91: 219–227.

Lozano, S.L., S.L. O'Halloran, K.W. Sargent, and J.C. Brazner. 1992. Effects of esfenvalerate on aquatic organisms in littoral enclosures. *Environ. Toxicol. Chem.* 11: 35–47.

Lundgren, A. 1985. Model ecosystems as a tool in freshwater and marine research. *Arch. Hydrobiol. Suppl.* 70: 157–196.

Lye, C.M., C.L.J. Frid, M.E. Gill, D.W. Cooper, and D.M. Jones. 1999. Estrogenic alkylphenols in fish tissues, sediments, and waters from the U.K. Tyne and Tees estuaries. *Environ. Sci. Technol.* 1009–1014.

Mac, M.J. and C.C. Edsall. 1991. Environmental contaminants and the reproductive success of lake trout in the Great Lakes: an epidemiological approach. *J. Toxicol. Environ. Health* 33: 375–394.

Malins, D.C, M.M. Krahn, M.S. Myers, L.D. Rhodes, D.W. Brown, C.A. Krone, B.B. McCain, and S.-L. Chan. 1985. Toxic chemicals in sediments and biota from a creosote-polluted harbor: relationships with hepatic neoplasms and other hepatic lesions in English sole (*Parophrys vetulus*). *Carcinogenesis* 6: 1463–1469.

Malins, D.C., B.B. McCain, D.W. Brown, M.S. Myers, M.M. Krahn, and S.-L. Chan. 1987. Toxic chemicals, including aromatic and chlorinated hydrocarbons and their derivatives, and liver lesions in white croaker (*Genyonemus lineatus*) from the vicinity of Los Angeles. *Environ. Sci. Technol.* 21: 765–770.

Maltby, L. 1992. The use of the physiological energetics of *Gammarus pulex* to assess toxicity: a study using artificial streams. *Environ. Toxicol. Chem.* 11: 79–85.

Maltby, L. and P. Calow. 1989. The application of bioassays in the resolution of environmental problems; past, present and future. *Hydrobiologia* 188/189: 65–76.

Maltby, L., C. Naylor, and P. Calow. 1990a. Effect of stress on a freshwater benthic detrivore: scope for growth in *Gammarus pulex*. *Ecotoxicol. Environ. Saf.* 19: 285–291.

Maltby, L., C. Naylor, and P. Calow. 1990b. Field deployment of a scope for growth assay involving *Gammarus pulex*, a freshwater benthic invertebrate. *Ecotoxicol. Environ. Saf.* 19: 292–300.

Maron, D.M. and B.N. Ames. 1983. Revised methods for the Salmonella mutagenicity test. *Mutat. Res.* 113: 173–215.

McCahon, C.P. and D. Pascoe. 1988. Use of *Gammarus pulex* (L.) in safety evaluation tests: culture and selection of a sensitive life stage. *Ecotoxicol. Environ. Saf.* 15: 245–252.

McCahon, C.P., S.F. Barton, and D. Pascoe. 1990. The toxicity of phenol to the freshwater crustacean *Aselus aquaticus* (L.) during periodic exposure—relationship between sub-lethal responses and body phenol concentrations. *Arch. Environ. Contam. Toxicol.* 19: 926–929.

McElroy, A.E. 1990. Polycyclic aromatic hydrocarbon metabolism in the polychaete *Nereis virens*. *Aquat. Toxicol.* 18: 35–50.

McElroy, A.E. and J.C. Means. 1988. Uptake, metabolism, and depuration of PCBs by the polychaete *Nereis virens*. *Aquat. Toxicol.* 11: 416–417.

McGee, B.L., C.E. Schlekat, and E. Reinharz. 1993. Assessing sublethal levels of sediment contamination using the estuarine amphipod *Leptocheirus plumulosus*. *Environ. Toxicol. Chem.* 12: 577–587.

McKim, J.M. 1977. Evaluation of tests with early life stages of fish for predicting long-term toxicity. *J. Fish. Res. Bd. Can.* 34: 1148–1154.

McKim, J.M., G.F. Olson, G.W. Holcombe, and E.P. Hunt. 1976. Long-term effects of methylmercuric chloride on three generations of brook trout (*Salvelinus fontinalis*): toxicity, accumulation, distribution and elimination. *J. Fish. Res. Bd. Can.* 33: 1726–2739.

McMaster, M.E., C.B. Portt, K.R. Munnkittrick, and D.G. Dixon. 1992. Milt characteristics, reproductive performance, and larval survival and development of white sucker exposed to bleached kraft mill effluent. *Ecotoxicol. Environ. Saf.* 23: 103–117.

Mersch-Sundermann, V., U. Schneider, G. Klopman, and H.S. Rosenkranz. 1994. SOS induction in *Escherichia coli* and *Salmonella* mutagenicity: a comparison using 330 compounds. *Mutagenesis* 9: 205–224.

Meyerhoff, R.D., D.W. Grothe, S. Sauer, and G.K. Dorulla. 1985. Chronic toxicity of tebuthiuron to an alga (*Selenastrum capricornutum*), a cladoceran (*Daphnia magna*), and the fathead minnow (*Pimephales promelas*). *Environ. Toxicol. Chem.* 4: 695–701.

Michel, X.R., P.M. Cassand, D.G. Ribera, and J.-F. Narbonne. 1992. Metabolism and mutagenic activation of benzo(a)pyrene by subcellular fractions from mussel (*Mytilus galloprovincialis*) digestive gland and sea bass (*Discenthrarcus labrax*) liver. *Comp. Biochem. Physiol.* 103C: 43–51.

Monosson, E. and J.J. Stegeman. 1991. Cytochrome P450E (P4501A) induction and inhibition in winter flounder by 3,3′,4,4′-tetrachlorobiphenyl: comparison of response in fish from Georges Bank and Narragansett Bay. *Environ. Toxicol. Chem.* 10: 765–774.

Moors, P.J. 1986. Decline in numbers of rockhopper penguins at Campbell Island. *Polar. Rec.* 23: 69–73.

Moreau, P., A. Bailone, and R. Devoret. 1976. Prophage l induction in *Escherichia coli* K12 *env* A *urv* B: a highly sensitive test for potential carcinogens. *Proc. Natl. Acad. Sci. U.S.A.* 73: 3700–3704.

Muir, D.C.G., G.P. Rawn, and N.P. Grift. 1985. Fate of the pyrethroid insecticide deltamethrin in small ponds: a mass balance study. *J. Agric. Food Chem.* 33: 603–609.

Muir, D.C.G., A.L. Yarechewski, and A. Knoll. 1986. Bioconcentration and disposition of 1,3,6,8-tetrachlorodibenzo-*p*-dioxin and octachlorodibenzo-*p*-dioxin by rainbow trout and fathead minnows. *Environ. Toxicol. Chem.* 5: 261–272.

Muir, D.C.G., M.D. Segstro, P.M. Welbourn, D. Toom, S.J. Eisenreich, C.R. Macdonald, and D.M. Whelpdale. 1993. Pattern of accumulation of airborne organochlorine contaminants in lichens from the upper Great Lakes region of Ontario. *Environ. Sci. Technol.* 27: 1201–1210.

Munkittrick, K.R., C.B. Portt, G.J. Van Der Kraak, I.R. Smith, and D. A. Rokosh. 1991. Impact of bleached kraft mill effluent on population characteristics, liver MFO activity, and serum steroid levels of a Lake Superior white sucker (*Catostomus commersoni*) population. *Can. J. Fish. Aquat. Sci.* 48: 1371–1380.

Myers, M.S., J.T. Landahl, M.M. Krahn, and B.B. McCain. 1991. Relationships between hepatic neoplasms and related lesions and exposure to toxic chemicals in marine fish from the U.S. West Coast. *Environ. Health Perspect.* 90: 7–15.

Nagel, R., H. Bresch, N. Caspers, P.D. Hansen, M. Markert, R. Munk, N. Scnholz, and B.B. ter Höfte. 1991. Effect of 3,4-dichloroaniline on the early life stages of the zebra fish (*Brachydanio rerio*): results of a comparative laboratory study. *Ecotoxicol. Environ. Saf.* 21: 157–164.

Nakajima, D., E. Kojima, S. Iwaya, J. Suzuki, and S. Suzuki. 1996. Presence of 1-hydroxypyrene conjugates in woody plant leaves and seasonal changes in their concentration. *Environ. Sci. Technol.* 30: 1675–1679.

Nebeker, A.V., M.A. Cairns, J.H. Gakstatter, K.W. Malueg, G.S. Schuytema, and D.F. Krawczyk. 1984. Biological methods for determining toxicity of contaminated freshwater sediments to invertebrates. *Environ. Toxicol. Chem.* 3: 617–630.

Nebeker, A.V., S.T. Onjukka, and M.A. Cairns. 1988. Chronic effects of contaminated sediment on *Daphnia magna* and *Chironomus tentans*. *Bull. Environ. Contam. Toxicol.* 41: 574–581.

Nebeker, A.V., G.S. Schuytema, W.L. Griffis, J.A. Barbitta, and L.A. Carey. 1989. Effect of sediment organic carbon on survival of *Hyalella azteca* exposed to DDT and endrin. *Environ. Toxicol. Chem.* 8: 705–718.

Neilson, A.H. and T. Larsson. 1980. The utilization of organic nitrogen for growth of algae: physiological aspects. *Physiol. Plant.* 48: 542–553.

Neilson, A.H., A.-S. Allard, S. Reiland, M. Remberger, A. Tärnholm, T. Viktor, and L. Landner. 1984. Tri- and tetra-chloroveratrole, metabolites produced by bacterial O-methylation of tri- and tetra-chloroguaiacol: an assessment of their bioconcentration potential and their effects on fish reproduction. *Can. J. Fish. Aquat. Sci.* 41: 1502–1512.

Neilson, A.H., H. Blanck, L. Förlin, L. Landner, P. Pärt, A. Rosemarin, and M. Söderström. 1989. Advanced hazard assessment of 4,5,6-trichloroguaiacol in the Swedish environment. In *Chemicals in the Aquatic Environment*, pp. 329-374. Ed. L. Landner, Springer-Verlag, Berlin.

Neilson, A.H., A.-S. Allard, S. Fischer, M. Malmberg, and T. Viktor. 1990. Incorporation of a subacute test with zebra fish into a hierarchical system for evaluating the effect of toxicants in the aquatic environment. *Ecotoxicol. Environ. Saf.* 20: 82–97.

Niederlehner, B.R., K.W. Pontasch, J.R. Pratt, and J. Cairns. 1990. Field evaluation of predictions of environmental effects from a multispecies-microcosm toxicity test. *Arch. Environ. Contam. Toxicol.* 19: 62–71.

Niimi, A.J., H.B. Lee, and G.P. Kissoon. 1989. Octanol/water partition coefficients and bioconcentration factors of chloronitrobenzenes in rainbow trout (*Salmo gairdneri*). *Environ. Toxicol. Chem.* 8: 817–823.

Nimmo, I.A. 1987. The glutathione S-transferases of fish. *Fish Physiol. Biochem.* 3: 163–172.

Norberg, T.J. and D.I. Mount. 1985. A new fathead minnow (*Pimephales promelas*) subchronic toxicity test. *Environ. Toxicol. Chem.* 4: 711–718.

Norberg-King, T.J. 1989. An evaluation of the fathead minnow seven-day subchronic test for estimating chronic toxicity. *Ecotoxicol. Environ. Saf.* 8: 1075–1089.

Nosek, J.A., S.R. Craven, J.R. Sullivan, J.R. Olsen, and R.E. Peterson 1992. Metabolism and disposition of 2,3,7,8-tetrachlorodibenzo-p-dioxin in ring-necked pheasant hens, chicks, and eggs. *J. Toxicol. Environ. Health* 35: 153–164.

Nyholm, N. 1985. Response variable in algal growth inhibition tests—biomass or growth rate? *Water Res.*19: 273–279.

Oikari, A. and T. Kunnamo-Ojala. 1987. Tracing of xenobiotic contamination in water with the aid of fish bile metabolites: a field study with caged rainbow trout (*Salmo gairdneri*). *Aquat. Pollut.* 9: 327–341.

Olson, L.E., J.L. Allen, and J.W. Hogan. 1977. Biotransformation and elimination of the herbicide dinitramine in carp. *J. Agric. Food Chem.* 25: 554–556.

Opperhuizen, A., and D.T.H.M. Sijm. 1990. Bioaccumulation and biotransformation of polychlorinated dibenzo-p-dioxins and dibenzofurans in fish. *Environ. Toxicol. Chem.* 9: 175–186.

Ottoboni, M.A. 1984. *The Dose Makes the Poison*. Vincente Books, Berkeley, CA.

Oulmi, Y. and T. Braunbeck. 1996. Toxicity of 4-chloroaniline in early life-stages of zebra fish *Brachydanio rerio*: I. Cytopathology of liver and kidney after microinjection. *Arch. Environ. Contam. Toxicol.* 30: 390–402.

Palm, H., Knuutinen, J., Haimi, J., Salminen, J., and Huhta, V. 1991. Methylation products of chorophenols, catechols and hydroquinones in soil and earthworm of sawmill environments. *Chemosphere* 23: 263–267.

Pechmann, J.H.K. and H.M. Wilbur. 1994. Putting declining amphibian populations in perspective: natural fluctuations and human impacts. *Herpetologica* 50: 65–84.

Peters, R. 1952. Lethal synthesis. *Proc. R. Soc. (London)* B 139: 143–170.

Petersen, L. B.-M. and R.C. Petersen. 1984. Effect of kraft pulp mill effluent and 4,5,6-trichloroguaiacol on the net spinning behaviour of *Hydropsyche angustipennis* (Trichoptera). *Ecol. Bull.* 36: 68–74.

Petersen, R.C. and L.B.-M. Petersen. 1989. Ecological concepts important for the interpretation of effects of chemicals on aquatic systems pp. 165–196. In *Chemicals in the Aquatic Environment* (L. Landner Ed.). Springer-Verlag, Berlin.

Petit, F., Y. Valotaire, and F. Pakdel. 1995. Differential functional activities of rainbow trout and human estrogen receptors expressed in the yeast *Saccaharomyces cerevisiae*. *Eur. J. Biochem.* 233: 584–592.

Petit, F., P. Le Goff, J.-P. Cravédi, Y. Valotaire, and F. Pakdel. 1998. Two complementary bioassays for screening the estrogenic potency of xenobiotics: recombinant yeast for trout estrogen receptor and trout hepatocyte cultures. *J. Mol. Endocrinol.* 19: 321–335.

Phipps, G.L., G.T. Ankley, D.A. Benoit, and V.R. Mattson. 1993. Use of the aquatic oligochaete *Lumbriculus variegatus* for assessing the toxicitiy and bioaccumulation of sediment-associated contaminants. *Environ. Toxicol. Chem.* 12: 269–279.

Pillai, M.C., H. Blethro, R.M. Higashi, C.A. Vines, and G.N. Cherr. 1997. Inhibition of the sea urchin sperm acrosome reaction by a lignin-derived macromolecule. *Aquat. Toxicol.* 37: 139–156.

Plewa, M.J., T. Gichner, H. Xin, K.-Y. Seo, S.R. Smith, and E.D. Wagner. 1993. Biochemical and mutagenic characterization of plant-activated aromatic amines. *Environ. Toxicol. Chem.* 12: 1353–1363.

Pochin, E.E. 1983. Sizewell Inquiry Transcripts, Day 151: 101 problems; past, present and future. *Hydrobiologia* 188/189: 65–76.

Pontasch, K.W., B.R. Niederlehner, and J. Cairns. 1989. Comparison of single-species microcosm and field responses to a complex effluent. *Environ. Toxicol. Chem.* 8: 521–532.

Powell, D.C., R.J. Aulerich, J.C. Meadows, D.E. Tillitt, M.E. Kelly, K.L. Stromborg, M.J. Melancon, S.D. Fitzgerald, and S.J. Bursian. 1998. Effects of 3,3′4,4′,5-pentachlorobiphenyl and 2,3,7,8-tetrachlorodibenzo-*p*-dioxin injected into the yolks of double-crested cormorant (*Phalacocorax auritus*) eggs prior to incubation. *Environ. Toxicol. Chem.* 17: 2035–2040.

Purvis, I.J., D. Chotai, C.W. Dykes, D.B. Lubahn, F.S. French, E.M. Wilson, and A.N. Hobden. 1991. An androgen-inducible expression system for *Saccharomyces cerevisiae*. *Gene* 106: 35–42.

Quillardet, P. and M. Hofnung. 1985. The SOS Chromotest, a colorimetric bacterial assay for genotoxins: procedures. *Mutat. Res.* 147: 65–78.

Quillardet, P. and M. Hofnung. 1993. The SOS Chromotest: a review. *Mutat. Res.* 297: 235–279.

Remberger, M., P.-Å. Hynning, and A.H. Neilson. 1991. 2,5-Dichloro-3,6-dihydroxybenzo-1,4-quinone: identification of a new organochlorine compound in kraft mill bleachery effluents. *Environ. Sci. Technol.* 25: 1903–1907.

Reynoldson, T.B., S.P. Thompson, and J.L. Bamsey. 1991. A sediment bioassay using the tubificid oligochaete worm *Tubifex tubifex*. *Environ. Toxicol. Chem.* 10: 1061–1072.

Robertson, B.K. and M. Alexander. 1998. Sequestration of DDT and dieldrin in soil: disappearance of acute toxicity but not the compounds. *Environ. Toxicol. Chem.* 17: 1034–1038.

Rodricks, J.V. 1992. *Calculated Risks*. Cambridge University Press, Cambridge, U.K.

Rosemarin, A., J. Mattsson, K.-J. Lehtinen, M. Notini, and E. Nylén. 1986. Effects of pulp mill chlorate (ClO_3) on *Fucus vesiculosus*—a summary of projects. *Ophelia* Suppl. 4: 219–224.

Rosenthal, H. and D.F. Alderdice. 1976. Sublethal effects of environmental stressors, natural and pollutional, on marine fish eggs and larvae. *J. Fish. Res. Bd. Can.* 33: 2047–2065.

Routledge, E.J. and J.P. Sumpter. 1996. Estrogenic activity of surfactants and some of their degradation products assessed using a recombinant yeast screen. *Environ. Toxicol. Chem.* 15: 241–248.

Routledge, E.J., D. Sheahan, C. Desbrow, G.C. Brighty, M. Waldock, and J.P. Sumpter. 1998. Identification of estrogenic chemicals in STW effluent. 2. *In vivo* responses in trout and roach. *Environ. Sci. Technol.* 32: 1559–1565.

Rutgers, M., I.M. van't Verlaat, B. Wind, L. Posthuma, and A.M. Breure. 1998. Rapid method for assessing pollution-induced community tolerance in contaminated soil. *Environ. Toxicol. Chem.* 17: 2210–2213.

Safe, S., Ed. 1987. *Mammalian Biologic and Toxic Effects of PCBs.* Springer-Verlag, Berlin.

Safe, S., K.W. Brown, K.C. Donnelly, C.S. Anderson, K.V. Markiewicz, M.S. McLachlan, A. Reischi, and O. Hutzinger. 1992. Polychlorinated dibenzo-*p*-dioxins and dibenzofurans associated with wood-preserving chemical sites: biomonitoring with pine needles. *Environ. Sci. Technol.* 26: 394–396.

Sandhu, S.S., B.S. Gill, B.C. Casto, and J.W. Rice. 1991. Application of *Tradescantia micronucleus* assay for *in situ* evaluation of potential genetic hazards from exposure to chemicals at a wood-preserving site. *Hazardous Waste Mater.* 8: 257–262.

Sarokin, D. and J. Schulkin. 1992. The role of pollution in large-scale population disturbances. Part 1: aquatic populations. *Environ. Sci. Technol.* 26: 1476–1484.

Saski, E.K., R. Mikkola, J.V.K. Kukkonen, and M.S. Salkinoja-Salonen. 1997. Bleached kraft pulp mill discharged organic matter in recipient lake sediment. *Environ. Sci. Pollut. Res.* 4: 194–202.

Schäfers, C. and R. Nagel. 1991. Effects of 3,4-dichloroaniline on fish populations. Comparison between r- and K-strategists: a complete life cycle test with the guppy (*Poecilia reticulata*). *Arch. Environ. Contam. Toxicol.* 21: 297–302.

Schäfers, C. and R. Nagel. 1993. Toxicity of 3,4-dichloroaniline to perch (*Perca fluviatilis*) in acute and early stage life stage exposures. *Chemosphere* 26: 1641–1651.

Scharenberg, W. 1991 Cormorants (*Phalacrocorax carbo sinensis*) as bioindicators for polychlorinated biphenyls. *Arch. Environ. Contam. Toxicol.* 21: 536–540.

Schlenk, D. and D.R. Buhler. 1991. Role of flavin-containing monooxygenase in the *in vitro* biotransformation of aldicarb in rainbow trout (*Oncorhyncus mykiss*). *Xenobiotica* 21: 1583–1589.

Sessions, S.K., R.A. Franssen, and V.L. Homer. 1999. Mophological clues from multilegged frogs: are retinoids to blame? *Science* 284: 800–802.

Shenker, J.M. and G.N. Cherr. 1990. Toxicity of zinc and bleached kraft mill effluent to larval English sole (*Parophrys vetulus*) and topsmelt (*Atherinops affinis*). *Arch. Environ. Contam. Toxicol.* 19: 680–685.

Sijm, D.T.H.M. and A. Opperhuizen. 1989. Biotransformation of organic chemicals by fish: enzyme activities and reactions. In *The Handbook of Environmental Chemistry* Vol. 2E (Ed. O. Hutzinger), pp. 164–235, Springer-Verlag, Berlin.

Simoneit, B.T.R. 1998. Biomarker PAHs in the environment pp. 176–221. In *PAHs and Related Compounds*, Vol. 3.J (A.H. Neilson Ed.) Springer-Verlag, Berlin.

Snell, T.W. and B.D. Moffat. 1992. A 2-d life cycle test with the rotifer *Brachionus calyciflorus*. *Environ. Toxicol. Chem.* 11: 1249–1257.

Sorvari, J. and M. Sillanpää. 1996. Influence of metal complex formation of heavy metal and free EDTA and DTPA acute toxicity determined by *Dapnia magna*. *Chemosphere* 33: 1119–1127.

Soto, A.M., C. Sonnenschein, K.L. Chung, M.F. Fernandez, N. Olea, and F.O. Serrano. 1995. The E-SCREEN assay as a tool to identify estrogens: an update on estrogenic environmental pollutants. *Environ. Health Perspect.* 103, Suppl. 7: 113–122.

Southworth, G.R., C.C. Keffer, and J.J. Beauchamp. 1980. Potential and realized bioconcentration. A comparison of observed and predicted bioconcentration of azaarenes in the fathead minnow (*Pimephales promelas*). *Environ. Sci. Technol.* 14: 1529–1531.

Spitsbergen, J.M., J.M. Kleeman, and R.E. Peterson. 1988. Morphological lesions and acute toxicity in rainbow trout (*Salmo gairdneri*) treated with 2,3,7,8-tetrachlorodibenzo-*p*-dioxin. *J. Toxicol. Environ. Health* 23: 333–358.

Spitsbergen, J.M., M.K. Walker, J.R. Olson, and R.E. Peterson. 1991. Pathologic alterations in early life stages of lake trout, *Salvelinus namaycushi*, exposed to 2,3,7,8-tetrachlorodibenzo-*p*-dioxin as fertilized eggs. *Aquat. Toxicol.* 19: 41–72.

Sprague, J.B. 1971. Measurement of pollutant toxicity to fish—III. Sublethal effects and "safe" concentrations. *Water Res.* 5: 24–266.

Stay, F.S., A. Kateko, C.M. Rohm, M.A. Fix, and D.P. Larsen. 1988. Effects of fluorene on microcosms developed from four natural communities. *Environ. Toxicol. Chem.* 7: 635–644.

Steadman, B.L., A.M. Farag, and H.L. Bergman. 1991. Exposure-related patterns of biochemical indicators in rainbow trout exposed to no. 2 fuel oil. *Environ. Toxicol. Chem.* 10: 365–374.

Stroomberg, G.J., C. Reuther, I. Konin, T.C. van Brummelen, C.A.M. van Gestel, C. Gooijer, and W.P. Cofino. 1996. Formation of pyrene metabolites by the terrestrial isopod *Porcello scaber*. *Chemosphere* 33: 1905–1914.

Sumpter, J.P. and S. Jobling. 1995. Vitellogenesis as a biomarker for estrogenic contamination of the aquatic environment. *Environ. Health Perspect.* 103, Suppl. 7: 173–178.

Sundelin, B. 1983. Effects of cadmium on *Pontoporeia affinis* (Crustacea: Amphipoda) in laboratory soft-bottom microcosms. *Mar. Biol.* 74: 203–212.

Susser, M. 1986. Rules of interference in epidemiology. *Reg. Toxicol. Pharmacol.* 6: 116–128.

Suter, G.W., A.E. Rosen, E. Linder, and D.F. Parkhurst. 1987. Endpoints for responses of fish to chronic toxic exposures. *Environ. Toxicol. Chem.* 6: 793–809.

Svanberg, O. and L. Renberg. 1988. Biological-chemical characterization of effluents for the evaluation of the potential impact on the aquatic environment pp. 244–255. In *Organic Micropollutants in the Aquatic Environment* (Eds. G. Angeletti and A. Björseth). Kluwer Academic Publishers, Dordrecht, The Netherlands.

Swain, W.R. 1991. Effects of organochlorine chemicals on the reproductive outcome of humans who consumed contaminated Great Lakes fish: an epidemiologic consideration. *J. Toxicol. Environ. Health* 33: 587–639.

Swartz, R.C., P.F. Kemp, D.W. Schults, G.R. Ditsworth, and R.J Ozretich. 1989. Acute toxicity of sediment from Eagle Harbor, Washington, to the infaunal amphipod *Rhepoxynius abronius*. *Environ. Toxicol. Chem.* 8: 215–222.

Sweeney, B.W., D.H. Funk, and L.J. Standley. 1993. Use of the stream mayfly *Cloeon triangulifer* as a bioasssy organism: life history response and body burden following exposure to technical chlordane. *Environ. Toxicol. Chem.* 12: 115–125.

Taraldsen, J.E. and Norberg-King, T.J. 1990. New method for determining effluent toxicity using duckweed (*Lemna minor*). *Environ. Toxicol. Chem.* 9: 761–767.

Taylor, E.J., S.J. Maund, and D. Pascoe. 1991. Toxicity of four common pollutants to the freshwater macroinvertebrates *Chironomus riparius* (Meigen) (Insecta: diptera) and *Gammarus pulex* (L.) (Crustacea: amphipoda). *Arch. Environ. Contam. Toxicol.* 21: 371–376.

Taylor, E.J., S.J. Blockwell, S.J. Maund, and D. Pascoe. 1993. Effects of lindane on the life-cycle of a freshwater macroinvertebrate *Chironomus riparius* Meigen (Insecta: diptera). *Arch. Environ. Contam. Toxicol.* 24: 145–150.

ten Hulscher, B.A. Vrind, H. van den Heuvel, L.E. van der Velde, P.C.M. van Noort, J.E.M. Beurskens, and H.A.J. Govers. 1999. Triphasic desorption of highly resistant chlorobenzenes, polychlorinated biphenyls, and polycyclic aromatic hydrocarbons in field contaminated sediment. *Environ. Sci. Technol.* 33: 126–132.

Thompson, A.R. and F.L. Gore. 1972. Toxicity of twenty-nine insecticides to *Folsomia candida*: laboratory studies. *J. Econ. Entomol.* 65: 1255–1260.

Tsushimoto, G., F. Matsumara, and R. Sago. 1982. Fate of 2,3,7,8-tetrachlorodibenzo-p-dioxin (TCDD) in an outdoor pond and in model aquatic ecosystems. *Environ. Toxicol. Chem.* 1: 61–68.

Tyler, C.R., R. van Aerle, T.H. Hutchinson, S. Maddix, and H. Trip. 1999. An *in vivo* testing system for endocrine disruptors in fish early life stages using induction of vitellogenin. *Environ. Toxicol. Chem.* 18: 337–347.

Ugrekhelidze, D., F. Korte, and G. Kvesitadze. 1997. Uptake and transformation of benzene and toluene by plant leaves. *Ecotoxicol. Environ. Saf.* 37: 24–29.

Underdal, B., O.M. Skulberg, E. Dahl, and T. Aune. 1989. Disastrous bloom of *Chrysochromulina polylepis* (Prymnesiophyceae) in Norwegian coastal waters 1988—mortality in marine biota. *Ambio* 18: 265–270.

van Brummelen, T.C. and S.C. Stuijfzand. 1993. Effects of benzo(a)pyrene on survival, growth and energy reserves in the terrestrial isopods *Oniscus asellus* and *Porcellio scaber*. *Sci. Total Environ.* Suppl., 921–930.

van Brummelen, T.C. and N.M. van Straalen. 1996. Uptake and elimination of benzo[a]pyrene in the terrestrial isopod *Porcello scaber*. *Arch. Environ. Contam. Toxicol.* 31: 277–285.

van Brummelen, T.C., C.A.M. Gestel, and R.A. Verweij. 1996. Long-term toxicity of five polycyclic aromatic hydrocarbons for the terrestrial isopods *Oniscus asellus* and *Porcellio scaber*. *Environ. Toxicol. Chem.* 15: 1199–1210.

Van Brummelen, T.V., B. van Hattum, T. Crommentuijn, and D.F. Kalf. 1998. Bioavailability and ecotoxicity of PAHs pp.203–263. In *PAHs and Related Compounds*, Vol. 3.J (A.H. Neilson Ed.) Springer-Verlag, Berlin.

van Dam, R.A., M.J. Berry, J.T. Ahokas, and D.A. Holdway. 1996. Comparative acute and chronic toxicity of diethylenetriamine pentaacetic acid (DTPA) and ferric-complexed DTPA to *Daphnia carinata*. *Arch. Environ. Contam. Toxicol.* 31: 433–443.

Van den Avyle, M.J., S.J. Garvick, V.S. Blazer, S.J. Hamilton, and W.G. Brumbaugh. 1989. Skeletal deformities in smallmouth bass (*Micropterus dolomieui*) from southern Appalachian reservoirs. *Arch. Environ. Contam. Toxicol.* 18: 688–696.

Van der Kraak, K.R. Munkittrick, M.E. McMaster, C.B. Portt, and J.P. Chang. 1992. Exposure to bleached kraft pulp mill effluent disrupts the pituitary-gonadal axis of white sucker at multiple sites. *Toxicol. Appl. Pharmacol.* 115: 224–233.

Van der Welden, M.E.J., J. van der Kolk, R. Bleuminck, W. Seinen, and M. van der Berg. 1992. Concurrence of P450 1A1 induction and toxic effects after administration of a low dose of 2,3,7,8-tetrachlorodibenzo-p-dioxin (TCDD) to the rainbow trout (*Oncorhynchus mykiss*). *Aquat. Toxicol.* 24: 123–142.

van Gestel, C.A.M. 1992. Validation of earthworm toxicity tests by comparison with field studies: a review of benomyl, carbendazin, carbofuran, and carbaryl. *Ecotoxicol. Environ. Saf.* 23: 221–236.

van Gestel, C.A.M., and W.-C. Ma. 1988. Toxicity and bioaccumulation of chlorophenols in earthworms in relation to bioavailability in soil. *Ecotoxicol. Environ. Saf.* 15: 289–297.

van Gestel, C.A.M. and W.-C. Ma. 1993. Development of QSARs in soil ecotoxicology: earthworm toxicity and soil sorption of chlorophenols, chlorobenzenes and chloroanilines. *Water Air Soil Pollut.* 69: 265–276.

van Gestel, C.A.M., D.M.M. Adema, and E.M. Dirven-van Breeman. 1996. Phytotoxicity of some chloroanilines and chorophenols, in relation to bioavailability in soil. *Water Air Soil Pollut.* 88: 119–132.

van Urk, J.M., F.C.M. Kerkum, and H. Smit. 1992. Life cycle patterns, density, and frequency of deformities in *Chironomus larvae* (Diptera: Chironomidae) over a contaminated sediment gradient. *Can. J. Fish. Aquat. Sci.* 49: 2291–2299.

Varanasi, U., J.E. Stein, M. Nishimoto, W.L. Reichert, and T.K. Collier. 1987. Chemical carcinogenesis in feral fish: uptake, activation, and detoxication of organic xenobiotics. *Environ. Health. Perspect.* 71: 155–170.

Veith, G.D. and V.M. Comstock. 1975. Apparatus for continuously saturating water with hydrophobic organic chemicals. *J. Fish. Res. Bd. Can.* 32: 1849–1851.

Visser, I.K.G., V.P. Kumarev, C. Örvell, P. de Vries, H.W.J. Broeders, M.W.G. van de Bildt, J. Groen, J.S. Teppema, M.C. Burger, F.G.C.M. UytdeHaag, and A.D.M.E. Osterhaus. 1990. Comparison of two morbilliviruses isolated from seals during outbreaks of distemper in North West Europe and Siberia. *Arch. Virol.* 111: 149–164.

Wachtmeister, C.A., L. Förlin, K:C. Arnoldsson, and J. Larsson. 1991. Fish bile as a tool for monitoring aquatic pollutants: studies with radioactively labelled 4,5,6-trichloroguaiacol. *Chemosphere* 22: 39–46.

Walker, C.H. 1983. Pesticides and birds—mechanisms of selective toxicity. *Agric. Ecosyst. Environ.* 9: 211–226.

Walker, M.K. and R.E. Peterson. 1991. Potencies of polychlorinated dibenzo-*p*-dioxin, dibenzofuran, and biphenyl congeners, relative to 2,3,7,8-tetrachlorodibenzo-*p*-dioxin, for producing early life stage mortality in rainbow trout (*Oncorhynchus mykiss*). *Aquat. Toxicol.* 21: 219–238.

Walker, M.K., L.C. Hufnagle, M.K. Clayton, and R.E. Peterson. 1992. An egg injection method for assessing early life stage mortality of polychlorinated dibenzo-*p*-dioxins, dibenzofurans, and biphenyls in rainbow trout (*Oncorhynchus mykiss*). *Aquat. Toxicol.* 22: 15–38.

Walsh, G.E., M.J. Yoder, L.L. McLaughlin, and E.M. Lores. 1987. Responses of marine unicellular algae to brominated organic compounds in six growth media. *Ecotoxicol. Environ. Saf.* 14: 215–222.

Wang, W. 1987. Root elongation method for toxicity testing of organic and inorganic pollutants. *Environ. Toxicol. Chem.* 6: 409–414.

Wang, W. 1991. Literature review on higher plants for toxicity testing. *Water Air Soil Pollut.* 59: 381–400.

Wei, L.Y. and C.R. Vossbrinck. 1992. Degradation of alachlor in chironomid larvae (Diptera: Chironomidae). *J. Agric. Food Chem.* 40: 1695–1699.

White, R., S. Jobling, S.A. Hoare, J.P. Sumpter, and M.G. Parker. 1994. Environmentally persistent alkylphenolic chemicals are estrogenic. *Endocrinology* 135: 175–182.

Wiederholm, T., A.-M. Wiederholm, and G. Milbrink. 1987. Bulk sediment bioassays with five species of fresh-water oligochaetes. *Water Air Soil Pollut.* 36: 131–154.

Winner, R.W. 1988. Evaluation of the relative sensitivities of a 7-d *Daphnia magna* and *Ceriodaphnia dubia* toxicity tests for cadmium and sodium pentachlorophenolate. *Environ. Toxicol. Chem.* 7: 153–159.

Woods, P.E., J.D. Paulauskis, L.A. Weight, M.A. Romano, and S.I. Guttman. 1989. Genetic variation in laboratory and field populations of the midge, *Chironomus tentans* Fab.: Implications for toxicology. *Environ. Toxicol. Chem.* 8: 1067–1074.

Yunker, M.B. and R.W. Macdonald. 1995. Composition and origins of polycyclic aromatic hydrocarbons in the Mackenzie River and on the Beaufort Sea Shelf. *Arctic* 48: 118–129.

Zacharewski, T. 1997. *In vitro* bioassays for assessing estrogenic substances. *Environ. Sci. Technol.* 31: 613–623.

Zieris, F.-J., D. Feind, and W. Huber 1988. Long-term effects of 4-nitrophenol in an outdoor synthetic aquatic system. *Arch. Environ. Contam. Toxicol.* 17: 165–175.

8

Microbiological Aspects
of Bioremediation*
In Collaboration with Ann-Sofie Allard

SYNOPSIS An overview is given of the potential for bioremediation of a range of contaminated sites. Most of the contaminants found belong to widely dispersed groups that are widespread and are generally persistent or toxic. Although attention has been focused on microbiological aspects of their application, a brief account of the application of higher plants and the role of bacteria in the rhizosphere is included. It is pointed out that a successful program requires integrated input from geologists, engineers, chemists, and microbiologists. It is emphasized that a protocol must be available for evaluating the success of the procedures that have been implemented. A discussion is presented on phytoremediation since this interfaces closely with microbial processes in the soil. Attention is directed to the effect of exposure on contaminants after deposition and to critical issues, including partial degradation, formation of metabolites, and recalcitrance of specific components in complex mixtures, and the cardinal issue of bioavailability after aging. An attempt is made to discuss the basic aspects of the biodegradation of components specific to the various sites and to illustrate the outcome of experiments in bioremediation in laboratory-based, pilot-scale, or full-scale field operations. Brief discussion is given of some less commonly perceived contaminants that may simultaneously be present.

8.1 Introduction

8.1.1 Orientation

This chapter attempts to provide an overview of the application of principles outlined in previous chapters to the bioremediation of contaminated terres-

---

* This chapter is an updated and considerably extended version of a review by the authors "Bioremediation of organic waste sites: a critical review of microbiological aspects" published in *International Biodeterioration* and *Biodegradation*, Volume 39, 1997, and is published here with the kind consent of the publishers Elsevier Science Limited.

trial sites. Cardinal microbiological processes will be briefly addressed but, for metabolic details, references will be made as far as possible to the chapters. Whereas Chapter 6 is devoted to a range of structural groups widely found among xenobiotics, this chapter is arranged according to the nature of the contaminated sites. These involve a wide spectrum of compounds that have been deposited onto or leaked into terrestrial sites. Not all of these compounds have been discussed in Chapter 6, and an attempt is made to describe some relevant experiments on bioremediation by using a selection of examples as illustration. It should be emphasized that many aspects of the bioremediation of terrestrial sites impinge on the aquatic environment due to (1) leaching of compounds from the terrestrial environment and (2) the facility of some with appreciable water solubility to be disseminated in groundwater.

There are several reasons for the increasing concern over the increasing volume of solid waste — both industrial and domestic.

1. Some sites occupy valuable land in urban areas, and it is therefore imperative that its reuse for commercial or domestic building must be insured against present or future health hazards.

2. There is widespread environmental concern over adverse long-term effects of landfills used for industrial waste because of potential hazard from the transformation of the components and discharge of leachate into water courses.

3. Although the volume of domestic waste is huge, its heterogeneity will be diminished with adoption of source-separation by households. It should also be emphasized that the sludge from industrial water treatment facilities may be sent to landfills that may therefore contain both undegraded — and presumably more recalcitrant — substrates and include terminal metabolites that may have been produced.

The range of compounds in nondomestic landfills is extremely wide and includes not only industrial waste from the manufacture of chemicals, but also substantial volumes of military waste containing a number of explosives and chemical warfare agents. It should also be appreciated that continuous atmospheric precipitation will have occurred, and therefore virtually any established atmospheric contaminant may also be present. Indeed, the atmospheric input into Lake Michigan has been found to exceed that from landfills (Hornbuckle et al. 1995), so that important interphase partitions should be taken into account. The range of relevant compounds to which attention should be directed is dependent on the availability of analytical procedures both for the original compounds and for their possible transformation products. An illustrative example is provided by the development of chromatographic procedures that have been applied to carbon black and coal tar and that have revealed the presence of PAHs with up to 9 rings (Bemgård et al. 1993).

Because of the toxicity of most of the compounds that are discussed here, there is increasing demand for remediation of such sites that involves removal of the offending compounds. In this overview, the concept will be enlarged to include remediation of riverine sediments in the vicinity of industrial production, and contamination of groundwater by low-molecular-mass compounds that have penetrated the overlying cover. The emphasis is placed on microbiological aspects, whereas engineering considerations that are of cardinal importance lie beyond the expertise of the authors. An attempt is made to summarize (1) a range of sites that have attracted attention, (2) the contaminants involved, (3) the results of full-scale and laboratory model experiments, and (4) the relevant microbiological aspects including the organisms, the pathways of biodegradation, and biotransformation, and (5) determinants, particularly bioavailability (Chapter 3, Sections 3.2.2 and 3.2.3, and Chapter 4, Section 4.6.3), and oxygen concentration (Chapter 4, Section 4.6.1.2). The review by Bollag (1992) addresses the important issue of fungus-mediated polymerization reactions of phenols and anilines that are noted again in Sections 8.2.3 and 8.2.4.4. Biological treatment of wastewater and of industrial exhaust gases is excluded: there is moreover an increasing tendency for recovery rather than destruction of valuable components of these discharges. Although this chapter is self-contained and reference is made to sections of preceding chapters for additional details, some degree of duplication is unavoidable.

It may seem that undue prominence has been given to PAHs and related compounds. This illustrates clearly the interaction of different, yet collectively important, issues.

1. *History* — The interest attached to sites containing these compounds that antedate the introduction of modern synthetic chemicals. Gas works and the simultaneous production of tar residues are common in all countries that used coal as a source of gas lighting and heating. Many of these are now abandoned, and the industry has been displaced by dominance of electricity and oil.

2. *Urban planning* — For convenience in the distribution of gas, the production sites were often placed close to areas of urban populations. As cities expand, these sites are therefore of substantial interest for new building, and the existence of these gas works residues presents a potentially serious problem.

3. *Recalcitrance* — It has been demonstrated that although some of the low molecular mass PAHs are biodegradable, those with more than four rings generally present a serious degree of recalcitrance.

4. *Health hazard* — it was established in the 1920s and 1930s that many components of coal tar were carcinogenic and that this was due to the presence of PAHs containing four to six aromatic rings (Chapter 7, Section 7.7.1). Their existence therefore presents a potential threat to residents, and office and factory users.

Previous chapters have dealt with important principles governing the distribution, the fate, and the toxicity of organic compounds in the aquatic environment. These may, with appropriate modification, be extended also to terrestrial systems. Substantial effort has been directed to a wide range of agrochemicals, and a few of these have already been used as illustration in these earlier chapters. Some important conclusions from these studies have a direct bearing on the subject of this chapter.

1. After successive applications of an agrochemical, a population develops that has enhanced capability for their degradation. This is negative for the functioning of the chemicals, but would be positive from the view of bioremediation. This is discussed in Chapter 4, Section 4.6.4, and an illustration of this positive effect in attenuating the concentration of chlorobenzene at a contaminated site is given in Section 8.2.4.2.

2. Many organic compounds partition into the organic matrix of soils and sediments, and these associated residues may be both less toxic and less amenable to biodegradation than the free compounds. This presents an adverse prognosis for bioremediation. This has been discussed in Chapter 3, Sections 3.2.2 and 3.2.3, and in Chapter 4, Section 4.6.3.

3. The composition of the original contaminants in sediments may be altered during deposition on land sites, either by biotic or abiotic transformation, or by loss through volatilization of the more volatile components (Chapter 3, Section 3.4.2), for example, alkylnaphthalenes in PAHs or partially dechlorinated PCBs that have been subjected to drying (Bushart et al. 1998). In addition, bioavailability may be altered by cycles of wetting and drying (White et al. 1998), so that the bioremediation of dredged material may be adversely affected.

In this overview, no account is given of procedure for genetic manipulation to improve the effectiveness of bacteria, and the account is devoted to indigenous organisms or to inoculation with naturally occurring organisms of established metabolic relevance. Procedures for the improvement of microorganisms for their applicability to treatment programs have been presented (Timmis et al. 1994). In many situations, the waste will have lain for considerable lengths of time with the result that a natural flora of degradative microorganisms will have developed through enrichment under natural conditions. This is well established for agrochemicals (Racke and Coats 1990), and it is, therefore, highly desirable to optimize the activity of such organisms in bioremediation programs.

There is an enormous literature in this area so that only a small fraction of this is cited: only primary publications have been cited, although as far as possible, references to review articles have been given where extensive literature citations may be found. Attention is drawn to overviews that offer valuable and stimulating general perspectives (Errampalli et al. 1997; Strauss 1997).

A number of contaminated sites have been used for illustration, though the extent to which bioremediation has received practical application is variable. For a number of reasons, emphasis is placed on basic microbiological issues that have emerged from controlled laboratory experiments.

1. These illustrate many cardinal issues that should be taken into consideration in scale-up to field exploitation.
2. They reveal fundamental issues such as the reasons for using biphenyl to facilitate degradation of PCBs, or of phenol or toluene to induce enzymes for the degradation of trichloroethene.
3. Mixtures of a number of compounds are often involved, and among groups such as PAHs and PCBs there are wide variations in the biodegradability of individual components.

For the sake of completeness, attention is drawn to some alternative procedures.

- Supercritical carbon dioxide extraction of hydrocarbon contaminants (Schleussinger et al. 1996)
- Extraction by vapor stripping (Siegrist et al. 1995)
- Thermal blanket *in situ* removal and destruction of PCBs (Iben et al. 1996)
- The application of sorbents (Verstraete and Devlieger 1996)
- An electrokinetic process combined with *in situ* chemical degradation (Ho et al. 1999).
- The application of enzymes (Bollag 1992).

A valuable review (Hamby 1996) has been given that summarizes chemical and physical treatments of soils, and contaminated ground and surface waters. Some procedures for chemical destruction of selected xenobiotics have been given in Chapter 4, Section 4.1.1 and 4.1.3, comments on sequential microbial and chemical reactions in Chapter 4, Section 4.2.3, and an example of combined biological and chemical remediation is given in Section 8.2.1.

8.1.2 Microbiological Considerations

Essential aspects of the metabolism of xenobiotics have been given in Chapters 4 and 6. Only cursory discussion is therefore given of classes of compounds that have already been reviewed; for others, some additional comments are given. Although a valuable review on microbiological aspects of bioremediation of sites contaminated with organochlorine compounds has been given

(Morgan and Watkinson 1989b), it is convenient to provide a short summary of some cardinal issues that are equally significant for all xenobiotics.

Natural Ecosystems

Many different microorganisms will invariably be present, and the degradation of a xenobiotic is frequently dependent upon the activity of several microorganisms, even though individual reactions may be carried out by only a single member of the consortium. Illustrative examples include the degradation of parathion (Daughton and Hsieh 1977) and chloropropionamides (Reanney et al. 1983), and the important hydrogen-transfer reactions and syntrophic associations among anaerobic bacteria (Schink 1991; 1997).

It is very seldom that in natural ecosystems only a single substrate is present. It is therefore important to examine how the regulation of degradative pathways may be affected. The regulatory pathways for monocyclic aromatic compounds such as phenol, benzoate, and hydroxybenzoates have been reviewed in detail (Ornston and Yeh 1982). Whereas biodegradation of a substrate is frequently induced by growth with an early metabolite, for example, of biphenyl (Furukawa et al. 1983) and naphthalene (Barnsley 1976) by salicylate, the metabolites in the lower pathways may suppress induction of essential degradative enzymes. For example, although the enzymes for the degradation of toluene in *Pseudomonas putida* strain pWWO are induced during succinate-limited growth, they are suppressed when succinate is in excess (Austen and Dunn 1980; Duetz et al. 1994). Where plausibly comparable degradation pathways exist, however, e.g., in the biodegradation of PAHs, a single organism may display wide versatility. For example, *P. paucimobilis* strain EPA505 is capable of degrading fluoranthene, pyrene, benz[*a*]anthracene, chrysene, benzo[*a*]pyrene, and benzo[*b*]fluoranthene (Ye et al. 1996). For halogenated substrates, the situation is more complex. Whereas degradation of PCBs is often carried out by organisms enriched by growth with biphenyl itself (Hernandez et al. 1995), naphthalene is a much less effective inducer (Pellizari et al. 1996). The degradation of toluene and of chlorobenzene generally proceeds by respective *extradiol* (2,3) and *intradiol* (1,2) fission of the initially formed catechols. Since, however, 3-chlorocatechol inhibits the activity of the 2,3-dioxygenase, the degradation of toluene and chlorobenzene are generally incompatible except in strains such as *Pseudomonas* sp. strain JS150 that have developed a strategy for overcoming this limitation (Haigler et al. 1992). Further discussion of this issue with further details of how this incompatibility may be overcome has been given in Chapter 4, Section 4.8.2. A strain of *Comamonas testosteroni* was, however, able to degrade both 4-chlorophenol and 4-methylphenol (Hollender et al. 1994), and *Phanerochaete chrysosporium* to degrade simultaneously chlorobenzene and toluene (Yadav et al. 1995). Comparable incompatibility due to the synthesis of toxic metabolites mayis also displayed in the degradation of chlorobenzoates (Reineke et al. 1982). An additional example is the synthesis of protoanemonin by organisms that degrade 4-chlorobiphenyl by *intradiol* fission of 4-chloroatechol (Blasco et al. 1995): protoanemonin may then inhibit the growth of other PCB-degrading organisms in terrestrial systems (Blasco et al. 1997).

In addition, some of the components may directly inhibit degradation by toxification of the relevant organism. The example of azaarenes in groundwater at a wood preservation site that inhibit PAH degradation (Lantz et al. 1997) is noted again in Section 8.2.1.

Alternative Electron Acceptors

In the absence of molecular oxygen, a number of alternative electron acceptors may be used: these include nitrate, sulfate, selenate, carbonate, chlorate, Fe(III), Cr(VI), and U(VI), and this has been discussed in Chapter 4, Section 4.3.32. In the succeeding sections, attention is directed primarily to the role of nitrate, sulfate, and Fe(III) with only parenthetical remarks on Cr(VI) and U(VI). The role of nitrate and sulfate as electron acceptors for the degradation of monocyclic aromatic compounds is discussed in Section 8.2.1, and the particularly broad metabolic versatility of sulfate-reducing bacteria should be especially noted.

Aging

This is one of the cardinal issues is that of aging (Hatzinger and Alexander 1995), and the term is applied to processes — whose mechanisms are only incompletely established — whereby organic compounds become associated with polymeric components of the soil matrix with the result that these initially monomeric compounds are no longer accessible to the relevant degradative microorganisms. This implies that evaluations of biodegradability that do not take this phenomenon into account may seriously underestimate the persistence of the compound. A critical issue is the degree of reversibility of these complexes since slow release by leaching could introduce the xenobiotic into the aqueous phase and thereby contaminate groundwater. Although this process has been suggested as a mechanism whereby in the short run their adverse effect may be diminished, it seems unwise to conclude that this is an acceptable long-term solution to diminishing their environmental impact (Alexander 1995b). This issue has been discussed in Chapter 3, Section 3.2 for partitioning between the water and/sediment/soil phases partitioning, the principles of which may readily be extrapolated to the terrestrial environment for soil/water partitioning. Important examples of this problem occur in the biorestoration of PAH-contaminated soil that is discussed in Section 8.2.1.

Biofilms

These are ubiquitous and only a few brief comments are given here in the context of biodegradation. Both in aquatic and terrestrial systems, bacteria may be present in the form of biofilms that are formed by production of polymeric products from the organisms themselves, and this may serve both to increase their penetration of the matrix and to localize the organisms and thereby prevent their loss from the system (Wolfaardt et al. 1995). Their role in wastewater trickling filter treatment systems is, for example, well established, and an example of their possible significance in the biorestoration of TCE-contaminated groundwater is given in Section 8.2.6.2 and the biodegradation of toluene in a contaminated stream in Section 8.2.6.1. Attention has been drawn in Chapter 4, Section 4.6.1.4 to the potential role of associations between bacteria and particulate matter in aquifers.

Metabolites

The significance of metabolites is a recurring theme in this overview, and their formation may limit the valuecompromise, the use of analyses for only the original substrate as a measure of biodegradation. Examples include the formation of chrysene dicarboxylate from benzo[*a*]pyrene (Schneider et al. 1996), and phenanthrene carboxylates from benz[*a*]anthracene (Mahaffey et al. 1988). In addition, their formation may be adverse either for the organism itself or for other organisms in the ecosystem. Attention has been drawn above to the role of potentially toxic metabolites in the regulation of biodegradation (Chapter 4, Section 4.8.2), but metabolites may undergo further purely chemical reactions to compounds that are terminal products. Examples are given in Chapter 4, Section 4.2.3, and include the formation of 5-hydroxyquinoline-2-carboxylate from 5-aminonaphthalene-2-sulfonate (Nörtemann et al. 1993), or of benzo[*b*]naphtho(1,2-*d*]thiophene from benzothiophene (Kropp et al. 1994). Microbial metabolites may be toxic to both bacteria and higher organisms: illustrative examples include (1) the toxicity to PCB-degrading bacteria of protoanemonin synthesized during the biodegradation of 4-chlorobiphenyl (Blasco et al. 1997), (2) of 3,4,3′,4′-tetrachloroazoxybenzene produced from 3,4-dichloroaniline (Lee and Kyung 1995), (3) the greater carcinogenicity of benzidine produced from 3,3′-dichlorobenzidine (Nyman et al. 1997), itself a reduction product of azo dyes and pigments that were formerly used, and (4) of products formed by microbial methylation of the chloroguaiacols to sensitive life stages of fish (Neilson and Allard 1984). Metabolites may also play a role in the association of the substrate with humic and fulvic acid components. For example, naphth-1-ol that is an established fungal metabolite of naphthalene may play a role in the association of naphthalene with humic material (Burgos et al. 1996), and ^{13}C-labeled metabolites of (9-^{13}C)-anthracene including 2-hydroxyanthracene-3-carboxylate and phthalate that were not extractable with acetone or dichloromethane could be recovered after alkaline hydrolysis (Richnow et al. 1998).

As illustration of unusual transformation products, the demonstration of volatile compounds of both Mo and W in gases evolved from a municipal landfill (Feldmann and Cullen 1997) may be given. These compounds were tentatively identified as $Mn(CO)_6$ and $W(CO)_8$ but neither the mechanism of their formation nor their potential health hazards have been resolved.

Utilization of N, P, S

Many substrates contain N, P, or S, and degradative organisms may utilize one or more of these, leaving the major part of the substrate intact. This is particularly important for munitions-related compounds with a high N/C ratio, while the addition of carbon sources in these circumstances may lead to the favorable development of anaerobic or facultatively anaerobic microorganisms.

Organisms

There are no organisms or groups of organisms universally applicable to bioremediation. Their compatibility with the site of application should be considered, as well as their physiology and biochemical potential. It should, however, be pointed out that some groups of organisms are metabolically highly versatile and are capable of degrading a wide spectrum of substrates. Examples of these include (1) organisms formerly classified within the genus *Pseudomonas* (Chapter 5, Section 5.9), (2) bacteria using C_1 substrates such as methane and methanol: and examples of their application to trichlororethene degradation are cited in Section 8.2.6.2, (3) Gram-positive bacteria belonging to the genera *Rhodococcus* and *Mycobacterium* to which increasing attention has also been directed (Chapter 4, Section 4.3.2).

Whereas in aquatic systems, bacteria are generally considered to be the major agents of biodegradation, the role of fungi in terrestrial systems may be equal or greater. Among fungi, considerable attention has been devoted to the metabolically versatile white-rot fungi (Shah et al. 1992; Barr and Aust 1994), and several examples are cited later. It is important to appreciate the wide range of organisms belonging to this group even though hitherto greatest attention has hitherto probably been devoted to *Phanerochaete chrysosporium*.

Substrate concentration

There is substantial evidence from the aquatic environment for the existence of threshold concentrations below which rates of degradation of xenobiotics are slow or even negligible (Alexander 1985). A number of hypotheses have been put forward including the critical concentrations required for induction and maintenance of the degradative enzymes (Janke 1987). Whether this also occurs in the terrestrial environment has been only superficially explored but may be important when only low concentrations of the substrate are available by desorption. Organisms may, however, be present that are adapted to such concentrations, and examples are given in Chapter 4, Section 4.6.2. Studies with *Burkholderia* sp. strain PS14 are instructive: 1,2,4,5-tetrachloro- and 1,2,4-trichlorobenzenes at concentrations of 500 nM were 70% mineralized in mineral liquid medium within 1 h and in spiked sterilized soil to ca. 80% after 100 min (Rapp and Timmis 1999). Although in this experiment there was no evidence for a threshold concentration >0.5 nM, the situation in aged soils in which mass transfer is limiting may be significantly less positive.

Temperature

Many contaminated sites are located in areas with temperatures considerably less than 15°C, and investigations into the role of temperature have therefore been carried out. Some of the cardinal results are discussed in Chapter 4, Section 4.6.1.1, and it may be concluded that although the rates at 4 to 5°C may be lower, selection of organisms for adaptation to the ambient temperature will generally ensure that degradative activity is retained. There may, however, be selective degradation of specific groups of components in mixtures such as PCBs.

The Role of Higher Plants: Phytoremediation

The use of higher plants (Schnoor et al. 1995; McIntyre and Lewis 1997; Salt et al. 1998) has been discussed: in the light of the role of both the plants themselves and rhizosomal bacteria that make use of plant exudates for growth, it is appropriate to discuss the problem in a few paragraphs in terms of the role of bacteria in the rhizosphere.

Higher plants may play different roles in remediation (Salt et al. 1998): they may carry out degradation of the contaminant themselves, or they may excrete compounds into the soil that promote microbial growth and thus indirectly mediate degradation. Although the structures of these have not been identified, it has been demonstrated that higher plants excrete a range of phenolic compounds (Fletcher and Hegde 1995; Hegde and Fletcher 1996), and some of these potential exudates are able to support the growth and metabolism of bacteria that degrade a range of PCB congeners (Donnelly et al. 1994; see Chapter 6, Section 6.5.1.1). There is therefore an intimate plausible relation between metabolites excreted by higher plants and their utilization to support bacterial degradation of xenobiotics in the rhizosphere. In addition, it has been shown that higher plants themselves may excrete enzymes that are able to degrade xenobiotics: for example degradation of 2-chlorobenzoate by exudates from Dahurian wild rye (*Elymus dauricus*) (Siciliano et al. 1998). It has also been shown that concentrations of 2-chlorobenzoate aged in soil for 2 years could be reduced by growth of Dahurian wild rye in the presence of bacterial inoculants that did not adversely effect the indigenous microflora (Siciliano and Germida 1999).

Mycorrhizal fungi play an important role in the nutrition and survival of higher plants, and the importance of their associated bacteria has been recognized. In a study of the bacterial communities in the mycorrhizospheres of *Pinus sylvestris–Suillus bovinus* and *P. sylvestris–Paxillus involutis*, a wide range of both Gram-negative and Gram-positive bacteria was recognized (Timonen et al. 1998). Among these were taxa such as *Burkholderia cepacia* and *Paenibacillus* sp. in which strains with established biodegradative capability have been recognized. The relevance of rhizosphere bacteria was illustrated by the demonstration of a biofilm community at the interface of the petroleum-contaminated soil/mycorrhizosphere, and the isolation from this of strains of *Pseudomonas fluorescens* biovars that were able to grow at the expense of 1,3-dimethyl benzene and 3-methylbenzoate, brought about fission of catechol, and harbored the plasmid-borne genes *xylE* and *xylMA* (Sarand et al. 1998).

1. Application of higher plants should, however, be devoted to a more thorough understanding of their metabolic capacity for xenobiotics, and the extent to which they may contribute to discharge of volatile metabolites to the atmosphere. A classic review (Sandermann 1994) provides references to a number of important biotransformations carried out by plants that involve successive detoxification by metabolism followed by conjugation. Plant cytochrome P-450 isoenzymes are discussed, and biotransformations illustrated by those of DDT, benzo[*a*]pyrene, and di-*sec*-octylphthalate. There are complex relations between the operation of cytochrome P-450 in biosyntheisis and detoxification, and the replacement of monooxygenase activity with peroxidase activity during aging (Khatisashvili et al. 1997): these should be taken into consideration. Plants can mediate both the uptake and the biotransformation of xenobiotics; examples including benzene in the leaves of spinach (*Spinacia oleracea*) have been noted in Chapter 3, Section 3.1.3 (Ugrekhelidze et al. 1997) and Chapter 4, Section 4.3.7, and in the latter, attention was drawn to the unresolved details in the reduction and conjugation of 2,4,6-trinitrotoluene (Bhadra et al. 1999). On the other hand, uptake and translocation may not be accompanied by transformation, for example, for hexahydro-1,3,5-trinitro-1,3,5-triazine in hybrid poplars (*Populus deltoides* × *nigra* DN34) grown in hydroponic solutions (Thompson et al. 1999).

2. As illustration of their positive indirect influence, it has been shown that (a) increased degradation of trichloroethene to CO_2 took place in soil slurries from a site with a cover of *Paspalum notatum*, *Lespdeza cuneata*, *Solidago* sp., and *Pinus taeda* (Walton and Anderson 1990) and (b) prairie grasses may stimulate treatment of PAH-contaminated soil (Aprill and Sims 1990). These results provide substance for the positive effect of plants in promoting a rhizobial flora which may contribute to the metabolism of contaminants. A wide range of compounds is excreted by higher plants, and, as noted above, considerable attention has been directed to phenolics. It has also been noted in Chapter 4, Section 4.8.1 that enzymes for the biodegradation of some PCB congeners could be induced by a number of terpenoids (Gilbert and Crowley 1997) that probably originate from higher plants, and in Chapter 4, Sections 4.6.2 and Chapter 6, Section 6.5.1.1 that some coumarin, flavone, flavanol, and flavanone plant metabolites were able to mediate the bacterial dechlorination of PCBs.

3. Experiments have been carried out on the mineralization of atrazine under various conditions (Burken and Schnoor 1996). Although in the absence of plants of poplar hybrids (*Populus deltoides nigra* DN34), root exudate resulted in only a slight stimulation on the mineralization of ^{14}C-labeled atrazine, addition of crushed roots provided a greater positive effect.

Some important conclusions may be drawn from the results of these studies.

- Microbial degradation in the root system of higher plants can be significant.
- A cardinal role of higher plants is the provision of organic carbon and of exudates.
- The effectiveness depends on the nature of the soil, and, in particular, its organic content.

In a more general context, bacteria associated with the roots of plants have important metabolic potential for the biodegradation of biogenic methane and CO: for example, methanotrophs associated with aquatic vegetation (King 1994) and the oxidation of CO by bacteria associated with freshwater macrophytes (Rich and King 1998).

Alternative Strategies

Three rather different strategies for bioremediation have been employed: these involve treatment *in situ*, on-site, or off-site, and each has both advantages and disadvantages. The last two offer a more controlled approach that may be based upon scale-up of procedures developed in simulation experiments in microcosms, although all of them necessitate maximal appreciation of the underlying microbiological issues. It should be emphasized that all procedures will of necessity be site specific since they must take into account both the physical environment and basic microbiological issues. An illustrative example of a problem that may arise in pump-and-treat systems for treating contaminated aquifers is the reduced permeability at the delivery site that was due to deposition of colloidal material that was mobilized during the treatment (Wiesner et al. 1996) and was apparently enhanced by gas formation from the hydrogen peroxide used to increase the oxygen concentration in the aquifer. Strategies will also be contaminant specific, and take into account both the pathways and the regulation mechanisms for degradation of the — generally complex — range of contaminants. For a given substrate, the induction and maintenance of the degradative enzymes will depend on the environmental conditions (Guerin and Boyd 1992a): the presence or addition of readily degradable substrates may enhance the bacterial population although not necessarily those required for degradation of recalcitrant compounds; an example is given in Section 8.2.6.2 in which addition of toluene to an aquifer to facilitate the degradation of trichloroethene by indigenous bacteria may result in dominance of populations that degrade trichloroethene only poorly. In general, the principles that have emerged from the study of aquatic systems that are outlined in previous chapters dealing with aquatic systems, can be adapted to the terrestrial environment — with some important change of emphasis: the cardinal role of association of xenobiotics with organic components of the soil (Bollag and Loll 1983), the concentration of water (water potential), and the potential role of fungi.

Compounds as Objectives in Bioremediation

Few of the compounds that are the object of bioremediation programs are readily degradedable, and although many have been shown to be *degradable* in controlled laboratory experiments, the cardinal question is whether they *are* degraded in the specific environment. Biodegradation implies an appreciable degree of mineralization and not merely biotransformation. General aspects of bioremediation have been discussed by Thomas and Ward (1989), by Morgan and Watkinson (1989a,b), by Edgehill (1992), and in a book (Alexander 1995a). Discussion of the pathways followed during the biodegradation and biotransformation of a wide range of xenobiotics by both bacteria and fungi has been given in Chapter 6, while biochemical aspects of the biodegradation of PAHs and related heteroarenes haves been reviewed (given by Neilson and Allard (1998).

Evaluation of Effectiveness of Bioremediation

An essential component of any bioremediation program is the evaluation of its success. A number of strategies have been proposed as potentially attractive solutions to evaluate the number of specific degradative organisms, or of the appropriate degradative enzymes. These have been discussed in detail in Chapter 5, Section 5.8 and no further comments are needed here.

Evaluation is not as readily accomplished as might be assumed for a number of reasons (Madsen 1991; Heitzer and Sayler 1993; Macdonald and Rittman 1993; Shannon and Unterman 1993). A valuable and critical review (Dott et al. 1995) discusses the need for an assessment that takes into account all the phases involved, and suggests that the difficulty of demonstrating *in situ* bioremediation requires extensive long-term studies. An illustrative example is provided by bioremediation of an aquifer in Switzerland that was contaminated with diesel fuel. Oxidants were provided in the form of O_2 and NO_3^- but measurements of a carbon and nitrogen balance were not sufficiently precise to assess the spatial effectiveness of the treatment. A laboratory microcosm using material from the aquifer was constructed: the use of gene probes revealed the presence of bacteria belonging to different phylogenetic groups, and bacteria able to degrade toluene and 1,3-dimethylbenzene were isolated (Hess et al. 1997). It was, however, pointed out that, because of the greater detail in organisms in the aquifer, there are limitations in the extent to which the results of the laboratory experiments may be extrapolated to the field situation. Procedures for examining natural communities and their metabolic activity are discussed in Chapter 5, Section 5.8.

Some of the cardinal issues for assessing the effectiveness of bioremediation include the following:

1. Diminution in the concentration of the substrates is not alone an acceptable measure of degradation, since loss may occur by volatilization or by transformation with the formation of transient or terminal metabolites. In addition there may be a continuous input, for example, by leaching.

2. The environment is highly heterogeneous so that representative sampling generally presents a serious problem, and evaluation of the spatial effectiveness is difficult (Hess et al. 1997).

3. The extent of leaching may be difficult to evaluate since few of the systems are sufficiently enclosed to make a convincing balance of the concentrations of the substrates and their metabolites — including CO_2 or CH_4 — that are lost to the atmosphere.

The collective evaluation of biodegradability, bioavailability, toxicity to plants and earthworms, and leaching has been demonstrated (Salanitro et al. 1997).

A number of strategies have been proposed as potentially attractive solutions to evaluate the number of specific degradative organisms and of the appropriate degradative enzymes.

1. Application of PCR to specific degradative organisms has been reviewed (Steffan and Atlas 1991), and it has been applied to hydrocarbon-degrading *Mycobacterium* sp. (Wang et al. 1996b).

2. Probes for toluene-2-monooxygenase have been used to evaluate the potential number of trichloroethene-degrading organisms in an aquifer (Fries et al. 1997b). In this study, repetitive extragenic palindromic PCR (REP-PCR) of isolates was used to classify their metabolic capability.

3. The reverse sample genome probe procedure has been used to monitor sulfate-reducing bacteria in oil-field samples (Voordouw et al. 1991), further extended to include 16 heterotrophic bacteria (Telang et al. 1997), and has been applied to evaluating the effect of nitrate on an oil-field bacterial community.

4. Metabolically-related parameters have also been used, and these include establishing any of the following:

   a. The presence of metabolites determined from laboratory experiments of degradation pathways: examples include *cis*-dihydrodiols of PAHs in a marine sediment (Li et al. 1996) and of naphthalene in leachate from a contaminated site (Wilson and Madsen 1996); chlorobenzoates from the degradation of PCBs (Flanagan and May 1993), Chapter 2, Section 2.4.2, and benzyl succinates from BTEX (Beller et al. 1995), Chapter 2, Section 2.6.1.

b. Some examples have been given in Chapter 5, Section 5.6 of enzymes with a low substrate specificity that are able to accept analogue substrates with the formation of colored metabolites. This may be used to detect enzymatic activity towards the desired substrates when a large number of colonies have to be examined: examples for PAHs include toluene 2,3-dioxygenase (Eaton and Chapman 1995), naphthalene dioxygenase (Ensley et al. 1983) and manganese peroxidase (Bogan et al. 1996), for trichloroethene cometabolism toluene-2-monooxygenase (Fries et al. 1997b), and for degradation of 2,4-dichlorophenoxyacetate, 2,4-dichlorophenoxyacetate/2-ketoglutarate dioxygenase that can accept 4-nitrophenoxyacetate as a substrate with the formation of colored 4-nitrophenol (Sassenella et al. 1997). Care ought to be exercised however, since it was also shown in the last example that not all strains with established degradative capability for 2,4-dichlorophenoxyacetate were able to form 4-nitrophenol from the surrogate substrate (Sassenella et al. 1997).

8.2 Representative Sites

Various sites that involve a range of contaminants and environments have been used as illustration. These sites have been the subject of either full-scale operation or microcosm experiments that simulate the conditions that will be encountered. For each site, an account is also given of microbiological issues with the emphasis on metabolic aspects that are relevant both to appreciating the design of the strategies and also for assessing their success. It is worth pointing out that agrochemicals frequently may contain structures that are not discussed here: for example, the diphenyl ether acifluorfen contains both aromatic nitro and trifluoromethyl groups. By-products from the manufacture of such compounds could present a potentially serious problem. An illustrative example of by-products from the production of 4-chloro-trifluoromethylbenzene is given in Section 8.2.4.3.

8.2.1 Coal-Distillation Products

Introduction

Coal gas for illumination — and later for heating — was produced by the destructive distillation of coal and resulted in the production of large volumes of tar. This is used in the form of creosote for wood preservation, so that both gas works sites and impregnation facilities became heavily contaminated with coal-tar products. The range of compounds involved is enormous, but attention has been directed to groups with established toxicity or persistence: PAHs, azaarenes, thiaarenes, and phenolic compounds — mainly cresols, xylenols, and 2,3,5-trimethylphenol (Mueller et al. 1989). It should be emphasized that as a result of analytical problems due to lack of authentic samples (Biggs and Fetzer 1996), the frequency and concentrations of PAHs with more than seven rings have probably been seriously underestimated. The principal reasons for attempting remediation of such sites is the established human carcinogenicity of important PAHs including benzo[a]pyrene, dibenz[a,h]anthracene, and the dibenzopyrenes — dibenzo[a,h]pyrene-, dibenzo[a,i]pyrene, and dibenzo[a,l]pyrenes. PAHs with five and six rings should therefore be the primary target in bioremediation programs (Chapter 7, Section 7.7.1). It is important to emphasize that, even under favorable conditions, complete removal of PAHs — particularly those with more than five rings will seldom be accomplished; it is therefore important to predetermine an acceptable limit for the levels.

Several reviews that cover various important aspects have been devoted to this topic. All of these provide valuable summaries of the various technologies that have been exploited, cover important engineering and geologic aspects, and include the following:

1. A discussion of the design of different systems and factors such as temperature, pH, nutrients, and oxygen supply (Thomas and Lester 1993);

2. A summary of the maximum concentrations that must be attained at contaminated sites, and discussion and exemplification of the various strategies that may be used — *in situ*, on-site, and bioreactors (Wilson and Jones 1993);

3. An evaluation of a range of available strategies for remediation (Ram et al. 1993);

4. A discussion of factors determining the effectiveness of bioremediation including the effect of complex mixtures of contaminants, the limitation of bioavailability, the value of surfactants, and of nutrients and inoculation (Hughes et al. 1997);

5. A procedure using extrapolation of data from laboratory experiments has been given (Findlay et al. 1995).

The mobility of bacteria and substrates at a coal-tar-contaminated site has been described (Madsen et al. 1996a) and involved the ingenious use of sorbents to trap both the bacteria and the analytes. It also drew attention to the important issue of the extent to which microorganisms penetrate the contaminated matrix. The extensive background on the biodegradation and biotransformation of PAHs has been reviewed and should be consulted for more-detailed metabolic and enzymatic aspects (Neilson and Allard 1998). The regulation of metabolism has been noted above in Section 8.1.2, and the introduction of salicylate as inducer for the degradation of naphthalene (Ogunseitan and Olson 1993) has been examined. Cells of *Pseudomonas saccharophila* P15 could not grow with fluoranthene, pyrene, benz[*a*]anthracene, chrysene, or benzo[*a*]pyrene, although they were able to metabolize them, and showed considerably enhanced rates of removal and of mineralization after induction with salicylate (Chen and Aitken 1999) although the deliberate introduction of inducers such as salicylate to enhance biodegradation of PAHs such as naphthalene (Ogunseitan and Olson 1993) does not seem attractive as a generally applicable strategy.

An attempt is made to illustrate some of the complex of factors that influence the degree of success of bioremediation programs.

On-Site or Off-Site Treatment

On-site treatment in batch reactors using refinery and wood-preserving waste added to soil resulted in highly variable loss of PAHs over a year (Aprill et al. 1990). An extensive study (Ellis et al. 1991) examined a number of design features and the relative merits of on-site and off-site treatments of a creosote-contaminated site. Preliminary experiments were carried out in liquid culture and in soil pots, and showed only limited advantage of adding strains isolated from the site, and rather variable success in removal of most PAHs except naphthalene. Two kinds of large-scale experiments were also carried out: (1) *on-site* by recirculation of groundwater and addition of surfactants, hydrogen peroxide, inorganic nutrients and (2) *off-site* in concrete-lined basins with recirculation of water and mixing by surface plowing. There were differences between *in situ* and *on-site* removal, and for most PAHs studied the degree of removal was variable for PAHs with up to four rings (Table 8.1). Data for the important group of PAHs with more than four rings were unfortunately not reported.

TABLE 8.1
Comparison of the Efficiency of *in Situ* and Off-Site
Procedures for Removal of Selected PAHs

	Mean % Removal	
Substrate	In Situ	Off-Site
Naphthalene	100	100
Acenaphthalene	3	87
Phenanthrene	8	74
Anthracene	36	69
Fluoranthene	53	57
Pyrene	37	53
Chrysene	76	50

A system that combined biological and chemical treatment for destruction of PAHs in contaminated soil has been given (Pradahn et al. 1997). This takes advantage of the activity of the indigenous microflora in the contaminated soil in bioreactors with an intermediate chemical reactor that used a mixture of ferrous sulfate and hydrogen peroxide (Fenton's reagent; see Chapter 4, Section 4.1.1 to carry out oxidation to more readily degradable substrates. The efficiency in removal of a range of PAHs by this process compared with conventional biological treatment is clearly shown by the results in Table 8.2.

TABLE 8.2
Comparison of the Removal of PAHs from
Contaminated Soil by Conventional Biological
Treatment and by Combined Biological and
Chemical Treatment

PAH: No. of Rings	Conventional	Combined
2–	31	96
3–	76	97
4–	69	95
5–	22	87
6–	0	88

The Physical State of the Contaminant

Although the bioavailability of xenobiotics and its decrease with aging have been considered as limiting factors in their biodegradation (Hatzinger and Alexander 1995), laboratory studies on the rate of degradation of phenanthrene using samples from a contaminated site showed that this depended critically on the source of the inoculum within the site (Sandoli et al. 1996). Most contaminated sites are highly heterogeneous in the concentration of the contaminant, and it has been shown that for naphthalene contained in coal-tar globules, the area-dependent mass transfer coefficient for globules was 10^3 greater than when the substrate was coated on microporous silica beads, and that this was an important factor in determining the rate of mineralization (Ghoshal and Luthy 1996). The sorption–desorption of PAHs has been extensively investigated, and the role of desorption in determining their biodegradability in aged sediments has been widely accepted (references in Carmichael et al. 1997). A definitive study using ^{14}C-phenanthrene and ^{14}C-chrysene showed that in contaminated soils their rates of mineralization were much lower than the rates of desorption from spiked sediments. By contrast, for aged substrates desorption rates were essentially comparable to rates of mineralization. This suggests that the indigenous microflora may have become adapted to the low substrate concentrations available by desorption (Carmichael et al. 1997). This would be consistent with the existence of bacteria capable of using low substrate concentrations, but would imply the limited effectiveness of adding bacteria that, although metabolically active, were not adapted to use low substrate concentrations. The bioavailability of xenobiotics and its decrease with aging have been considered as limiting factors in their biodegradation (Hatzinger and Alexander 1995), and this is discussed below in the context of the role of surfactants. The results that are discussed in Chapter 3, Section 3.2.2 (Kan et al. 1998) provide a graphic illustration of the differences in desorption predicted from a reversible and an irreversible model, and the lack of correlation between chemical extractability and bioavailability (Chung and Alexander 1998) has been noted in Chapter 3, Section 3.2.3 and Chapter 4, Section 4.6.3.

An important issue emerged from the results of experiments with ^{14}C-labeled pyrene added to a pristine forest soil (Guthrie and Pfaender 1998): (1) extensive mineralization took place only in samples amended with a pyrene-degrading microbial community, (2) there was a substantially greater nonextractable fraction of label in soils containing either the natural or introduced microflora compared with an azide-treated control, (3) metabolites that could be released by acid and base extraction remained in the soil after 270 days of incubation.

These results clearly indicate the existence of a number of issues that must be resolved in the application of bioremediation to PAH-contaminated soils, and that great care should be exercised in making generalizations.

Environmental Parameters

Two aspects of general significance have been examined: the water concentration and the oxygen concentration.

Extensive studies using microcosm experiments with creosote-contaminated soil revealed losses in surface soil samples that exceeded those from the sediment, and that the use of soil slurries was more effective than solid-phase systems. The removal of benz[*a*]anthracene, benzo[*b/k*]fluoranthene, and indeno[1,2,3-*c,d*]pyrene was poor even using the slurries, although the heteroarenes — quinolines, acridine, and carbazole — were partially removed (Mueller et al. 1991a,b). Laboratory experiments using spiked soil indicated that biodegradation by inoculation with bacterial strains that degrade anthracene and pyrene was inhibited by introduction of a mineral medium, and occurred effectively only by addition of water (Kästner et al. 1998). These observations are in general agreement with the results of a study with *Pseudomonas putida* mt-2 during degradation of toluene (Holden et al. 1997). Matric water potential results from the interaction of soil water with solid surfaces and is generally the major component of total water potential: growth rates showed a shallow maximum at a matric water potential of –0.25 MPa, thereafter decreasing steadily to –1.5 MPa.

The physical design of any system therefore requires careful consideration. Whereas oxygen must be accessible, it has been shown that the degradation of pyrene by a strain of *Mycobacterium* sp. can occur at low oxygen concentration (Fritsche 1994), and this is consistent with the observation that toluene-degrading bacteria isolated from sites such as aquifer sand or groundwater with low concentrations of oxygen have lower values of K_m for O_2 and for

3-methylcatechol than do other strains (Kukor and Olsen 1996). Although the biodegradation of naphthalene in a subsurface site was oxygen limited and was not facilitated by addition of nitrate as alternative electron acceptor (Madsen et al. 1996b), degradation of anthracene, phenanthrene, and pyrene has been shown in three strains of denitrifying pseudomonads under both aerobic and denitrifying conditions (McNally et al. 1998). This is in contrast to the effectiveness of nitrate as electron acceptor for the degradation of monocyclic aromatic hydrocarbons in groundwater (Section 8.2.6.1). It is important to note that, under oxygen limitation, cultures of even strictly anaerobic methanogenic bacteria may coexist with aerobic bacteria (Gerritse and Gottschal 1993). An important observation is the toxicity of naphthalene to a naphthalene-degrading strain of *P. putida* G7 under conditions of oxygen or nitrogen limitation since both of these conditions will generally be prevalent at contaminated sites (Ahn et al. 1998).

The physical design of any system therefore requires careful consideration, and aeration may be expensive, but obligatory.

Supplementation with Fungi

There has been considerable interest in the use of fungi in bioremediation programs, although quinones may be produced as transient or terminal metabolites. For example, although anthracene-9,10-quinone is formed from anthracene is a terminal metabolite with *Pleurotus ostreatus* strain MUCL 29257, it is only transiently formed by *Phanerochaete chrysosporium* strain DSM 1556 (Andersson and Henrysson 1996). White-rot fungi belonging to other taxa may, however, be more effective in mineralization. For example, *Nematoloma frowardii* mineralized 8.6% of the ^{14}C-labeled phenanthrene supplied, although a substantial portion (46%) was associated with the mycelia (Sack et al. 1997; Hofrichter et al. 1998). A few examples of bioremediation using fungi are given as illustration. The supplementation of indigenous bacteria with *Ph. chrysosporium* increased the loss of phenanthrene in a former oil gasification site, although there was apparently a significant contribution from "polar metabolites" (Brodkorb and Legge 1992). Phenanthrene is not, however, representative of PAHs of toxicological significance, and the presence of unidentified metabolites may plausibly be interpreted as transformation products. The application of the white-rot fungus *Ph. sordida* to treat creosote-contaminated soil has been described, but again had limited success with PAHs having more than four rings (Davis et al. 1993). In addition to white-rot fungi, experiments have been carried out in microcosms supplemented with inorganic N and P sources using the nonlignolytic fungus *Cunninghamella echinulata* var. *elegans*. Not only the expected loss of three- and four-ring PAHs was observed, but also partial loss of benzo[*b*]fluoranthene and benzo[*g,h,i*]perylene (Cutright 1995). In this study hydrogen peroxide was used as source of oxygen, although in natural systems the presence of bacteria with catalase activity would probably effectively destroy the hydrogen peroxide.

An important series of observations with a PAH-contaminated soil (May et al. 1997) showed that although the fraction that was mineralized was low (~2.5%), the PAHs were lost by polymerization processes mediated by the fungus. It was suggested that this was consistent with (1) the formation by this organism of pyrene-1,6-quinone and pyrene-1,8-quinone from pyrene (Hammel et al. 1986) and the 1,6-, 3,6-, and 6,12-quinones from benzo[a]pyrene (Haemmerli et al. 1986), and (2) the tendency of quinones to polymerize and associate with soil humic material. The loss of the larger PAHs was highly variable ranging from ~70% for benzo[b]fluoranthene to ~5.5% for benzo[a]pyrene.

It is worth drawing attention to the significance of other issues. In natural ecosystems, other microorganisms including bacteria are almost always present and it has been shown that, in experiments using [7,10-^{14}C]benzo[a]pyrene, incubation for 215 days with *Bjerkandera* sp. strain BOS55 alone resulted in the formation of 13.5% $^{14}CO_2$ and 61% of labeled metabolites in the water phase. Addition of microbial cultures from forest soil at Day 15 increased the $^{14}CO_2$ evolution to 32% and reduced the water-soluble metabolites to 18% (Kottermann et al. 1998).

There a number of important issues to which attention is drawn, and some of these have been only incompletely resolved.

The Range of Substrate Contaminants

It should be emphasized that the development of suitable chromatographic columns for GC analysis has begun to reveal the presence of PAHs with molecular masses greater than 300. Those with more than seven rings are probably ubiquitous, although their determination is made difficult by the inaccessibility of reference compounds and by their long retention times on GC columns (Biggs and Fetzer 1996). PAHs with molecular mass >300 that occur in both carbon black and the NIST SRM 1597 coal-tar extract include dibenzo[a,j]coronene (M_r 400) with nine rings, dinaphtho[2,1-a : 2,1-h]anthracene (M_r 378) and benzo[a]coronene (M_r 350) with eight rings (Bemgård et al. 1993). The degradability of such compounds is unknown, and they may possibly contain carcinogenic representatives. Alkylated phenols are major components of creosote, and their aerobic degradation is accomplished by either of two pathways: (1) hydroxylation to catechols followed by *extradiol* ring fission or (2) oxidation of the methyl group to carboxyl followed by hydroxylation to hydroquinones and fission by the gentisate pathway (Chapter 6, Section 6.2.1; Bayly et al. 1988).

The Range of Degradative Bacteria

Bacteria vary considerably in their capacity to degrade a range of PAHs. Some degrade only the more readily degradable naphthalene and phenanthrene, but phenanthrene, pyrene, and fluoranthene do not seem to be uniformly recalcitrant (Chapter 6, Section 6.2.3). *Pseudomonas paucimobilis* strain EPA 505 can degrade fluoranthene, pyrene, benz[a]anthracene, chrysene, benzo[a]pyrene, and benzo[b]fluoranthene (Ye et al. 1996), and Gram-positive mycobacteria have attracted increasing attention: for example, a strain of *Mycobacterium* sp. degrades pyrene, benz[a]anthracene, and benzo[a]pyrene (Schneider et al. 1996). Extrapolation of results from the more-readily degraded naphthalene and phenanthrene should, however, be carried out with caution.

Bioavailability and the Use of Surfactants

This is a cardinal issue with many sometimes contradictory facets that merits more-detailed consideration than can be given here. Strategies used by cells for substrates with low or negligible water solubility have been discussed in Chapter 4, Section 4.6.2., and many of these issues are relevant to strategies for bioremediation. The accessibility of the substrates at sites with a long history of contamination is a serious limitation to bioremediation and attempts have been made to increase this by addition of surfactants (Grimberg et al. 1996). The additional role of bacteria associated with particulate material, and the possible role of surfactants in promoting the adhesion of bacteria to surfaces has been noted briefly in Chapter 4, Section 4.6.1.4. A review (Volkering et al. 1998) has summarized data from experiments with alkanes and PAHs that used a range of surfactants. A few of the sometimes contradictory aspects of the role of surfactants will be summarized briefly.

1. This is a complex issue and significant differences have been found among different organisms (Guerin and Boyd 1992b; Crocker et al. 1995), and indeed incorporation of PAHs into micelles may make them less accessible (Volkering et al. 1995).

2. It has been shown in biodegradation experiments using a mixed bacterial culture, that components in a mixture of naphthalene, phenanthrene, and pyrene were bioavailable from the micellar phase of Triton X-100, though the degree of bioavailability decreased with increasing surfactant concentration (Guha et al. 1998). A model was constructed that could be used to determine optimal surfactant concentration for criteria including maximum rates of biodegradation.

3. There is also evidence that, as for alkanes, biosurfactants may be produced during growth with PAHs (Déziel et al. 1996), although the practical implementation of this may be limited by the requirement of a suitable monomer such as glucose or mannitol for their synthesis. *P. aeruginosa* GL1 that was part of a community that degraded PAHs produced a rhamnolipid in the late growth stage of growth with glycerol or *n*-hexadecane that inhibited the growth of *Bacillus cereus* and *Rhodococcus erythropolis* (Arino et al. 1998).

4. The biodegradation of pyrene, chrysene, fluoranthene, benz[a]anthracene, dibenz[a,h]anthracene, benzo[a]pyrene, and

coronene by *Stenotrophomonas maltophilia* has been studied in the presence of a range of synthetic surfactants (Boonchan et al. 1998). Non-neutral surfactants were toxic, biodegradation was inhibited by the neutral Igepal CA-630, and the positive enhancement of removal of substrates was generally low — in the range of 10%.

5. A comparison was made of the mineralization of ^{14}C-fluoranthene by two strains each of *Mycobacterium* sp. and *Sphingomonas* sp. (Willumsen et al. 1998). Triton X-100 decreased the rate of mineralization of the *Mycobacterium* strains, inhibited mineraliztion by *Sphingomonas* sp. strain EPA505, and diminished the survival of both Gram-negative strains during 50-h incubation. It is worth noting that the *Sphingomonas* strain EPA 505 is able to degrade a wide range of PAHs (Ye et al. 1996), and would therefore otherwise be a potential candidate for the bioremediation of PAH-contaminated sites. The differential effects of the neutral surfactants 2,2,4,4,6,8,8-heptamethylnonane and Triton X-100 on the formation of *cis*-dihydrodiols of naphthalene and phenanthrene by *Pseudomonas* sp. strain 9816/11 and *S. yanoikuyae* strain B8/36 further illustrated the significance of both the organism and the structure of the surfactants (Allen et al. 1999).

6. A lipoprotein surfactant is produced during growth of *P. marginalis* strain PD-14B with succinate, and prevented flocculation of cells in media containing anthracene, acenaphthylene, naphthalene, and chrysene (Burd and Ward 1996).

7. Increased removal of phenanthrene from soil columns spiked with the rhamnolipid mixture synthesized by *P. aeruginosa* UG2 has been demonstrated and shown to depend both on increased desorption of the substrate and on partitioning into micelles (Noordman et al. 1998). On the other hand, the addition of the biosurfactant from the same strain of *P. aeruginosa* strain UG2 or of sodium dodecyl sulfate had no effect on the rate of biodegradation of anthracene and phenanthrene from a chronically contaminated soil, though at concentrations of sodium dodecyl sulfate >100 mg/kg there was a marked inhibition of the degradation of fluoranthene, pyrene, benz[*a*]anthracene, and chrysene (Deschênes et al. 1996). It was suggested that this was due to competition with increasingly dominant surfactant-degrading microorganisms, and this is supported by evidence that the addition of nonionic surfactants (alcohol ethoxylates) was more effective in enhancing biodegradation of PAHs in coal-tar-contaminated soil than the more readily degradable glycoside surfactants (Madsen and Kristensen 1997).

8. During degradation of artificial surfactants, the consumption of oxygen may, however, result in diminished PAH degradation (Tiehm et al. 1997), so that surfactants with lower biodegradability would be advantageous from this point of view.

The relation between the type and concentration of surfactant and the stimulation of biodegradation is therefore complex and not so far completely understood (Burd and Ward 1996), although there seems to be general agreement that supplementation with surfactants is one of the factors that may enhance bioremediation of contaminated sites (Walter et al. 1997). Some cardinal issues appear to include the following (Willumsen et al. 1998):

- Diminished bioavailability of substrates from micelles;
- Toxicity or inhibition of biodegradation by some non-neutral synthetic surfactants as a result of surfactant-induced permeabilization of the cells;
- Toxicity as a result of the surfactant-enhanced PAH concentration in the aqueous phase;
- Adverse effect of surfactants on bacterial adhesion to surfaces in the matrix;
- Adverse effect of readily degraded surfactants that diminish the oxygen concentration available to the substrate-degrading organisms;
- The need to maintain a high density of the substrate-degrading organisms.

Heteroarenes

It is important to appreciate that not only PAHs may be present, but also the corresponding analogues containing nitrogen (azaarenes) and sulfur (thiaarenes): this important group of contaminants has been reviewed in detail by Herod (1998), and some azaarenes such as dibenzo[*a,h*]acridine and 7*H*-dibenzo[*c,g*]carbazole are carcinogenic. The biodegradation — and therefore the possibility of remediating sites contaminated with heteroarenes — presents a number of unresolved issues. The fate of azaarenes seems not to have been examined directly in this context, while that of thiaarenes has attracted attention mainly in connection with the desulfphurization of coal and coal oil (Wang et al. 1996a; Macpherson et al. 1998). Azaarenes do, however, have an inhibitory effect on the biodegradation of PAHs even in a versatile strain such as *Sphingomonas paucimobilis* strain EPA505 (Lantz et al. 1997). Details of their metabolism may be found in Sections 6.3 and 6.7.4, and reference may also be made to the review (by Neilson and Allard 1988). By comparison with their carbocyclic analogues, much less is known about the biodegradation of azaarenes, although the biochemistry of the aerobic degradation of quinolines (including methyl- and carboxyquinolines) has been extensively investigated by Lingens and his colleagues (Schwartz and Lingens 1994) (Chapter 6, Section 6.3.1). As a broad generalization, it may be stated that the initial reaction frequently involves formal hydroxylation (addition of the elements of H_2O followed by dehydrogenation) in the hetero ring, whereas fission of the carbocyclic ring is accomplished by dioxygenation. It is worth noting that

some of these organisms may be quite restricted in their ability to use other
N-heterocyclic compounds: for example, a strain of *Rhodococcus* sp. that was
isolated by enrichment of pristine samples with quinoline was unable to use
isoquinoline or acridine, and supported only poor growth with a series of
pyridines (O'Loughlin et al. 1996). Although the degradation of acridine and
phenanthridine seems, however, to have been examined only parenthetically,
the degradation of carbazole has been examined more extensively. An
exhaustive study has revealed the production of several terminal metabolites
formed by cyclization of the fission products from the carbocyclic ring (Gieg
et al. 1996). On the other hand, the relevant benzocarbazoles seem not to have
been examined. The mechanism of the degradation of dibenzofurans has
been established in detail and is accomplished by angular dioxygenation
with fission of the furan ring, while the biodegradation of benzothiophenes
is complicated by oxidation at the sulfur atom (Chapter 6, Section 6.3.1). The
biotransformation of dimethylbenzothiophenes was studied in three strains
of *Pseudomonas* sp. growing with 1-methylnaphthalene or glucose (Kropp et
al. 1996). A variety of transformations for different isomers was observed: (1)
oxidation of the methyl groups to aldehyde groups and carboxylic acids, (2)
oxidation at the S atom to produce sulfoxides and sulfones, and (3) oxidation
in the thiophene ring to form the 2- and 3- hydroxy compounds [2(3*H*) ones,
3(2*H*) ones], and 2,3-diones. The same pseudomonads degraded 3,4-dime-
thyl- and 2,7-dimethyldibenzothiophenes by oxidative fission of one of the
benzene rings and further oxidation of the thiophene ring.

The Role of Fungi

It is important to emphasize that, in contrast to bacterial degradation of
PAHs, many fungi carry out only transformation *without* ring fission. An
important consequence is that demonstration of loss during bioremediation
experiments is equivocal. Generally a monooxygenase produces an epoxide
which is either hydrolyzed to a *trans*-dihydrodiol or rearranged to a phenol
that is often conjugated. The most comprehensive studies have been made
using *Cunninghamella* sp. (references in Sutherland et al. 1995). It has been
suggested that naphth-1-ol which is one of the expected transformation products of naphthalene is the form in which at least part of naphthalene is associated with soil matrices (Burgos et al. 1996).

Although considerable attention has been directed to the application of
white-rot fungi, several important issues should be noted, in particular their
quantitative role in the mineralization of PAHs. Some brief comments on the
oxidative enzymes of these organisms is given in Chapter 4, Sections 4.3.6
and 4.4.4.

1. Most attention has hitherto been given to species of *Phanerochaete*
    especially *Ph. chrysosporium* and *Ph. sordida*, although other species
    and other taxa (Field et al. 1992) may be of greater interest in view
    of their greater biodegradative capability.

2. Whereas transformation has generally been demonstrated — often to quinones — only relatively low levels of mineralization to CO_2 have frequently been observed, for example in *Pleurotus ostreatus* (Bezalel et al. 1996a; Hofrichter et al. 1998), and it has been shown that a substantial fraction of phenanthrene was nonextractable and presumably bound to fungal mycelia (Bezalel et al. 1996b).

Anoxic or Anaerobic Environments

In many circumstances, the subsurface environment will be anoxic or even anaerobic due to the activity of aerobic and facultatively anaerobic bacteria in the surface layers of the soil. It is therefore essential to take into consideration the extent to which anaerobic degradation may be expected to be significant. Reactions may take place under sulfidogenic or methanogenic conditions, and the occurrence of sulfate in contaminated sites containing building-material waste and the metabolic versatility of sulfate-reducing bacteria makes them particularly attractive. Details of these reactions have been given in Chapter 6, Section 6.7.

1. **Hydrocarbons** — Up until now, degradation of aromatic hydrocarbons has been limited to monocyclic representatives — particularly toluene, and there is only circumstantial evidence so far for the degradation of naphthalene and phenanthrene coupled to sulfate reduction under anaerobic conditions (Coates et al. 1996a) and the degradation of naphthalene, phenanthrene, and pyrene under denitrifying conditions (McNally et bal. 1998). This issue has been discussed more fully in Chapter 6, Section 6.7.3.2, and in the context of BTEX remediation in Section 8.2.6.1.

2. **Phenols** — The anaerobic degradation of phenol generally proceeds by carboxylation, followed by dehydroxylation, and fission of the ring after partial reduction (Brackmann and Fuchs 1993). Degradation of *o*-cresol (Bisaillon et al. 1991) and *m*-cresol (Roberts et al. 1990) has been observed under methanogenic conditions, and a summary of the possibilities under denitrifying conditions has been given (Rudolphi et al. 1991). An alternative pathway involving oxidation of the methyl groups under denitrifying conditions has also emerged (Bonting et al. 1995).

3. **Heteroarenes** — Although the anaerobic degradation of azaarenes containing several N atoms — particularly pyrimidines and purines — has been extensively studied (Chapter 6, Section 6.7.4.1), less is known about those with only a single hetero atom. Evidence for the loss of pyridine in sulfate-reducing and methanogenic aquifer slurries and of 2-methylpyridine in the former has been presented (Kuhn and Suflita 1989), although no details of the degradative pathways are available. Whereas indole and quinoline

were degraded anaerobically under nitrate-reducing, sulfate-reducing, and methanogenic conditions, neither benzofuran nor benzothiophene were degraded (Licht et al. 1996).

Abiotic Transformation

Although this issue has not been frequently addressed, it is relevant to evaluating the occurrence of *bio*remediation. Several important issues emerged from a study (Wischmann and Steinhart 1997) using a range of three-ring and four-ring PAHs.

1. In contrast to the three-ring compounds, residues of benz[*a*]anthracene, chrysene, and benzo[*a*]pyrene were found after 15 weeks incubation in compost-amended soil.

2. Neither dihydrodiols formed by bacterial dioxygenation nor phenols from fungal monooxygenation followed by rearrangement or hydrolysis and elimination were found.

3. Whereas plausible fungal oxidation products (Chapter 6, Section 6.2.2; Neilson and Allard 1998) of anthracene, acenaphthylene, fluorene, and benz[*a*]anthracene — anthracene-9,10-quinone, acenaphthene-9,10-dione, fluorene-9-one, and benz[*a*]anthracene-7,12-quinone — were found transiently in compost-amended soil, these were formed even in sterile controls by abiotic reactions.

These results clearly illustrate the care that must be exercised in interpreting the occurrence of PAH oxidation products in bioremediation experiments as evidence of biological activity.

All these factors should be carefully considered in the design and implementation of any bioremediation program, and the rather negative views presented above are supported by quoting the succinct conclusions of a study on PAH loss in laboratory microcosms using soil from a site contaminated with PAHs from a previous gas-manufacturing facility (Erickson et al. 1993). (1) PAHs were resistant to mineralization, (2) the soils contained significant populations of bacteria that were not adversely affected by the toxicity of the contaminants, (3) addition of readily degradable naphthalene or phenanthrene did not improve degradation of the indigenous PAHs, (4) PAHs were unavailable for microbial degradation, and (5) PAHs were not leached into aqueous extracts from the soil.

Since the indigenous microflora will have been exposed to the substrates, the relevant degradative pathways will often be present in populations of these microorganisms and induction of the appropriate degradative enzymes will have taken place. Indeed, many of the most versatile organisms have been isolated from such contaminated sites. If therefore organisms with the relevant degradative capacity have been demonstrated, the prognosis for remediation is good. On the other hand, although a mixed flora obtained by enrichment with creosote was able to degrade two- and three-ring PAHs, those with four or five more rings including fluoranthene that is a dominant component of creosote were unaltered (Selifonov et al. 1998). Additional readily degradable carbon sources are probably counterproductive to maintaining the required degradative capacity, although addition of inorganic nutrients and provision of suitable oxygen concentrations and of water are necessary. There are therefore several potentially serious problems that merit further critical investigation.

All these factors should be carefully considered in the design and implementation of any bioremediation program.

Conclusion

There are several important conclusions that may be drawn from the results of these studies:

- Determination of degradation, as opposed to transformation to metabolites or loss by other processes such as volatilization, is difficult to assess.

- The collective evidence suggests that although PAH components with up to four rings are probably lost, those with five or more rings may be recalcitrant.

- The presence of azaarenes — primarily quinolines, acridine, and carbazole — in creosote may inhibit or be incompatible with the degradation of PAHs.

- The presence of high concentrations of phenols may have an adverse effect of the microflora as a result of their toxicity.

- The extent to which bioavailability is a limiting factor and may be circumvented by addition of surfactants remains unresolved.

- The most serious limitation is probably the degree to which the substrates are accessible to the relevant microorganisms, while the advantage of adding surfactants is equivocal.

- Oxygen limitation may be a seriously limiting factor in deep subsurface sites, and there is evidence that degradation in the "solid" phase is less effective than in slurries.

On the basis of these limitations, a cautious view of the effectiveness of bioremediation of PAHs seems justified.

8.2.2   Refinery Waste and Stranded Oil: Petroleum Hydrocarbons

There are several different situations in which petroleum hydrocarbons pose a problem that has attracted solution by bioremediation. These include (1) oil-refinery waste and contamination of the surrounding soil, (2) leakage from oil pipelines and underground storage tanks or basins, and (3) spills of crude oil in the marine environment after accidents at sea.

In many important respects, the basic issues are similar and the strategy for all experiments on bioremediation may be viewed against the background provided in a valuable review by Atlas (1981). The broad classes of compounds involved (alkanes, cycloalkanes, aromatic hydrocarbons, and heteroarenes) that are formed by transformation of plant and algal residues were summarized together with the range of microorganisms capable of carrying out their degradation. Important environmental determinants of biodegradation were outlined including the addition of nutrients, the concentration of oxygen, the temperature, the salinity, and the role of the hydrocarbon–water interface as the site of active degradation. It is worth pointing out (1) that the biodegradability of cycloalkanes has attracted less attention that the other components and that some present potential recalcitrance and (2) that mass spectrometric methods using acetone chemical ionization (CI) have been developed to distinguish alkenes from their isomeric cycloalkanes (Roussis and Fedora 1997).

The metabolism of alkanes is discussed in Chapter 6, Section 6.1.1 and of cycloalkanes in Chapter 6, Section 6.1.2. Reviews on the microbial metabolism including biochemical aspects are available, and include those of Britton (1984) on alkanes and of Morgan and Watkinson (1994) that includes also cycloalkanes and some aromatic compounds. Virtually all the issues that are discussed in these recur in the examples that will be used as illustration. Some broad generalizations are therefore justified.

1. Biodegradation of all hydrocarbons requires access to electron acceptors, generally oxygen in natural situations, added hydrogen peroxide in terrestrial systems, or nitrate or sulfate under anaerobic conditions that prevail at deeper levels of the soil.

2. The biodegradation of *n*-alkanes and the pathways whereby it is accomplished are well established. The range of compounds includes branched-chain compounds such as pristane, and although branching may present an obstacle (Pirnick 1977), this may be overcome by carboxylation (Rontani et al. 1997). The biodegradation of aromatic hydrocarbons has been noted above, and it is necessary only to repeat that, whereas naphthalene and phenanthrene are readily degradable, PAHs with more than five rings may be more recalcitrant.

3. Many bacteria produce surfactants in response to exposure to hydrocarbons, and these have been demonstrated both for those that degrade alkanes and for those that degrade PAHs (Déziel et al. 1996). The positive effect of adding surfactants is, however, equivocal (Deschênes et al. 1996).

4. Occasionally, biotransformation in the absence of substantial biodegradation may be acceptable. Dicyclopentadiene is produced
pyrolytically in petrochemical plants and has a nauseating and
penetrating odor. Although it was degraded by a mixed bacterial
culture, the major part was transformed into a series of oxygenated
compounds that were presumed to be less malodorous (Stehmeier
et al. 1996; Shen et al. 1998).

8.2.2.1   Terrestrial Habitats

A number of reviews cover different aspects of the bioremediation of soil
contaminated with hydrocarbon refinery waste. The relative merits of *in situ*
and on-site procedures, the role of electron acceptors, and the addition of
inorganic nutrients have been discussed by Dott et al. (1995), who also
showed that inoculation even with organisms isolated from the sites — and
putatively able to degrade the contaminants — had only very limited effect.
The important determinants were, therefore, provision of oxygen as O_2 or
H_2O_2 and of an inorganic source of nitrogen. This conclusion is supported by
the results of attempts to improve the effectiveness of remediation of diesel-
contaminated alpine soils by adding inorganic nutrients or by inoculation
with a psychrophilic culture of oil-degrading microorganisms (Margesin and
Schinner 1997): the former was more effective, although a substantial loss
occurred by abiotic processes, and only moderate remediation was achieved.
The improvements induced by adding inorganic nutrients are, however,
determined by a number of factors including the level of fertilization (Braddock et al. 1997). For example, the effect of spraying nitrate solutions to plots
that had been contaminated with both BTEXTMB (TMB = trimethylbenzene)
and jet fuel and from which the surface vegetation had been removed
(Hutchins et al. 1998) was equivocal. Although hydrocarbon removal could
be demonstrated, it was not possible to assign success solely to the addition
of nitrate due to the intervention of soil leaching. In microcosms prepared
from core samples, however, removal occurred in nitrate and/or iron-reducing conditions, but not sulfate-reducing conditions. The modes of dissipation
and a summary of remedial techniques including designs for solid-phase systems and bioreactors "biopiles" have been given (Bossert and Compeau
1995): the authors conclude that the potential of bioremediation has not been
fully exploited and that the important issues of bioavailabilty and means of
evaluating effectiveness should be addressed. An investigation into the
design of laboratory reactors for the degradation of diesel-contaminated soil
showed (1) the advantage of periodic slurry systems with recycling at a rate
that is determined by the degree of contamination of the soil and (2) that periodic aeration was as effective as continuous aeration but resulted in significantly less loss by volatilization (Cassidy and Irvine 1997).

There is considerable interest in the removal of contaminants from former
gas filling stations.The surrounding soil is contaminated not only with fuel
hydrocarbon residues but also with alkyl lead compounds that were

previously used as gasoline additives. It has been shown that while the latter are readily degradable, this is inhibited by the presence of the concomitant hydrocarbons (Mulroy and Ou 1998).

8.2.2.2 Marine Habitats

The Baffin Island Oil Spill Project (BIOS) — General aspects of the degradation of oil in the marine environment have been given (van der Linden 1978), who discusses important issues including the effect of adding nutrients and the significance of temperature. Attention is drawn to a unique investigation that involved the deliberate discharge of oil and oil dispersants into a geographically remote and restricted environment in the Canadian Arctic. The reasons for adopting this strategy were themselves important: it was felt that, in spite of the effort already devoted to the persistence and toxicity of hydrocarbons in the marine environment, much of this knowledge could not validly be extrapolated to an arctic environment with low water temperatures, and a restricted — and possibly sensitive — pristine biota. It is not possible to review in detail the conclusions from this study which have been published in a series of comprehensive articles in a supplement to *Arctic* (1987, vol. 40) (Sergy 1987; Owens et al. 1987): this idea alone should be warmly welcomed as an alternative to the publication of details in miscellaneous reports that may not be readily accessible to interested members of the scientific community. This study is also important since, although it was designed to provide a predictive basis for future activity in the event of an oil spill, it also clearly illustrated the kinds of data that provide the basis for epidemiological studies. Possibly of most direct relevance in the present context are those aspects of its design that could profitably be incorporated into a wider range of hazard assessment procedures, and an attempt is therefore made to summarize these features.

1. Comprehensive chemical analyses of samples of water, sediment, and of biota were carried out both *before* and *after* the spill: this cannot, of course, be done in most cases, and this illustrates a serious limitation in field studies where lack of background data or difficulty in finding an uncontaminated control locality is frequently encountered. Sum parameters were sparingly employed in BIOS, and emphasis was placed on the analysis of specific compounds: attention was directed not only to PAHs, but also to azaarenes, dibenzothiophenes, and hopanes. Therefore, a clear distinction could be made between the input from the oil deliberately discharged and that arising from natural biological reactions or mediated by atmospheric transport. The importance of both these issues has been noted earlier in appropriate parts of this book.

2. Ice scouring of the intertidal zone in arctic waters makes this virtually sterile — a fact that was noted more than 160 years ago by Keilhau (1831) — so that attention was directed to components of the subtidal zone to which little attention had previously been

directed, and which was expected to be particularly sensitive to oil spills. Changes in the components of the macrobenthos including infauna, epibenthos, and macroalgae were examined, and attention was also directed to the histopathological and biochemical responses of bivalve mollusks that were affected in different ways by exposure to the dispersed and the undispersed oil.

The original reports should be consulted for details of the conclusions drawn from this comprehensive investigation, but the following seem of particular importance for this specific class of pollutant. The study provided:

1. A basis for determining remedial action in the event of an oil spill;
2. An evaluation of the advantage or disadvantage of using oil dispersants;
3. An estimate of the persistence of oil washed ashore;
4. An assessment of the effectiveness of procedures for removing stranded oil from beaches.

Experience with some major spills of crude oil has been summarized (Atlas 1995; Swannell et al. 1996) and attention drawn to a number of important general issues.

1. Rates of biodegradation are generally low compared with the speed and ease of mechanical removal, and are strongly dependent on whether the beaches are sandy or rocky.
2. The form in which nutrients are administered, e.g., granulated or slow release is important, and a periodic monitoring program should be implemented to determine the need for addition of nutrients in view of the possible adverse effects resulting from excess addition of nutrients, e.g., abnormally extensive algal growth and high bacterial activity leading to oxygen limitation.
3. It is important to be able to assess biodegradation of the oil relative to that of its most recalcitrant components.

A review (Sugai et al. 1997) of the remediation following the oil spill from the *Exxon Valdez* that includes a retrospective analysis revealed a change in the populations of alkane-degrading and aromatic hydrocarbon-degrading bacteria: this could be attributed to the enrichment effect of oleophilic fertilizers and terrestrial alkane waxes. It is worth drawing attention to marine bacteria of the genus *Marinobacter* (Gauthier et al. 1992) that are able to degrade aliphatic hydrocarbons and related compounds.

8.2.2.3 Conclusion

There seems little doubt that biodegradation of many components of petroleum may be facilitated by addition of inorganic nutrients — primarily nitrogen — and provision of suitable oxidant(s). It seems that inoculation with exogenous organisms — even when these have been isolated from the same site — is not generally effective and that indigenous organisms are effective provided that suitable nutrients and a supply of oxygen (as O_2 or H_2O_2) are available. Greatest effectiveness would probably be achieved by optimizing the capacity of these indigenous hydrocarbon-degrading organisms that are virtually ubiquitous. The limitations that have already been noted for PAHs are equally relevant here.

8.2.3 Wood Preservation Sites: Chlorophenolic Compounds

Creosote contamination has been noted in Section 8.2.1, and in this section only contamination with chlorophenolic compounds — particularly pentachlorophenol — will be discussed. It should, however, be noted that commercial samples of pentachlorophenol also contain lower chlorinated congeners and possibly chlorinated dibenzodioxins and dibenzofurans. Concern with their production, use, and disposal of chlorophenols is motivated by their widely distributed toxic effects (Ahlborg and Thunberg 1980) together with their possible persistence. The degradation of chlorophenols has been extensively investigated by a number of workers, the pathways have been reviewed (Häggblom 1990), and the application of specific strains to remediation of a wood-preservation site in Finland and elsewhere has been presented in detail (Häggblom and Valo 1995). Contamination of surrounding groundwater may also have occurred, and fluidized-bed reactors inoculated with activated sludge from pulp mill producing chemothermomechanical pulp have been used to treat groundwater highly contaminated with chlorophenols. An important aspect of this study is that the organisms were fully functional at the relatively low temperature of 5 to 7°C (Järvinen et al. 1994). Evaluation of white-rot fungi for treatment of pentachlorophenol-contaminated soil has been made (Lamar and Dietrich 1990; Lamar et al. 1993), and two issues have emerged: the production of the highly lipophilic pentachloroanisole and the association of the substrate with the soil matrix that presents a potentially serious problem, which is noted below.

Some general microbiological issues are briefly summarized.

Aerobic Bacteria

The aerobic bacterial degradation of phenols with fewer than three chlorine atoms occurs by initial hydroxylation, followed by ring fission and elimination of chloride from the acyclic intermediates. For compounds with more than three chlorine atoms, a different mechanism has been elucidated for the degradation of pentachlorophenol by *Mycobacterium chlorophenolicus* strain PCP1 (Apajalahti and Salkinoja-Salonen 1987a,b), and for a strain of *Flavobacterium* sp. (Steiert and Crawford 1986; Xun et al. 1992): this involves hydrolytic loss of chloride, followed by reductive and hydrolytic elimination to produce 1,2,4-trihydroxybenzene or 2,6-dichlorohydroquinone. This is discussed in Chapter 6, Section 6.5.1.2, and further details may be found in a review (Häggblom 1990). Several other pentachlorophenol-degrading organisms including the Gram-negative *Sphingomonas chlorophenolica* strain RA2 have also been examined. In the application of such strains to bioremediation of contaminated sites, a number of important issues must be taken into consideration.

1. The tolerance of the strains to high concentrations of pentachlorophenol — *S. chlorophenolica* appears to be less sensitive than *M. chlorophenolicus* (Miethling and Karlson 1996). This may be attributed to the ability of the cells to adapt their metabolism to avoid synthesis of toxic concentrations of chlorinated hydroquinones, and this is consistent with the low levels of these metabolites measured in the cytoplasm of cells metabolizing pentachlorophenol (McCarthy et al. 1997). Inocula have also been immobilized on polyurethane which, in addition, ameliorates the toxicity of chlorophenols (Valo et al. 1990).

2. The longevity of the strains in the environment — This is dramatically illustrated by the fact that an initial inoculum of 10^8 cells/g of *S. chlorophenolica* fell to zero within 7 months whereas for *M. chlorophenolicum*, 10^6 to 10^7 cells/g could be recovered from a similar inoculum (Miethling and Karlson 1996).

3. There are important requirements that may be necessary — An additional substrate may be required either to promote cell growth or to serve as a reductant, and degradation is functional under low oxygen concentrations and may be facilitated by sorption of the cells onto polyurethane films by the presence of straw compost (Laine and Jörgensen 1996).

4. Microbial *O*-methylation to chloroanisoles may occur, and under aerobic conditions these may be terminal metabolites which, because of their lipophilic character, may be transported into higher biota and hence into the food chain via earthworms (Palm et al. 1991).

The detection and persistence of *M. chrorophenolicum* in soil that is a prerequisite for demonstrating its positive effect on biodegradation has used 16S rRNA combined with PCR (Briglia et al. 1996) that had a detection limit of ~3 $\times 10^2$ cells/g. It should be noted that extensive investigations by Laine have shown the effectiveness of additives in facilitating the biodegradation of chlorophenols, and that for contaminated soils that already provide degradative microorganisms, the addition of an inoculum, however, provides little or no additional advantage (Laine and Jörgensen 1997).

Degradation by Fungi

Both 2,4-dichlorophenol and 2,4,5-trichlorophenol may be degraded by the white-rot fungus *Phanerochaete chrysosporium* by a complex pathway

(Chapter 6, Section 6.5.2) involving oxidative displacements of chloride and O-methylation with the formation of 1,2,4,5-tetrahydroxybenzene before ring fission (Valli and Gold 1991). Although pentachloroanisole is formed from pentachlorophenol by a number of species of *Phanerochaete*, it is subsequently mineralized. The formation of products by oxidative coupling with loss of chloride mediated by laccase or peroxidase should also be noted (Dec and Bollag 1994), since these reactions may be the basis of associations between chlorophenols and humic substances (Arjmand and Sandermann 1985). Fungal activity may also result in the formation of stable associations by reaction of xenobiotics and their metabolites with humic acids (Section 3.2.4).

Anaerobic Dechlorination

Anaerobic dechlorination of both aliphatic and aromatic organochlorine compounds is discussed in detail in Chapter 6, Sections 6.4.4 and 6.6. It remains only to note that although dechlorination is a widespread reaction, this may represent merely a biotransformation and not ultimate degradation. Indeed, partial dechlorination is extremely common. A review of anaerobic dechlorination has been given by Mohn and Tiedje (1992). Under these conditions — and in contrast to aerobic conditions — the more highly substituted compounds are more readily transformed. Partial dechlorination of pentachlorophenol has been observed, although the resulting chlorophenols are degradable under aerobic conditions, so that under suitable conditions complete mineralization may be accomplished. Pure cultures of *Desulfomonile tiedjei* are able to carry out dechlorination: strain DCB-1 of polychlorinated phenols (Mohn and Kennedy 1992) and strain DCB-2 of 2,4,6-trichlorophenol and dichlorophenols (Madsen and Licht 1992).

Phenoxyalkanoic Acids

It is convenient to add some brief comments on chlorophenoxyalkanoic acids since the pathway for their degradation that involves the intermediate formation of chlorophenols is well established, and several groups of organisms able to do so have been isolated. Kinetic parameters for the rate of degradation and for sorption of 2,4-D (Estrella et al. 1993) have been carried out and showed higher values for the maximum growth rate constants in column experiments than in batch experiments. Experiments on the bioremediation of 2,4-dichlorophenoxyacetic acid-contaminated soil using *Pseudomonas cepacia* strain BRI 6001 were promising and indicated the critical role of the cell density of the inoculum added to nonsterile soil (Comeau et al. 1993). On the other hand, the bacterial degradation of 2,4-D sorbed to sterile soil has been shown to be extremely slow (Ogram et al. 1985), and the rates in soil with a low organic content were greater than in an organic-rich soil (Greer and Shelton 1992). The biodegradation of phenoxypropion-2-oic acids is more complex, and mecoprop-2-(2-methyl-4-chlorophenoxy)propionic acid has been consistently found in municipal landfill leachates (Gintautas et al. 1992).

Whereas in pure cultures of bacteria, only one enantiomer of 2-(2-methyl-4-chlorophenoxy)propionic acid — the (S)-(–)-enantiomer — may be degraded (Tett et al. 1994), in agricultural soil treated with a herbicide mixture containing 2-(2,4-dichlorophenoxy)propionic acid both enantiomers were degraded with the opposite preference (Garrison et al. 1996). The same preference for degradation of racemic 2-(2-methyl-4-chlorophenoxy)propionic acid was observed in *Sphingomonas herbicidovorans* by Zipper et al. (1996), who made the important observation that cells grown with *either* of the pure enantiomers preferentially degraded that enantiomer. Current formulations, however, contain only the biologically active and more readily biodegradable enantiomer. Additional comments on the enantiomerization of 2-phenoxypropionic acids (Buser and Müller 1997) has been given in Chapter 4, Section 4.2.2.

Conclusion

There is evidence for the success of bioremediation of chlorophenol-contaminated sites by on-site processes using a number of aerobic microorganisms as inocula, although their activity depends on a number of factors that should be investigated in each case (Miethling and Karlson 1996). On the other hand, advantage should be taken of indigenous microflora together with the addition of suitable sorbents. A relatively high degree of mineralization may be attained, although the possibly undesired formation of chloroanisoles might present a potential hazard, together with the fungal-catalyzed polymerization of chlorophenols. The extent to which this is applicable to chlorophenols containing methyl or nitro substituents is apparently unresolved.

8.2.4 Chemical Production Waste

The range of compounds in waste or in by-products from chemical production is enormous, and includes compounds containing fluorine, chlorine, bromine, iodine, and nitro, sulfonate, and thiophosphate groups. A useful structural listing of a wide range of chemicals has been compiled (Hartter 1985; Swoboda-Colberg 1995), and attention is drawn to two less-prominent groups containing the CF_3 group (Section 8.2.4.3), and organic iodo compounds such as the herbicide ioxynil and X-ray contrast agents. Chlorophenols and phenoxyalkanoic acid herbicides have been discussed in Section 8.2.3, and the commercial chlorophenols products generally contain impurities including chlorinated dioxins and furans. Nitro compounds will be discussed in Section 8.2.5.1, and phosphorothioate esters briefly noted in Section 8.2.5.2. The present discussion will therefore be concentrated on halogenated aromatic and alicyclic hydrocarbons, although attention is directed to other structural groups such as benzothiazoles (Reemtsma et al. 1995; Kumata et al. 1996) that have wide industrial application but about which little is known of their biodegradability (De Vos et al. 1993). It should be emphasized that accumulation of by-products may present problems so that attention should not be confined to the final product. Illustrative examples include diphenyl sulfone from the production of phenol by sulfonation of benzene followed by hydrolysis (Wick and Gschwend 1998), and the diverse compounds containing chloro and CF_3 substituents resulting from the production of 4-chloro-(trifluoromethyl) (Jaffe and Hites 1985) that is noted later in Section 8.2.4.3.

8.2.4.1 Chlorinated Alicyclic Hydrocarbons

Hexachlorocyclohexanes (HCHs) are among the organochlorine compounds at sites contaminated with former pesticide production (Lang et al. 1992), and although evidence has been presented for the existence of a rich soil microflora at such sites, it appears that many of the substrates remain even after 30 years (Feidieker et al. 1994). This indicates the existence of one or more limiting factors, of which the degree of bioavailability is probably of major significance, and: this is consistent with the established difficulty of chemical recovery of HCH from aged contaminated soil (Westcott and Worobey 1985). The situation is additionally complicated by the existence of various isomers of HCH ($\alpha,\beta,\gamma,\delta$) each of which behaves differently.

The aerobic degradation of γ-hexachloro[*aaaeee*]cyclohexane (HCH) has been accomplished with a strain of *Pseudomonas paucimobilis* strain UT26 (Nagasawa et al. 1993), and involves a complex chain of elimination, hydrolytic displacement of chlorine, and ring fission reactions (Chapter 6, Sections 6.4.1 and 6.4.2). It should be noted, however, that the biotransformation to chlorobenzenes as terminal metabolites presents an undesirable alternative. The aerobic degradation of α-hexachloro[*aaaaee*]cyclohexane (HCH) by indigenous bacteria in soil slurries has been demonstrated (Bachmann et al. 1988), and in soil is apparently limited by rates of soil desorption and intraparticle mass transfer (Rijnaarts et al. 1990). Although the β-isomer (*eeeeee*) is highly persistent, its susceptibility to degradation under methanogenic conditions in a laboratory column reactor has been studied (Middeldorp et al. 1996): chlorobenzene and benzene were terminal metabolites plausibly produced from an intermediate δ-2,3,4,5-tetrachlorocyclohex-1-ene. The anaerobic transformation of the structurally related γ-hexachlorocyclohexene and γ-1,3,4,5,6-pentachlorocyclohexene by *Clostridium rectum* produced 1,2,4-trichlorobenzene and 1,4-dichlorobenzene, respectively (Ohisa et al. 1980).

It should be also noted that considerable attention has been directed to the highly toxic polychlorinated bornanes (toxaphene) and derivatives of hexachlorocyclopentadiene (chlordane) that were used as pesticides. Chlordane has been recovered from biota (Buser et al. 1992), and toxaphene components from sediments (Stern et al. 1996), so that these are highly persistent contaminants. For both, the commercial products contained a large number of congeners, and the analysis of chlordane is complicated additionally by the occurrence of enantiomers, some of which have undergone degradation and transformation after discharge. Nevertheless, only limited attention appears to have been directed to waste from locations at which these products were manufactured.

8.2.4.2 Chlorinated and Brominated Aromatic Hydrocarbons

The structural range of industrially important representatives of these groups is enormous, and includes chlorobenzenes (solvents), to PCBs (hydraulic and insulating fluids), and polybrominated biphenyls (PBBs) and polybrominated diphenyl ethers (PBDE) ethers (flame retardants). There is widespread concern over both the persistence and the potential toxicity of all these compounds, so that sites contaminated during their production represent a threat both to the environment and to human health. Comments have already been made in Chapter 6, Section 6.5.1.1 on the pathways for the aerobic bacterial degradation of chlorobenzenes and chlorobiphenyls.

Chlorobenzenes

Although the possible incompatibility of pathways for the degradation of chlorobenzene and toluene has been noted in Chapter 4, Section 4.8.2, a strain of *Ralstonia* sp. from a site contaminated with chlorobenzene contained the genes for both chlorocatechol degradation and the dioxygenase system for the degradation of benzene/toluene (van der Meer et al. 1998). The evolution of this strain resulted in a natural lowering in the groundwater concentration. Aerobic organisms capable of degrading chlorobenzenes with fewer than five chlorine atoms have been described and the pathways elucidated (Chapter 4, Section 4.4.2 and Chapter 6, Section 6.5.1.2): the metabolism may be initiated by dioxygenation without elimination of a chlorine atom from 1,2,4-trichlorobenzene or, with elimination of a single chlorine atom, from 1,2,4,5-tetrachlorobenzene (Sander et al. 1991). The possible application of the soluble methane monooxygenase of *Methylosinus trichosporium* to the bioremediation of 1,2,3-trichlorobenzene is attended by three negative factors: (1) the formation of the toxic 2,3,4- and 3,4,5-trichlorophenol, (2) the requirement of formate, and (3) the permanent inactivation of the cells by the substrate (Sullivan and Chase 1996). The anaerobic dechlorination of hexachlorobenzene (HCB) has been described in anaerobic mixed cultures supplemented with electron donors including lactate, ethanol, or glucose (Holliger et al. 1992): successive and partial dechlorination produced 1,2,4- and 1,3,5-trichlorobenzenes, whereas the 1,2,3-trichlorobenzene was further dechlorinated. The partial dechlorination of 1,2,3,4-tetra-, 1,2,3,5-tetra-, and pentachlorobenzene has been examined in a methanogenic mixed culture using lactate as electron donor (Middeldorp et al. 1997), and, for undisclosed reasons, sterile Rhine River sand was needed to maintain dechlorination activity.

Polychlorinated Biphenyls PCBs

There is an enormous literature on the degradation of PCBs, and this chapter does not attempt to cover this topic entirely. The degradation of PCBs is exacerbated not only by the number of congeners but by the fact that heavily *ortho*-substituted congeners occur in enantiomeric forms (Chapter 2, Section 2.4.2). Attention is directed to reviews on anaerobic dechlorination (Bedard and Quensen 1995) and to those that include a summary of aerobic degradation (Bedard 1990; Sylvestre 1995). The last also includes valuable comments on the regulation of PCB degradation and the significance of chlorobenzoate degradation. As background to a discussion of bioremediation technologies, a few simplistic remarks are given here on the metabolic pathways used for the degradation of PCBs that is discussed in detail in Chapter 6, Section 6.5.1.1. It should, however, be emphasized that there are considerable differences among the congeners both in their susceptibility to degradation and in the details of the mechanisms whereby this is accomplished, and among the various organisms. No attempt, however, is made to discuss here the details of these important differences. One important parameter is temperature, which is of particular significance for naturally occurring mixed cultures of organisms in the natural environment, since temperature may result in important changes in the composition of the microbial flora as well as on the rates. Some illustrative examples of the importance of temperature under both aerobic and anaerobic situations include the following.

1. An anaerobic sediment sample was incubated with 2,3,4,6-tetrachlorobiphenyl at various temperatures between 4 and 66°C (Wu et al. 1997). The main products were 2,4,6- and 2,3,6-trichlorobiphenyl and 2,6-dichlorobiphenyl: the first was produced maximally and discontinuously at 12 and 34°C, the second maximally at 18°C, and the third was dominant from 25 to 30°C. Dechlorination was not observed above 37°C.

2. Sediment samples from a contaminated site were spiked with Arochlor 1242 and incubated at 4°C for several months (Williams and May 1997). Degradation by aerobic organisms in the upper layers of the sediment — but not in those >15 mm from the surface — occurred with the selective production of di- and trichlorobiphenyls. Some congeners were not found including the 2,6- and 4,4'-dichlorobiphenyls and a wider range of trichlorobiphenyls which were presumably further degraded.

A number of aerobic organisms are able to degrade various congeners of PCB. Most have been isolated by enrichment with biphenyl as sole carbon and energy source. These include both Gram-negative strains belonging to the genera *Alcaligenes* and *Pseudomonas*, and Gram-positive strains of the genus *Rhodococcus*. It has been noted in Chapter 4, Section 4.6.2 and Chapter 6, Section 6.5.1.1 that growth and dechlorination may be accomplished by bacteria growing not only at the expense of biphenyl that was used for enrichment of the organisms, but also at the expense of typical plant metabolites. Aerobic metabolism is initiated by dioxygenation of the ring bearing least chlorine substituents, followed by ring fission either to chlorobenzoates or to chloroacetophenones (Bedard et al. 1987). Either biphenyl-2,3-dioxygenase or the 3,4-dioxygenase may be involved. In an ingenious application, the recovery of chlorobenzoates has been used to demonstrate aerobic degradation under natural conditions (Flanagan and May 1993). The formation of protoanemonin from 4-chlorocatechol that is an intermediate in the degradation of 4-chlorobiphenyl (Blasco et al. 1995), and its adverse effect on PCB-degrading bacteria in soil (Blasco et al. 1997) (Chapter 4, Section 4.8.2 and Chapter 6, Section 6.5.1.1) may present a significant problem in the bioremediation of PCB-contaminated sites even though it may be possible to develop strains that overcome this pathway. Conversely, naturally occurring compounds may be able to induce the enzymes for PCB metabolism. For example, effective degradation by cometabolism of a number of congeners in Arochlor 1242 was induced in *Arthrobacter* sp. strain B1B by carvone that was not used as a growth substrate, although it was toxic at high concentrations (>500 mg/l) (Gilbert and Crowley 1997). Other structurally related compounds including limonene, *p*-cymene, and isoprene were also effective.

A valuable overview (Adriaens and Vogel 1995) includes a range of chlorinated aromatic compounds and provides illustrative engineering designs. In the light of the microbiological background that has been presented, a number of different strategies may be distinguished and all of these have been applied to and evaluated in contaminated soils or sediments.

1. Contaminated soil has been inoculated with bacteria of established capacity for degradation of chlorobenzoates (Hickey et al. 1993). In the presence of added biphenyl, mineralization of PCBs was shown, although it was pointed out that there may exist incompatibility between the production of chlorocatechols from chlorobenzoates and their inhibition of dihydroxybiphenyl-2,3-dioxygenase that catalyzes the ring fission of many PCBs.

2. Experiments using *P. testosteroni* strain B-356 in microcosm systems revealed the necessity of adding biphenyl to encourage degradation of the tetrachlorinated congeners of Arochlor 1242, and the effectiveness of repeated inoculation (Barriault and Sylvestre 1993).

3. *In situ* stimulation of aerobic PCB degradation has been shown in Hudson River sediments (Harkness et al. 1993). Biodegradative capacity of the indigenous organisms was enhanced by the addition of biphenyl and inorganic nutrients, although repeated inoculation with an organism having established capability to degrade PCBs did not apparently improve the degree of biodegradation.

4. Anaerobic strategies are complicated by the different susceptibility of the various congeners to dechlorination and by the effect of electron acceptors, especially sulfate. Ingenious alternatives have used addition of specific PCB congeners that are more readily dechlorinated to "prime" dechlorination at specific positions (Bedard and Quensen 1995), and the use of dibrominated biphenyls in the

presence of malate to stimulate dechlorination of the hexachloro- to nanochlorobiphenyls (Bedard and Quensen 1995). The use of brominated biphenyls to induce dechlorination of highly chlorinated biphenyls has been examined. Di- and tribromobiphenyls were the most effective in dechlorinating the hepta, hexa, and pentachloro congeners and were themselves reduced to biphenyl (Bedard et al. 1998), and 2,6-dibromobiphenyl stimulated the growth of anaerobes that effectively dechlorinated hexa-, hepta-, octa-, and nanochlorobiphenyls over the temperature range from 8 to 30°C (Wu et al. 1999). Other compounds have been examined as "primers" for dechlorination of hexa- to nonachloro congeners (Deweerd and Bedard 1999): a number of substituted brominated monocyclic aromatic compounds were examined, and 4-bromobenzoate was effective — though less so than 2,6-dibromobiphenyl. The chlorobenzoates that are metabolites of aerobic degradation were ineffective. The positive effect of brominated biphenyls in "priming" the anaerobic dechlorination of PCBs is also encountered in the dechlorination of octachlorodibenzo[1,4]dioxin to the 2,3,7,8 congener induced by 2-bromodibenzo[1,4]dioxin (Albrecht et al. 1999; Chapter 3, Section 3.6.3).

5. As for PAHs, attempts have been made to increase bioavailability by use of surfactants, and once again a complex situation has emerged (Fava and Di Gioia 1998). Triton 100 exerted both positive and negative effects: in soil slurries, although it was not metabolized by the soil microflora, it adversely affected the degradation of chlorobenzoate intermediates, whereas in fixed-bed reactors depletion of PCBs was enhanced.

Polybrominated Biphenyls (PBBs) and Polybrominated Diphenyl Ethers (PBDEs)

Highly brominated biphenyls and diphenyl ethers have been manufactured as flame-retardants, and the diphenyl ethers have apparently become environmental contaminants (references in Sellström et al. 1998; see also Chapter 3, Section 3.6.3). One of the reasons for concern is the formation of polybrominated dibenzofurans and dibenzo[-1,4]-dioxins (although *not* 2,3,7,8-tetrabromdibenzo[-1,4]-dioxin) during subsequent incineration (Buser 1986). Mixed cultures of organisms isolated from sediments contaminated with PCBs and PBBs were shown to debrominate PBBs under anaerobic conditions (Morris et al. 1992), and the dominant congener — 2,4,5,2',4',5'-hexabromobiphenyl — could be successively debrominated to 2,2'-dibromobiphenyl. On the other hand, in sediments from the most heavily contaminated site containing contaminants in addition to PBBs, very little debromination occurred and the recalcitrance was attributed to the toxicity of the other contaminants (Morris et al. 1993).

Chlorinated Dibenzodioxins

These are probably ubiquitous in the environment, and their biodegradability has been extensively examined. This will not be discussed in any detail here, and attention will be drawn merely to a few investigations that illustrate important new aspects.

Chlorinated dioxins have been recovered from sediments contaminated with both industrial discharge and atmospheric deposition (Macdonald et al. 1992; Götz et al. 1993; Evers et al. 1993), and attention has therefore been directed to anaerobic dechlorination processes (Beurskens et al. 1995; Adriaens et al. 1995). It has been suggested that microbial dechlorination of cell-partitioned 2,3,7,8-tetrachlorodibenzo[1,4]dioxin in aged sediments was as effective as that in freshly spiked sediment (Barkovskii and Adriaens 1996). This suggests that the kinetics of transport into the cells is a cardinal determinant, although the extent to which hydrophobic xenobiotics in sediments are partitioned into the indigenous microflora has not been generally established.

Chlorinated dioxins occur in atmospheric deposition (Koester and Hites 1992) and will thereby enter the terrestrial environment. The degradation of tetrachloro- to octachlorodibenzo[1,4]dioxins has been examined in low-nitrogen medium by the white-rot fungus *Phanerochaete sordida* YK-624 (Takada et al. 1996). All the compounds were extensively degraded, and the ring-cleavage of 2,3,7,8-tetra- and octachlorodibenzo[1,4]dioxin produced 4,5-dichlorocatechol and tetrachlorocatechol. These results establish important new evidence for the biodegradability of even highly chlorinated dibenzodioxins. High cell densities of a strain of *Sphingomonas* sp. that is able to degrade chlorinated dibenzofurans and dibenzo[1,4]dioxins (Wilkes et al. 1996) were introduced into soil microcosms spiked with 2-chloro-dibenzo[1,4]dioxin (Halden et al. 1999). Although this was removed relatively rapidly, the toxicity of 2-chloro and the adverse effect of soil organic material present serious limitation to its application in bioremediation.

8.2.4.3   Organofluoro Compounds

These have found a wide range of application including use as, for example, agrochemicals (Cartwright 1994), liquid crystals (Inoi 1994), fluorine-containing dyes (Herd 1994; Engel 1994), plastics (Feiring 1994), elastomers (Logothetis 1994), and chemotherapeutic agents (Edwards 1994). The plethora of structures is a tribute to the ingenuity of the synthetic chemist, and both aliphatic and aromatic structures are widely represented. Attention is directed primarily to groups of compounds containing the CF_3 group such as that including agrochemicals, such as the diphenyl ether herbicides (Lee et al. 1995) and fluoroaromatic compounds used as pharmaceuticals such as the fluoroquinolone carboxylic acids that are used in substantial amounts in large-scale animal husbandry (Nowara et al. 1997).

Two examples of the existence of fluorinated organic compounds are used as illustration of a problem area.

1. Sediments containing waste from the production of 4-chloro-(trifluoromethyl)benzene contained a number of compounds containing chloro and CF_3 substituents: these included benzophenone, a difluorodiphenylmethane, and several biphenyls (Jaffe and Hites 1985), although the more widespread occurrence of these compounds is unknown. This is important since, whereas pathways for the degradation of fluorophenols and fluorobenzoates are well established, the CF_3 group appears to be resistant to both chemical and microbiological degradation. A combination of photolytic and microbiological reactions, however, brings about the degradation of 3- and 4-trifluoromethylbenzoate via 7,7,7-trifluoro-hepta-2,4-diene-6-one carboxylate that was subsequently degraded photochemically with the loss of fluoride (Taylor et al. 1993).

2. A series of dibenzo[1,4]dioxins bearing nitro and trifluoromethyl substituents in one ring and chloro in the other was synthesized for comparison with impurities in commercial samples of the lampricide 3-trifluoromethyl-4-nitrophenol, and to assess their ability to induce MFO activity (Hewitt et al. 1998). Although they appeared to be relatively weak inducers, there were other unidentified components that were highly active.

8.2.4.4 Chlorinated Anilines

These are the starting compounds for production of a range of agrochemicals (insecticides, herbicides, fungicides) including anilides, carbamates, and ureas. Their hydrolysis products — halogenated anilines, particularly chloroanilines — are widely distributed and notorious for association with, and incorporation into humic substances (Bollag and Loll 1983). The mechanism of this interaction has been examined in model enzyme-mediated reactions with ferulic acid (Tatsumi et al. 1994), and it has been shown that oligomerization of 4-chloroaniline mediated by oxidoreductases may produce 4,4′-dichloroazobenzene and 4-chloro-4′-aminodiphenyl as well as trimers and tetramers (Simmons et al. 1987). From a purely microbiological view, there are some important issues.

1. Although aerobic bacteria that degrade monochloroanilines have been isolated (Chapter 6, Section 6.2.1), their effectiveness in the biodegradation of dichloroanilines appears to have been less explored (Brunsbach and Reineke 1993). The regulation of chloroaniline degradation that is discussed in Chapter 6, Section 6.2.1 should be taken into account, in particular the possible requirement for additional carbon sources. The anaerobic dechlorination of chloroanilines has been discussed in Chapter 6, Section 6.6, and the pathway for the anaerobic degradation of aniline itself by carboxylation followed by deamination is discussed in Chapter 6, Section 6.7.3.1.

2. Biotransformation to the acetanilides has been demonstrated, although a strain *of P. putida* was able to degrade 3,4-dichloroaniline (You and Bartha 1982). The chloroaniline component of lignin complexes is mineralized by *Phanerochaete chrysosporium* slowly — although as effectively as the free compound (Arjmand and Sandermann 1985). The mineralization of 3,4-dichloroaniline by *P. chrysosporium* proceeds sequentially by conjugation with 2-ketoglutaryl CoA to produce the amide and N-(3,4-dichloro)succinamide before mineralization (Sandermann et al. 1998). Negative aspects of fungal metabolism include (a) the transformation of 3,4-dichloroaniline by white-rot fungi (Pieper et al. 1992) which may produce the toxic 3,3',4,4'-tetrachloroazobenzene (Lee and Kyung 1995) and (b) association with humic and fulvic acids (Section 3.2.4) (Arjmand and Sandermann 1985).

The prognosis for bioremediation of residues containing polychlorinated anilines and their impurities is therefore somewhat discouraging, although removal from soil of 2-chloro-N-isopropylacetanilide by a strain able to degrade this has been reported (Martin et al. 1995). This strain was able to dechlorinate the substrate, although it was unable to grow with either aniline or phenol. Evidence for bioremediation based on loss of the initial substrate should therefore be evaluated critically, even when production of CO_2 — presumably from the alkyl substituents — has been demonstrated.

8.2.4.5   Conclusion

There is therefore unequivocal evidence for the success of *in situ* strategies for bioremediation of PCB in soils and sediments using indigenous microorganisms. The major problem arises from the complex mixture of compounds in PCB commercial preparations that include not only a range of congeners but also polychlorinated terphenyls (Wester et al. 1996). Current research is therefore directed to understanding the basis of the biodegradability of specific congeners, and to the possibility of developing strategies to overcome this. Before implementation of bioremediation of sites contaminated with PBBs, PCDDs, aromatic compounds containing CF_3 groups, and chloroanilines can be successfully implemented, however, important unresolved issues remain.

8.2.5   Military Waste

8.2.5.1   Explosives

A range of nitro compounds has been used as military explosives, including nitrotoluenes, nitrate esters, and nitramines. There is concern over the destruction of these toxic, dangerous and potentially explosive products, so that considerable effort has been given to the development of microbiological processes of destruction (Kaplan 1990): a review gives details of the degradation pathways of the relevant compounds has been given (Walker and

Kaplan 1992). Specific details on the biodegradation of aromatic nitro compounds may be found in Chapter 6, Section 6.8.2, in a review (Spain 1995a), and in a book (Spain 1995b). Compared with the wastes that have been considered above, these compounds are distinguished by a higher N/C ratio, so that, although they may serve as sources of both C and N, only the N is used by some organisms: in this case, an additional carbon source must obligatorily be added.

Only limited data are available on full-scale or pilot-scale operations and may present an overly simplified view. In general, reduction to amines is the initial — and sometimes the only — transformation.

1. Indigenous microorganisms from a munitions-contaminated site reduced a single nitro group of 2,4-dinitrotoluene, 2,6-dinitrotoluene (DNTs) and 2,4,6-trinitrotoluene (TNT) to amines, while TNT was mineralized to ca. 10% (Bradley et al. 1994). A preliminary experiment using soil contaminated with both TNT and nitramines was carried out by adding phosphate and starch to the surface: the anaerobic flora that developed reduced the nitro groups of TNT, and subsequently the reduced compounds were removed though their fate was not established (Funk et al. 1993). One important additional observation was that under aerobic conditions, or under anaerobic conditions at alkaline pH, polymerization of the intermediate amines took place. This experiment has since been extended to an evaluation of an *in situ* bioreactor (Funk et al. 1995).

2. Bioreactors containing an undefined anaerobic consortium reduced TNT to 2,4,6-triaminotoluene (TAT) in the presence of glucose (Daun et al. 1998). The sorption of TAT to montmorillonite clay was irreversible, and the substrate could not be released by solvent extraction or by acid or alkaline treatment. Similar results were obtained with humic acids in which covalent reaction with carbonyl or activated C=C bonds bonding presumably occurs (Chapter 3, Section 3.2.4). Results from laboratory experiments using ^{14}C-labeled TNT in reactors to which molasses was added as carbon source showed that after 9 weeks, 83% of the radioactivity was recovered in soil components (humin, humic acids, and fulvic acids) (Drzyzga et al. 1998). These results illustrate the important issue of the association of metabolic products from TNT with soil components which in turn may reduce the effectiveness of remediation.

3. Simultaneous degradation of 2,4- and 2,6-dinitrotoluene has been achieved in a fluidized-bed biofilm reactor (Lendemann et al. 1998), and was successfully operated with contaminated groundwater containing 2- and 4-nitrotoluenes and 2,4- and 2,6-dinitrotoluenes. Nitrite formed during degradation was accounted for as nitrate.

The effectiveness of clostridia in the anaerobic digestion of a number of nitroaromatics prompted development of a procedure for the production of spores of *Clostridium bifermentans*, and a medium for their effective production has been developed (Sembries and Crawford 1997).

Nitrotoluenes

Reviews of the biodegradation and biotransformation of nitroaroamatic compounds have been given (Spain 1995a,b; Crawford 1995) to which reference should be made for details, so that only a very brief summary is justified here. Three principal reactions are found among aerobic bacteria: (1) oxidative elimination of nitrite, (2) partial or complete reduction of the nitro group, and (3) reduction of the aromatic ring (Chapter 6, Section 6.8.2).

1. *Oxidative elimination of nitrite* — Dioxygenase-mediated loss of nitrite with formation of a catechol. For 2,4-DNT, degradation by *Pseudomonas* sp. strain TNT is initiated by formation of 4-methyl-5-nitrocatechol, whereas loss of the second nitro group results in the formation of an intermediate 1,4-quinone before ring fission occurs (Haigler et al. 1994). Alternative pathways for ring fission have been presented (Spain 1995a,b).

2. *Reduction of the nitro group* — This has been observed under aerobic conditions, and a number of alternative pathways have emerged. The biodegradation of nitrobenzene by a strain of *P. pseudoalcaligenes* proceeds by rearrangement of the initially formed hydroxylamine to 2-aminophenol before degradation by *extradiol* ring cleavage (Nishino and Spain 1993). The situation for more highly nitrated toluene is more complex. A range of metabolites is formed from TNT by a strain of *Pseudomonas* sp., and includeds both hydroxylamino-dinitrotoluenes and tetranitroazoxytoluenes formed by condensation of the partially reduced hydroxylamine and nitroso compounds (Haïdour and Ramos 1996). The metabolism of TNT by *P. pseudoalcaligenes* strain JS 52 produces dihydroxylamino compounds that are converted into metabolites of unestablished structure, and into 4-amino-2-hydroxylamino-6-nitrotoluene that is also metabolized further. The biodegradation of 2,4-DNT by *Phanerochaete chrysosporium* involves initial reduction of one of the nitro groups followed by a series of reactions (Valli and Gold 1991) similar to those involved in the degradation of 2,4-dichlorophenol (Valli et al. 1992) — oxidation, O-methylation, and further oxidation to 1,2,4-trihydroxybenzene. TNT is also transformed by this organism and partially mineralized, although it was shown that 4-hydroxylamino-2,6-dinitrotoluene was a potent inhibitor of lignin peroxidase activity (Bumpus and Tatarko 1994). Partial reduction of 2,4-DNT to the dinitroazo compound has been observed in a strain of the fungus *Mucrosporium* (McCormick et al. 1978), and in a strain

of *Pseudomonas* sp. (Duque et al. 1993). A range of nonlignolytic fungi has been examined for their ability to transform TNT (Bayman and Radkar 1997): tetranitroazoxytoluene, 4-amino-2,6-dinitrotoluene, and hydroxylaminodinitrotoluenes were the major metabolites and none of the fungi was able to mineralize the substrate.

Consistent with the presence of three strongly electron-withdrawing groups and the established formation of charge–transfer complexes involving trinitroarenes, a strain of *Mycobacterium* sp. forms a hydride–TNT complex in addition to products from reduction of the nitro groups (Vorbeck et al. 1994). Although of minor current interest as an explosive, it is noted that a similar complex is formed from 2,4,6-trinitrophenol that was a component of the explosive lyddite although the intermediate is further degraded (Lenke and Knackmuss 1996).

Reduction by anaerobic bacteria has been widely observed, and a sulfate-reducing organism that uses TNT as an N-source carried out sequential reduction to TAT (Preuss et al. 1993), although the pathway for the further metabolism of this compound from which NH_4^+ is released was not elucidated. Incubation of TNT under anaerobic conditions was shown to produce TAT and methylphloroglucinol (Funk et al. 1993), and it is attractive to hypothesize that further degradation of the latter is analogous to that used for the degradation of 3,4,5-trihydroxybenzoate by *Pelobacter acidigallici* that proceeds via 1,3,5-trihydroxybenzene before ring cleavage (Brune and Schink 1992). Transformation of TNT by *C. bifermentens* produced TAT via a number of partially reduced amino- and hydroxylamino compounds (Lewis et al. 1996), together with a compound produced by condensation of TAT with pyruvaldehyde.

3. *Reduction of the aromatic ring* — The degradation of phenols containing several nitro groups has revealed an unusual reaction. Although this occurs with the elimination of nitrite — not apparently by any of the pathways illustrated above — terminal metabolites are formed by reduction of the aromatic ring. 4,6-Dinitrohexanoate is produced from 2,4-dinitrophenol (Lenke et al. 1992) and 2,4,6-trinitrocyclohexanone from 2,4,6-trinitrophenol (Lenke and Knackmuss 1992).

Nitrodiphenylamines

Although these are no longer used as explosives, they are formed from diphenylamine that is used as a stabilizer for nitrocellulose explosives. A number of nitrated diphenylamines are then formed by reaction with the NO_x that is produced. The metabolism of nitrodiphenylamines has been examined under anaerobic conditions, and phenazine and 4-aminoacridine that are cyclization products of the initially formed 2-aminodiphenylamine have been identified (Drzyzga et al. 1996).

Nitrate Esters

A review (White and Snape 1993) has been devoted to this topic, and only very brief comments can be added to this relatively unexplored area. The biotransformation of gylcerol trinitrate by strains of *Bacillus thuringiensis/cereus* or *Enterobacter agglomerans* involves the expected successive loss of nitrite with formation of glycerol (Meng et al. 1995). The transformation of pentaerythritol tetranitrate by *E. cloacae* that uses this as a source of N proceeds comparably with further metabolism of the two hydroxymethyl groups produced by hydrolysis to the dialdehyde (Binks et al. 1996). The enzymes that produce nitrite from the nitrate esters are apparently reductases, and the enzyme from *E. cloacae* is strongly inhibited by steroids and is capable of the reduction of cyclohex-2-ene-1-one (French et al. 1996). In a medium containing glucose and ammonium nitrate, glyceryl trinitrate is transformed by *Penicillium corylophilum* to the dinitrate and the mononitrate before complete degradation (Zhang et al. 1997). On the other hand, the metabolism of glyceryl trinitrate by *Phanerochaete chrysosporium* involves the production of the physiologically important nitric oxide (Servent et al. 1991).

Nitramines

Of the nitramines used as explosives, greatest attention has been directed to the degradation of 1,3,5-trinitro-hexahydro-1,3,5-triazine (RDX).This is degraded by several species of enterobacteriacaea (Kitts et al. 1994), and is used by *Stenotrophomonas* sp. strain PB1 as a source of N, with the production of metabolites formed by loss of two nitro groups and a single N-atom from the triazine: their structures were tentatively put forward (Binks et al. 1995), although an alternative anaerobic pathway involving reduction of the nitro groups to nitroso and a series of transformations yielding hydrazine, dimethylamine, and 1,2-dimethylhydrazine has been presented (McCormick et al. 1981).

   In addition to military uses, it should be noted that aromatic nitro compounds are of considerable industrial value for the production of anilines (Hartter 1985) that are components of agrochemicals, some of which have been briefly noted in Section 8.2.4.4.

Conclusions

Although the possibility of bioremediation of sites contaminated with nitrotoluene waste is clearly possible, important issues should, however, be clearly appreciated. These include (1) that addition of additional carbon sources may be necessary to accomplish partial or complete reduction of nitro groups, (2) under aerobic conditions, dimeric azo compounds may be formed as terminal metabolites, and (3) that aromatic amines may be incorporated into humic material by covalent bonding and thereby resist further degradation.

8.2.5.2 Chemical Warfare Agents

A few brief notes are added on phosphorofluoridates even though their destruction by microbial activity — although clearly possible — may be limited by their toxicity to the requisite microorganisms. One of the motivations for their inclusion is the fact that the hydrolytic enzyme(s) responsible for defluorination — organophosphorous acid anhydrase (OPA) — is widespread, and is found in a number of bacteria (Landis and DeFrank 1990). The microbial hydrolysis of organophosphorus pesticides and cholinesterase inhibitors is accomplished by several distinct enzymes, all of which are termed organophosphorus acid anhydrolases. These have been reviewed (DeFrank 1991) so that only a few additional comments are justified.

The enzymes from *Pseudomonas putida* strain MG and *Flavobacterium* sp. [ATCC 27551] are carried on plasmids (Mulbry et al. 1986), and are membrane bound. Although they hydrolyze a range of phosphorothioate esters at rates greatly exceeding those for chemical hydrolysis and have been proposed for application to destruction of these (Munnecke 1976), they hydrolyze phosphorofluoridates much more slowly (Dumas et al. 1989). This, together with the fact that they are membrane bound, makes them of somewhat limited attraction for the destruction of phosphorofluoridates. Two other enzymes have attracted considerable attention for the hydrolysis of phosphorofluoridates: one from *B. stearothermophilus* is also membrane bound, whereas the other from *Alteromonas* sp. strain JD 6.5 is a soluble enzyme. OPA anhydrolase activity has also been observed in a number of other organisms (DeFrank et al. 1993). The enzyme designated OPA-2 has been purified from *Alteromonas* strain JD 6.5, and has a molecular mass of 60 kDa (DeFrank and Cheng 1991). There has been considerable speculation on the natural substrate for OPA anhydrolase activity, and a surprising fact emerged from a detailed comparison of the amino acid sequence. This revealed a high degree of homology with an aminopeptidase from *Escherichia coli* and that the OPA anhydrolase had high prolidase activity (Cheng et al. 1996).

All of these observations underscore the potential for application of appropriate OPAs to the destruction of organophosphorus compounds with anticholinesterase activity (Cheng and Calomiris 1996). Since, however, hydrolysis results in release of fluoride, the possibility of its subsequent incorporation into organic substrates to produce fluoroacetate and 4-fluorothreonine (Reid et al. 1995) may be worth consideration.

8.2.6 Groundwater Contamination

8.2.6.1 Benzene/Toluene/Ethylbenzene/Xylenes (BTEX)

These compounds along with trimethylbenzenes are components of automobile gasoline, and may enter aquatic systems and eventually groundwater through spillage at filling stations, leakage from underground storage tanks, or during transport.

Aerobic Conditions

All these substrates can be degraded under aerobic conditions, and although there appear to be significant differences among the xylene isomers, mutant strains have been isolated that can degrade all three isomers (Di Lecce et al. 1997). Reviews have covered various different aspects of this problem: (1) biodegradation in gas condensate-contaminated groundwater with emphasis on the role of inorganic nutrients and oxygen (Morgan et al. 1993), (2) the determinative factors for treating aquifer plumes after accidental spillage of automobile fuel with discussions of the design of systems and the role of the indigenous microflora (Salanitro 1993), and (3) general procedures including provision of oxygen and comments on anaerobic treatment (Bowlen and Kosson 1995).

The pathways for the aerobic biodegradation of alkylated benzenes have been elucidated in extensive investigations, and have been discussed in Chapter 6, Section 6.2.1, so that only some of the salient features need be briefly summarized here. The genes for the degradation of toluene may be either chromosomal or plasmid borne. The latter may, therefore, be lost in the absence of selection pressure. The degradation of toluene may take a number of different pathways: (1) involving either dioxygenation or successive monooxygenation of the substrate with subsequent ring cleavage leaving the alkyl group intact and (2) involving oxidation of the alkyl group to a carboxylic acid before ring cleavage. For xylenes, it appears that there are significant differences in the degradability of the isomers, *o*-xylene being apparently more recalcitrant.

The results of laboratory experiments with fluidized-bed reactors treating toluene-contaminated influent (Massol-Deyá et al. 1997) showed that (1) established toluene-degrading strains in the biofilm were displaced by groundwater organisms from a putatively pristine source, (2) these organisms were able to establish a stable microbial community, and (3) these indigenous organisms were able to bring about a decrease in toluene concentration in the effluent from 0.140 to 0.063 mg/l. These results clearly show the advantage and desirability of using indigenous organisms under appropriate conditions, and that they may effectively degrade low substrate concentrations. The latter is consistent with the ability of bacteria in natural aquatic systems to utilize low substrate concentrations (Chapter 4, Section 4.65.2). An experiment on toluene biodegradation under field and laboratory conditions provided results of value in the design of laboratory simulation experiments and illustrated the caution required in assessing the fate of contaminants. Experiments were carried out under three conditions: (1) in flow-through horizontal columns containing sediment and rocks; (2) in shaking cultures containing sediments, rocks, or plant material; and (3) *in situ* in a contaminated stream (Cohen et al. 1995). There were a number of important conclusions:

1. The rates of biodegradation observed in the shake-flask experiment more closely paralleled those from the *in situ* experiment than those using the flow-though column system.

2. Biodegradation took place on streambed surfaces particularly rocks so that the formation of biofilms was essential.
3. Conversion to CO_2 was incomplete, and 15% of the substrate was converted to soluble metabolites and 62% to biomass or insoluble material.

Consistent with this, it was observed during the metabolism of ^{14}C-labeled BTEX that, although the substrates were mineralized to ~70%, ~20% of the label in toluene and ~30% in *o*-xylene were found in humus. It was suggested that the alkylated catechol metabolites might plausibly be responsible for this association (Tsao et al. 1998).

The effect of increased oxygen was examined in contaminated groundwater (Gibson et al. 1998). Oxygen was supplied by diffusion from silicone tubing to areas within the plume that was oxygen-deficient due to the activity of indigenous bacteria. The rate of aerobic BTEX degradation increased at the expense of anaerobic degradation, and resulted in the degradation of both benzene and ethylbenzene that are less readily degraded under anaerobic conditions.

Anoxic or Anaerobic Conditions

In many situations, such as in deep groundwater, however, oxygen concentration may be severely limiting due to its consumption by surficial aerobic organisms or its low rate of transport into the system. There has therefore been great interest in the anaerobic degradation of BTEX. Although this has been demonstrated for benzene under anaerobic conditions by methanogenic consortia, sulfate-reducing bacteria, and Fe(III)-reducing bacteria, greatest attention has been directed in laboratory studies to the anaerobic degradation of toluene in the presence of nitrate. This is supported by the fact that the appropriate organisms are widely distributed (Fries et al. 1994).

Some of the important principles may be illustrated from two groups of experiments. The first used a laboratory microcosm and a field experiment with a contaminated aquifer (Barbaro et al. 1992). Although this clearly showed the effectiveness of adding nitrate, there were significant differences in the extent to which the substrates were degraded: for example, benzene appeared resistant, whereas the xylenes and ethyl benzene were less degradable than toluene. Nitrate was apparently consumed by indigenous organisms at the expense of other carbon substrates, so that this was effectively removed from the system. In this experiment — and possibly in general — sorption of these substrates to organic carbon was not a limiting factor. In the second example, the effect of nitrate or sulfate additions was examined by adding either of these together with samples of BTEX to a contaminated aquifer in the form of "slugs" of purified groundwater (Reinhard et al. 1997). For nitrate additions, the rates of removal of the substrates were in the order benzene < *o*-xylene < ethylbenzene < toluene = *m*-xylene, the last three being removed within 6 days. For sulfate additions, removal of toluene and all of

the xylenes occurred within 50 days. Metabolites characteristic for degradation (Beller et al. 1995; Chapter 6, Section 6.7.3) were isolated as confirmation
of biotransformation. The important observation was made that, although
the results were broadly in agreement with those from laboratory microcosm
experiments, the rates of degradation of toluene and *m*- and *p*-xylenes
seemed to be lower. Benzene and *o*-xylene seem to be considerably more
recalcitrant than other components of BTEX. A review (Lovley 1997) has summarized the various strategies and suggests that uncertainties, particularly in
the bioremediation of benzene, can only be resolved by greater emphasis on
field-oriented studies and a better understanding of the reactions involved
and the factors that limit the rates of degradation.

These experiments should be viewed against the results of extensive laboratory investigations of the anaerobic degradation and transformation of the
individual components of BTEX. The degradation of ethylbenzene and
n-propylbenzene — though not xylenes — has been shown in pure cultures of
denitrifying bacteria (Rabus and Widdel 1995). In crude oil, toluene, ethylbenzene, and *m*-xylene were not degraded through *o*-xylene or *p*-xylene (Rabus
and Widdel 1996). Degradation of toluene by the sulfate-reducing anaerobe
Desulfobacula toluolica has also been shown (Rabus et al. 1993) and by another
sulfate-reducing bacterium with a novel 16S rRNA gene sequence (Beller et al.
1996). The mechanism of these anaerobic degradations is complex and only a
very brief summary using toluene as illustration is justified, since reviews
have been given by the pioneers in this area (Evans and Fuchs 1988; Fuchs et
al. 1994), and detailed discussion has been given in Chapter 6, Section 6.7.3.1.
In summary, there are three cardinal reactions:

1. Oxidation of the methyl group to carboxyl by a number of different
   pathways.
2. Partial reduction of the aromatic ring of the CoA ester of benzoate
   before ring fission.
3. Fission of the ring of cyclohexa-1,4-diene carboxylate (or possibly
   also the cyclohex-1-enecarboxylate) by a series of hydroxylation,
   and reductive and elimination reactions to produce 3-hydroxyp-
   imelate (or pimelate) that then enters the metabolic system.

It is worth noting that benzylsuccinates and benzylfumarates have been
proved to be intermediates during the oxidation of methyl to carboxylate
(Biegert et al. 1996; Beller and Spormann 1997), and that, as noted above,
these compounds together with the analogous compounds that would be
produced from xylenes have been used in support of the active *in situ* anaerobic degradation of toluene and xylenes (Beller et al. 1995).

8.2.6.2 Choroethenes

Trichloroethene (TCE) has been used extensively as a solvent and degreasing
agent, and substantial effort has been devoted to the bioremediation of

groundwater contamined with this. Determinative issues are the relatively low substrate concentrations combined with the large volumes of groundwater that are involved. There has therefore been greater interest in the application of *in situ* procedures than in more conventional "pump-and-treat" on-site methods. In order to appreciate the microbiology of the procedures, a brief summary of the relevant aspects is given. Reviews have been given that provide additional details (Ensley 1991; Arp 1995), and relevant details of the reactions have been given in Chapter 6, Section 6.4. A valuable review (Lee et al. 1998) includes a discussion of microbial processes and provides illustrative examples of both *in situ* and natural attenuation.

Aerobic Conditions

TCE is not able to support the growth of a single organism, but it is susceptible to cooxidation by different oxygenases elaborated by organisms during growth with structurally unrelated substrates. A review of methanotrophic bacteria (Hanson and Hanson 1996) contains a useful account of their application to bioremediation of TCE-contaminated sites.

There are several bacterial oxygenases that can bring about cooxidation of TCE after induction on the relevant substrates: methane monooxygenase in *Methylosinus trichosporium* strain OB3b (Jahng and Wood 1994), propane monooxygenase in *Mycobacterium vaccae* (Vanderberg et al. 1995), and ammonia monooxygenase (Arciero et al. 1989; Vannelli et al. 1990). Although the methane monooxygenase system involves an epoxide as intermediate that may form CO which then toxifies the organism, this may be overcome by addition of a suitable reductant such as formate (Henry and Grbic-Galic 1991). An alternative strategy is the intermittent provision of substrate TCE (Walter et al. 1997).

Pseudomonas putida strain F1 elaborates a dioxygenase for the degradation of toluene, and this is able to cooxidize TCE that may also be carried out by growth with phenol (Nelson et al. 1988). A similar situation prevails for the *iso*propylbenzene-degrading *Rhodococcus erythropolis* strain BD-2 (Dabrock et al. 1994). Increased biodegradation of TCE has been observed in a constructed strain *of Escherichia coli* carrying the genes for the large subunit of toluene terminal oxygenase (from *P. putida* F1), and the small subunit of biphenyl terminal oxygenase, and the ferredoxin and ferredoxin reductase (from *P. pseudoalcaligenes* KF707) (Furukawa et al. 1994). Monooxygenase systems may also be involved in the degradation of toluene and bring about the cooxidation of TCE under suitable growth conditions: in *P. cepacia* strain G4 that carries out *ortho*-hydroxylation, in *P. mendocina* KR-1 that carries out *para*-hydroxylation, and *P. pickettii* that carries out *meta*-hydroxylation.

The complexity introduced by exposure of an established mixed culture growing with a single substrate to an alternative cosubstrate is illustrated by the following. A stable mixed culture of *P. putida* mt-2, *P. putida* F1, *P. putida* GJ31, and *Burkholderia cepacia* G4 growing with limited concentrations of toluene was established. Exposure to trichloroethene for a month resulted in

loss of viability of the last three organisms and resulted in a culture dominated by *P. putida* mt-2 from which mutants had fortuitously arisen (Mars et
al. 1998).

Two different types of experiments on bioremediation of sites contaminated
with TCE have been carried out, and they have been preceded by and have
taken advantage of the valuable results obtained in microcosm experiments.

Application of Indigenous Bacteria

The first procedure used an indigenous microflora enriched by the addition
of substrates that induced the appropriate oxygenases. Microcosm experiments (Hopkins et al. 1993b) demonstrated the differential effects of phenol,
toluene, methane, and ammonia: whereas the first two substrates were
equally effective in removing both TCE and *trans*-1,2-dichloroethene, methane was only marginally effective in removing TCE but more effective in
removing *trans*-1,2-dichloroethene, while ammonia was less effective in
removing of all the substrates. It was shown that in bioreactors fed with phenol under various conditions, maximum removal of TCE occurred when phenol was supplied as a pulse every 24 h: in this reactor, there was a greater
diversity of microorganisms and higher numbers that degraded phenol and
TCE (Shih et al. 1996). Field evaluation of this procedure has been carried out,
using either phenol or toluene or methane as oxygenase inducers (Hopkins
et al. 1993a; Hopkins and Mccarty 1995). The effectiveness of the various substrates for removal of a number of chloroethenes is given in the Table 8.3 from

TABLE 8.3
Influence of Growth Substrates on the Removal of Chloroethenes

Growth Substrate	Concentration (mg/l)	% Removal of Substrate				
		TCE	1,1-DCE	*cis*-DCE	*trans*-DCE	MCE
Methane	6.6	19	—	43	90	95
Phenol	12.5	94	54	92	73	>98
Toluene	9	93	—	>98	75	—

TCE = trichloroethene; 1,1-DCE = 1,1-dichloroethene; *cis-cis*-DCE = *cis-cis*-1,2-dichloro-
ethene; *trans-trans*-DCE = *trans-trans*-1,2-dichloroethene; MCE = chloroethene (vinyl
chloride).

which it is clear that all of the chloroethenes except 1,1-dichloroethene were
effectively removed. There remains no serious concern over the recalcitrance
of these — and particularly of vinyl chloride — produced by partial anaerobic dechlorination of TCE. In this experiment, the concentrations of residual
phenol and toluene were below the standards set for drinking water. A full-
scale treatment of TCE-contaminated groundwater used pulsed addition of
toluene and supply of oxygen both as O_2 and H_2O_2, and a high level of TCE
removal was achieved over a 410-day period (McCarty et al. 1998).

Subsequent microbiological investigations after treatment revealed several important additional facts:

1. The adverse effect of 1,1-dichloroethene on the populations which could be restored after its removal, although the original diversity was recovered only after treatment with toluene and TCE. The dominant Gram-negative taxa belonged to the genera *Comamonas-Variovorax*, *Azoarcus*, and *Burkholderia* (Fries et al. 1997a).

2. Although many Gram-negative strains hybridized to a toluene 2-monooxygenase probe, not all positive strains effectively degraded trichloroethene, and none were as effective as *B. cepacia* strain G4. Since many indigenous strains were able to degrade toluene — and not all were able to metabolize trichloroethene — such strains would dominate populations in the absence of TCE (Fries et al. 1997b). This is consistent with the results of a laboratory study using *B. cepacia* strain G4 (Mars et al. 1996) to which reference has already been made.

Application of Exogenous Bacteria

In the second procedure, organisms with an established degradation potential for TCE were injected at the site.

1. Cells of *Methylosinus trichosporium* strain OB3b were added and became attached to the sediments to form an *in situ* "bioreactor:" the concentration of TCE in samples withdrawn through the bio-filter region fell rapidly, although after 40 days the concentrations had reverted to background levels (Duba et al. 1996). The major remaining unresolved issue is therefore how the degradative activity can be maintained during the time required for effective bioremediation.

2. In light of possible concern over the introduction of phenol into aquifers, mutant strains of *Burkholderia* (*Pseudomonas*) *cepacia* strain G4 in which toluene 2-monooxygenase is constitutive have been examined in microcosm experiments. Effective removal of TCE could be demonstrated (Krumme et al. 1993; Munakata-Marr et al. 1996), although in the first experiment, serious loss of viability of the cells occurred during the 10-week incubation, and this therefore presents an additional challenge.

Experiments have been conducted in a contaminated aquifer to examine the effect of Triton X-100 on increasing the release of TCE from sediments (Sahoo et al. 1998), but the increase was rather modest, ~30%, and is consistent with the limited effect of surfactants in increasing the availability of PAHs (Section 8.2.1).

Anoxic and Anaerobic Conditions

The anaerobic dechlorination of chloroethenes has received attention that has been encouraged by concern that partial dechlorination could result in the production of vinyl chloride as a terminal metabolite. After induction by growth *of Desulfomonile tiedjei* DCB-1 with 3-chlorobenzoate, partial dechlorination of tetrachloroethene to TCE and *cis*-1,1-dichloroethene has been observed (Cole et al. 1995; Townsend and Suflita 1996). Dehalogenases for both TCE and tetrachloroethene have been examined in a strain of *Dehalospirillum multivorans* (Neumann et al. 1994), and it has been shown that the dechlorination of tetrachloroethene in anaerobic aquifer slurries was stimulated by C_3 carboxylic acids or ethanol (Gibson and Sewell 1992).

The anaerobic degradation of chloroalkanes and chloroalkenes has been discussed in detail in Chapter 6, Section 6.4.4, and the application of reductive processes to bioremediation has been examined under denitrifying, sulfidogenic, and methanogenic conditions. References to practical implementation including many from conference proceedings and reports are given in the review by Lee et al. (1998).

Conclusion

In situ procedures using indigenous organisms induced for the aerobic degradation of chloroethenes have been successful, particularly in broadening the range of substrates to include those which might be produced by partial anaerobic dechlorination. Only limited use has been made of anaerobic processes, although these may be effectively combined with aerobic reactions. The application of introduced organisms is currently limited by the relatively short time during which their activity can be maintained, but this could possibly be achieved by designing more effective *"in situ* bioreactors."

8.2.6.3　Methyl t-Butyl Ether (MTBE)

Methyl *t*-butyl ether (MTBE) has replaced tetraethyl lead as an octane booster in gasoline and to minimize automobile discharge of CO and NO in cold weather. It is produced on a large scale, and there has been concern over its possible adverse effects. Its occurrence in groundwater could be associated with gasoline spillage or leakage from underground tanks that store oxygenated fuel. This has already been discussed as a source for BTEX. A survey has revealed the widespread distribution of MTBE in water in the United States, although in no case did concentrations exceed those set for drinking-water standards (Squillace et al. 1996). Contrary to initial fears, this compound was not associated with the simultaneous presence of BTEX so that its source remains unestablished. Attention has been directed to its toxicity, and a somewhat complex picture has emerged (references in Kado et al. 1998; Nihlén et al. 1998). For the sake of completeness, attention is drawn here only to various aspects of its biodegradability.

Experiments have been carried out under different conditions with apparently conflicting conclusions.

1. Under aerobic conditions, a mixed bacterial culture was able to degrade the substrate with intermediate formation of *t*-butanol (Salanitro et al. 1994).

2. Evidence from field measurements indicated slow decay near the source of contamination but persistence further downstream, and this was consistent with the results of laboratory experiments that showed only slow biodegradation of MTBE under aerobic condition (Borden et al. 1997).

3. In a simulated spill, groundwater was spiked with BTEX, MTBE, and chloride as marker, and injected into a sandy aquifer (Schirmer and Barker 1998). Whereas the level of BTEX fell, loss of MTBE was much slower and only ca. 3% remained after 8 years, although the mechanism of its loss was not resolved.

4. Degradation of [U-^{14}C]-MTBE and [U-^{14}C]-*t*-butanol has been demonstrated within 105 days in laboratory micrososms containing stream sediments in a contaminated area (Bradley et al. 1999).

5. In prolonged incubation with a variety of anaerobic conditions, the substrate was recalcitrant under sulfate-reducing conditions and very poorly degraded under nitrate-reducing conditions (Mormille et al. 1994).

The collective evidence suggests the variable recalcitrance of MTBE under aerobic conditions, its recalcitrance anaerobic conditions, and the presence of the putative degradation product *t*-butanol cannot necessarily be construed as evidence for biodegradation (Landmeyer et al. 1998). Bioremediation of groundwater contaminated with MTBE therefore seems problematic.

8.2.7 Metabolic Interaction of Metal Cations and Organic Compounds

Although discussion of strategies for remediation of base-metal-contaminated sites lies beyond the scope of this review, it has already been noted in Chapter 4, Section 4.3.3 that Cr(VI) and U(VI) may serve as alternative electron acceptors to oxygen: and it therefore seems appropriate to mention examples in which their metabolic interaction with organic degradation has been utilized. Whereas the reduction of metals at higher oxidation levels can be gratuitously carried out by many bacteria, in the present context, the cardinal issue is its coupling to the oxidation of an organic substrate under anaerobic conditions.

1. Whereas the gratuitous aerobic reduction of chromate by bacteria is widely distributed (Lovley 1993), the reduction of chromate

Cr(VI) to Cr(III) anaerobically at the expense of the oxidation of organic compounds is uncommon. The oxidation of benzoate to CO_2 using Cr(VI) as oxidant has been accomplished in cultures that were initially enriched using nitrate as an additional electron acceptor (Shen et al. 1996). In the absence of nitrate, there was good agreement between the removal of benzoate and the reduction of Cr(VI), and this could be an attractive procedure in view of the low toxicity and ready degradability of benzoate if this were used in excess.

2. The organism now known as *Geobacter metallireducens* is a strictly anaerobic organism that can couple the oxidation of acetate, toluene, or phenol to the reduction of Fe(III) supplied as an insoluble oxide (Lovley 1991), and strains of the genus have been isolated from a variety of sediments (Coates et al. 1996b). In addition, this organism can couple the oxidation of acetate to the reduction of U(VI) to insoluble U(IV) (Lovley et al. 1991). This opens the possibility of simultaneously remediation of sites containing organic contaminants and U(VI) residues. For the sake of completeness, it is noted that the acidophilic organisms *Thiobacillus ferrooxidans* and *Th. thiooxidans* are able to carry out a formally equivalent reduction of pertechnate (Tc^{VII}) to insolube Tc^V and Tc^{IV} (Lyalikova and Khizhnyak 1996).

3. A mixed bacterial culture obtained by enrichment with tributyl phosphate as sole source of carbon and phosphorus was immobilized in a polyurethane foam matrix (Thomas and Macaskie 1996). These nongrowing cells produced phosphate by hydrolysis of the substrate, and this could be coupled to the formation of uranyl phosphate that was washed out of the column and then precipitated.

8.2.8 Municipal Waste

Only a few parenthetical remarks are given for completeness. In view of the volumes that may be involved in municipal landfills (Suflita et al. 1992), it is not generally practical to attempt remediation of this highly heterogeneous solid phase. The potential adverse effect of leachate that may reach water courses justifies serious concern, and this is underscored by the range of compounds that have been recovered in leachates from municipal landfills. These may include virtually any organic compound together with their transformation products: some illustrative examples include 2-(4-chloro-2-methylphe-noxy)propionic acid, *N-n*-butylbenzenesulphonamide, camphor, 2-hydroxy-benzothiazole, and 2,2'- and 4,4'-dihydroxydiphenylmethane (Öman and Hynning 1993). Phthalate esters are virtually ubiquitous and are therefore to be expected in municipal landfills. On the basis of the collective evidence, however, they are degradable by a range of aerobic bacteria (Ribbons et al. 1984), although details of the mechanism under anaerobic conditions are less fully established, and the relevance of results from experiments under anaerobic denitrifying conditions (Nozawa and Maruyama 1988) may be questioned in this context. Attention has already been directed to the biodegradability of phenoxyalkanoic acids in Section 8.2.3. A range of wood-related products may be expected, and considerable attention has been directed to the biodegradation of lignin and hemicellulose that are complex polymers ubiquitous in higher plants. This has been accomplished particularly by fungi (Kirk and Farrell 1987; Jeffries 1994), although it appears that the variability in the structures of these polymers may result in significant recalcitrance, and that there are significant incompletely resolved issues on the role of bacteria and the regulation of "ligninases" in white-rot fungi. The biodegradability of leachate under methanogenic conditions has been examined, although there was evidence of toxicity at high concentrations (O'Connor et al. 1990).

Landfills containing municipal waste are extremely heterogeneous both in texture and in the range of compounds that have been deposited, and they may be used to illustrate a potentially serious problem of wider significance — the chemical interaction of reactive microbial metabolites from different substrates. Such metabolites have diverse structures including, for example, aromatic amines, benzothiazole sulfones, quinones, and aldehydes. These may react with each other, or with reactive substituents of humic and fulvic acids to produce stable compounds that may be recalcitrant to microbial degradation or of limited bioavailability. Indeed, these condensation products may not be readily detected using conventional analytical protocols, particularly when these are limited to a restricted list such as those designated "priority pollutants." They may also contain transformation products, and the example of 3,3'-dichlorobenzidine and its dechlorination product benzidine has already been noted in Section 8.1.2: this example also reveals the presence of persistent degradation products of azo pigments based on 3,3'-dichlorobenzidine even though its use was prohibited many years ago.

Attention is drawn to two broad classes of compounds that may be encountered.

1. *Monomeric colorants and pigments* — These encompass a wide range of structures (Grimmel 1953), but will seldom be encountered in the free state. Although these compounds are widely used in fabrics, plastics, coatings, and color printing, their volume is low compared with that of the matrices that will generally present greater problems. The degradation of azo compounds has been discussed in Chapter 6, Section 6.8.3, while some of the potential complexities of the aerobic metabolism of anilines have been noted in Chapter 6, Section 6.2.1. It should be noted that sulfanilic acid formed as reduction product of azo dyes may present a problem since, after further oxidation, it may be polymerized to terminal products (Kulla et al. 1983).

2. *Polymeric products* — There may be substantial volumes of these which include both naturally occurring structures such as cellulose, rubber, polypeptides (proteins), and a much wider range of synthetic

products including polyesters, polyamides, polyurethane, polystyrene, polyethylene and polypropylene, polyvinyl chloride, and polyacrylamide. Most of the latter have been developed with the specific aim of resistance to degradation. Some, including types of polyurethane, are, however, degradable by extracellular enzymes from both fungi and bacteria (Howard et al. 1999).

8.2.9 Miscellaneous Contaminants

Brief attention is drawn to a number of monomeric compounds that have received less publicity. These are not necessarily directly related to specific industrial activity, but their presence may be revealed by careful analysis of samples from landfills and sediments. In view of their general recalcitrance, their removal during application of conventional procedures is questionable.

1. Linear styrene dimers have a wide spectrum of use including lubricants and plasticizers. Those without methyl substituents on the rings, or with only one, were biodegradable whereas those with methyl substituents in both rings were recalcitrant (Higashimura et al. 1983).

2. Detergents comprise surfactants combined in commercial preparations with compounds that function as chelating agents, and both surfactants and their degradation products may be encountered in domestic waste. Attention is directed to a review of the bacterial degradation of anionic surfactants comprising alkyl sulfates, aromatic sulfonates, and alkyl ethoxy sulfates (White and Russell 1994) from which it is clear that complete mineralization of some groups may be difficult to achieve. There has been increasing interest in thermally stable fluorinated surfactants (Dams 1993) and in perfluoropolyethers (Gurarini et al. 1993). The biodegradability of some surfactants has been examined (Remde and Debus 1996), and the apparent recalcitrance of an anionic fluorinated sulfonic acid is noteworthy.

3. 2-Mercaptobenzothiazoles have wide industrial use as surface-active biocides (Reemtsma et al. 1995), and (4-morpholino)-2-benzothiazole is used in rubber vulcanization and has been encountered in a range of environmental samples (Kumata et al. 1996). Benzothiazole, 2-hydroxybenzothiazole, and (4-morpholino)-2-benzothiazole have been found in urban runoff, in urban particulate matter, and in road dust, all of which are attributable to their leaching from rubber material (Reddy and Quinn 1997).

4. Silicones have found diverse and increasing use and are generally considered highly stable. Although the levels in effluents from municipal treatment plants were generally below the limits of quantification, they have been found in substantial amounts in the sludges, in sediments in the vicinity of the outfalls, and in agricultural

soil that was amended with sewage sludge (Fendinger et al. 1997). Polydimethylsiloxanes are, however, at least partially degraded in soil to dimethylsilanediol (Carpenter et al. 1995; Fendinger et al. 1997), and the mineralization in soil of the monomer — dimethyl-silanediol — has been shown (Sabourin et al. 1996). The interesting observations were made that in liquid cultures degradation could be obtained with *Fusarium oxysporium* growing concurrently with propan-2-ol or acetone, and by a strain of *Arthrobacter* sp. growing concurrently with dimethylsulfone. The hydrolysis and degradation of tetraalkoxysilanes that have found various applications including use as heat-exchange fluids has been noted in Chapter 6, Section 6.9.6.

5. Aliphatic organobromine compounds are used as agrochemicals and may penetrate water courses. Brief note is made of the biodegradation of methyl bromide by methanotrophic bacteria (Oremland et al. 1994), and in the alkaline borate Lake Mono by trimethylamine-utilizing methylotrophs (Connell et al. 1997). The degradation of 1,2-dibromoethane by mixed cultures of aerobic organisms has also been reported (Freitas dos Santos et al. 1996), and its anaerobic debromination to ethene by methanogenic bacteria has been observed (Belay and Daniels 1987).

6. The degradation of important fluorohydrocarbons is discussed in Chapter 6, Sections 6.4.4 and 6.10.1, and attention is directed to the possible role of bacteria in the destruction of chlorofluorohydrocarbons (Oremland et al. 1996).

7. Dicyclopentadiene has a nauseating smell and has been found as a contaminant in an aquifer used for drinking water (Ventura et al. 1997): it was speculated that it might have originated from an illegal landfill.

8.3 A Hierarchical Strategy

It is obvious that there are several basic questions that must be answered before considering bioremediation:

- What compounds are present and which have to be removed?
- By what means is success to be evaluated?
- Are indigenous or exogenous organisms to be used?

In virtually all the examples that have been used as illustration, three stages may be discerned, and it is convenient to assemble these into a hierarchical system: I — basic laboratory studies, II — microcosm studies using site material, III — evaluation in large-scale systems: on-site pilot plants or *in situ* installations. It is essential that geologic, engineering, microbiological, and chemical interaction be set up at the beginning and maintained throughout. The principles of the design of laboratory experiments have been discussed in Chapter 5 that includes illustrative examples of both microcosm (Section 5.3.2), and mesocosm (Section 5.3.3) systems, and several examples of the use of microcosms have been given in this chapter.

I. Laboratory Experiments

For both exogenous and indigenous organisms:

- Examination of pathways including formation of intermediate and terminal metabolites;
- Use of the substrates as sources of C, N, P, S;
- Physiological optima of temperature, pH, oxygen concentration;
- Toxicity of substrates;
- Induction of catabolic enzymes by growth with a structurally related substrate;
- Degradation of a compound during growth with a metabolically compatible substrate (cometabolism or concurrent metabolism);
- The existence of metabolically incompatible mixtures of substrates;
- Evaluation of relevant probes for the presence of the relevant microorganisms or their activity.

For many exogenous organisms, this will already have been established so that only complementation will be required for indigenous organisms: determination of their presence, number, and metabolic activity by conventional laboratory procedures. It may be valuable to set up enrichment cultures using site material to provide suitable material for laboratory and microcosm studies.

II. Microcosm Experiments

These should be constructed from material collected from the relevant site and simulate aerobic, microaerophilic or anaerobic conditions. They should provide the kinetic data necessary for large-scale operation, and specifically address the following:

- Rates of substrate loss;
- Data on existence and stability of metabolites, and their toxicity to other biota;
- Existence of threshold substrate concentration below which rates of degradation are low or negligible;
- Effectiveness of analogue substrates and metabolites in promoting degradation;

- Stability of the system under prolonged operation;

It is appropriate at this stage to evaluate procedures that will be used to assess the effectiveness of bioremediation. These may include (a) use of radiolabelled substrates (that will not generally be permitted in field operation), (b) evaluation of metabolites and (c) evaluation of markers such as specific enzymes.

III. Large-Scale Operations

Apart from the preceding issues, it is particularly at this stage that very close collaboration with geologists and engineers is maintained: putatively optimal solutions may not be realistic for the site, and alternatives may have to be adopted. The choice may then depend on returning to Stage II — or even Stage I to evaluate these new constraints. A good example is provided by the reduced permeability of substrata encountered during application of a pump-and-treat system: this was due to several causes including mobilization of colloidal material in the system and subsequent deposition and fouling at injection sites and infiltration zones (Wiener et al. 1996). It is probably inadvisable to evaluate alternative strategies in large-scale systems at initial stages of the program. All the intrinsic problems of scale-up will clearly have to be addressed and solved and these include: (1) mobility of microorganisms, substrates, nutrients, and metabolites; (2) oxygen transport, or maintenance of other electron acceptor, e.g., nitrate; (3) stability of the system.

8.4 Concluding Comments

Whereas the potential of bioremediation is substantial, its application has important limitations that are apparent from the many examples used here as illustration. The authors feel that these limitations can be overcome only when adequate attention is directed to fundamental microbiological, chemical and engineering issues. Only then can the barriers to achieving success in full-scale operation be surmounted, since there are significant problems with *in situ* or on-site programs that have not taken advantage of experiments under the controlled conditions in microcosms or small-scale bioreactors. Of the several unresolved microbiological issues, it is suggested that the following deserve particular consideration:

- The degree to which the compounds are accessible to the appropriate microorganisms;
- The extent to which the population of degradative organisms can be maintained and increased;

- The relative biodegradability of components of complex mixtures such as PAHs and PCBs, since some of the less readily degraded components may be the least desirable from an environmental and toxicological viewpoint;
- Even when removal of, say, 90% of the contaminant has been achieved, it may not be possible to attain 99% removal due to the possible existence of threshold concentrations below which rates of degradation are slow or even negligible;
- Evaluation of success is of primary importance. Loss of substrates is a *necessary but not sufficient* condition, in the light of the frequency of biotransformation reactions and the formation of terminal and possibly toxic metabolites.

1. Application of PCR to specific degradative organisms has been reviewed (Steffan and Atlas 1991), and it has been applied to hydrocarbon-degrading *Mycobacterium* sp. (Wang et al. 1996b).

2. Probes for toluene-2-monooxygenase have been used to evaluate the potential number of TCE-degrading organisms in an aquifer (Fries et al. 1997b). In this study, repetitive extragenic palindromic PCR (REP-PCR) of isolates was used to classify their metabolic capability.

3. The reverse-sample genome probe procedure has been used to monitor sulfate-reducing bacteria in oil-field samples (Voordouw et al. 1991), further extended to include 16 heterotrophic bacteria (Telang et al. 1997), and has been applied to evaluating the effect of nitrate on an oil-field bacterial community.

4. On the basis of examination of a collection of strains of *Sphingomonas* sp. from a range of localities, it has been suggested that the glutathione *S*-transferase-encoding gene might be used as a marker for PAH-degrading bacteria (Lloyd-Jones and Lau 1997).

5. Metabolically related parameters have also been used, and these include establishing any of the following.

   a. The presence of metabolites determined from laboratory experiments of degradation pathways: examples include *cis*-dihydrodiols of PAHs in a marine sediment (Li et al. 1996) and of naphthalene in leachate from a contaminated site (Wilson and Madsen 1996); chlorobenzoates from the degradation of PCBs (Flanagan and May 1993; Section 8.2.4.2), and benzyl succinates from BTEX (Beller et al. 1995; Section 8.2.6.1).

   b. Some examples have been given in Section 5.7 of enzymes with a low substrate specificity that are able to accept analogue substrates with the formation of colored metabolites. This may be used to detect enzymatic activity toward the desired substrates when a large number of colonies have to be examined: examples for PAHs include toluene 2,3-dioxygenase (Eaton and Chapman 1995), naphthalene dioxygenase (Ensley et al. 1983), and manganese peroxidase (Bogan et al. 1996), for TCE cometabolism toluene-2-monooxygenase (Fries et al. 1997b), and for degradation of 2,4-dichlorophenoxyacetate, 2,4-dichlorophenoxyacetate/2-ketoglutarate dioxygenase that can accept 4-nitrophenoxyacetate as a substrate with the formation of colored 4-nitrophenol (Sassenella et al. 1997). Care ought, however, to be exercised since it was also shown in the last example that not all strains with established degradative capability for 2,4-dichlorophenoxyacetate were able to form 4-nitrophenol from the surrogate substrate (Sassenella et al. 1997).

References

Adriaens, P. and T.M. Vogel. 1995. Biological treatment of chlorinated organics. pp. 435–486. In *Microbial Transformation and Degradation of Toxic Organic Chemicals* (Eds. L.Y. Young and C.E. Cerniglia). Wiley-Liss, New York.

Adriaens, P., Q. Fu, and D. Grbic-Galic. 1995. Bioavailability and transformation of highly chlorinated dibenzo-*p*-dioxins and dibenzofurans in anaerobic soils and sediments. *Environ. Sci. Technol.* 29, 2252–2260.

Ahlborg, U.G. and T.M. Thunberg. 1980. Chlorinated phenols: occurrence, toxicity, metabolism, and environmental impact. *CRC Crit. Rev. Toxicol.* 7, 1–35.

Ahn, I.-S., W.C. Ghiorse, L.W. Lion, and M.L. Shuler. 1998. Growth kinetics of *Pseudomonas putida* G7 on naphthalene and toxicity during nutrient deprivation. *Biotechnol. Bioeng.* 59: 587–594.

Albrecht, I.D., A.L. Barkovskii, and P. Adriaens. 1999. Production and dechlorination of 2,3,7,8-tetrachlorodibenzo-*p*-dioxin in historically contaminated estuarine sediments. *Environ. Sci. Technol.* 33: 737–744.

Alexander, M. 1985. Biodegradation of organic chemicals. *Environ. Sci. Technol.* 18: 106–111.

Alexander, M. 1995a. *Biodegradation and Bioremediation.* Academic Press, San Diego, CA.

Alexander, M. 1995b. How toxic are toxic chemicals in soil? *Environ. Sci. Technol.* 29: 2713–2717.

Allen, C.C.R., D.R. Boyd, F. Hempenstall, M.J. Larkin, and N.D. Sharma. 1999. Contrasting effects of a nonionic surfactant on the biotransformation of polycyclic aromatic hydrocarbons to *cis*-dihydrodiols by soil bacteria. *Appl. Environ. Microbiol.* 65: 1335–1339.

Andersson, B.E. and T. Henrysson. 1996. Accumulation and degradation of dead-end metabolites during treatment of soil contaminated with polycyclic aromatic hydrocarbons with five strains of white-rot fungi. *Appl. Microbiol. Biotechnol.* 46: 647–652.

Apajalahti, J.H.A. and M.S. Salkinoja-Salonen. 1987a. Dechlorination and *para*-hydroxylation of polychlorinated phenols by *Rhodococcus chlorophenolicus. J. Bacteriol.* 169: 675–681.

Apajalahti, J.H.A. and M.S. Salkinoja-Salonen. 1987b. Complete dechlorination of tetrachlorohydroquinone by cell extracts of pentachloro-phenol induced *Rhodococcus chlorophenolicus. J. Bacteriol.* 169, 5125–5130.

Aprill, W. and R.C. Sims. 1990. Evaluation of the use of prairie grasses for stimulating polycyclic aromatic hydrocarbon treatment in soil. *Chemosphere* 20, 253–265.

Aprill, W., R.C. Sims, J.L. Sims, and J.E. Mattews. 1990. Assessing detoxification and degradation of wood preserving and petroleum wastes in contaminated soil. *Waste Manage. Res.* 8: 45–65.

Arciero, D., T. Vannelli, N. Logan, and A.B. Hooper. 1989. Degradation of trichloroethylene by the ammonia-oxidizing bacterium *Nitrosomonas europaea. Biochem. Biophys. Res. Commun.* 159: 640–642.

Arjmand, M. and H. Sandermann. 1985. Mineralization of chloroaniline/lignin conjugates and of free chloroanilines by the white-rot fungus *Phanerochaete chrysosporium. J. Agric. Food Chem.* 33: 1055–1060.

Arp, D.J. 1995. Understanding the diversity of trichloroethylene co-oxidations. *Curr. Opinion Biotechnol.* 6: 352–358.

Atlas, R.M. 1981. Microbial degradation of petroleum hydrocarbons: an environmental perspective. *Microbiol. Rev.* 45: 180–209

Atlas, R.M. 1995. Bioremediation of petroleum pollutants. *Int. Biodeterior. Biodeg.* 35: 317–327.

Austen, R.A. and N.W. Dunn. 1980. Regulation of the plasmid-specified naphthalene catabolic pathway of *Pseudomonas putida. J. Gen. Microbiol.* 117: 521–528.

Bachmann, A., P. Walet, P. Wijnen, W. de Bruin, J.L.M. Huntjens, W. Roelofsen, and A.J.B. Zehnder. 1988. Biodegradation of alpha- and beta-hexachlorocyclohexane in a soil slurry under different redox conditions. *Appl. Environ. Microbiol.* 54: 143–149.

Barbaro, J.R., J.F. Barker, L.A. Lemon, and C.I. Mayfield. 1992. Biotransformation of BTEX under anaerobic, denitrifying conditions: field and laboratory observations. *J. Contam. Hydrol.* 11: 245–272.

Barkovskii, A.I. and P. Adriaens. 1996. Microbial dechlorination of historically present and freshly spiked chlorinated dioxins and diversity of dioxin-dechlorinating populations. *Appl. Environ. Microbiol.* 62: 4556–4562.

Barnsley, E.A. 1976. Role and regulation of the *ortho* and *meta* pathways of catechol metabolism in pseudomonads metabolizing naphthalene and salicylate. *J. Bacteriol.* 125: 404–408

Barr, D.P. and S.D. Aust. 1994. Mechanisms white-rot fungi use to degrade pollutants. *Environ. Sci. Technol.* 28: 78A–87A.

Barriault, D. and M. Sylvestre. 1993. Factors affecting PCB degradation by an implanted bacterial strain in soil microcosms. *Can. J. Microbiol.* 39: 594–602.

Bayly, R., R. Jain, C.L. Poh, and R. Skurray. 1988. Unity and diversity in the degradation of xylenols by *Pseudomonas* spp.: a model for the study of microbial evolution. In *Microbial Metabolism and the Carbon Cycle* pp. 359–379 (Eds. S.R. Hagedorn, R.S. Hanson, and D.A. Kunz). Harwood Academic Publishers, Chur, Switzerland.

Bayman, P. and G.V. Radkar. 1997. Transformation and tolerance of TNT (2,4,6-trinitrotoluene) by fungi. *Int. Biodeterior. Biorem.* 39: 45–53.

Bedard, D.L. 1990. Bacterial transformation of polychlorinated biphenyls pp. 369–388. In *Biotechnology and Biodegradation*, Vol. 4 (Eds. D. Kamely, A. Chakrabarty, and G.S. Omenn). Gulf Publishing Company, Houston.

Bedard, D.L. and J.F. Quensen III. 1995. Microbial reductive dechlorination of polychlorinated biphenyls pp. 127–216. In *Microbial Transformation and Degradation of Toxic Organic Chemicals* (Eds. L.Y. Young and C.E. Cerniglia). Wiley-Liss, New York.

Bedard, D.L., M.L. Haberl, R.J. May and M.J. Brennan. 1987. Evidence for novel mechanisms of polychlorinated biphenyl metabolism in *Alcaligenes eutrophus* H 850. *Appl. Environ. Microbiol.* 53: 1103–1112.

Bedard, D.L., H. van Dort, and K.A. Deweerd. 1998. Brominated biphenyls prime extensive microbial reductive dehalogenation of Arochlor 1260 in Housatonic River sediment. *Appl. Environ. Microbiol.* 64: 1786–1795.

Belay, N. and L. Daniels. 1987. Production of ethane, ethylene, and acetylene from halogenated hydrocarbons by methanogenic bacteria. *Appl. Environ. Microbiol.* 53: 1604–1610.

Beller, H.R. and A.M. Spormann. 1997. Anaerobic activation of toluene and *o*-xylene by addition to fumarate in denitrifying strain T. *J. Bacteriol.* 179: 670–676.

Beller, H.R., W.-H. Ding, and M. Reinhard. 1995. Byproducts of anaerobic alkylbenzene metabolism useful as indicators of *in situ* bioremediation. *Environ. Sci. Technol.* 29: 2864–2870.

Beller, H.R., A.M. Spormann, P.K. Sharma, J.R. Cole, and M. Reinhard. 1996. Isolation and characterization of a novel toluene-degrading sulfate-reducing bacterium. *Appl. Environ. Microbiol.* 62: 1188–1196.

Bemgård, A., A. Colmsjö, and B-O. Lundmark. 1993. Gas chromatographic analysis of high-molecular-mass polycyclic aromatic hydrocarbons II. Polycyclic aromatic hydrocarbons with relative molecular masses exceeding 328. *J. Chromatogr.* 630: 287–295.

Beurskens, J.E.M., M. Toussaint, J. de Wolf, J.M.D. van der Steen, P.C. Slot, L.C.M. Commandeur, and J.R. Parsons. 1995. Dehalogenation of chlorinated dioxins by an anaerobic consortium from sediment. *Environ. Toxicol. Chem.* 14: 939–943.

Bezalal, L., Y. Hsadar, and C.E. Cerniglia. 1996a. Mineralization of polycyclic aromatic hydrocarbons by the white-rot fungus *Pleurotus ostreatus*. *Appl. Environ. Microbiol.* 62. 292–295.

Bezalel, L., Y. Hadar, P.P. Fu, J.P. Freeman, and C.E. Cerniglia. 1996b. Metabolism of phenanthrene by white-rot fungus *Pleurotus ostreatus*. *Appl. Environ. Microbiol.* 62: 2547–2553

Bhadra, R., D.G. Wayment, J.B. Hughes, and J.V. Shanks. 1999. Confirmation of conjugation processes during TNT metabolism by axenic plant roots. *Environ. Sci. Technol.* 33: 446–452.

Biegert, T., G. Fuchs, and J. Heider. 1996. Evidence that anaerobic oxidation of toluene in the denitrifying bacterium *Thauera aromatica* is initiated by formation of benzylsuccinate from toluene and fumarate. *Eur. J. Biochem.* 238: 661–668.

Biggs, W.R. and J.C. Fetzer. 1996. Analytical techniques for large polycyclic aromatic hydrocarbons: a review. *Trends Anal. Chem.* 15: 196–205.

Binks, P.R., S. Nicklin, and N.C. Bruce. 1995. Degradation of hexahydro-1,3,5-trinitro-1,3,5-triazine RDX by *Stenotrophomonas maltophilia* PB1. *Appl. Environ. Microbiol.* 61: 1318–1322.

Binks, P.R., C.E. French, S. Nicklin, and N.C. Bruce. 1996. Degradation of pentaerythritol tetranitrate by *Enterobacter cloacae* PB2. *Appl. Environ. Microbiol.* 62: 1214–1219.

Bisaillon, J.-G., F. Lépine, R. Beaudet, and M. Sylvestre. 1991. Carboxylation of *o*-cresol by an anaerobic consortium under methanogenic conditions. *Appl. Environ. Microbiol.* 57: 2131–2134.

Blasco, R., R.-M. Wittich, M. Mallavarapu, K.N. Timmis, and D.H. Pieper. 1995. From xenobiotic to antibiotic, formation of protoanemonin from 4-chlorocatechol by enzymes of the 3-oxoadipate pathway. *J. Biol. Chem.* 270: 29229–29235.

Blasco, R., M. Mallavarapu, R.-M. Wittich, K.N. Timmis, and D.H. Pieper. 1997. Evidence that formation of protoanemonin from metabolites of 4-chlorobiphenyl degradation negatively affects the survival of 4-chlorobiphenyl-cometabolizing microorganisms. *Appl. Environ. Microbiol.* 63: 434.

Bogan, B.W., B. Schoenike, R.T. Lamer, and D. Cullen. 1996. Manganese peroxidase mRNA and enzyme activity levels during bioremediation of polycyclic aromatic hydrocarbon-contaminated soil with *Phanerochaete chrysosporium*. *Appl. Environ. Microbiol.* 62: 2381–2386.

Bollag, J.M. 1992. Decontaminating soil with enzymes. *Environ. Sci. Technol.* 26: 1876–1881.

Bollag, J.-M. and M.J. Loll. 1983. Incorporation of xenobiotics into soil humus. *Experientia* 39: 1221–1231.

Bonting, C.F.C., S. Schneider, G. Schmidtberg, and G. Fuchs. 1995. Anaerobic degradation of *m*-cresol via methyl oxidation to 3-hydroxybenzoate by a denitrifying bacterium. *Arch. Microbiol.* 164: 63–69.

Boonchan, S., M.L. Britz, and G.A. Stanley. 1998. Surfactant-enhanced biodegradation of high molecular weight polycyclic aromatic hydrocarbons by *Stenotrophomonas maltophilia*. *Biotechnol. Bioeng.* 59: 480–494.

Borden, R.C., R.A. Daniel, L.E. LeBrun, and C.W. Davis. 1997. Intrinsic biodegradation of MTBE and BTEX in a gasoline-contaminated aquifer. *Water Resour. Res.* 33: 1105–1115.

Bossert, I.D. and Compeau, G.C. 1995. Cleanup of petroleum hydrocarbon contamination in soil pp. 77–125. In *Microbial Transformation and Degradation of Toxic Organic Chemicals* (Eds. L.Y. Young and C.E. Cerniglia). Wiley-Liss, New York.

Bowlen, G.F. and D.S. Kosson. 1995. *In situ* processes for bioremediation of BTEX and petroleum fuel products pp. 515–545. In *Microbial Transformation and Degradation of Toxic Organic Chemicals* (Eds. L.Y. Young and C.E. Cerniglia). Wiley-Liss, New York.

Brackmann, R. and G. Fuchs. 1993. Enzymes of anaerobic metabolism of phenolic compounds. 4-Hydroxybenzoyl-CoA reductase dehydroxylating from a denitrifying *Pseudomonas* sp. *Eur. J. Biochem.* 213: 563–571.

Braddock, J.F., M.L. Ruth, P.H. Catterall, J.L. Walworth, and K.A. McCarthy. 1997. Enhancement and inhibition of microbial activity in hydrocarbon-contaminated arctic soils: implications for nutrient-amended bioremediation. *Environ. Sci. Technol.* 31: 2078–2084.

Bradley, P.M., F.H. Chapelle, J.E. Landmeyer, and J.G. Schumacher. 1994. Microbial transformation of nitroaromatics in surface soils and aquifer materials. *Appl. Environ. Microbiol.* 60: 2170–2175.

Briglia, M., R.I.L. Eggen, W.M. de Vos et al. 1996. Rapid and sensitive method for the detection of *Mycobacterium chlorophenolicum* PCP-1 in soil based on 16S rRNA gene-targeted PCR. *Appl. Environ. Microbiol.* 62: 1478–1480.

Brodkorb, T.S. and R.L. Legge. 1992. Enhanced biodegradation of phenanthrene in soil tar-contaminated soils supplemented with *Phanerochaete chrysosporium*. *Appl. Environ. Microbiol.* 58: 3117–3121.

Brune, A. and B. Schink. 1992. Phloroglucinol pathway in the strictly anaerobic *Pelobacter acidigallici* fermentation of trihydroxybenzenes to acetate via triacetic acid. *Arch. Microbiol.* 157: 417–424.

Brunsbach, F.R. and W. Reineke. 1993. Degradation of chloroanilines in soil slurry by specialized organisms. *Appl. Microbiol. Biotechnol.* 40: 402–407.

Bumpus, J.A. and M. Tatarko. 1994. Biodegradation of 2,4,6-trinitrotoluene by *Phanerochaete chrysosporium*: identification of initial degradation products and the discovery of a metabolite that inhibits lignin peroxidases. *Curr. Microbiol.* 28: 185–190.

Burd, G. and O.P. Ward. 1996. Involvement of a surface-active high molecular weight factor in degradation of polycyclic aromatic hydrocarbons by *Pseudomonas marginalis. Can. J. Microbiol.* 42: 791–797.

Burgos, W.D., J.T. Novak, and D.F. Berry. 1996. Reversible sorption and irreversible binding of naphthalene and α-naphthol to soil: elucidation of processes. *Environ. Sci. Technol.* 30: 1205–1211.

Burken, J.G. and J.L. Schnoor. 1996. Phytoremediation: plant uptake of atrazine and role of root exudates. *J. Environ. Eng.* 122: 958–963.

Buser, H.-R. 1986. Polybrominated dibenofurans and dibenzo-*p*-dioxins: thermal reaction products of polybrominated diphenyl ether flame retardants. *Environ. Sci. Technol.* 20: 404–408.

Buser, H.-R. and M.D. Müller. 1997. Conversion reactions of various phenoxyalkanoic acid herbicides in soil. 2. Elucidation of the enantiomerization process of chiral phenoxy acids from incubation in a D_2O / soil system. *Environ. Sci. Technol.* 31: 1960–1967.

Buser, H.-R., M.D. Müller, and C. Rappe. 1992. Enantioselective determination of chlordane components using chiral high-resolution gas chromatography-mass spectrometry with application to environmental samples. *Environ. Sci. Technol.* 26: 1533–1540.

Bushart, S.P., B. Bush, E.L. Barnard, and A. Bott. 1998. Volatilization of extensively dechlorinated polychlorinated biphenyls from historically contaminated sediments. *Environ. Toxicol. Chem.* 17: 1927–1933.

Carmichael, L.M., R.F. Christman, and F.K. Pfaender. 1997. Desorption and mineralization kinetics of phenanthrene and chysene in contaminated soils. *Environ. Sci. Technol.* 31: 126–132.

Carpenter, J.C., J.A. Cella, and S.B. Dorn. 1995. Study of the degradation of polydimethylsiloxanes on soil. *Environ. Sci. Technol.* 29: 864–868.

Cartwright, D. 1994. Recent developments in fluorine-containing agrochemicals pp.237–262. In *Organofluorine Chemistry Principles and Commercial Applications* (Eds. R.E. Banks, B.E. Smart, and J.C. Tatlow). Plenum Press, London.

Cassidy, D.P. and R.L. Irvine. 1997. Biological treatment of a soil contaminated with diesel fuel using periodically operated slurry and solid phase reactors. *Water Sci. Technol.* 35 (1): 185–192.

Chen, S.-W. and M.D. Aitken. 1999. Salicylate stimulates the degradation of high-molecular-weight polycyclic aromatic hydrocarbons by *Pseudomonas saccharophila* P15. *Environ. Sci. Technol.* 33: 435–439.

Cheng, T.-C. and J.J. Calomiris. 1996. A cloned bacterial enzyme for nerve agent decontamination. *Enz. Microbiol. Technol.* 18: 597–601.

Cheng, T.-C., S.P. Harvey, and G.L. Chen. 1996. Cloning and expression of a gene encoding a bacterial enzyme for decontamination of organophosphorus nerve agents and nucleotide sequence of the enzyme. *Appl. Environ. Microbiol.* 62: 1636–1641.

Chung, N. and M. Alexander. 1998. Differences in sequestration and bioavailability of organic compounds aged in different soils. *Environ. Sci. Technol.* 32: 855–860.

Coates, J.D., R.T. Anderson, and D.R. Lovley. 1996a. Oxidation of polycyclic aromatic hydrocarbons under sulfate-reducing conditions. *Appl. Environ. Microbiol.,* 62: 1099–1101

Coates, J.D., E.J.P. Phillips, D.J. Lonergan, H. Jenter, and D.R. Lovley. 1996b. isolation of *Geobacter* species from diverse sedimentary environments. *Appl. Environ. Microbiol.* 62: 1531–1536.

Cohen, B.A., L.R. Krumholz, H. Kim, and H.E. Hemond. 1995. *In situ* biodegradation of toluene in a contaminated stream. 2. Laboratory studies. *Environ. Sci. Technol.* 29: 117–125.

Cole, J.R., B.Z. Fathepure, and J.M. Tiedje. 1995. Tetrachloroethene and 3-chlorobenzoate dechlorination activities are co-induced in *Desulfomonile tiedjei* DCB-1. *Biodegradation* 6: 167–172.

Comeau, Y., C.W. Greer, and R. Samson. 1993. Role of inoculum preparation and density on the bioremediation of 2,4-D-contaminated soil by bioaugmentation. *Appl. Microbiol. Biotechnol.* 38: 681–687.

Connell, T.L., S.B. Joye, L.G. Miller, and R.S. Oremland. 1997. Bacterial oxidation of methyl bromide in Mono Lake, California. *Environ. Sci. Technol.* 31: 1489–1495.

Crawford, R.L. 1995. The microbiology and treatment of nitroaromatic compounds. *Curr. Opinion Biotechnol.* 6: 329–336.

Crocker, F.H., W.F. Guerin, and S.A. Boyd. 1995. Bioavailability of naphthalene sorbed to cationic surfactant-modified smectite clay. *Environ. Sci. Technol.* 29: 2953–2958

Cutright, T.J. 1995. Polycyclic aromatic hydrocarbon degradation and kinetics using *Cunninghamelle echinulata* var. *elegans. Int. Biodeterior. Biodeg.* 35: 397–408.

Dabrock, B., M. Keßeler, B. Averhoff, and G. Gottschalk. 1994. Identification and characterization of a transmissible linear plasmid from *Rhodococcus erythropolis* BD2 that encodes isopropylbenzene and trichloroethylene catabolism. *Appl. Environ. Microbiol.* 60: 853–860.

Dams, R. 1993. Fluorochemical surfactants for hostile environments. *Spec. Chem.* 13: 4–6.

Daughton, C.G. and D.P.H. Hsieh. 1977. Parathion utilization by bacterial symbionts in a chemostat. *Appl. Environ. Microbiol.* 34: 175–184.

Davis, M.W., J.A. Glaser, J.W. Evans, and R.T. Lamer. 1993. Field evaluation of the lignin-degrading fungus *Phanerochaete sordida* to treat creosote-contaminated soil. *Environ. Sci. Technol.* 27: 2572–2576.

De Vos, D.E., H. De Wever, G. Bryon, and H. Verachtert. 1993. Isolation and characteristics of 2-hydroxybenzothiazole-degrading bacteria. *Appl. Microbiol. Biotechnol.* 39: 377–381.

Dec, J. and J.-M. Bollag. 1994. Dehalogenation of chlorinated phenols during oxidative coupling. *Environ. Sci. Technol.* 28: 484–490.

DeFrank, J.J. 1991. Organophosphorus cholinesterase inhibitors: detoxification by microbial enzymes pp. 165–180. In *Applications of Enzyme Biotechnology* (Ed. J.W. Kelly and T.O. Baldwin). Plenum Press, New York.

DeFrank, J.J. and T.-C. Cheng. 1991. Purification and properties of an organophosphorus acid anhydrase from a halophilic bacterial isolate. *J. Bacteriol.* 173: 1938–1943.

DeFrank, J.J., W.T. Beaudry, T.-C. Cheng, S.P. Harvey, A.N. Stroup, and L.L. Szafraniec. 1993. Screening of halophiliuc bacteria and *Alteromonas* species for organophosphorus hydrolyzing enzyme activity. *Chem.-Biol. Interact.* 87: 141–148.

Deschênes, L., P. Lafrance, J.-P. Villeneuve, and R. Samson. 1996. Adding sodium dodecyl sulfate and *Pseudomonas aeruginosa* UG2 biosurfactants inhibits polycyclic hydrocarbon biodegradation in a weathered creosote-contaminated soil. *Appl. Microbiol. Biotechnol.* 46: 638–646.

Déziel, É., G. Paquette, R. Villemur, F. Lépine, and J-G. Bisaillon. 1996. Biosurfactant production by a soil *Pseudomonas* strain growing on polycyclic aromatic hydrocarbons. *Appl. Environ. Microbiol.* 62, 1908–1912.

Di Lecce, C., M. Accarino, F. Bolognese, E. Galli, and P. Barbieri. 1997. Isolation and metabolic characterization of a *Pseudomonas stutzeri* mutant able to grow on all three isomers of xylene. *Appl. Environ. Microbiol.* 63: 3279–3281.

Donnelly, P.K., R.S. Hegde, and J.S. Fletcher. 1994. Growth of PCB-degrading bacteria on compounds from photosynthetic plants. *Chemosphere* 28: 981–988.

Dott, W., D. Feidieker, M. Steiof, P.M. Beckerm, and P. Kämpfer. 1995. Comparison of *ex situ* and *in situ* techniques for bioremediation of hydrocarbon-polluted soils. *Int. Biodeterior. Biodeg.* 301–316.

Drzyzga, O., A. Schmidt, and K.-H. Blotevogel. 1996. Cometabolic transformation and cleavage of nitrodiphenylamines by three newly isolated sulfate-reducing bacterial strains. *Appl. Environ. Microbiol.* 62: 1710–1716.

Drzyzga, O., D. Bruns-Nagel, T. Gorontzy, K.-H. Blotevogel, and E. von Löw. 1998. Incorporation of ^{14}C-labeled 2,4,6-trinitrotoluene metabolites into different soil fractions after anaerobic and anaerobic-aerobic treatment of soil/molasss mixtures. *Environ. Sci. Technol.* 32: 3529–3535.

Duba, A.G., K.J. Jackson, M.C. Jovanovich, R.B. Knapp, and R. T. Taylor. 1996. TCE remediation using *in situ* resting-state bioaugmentation. *Environ. Sci. Technol.* 30: 1982–1989.

Duetz, W.A., S. Marqués, C. de Jong, J.L. Ramos, and J.G. van Ande. 1994 Inducibility of the TOL catabolic pathway in *Pseudomonas putida* pWW0 growing on succinate in continuous culture: evidence for carbon catabolite repression control. *J. Bacteriol.* 176: 2354–2361.

Dumas, D.P., J.R. Wild, and F.M. Raushel. 1989. Diisopropylfluorophosphate hydrolysis by an organophosphate anhydrase from *Pseudomonas diminuta*. *Biotechnol. Appl. Biochem.* 11: 235–243.

Duque, E., A. Haïdour, F. Godoy, and J.L. Ramos. 1993. Construction of a *Pseudomonas* hybrid strain that mineralizes 2,4,6-trinitrotoluene. *J. Bacteriol.* 175: 2278–2283.

Eaton, R.W. and P. Chapman. 1995. Formation of indigo and related compounds from indolecarboxylic acids by aromatic acid-degrading bacteria: chromogenic reactions for cloning genes encoding dioxygenases that act on aromatic acids. *J. Bacteriol.* 177: 6983–6988.

Edgehill, R. 1992. Factors influencing the success of bioremediation. *Biotechnology* 2: 297–301.

Edwards, P.N. 1994. Uses of fluorine in chemotherapeutics pp. 501–542. In *Organofluorine Chemistry Principles and Commercial Applications* (Eds. R.E. Banks, B.E. Smart, and J.C. Tatlow). Plenum Press, London.

Ellis, B., P. Harold, and H. Kronberg. 1991. Bioremediation of a creosote-contaminated site. *Environ. Sci. Technol.* 12: 447–459.

Engel, A. 1994. Fluorine-containing dyes. B. Other fluorinated dyestuffs pp. 321–338. In *Organofluorine Chemistry. Principles and Commercial Applications* (Eds. R.E. Banks, B.E. Smart, and J.C. Tatlow). Plenum Press, New York.

Ensley, B.D. 1991. Biochemical diversity of trichloroethylene metabolism. *Annu. Rev. Microbiol.* 45: 283–299.

Ensley, B.D., B.J. Ratzken, T.D. Osslund, M.J. Simon, L.P. Wackett, and D.T. Gibson. 1983. Expression of naphthalene oxidation genes in *Escherichi coli* results in the biosynthesis of indigo. *Science* 222: 167–169.

Erickson, D.C., R.C. Loehr, and E.F. Neuhauser. 1993. PAH loss during bioremediation of manufactured gas plant site soils. *Water Res.* 27: 911–919.

Errampalli, D., J.T. Trevors, H. Lee, K. Leung, M. Cassidy, K. Knoke, T. Marwood, K. Shaw, M. Blears, and E. Chung. 1997. Bioremediation: a perspective. *J. Soil Contam.* 6: 207–218.

Estrella, M.R., M.L. Brusseau, R.S. Maier, I.L. Pepper, P.J. Wierenga, and R.M. Miller. 1993. Biodegradation, sorption, and transport of 2,4-dichlorophenoxyacetic acid in saturated and unsaturated soils. *Appl. Environ. Microbiol.* 59: 4266–4273.

Evans, W.C. and G. Fuchs. 1988. Anaerobic degradation of aromatic compounds. *Annu. Rev. Microbiol.* 42: 289–317.

Evers, E.H.G., H.J.C. Klamer, R.W.P.M. Laane, and H.A.J. Govers. 1993. Polychlorinated dibenzo-*p*-dioxin and dibenzofuran residues in estuarine and coastal North Sea sediments. Sources and distribution. *Environ. Toxicol. Chem.* 12: 1583–1598.

Fava, F. and D. Di Gioia. 1998. Effects of Triton X-100 and quillaya saponin on the *ex situ* bioremediation of a chronically polychlorobiphenyl-contaminated soil. *Appl. Microbiol. Biotechnol.* 50: 623–630.

Feidieker, D., P. Kämpfer, and W. Dott. 1994. Microbiological and chemical evaluation of a site contaminated with chlorinated aromatic compounds and hexachlorocyclohexanes. *FEMS Microbiol. Ecol.* 15: 265–278.

Feiring, A.E. 1994. Fluoroplastics pp. 339–372. In *Organofluorine Chemistry Principles and Commercial Applications* (Eds. R.E. Banks, B.E. Smart, and J.C. Tatlow). Plenum Press, London.

Feldmann, J. and W.R. Cullen. 1997. Occurrence of volatile transition metal compounds in landfill gas: synthesis of molybdenum and tungsten carbonyls in the environment. *Environ. Sci. Technol.* 31: 2125–2129.

Fendinger, N.J., D.C. Mcavoy, W.S. Eckhoff, and B.B. Price. 1997. Environmental occurrence of polymethylsiloxane. *Environ. Sci. Technol.* 31: 1555–1563.

Ferring, A.E. 1994. Fluoroplastics pp.339–372. In *Organofluorine Chemistry. Principles and Commercial Applications* (Eds. R.E. Banks, B.E. Smart, and J.C. Tatlow). Plenum Press, New York.

Field, J.A., E. de Jong, G.F. Costa, and J.A.M. de Bont. 1992. Biodegradation of polycyclic aromatic hydrocarbons by new isolates of white-rot fungi. *Appl. Environ. Microbiol.* 58: 2219–2226.

Findlay, M., S. Fogel, L. Conway, and A. Taddeo. 1995. Field treatment of coal tar-contaminated soil based on results of laboratory treatment studies. pp. 487–513. In *Microbial Transformation and Degradation of Toxic Organic Chemicals* (Eds. L.Y. Young and C.E. Cerniglia). Wiley-Liss, New York.

Fiorella, P.D. and J.C. Spain. 1997. Transformation of 2,4,6-trinitrotoluene by *Pseudomonas pseudoalkaligenes* JS 52. *Appl. Environ. Microbiol.* 63: 2007–2015.

Flanagan, W.P. and R.J. May. 1993. Metabolite detection as evidence for naturally occurring aerobic PCB degradation in Hudson River sediments. *Environ. Sci. Technol.* 27: 2207–2212.

Fletcher, J.S. and R.S. Hegde. 1995. Release of phenols by perennial plant roots and their potential importance in bioremediation. *Chemosphere* 31: 3009–3016.

Freitas dos Santos, L.M., D.J. Leak, and A.G. Livingstone. 1996. Enrichment of mixed cultures capable of aerobic degradation of 1,2-dibromoethane. *Appl. Environ. Microbiol.* 62: 4675–4677.

French, C.E., S. Nicklin, and N.C. Bruce. 1996. Sequence and properties of pentaerythritol tetranitrate reductase from *Enterobacter cloacae* PB2. *J. Bacteriol.* 178: 6623–6627.

Fries, M.R., J. Zhou, J. Chee-Sandford, and J.M. Tiedje. 1994. Isolation, characterization, and distribution of denitrifying toluene degraders from a variety of habitats. *Appl. Environ. Microbiol.* 60: 2802–2810.

Fries, M.R., G.D. Hopkins, P.L. McCarty, L.J. Forney, and J.M. Tiedje. 1997a. Microbial succession during a field evaluation of phenol and toluene as the primary substrates for trichloroethene cometabolism. *Appl. Environ. Microbiol.* 63: 1515–1522.

Fries, M.R., L.J. Forney, and J.M. Tiedje. 1997b. Phenol- and toluene-degrading microbial populations from an aquifer in which successsful trichloroethene cometabolism occurred. *Appl. Environ. Microbiol.* 63: 1523–1530.

Fritzsche, C. 1994. Degradation of pyrene at low defined oxygen concentrations by *Mycobacterium* sp. *Appl. Environ. Microbiol.* 60: 1687–1689

Fuchs, G., M.E.S. Mohame, U. Alenschmidt, J. Koch, A. Lack, R. Brackmann, C. Lochmeyer, and B. Oswald. 1994. Biochemistry of anaerobic biodegradation of aromatic compounds. pp. 513–553. In *Biochemistry of Microbial Degradation* (Ed. C. Ratledge). Kluwer Academic Publishers, Dordrecht, the Netherlands.

Funk, S.B., D.J. Roberts, D.L. Crawford, and R.L. Crawford. 1993. Initial-phase optimization for bioremediation of munition compound-contaminated soils. *Appl. Environ. Microbiol.* 59: 2171–2177.

Funk, S.B., D.L. Crawford, R.L. Crawford, G. Mead, and W. Davis-Hoover. 1995. Full-scale anaearobic bioremediation of tritritrotoluene TNT contaminated soil. *Appl. Biochem. Biotechnol.* 51/52: 625–633.

Furukawa, K., J.R. Simon, and A.M. Chakrabarty. 1983. Common induction and regulation of biphenyl, xylene / toluene, and salicylate catabolism in *Pseudomonas paucimobilis. J. Bacteriol.* 154: 1356–1362.

Furukawa, K., J. Hirose, S. Hayashida, and K. Nakamura. 1994. Efficient degradation of trichloroethylene by a hybrid aromatic ring dioxygenase. *J. Bacteriol.* 176: 2121–2123.

Garrison, A.W., P. Schmitt, D. Martens, and A. Kettrup. 1996. Enantiomeric selectivity in the environmental degradation of dichloroprop as determined by high-performance capillary electrophoresis. *Environ. Sci. Technol.* 30: 2449–2455.

Gauthier, M.J., B. Lafay, R. Christen, L. Fernandez, M. Acquaviva, P. Bonin, and J.-C. Bertrand. 1992. *Marinobacter hydrocarbonoclasticus* gen. nov., sp. nov., a new extremely halotolerant, hydrocarbon-degrading marine bacterium. *Int. J. Syst. Bacteriol.* 42: 568–576.

Gerritse, J. and J.C. Gottschal. 1993. Two-membered mixed cultures of methanogenic and aerobic bacteria in O_2-limited chemostats. *J. Gen. Microbiol.* 139: 1853–1860.

Ghoshal, S. and R.G. Luthy. 1996. Bioavailability of hydrophobic organic compounds from nonaqueous-phase liquids: the biodegradation of naphthalene from coal tar. *Environ. Toxicol. Chem.* 15: 1894–1900

Gibson, S.A. and G.W. Sewell. 1992. Stimulation of reductive dechlorination of tetrachloroethene in anaerobic aquifer microcosms by addition of short-chain organic acids or alcohols. *Appl. Environ. Microbiol.* 58: 1392–1393.

Gibson, T.L., A.S. Abdul, and P.D. Chalmer. 1998. Enhancement of *in situ* bioremediation of BTEX-contaminated ground water by diffusion from silicone tubing. *Ground Water Monit. Remed.* 18: 93–104.

Gieg, L.M., A. Otter. and P.M. Fedorak. 1996. Carbazole degradation by *Pseudomonas* sp. LD2, metabolic characteristics and identification of some metabolites. *Environ. Sci. Technol.* 30: 575–585

Gilbert, E.S. and D.E. Crowley. 1997. Plant compounds that induce polychlorinated biphenyl degradation by *Arthrobacter* sp. strain B1B. *Appl. Environ. Microbiol.* 63: 1933–1938.

Gintautas, P.A., S.R. Daniel. and D.L. Macalady. 1992. Phenoxyalkanoic acid herbicides in municipal landfill leachates. *Environ. Sci. Technol.* 26: 517–521.

Götz, R., P. Friesel, K. Roch, O. Päpke, M. Ball, and A. Lis. 1993. Polychlorinated-*p*-dioxins PCDDs, dibenzofurans PCDFs, and other chlorinated compounds in the River Elbe: results on bottom sediments and fresh sediments collected in sedimentation chambers. *Chemosphere* 27: 105–111.

Greer, L.E. and D.R. Shelton. 1992. Effect of inoculant strain and organic matter content on kinetics of 2,4-dichlorophenoxyacetic acid degradation in soil. *Appl. Environ. Microbiol.* 58: 1459–1465.

Grimberg, S.J., W.T. Stringfellow, and M.D. Aitken. 1996. Quantifying the biodegradation of phenanthrene by *Pseudomonas stutzeri* P 16 in the presence of a nonionic surfactant. *Appl. Environ. Microbiol.* 62: 2387–2392.

Grimmel, H.W. 1953. Organic dyes pp. 243–391. In *Organic Chemistry. An Advanced Treatise.* Vol. III. (Ed. H. Gilman). John Wiley & Sons, Inc., New York.

Guerin, W.F. and S.A. Boyd. 1992a. Maintenance and induction of naphthalene degradation activity in *Pseudomonas putida* and an *Alcaligenes* sp. under different culture conditions. *Appl. Environ. Microbiol.* 61: 4061–4068.

Guerin, W.F. and S.A. Boyd. 1992b. Differential bioavailability of soil-sorbed naphthalene to two bacterial species. *Appl. Environ. Microbiol.* 58: 1142–1152.

Guha, S., P.R. Jaffé, and C.A. Peters. 1998. Bioavailability of mixtures of PAHs partitioned into the micellar phase of a nonionic surfactant. *Environ. Sci. Technol.* 32: 2317–2324.

Gurarini, A., G. Guglielmetti, M. Vincenti, P. Guarda, and G. Marchionni. 1993. Characterization of perfluoropolyethers by desorption chemical ionization and tandem mass spectrometry. *Anal. Chem.* 65: 970–975.

Guthrie, E.A. and F.K. Pfaender. 1998. Reduced pyrene bioavailability in microbially active soils. *Environ. Sci. Technol.* 32: 501–508.

Haemmerli, S.D., M.S.A Leisola, D. Sanglard, and A. Fiechter. 1986. Oxidation of benzo[a]pyrene by extracellular ligninases of *Phanerochaete chrysosporium. J. Biol. Chem.* 261: 6900–6903

Häggblom, M. 1990. Mechanisms of bacterial degradation and transformation of chlorinated monoaromatic compounds. *J. Basic Microbiol.* 30: 115–141.

Häggblom, M.M., and R.J. Valo. 1995. Bioremedation of chlorophenol wastes. pp. 389–434. In *Microbial Transformation and Degradation of Toxic Organic Chemicals* (Eds. L.Y. Young and C.E. Cerniglia). Wiley-Liss, New York.

Haïdour, A., and J.L. Ramos. 1996. Identification of products resulting from the biological reduction of 2,4,6-trinitrotoluene, 2,4-dinitrotoluene, and 2,6-dinitrotoluene by *Pseudomonas* sp. *Environ. Sci. Technol.* 30: 2365–2370.

Haigler, B.E., C.A. Pettigrew, and J.C. Spain. 1992. Biodegradation of mixtures of substituted benzenes by *Pseudomonas* sp. strain JS150. *Appl. Environ. Microbiol.* 58: 2237–2244.

Haigler, B.E., S.F. Nishino, and J.C. Spain. 1994. Biodegradation of 4-methyl-5-nitrocatechol by *Pseudomonas* sp. strain DNT. *J. Bacteriol.* 176: 3433–3437.

Hamby, D.M. 1996. Site remediation techniques supporting environmental restoration activities — a review. *Sci. Total. Environ.* 191: 203–224.

Hammel, K.E., B. Kalyanaraman, and T.K. Kirk. 1986. Oxidation of polycyclic aromatic hydrocarbons and dibenzo[p]-dioxins by *Phanerochaete chrysosporium*. *J. Biol. Chem.* 261: 16948–16952

Hanson, R.S., and T.E. Hanson. 1996. Methanotrophic bacteria. *Microbiol. Rev.* 60: 439–471.

Harkness, M.R., J.B. McDermott, D.A. Abramowicz, J.J. Salvo, W.P. Flanagan, M.L. Stephens, F.J. Mondello, R.J. May, J.H. Lobos, K.M. Carroll, M.J. Brennan, A.A. Bracco, K.M. Fish, G.L. Warner, P.R. Wilson, D.K. Dietrich, D.T. Lin, C.B. Morgan, and W.L. Gately. 1993. *In situ* stimulation of aerobic PCB biodegradation in Hudson River sediments. *Science* 259: 503–507.

Hartter, D.R. 1985. The use and importance of nitroaromatic compounds in the chemical industry. In *Toxicity of Nitroaromatic Compounds*, pp.1–13. Hemisphere Publishing Corporation, Washington, D.C.

Hatzinger, P.B. and M. Alexander. 1995. Effect of ageing of chemicals in soil on their biodegradability and extractability. *Environ. Sci. Technol.* 29: 537–545.

Hegde, R.S. and J.S. Fletcher. 1996. Influence of plant growth stage and season on the release of root phenolics by mulberry as related to development of phytoremediation technology. *Chemosphere* 32: 2471–2479.

Heitzer, A. and G.S. Sayler. 1993. Monitoring the efficacy of bioremediation. *Trends Biol. Technol.* 11: 334–343.

Henry, S.M. and D. Grbic-Galic. 1991. Influence of endogenous and exogenous electron donors and trichloroethylene oxidation toxicity on trichloroethylene oxidation by methanotrophic cultures from a groundwater aquifer. *Appl. Environ. Microbiol.* 57: 236–244.

Herd, K.J. 1994. Fluorine-containing dyes A. Reactive dyes pp. 287–314. In *Organofluorine Chemistry Principles and Commercial Applications* (Eds. R.E. Banks, B.E. Smart, and J.C. Tatlow). Plenum Press, London.

Hernandez, B.S., J.J. Arensdorf, and D.D. Focht. 1995. Catabolic characteristics of biphenyl-utilizing isolates which cometabolize PCBs. *Biodegradation* 6: 65–72.

Herod, A. A. 1998. Azaarenes and thiaarenes pp. 271–323. In *Handbook of Environmental Chemistry* (Ed. A.H. Neilson) Vol. 3, Part I. Springer-Verlag, Berlin.

Hess, A., B. Zarda, D. Hahn, A. Häner, D. Stax, P. Höhener, and J. Zeyer. 1997. *In situ* analysis of denitrifying toluene- and *m*-xylene-degrading bacteria in a diesel fuel-contaminated laboratory aquifer column. *Appl. Environ. Microbiol.* 63: 2136–2141.

Hewitt, L.M., J.H. Carey, K.R. Munkittrick, J.L. Parrott, K.R. Solomon, and M.R. Servos. 1998. Identification of chloro-nitro-trifluoromethyl-substituted dibenzo-p-dioxins in lampricide formulations of 3-trifluoromethyl-4-nitrophenol: assessment to induce mixed function oxidase activity. *Environ. Toxicol. Chem.* 17: 941–950.

Hickey, W.J., D.B. Searles, and D.D. Focht. 1993. Enhanced mineralization of polychlorinated biphenyls in soil inoculated with chlorobenzoate-degrading bacteria. *Appl. Environ. Microbiol.* 59: 1194–1200.

Higashimura, T., M. Sawamoto, T. Hiza, M. Karaiwa, T. Tsuchii, and T. Suzuki. 1983. Effect of methyl substitution on microbial degradation of linear styrene dimers by two soil bacteria. *Appl. Environ. Microbiol.* 46: 386–391.

Ho, S.V. and 13 coauthors. 1999. The lasagna technology for *in situ* soil remediation. 2. Large field test. *Environ. Sci. Technol.* 33: 1092–1099.

Hofrichter, M., K. Scheibner, I. Schneegaß, and W. Fritsche. 1998. Enzymatic combusion of aromatic and aliphatic compounds by manganese peroxidase from *Nematoloma frowardii*. *Appl. Environ. Microbiol.* 64: 399–404.

Holliger, C., G. Schraa, A.J.M. Stams, and A.J.B. Zehnder. 1992. Enrichment and properties of an anaerobic mixed culture reductively dechlorinating 1,2,3-trichlorobenzene to 1,3-dichlorobenzene. *Appl. Environ. Microbiol.* 58: 1636–1644.

Hopkins, G.D. and P.L. Mccarty. 1995. Field evaluation of *in situ* aerobic cometabolism of trichloroethylene and three dichloroethylene isomers using phenol and toluene as the primary substrates. *Environ. Sci. Technol.* 29: 1628–1637.

Hopkins, G.D., J. Munakata, L. Semprini, and P.L. McCarty. 1993a. Trichloroethylene concentration effects on pilot-scale *in situ* groundwater bioremediation by phenol-oxidizing microorganisms. *Environ. Sci. Technol.* 27: 2542–2547.

Hopkins, G.D., L. Semprini, and P.L. McCarty. 1993b. Microcosm and *in situ* field studies of enhanced biotransformation of trichloroethylene by phenol-oxidizing microorganisms. *Appl. Environ. Microbiol.* 59: 2277–2285.

Hornbuckle, K.C., C.W. Sweet, D.L. Swackhamer, and S.J. Eisenreich. 1995. Assessing annual water-air fluxes of polychlorinated biphenyls in Lake Michigan. *Environ. Sci. Technol.* 29: 869–877.

Hughes, J.B., D.M. Beckles, S.D. Chandra, and C.H. Ward. 1997. Utilization of bioremediation processes for the treatment of PAH-contaminated sediments. *J. Ind. Microbiol. Biotechnol.* 18: 152–160.

Hurst, G.B., K. Weaver, M.J. Doktycz, M.V. Buchanan, A.M. Costello, and M.E. Lidstrom. 1998. MALDI-TOF analysis of polymerase chain reaction products from methanotrophic bacteria. *Anal. Chem.* 70: 2693–2698.

Hutchins, S.R., D.E. Miller, and A. Thomas. 1998. Combined laboratory/field study on the use of nitrate for *in situ* bioremediation of a fuel-contaminated aquifer. *Environ. Sci. Technol.* 32: 182–1840

Iben, I.E.T. and 10 coauthors. 1996. Thermal blanket for *in situ* remediation of surficial contamination: a pilot test. *Environ. Sci. Technol.* 30: 3144–3154

Inoi, T. 1994. Fluorinated liquid crystals pp. 263–286. In *Organofluorine Chemistry Principles and Commercial Applications* (Eds. R.E. Banks, B.E. Smart, and J.C. Tatlow). Plenum Press, London.

Jaffe, R. and R.A. Hites. 1985. Identification of new, fluorinated biphenyls in the Niagara River Lake Ontario area. *Environ. Sci. Technol.* 19: 736–740.

Jahng, D. and T.K. Wood. 1994. Trichloroethylene and chloroform degradation by a recombinant pseudomonad expressing soluble methane monooxygenase from *Methylosinus trichosporium* OB3b. *Appl. Environ. Microbiol.* 60: 2473–2482.

Janke, D. 1987. Use of salicylate to estimate the threshold inducer level for *de novo* synthesis of the phenol-degrading enzymes in *Pseudomonas putida* strain H. *J. Basic Microbiol.* 27: 83–89.

Järvinen, K.T., E.S. Melin, and J.A. Puhakka. 1994. High-rate bioremediation of chlorophenol-contaminated groundwater at low temperatures. *Environ. Sci. Technol.* 28: 2387–2392.

Jeffries, T.W. 1994. Biodegradation of lignin and hemicelluloses pp. 233–277. In *Biochemistry of Microbial Degradation* (Ed. C. Ratledge). Kluwer Academic Publishers, Dordrecht, the Netherlands.

Kado, N.Y., P.A. Kuzmicky, G. Loarca-Piña, and M.M. Mumtaz. 1998. Genotoxicity testing of methyl tertiary-butyl ether (MTBE) in the Salmonella microsuspension assay and mouse bone marrow micronucleus test. *Mutat. Res.* 412: 131–138.

Kan, A.T., G. Fu., M. Hunter, and M.B. Tomson. 1998. Irreversible sorption of neutral hydrocarbons to sediments: experimental observations and model predictions. *Environ. Sci. Technol.* 32: 892–902.

Kaplan, D.L. 1990. Biotransformation pathways of hazardous energetic organo-nitro compounds. pp.155–181. In *Biotechnology and Biodegradation* Vol. 4 (Eds. D. Kamely, A. Chakrabarty, and G.S. Omenn). Gulf Publishing Company, Houston, TX.

Kästner, M., M. Breuer-Jammali, and B. Mahro. 1998. Impact of inoculum protocols, salinity, and pH on the degradation of polycyclic aromatic hydrocarbons (PAHs) and survival of PAH-degrading bacteria introduced into soil. *Appl. Environ. Microbiol.* 64: 359–362.

Keilhau, B.M. 1831. Reise i Øst- og Vest-Finnmarken samt til Beeren-Eiland og Spitsbergen i Aarene 1827 og 1828. Christiana.

Khatisashvili, G., M. Gordenziani, G. Kvesitadze, and F. Korte. 1997. Plant monooxygenases: participation in xenobiotic oxidation. *Ecotoxicol. Environ. Saf.* 36: 118–122.

King, G.M. 1994. Association of methanotrophs with the roots and rhizomes of aquatic vegetation. *Appl. Environ. Microbiol.* 60: 3220–3227.

Kirk, T.K. and R.L. Farrell. 1987. Enzymatic "combustion": the microbial degradation of lignin. *Annu. Rev. Microbiol.* 41: 465–505.

Kitts, C.L., D.P. Cunningham, and P.J. Unkefer. 1994. Isolation of three hexahydro-1,3,5-trinitro-1,3,5-triazine-degrading species of the family Enterobacteriaceae from nitramine explosive-contaminated soil. *Appl. Environ. Microbiol.* 60: 4608–4711.

Koester, C.J. and R.A. Hites. 1992. Wet and dry deposition of chlorinated dioxins and furans. *Environ. Sci. Technol.* 26: 1375–1382.

Kotterman, M.J.J., E.H. Vis, and J.A. Field. 1998. Successive mineralization and detoxification of benzo[*a*]pyrene by the white-rot fungus *Bjerkandera* sp. strain BOS55 and indigenous microflora. *Appl. Environ. Microbiol.* 64: 2853–2858.

Kropp, K.G., J. A. Goncalves, J.T. Anderson, and P.M. Fedorak. 1994. Microbially mediated formation of benzonaphthothiophenes from benzo[*b*]thiophenes. *Appl. Environ. Microbiol.* 60: 3624–3631.

Kropp, K.G., S. Saftic, J.T. Andersson, and P.M. Fedorak. 1996. Transformations of six isomers of dimethylbenzothiophenes by three *Pseudomonas* strains. *Biodegradation* 7: 203–221.

Krumme, M.L., K.N. Timmis, and D.F. Dwyer. 1993. Degradation of trichloroethylene by *Pseudomonas cepacia* G4 and the constitutive mutant strain G4 52223 PR1 in aquifer systems. *Appl. Environ. Microbiol.* 59: 2746–2749.

Kuhn, E.P. and J.M. Suflita. 1989. Microbial degradation of nitrogen, oxygen and sulfur heterocyclic compounds under anaerobic conditions: studies with aquifer samples. *Environ. Toxicol. Chem.* 8: 1149–1158.

Kukor, J.J. and R.H. Olsen. 1996. Catechol 2,3-dioxygenases functional in oxygen-limited hypoxic environments. *Appl. Environ. Microbiol.* 62: 1728–1740.

Kulla, H.G., F. Klausener, U. Meyer, B. Lüdeke, and T. Leisinger. 1983. Interference of aromatic sulfo groups in the microbial degradation of the aza dyes Orange I and Orange II. *Arch. Microbiol.* 135: 1–7.

Kumata, H., H. Takada, and N. Ogura. 1996. Determination of 2-4-morpholinyl benzothiazole in environmental samples by a gas chromatograph equipped with a flame photometric detector. *Anal. Chem.* 68: 1976.

Laine, M.M. and K.S. Jörgensen. 1996. Straw compost and bioremediated soil as inocula for the biorermediation of chlorophenol contaminated soil. *Appl. Environ. Microbiol.* 62: 1507–1513.

Laine, M.M. and K.S. Jörgensen. 1997. Effective and safe composting of chlorophenol contaminated soil in pilot scale. *Environ. Sci. Technol.* 30: 371–378.

Lamar, R.T. and D.M. Dietrich. 1990. *In situ* depletion of pentachlorophenol from contaminated soil by *Phanerochaete* spp. *Appl. Environ. Microbiol.* 56: 3093–3100.

Lamar, R.T., J.W. Evans, and J. Glaser. 1993. Solid-phase treatment of a pentachlorophenol-contaminated soil using lignin-degrading fungi. *Environ. Sci. Technol.* 27: 2566–2571.

Landis, W.G. and J.J. DeFrank. 1990. Enzymatic hydrolysis of toxic organofluorophosphate compounds. pp. 183–201. In *Biotechnology and Biodegradation* Vol. 4 (Eds. D. Kamely, A. Chakrabarty, and G.S. Omenn). Gulf Publishing Company, Houston, TX.

Landmeyer, J.E., F.H. Chapelle, P.M. Bradley, J.F. Pamkow, C.D. Church, and P.G. Tratnyek. 1998. Fate of MTBE relative to benzene in a gasoline-contaminated aquifer. *Ground Water Monit. Remed.* 18: 93–102.

Lang, E., H. Viedt, J. Egestorff, and H.H. Hanert. 1992. Reaction of the soil microflora after contamination with chlorinated aromatic compounds and HCH. *FEMS Microbiol. Ecol.* 86: 275–282.

Lantz, S.E., M.T. Montgomery, W.W. Schultz, P.H. Pritchard, B.J. Spargo, and J.G. Mueller. 1997. Constituents of an organic wood preservative that inhibit the fluoranthene-degrading activity of *Sphingomonas paucimobilis* strain EPA505. *Environ. Sci. Technol.* 31: 3573–3580.

Lee, H.J., M.V. Duke, J.H. Nirk, M. Yamamoto, and S.O. Duke. 1995. Biochemical and physiological effects of benzheterocycles and related compounds. *J. Agric. Food Chem.* 43: 2722–2727.

Lee, M.D., J.M. Odom, and R.J. Buchanan. 1998. New persectives on microbial dehalogenation of chlorinated solvents: insights from the field. *Annu. Rev. Microbiol.* 52: 423–452.

Lendemann, U., J.C. Spain, and B.F. Smets. 1998. Simultaneous biodegradation of 2,4-dinitrotoluene and 2,6-dinitrotoluene in an anerobic fluidized-bed biofilm reactor. *Environ. Sci. Technol.* 32: 82–87.

Lenke H. and H.-J. Knackmuss. 1996. Initial hydrogenation and extensive reduction of substituted 2,4-dinitrophenols. *Appl. Environ. Microbiol.* 62: 784–790.

Lewis, T.A., S. Goszczynski, R.L. Crawford, R.A. Korus, and W. Admassu, 1996. Products of anaerobic 2,4,6-trinitrotoluene TNT transformation *by Clostridium bifermentans. Appl. Environ. Microbiol.* 62: 4669–6774.

Li, X.-F., X-C. Le, C.D. Simpson, W.R. Cullen, and K.J. Reimer. 1996. Bacterial transformation of pyrene in a marine environment. *Environ. Sci. Technol.* 30: 1115–1119

Licht, D., B.K. Ahring, and E. Arvin, 1996. Effects of electron acceptors, reducing agents, and toxic metabolites on anaerobic degradation of heterocyclic compounds. *Biodegradation* 7: 83–90.

Lloyd-Jones, G. and P.C.K. Lau. 1997. Glutathione *S*-transferase-encoding gene as a potential probe for environmental bacterial isolates capable of degrading polycyclic aromatic hydrocarbons. *Appl. Environ. Microbiol.* 63: 3286–3290.

Logothetis, A.L. 1994. Fluoroelastomers pp. 373–396. In *Organofluorine Chemistry. Principles and Commercial Applications* (Eds. R.E. Banks, B.E. Smart, and J.C. Tatlow). Plenum Press, New York.

Lovley, D.R. 1991. Dissimilatory FeIII and MnIV reduction. *Microbiol. Rev.* 55: 259–287.

Lovley, D.R. 1993. Dissimilatory metal reduction. *Annu. Rev. Microbiol.* 47: 263–290.

Lovley, D.R. 1997. Potential for anaerobic bioremediation of BTEX in petroleum-contaminated aquifers. *J. Ind. Microbiol. Biotechnol.* 18: 75–81.

Lovley, D.R., E.J.P. Phillips, Y. A. Gorby, and E.R. Landa. 1991. Microbial reduction of uranium. *Nature* 350: 413–416.

Lyalikova, N.N. and T.V. Khizhnyak. 1996. Reduction of heptavalent Technetium by acidophilic bacteria of the genus *Thiobacillus*. *Microbiology* 65: 468–473. Reprinted from *Mikrobiologiya* 65: 533–538.

Macdonald, J.A. and B.E. Rittman. 1993. Performance standards for *in situ* bioremediation. *Environ. Sci. Technol.* 27: 1974–1979.

Macdonald, R.W., W.J. Cretney, N. Crewe, and D. Paton. 1992. A history of octachlorodibenzo-*p*-dioxin, 2,3,7,8-tetrachlorodibenzofuran, and 3,3′,4,4′-tetrachlorobiphenyl contamination in Howe Sound, British Columbia. *Environ. Sci. Technol.* 26: 1544–1550.

Madsen, E.L. 1991. Determining *in situ* biodegradation. Facts and challenges. *Environ. Sci. Technol.* 25: 1663–1673.

Madsen, T. and P. Kristensen. 1997. Effects of bacterial inoculation and nonionic surfactants on degradation of polycyclic aromatic hydrocarbons in soil. *Environ. Toxicol. Chem.* 16: 631–637.

Madsen, T. and D. Licht. 1992. Isolation and characterization of an anaerobic chlorophenol-transforming bacterium. *Appl. Environ. Microbiol.* 58: 2874–2878.

Madsen, E.L., C.L. Mann, and S.E. Bilotta. 1996a. Oxygen limitations and aging as explanations for the field persistence of naphthalene in coal tar-contaminated surface sediments. *Environ. Toxicol. Chem.* 15: 1876–1882.

Madsen, E.L., C.T. Thomas, M.S. Wilson, R.L. Sandoli, and S.E. Bilotta. 1996b. *In situ* dynamics of aromatic hydrocarbons and bacteria capable of PAH metabolism in a coal tar-waste-contaminated field site. *Environ. Sci. Technol.* 30: 2412–2416.

Mahaffey, W.R., D.T. Gibson, and C.E. Cerniglia. 1988. Bacterial oxidation of chemical carcinogens: formation of polycyclic aromatic acids from benz[*a*]anthracene. *Appl. Environ. Microbiol.* 54: 2415–2423.

Margesin, R. and F. Schinner. 1997. Efficiency of indigenous and inoculated cold-adapted soil microorganisms for biodegradation of diesel oil in alpine oils. *Appl. Environ. Microbiol.* 63: 2660–2664.

Mars, A.E., G.T. Prins, P. Wietzes, W. de Konig, and D.B. Janssen. 1998. Effect of trichloroethylene on the competitive behaviour of toluene-degrading bacteria. *Appl. Environ. Microbiol.* 64: 208–215.

Mars, A.E., J. Houwing, J. Dfolfing, and D.B. Janssen. 1996 Degradation of toluene and trichloroethylene by *Burkholderia cepacia* G4 in growth-limited fed-batch culture. *Appl. Environ. Microbiol.* 62: 886–891.

Martin, M., E. Ferrer, R. Alonso, and J. Fernández. 1995. Bioremediation of soil contaminated by propachlor using native bacteria. *Int. Biodeterior. Biodeg.* 213–225.

Massol-Deyá, A., R. Weller, L. Ríos-Hernández, J.-Z. Zhou, R.F. Hickey, and J.M. Tiedje. 1997. Succession and convergence of biofilm communities in fixed-film reactors treating aromatic hydrocarbons in groundwater. *Appl. Environ. Microbiol.* 63: 270–276.

May, R., P. Schröder, and H. Sandermann. 1997. *Ex situ* process for treating PAH-contaminated soil with *Phanerochaete chrysosporium*. *Environ. Sci. Technol.* 31: 2626–2633.

McCarthy, D.L., A.A. Claude, and S.D. Copley. 1997. *In vivo* levels of chlorinsted hydroquinones in a pentachlorophenol-degrading bacterium. *Appl. Environ. Microbiol.* 63: 1883–1888.

McCarty, P.L., M.N. Goltz, G.D. Hopkins, M.E. Dolan, J.P. Allan, B.T. Kawakami, and T.J. Carrothers. 1998. Full-scale evaluation of *in situ* cometabolic degradation of trichloroethylene in groundwater through toluene injection. *Environ. Sci. Technol.* 32: 88–100.

McCormick, N.G., J.H. Cornell, and A.M. Kaplan. 1978. Identification of biotransformation products from 2,4-dinitrotoluene. *Appl. Environ. Microbiol.* 35: 945–948.

McCormick, N.G., J.H. Cornell, and A.M. Kaplan. 1981. Biodegradation of hexahydro-1,3,5-trinitro-1,3,5-triazine. *Appl. Environ. Microbiol.* 42: 817–823.

McIntyre, T. and G.M. Lewis. 1997. The advancement of phytoremediation as an innovative environmental technology for stabilization, remediation, or restoration of contaminated sites in Canada: a discussion paper. *J. Soil Contam.* 6: 227–241.

McNally, D.L., J.R. Mihelcic, and D.R. Lueking. 1998. Biodegradation of three- and four-ring polycyclic aromatic hydrocarbons under aerobic and denitrifying conditions. *Environ. Sci. Technol.* 32: 2633–2639.

Meng, M., W.-Q. Sun, L.A. Geelhaar, G. Kumar, A.R. Patel, G.F. Payne, M.K. Speedie, and J.R. Stacy. 1995. Denitration of glycerol trinitrate by resting cells and cell extracts of *Bacillus thuringiensis/cereus* and *Enterobacter agglomerans*. *Appl. Environ. Microbiol.* 61: 2548–2553.

Middeldorp, P.J., M. Jaspers, A.J.B. Zehnder, and G. Schraa. 1996. Biotransformation of α–, β–, γ–, and δ-hexachlorocyclohexane under methanogenic conditions. *Environ. Sci. Technol.* 30: 2345–2349.

Middeldorp, P.J.M., J. de Wolf, A.J.B. Zehnder, and G. Schraa. 1997. Enrichment and properties of a 1,2,4-trichlorobenzene-dechlorinating microbial consortium. *Appl. Environ. Microbiol.* 63: 1225–1229.

Miethling, R. and U. Karlson. 1996. Accelerated mineralization of pentachlorophenol in soil upon inoculation with *Mycobacterium chlorophenolicum* PCP1 and *Sphingomonas chlorophenolica* RA2. *Appl. Environ. Microbiol.* 62: 4361–4366.

Mohn, W.W. and K.J. Kennedy. 1992. Reductive dehalogenation of chlorophenols by *Desulfomonile tiedjei* DCB-1. *Appl. Environ. Microbiol.* 58: 1367–1370.

Mohn, W.W. and J.M. Tiedje. 1992. Microbial reductive dehalogenation. *Microbiol. Rev.* 56: 482–507.

Morgan, P. and R.J. Watkinson. 1989a. Hydrocarbon degradation in soils and methods for soil biotreatment. *CRC Crit. Rev. Biotechnol.* 8: 305–333.

Morgan, P. and R.J. Watkinson. 1989b. Microbiological methods for the cleanup of soil and ground water contaminated with halogenated organic compounds. *FEMS Microbiol. Rev.* 63: 277–300.

Morgan, P., S.T. Lewis, and R.J. Watkinson. 1993. Biodegradation of benzene, toluene, ethylbenzene and xylenes in gas-condensate-contaminated water. *Environ. Pollut.* 82: 181–190.

Mormille, M.R., S. Liu, and J.M. Suflita. 1994. Anaerobic biodegradation of gasoline oxygenates: extrapolation of information to multiple sites and redox conditions. *Environ. Sci. Technol.* 28: 1727–1732.

Morris, P.J., J.F. Quensen, III, J.M. Tiedje, and S.A. Boyd. 1992. Reductive debromination of the commercial polybrominated biphenyl mixture Firemaster BP6 by anaerobic microorganisms from sediments. *Appl. Environ. Microbiol.* 58: 3249–3256.

Morris, P.J., J.F. Quensen, III, J.M. Tiedje, and S.A. Boyd. 1993. An assessment of the reductive debromination of polybrominated biphenyls in the Pine River reservoir. *Environ. Sci. Technol.* 27: 1580–1586.

Mueller, J.G., P.J. Chapman, and P.H. Pritchard. 1989. Creosote-contaminated sites. Their potential for bioremediation. *Environ. Sci. Technol.* 23. 1197–1201.

Mueller, J.G., S.E. Lantz, B.O. Blattmann, and P.J. Chapman. 1991a. Bench-scale evaluation of alternative biological treatment processes for the remediation of pentachlorophenol- and creosote-contaminated materials: solid-phase bioremediation. *Environ. Sci. Technol.* 25: 1045–1055.

Mueller, J.G., S.E. Lantz, B.O. Blattmann, and P.J. Chapman. 1991b. Bench-scale evaluation of alternative biological treatment processes for the remediation of pentachlorophenol- and creosote-contaminated materials: slurry-phase bioremediation. *Environ. Sci. Technol.* 25: 1055–1061.

Mulbry, W.W., J.S. Karns, P.C. Kearney, J.O. Nelson, C.S. McDaniel, and J.R. Wild. 1986. Identification of a plasmid-borne parathion hydrolase gene from *Flavobacterium* sp. by Southern hybridization with *opd* from *Pseudomonas diminuta. Appl. Environ. Microbiol.* 51: 926–930.

Mulroy, P.T. and L.-T. Ou. 1998. Degradation of tetraethyllead during the degradation of leaded gasoline hydrocarbons in soil. *Environ. Toxicol. Chem.* 17: 777–782.

Munakata-Marr, J., P.L. Mccarty, M.S. Shields, M. Reagin, and S.C. Francesconi. 1996. Enhancement of trichloroethylene degradation in aquifer microcosms bioaugmented with wild-type and genetically altered *Burkholderia Pseudomonas cepacia* G4 and PR1. *Environ. Sci. Technol.* 30: 2045–2052.

Munnecke, D.M. 1976. Enzymatic hydrolysis of organophosphate insecticides, a possible pesticide disposal method. *Appl. Environ. Microbiol.* 32: 7–13.

Nagasawa, S., R. Kikuchi, Y. Nagata, M. Takagi, and M. Matsuo. 1993. Aerobic mineralization of γ-HCH by *Pseudomonas paucimobilis* UT26. *Chemosphere* 26: 1719–1728.

Nazina, T.N., A.E. Ivanova, M. Wagner, B. Ziran, R.R. Ibatullin, G.F. Kandaurova, Y.M. Miller, Y. Belyaev, and M.V. Ivanov. 1996. Introduction of *Clostridium tyrobutyricum* and molasses into the oil stratum of the Romashkinskoe oil field and its influence on microbial processes in the stratum. *Microbiology* 65: 355–359 Reprinted from Mikrobiologiya 65: 403–408.

Neilson, A.H. and A.-S. Allard. 1998. Microbial metabolism of PAHs and heteroarenes, pp. 1–80. In *Handbook of Environmental Chemistry* (Ed. A.H. Neilson). Vol. 3, Part J, Springer-Verlag, Berlin.

Nelson, M.J.K., S.O. Montgomery, and P.H. Pritchard. 1988. Trichloroethylene metabolism by microorganisms that degrade aromatic compounds. *Appl. Environ. Microbiol.* 54: 604–606.

Neumann, A., H. Scholz-Muramatsu, and G. Diekert. 1994. Tetrachloroethene metabolism of *Dehalospirillum multivorans. Arch. Microbiol.* 162: 295–301.

Nihlén, A., R. Wålindrer, A. Löf, and G. Johanson. 1998. Experimental exposure to methyl *tertiary*-butyl ether II. Acute effects in humans. *Toxicol. Appl. Pharmacol.* 148: 281–287.

Nishino, S.F. and J.C. Spain. 1993. Degradation of nitrobenzene by a *Pseudomonas pseudoalcaligenes. Appl. Environ. Microbiol.* 59: 2520–2525.

Noordman, W.H., W. Ji, M.L. Briusseau, and D.B. Janssen. 1998. Effects of rhamnolipid biosurfactants on removal of phenanthrene from soil. *Environ. Sci. Technol.* 32: 1806–1812

Nörtemann, B., A. Glässer, R. Machinek, G. Remberg, and H.-J. Knackmuss. 1993. 5-Hydroxyquinoline-2-carboxylic acid, a dead-end metabolite from the bacterial oxidation of 5-aminonaphthalene-2-sulfonic acid. *Appl. Environ. Microbiol.* 59: 1898–1903.

Nowara, A., J. Burhenne, and M. Spiteller. 1997. Binding of fluoroquinolone carboxylic acid derivatives to clay minerals. *J. Agric. Food Chem.* 45: 1459–1463.

Nozawa, T. and Y. Maruyama. 1988. Anaerobic metabolism of phthalate and other aromatic acids by a denitrifying bacterium. *J. Bacteriol.* 170: 5778–5784.

Nyman, M.C., A.K. Nyman, L.S. Lee, L.F. Nies, and E.R. Blatchley. 1997. 3,3'-Dichlorobenzidine transformation processes in natural sediments. *Environ. Sci. Technol.* 31: 1068–1073

O'Connor, O.A., R. Dewan, P. Galuzzi, and L.Y. Young. 1990. Landfill leachates: a study of its anaerobic mineralization and toxicity to methanogenesis. *Arch. Environ. Contam. Toxicol.* 18: 143–147.

O'Loughlin, E.J., S.R. Kehrmeyer, and G.K. Sims. 1996. Isolation, characterization, and substrate utilization of a quinoline-degrading bacterium. *Int. Biodeterior. Biodeg.* 38: 107–118.

Ogram, A.V., R.E. Jessup, L.T. Ou, and P.S.C. Rao. 1985. Effects of sorption on biological degradation rates of 2,4-dichlorophenoxyacetic acid in soils. *Appl. Environ. Microbiol.* 49: 582–587.

Ogunseitan, O.A. and B.H. Olson. 1993. Effect of 2-hydroxybenzoate on the rate of naphthalene mineralization in soil. *Appl. Microbiol. Biotechnol.* 38: 799–807.

Ohisa, N., M. Yamaguchi, and N. Kurihara. 1980. Lindane degradation by cell-free extracts of *Clostridium rectum*. *Arch. Microbiol.* 125: 221–225.

Öman, C. and P.-Å. Hynning. 1993. Identification of organic compounds in municipal landfill leachates. *Environ. Pollut.* 80: 265–271.

Oremland, R.S., D.J. Lonergan, C.W. Culbertson, and D.R. Lovley. 1996. Microbial degradation of hydrochlorofluorocarbons ($CHCl_2F$ and $CHXCl_2CF_3$) in soils and sediments. *Appl. Environ. Microbiol.* 62: 1818–1821.

Oremland, R.S., L.C. Miller, C.W. Culbertson, T.L. Connell, and L. Jahnke. 1994. Degradation of methyl bromide by methanotrophic bacteria in cells suspensions and soils. *Appl. Environ. Microbiol.* 60: 3640–3646.

Ornston, L.N. and W.-K. Yeh. 1982. Recurring themes and repeated sequences in metabolic evolution pp.105–126. In *Biodegradation and Detoxification of Environmental Pollutants* (Ed. A.M. Chakrabarty). CRC Press, Boca Raton, FL.

Owens, E.H., J.R. Harper, W. Robson, and P.D. Boehm. 1987. Fate and persistence of crude oil stranded on a sheltered beach. *Arctic* 40, Suppl. 1, 109–123.

Palm, H., J. Knuutinen, J. Haimi, J. Salminen, and V. Huhta. 1991. Methylation products of chorophenols, catechols and hydroquinones in soil and earthworm of sawmill environments. *Chemosphere* 23: 263–267.

Pellizari, V.H., S. Bezborodnikov, J.F. Quensen, and J.M. Tiedje. 1996 Evaluation of strains isolated by growth on naphthalene and biphenyl for hybridization of genes to dioxygenase probes and polychlorinated biphenyl-degrading ability. *Appl. Environ. Microbiol.* 62: 2053–2058

Pieper, D.H., R. Winkler, and H. Sandermann. 1992. Formation of a toxic dimerization product of 3,4-dichloroaniline by lignin peroxidase from *Phanerochaete chrysosporium*. *Angew. Chem.* 104: 60–61.

Pirnik, M.P. 1977. Microbial oxidation of methyl branched alkanes. *Crit. Rev. Microbiol.* 5: 413–422.

Pradhan, S.P., J.R. Paterek, B.Y. Liu, J.R. Conrad, and V.J. Srivastava. 1997. Pilot-scale bioremediation of PAH-contaminated soils. *Appl. Biochem. Biotechnol.* 63–65: 759–773.

Preuss, A., J. Fimpel, and G. Diekert. 1993. Anaerobic transformation of 2,4,6-trinitrotoluene (TNT). *Arch. Microbiol.* 159: 345–353.

Rabus, R. and F. Widdel. 1995. Anaerobic degradation of ethylbenzene and other aromatic hydrocarbons by new denitrifying bacteria. *Arch. Microbiol.* 163: 96–103.

Rabus, R. and F. Widdel. 1996. Utilization of alkylbenzenes during anaerobic growth of pure cultures of denitrifying bacteria on crude oil. *Appl. Environ. Microbiol.* 62: 1238–1241.

Rabus, R., R. Nordhaus, W. Ludwig, and F. Widdel. 1993. Complete oxidation of toluene under strictly anoxic conditions by a new sulfate-reducing bacterium. *Appl. Environ. Microbiol.* 59: 1444–1451.

Racke, K.D. and J.R. Coats 1990. *Enhanced Biodegradation of Pesticides in the Environment* American Chemical Society Symposium Series 426. American Chemical Society, Washington. D.C.

Ram, N.M., D.H. Bass, R. Falotico, and M. Leahy. 1993. A decision framework for selecting remediation technologies at hydocarbon-contaminated sites. *J. Soil Contam.* 2: 167–189.

Reanney, D.C., P.C. Gowland, and J.H. Slater. 1983. Genetic interactions among microbial communities. *Symp. Soc. Gen. Microbiol.* 34: 379–421.

Reddy, C.M. and J.G. Quinn. 1997. Environmental chemistry of benzothiazoles derived from rubber. *Environ. Sci. Technol.* 31: 2847–2853.

Reemtsma, T., O. Fiehn, G. Kalnowski, and M. Jekel. 1995. Microbial transformations and biological effects of fungicide-derived benzothiazoles determined in industrial wastewater. *Environ. Sci. Technol.* 29: 478–485.

Reid, K.A., J.T.G. Hamilton, R.D. Bowden, D. O'Hagan, L. Dasaradhi, M.R. Amin, and D.B. Harper. 1995. Biosynthesis of fluorinated secondary metabolites by *Streptomyces cattleya*. *Microbiology* (U.K.) 141: 1385–1393.

Reinecke, W., D.J. Jeenes, P.A. Williams, and H.-J. Knackmuss. 1982. TOL plasmid pWWO in constructed halobenzoate-degrading *Pseudomonas* strains: prevention of meta pathway. *J. Bacteriol.* 150: 195–201.

Remde, A. and R. Debus. 1996. Biodegradability of fluorinated syrfactants under aerobic and anaerobic conditions. *Chemosphere* 32: 1563–1574.

Ribbons, D.W., P. Keyser, D.A. Kunz, B.F. Taylor, R.W. Eaton, and B.N. Anderson. 1984. Microbial degradation of phthalates pp. 371–395. In *Microbial Degradation of Organic Compounds* (D.T. Gibson ed.). Marcel Dekker, New York.

Rich, J.J. and G.M. King. 1998. Carbon monoxide oxidation by bacteria associated with the roots of freshwater macrophytes. *Appl. Environ. Microbiol.* 64: 4939–4943.

Richnow, H.H., A. Eschenbach, B. Mahro, R. Seifert, P. Wehrung, P. Albrecht, and W. Michaelis. 1998. The use of ^{13}C-labelled polycyclic aromatic hydrocarbons for the analysis of their transformation in soil. *Chemosphere* 36: 2211–2224.

Rijnaarts, H.H.M., A. Bachmann, J.C. Jumelet, and A.J.B. Zehnder. 1990. Effect of desorption and intraparticle mass transfer on the aerobic biomineralization of alpha-hexachlorocyclohexane in a contaminated calcareous soil. *Environ. Sci. Technol.* 24: 1349–1354.

Roberts, D.J., P.M. Fedorak, and S.E. Hrudey. 1990. CO_2 incorporation and 4-hydroxy-2-methylbenzoic acid formation during anaerobic metabolism of *m*-cresol by a methanogenic consortium. *Appl. Environ. Microbiol.* 56: 472–478.

Rontani, J.-F., M.J. Gilewicz, V.D. Micgotey, T.L. Zheng, P.C. Bonin, and J.-C. Bertrand. 1997. Aerobic and anaerobic metabolism of 6,10,14-trimethylpentadecan-2-one by a denitrifying bacterium isolated from marine sediments. *Appl. Environ. Microbiol.* 63: 636–643.

Roussis, S.G. and J.W. Fedora. 1997. Determination of alkenes in hydrocarbon matrices by acetone chemical ionization mass spectrometry. *Anal. Chem.* 69: 1550–1556.

Rudolphi, A., A. Tschech, and G. Fuchs. 1991. Anaerobic degradation of cresols by denitrifying bacteria. *Arch. Microbiol.* 155: 238–248.

Rüttimann-Johnson, C. and R.T. Lamar. 1996. Polymerization of pentachlorophenol and ferulic acid by fungal extracellular lignin-degrading enzymes. *Appl. Environ. Microbiol.* 62: 3890–3893.

Sabourin, C.L., J.C. Carpenter, T.K. Leib, and J.L. Spivack. 1996. Biodegradation of dimethylsilanediol in soils. *Appl. Environ. Microbiol.* 62: 4352–4360.

Sack, U., M. Hofrichter, and W. Fritsche. 1997. Degradation of phenanthrene and pyrene by *Nematoloma frowardii*. *J. Basic Microbiol.* 4: 287–293.

Sahoo, D., J.A. Smith, T.E. Imbrigiotta, and H.M. Mclallan. 1998. Surfactant-enhanced remediation of a trichloroethene-contaminated aquifer. 2. Transport of TCE. *Environ. Sci. Technol.* 32: 1686–1693.

Salanitro, J.P. 1993. The role of bioattenuation in the management of aromatic hydrocarbon plumes in aquifers. *Ground Water Monit. Remed.* 13: 150–161.

Salanitro, J.P., L.A. Diaz, M.P. Williams, and H.L. Wisniewski. 1994. Isolation of a bacterial culture that degrades methyl *t*-butyl ether. *Appl. Environ. Microbiol.* 60: 2593–2596.

Salanitro, J.P., P.B. Dorn, M.H. Huesmann, K.O. Moore, I.A. Rhodes, L.M.R. Jackson, T.E. Vipond, M.M. Western, and H.L. Wisniewski. 1997. Crude oil hydrocarbon bioremediation and soil ecotoxicity assessment. *Environ. Sci. Technol.* 31: 1769–1776.

Salt, D.E., R.D. Smith, and I. Raskin. 1998. Phytoremediation. *Annu. Rev. Plant Physiol.* 49: 643–668.

Sander, P., R.-M. Wittich, P. Fortnagel, H. Wilkes, and W. Francke. 1991. Degradation of 1,2,4-trichloro- and 1,2,4,5-tetrachlorobenzene by *Pseudomonas* strains. *Appl. Environ. Microbiol.* 57: 1430–1440.

Sandermann, H. 1994. Higher plant metabolism of xenobiotics: the "green liver" concept. *Pharmacogenetics* 4: 225–241.

Sandoli, R.L., W.C. Ghiorse, and E.L. Madsen. 1996. Regulation of microbial phenanthrene mineralization in sediment samples by sorbent-sorbate contact time, inocula and gamma irradiation-induced sterilization artifacts. *Environ. Toxicol. Chem.* 15: 1901–1907.

Sarand, I., S. Timonen, E.-L. Nurmiaho-Lassila, T. Koivula, K. Haatela, M. Romantschuk, and R. Sen. 1998. Microbial biofilms and catabolic plasmid harbouring degradative fluorescent pseudomonads in Scots pine mycorrhizospheres developed on petroleum contaminated soil. *FEMS Microbiol. Ecol.* 27: 115–126.

Sassanella, T.M., F. Fukumori, M. Bagdasarian, and R.P. Hausinger. 1997. Use of 4-nitrophenoxyacetic acid for detection and quantification of 2,4-dichlorophenoxyacetic acid 2,4-D/α-ketoglutarate dioxygenase activity in 2,4-D-degrading microorganisms. *Appl. Environ. Microbiol.* 63: 1189–1191.

Schink, B. 1991. Syntrophism among prokaryotes pp. 276–299. In *The Prokaryotes* (Eds. A. Balows, H.G. Trüper, M. Dworkin, W. Harder, and K.-H. Schleifer). Springer-Verlag, Heidelberg.

Schink, B. 1997. Energetics of syntrophic cooperation in methanogenic degradation. *Microbiol. Mol. Biol. Rev.* 61: 262–280.

Schirmer, M. and J.F. Barker. 1998. A study of long-term MTBE attenuation in the Borden aquifer, Ontario, Canada. *Groundwater Monit.*: 113–122.

Schleussinger, A., B. Ohlmeier, I. Reiss, and S. Schulz. 1996. Moisture effects on the cleanup of PAH-contaminated soil with dense carbon dioxide. *Environ. Sci. Technol.* 30: 3199–3204.

Schneider, J., R. Grosser, K. Jayasimhulu, W. Xue, and D. Warshawsky. 1996. Degradation of pyrene, benz[a]anthracene, and benzo[a]pyrene by *Mycobacterium* sp. strain RGHII-135, isolated from a former coal gasification site. *Appl. Environ. Microbiol.* 62: 13–19

Schnoor, J.L., L.A. Licht, S.V. McCutcheon, N.L. Wolfe, and L.H. Carreiora. 1995. Phytoremediation of organic and nutrient contaminants. *Environ. Sci. Technol.* 29: 318A–323A.

Schwartz, G. and F. Lingens. 1994. Bacterial degradation of N-heterocyclic compounds. pp.59–486. In *Biochemistry of Microbial Degradation* (Ed. C. Ratledge). Kluwer Academic Publishers, Dordrecht, The Netherlands.

Selifonov, S.A., P.J. Chapman, S.B. Akkerman, J.E. Gurst, J.M. Bortiatynski, M.A. Nanny, and P.G. Hatcher. 1998. Use of ^{13}C nuclear magnetic resonance to assess fossil fuel biodegradation: fate of $[1-^{13}C]$acenaphthene in creosote polycyclic aromatic compound mixtures degraded by bacteria. *Appl. Environ. Microbiol.* 64: 1447–1453.

Sellström, U., A. Kierkegaard, C. de Witt, and B. Jansson. 1998. Polybrominated diphenyl ethers and hexabromocyclododecane in sediment and fish from a Swedish river. *Environ. Toxicol. Chem.* 17: 1065–1072.

Sembries, S. and R.L. Crawford. 1997. Production of *Clostridium bifermentans* spores as inoculum for bioremediation of nitroaromatic contaminants. *Appl. Environ. Microbiol.* 63: 2100–2104.

Sergy, G.A., Ed. 1987. The Baffin Island Oil Spill (BIOS) Project. *Arctic* 40: Suppl. 1: 1–279.

Servent, D., C. Ducrorq, Y. Henry, A. Guissani, and M. Lenfant. 1991. Nitroglycerin metabolism by *Phanerochaete chrysosporium*: evidence for nitric oxide and nitrite formation. *Biochim. Biophys. Acta* 1074: 320–325.

Shah, M.M., D.P. Barr, N. Chung, and S.D. Aust. 1992. Use of white-rot fungi in the degradation of environmental chemicals. *Toxicol. Lett.* 64: 493–501.

Shannon, M.J.R and R. Unterman. 1993. Evaluating bioremediation, distinguishing fact from fiction. *Annu. Rev. Microbiol.* 47: 715–738.

Shen, H., P.H. Pritchard, and G.W. Sewell. 1996. Microbial reduction of CrVI during anaerobic degradation of benzoate. *Environ. Sci. Technol.* 30: 1667–1674.

Shen, Y., L.G. Stehmeier, and G. Voordouw. 1998. Identification of hydrocarbon-degrading bacteria in soil by reverse sample gene probing. *Appl. Environ. Microbiol.* 64: 637–645.

Shih, C.-C., M.E. Davey, J. Zhou, J.M. Tiedje, and C.S. Criddle. 1996. Effects of phenol feeding pattern on microbial community structure and cometabolism of trichloroethylene. *Appl. Environ. Microbiol.* 62: 2953–2960.

Siciliano, S.D. and J.J. Germida. 1999. Enhanced phytoremediation of chlorobenzoates in rhizosphere soil. *Soil. Biol. Biochem.* 31: 299–305.

Siciliano, S.D., H. Holdie, and J.J. Germida. 1998. Enzymatic activity in root exudates of Dahurian wild rye (*Elymus daurica*) that degrades 2-chlorobenzoic acid. *J. Agric. Food Chem.* 46: 5–7.

Siegrist, R.L., O.R. West, M.I. Morris, D.A. Pickering, D.W. Greene, C.A. Muhr, D.D. Davenport, and J.S. Gierke. 1995. *In situ* mixed region vapor stripping in low-permeability media. 2. Full-scale field experiments. *Environ. Sci. Technol.* 29: 2198–2207.

Simmons, K.E., R.D. Minard, and J.-M. Bollag. 1987. Oligomerization of 4-chloroaniline by oxidoreductases. *Environ. Sci. Technol.* 21: 999–1003.

Spain, J.C. 1995a. Biodegradation of nitroaromatic compounds. *Annu. Rev. Microbiol.* 49: 523–555.

Spain, J.C., Ed. 1995b. *Biodegradation of Nitroaromatic Compounds.* Plenum Press, New York.

Squillace, P.I., J.S. Zogorski, W.G. Wilber, and C.V. Price. 1996. Preliminary assessment of the occurrence and possible sources of MTBE in groundwater in the United States, 1993–1994. *Environ. Sci. Technol.* 30: 1721–1730.

Steffan, R.J. and R.M. Atlas. 1991. Polymerase chain reactions, applications in environmental microbiology. *Annu. Rev. Microbiol.* 45: 137–161.

Stehmeier, L.G., T.R. Jack, and G. Voordouw. 1996. *In vitro* degradation of dicyclopentadiene by microbial consortia isolated from hydrocarbon-contaminated soil. *Can. J. Microbiol.* 42: 1051–1060.

Steiert, J.G. and R.L. Crawford. 1986. Catabolism of pentachlorophenol by a *Flavobacterium* sp. *Biochem. Biophys. Res. Commun.* 141: 1421–1427.

Stern, G.A., M.D. Loewen, B.M. Miskimmin, D.C.G. Muir, and J.B. Westmore. 1996. Characterization of two major toxaphene components in treated lake sediment. *Environ. Sci. Technol.* 30: 2251–2258.

Strauss, H.S. 1997. Is bioremediation a green technology? *J. Soil Contam.* 6: 219–225.

Suflita, J.M., C.P. Gerba, R.K. Ham, A.C. Palmisano, W.L. Rathje, and J.A. Robinson. 1992. The world's largest landfill. A multidisciplinary investigation. *Environ. Sci. Technol.* 26: 1486–1495.

Sugai, S.F., J.E. Lindstrom, and J.F. Braddock. 1997. Environmental influences on the microbial degradation of *Exxon Valdez* oil on the shorelines of Prince William Sound, Alaska. *Environ. Sci. Technol.* 31: 1564–1572.

Sullivan, J.P. and H.A. Chase. 1996. 1,2,3-Trichlorobenzene transformation by *Methylosinus trichosporium* OB3b expressing soluble methane monooxygenase. *Appl. Microbiol. Biotechnol.* 45: 427–433.

Sutherland, J.B., F. Rafii, A.A. Kahn, and C.E. Cerniglia. 1995. Mechanisms of polycyclic aromatic hydrocarbon degradation pp. 269–306 In *Microbial Transformation and Degradation of Toxic Organic Chemicals* (Eds. L.Y. Young and C.E. Cerniglia). Wiley-Liss, New York.

Swannell, R.P.J., K. Lee, and M. McDonagh. 1996. Field evaluations of marine oil spill bioremediation. *Microbiol. Revs.* 60: 342–365.

Swoboda-Colberg, N.G. 1995. Chemical contamination of the environment: sources, types, and fate of synthetic organic chemicals pp. 27–74. In *Microbial Transformation and Degradation of Toxic Organic Chemicals* (Eds. L.Y.Young and C.E. Cerniglia). Wiley-Liss, New York.

Sylvestre, M. 1995. Biphenyl/chlorobiphenyls catabolic pathway of *Comamonas testosteroni* B-356: prospect for use in bioremediation. *Int. Biodeterior. Biodeg.* 35: 189–211.

Takada, S., M. Nakamura, T. Matsueda, R. Kondo, and K. Sakai. 1996. Degradation of polychlorinated dibenzo-*p*-dioxins and polychlorinated dibenzofurans by the white-rot fungus *Phanerochaete sordida* YK-624. *Appl. Environ. Microbiol.* 62: 4323–4328.

Tatsumi, K., A. Freyer, R.D. Minard, and J.-M. Bollag. 1994. Enzyme-mediated coupling of 3,4-dichloroaniline and ferulic acid: a model for pollutant binding to humic materials. *Environ. Sci. Technol.* 28: 210–215.

Telang, A.J., S. Ebert, L.M. Focht, D.W. Westlake, G.E. Jenneman, D. Gevertz, and G. Voordouw. 1997. Effect of nitrate injection on the microbial community in an oil field monitored by reverse sample genome probing. *Appl. Environ. Microbiol.* 63: 1785–1793.

Tett, V.A., A.J. Willetts, and H.M. Lappin-Scott. 1994. Enantioselective degradation of the herbicide mecoprop [2-2-methyl-4-chlorophenoxypropionic acid] by mixed and pure bacterial cultures. *FEMS Microbiol. Ecol.* 14: 191–200.

Thomas, A.O. and J.M. Lester. 1993. The microbial remediation of former gasworks sites, a review. *Environ. Technol.* 14: 1–24.

Thomas, J.M. and C.H. Ward. 1989. *In situ* biorestoration of organic contaminants in the subsurface. *Environ. Sci. Technol.* 23: 760–766.

Thomas, R.A.P. and L.E. Macaskie. 1996. Biodegradation of tributyl phosphate by naturally occurring microbial isolates and coupling to the removal of uranium from aqueous solution. *Environ. Sci. Technol.* 30: 2371–2375.

Thompson, P.L., L.A. Ramer, and J.L. Schnoor. 1999. Hexahydro-1,3,5-trinitro-1,3,5-triazine translocation in poplar trees. *Environ. Toxicol. Chem.* 18: 279–284.

Tiehm, A., M. Stieber, P. Werner, and F.M. Frimmel. 1997. Surfactant-enhanced mobilization and biodegradation of polycyclic aromatic hydrocarbons in manufactured gas plant soil. *Environ. Sci. Technol.* 31: 2570–2576.

Timmis, K.N., R.J. Steffan, and R. Unterman. 1994. Designing microorganisms for the treatment of toxic wastes. *Annu. Rev. Microbiol.* 48: 525–557.

Timonen, S., K.S. Jörgensen, K. Haahtela, and R. Sen. 1998. Bacterial community structure at defined locations of *Pinus sylvestris–Suillus bovinus* and *Pinus sylvestris–Paxillus involutus* mycorrhizospheres in dry pine forest humus and nursery peat. *Can. J. Microbiol.* 44: 499–513.

Townsend, G.T. and J.M. Suflita. 1996. Characterization of chloroethylene dehalogenation by cell extracts of *Desulfomonile tiedjei* and its relationship to chlorobenzoate dehalogenation. *Appl. Environ. Microbiol.* 62: 2850–2853.

Tsao, C.-W., H.-G. Song, and R. Bartha. 1998. Metabolism of benzene, toluene, and xylene hydrocarbons in soil. *Appl. Environ. Microbiol.* 64: 4924–4929.

Valli, K. and M.H. Gold. 1991. Degradation of 2,4-dichlorophenol by the lignin-degrading fungus *Phanerochaete chrysosporium*. *J. Bacteriol.* 173: 345–352.

Valli, K., B.J. Brock, D.K. Joshi, and M.H. Gold. 1992. Degradation of 2,4-dinitrotoluene by the lignin-degrading fungus *Phanerochaete chrysosporium*. *Appl. Environ. Microbiol.* 58: 221–228.

Valo, R.J., M.M. Häggblom, and M.S. Salkinoja-Salonen. 1990. Bioremediation of chlorophenol containing simulated ground water by immobilized bacteria. *Water Res.* 24: 253–258.

van der Linden, A.C., 1978. Degradation of oil in the marine environment pp.165–200. In *Development in Biodegradation of Hydrocarbons* — 1. (Ed. R.J. Watkinson). Applied Science Publishers., London.

Van der Meer, J.R., C. Werlen, S.F. Nishino, and J.C. Spain. 1998. Evolution of a pathway for chlorobenzene metabolism leads to natural attenuation in contaminated groundwater. *Appl. Environ. Microbiol.* 64: 4185–4193.

Vanderberg, L.A., B.L. Burback, and J.J. Perry. 1995. Biodegradation of trichloroethylene by *Mycobacterium vaccae*. *Can. J. Microbiol.* 41: 298–301.

Vannelli, T., M. Logan, D.M. Arciero, and A.N. Hooper. 1990. Degradation of halogenated aliphatic compounds by the ammonia-oxidizing bacterium *Nitrosomonas europaea*. *Appl. Environ. Microbiol.* 56: 1169–1171.

Ventura, F., J. Romero, and J. Pares. 1997. Determination of dicyclopentadiene and its derivatives as compounds causing odors in groundwater supplies. *Environ. Sci. Technol.* 31: 2368–2374.

Verstraete, W. and W. Devlieger. 1996. Formation of non-bioavailable organic residues in soil: perspectives for site remediation. *Biodegradation* 7: 471–485.

Volkering, F., A.M. Breure, J.G. van Andel, and W.H. Rulkens. 1995. Influence of nonionic surfactants on bioavailability and biodegradation of polycyclic hydrocarbons. *Appl. Environ. Microbiol.* 61: 1699–1705.

Volkering, F., A.M. Breure, and W.H. Rulkens, W.H. 1998. Microbiological aspects of surfactant use for biological soil remediation. *Biodegradation* 8: 401–417.

Voordouw, G., J.K. Voordouw, R.R. Karkhoff-Schweiser, P.M. Fedorak, and D.W.S. Westlake. 1991. Reverse sample genome probing, a new technique for identification of bacteria in environmental samples by DNA hybridizatiin, and its application to the identification of sulfate-reducing bacteria in oil field samples. *Appl. Environ. Microbiol.* 57: 3070–3078.

Vorbeck, C., H. Lenke, P. Fischer, and H.-J. Knackmuss. 1994. Identification of a hydride-Meisenheimer complex as a metabolite of 2,4,6-trinitrotoluene by a *Mycobacterium strain*. *J. Bacteriol.* 176: 932–934.

Walker, J.E. and D.L. Kaplan. 1992. Biological degradation of explosives and chemical agents. *Biodegradation* 3: 369–385.

Walter, G.A., S.E. Strand, R.P. Herwig, T.P. Treat, and H.D. Stensel. 1997. Trichloroethylene and methane feeding strategies to sustain degradation by methanotrophic enrichments. *Water Environ. Res.* 69: 1066–1074.

Walter, M.V., E.C. Neilson, G. Firmstone, D.G. Martin, M.J. Clayton, S. Simpson, and S. Spaulding. 1997. Surfactant enhances biodegradation of hydrocarbons: microcosm and field study. *J. Soil Contam.* 6: 61–77.

Walton, B.T. and T.A. Anderson. 1990. Microbial degradation of trichloroethylene in the rhizosphere: potential application to biological remediation of waste sites. *Appl. Environ. Microbiol.* 56: 1012–1016.

Wang, P., A.E. Humphrey, and S. Krawiec. 1996a. Kinetic analysis of desulfurization of dibenzothiophene by *Rhodococcus erythropolis* in continuous cultures. *Appl. Environ. Microbiol.* 62: 66–3068.

Wang, R.-F., A. Luneau, W.-W. Cao, and C.E. Cerniglia. 1996b. PCR detection of polycyclic hydrocarbon-degrading mycobacteria. *Environ. Sci. Technol.* 30: 307–311.

Westcott, N.D. and B.L. Worobey. 1985. Novel solvent extraction of lindane from soil. *J. Agric. Food Chem.* 33: 58–60.

Wester, P.G., J. de Boer, and U.A.T. Brinkman. 1996. Determination of polychlorinated terphenyls in aquatic biota and sediment with gas chromatography/mass spectrometry using negative chemical ionization. *Environ. Sci. Technol.* 30: 473–480.

White, G.F. and N.J. Russell. 1994. Biodegradation of anionic surfactants and related molecules. pp. 143–177. In *Biochemistry of Microbial Degradation* (Ed. C. Ratledge). Kluwer Academic Publishers, Dordrecht, The Netherlands.

White, G.F. and J.R. Snape. 1993. Microbial cleavage of nitrate esters, defusing the environment. *J. Gen. Microbiol.* 139: 1947–1957.

White, J.C., A. Quiñones-Rivera, and M. Alexander. 1998. Effect of wetting and drying on the bioavailability of organic compounds sequestered in soil. *Environ. Toxicol. Chem.* 17: 2378–2382.

Wick, and P.M. Gschwend. 1998. Source and chemodynamic behaviour of diphenyl sulfone and *ortho*- and *para*-hydroxybiphenyl in a small lake receiving discharges from an adjacent superfund site. *Environ. Sci. Technol.* 32: 1319–1328.

Wiesner, M.R., M.C. Grant, and S.R. Hutchins. 1996. Reduced permeability in groundwater remediation systems: role of mobilized colloids and injected chemicals. *Environ. Sci. Technol.* 30: 3184–3191.

Williams, W.A. and R.J. May. 1997. Low-temperature microbial aerobic degradation of polychlorinated biphenyls in sediment. *Environ. Sci. Technol.* 31: 3491–3496.

Willumsen, P.A., U. Karlson, and P.H. Pritchard. 1998. Response of fluoranthene-degrading bacteria to surfactants. *Appl. Microbiol. Biotechnol.* 50: 475–483.

Wilson, M.S. and E.L. Madsen. 1996. Field extraction of a transient intermediary metabolite indicative of real time *in situ* naphthalene biodegradation. *Environ. Sci. Technol.* 30: 2099–2103

Wilson, S.S. and K.C. Jones. 1993. Bioremediation of soil contaminated with polynuclear aromatic hydrocarbons PAHs, a review. *Environ. Pollut.* 81: 229–249.

Wischmann, H. and H. Steinhart. 1997. The formation of PAH oxidation products in soils and soil/compost mixtures. *Chemosphere* 35: 1681–1698.

Wolfaardt, G.M., J.R. Lawrence, R.D. Robarts, and D.E. Caldwell. 1995. Bioaccumulation of the herbicide diclofop in extracellular polymers and its utilization by a biofilm community during starvation. *Appl. Environ. Microbiol.* 61: 152–158.

Wu, Q., D.L. Bedard, and J. Wiegel. 1997. Effect of incubation temperature on the route of microbial reductive dechlorination of 2,3,4,6-tetrachlorobiphenyl in polychlorinated biphenyl (PCB)- contaminated and PCB free freshwater sediments. *Appl. Environ. Microbiol.* 63: 28366–2843.

Wu, Q., D.L. Bedard, and J. Wiegel. 1999. 2,6-Dibromobiphenyl primes extensive dechlorination of Arochlor 1260 in contaminated sediment at 8–30°C by stimulating growth of PCB-dehalogenating microorganisms. *Environ. Sci. Technol.* 33: 595–602.

Xun, L., E. Topp, and C.S. Orser. 1992. Confirmation of oxidative dehalogenation of pentachlorophenol by a *Flavobacterium* pentachlorophenol hydroxylase. *J. Bacteriol.* 174: 5745–5747.

Yadav, J.S., R.E. Wallace, and C.A. Reddy. 1995. Mineralization of mono- and dichlorobenzenes and simultaneous degradation of chloro- and methyl-substituted benzenes by the white-rot fungus *Phanerochaete chrysosporium*. *Appl. Environ. Microbiol.* 61: 677–690.

Yamabe, M. 1994. Fluoropolymer coatings pp. 397–402. In *Organofluorine Chemistry Principles and Commercial Applications* (Eds. R.E. Banks, B.E. Smart, and J.C. Tatlow). Plenum Press, London.

Ye, D., M.A. Siddiq, A. Maccubbin, S. Kumar, and H.C. Sikka. 1996. Degradation of polynuclear aromatic hydrocarbons by *Sphingomonas paucimobilis*. *Environ. Sci. Technol.* 30: 136–142

Zipper, C., K. Nickel, W. Angst, and H.-P. E. Kohler. 1996. Complete microbial degradation of both enantiomers of the chiral herbicide mecoprop [2-4-chloro-2-methylphenoxypropionic acid] in an enantioselective manner by *Sphingomonas herbicidovorans*. sp. nov. *Appl. Environ. Microbiol.* 62: 4318–4322.

Index